THE CARBOHYDRATES

Chemistry and Biochemistry

SECOND EDITION

VOLUME IB

CONTRIBUTORS

Don C. DeJongh

H. S. El Khadem

R. J. Ferrier

John W. Green

Laurance D. Hall

Anthony Herp

Martin I. Horowitz

Derek Horton

L. Mester

Frank S. Parker

H. Paulsen

A. S. Perlin

K.-W. Pflughaupt

Glyn O. Phillips

Olof Theander

R. Stuart Tipson

Joseph D. Wander

Neil R. Williams

THE CARBOHYDRATES

Chemistry and Biochemistry

SECOND EDITION

EDITED BY

Ward Pigman

Department of Biochemistry
New York Medical College
New York, New York

Derek Horton

Department of Chemistry
The Ohio State University
Columbus, Ohio

ASSISTANT EDITOR

Joseph D. Wander

Department of Chemistry
University of Georgia
Athens, Georgia

VOLUME IB

1980

ACADEMIC PRESS

A Subsidiary of Harcourt Brace Jovanovich, Publisher

New York   London   Toronto   Sydney   San Francisco

572.56 CAR

ACADEMIC PRESS, INC.
111 Fifth Avenue, New York, New York 10003

United Kingdom Edition published by
ACADEMIC PRESS, INC. (LONDON) LTD.
24/28 Oval Road, London NW1 7DX

LIBRARY OF CONGRESS CATALOG CARD NUMBER: 68-26647

ISBN 0-12-556351-5

PRINTED IN THE UNITED STATES OF AMERICA

80 81 82 83   9 8 7 6 5 4 3 2 1

We dedicate this work to the persons most responsible for our professional development

HORACE S. ISBELL AND THE LATE MELVILLE L. WOLFROM

CONTENTS

16. Amino Sugars

Derek Horton and Joseph D. Wander

17.   Deoxy and Branched-Chain Sugars

Neil R. Williams and Joseph D. Wander

18.   Thio Sugars and Derivatives

Derek Horton and Joseph D. Wander

19.   Unsaturated Sugars

R. J. Ferrier

20. Glycosylamines

H. Paulsen and K.-W. Pflughaupt

21. Hydrazine Derivatives and Related Compounds

L. Mester and H. S. El Khadem

22.   Reduction of Carbohydrates

John W. Green

23.   Acids and Other Oxidation Products

Olof Theander

24. Oxidative Reactions and Degradations

John W. Green

25. Glycol-Cleavage Oxidation

A. S. Perlin

26. The Effects of Radiation on Carbohydrates

Glyn O. Phillips

27. Physical Methods for Structural Analysis

28.   Separation Methods: Chromatography and Electrophoresis

Martin I. Horowitz

LIST OF CONTRIBUTORS

Numbers in parentheses indicate the pages on which the authors' contributions begin.

Don C. DeJongh (1327), Finnigan Institute, Atkinson Square Building 5, Cincinnati, Ohio 45246

H. S. El Khadem (929, 1376), Department of Chemistry, Michigan Technological University, Houghton, Michigan 49931

R. J. Ferrier (843, 1354), Department of Chemistry, Victoria University of Wellington, Private Bag, Wellington, New Zealand

John W. Green (959, 1101), The Institute of Paper Chemistry, Appleton, Wisconsin 54911

Laurance D. Hall (1299), Department of Chemistry, University of British Columbia, Vancouver, British Columbia V6T 1W5, Canada

Anthony Herp (1276), Department of Biochemistry, New York Medical College, Valhalla, New York 10595

Martin Horowitz (1445), Department of Biochemistry, New York Medical College, Valhalla, New York 10595

Derek Horton (644, 799), Department of Chemistry, The Ohio State University, Columbus, Ohio 43210

L. Mester (929), Institut de Chimie des Substances Naturelles, C.N.R.S., 91 Gif-sur-Yvette, France

Frank S. Parker (1376, 1394), Department of Biochemistry, New York Medical College, Valhalla, New York 10595

H. Paulsen (881), Institut für Organische Chemie, Universität Hamburg, 2000 Hamburg 13, Federal Republic of Germany

A. S. Perlin (1169), Department of Chemistry, McGill University, Montreal, Quebec H3A 2K6, Canada

K.-W. Pflughaupt (881), Neurologische Universitäts Klinik der Universität Würzburg, 8700 Würzburg, Federal Republic of Germany

Glyn O. Phillips (1217), The North Wales Institute of Higher Education, Kelsterton College, Deeside, Clwyd, North Wales

Olof Theander (1013), Department of Chemistry, Swedish University of Agricultural Sciences, Uppsala S-750 07, Sweden

R. Stuart Tipson (1394), 10303 Parkwood Dr., Box 143, Kensington, Maryland 20795

Joseph D. Wander (644, 761, 749), Department of Chemistry, University of Georgia, Athens, Georgia 30602

Neil R. Williams (761), Department of Chemistry, University of London, London WC1E 7HX, England

PREFACE

This edition of "The Carbohydrates" is a complete revision of the 1957 work, which was based on "The Chemistry of the Carbohydrates" (1948). Because of its size, it has been divided into four volumes (IA and B and IIA and B). The considerably greater length of this edition is a reflection of the rapid growth of research in the field.

In retrospect, the previous edition contained very little that has needed correction, but new fields of knowledge have developed. Thus, conformational analysis has made spectacular advances with the development of nuclear magnetic resonance methods, and X-ray crystallography has become widely used as a general structural tool. Amino sugars and uronic acids have attained great significance because of their widespread occurrence in important biological substances. Unsaturated sugars and dicarbonyl sugars have become especially important with the development of newer methods of synthesis, separation, and structural characterization. A chapter has been added on the effects of ionizing radiations and of autoxidation reactions. The major physical methods of characterization and greatly advanced methods of separation are described in additional chapters. The literature on nucleosides and carbohydrate-containing antibiotics has expanded to the extent that these subjects have necessitated full chapters. With the discovery of transglycosylation reactions, the number of known oligosaccharides and enzymes acting on carbohydrates has vastly increased. A new chapter on the biosynthesis of sugars and complex saccharides was required to cover this rapidly growing field.

In the previous edition, the discussion of polysaccharides was reduced to two chapters because of the prior appearance of "The Polysaccharides" by Whistler and Smart. In the present edition, the original practice of having separate chapters for the main types of polysaccharide has been restored. Chapters on the rapidly growing fields of glycolipids and glycoproteins are incorporated.

The two final chapters in Volume IIB cover the nomenclature for carbohydrates and for enzymes having carbohydrates as substrates. The enzyme names were extracted from the official report, but the names were modified to conform as much as possible to standard carbohydrate nomenclature. In the other chapters, official carbohydrate nomenclature has been used, but both old and new enzyme names are given.

The conformational terminology used in Volume IB for five- and six-membered ring sugars accords with the British–American Rules [*Chem. Commun.* 505 (1973)] that were promulgated in replacement of the older Reeves system employed in the other volumes of this edition. For the general nomenclature of carbohydrates and their derivatives, the established system as set out in Chapter 46 has been used throughout this treatise in conjunction with the most widely accepted working drafts on official terminology for unsaturated and branched-chain sugars, oligosaccharide chains, and polysaccharides. A set of definitive, international recommendations for carbohydrate nomenclature had not been made final at the time Volume IB went to press. When these recommendations are eventually published, little difference in substance from the usage in this edition is to be expected, except that the highest-numbered asymmetric center in a monosaccharide chain will be used uniformly as a basis for assigning the anomeric α and β designators.

As in the previous edition, the chapter authors were encouraged to be selective rather than exhaustive in their citation of the literature. The objective has been to achieve a proper balance, in correct historical perspective, between the important early papers and the more recent developments. Even in a work the size of this one, it is possible to cite only a small fraction of the total published literature on the subject, and the material selected reflects the collective judgment of the chapter authors and the editors.

For the chapters in Volume IB, selective coverage of articles published through 1979 has been made. For more detailed treatment of individual subject areas, the reader is referred especially to the annual series *Advances in Carbohydrate Chemistry and Biochemistry* for authoritative, in-depth articles. For detailed listing of all articles and patents published in the carbohydrate field, both before and since the date of publication of this book, the reader should consult the Carbohydrates section of *Chemical Abstracts* (currently Section 33) and also the Cellulose and Industrial Carbohydrates sections (currently Sections 43 and 44). For new research articles on the carbohydrates, the international journal *Carbohydrate Research*, inaugurated in 1965, provides a prime source. The series of *Specialist Periodical Reports on Carbohydrate Chemistry* (Chemical Society, London) serves to catalogue, on an annual basis starting in 1967, a major proportion of the work published on the carbohydrates during each year; Volume 11 covering the 1977 literature was published in late 1979. The *Advances*, *Carbohydrate Research*, the Carbohydrates section of *Chemical Abstracts*, and the *Specialist Periodical Reports on Carbohydrate Chemistry* serve to complement this treatise and with it provide a complete bibliographic core on the subject, including detailed foundations in the past literature, and continuing developments subsequent to the publication of this edition.

For extracting from the literature new articles on carbohydrates as they appear, a most valuable tool is the computer-assisted search of current journals and patents. *CA Search,* produced by Chemical Abstracts Service, covers over 14,000 periodicals and is issued weekly on magnetic tape. Titles and literature citations are printed out automatically in response to a given search profile. Search profiles for selecting titles dealing with carbohydrates have been devised and may be modified appropriately to reflect individual interests and emphasis.

This book is an international collaborative effort, and sixty-four authors were involved in the writing of the various chapters. They reside in Australia, the British Isles, Canada, France, Germany, Japan, and the United States. Most of the chapters were read by other workers in the field, and warm thanks are extended to the many colleagues who gave valued assistance in this way.

The editors owe special appreciation for the help of Drs. Anthony Herp, Joseph D. Wander, the late Hewitt G. Fletcher, Jr., and Leonard T. Capell. Dr. Wander acted as assistant editor for Volume IB, and helped in rewriting major sections of several chapters and in verifying numerous references; Dr. Herp served in a similar capacity for Volumes IIA and B. Dr. Fletcher read initial drafts and galley proofs of many of the chapters, and Dr. Capell was responsible for the Subject Indexes.

The New York Medical College and The Ohio State University gave support and encouragement to the editors in the preparation of these volumes, and Academic Press gave the expected hearty cooperation. The dedicated secretarial support of Mary LaSalette Sayre throughout the long endeavor in producing this edition is especially appreciated.

Ward Pigman died suddenly and unexpectedly on September 30, 1977. A brief account of his life and work has been published [D. Horton, *Nature,* **272,** 476 (1978)] and a fuller article on this major personage in carbohydrate chemistry has been compiled by A. Herp [*Advan. Carbohydr. Chem. Biochem.* **37,** 1 (1980)]. The present set of volumes comprising the second edition of "The Carbohydrates" stands as a continuing monument to this outstanding and unusual scientist.

Derek Horton

CONTENTS OF OTHER VOLUMES

VOLUME IA

16. AMINO SUGARS

DEREK HORTON AND JOSEPH D. WANDER

I. INTRODUCTION

The designation *amino sugar* is applied to carbohydrate derivatives in which one (or more) of the hydroxyl groups attached to the carbon backbone of a sugar molecule is replaced by a free or substituted amino group. Nitrogen normally adopts valence states different from those of the chalcogens oxygen and sulfur, so that its introduction into a sugar constitutes a functional-group replacement rather than the electronic homologation that relates the thio sugars (Chapter 18) to their oxygenated counterparts; the accepted nomenclature (Vol. IIB, Chapter 46) is based on replacement terminology—for example, 2-amino-2-deoxy-α-D-glucopyranose (**1**).

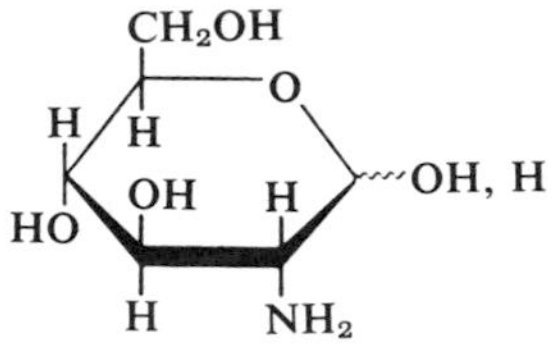

$$CH_2OH$$

1

The first amino sugar was discovered in 1875 by Ledderhose,[1] who isolated **1** (earlier termed *glucosamine*) from an acid hydrolyzate of lobster shell; however, sustained interest in this class of sugar was initiated only at the beginning of the 1950's, partly as a consequence of the observation[2] (in 1946) that 2-deoxy-2-(methylamino)-L-glucose is a component of the antibiotic streptomycin. The literature prior to 1966 on monosaccharide amino sugars has been the subject of a comprehensive article[3] as part of a treatise dealing with the multifarious aspects of the chemistry and biology of amino sugars. Other aspects of the chemistry of nitrogenous carbohydrate derivatives are treated elsewhere in this monograph as follows: aminocyclitols (Vol. IA, Chapter 15, Section VIII), glycosylamines (Chapter 20), osazones and other carbonyl-group condensation products (Chapter 21), aminonucleosides (Vol. IIA, Chapter 29), antibiotic substances (Vol. IIA, Chapter 31; see also subsequent reviews of aminoglycoside antibiotics and mechanisms of resistance to them[4]), biosynthesis of amino sugars[5] (Vol. IIA, Chapter 34, Section IX), and the occurrence of amino sugars as components of biopolymers (chitin, Vol. IIA, Chapter 36, Section XI; glycosaminoglycans, Vol. IIB, Chapter 42; glycoproteins, Vol. IIB, Chapter 43; and glycolipids, Vol. IIB, Chapter 44); methods for isolation and for chemical analysis of amino sugars are treated in Chapter 28 and in Vol. IIB, Chapter 45, respectively. This chapter will discuss the occurrence, preparation, and reactions of monosaccharide amino sugars. Methods for chemical synthesis are treated first.

II. SYNTHESIS OF MONOSACCHARIDE AMINO SUGARS

A. ADDITION OF CARBON NUCLEOPHILES TO CARBONYL GROUPS IN THE PRESENCE OF AMINES

1. *Hydrogen Cyanide*

The classic synthesis of 2-amino-2-deoxy-D-glucose (1) by Fischer and Leuchs[6] was accomplished by the sequential condensation of D-arabinose with ammonia and subsequently with hydrogen cyanide to produce an epimeric mixture of 2-amino-2-deoxy-D-gluconitrile and a lower proportion of 2-amino-2-deoxy-D-mannononitrile; hydrolysis of the former afforded 2-amino-2-deoxy-D-gluconic acid, which was lactonized and the lactone reduced with sodium amalgam to afford 1 in low yield.

Levene and co-workers[7,8] demonstrated the generality of this reaction by preparing all eight of the 2-amino-2-deoxy-D-hexonic acids from the four D-aldopentoses; in each reaction, the epimeric product having the 2,3-*threo* arrangement preponderates. 2-Amino-2-deoxy-D-mannonic acid, the minor product from the reaction of D-arabinose, was prepared in useful quantities by epimerization[7,9] of 2-amino-2-deoxy-D-gluconic acid in aqueous pyridine at 100°.

An important, practical improvement of this reaction was developed by Kuhn and Kirschenlohr,[10] who effected controlled hydrogenation of the aminonitriles directly to the 2-aminoaldoses in good yields by use of palladium on barium sulfate as the catalyst.

The introduction of a primary amine instead of ammonia in this reaction sequence was first demonstrated by Votoček and Lukeš,[11] who prepared 2-deoxy-2-(methylamino)-D-gluconic acid from D-arabinose by the sequential action of methylamine and hydrogen cyanide. Wolfrom and co-workers[12] employed the analogous reaction of L-arabinose in preparing 2-deoxy-2-(methylamino)-L-glucose, which had been isolated[2] shortly before as a constituent of streptomycin. Kuhn and co-workers[10,13] substituted benzylamine or aniline for ammonia in this reaction; the stereoselectivity of the initial addition was not altered significantly, but subsequent epimerization of the nitriles at the new center occurred readily in warm alcohol. Using this modification, Kuhn and Jochims[13] reported that yields of 2-amino-2-deoxy-D-allose (4) [from D-ribose (2)] and 2-amino-2-deoxy-D-talose (from D-lyxose) could be increased from less than 10% to more than 50%. Addition of benzylamine to 2 affords a glycosylamine intermediate which undergoes attack by cyanide to afford preponderantly the D-*altro* adduct 5; however, 5

References start on p. 737.

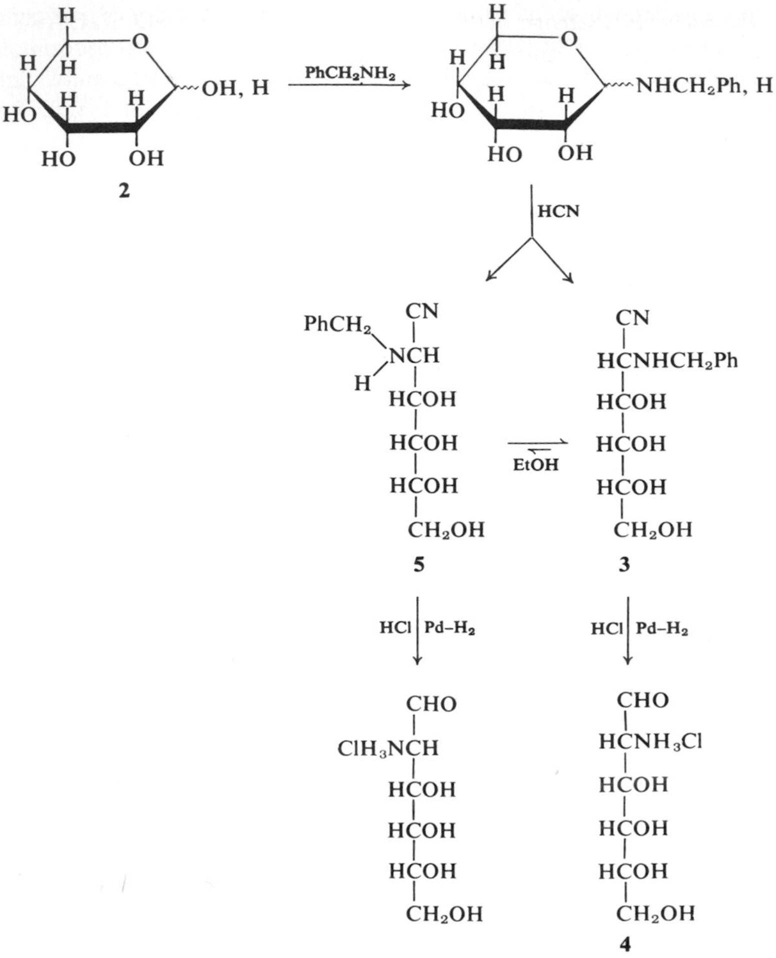

epimerizes in warm, alcoholic solution to give the less-soluble isomer **3**, which crystallizes out, thus displacing the equilibrium and drawing the inversion to completion. In the final step, reduction of the nitrile to the aldehyde (by way of a presumed imine) is accompanied by hydrogenolysis of the benzyl–nitrogen bond; N-alkyl groups are not hydrogenolyzed by this procedure. Specific, isotopic labels have been introduced[14] into 2-amino-2-deoxy-D-glucose (**1**) by the use of [^{15}N]benzylamine and also by the use of D-[1-^{14}C]-arabinose; ^{13}C-labeling at C-1 has also been effected.[14a]

The glycosylamine is not an obligatory intermediate in this reaction, as 1,2:3,4-di-O-isopropylidene-α-D-*galacto*-hexodialdo-1,5-pyranose reacts with

benzylamine and hydrogen cyanide to afford a 6-amino-6-deoxyhepturono-
nitrile derivative of unspecified configuration.[15] Paulsen and Budzis[16]
condensed 1,2-O-isopropylidene-α-D-*xylo*-pentodialdo-1,4-furanose with *N*-
methylhydroxylamine; the resulting imine oxide **6** incorporated hydrogen
cyanide very efficiently, forming a 7:2 mixture of **8** and **7** in 90% net yield.
After reduction of the mixture with hydroxylamine, and hydrolysis of the
nitrile, the amide **9** crystallized in 61% yield.

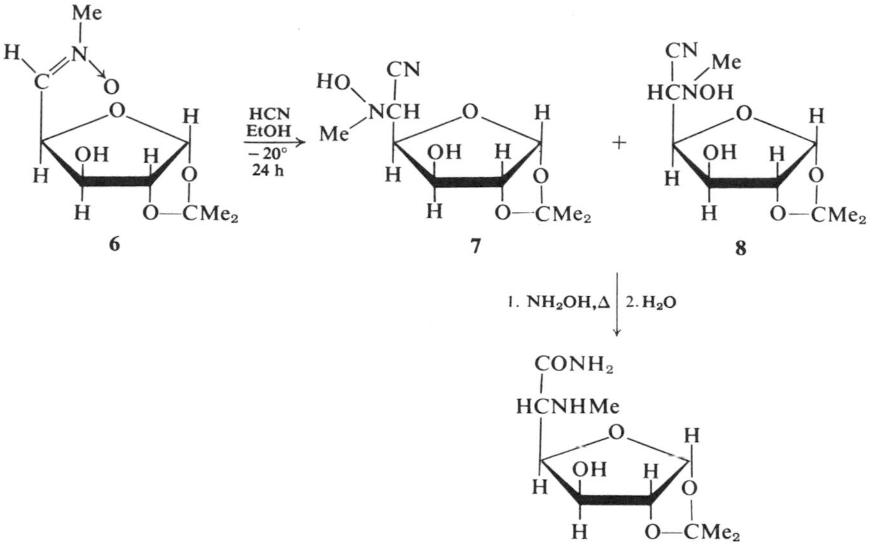

A similar process of addition can be effected with glycosuloses. 5-O-
Benzyl-1,2-O-isopropylidene-α-D-*erythro*-pentofuranos-3-ulose (**10**) reacts
with potassium cyanide and ammonium carbonate to afford[17] the branched-
chain derivative **11**; saponification and acetylation of **11** are accompanied by
lactonization to form compound **12**, which sequence defines the stereochemistry of the new asymmetric center.

3-C-Cyano-1,2:5,6-di-O-isopropylidene-α-D-allofuranose (**12a**) combines
with ammonium chloride in the presence of potassium hydroxide and cyanide
to afford[17a] the corresponding diacetal (**12b**) of 3-amino-3-C-cyano-3-deoxy-
D-allose, which may be hydrolyzed to form an amino acid.

2-Aminonitriles were shown by Kuhn, Weiser, and H. Fischer[18] to undergo
1,2 elimination of water under alkaline conditions, thereby forming α,β-
unsaturated nitriles. Thus, nitriles **3** and **5**, as well as their D-*gluco* and

---

References start on p. 737.

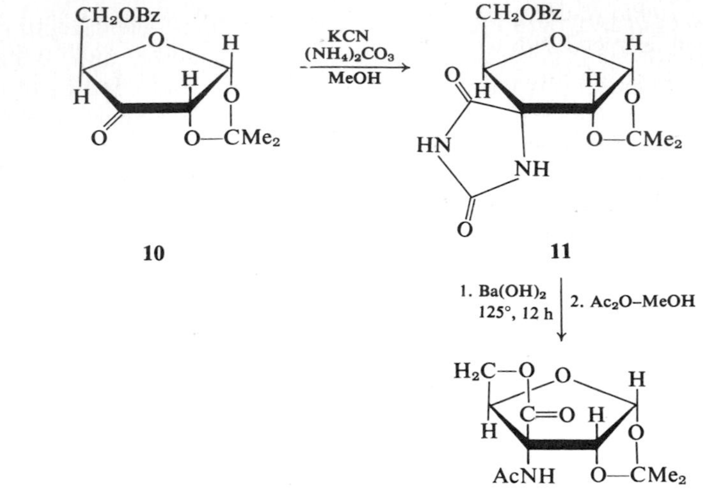

10 → **11**

KCN
(NH₄)₂CO₃
MeOH

1. Ba(OH)₂ 2. Ac₂O–MeOH
125°, 12 h

12

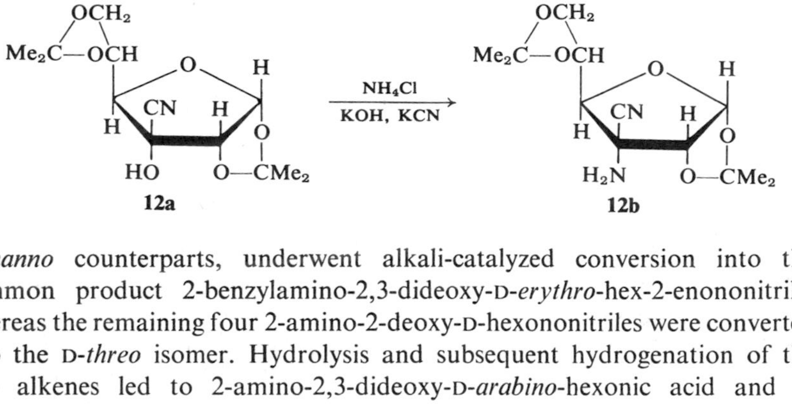

12a → **12b**

NH₄Cl
KOH, KCN

D-*manno* counterparts, underwent alkali-catalyzed conversion into the common product 2-benzylamino-2,3-dideoxy-D-*erythro*-hex-2-enononitrile, whereas the remaining four 2-amino-2-deoxy-D-hexononitriles were converted into the D-*threo* isomer. Hydrolysis and subsequent hydrogenation of the two alkenes led to 2-amino-2,3-dideoxy-D-*arabino*-hexonic acid and 2-amino-2,3-dideoxy-D-*xylo*-hexonic acid, respectively, and further reduction by sodium amalgam produced the corresponding aldoses, although in low yield.

By the Kiliani procedure, Levene and Matsuo[19] converted 2-amino-2-deoxy-D-glucose (**1**) into 3-amino-3-deoxy-D-*glycero*-D-*ido*-heptonic acid and the D-*glycero*-D-*gulo* epimer; the aldoses were prepared later.[20] 3-Amino-3-deoxy-D-*glycero*-L-*gluco* (and L-*manno*)-heptonic acids were similarly prepared from 2-amino-2-deoxy-D-galactose.

2. *Other Carbon Nucleophiles*

In the presence of ammonium acetate, cyclohexyl isocyanate converts 2,3:4,5-di-O-isopropylidene-*aldehydo*-L-arabinose into a mixture of 2-(N-cyclohexylacetamido)-2-deoxy-3,4:5,6-di-*O*-isopropylidene-L-gluconamide and its L-*manno* isomer; similar treatment of 2,3:4,5-di-*O*-isopropylidene-*aldehydo*-D-xylose affords[21] a presumed single isomer of unspecified configuration.

Addition of a two-carbon fragment has been accomplished by the reaction of *aldehydo*-aldoses with *N*-pyruvylideneglycinatoaquocopper(II) (13); treatment with 13 and subsequently with aqueous ammonia converted[22] 2,3-*O*-isopropylidene-D-erythrose into 2-acetamido-2-deoxy-D-gluconic acid in 60% yield.

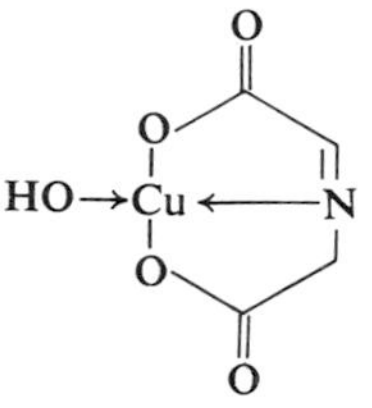

13

Satoh and Kiyomoto[23] converted D-mannose directly into a mixture of 2-acetamido-1,2-dideoxy-1-nitro-D-*glycero*-D-*talo*-heptitol plus a small proportion of the corresponding D-*glycero*-D-*galacto* analogue by the action of nitromethane and ammonia in acetic anhydride (see Section II,H).

B. CARBON–CARBON BOND CLEAVAGE OF HIGHER HOMOLOGUES

Conversion of 2-acetamido-2-deoxy-D-glucose (14) into 2-amino-2-deoxy-D-xylose hydrochloride (17) was accomplished by converting the former into the diethyl dithioacetal and this into the thiofuranoside 15; cleavage of the C-5–C-6 bond by periodic acid, and reduction of the resultant aldehyde, yielded the lower homologue 16, which underwent hydrolysis to afford 17 in an overall 50% yield.[24] Although the low net yields and the requirements for suitable starting materials limit the general synthetic utility of this reaction, it is stereochemically definitive and thus potentially valuable in configurational assignment. A similar sequence of reactions was used to degrade 2-acetamido-2-deoxy-D-galactose into 2-amino-2-deoxy-D-arabinose.[25]

3-Acetamido-3-deoxy-D-*glycero*-D-*galacto*-heptose forms both methyl furanosides and methyl pyranosides. Periodate cleavage and subsequent reduction and hydrolysis degraded the latter by one carbon atom to 3-amino-3-deoxy-D-galactose, whereas the same sequence of reactions converted both of the former[20] and 3-acetamido-3-deoxy-1,2-O-isopropylidene-α-D-galacto-furanose[26] into 3-amino-3-deoxy-L-arabinose. 3-Acetamido-3-deoxy-1,2-O-isopropylidene-α-D-allofuranose,[27] its 1,2-O-cyclohexylidene[28] analogue, and its 3-azido-3-deoxy precursor[29] have likewise been converted into 3-acet-amido-3-deoxy-D-ribose; degradation of 3-acetamido-3-deoxy-1,2-O-iso-propylidine-α-D-gulofuranose similarly afforded[30] 3-acetamido-3-deoxy-L-lyxose.

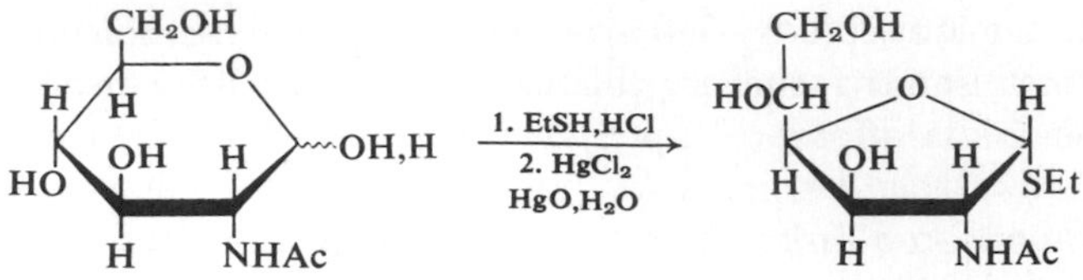

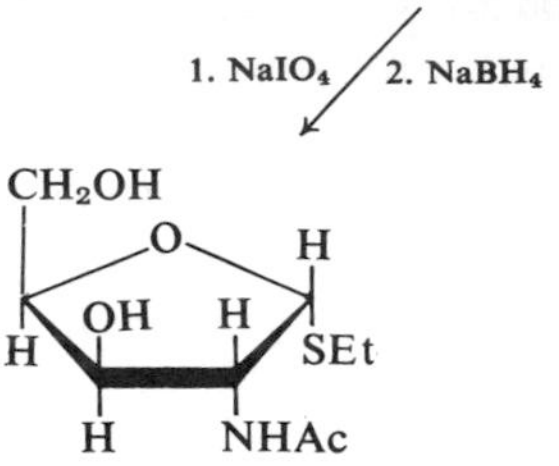

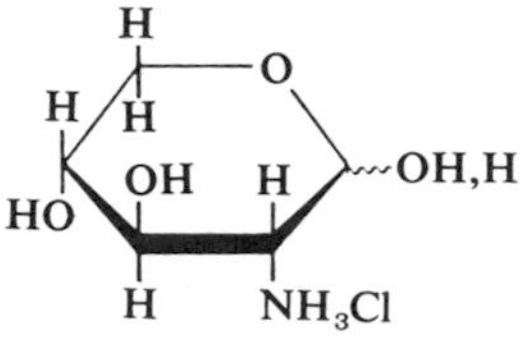

The 3,4-isopropylidene acetals of 2-acetamido-2-deoxy-D-glucitol[31] and its 2-acetamido-1,2-dideoxy[32] and 1,2-diacetamido-1,2-dideoxy[33] counter-parts undergo C-5–C-6 scission by periodate to afford, after hydrolysis, 4-acetamido-4-deoxy-L-xylose (C-5 becoming C-1 after the oxidation), 4-acetamido-4,5-dideoxy-L-xylose, and 4,5-diacetamido-4,5-dideoxy-L-xylose, respectively.

Chain-descent at the reducing terminus of suitably protected 3(or higher)-aminoaldoses has been accomplished by periodate oxidation of reducing aldoses. In this way, 2-acetamido-2-deoxy-D-arabinose has been prepared both from 3-acetamido-3-deoxy-D-mannopyranose[34] and from the D-*gluco*[35] isomer, 2-acetamido-2-deoxy-D-lyxose[36] from 3-acetamido-3-deoxy-D-talopyranose, 5-acetamido-2,5-dideoxy-D-*threo*-pentose[37] from 6-acetamido-3,6-dideoxy-D-*xylo*-hexose, and 2-acetamido-2,3,5-trideoxy-D-*erythro*-pentose[38] from 3-acetamido-3,4,6-trideoxy-D-*arabino*-hexose. Deane and Inch[39] so degraded 1-azido-1,3,4-trideoxy-D-*threo*-hexitol in the course of preparing 3(S)-piperidinol.

3-Acetamido-3-deoxy-D-altrose, which is resistant[40] to periodate, was degraded[41] by the MacDonald–Fischer procedure to 2-acetamido-2-deoxy-D-ribose.

C. INTRAMOLECULAR REARRANGEMENTS

The reaction of D-glucose with *p*-toluidine was found by Amadori[42] to afford, in addition to the anticipated glycosylamine, a product that Kuhn and Weygand later identified as 1-deoxy-1-(*p*-toluidino)-D-fructose (**18**), which may be presumed,[43a] by analogy to the *N*,*N*-dibutyl analogue, to favor the β-D $^2C_5$ arrangement[43b] illustrated. The early chemistry of these so-called Amadori compounds has been reviewed by Hodge,[44] who has generalized[45] that sterically unhindered primary and secondary amines undergo this acid–base-catalyzed condensation and rearrangement with aldoses; steric factors make such bis(secondary alkyl)amines as 2,6-dimethylpiperidine unreactive, and electronic effects depress the basicity of acylamines, and some diarylamines and haloanilines,[46] sufficiently to prevent their reaction. Enolic forms appear to be general intermediates.

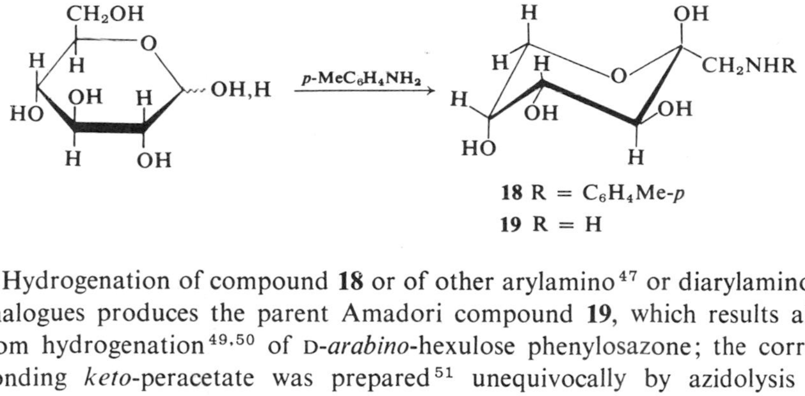

18 R = C$_6$H$_4$Me-*p*
19 R = H

Hydrogenation of compound **18** or of other arylamino[47] or diarylamino[48] analogues produces the parent Amadori compound **19**, which results also from hydrogenation[49,50] of D-*arabino*-hexulose phenylosazone; the corresponding *keto*-peracetate was prepared[51] unequivocally by azidolysis of 3,4,5,6-tetra-*O*-acetyl-1-bromo-1-deoxy-*keto*-D-fructose followed by reduction.

Reactions of aldoses with amino acids have been implicated in nonenzymic browning (or Maillard) reactions of various foodstuffs,[52] and *N*-(1-deoxy-ketos-1-yl)amino acids have been characterized[53] as products from such reactions. Whereas early studies of Maillard processes employed color

---

reactions and paper chromatography to monitor progress of the reaction, improved techniques of trimethylsilylation permit rapid quantitation[54] by gas chromatography. Other amino acids have also been used in Amadori reactions. 2-Aminoethanephosphonic acid reacts with D-glucose to afford[55] 2-(1-deoxy-D-fructos-1-ylamino)ethanephosphonic acid, whereas 2-amino-ethanesulfonic acid is unreactive; however, 4,6-O-benzylidene-D-glucose reacts with the sodium salt of the acid to form a glycosylamine that has been converted[56] into the Amadori product, 2-(4,6-O-benzylidene-1-deoxy-D-fructos-1-ylamino)ethanesulfonic acid. 2-(D-*threo*-1,2,3-Trihydroxypropyl)-5(and 6)-(D-*glycero*-2,3-dihydroxypropyl)pyrazine have been identified as products of the nonenzymic browning reaction of D-xylose by ammonium formate.[63a]

5-Amino-5-deoxy-D-xylopyranose (**20**), which is formally a glycosylamine, rearranges[57] slowly in the presence of dilute, mineral acid to the anhydro-pentulose derivative **21**. One of the products[58] of the reaction of 1,2:5,6-di-O-isopropylidene-D-*arabino*-3-hexulose (**22**) with aniline in refluxing toluene is compound **23**, and the authors suggested that steps of the Amadori and

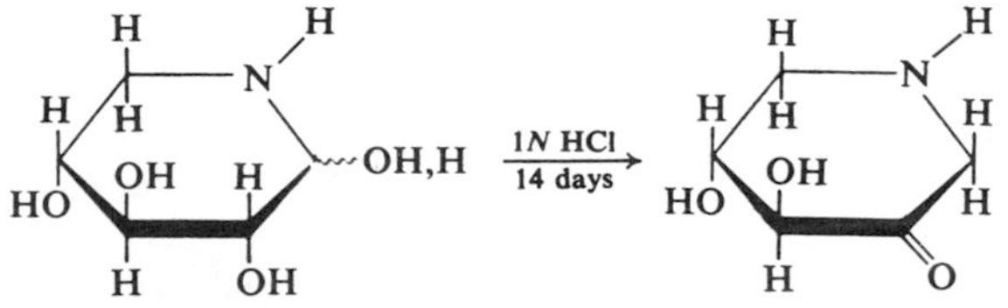

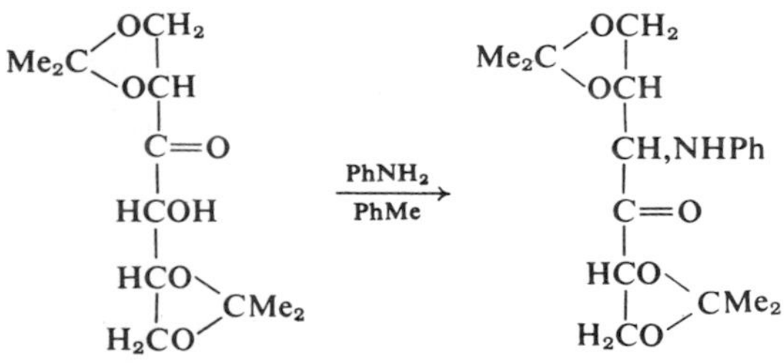

subsequent browning reactions of **22** are acid–base-catalyzed, but that each has a characteristic pH-dependence.

The (formally identical) rearrangement of N-benzyl-D-fructopyranosylamine into 2-benzylamino-2-deoxy-D-glucose and -D-mannose was demonstrated by Carson,[59] who converted the former into 2-amino-2-deoxy-D-glucose (**1**) by hydrogenolysis. Heyns and his co-workers[60] extended this reaction to the synthesis of *N*-(1-carboxyalkyl) derivatives of **1** and 2-amino-2-deoxy-D-mannose by rearrangement of the products of reaction of D-fructose with amino acids. A comparative study[61] revealed that D-tagatose reacts readily with ammonia to afford 2-amino-2-deoxy-D-galactose plus a trace of 2-amino-2-deoxy-D-talose, whereas D-fructose and D-psicose react more slowly, and D-sorbose reacts the least readily; each reaction produces an epimeric mixture, the stereoselectivity decreasing with the reaction rate. Ammonolysis of 1-amino-1-deoxy-D-fructose at elevated pressure has been suggested[62] as a manufacturing process for 2-amino-2-deoxy-D-glucose; this suggests that the 2-aminoaldose is a more stable entity than the 1-aminoketose.

The lower homologue D-*threo*-pentulose combines with 3-aminopropanoic acid to afford,[63] in 32% crude yield, a mixture of 2-(2-carboxyethylamino)-2-deoxy-D-xylose and -D-lyxose, from which the former was isolated pure in 50% yield. The generally difficult separations and low yields detract from the synthetic utility of this process in most preparations for which alternative routes are available.

D. DISPLACEMENT BY NITROGEN NUCLEOPHILES

1. *Opening of Epoxides*

The attack of nitrogen nucleophiles upon terminal epoxides occurs at the primary center. Thus, 5,6-anhydro-1,2-*O*-isopropylidene-3-*O*-(methylsulfonyl)-β-L-idofuranose reacts[64] with liquid ammonia to form 6-amino-6-deoxy-1,2-*O*-isopropylidene-3-*O*-(methylsulfonyl)-β-L-idofuranose, and 5,6-anhydro-3-azido-3-deoxy-1,2-*O*-isopropylidene-α-D-glucofuranose (**24**) is attacked[65] by phenothiazine in the presence of sodium amide to form the *N*-(6-deoxyaldos-6-yl)phenothiazine derivative **25**.

In conformationally predictable epoxypyranose systems, the favored mode of opening is that leading to a *trans*-diaxial arrangement of the nucleophile

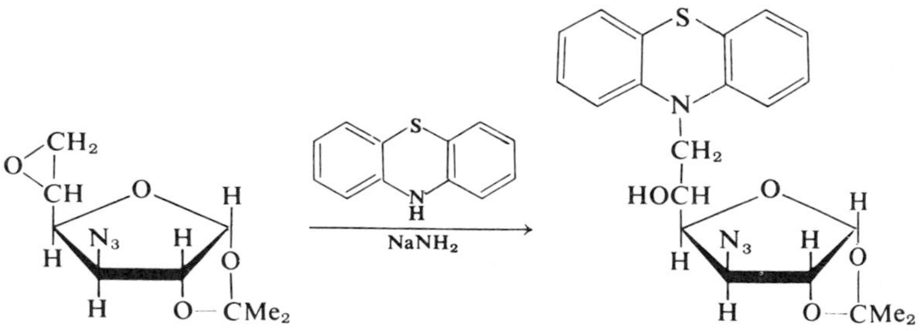

24   25

References start on p. 737.

and the oxygen atom of the oxirane precursor. Thus, ammonolysis of 1,6:2,3-dianhydro-β-D-talopyranose, and hydrolysis of the major adduct, provided[66] a constitutional synthesis of 2-amino-2-deoxy-D-galactose, and similar opening of 1,6:3,4-dianhydro-β-D-talopyranose produced[66a] 4-amino-1,6-anhydro-β-D-mannopyranose, together with a trace of the product of diequatorial opening, 3-amino-1,6-anhydro-3-deoxy-β-D-idopyranose. 1,6:2,3-Dianhydro-4-O-benzyl-β-D-mannopyranose is ammonolyzed to the all-axial 2-amino-1,6-anhydro-4-O-benzyl-2-deoxy-β-D-glucopyranose, which was isolated[67] as the diacetate; similar specificity of opening was observed[68] in the methanolic ammonolysis of the corresponding 4-deoxy-4-fluoro derivative, and 1,6-anhydro-2-deoxy-2-fluoro-β-D-galactopyranose reacts under similar conditions to afford 4-amino-1,6-anhydro-2,4-dideoxy-2-fluoro-β-D-glucopyranose. 1,6:2,3-Dianhydro-β-D-allopyranose undergoes ammonolysis[69] at 110° to produce 3-amino-1,6-anhydro-3-deoxy-β-D-glucopyranose in 70% yield.

Conversely, methyl 2,3-anhydro-4,6-O-benzylidene-α-D-talopyranoside (26), for which the $^4C_1$(D) conformation is strongly favored, undergoes ammonolysis to yield[70,71] (after acetylation) mainly methyl 3-acetamido-4,6-O-benzylidene-3-deoxy-α-D-idopyranoside (27), the product of diaxial opening. The same stereochemistry results[71] from the ammonolysis of the D-*gulo* analogue (28) of 26 to form 29, and ammonolysis of methyl 4,6-O-benzylidene-2,3-di-O-p-tolylsulfonyl-α-D-galactopyranoside in the presence of methoxide ion affords a mixture of compounds 27 and 29, suggesting the intermediacy of 26 and 28 in the displacement process.

Ammonolysis of methyl 2,3-anhydro-4,6-O-benzylidene-α-D-allopyranoside, which is constrained to exist in the $^4C_1$(D) conformation, is known[72] to give mainly methyl 2-amino-4,6-O-benzylidene-2-deoxy-α-D-altropyranoside; methylamine and guanidine also attack C-2 to form[73] the corresponding 2-methylamino and 2-guanidino analogues, whereas the reaction[74] with

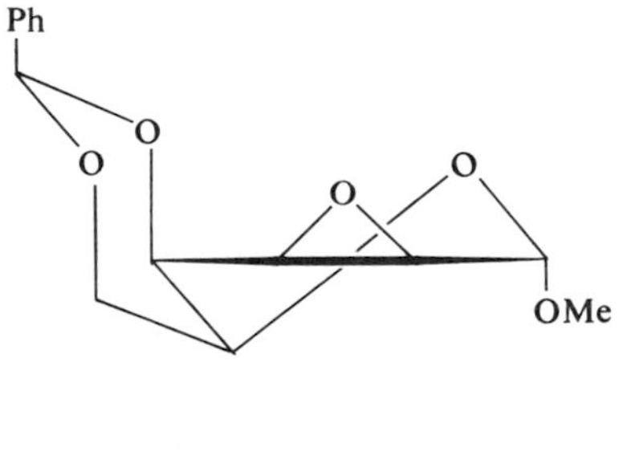

26 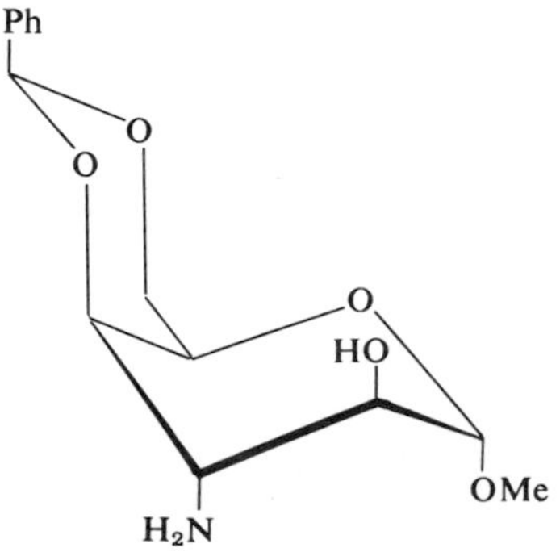   27

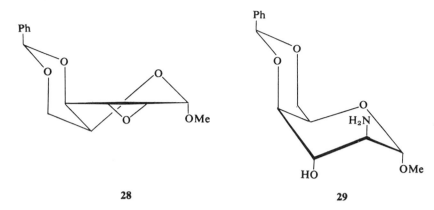

28   29

dimethylamine is less selective, allowing the isolation of methyl 3-deoxy-3-(dimethylamino)-D-glucose, which was converted, by hydrogenolysis of the 6-O-p-tolylsulfonyl derivative and subsequent hydrolysis, into 3,6-dideoxy-3-(dimethylamino)-D-glucose (*mycaminose*), a constituent of the antibiotic magnamycin.

Ammonolysis of methyl 2,3-anhydro-4,6-O-benzylidene-α-D-mannopyranoside in liquid ammonia gives mainly the 3-amino-3-deoxy-D-*altro* product; addition of methanol to the solvent lowers the selectivity somewhat, and the corresponding reaction with ammonium carbonate gives a 1:1 mixture of the isomeric products.[74a]

Azidolysis of methyl 2,3-anhydro-4,6-O-benzylidene-D-hexopyranosides parallels ammonolysis, and Guthrie and Liebmann[75] have interpreted approximate kinetic data for the entire series of stereoisomeric epoxides as signifying that the kinetic effects directing the point of attack are tempered by differences in electronic stability of intermediates.

Stereochemical directing of conformationally mobile epoxyhexopyranoses is less certain, but generally appears to lead mainly to the product of diaxial opening from the favored conformation of the starting materials. Thus, ammonolysis[76] of methyl 2,3-anhydro-4,6-di-O-methyl-β-D-mannopyranoside (**31**) afforded 3-amino-D-*altro* (**30**) and 2-amino-D-*gluco* (**32**) products in a ratio of 9:1; N-acetylation and subsequent methylation of **32** completed the first constitutional synthesis of a derivative of 2-amino-2-deoxy-D-glucose (**1**), whereas similar derivatization of **30** and of the product of ammonolysis[77] of methyl 2-chloro-2-deoxy-β-D-glucopyranoside (**35**) afforded **33**, implicating[76] the D-*manno* epoxide **34** as an intermediate in the latter reaction.

Methyl 2,3-anhydro-β-D-talopyranoside (**36**) underwent attack by ammonia preponderantly[66a,70a] at C-3, as did[38] the related 4,6-dideoxy sugar **37**, and methyl 2,6-di-*O*-benzoyl-3,4-di-*O*-(methylsulfonyl)-α-D-glucopyranoside is converted by hydrazinolysis into a D-*allo* 2,3-epoxide, which undergoes attack[78] at C-2 to yield, after reduction and acetylation, methyl 2,4-di-acetamido-3,6-di-*O*-acetyl-2,4-dideoxy-α-D-idopyranoside. Partial epoxide migration occurs in the azidolysis[79] of methyl 2,3-anhydro-6-deoxy-4-*O*-(methylsulfonyl)-α-L-gulopyranoside (**38**), resulting in the formation of **39** in 10.8% yield by direct displacement, and of **40** in 6.5% yield by opening of an intermediate, the L-*galacto* 3,4-epoxide; the benzyl β-L-pyranoside analogue **41** forms[80] compounds **42** and **43** in the same general proportions, together with a minor proportion[81] of **44**, which is formed by initial attack at C-3. Methyl 2,3-anhydro-4-*O*-benzoyl-6-deoxy-α-D-allopyranoside reacts with methanolic ammonia to produce[81a] methyl 2-amino-4-*O*-benzoyl-2,6-dideoxy-α-D-altropyranoside and methyl 3-amino-4-*O*-benzoyl-3,6-dideoxy-α-D-glucopyranoside in 24% and 52% yield, respectively.

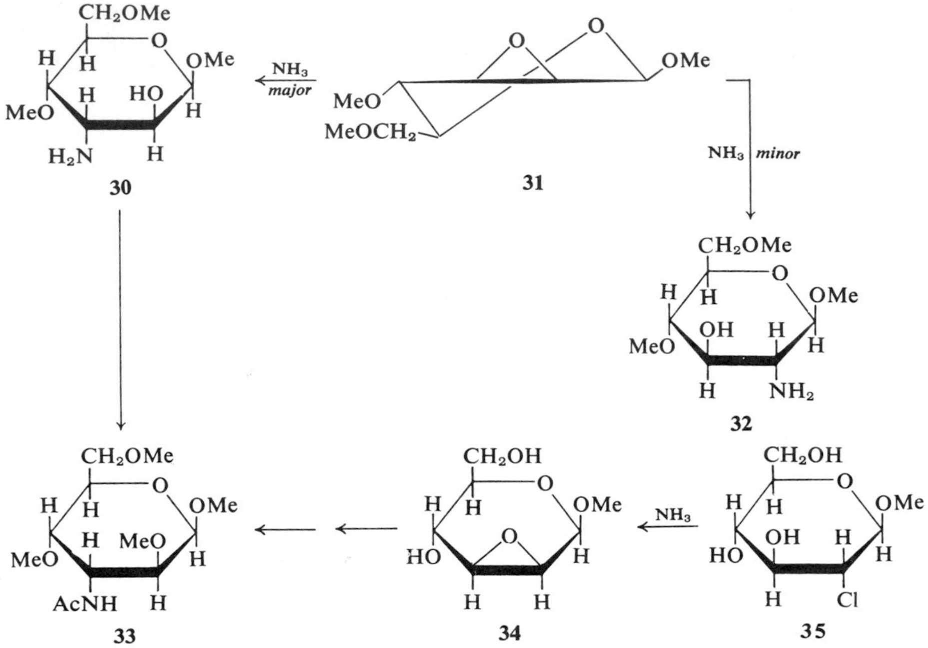

Ammonolysis[82] and, to a smaller extent, azidolysis[83] of methyl 3,4-anhydro-6-deoxy-β-D-talopyranoside (**45**) resulted in substitution at C-3 to afford the 3-amino(or 3-azido)-3-deoxy-D-*ido* product; however, azidolysis of the 2-*O*-benzoyl derivative (**46**) of **45** affords methyl 4-azido-4,6-dideoxy-α-D-mannopyranoside, and it was suggested[84] that the (larger) benzoyloxy group in the quasi-axial position of the idealized conformation depicted for

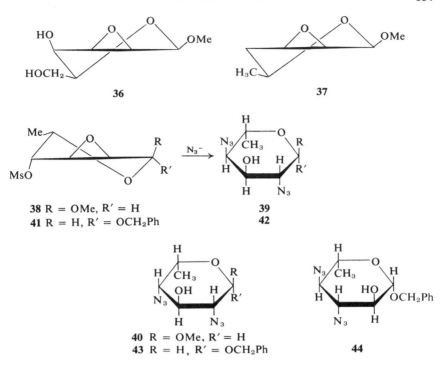

36

37

38 R = OMe, R' = H
41 R = H, R' = OCH₂Ph

N_3^- →

39
42

40 R = OMe, R' = H
43 R = H, R' = OCH₂Ph

44

45 might relatively more favor the general conformation illustrated for **46**, from which the observed major product would be formed by diaxial opening. Methyl 3,4-anhydro-6-deoxy-2-O-(methylsulfonyl)-α-D-galactopyranose (**47**) undergoes favored attack[85] by azide ion at C-4, even though the conformation depicted has all three of its substituents unfavorably disposed.

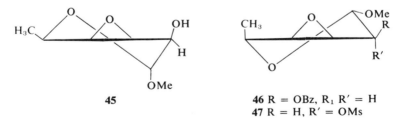

45

46 R = OBz, R₁ R' = H
47 R = H, R' = OMs

Ammonolysis of methyl 3,4-anhydro-6-deoxy-L-*arabino*-hex-5-enopyranosides gives the products of attack at C-4; however, an allylic shift subsequent to azidolysis results in formation of methyl 6-azido-4,6-dideoxy-L-*threo*-hex-4-enopyranosides.[85a]

References start on p. 737.

Methyl 2,3-anhydro-β-L-ribopyranoside underwent reaction with aqueous ammonia at 100° to form methyl 3-amino-3-deoxy-β-L-xylopyranoside[86] in 65% yield; similar addition of methylamine to the enantiomorphic epoxide produced methyl 3-deoxy-3-(methylamino)-β-D-xylopyranoside, which is identical[87] with methyl gentosaminide.

Opening of 2,3-anhydropentofuranoses does not follow a simple generalization, although attack by ammonia at C-3 is most frequently favored. Methyl 2,3-anhydro-α- and -β-D-lyxofuranoside react[88] with aqueous ammonia to form the corresponding methyl 3-amino-3-deoxy-D-arabinofuranosides, and the reaction of 9-(2,3-anhydro-β-D-lyxofuranosyl)adenine with azide ion was later found[89] to afford 9-(3-azido-3-deoxy-β-D-arabinofuranosyl)adenine in 78% yield and the 2-azido-2-deoxy-D-*xylo* isomer in less than 5% yield. Methyl 2,3-anhydro-α-D-ribofuranoside reacts[90] with ammonia to form a 3:2 mixture of methyl 2-amino-2-deoxy-α-D-arabinofuranoside and methyl 3-amino-3-deoxy-α-D-xylofuranoside, whereas the 5-deoxy analogue reacts under similar conditions to form[90,91] methyl 2-amino-2,5-dideoxy-α-D-xylofuranoside and methyl 3-amino-3,5-dideoxy-α-D-arabinofuranoside in approximately equal amounts.

Essentially nonspecific opening occurs with nonterminal, acyclic epoxides. Thus, ammonolysis of 3,4-anhydro-1,2:5,6-di-*O*-isopropylidene-D-altritol produces[92] the 1,2:5,6-diisopropylidene acetals of 3-amino-3-deoxy-D-mannitol (attack at C-3) and 3-amino-3-deoxy-D-iditol (attack at C-4).

The oxetane derivative 3,5-anhydro-1,2-*O*-isopropylidene-β-L-idofuranose is attacked at C-5 by azide ion in boiling *N,N*-dimethylformamide, giving 5-azido-5-deoxy-1,2-*O*-isopropylidene-α-D-glucofuranose in 40% yield; however, heating of the same reactants in aqueous 2-methoxyethanol affords the isomeric 6-azido-6-deoxy-1,2-*O*-isopropylidene-α-D-glucofuranose plus a trace of 3,6-anhydro-1,2-*O*-isopropylidene-α-D-glucofuranose, the presumed intermediate.[93]

2. Direct Replacement of Sulfonic Esters

a. Sulfonic Esters of Primary Alcohols.—Direct conversion of a single, unprotected, primary hydroxyl group into an amine derivative has been accomplished[94] by the combined action of phthalimide, ethyl azodicarboxylate, and triphenylphosphine; thus, 1,2:3,4-di-*O*-isopropylidene-α-D-galactopyranose (**48**) was converted into 6-deoxy-1,2:3,4-di-*O*-isopropylidene-6-phthalimido-α-D-galactopyranose (**49**) in 66% yield. More-traditional methods, however, involve initial conversion of the primary hydroxyl group into a better leaving group, commonly a halide or a sulfonic ester; ammonolysis of the 6-*p*-toluenesulfonate (**50**) of **48** gave[95] the aminodeoxy sugar **51**, and reactions of compound **50** with hydrazine[96] and dimethylamine[97] produced **52** and **53**, respectively, the former reaction being complicated by the formation of bis(6-deoxyhexos-6-yl)hydrazines. Formation of secondary and tertiary amines can compete with ammonolysis reactions; the iodide **54** affords **51**, but only in low yield owing to the extensive formation[97a] of bis- and tris-(6-deoxyhexos-6-yl)amines.

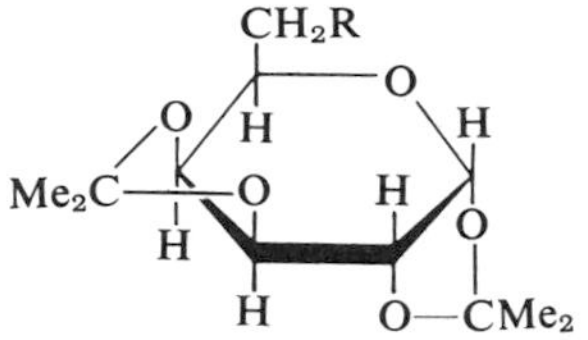

48 R = OH

49 R =

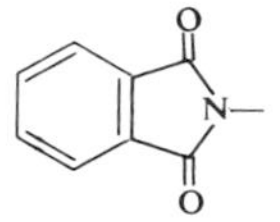

50 R = OTs

51 R = NH$_2$
52 R = NHNH$_2$
53 R = NMe$_2$
54 R = I

E. Fischer and Zach[98] obtained methyl 6-amino-6-deoxy-α-D-glucopyranoside by ammonolysis of methyl 2,3,4-tri-O-acetyl-6-bromo-6-deoxy-α-D-glucopyranoside, and the same product was obtained[99] in similar yield by azidolysis of the corresponding 6-O-p-tolylsulfonyl derivative and subsequent hydrogenation; the latter method was recommended[99] as more certain than direct ammonolysis, although ammonolysis offers the higher yield for that particular example. Azidolysis of methyl 2,3-di-O-acetyl-4-O-benzoyl-6-bromo-6-deoxy-α-D-mannopyranoside and hydrogenation of the deacylated azido sugar afforded[100] methyl 6-amino-6-deoxy-α-D-mannopyranoside.

6-Aminohexose derivatives have also been prepared by reactions of furanoid derivatives. 1,2-O-Isopropylidene-6-O-p-tolylsulfonyl-α-D-glucofuranose[101] and the 1,2:3,5-di-O-isopropylidene[102] and 3,5-O-benzylidene-1,2-O-isopropylidene[103] analogues undergo ammonolysis to form the corresponding 6-amino-6-deoxy-D-glucose derivatives. Reaction of the diisopropylidene sulfonate with [15N]phthalimide ion and subsequent hydrazinolysis of the imide linkage provides[104] an efficient route to 6-[15N]amino-6-deoxy-D-glucose. Although intramolecular participation does not occur in the reaction of the monoisopropylidene acetal, attempted intermolecular displacement reactions failed because of participation in 3-benzamido-3-deoxy-1,2-O-isopropylidene-6-O-p-tolylsulfonyl-α-D-glucofuranose[105] and in 3-acetamido-3-deoxy-1,2-O-isopropylidene-5-O-(methylsulfonyl)-α-D-xylofuranose[106]; however, successful syntheses were realized by introducing azido groups and reducing and acetylating them to the acylamido functions after the displacement step.

1,2-O-Isopropylidene-5-O-p-tolylsulfonyl-α-D-xylofuranose has been converted into 5-amino-5-deoxy-1,2-O-isopropylidene-α-D-xylofuranose by direct ammonolysis[107]; a similar transformation was effected[108] by hydrazinolysis and subsequent hydrogenation.

6-Formamido-9-(2,3-O-isopropylidene-5-O-p-tolysulfonyl-β-D-ribofuranosyl) purine reacts with azide ion in methyl sulfoxide to form, after saponification, 5'-azido-5'-deoxyadenosine, which can be hydrogenated[109] to 5'-amino-5'-deoxyadenosine, or photolyzed in benzene as a route to[110] 9-(2,3-O-isopropylidene-5-*aldehydo*-β-D-*ribo*-pentodialdo-1,4-furanosyl)adenine. 9-(5-O-p-Tolylsulfonyl-β-D-ribofuranosyl)guanine has been converted by direct aminolysis into a number of 5-alkylamino-5-deoxy derivatives, isolated in excellent yields.[111]

Selective reactions of primary sulfonates are possible because of their greater reactivity toward nucleophiles. Thus, 1,2-O-isopropylidene-5,6-di-O-p-tolylsulfonyl-α-D-glucofuranose undergoes selective azidolysis[112] of the 6-sulfonyloxy group in methyl sulfoxide during 45 minutes at 90°. Similarly, brief azidolysis of methyl 2,3,5-tri-O-p-tolylsulfonyl-β-D-ribopyranoside[113] and of 1,4-anhydro-2,3,5-tri-O-p-tolylsulfonyl-L-xylitol[114] in N,N-dimethylformamide results in displacement of the 5-sulfonyloxy group.

b. Sulfonic Esters of Secondary Alcohols.—The replacement of isolated, secondary sulfonyloxy groups proceeds stereospecifically[115] with configurational inversion, as discussed in general in Vol. IA, Chapter 3, Section V, B; however, participation by proximal functional groups may alter the course of such reactions.

Direct displacement of a secondary hydroxyl group has, in principle, been demonstrated[94] in the conversion of compound **55** into **56**; however, it is known[116] that, whereas sulfonyloxy groups adjacent to the anomeric center are somewhat resistant to displacement, methyl 4,6-O-benzylidene-3-deoxy-2-O-p-tolylsulfonyl-α-D-*ribo*-hexopyranoside (but not the *arabino* isomer) undergoes substitution by azide ion, so that the utility of the former reaction remains to be demonstrated. Displacement reactions by halide may be practicable when suitable substrates are available. Azidolysis of methyl 4-O-benzoyl-3-chloro-3,6-dideoxy-β-D-allopyranoside affords[117] methyl 3-azido-4-O-benzoyl-3,6-dideoxy-β-D-glucopyranoside in 88% yield; however, hydrazinolysis of methyl 2-chloro-2-deoxy-D-glucopyranoside resulted[118] in formation of the 1,2-diazole derivative **57**.

(i) Simple displacement reactions of sulfonyloxy groups. 1,2:5,6-Di-O-isopropylidene-3-O-p-tolylsulfonyl-α-D-glucofuranose was ammonolyzed[118] and hydrazinolyzed[119] to the corresponding 3-amino- and 3-hydrazino-D-allose[120] derivatives by Freudenberg and co-workers. This sulfonate is highly resistant to displacement (the corresponding p-bromobenzenesulfonate suffers bromide[121] displacement by dimethylamine), particularly[118] by ionic nucleophiles, and azidolysis can be effected only under forcing conditions in such solvents as N,N-dimethylformamide[122] and hexamethylphosphoric

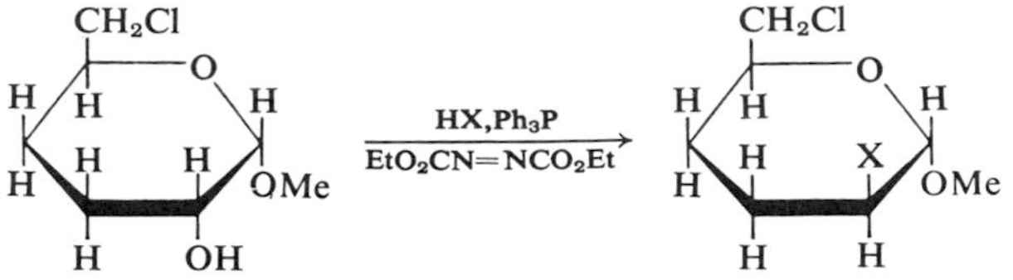

triamide,[123] which solvate cations well but not anions. The epimeric D-*allo* (*endo*) sulfonate, however, reacts more readily[124] with sodium azide to afford 3 - azido - 3 - deoxy - 1,2 : 5,6 - di - O - isopropylidene - α - D - glucofuranose; 3,6-anhydro-1,2-O-isopropylidene-5-O-p-tolylsulfonyl-α-D-glucofuranose[125] and 1,2 : 5,6-di-O-isopropylidene-3-O-p-tolylsulfonyl-β-L-talofuranose,[126] in which the sulfonyloxy groups likewise occupy an *endo* position, react similarly to form the 5-azido- and 3-azido-L-*ido* products, respectively. 3-Sulfonic esters of 1,2-O-isopropylidene-α-D-xylo-[127] and -α-D-ribofuranose[128] readily undergo azidolysis to yield, after hydrogenation, derivatives of 3-amino-3-deoxy-D-ribose and -D-xylose, respectively.

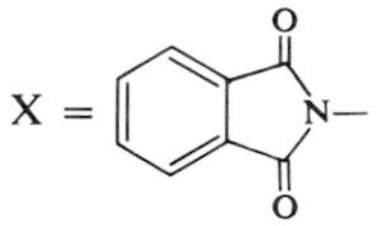

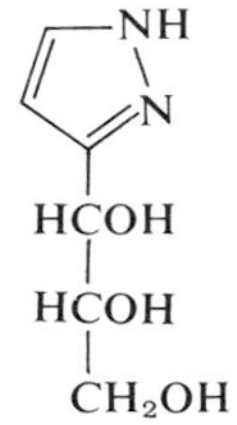

Methyl 5-azido-5-deoxy-2,3-di-O-p-tolylsulfonyl-β-D-ribofuranoside[129] and methyl 5-O-(tetrahydropyran-2-yl)-2,3-di-O-p-tolylsulfonyl-β-D-ribofurano-side[130] undergo selective displacement by azide ion at the more-reactive 3 position, whereas steric effects appear to direct the selective azidolysis[131] at C-4 of 2,5-anhydro-3,4-di-O-p-tolylsulfonyl-D-ribose di-isobutyl dithioacetal. Steric effects evidently also account for the observation[132] that methyl 3,5,6-tri-O-methyl-2-O-p-tolylsulfonyl-β-D-glucofuranoside is converted into the corresponding 2-amino-2-deoxy-D-*manno* derivative by hydrazinolysis and subsequent reduction, whereas the 3,5,6-tri-O-benzyl analogue does not react with hydrazine.

References start on p. 737.

Hydrazinolysis and subsequent reduction have been used to convert methyl 3,4-O-isopropylidene-2-O-p-tolylsulfonyl-β-L-arabinopyranoside   into methyl 2-amino-2-deoxy-3,4-O-isopropylidene-β-L-ribopyranoside[133]   and methyl 6-deoxy-2,3-O-isopropylidene-4-O-(methylsulfonyl)-α-L-mannopyranoside into methyl 4-amino-4,6-dideoxy-2,3-O-isopropylidene-α-L-talopyranoside.[134] Displacements by azide are, however, generally preferred as a means of introducing a nitrogenous substituent, because of the decreased probability of side reactions subsequent to the substitution reaction. By this procedure (and a reduction step), methyl 2,3,6-tri-O-benzoyl-4-O-(methylsulfonyl)-α-D-galactopyranoside was converted[135] into a 4-amino-4-deoxy-D-glucose derivative, methyl 2-azido-4,6-O-benzylidene-2-deoxy-3-O-p-tolylsulfonyl-α-D-altropyranoside into methyl 2,3-diazido-4,6-O-benzylidene-2,3-dideoxy-α-D-mannopyranoside,[136] and ethyl 2,3,6-trideoxy-4-O-p-tolylsulfonyl-α-D-$threo$-hexopyranoside into ethyl 4-amino-2,3,4,6-tetradeoxy-α-D-$erythro$-hexopyranoside.[137] A sequence of azidolysis, deprotection, and hydrogenation has also been used as a means of amination, as in the conversion of 1-[2,3,6-tri-O-benzoyl-4-O-(methylsulfonyl)-β-D-galactopyranosyl]uracil into 1-(4-amino-4-deoxy-β-D-glucopyranosyl)uracil[138] and of methyl 2,4,6-tri-O-benzyl-3-O-p-tolylsulfonyl-α-D-glucopyranoside into 3-amino-3-deoxy-D-allose.[139] The reaction sequence of azidolysis, hydrogenation, and saponification converted[140] methyl 2-benzamido-3,6-di-O-benzoyl-2-deoxy-4-O-(methylsulfonyl)-α-D-glucopyranoside into methyl 2,4-dibenzamido-2,4-dideoxy-α-D-galactopyranoside, the 4-benzamido group arising by O → N migration during the saponification step.

Methyl 2,6-di-O-benzoyl-3,4-di-O-(methylsulfonyl)-α,β-D-galactopyranoside was recovered unchanged after extended treatment with sodium azide in hot hexamethylphosphoric triamide,[141] whereas methyl 2,3,4,6-tetra-O-(methylsulfonyl)-D-glucopyranoside was converted[142] into methyl 4,6-diazido-4,6-dideoxy-2,3-di-O-(methylsulfonyl)-D-galactopyranoside under less-forcing conditions. A similar reaction with methylamine also gave an azetidine.[142a]

In general, nucleophilicity cannot be separated from basicity. Incipient conjugation favors the conversion[143] of methyl [methyl 2,3-di-O-benzoyl-4-O-(methylsulfonyl)-α-D-galactopyranosid]uronate (58) into the α,β-unsaturated ester—namely methyl (methyl 2,3-di-O-benzoyl-4-deoxy-β-L-$threo$-hex-4-

58   59

enopyranosid)uronate (**59**); however, in the azidolysis of methyl 6-deoxy-2,3-
O-isopropylidene-4-*O*-(methylsulfonyl)-α-L-talopyranoside, elimination
of methanesulfonic acid to produce the corresponding 4,5-alkene preponderates[144,155] over the displacement reaction by a factor of three. Participation
reactions may also compete (see Section II,D,2,*b,ii*).

4-*O*-Benzoyl-1,2:5,6-di-*O*-isopropylidene-3-*O*-(methylsulfonyl)-D-mannitol
reacts with azide ion in *N,N*-dimethylformamide to afford[92] the product of
direct displacement (3-azido-4-*O*-benzoyl-3-deoxy-1,2:5,6-di-*O*-isopropyl-
idene-D-altritol) despite the adjacent benzoyloxy group. Similarly, reaction of
5-deoxy-2,3-*O*-isopropylidene-4-*O*-*p*-tolylsulfonyl-D-arabinose diethyl acetal
with azide ion[146] or methylhydrazine[147] yields, after hydrogenation in the
presence of a nickel catalyst, 4-amino-4,5-dideoxy-2,3-*O*-isopropylidene-L-
xylose diethyl acetal and its *N*-methyl analogue, respectively. Azidolysis
of 4,5-di-*O*-benzyl-1,3-*O*-benzylidene-2-*O*-(methylsulfonyl)-6-*O*-trityl-L-galac-
titol and reduction by lithium aluminum hydride afforded 2-amino-4,5-
di-*O*-benzyl-1,3-*O*-benzylidene-2-deoxy-6-*O*-trityl-D-talitol[148] in 50% yield.
Azidolysis of 2,3:5,6-di-*O*-isopropylidene-4-*O*-*p*-tolylsulfonyl-D-glucose
dimethyl acetal primarily effects elimination of the sulfonate to form[149]
4-deoxy-2,3:5,6-di-*O*-isopropylidene-D-*erythro*-hex-3-enose dimethyl acetal.

Azidolysis of methyl 6-deoxy-2,3-*O*-isopropylidene-4-*O*-(methylsulfonyl)-
α-L-mannopyranoside (**60**) has been reported[150] to proceed through participation of the ring-oxygen atom, the major product being methyl 5-azido-5,6-
dideoxy-2,3-*O*-isopropylidene-α-L-talofuranoside (**61**); the same observation[151]
has been made in the D series. Hydrazinolysis of **60**, however, proceeded
primarily by direct displacement[150] to produce, after hydrogenation, methyl
4-amino-4,6-dideoxy-2,3-*O*-isopropylidene-α-L-talopyranoside, together with
smaller proportions of rearranged products. Azidolysis of the 4-deoxy-4-iodo
analogue (**62**) of **60** affords principally **61**, whereas azidolysis of the 4-epimer
(**63**) of **62** produces equal amounts of **64** and **65**, plus a trace of **61**; the
authors[152] postulated that separation of the iodide ion from **62** is sterically

60 R = OMs
62 R = I

61

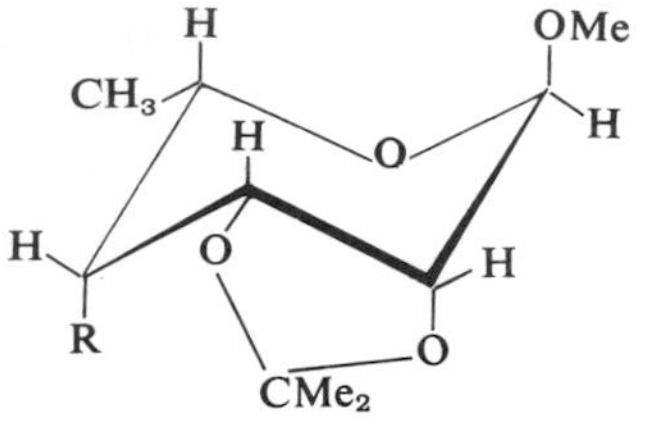

63 R = I
65 R = N$_3$

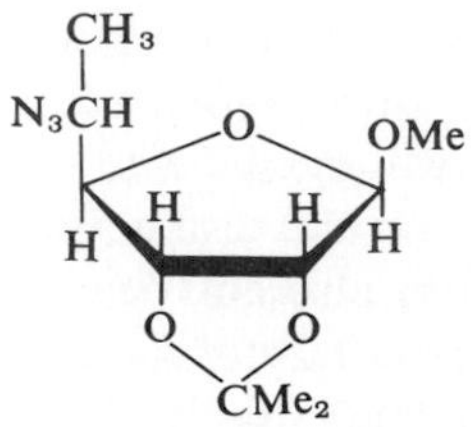

64

accelerated to favor an SN1 process, but that the (less-crowded) 4-epimer reacts by bimolecular displacement, with or without participation.

(*ii*) *Intramolecular displacements by nitrogen functionalities to form five- and six-membered rings.* The presence of a basic nitrogen atom in a molecule that has an accessible sulfonyloxy group commonly results in intramolecular displacement to form an epimine (aziridine). Methyl 2,3-dibenzamido-2,3-dideoxy-6-O-(methylsulfonyl)-β-D-galactopyranoside is partially converted[153] into methyl 2-benzamido-(3,6-benzoylimino)-2,3,6-trideoxy-β-D-galactopyranoside in the presence of sodium acetate or sodium ethoxide. Hydrogenolysis of 6-azido-6-deoxy-1,2-O-isopropylidene-3-O-(methylsulfonyl)[104]- or 3-O-p-tolylsulfonyl-α-D-allofuranose[154] produces, after brief treatment with sodium acetate, 3,6-dideoxy-3,6-imino-1,2-O-isopropylidene-α-D-glucofuranose (**66**); the 1,2-O-cyclohexylidene-5-O-(methylsulfonyl) analogue (**67**) was produced[155] (and not the 5,6-alkene) by the action of zinc and sodium iodide in N,N-dimethylformamide[155a] on 3-azido-1,2-O-cyclohexylidene-3-deoxy-5,6-di-O-(methylsulfonyl)-α-D-glucofuranose.

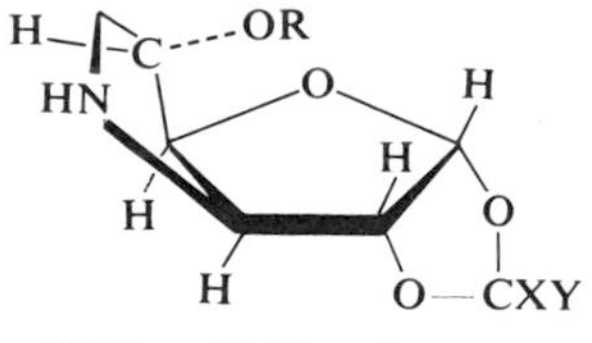

66 R = H, X = Y = Me
67 R = Ms, XY = (CH$_2$)$_5$

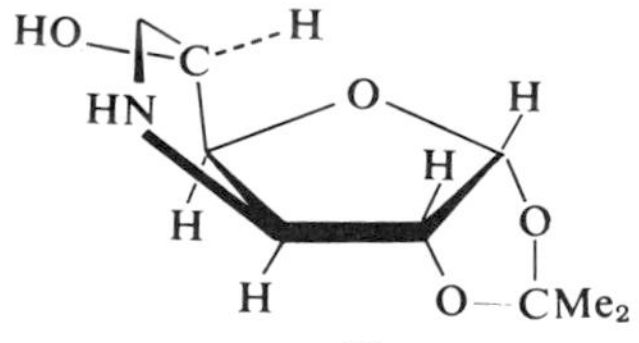

68

Two types of participation occur sequentially in the conversion[156] of 3-acetamido-3-deoxy-1,2-O-isopropylidene-5,6-di-O-(methylsulfonyl)-α-D-glucofuranose (by way of an intermediate 3,5-oxazine intermediate; see Section II,J) into **68**, whose structure has been proved by inversion of configuration[154] at C-5 of **66** and by X-ray crystallography.[157]

Assisted displacement of the sulfonyloxy group occurs during treatment of methyl 3-acetamido-3,6-dideoxy-4-*O*-methyl-2-*O*-(methylsulfonyl)-α-L-gluco-pyranoside with ethoxide ion, whereas both the 4-unsubstituted analogue (**68a**) and its 4-epimer undergo displacement with ring contraction to afford a common product, methyl 2-*C*-(1-acetamido-1-ethoxymethyl)-2,5-dideoxy-α-L-arabinofuranoside[157a] (**68b**).

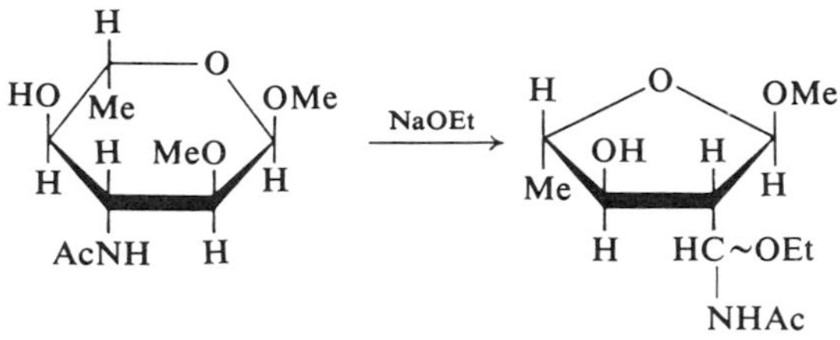

68a   68b

Ammonolysis of 3-*O*-methyl-2,4-*O*-methylene-1,5-di-*O*-*p*-tolylsulfonylxylitol produces only 1,5-dideoxy-1,5-imino-3-*O*-methyl-2,4-*O*-methylenexylitol, whereas displacement by potassium phthalimide affords an acyclic diamine derivative.[158] Spontaneous cyclization also occurs[159] after hydrogenation of 1-azido-2,5-di-*O*-benzyl-4-*O*-*p*-tolylsulfonyl-L-ribitol and 2-azido-2-deoxy-3,4-*O*-isopropylidene-1-*O*-(tetrahydropyran-2-yl)-5-*O*-*p*-tolylsulfonyl-L-arabinitol to produce, after *N*-sulfonylation, the all-*cis*[160] derivatives **69** and **70**, respectively.

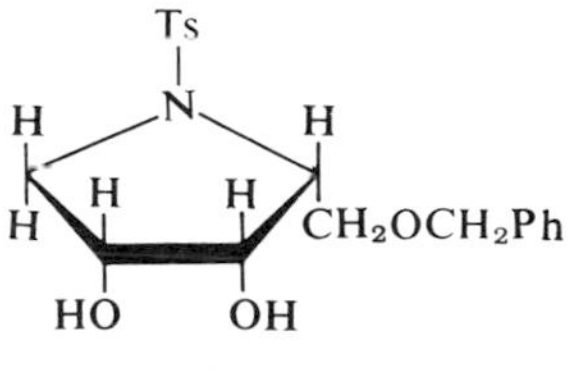

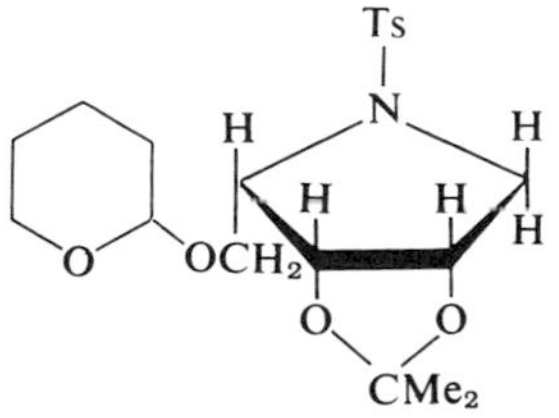

69   70

Careful hydrogenation of methyl 5-azido-5-deoxy-2,3-di-*O*-*p*-tolylsulfonyl-β-D-ribofuranoside produces a 5-amino derivative that reacts[109] with 6-chloropurine to form **71**. 2,5-Anhydro-1-azido-1-deoxy-3,4-di-*O*-*p*-tolylsulfonyl-D-xylitol was hydrogenated, *N*-sulfonylated, and subsequently cyclized by methoxide ion, to afford the bicyclo[2.2.1] product **72** in high yield; the epimer **73** resulted from spontaneous cyclization during reduction of 1,4-anhydro-2-azido-2-deoxy-5-*O*-(methylsulfonyl)-3-*O*-*p*-tolylsulfonyl-L-lyxitol.[110] Platinum-catalyzed hydrogenation of either 2-azido-1,2,3,4-

tetradeoxy-5-O-p-tolylsulfonyl-1,4-p-tolylsulfonylimino-D-threo-pentitol or 1-azido-1,2,3,5-tetradeoxy-4-O-p-tolylsulfonyl-2,5-p-tolylsulfonylimino-L-erythro-pentitol resulted in cyclization to produce, after N-sulfonylation, the common[161] product **74**.

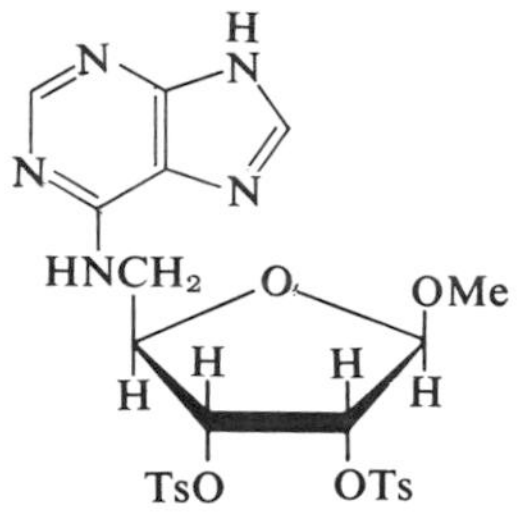

71

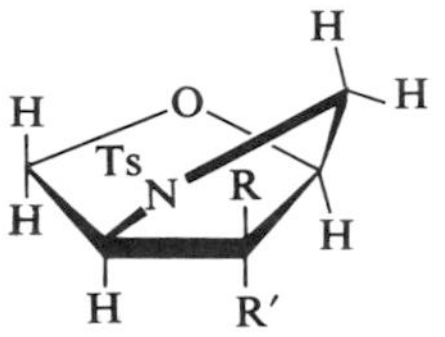

72 R = OTs, R′ = H
73 R = H, R′ = OTs

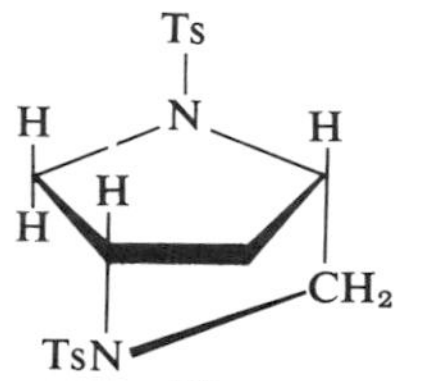

74

Methyl 4-O-benzoyl-6-bromo-2,6-dideoxy-2-p-toluenesulfonamido-α-D-altropyranoside is gradually converted into a 2,6-p-tolylsulfonylimino derivative, whereas the corresponding 3-O-p-tolylsulfonyl derivative reacts more readily, affording only methyl 4-O-benzoyl-6-bromo-2,3,6-trideoxy-2,3-p-tolylsulfonylimino-α-D-mannopyranoside.[162] (Reactions to form aziridines are discussed in Section II,D,2,b,iii.)

Hydrazine is a bidentate nucleophile, and it may participate in several ways, depending on the architecture of the target molecule. Hydrazinolysis of methyl 3,4-O-isopropylidene-2,6-di-O-p-tolylsulfonyl-α-D-galactopyranoside presumably proceeds by initial formation of a 6-hydrazino derivative in which the substituted nitrogen atom displaces the 2-sulfonyloxy group as well; reduction of the α,α-disubstituted hydrazine product gives methyl 2,6-dideoxy-2,6-imino-3,4-O-isopropylidene-α-D-talopyranoside.[163] Vicinal disulfonates that are not trans-related produce 1-aminoaziridines (Section II,D,2,b,iii). Methyl 2,3-di-O-acetyl-4,6-di-O-(methylsulfonyl)-α-D-glucopyranoside reacts with hydrazine to form[78] the α,β-disubstituted hydrazine **75**, which has been catalytically hydrogenated to a 4,6-diamino sugar derivative; the analogous 2,3-di-O-benzyl analogue[164] was similarly prepared, milder reduction being effected by hydrazine in the presence of Raney nickel. 1,2-O-Isopropylidene-3,5-di-O-(methylsulfonyl)-6-O-trityl-α-D-allofuranose reacts with hydrazine in like manner, forming[164] the cyclic hydrazine **76**, and the related compound **77** was prepared by hydrazinolysis[26] of 1,2-O-isopropylidene-3,5-di-O-p-tolylsulfonyl-β-L-arabinofuranose; catalytic hydrogenation of the N–N bond in each example afforded the respective diamino sugars.

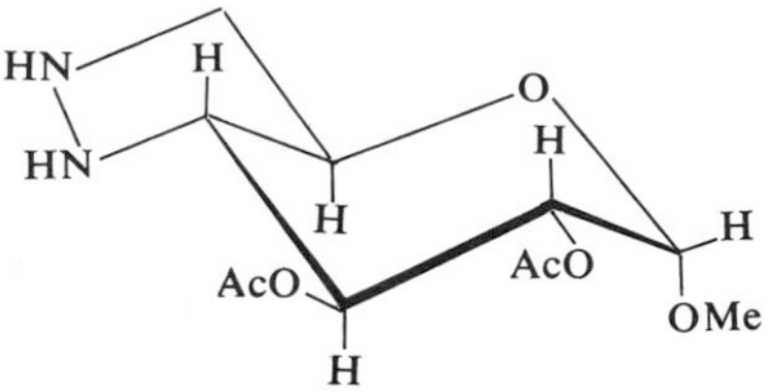

75

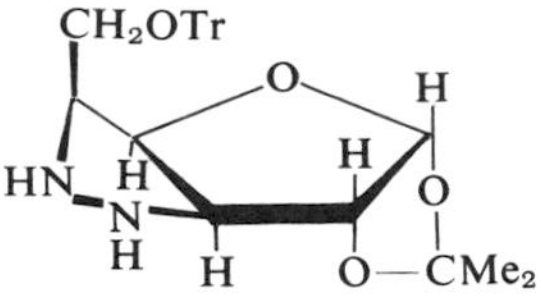

76

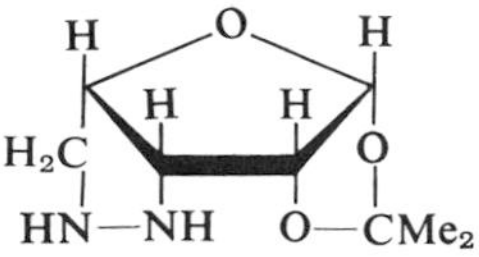

77

Reduction of the nitro group of the branched-chain derivative 1,2-*O*-isopropylidene-3-*C*-nitromethyl-5-*O*-*p*-tolylsulfonyl-α-D-ribofuranose caused cyclization[165] to **78**, thereby defining the stereochemistry of the branch point. Trifluoroacetic acid-catalyzed reaction of 3,4,6-tri-*O*-acetyl-1,5-anhydro-2-deoxy-D-*arabino*-hex-1-enitol (*tri-O-acetyl-*D-*glucal*) with **79** was shown[166] (by X-ray crystallography) to produce **80**. 2-Aminophenyl 3,4,6-tri-*O*-acetyl-2-*O*-(methylsulfonyl)-β-D-glucopyranoside was cyclized in the presence of acetate ion to 3′,4′,6′-tri-*O*-acetyl-β-D-mannopyrano[1′,2′:*b*]-2,3-dihydrobenzo[*e*]-thiazine[167] (**81**); the 6′-*O*-*p*-tolylsulfonyl analogue of **81** underwent further cyclization[168] to afford **82**.

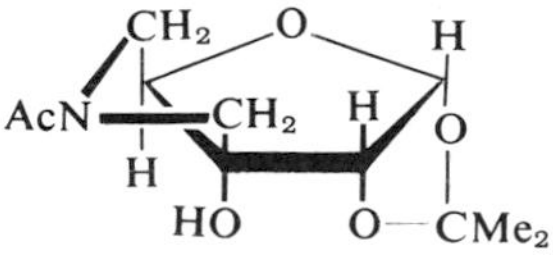

78

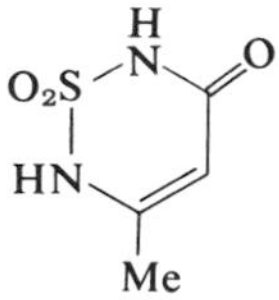

79

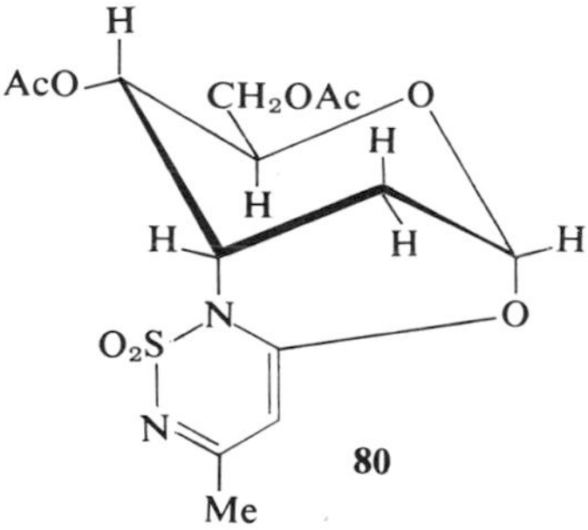

80

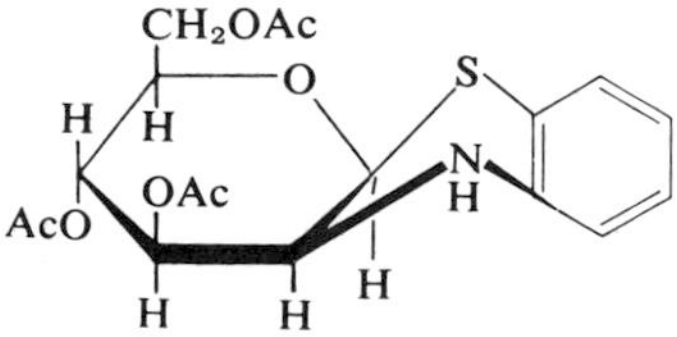

81

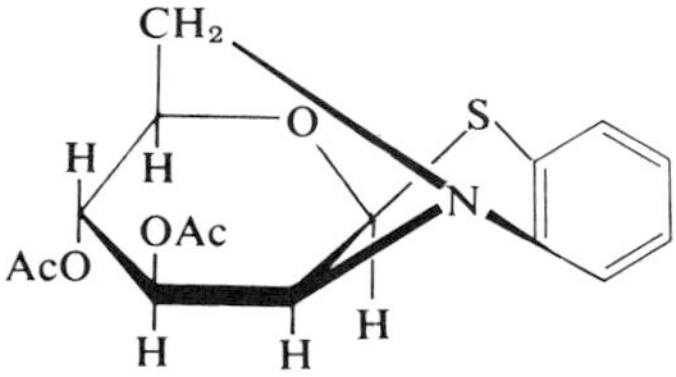

82

References start on p. 737.

A six-membered-ring intermediate may be formed in the rearrangement[169] of **83** to **84** in refluxing toluene.

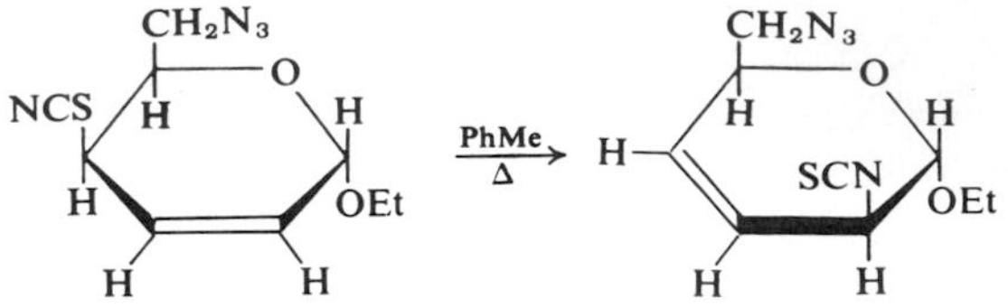

83   **84**

cis-Diamino compounds have on occasion been prepared by participation reactions involving attack by imines. The action of benzylamine on compound **85** afforded[170] **86** via an *N*-benzylguanidine intermediate, and compound **110** (see p. 670) was converted[171] by the action of aniline in refluxing ethanol into the imidaz᷉ ᷉idine **87**; ammonolysis of the *N*-benzyl derivative of **110** yielded[170] the related derivative **88**. In general, however, the product of similar reactions is an aziridine (see Section II,D,2,*b,iii*). Methyl 4,6-*O*-

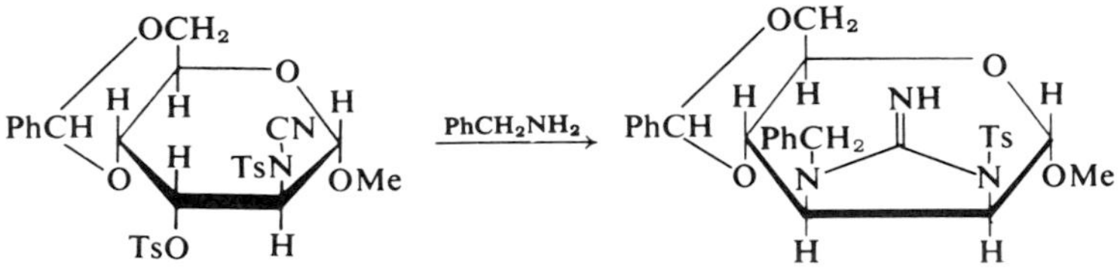

85   **86**

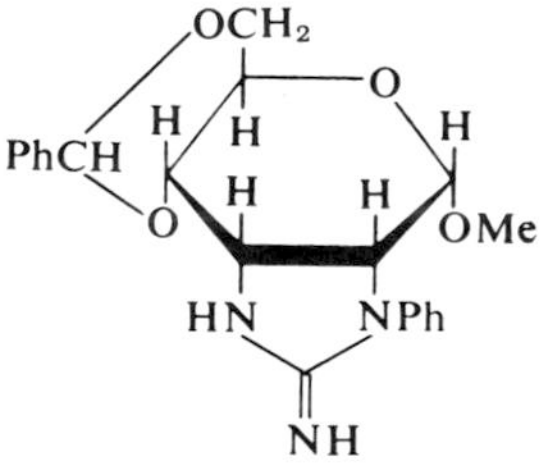

87

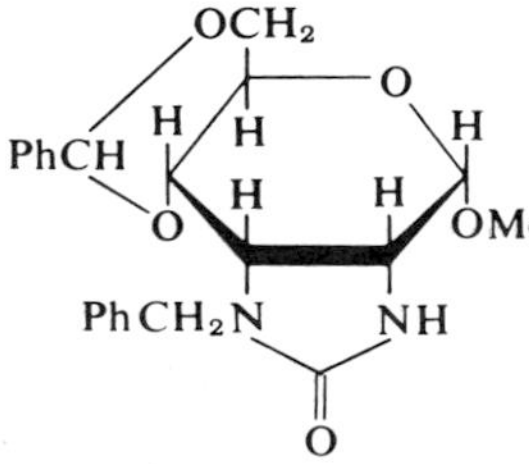

88 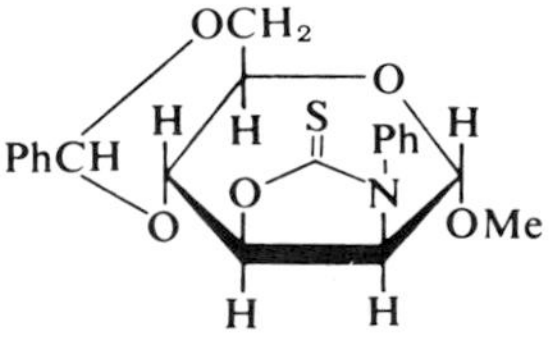   **89**

benzylidene-2-*O*-(methylsulfonyl)-3-*O*-(*N*-phenylthioureido)-α-D-glucopyran-
oside was converted into the oxazolidinethione **89** by treatment with sodium
methoxide at room temperature.[172]

(*iii*) *Formation of aziridine derivatives.* The conformationally predictable
methyl 4,6-*O*-benzylidene-α-D-altropyranoside system has provided the most
intensively studied substrate for examining this type of participation, and a
number of different *N*-substituents were found to react in the presence of
strong bases to afford aziridines. The conversions (**90** → **91**),[173] (**93** → **94**),[170]
(**98** → **99**),[173] (**100** → **101**),[174] (**102** → **103**),[175] (**104** → **105**),[176] and (**106** →
99)[177] were effected by heating in the presence of an alkoxide; the conversions
(**90** → **92**),[173] (**95** → **96**),[178] (**106** → **107**),[177] and (**108** → **101** + **109**)[171]
occurred at room temperature in the presence of alkoxides. Ethanolic ammonia
converted[171] **110** into **111**, whereas refluxing aqueous sodium hydroxide
converted[174] **110** into **99**. Reduction by hydrazine in Raney nickel was
reported[173,179] to accomplish the conversions **97** → **91**, and **112** → **99**,
although further study[180] suggested that the reaction is less simple and
reliable than was originally thought and that conformational factors[181] exert
an effect on the course of the reaction. Transformation of **90** into **92** was also
catalyzed[182] by (the weak base) potassium cyanide, although formation of an
oxazoline occurred as a competing process. (See Section II,J and the definitive
review by Goodman.[183])

The less-favored, cyclization reaction of the *trans*-diequatorial isomers
methyl 4,6-*O*-benzylidene-3-deoxy-3-*p*-toluenesulfonamido-2-*O*-*p*-tolyl-
sulfonyl-α-D-glucopyranoside[178] and methyl 2-benzamido-4,6-*O*-benzylidene-
2-deoxy-3-*O*-(methylsulfonyl)-α-D-glucopyranoside[184] to form compounds **96**
and **99**, respectively, occurs under forcing (strongly basic) conditions, possibly
through conformationally perturbed forms. Methyl 4,6-*O*-benzylidene-
2,3-dideoxy-2,3-imino-α(and *β*)-D-gulopyranosides have been prepared[185]
from the corresponding D-*talo* epoxides by successive azidolysis, sulfonylation,
and reduction by lithium aluminum hydride.

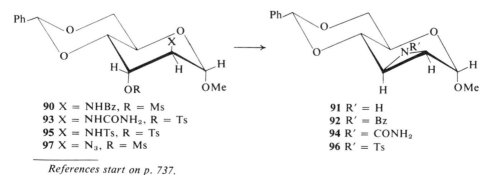

90 X = NHBz, R = Ms	91 R′ = H
93 X = NHCONH₂, R = Ts	92 R′ = Bz
95 X = NHTs, R = Ts	94 R′ = CONH₂
97 X = N₃, R = Ms	96 R′ = Ts

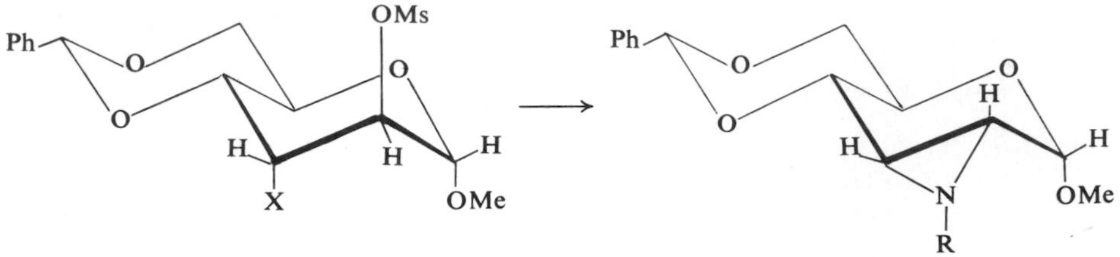

98 X = NHBz
100 X = NHCONH$_2$
102 X = NHCSNH$_2$
104 X = NHC(=S)SMe
106 X = NHC(=NNO$_2$)NH$_2$
108 X = NHC(=NH)NHOH
110 X = NHCN
112 X = N$_3$

99 R = H
101 R = CONH$_2$
103 R = CSNH$_2$
105 R = C(=S)SMe
107 R = C(=NNO$_2$)NH$_2$
109 R = C(=NOH)NH$_2$
111 R = CN

1,6-Anhydro-4-O-benzyl-3-deoxy-2-O-(methylsulfonyl)-3-p-nitrobenzamido-β-D-glucopyranose is converted[69] into 1,6-anhydro-4-O-benzyl-2,3-dideoxy-2,3-imino-β-D-mannopyranose in 90% yield by warming in the presence of isopropoxide ion; lower yields resulted from the use of different amides or sulfonates. Action of lithium aluminum hydride on methyl 3-benzamido-3,6-dideoxy-2,4-di-O-(methylsulfonyl)-α-L-glucopyranoside produced[186] methyl 3,4,6-trideoxy-3,4-imino-α-L-galactopyranoside to the exclusion of the D-*manno* 2,3-imine; the latter was, however, prepared[187] (in the D series) by the action of lithium aluminum hydride on methyl 2-benzamido-2-deoxy-3,4,6-tri-O-(methylsulfonyl)-α-D-glucopyranoside.

A number of 2,3-imino derivatives of 1,4-anhydropentitols and pentofuranosides have been prepared[129–131] by azidolysis and reduction of vicinal, *cis* disulfonates; thus, methyl 5-deoxy-2,3-di-O-p-tolylsulfonyl-β-D-ribofuranoside was converted[130] into methyl 2,3,5-trideoxy-2,3-imino-β-D-lyxofuranoside.

Terminal aziridine derivatives were prepared by Saeki et al.,[188] who reduced 6-azido-3-O-benzyl-6-deoxy-1,2-O-isopropylidene-5-O-p-tolylsulfonyl-α-D-glucofuranose with lithium aluminum hydride and obtained 113, which was also prepared[189] by similar reduction of 3-O-benzyl-1,2-O-isopropylidene-5-O-(phenylsulfonyl)-α-D-glucofuranurononitrile. Failure to protect the 3-hydroxyl group resulted[190] in decreased yields. The aziridine 114 resulted[191] from the action of sodium methoxide upon 2-benzamido-1-chloro-1,2-dideoxy-3,5:4,6-di-O-ethylidene-L-iditol. Hydrazinolysis of vicinal di(methanesulfonates) afforded compounds 115 (ref. 192) and 116 (ref. 193) in a single step.

Reduction of 3-C-cyano-1,2:5,6-di-O-isopropylidene-3-O-p-tolylsulfonyl-α-D-allofuranose by treatment with lithium aluminum hydride or with

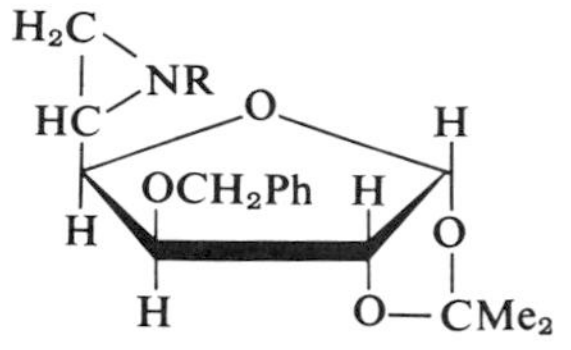

113 R = H
115 R = NH₂

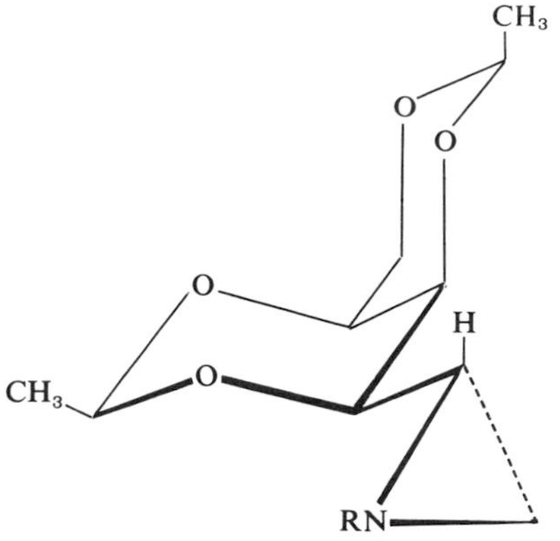

114 R = H
116 R = NH₂

methylmagnesium bromide affords a *spiro*-aziridine and its *C*-dimethyl homologue, respectively, displacement occurring at the tertiary center[193a] with inversion of configuration.[193b]

Hydrogenation in the presence of platinum converted[194] 1,2-*O*-isopropylidene-3,5-di-*O*-(methylsulfonyl)-α-D-glucofuranurononitrile into 6-amino-5,6-dideoxy-1,2-*O*-isopropylidene-3-*O*-(methylsulfonyl)-α-D-*xylo*-hexofuranose; an elimination–hydrogenation mechanism was proposed. The action of hydrazine in the presence of Raney nickel converts vicinal azidosulfonates and *N*-aminoaziridines into aminodideoxy structures, apparently by reductive opening of an intermediate aziridine ring. 6-Azido-6-deoxy-1,2-*O*-isopropylidene-3,5-di-*O*-(methylsulfonyl)-α-D-glucofuranose was thus converted[194,195] into 5-amino-5,6-dideoxy-1,2-*O*-isopropylidene-3-*O*-(methylsulfonyl)-β-L-idofuranose (117), as was the hydrazine[193] 118. Under the same conditions, compound 116 was reduced further, yielding[192] 5,6-dideoxy-1,3:2,4-di-*O*-ethylidene-D-*xylo*-hexitol as the major (68% yield) product, together with the expected 2-amino-2-deoxy-L-iditol derivatives (19%).

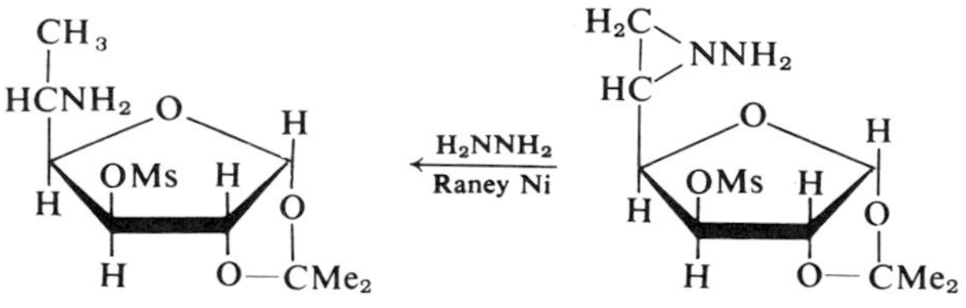

117

118

$\xleftarrow{\text{H}_2\text{NNH}_2}{\text{Raney Ni}}$

E. Reactions of Aziridines with Nucleophiles

N-Acylated aziridines are susceptible to nucleophilic ring-opening in much the same way as epoxides; the unsubstituted aziridines require more vigorous conditions.[195a] Warm acetic acid converted 5,6-acetylimino-3-*O*-benzyl-5,6-dideoxy-1,2-*O*-isopropylidene-α-D-glucofuranose into its 5-acetamido-6-*O*-acetyl[196] analogue, and the action of benzoyl chloride in pyridine and of azide ion in 2-methoxyethanol converted[197] the imine **114** into 2-benzamido-1-chloro-1,2-dideoxy-3,5:4,6-di-*O*-ethylidene-D-iditol, and its 1-azido analogue, respectively, attack occurring at the primary position in all examples.

The configurational direction of this reaction was confused for a while. Methyl 2,3-benzoylimino-α-D-allopyranoside (**119**) reacted with azide ion to afford a crude product that was identified[198] as methyl 3-azido-2-benzamido-4,6-*O*-benzylidene-2,3-dideoxy-α-D-glucopyranoside, the product of diequatorial opening; Guthrie and Williams[198a] subsequently purified this product and found it to contain the 2-azido-3-benzamido (diaxial) product (56% yield) and only a small (18%) amount of the diequatorial product. The α-D-*manno* stereoisomer also underwent attack at C-3 by azide[198] (in the presence of ammonium chloride[198a]) or thiocyanate ion,[199] affording the 3-azido- and 3-thiocyanato-D-*altro* products (diaxial opening). The *N*-methyl analogue (**120**) of **119** reacts with hydrochloric acid in acetone to form[200]

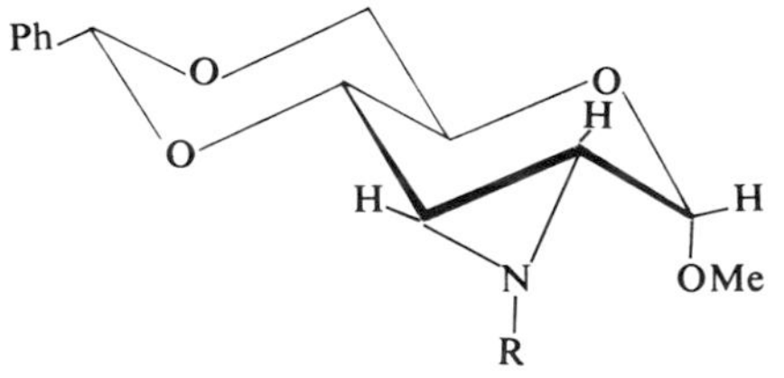

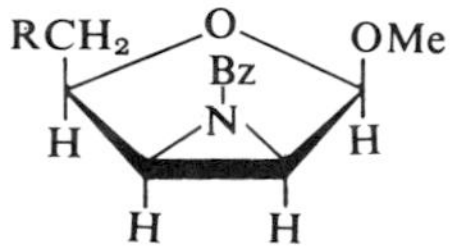

119 R = Bz

120 R = Me

121 R = –O–⟨ ⟩

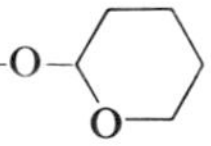

122 R = NHBz

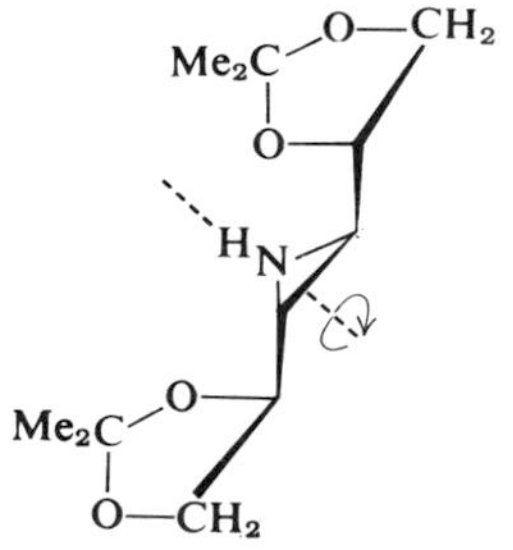

123

methyl 4,6-O-benzylidene-2-chloro-2,3-dideoxy-3-(methylamino)-α-D-altro-pyranoside, the diaxial product. Methyl 2,3-benzoylimino-2,3-dideoxy-5-O-(tetrahydropyran-2-yl)-β-D-lyxofuranoside[130] (121) and its 5-benzamido-5-deoxy analogue[129] (122) react with azide ion principally by attack at C-3, the 2-azido product being formed in about half the yield.

1,6-Anhydro-2,3-dideoxy-2,3-imino-β-D-mannopyranose undergoes base-catalyzed intramolecular opening at C-3 to afford[69] 2-amino-1,6-anhydro-2-deoxy-β-D-mannopyranose via a 1,6:3,4-dianhydro intermediate. The effect of a directed attack is also seen[201] in the reaction of the aziridine 123 with azide ion; because the two electrophilic sites are symmetrically equivalent, the reagent may attack at either site, giving 3-amino-4-azido-3,4-dideoxy-1,2:5,6-di-O-isopropylidene-D-altritol. 2,3-Dideoxy-2,3-imino-4,5:6,7-di-O-isopropylidene-L-*glycero*-L-*galacto*-heptonamide suffers attack at C-3 (the site remote from the deactivating group) by several[202] different reagents, catalytic hydrogenation subsequently affording 2-amino-2,3-dideoxy-4,5:6,7-di-O-isopropylidene-L-*gluco*-heptonamide.

Opening of N-acylaziridines by iodide ion (also a good leaving group) results in a second displacement to form an oxazoline; the position of initial attack parallels the preceding examples, and the (carbonyl) oxygen atom effects the second displacement. Thus, compounds 124 (ref. 189), 125 (ref. 199) and 126 (ref. 203) were formed from the corresponding N-acylaziridine derivatives by the action of sodium iodide; the reaction to give 125 also occurred[199] to a small extent in refluxing N,N-dimethylformamide in the absence of an added nucleophile. In the absence of acidic buffers, azidolysis of methyl 2,3-benzoylimino-4,6-O-benzylidene-2,3-dideoxy-α-D-mannopyran-oside affords mainly[198a] 125.

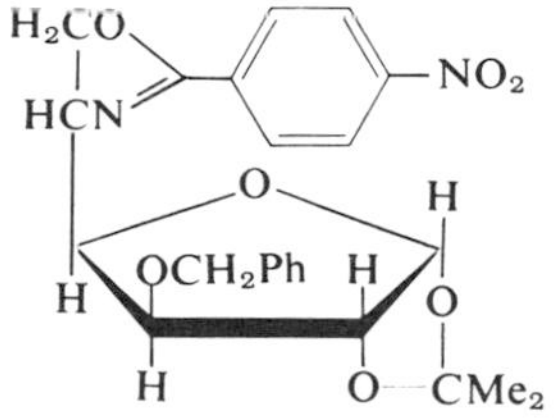

124 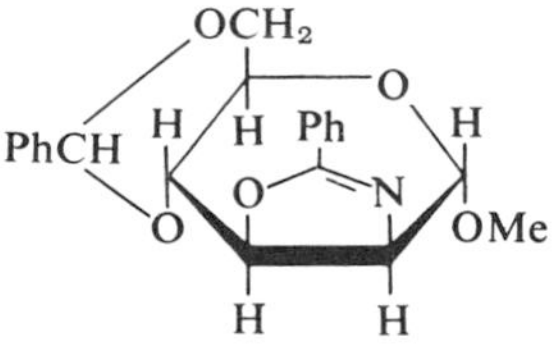   125 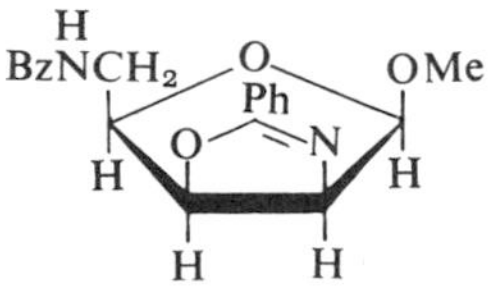   126

F. HYDROGENATION OF CARBON–NITROGEN MULTIPLE BONDS

Reduction of nitriles offers a route to terminal amino sugar derivatives. Catalytic hydrogenation of benzyl 3,4-di-O-acetyl-2-(benzyloxycarbonyl)-amino-2-deoxy-α-D-glucopyranosidurononitrile in acidic methanol affords[204]

References start on p. 737.

2,6-diamino-2,6-dideoxy-D-glucose dihydrochloride (*neosamine C*); the D-*galacto* isomer has also been prepared[205] via such a reduction. Hydrogenation in the presence of Raney nickel converted 5-benzylamino-5-deoxy-1,2-*O*-isopropylidene-β-L-idofuranurononitrile into 6-amino-5-benzylamino-5,6-dideoxy-1,2-*O*-isopropylidene-β-L-idofuranose[206] (however, see Section II,D,2, *e,iii*), and treatment of tri-*O*-benzyl-D-ribofuranosyl bromide with mercuric cyanide and the product with lithium aluminum hydride gave[207] 1-amino-2,5-anhydro-3,4,6-tri-*O*-benzyl-1-deoxy-D-altritol and -D-allitol in 38% and 18% yields, respectively.

Hydrogenation of 1,2-*O*-isopropylidene-α-D-*xylo*-pentodialdo-1,4-furanose 5-phenylhydrazone produced[107,208,209] 5-amino-5-deoxy-1,2-*O*-isopropylidene-α-D-xylofuranose; similar conversion of aldose phenylhydrazones was demonstrated to be a general means[107,208,209] to prepare 1-amino-1-deoxyalditols. D-*arabino*-Hexulose phenylosazone has been converted into 1-amino-1-deoxy-D-fructose by zinc and acetic acid[49] or, better, by palladium-catalyzed hydrogenation.[50] L-*threo*-Hexos-2,3-diulosono-1,4-lactone 2,3-bis(phenylhydrazone) has been converted into the enediamine[210] **127** in 90% yield by catalytic hydrogenation. The amino reductone **128** cyclizes readily to form **129**, and hydrogenation of either form in the presence of platinum oxide affords 2,4-diamino-1,6-anhydro-2,4-dideoxy-β-D-talopyranose.[211] The carbonyl derivative of choice for introduction of amino groups by reduction is the oxime, and the development of mild oxidation procedures (discussed in Chapter 24) has caused this route to become a versatile source of amino sugars. Lindberg and Theander[212] reduced the oxime **131** both catalytically

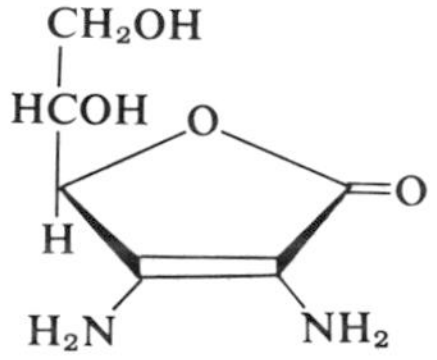

127

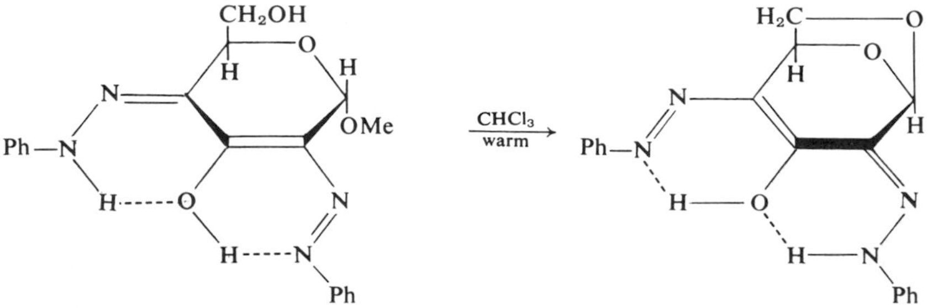

128

129

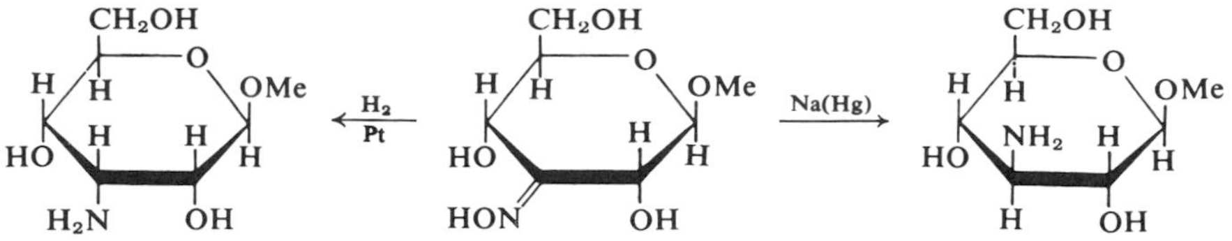

130   **131**   **132**

and by the action of sodium amalgam, obtaining the epimeric products **130** and **132**, respectively. Lithium aluminum hydride reduces methyl 6-deoxy-3,4-*O*-isopropylidene-β-L-*lyxo*-hexopyranoside-2-ulose oxime to methyl 2-amino-2,6-dideoxy-3,4-*O*-isopropylidene-β-L-galactopyranoside plus a trace of the L-*talo* epimer.[213]

Lemieux and co-workers examined the catalytic hydrogenation of D-*arabino*-hexopyranosulose oxime and a number of its derivatives, observing that reduction of the penta-*O*-acetyl derivative affords an optimal yield (near 80%) of the 2-amino product having the D-*manno*[214] stereochemistry, whereas hydrogenation of the isopropyl tetra-*O*-acetyl-α-D-glycoside afforded the corresponding D-*gluco*[215] product in optimal (73%) yield. The rigid, bicyclic oximes **133** and **134** undergo stereospecific, catalytic reduction from the less-hindered side (in accordance with principles discussed in Chapter 22) to afford products having the D-*talo*[216] configuration.

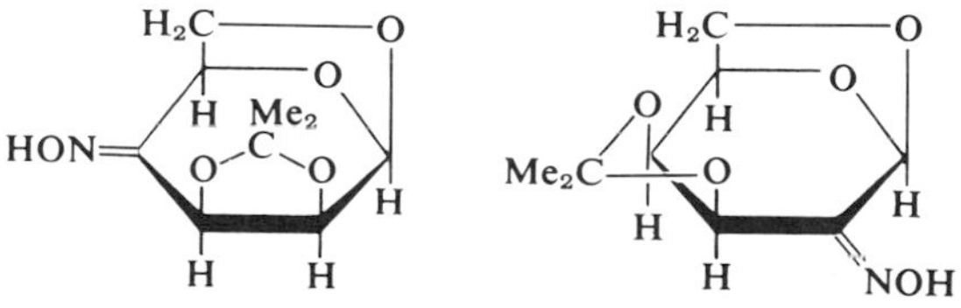

133   **134**

Methyl 2,3-*O*-isopropylidenehexopyranosid-4-ulose oximes of several stereochemical series react with lithium aluminum hydride to generate, in each example, the epimer having the amino group *cis* to the isopropylidene acetal grouping[217]; methyl 2,3,6-trideoxy-α-D-*glycero*-hexopyranosid-4-ulose oxime, however, was hydrogenated to a 3:2 mixture[218] of the 4-amino-D-*threo* and -D-*erythro* products. Reduction of 1,2-*O*-alkylidenealdofuranos-3-ulose oximes by lithium aluminum hydride likewise produced the epimer having the 3-amino group *cis* to the isopropylidene (or cyclohexylidene) acetal grouping in a number[28,30,219] of examples. Thus, 1,2-*O*-isopropylidene-5-*O*-trityl-α-D-*erythro*-pentofuranos-3-ulose oxime (**135**) was converted[220] into the 3-amino-3-deoxy-D-ribose derivative **136**; however, oxidation of the 5-*O*-benzoyl

analogue (**137**) of **135** with a peroxy acid afforded a 2:1 mixture of the
3-nitro-D-*xylo* product **139** and the D-*ribo* isomer[221] (**138**).

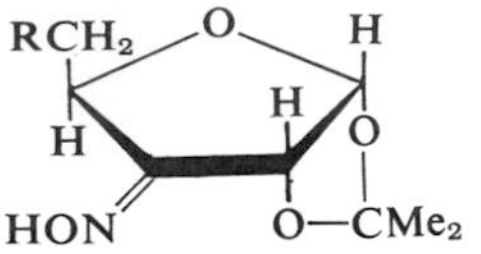

135 R = OCPh$_3$ **136** R = OCPh$_3$, R' = NH$_2$ **139**
137 R = OBz   **138** R = OBz, R' = NO$_2$

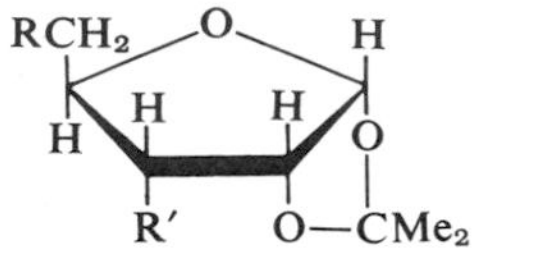

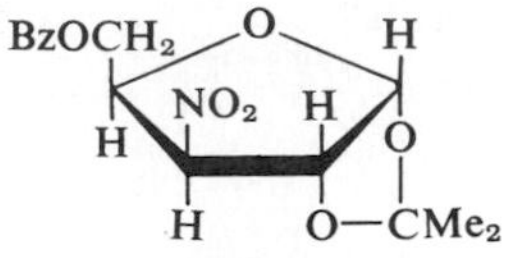

The *E*-oxime of methyl 4,6-*O*-benzylidene-2-deoxy-α-D-*erythro*-hexo-
pyranosid-3-ulose is reported[222] to afford the 3-amino-D-*ribo* product in 98%
yield by reduction with lithium aluminum hydride; catalytic hydrogenation
of this compound or reduction of the *E,Z* mixture with lithium aluminum
hydride produced a mixture of the two 3-epimers. Catalytic reduction of
D-fructose oxime produced[223] 2-amino-2-deoxy-D-mannitol, whereas the
action of lithium aluminum hydride on 8-deoxy-1,2:3,4-di-*O*-isopropylidene-
α-D-*glycero*-D-*galacto*-octopyranos-6-ulose oxime gave, after *N*-acetylation,
a 1:1 mixture[224] of 6-acetamido-6,8-dideoxy-1,2:3,4-di-*O*-isopropylidene-α-
D-*erythro*-D-*galacto*- and -D-*threo*-D-*galacto*-octopyranose.

G. ADDITION OF NITROGEN NUCLEOPHILES TO ALKENES

Nucleophilic addition to carbon–carbon double bonds activated by
conjugation to a polar, unsaturated group provides a general means of introducing nitrogen substitution; for nitro compounds, subsequent reduction to
the corresponding amine, or the Nef reaction (see Vol. IA, Chapter 3,
Section III,B), is commonly used to generate a new amino group or a carbonyl
group, respectively.

Acylated, vicinal nitroalcohols readily undergo base-catalyzed elimination
to form nitroalkenes. 3,4,5,6-Tetra-*O*-acetyl-1,2-dideoxy-1-nitro-D-*arabino*-
hex-1-enitol was found[225] to add ammonia, forming principally 2-acetamido-
1,2-dideoxy-D-mannitol, from which 2-amino-2-deoxy-D-mannose was
prepared. This sequence has been shown to be a useful, general[23,226] route to
2-aminoaldoses having the 2,3-*erythro* configuration, the *threo* epimer being
formed in minor amount for many examples. Addition of ammonia (and
amines) to methyl 3-deoxy-3-nitroaldopyranosides, is, apparently, thermodynamically controlled; thus, methyl 2,4,6-tri-*O*-acetyl-3-deoxy-3-nitro-β-D-
glucopyranoside reacts with methanolic ammonia by sequential elimination
and addition to give, after reduction and acetylation, methyl 2,3,4-triacet-
amido-6-*O*-acetyl-2,3,4-trideoxy-β-D-glucopyranoside[227]; the 4,6-*O*-benzyl-
idene analogue **140** reacts (via the intermediate alkene **141**) with ammonium

hydroxide in tetrahydrofuran to afford the 2-substituted, D-*gluco* derivative[228] **142** as the preponderant product, whereas the 6-deoxy-L analogue gave a mixture of 2-epimeric triamines, the composition varying with the reaction conditions.[229] Addition of **141** to **142** in hot benzene produced[230] the secondary amine 2,2′-iminobis(methyl 4,6-*O*-benzylidene-2,3-dideoxy-3-nitro-α-D-glucopyranoside).

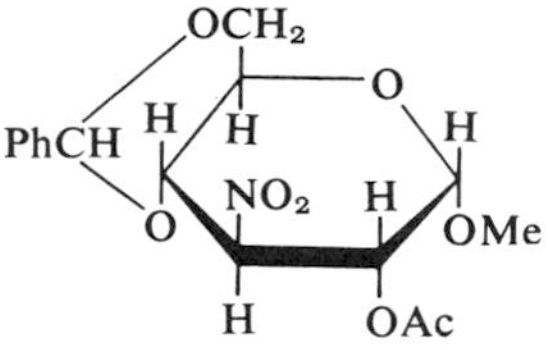

140 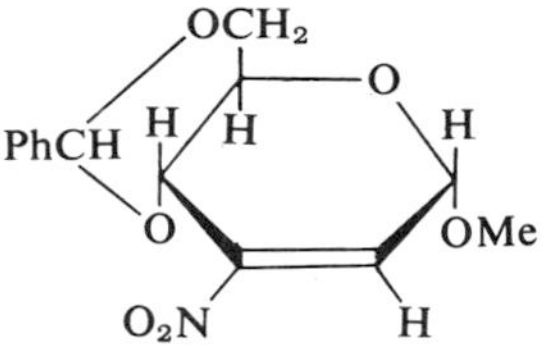   **141** 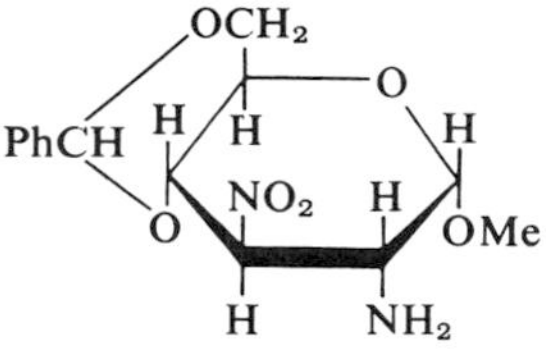   **142**

Addition of *N*-bromoacetamide to **141** gave **143**, which was reduced, the product deprotected, and the product acetylated to afford methyl 2,3-diacetamido-4,6-di-*O*-acetyl-2,3-dideoxy-α-D-mannopyranoside; treatment of the other three stereoisomers in the D series with this reagent yielded, for each example, the stereoisomer having the nitro group equatorially disposed and the 2-acetamido group *trans* to the aglycon.[231] The same reagent converted both the primary alkene 3-*O*-acetyl-5,6-dideoxy-1,2-*O*-isopropylidene-6-nitro-α-D-*xylo*-hex-5-enofuranose (**144**) and 5-acetamido-3-*O*-acetyl-5,6-dideoxy-1,2-*O*-isopropylidene-6-nitro-α-D-glucofuranose (**145**) into the dibromide[232] **146**.

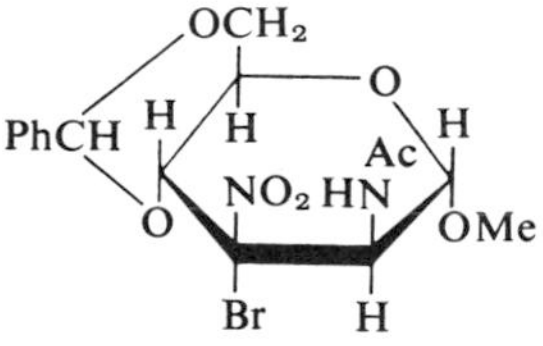

143 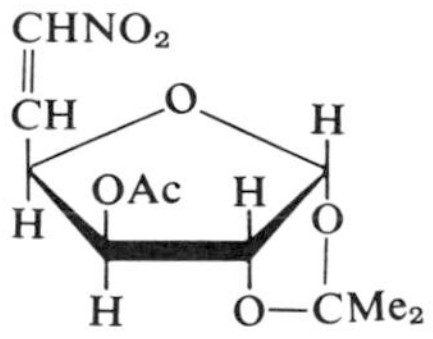   **144** 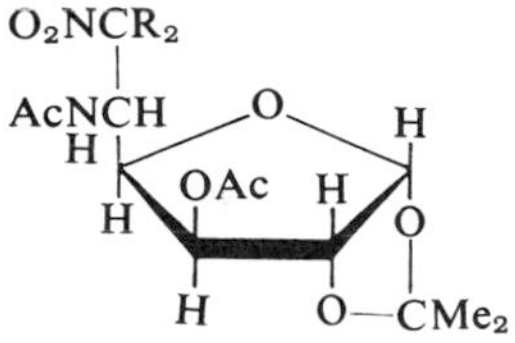   **145** R = H
                                              146 R = Br

Addition of ammonia to the Wittig product **147** gave a mixture[233] of the epimeric 3-acetamido-2,3-dideoxy-4,5:6,7-di-*O*-isopropylidene-L-*manno*- and -L-*gluco*-heptonamides, the former (*erythro* adduct) preponderating; similar reaction of the 2-bromo derivative **148** proceeded further by cyclization of the L-*manno* adduct to afford the 2,3-*trans* aziridine, namely, ethyl 2,3-dideoxy-2,3-imino-4,5:6,7-di-*O*-isopropylidene-L-*glycero*-L-*galacto*-heptonate,[234] in 90% yield.

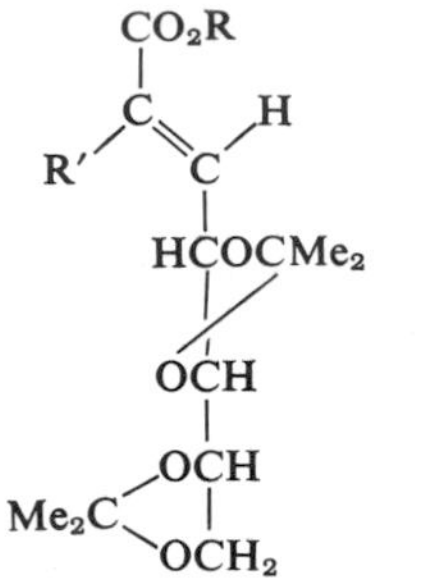

147 R = Me, R' = H
148 R = Et, R' = Br

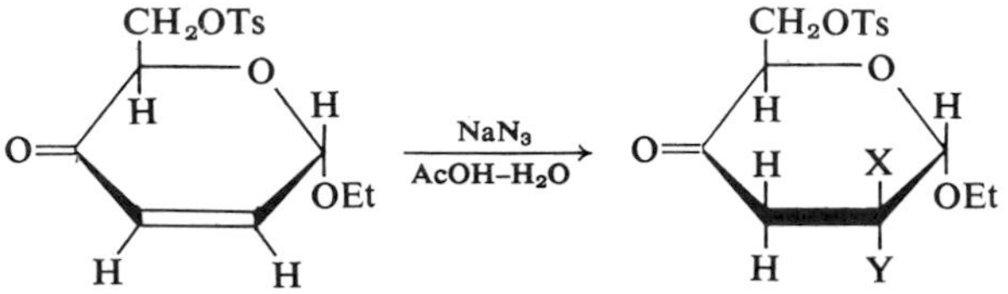

149

150 X = N_3, Y = H
151 X = H, Y = N_3

The enone **149** reacted with azide ion to form an initial adduct **150** (isolated after 5 minutes) that was gradually converted (5 hours) into the (presumably) thermodynamic product[235] **151**.

Mercury(II) acetate-catalyzed addition of sodium azide to 3-deoxy-1,2:5,6-di-O-isopropylidene-3-C-methylene-α-D-*ribo*-hexofuranose afforded a 3-azido-3-deoxy-3-C-methylhexose; D-*gluco* stereochemistry was demonstrated by reduction of the azide functionality, graded hydrolysis of the 5,6-acetal, and methanesulfonylation of O-5 and O-6, the product containing a 3,6-imino grouping that indicates[235a] *cis* disposition of C-5 and N-3 (compare Section II,D,2,*b*,*ii*).

3,4,6-Tri-O-acetyl-1,5-anhydro-2-deoxy-D-*arabino*-hex-1-enitol (*tri-O-acetyl*-D-*glucal*, **152**) reacts with nitrosyl chloride to form[236] (dimeric) 3,4,6-tri-O-acetyl-2-deoxy-2-nitroso-α-D-glucopyranosyl chloride (**153**), which reacts with alcohols[236,237] to give 2-oximino glycosides, and with silver acetate, followed by reduction by an active metal, to give[235] 1,3,4,6-tetra-O-acetyl-2-amino-2-deoxy-β-D-glucopyranose. The nitrosyl chloride adduct of 1,5-anhydro-2,6-dideoxy-3,4-di-O-methyl-L-*arabino*-hex-1-enitol underwent reduction by zinc–copper mixtures in acetic acid to afford[238] a mixture of 2-acetamido-2,6-dideoxy-4-O-methyl- and -3,4-di-O-methyl-L-glucopyranose, the former (major) product presumably arising by an elimination–addition side process. Addition of nitrosyl chloride to 3,4-di-O-acetyl-1,5-anhydro-2,6-dideoxy-L-*lyxo*-hex-1-enitol (*diacetyl-L-fucal*) affords, after glycosidation and reduction, a 7:3 mixture of L-*galacto* and L-*talo* products.[238a] Chromous chloride catalyzes the addition of *N*-(chloroethyl)carbamate to **152** to give[239] a mixture of four stereoisomeric products in which compound **154** preponderates. By virtue of the reluctance of the 3-substituent to act as a leaving group (see Chapter 19), the 3-acetamidoglycal **155** undergoes acid-catalyzed 1,2-addition of methanol (directed by participation of the acetamido group) to afford the 2-deoxy-α-glycoside[240] **156**.

The glycal **156a** was isolated[240a] from the reaction of ethyl 4,6-diacetamido-2,3-dideoxy-D-*erythro*-hex-2-enopyranoside with chlorosulfonyl isocyanate in ether.

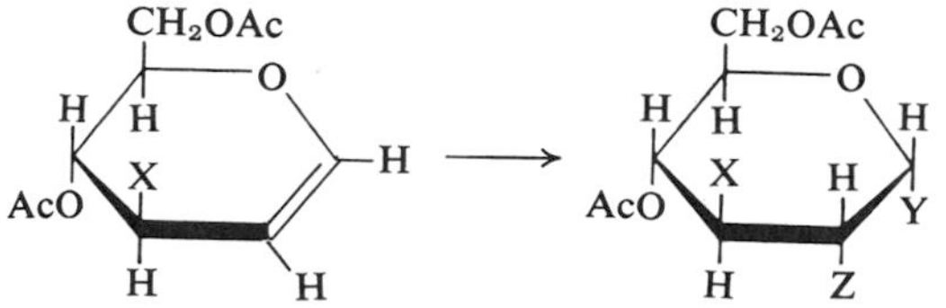

152 X = OAc
155 X = NHAc
156a X = NHCO$_2$Et

153 X = OAc, Y = Cl, Z = —N=O)$_2$
154 X = OAc, Y = Cl, Z = —NHCO$_2$Et
156 X = NHAc, Y = OMe, Z = H

H. NITROMETHANE CONDENSATIONS AND RELATED REACTIONS

The reaction sequence: aldose → *C*-nitroalditol → higher aldose by use of nitromethane is discussed in Vol. IA, Chapter 3,III,B; however, the nitro sugar may also be reduced to the amino sugar analogue or be aminated further (Section II,G). Thus, methyl 2,3,4-tri-*O*-acetyl-6-deoxy-6-nitro-1-thio-α-D-galactopyranoside was condensed with acetaldehyde in the presence of sodium, and the product treated with lithium aluminum hydride to give a mixture of four diastereoisomeric methyl 6-amino-6,8-dideoxy-1-thio-octopyranosides, of which the D-*erythro*-D-*galacto* isomer[241] was identical with methyl 1-thiolincosaminide.

Baer and H. O. L. Fischer[242] treated **157,** the "dialdehyde" produced by periodate oxidation of methyl α-D-glucopyranoside, with nitromethane in the presence of sodium methoxide, and obtained a mixture of methyl 3-deoxy-3-nitro-α-D-mannopyranoside (**159**) and the D-*gluco* and D-*talo* stereoisomers as major, minor, and trace constituents, respectively; some of the D-*galacto* isomer was produced[36] by extended exposure of the initial mixture to aqueous base. This stereoselectivity conforms[3] to Cram's rule, wherein the favored product corresponds to that resulting from selective attack by the nitroalkylate anions on the carbonyl groups from the less-hindered side to give, as the major intermediate in the basic medium, the *aci*-nitro salt **158**, in which the oxygen atoms at the newly created asymmetric centers are *trans* to the adjacent substitutent; the 2-epimer and, less, the 4-epimer of **158** are also formed initially, and the aqueous base catalyzes formation of the fourth stereoisomer. Neutralization results in exclusive formation of the equatorial nitro group, which is present in compound **159** and the three other isomers. The β-D anomer of **157** reacted similarly[35,243] to afford methyl 3-deoxy-3-nitro-β-D-glucopyranoside (1,2-*trans*) as the major product.

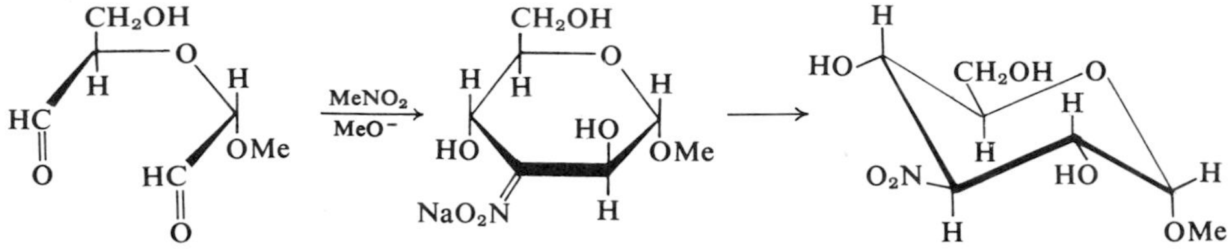

157   **158**   **159**

Periodate oxidation, subsequent condensation with nitromethane, and hydrogenation in the presence of Raney nickel converts methyl 6-deoxy-α-D-glucopyranoside into a mixture[243a] of 3-amino-3,6-dideoxy-D-hexoses having the *gluco, galacto, manno, talo,* and *ido* stereochemistry in the ratio 25:6.8:6.1:1:1.5. The same sequence has been applied to dextran.[243b]

The "dialdehyde" **160** (prepared by periodate oxidation of 1,6-anhydro-β-D-glucopyranose) was condensed with nitromethane to afford, after reduction, 3-amino-1,6-anhydro-3-deoxy-β-D-altropyranose (**161**) and the D-*gulo* isomer as major products, together with a smaller proportion of the all-equatorial D-*ido* compound.[244] As these anhydrides are constrained to exist in the

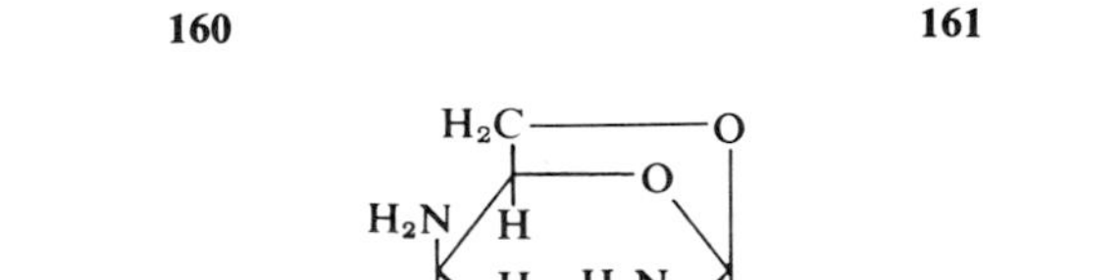

160   **161**

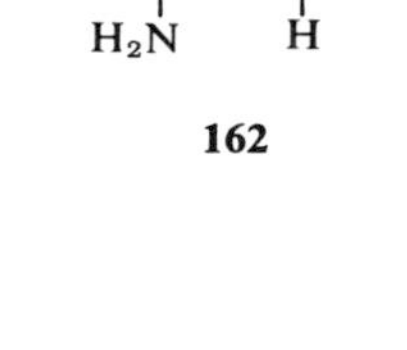

162

$^1C_4(D)$ conformation, the nitro group is equatorial in all the products. Similar cyclization of **160** in an excess of benzylamine afforded 1,6-anhydro-2,4-dibenzamido-2,3,4-trideoxy-3-nitro-β-D-idopyranose[245] as the major product, whence the D-*ido* triamine **162** was prepared by catalytic hydrogenation.

Periodate oxidation of any methyl β-D(or α-L)-pentopyranoside produces the "dialdehyde" **163**, which condenses with nitromethane to afford mainly[246] the β-D-*erythro* nitronate **164**; acidification gives mainly methyl 3-deoxy-3-nitro-β-D-ribopyranoside plus a small proportion of the D-*xylo* isomer. Periodate oxidation and nitromethane cyclization converted methyl 4,6-*O*-benzylidine-α-D-glucopyranoside into a mixture[247] of four stereoisomeric methyl 5,7-*O*-benzylidene-3-deoxy-3-nitro-β-D-heptoseptanosides (**165**); the 4,6-*O*-ethylidene analogue was earlier[248] reported to afford the D-*glycero*-D-*manno* septanoside in 41% yield.

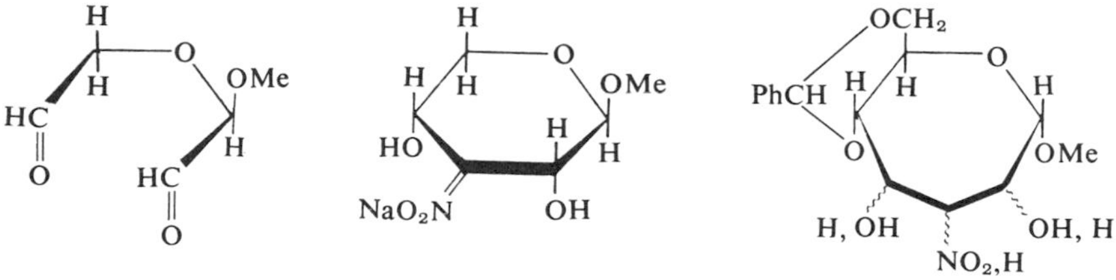

163   164   165

The identity of the cyclization product is affected only by features still remaining in the dialdehyde after cyclization. Thus, 1-(β-D-ribofuranosyl)-uracil[249] and 1-(β-D-glucopyranosyl)thymine[250] are converted by sequential reaction with periodate, nitromethane, and hydrogen (catalytic) into the respective 3-amino-3-deoxy-β-D-glucopyranosyl nucleosides. Access to a nitrofuranose was gained[251] through periodate oxidation of methyl 6-deoxy-6-nitro-α-D-glucopyranoside, base-catalyzed cyclization of the resultant dialdehyde **166**, and borohydride reduction, which led to an 8:1 mixture of methyl 3-deoxy-3-nitro-β-L-ribofuranoside and its L-*arabino* epimer, whence the 3-amino-3-deoxypentosides were obtained by catalytic hydrogenation. Base-catalyzed cyclization of a 6-deoxy-6-nitroaldose has been employed as a synthetic route to deoxynitroinositols.[251a] Baer[252] has presented a detailed discussion of the chemistry of nitro sugars.

Cyclization of periodate-oxidized methyl 4,6-*O*-benzylidene-α-D-glucopyranoside to methyl 4,6-*O*-benzylidene-3-deoxy-3-phenylazo-α-D-glucopyranoside was accomplished by the action of aqueous phenylhydrazine[253];

subsequent reduction afforded the 3-amino-3-deoxy-D-glucose derivative[254] **167.** Use of *N,N*-dimethylformamide as the solvent for the phenylhydrazonation led to the isolation[255] of the 2,3-diamino-D-*manno* product[255] **167a,** whereas use of methylhydrazine in the same solvent ultimately afforded[256] compound **167.**

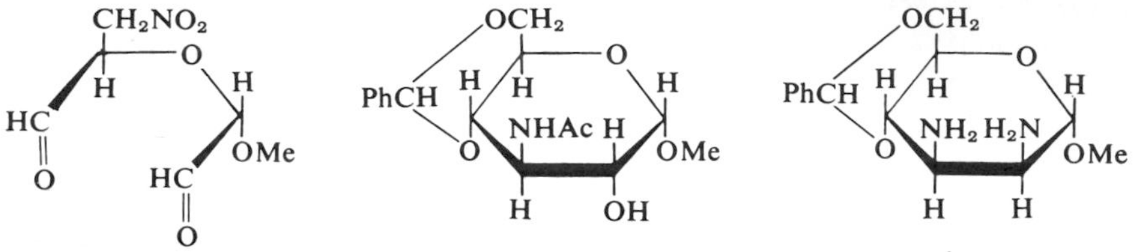

166   167   167a

I. EPIMERIZATION OF ACETAMIDO SUGARS

2-Acetamido-2-deoxyaldoses readily epimerize in dilute, aqueous alkali, the acylamido grouping serving the dual function of promoting detachment of H-2 and of prohibiting ketonization. 2-Acetamido-2-deoxy-D-glucose equilibrates[257] in about two days at pH 11, to a 4:1 mixture with the D-*manno* epimer from which the latter has been isolated[258] preparatively; nickel(II) carbonate is reported[259] to displace the equilibrium to 3:1. Epimerization of 2-amino-2-deoxy-D-gluconic acid in aqueous pyridine offers an alternative[7,9] route. The corresponding equilibrium between 2-acetamido-2-deoxy-D-ribose and the D-*arabino* epimer favors the latter[260] by about 2:1, and 2-acetamido-2-deoxy-L-xylose has been prepared[261] by epimerization of the L-*lyxo* isomer. *O*-Benzylation of 2-acetamido-2-deoxy-D-mannose afforded only[262] benzyl 2-acetamido-3,4,6-tri-*O*-benzyl-2-deoxy-β-glucopyranoside. Action of deuteroxide ion forms [2-²H] analogues from 2-acetamido-2-deoxy sugars.[262a]

Oxidation of 2-acetamido-2-deoxy-D-glucose by bromine–water resulted in an epimeric mixture[263] of D-*gluco* and D-*manno* lactones. Treatment of 2-acetamido-2-deoxy-D-mannono-1,4-lactone with aqueous cyclohexylamine produced a mixture of the cyclohexylammonium 2-amino-2-deoxy-D-hexonates having D-*gluco* and D-*manno* configurations, whereas the 2-acetamido-2-deoxy-D-mannonate ion was stable under the same conditions.[264]

Oxidation of methyl 2-azido-4,6-*O*-benzylidene-2-deoxy-α-D-altropyranoside and of the 3-azido-3-deoxy analogue with methyl sulfoxide–*N,N*-dicyclohexylcarbodiimide–pyridinium phosphate resulted in inversion of configuration[265] of the carbon atom bearing the azido group. In the presence of sodium hydrogencarbonate in acetone, 3-deoxy-1,2:5,6-di-*O*-isopropylidene-3-nitro-α-D-glucofuranose slowly equilibrates[266] to an epimeric mixture containing 55% of the 3-nitro-D-*allo* isomer and 45% of the starting material; nitropyranosides are, however, stable under these conditions.

J. CONFIGURATIONAL INTERCONVERSION OF HYDROXYL GROUPS

Because (a) the number of starting materials available is finite, and (b) most of the synthetic conversions discussed in the preceding sections afford only one of two (or more) isomeric products that could, at least in principle, be thus prepared, methods for specific, stereochemical inversion of hydroxyl groups in an epimer (or another disastereoisomer) constitute the synthesis of choice for some amino sugars that might otherwise be virtually inaccessible. Such routes also present a means for determining the configuration of unidentified compounds by predictable conversion into analogues of known stereochemistry.

Direct interconversions are known. Chromic acid–pyridine oxidation[267] of 6-acetamido-6,8-dideoxy-1,2:3,4-di-O-isopropylidene-β-L-threo-D-galacto-octopyranose, and borohydride reduction[267,268] of the resultant glycos-7-ulose, afforded a 7:3 mixture of the 6-epimer and the starting material, and oxidation of the hydroxyl group followed by reduction with borohydride converted methyl 3-acetamido-4,6-O-benzylidene-3-deoxy-α-D-manno-pyranoside into the D-gluco[269] epimer.

A net inversion at C-5 was accomplished by treatment of methyl 3,4-di-O-acetyl-6-bromo-2,6-dideoxy-2-(methoxycarbonyl)amino-α-D-glucopyranoside with silver fluoride to form the 5,6-alkene, hydrogenation of which produced methyl 3,4-di-O-acetyl-2,6-dideoxy-2-(methoxycarbonyl)amino-β-L-idopyran-oside[270] in nearly 60% yield and about half as much of the α-D-gluco isomer. Nucleophilic displacement of the sulfonyloxy group of methyl 3-azido-2,3,6-trideoxy-4-O-(methylsulfonyl)-α-L-arabino-hexopyranoside and subsequent hydrogenation and deprotection afforded the first constitutional synthesis of 3-amino-2,3,6-trideoxy-L-lyxo-hexose[271] (daunosamine).

Intramolecular reactions are generally much favored over their inter-molecular counterparts, and assisted displacements of sulfonates occur considerably more readily than related, bimolecular reactions (see also Section II,D,2,b,ii and iii, Chapter 18, Section III,A,1,b, and the perspicacious review[183] by Goodman).

The utility of epoxide amination reactions (Section II,D,1) was greatly enhanced when Baker and Schaub[86] extended the concepts of participation by neighboring acyloxy groups to the conversion of methyl 3-acetamido-3-deoxy-2,4-di-O-(methylsulfonyl)-β-L-xylopyranoside (168, from ammonolysis of methyl 2,3-anhydro-β-L-ribopyranoside) into methyl 3-acetamido-3-deoxy-D-ribopyranoside (173) by the action of (the necessarily weak base) sodium acetate in 2-methoxyethanol at 120°. The reaction probably proceeded by assisted displacement of the 4-sulfonyloxy group (the 2-substituent was presumed by the authors to react first), forming an oxazolinium ion (**169**) that was subsequently hydrated to **170** and opened to liberate compound **171** having a configurationally inverted alcoholic group; a similar sequence of steps effected displacement of the remaining sulfonate group via the oxazolinium ion **172**, affording **173** as the final product, whereas a similar reaction in ethanol afforded a monosulfonate (**171**), an enantiomorphic analogue of which was identified,[272] by conversion into 2,3-diamino-2,3-dideoxy-L-ribose. 3-Amino-3-deoxy-D-ribose (**176**) was prepared by deprotection of **173**; an alternative synthesis[88] proceeded by ammonolysis of methyl

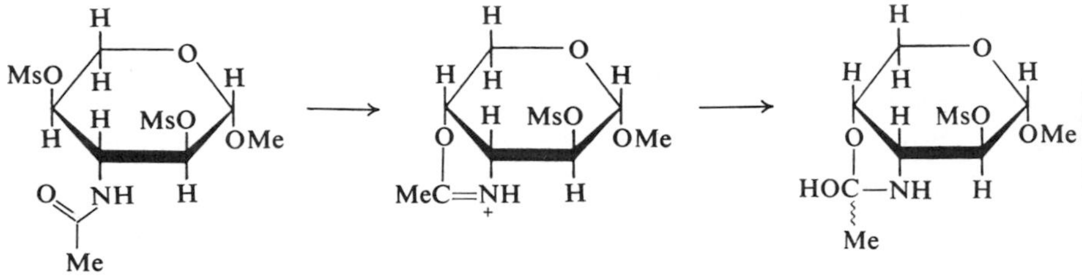

168   169   170

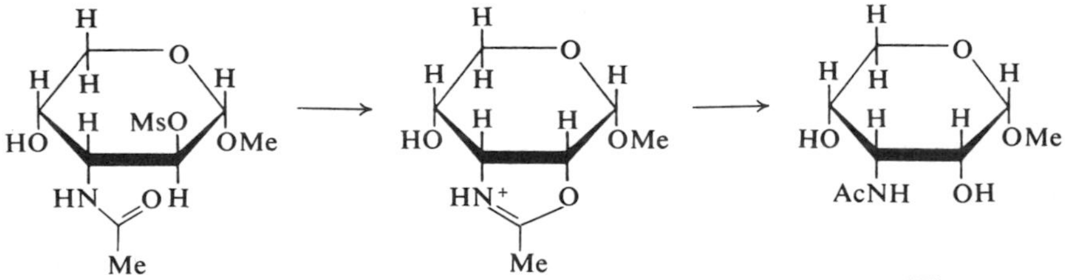

171   172   173

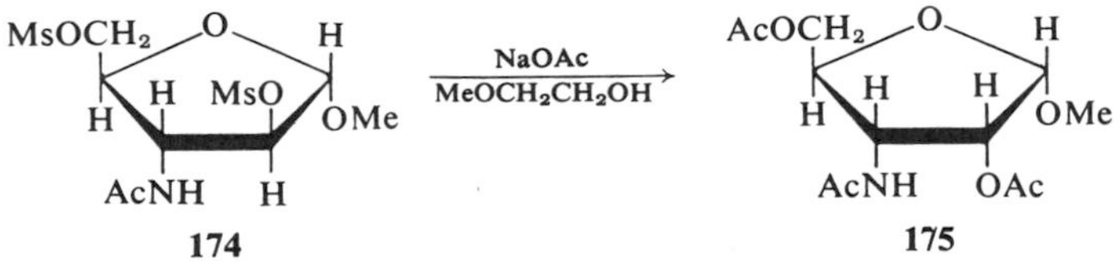

174   175

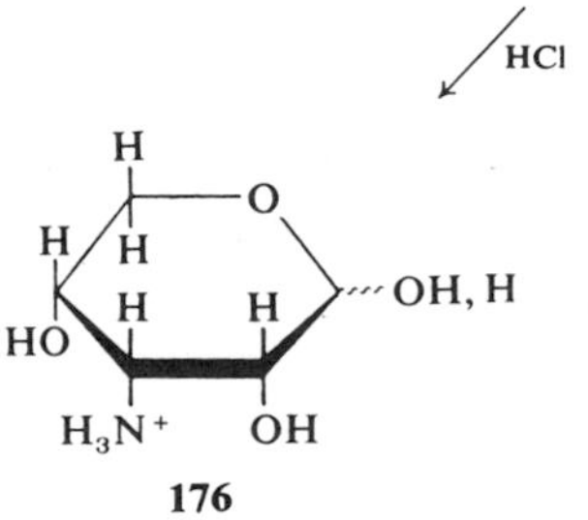

176

2,3-anhydro-α-D-lyxofuranoside, *N*-acetylation, and sulfonylation, to give **174**, which was converted into **175** by the action of sodium acetate in hot 2-methoxyethanol, and acetylation, hydrolysis then affording **176** and thus confirming that inversion had occurred.

The greater reactivity of C-4 allows of selective acylation and sulfonylation procedures (see ref. 115), as in the conversion of methyl 3-acetamido-3,6-dideoxy-α-L-glucopyranoside (**177**) into 4-*O*-acyl-2-*O*-*p*-tolylsulfonyl (**178**) and 2-*O*-acyl-4-*O*-*p*-tolylsulfonyl derivatives (**179**). Sodium acetate in refluxing 2-methoxyethanol converted the former into an L-*manno* product **180**, whereas the same conditions converted the latter into an L-*galacto* product[88,273,274] **181** related to the monosulfonate (**183**) obtained by short (2.5 hours) treatment of the 2,4-di-*O*-(methylsulfonyl) derivative (**182**) of **177** under similar[275] conditions.

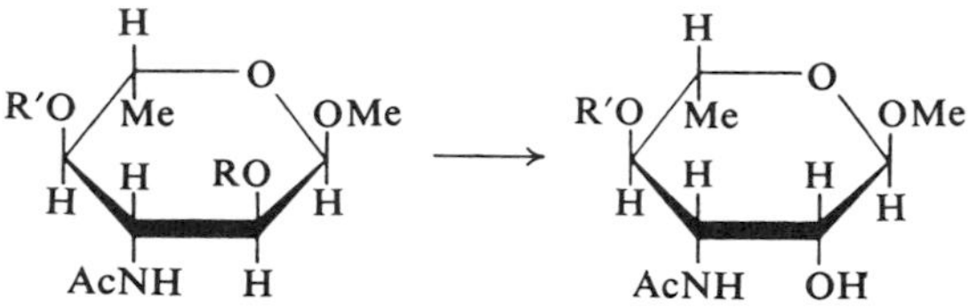

177 R = R' = H
178 R = Ts, R' = Ac or Bz
179 R = Ac or Bz, R' = Ts
182 R = R' = Ms

180 R' = Ac or Bz

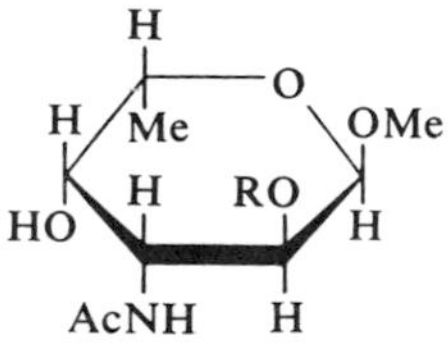

181 R = Ac or Bz
183 R = Ms

Methyl 3-acetamido-4,6-*O*-benzylidene-3-deoxy-2-*O*-(methylsulfonyl)-α-D-altropyranoside was readily converted into a D-*allo*[86] derivative, whence 3-amino-3-deoxy-D-allose was liberated by acid hydrolysis; by similar, internal displacement, Jeanloz[276] prepared 2-amino-2-deoxy-D-allose (**187**) (and -D-talose[71,277]) from methyl 2-acetamido-4,6-*O*-benzylidene-2-deoxy-3-*O*-(methylsulfonyl)-α-D-glucopyranoside (**184**) (and -α-D-idopyranoside), whereas the corresponding D-*galacto* acetamidosulfonate underwent displacement only after hydrolysis of the benzylidene acetal, affording a derivative[278] of 2-amino-2-deoxy-D-gulose.

The oxazoline **186** is sufficiently stable that it was isolated[279] from a similar reaction of the benzamido analogue (**185**) of **184**.

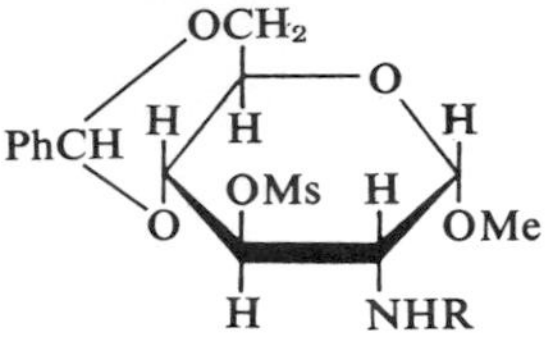

184 R = Ac
185 R = Bz

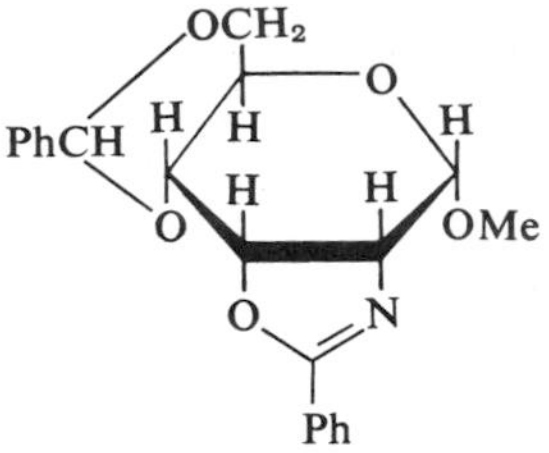

186

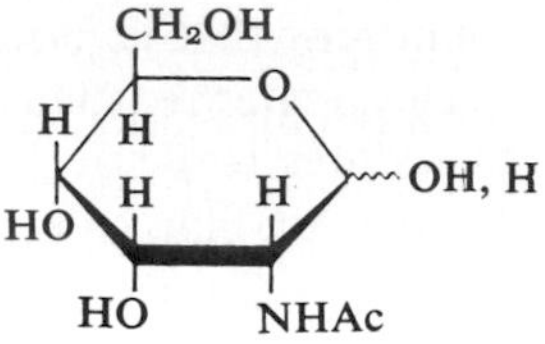

187

Displacement of an epoxide by participation occurs in a directed orientation. Benzyl 3,4-anhydro-2-(benzyloxycarbonyl)amino-2-deoxy-α-[280] and -β-D-galactopyranoside[281] rearrange in hot, aqueous acetic acid by opening at C-3 to afford benzyl α(and β)-D-gulopyranosido[2′,3′:4,5]-4,5-dihydro-2-oxazolinone. A 3,4-epoxide is also probably intermediate in the conversion[282] of methyl 2-benzamido-3,6-di-*O*-benzoyl-2-deoxy-4-*O*-(methylsulfonyl)-α-D-glucopyranoside into methyl (or benzyl) β-D-gulopyranosido[2′,3′:4,5]-4,5-dihydro-2-phenyloxazoline in the presence of methoxide ion.

Nonvicinal participating groups may assist in configurational inversion if they are sterically available. Meyer zu Reckendorf[283] prepared 2,6-diamino-2,6-dideoxy-D-galactose by internal displacement of the sulfonyloxy group from methyl 2,6-dibenzamido-2,6-dideoxy-3-*O*-methyl-4-*O*-(methylsulfonyl)-β-D-glucopyranoside through participation of the 6-benzamido group in a six-membered [5,6-dihydro-2-phenyl-4*H*-1,3-oxazine] ring intermediate. Oxazine and oxazoline intermediates are presumably formed in the conversion[284] of methyl 2,6-diacetamido-2,6-dideoxy-3,4-di-*O*-(methylsulfonyl)-α-D-glucopyranoside into methyl 2,6-diacetamido-2,6-dideoxy-α-D-gulopyranoside. The dihydroöxazine derivative **188** was isolated[285] in high yield by treatment of 5-benzamido-5-deoxy-1,2-*O*-isopropylidene-3-*O*-(methylsulfonyl)-β-D-xylofuranose with sodium benzoate in boiling *N,N*-dimethylformamide, whereas the corresponding *N*-acetyl derivative was converted further into 5-acetamido-5-deoxy-1,2-*O*-isopropylidene-β-D-lyxofuranose.

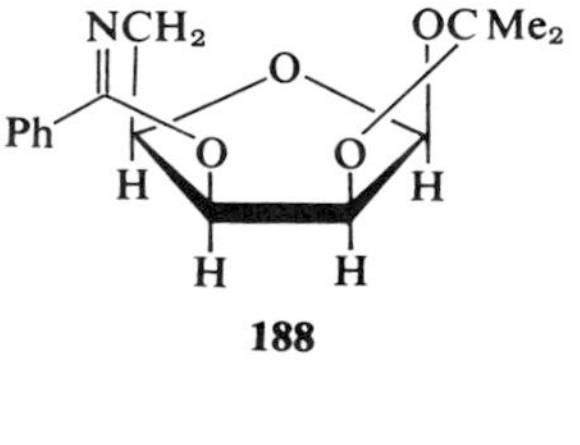

188

Sequential displacements of dioxolanium and oxazolinium ions have been invoked[286] to account for the acetate-catalyzed conversion of benzyl 2-acetamido-3-O-benzoyl-2-deoxy-4,6-di-O-(methylsulfonyl)-α-D-glucopyranoside (189) in 2-methoxyethanol into a D-*gulo* product from which 190 was prepared by saponification; however, 191 in potassium acetate–acetic acid,[287] and 192 in N,N-dimethylformamide containing sodium benzoate,[288] underwent inversion at C-4 only, affording products having the D-*galacto* configuration. Paulsen and co-workers found that extended exposure of 3-acetamido-1,2,4,6-tetra-O-acetyl-3-deoxy-β-D-glucopyranose to antimony pentachloride resulted, at equilibrium between the various acetoxonium and oxazolinium ions, in formation of a single, bridged onium ion that could be converted either into 3-deoxy-D-mannopyrano[2′,3′:5,4]-4,5-dihydro-3-methyloxazoline (193) (by sodium methoxide) or into 3-amino-3-deoxy-D-mannose (by 6M hydrochloric acid), whereas similar treatment of the per-O-acetyl derivative of 6-benzamido-6-deoxy-β-D-glucopyranose afforded, after treatment with

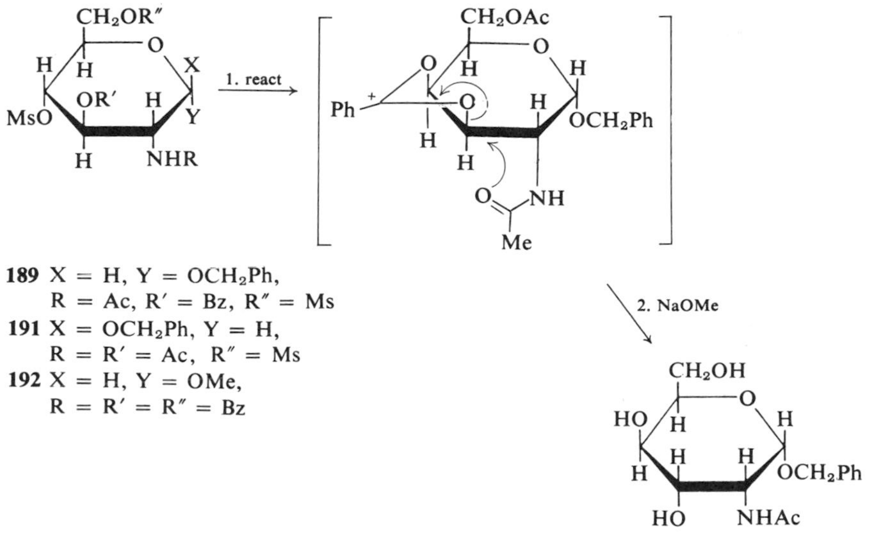

189 X = H, Y = OCH₂Ph,
    R = Ac, R′ = Bz, R″ = Ms
191 X = OCH₂Ph, Y = H,
    R = R′ = Ac, R″ = Ms
192 X = H, Y = OMe,
    R = R′ = R″ = Bz

190

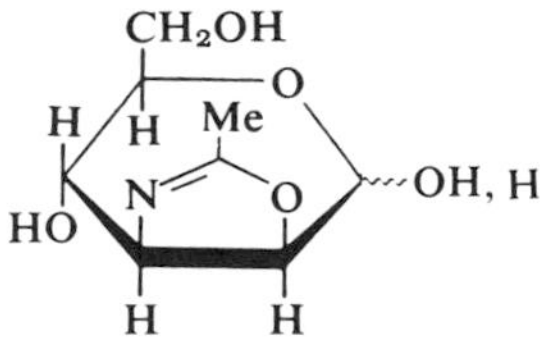

193 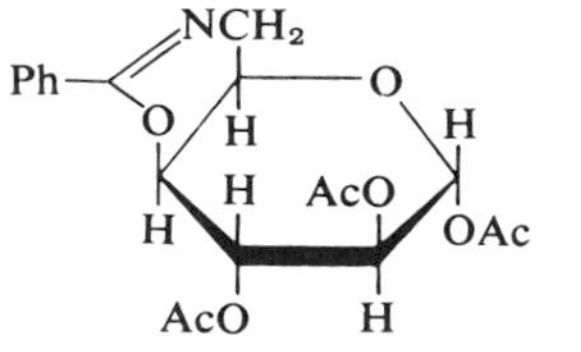   194 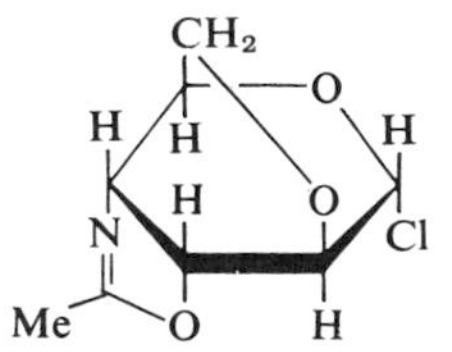   195

sodium hydrogencarbonate, 1,2,3-tri-*O*-acetyl-α-D-idopyrano[4′,5′:6,5]-5,6-dihydro-2-phenyl-4*H*-oxazine[289] (**194**); 4-benzamido-1,2,3,6-tetra-*O*-benzoyl-4-deoxy-D-glucopyranose likewise forms an oxazolinium ion in the presence of antimony pentachloride, but a subsequent, transannular displacement-reaction resulted in isolation[290] of the anhydride **195**. In all cases, however, the nitrogen-containing heterocyclic ions were much more stable than their dioxa counterparts.

In acyclic systems, free rotation about carbon–carbon bonds permits hydroxyl inversion from either *erythro* or *threo*, vicinal, acylamino alcohols. Thus, Baker and Haines[92] converted 3-benzamido-3-deoxy-1,2:5,6-di-*O*-isopropylidene-4-*O*-(methylsulfonyl)-D-altritol into 3-benzamido-3-deoxy-1,2:5,6-di-*O*-isopropylidene-D-iditol by way of the isolable oxazoline **196**, whereas Gigg and Warren[148] accomplished the reverse by transforming 2-benzamido-6-*O*-benzoyl-1,4,5-tri-*O*-benzyl-2-deoxy-3-*O*-(methylsulfonyl)-L-talitol into the stable oxazoline **197**. 6-Benzamido-6-deoxy-1,2-*O*-isopropylidene-5-*O*-*p*-tolylsulfonyl-α-D-glucofuranose was likewise converted into a 5,6-oxazoline with inversion[108] at C-5.

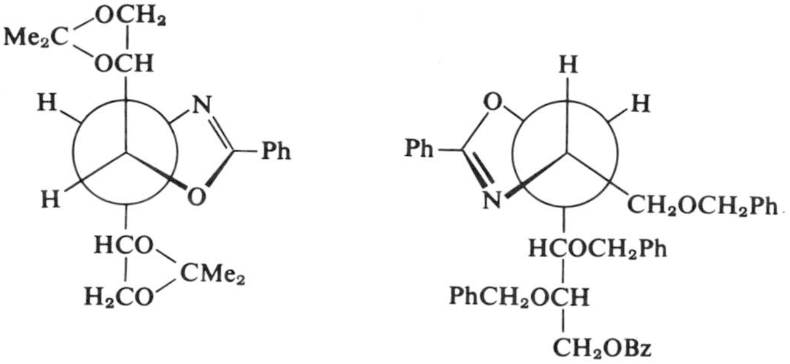

196

197

K. Racemic, Total Syntheses

The first stereospecific synthesis of a racemic, aminodeoxy sugar derivative from aliphatic starting materials was accomplished by treatment of 1-chloro-1-nitrosocyclohexane with methyl *trans,trans*-2,4-hexadienoate (*methyl sorbate*) to afford, after *N*-benzoylation, the hydroxylamine derivative **198**; *cis*-hydroxylation of **198** occurred on the less-hindered side to form, after deprotection and hydrogenolysis, 5-amino-5,6-dideoxy-DL-allonic acid, whereas epoxidation of **198** and the subsequent steps of epoxide opening, deprotection, and hydrogenolysis gave[291] 5-amino-5,6-dideoxy-DL-gulonic acid. Mochalin *et al.*[292] prepared the racemic epoxides **199** (ref. 292a) and **200** (ref. 292b) from the corresponding 5,6-dihydro-2*H*-pyran derivatives;

aminolysis (with a number of different amines) resulted in (presumably axial) attack at the position remote from the ring-oxygen atom to produce **201** and **202**, respectively.

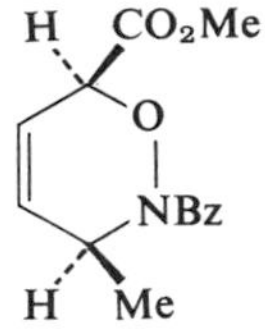

198

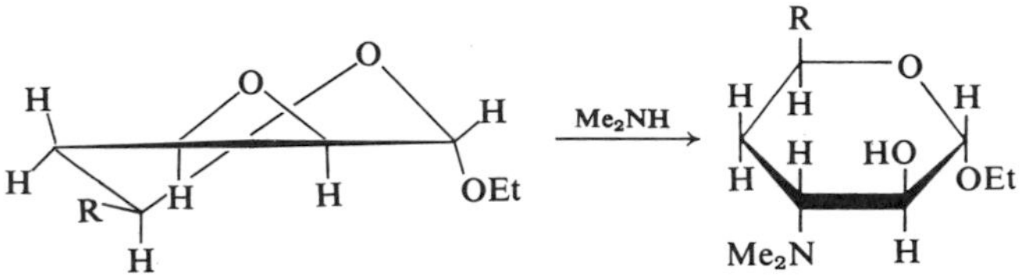

199 R = H
200 R = CCl$_3$

201 R = H
202 R = CCl$_3$

Racemic 6-acetoxymethyl-4,5-dihydro-6*H*-pyran reacted with nitrosyl chloride, and, in a second step, methanol, to produce the oxime **203**, which was hydrogenated and the product acetylated to a mixture of methyl 2-acetamido-6-*O*-acetyl-2,3,4-trideoxy-α-DL-*erythro*(and *threo*)-hexopyranosides (**204** and **205**, respectively); azidolysis of the corresponding 6-*O*-*p*-tolylsulfonyl

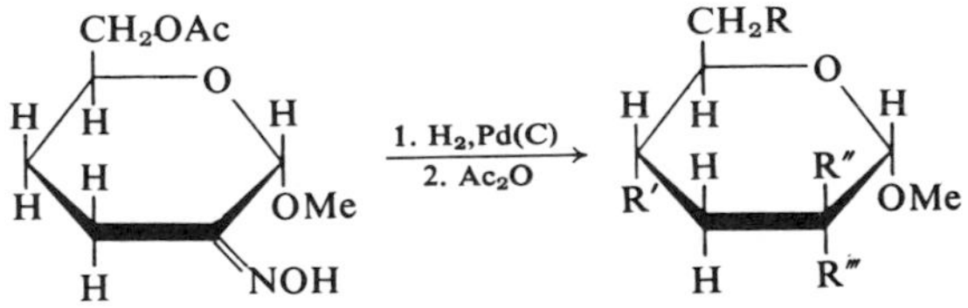

203

1. H$_2$,Pd(C)
2. Ac$_2$O

204 R = OAc, R′ = R″ = H, R‴ = NHAc
205 R = OAc, R′ = R″ = H, R‴ = NHAc
206 R = R‴ = NHAc, R′ = R″ = H
207 R = R″ = NHAc, R′ = R‴ = H
210 R = R″ = H, R′ = R‴ = NH$_2$

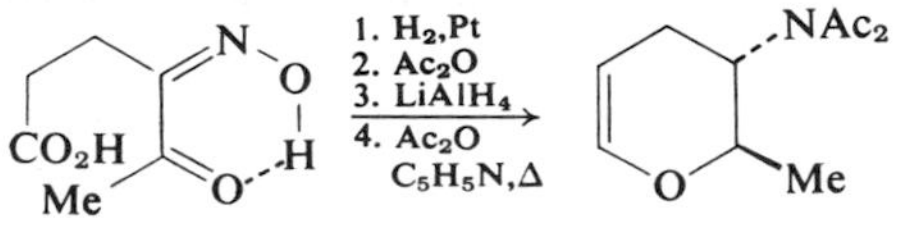

208

1. H$_2$,Pt
2. Ac$_2$O
3. LiAlH$_4$
4. Ac$_2$O
C$_5$H$_5$N,Δ

209

derivatives and reduction afforded derivatives of 2,6-diamino-2,3,4,6-tetra-deoxy-DL-*erythro*-hexose (**206**, DL-*purpurosamine*) and the DL-*threo* analogue **207**, DL-*epipurpurosamine*).[293a,b] The action of nitrosyl chloride on 6-methyl-3,4-dihydro-2*H*-pyran-2-one gave an adduct that reacted with water to yield the keto-oxime **208** by spontaneous opening of the tautomeric hemiacetal–ester; hydrogenation afforded a racemic amino alcohol that lactonized, and *N*-acetylation, reduction of the carboxyl group to a hemiacetal, and pyrolysis of the corresponding *O*-acetyl derivative produced 3-(*N*-acetylacetamido)-3,4-dihydro-2-methyl-2*H*-pyran (**209**), which was treated sequentially with (*a*) nitrosyl chloride, (*b*) methanol, (*c*) hydrogen, and (*d*) alkali to produce mainly methyl 2,4-diamino-2,3,4,6-tetradeoxy-α-DL-*arabino*-hexopyranoside (**210**, *methyl* DL-*kasugaminide*).[294]

An *N*-benzoylated methyl glycoside of DL-nojirimycin has been prepared[294a] from 3-cyanopyridine by a multistep sequence of reactions that included photochemical addition of methanol and oxidative cleavage of the C-2–C-3 bond.

The synthesis of 1,4-dideoxy-1,4-imino-DL-ribitol from pyrrole-2-carboxylic acid was accomplished[295] by 2,5-hydrogenation of the 2,3- and 4,5-double bonds and reduction of the carboxyl group (as an ester), with subsequent *cis*-hydroxylation from the less-hindered side of the resulting 3,4-double bond.

L. Synthetic Mono-, Di-, Tri-, and Tetra-Amino Sugars

A brief compilation of synthetic aminodeoxy aldoses is presented in Table I. Although diaminodideoxy sugars were discovered in Nature much later than the monoamino analogues, a substantial group of synthetic examples (see Table II) has been prepared, generally according to the methods described in Sections II, D, G, H, and J. Tri- and tetra-aminoaldoses, which offer only theoretical interest at present, have been similarly prepared; derivatives (all equatorial substituents) of 2,3,4-triamino-2,3,4-trideoxy-D-glucopyranose[227,228] and of 2,3,4-triamino-1,6-anhydro-2,3,4-trideoxy-β-D-idopyranose[245] are prepared readily from 3-deoxy-3-nitrohexose derivatives, although methyl 2,3,4-triacetamido-2,3,4,6-tetradeoxy-α-L-mannopyranoside was isolated as a minor product[229] from such a reaction. Displacement reactions afforded methyl 3,4,6-tribenzamido-3,4,6-trideoxy-β-D-allopyranoside[197] and a mixture[129] of methyl 2,3,5-tribenzamido-2,3,5-trideoxy-β-D-arabino- and -xylo-furanosides. Methyl 2,3,4,6-tetraacetamido-2,3,4,6-tetradeoxy-α-D-pyranosides have been prepared in the *gluco*,[324] *ido*,[325] and *galacto*[325] series, and syntheses of unacetylated analogues [as the tetrakis(hydrochlorides)] having the D-*gluco*[326] and D-*galacto*[327] configurations were reported later. Derivatives of 2,4-diamino-2,3,4-trideoxy-D-*ribo*-hexose have also been prepared.[327a]

TABLE I

References[a] to Synthetic Aminodeoxy-Aldopentoses and -Aldohexoses, and Their Hydrochlorides, or N-Acetyl Derivatives

Hydrochloride of aminodeoxy derivative	Position of substitution by nitrogen				
	2	3	4	5	6
Hexoses					
D-Allose	276, 296	139, 297	—	—	297
D-Altrose	296	298[b]	—	—	—
D-Glucose	299	300	135, 301[b]	302	303
D-Mannose	304	34	—	—	305
D-Gulose	278, 306	307[c]	—	—	—
D-Idose	306	307[c]	—	—	—
D-Galactose	308	20	309	—	—
D-Talose	310, 277	36	—	—	—
Pentoses					
D-Ribose	108	86, 200	—	57	
D-Arabinose	311	26, 312	313	57	
D-Xylose	311	314	—	57, 315	
D-Lyxose	108	30[b–d]	316[d]	57, 317[b]	

[a] Selected references only; for a more-extensive compilation, see ref. 3.
[b] N-Acetyl derivative.
[c] Various simple derivatives.
[d] L Enantiomorph.

TABLE II

References to Synthetic Diaminodideoxy-Pentoses and -Hexoses

Dihydrochloride of diaminodideoxy derivative	Positions of substitution by nitrogen					
	2,3	2,4	2,6	3,5	3,6	4,6
D-Allose	197	—	197	—	—	—
D-Altrose	—	—	317a[a]	—	—	—
D-Glucose	197, 318	319	204, 319a	—	104	320
D-Mannose	—	—	321, 322	—	—	—
D-Gulose	—	—	282	—	—	—
L-Idose	—	—	323[b]	—	—	—
D-Galactose	—	319	281	—	—	320
L-Ribose	273	—	—	—		
L-Arabinose	—	—	—	26		
L-Lyxose	—	—	—	26		

[a] Di-N-acetyl derivative.
[b] Dipicrate.

References start on p. 737.

III. REACTIONS OF AMINO SUGARS

In general, amino sugars undergo many of the same reactions as their oxygenated analogues, but the nitrogenous substitutent introduces additional characteristics. The free amino group is relatively basic, experiencing only a slight (although characteristic) dependence[328] of its properties upon structure, and protonation of the nitrogen atom occurs readily, especially under acidic conditions, profoundly altering the polar character of the amino sugar. Thus, the anomeric equilibria of amino sugars differ somewhat from those of their oxygenated analogues, and internal proton-transfer processes may accelerate the attainment of equilibrium by the former[329]; furthermore, the conversion of 2-amino-2-deoxy-D-glucose into its diethyl dithioacetal[330] by the action of ethanethiol plus hydrochloric acid requires strenuous conditions, whereas the analogous transformation[331] of 2-acetamido-2-deoxy-D-glucose, in which the amide nitrogen atom is far less basic, occurs readily. The amino group may also serve as an effective nucleophile, generally entering into displacement reactions much more readily than hydroxyl groups; this property, however, facilitates specific protection of amino groups, as by acylation with acetic (or other) anhydrides in hydroxylic[332] solvents or with benzyl chloroformate.[333] The N-acetyl protecting group is generally stable, under all but extreme (strongly acidic or basic) reaction[334] conditions, whereas the N-benzyloxycarbonyl group may be removed under mild conditions by hydrogenolysis; however, either may participate in displacement reactions (see Section II,D,2 and II,J).

Condensation of amino sugars with 1-halo-2,4-dinitrobenzenes or with aromatic aldehydes affords N-(2,4-dinitrophenyl) or substituted N-arylidene derivatives, respectively, which are useful, nonparticipating, protecting groups. Use of two frequently used N-protecting groups is illustrated by the acetylation of **211** and of **212**, with subsequent hydrogenolysis,[335] or mild[336] hydrolysis, respectively, the product being 1,3,4,6-tetra-O-acetyl-2-amino-2-deoxy-D-glucopyranose from both sequences.

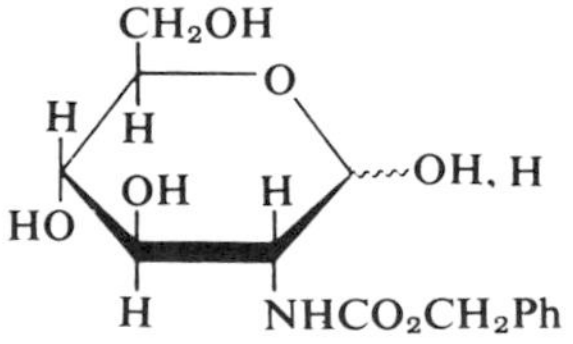

211 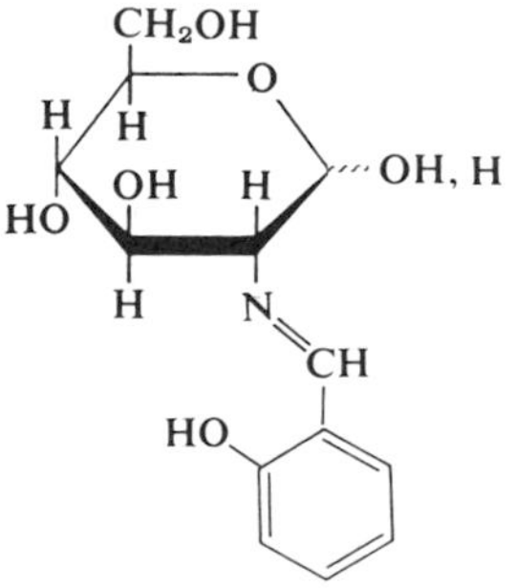   **212**

When the amino group is suitably deactivated, the hydroxyl groups of amino sugars display reactions typical of simple sugars; acetals, esters, and ethers are formed in a predictable manner. Methyl ethers are particularly useful as reference compounds in structural work on complex carbohydrates, and numerous partially O-methylated amino sugars have been prepared and catalogued[337]; the N-methyl-O-methyl derivatives are also useful in this regard,[337a] and further conversions to protect unsubstituted polar groups permit the use of gas-chromatographic methods.[337b]

Reactions unique to the amino sugars are treated under three categories: sugars having a nitrogen atom in the ring, nitrosation and deamination reactions, and formation of nitrogenous heterocycles and related derivatives.

A. SUGARS HAVING A NITROGEN ATOM IN THE RING

4-Amino-4-deoxy- and 5-amino-5-deoxy-aldoses (plus homologous amino-deoxyketoses and a few related structures) and their mono-N-substituted derivatives exist as an equilibrium of ring forms that include structures having a nitrogen atom as a constituent of the ring. In general, (a) the N-ring tautomers are relatively favored (in comparison to oxygenated analogues) unless the amino group is strongly deactivated, as by N-acetylation, and (b) 5-aminopyranose forms are somewhat more stable than 4-aminofuranose structures; however, the configuration exerts some influence upon the position of the equilibrium, and judicious use of protecting groups may dictate a selected ring closure. Paulsen and Todt[338] have presented a detailed discussion of this subject.

1. Ring Closure Involving Nonacylated Amino Groups

a. Piperidine Derivatives.—Acid hydrolysis of derivatives of 5-amino-5-deoxy-D-xylofuranose generates the Mannich base 20; the observed[57,339,340] product is, however, the thermodynamically favored 3-pyridinol (216), formed by dehydration of 20 to the Schiff base 213, 1,2 migration of the double bond to generate an enol, and acid-catalyzed eliminations. Under milder conditions, part of the enol undergoes hydration, forming[339] 214, formally an Amadori rearrangement product (see Section II,B). N-Alkylation is qualitatively without effect upon the course of this reaction, as N-alkyl-pyridinium salts (for example,[341] 217) have been isolated in good yield after attempted hydrolysis of 5-alkylamino-5-deoxyaldose derivatives.

Controlled hydrolysis by the action of sulfurous acid was found to result in formation of a stable, bisulfite adduct (215), from which 20 could be liberated

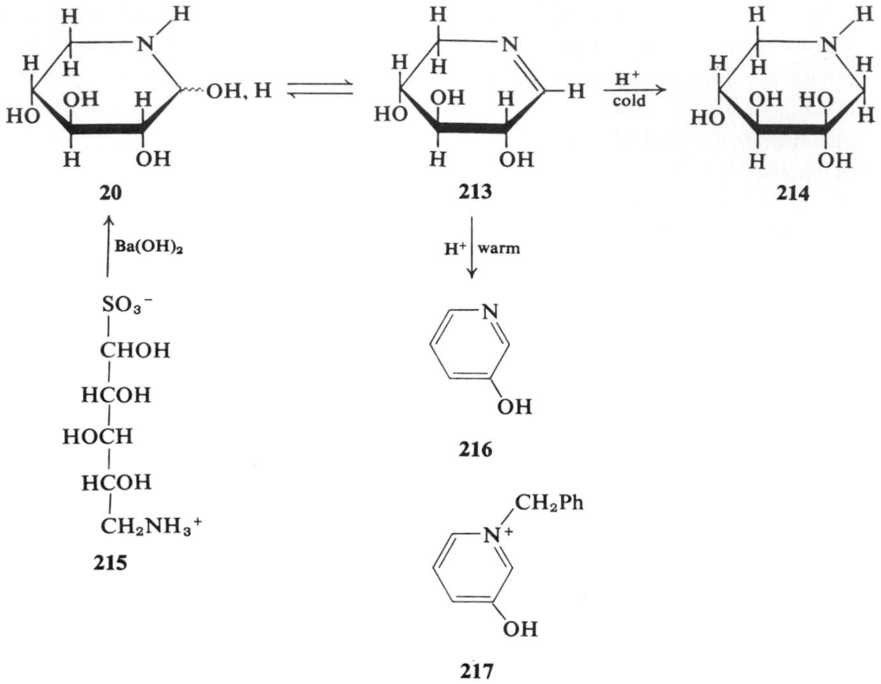

20   **213**   **214**

Ba(OH)₂

H⁺ | warm

SO_3^-
|
CHOH
|
HCOH
|
HOCH
|
HCOH
|
$CH_2NH_3^+$

215

216

217

by treatment with barium hydroxide.[342] Even under optimal conditions, compound **20** exists in equilibrium with the dehydrated form **213**, and the results of circular-dichroism studies indicate the almost ubiquitous presence of the dehydration product **213** as a contaminant; mild acidification (pH 6 to 6.8) results in rapid conversion of **20** into **214**, stronger acidic treatment causing aromatization[342] to 3-pyridinol (**216**). Reduction of such precursors as **218** or **219** under neutral conditions is complicated[343] by the susceptibility of **20** (and related molecules) to hydrogenolysis, producing such iminoalditols as **220**; the strongly nucleophilic cyanide ion reacts with **20** ⇌ **213** to form 2,6-dideoxy-2,6-imino-D-idononitrile (**221**), whereas reactions with simple amines or alcohols afforded[343] only **214** and **216**, in accordance with the intrinsic lability[147] of aglycons in nonacylated N-ring sugars.

Although configuration exerts only a slight, quantitative effect upon the course of these reactions, a potential leaving group at C-3 is an absolute requirement[57] for aromatization; thus, acid hydrolysis of 5-amino-3,5-dideoxy-1,2-O-isopropylidene-α-D-erythro-pentofuranose afforded only an Amadori rearrangement product. Replacement of the anomeric hydrogen atom by a hydroxymethyl group, as in 6-amino-6-deoxy-L-sorbose (**222**), acts to stabilize the 2-piperidinol form; thus, the action of 2M hydrochloric acid

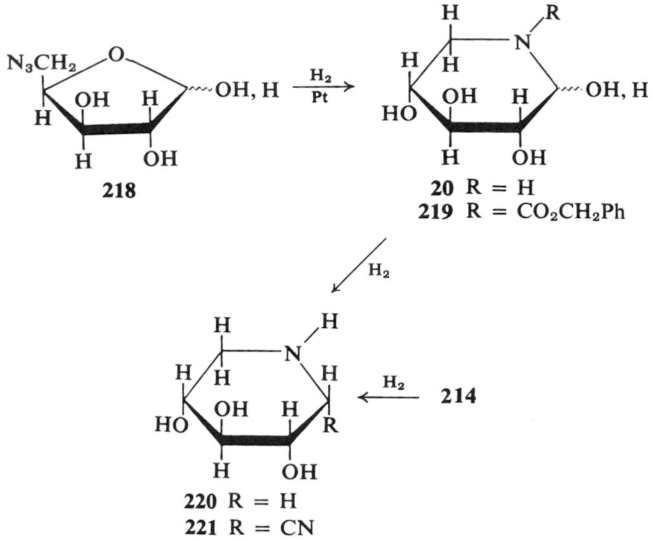

218

20 R = H
219 R = CO$_2$CH$_2$Ph

220 R = H
221 R = CN

at 65° upon 6-amino-6-deoxy-2,3-*O*-isopropylidene-α-L-*xylo*-hexulofuranose (**224**) gives 3-hydroxy-2-pyridinemethanol (**225**), whereas the same reagents at 22° produce the crystalline, deacetalated furanose hydrochloride (**223**), in 70% yield, which rearranges reversibly in dilute base to form[344] **222**. The Amadori product is not observed, but the Schiff-base form is present in equilibrium; mild acidification results in formation of **225**, whereas strongly acidic conditions cause **222** to revert[344] to **223**.

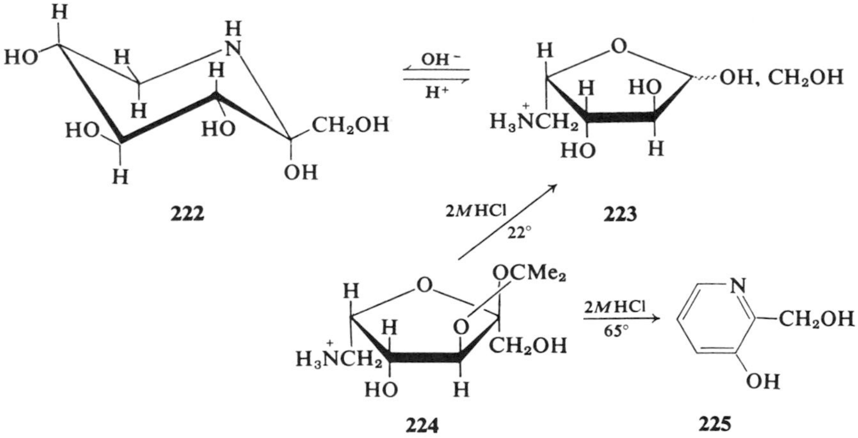

222

223

224

225

The isomeric 5-amino-5-deoxyhexose forms are likewise quite stable; 5-amino-5-deoxy-D-glucopyranose (*nojirimycin*) occurs[301] as a natural product (see Section IV,D). 5,6-Diamino-5,6-dideoxy-D-glucopyranose (**226**), which was prepared by saponification of its bisulfite adduct, is partially protected from the usual degradation processes by extensive (20–40% at equilibrium[345]; see Vol. IA, Chapter 13,II,D) conversion into the 1,6-anhydro-β-D-pyranose (**227**).

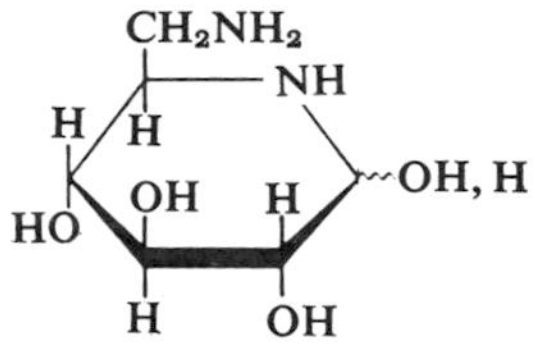

226 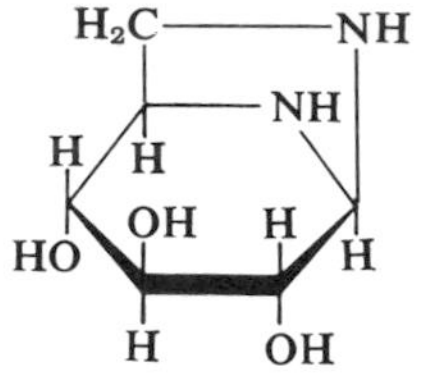   **227**

b. Pyrrolidine Derivatives.—4-Amino-4-deoxy-L-lyxose (or any other 4-amino-4-deoxyaldose), could, in principle, exist as an equilibrium mixture of four ring forms (**228–231**); however, the pyranose (**228**) form, which should be favored on stereochemical grounds, is not observed (except under strongly acidic conditions[346]), presumably as a consequence of the nucleophilicity of the 4-amino group. The unsymmetrical dimer **231** is destabilized by the all-*cis* arrangement of substituents, but **231** nonetheless dominates the equilibrium at high concentration in an anhydrous medium; the Schiff base **230** preponderates[316] over **229**. The 2-epimeric 4-amino-4,5-dideoxy-L-*xylo* dimer (**232**), in which the exocyclic carbon substituent[146,149] occupies the *exo* position, is the principal[347] form observed. 4,5-Dideoxy-4-(methyl-amino)-L-lyxose, which cannot form an unsymmetrical dimer analogous to **232**, adopts[147] a symmetrical, 1,2′:2,1′-dianhydride form (**233**). Methyl 5-amino-5-deoxy-D-*glycero*-D-*galacto*-nonulosonate (*methyl neuraminate*, see Section IV,D) exists[348] almost solely in the Schiff-base form, suggesting that ketoses may dimerize less readily than isomeric aldoses. The presence of a 2-amino group also appears to disfavor dimerization.[324b,349]

In contrast to the six-membered-ring analogues, the 4-amino-4-deoxyaldoses are virtually unaffected[146,347] by treatment with acids; an 80% recovery[146] (as the bisulfite adduct) of starting material was achieved following exposure of **232** to 2*M* hydrochloric acid for 2 hours at 80°. This observation has synthetic utility: 2-amino-4-azido-2,4-dideoxy-D-galactose was catalytically hydrogenated in the presence of 6*M* hydrochloric acid to afford[349] 2,4-diamino-2,4-dideoxy-D-galactose instead of a 1,4-iminoalditol (which was formed in 1*M* acid in the presence of palladium and hydrogen).

c. Azepin Derivatives.—In practice, only pyranose and furanose forms appear to be favored. 6-Amino-6-deoxy-L-idose spontaneously adopts[345] the all-

228 229 230

231

232

233

equatorial 1,6-anhydropyranose form, which is also a 1,5-anhydroseptanose, and 6-amino-5,6-dideoxy-D-*xylo*-hexose, which cannot form a pyranose ring, exists[309] as the 1,6-anhydrofuranose (1,4-anhydroseptanose); however, 1,6-dideoxy-1,6-iminoalditols are the product of reduction[309,345] of 6-aminoaldoses.

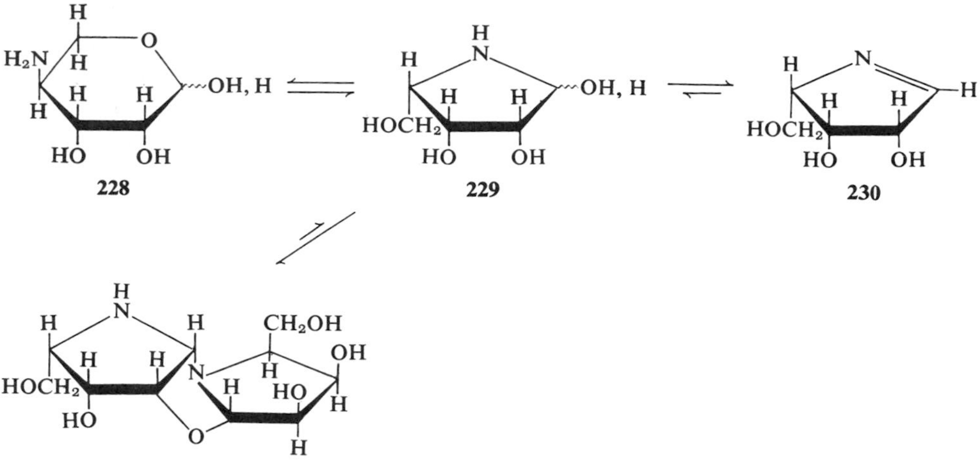

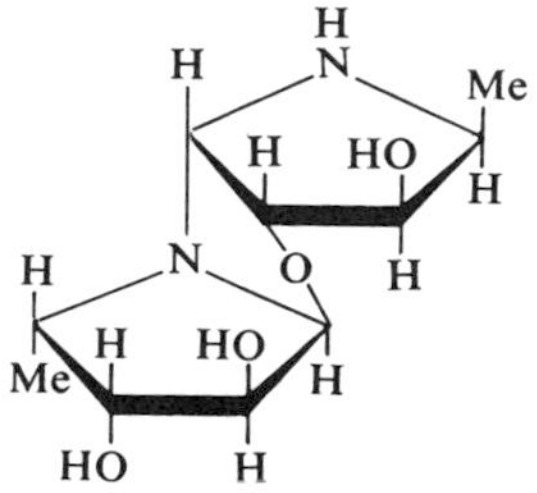

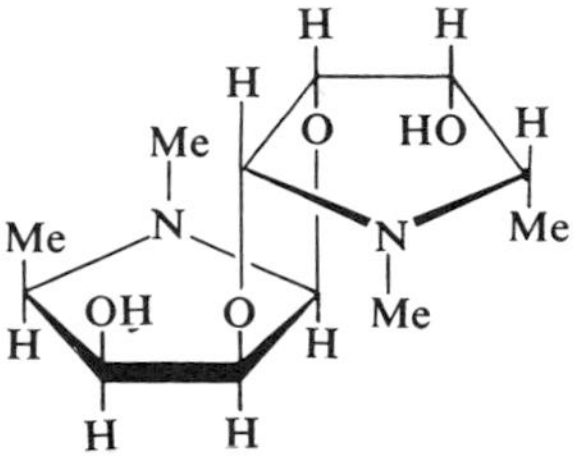

d. Hydrazines.—Only a five-membered ring can form on cyclization of a 3-deoxy-3-hydrazinoaldose; thus, acid hydrolysis of 3-deoxy-3-hydrazino-1,2:5,6-di-*O*-isopropylidene-α-D-allofuranose afforded[350] 3-(D-*erythro*-1,2,3-trihydroxypropyl)-2*H*-pyrazole (**57**) by dehydration of the initial hydrazone **234**. 4-Deoxy-4-hydrazinoaldoses and 5-deoxy-5-hydrazinoaldoses undergo a cyclization through whichever of the two nitrogen atoms allows formation of a six-membered ring, unless that nitrogen atom is unavailable. Thus, 5-deoxy-5-hydrazino-D-xylose cyclizes spontaneously in aqueous barium

References start on p. 737.

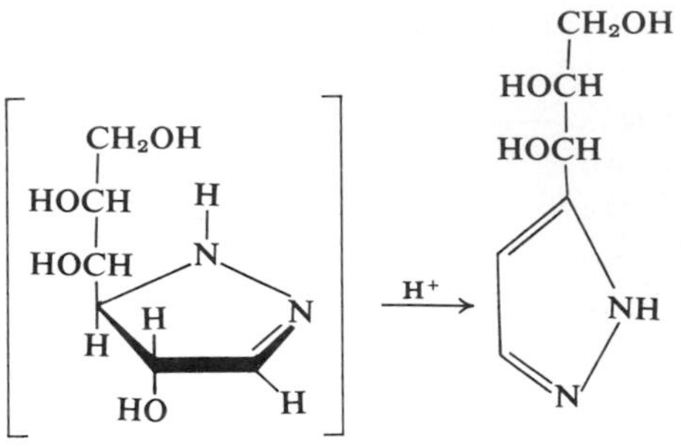

234   57

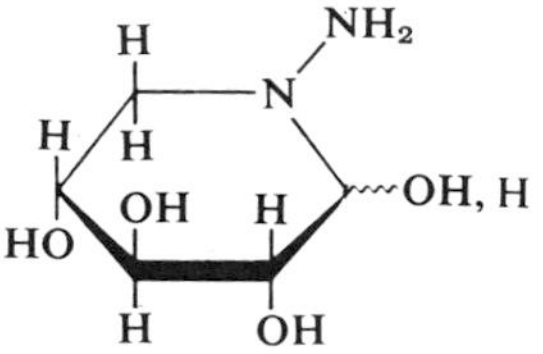

236 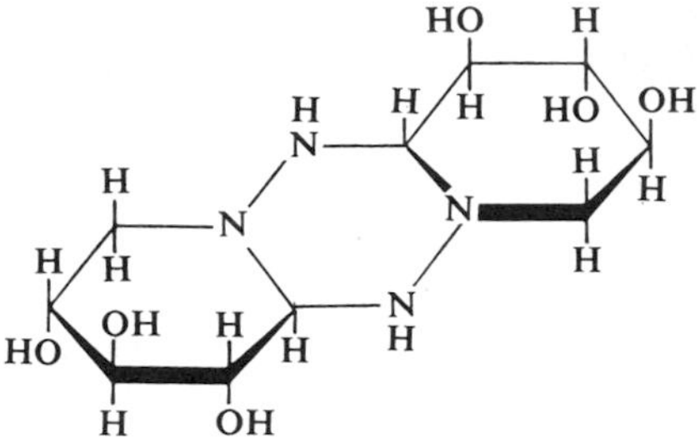   235

hydroxide to give N-amino-(5-amino-5-deoxy-D-xylopyranose) (**236**), which forms the symmetrical dimer **235** upon neutralization[351]; compound **236** ⇌ **235** is slightly more stable to acid than its N-unsubstituted analogue **20**, because of deactivation by the N-substituent. Forcing hydrazinolysis of 1,2:5,6-di-O-isopropylidene-3-O-p-tolylsulfonyl-α-D-glucofuranose is presumed to proceed via the sequence: (a) 3,4-elimination of toluenesulfonic

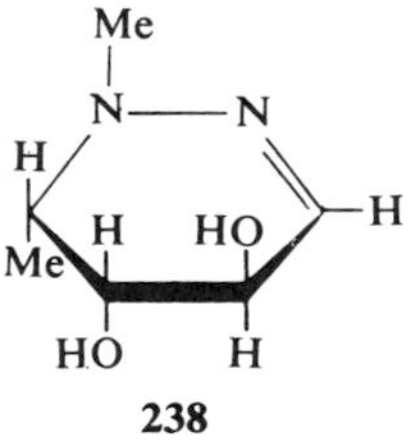

238

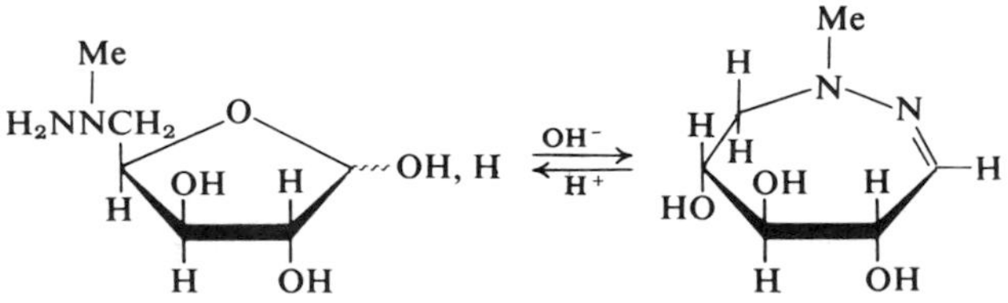

239   237

acid, (b) 4,3-addition of hydrazine, (c) ring-opening to a glycos-4-ulose 4-hydrazone, (d) cyclization by hydrazonation of C-1 (with loss of one isopropylidene group), and (e) 2,3 elimination of water to afford 3-[2,2-dimethyl-1,3-dioxolane-4(S)-yl)]pyridazine.[351a]

4,5-Dideoxy-4-(1-methylhydrazino)-L-xylose cyclized[352] immediately upon formation, giving the hydrazone **238**, which was converted into 1,6-dimethyl-pyridazine by the action of 2M hydrochloric acid at 80°; however, 5-deoxy-5-(1-methylhydrazino)-D-xylofuranose (**239**) undergoes reversible, base-catalyzed formation of the seven-membered-ring[353] hydrazone **237**, owing to the unavailability of the tertiary nitrogen atom.

2. Ring Closure Involving Acylated Amino Groups

Whereas the amino group is able to compete very effectively for the aldehydogenic center, the depletion of electron density wrought by N-acylation renders the acylamido group inferior to the hydroxyl group as a nucleophile; accordingly, stereochemical factors and the protecting groups present dictate the circumstances under which acylamido groups can participate in ring formation. N-Acylation also decreases the basicity of the nitrogen atom, which, in turn, increases the stability of the N-ring toward acids.

a. Piperidine Derivatives.—Acid-catalyzed hydrolysis of 5-acetamido-5-deoxy-1,2-O-isopropylidene-α-D-xylofuranose produces a mixture[340,354,355] of crystalline 5-acetamido-5-deoxy-α-D-xylopyranose (**241**) and the syrupy furanose (**240**); compound **241** is also formed upon N-acetylation[343,356] of **20**. Glycosidation of **241** proceeded[354] straightforwardly in acidified methanol, the glycosidic bond being stabilized by the acyl group, whereas attempted[357] preparation of the corresponding acetylated glycosyl halides was unsuccessful.

Equilibration between compounds **241** and **240** is extremely slow at neutral pH. Heat,[354] acid,[340,355] and base[350,355] accelerate the interconversion, although basic conditions also foster[340] extensive decomposition. Vigorous acidic hydrolysis causes[339,340] N-deacylation and subsequent aromatization.

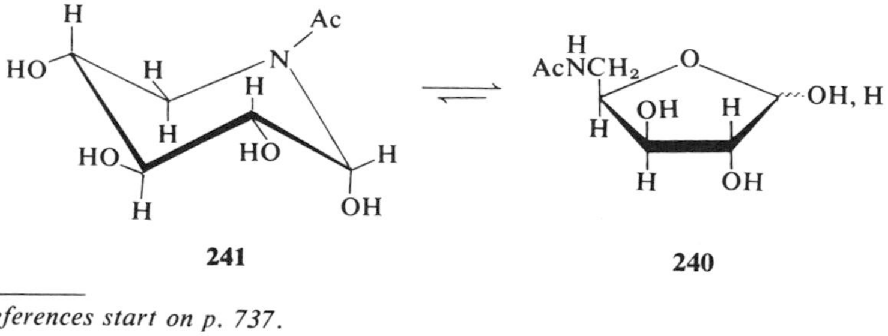

241   **240**

A singularly strong, anomeric effect[357a] (Vol. IA, Chapter 5) appears to operate[358] in such acylated N-ring compounds as **241**, and the doubling of the nuclear magnetic resonance signals[359] is a consequence of hindered rotation about the (conjugated) N—acyl bond. For **241** ⇌ **240**, the pyranose:furanose ratio[340,355] is ~ 2:1; the corresponding ratio for the N-benzoyl analogue[357] is ~ 3:1, and 5-(benzyloxycarbonyl)amino-5-deoxy-D-xylopyranose has only traces[343] of the furanose forms at equilibrium, reflecting the progressively diminishing extent of deactivation of the respective amide-nitrogen atoms.

5-Acetamido-5-deoxy-D-lyxose forms approximately equal[283,317] amounts of the pyranose and furanose tautomers, whereas the L-*arabino*[355,360] and D-*ribo*[355] isomers favor the furanose forms to the extent of 4:1 and 10:1, respectively; more-favorable yields of the pyranose forms have been obtained by N-acetylation of the respective[340,356] 5-amino-5-deoxyaldopyranoses. The corresponding 5-(benzyloxycarbonyl)amino-5-deoxypentoses exist[361] almost exclusively in the pyranose form. 5-(Benzyloxycarbonyl)amino-5,6-dideoxy-3-O-(methysulfonyl)-L-idopyranose, which is destabilized by an axial C-methyl group, preponderates[362] over the furanose tautomer by a factor of only 4:1 at equilibrium. 5-Acetamido-5-deoxy-L-idose adopts the pyranose[345] form to some extent, presumably because of a favored conversion into the 1,6-anhydroaldopyranose form; 5,6-diacetamido-5,6-dideoxy-L-idose is a typical[345] 5-acetamido-5-deoxyaldohexose, in that the furanose tautomers are formed exclusively. 6-Acetamido-6-deoxy-D-fructose[363] and -L-sorbose[344] both exist in the furanose forms only.

The members of the series of 3,5-diacetamido-3,5-dideoxypentoses favor the pyranose form somewhat more than their 3-hydroxy analogues; the D-*xylo*[105] and L-*lyxo*[26] isomers exist almost exclusively as the pyranose forms, whereas the D-*ribo*[105] and L-*arabino*[26] isomers exhibit pyranose : furanose ratios of 5:1 and 2:7, respectively. A 2,5-diacetamido-2,5-dideoxy-D-xylofuranose is reportedly[364] the exclusive tautomer.

The greater stability of the six-membered ring is evident in the observation[33,356] that 4,5-diacetamido-4,5-dideoxy-L-xylose adopts a pyranose tautomer, the furanose forms occurring[365] as minor constituents.

b. Pyrrolidine Derivatives.—4-Acetamido-4-deoxy-L-erythrose[363] and -D-threose[366] (**243**), and 4-acetamido-4,5-dideoxy-L-xylose[31,32] (**244**) adopt furanose forms rather than acyclic tautomers. 4-Acetamido-4,5-dideoxy-2,3-O-isopropylidene-L-xylose[32] and the acetal[347,367] **245** exist as acyclic forms, because cyclization through the acetamido group is insufficiently stable to support a *trans*-5,5-ring fusion; mild, acid hydrolysis of the isopropylidene groups, however, affords the cyclic products **244** and **242**, respectively.

4-Acetamido-4-deoxy-D-ribose, for which the furanose structures would most probably be favored (see Vol. IA, Chapter 4, Section I,C), was found to

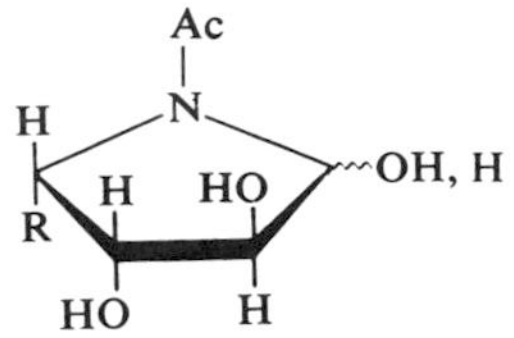

243 R = H
244 R = Me

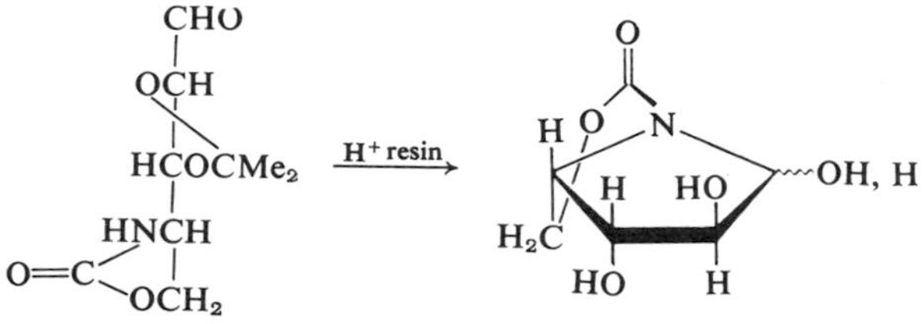

245

242

react (as a tetraacetate) with methanolic hydrochloric acid to yield the methyl 4-acetamido-4-deoxy-D-ribofuranosides; 4-acetamido-4-deoxy-L-xylose, however, exists exclusively[368] in the pyranose form, which results spontaneously[347] from N-acetylation of 4-amino-4-deoxy-L-xylofuranose.

In contrast to the aldopentose derivatives, numerous[338] studies on 4-acetamido-4-deoxyhexoses have demonstrated no example of a furanose tautomer. The same principle holds true of the (less-reactive) ketoses, as 5-acetamido-5-deoxy-D-*threo*- and -L-*erythro*-pentulose[369] and 5-acetamido-5,6-dideoxy-L-*xylo*-hexulose[370] have been reported to adopt acyclic forms instead of pyrrolidine structures.

Methanolysis of the apiose analogue 3-(acetamidomethyl)-1,2-O-isopropylidene-β-L-threofuranose (246a) affords the corresponding methyl furanosides in 55% yield, together with methyl 4-acetamido-4-deoxy-3-C-(hydroxymethyl)-2,3-O-isopropylidene-β-D-erythrofuranoside (246) in 25% yield; however, acetolysis of 246a produces[370a] only acetylated 4-amino-4-deoxy-2,3-O-isopropylidene-D-erythrofuranoses, presumably because acetal migration occurring subsequent to furanose-pyrrolidine interchange, which is subject to competition by transacetalation in acidic methanol, proceeds to completion under conditions of acetolysis.

c. *Lactams.*—In aminodeoxyaldonic acids, competition for the carboxyl group may occur between suitably situated amino and hydroxyl groups. The 1,4-lactam [for example, 4-amino-4-deoxy-2,3-O-isopropylidene-D-erythrono-1,4-lactam[371,372] (247)] is sterically favored and is, therefore, inert to acid

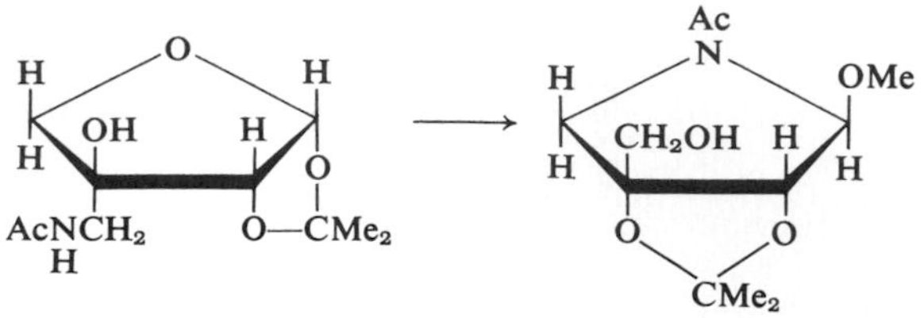

246a   **246**

hydrolysis of the lactam group; 1,5- and 1,6-lactams, however, may be converted into 1,4-lactones under acidic conditions. Thus, 5-amino-5-deoxy-2,3-*O*-isopropylidene-D-ribono-1,5-lactam (**248**), which is formed spontaneously upon hydrogenation of 5-azido-5-deoxy-2,3-*O*-isopropylidene-D-ribono-1,4-lactone (**249**), reverts to the sterically favored 1,4-lactone tautomer **250** in acid solution.[373] 6-Amino-6-deoxy-D-gulono-1,6-lactam (**251**) resulted from exposure of the tautomeric 1,4-lactone to alkali at room temperature; however, the ring in compound **251** reopened in hot water.[374]

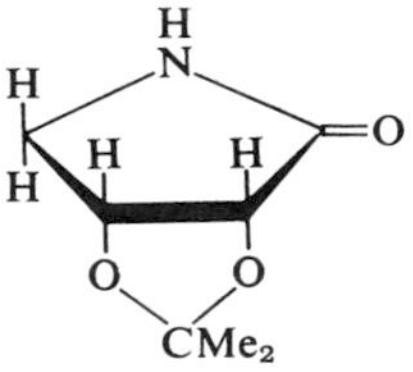

247   **248**   **249** R = N$_3$
                                            250 R = NH$_3$$^+$

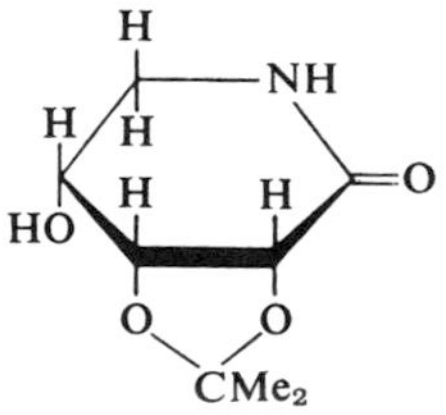

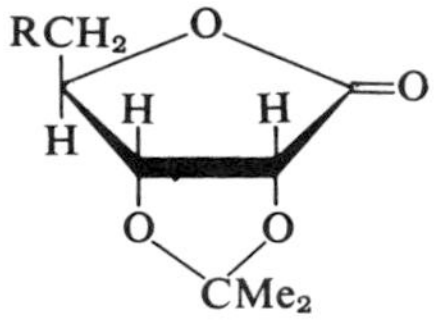

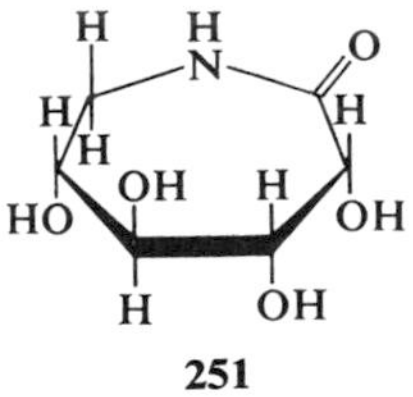

251

B. NITROSATION AND DEAMINATION OF AMINO SUGARS

1. *Reactions Leading to Isolable* N-*Nitroso and Diazo Derivatives*

The reaction of primary or secondary amino groups in sugars and their derivatives with nitrosating agents (for example, nitrosyl chloride, dinitrogen tetraoxide, alkyl nitrites, or solutions of sodium nitrite in aqueous acid) involves the initial formation of an *N*-nitrosoamine, which may be stable enough to be isolated. Nitrosation of the urea derivative[375] **252** and of the bicyclic N-ring[376] derivative **253** by nitrites in aqueous acid resulted in formation of the isolable *N*-nitrosoamines **254** (analogues of the antibiotic streptozotocin, Section IV,A,3) and **255**, respectively. The nitrosoamide **257** [prepared by treating 2-acetamido-1,3,4,6-tetra-*O*-acetyl-2-deoxy-β-D-glucopyranose (**256**) with nitrosyl chloride[377] or dinitrogen tetraoxide[378]] spontaneously decomposes (during several days) to afford a variety[377,378] of nitrogen-free products, the rate of decomposition being accelerated[378] by impurities; photolysis[378] reconverts **257** into the original acetamido derivative **256**. Potassium hydroxide in hot isopropyl alcohol converted[378] compound **257** into a C_5 acetylenic sugar, namely 1,2-dideoxy-D-*erythro*-pent-1-ynitol; the 4-epimer of **256** was likewise converted[379] into 1,2-dideoxy-D-*threo*-pent-1-ynitol, indicating the loss of C-1, and generation of a 2,3-triple bond with retention of the stereochemistry at C-4 and C-5 of the original hexose. Conjugation within the functional group hinders rotation about the N—N bond of nitrosoamines, and the anisotropic, magnetic properties of the N—O double bond[207,380] allow differentiation between the *syn* and *anti* (geometrical) isomers on the basis of chemical shifts measured in the n.m.r. spectrum.

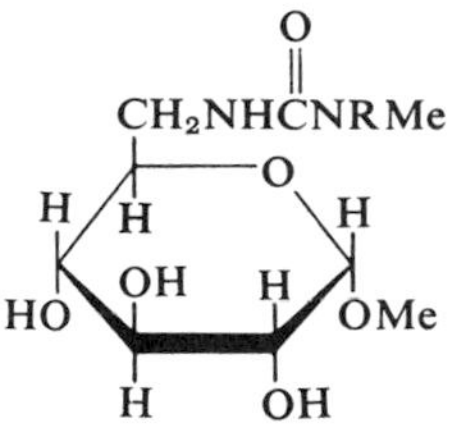

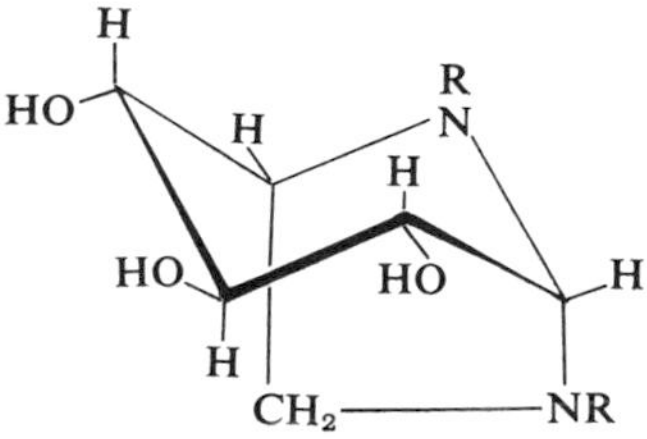

| 252 R = H | 253 R = H | 256 R = H |
| 254 R = N=O | 255 R = N=O | 257 R = N=O |

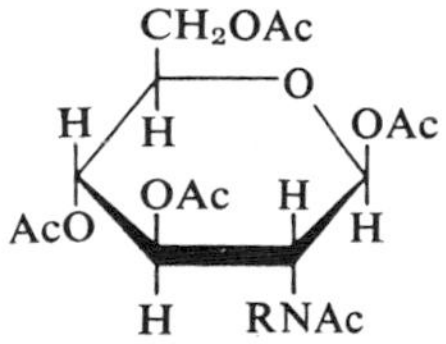

Nitrosation of primary amines is considered to begin similarly, but the initial *N*-nitrosoamine usually decomposes rapidly to generate a diazonium ion. Ethyl 2-amino-4,6-*O*-benzylidene-2-deoxy-D-gluconate (**258**) reacted with nitrous acid to give a stable, nitrogen-containing product[8a] that was later shown[381] to be the diazo sugar **259**. This unusual example of retention of the nitrogen atoms by a carbohydrate-derived, aliphatic diazonium ion accords with the classical observation that electron-withdrawing groups (for example, carbonyl groups[382]) adjacent to the nitrogenous center act to stabilize the diazonium ion and to encourage deprotonation to form diazo compounds.

Photolysis of methyl 3,4,5,6-tetra-*O*-acetyl-2-deoxy-2-diazo-D-*arabino*-hexonate in 2-propanol afforded a 61% yield of methyl 3,4,5,6-tetra-*O*-acetyl-2-deoxy-D-*arabino*-hexonate, whereas thermolysis of the diazo sugar gave[381] mainly the isomeric 2-deoxy-2,3-unsaturated analogue **260**. The diazo sugar **261** was prepared[383] by base-catalyzed decomposition of the 4-[2-(2,4,5-trichlorophenylsulfonyl)hydrazone] of a 4-keto precursor, and illustrates an alternative preparative route to stable diazo sugars. This compound may also be prepared by the action of lead tetraacetate on a hydrazone of the 4-ketone.[383a]

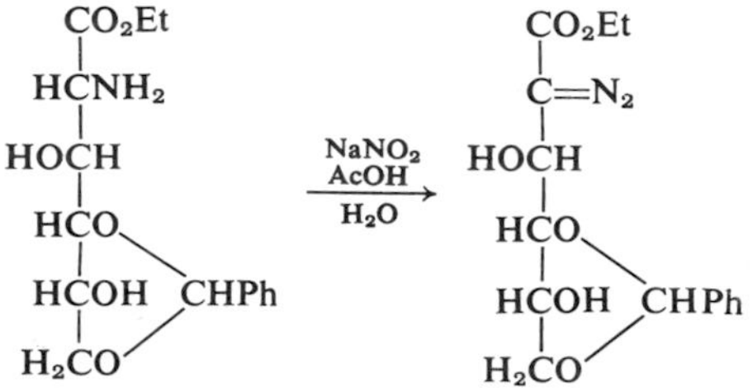

In general, however, the sequences of reactions subsequent to nitrosation of a primary, aliphatic amine include a rapid-decomposition step, in which molecular nitrogen departs and the developing positive charge is dispelled by the attack of a nucleophile or by an elimination step. In practice, a number of different processes often compete in this final step, giving rise to complex mixtures of products, many of which are formed by skeletal rearrangement. As (*a*) a complete accounting has seldom been reported for all the products of deamination reactions, and (*b*) single products are often reported in low yields, the true[384] degree of complexity of most deamination processes is not yet known. The activation energies for reactions subsequent to the slow step of nitrosation are supposedly[385] small, so that conformational factors in the ground state exert a profound effect on the net course of the reaction.

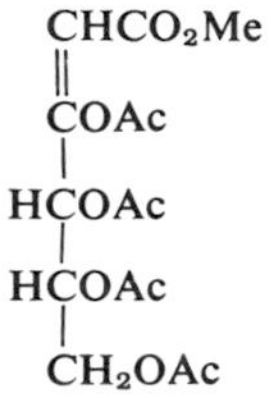

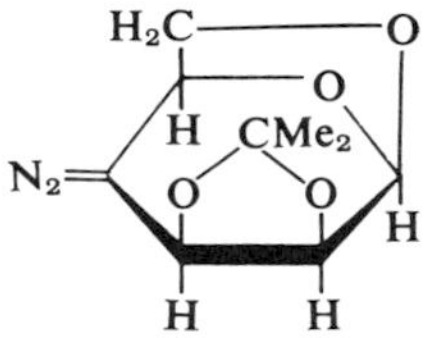

2. *Deamination of Amino Groups Substituted on Sugar Rings*

Ledderhose[386] was the first to treat 2-amino-2-deoxy-D-glucose (**1**) with nitrous acid; the major product of this reaction, 2,5-anhydro-D-mannose (**262**), was also isolated from similar deamination of the methyl α-[387–389] and β-glycosides[388–390] of **1**, and of chitosan,[391] a β-D-(1 → 4)-linked polymer of **1**. Reduction[392] of the unstable anhydro sugar **262**, preferably[387] in the presence of carbon dioxide as a buffer, afforded crystalline 2,5-anhydro-D-mannitol, which was converted[392] by conventional techniques, including glycol-cleavage oxidation, into the symmetrical compound **263** and, in a separate reaction sequence, into the optically active compound **264**; the dissymmetry of **264** revealed that the asymmetry of C-2 and C-5 in **262** is of the same sense (*R* by the Cahn–Ingold–Prelog convention), in accord with the initial, stereochemical assignment by Levene and LaForge.[393]

1

262

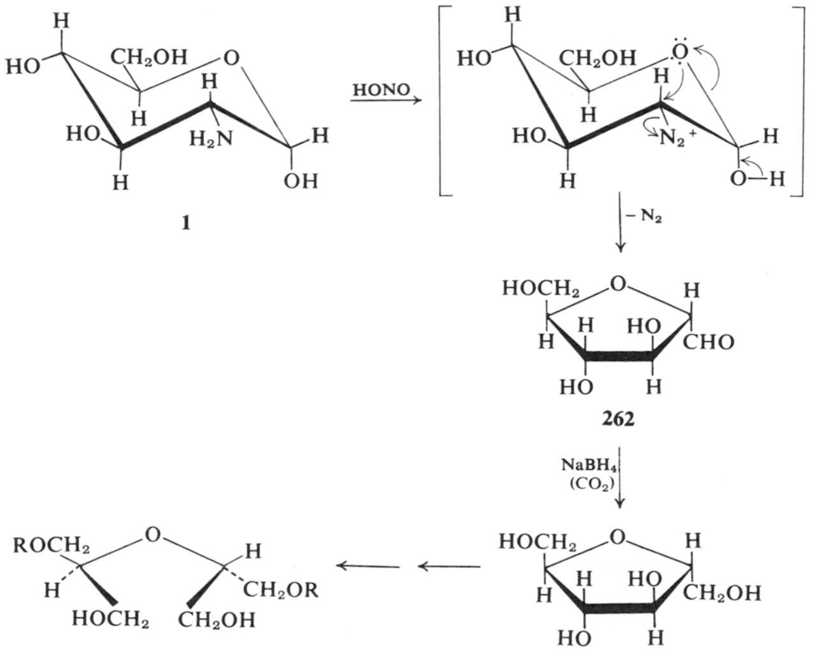

263 R = H
264 R = *p*-MeC$_6$H$_4$SO$_2$

The anhydride **262** is formed by migration of the ring-oxygen atom from its vicinal, antiparallel relationship to the developing vacant orbital, with subsequent deprotonation at O-1 to generate the aldehyde group containing C-1. Much later, it was found[389] that, after nitrosation, the alternative, vicinal, antiparallel group (C-4) of the methyl β-pyranoside of **1** migrates

References start on p. 737.

likewise, affording, as a minor product, the ring-contracted, chain-branched derivative **265**, which, after reduction, was isolated as **266**.

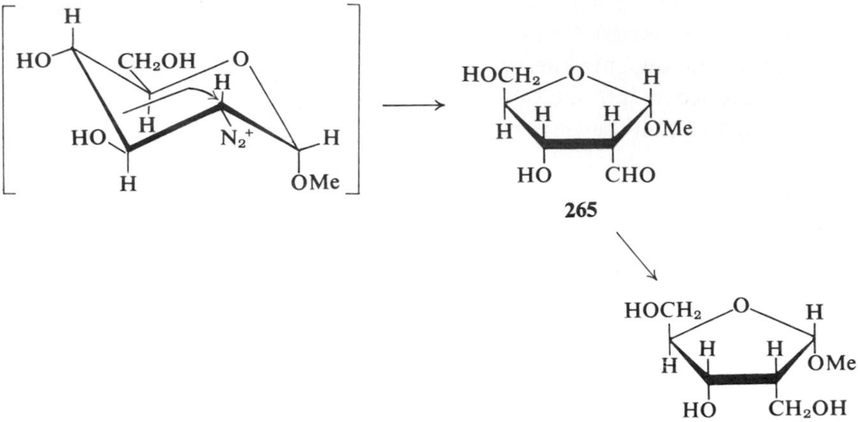

265

266

Upon deamination, 2-amino-2-deoxy-D-galactose similarly gives[394,395] an unstable 2,5-anhydrohexose in good yield. Formation of 2,5-anhydroaldoses in the presence of indole produces a colored species whose absorption at 492 nm has been used to determine[396] (see Vol. IIB, Chapter 45) small amounts of **1** and its 4-epimer. The importance of the equatorial disposition of the amino group for the ring-contraction process was demonstrated in the observation[25,108,133] that 2-amino-2-deoxyaldopentopyranoses, which may exist as equilibrium mixtures of conformers having the amino group respectively axial and equatorial, undergo the latter reaction to afford solutions whose absorptions at 492 nm were interpreted as being proportional to the fraction of the amine present in the "equatorial form" at conformational equilibrium. Methyl 3-amino-3-deoxy-α-D-glucopyranoside[397] (**267**) and its (D-*manno*) 2-epimer[395] **268** undergo deamination with concomitant migration of the vicinal, antiparallel group (C-5), to afford, principally, the branched-chain dialdose derivatives **269** and **270**, respectively, the latter in 60% yield.

As a general rule, factors that act to diminish localization of charge, as nitrogen departs, appear to suppress rearrangements. Thus, deamination of 2-amino-1,5-anhydro-2-deoxy-D-glucitol gave[398] 1,5-anhydro-D-glucitol as the major product, presumably by initial participation of O-5. Had migration occurred, an unstabilized, primary, carbonium ion would have developed at C-1. Likewise, deamination of 1,2,4,6-tetra-O-acetyl-3-amino-3-deoxy-α-D-glucopyranose by sodium nitrite in aqueous perchloric acid led to the formation[399] of isolable amounts of α-D-glucopyranose pentaacetate; re-

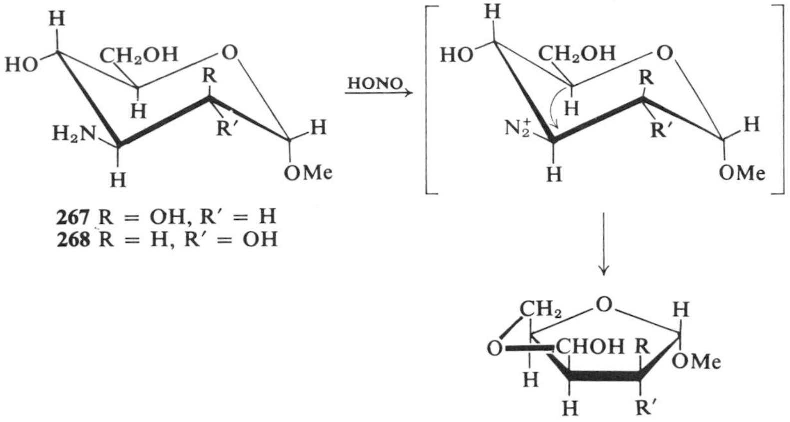

267 R = OH, R′ = H
268 R = H, R′ = OH

269 R = OH, R′ = H
270 R = H, R′ = OH

arrangement is, apparently, suppressed in this example, because of meso-meric withdrawal of electron density from (acylated) O-4, which destabilizes the incipient oxonium ion that would form at C-4 if C-5 were to migrate to the cationic center.

The action of ethyl nitrite in aqueous, ethanolic acetic acid, however, converted[249] 1-(3-amino-3-deoxy-β-D-glucopyranosyl)uracil into 1-(β-D-allo-pyranosyl)uracil, replacement in this example occurring with inversion of configuration, presumably through direct displacement by water in the solvent mixture. A detailed study of the products of deamination of methyl 4-amino-4-deoxy-α-D-glucopyranoside indicated[398,400] that the principal intermediate is the ion **272**, which is formed by participation of the vicinal, antiparallel, ring-oxygen atom; hydrolytic opening of the C-4–O-5 bond of **272** (to form **271**), initial attack of C-2 at C-4 (to form, after epimerization at C-3, **273**), hydrolytic opening of the C-1–O-5 bond of **272** (to form the unstable **274**, whose presence was inferred from its decomposition products), and hydrolytic opening of the C-5–O-5 bond of **272** (to form **275**) were invoked to account for 37%, 14%, ~10%, and 7% yields, respectively, of the starting material.

Deamination of the *trans*-diaxial amino alcohols methyl 2(or 3)-amino-4,6-O-benzylidene-2(or 3)-deoxy-α-D-altropyranoside gave[401] high yields of methyl 2,3-anhydro-4,6-O-benzylidene-α-D-allo(or manno)pyranoside by attack of the vicinal, antiparallel, hydroxyl group; however, a similar reaction of 2-amino-1,6-anhydro-2-deoxy-β-D-glucopyranose led to the isolation of tautomeric forms of 2,6-anhydro-D-mannose, which arise by migration of the *trans*, antiparallel, ring-oxygen atom (O-5) rather than of the vicinal hydroxyl

References start on p. 737.

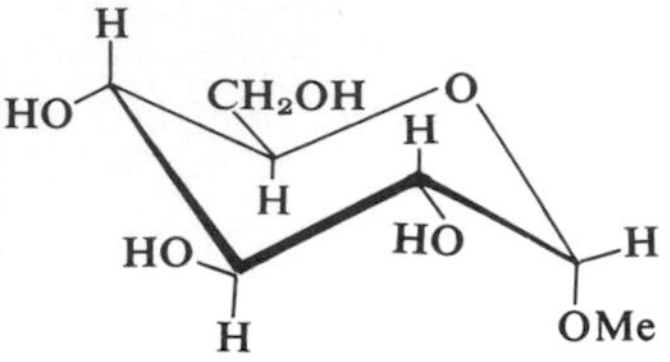

271

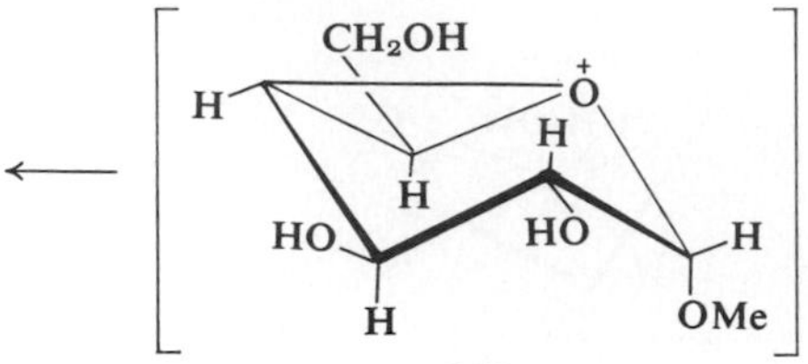

272

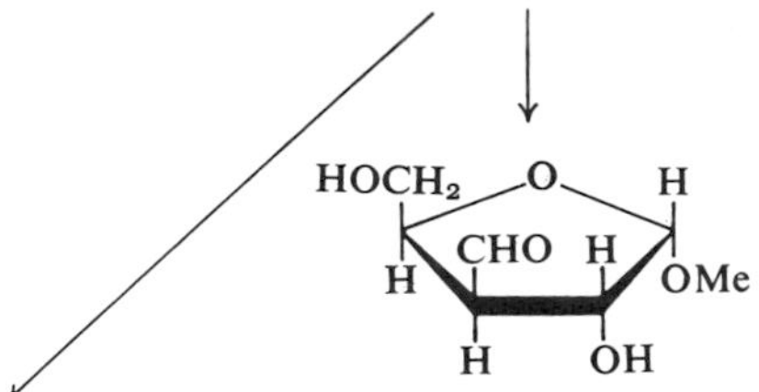

273

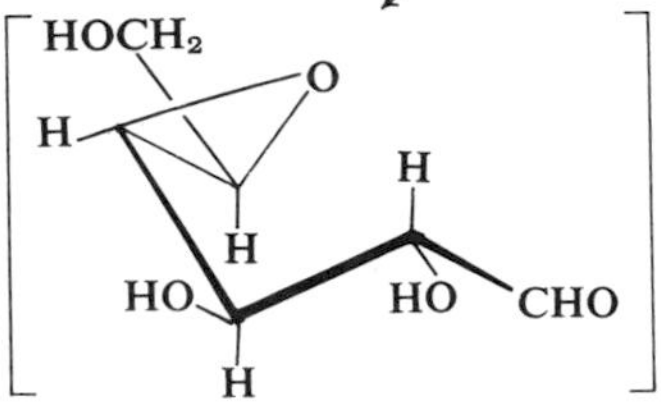

274

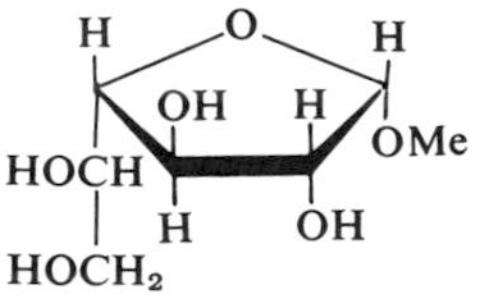

275

group (O-3), suggesting that migration of O-5 entails the development of less strain in the reaction. Deamination of methyl 4-amino-4-deoxy-α-D-galacto-pyranoside (**276**) in dilute, aqueous, acetic acid gave[402] as the major consti-tuent (40% yield) the "normal"[364,403] product of elimination from axial-equatorial amino alcohols–namely, methyl 4-deoxy-α-D-*erythro*-hexopyrano-sid-3-ulose (**277**), together with minor proportions of at least five other products; under similar conditions, however, 2-amino-2-deoxy-D-manno-pyranose gave[304,402,404,405] D-glucose (72% yield[404]), and methyl 3-amino-3-deoxy-β-D-allopyranoside gave[402] mainly (30% yield) methyl β-D-gluco-pyranoside (displacement with inversion), together with at least six minor components. Deamination of 2-amino-2-deoxy-D-mannose in ^{18}O-labeled water affords D-glucose labeled at both O-1 and O-2; from this, Llewellyn and Williams[405a] concluded that a 1,2-anhydride forms as an intermediate subsequent to dissociation of the carbon–nitrogen bond in the diazonium ion. Methyl 2-amino-2-deoxy-α-D-mannopyranoside[405b] and 2-amino-1,5-anhydro-2-deoxy-D-mannitol[405c] give "normal" 2-deoxy-3-ketosugars as major products of deamination.

Deamination of methyl 4-amino-4,6-dideoxy-2,3-O-isopropylidene-α-L-talopyranoside (278) in 90% aqueous acetic acid results in almost complete conversion[406] of the sugar into a mixture of methyl 6-deoxy-α-L-mannopyranoside (279) and its 4-acetate (280); the analogous conversion of the 4-epimer (281) of 278 afforded 279 and 280 as minor (30% total) products

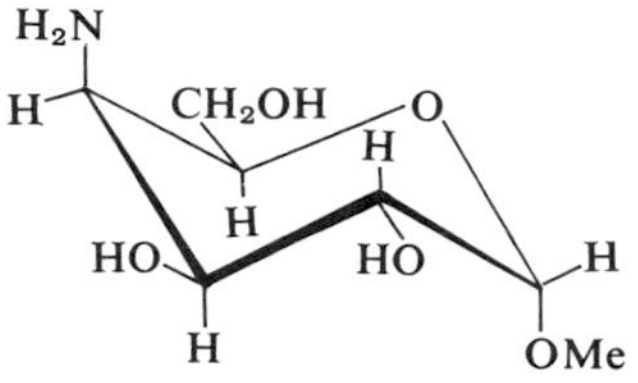

276

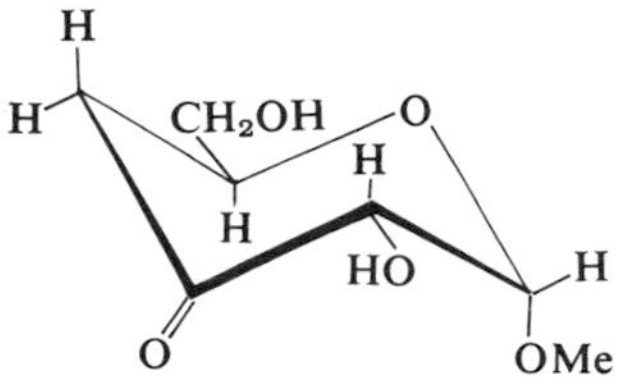

277

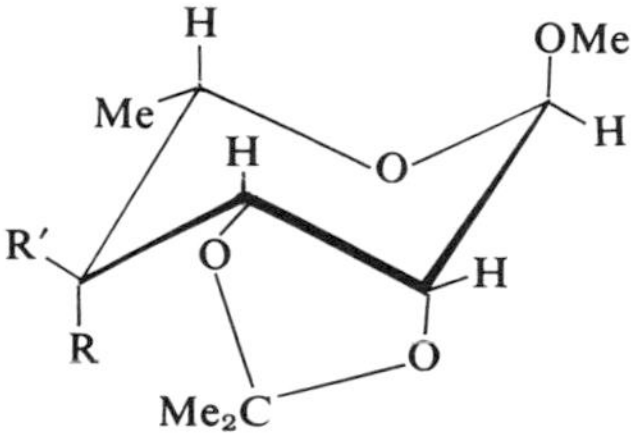

278 R = NH₂, R′ = H
279 R = H, R′ = OH
280 R = H, R′ = OAc
281 R = H, R′ = NH₂

and the ring-contracted products 282 and 283 as major constituents, presumably through displacement reactions, by the solvent, at C-5 or C-4 of the proposed[406] intermediate 284. In the corresponding reaction of methyl 4-amino-4-deoxy-2,3-O-isopropylidene-α-D-lyxopyranoside (286), attack at the primary position preponderates to give[407] 287 and 288 in a net yield of 61%, together with methyl 2,3-O-isopropylidene-β-L-ribofuranoside and its 4-acetate in 39% net yield. Attempted generation of the intermediate 284 by deamination of 285 and of its 5-epimer gave[408] only furan-derived products, reinforcing the impression that the stereochemical requirements for participation are rather exacting.

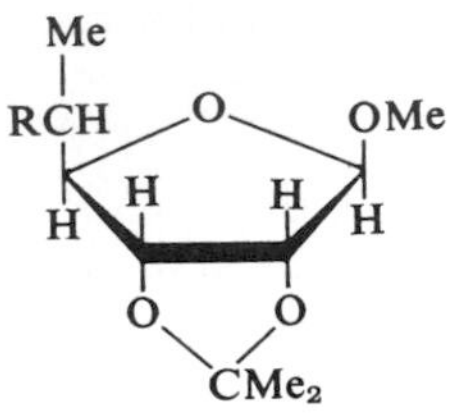

282 R = OH   **284** 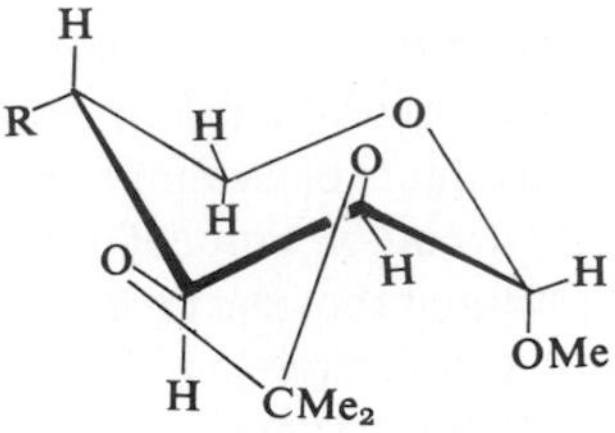   **286** R = NH$_2$
283 R = OAc   **287** R = OH
285 R = NH$_2$ **288** R = OAc

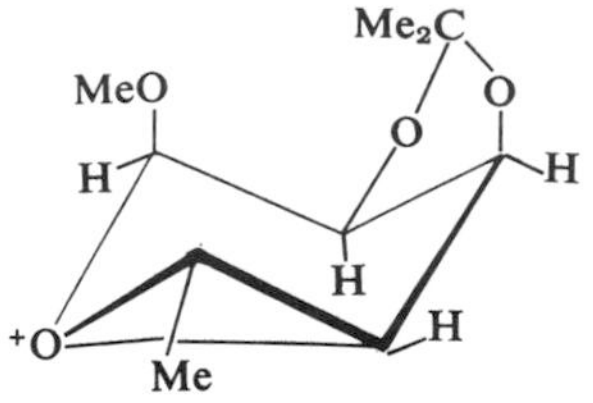

Deamination of both 2,5-diamino-1,4:3,6-dianhydro-2,5-dideoxy-D-gluc-itol and -D-mannitol, in which the amino groups are attached to *cis*-fused, five-membered rings, resulted in formation[409] of 1,4:3,6-dianhydro-L-iditol, presumably by sterically directed, *exo* attack of solvent water on carbonium ion intermediates.

3. *Deamination of Amino Groups Present in Acyclic Sugars or in Side Chains*

The action of nitrous acid on 1-amino-1-deoxy-D-glucitol[410] and -D-mannitol[411] afforded the respective 1,4-anhydroalditols of the same configuration. R. Barker[412] subsequently determined that the 1,4-anhydride is the preponderant product of deamination of 1-amino-1-deoxyhexitols in which the 3- and 5-hydroxyl groups are *threo*-related, whereas substantial proportions of the free alditols are formed, in competition with the anhydride, in deaminations of those isomers having the 3- and 5-hydroxyl groups in the *erythro* relationship; the latter diol arrangement introduces conformational instability[413] about the C-3–C-4 and C-4–C-5 bonds, and presumably favors substantial population of a ground-state conformation from which anhydride formation does not occur.

Detailed investigation[414] of the analogous reaction of 1-amino-1-deoxy-pentitols revealed that the ratio of 1,4-anhydride to acyclic pentitol in the products increases in the order *lyxo* < *arabino* < *ribo* < *xylo*; this order correlates qualitatively with decreasing conformational stability as predicted from intramolecular interactions (see ref. 413 and Vol IA, Chapter 5, Section VII), and it was subsequently demonstrated[415] by 250-MHz n.m.r. spectroscopy that extended conformations are adopted in aqueous solutions by 1-amino-1-deoxy-D-lyxitol hydrochloride (**289**) and its D-*arabino* analogue, but that twisted (sickle) arrangements prevail for the D-*ribo* and D-*xylo* (**290**) isomers. A third product from deamination of the amino pentitols, an isomeric 1,4-anhydride that is formed to the extent of 10 to 20%, was thought[414] to arise by initial attack of O-2 on C-1, with subsequent attack of O-5 on C-2 to displace O-2 with inversion of configuration; no correlation of the latter product with configuration is yet evident.

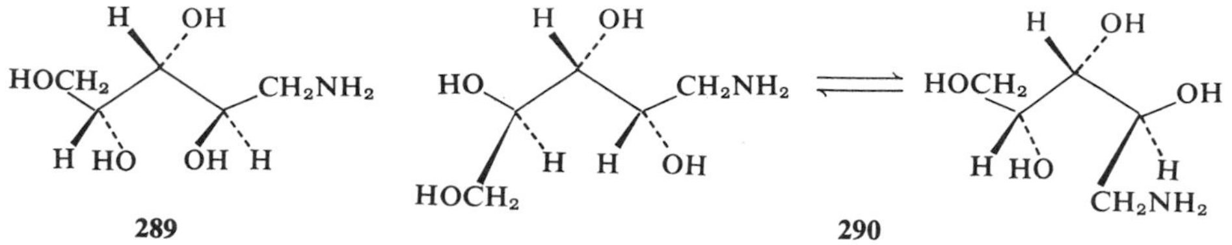

289   **290**

Deamination of 2-amino-2-deoxy-D-glucitol did not yield the anticipated[410] anhydride, but, instead, gave 2-deoxy-D-*arabino*-hexose[416] (**291**, 46% yield), together[417] with smaller proportions of the isomeric 2-deoxy-D-*erythro*-3-hexulose (**292**, 23%), D-mannitol (5%), an impure syrup [tentatively identified as 2-deoxy-2-*C*-(hydroxymethyl)-D-ribo(or arabino)furanose (**293**)], and a fifth, unidentified constituent; α elimination from the 2-cation to produce an enol that subsequently tautomerizes was suggested[418] as the genesis of **291**, and it could account for the formation of **292** as well.

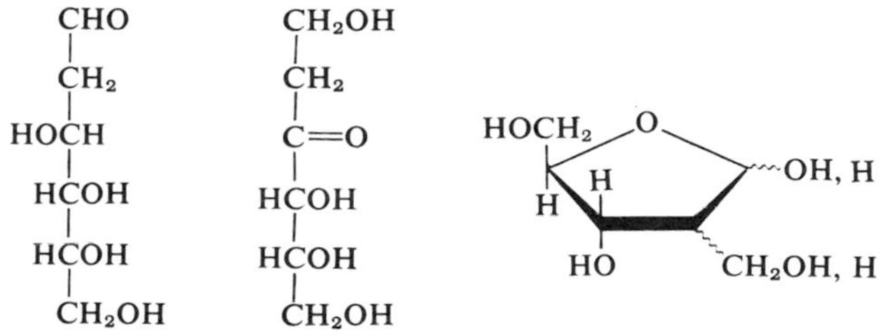

291   **292**   **293**

In contrast, 2-amino-2-deoxyhexonic acids undergo deamination to afford 2,5-anhydrohexonic acids of the same[393,394,419,420] configuration, and Foster[418] proposed initial attack of the carboxyl oxygen atom on C-2 (with inversion) to form an unstable α-lactone that is subsequently opened by attack of O-5 on C-2 (restoring the original configuration) to form the product observed. The apparently anomalous conversion[420] of 2-amino-2-deoxy-D-idono-1,4-lactone into 2,5-anhydro-D-gulonic acid (inversion at C-2) was rationalized[421] as resulting from direct, initial attack by O-5 on C-2 owing to its favorable location at the rear of C-2.

The potentially critical dependence of the reaction course upon conditions is illustrated in the deamination of 2-amino-2-deoxy-D-glucose diethyl dithioacetal (**295**); this compound reacts with nitrous acid in water[422] at pH 0 to form[423] 2-*S*-ethyl-2-thio-D-glucose (**296**). The latter is produced readily[424] by isomerization of the D-*manno* epimer (**298**). In contrast, deamination of **295** in aqueous acetic acid[425] at pH 5 gave[426] the dithiofuranoside **294**. Initial attack at C-2 by a sulfur atom, forming an intermediate episulfonium ion (**297**), would generate the 2-thio-D-*manno* skeleton, and subsequent attack by O-4 would then lead[423,426] to the thiofuranoside **294**. Although attack by water on **297** (to give **298**) with subsequent epimerization to **296** appears a plausible route to **296**, the intermediacy of **298** has not been detected; solvent attack on **297** may be preceded by a second C-2 inversion[426b] in the route to **296**.

299 R = NH_2, R′ = H, R″ = OH
300 R = NH_2, R′ = OH, R″ = H
301 R = NH_2, R′ = SCH_2Ph, R″ = H
302 R = OH, R′ = SCH_2Ph, R″ = H

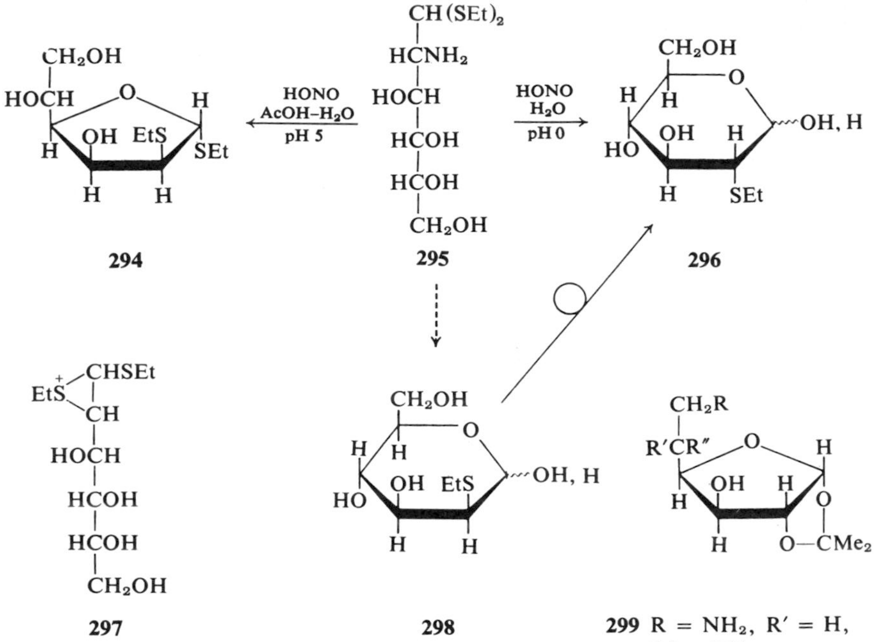

6-Amino-6-deoxy-1,2-*O*-isopropylidene-β-L-idofuranose (**299**) cyclized to the corresponding 3,6-anhydrofuranose[427] upon deamination, whereas similar deamination of the D-*gluco* isomer (**300**) afforded[410] 5,6-anhydro-1,2-*O*-isopropylidene-α-D-glucofuranose instead; presumably formation of the

3,6-anhydride from **300** is disfavored, because the product would contain a 5-*endo* substituent. As both of these reactions proceed by participation of oxygen atoms, it remains to be disproved that deamination of the 5-thio sugar derivative **301** to afford the product of replacement (**302**) occurs by formation and reopening of an episulfonium ion, rather than without any participation of the powerfully nucleophilic sulfur atom, as proposed[428] (see also Chapter 18, Section III,A,4,*a*).

Deamination of 1-amino-1-deoxy-D-fructose[429] and -D-tagatose[430] (**303**) gave D-fructose and D-tagatose (**304**), respectively, in fair yield; 6-amino-6-deoxy-1,2:3,4-di-*O*-isopropylidene-α-D-galactopyranose[95] and 1,2,3,4-tetra-*O*-acetyl-6-amino-6-deoxy-D-glucopyranose[431] were, likewise, readily converted into the corresponding 6-hydroxy derivatives by the action of nitrous acid. In contrast, deamination of an unsubstituted example (6-amino-6-deoxy-D-galactose) gave[95] a syrupy mixture of products, and similar reaction[432] of 1-amino-2,6-anhydro-2-deoxy-D-*glycero*-D-*galacto*-heptitol (**305**) gave the replacement product **306** (58%), together with both anomers of **307**. The authors[432] interpreted the latter product mixture as arising from reactions of diazotized **305** as different rotamers about the C-1–C-2 bond.

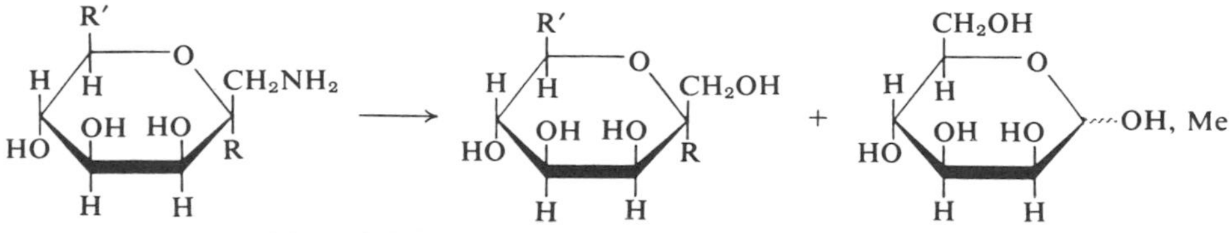

303 R = OH, R′ = H
305 R = H, R′ = CH₂OH

304 R = OH, R′ = H
306 R = H, R′ = CH₂OH

307

4. Reactions of Miscellaneous Amines with Nitrous Acid

Nitrosation of methyl 4,6-*O*-benzylidene-2,3-imino-β-D-allo(and D-manno)-pyranoside gave[433] methyl 4,6-*O*-benzylidene-2,3-dideoxy-β-D-*erythro*-hex-2-enopyranoside, and the 4-epimeric alkene was similarly obtained[185] from the D-*gulo* and D-*talo* imines; the D-*allo* N-nitrosoimine was isolable at low temperatures, but decomposed rapidly to the alkene. 2,5-Anhydro-3,4-dideoxy-3,4-imino-L-arabinose diisobutyl dithioacetal was converted into 2(*R*)-[di(isobutylthio) methyl]-2,5-dihydrofuran[131] by nitrosation in aqueous acetic acid. 2-(3-Deoxy-1,2:5,6-di-*O*-isopropylidene-α-D-allofuranos-3-ylidene)aziridine underwent deaminative elimination to form 3-deoxy-1,2:5,6-di-*O*-isopropylidene-3-*C*-methylene-α-D-*ribo*-hexofuranose.[193a]

Selective nitrosations have been demonstrated. 3'-Amino-3'-deoxyadenosine (308) gave[434] an 8% yield (based on 50% recovery of starting material) of 3'-amino-3'-deoxyinosine (309) upon partial deamination under controlled conditions, indicating selective attack on the aryl amino group. 1-Amino-1,2,6-trideoxy-2,6-imino-D-iditol (310) reacted[207] with silver nitrite in $2M$ hydrochloric acid at the imino group only, to afford the nitrosoimino amine 311, whereas the same reagent in 1% acetic acid caused deaminative cyclization of 310 to form 2,6-anhydro-1,5-dideoxy-1,5-(nitrosoimino)-D-iditol (312).

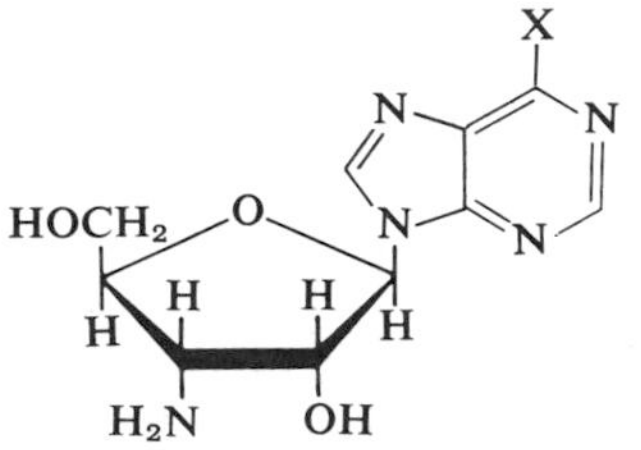

308 X = NH$_2$
309 X = OH

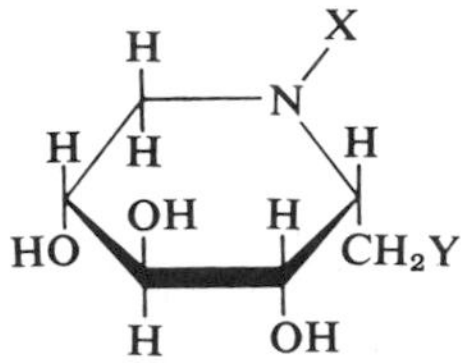

310 X = H, Y = NH$_2$
311 X = NO, Y = NH$_3$Cl

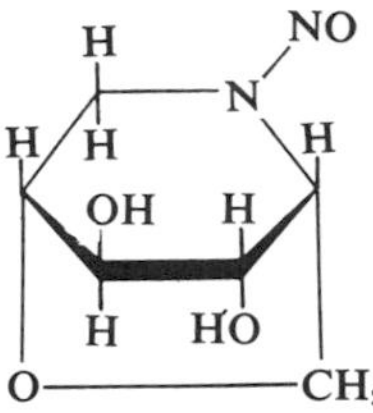

312

Hydrolytic cleavage of the aglycon from 2-amino-2-deoxyglycosides proceeds only[435] with difficulty; nitrous acid deamination offers a mild, alternative method whereby relatively sensitive aglycons may be recovered intact and selective glycosidic cleavages may be accomplished, as in the partial degradation[436] of heparin to yield oligosaccharide derivatives. In sugar derivatives containing other sensitive features, unsubstituted amide groups (—CONH$_2$) have been converted into carboxyl groups by treatment[437] with nitrous acid.

5. Other Reactions in Which Deamination Occurs

The action of mercuric oxide in hot water has been stated to convert[304] 2-amino-2-deoxy-D-mannose into 2,5-anhydro-D-glucose. Oxidation of 2-amino-2-deoxy-D-glucose (1) by nitric acid is accompanied by deamination, the product isolated[438] being 2,5-anhydro-D-mannaric acid (313). An excess of phenylhydrazine effects oxidative deamination of 2-amino-2-deoxyaldoses, as in the conversion[439] of 1 into D-*arabino*-hexulose phenylosazone, and treatment of 1 with *o*-phenylenediamine in the presence of copper(II) acetate gave[440] the formally related quinoxaline derivative 314.

Other oxidative deamination reactions of 2-amino-2-deoxyaldoses result in chain shortening. Storage of 2-amino-2-deoxy-D-glucose (free base) in aqueous solution results in gradual, spontaneous[441] decomposition to a mixture of products including D-arabinose and 5-(hydroxymethyl)-2-furaldehyde; the degradation is accelerated[442] in alkaline solution. The action of hydrogen peroxide on 2-amino-2-deoxy-D-glucose (1) gives a mixture of acidic products[443] plus D-arabinose; better yields of D-arabinose result from treatment of 1 with hypochlorite[444] ion, N-chloro-p-toluenesulfon-amide[445] (*chloramine T*), chlorine,[446] and ninhydrin.[447] The blue color produced in the last reaction has found use in the analysis[431,448] (qualitative and quantitative) of amino sugars (see Vol. IIB, Chapter 45).

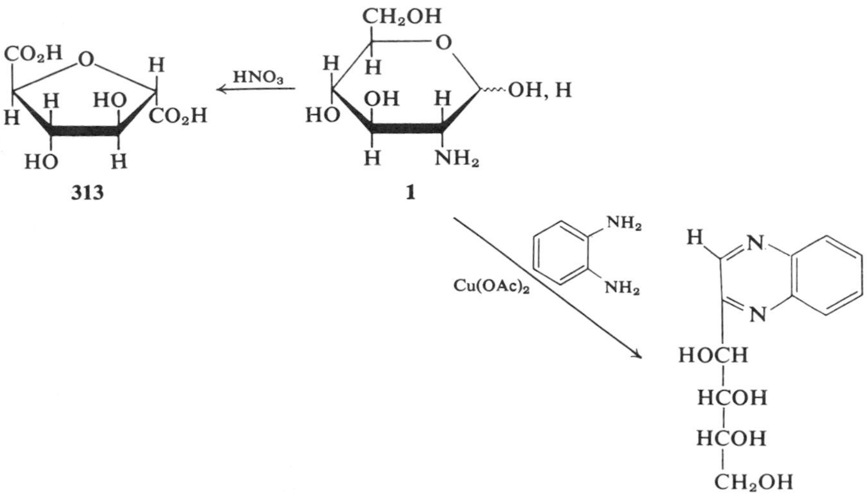

In the presence of an aqueous solution of barium hydroxide, gaseous oxygen[449] converts 1 into D-arabinonic acid. More-extensive degradation of 1 to levulinic and formic acids takes place[450] upon heating the amino sugar in dilute sulfuric acid, and the action of hot, dilute sulfuric acid and lead oxide converts 1 into a mixture[451] of formaldehyde and formic acid.

C. OTHER REACTIONS FOR CONVERTING AMINO SUGARS INTO HETEROCYCLES

1. *1,2-Oxazolines and Glycosyl Substitution Reactions*

The usual preparative sequences for glycosyl halides are generally applicable to amino sugars as well. Treatment with acetyl chloride converts[452] 2-acet-amido-2-deoxy-D-glucose into 2-acetamido-3,4,6-tri-O-acetyl-2-deoxy-α-D-glucopyranosyl chloride (315); however, 3-acetamido-2,4,6-tri-O-acetyl-3-deoxy-D-galactopyranosyl bromide, which is prepared by treatment of the corresponding peracetate with hydrogen bromide–acetic acid, reacts further if not isolated at the appropriate time, yielding[453] a 1,6-dibromide. Such chlorides as **315** are generally to be preferred as synthetic intermediates, because the fluoride **316** is rather[454] inert, whereas isolation of the bromide **317**, which is activated by extremely efficient participation of the acylamido group, is possible only under anhydrous[455] conditions. Compound **317** was found to undergo further reaction to form[455] the saturated heterocycle **318**, whereas the 2-benzamido analogue **319** underwent facile[456] conversion into the isolable oxazolinium bromide **320**; a subsequent study[457] established that the three forms are present in an equilibrium in which (*a*) electron-withdrawing *R*-groups (such as CF_3 or p-$C_6H_4NO_2$) favor formation of the glycosyl halide, (*b*) *R*-groups (such as p-C_6H_4OMe) that donate by resonance tend to favor adoption of the conjugated oxazoline form, and (*c*) inductively donating groups (such as Me) direct quaternization of the nitrogen atom in the oxazoline ring, although solvent effects may also influence the composition of the equilibrium[458] mixture. General preparations of oxazolines have been accomplished by treating 2-acylamino-2-deoxyaldoses with hot[459] acetic anhydride–zinc chloride, by treating acylated glycosyl halides with silver(I) salts of strong acids in pyridine under aprotic[460] conditions, and, perhaps best, by treating peracyl derivatives with ferric chloride[461] in dichloromethane. The action of acidic acetone on 2-benzamido-2-deoxy-D-glucose gave[462] the furano[2′,1′:4,5]-2-phenyl-2-oxazoline **321**. (Nonanomeric oxazolines are discussed in Section II,E.)

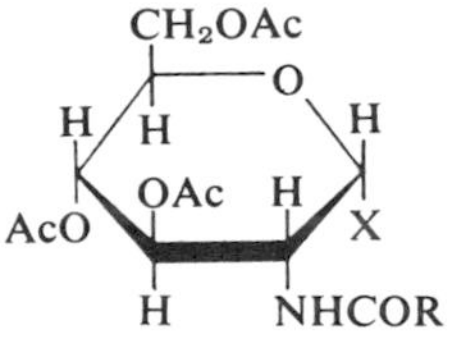

315 R = Me X = Cl
316 R = Me X = F
317 R = Me X = Br
319 R = Ph X = Br

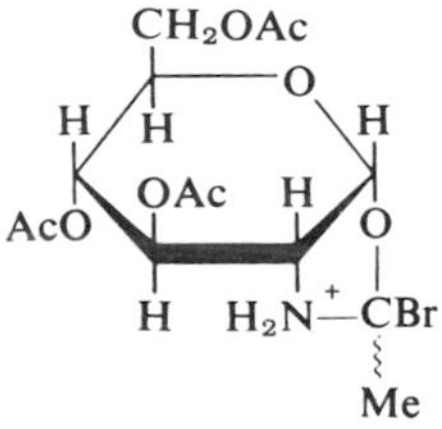

318

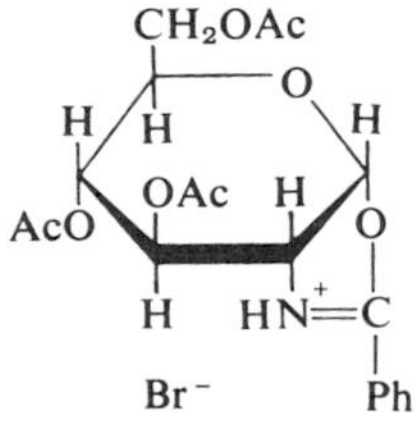

320

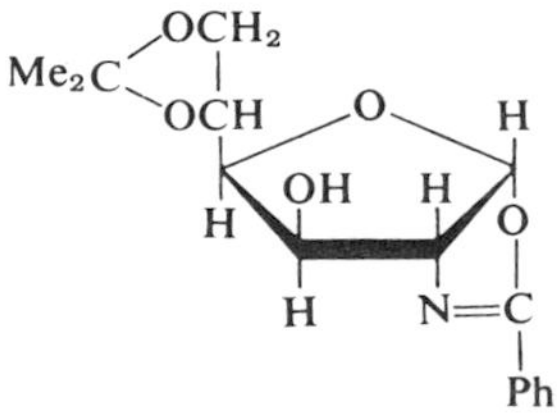

321

2-Acetamido-2-deoxy-D-glucose undergoes condensation, under acid catalysis, with alcohols, arylamines, and thiols, to afford the corresponding glycosides,[435] glycosylamines,[463] and thioglycosides.[464] The last reaction produces only the β anomer, and it was suggested that an oxazolinium intermediate occurs as an intermediate form. Direct glycosidation of 2-acetamido-2-deoxy-D-mannose gave[260] all four possible methyl glycosides. By analogy with the Koenigs–Knorr reaction of O-acylglycosyl halides[464a] (Vol. IA, Chapter 9, Section II,C,1), condensation of 315 with nucleophiles [for instance, 9-(chloromercuri)-6-(dimethylamino)purine[465] or silver thiocyanate[466]] affords the 1,2-*trans* product (322 and 323, respectively); the idea of closed-ion intermediates is consistent with the observations that methanol adds to 320 to give[456] the β-glycoside, and that 1,3,4,6-tetra-O-acetyl-2-amino-2-deoxy-α-D-glucopyranose hydrobromide undergoes acetyl migration during reaction with silver thiocyanate to form[467] 323. In a later paper[468] it was asserted that glycopyrano[2′,1′:4,5]-2-oxazolines are more reactive than glycosyl halides to oxygen nucleophiles. The extension to furanosyl derivatives is uncertain, because, in acidic methanol, compound 321 and its 3-epimer underwent rapid ring expansion[469] to form the methyl β-pyranosides as preponderant, and major, products, respectively.

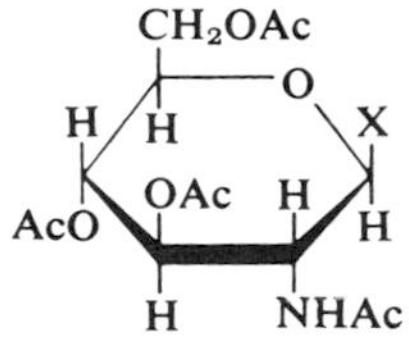

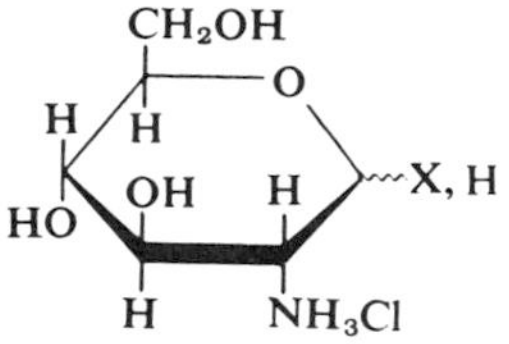

322 X = [purine structure with NMe₂]

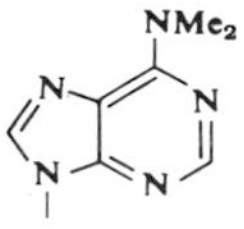

323 X = NCS

324 X = OMe

325 X = NMe₃Cl

Reactions proceeding through closed-ion intermediates have the feature that only one anomer of the product is formed in appreciable proportion. Access to the second anomer is sometimes gained by the use of glycosyl halide derivatives in which a (powerfully deactivating) nonparticipating group protects the nitrogen atom. In the presence of an acid acceptor, 3,4,6-tri-O-acetyl-2-amino-2-deoxy-α-D-glucopyranosyl bromide hydrobromide condenses with methanol[470] or trimethylamine[471] to afford, after acidification, salts of 324 and 325, respectively. Reaction of 3,4,6-tri-O-acetyl-2-deoxy-2-(2,4-dinitroanilino)-α-D-glucopyranosyl bromide with 9-benzamido-6-(chloromercuri)purine gave[472] an anomeric mixture of nucleosides from which the α-D-pyranose was isolated as a minor product; however, a similar condensation of the 2-trifluoroacetamido analogue gave[473] only the β product, from which the protecting groups were removed in cold, methanolic ammonia. Glycosidation of Schiff-base derivatives—such as 3,4,6-tri-O-acetyl-2-deoxy-2-salicylideneamino-α-D-glucopyranosyl bromide, which was condensed with methanol to afford[299,474] the corresponding β-D-glucoside **326**—can be accomplished without cleavage of the imino group. Pyrolysis of the 2-methyl analogue of **320** in a mixture of benzene and tetra-N-methylurea containing p-toluenesulfonic acid gave[475] a good yield of the glycal derivative **327**.

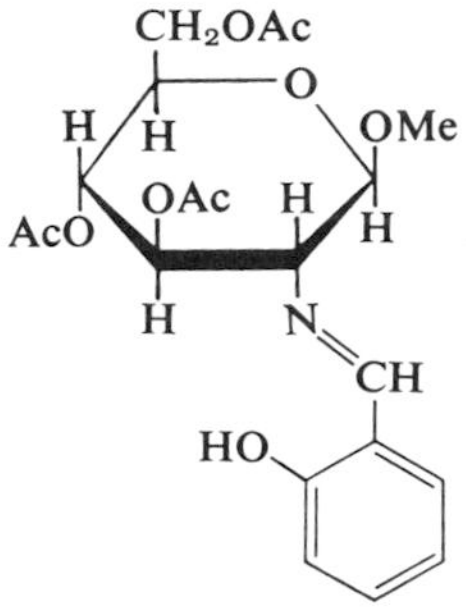

326 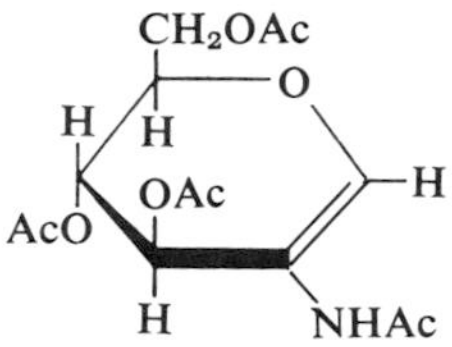   **327**

Glycosidation (as well as other acid-catalyzed reactions) of amino sugars is inhibited by protonation[435] of the amino group. Hydrolysis of 2-aminoglycopyranosides[435,476] is extremely slow, whereas the 2-acetamido analogues react faster than the corresponding nitrogen-free glycosides until N-deacetylation has proceeded to a large extent; N-(2,4-dinitrophenyl) derivatives are hydrolyzed at a constant[476] rate, slightly lower than the initial rate for the N-acetyl derivatives. Likewise, alkyl 2-amino-2-deoxy-D-glucofuranosides are strongly resistant to acid hydrolysis, whereas the corresponding N-acetyl derivatives are hydrolyzed very readily.[477]

2. Other Heterocycles Containing Nitrogen Atoms

2-Amino-2-deoxy-D-glucose (**1**) and its derivatives have long been known to react with phenyl isocyanate[478] and phenyl isothiocyanate[479] to give heterocyclic products; controversy about the tautomeric identity (detailed in ref. 3, pp. 114–116) of these and related molecules was resolved by periodate-oxidation studies, which showed[480] the product of reaction with 3,4-dichlorophenyl isothiocyanate to be the furanose **328**. Reaction of **1** with cyanamide gave a mixture[481] of **329** plus an acyclic isomer **330**, both of which may be derived through dehydration of a monocyclic imidazolin-4-ol[482] intermediate, whereas 1-amino-1-deoxy-D-fructose reacted with phenyl (as well as other) isothiocyanates to form[483] only the acyclic product **331**; however, the action of ethyl iminoacetate in *N*,*N*-dimethylformamide converted[484] **1** into α-D-glucopyrano[2′,1′:4,5]-2-methyl-1*H*-imidazoline (**332**). The related structure **333** was assigned[485] to the product from reaction of D-glucose with thiocyanate ion, whereas the adduct of **1** and thiocyanate appears[486] to have a structure analogous to **331**. 2-Amino-2-deoxy-D-gluconic acid combines[487] with phenyl isocyanate to form the hydantoin **334**. Acetolysis of 2-acetamido-2-deoxy-*N*-*p*-tolyl-β-D-glucopyranosylamine affords 2-methyl-4-(D-*arabino*-1,2,3,4-tetraacetoxybutyl)-1-*p*-tolylimidazole in 12% yield.[487a] Carbon disulfide slowly condenses with **1** (in methanol at 5°) to form

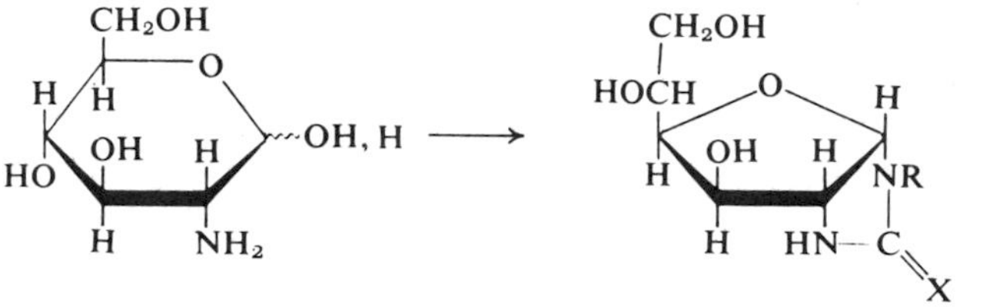

1

328 R = 3,4-$C_6H_3Cl_2$, X = S
329 R = H, X = NH

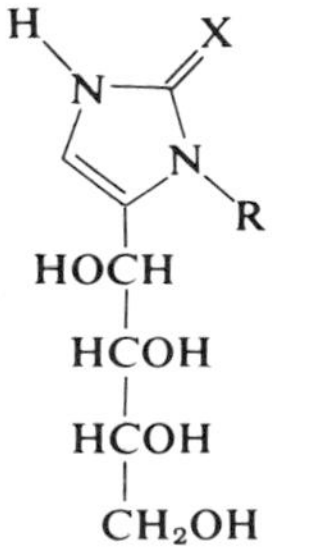

330 R = H, X = NH
331 R = Ph, X = S

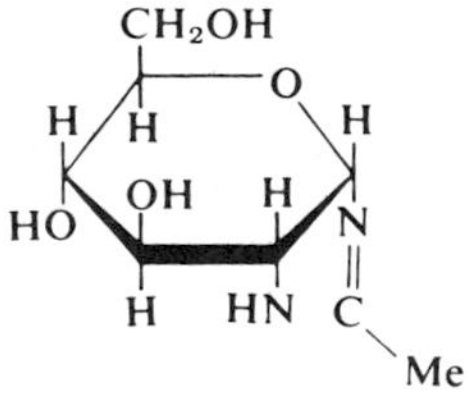

332

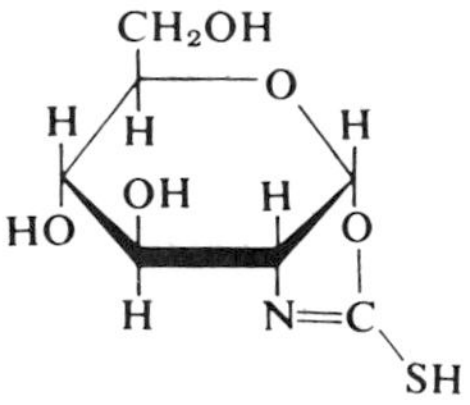

333

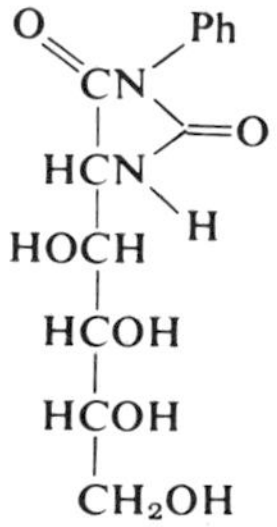

334

References start on p. 737.

5-hydroxy-4-(D-*arabino*-tetrahydroxybutyl)-2,3-dihydrothiazoline-2-thione.[487b] The tetra-*O*-acetyl derivative of **1** reacts with isocyanates to form urea derivatives.[487c] Treatment of 2-amino-2-deoxy-D-gluconic acid with hot sodium acetate–acetic anhydride gives small amounts of 3-acetamido-6-(acetoxymethyl)-2*H*-pyran-2-one[487d,e] plus far larger amounts of (*E*)- and (*Z*)-3-acetamido-5-(2-acetoxyethylidene)-2,5-dihydrofuran-2-one.[487e]

Vicinal amino alcohols react with phosgene, either directly or indirectly, to form oxazolidinone derivatives, which are suitable for protection. Thus, the action of Lewis acids converted[462] 1,3,4,6-tetra-*O*-acetyl-2-(benzyloxycarbonyl)amino-2-deoxy-β-D-glucopyranose into 3,4,6-tri-*O*-acetyl-α-D-glucopyrano[2′,1′:4,5]-2-oxazolidinone, whereas *trans*-[2′,3′:4,5]-2-oxazolidinones arose from treatment of benzyl 2-amino-4,6-*O*-benzylidene-2-deoxy-β-D-glucopyranoside with phosgene[488] in the presence of pyridine, and from the reactions of methyl 2-amino-2-deoxy-α-D-glucopyranoside[489] and of a 2-deoxy-2-methylamino-α-L-glucopyranoside[490] with *p*-nitrophenyl chloroformate in the presence of hydroxide ion; the same conditions introduced[489] a 4,6-carbonyl bridge into methyl 6-amino-6-deoxy-α-D-glucopyranoside. Sodium hydroxide in aqueous ethanol acted[491] to convert methyl 2-(benzyloxycarbonyl)amino-2-deoxy-6-*O*-*p*-tolylsulfonyl-α-D-glucopyranoside into the tricyclic anhydride **335**, presumably by initial closure of the 3,6-anhydro ring, with subsequent formation of the tetrahydro-1,3-oxazin-2-one ring. 6-Amino-6-deoxy-1,2:3,4-di-*O*-isopropylidene-α-D-galactopyranose reacted with phosgene to afford the monomeric isocyanate[492] derivative **336**, and with cyanamide to form the guanidino[493] analogue **337**; the carbamate formed by treating **336** with propargyl alcohol was prepared as a potential therapeutic agent.[492] Sequential treatment of 3-amino-3-deoxy-1,2:5,6-di-*O*-isopropylidene-α-D-allofuranose with (*a*) cyanamide, (*b*) dilute hydrochloric acid, and (*c*) acetic anhydride gave[493] **338**. Acid-catalyzed condensation of 2-acetamido-2-deoxy-D-mannose with 2,2-dimethoxyethane gave 2-acetamido-2-deoxy-2-*N*,3-*O*-isopropylidene-5,6-*O*-isopropylidene-D-mannofuranose and the methyl α-furanoside; 2,3-*trans* analogues did not react similarly.[494]

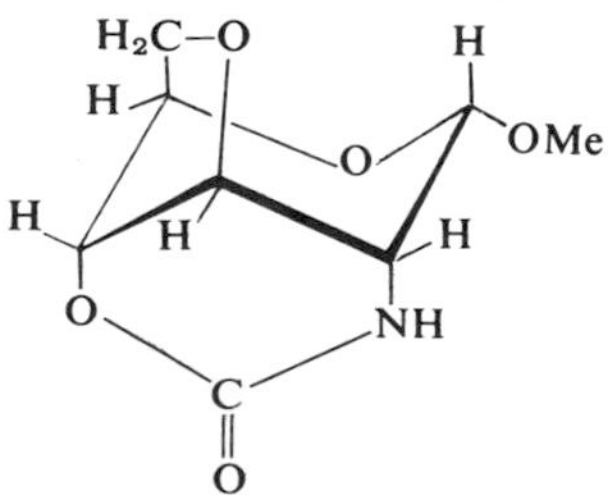

335 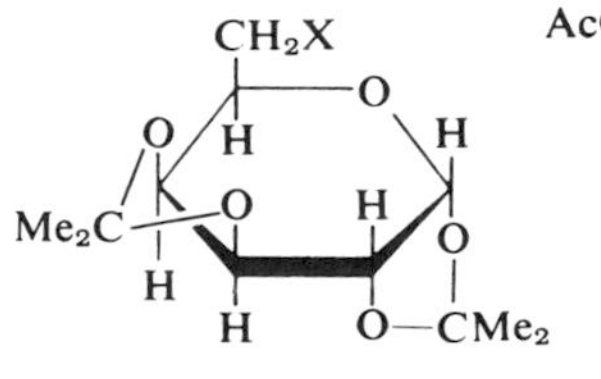   **336** X = NCO   **338**
              337 X = NHC(=NH)NH₂

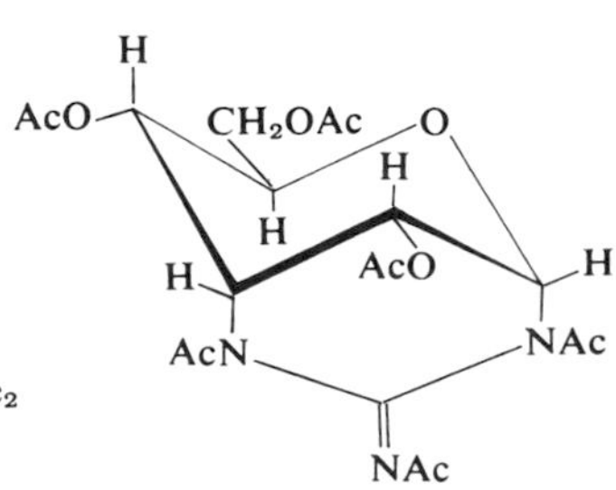

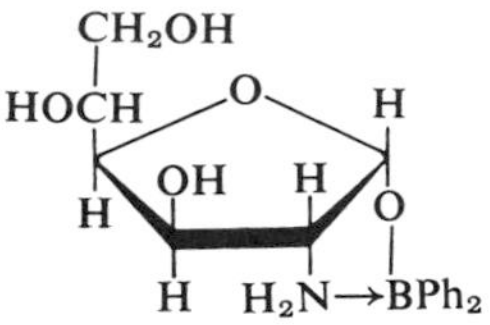

339

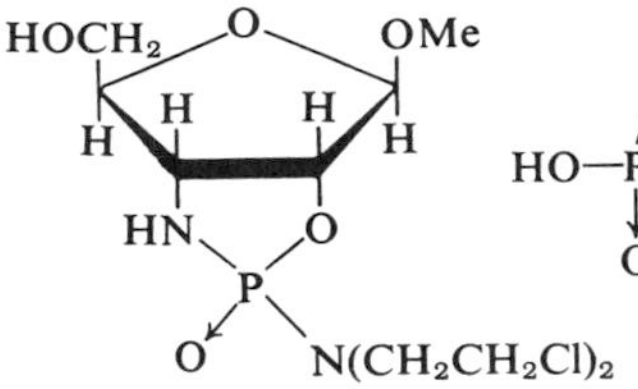

340

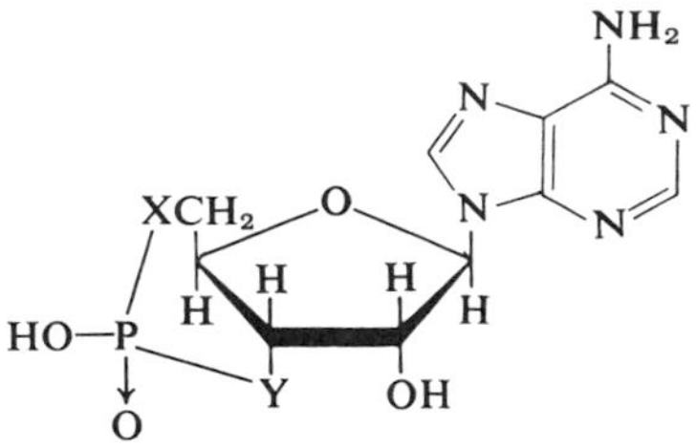

341 X = NH, Y = O
342 X = O, Y = NH

Reaction of **1** with isobutyl diphenylborinate gave[495] **339**, and condensation of methyl 3-amino-3-deoxy-β-D-ribofuranoside with bis(2-chloroethyl)-phosphoramido dichloride gave[496] the potential antineoplastic agent **340**. The isomeric adenosine 3′,5′-cyclic phosphate analogues **341** (ref. 497) and **342** (ref. 498) compete with the natural nucleotide in activating the protein kinase of bovine muscle.

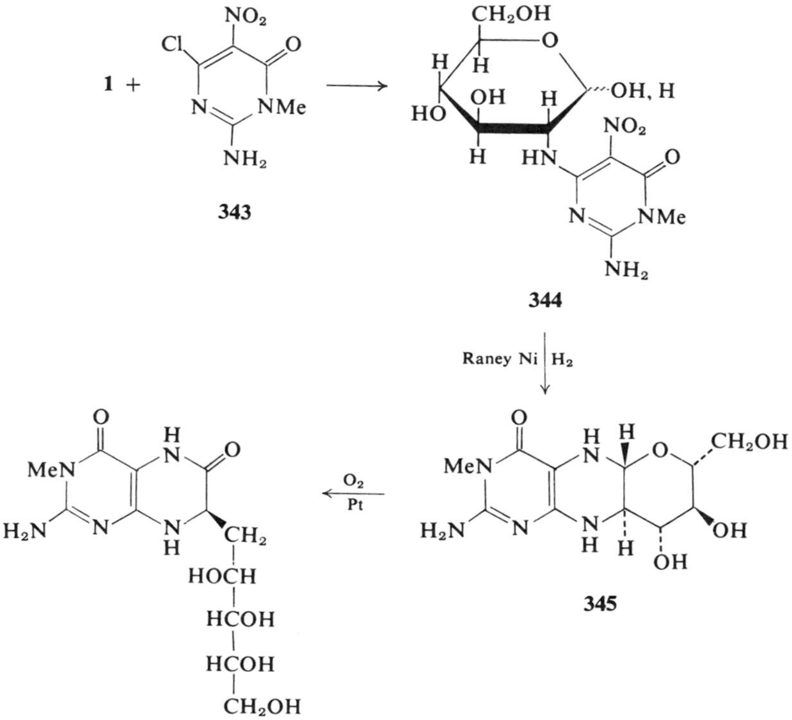

343, **344**, **345**, **346**

References start on p. 737.

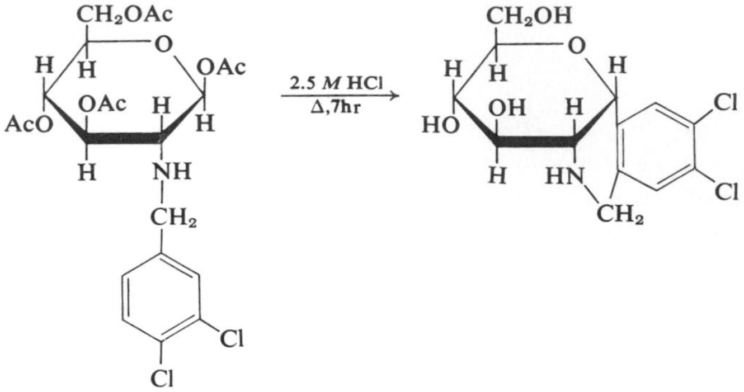

347   348

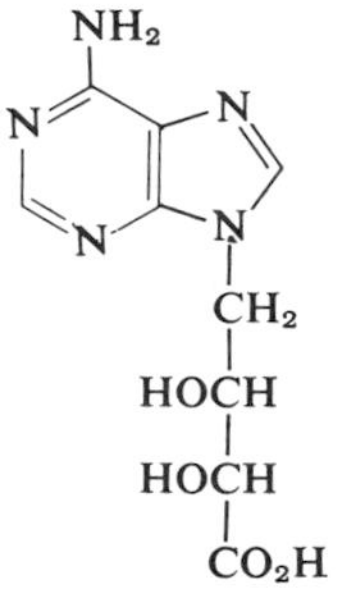

349

Amino sugars have functioned as starting materials in syntheses of heterocycles. Nucleophilic, aromatic substitution on **343** by 2-amino-2-deoxy-D-glucose (**1**) gave **344**, which cyclized spontaneously after reduction of the nitro group to form the pteridine derivative **345**; catalytic oxidation gave[499] **346**. Internal, electrophilic substitution to form **348** resulted[500] from acidification of **347** and subsequent deprotonation of the product. The natural hypocholesterolemic agent **349** was prepared by treatment of 4-amino-4-deoxy-2,3-O-isopropylidene-D-erythronic acid with 4-amino-6-chloro-5-nitro-pyrimidine, followed by reduction, and subsequent N-formylation of the 5-substituent, and base-catalyzed cyclization.[501]

3. *Pyrroles and Other Products of Aldol-Type Reactions of Amino Sugars*[502]

Pauly and Ludwig[503] condensed 2-amino-2-deoxy-D-glucose (**1**) with 2,4-pentanedione and with ethyl acetoacetate, obtaining initial products that were subsequently treated with *p*-dimethylaminobenzaldehyde (*Ehrlich's reagent*) to generate an intensely red compound; the initial product (**350**) of the former reaction was afterward isolated[504] in pure form, but the analogous initial product (**351**), which was subsequently[505] characterized from the condensation with ethyl acetoacetate, had physical constants in disaccord with those initially[503] reported. Cessi and Serafini-Cessi[506] identified the Schiff base **352** in the product mixture from the reaction of **1** with 2,4-pentanedione, and the tetra-*O*-acetyl analogue **353** was formed in good yield[507] from reaction of the tetra-*O*-acetyl-β-D derivative of **1** with the same diketone. Deacetylation and cyclization of **353** by the action of aqueous barium hydroxide afforded[507] **350**, suggesting the intermediacy of **352** in the formation of the pyrrole ring. Serafini-Cessi and Cessi[508] reported the isolation of the 4-epimer of **352** from treatment of 2-amino-2-deoxy-D-galactose with 2,4-pentanedione; in the presence of triethylamine, the same reactants afford a 1:2 adduct, identified as **354**, which presumably arises by nucleophilic addition to the carbonyl group of an initially formed enamine. Base-catalyzed liberation of the amino group was proposed as the final step preceding cyclization to form the pyrrole ring.

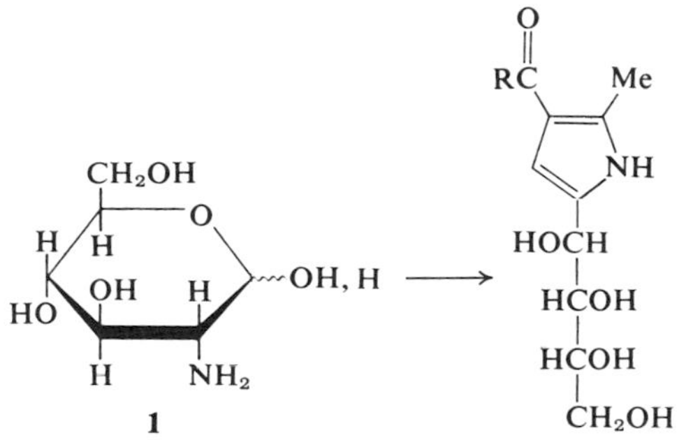

350 R = Me
351 R = OEt

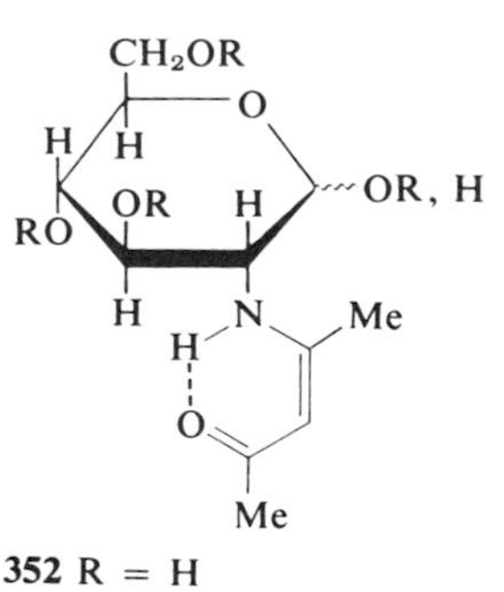

352 R = H
353 R = Ac (β-anomer)

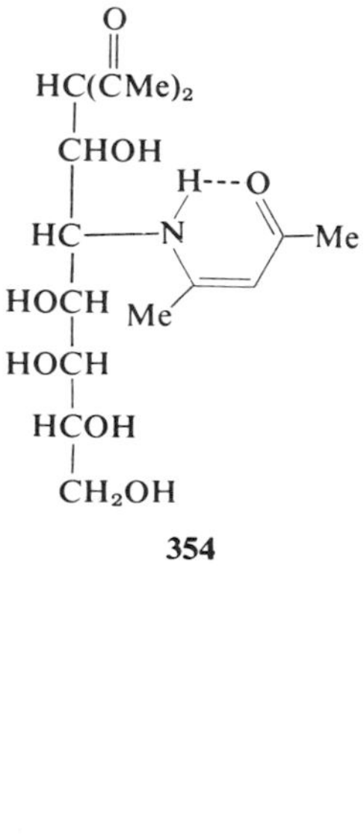

354

The fundamental reaction sequence appears to be quite general[502,509] for β-diketones, β-ketoesters, and β-ketoaldehydes, exhibiting only minor, quantitative variations in the proportion of eneamine to pyrrole isolated from different dicarbonyl compounds. The related enamines 356 and 357 resulted from reactions of 1 with[510] 355 and dimethyl butynedicarboxylate,[511] respectively, the latter enamine undergoing conversion (as the 3,4,6-triacetate) into the monoester 358 in a subsequent reaction employing more-stringent conditions. Isomeric enamines and pyrroles result from the analogous condensation of 1-amino-1-deoxy-2-ketoses with dicarbonyl compounds; thus, 1-amino-1-deoxy-D-fructose reacted during 30 minutes with 1-phenyl-1,3-butanedione in methanolic triethylamine to afford 359 in 47% yield, whereas the same reagents after 24 hours at room temperature produced[512] the pyrrole[513] 360.

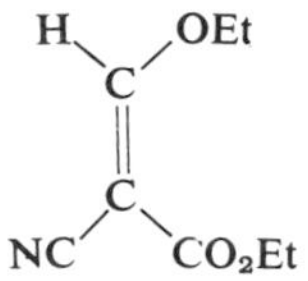

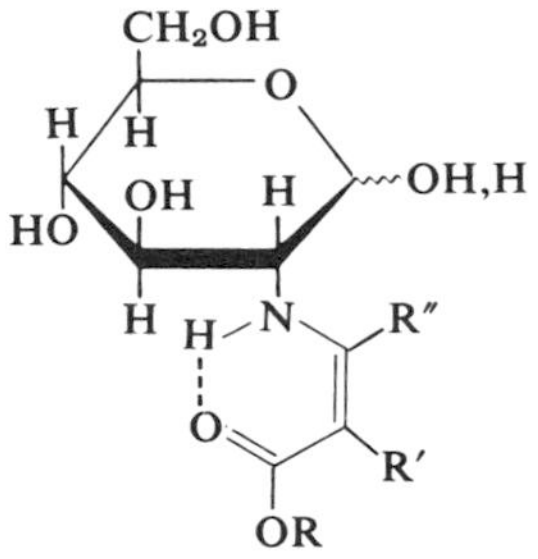

356 R = Et, R′ = CN, R″ = H
357 R = Me, R′ = H, R″ = CO₂Me

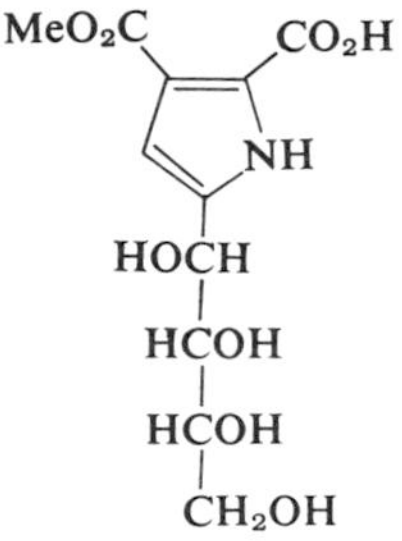

358

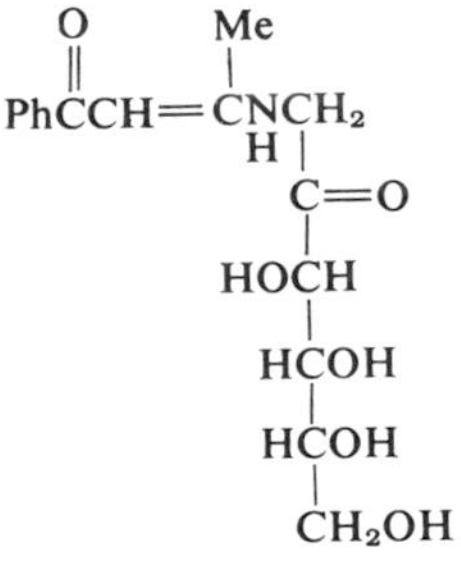

359

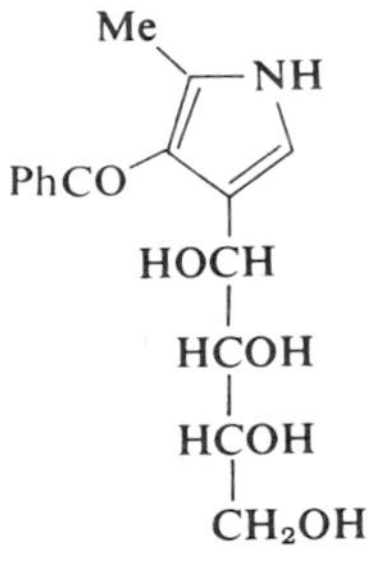

360

N-Acetylation prevents cyclization to form the pyrrole ring, limiting the reaction of the active methylene compound to nucleophilic attack at the "carbonyl" group. Reaction of 2-acetamido-2-deoxy-D-mannose (361) (or -D-glucose) with 2-oxobutanedioic (oxaloacetic) acid[257] or 2-oxopropanoic (pyruvic) acid[514] in water at pH 10 to 11 afforded an epimeric mixture of products from which 5-acetamido-3,5-dideoxy-D-*glycero*-D-*galacto*-nonulo-sonic acid (**362**, *N-acetylneuraminic acid*; see Section IV,D) was isolated[515] in low yield. Short reaction times minimize epimerization, so that, under optimal conditions, ~ 10% yields[516] of **362** are available from **361** by this method; complexation by sodium tetraborate (compare Section II,I) alters the course of epimerization, raising[514] the yields to above 20%. Kuhn and Baschang[517] treated **361** with the di(*tert*-butyl) ester of 2-oxobutanedioic acid to form a mixture of lactones related to **363**, which were decarboxylated in warm water to form **364** (and isomers and tautomers), and **364** was subsequently saponified to afford **362** in 34% yield. The analogous reaction sequence starting with the more readily available 2-acetamido-4,6-*O*-benzylidene-2-deoxy-D-gluco-pyranose afforded **362** in ~ 30% yield after hydrolysis of the acetal function, offering[517] the most satisfactory route to **362**.

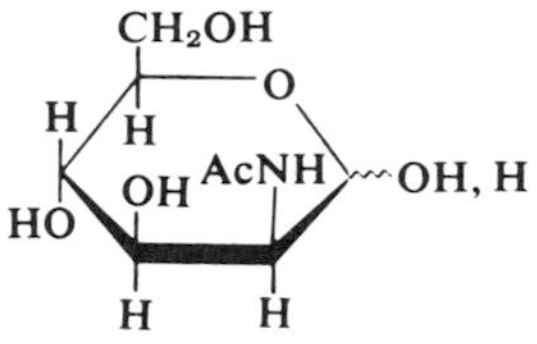

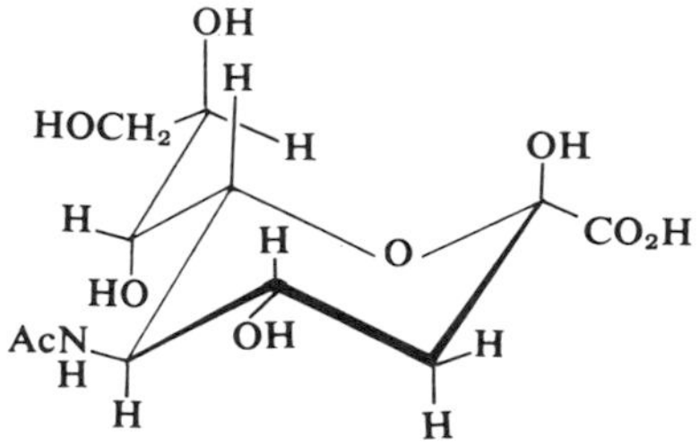

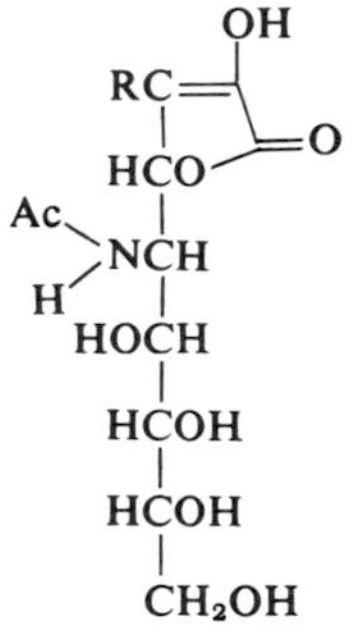

D-Fructosylamine reacts with 2,4-pentanedione to form[518] **350**, presumably by an initial, Amadori type of rearrangement (Section II,C); *N*-alkylated analogues were similarly converted[519] into *N*-alkylpyrroles, which also arise[520] from condensation reactions of 2-(alkylamino)-2-deoxyaldoses.

Elson and Morgan[521] instituted the use of the red color produced on sequential reaction of a 2-amino-2-deoxyaldose with 2,4-pentanedione and then *p*-dimethylaminobenzaldehyde as a quantitative, analytical method for those amino sugars (see Vol. IIB, Chapter 45, p. 745 and p. 753, and refs. 3 and 522); interference may, however, arise from other reactions cited in this section, and from spurious[504,523] Amadori products formed by interaction of non-amino sugars with amino acids. The influence of substitution is also uncertain as 4- or 6-*O*-methyl groups do not appear to affect quantitation of **1**, whereas a 3-*O*-methyl group may[435b,524] interfere in the determination.[525] Furthermore, the reaction is sensitive[3] to slight alterations in the conditions, and less-extensively substituted pyrroles, which occur[526] in varying proportions in the product mixtures from the first step, react with p-dimethylaminobenzaldehyde to form compounds having slightly different properties of light absorption, so that reliable quantitation requires sedulous control of conditions employed and of sample purity. With proper[527] verification, the Elson–Morgan procedure is useful as a spray reaction for making visible those spots on paper chromatograms that correspond to amino sugars.[528] Kuhn's[528a] identification of Chromogen I from this reaction as methyl 2-acetamido-2-deoxy-α,β-D-erythro-hex-2-enofuranoside has been corroborated by synthesis.[528b]

IV. NATURALLY OCCURRING AMINO SUGARS

In this section are discussed various amino sugars that have been demonstrated to occur as natural products in free or combined form.

A. 2-AMINO-2-DEOXY SUGARS

1. 2-Amino-2-deoxy-D-galactose

2-Amino-2-deoxy-D-galactose (chondrosamine or D-galactosamine), the second amino sugar found in Nature, was isolated from chondroitin sulfate preparations by Levene and LaForge[529] in 1914; the isolation is tedious,[530] and synthesis, either by ammonolysis[66] of 1,6:2,3-dianhydro-β-D-talopyranose, by the amino nitrile method (Section II,A) from D-lyxose,[10,308] or by nucleophilic inversion[286] at C-4 of 1 (Section II,J), is a practicable alternative. The anomeric pyranose hydrochlorides, which were separated by Levene[531] by fractional recrystallization from methanol, are present in almost equal amounts[532] at mutarotational equilibrium in water; the β-D anomer has been shown by n.m.r. spectroscopy[532] and X-ray crystallography[533] to favor the $^4C_1$(D) conformation. Nuclear magnetic resonance spectroscopy indicated the same conformation for the N-acetyl[532,534] and peracetyl[535] derivatives in solution. This sugar is also a constituent[535a] of the antibiotic racemomycin and of numerous bacterial polysaccharides.

2-Amino-2-deoxy-D-galacturonic acid, a component[536] of the Vi antigen from Escherichia coli, has been prepared[537] by oxidation of benzyl 2-(benzyloxycarbonyl)amino-2-deoxy-α-D-galactopyranoside with oxygen in the presence of platinum oxide and subsequent deprotection of the product.

2. 2-Amino-2,6-dideoxy-D- and -L-galactose

The D-enantiomorph (D-fucosamine) was isolated[538] from a hydrolyzate of the specific lipopolysaccharide of Chromobacterium violaceum (strain NCTC 7917), and synthesized[539] by hydrogenolysis of 2-amino-2,6-dideoxy-6-iodo-

D-galactose. 2-Amino-2,6-dideoxy-L-galactose (L-*fucosamine*), which occurs[540] as a constituent of antigenic Type V pneumococcal capsular polysaccharide, was constitutionally synthesized[541] by the amino nitrile method (Section II,A). Both occur in various lipopolysaccharide antigens of *Pseudomonas aeruginosa* (together with the D-*gluco* isomer) in characteristically different amounts.[541a]

3. *2-Amino-2-deoxy-D-glucose*

2-Amino-2-deoxy-D-glucose (**1**, D-*glucosamine, chitosamine*), the first (1876) amino sugar to be found[1] in Nature, is one of the most abundant monosaccharides. It occurs as a major constituent in the hard shells of crustaceans and other arthropods, in many fungi, and it is ubiquitously distributed, in higher animals (for example, as a constitutent of numerous[542] glycosaminoglycans and -glycuronans). It is, therefore, not only available and inexpensive but an intrinsically important molecule as well, so that no wonder attaches to the fact that almost half of the literature on amino sugars has been concerned with **1** and its derivatives.

Synthesis of **1** by the cyanohydrin reaction of D-arabinosylamine was reported[6] 37 years before the stereochemistry at C-2 was elucidated by chemical conversions[76] and by X-ray crystallography,[543] but hydrolysis of crustacean shells remains the preparative method of choice for **1**. Wolfrom *et al.*[330] subsequently converted **1** into a derivative of L-alanine, which provided the first correlation between the configurational conventions in use for sugars and for amino acids.

Compound **1** is stable as the crystalline hydrochloride or *N*-acetyl derivatives; however, thermolysis[544] of these derivatives occurs more readily (but less efficiently) than that of nonaminated analogues, presumably owing to catalysis by the basic nitrogen atom. The free base of **1** is degraded, especially in alkaline, aqueous solution, forming decomposition products[332,441,442,545] including 2,5-bis(D-*arabino*-tetrahydroxybutyl)pyrazine (**365**). The $^4C_1(D)$ conformation (of the α-anomer) has been demonstrated by n.m.r. spectroscopy to be the favored conformer of[532] **1**, its *N*-acetyl derivative,[532,534] and its peracetate[535,546]; specific introduction of trideuterioacetyl groups into **1** allowed detailed interpretation of the n.m.r.[546,547] and mass spectra[547] of the peracetylated α-pyranose form.

Compound **1** also occurs as important derivatives. The 3-*O*-(D-1-carboxyethyl)[446b] derivative (**366**, *muramic acid*), as its *N*-acetyl derivative, occurs[548] widely in bacterial[549] cell walls; the 6-phosphate of **366** has been identified[550] as a constituent of several gram-positive, bacterial cell walls, and the *N*-glycolyl derivative **367** was identified[551] as a cell-wall component of *Mycobacterium smegmatis*. Considerable stereoselectivity attends the reaction

---

References start on p. 737.

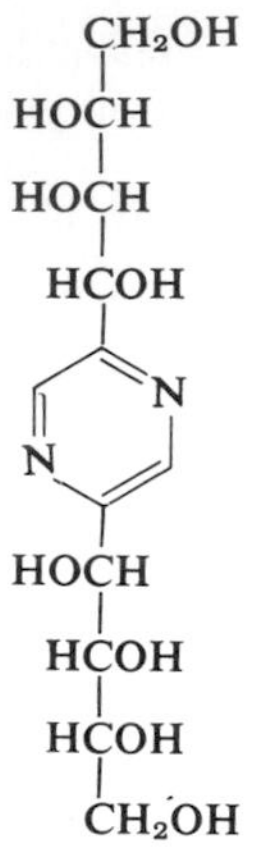

CH₂OH
|
HOCH
|
HOCH
|
HCOH

(pyrazine ring structure)

HOCH
|
HCOH
|
HCOH
|
CH₂OH

365

CH₂OH
(pyranose ring structure with OR, HO, H, OH,H, NHR')

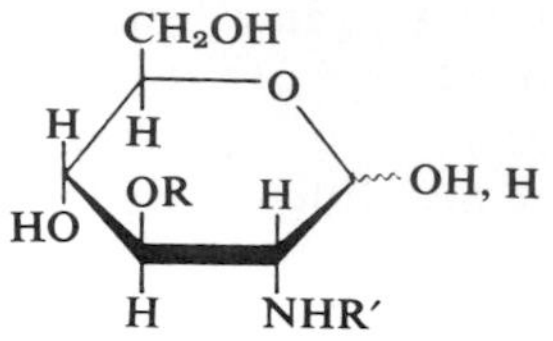

1 R = R' = H
366 R = (R)-CHMeCO₂H, R' = H
367 R = (R)-CHMeCO₂H, R' = COCH₂OH
368 R = H, R' = CON(Me)N=O

of various 2-acetamido-4,6-O-benzylidene-2-deoxy-β-D-glucopyranosides with racemic 2-chloropropanoic acid, the natural (R) configuration preponderating[552] in the product, which is a useful, synthetic precursor of **366**.

The derivative **368** (*streptozotocin*), which was isolated[553] from cultures of *Streptomyces achromogenes* and later synthesized[554] in low yield (also see Section II,B), exerts a cytotoxic[555] effect on pancreatic beta-cells. The antibiotic substance *diamycin* has been shown[556] to contain residues of **1** as its only amino sugar constituent. (Synthetic) 2-acetamido-2-deoxy-D-glucono-1,4-lactone inhibits[557] 2-acetamido-2-deoxy-β-D-glucosidase in bull epididymis. 2-Amino-2-deoxy-D-glucuronic acid has been identified as a constituent of type-specific substance *d* of *Haemophilus influenzae*[558] and of a *Staphylococcus* polysaccharide[559] antigen.

4. 2-Amino-2,6-dideoxy-D-glucose and 2-Deoxy-2-(methylamino)-L-glucose

Syntheses of both of these sugars were undertaken in advance of their discovery in Nature. The former (D-quinovosamine) was prepared[560] from **1** by conversions, and ultimate hydrogenolysis of the 6-substituent; its isolation[561] from a bacterial polysaccharide followed six years later (compare ref. 541a). 2-Deoxy-2-(methylamino)-L-glucose, which Wolfrom *et al.*[12] prepared by a cyanohydrin sequence (Section II,A) from *N*-methyl-L-arabinosylamine, had been reported[2,562] earlier the same year (1946) as a constituent of the antibiotic streptomycin; this was the third amino sugar to be found in Nature, and the first one in an economically significant product.

5. 2-Amino-2-deoxy-D-gulose and 2-Deoxy-2-(methylamino)-D-gulose

2-Amino-2-deoxy-D-gulose was recognized as a component residue of the antibiotics streptothricin and streptolin B by van Tamelen *et al.*[563] It was synthesized from D-xylose in low yield by the amino nitrile procedure[564] (Section II,A), and from methyl 2-acetamido-4,6-O-benzylidene-2-deoxy-3-O-(methylsulfonyl)-α-D-galactopyranoside by solvolysis[278] of the 3-substituent and subsequent deprotection. The N-methyl derivative was identified[565] as a constituent of streptothricin analogues LL-AC541 and LL-AB644.

A 2-amino-2-deoxyguluronic acid (chirality not specified) has been identified as a constituent in cell walls of *Halococcus* sp., strain 24 (ref. 565a) and the antigenic principle of *Vibrio parahaemolyticus* K15 (ref. 565b).

6. *2-Amino-2-deoxy-D-mannuronic Acid and 2-Amino-2,6-dideoxy-L-mannose*

The former has been identified as a probable[566] constituent of the cell-wall polysaccharide of *Micrococcus lysodeikticus*; the latter (also known as L-*rhamnosamine*) was prepared by hydrogenation[567] of the oxime of methyl 6-deoxy-L-*arabino*-hexopyranosidulose and subsequent deprotection, five years prior to its characterization[568] in hydrolyzates of the lipopolysaccharide from *Escherichia coli* U 41/14.

7. *2-Amino-2-deoxy-D-talose and 2-Amino-2,6-dideoxy-D-talose*

2-Amino-2-deoxy-D-talose has been synthesized by a modified amino nitrile procedure[310] (Section II, A) and by configurational inversion[277] of 2-amino-2-deoxy-D-idose at C-3. It (or its enantiomorph) has been identified as a (probable) minor constituent of ovine[569] and bovine[570] cartilage, whereas a reported[571] product isolated from hyaline cartilage chondroitin sulfate had physical constants in disagreement with those of the synthetic sugar. The occurrence[572] of 2-amino-2,6-dideoxy-D-talose (*pneumosamine*) in Type V pneumococcal capsular polysaccharide was proved by its unequivocal synthesis.[213,573]

8. *2-Amino-2,3-dideoxy-D-ribo-hexose*

2-Amino-2,3-dideoxy-D-*ribo*-hexose has been shown[574] to be a constituent of the antibiotics lividomycin A and B; its structure has been verified by syntheses of the natural form[574a] and of the enantiomorph.[574b]

9. *2-Amino-2-deoxypentoses*

The 5-O-carbamoyl derivative of 2-amino-2-deoxy-L-xylonic acid has been identified[575] as a constituent of several antifungal agents, proliferated by *Streptomyces cacoi*, which show activity against *Pellicularia filamentosa f. sasakii*; two other, active agents in the same broths contained the 3-deoxy

---

References start on p. 737.

analogue 2-amino-2,3-dideoxy-L-*erythro*-pentonic acid. Both were identified by spectroscopic methods, and the former has been synthesized.[575a]

A 9-(2-amino-2-deoxypentofuranosyl)guanine of unspecified configuration, produced by *Aerobacter* KY 3071, exhibits limited antibacterial and antitumor activity.[575b]

B. 3-Amino-3-deoxy Sugars

1. 3-Amino-3,6-dideoxy-D-galactose

3-Amino-3,6-dideoxy-D-galactose occurs as a constituent of cells of *Xanthomonas campestris*,[575c] and *Citrobacter* and *Salmonella* strains.[576]

2. 3-Amino-3-deoxy-D-glucose

This sugar (also called *kanosamine*[399]) was isolated[431] from the antibiotic kanamycin in 1958, and was later identified[577] as a constituent of another antibiotic, hikizimycin; it occurs[578] as the free sugar in fermentation broths of *Bacillus aminoglucosidicus*. Numerous different syntheses have been reported.[35,36,124,212,243,254,579]

3-Amino-3,6-dideoxy-D-glucose was found to be a constituent of lipopolysaccharides isolated from[580] *Escherichia coli* O71 and from[581] *Citrobacter freundii*; the structure of 3,6-dideoxy-3-(dimethylamino)-D-glucose (*mycaminose*), which is a component of the antibiotics[582] magnamycin and magnamycin B, the veterinary antibiotic tylosin,[583a] and several components of the platenomycin complex of antibiotics produced by *Streptomyces platensis*,[583b] was proved by two independent, direct syntheses.[74,583]

3. 3-Amino-3,6-dideoxy-D-mannose

The fluvamycin A[583c] hexose (*mycosamine*) isolated from the antibiotics nystatin,[584] amphotericin B,[585] pimaricin,[586] and rimocidin[586a] was shown[587,588] to be 3-amino-3,6-dideoxy-D-mannose; synthetic proof of structure was provided[587] by reductive displacement of the 6-O-p-toluenesulfonyloxy group from a partially protected derivative of 3-amino-3-deoxy-D-mannose.

4. 3-Amino-3-deoxypentoses

The 3-amino-3-deoxy-D-ribose residue has been identified as a component of the antibiotic puromycin,[589] and also isolated[590] in the form of 3'-amino-3'-deoxyadenosine from cultures of *Helminthosporium* species. The constitutional[86,312,591] synthesis involved epoxide scission by ammonia, but superior yields were later reported by methods involving reduction of an oxime,[220] or chain-shortening[29] of 3-azido-3-deoxy-1,2-O-isopropylidene-α-D-allofuranose by oxidation with periodic acid.

3-Deoxy-3-methylamino-D-xylose (*gentosamine*) was identified[87] as a constituent of the antibiotics gentamicin A and 66-40B (from *Micromonospora inyoensis*, ref. 591a). The constitutional synthesis involved opening of a 2,3-epoxide[87] by methylamine; a later preparation proceeded by periodate-oxidative[592] chain-descent. 3-Deoxy-3-methylamino-L-arabinose was identified[591a] as a constituent of antibiotic 66-40D from *M. inyoensis*, and its 4-*C*-methyl homologue was isolated from the gentamicin antibiotic[591b] XK-62-2 and from sisomycin.[591c]

The branched chain sugar 3-deoxy-4-*C*-methyl-3-(methylamino)-L-arabinose (**370**, *garosamine*) was isolated as a methyl glycoside[592a] upon methanolysis of gentamicins C_1, C_2, or C_{1a}; it was later found[592b] to be a constituent of the aminocyclitol antibiotic G-52. Reaction of 3-deoxy-1,2-*O*-isopropylidene-3-methylamino-L-*threo*-pentopyranosid-4-ulose (**369**) with methylmagnesium iodide gave, after methanolysis, a 4-epimeric mixture from which **370** could be isolated. As the *N*-benzyloxycarbonyl derivative (**371**) of the natural isomer formed an oxazolidinone derivative **372**, whereas the 4-epimer did not, the L-*arabino* stereochemistry was established.[592]

5. *3-Aminodi- and -tri-deoxyhexoses*

3-Amino-2,3,6-trideoxy-L-*lyxo*-hexose (*daunosamine*) is the sugar component of the clinically important, cytotoxic antibiotics daunorubicin,[593] and carminomycin[593a] and adriamycin[594]; the proper stereochemistry was attained[271] synthetically from L-rhamnose by an epoxide-opening reaction followed by an inversion step; a preparative synthesis from D-mannose has also been recorded,[594a] and the 5-epimer has been prepared.[594b] 3-Amino-2,3,6-trideoxy-L-*arabino*-hexose and its 4-methyl ether (L-*actinosamine*) have been identified as constituents of actinoidins A and B.[594c] 2,3,6-Trideoxy-3-(dimethylamino)-D-*lyxo*-hexose (D-*rhodosamine*) was identified as a constituent of megalomycins A, B, C_1, and C_2,[595] and β-rhodomycins S-1b, S-2, S-3, and S-4 (ref. 595a); an amino sugar liberated[596] from the broad-spectrum antibiotic vancomycin was identified[597] as 3-amino-2,3,6-trideoxy-3-*C*-methyl-L[598]-*lyxo*-hexose (*vancosamine*).

3-Amino-2,3,6-trideoxy-L-*ribo*-hexose (*ristosamine*) was shown by degradation[599] and later by synthesis[599a] to be a constituent of ristomycin A; other syntheses of this sugar[599b] and its enantiomorph[594b] followed. The L-*arabino* stereoisomer (*acosamine*), isolated from actinoidin,[599c] was prepared[599d] from an intermediate in the synthesis of daunosamine. 3-(Dimethylamino)-2,3,6-trideoxy-D-*xylo*-hexose (*angolosamine*) occurs in the antibiotic angolamycin.[599e]

Synthesis[599f] verified identification[599g] of 3,4,6-trideoxy-3-(dimethylamino)-

---

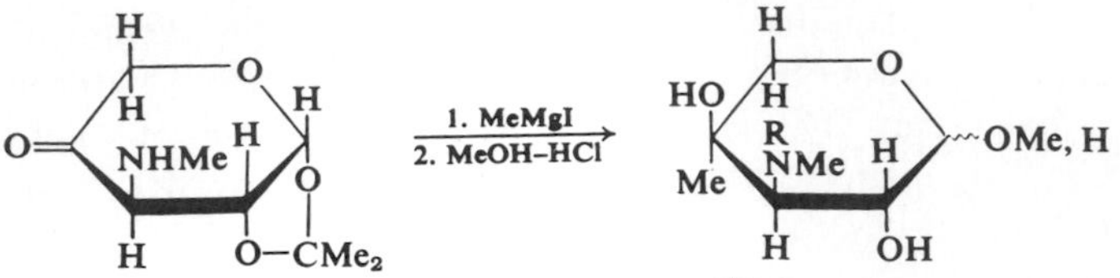

369

370 R = H
371 R = CO$_2$CH$_2$Ph

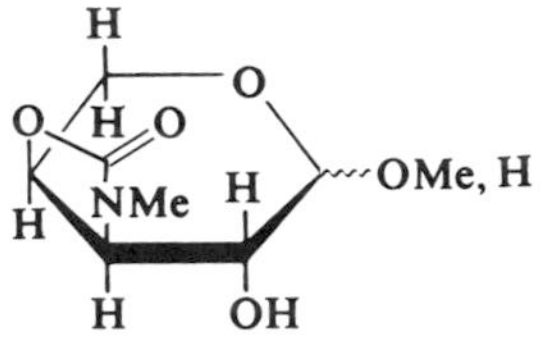

372

D-*xylo*-hexose (*desosamine*), a constituent residue of the antibiotics picromycin,[599h] methymycin,[599i] neomethymycin,[599j] narbomycin,[599k] erythromycin,[599l] and oleandomycin.[599m]

3-Amino-3,4-dideoxy-D-*xylo*-hexopyranuronic acid (*ezoaminuroic acid*) has been identified[600] spectroscopically as a constituent of antifungal agents known as ezomycins A$_1$ and A$_2$, and it has been synthesized.[600a]

C. 4-Amino-4-deoxy Sugars

The isolation of 4-amino-4-deoxyaldoses after acid hydrolysis of natural products is complicated owing to their facile decomposition by way of pyrrole derivatives (see Section III,A).

1. *4-Amino-4-deoxyaldoses*

4-Amino-4-deoxy-D-glucose has been identified[601] as a constituent of the antibiotics P-2563 (P) and (A), from *Pseudomonas fluorescens,* and[601a] the antibiotic apramycin. 4-Amino-4-deoxy-L-arabinose has been characterized as a constituent of lipopolysaccharides[602] isolated from two *Salmonella* R mutant strains. 4-Amino-4-deoxy-(D or L)-*glycero*-D-*gluco*-heptose was identified[603] as a component residue of the cytotoxic, antifungal agent septacidin, which is produced[604] by *Streptomyces fimbriatus.* 4-Amino-4-deoxy-D-*glycero*-D-*galacto*-D-*gluco*-undecanose occurs as a component residue of the nucleoside antibiotic hikizimycin *(anthelmycin).*[577b]

4-Amino-4-deoxy-D-glucuronamide has been identified[605] as the carbohydrate moiety of the pyrimidine nucleoside antibiotic[606] gougerotin and as a constituent[606a] of aspiculamycin; 1-(4-amino-4-deoxy-β-D-glucopyranosyluronic acid) cytosine ("*C-substance*"), a degradation product of gougerotin, has been prepared[141,607] by direct synthesis. The identity of 4-amino-4,6-

dideoxy-D-galactose, isolated from lipopolysaccharides[608] of *Escherichia coli* strain O10:K5(L):H4 and as a thymidine 5′-pyrophosphate derivative,[609] from *E coli* Y-10, has been verified by synthesis.[610]

4-Amino-4,6-dideoxy-D-glucose (*viosamine*) has been identified[611] as a component residue of the lipopolysaccharides of *Escherichia coli* strains[609,610] O7:K1:(L):H⁻, and of[612] *Chromobacterium violaceum*; this amino sugar also occurs in extracts of *E. coli* strain B as a thymidine 5′-(glycosyl pyrophosphate)[613] derivative. 4,6-Dideoxy-4-methylamino-D-glucose (*bamosamine*) and 4,6-dideoxy-4-(dimethylamino)-D-glucose (*amosamine*) were isolated as constituents of the *Streptomyces*-generated antibiotics bamicetin[614] and amicetin,[615] respectively; amosamine is also reported[615a] to be a constituent of the macrolide antibiotics maridomycin I, III, IV, V, and VI. All three of these sugars were prepared by Stevens *et al.*,[616] the last two by specific *N*-mono- and di-alkylation of the first one.

4-Amino-4,6-dideoxy-D-mannose (*perosamine*) was identified[617] as the sugar component of perimycin, a heptaenic, antifungal agent produced by cultures of *Streptomyces coelicolor* var. *aminophilus*; synthesis of this sugar by epoxide opening with ammonia was reported shortly thereafter.[84]

Sibiromycin, an antitumor principle elaborated by *Streptoporangium sibiricum*, is reported[617a] to be a glycoside of 4,6-dideoxy-3-C-methyl-4-methylamino-D-altrose.

2. 4-Amino-trideoxy- and -tetradeoxy Sugars

2,4,6-Trideoxy-3-*O*-methyl-4-methylamino-D-*ribo*-hexose has been identified as the sugar of the steroidal cardiac glycosides holacurtin[618] and mitiphylline.[619] 4-Amino-2,4,6-trideoxy-3-*O*-methyl-D-*ribo*-hexose (D-*holosamine*) is present in the related glycoside holantosine B[619]; synthesis of these sugars was accomplished[620,621] by a double-inversion sequence starting from methyl 4,6-*O*-benzylidene-2-deoxy-α-D-*ribo*-hexopyranoside. The 4-epimer, 4-amino-2,4,6-trideoxy-3-*O*-methyl-D-*xylo*-hexose (*holacosamine*), has been shown[622] to occur as a constituent of the steroidal glycoside holarosine B, isolated from *Holarrhena antidysenterica*.

The nucleoside antibiotic blasticidin S[623] has been shown[624] to contain as the sugar portion the unusual, unsaturated sugar 4-amino-2,3,4-trideoxy-D-*erythro*-hex-2-enosyluronic acid; the synthesis[141,625] of several derivatives of this sugar has been reported.

2,3,4,6-Tetradeoxy-4-(dimethylamino)-D-*erythro*-hexose (*forosamine*) is one[626] of three sugars found in the spiramycins (A, B, and C) isolated[627] from *Streptomyces ambofaciens*; the configuration of this sugar was ascertained[137] by synthesis, and both it and the D-*threo* epimer (*ossamine*, identified[627a] as a constituent of a fungal metabolite) have been prepared in good yield by reduction of the corresponding 4-oxime[218] (see Section II,F). Forosamine has also been prepared from 2,3,4,6-tetradeoxy-4-(dimethylamino)-D-*erythro*-hex-2-enonic acid by reduction.[627b] 4-Amino-2,3,4,6-tetradeoxy-L-*erythro*-hexose (*tolyposamine*) occurs in the antibiotic tolypomycin.[627c]

D. 5-Amino-5-deoxy Sugars

5-Amino-5-deoxy-D-glucose (*nojirimycin*) is an antibiotic produced by *Streptomyces roseochromogenes* R-468 (ref. 628), *S. lavendulae* SF-425 (ref. 629), and *S. nojiriensis* n. sp. SF-426 (ref. 630); its existence in the 5-aminopyranose form (the first example in Nature, see Section III,A) was inferred[629] from chemical and spectroscopic evidence, and the stereochemical identification was verified[302] by stereoselective reduction of the 5-oxime of 3-*O*-benzoyl-1,2-*O*-isopropylidene-6-*O*-trityl-α-D-*xylo*-hexofuranos-5-ulose and by the properties of the product of hydrolysis of the known[631] 5-amino-5-deoxy-1,2-*O*-isopropylidene-α-D-glucofuranose.

5-Amino-5-deoxy-D-alluronic acid is a constituent of the series of polyoxins (A–L), bound as a 5'-substituted 1-(β-D-glycofuranosyl)uracil; the amino group is incorporated into an amide residue in ten of the twelve polyoxins identified. The configuration of the sugar moiety was proved[575] by X-ray crystallographic analysis of the *N-p*-bromobenzenesulfonyl derivative of polyoxin C.

5-Acetamido-3,5-dideoxy-D-*glycero*-D-*galacto*-nonulosonic acid (**362**, N-*acetylneuraminic acid*) was probably first isolated[632] in 1899 as an impure copper chelate prepared from the contents of an ovarian cyst; the pure acid was isolated much later from such diverse sources as bovine submaxillary mucin,[633] cow colostrum,[634] human milk,[635a] horse serum,[636] and the gelatinous coating[637] of sea-urchin eggs. 4-*O*-Acetyl,[635] 7-*O*-acetyl[635,638] [the first example of a crystalline sialic (*N*-acylnonulosaminic) acid], and 7,8-(or 7,9-)di-*O*-acetyl[639] derivatives of **362** have also been found in Nature;

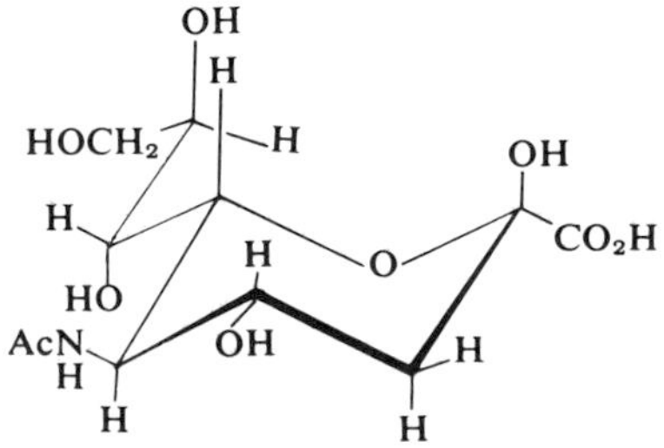

362

the *N*-glycolyl analogue and its 8-methyl ether occur as constituents of porcine submaxillary mucin[635] and of the starfish[640] *Asterias forbesi*, respectively. An elevated amount of **362** in the cerebrospinal fluid of patients who have meningitis is claimed[640a] to be pathognomonic of infection by a pyogenic bacterium. Edible bird's nest is a non-tissue source[640b] of **362**.

Gottschalk[641] first proposed that the sialic acids are nonulosaminic acids, and it was subsequently found[642] that an enzyme present in *Vibrio cholerae* degrades **362** to give 2-acetamido-2-deoxy-D-glucose plus pyruvic acid, the exact reverse of the initial synthesis by Cornforth *et al.*[515] The correct stereochemistry of **362** followed from the observation[643] that an enzyme from *Clostridium perfringens* reversibly catalyzes the conversion: **362** $\rightleftharpoons$ 2-acetamido-2-deoxy-D-mannose plus pyruvic acid, but fails to effect similar condensation of 2-acetamido-2-deoxy-D-glucose. (Epimerization accompanying base-catalyzed, active methylene reactions leading to **362** is evaluated in Section II,C,3; an involved, alternative synthesis, based on a Wittig condensation, has been described.[644]) The $^2C_5(D)$ conformation depicted for **362** is supported by p.m.r.-spectroscopic[645] data, and corresponds to the structure determined[646] by X-ray crystallography; the *N*-deacylated analogue exists,[348] however, as a five-membered, cyclic Schiff-base.

The sialic acids have been reviewed in detail (see ref. 548).

E. 6-AMINO-6-DEOXY SUGARS

6-Amino-6-deoxy-D-glucose has been identified[431] as a constituent of the antibiotic kanamycin A; the facility of displacements at primary centers makes this a highly accessible molecule, and an efficient synthesis of the ^{15}N-labeled compound has been reported.[103]

6-Amino-6,8-dideoxy-D-*erythro*-D-*galacto*-octose (*lincosamine*) has been identified[647] as the sugar portion of the antibiotic lincomycin, which is produced[648,649] by *Streptomyces lincolnensis* var. *lincolnensis*; this sugar has been prepared by three approaches: (*a*) reduction of an oxime,[224,268] (*b*) stepwise chain-ascent by an aminonitrile condensation and a subsequent Grignard reaction,[267] and (*c*) an aldol reaction of a 6-deoxy-6-nitro sugar.[241] Treatment of lincomycin with thionyl chloride resulted[650] in replacement, with configurational inversion, of the 7-hydroxyl group by a chlorine atom, to afford a superior antibiotic, clindamycin. The 7-*O*-methyl analogue of lincosamine has been identified[648,651] as the sugar residue of the related antibiotic celesticetin, and prepared by diazoethylation of 1,2:3,4-di-*O*-isopropylidene-α-D-galactopyranuronyl chloride and selective reduction of the 6-oxime thus prepared.[652]

A 6-amino-6-deoxyheptonic acid of undetermined configuration has been found to occur as part of an unusual cyclic orthoester group in the antibiotics hygromycin B,[653] A-396-I,[654] and SS-56C[655]; the antihelminthic agents destomycin A and B contain 6-amino-6-deoxy-L-*glycero*-D-*galacto*- and -L-*glycero*-D-*gluco*-heptonic acid, respectively.[655a]

F. DIAMINO SUGARS

1. *Diaminodideoxyaldoses*

2,6-Diamino-2,6-dideoxy-D-glucose (*neosamine C*) has been characterized as a constituent of the antibiotics neomycin C,[656] zygomycin A,[657] hybrimycins A and B,[658] kanamycin B,[659] ribostamycin,[660] and butirosins A and B[661]; the stereochemistry of this sugar was determined by comparison with a synthetic sample.[662] The 5-epimer, 2,6-diamino-2,6-dideoxy-L-idose (*neosamine B, paromose*), was identified as a component residue of neomycin B,[663] paromomycin,[323,663a] hybrimycins A₁ and B₁,[658] lividomycins A and B,[664] "D-mannosylparomomycin,"[665] and framycetin[666]; the L-*ido* configuration was deduced from degradation studies,[667] and verified by synthesis.[668]

2,3-Diamino-2,3-dideoxy-D-glucose has been identified[668a] as a constituent of the lipid A component from lipopolysaccharides of *Rhodopseudomonas viridis* and *Rh. palustris*. The antibiotic prumycin was shown[668b] to be 4-(D-alanylamino)-2-amino-2,4-dideoxy-L-arabinose. A 1,4-diamino-1,4-dideoxyhexitol reportedly[601] occurs in the antibiotics P-2563 (P) and (A).

2. *Diaminotrideoxyaldoses*

2,4-Diamino-2,4,6-trideoxy-D-glucose (*bacillosamine*) was isolated[669] from *Bacillus lichenformis* ATCC 9945 (formerly *B. subtilis*), and subsequently identified by physicochemical methods[670] and by synthesis.[671] An unidentified stereoisomer occurs in *Diplococcus pneumoniae* as a uridine 5'-pyrophosphate derivative[672] and as a constituent of "C substance."[673]

2,6-Diamino-2,3,6-trideoxy-D-*ribo*-hexose (*nebrosamine*) was identified[674] by p.m.r. spectroscopy as a constituent residue of tobramycin; the stereochemical assignment was verified by two independent[235,675] syntheses. 2,6-Diamino-2,4,6-trideoxy-D-*xylo*-hexose has been found[675a] as a constituent of the 4'-deoxybutirosins Bu-1975c₁ and Bu-1975c₂.

2-Amino-2,3,7-trideoxy-7-methylamino-D-*glycero*-D-*allo*-octodialdose occurs[601a] in the antibiotic apramycin, elaborated by *Streptomyces tenebrarius*.

3. *Diamino-tetra- and -penta-deoxyaldoses*

2,4-Diamino-2,3,4,6-tetradeoxy-D-*arabino*-hexose (*kasugamine*) was identified[676] by physicochemical methods as a component of kasugamycin;

stereospecific syntheses of the methyl α-glycoside of this sugar have been reported for the racemic[294] and the D sugar,[116] which was later also found in the antibiotic minosaminomycin.[676a]

2,6-Diamino-2,3,4,6-tetradeoxy-D-*erythro*-hexose (*purpurosamine C*) was identified as a constituent of gentamicin C_{1a} on the basis of n.m.r.[677] and mass-spectral[678] data; its stereochemistry was proved by its nonidentity[679] with the corresponding derivative of synthetic 2,6-diamino-2,3,4,6-tetra-deoxy-D-*threo*-hexose, and subsequently by synthesis.[680] D-Ribo-2,6-diamino-2,3,4,6,7-pentadeoxyheptose (*purpurosamine B*)[293b] and a 2-amino-2,3,4,6,7-pentadeoxy-6-methylaminoheptose (*purpurosamine A*) of unknown stereochemistry have been identified by physicochemical methods[677,678] as constituents of gentamicins C_2 and C_1, respectively; 2,6-diamino-2,3,4,6-tetradeoxy-D-*glycero*-hex-4-enopyranose was identified[681] as a component residue of sisomicin by its hydrogenation to give the 5-epimer of purpurosamine C, and the assignment has been verified by synthesis.[680] 2-Amino-2,3,4,6-tetradeoxy-6-methylamino-D-*glycero*-hex-4-enopyranose occurs[592b] as a constituent of the aminocyclitol antibiotic G-52.

3,6-Diamino-2,3,4,6-tetradeoxy-L-*threo*-hexono-1,5-lactone was shown[240] by direct synthesis to be a component residue of negamycin.

4. Triaminotrideoxyaldonic Acid

2,3,5-Triamino-2,3,5-trideoxy-D-arabinonic acid occurs as a constituent of 2-amino-5(S)-carboxy-4(S)-[1(R)-hydroxy-2-aminoethyl]-4,5-dihydroimidazole (**373**, *streptolidine, roseonine,* or *geamine*), which has been isolated from hydrolyzates of streptothricin,[682] roseothricin,[683] geomycin,[684] racemomycin,[685] and mycothricin.[686] After the structure of **373** had been elucidated by chemical methods,[687] the absolute configuration was established by X-ray crystallography[688]; a constitutional synthesis of **373** from D-ribose was subsequently reported.[689]

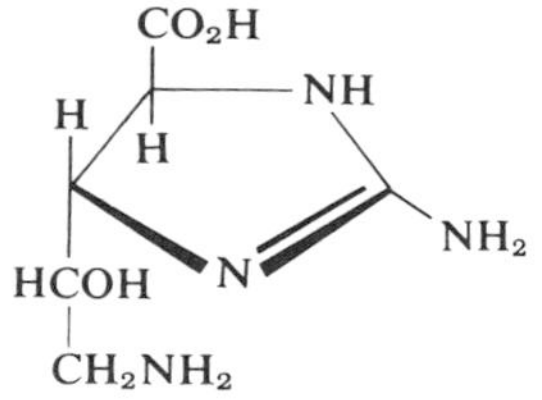

373

REFERENCES

1. G. Ledderhose, *Ber.*, **9**, 1200 (1876); *Z. Physiol. Chem.*, **2**, 213 (1878).
2. F. A. Kuehl, Jr., E. H. Flynn, F. W. Holly, R. Mozingo, and K. Folkers, *J. Amer. Chem. Soc.*, **68**, 536 (1946); **69**, 3032 (1947).

3. D. Horton, in "The Amino Sugars" (R. W. Jeanloz and E. A. Balazs, eds.), Vol. IA, Academic Press, New York, 1969, pp. 1–211.
4. S. Umezawa, *Advan. Carbohyd. Chem. Biochem.*, **31**, 111 (1974); H. Umezawa, *ibid.*, **31**, 183 (1974).
5. H. Grisebach, *Advan. Carbohyd. Chem. Biochem.*, **35**, 81 (1978).
6. E. Fischer and H. Leuchs, *Ber.*, **35**, 3787 (1902); **36**, 24 (1903).
7. P. A. Levene, *J. Biol. Chem.*, **36**, 73 (1918).
8. (a) P. A. Levene, "Hexosamines and Mucoproteins," Longmans, Green, London, 1925; (b) P. A. Levene and E. P. Clark, *J. Biol. Chem.*, **46**, 19 (1921).
9. P. A. Levene, *J. Biol. Chem.*, **26**, 367 (1916).
10. R. Kuhn and W. Kirschenlohr, *Angew. Chem.*, **67**, 786 (1955); R. Brossmer, *Methods Carbohyd. Chem.*, **1**, 216 (1962).
11. E. Votoček and R. Lukeš, *Collect. Czech. Chem. Commun.*, **7**, 424 (1935).
12. M. L. Wolfrom, A. Thompson, and I. R. Hooper, *J. Amer. Chem. Soc.*, **68**, 2343 (1946); M. L. Wolfrom and A. Thompson, *ibid.*, **69**, 1847 (1947).
13. R. Kuhn and J. C. Jochims, *Ann.*, **628**, 172 (1961).
14. U. Hornemann, *Carbohyd. Res.*, **28**, 171 (1973).
14a. T. E. Walker and R. Barker, *Carbohyd. Res.*, **64**, 266 (1978).
15. H. Saeki, T. Iwashige, E. Ohki, K. Furuya, and M. Shirasaka, *Ann. Sankyo Res. Lab.*, **19**, 37 (1967); *Chem. Abstr.*, **68**, 96, 075 (1968).
16. H. Paulsen and M. Budzis, *Ber.*, **107**, 2009 (1974).
17. H. Yanagisawa, M. Kinoshita, S. Nakada, and S. Umezawa, *Bull. Chem. Soc. Japan*, **43**, 246 (1970).
17a. J.-M. Bourgeois, *Helv. Chim. Acta*, **58**, 369 (1975).
18. R. Kuhn, D. Wieser, and H. Fischer, *Ann.*, **628**, 207 (1959); **644**, 117 (1961).
19. P. A. Levene and I. Matsuo, *J. Biol. Chem.*, **39**, 105 (1919).
20. R. Kuhn and G. Baschang, *Ann.*, **636**, 164 (1960).
21. M. F. Shostakovskiĭ, K. F. Lavrova, N. N. Aseeva, and A. I. Polyakov, *Izv. Akad. Nauk SSSR, Ser. Khim.*, 1168 (1969).
22. S. Ohdan, T. Okamoto, S. Maeda, T. Ichikawa, Y. Araki, and Y. Ishido, *Bull. Chem. Soc. Japan*, **46**, 981 (1973).
23. C. Satoh and A. Kiyomoto, *Carbohyd. Res.*, **7**, 138 (1968).
24. M. L. Wolfrom and K. Anno, *J. Amer. Chem. Soc.*, **75**, 1038 (1953); M. L. Wolfrom and A. Thompson, *Methods Carbohyd. Chem.*, **1**, 209 (1962).
25. M. L. Wolfrom and Z. Yosizawa, *J. Amer. Chem. Soc.*, **81**, 3477 (1959).
26. J. S. Brimacombe and A. M. Mofti, *J. Chem. Soc. (C)*, 1634 (1971).
27. H. Ando and J. Yoshimura, *Bull. Chem. Soc. Japan*, **43**, 2966 (1970).
28. K. Onodera, N. Kashimura, and N. Miyazaki, *Carbohyd. Res.*, **21**, 159 (1972).
29. A. K. M. Anisuzzaman and R. L. Whistler, *J. Org. Chem.*, **37**, 3187 (1972).
30. J. S. Brimacombe, A. M. Mofti, and M. Stacey, *Carbohyd. Res.*, **16**, 303 (1971).
31. A. E. El-Ashmawy and D. Horton, *Carbohyd. Res.*, **3**, 191 (1966).
32. A. E. El-Ashmawy and D. Horton, *Carbohyd. Res.*, **1**, 164 (1965).
33. M. L. Wolfrom, J. L. Minor, and W. A. Szarek, *Carbohyd. Res.*, **1**, 156 (1965).
34. H. H. Baer and H. O. L. Fischer, *J. Amer. Chem. Soc.*, **82**, 3709 (1960).
35. H. H. Baer, *J. Amer. Chem. Soc.*, **83**, 1882 (1961).
36. H. H. Baer, *J. Amer. Chem. Soc.*, **84**, 83 (1962).
37. H. Weidmann and N. Wolf, *Monatsh. Chem.*, **102**, 747 (1971).
38. K. Čapek and J. Jarý, *Collect. Czech. Chem. Commun.*, **37**, 484 (1972).
39. C. C. Deane and T. D. Inch, *Chem. Commun.*, 813 (1969).
40. F. J. McEvoy, B. R. Baker, and M. J. Weiss, *J. Amer. Chem. Soc.*, **82**, 209 (1960).

41. B. Coxon and L. Hough, *J. Chem. Soc.*, 1643 (1961).
42. M. Amadori, *Atti Real. Acad. Naz. Lincei*, [6] **2**, 337 (1925); **9**, 68, 226 (1929); **13**, 72 (1931).
43. R. Kuhn and F. Weygand, *Ber.*, **70**, 769 (1937).
43a. S. B. Tjan and G. A. M. van der Ouweland, *Tetrahedron*, **30**, 2891 (1974).
43b. British–American Rules for Conformational Nomenclature, *Chem. Commun.*, 505 (1973).
44. J. E. Hodge, *Advan. Carbohyd. Chem.*, **10**, 169 (1955).
45. J. E. Hodge, and B. E. Fisher, *Methods Carbohyd. Chem.*, **2**, 99 (1963).
46. L. Rosen, J. W. Woods, and W. Pigman, *Ber.*, **90**, 1038 (1957).
47. R. Kuhn and H. J. Haas, *Ann.*, **600**, 148 (1956).
48. J. Druey and G. Huber, *Helv. Chim. Acta*, **40**, 342 (1957).
49. E. Fischer, *Ber.*, **19**, 1920 (1886).
50. M. L. Wolfrom, H. S. El Khadem, and J. R. Vercellotti, *J. Org. Chem.*, **29**, 3284 (1964).
51. H. Hřebabecký, J. Krupička, and J. Farkaš, *Collect. Czech. Chem. Commun.*, **38**, 3181 (1973).
52. For reviews, see: G. P. Ellis, *Advan. Carbohyd. Chem.*, **14**, 63 (1959); J. E. Hodge, *J. Agr. Food Chem.*, **1**, 928 (1953); J. P. Danechy and W. Pigman, *Advan. Food Res.*, **3**, 241 (1951).
53. P. A. Finot and J. Mauron, *Helv. Chim. Acta*, **52**, 1488 (1969); A. Gottschalk, *Biochem. J.*, **52**, 455 (1952); P. H. Lowy and H. Borsook, *J. Amer. Chem. Soc.*, **78**, 3175 (1956); K. Heyns and H. Paulsen, *Ann.*, **622**, 160 (1959).
54. M. L. Wolfrom, N. Kashimura, and D. Horton, *J. Agr. Food Chem.*, **22**, 791 (1974).
55. S. K. Das, J. L. Abernethy, and L. D. Quin, *Carbohyd. Res.*, **30**, 379 (1973).
56. K. Heyns, H. Behre, and H. Paulsen, *Carbohyd. Res.*, **5**, 225 (1967).
57. H. Paulsen and F. Leupold, *Ber.*, **102**, 2822 (1969).
58. J. Yoshimura, M. Funabashi, and H. Simon, *Carbohyd. Res.*, **11**, 276 (1969).
59. J. F. Carson, *J. Amer. Chem. Soc.*, **77**, 1881, 5957 (1955); **78**, 3728 (1956).
60. K. Heyns, H. Breuer, and H. Paulsen, *Ber.*, **90**, 1374 (1957); K. Heyns and H. Noack, *ibid.*, **95**, 720 (1962), and references therein.
61. K. Heyns, H. Paulsen, R. Eichstedt, and M. Rolle, *Ber.*, **90**, 2039 (1957).
62. J. Druey and G. Huber, U.S. Patent 2,918,462 (1959); *Chem. Abstr.*, **54**, 7579 (1960).
63. K. Heyns, K.-W. Pflughaupt, and H. Paulsen, *Ber.*, **101**, 2800 (1968).
63a. H. Tsuchida, S. Tachibana, and M. Komoto, *Agr. Biol. Chem.* (Tokyo), **40**, 1241 (1976).
64. J. Kovář and J. Jarý, *Collect. Czech. Chem. Commun.*, **33**, 549 (1968).
65. W. Meyer zu Reckendorf, *Arch. Pharm.*, **305**, 29 (1972).
66. (a) S. P. James, F. Smith, M. Stacey, and L. F. Wiggins, *Nature*, **156**, 308 (1945); *J. Chem. Soc.*, 625 (1946); (b) R. W. Jeanloz and P. J. Stoffyn, *Methods Carbohyd. Chem.*, **1**, 221 (1962).
67. F. Schmitt and P. Sinaÿ, *Carbohyd. Res.*, **29**, 99 (1973); M. Černý, O. Juláková, and J. Pacák, *Collect. Czech. Chem. Commun.*, **39**, 1391 (1974).
68. J. Pacák, P. Drašar, D. Štropová, M. Černý, and M. Buděšínsky, *Collect. Czech. Chem. Commun.*, **38**, 3936 (1973).
69. M. Černý, T. Elbert, and J. Pacák, *Collect. Czech. Chem. Commun.*, **39**, 1752 (1974).
70. (a) L. F. Wiggins, *J. Chem. Soc.*, 522 (1944); (b) J. G. Buchanan and K. J. Miller, *ibid.*, 3392 (1960).
71. R. W. Jeanloz, Z. Tarasiejska-Glazer, and D. A. Jeanloz, *J. Org. Chem.*, **26**, 532 (1961).

72. (a) S. Peat and L. F. Wiggins, *J. Chem. Soc.*, 1810 (1938); A. B. Foster, M. Stacey, and S. V. Vardheim, *Nature*, 180, 247 (1957); *Acta Chem. Scand.*, 12, 1605 (1958); (b) J. Jarý and K. Čapek, *Methods Carbohyd. Chem.*, 6, 238 (1972).
73. S. N. Danilov and I. S. Lyshanskiĭ, *Zh. Obshch. Khim.*, 24, 2106 (1955).
74. A. B. Foster, T. D. Inch, J. Lehmann, M. Stacey, and J. M. Webber, *Chem. Ind.* (London), 142 (1962).
74a. I. Ježo, *Chem. Zvesti*, 29, 124 (1975).
75. R. D. Guthrie and J. A. Liebmann, *Carbohyd. Res.*, 33, 355 (1974), and references therein.
76. W. N. Haworth, W. H. G. Lake, and S. Peat, *J. Chem. Soc.*, 271 (1939).
77. E. Fischer, M. Bergmann, and H. Schotte, *Ber.*, 53, 509 (1920); P. A. Levene and G. M. Meyer, *J. Biol. Chem.*, 55, 221 (1923).
78. T. Suami and T. Shoji, *Bull. Chem. Soc. Japan*, 43, 2948 (1970).
79. U. Zehavi and N. Sharon, *J. Org. Chem.*, 37, 2141 (1972).
80. A. Liav and N. Sharon, *Carbohyd. Res.*, 30, 109 (1973).
81. A. Liav and N. Sharon, *Carbohyd. Res.*, 37, 248 (1974).
81a. H. H. Baer and S.-H. Lee Chiu, *Carbohyd. Res.*, 28, 390 (1973).
82. J. Jarý, K. Čapek, and J. Kovář, *Collect. Czech. Chem. Commun.*, 28, 2171 (1963).
83. C. L. Stevens, S. K. Gupta, R. P. Glinski, K. G. Taylor, P. Blumbergs, C. P. Schaffner, and C.-H. Lee, *Carbohyd. Res.*, 7, 502 (1968).
84. C. L. Stevens, R. P. Glinski, K. G. Taylor, P. Blumbergs, and S. K. Gupta, *J. Amer. Chem. Soc.*, 92, 3160 (1970).
85. K. Čapek and J. Jarý, *Collect. Czech. Chem. Commun.*, 38, 2518 (1973).
85a. M. Brockhaus, W. Gorath, and J. Lehmann, *Ann.*, 89 (1976).
86. B. R. Baker and R. E. Schaub, *J. Amer. Chem. Soc.*, 75, 3864 (1953); *J. Org. Chem.*, 19, 646 (1954); also see D. J. Cooper, D. H. Davies, A. K. Mallams, and A. S. Yehaskel, *J. Chem. Soc. Perkin I*, 785 (1975).
87. H. Maehr and C. P. Schaffner, *J. Amer. Chem. Soc.*, 92, 1697 (1970).
88. B. R. Baker, R. E. Schaub, J. P. Joseph, and J. H. Williams, *J. Amer. Chem. Soc.*, 76, 4044 (1954).
89. A. P. Martinez, D. F. Calkins, E. J. Reist, W. W. Lee, and L. Goodman, *J. Heterocycl. Chem.*, 7, 713 (1970).
90. J. A. Montgomery, M. C. Thorpe, S. D. Clayton, and H. J. Thomas, *Carbohyd. Res.*, 32, 404 (1974).
91. H. Kuzuhara and S. Emoto, *Agr. Biol. Chem.* (Tokyo), 27, 689 (1963); 28, 184 (1964).
92. B. R. Baker and A. H. Haines, *J. Org. Chem.*, 28, 442 (1963).
93. U. G. Nayak and R. L. Whistler, *J. Org. Chem.*, 33, 3582 (1968).
94. A. Zamojski, W. A. Szarek, and J. K. N. Jones, *Carbohyd. Res.*, 23, 460 (1972).
95. K. Freudenberg and A. Doser, *Ber.*, 58, 294 (1925); V. I. Veksler, *Zh. Obshch. Khim.*, 31, 989 (1961).
96. K. Freudenberg and R. M. Hixon, *Ber.*, 56, 2119 (1923).
97. K. Freudenberg and K. Smeykal, *Ber.*, 59, 100 (1926).
97a. W. M. Corbett, *J. Chem. Soc.*, 2926 (1961).
98. E. Fischer and K. Zach, *Ber.*, 44, 132 (1911).
99. F. D. Cramer, H. Otterbach, and H. Springmann, *Ber.*, 92, 384 (1959); F. D. Cramer, *Methods Carbohyd. Chem.*, 1, 242 (1962).
100. D. Horton and A. E. Luetzow, *Carbohyd. Res.*, 7, 101 (1968).
101. H. Ohle and L. von Vargha, *Ber.*, 61, 1203 (1928).
102. H. Ohle and L. von Vargha, *Ber.*, 62, 2425 (1929).
103. B. Helferich and R. Mittag, *Ber.*, 71, 1585 (1938).

104. B. Coxon, *Carbohyd. Res.*, **11**, 153 (1969); **19**, 197 (1971).
105. W. Meyer zu Reckendorf, *Ber.*, **101**, 3802 (1968).
106. J. S. Brimacombe and A. M. Mofti, *Carbohyd. Res.*, **16**, 167 (1971).
107. S. Akiya and T. Osawa, *Yakugaku Zasshi*, **76**, 1280 (1956); B. Helferich and M. Burgdorf, *Tetrahedron*, **3**, 274 (1958).
108. M. L. Wolfrom, F. Shafizadeh, R. K. Armstrong, and T. M. Shen Han, *J. Amer. Chem. Soc.*, **81**, 3716 (1959).
109. M. G. Stout, M. J. Robins, R. K. Olsen, and R. K. Robins, *J. Med. Chem.*, **12**, 658 (1969).
110. D. C. Baker and D. Horton, *Carbohyd. Res.*, **21**, 393 (1972).
111. K. Schattka and B. Jastorff, *Ber.*, **105**, 3824 (1972).
112. W. Meyer zu Reckendorf and N. Wassiliadou-Micheli, *Ber.*, **101**, 2294 (1968).
113. J. Hildesheim, J. Cléophax, S. D. Gero, and R. D. Guthrie, *Tetrahedron Lett.*, 5013 (1967).
114. J. Cléophax, S. D. Gero, and A. M. Sepulchre, *Carbohyd. Res.*, **7**, 505 (1968).
115. R. S. Tipson, *Advan. Carbohyd. Chem.*, **8**, 107 (1953); D. H. Ball and F. W. Parrish, *Advan. Carbohyd. Chem. Biochem.*, **24**, 139 (1969).
116. K. Kitahara, S. Takahashi, H. Shibata, N. Kurihara, and M. Nakajima, *Agr. Biol. Chem.* (Tokyo), **33**, 748 (1969).
117. J. Staněk, Jr., K. Čapek, and J. Jarý, *Collect. Czech. Chem. Commun.*, **39**, 1479 (1974).
118. K. Freudenberg, G. Burkhart, and E. Braun, *Ber.*, **59**, 714 (1926).
119. K. Freudenberg and F. Brauns, *Ber.*, **55**, 3233 (1922).
120. R. U. Lemieux and P. Chü, *J. Amer. Chem. Soc.*, **80**, 4745 (1958).
121. D. Horton, J. S. Jewell, and H. S. Prihar, *Can. J. Chem.*, **46**, 1580 (1968).
122. U. G. Nayak and R. L. Whistler, *J. Org. Chem.*, **34**, 3819 (1969).
123. R. L. Whistler and L. W. Doner, *Methods Carbohyd. Chem.*, **6**, 215 (1972).
124. D. T. Williams and J. K. N. Jones, *Can. J. Chem.*, **45**, 7 (1967); W. Meyer zu Reckendorf, *Angew. Chem.*, **78**, 1023 (1966); J. S. Brimacombe, J. G. H. Bryan, A. Husain, M. Stacey, and M. S. Tolley, *Carbohyd. Res.*, **3**, 318 (1967); A. C. Richardson, *Methods Carbohyd. Chem.*, **6**, 218 (1972).
125. M. L. Wolfrom, J. Bernsmann, and D. Horton, *J. Org. Chem.*, **27**, 4505 (1962).
126. J. S. Brimacombe, P. A. Gent, and J. H. Westwood, *Carbohyd. Res.*, **12**, 475 (1970).
127. J. Defaye and A. M. Miquel, *Carbohyd. Res.*, **9**, 250 (1969).
128. T. Tsuchiya, K. Suo, and S. Umezawa, *Bull. Chem. Soc. Japan*, **43**, 531 (1970).
129. J. Cléophax, S. D. Gero, and J. Hildesheim, *Chem. Commun.*, 94 (1968); J. Hildesheim, J. Cléophax, A. M. Sepulchre, and S. D. Gero, *Carbohyd. Res.*, **9**, 315 (1969).
130. J. Cléophax, S. D. Gero, J. Hildesheim, A. M. Sepulchre, R. D. Guthrie, and C. W. Smith, *J. Chem. Soc.* (C), 1385 (1970).
131. J. Cléophax, S. D. Gero, R. D. Guthrie, *Tetrahedron Lett.*, 567 (1967); J. Cléophax, J. Hildesheim, A. M. Sepulchre, and S. D. Gero, *Bull. Soc. Chim. Fr.*, 153 (1969).
132. W. Roth and W. Pigman, *J. Org. Chem.*, **26**, 2455 (1961).
133. M. L. Wolfrom, F. Shafizadeh, and R. K. Armstrong, *J. Amer. Chem. Soc.*, **80**, 4885 (1958).
134. J. Jarý and P. Novák, *Collect. Czech. Chem. Commun.*, **34**, 1744 (1968); J. Jarý and A. Zobáčová, *Methods Carbohyd. Chem.*, **6**, 229 (1972).
135. E. J. Reist, R. R. Spencer, B. R. Baker, and L. Goodman, *Chem. Ind.* (London), 1794 (1962).
136. R. D. Guthrie and D. Murphy, *Chem. Ind.* (London), 1473 (1962).

137. (a) C. L. Stevens, G. E. Gutowski, K. G. Taylor, and C. P. Bryant, *Tetrahedron Lett.*, 5717 (1966); (b) C. L. Stevens and C. P. Bryant, *Methods Carbohyd. Chem.*, **6**, 225 (1972).

138. T. Kondo and T. Goto, *Agr. Biol. Chem.* (Tokyo), **35**, 625 (1971).

139. S. Koto, N. Kawakatsu, and S. Zen, *Bull. Chem. Soc. Japan*, **46**, 876 (1973).

140. M. W. Horner, L. Hough, and A. C. Richardson, *Carbohyd. Res.*, **17**, 209 (1971).

141. K. A. Watanabe, R. S. Goody, and J. J. Fox, *Tetrahedron*, **26**, 3883 (1970).

142. K. Hess and H. Stenzel, *Ber.*, **68**, 981 (1935).

142a. C. R. Hall and T. D. Inch, *Carbohyd. Res.*, **53**, 254 (1977).

143. P. L. Gill, M. W. Horner, L. Hough, and A. C. Richardson, *Carbohyd. Res.*, **17**, 213 (1971).

144. J. S. Brimacombe, O. A. Ching, and M. Stacey, *Carbohyd. Res.*, **8**, 498 (1968).

145. J. S. Brimacombe, O. A. Ching, and M. Stacey, *J. Chem. Soc.* (*C*), 1270 (1969).

146. H. Paulsen, K. Propp, and J. Brüning, *Ber.*, **102**, 469 (1969).

147. H. Paulsen, H. Koebernick, and H. Schönherr, *Ber.*, **105**, 1515 (1972).

148. R. Gigg and C. D. Warren, *J. Chem. Soc.* (*C*), 2661 (1968).

149. H. Paulsen, K. Steinert, and K. Heyns, *Ber.*, **103**, 1599 (1970).

150. J. Jarý, P. Novák, and Z. Šamek, *Ann.*, **740**, 98 (1970).

151. C. L. Stevens, R. P. Glinski, K. G. Taylor, and F. Sirokman, *J. Org. Chem.*, **35**, 592 (1970).

152. A. I. Usov, K. S. Adamyants, and N. K. Kochetkov, *Izv. Akad. Nauk SSSR, Ser. Khim.*, 2546 (1968).

153. W. Meyer zu Reckendorf and N. Wassiliadou-Micheli, *Ber.*, **103**, 37 (1970).

154. J. S. Brimacombe and A. M. Mofti, *Carbohyd. Res.*, **18**, 157 (1971).

155. H. Ohrui and S. Emoto, *Carbohyd. Res.*, **10**, 221 (1969).

155a. R. S. Tipson and A. Cohen, *Carbohyd. Res.*, **1**, 338 (1965).

156. J. S. Brimacombe and J. G. H. Bryan, *Carbohyd. Res.*, **6**, 423 (1968).

157. J. S. Brimacombe, J. Iball, and J. N. Low, *J. Chem. Soc. Perkin II*, 937 (1972).

157a. K. Čapek, J. Jarý, and Z. Samek, *Collect. Czech. Chem. Commun.*, **40**, 149 (1975).

158. A. N. Anikeeva, L. G. Revel'skaya, N. A. Khrenova, and S. N. Danilov, *Zh. Obshch. Khim.*, **37**, 997 (1967).

159. A. Gateau, A. M. Sepulchre, A. Gaudemer, and S. D. Gero, *Carbohyd. Res.*, **24**, 474 (1972).

160. J. Defaye and D. Horton, *Carbohyd. Res.*, **14**, 128 (1970).

161. A. M. Sepulchre, J. Cléophax, J. Hildesheim, and S. D. Gero, *Carbohyd. Res.*, **14**, 1 (1970).

162. T. L. Hullar and S. B. Siskin, *J. Org. Chem.*, **35**, 225 (1970).

163. A. Zobáčová and J. Jarý, *Collect. Czech. Chem. Commun.*, **29**, 2042 (1964).

164. H. Paulsen and D. Stoye, *Ber.*, **102**, 3833 (1969).

165. G. J. Lourens, *Carbohyd. Res.*, **17**, 35 (1971).

166. G. García-Muñoz, R. Madroñero, M. Stud, F. Florencio, C. Foces, S. García-Blanco, M. Rico, and P. Smith, *J. Heterocycl. Chem.*, **11**, 281 (1974).

167. M. Sekiya and S. Ishiguro, *Carbohyd. Res.*, **22**, 325 (1972).

168. M. Sekiya and S. Ishiguro, *Carbohyd. Res.*, **22**, 337 (1972).

169. R. D. Guthrie and G. J. Williams, *Chem. Commun.*, 923 (1971).

170. B. R. Baker and T. L. Hullar, *J. Org. Chem.*, **30**, 4038 (1965).

171. B. R. Baker and T. Neilson, *J. Org. Chem.*, **29**, 1063 (1964); T. L. Hullar and T. Neilson, *Methods Carbohyd. Chem.*, **6**, 260 (1972).

172. B. R. Baker, K. Hewson, L. Goodman, and A. Benitez, *J. Amer. Chem. Soc.*, **80**, 6577 (1958).

173. R. D. Guthrie, D. Murphy, D. H. Buss, L. Hough, and A. C. Richardson, *Proc. Chem. Soc.*, 84 (1963).
174. B. R. Baker and T. Neilson, *J. Org. Chem.*, **29**, 1057 (1964).
175. B. R. Baker and T. Neilson, *J. Org. Chem.*, **29**, 1051 (1964).
176. J. E. Christensen and L. Goodman, *J. Amer. Chem. Soc.*, **82**, 4738 (1960); L. Goodman and J. E. Christensen, *ibid.*, **83**, 3823 (1961).
177. B. R. Baker and T. Neilson, *J. Org. Chem.*, **29**, 1047 (1964).
178. B. R. Baker and T. L. Hullar, *J. Org. Chem.*, **30**, 4049 (1965).
179. R. D. Guthrie and D. Murphy, *J. Chem. Soc.*, 5288 (1963).
180. R. D. Guthrie and D. Murphy, *Carbohyd. Res.*, **4**, 465 (1967); R. D. Guthrie, R. D. Wells, and G. J. Williams, *ibid.* **10**, 172 (1969).
181. R. D. Guthrie and R. D. Wells, *Carbohyd. Res.*, **24**, 11 (1972).
182. W. Meyer zu Reckendorf, *Ber.*, **98**, 93 (1965).
183. L. Goodman, *Advan. Carbohyd. Chem.*, **22**, 109 (1967).
184. C. F. Gibbs, L. Hough, and A. C. Richardson, *Carbohyd. Res.*, **1**, 290 (1965).
185. R. D. Guthrie and J. A. Liebmann, *J. Chem. Soc. Perkin I*, 650 (1974).
186. A. D. Barford and A. C. Richardson, *Carbohyd. Res.*, **4**, 408 (1967).
187. C. F. Gibbs, L. Hough, A. C. Richardson, and J. Tjebbes, *Carbohyd. Res.*, **8**, 405 (1968).
188. H. Saeki, T. Iwashige, and E. Ohki, *Chem. Pharm. Bull.* (Tokyo), **16**, 188 (1968); H. Saeki and E. Ohki, *ibid.*, **16**, 2471 (1968).
189. K. Ichimura, *Bull. Chem. Soc. Japan*, **43**, 2501 (1970).
190. H. Saeki and E. Ohki, *Chem. Pharm. Bull.* (Tokyo), **17**, 1664 (1969).
191. A. D. Barford and A. C. Richardson, *Carbohyd. Res.*, **14**, 217 (1970).
192. H. Paulsen and M. Budzis, *Ber.*, **103**, 3794 (1970).
193. H. Paulsen and D. Stoye, *Ber.*, **102**, 820 (1969).
193a. J.-M. Bourgeois, *Helv. Chim. Acta*, **57**, 2553 (1974).
193b. J. S. Brimacombe, J. A. Miller, and U. Zakir, *Carbohyd. Res.*, **49**, 233 (1976).
194. H. Weidmann, E. Stieger, and H. Schwarz, *Monatsh. Chem.*, **101**, 871 (1970).
195. C. F. Gibbs and L. Hough, *Chem. Commun.*, 1210 (1969); *Carbohyd. Res.*, **15**, 29 (1970).
195a. Y. Ali, A. C. Richardson, C. F. Gibbs, and L. Hough, *Carbohyd. Res.*, **7**, 255 (1968).
196. H. Saeki and E. Ohki, *Chem. Pharm. Bull.* (Tokyo), **16**, 962, 2477 (1968).
197. W. Meyer zu Reckendorf, *Ber.*, **97**, 325, 1275 (1964).
198. R. D. Guthrie and D. Murphy, *J. Chem. Soc.*, 3828 (1965).
198a. R. D. Guthrie and G. J. Williams, *J. Chem. Soc. Perkin I*, 801 (1976).
199. Z. M. El Shafei and R. D. Guthrie, *J. Chem. Soc.(C)*, 843 (1970).
200. C. F. Gibbs and L. Hough, *Carbohyd. Res.*, **18**, 363 (1971).
201. A. D. Barford and A. C. Richardson, *Carbohyd. Res.*, **14**, 231 (1970).
202. B. A. Dmitriev, N. É. Bairamova, and N. K. Kochetkov, *Izv. Akad. Nauk SSSR, Ser. Khim.*, 650 (1970).
203. J. Hildesheim, E. Walczak, and S. D. Gero, *C. R. Acad. Sci., Ser. C*, **267**, 980 (1968); S. D. Gero, J. Hildesheim, E. Walczak, R. D. Guthrie, and C. W. Smith, *J. Chem. Soc. (C)*, 1402 (1970).
204. H. Weidmann and H. K. Zimmerman, Jr., *Angew. Chem.*, **72**, 750 (1960); *Ann.*, **641**, 138 (1961).
205. L. V. Smith, Jr., P. H. Gross, K. Brendel, and H. K. Zimmerman, Jr., *Ann.*, **681**, 228 (1965).
206. H. Paulsen and E. Mäckel, *Ber.*, **102**, 3844 (1969).

207. M. W. Winkley, *Carbohyd. Res.*, **31**, 245 (1973).
208. M. L. Wolfrom, F. Shafizadeh, J. G. Wehrmüller, and R. K. Armstrong, *J. Org. Chem.*, **23**, 571 (1958).
209. F. Shafizadeh, J. G. Wehrmüller, and M. L. Wolfrom, *Abstr. Papers Amer. Chem. Soc. Meeting*, **130**, 22D, (1956).
210. B. Gross, M. El Sekily, S. Mancy, and H. El Khadem, *Carbohyd. Res.*, **37**, 384 (1974).
211. W. Meyer zu Reckendorf, *Ber.*, **103**, 2424 (1970).
212. B. Lindberg and O. Theander, *Acta Chem. Scand.*, **13**, 1226 (1959).
213. P. M. Collins and W. G. Overend, *Chem. Ind.* (London), 375 (1963); J. S. Brimacombe, J. G. H. Bryan, and M. Stacey, *Carbohyd. Res.*, **1**, 258 (1965).
214. R. U. Lemieux and T. L. Nagabhushan, *Can. J. Chem.*, **46**, 401 (1968).
215. R. U. Lemieux, K. James, T. L. Nagabhushan, and Y. Ito, *Can. J. Chem.*, **51**, 33 (1973); see also, R. U. Lemieux and S. W. Gunner, *ibid.*, **46**, 397 (1968).
216. A. K. Chatterjee, D. Horton, J. S. Jewell, and K. D. Philips, *Carbohyd. Res.*, **7**, 173 (1968).
217. S. W. Gunner, W. G. Overend, and N. R. Williams, *Carbohyd. Res.*, **4**, 498 (1967); C. L. Stevens, R. P. Glinski, and K. G. Taylor, *J. Org. Chem.*, **33**, 1586 (1968); C. L. Stevens and K. K. Balasubramanian, *Carbohyd. Res.*, **21**, 166 (1972); B. A. Dmitriev, A. Ya. Chernyak, O. S. Chizhov, and N. K. Kochetkov, *Izv. Akad. Nauk SSSR, Ser. Khim.*, 2298 (1972): C. L. Stevens and C. P. Bryant, *Methods Carbohyd. Chem.*, **6**, 235 (1972).
218. E. L. Albano and D. Horton, *Carbohyd. Res.*, **11**, 485 (1969).
219. J. M. J. Tronchet, R. Graf, and J. Tronchet, *Helv. Chim. Acta*, **52**, 315 (1969); J. M. J. Tronchet and R. Graf, *ibid.*, **53**, 851 (1970); **55**, 2286 (1972).
220. W. Sowa, *Can. J. Chem.*, **46**, 1586 (1968).
221. T. Takamoto, R. Sudoh, and T. Nakagawa, *Tetrahedron Lett.*, 2053 (1971).
222. P. J. Beynon, P. M. Collins, and W. G. Overend, *J. Chem. Soc.* (C), 272 (1969).
223. W. Roth, W. Pigman, and I. Danishefsky, *Tetrahedron*, **20**, 1675 (1964).
224. G. B. Howarth, W. A. Szarek, and J. K. N. Jones, *J. Chem. Soc.* (C), 2218 (1970).
225. A. N. O'Neill, *Can. J. Chem.*, **37**, 1747 (1959); J. C. Sowden and M. L. Oftedahl, *J. Amer. Chem. Soc.*, **82**, 2303 (1960); *Methods Carbohyd. Chem.*, **1**, 235 (1962).
226. J. C. Sowden and M. L. Oftedahl, *J. Org. Chem.*, **26**, 2153 (1961); Y. Ito, Y. Ohashi, and T. Miyagishima, *Carbohyd. Res.*, **9**, 125 (1969); M. B. Perry and V. Daoust, *ibid.*, **31**, 131 (1973); *Can. J. Chem.*, **51**, 3039 (1973), and earlier papers; M. B. Perry and J. Furdová, *Methods Carbohyd. Chem.*, **7**, 25 (1976); M. B. Perry, *ibid.*, **7**, 29, 32 (1976).
227. F. W. Lichtenthaler, P. Voss, and N. Majer, *Angew. Chem.*, **81**, 221 (1969); see also, T. Nakagawa, Y. Sato, T. Takamoto, F. W. Lichtenthaler, and N. Majer, *Bull. Chem. Soc. Japan*, **43**, 3866 (1970).
228. H. H. Baer and F. Rajabalee, *Carbohyd. Res.*, **12**, 241 (1970); see also, H. H. Baer and C.-W. Chiu, *ibid.*, **31**, 347 (1973).
229. F. W. Lichtenthaler and W. Fischer, *Chem. Commun.*, 1081 (1970).
230. H. H. Baer and F. Rajabalee, *Can. J. Chem.*, **47**, 4086 (1969).
231. H. H. Baer and W. Rank, *Can. J. Chem.*, **52**, 2257 (1974).
232. W. Rank and H. H. Baer, *Carbohyd. Res.*, **35**, 65 (1974).
233 B. A. Dmitriev, N. É. Bairamova, A. A. Kost, and N. K. Kochetkov, *Izv. Akad. Nauk SSSR, Ser. Khim.*, 2491 (1967).
234. B. A. Dmitriev, N. É. Bairamova, and N. K. Kochetkov, *Izv. Akad. Nauk SSSR, Ser. Khim.*, 2691 (1967).

235. J. Cléophax, S. D. Gero, J. Leboul, and A. Forchioni, *Chem. Commun.*, 710 (1973).
235a. J. S. Brimacombe, J. A. Miller, and U. Zakir, *Carbohyd. Res.*, **41**, C3 (1975); **44**, C9 (1975); see Ref. 193b.
236. W. J. Serfontein, J. H. Jordaan, and J. White, *Tetrahedron Lett.*, 1069 (1964); R. U. Lemieux, T. L. Nagabhushan, and I. K. O'Neill, *ibid.*, 1909 (1964).
237. R. U. Lemieux, Y. Ito, K. James, and T. L. Nagabhushan, *Can. J. Chem.*, **51**, 7 (1973).
238. M. B. Perry and V. Daoust, *Can. J. Chem.*, **52**, 2425 (1974).
238a. P. J. Garegg, B. Lindberg, and T. Norberg, *Acta Chem. Scand.*, *Ser. B*, **28**, 1104 (1974).
239. J. Lessard, H. Driguez, and J. P. Vermes, *Tetrahedron Lett.*, 4887 (1970).
240. S. Shibahara, S. Kondo, K. Maeda, H. Umezawa, and M. Ohno, *J. Amer. Chem. Soc.*, **94**, 4353 (1972); see also, P. J. L. Daniels, A. K. Mallams, and J. J. Wright, *Chem. Commun.*, 675 (1973).
240a. R. H. Hall, A. Jordaan, and O. G. DeVilliers, *J. Chem. Soc. Perkin I*, 626 (1975).
241. B. J. Magerlein, *Tetrahedron Lett.*, 33 (1970).
242. H. H. Baer and H. O. L. Fischer, *Proc. Nat. Acad. Sci. U.S.*, **44**, 991 (1958).
243. H. H. Baer, *Ber.*, **93**, 2865 (1960); F. W. Lichtenthaler, *Methods Carbohyd. Chem.*, **6**, 250 (1972); See also, K. Čapek, J. Staněk, Jr., and J. Jarý, *Collect. Czech. Chem. Commun.*, **39**, 1462 (1974).
243a. K. Čapek, J. Staněk, Jr., and J. Jarý, *Collect. Czech. Chem. Commun.*, **39**, 1462 (1974).
243b. K. Maekawa, Y. Miyoshi, and K. Tsuru, *Agr. Biol. Chem.* (Tokyo), **40**, 1951 (1976).
244. A. C. Richardson and H. O. L. Fischer, *Proc. Chem. Soc.*, 341 (1960); *J. Amer. Chem. Soc.*, **83**, 1132 (1961).
245. F. W. Lichtenthaler, T. Nakagawa, and A. El-Scherbiney, *Angew. Chem.*, **79**, 530 (1967); F. W. Lichtenthaler and T. Nakagawa, *Ber.*, **101**, 1846 (1968).
246. H. H. Baer and H. O. L. Fischer, *J. Amer. Chem. Soc.*, **81**, 5184 (1959).
247. M. L. Wolfrom, U. G. Nayak, and T. Radford, *Proc. Nat. Acad. Sci. U.S.*, **58**, 1848 (1967).
248. G. Baschang, *Ann.*, **663**, 167 (1963).
249. K. A. Watanabe, J. Beránek, H. A. Friedman, and J. J. Fox, *J. Org. Chem.*, **30**, 2735 (1965).
250. F. W. Lichtenthaler and H. P. Albrecht, *Ber.*, **100**, 1845 (1967).
251. H. H. Baer and I. Furić, *J. Org. Chem.*, **33**, 3731 (1968).
251a. J. Kovář and H. H. Baer, *Carbohyd. Res.*, **45**, 161 (1975).
252. H. H. Baer, *Advan. Carbohyd. Chem. Biochem.*, **24**, 67 (1969).
253. R. D. Guthrie and L. F. Johnson, *J. Chem. Soc.*, 4166 (1961).
254. G. J. F. Chittenden and R. D. Guthrie, *J. Chem. Soc.*, 2358 (1963); G. J. F. Chittenden, R. D. Guthrie, and J. F. McCarthy, *Carbohyd. Res.*, **1**, 196 (1965).
255. B. E. Davidson, R. D. Guthrie, and D. Murphy, *Carbohyd. Res.*, **5**, 449 (1967).
256. E. O. Bishop, R. D. Guthrie, and J. E. Lewis, *Carbohyd. Res.*, **5**, 477 (1967).
257. S. Roseman and D. G. Comb, *J. Amer. Chem. Soc.*, **80**, 3166 (1958); R. Kuhn and R. Brossmer, *Ann.*, **616**, 221 (1958); J. Brug and G. B. Paerels, *Nature*, **182**, 1159 (1958).
258. C. T. Spivak and S. Roseman, *J. Amer. Chem. Soc.*, **81**, 2403 (1959).
259. B. Coxon and L. Hough, *J. Chem. Soc.*, 1577 (1961).
260. J. Yoshimura, H. Sakai, N. Oda, and H. Hashimoto, *Bull. Chem. Soc. Japan*, **45**, 2027 (1972).

261. M. L. Wolfrom, D. Horton, and A. Böckmann, *Chem. Ind.* (London), 41 (1963); *J. Org. Chem.*, **29**, 1479 (1964).

262. J. R. Plimmer, N. Pravdić, and H. G. Fletcher, Jr., *J. Org. Chem.*, **32**, 1982 (1967).

262a. W. L. Salo, M. Hamari, and L. Hallcher, *Carbohyd. Res.*, **50**, 287 (1976); M. Hanchak and W. Korytnyk, *ibid.*, **52**, 219 (1976).

263. N. Pravdić and H. G. Fletcher, Jr., *Carbohyd. Res.*, **19**, 339 (1971).

264. E. Zissis, H. W. Diehl, and H. G. Fletcher, Jr., *Carbohyd. Res.*, **28**, 327 (1973).

265. Y. Ali and A. C. Richardson, *Chem. Commun.*, 554 (1967); *Carbohyd. Res.*, **5**, 441 (1967).

266. J. Kovář and H. H. Baer, *Can. J. Chem.*, **49**, 3203 (1971).

267. H. Saeki and E. Ohki, *Chem. Pharm. Bull.* (Tokyo), **18**, 412, 789 (1970).

268. G. B. Howarth, W. A. Szarek, and J. K. N. Jones, *Chem. Commun.*, 1339 (1969).

269. H. Shibata, I. Takeshita, N. Kurihara, and M. Nakajima, *Agr. Biol. Chem.* (Tokyo), **32**, 1006 (1968).

270. D. Ikeda, T. Tsuchiya, and S. Umezawa, *Bull. Chem. Soc. Japan*, **44**, 2529 (1971).

271. J. P. Marsh, Jr., C. W. Mosher, E. M. Acton, and L. Goodman, *Chem. Commun.*, 973 (1967).

272. E. J. Reist and S. H. Cruse, *J. Org. Chem.*, **34**, 3029 (1969).

273. K. Čapek, J. Šteffková, and J. Jarý, *Collect. Czech. Chem. Commun.*, **31**, 1854 (1966).

274. A. C. Richardson, *Carbohyd. Res.*, **4**, 415 (1967); see also ref. 117.

275. A. C. Richardson and K. A. McLauchlan, *J. Chem. Soc.*, 2499 (1962).

276. R. W. Jeanloz, *J. Amer. Chem. Soc.*, **79**, 2591 (1957); *Methods Carbohyd. Chem.*, **1**, 212 (1962).

277. R. W. Jeanloz, *Methods Carbohyd. Chem.*, **1**, 238 (1962).

278. Z. Tarasiejska and R. W. Jeanloz, *J. Amer. Chem. Soc.*, **79**, 2660, 4215 (1957); R. W. Jeanloz, *Methods Carbohyd. Chem.*, **1**, 231 (1962).

279. W. Meyer zu Reckendorf and W. A. Bonner, *Ber.*, **95**, 1917 (1962).

280. P. H. Gross, K. Brendel, and H. K. Zimmerman, Jr., *Ann.*, **680**, 155, 159 (1964).

281. G. D. Shryock and H. K. Zimmerman, (Jr.), *Carbohyd. Res.*, **3**, 14 (1966).

282. M. W. Horner, L. Hough, and A. C. Richardson, *J. Chem. Soc.* (C), 99 (1971).

283. W. Meyer zu Reckendorf, *Ber.*, **96**, 2019 (1963); *Methods Carbohyd. Chem.*, **6**, 270 (1972).

284. J. S. Brimacombe, I. Da'Aboul, and L. C. N. Tucker, *Carbohyd. Res.*, **25**, 522 (1972).

285. S. Hanessian, *J. Org. Chem.*, **32**, 163 (1967).

286. M. Parquet and P. Sinaÿ, *Carbohyd. Res.*, **18**, 195 (1971).

287. P. H. Gross, F. du Bois, and R. W. Jeanloz, *Carbohyd. Res.*, **4**, 244 (1967).

288. M. W. Horner, L. Hough, and A. C. Richardson, *J. Chem. Soc.* (C), 1336 (1970).

289. H. Paulsen and C.-P. Herold, *Ber.*, **104**, 1311 (1971).

290. H. Paulsen and Ö. Kristinsson, *Ber.*, **105**, 3456 (1972).

291. B. Belleau and Y.-K. Au-Young, *J. Amer. Chem. Soc.*, **85**, 64 (1963).

292. (a) V. B. Mochalin, Yu. N. Porshnev, and G. I. Samokhvalov, *Zh. Obshch. Khim.*, **38**, 85 (1968); (b) *ibid.*, **38**, 427 (1968).

293. (a) J. S. Brimacombe, I. Da'Aboul, and L. C. N. Tucker, *J. Chem. Soc. Perkin I*, 263 (1974); compare J. S. Brimacombe, F. Hunedy, A. M. Mather, and L. C. N. Tucker, *Carbohyd. Res.*, **68**, 231 (1979); see also (b) M. Chmielewski, A. Konowal, and A. Zamojski, *ibid.*, **70**, 275 (1979) for synthesis of a C_7 analogue, racemic purpurosamine B.

294. Y. Suhara, F. Sasaki, K. Maeda, H. Umezawa, and M. Ohno, *J. Amer. Chem. Soc.*, **90**, 6559 (1968).

294a. M. Natsume and M. Wada, *Chem. Pharm. Bull.* (Tokyo), **23**, 2567 (1975).
295. V. Nair and R. H. Walsh, *Carbohyd. Res.*, **36**, 131 (1974).
296. R. Kuhn and H. Fischer, *Ann.*, **617**, 88 (1958).
297. J. Jarý, Z. Kefurtová, and J. Kovář, *Collect. Czech. Chem. Commun.*, **34**, 1452 (1969).
298. B. Coxon and L. Hough, *J. Chem. Soc.*, 1463 (1961).
299. J. C. Irvine and J. C. Earl, *J. Chem. Soc.*, **121**, 2370 (1922).
300. M. Murase, *J. Antibiot.*, *Ser. A*, **14**, 367 (1961).
301. R. W. Jeanloz and A. M. C. Rapin, *J. Org. Chem.*, **28**, 2978 (1963).
302. S. Inouye, T. Tsuruoka, T. Ito, and T. Niida, *Tetrahedron*, **23**, 2125 (1968).
303. E. Hardegger, G. Zanetti, and K. Steiner, *Helv. Chim. Acta*, **46**, 282 (1963).
304. P. A. Levene, *J. Biol. Chem.*, **39**, 69 (1919).
305. D. Horton and A. E. Luetzow, *Carbohyd. Res.*, **7**, 101 (1968).
306. R. Kuhn and W. Bister, *Ann.*, **617**, 92 (1958).
307. R. W. Jeanloz and D. A. Jeanloz, *J. Org. Chem.*, **26**, 537 (1961).
308. R. Kuhn and W. Kirschenlohr, *Ann.*, **600**, 126 (1956).
309. H. Paulsen and K. Todt, *Ber.*, **100**, 512 (1967).
310. R. Kuhn and H. Fischer, *Ann.*, **612**, 65 (1958).
311. R. Kuhn and G. Baschang, *Ann.*, **628**, 193 (1959).
312. B. R. Baker, R. E. Schaub, and J. H. Williams, *J. Amer. Chem. Soc.*, **77**, 7 (1955).
313. A. J. Dick and J. K. N. Jones, *Can. J. Chem.*, **46**, 425 (1968).
314. R. E. Schaub and M. J. Weiss, *J. Amer. Chem. Soc.*, **80**, 4683 (1958).
315. M. Lamchen and R. L. Whistler, *Carbohyd. Res.*, **16**, 309 (1971).
316. H. Paulsen, J. Brüning, and K. Heyns, *Ber.*, **103**, 1621 (1970).
317. (a) J. S. Brimacombe, F. Hunedy, and M. Stacey, *J. Chem. Soc.* (*C*), 1811 (1968); (b) A. Zobáčová, V. Heřmánková, and J. Jarý, *Collect. Czech. Chem. Commun.*, **32**, 3560 (1967).
318. W. Meyer zu Reckendorf, *Methods Carbohyd. Chem.*, **6**, 266 (1972).
319. W. Meyer zu Reckendorf and N. Wassiliadou-Micheli, *Ber.*, **105**, 2998 (1972).
319a. W. Meyer zu Reckendorf, L. Rolf, and N. Wassiliadou-Micheli, *Carbohyd. Res.*, **45**, 307 (1975).
320. J. Hill, L. Hough, and A. C. Richardson, *Carbohyd. Res.*, **8**, 7 (1968).
321. M. L. Wolfrom, P. Chakravarty, and D. Horton, *J. Org. Chem.*, **30**, 2728 (1965).
322. W. Meyer zu Reckendorf, *Methods Carbohyd. Chem.*, **6**, 274 (1972).
323. T. H. Haskell, J. C. French, and Q. R. Bartz, *J. Amer. Chem. Soc.*, **81**, 3481 (1959).
324. (a) Y. Ali and A. C. Richardson, *J. Chem. Soc.* (*C*), 320 (1969); (b) see also W. Meyer zu Reckendorf and N. Wassiliadou-Micheli, *Ber.*, **107**, 1188 (1974), and citations therein.
325. Y. Ali and A. C. Richardson, *J. Chem. Soc.* (*C*), 1764 (1968).
326. H. H. Baer and M. Bayer, *Carbohyd. Res.*, **14**, 114 (1970); W. Meyer zu Reckendorf, *Chimia*, **24**, 16 (1970); *Ber.*, **104**, 1976 (1971).
327. W. Meyer zu Reckendorf, *Tetrahedron Lett.*, 287 (1970).
327a. W. Meyer zu Reckendorf, U. Kamprath-Scholz, E. Bischof, and N. Wassiliadou-Micheli, *Ber.*, **108**, 3397 (1975).
328. See, for examples: C. L. Stevens, P. Blumbergs, and D. H. Otterbach, *J. Org. Chem.*, **31**, 2817 (1966); C. B. Barlow, R. D. Guthrie, and A. M. Prior, *Carbohyd. Res.*, **10**, 481 (1969); A. Neuberger and A. T. Fletcher, *J. Chem. Soc.* (*B*), 178 (1969); *Carbohyd. Res.*, **17**, 79 (1971).
329. D. Horton, J. S. Jewell, and K. D. Philips, *J. Org. Chem.*, **31**, 3843 (1966); M. L. Wolfrom, P. Chakravarty, and D. Horton, *ibid.*, **31**, 2502 (1966).

330. M. L. Wolfrom, R. U. Lemieux, and S. M. Olin, *J. Amer. Chem. Soc.*, **71**, 2870 (1949).
331. M. L. Wolfrom and K. Anno, *J. Amer. Chem. Soc.*, **74**, 6150 (1952).
332. R. Breuer, *Ber.*, **31**, 2193 (1898).
333. E. Chargaff and M. Bovarnick, *J. Biol. Chem.*, **118**, 421 (1937).
334. For a survey of methods for *N*-deacetylation, see: S. Hanessian, *Methods Carbohyd. Chem.*, **6**, 208 (1972); see also: B. Nilsson and S. Svensson, *Carbohyd. Res.*, **62**, 377 (1978).
335. W. H. Bromund and R. M. Herbst, *J. Org. Chem.*, **10**, 267 (1945).
336. M. Bergmann and L. Zervas, *Ber.*, **64**, 975 (1931).
337. R. W. Jeanloz, *Advan. Carbohyd. Chem.*, **13**, 189 (1958); see also, ref. 5.
337a. M. B. Perry and A. C. Webb, *Can. J. Chem.*, **47**, 4091 (1969).
337b. Discussed in detail in G. G. S. Dutton, *Advan. Carbohyd. Chem. Biochem.*, **28**, 11 (1973); **30**, 10 (1974); see also, J. Lönngren and S. Svensson, *ibid.*, **29**, 41 (1974).
338. H. Paulsen and K. Todt, *Advan. Carbohyd. Chem.*, **23**, 116 (1968).
339. H. Paulsen, *Ann.*, **683**, 187 (1965).
340. H. Paulsen, *Ann.*, **670**, 121 (1963).
341. H. Paulsen, K. Todt, and K. Heyns, *Ann.*, **679**, 168 (1964).
342. H. Paulsen, K. Todt, and F. Leupold, *Tetrahedron Lett.*, 567 (1965).
343. H. Paulsen, F. Leupold, and K. Todt, *Ann.*, **692**, 200 (1966).
344. H. Paulsen, I. Sangster, and K. Heyns, *Ber.*, **100**, 802 (1967).
345. H. Paulsen and K. Todt, *Ber.*, **99**, 3450 (1966).
346. H. Paulsen, K. Propp, and K. Heyns, *Tetrahedron Lett.*, 683 (1969).
347. H. Paulsen, J. Brüning, K. Propp, and K. Heyns, *Tetrahedron Lett.*, 999 (1968).
348. W. Gielen, *Z. Physiol. Chem.*, **348**, 329 (1967).
349. H. Paulsen and U. Grage, *Ber.*, **107**, 2016 (1974).
350. K. Freudenberg and A. Doser, *Ber.*, **56**, 1243 (1923).
351. H. Paulsen and G. Steinert, *Ber.*, **100**, 2467 (1967).
351a. P. Smit, G. A. Stork, and H. C. Vander Plas, *J. Heterocycl. Chem.*, **12**, 957 (1975).
352. H. Paulsen and G. Steinert, *Ber.*, **103**, 1834 (1970).
353. H. Paulsen and G. Steinert, *Ber.*, **103**, 475 (1970).
354. J. K. N. Jones and W. A. Szarek, *Can. J. Chem.*, **41**, 636 (1963).
355. S. Hanessian and T. H. Haskell, *J. Org. Chem.*, **28**, 2604 (1963).
356. S. Hanessian, *Chem. Ind.* (London), 1296 (1965).
357. M. S. Patel and J. K. N. Jones, *Can. J. Chem.*, **43**, 3105 (1965).
357a. W. A. Szarek and D. Horton (eds.), *Amer. Chem. Soc. Symp. Ser.*, **87** (1979).
358. H. Paulsen and F. Leupold, *Carbohyd. Res.*, **3**, 47 (1966).
359. W. A. Szarek, S. Wolfe, and J. K. N. Jones, *Tetrahedron Lett.*, 2743 (1964).
360. J. K. N. Jones and J. C. Turner, *J. Chem. Soc.*, 4699 (1962).
361. H. Paulsen and F. Leupold, *Ber.*, **102**, 2804 (1969).
362. H. Paulsen and M. Friedmann, *Ber.*, **105**, 731 (1972).
363. W. A. Szarek and J. K. N. Jones, *Can. J. Chem.*, **42**, 20 (1964).
364. H. Weidmann, *Ann.*, **687**, 250 (1965).
365. S. Hanessian, *Carbohyd. Res.*, **1**, 178 (1965).
366. W. A. Szarek and J. K. N. Jones, *Can. J. Chem.*, **43**, 2345 (1965).
367. H. Paulsen, J. Brüning, and K. Heyns, *Ber.*, **102**, 459 (1969).
368. A. J. Dick and J. K. N. Jones, *Can. J. Chem.*, **43**, 977 (1965).
369. J. K. N. Jones, M. B. Perry, and J. C. Turner, *Can. J. Chem.*, **40**, 503 (1962).
370. J. K. N. Jones, M. B. Perry, and J. C. Turner, *Can. J. Chem.*, **39**, 2400 (1961).
370a. D. H. Ball, F. H. Bissett, M. H. Halford, and L. Long, Jr., *Carbohyd. Res.*, **45**, 91 (1975).

371. S. Hanessian, *J. Org. Chem.*, **34**, 675 (1969).
372. S. Hanessian and T. H. Haskell, *J. Heterocycl. Chem.*, **1**, 57 (1964).
373. S. Hanessian and T. H. Haskell, *J. Heterocycl. Chem.*, **1**, 55 (1964).
374. H. Weidmann and E. Fauland, *Ann.*, **679**, 192 (1964).
375. T. Machinami and T. Suami, *Bull. Chem. Soc. Japan*, **46**, 1013 (1973).
376. H. Paulsen and U. Grage, *Ber.*, **102**, 3854 (1969).
377. J. W. Llewellyn and J. M. Williams, *Carbohyd. Res.*, **28**, 339 (1973); **42**, 168 (1975).
378. D. Horton and W. Loh, *Carbohyd. Res.*, **36**, 121 (1974).
379. D. Horton and W. Loh, *Carbohyd. Res.*, **38**, 189 (1974).
380. H. Paulsen, K. Todt. and H. Ripperger, *Ber.*, **101**, 3365 (1968).
381. D. Horton and K. D. Philips, *Carbohyd. Res.*, **22**, 151 (1972); compare Y. Gelas-Mialhe and D. Horton, *J. Org. Chem.*, **43**, 2307 (1978).
382. A. Angeli, *Ber.*, **26**, 1715 (1893); see also, R. Breslow and C. Yuan, *J. Amer. Chem. Soc.*, **80**, 5991 (1958).
383. D. Horton and E. K. Just, *Chem. Commun.*, 1116 (1969).
383a. M. Alexander and D. Horton, *Abstr. Papers Amer. Chem. Soc. Meeting*, **174**, CARB 12 (1977).
384. For a discussion in more detail, see: J. M. Williams, *Advan. Carbohyd. Chem. Biochem.*, **31**, 9 (1975).
385. E. L. Eliel, N. A. Allinger, S. J. Angyal, and G. A. Morrison, "Conformational Analysis," Wiley, New York, 1965, pp. 28–31.
386. G. Ledderhose, *Z. Physiol. Chem.*, **4**, 139 (1880).
387. D. Horton and K. D. Philips, *Carbohyd. Res.*, **30**, 367 (1973).
388. A. B. Foster, E. F. Martlew, and M. Stacey, *Chem. Ind.* (London), 825 (1953).
389. C. Erbing, B. Lindberg, and S. Svensson, *Acta Chem. Scand.*, **27**, 3699 (1973); see also G. O. Aspinall, E. Przybylski, R. G. S. Ritchie, and C. O. Wong, *Carbohyd. Res.*, **66**, 225 (1978) for microanalytical applications with amino sugar-containing oligosaccharides.
390. J. C. Irvine and A. Hynd, *J. Chem. Soc.*, **101**, 1128 (1912).
391. W. Ambrecht, *Biochem. Z.*, **95**, 108 (1919).
392. B. C. Bera, A. B. Foster, and M. Stacey, *J. Chem. Soc.*, 4531 (1956).
393. P. A. Levene and F. B. LaForge, *J. Biol. Chem.*, **21**, 345, 351 (1915); P. A. Levene, *Biochem. Z.*, **124**, 37 (1921).
394. P. A. Levene and F. B. LaForge, *J. Biol. Chem.*, **20**, 433 (1915).
395. P. W. Austin, J. G. Buchanan, and R. M. Saunders, *Chem. Commun.*, 146 (1965); *J. Chem. Soc.* (C), 372 (1967).
396. Z. Dische and E. Borenfreund, *J. Biol. Chem.*, **184**, 517 (1950); D. Exley, *Biochem. J.*, **67**, 52 (1957).
397. S. Inoue, personal communication, cited in ref. 384.
398. N. M. K. Ng Ying Kin, J. M. Williams, and A. Horsington, *J. Chem. Soc.* (C), 1578 (1971).
399. M. J. Cron, O. B. Fardig, D. L. Johnson, D. F. Whitehead, I. R. Hooper, and R. U. Lemieux, *J. Amer. Chem. Soc.*, **80**, 4115 (1958).
400. N. M. K. Ng Ying Kin, J. M. Williams, and A. Horsington, *Chem. Commun.*, 971 (1969).
401. L. F. Wiggins, *Nature*, **157**, 300 (1946).
402. N. M. K. Ng Ying Kin and J. M. Williams, *Chem. Commun.*, 1123 (1971).
403. M. Chérest, H. Felkin, J. Sicher, F. Šipoš, and M. Tichý, *J. Chem. Soc.*, 2513 (1965).
404. D. Horton, K. D. Philips, and J. Defaye, *Carbohyd. Res.*, **21**, 417 (1972).
405. S. Hase and Y. Matsushima, *J. Biochem.* (Tokyo), **69**, 559 (1971).

405a. J. W. Llewellyn and J. M. Williams, *Carbohyd. Res.*, **42**, 168 (1975).

405b. J. W. Llewellyn and J. M. Williams, *J. Chem. Soc. Perkin I*, 1997 (1973).

405c. J. A. Ballantine, G. Hutchinson, and J. M. Williams, *Carbohyd. Res.*, **50**, C9 (1976).

406. A. K. Al-Radhi, J. S. Brimacombe, and L. C. N. Tucker, *Chem. Commun.*,1250 (1070); *J. Chem. Soc. Perkin I*, 315 (1972).

407. J. S. Brimacombe, J. Minshall, and L. C. N. Tucker, *Carbohyd. Res.*, **32**, C7 (1974).

408. J. S. Brimacombe and J. Minshall, *Carbohyd. Res.*, **25**, 267 (1972); J. S. Brimacombe, J. Minshall, and L. C. N. Tucker, *ibid.*, **35**, 55 (1974).

409. V. G. Bashford and L. F. Wiggins, *J. Chem. Soc.*, 371 (1950).

410. V. G. Bashford and L. F. Wiggins, *Nature*, **165**, 566 (1950).

411. V. G. Bashford and L. F. Wiggins, *J. Chem. Soc.*, 299 (1948).

412. R. Barker, *J. Org. Chem.*, **29**, 869 (1964).

413. D. Horton, P. L. Durette, and J. D. Wander, *Ann. N. Y. Acad. Sci.*, **222**, 884 (1973), and references therein.

414. D. D. Heard, B. G. Hudson, and R. Barker, *J. Org. Chem.*, **35**, 464 (1970).

415. M. Blanc-Muesser, J. Defaye, and D. Horton, *Carbohyd. Res.*, **68**, 175 (1979).

416. Y. Matsushima, *Bull. Chem. Soc. Japan*, **24**, 144 (1951).

417. T. Bando and Y. Matsushima, *Bull. Chem. Soc. Japan*, **46**, 593 (1973).

418. A. B. Foster, *Chem. Ind.* (London), 627 (1955).

419. P. A. Levene, *J. Biol. Chem.*, **63**, 95 (1925).

420. P. A. Levene, *J. Biol. Chem.*, **36**, 89 (1918).

421. J. Defaye, *Advan. Carbohyd. Chem. Biochem.*, **25**, 181 (1970).

422. A. E. El Ashmawy, D. Horton, L. G. Magbanua, and J. M. J. Tronchet, *Carbohyd. Res.*, **6**, 299 (1968).

423. For an overview, see D. Horton, *Pure Appl. Chem.*, **42**, 301 (1975).

424. B. Berrang and D. Horton, *Chem. Commun.*, 1038 (1970); A. Ducruix, C. Pascard-Billy, D. Horton, and J. D. Wander, *Carbohyd. Res.*, **29**, 276 (1973).

425. J. Defaye, *Bull. Soc. Chim. Fr.*, 1101 (1967).

426. (a) J. Defaye, A. Ducruix, and C. Pascard-Billy, *Bull. Soc. Chim. Fr.*, 4514 (1970); (b) J. Defaye, T. Nakamura, D. Horton, and K. D. Philips, *Carbohyd. Res.*, **16**, 133 (1971).

427. H. Ohle and R. Lichtenstein, *Ber.*, **63**, 2905 (1930).

428. R. L. Whistler and R. E. Pyler, *Carbohyd. Res.*, **12**, 201 (1970).

429. E. Fischer and J. Tafel, *Ber.*, **20**, 2566 (1887).

430. R. Gruennagel and H. J. Haas, *Ann.*, **721**, 234 (1969).

431. M. J. Cron, O. B. Fardig, D. L. Johnson, H. Schmitz, D. F. Whitehead, I. R. Hooper, and R. U. Lemieux, *J. Amer. Chem. Soc.*, **80**, 2342 (1958).

432. J. C. Sowden, C. H. Bowers, and K. O. Lloyd, *J. Org. Chem.*, **29**, 130 (1964).

433. R. D. Guthrie and D. King, *Carbohyd. Res.*, **3**, 129 (1966).

434. N. N. Gerber, *J. Med. Chem.*, **7**, 204 (1964).

435. (a) R. C. G. Moggridge and A. Neuberger, *J. Chem. Soc.*, 745 (1938); (b) A. B. Foster, D. Horton, and M. Stacey, *ibid.*, 81 (1957).

436. A. B. Foster, R. Harrison, T. D. Inch, M. Stacey, and J. M. Webber, *J. Chem. Soc.*, 2279 (1963); M. L. Wolfrom, P. Y. Wang, and S. Honda, *Carbohyd. Res.*, **11**, 179 (1969); also, see ref. 384.

437. M. Ishidate and M. Matsui, *Yakugaku Zasshi*, **82**, 662 (1962); *Chem. Abstr.*, **58**, 4639g (1963); K. Ochi, Jap. Pat. 19,626 (1971); *Chem. Abstr.*, **75**, 64,218 (1971).

438. F. Tiemann, *Ber.*, **17**, 241 (1884); **27**, 118 (1894); F. Tiemann and R. Haarmann, *ibid.*, **19**, 1257 (1886).

439. F. Tiemann, *Ber.*, **19**, 49 (1886).
440. R. Lohmar and K. P. Link, *J. Biol. Chem.*, **150**, 351 (1943).
441. K. Heyns, C.-M. Koch, and W. Koch *Z. Physiol. Chem.*, **296**, 121 (1954).
442. M. I. Taha, *J. Chem. Soc.*, 2468 (1961).
443. M. R. Everett and F. Sheppard, *Univ. Oklahoma Med. School, Dept. Biochem.*, (1944); *Chem. Abstr.*, **40**, 547 (1946).
444. Y. Matsushima, *Bull. Chem. Soc. Japan*, **24**, 17 (1951).
445. R. M. Herbst, *J. Biol. Chem.*, **119**, 85 (1937).
446. (a) S. Gardell, F. Heijkenskjold, and A. Roch-Norlund, *Acta Chem. Scand.*, **4**, 970 (1950); (b) see also: A Veyrières and R. W. Jeanloz, *Biochemistry*, **9**, 4153 (1970).
447. Y. Matsushima, *Sci. Pap. Osaka Univ.*, **32**, 7 (1951).
448. P. J. Stoffyn and R. W. Jeanloz, *Arch. Biochem. Biophys.*, **52**, 373 (1954).
449. E. Hardegger, K. Kreis, and H. (S.) El Khadem, *Helv. Chim. Acta*, **35**, 618 (1952).
450. H. Hamburger, *Biochem. Z.*, **36**, 1 (1911).
451. Y. Matsushima, *Sci. Pap. Osaka Univ.*, **33**, 9 (1951).
452. D. Horton, *Methods Carbohyd. Chem.*, **6**, 282 (1972); *Org. Syn.*, **46**, 1 (1966).
453. F. W. Lichtenthaler, G. Bambach, and U. Scheidegger, *Ber.*, **102**, 986 (1969).
454. F. Micheel and H. Wulff, *Ber.*, **89**, 1521 (1956).
455. F. Micheel and H. Petersen, *Ber.*, **92**, 298 (1959).
456. F. Micheel, F.-P. van de Kamp, and H. Petersen, *Ber.*, **90**, 521 (1957).
457. H. Weidmann, D. Tartler, P. Stöckl, L. Binder, and H. Hönig, *Carbohyd. Res.*, **29**, 135 (1973); see: P. Stöckl, H. Hönig, and H. Weidmann, *J. Carbohyd. Nucleosides, Nucleotides*, **1**, 169 (1974).
458. H. Weidmann, D. Tartler, P. Stöckl, and H. Hönig, *Monatsh. Chem.*, **103**, 883 (1972).
459. N. Pravdić, T. D. Inch, and H. G. Fletcher, Jr., *J. Org. Chem.*, **32**, 1815 (1967).
460. A. Ya. Khorlin, M. L. Shul'man, S. E. Zurabyan, I. M. Privalova, and Yu. L. Kopaevich, *Izv. Akad. Nauk SSSR, Ser. Khim.*, 236, 2094 (1968).
461. K. L. Matta and O. P. Bahl, *Carbohyd. Res.*, **21**, 460 (1972); see also M. A. Nashed, *ibid.*, **71**, 299 (1979), and compare C. D. Warren, M. A. E. Shaban, and R. W. Jeanloz, *ibid.*, **59**, 427 (1977) and M. A. Nashed, C. W. Slife, M. Kiso, and L. Anderson, *ibid.*, **58**, C13 (1977) for applications of oxazolines in oligosaccharide synthesis.
462. S. Konstas, I. Photaki, and L. Zervas, *Ber.*, **92**, 1288 (1959); however, see F. Micheel, *ibid.*, **102**, 2880 (1969).
463. Y. Inouye, K. Onodera, and J. Nakatani, *Nippon Nogeikagaku Kaishi*, **25**, 550 (1951); *Chem. Abstr.*, **48**, 2002 (1954); Y. Inouye, K. Onodera, and S. Kitaoka, *ibid.*, **29**, 139, 143 (1955); *Chem. Abstr.*, **50**, 825, 826 (1956); R. H. Hackman, *Aust. J. Biol. Sci.*, **8**, 83 (1955).
464. L. Hough and M. I. Taha, *J. Chem. Soc.*, 2042 (1956).
464a. K. Igarashi, *Advan. Carbohyd. Chem. Biochem.*, **34**, 243 (1977).
465. B. R. Baker, J. P. Joseph, R. E. Schaub, and J. H. Williams, *J. Org. Chem.*, **19**, 1786 (1954).
466. F. Micheel, H. Petersen, and J. Köchling, *Ber.*, **93**, 1 (1960).
467. F. Micheel and W. Lengsfeld, *Ber.*, **89**, 1246 (1956).
468. J.-C. Jacquinet, S. E. Zurabyan, and A. Ya. Khorlin, *Carbohyd. Res.*, **32**, 137 (1974).
469. W. Meyer zu Reckendorf, N. Wassiliadou-Micheli, and N. Delevallee, *Ber.*, **102**, 1076 (1969).
470. J. C. Irvine, D. McNicoll, and A. Hynd, *J. Chem. Soc.*, **99**, 250 (1911).
471. F. Micheel and H. Micheel, *Ber.*, **65**, 253 (1932).

472. M. L. Wolfrom, H. G. Garg, and D. Horton, *J. Org. Chem.*, **30**, 1556 (1965).
473. M. L. Wolfrom and P. J. Conigliaro, *Carbohyd. Res.*, **11**, 63 (1969).
474. J. C. Irvine and J. C. Earl, *J. Chem. Soc.*, **121**, 2376 (1922).
475. W. L. Salo and H. G. Fletcher, Jr., *J. Org. Chem.*, **34**, 3189 (1969).
476. P. F. Lloyd and B. Evans, *J. Chem. Soc.* (*C*), 2753 (1968).
477. M. W. Whitehouse and P. W. Kent, *Tetrahedron*, **4**, 425 (1958).
478. H. Steudel, *Z. Physiol. Chem.*, **33**, 223 (1901); **34**, 352 (1902).
479. C. Neuberg and H. Wolff, *Ber.*, **34**, 3840 (1901).
480. H. Fritz, C. J. Morel, and O. Wacker, *Helv. Chem. Acta*, **51**, 569 (1968).
481. J. Yoshimura, T. Sekiya, and T. Iida, *Bull. Chem. Soc. Japan*, **45**, 1227 (1972).
482. J. E. Scott, *Carbohyd. Res.*, **14**, 389 (1970).
483. F. García González, J. Fernández-Bolaños, and J. Fuentes Mota, *Carbohyd. Res.*, **22**, 436 (1972).
484. M. H. Fischer and B. A. Lewis, *Chem. Ind.* (London), 192 (1967).
485. G. Zemplén, A. Gerecs, and M. Rados, *Ber.*, **69**, 748 (1936).
486. F. García González and J. Fernández-Bolaños, *An. Real Soc. Espan. Fis. Quim.*, *Ser. B.*, **45**, 1527 (1949); A. Paneque Guerrero, F. García González, and J. Fernández-Bolaños, *An. Inst. Farmacol. Espan.*, **5**, 309 (1956).
487. C. Neuberg, H. Wolff, and W. Neimann, *Ber.*, **35**, 4009 (1902).
487a. S. Hirano and R. Yamasaki, *Carbohyd. Res.*, **43**, 377 (1975).
487b. J. C. Jochims, *Ber.*, **108**, 2320 (1975).
487c. N. D. Heindel, H. D. Burns, T. Honda, V. R. Risch, and L. W. Brady, *Org. Prep. Proced. Int.*, **7**, 291 (1975).
487d. C. Neuberg, *Ber.*, **35**, 4009 (1902); M. Bergmann, L. Zervas, and E. Silberkweit, *Ber.*, **64**, 2428 (1931).
487e. C. T. Clarke, J. H. Jones, and R. Walker, *J. Chem. Soc. Perkin I*, 1001 (1976).
488. K. Miyai and P. H. Gross, *J. Org. Chem.*, **34**, 1638 (1969).
489. S. Umezawa, Y. Takagi, and T. Tsuchiya, *Bull. Chem. Soc. Japan*, **44**, 1411 (1971).
490. S. Umezawa, T. Tsuchiya, T. Yamasaki, H. Sano, and Y. Takahashi, *J. Amer. Chem. Soc.*, **96**, 920 (1974).
491. A. B. Foster, M. Stacey, and S. V. Vardheim, *Acta Chem. Scand.*, **13**, 281 (1959).
492. E. M. Bessell and J. H. Westwood, *Carbohyd. Res.*, **19**, 389 (1971).
493. J. Yoshimura, T. Sekiya, and Y. Ogura, *Bull. Chem. Soc. Japan*, **47**, 1219 (1974).
494. A. Hasegawa and H. G. Fletcher, Jr., *Carbohyd. Res.*, **29**, 223 (1973).
495. A. M. Yurkevich and O. N. Shevtsova, *Zh. Obshch. Khim.*, **42**, 1172 (1972).
496. A. N. Fujiwara, E. M. Acton, and L. Goodman, *J. Heterocycl. Chem.*, **7**, 891 (1970).
497. B. Jastorff and H. P. Bär, *Eur. J. Biochem.*, **37**, 497 (1973).
498. M. Morr, M.-R. Kula, G. Roesler, and B. Jastorff, *Angew. Chem.*, **86**, 308 (1974).
499. W. Pfleiderer, E. Bühler, and D. Schmidt, *Ber.*, **101**, 3794 (1968).
500. O. Wacker and H. Fritz, *Helv. Chim. Acta*, **50**, 2481 (1967).
501. T. Kamiya, Y. Saito, M. Hashimoto, and H. Seki, *Tetrahedron*, **28**, 899 (1972).
502. For a review, see: F. García González and A. Gómez Sánchez, *Advan. Carbohyd. Chem.*, **20**, 303 (1965).
503. H. Pauly and E. Ludwig, *Z. Physiol. Chem.*, **121**, 170 (1922).
504. R. Boyer and O. Fürth, *Biochem. Z.*, **282**, 242 (1935).
505. F. García González, *An. Real Soc. Espan. Fis. Quim.*, **32**, 815 (1934).
506. C. Cessi and F. Serafini-Cessi, *Biochem. J.*, **88**, 132 (1963).
507. A. Gómez Sánchez, A. Cert Ventulá, and U. Scheidegger, *Carbohyd. Res.*, **18**, 173 (1971).

508. F. Serafini-Cessi and C. Cessi, *Biochem. J.*, **120**, 873 (1970).
509. A. Gómez Sánchez, E. Toledano, and M. Gómez Guillén, *J. Chem. Soc. Perkin I*, 1237 (1974), and earlier papers cited therein; F. García González, J. Fernández-Bolaños, and F. Alcudia, *An. Quim.*, **67**, 383 (1971), and earlier papers cited therein.
510. A. Gómez Sánchez, A. Cert Ventulá, M. Gómez Guillén, and U. Scheidegger, *An. Quim.*, **67**, 545 (1971).
511. A. Gómez Sánchez, M. Gómez Guillén, E. Pando Ramos, and A. Cert Ventulá, *Carbohyd. Res.*, **35**, 39 (1974).
512. F. García González, A. Gómez Sánchez, M. Gómez Guillen, and M. Tena Aldave, *An. Quim.*, **67**, 389 (1971).
513. M. J. Crumpton, *Biochem. J.*, **72**, 479 (1959).
514. M. J. How, M. D. A. Halford, and M. Stacey, *Carbohyd. Res.*, **11**, 313 (1969).
515. J. W. Cornforth, M. E. Daines, and A. Gottschalk, *Proc. Chem. Soc.*, 25 (1957); J. W. Cornforth, M. E. Firth, and A. Gottschalk, *Biochem. J.*, **68**, 57 (1958).
516. H. Rinderknecht and T. Rebane, *Experientia*, **19**, 342 (1963).
517. R. Kuhn and G. Baschang, *Ann.*, **659**, 156 (1962).
518. A. Gómez Sánchez, F. García González, and J. Gasch Gómez, *Rev. Espan. Fisiol.*, **14**, 277 (1958).
519. F. García González, J. Fernández-Bolaños, and A. Paneque Guerrero, *An. Real Soc. Espan. Fis. Quim.*, Ser. B, **57**, 379 (1961).
520. J. V. Scudi, G. E. Boxer, and V. Jelinek, *Science*, **104**, 486 (1946).
521. L. A. Elson and W. T. J. Morgan, *Biochem. J.*, **27**, 1824 (1933).
522. A. B. Foster and M. Stacey, *Advan. Carbohyd. Chem.*, **7**, 247 (1952); A. B. Foster and D. Horton, *ibid.*, **14**, 213 (1959).
523. N. F. Boas, *J. Biol. Chem.*, **204**, 553 (1953), and references therein.
524. P. G. Johansen, R. D. Marshall, and A. Neuberger, *Biochem. J.*, **77**, 239 (1960).
525. R. Belcher, A. J. Nutten, and C. M. Sambrook, *Analyst*, **79**, 201 (1954).
526. B. Schloss, *Anal. Chem.*, **23**, 1321 (1951); J. W. Cornforth and M. E. Firth, *J. Chem. Soc.*, 1091 (1958).
527. D. Horton, J. Vercellotti, and M. L. Wolfrom, *Biochim. Biophys. Acta*, **50**, 358 (1961).
528. S. M. Partridge, *Biochem. J.*, **42**, 238 (1948).
528a. R. Kuhn and G. Krüger, *Ber.*, **89**, 1473 (1956).
528b. J.-M. Beau, P. Rollin, and P. Sinaÿ, *Carbohyd. Res.*, **53**, 187 (1977).
529. P. A. Levene and F. B. LaForge, *J. Biol. Chem.*, **18**, 123 (1914).
530. R. Heyworth, D. H. Leaback, and P. G. Walker, *J. Chem. Soc.*, 4121 (1959), and references therein.
531. P. A. Levene, *J. Biol. Chem.*, **57**, 337 (1923).
532. D. Horton, J. S. Jewell, and K. D. Philips, *J. Org. Chem.*, **31**, 4022 (1966).
533. M. Takai, S. Watanabe, T. Ashida, and M. Kakudo, *Acta Crystallogr.*, **B28**, 2370 (1972).
534. See D. G. Steefkerk, M. J. A. de Bie, and J. F. G. Vliegenthart, *Carbohyd. Res.*, **33**, 339 (1974).
535. T. D. Inch, J. R. Plimmer, and H. G. Fletcher, Jr., *J. Org. Chem.*, **31**, 1825 (1966).
535a. R. Taniyama, Y. Suwada, and J. Kitigawa, *J. Antibiot.*, **24**, 662 (1971); R. Taniyama, Y. Suwada, and S. Tanaka, *Chem. Pharm. Bull.* (Tokyo), **22**, 337 (1974).
536. W. R. Clark, J. McLaughlin, and M. E. Webster, *J. Biol. Chem.*, **230**, 81 (1958); K. Heyns, G. Kiessling, W. Lindberg, H. Paulsen, and M. E. Webster, *Ber.*, **92**, 2435 (1959).
537. K. Heyns and M. Beck, *Ber.*, **90**, 2443 (1957).
538. M. J. Crumpton and D. A. L. Davies, *Biochem. J.*, **64**, 22P (1956); **70**, 729 (1958).

539. U. Zehavi and N. Sharon, *J. Org. Chem.*, **29**, 3654 (1964).
540. S. A. Barker, J. S. Brimacombe, M. J. How, M. Stacey, and J. M. Williams, *Nature*, **189**, 303 (1961).
541. R. Kuhn, W. Bister, and W. Dafeldecker, *Ann.*, **628**, 186 (1959).
541a. T. H. Haskell, D. Horton, and G. Rodemeyer, *Carbohyd. Res.*, **55**, 35 (1977); D. Horton, G. Rodemeyer, and R. Rodemeyer, *ibid.*, **56**, 129 (1977).
542. See: "The Amino Sugars," (R. W. Jeanloz and E. A. Balazs, eds.), Vol. IIA. Academic Press, New York, 1965.
543. E. G. Cox and G. A. Jeffrey, *Nature*, **143**, 894 (1939); S. C. Chu and G. A. Jeffrey, *Proc. Roy. Soc.* (London), **A285**, 470 (1965).
544. F. Shafizadeh, G. D. McGinnis, R. A. Susott, and M. H. Meshreki, *Carbohyd. Res.*, **33**, 191 (1974).
545. C. A. Lobry de Bruyn and W. Alberda van Ekenstein, *Rec. Trav. Chim.* (Pays-Bas), **18**, 77 (1899); K. Stolte, *Beitr. Chem. Physiol. Pathol.*, **11**, 19 (1907); *Chem. Zentr.*, *I*, 224 (1908).
546. D. Horton, J. B. Hughes, J. S. Jewell, K. D. Philips, and W. N. Turner, *J. Org. Chem.*, **32**, 1073 (1967); D. Horton, W. E. Mast, and K. D. Philips, *ibid.*, **32**, 1471 (1967).
547. R. C. Dougherty, D. Horton, K. D. Philips, and J. D. Wander, *Org. Mass Spectrom.*, **7**, 805 (1973).
548. For a review, see: G. Blix and R. W. Jeanloz, in "The Amino Sugars" (R. W. Jeanloz and E. A. Balazs, eds.), Vol. IA. Academic Press, New York, 1969, pp. 213–265.
549. R. E. Strange and L. H. Kent, *Biochem. J.*, **71**, 333 (1959); L. H. Kent and R. E. Strange, *Methods Carbohyd. Chem.*, **1**, 750 (1962).
550. G. Ågren and C. H. de Verdier, *Acta Chem. Scand.*, **12**, 1927 (1958); Y. Konami, T. Osawa, and R. W. Jeanloz, *Biochemistry*, **10**, 192 (1971); S. Hase and Y. Matsushima, *Bull. Chem. Soc. Japan*, **47**, 1190 (1974).
551. P. Sinaÿ, *Carbohyd. Res.*, **16**, 113 (1971).
552. R. W. Jeanloz, E. Walker, and P. Sinaÿ, *Carbohyd. Res.*, **6**, 184 (1968).
553. J. J. Vavra, C. DeBoer, A. Dietz, L. J. Hanka, and W. T. Sokoloski, *Antibiot. Annu.*, 230 (1959–1960).
554. R. R. Herr, H. K. Jahnke, and A. D. Argoudelis, *J. Amer. Chem. Soc.*, **89**, 4808 (1967); E. Hardegger, A. Meier, and A. Stoos, *Helv. Chim. Acta*, **52**, 2555 (1969).
555. N. Rakieten, M. L. Rakieten, and M. V. Nadkarni, *Cancer Chemother. Rep.*, **29**, 91 (1963).
556. E. Meyers, D. M. Slusarchyk, J. L. Bouchard, and F. L. Weisenborn, *J. Antibiot.*, **22**, 490 (1969).
557. N. Pravdić, E. Zissis, M. Pokorny, and H. G. Fletcher, Jr., *Carbohyd. Res.*, **32**, 115 (1974).
558. A. R. Williamson and S. Zamenhof, *J. Biol. Chem.*, **238**, 2255 (1963).
559. S. Hanessian and T. H. Haskell, *J. Biol. Chem.*, **239**, 2758 (1964).
560. C. J. Morel, *Helv. Chim. Acta*, **41**, 1501 (1958).
561. E. J. Smith, *Biochem. Biophys. Res. Commun.*, **15**, 593 (1964).
562. R. U. Lemieux and M. L. Wolfrom, *Advan. Carbohyd. Chem.*, **3**, 337 (1948).
563. E. E. van Tamelen, J. R. Dyer, H. E. Carter, J. V. Pierce, and E. E. Daniels, *J. Amer. Chem. Soc.*, **78**, 4817 (1956).
564. R. Kuhn, W. Kirschenlohr, and W. Bister, *Angew. Chem.*, **69**, 60 (1957).
565. D. B. Borders, K. J. Sax, J. E. Lancaster, W. K. Hausmann, L. A. Mitscher, E. R. Wetzel, and E. L. Patterson, *Tetrahedron*, **26**, 3123 (1970).

565a. R. Reistad, *Carbohyd. Res.*, **36**, 420 (1974).

565b. M. Torii, K. Sakakibara, and K. Kuroda, *Eur. J. Biochem.*, **37**, 401 (1973).

566. H. R. Perkins, *Biochem. J.*, **86**, 475 (1963).

567. J. S. Brimacombe and M. C. Cook, *Chem. Ind.* (London), 1281 (1963); *J. Chem. Soc.*, 2663 (1964).

568. B. Jann and K. Jann, *Eur. J. Biochem.*, **5**, 173 (1968).

569. M. J. Crumpton, *Nature*, **180**, 605 (1957).

570. R. Heyworth and P. G. Walker, *Proc. Intern. Congr. Biochem. 4th Vienna*, **1**, 7 (1958).

571. H. Muir, *Biochem. J.*, **65**, 33P (1957).

572. S. A. Barker, M. Stacey, and J. M. Williams, *Bull. Soc. Chim. Biol.*, **42**, 1611 (1960); S. A. Barker and M. Stacey, *Biochem. J.*, **82**, 37P (1962).

573. J. S. Brimacombe and M. J. How, *J. Chem. Soc.*, 3886 (1963).

574. W. Meyer zu Reckendorf and W. A. Bonner, *Tetrahedron*, **19**, 1711 (1963).

574a. E. Jegou, J. Cléophax, J. Leboul, and S. D. Gero, *Carbohyd. Res.*, **45**, 323 (1975).

574b. H. Sano, T. Tsuchiya, Y. Ban, and S. Umezawa, *Bull. Chem. Soc. Japan*, **49**, 313 (1976).

575. K. Isono, K. Asahi, and S. Suzuki, *J. Amer. Chem. Soc.*, **91**, 7490 (1969).

575a. H. Kuzuhara and S. Emoto, *Tetrahedron Lett.*, 5051 (1973).

575b. T. Nakanishi, F. Tomita, and T. Suzuki, *Agr. Biol. Chem.* (Tokyo), **38**, 2465 (1974).

575c. G. Ashwell and W. A. Volk, *J. Biol. Chem.*, **240**, 4549 (1965).

576. O. Lüderitz, E. Ruschmann, O. Westphal, R. A. Raff, and R. Wheat, *J. Bacteriol.*, **93**, 1681 (1967).

577. (a) B. C. Das, J. Defaye, and K. Uchida, *Carbohyd. Res.*, **22**, 293 (1972); (b) M. Vuilhorgne, S. Ennifar, B. P. Das, J. W. Paschall, R. Nagarajan, E. W. Hagaman, and E. Wenkert, *J. Org. Chem.*, **42**, 3289 (1977).

578. S. Umezawa, K. Umino, S. Shibahara, and S. Omoto, *Bull. Chem. Soc. Japan*, **40**, 2419 (1967); S. Umezawa, K. Umino, S. Shibahara, M. Hamada, and S. Omoto, *J. Antibiot.*, *Ser. A*, **20**, 355 (1967).

579. R. Kuhn and G. Baschang, *Ann.*, **628**, 206 (1959); T. Trnka, M. Černý, M. Buděšínsky, and J. Pacák, *Collect. Czech. Chem. Commun.*, **40**, 3038 (1975).

580. B. Jann, K. Jann, and E. Müller-Seitz, *Nature*, **215**, 170 (1967).

581. R. A. Raff and R. W. Wheat, *J. Biol. Chem.*, **242**, 4610 (1967).

582. F. A. Hochstein and P. P. Regna, *J. Amer. Chem. Soc.*, **77**, 3353 (1955).

583. A. C. Richardson, *Proc. Chem. Soc.*, 430 (1961); A. B. Foster, T. D. Inch, J. Lehmann, M. Stacey, and J. M. Webber, *J. Chem. Soc.*, 2116 (1962).

583a. H. Achenbach, W. Regel, and W. Karl, *Ber.*, **108**, 2481 (1975).

583b. A. Kinumaki, I. Takamori, Y. Sugawara, Y. Seki, M. Suzuki, and T. Okuda, *J. Antibiot.*, **27**, 117 (1974); A. Kinumaki, I. Takamori, Y. Sugawara, M. Suzuki, and T. Okuda, *ibid.*, **27**, 107 (1974).

583c. L. F. Kruglikova and Yu. D. Shenin, *Antibiotiki*, **21**, 407 (1976); *Chem. Abstr.*, **85**, 108, 914 (1976).

584. D. R. Walters, J. D. Dutcher, and O. Wintersteiner, *J. Amer. Chem. Soc.*, **79**, 5076 (1957).

585. J. D. Dutcher, M. B. Young, J. H. Sherman, W. E. Hibbits, and D. R. Walters, *Antibiot. Annu.*, 866 (1956–1957).

586. J. B. Patrick, R. P. Williams, C. F. Wolf, and J. S. Webb, *J. Amer. Chem. Soc.*, **80**, 6688 (1958).

586a. L. Falkowski, J. Golik, J. Zielinski, and E. Borowski, *J. Antibiot.*, **29**, 197 (1976).

587. M. H. von Saltza, J. Reid, J. D. Dutcher, and O. Wintersteiner, *J. Amer. Chem. Soc.*, **83**, 2785 (1961).
588. J. D. Dutcher, D. R. Walters, and O. Wintersteiner, *J. Org. Chem.*, **28**, 995 (1963); M. H. von Saltza, J. D. Dutcher, J. Reid, and O. Wintersteiner, *ibid.*, **28**, 999 (1963); O. Ceder, G. Eriksson, J. M. Waisvisz, and M. G. van der Hoeven, *Acta Chem. Scand.*, **18**, 98 (1964).
589. C. W. Waller, P. W. Fryth, B. L. Hutchings, and J. H. Williams, *J. Amer. Chem. Soc.*, **75**, 2025 (1953).
590. N. N. Gerber and H. A. Lechevalier, *J. Org. Chem.*, **27**, 1731 (1962).
591. B. R. Baker and R. E. Schaub, *J. Amer. Chem. Soc.*, **75**, 3864 (1953).
591a. D. H. Davies, D. Greeves, A. K. Mallams, J. B. Morton, and R. W. Tkach, *J. Chem. Soc. Perkin I*, 814 (1975).
591b. R. S. Egan, R. L. DeVault, S. L. Mueller, M. I. Levenberg, A. C. Sinclair, and R. S. Stanaszek, *J. Antibiot.*, **28**, 29 (1975).
591c. H. Riemann, D. J. Cooper, A. K. Mallams, R. Jaret, A. Yehaskel, M. Kigelman, H. F. Vernay, and D. Schumacher, *J. Org. Chem.*, **39**, 1451 (1974).
592. W. Meyer zu Reckendorf and E. Bischof, *Ber.*, **105**, 2546 (1972).
592a. D. J. Cooper, M. D. Yudis, R. D. Guthrie, and A. M. Prior, *J. Chem. Soc. (C)*, 960 (1971).
592b. P. J. L. Daniels, R. S. Jaret, T. L. Nagabhushan, and W. N. Turner, *J. Antibiot.*, **29**, 488 (1976).
593. F. Arcamone, G. Franceschi, P. Orezzi, G. Cassinelli, W. Barbieri, and R. Mondelli, *J. Amer. Chem. Soc.*, **86**, 5334 (1964).
593a. M. G. Brazhkinova, V. B. Zbarskii, D. Tresselt, and K. Eckardt, *J. Antibiot.*, **29**, 469 (1976).
594. F. Arcamone, G. Cassinelli, G. Franceschi, S. Penco, C. Pol, S. Rendaelli, and A. Selva, *Proc. Int. Symp. Adriamycin*, 9 (1971).
594a. D. Horton and W. Weckerle, *Carbohyd. Res.*, **44**, 227 (1975).
594b. D. Horton and W. Weckerle, *Carbohyd. Res.*, **46**, 227 (1976).
594c. I. A. Spiridonova, M. S. Yurina, N. N. Lomakina, F. Sztaricskai, and R. Bognár, *Antibiotiki*, **21**, 304 (1976); *Chem. Abstr.*, **85**, 108, 925 (1976).
595. A. K. Mallams, *J. Amer. Chem. Soc.*, **91**, 7505 (1969); *J. Chem. Soc. Perkin I*, 1369 (1973).
595a. H. Brockmann and H. Greve, *Tetrahedron Lett.*, 831 (1975).
596. N. N. Lomakina, I. A. Spiridonova, R. Bognár, M. Puksás, and F. Sztaricskai, *Antibiotiki*, **13**, 975 (1968).
597. W. D. Weringa, D. H. Williams, J. Feeney, J. P. Brown, and R. W. King, *J. Chem. Soc. Perkin I*, 443 (1972).
598. R. M. Smith, A. W. Johnson, and R. D. Guthrie, *Chem. Commun.*, 361 (1972); A. W. Johnson, R. M. Smith, and R. D. Guthrie, *J. Chem. Soc. Perkin I*, 2153 (1972).
599. R. Bognár, F. Sztaricskai, M. E. Munk, and J. Tamas, *Magyar Kem. Foly.*, **80**, 385 (1974); I. A. Spiridonova, N. N. Lomakina, F. Sztaricskai, and R. Bognár, *Antibiotiki*, **19**, 400 (1974).
599a. W. W. Lee, H. Y. Wu, J. J. Marsh, Jr., C. W. Mosher, E. M. Acton, L. Goodman, and D. W. Henry, *J. Med. Chem.*, **18**, 767 (1975); R. Bognár, F. Sztaricskai, M. E. Munk, and J. Tamas, *J. Org. Chem.*, **39**, 2971 (1974); F. Sztaricskai, I. Pelyvás, L. Szilágyi, R. Bognár, J. Tamás, and A. Neszméyli, *Carbohyd. Res.*, **65**, 193 (1978).
599b. F. Sztaricskai, L. Pelyvas, R. Bognár, and G. Bujtas, *Tetrahedron Lett.*, 1111 (1975); F. Arcamone, A. Bargiotti, G. Cassinelli, S. Penco, and S. Hanessian, *Carbohyd. Res.*, **46**, C3 (1976); H. H. Baer and F. F. Z. Georges, *ibid.*, **55**, 251 (1977).

599c. N. N. Lomakina, I. A. Spiridonova, I. Yu, N. Sheinker, and T. F. Vlasova, *Khim. Prir. Soedin.*, **9**, 101 (1973); *Chem. Abstr.*, **78**, 148, 170 (1973).

599d. W. W. Lee, H. Y. Wu, J. E. Christensen, L. Goodman, and D. W. Henry, *J. Med. Chem.*, **18**, 768 (1975).

599e. H. Brockmann, H. B. König, and R. Oster, *Ber.*, **87**, 856 (1954); C. H. Bolton, A. B. Foster, M. Stacey, and J. M. Webber, *J. Chem. Soc.*, 4831 (1961); *Chem. Ind.* (London), 1945 (1962); W. Hofheintz and H. Grisebach, *Tetrahedron Lett.*, 377 (1962).

599f. M. Brufani and W. Keller-Schierlein, *Helv. Chim. Acta*, **49**, 1962 (1966).

599g. F. Korte, A. Bilow, and R. Heinz, *Tetrahedron*, **18**, 657 (1962); H. Newman, *J. Org. Chem.*, **29**, 1461 (1964).

599h. H. Brockmann and W. Henkel, *Ber.*, **84**, 284 (1951); H. Muxfeldt, S. Shrader, P. Hansen, and H. Brockmann, *J. Amer. Chem. Soc.*, **90**, 4748 (1968).

599i. S. E. DeVoe, H. B. Renfroe, and W. K. Hausmann, *Antibiot. Ag. Chemother.*, 125 (1963).

599j. R. Anliker, D. Dvornik, K. Gubler, H. Heusser, and V. Prelog, *Helv. Chim. Acta*, **39**, 1785 (1956).

599k. V. Prelog, A. M. Gold, G. Talbot, and A. Zamojski, *Helv. Chim. Acta*, **45**, 4 (1962).

599l. P. F. Wiley, K. Gerzon, E. H. Flynn, M. V. Sigal, Jr., O. Weaver, J. C. Quarck, R. R. Chauvette, and R. Monahan, *J. Amer. Chem. Soc.*, **79**, 6062 (1957).

599m. F. Blindenbacher and T. Reichstein, *Helv. Chim. Acta*, **31**, 2061 (1948).

600. K. Sakata, A. Sakurai, and S. Tamura, *Tetrahedron Lett.*, 1533 (1974).

600a. T. Ogawa, M. Akatsu, and M. Matsui, *Carbohyd. Res.*, **44**, C22 (1975).

601. K. Nara, Y. Sumino, K. Katamoto, S. Akiyama, and M. Asai, *Chem. Lett.*, 33 (1977); see also M. M. Ponpipom, R. Bugianesi, E. Walton, and T. Y. Shen, *Carbohyd. Res.*, **65**, 121 (1978) for synthetic studies on sorbistin A_1 (GlA_1), 1,4-diamino-1,4-dideoxy-3-O-(4-deoxy-4-propionamido-α-D-glucopyranosyl)-D-glucitol, first synthesized by T. Ogawa, K. Katano, and M. Matsui, *ibid.*, **60**, C1 (1978).

601a. S. O'Connor, L. K. T. Lam, N. D. Jones, and M. O. Chaney, *J. Org. Chem.*, **41**, 2087 (1976).

602. W. A. Volk, C. Galanos, and O. Lüderitz, *FEBS Lett.*, **8**, 161 (1970); *Eur. J. Biochem.*, **17**, 223 (1970).

603. M. H. von Saltza, J. D. Dutcher, and J. Reid, *Abstr. Papers Amer. Chem. Soc. Meeting*, **148**, 15Q (1964); H. Agahigian, G. D. Vickers, M. H. von Saltza, J. Reid, A. I. Cohen, and H. Gauthier, *J. Org. Chem.*, **30**, 1085 (1965).

604. J. D. Dutcher, M. H. von Saltza, and F. E. Pansy, *Antimicrob. Ag. Chemother.*, 83 (1963).

605. J. J. Fox, Y. Kuwada, and K. A. Watanabe, *Tetrahedron Lett.*, 6029 (1968).

606. T. Kanzaki, E. Higashide, H. Yamamoto, M. Shibata, K. Nakazawa, H. Iwasaki, T. Takewaka, and A. Miyake, *J. Antibiot., Ser. A.*, **15**, 93 (1962); J. J. Fox, Y. Kuwada, K. A. Watanabe, T. Ueda, and E. B. Whipple, *Antimicrob. Ag. Chemother.*, 518 (1964).

606a. T. Haneishi, A. Terehashi, and M. Arai, *J. Antibiot.*, **27**, 338 (1974); F. W. Lichtenthaler, T. Morino, and H. M. Menzel, *Tetrahedron Lett.*, 665 (1975).

607. K. A. Watanabe, M. P. Kotick, and J. J. Fox, *J. Org. Chem.*, **35**, 231 (1970), see also: *idem.*, *Chem. Pharm. Bull.* (Tokyo), **17**, 416 (1969); M. P. Kotick, R. S. Klein, K. A. Watanabe, and J. J. Fox, *Carbohyd. Res.*, **11**, 369 (1969).

608. B. Jann and K. Jann, *Eur. J. Biochem.*, **2**, 26 (1967).

609. J. L. Strominger and S. S. Scott, *Biochim. Biophys. Acta*, **35**, 552 (1959); T. Okazaki, R. Okazaki, J. L. Strominger, and S. Suzuki, *Biochem. Biophys. Res. Commun.*, **7**, 300 (1962).

610. C. L. Stevens, P. Blumbergs, D. H. Otterbach, J. L. Strominger, M. Matsuhashi, and D. N. Dietzler, *J. Amer. Chem. Soc.*, **86**, 2937 (1964); C. L. Stevens, P. Blumbergs, and D. H. Otterbach, *J. Org. Chem.*, **31**, 2817 (1966).

611. C. L. Stevens, P. Blumbergs, F. A. Daniher, J. L. Strominger, M. Matsuhashi, D. N. Dietzler, S. Suzuki, T. Okazaki, K. Sugimoto, and R. Okazaki, *J. Amer. Chem. Soc.*, **86**, 2939 (1964); C. L. Stevens, P. Blumbergs, F. A. Daniher, D. H. Otterbach, and K. G. Taylor, *J. Org. Chem.*, **31**, 2822 (1966).

612. R. W. Wheat, E. L. Rollins, and J. M. Leatherwood, *Biochem. Biophys. Res. Commun.*, **9**, 120 (1962).

613. R. Okazaki, T. Okazaki, and Y. Kuriki, *Biochim. Biophys. Acta*, **38**, 384 (1960).

614. J. W. Hinman, E. L. Caron, and C. DeBoer, *J. Amer. Chem. Soc.*, **75**, 5864 (1953).

615. T. H. Haskell, *J. Amer. Chem. Soc.*, **80**, 747 (1958).

615a. M. Muroi, M. Izawa, and T. Kishi, *Chem. Pharm. Bull.* (Tokyo) **24**, 450, 463 (1976).

616. C. L. Stevens, P. Blumbergs, F. A. Daniher, D. H. Otterbach, and K. G. Taylor, *J. Org. Chem.*, **31**, 2822 (1966).

617. C.-H. Lee and C. P. Schaffner, *Tetrahedron Lett.*, 5837 (1966); *Tetrahedron*, **25**, 2229 (1969).

617a. A. S. Mesentsev, V. V. Kulyaeva, and L. M. Rubasheva, *J. Antibiot.*, **27**. 866 (1974).

618. M.-M. Janot, P. Devissaguet, Q. Khuong-Huu, J. Parello, N. G. Bisset, and R. Goutarel, *C.R. Acad. Sci.*, *Ser. C.*, **266**, 388 (1968).

619. M.-M. Janot, M. Lebœuf, A. Cave, R. Wijesekera, and R. Goutarel, *C.R. Acad. Sci.*, *Ser. C*, **267**, 1050 (1968).

620. M.-M. Janot, Q. Khuong-Huu, C. Monneret, I. Kaboré, J. Hildesheim, S. D. Gero, and R. Goutarel, *Tetrahedron*, **26**, 1695 (1970).

621. J. Hildesheim, S. D. Gero, Q. Khuong-Huu, and C. Monneret, *Tetrahedron Lett.*, 2849 (1969).

622. R. Goutarel, C. Monneret, P. Choay, I. Kaboré, and Q. Khuong-Huu, *Carbohyd. Res.*, **24**, 297 (1972).

623. K. Fukuhara, T. Misato, I. Ishii, and M. Asakawa, *Bull. Agr. Chem. Soc. Japan*, **19**, 181 (1955); S. Takeuchi, K. Hirayama, K. Ueda, H. Sakai, and K. Yonehara, *J. Antibiot.*, *Ser. A*, **11**, 1 (1958).

624. N. Ōtake, S. Takeuchi, T. Endō, and H. Yonehara, *Tetrahedron Lett.*, 1404, 1411 (1965); *Agr. Biol. Chem.* (Tokyo), **30**, 126, 132 (1966); J. J. Fox and K. A. Watanabe, *Tetrahedron Lett.*, 897 (1966); H. Yonehara and N. Ōtake, *ibid.*, 3785 (1966).

625. K. A. Watanabe, I. Wempen, and J. J. Fox, *Chem. Pharm. Bull.* (Tokyo), **18**, 2368 (1970); K. A. Watanabe, T. M. K. Chiu, U. Reichman, C. K. Chu, and J. J. Fox, *Tetrahedron*, **32**, 1493 (1976).

626. R. Paul and S. Tchelitcheff, *Bull. Soc. Chim. Fr.*, 443, 734, 1059 (1957).

627. S. Pinnert-Sindico, L. Ninet, J. Preud'homme, and C. Cosar, *Antibiot. Annu.*, 724 (1954); R. Corbaz, L. Ettlinger, E. Gäumann, W. Keller-Schierlein, F. Kradolfer, E. Kyburz, L. Neipp, V. Prelog, W. Wettstein, and H. Zahner, *Helv. Chim. Acta*, **39**, 304 (1956); M. E. Keuhne and B. W. Benson, *J. Amer. Chem. Soc.*, **87**, 4660 (1965).

627a. C. L. Stevens, G. E. Gutowski, C. P. Bryant, R. P. Glinski, O. E. Edwards, and G. M. Sharma, *Tetrahedron Lett.*, 1181 (1969).

627b. I. Dyong, R. Knollman, and N. Jersch, *Angew. Chem.*, **88**, 301 (1976).

627c. T. Kishi, S. Harada, M. Asai, M. Muroi, and K. Mizuno, *Tetrahedron Lett.*, 97 (1969).

628. T. Nishikawa and N. Ishida, *J. Antibiot.*, *Ser. A*, **18**, 132 (1965).

629. S. Inouye, T. Tsuruoka, and T. Niida, *J. Antibiot.*, *Ser. A*, **19**, 288 (1966).

630. N. Ishida, K. Kumagai, T. Niida, T. Tsuruoka, and H. Yumoto, *J. Antibiot., Ser. A,* **20,** 66 (1967).
631. R. L. Whistler and R. E. Gramera, *J. Org. Chem.,* **29,** 2609 (1964).
632. J. B. Leathes, *Arch. Exptl. Pathol. Pharmakol.,* **43,** 245 (1899).
633. E. Klenk and H. Faillard, *Z. Physiol. Chem.,* **298,** 230 (1954).
634. F. Zilliken, G. A. Braun, and P. György, *Arch. Biochem. Biophys.,* **54,** 564 (1955).
635. (a) G. Blix, E. Lindberg, L. Odin, and I. Werner, *Nature,* **175,** 340 (1955); (b) *idem., Acta Soc. Med. Upsalien.,* **61,** 1 (1956).
636. T. Yamakawa and S. Suzuki, *J. Biochem.* (Tokyo), **42,** 727 (1955).
637. G. Blix, *Z. Physiol. Chem.,* **240,** 43 (1936).
638. G. Blix and E. Lindberg, *Acta Chem. Scand.,* **14,** 1809 (1960).
639. J. Immers, *Acta Chem. Scand.,* **22,** 2046 (1968).
640. L. Warren, *Biochim. Biophys. Acta,* **83,** 129 (1964).
640a. A. S. Balasubramanian, P. T. Raman, and G. M. Taori, *Ind. J. Med. Res.,* **62,** 781 (1974) and references therein.
640b. J. E. Martin, S. E. Tanenbaum, and M. Flashner, *Carbohyd. Res.,* **56,** 423 (1977); *see:* G. Blix, *Methods Carbohyd. Chem.,* **1,** 246 (1962).
641. A. Gottschalk, *Nature,* **176,** 881 (1955).
642. R. Heimer and K. Meyer, *Proc. Nat. Acad. Sci. U.S.,* **42,** 728 (1956).
643. D. G. Comb and S. Roseman, *J. Amer. Chem. Soc.,* **80,** 497 (1958).
644. M. N. Mirzayanova, L. P. Davydova, and G. I. Samokhvalov, *Zh. Obshch. Khim.,* **40,** 693 (1970).
645. P. Lutz, W. Lochinger, and G. Taigel, *Ber.,* **101,** 1089 (1968).
646. J. L. Flippen, *Acta Crystallogr.,* **B29,** 1881 (1973).
647. H. Hoeksema, B. Bannister, R. D. Birkenmeyer, F. Kagan, B. J. Magerlein, F. A. MacKellar, W. Schroeder, G. Slomp, and R. R. Herr, *J. Amer. Chem. Soc.,* **86,** 4223 (1964).
648. H. Hoeksema, *J. Amer. Chem. Soc.,* **86,** 4224 (1964).
649. D. J. Mason, A. Dietz, and C. DeBoer, *Antimicrob. Ag. Chemother.,* 554 (1962).
650. B. J. Magerlein, R. D. Birkenmeyer, and F. Kagan, *Antimicrob. Ag. Chemother.,* 727 (1966).
651. H. Hoeksema and J. W. Hinman, *J. Amer. Chem. Soc.,* **86,** 4979 (1964).
652. S. M. David and J.-C. Fischer, *Carbohyd. Res.,* **38,** 147 (1974).
653. R. L. Mann and W. W. Bromer, *J. Amer. Chem. Soc.,* **80,** 2714 (1958).
654. J. Shoji and Y. Nakagawa, *J. Antibiot.,* **23,** 569 (1970).
655. S. Inouye, T. Shomura, H. Watanabe, T. Totsugawa, and T. Niida, *J. Antibiot.,* **26,** 374 (1973).
655a. S. Kondo, T. Iinuma, H. Naganawa, M. Shimura, and Y. Sekizawa, *J. Antibiot.,* **28,** 79 (1975).
656. J. D. Dutcher, N. Hosansky, M. N. Donin, and O. Wintersteiner, *J. Amer. Chem. Soc.,* **73,** 1384 (1951).
657. S. Horii, T. Yamaguchi, H. Hitomi, and A. Miyake, *Chem. Pharm. Bull.* (Tokyo), **9,** 340 (1961); K. L. Rinehart, Jr., M. Hichens, K. Streigler, K. R. Rover, T. P. Culbertson, S. Tatsuoka, S. Horii, T. Yamaguchi, H. Hitomi, and A. Miyake, *J. Amer. Chem. Soc.,* **83,** 2964 (1961).
658. W. T. Shier, K. L. Rinehart, Jr., and D. Gottlieb, *Proc. Nat. Acad. Sci. U.S.,* **63,** 198 (1969).
659. T. Ito, M. Nishio, and H. Ogawa, *J. Antibiot., Ser. A.,* **17,** 189 (1964).
660. E. Akita, T. Tsuruoka, N. Ezaki, and T. Niida, *J. Antibiot.,* **23,** 173 (1970).
661. P. W. K. Woo, H. W. Dion, and Q. R. Bartz, *Tetrahedron Lett.,* 2617 (1971).
662. H. Weidmann and H. K. Zimmerman, Jr., *Ann.,* **639,** 198 (1961); **641,** 132 (1961).

663. K. L. Rinehart, Jr., P. W. K. Woo, A. D. Argoudelis, and A. M. Giesbrecht, *J. Amer. Chem. Soc.*, **79**, 4567 (1957); K. L. Rinehart, Jr., P. W. K. Woo, and A. D. Argoudelis, *ibid.*, **79**, 4568 (1957); **80**, 6461 (1958).

663a. T. H. Haskell, J. C. French, and Q. R. Bartz, *J. Amer. Chem. Soc.*, **81**, 3482 (1959).

664. T. Oda, T. Mori, and Y. Kyotani, *J. Antibiot.*, **24**, 511 (1971).

665. T. Mori, Y. Kyotani, I. Watanabe, and T. Oda, *J. Antibiot.*, **25**, 317 (1972).

666. M.-M. Janot, H. Pénau, D. Van Stolk, G. Hagemann, and L. Pénasse, *Bull. Soc. Chim. Fr.*, 1458 (1954).

667. T. H. Haskell and S. Hanessian, *J. Org. Chem.*, **28**, 2598 (1963).

668. W. Meyer zu Reckendorf, *Angew. Chem.*, **75**, 573 (1963); *Tetrahedron*, **19**, 2033 (1963).

668a. J. Roppel, H. Mayer, and J. Weckesser, *Carbohyd. Res.*, **40**, 31 (1975); G. Keilich, J. Roppel, and H. Mayer, *ibid.*, **51**, 129 (1976).

668b. H. Kuzuhara and S. Emoto, *Tetrahedron Lett.*, 1853 (1975); S. Ōmura, M. Katagiri, K. Atsumi, T. Hata, A. A. Jakubowski, E. B. Springs, and M. Tischler, *J. Chem. Soc. Perkin I*, 1627 (1974); H. Hashimoto, T. Nishide, F. Chiba, and J. Yoshimura, *Carbohyd. Res.*, **60**, 75 (1978).

669. N. Sharon and R. W. Jeanloz, *Biochim. Biophys. Acta*, **31**, 277 (1959); *J. Biol. Chem.*, **235**, 1 (1960).

670. U. Zehavi and N. Sharon, *J. Biol. Chem.*, **248**, 433 (1973).

671. A. Liav, J. Hildesheim, U. Zehavi, and N. Sharon, *Carbohyd Res.*, **33**, 217 (1974); compare A. Liav, I. Jacobsen, M. Sheinblatt, and N. Sharon, *ibid.*, **66**, 95 (1978).

672. J. Distler, B. Kaufman, and S. Roseman, *Arch. Biochem. Biophys.*, **116**, 466 (1966).

673. D. E. Brundish and J. Baddiley, *Biochem. J.*, **110**, 573 (1968).

674. K. F. Koch and J. A. Rhoades, *Antimicrob. Ag. Chemother.*, 309 (1970).

675. C. L. Brewer and R. D. Guthrie, *J. Chem. Soc. Perkin I*, 657 (1974).

675a. M. Konishi, K. Numata, K. Shimota, H. Tsukiura, and H. Kawaguchi, *J. Antibiot.*, **27**, 471 (1974).

676. Y. Suhara, K. Maeda, H. Umezawa, and M. Ohno, *J. Antibiot., Ser. A*, **18**, 184 (1965); *Tetrahedron Lett.*, 1239 (1966); *Advan. Chem. Ser.*, **74**, 15 (1968).

676a. K. Iinuma, S. Kondo, K. Maeda, and H. Umezawa, *J. Antibiot.*, **28**, 613 (1975).

677. D. J. Cooper, M. J. Yudis, H. M. Marigliano, and T. Traubel, *J. Chem. Soc. (C)*, 2876 (1971).

678. D. C. DeJongh and S. Hanessian, *J. Amer. Chem. Soc.*, **87**, 3744 (1965).

679. R. D. Guthrie and G. J. Williams, *J. Chem. Soc. Perkin I*, 2619 (1972).

680. J. Cléophax, S. D. Gero, E. Jagou-Aumont, J. Leboul, D. Mercier, and A. Forchioni, *Chem. Commun.*, 11 (1975).

681. D. J. Cooper, R. S. Jaret, and H. Reinmann, *Chem. Commun.*, 285 (1971); H. Reinmann, R. S. Jaret, and D. J. Cooper, *ibid.*, 924 (1971).

682. H. E. Carter, R. K. Clark, Jr., P. Kohn, J. W. Rothrock, W. R. Taylor, C. A. West, G. B. Whitfield, and W. G. Jackson, *J. Amer. Chem. Soc.*, **76**, 566 (1954).

683. K. Nakanishi, T. Ito, M. Ohashi, I. Morimoto, and Y. Hirata, *Bull. Chem. Soc. Japan*, **27**, 539 (1954).

684. H. Brockmann and H. Musso, *Ber.*, **88**, 648 (1955).

685. H. Taniyama and S. Takemura, *Yakugaku Zasshi*, **77**, 1215 (1957).

686. G. Rangaswami, C. P. Schaffner, and S. A. Waksman, *Antibiot. Chemother.*, **6**, 675 (1956).

687. H. E. Carter, C. C. Sweeley, E. E. Daniels, J. E. McNary, C. P. Schaffner, C. A. West, E. E. van Tamelen, J. R. Dyer, and H. A. Whaley, *J. Amer. Chem. Soc.*, **83**, 4296 (1961).

688. B. W. Bycroft and T. J. King, *Chem. Commun.*, 652 (1952).

689. S. Kusumoto, T. Tsuji, and T. Shiba, *Bull. Chem. Soc. Japan*, **47**, 2690 (1974).

17. DEOXY AND BRANCHED-CHAIN SUGARS

NEIL R. WILLIAMS AND JOSEPH D. WANDER

I. INTRODUCTION

Deoxy and branched-chain sugars occur widely in a range of plants, fungi, and bacteria, and their increasing importance is reflected in the current interest in their chemistry. Available elsewhere are a number of articles and reviews covering deoxy and branched-chain sugars generally,[1,2] 2-deoxy sugars,[3,4] branched-chain sugars of natural occurrence,[5-8] 3,6-dideoxy sugars in bacterial lipopolysaccharides,[9] and the deoxy sugars of cardiac glycosides.[10] This chapter focuses attention on their natural occurrence, structural types, methods of synthesis, and their general properties compared with those of the more common monosaccharides, with the emphasis on aldopentoses and aldohexoses, since very little is as yet known of other classes.

The deoxy sugars are formally derived from normal sugars by replacement of one or more hydroxyl groups present in normal sugars (excluding the glycosidic hydroxyl group of the hemiacetal form). Branched-chain sugars, as the name implies, have a branched as opposed to a straight carbon-chain backbone. They are formally derived from a normal sugar by replacement of either a hydrogen or a hydroxyl group on a secondary carbon atom by a carbon side chain. Most of the known, naturally occurring, branched-chain sugars have the carbon side chain replacing a hydrogen atom, making the branch point a tertiary alcohol. It happens that most naturally occurring branched-chain sugars are also deoxy sugars, but for the purposes of this chapter they will be considered separately.

The deoxy sugars are named according to the generally accepted system of nomenclature (Vol. IIB, Chapter 46). For the branched-chain sugars a convention will be followed in which the carbon side chain is considered to be a substituent (replacing a hydrogen atom) on the unsubstituted monosaccharide having the same configuration of hydroxyl groups; that is, the configurational designation is controlled by the arrangement of oxygen atoms attached to the main carbon chain. When the side chain replaces an hydroxyl group, making the branch point a deoxy position, the orientation of the branching group (in the Fischer projection) is used with the other centers to define the total configurational designation. The key example of this class in Nature is the 2-C-butyl lactone, blastmycinone. Based on their source of origin, most naturally occurring branched-chain sugars have trivial names that are commonly used. These names are convenient while the sugars are not too numerous and may possess cumbersome systematic names. However, this practice has unfortunately led to different names being used for the same sugar, or different derivatives of the same sugar, a confusion the more regrettable for the more-numerous deoxy sugars, where trivial names are less excusable.

II. DEOXY SUGARS

A. Occurrence

The majority of naturally occurring deoxy sugars are 6-deoxyhexoses, with or without deoxy positions elsewhere in the carbon chain, the outstanding exception being 2-deoxy-D-erythro-pentose ("2-deoxy-D-ribose"), occurring in deoxyribonucleic acid. The simple 6-deoxy sugars are considered in Chapter 2. Table I lists the other naturally occurring deoxy sugars. As the table indicates, although most of the 2,6-dideoxyhexoses occur in cardiac glycosides, some are to be found in antibiotic compounds produced by microorganisms. The 3,6-dideoxyhexoses are, with one exception, confined to the cell-wall lipopolysaccharides of Salmonella and other gram-negative bacteria. 3-Deoxy-D-erythro-pentose (cordycepose) is of special interest, since it constitutes a naturally occurring 3-deoxy sugar; it is the sugar component of the antibiotic nucleoside cordycepin (3'-deoxyadenosine), and it was originally thought to possess a branched chain.

TABLE I

PRINCIPAL DEOXY SUGARS OF NATURAL OCCURRENCE[a]

Sugar	Trivial name	Source	References
2-Deoxy-D-*erythro*-pentose	2-Deoxy-D-ribose	Deoxyribonucleic acid	3
2-Deoxy-D-*arabino*-hexose	2-Deoxy-D-glucose	Cardiac glycosides	10
2-Deoxy-D-*xylo*-hexose		Cardiac glycosides	10
3-Deoxy-D-*erythro*-pentose	Cordycepose	Cordycepin	11, 12
4-Deoxy-D-*arabino*-hexose		*Citrobacter* species	13
6-Deoxy-D-allose			
2,3-dimethyl ether		Tylosin	14
6-Deoxy-D-altrose			
4-methyl ether		Sodarin	15
6-Deoxy-L-altrose			
3-methyl ether	L-Vallarose		16
6-Deoxy-D-galactose			
3-methyl ether	Digitalose	Cardiac glycosides	16
4-methyl ether		Flambamycin	17
		Everninomycins	
		B;C;D;-2	18
		Flambeurekanose	19a
6-Deoxygalactose			
2-methyl ether		Seaweed	20
3-methyl ether		Seaweed	20
6-Deoxy-L-mannose			
3-methyl ether	L-Acofriose	*Klebsiella* K73:O10	21
		Gram-negative	
		bacteria	22
		Mycoside G	23
6-Deoxy-D-talose			
3-methyl ether		*Rhodopseudomonas*	
		palustris	24
6-Deoxy-D-*manno*-heptose		Bacterial lipopoly-	
		saccharide	25
2,6-Dideoxy-D-*arabino*-hexose	Canarose	Cardiac glycosides	26
	Chromose C	Chromomycin	27
	Olivose	Olivomycin	28
		Flambamycin	17b
		Everninomycins	
		B;C;D;-2	18
		Chromocyclomycin	29
		Oxamicetin	29a

(Continued)

TABLE I— *Continued*

Sugar	Trivial name	Source	References
3-methyl ether	D-Oleandrose	Cardiac glycosides	10, 30
		Cynanchum sibiricum	30a
2,6-Dideoxy-L-*arabino*-hexose			
3-methyl ether	L-Oleandrose	Cardiac glycosides	10
2,6-Dideoxy-D-*lyxo*-hexose	Oliose	Oleandomycin	31
		Olivomycin	28
		Chromomycin[b]	27
		Cardiac glycosides[c]	10
		Chromocyclomycin	29
3-methyl ether	D-Diginose	Cardiac glycosides	10
4-methyl ether	Olivomose	Olivomycin	28
	Chromose A	Chromomycin	32
		α-Rhodomycin	33
		β-Rhodomycins	
		S-1b,2,3,4	34
2,6-Dideoxy-L-*lyxo*-hexose			
3-methyl ether	L-Diginose	Cardiac glycosides	10
2,6-Dideoxy-D-*ribo*-hexose	D-Digitoxose	Cardiac glycosides	10
		Lipomycin	35
		Erysimum marschallianum	35a
3-methyl ether	D-Cymarose	Cardiac glycosides	10, 30
		Cynanchum sibiricum	30a
2,6-Dideoxy-(D-*ribo* or L-*lyxo*)-hexose			
3-methyl ether	Variose	Variomycin	36
2,6-Dideoxy-D-*ribo*-hexonic acid		Flambamycin	17b
		Everninomycins	
		B;C;D;-2	18
2,6-Dideoxy-D-*xylo*-hexose	D-Boivinose	Cardiac glycosides	10
		Coloroside	36a
3-methyl ether	D-Sarmentose	Cardiac glycosides	10
3,6-Dideoxy-D-*arabino*-hexose	Tyvelose	Bacterial lipopoly-saccharides	32
3,6-Dideoxy-L-*arabino*-hexose	Ascarylose	Bacterial lipopoly-saccharides	32
		Glycolipid	37a, b
3,6-Dideoxy-D-*ribo*-hexose	Paratose	Bacterial lipopoly-saccharides	32
3,6-Dideoxy-D-*xylo*-hexose	Abequose	Bacterial lipopoly-saccharides	32
		Acanthophyllum gypsophiloides	37c
3,6-Dideoxy-L-*xylo*-hexose	Colitose	Bacterial lipopoly-saccharides	32

(*Continued*)

TABLE I—*Continued*

Sugar	Trivial name	Source	References
4,6-Dideoxy-D-*xylo*-hexose			
3-methyl ether	Chalcose	Chalcomycin	38
	Lankavose	Lankamycin	39
2,3,6-Trideoxy-D-*erythro*-			
hexose	Amicetose	Amicetin	40
4-methyl ether		Dianemycin	41
		Septamycin	42
		A 204 A	43
		Dihydro A-204	43a
		RO 21-6150	44
2,3,6-Trideoxy-L-*threo*-hexose	Rhodinose	Rhodomycin	34, 40
		Streptolydigin	45

[a] Excluding the 6-deoxyhexoses (see Vol. IA, Chapter 2).

[b] In chromomycin, the 3-acetate is called Chromose B.

[c] The D or L configuration of the sugar from cardiac glycosides has not been established.

B. SYNTHESIS

A wide range of methods has been developed for the synthesis of deoxy sugars. Certain of these methods, although in principle generally applicable, have hitherto found only occasional use for the synthesis of specific sugars. It is convenient to consider the utility of the methods in relation to the position in the sugar chain where the deoxy group is required.

1. *2-Deoxyaldoses*

The principal synthetic methods used have been (*a*) the direct or indirect hydration of glycals (see also Chapter 19), (*b*) the direct or indirect reduction of sugar 2,3-epoxides, (*c*) the condensation of pentoses with nitromethane, Wittig reagents, 2-(1,3-dithianyl)lithium, and active methylene compounds leading to 2-deoxyhexoses, and (*d*) reduction of deoxyhalogeno sugars or sulfonic esters.

a. From Glycals.—Although the yields are sometimes poor, the original Fischer method for the direct hydration of glycals by using dilute sulfuric acid[46] has been used frequently[3]; for example, D-glucal (**1**) yields 2-deoxy-D-*arabino*-hexose (**2**). In attempts to improve the yields obtained, particularly in the conversion of D-arabinal into 2-deoxy-D-*erythro*-pentose, many indirect methods for the addition of water to the double bond have been proposed.

References start on p. 791.

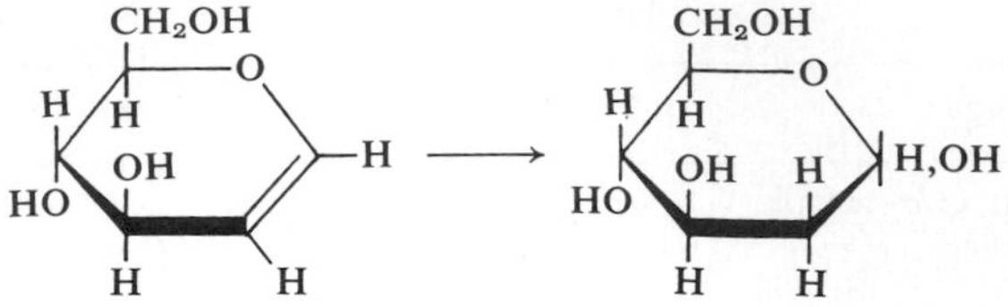

Treatment of glycals with hydrogen chloride or bromide in alcohol,[47] or in benzene followed by methanolysis in the presence of silver carbonate,[48] gives the 2-deoxyglycoside (3), which is readily hydrolyzed to the free sugar. Addition of bromine,[46] chlorine,[49] or Prévost's reagent[50] yields 2-deoxy-2-halo derivatives (4), the halogen atom of which can be removed by catalytic reduction to furnish the deoxy sugar. This method can provide[51] a route to the deoxy sugar isotopically labeled specifically at C-2. The methoxy-mercuration of glycals, furnishing the methyl 2-acetoxymercuri glycoside (5), has also been recommended,[52-54] since these intermediates are readily reduced by potassium borohydride or by photolysis[55] to the deoxy sugar in good yield. Hydrogenolysis of 1,2-*cis* glycopyranosyl bromides may also be used.[55a]

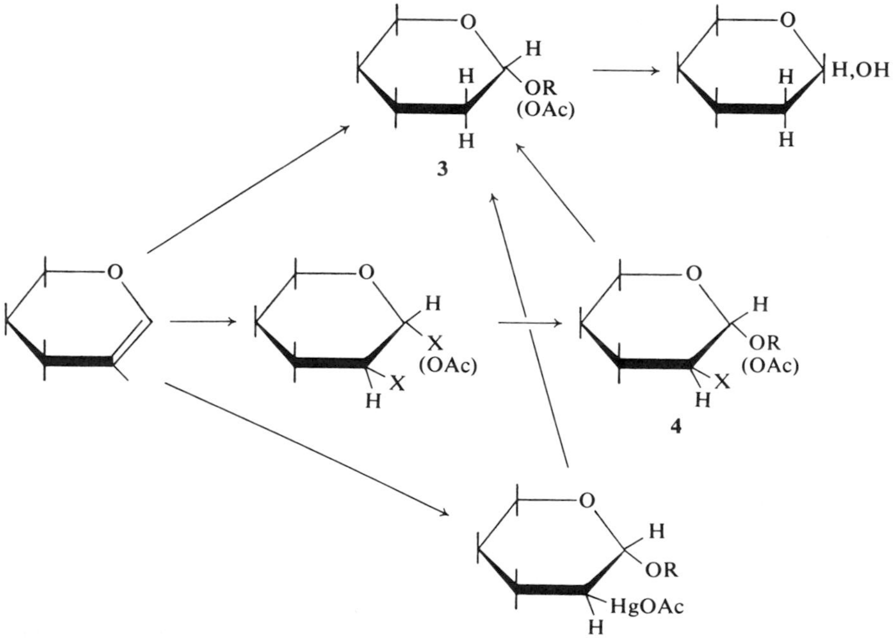

b. From Sugar 2,3-Epoxides.—The epoxide ring in these sugars may be reduced directly, by using lithium aluminum hydride or catalytic hydrogenation, or indirectly, by first treating the epoxide with sodium thiolates or thiocyanate, or with hydrogen halides, to furnish thio- or halo-sugar inter-mediates, which are then reduced catalytically to the deoxy sugar. The preferred method is determined by the ratio of the two possible isomeric products formed in any particular case through the two alternative modes of epoxide cleavage, and by the ease of separation of the isomers. In the hexose series, preferential formation of the diaxially oriented isomer in accordance with the Fürst–Plattner rule means that the 2,3-anhydro alloside (or guloside) **6** yields the 2-deoxy sugar **7** as the major product, whereas the 2,3-anhydro mannoside (or taloside) **8** gives[56] mainly the 3-deoxy isomer **9**. There are

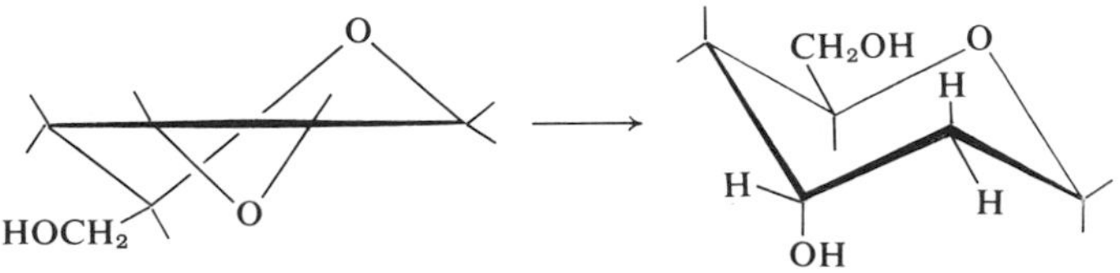

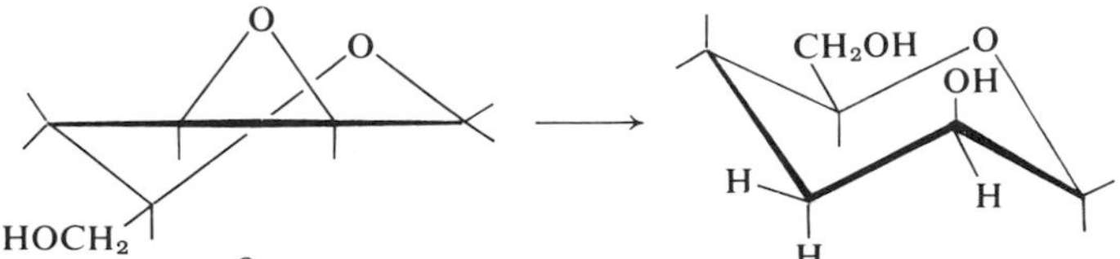

exceptions, however, notably when catalytic hydrogenation is employed; thus, for example, methyl 2,3-anhydro-4,6-*O*-benzylidene-α-D-allopyranoside (**10**) on nickel-catalyzed hydrogenation yields primarily the 3-deoxy-D-*ribo*-hexose isomer (**11**), whereas reduction with lithium aluminum hydride furnishes[57] the 2-deoxy-D-*ribo*-hexose derivative (**12**).

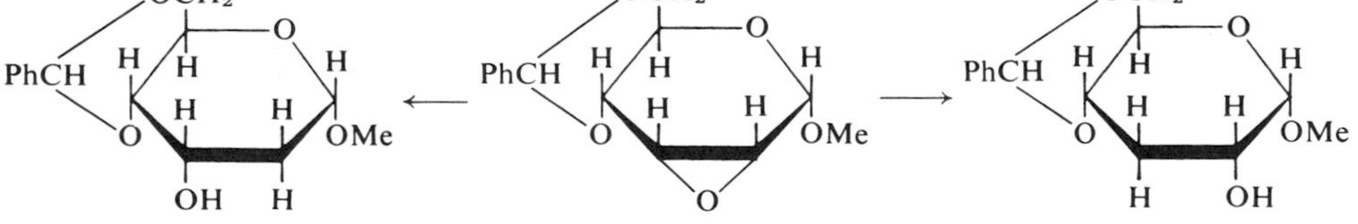

Reduction of 2,3-anhydro sugars of the pentose series usually leads to the preferential formation of the 3-deoxy isomer, although again exceptions have been noted among furanoside derivatives[58,59]; for example, reduction of the 2,3-anhydro riboside **13** with lithium aluminum hydride gives predominantly the 2,5-dideoxy-*erythro*-pentoside **14**.

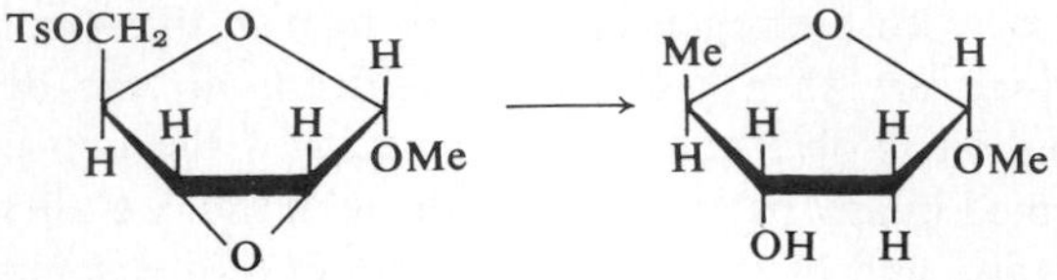

13　　　　　　　**14**

c. Condensation Reactions.—The nitromethane synthesis,[60] developed by Fischer and Sowden,[61] has been used to convert pentoses into the corresponding 2-deoxyhexoses and is particularly useful for preparing such sugars labeled at C-1. Thus, 2-deoxy-D-*arabino*-hexose may be synthesized from D-arabinose,[61] by successively forming the deoxynitroalditol **15** by condensation with nitromethane in alkali, the nitrohexene **16** by acetylation and hydrogen carbonate-catalyzed elimination of acetic acid, the dideoxynitroalditol **17** by reduction, and finally the free sugar **18** by Nef degradation (improved by preliminary deacetylation[62]). Wittig reaction[63] of aldehydo sugars with methoxy-[64] and (methylthio)-methylenetriphenylphosphorane[65] affords aldenitol derivatives analogous to **16**, and condensation with the active methylene group of ethyl oxaloacetate[65a] gives a diester; any of these products may be hydrolyzed directly to the aldose. Reaction of 2,4-*O*-benzylidene-1-deoxy-1-iodo-D-erythritol with 2-(1,3-dithianyl)lithium gave the trimethylene dithioacetal of 2-deoxy-D-*erythro*-pentose.[66]

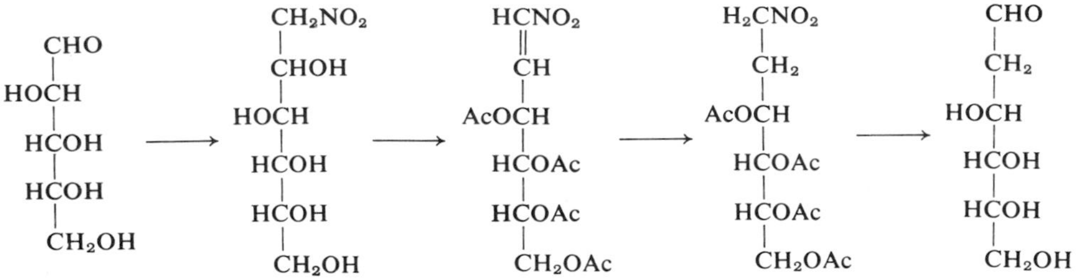

15　　　　　　**16**　　　　　　**17**　　　　　　**18**

2-Deoxy-D-*erythro*-pentose, labeled at C-1, has been synthesized by a chain-extension method, starting from 2,4-*O*-ethylidene-D-erythritol, following the sequence[67]:

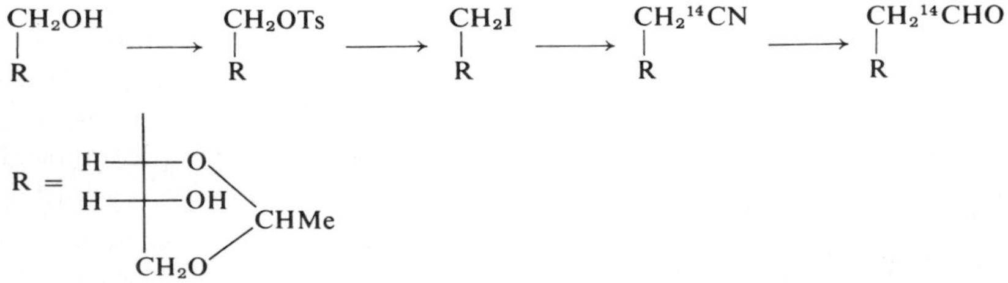

Special attention, of course, has been focused on the synthesis of 2-deoxy-D-*erythro*-pentose. The most convenient method for this would appear to be either the alkaline degradation of 3-*O*-(methylsulfonyl)-D-glucose (**19**),[68] the Ruff degradation of the "glucosaccharinic" acids (see Chapter 23) formed by alkaline treatment of D-glucose,[69] or sequential elimination, reduction, and oxidation, starting from D-glucono-1,4-lactone.[70]

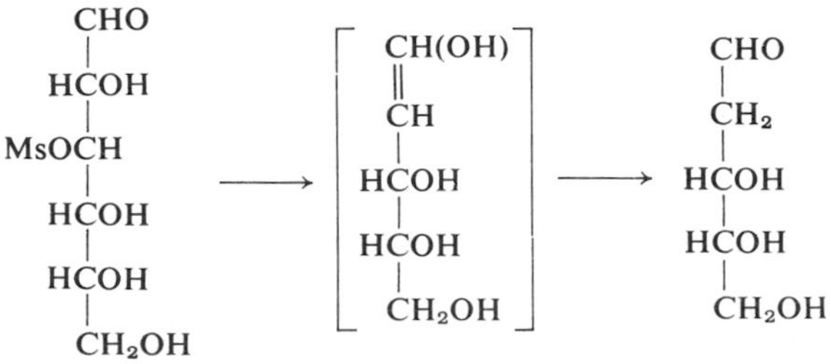

$$
\begin{array}{ccc}
\begin{array}{c}
\text{CHO} \\
| \\
\text{HCOH} \\
| \\
\text{MsOCH} \\
| \\
\text{HCOH} \\
| \\
\text{HCOH} \\
| \\
\text{CH}_2\text{OH} \\
\textbf{19}
\end{array}
& \longrightarrow &
\left[\begin{array}{c}
\text{CH(OH)} \\
\| \\
\text{CH} \\
| \\
\text{HCOH} \\
| \\
\text{HCOH} \\
| \\
\text{CH}_2\text{OH}
\end{array}\right]
& \longrightarrow &
\begin{array}{c}
\text{CHO} \\
| \\
\text{CH}_2 \\
| \\
\text{HCOH} \\
| \\
\text{HCOH} \\
| \\
\text{CH}_2\text{OH}
\end{array}
\end{array}
$$

Other methods for 2-deoxy sugars occasionally used include the deamination of 2-amino-2-deoxyalditols with nitrous acid,[71] the pyrolytic rearrangement of 2-*O*-(methylthio)thiocarbonyl derivatives followed by Raney nickel desulfurization[72] (see synthesis of 3-deoxy-D-*ribo*-hexose below), the degradation of 3-deoxyhexoses to 2-deoxypentoses with periodic acid[73] or lead tetraacetate,[74] the degradation of 3-deoxyhexose oximes with 2,4-dinitrofluorobenzene,[75] the Ruff degradation of calcium 3-deoxyaldonates,[76] photolysis of sulfonic esters,[77] hydrogenolysis of halogenated sugars,[78] and reaction of 2,3-benzylidene acetals with butyllithium.[79] 2′-Deoxynucleotides have been prepared by hydrogenation of 2′-thionucleosides,[79a] and racemic syntheses have employed *cis*-hydroxylation of dihydropyrans by permanganate.[79b]

2. *3-Deoxyaldoses*

The method generally used for these sugars is the direct or indirect reduction of 2,3- or 3,4-anhydro sugar derivatives. Analogous considerations apply, as in the use of epoxides for preparing 2-deoxy sugars. In this case, direct reduction with lithium aluminum hydride or by Raney nickel-catalyzed hydrogenation has been preferred, the use of thio or halo sugar intermediates occasionally being advocated. All of the 3-deoxyhexoses are available by this method—for example, 3-deoxy-D-*arabino*-hexose (**20**) from the 2,3-anhydro-D-mannoside **21**, by using lithium aluminum hydride,[57] and synthesis of the 3-deoxy-D-*ribo*-hexose derivative **11** mentioned above. 3-Deoxy-D- and

L-*erythro*-pentose (**22**; D form shown) have been synthesized from the 2,3-anhydro-riboside **23** by using sodium methanethioxide[80] or hydrobromic acid[81] to furnish the 3-thio (**23a**) and the 3-bromo-3-deoxy analogue (**23b**), respectively; each was then reduced catalytically to the methyl glycoside of the deoxy sugar **22**. Similarly, a 3-deoxy-D-*erythro*-pentoside is obtained almost exclusively by catalytic hydrogenation of methyl 2,3-anhydro-β-D-ribofuranoside.[82]

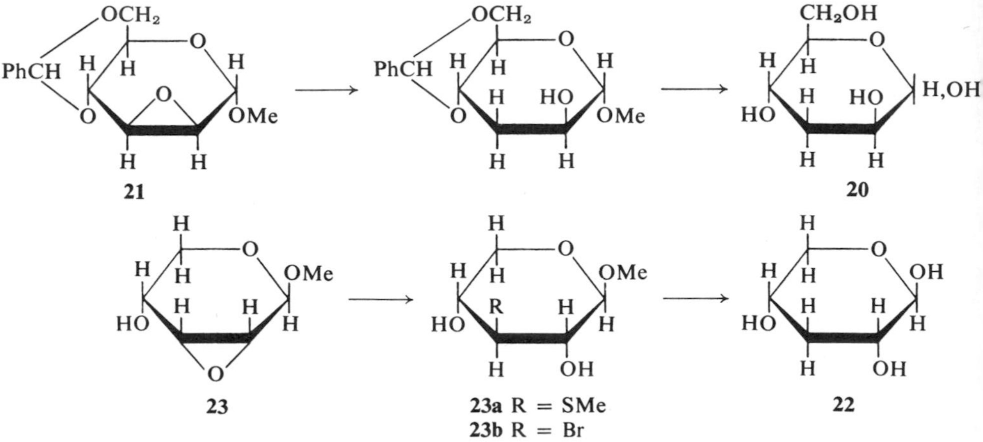

A convenient synthesis of 3-deoxy-D-*ribo*-hexose has been developed, involving the thermal rearrangement of 1,2:5,6-di-O-isopropylidene-3-O-(methylthio)thiocarbonyl-α-D-glucofuranose (**24**) to the 3-S-(methylthio)-carbonyl-3-thio derivative **25**, which on desulfurization with Raney nickel gives[83,84] the 3-deoxy-D-*ribo*-hexose derivative **26**. Photolysis of the 3-O-(dimethylthiocarbamoyl) analogue of **24** also yields[85] the 3-deoxy sugar **26**.

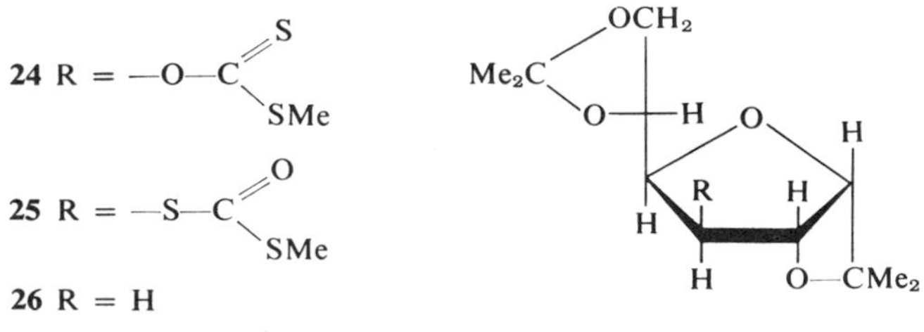

The stereospecific hydrogenation of the readily available unsaturated sugar **27** provides a useful synthesis of the 3-deoxy-D-*lyxo*-hexose derivative **28**, from which the free sugar is readily obtained on hydrolysis.[75]

3-Deoxyhexoses have also been prepared from 2-deoxypentoses by chain extension, via either the Kiliani synthesis[86] or the Fischer–Sowden synthesis.[87]

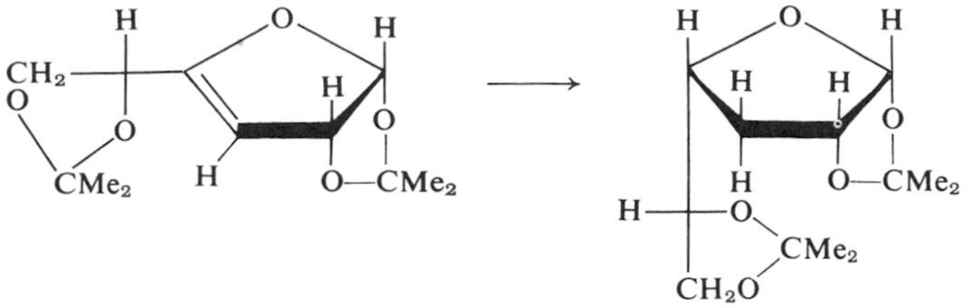

27   28

Halogenolysis and reduction of kanamycin B 3′-phosphate gave 3′-deoxy-kanamycin.[87a]

3. 4-Deoxyaldoses

The synthesis of 4-deoxy sugars from 3,4-anhydro sugars has not proved a very satisfactory route, since the 3-deoxy isomer tends to be the major product from accessible 3,4-anhydrohexosides. However, conditions have been found for the synthesis of 4-deoxy-D-xylo-hexose from the derivatives **29** (ref. 88) and **30** (ref. 89) in reasonable yields, by using catalytic hydrogenation.

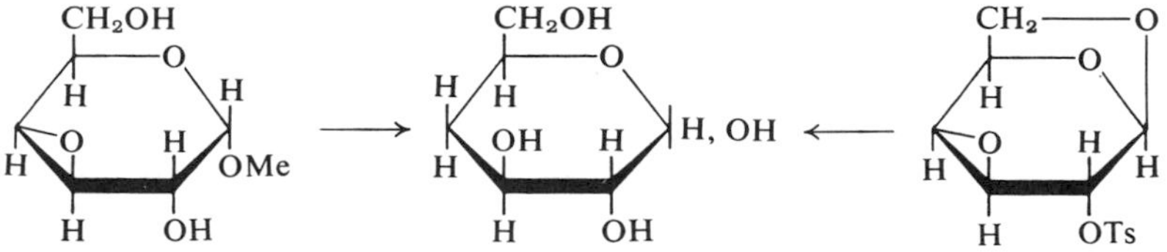

29   30

In the pentose series, treatment of methyl 3,4-anhydro-β-L-ribopyranoside (**31**) with hydrobromic acid yields the 4-bromo-4-deoxy isomer (**32**) stereospecifically, which on reduction gives[90] the 4-deoxy-L-erythro-pentose glycoside **33**.

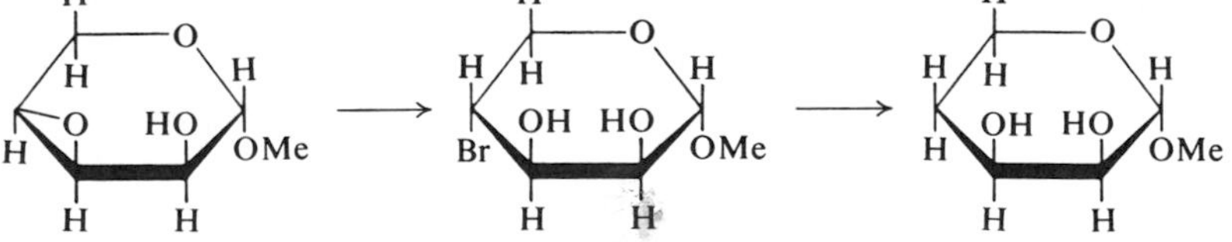

31   32   33

Two potentially general methods have been used to introduce a 4-deoxy group into a sugar. In both cases the terminal carbon atom was simultaneously converted into a deoxy group. The first method, developed by Kochetkov and his co-workers,[91] involved the conversion of the 3-O-methyl-D-glucose derivative **34** into the diiodo sugar **35** in one step with triphenyl phosphite methiodide under mild conditions. Catalytic reduction of **35** then gave the dideoxy sugar **36**, which could be transformed into 3-*O*-methyl-4,6-dideoxy-D-*xylo*-hexose (*chalcose*) (**37**).

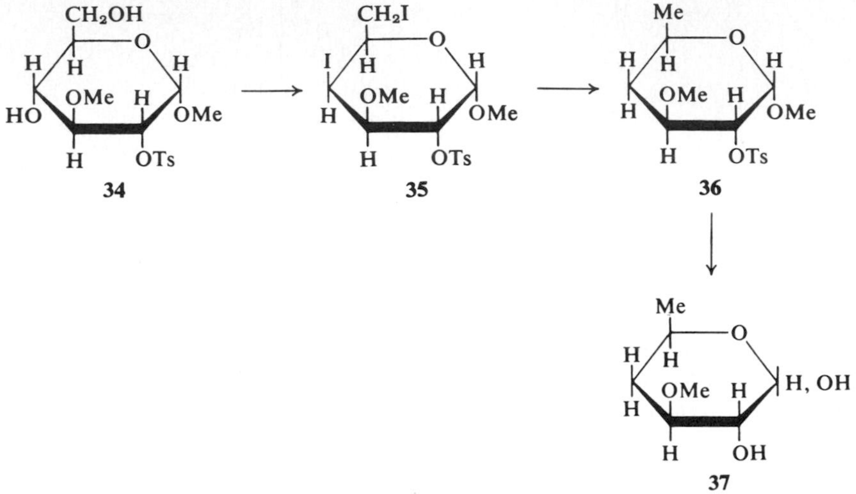

Chalcose was also prepared[92] by chlorosulfonation and subsequent hydrogenation of methyl 3-*O*-methyl-D-glucopyranoside. Deoxygenation of C-4 in hexopyranosides may be accomplished by using thionyl chloride followed by aluminum chloride,[93] sulfuryl chloride,[94] or methanesulfonyl chloride in *N*,*N*-dimethylformamide,[95] reduction being effected by action of Raney nickel[93] or tributyltin hydride[94]; reduction of 4,6-dichloro-4,6-dideoxy sugars occurs at the 4-position somewhat faster than at the primary center.[93]

In the other method, the sulfonic ester groups in the D-glucose derivative **38** were displaced by thiocyanate, by using potassium thiocyanate in *N*,*N*-dimethylformamide, and the resulting dithio compound could be desulfurized with Raney nickel[96] to the 4,6-dideoxy-D-*xylo*-hexoside **39**.

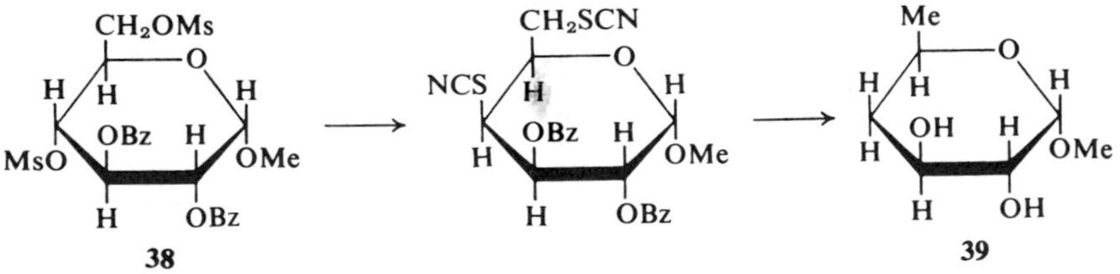

4. *5-Deoxyhexoses*

5-Deoxy-D-*xylo*-hexose has been synthesized via its 1,2-*O*-isopropylidene derivative **40** from the unsaturated sugars **41** and **42**, the former being hydrated by the hydroboration procedure,[97] the latter catalytically hydrogenated.[98] The derivative **40** has also been prepared from the 5,6-anhydro-D-glucose derivative **43** by catalytic hydrogenation with alkali-free Raney nickel, which causes abnormal cleavage of the epoxide ring to give the primary rather than the secondary alcohol.[99,100] The synthesis of **40** was achieved before these reports, but at the time it was mistaken for a derivative of 6-deoxy-L-idose.[101] A convenient preparation of **40** proceeds by reduction of 6-O-benzoyl-1,2-O-isopropylidene-5-O-p-tolylsulfonyl-α-D-glucopyranose.[100]

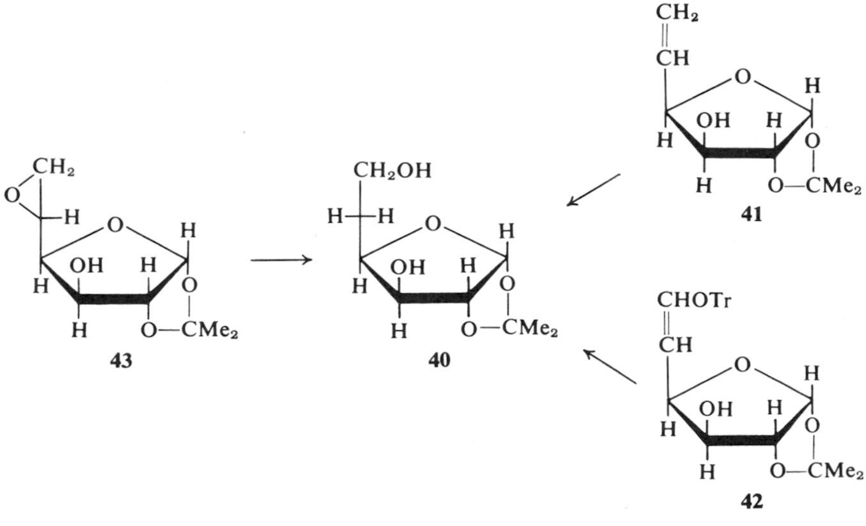

5-Deoxy-D-*ribo*-hexose ("*homoribose*") has been prepared[102] analogously from the C-3 epimer of **42**, but the authors suggested that a more practical synthesis is by isomerization of the D-*xylo* isomer (**40**) via its 3-methanesulfonate by using sodium benzoate in N,N-dimethylformamide. Alternatively, the 5-cyano-5-deoxy-D-ribose derivative **44** may be reduced and then deaminated with nitrous acid in the usual way to give[103] the 5-deoxy-D-*ribo*-hexose derivative **45**.

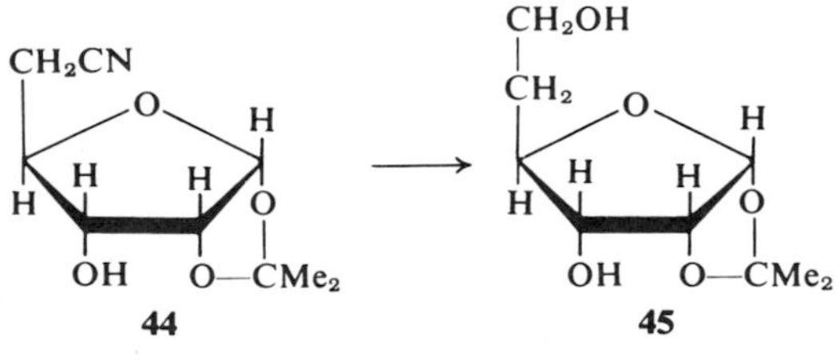

Other routes, potential and realized, to ($\omega - 1$)-deoxy sugars include the addition of such carbon nucleophiles as 2-(1,3-dithianyl)lithium to terminal epoxides[104] or deoxyhalogeno sugars,[66] Wittig reactions,[63,105] or condensation of active methylene reagents[106] with aldehydo derivatives of dialdoses.

5. *5-Deoxypentoses and 6-Deoxyhexoses*

The standard method for converting the terminal group on a sugar chain into a methyl group has been to reduce the methanesulfonate or *p*-toluenesulfonate of the primary alcohol, either with lithium aluminum hydride directly or, in cases where this gives poor yields, indirectly by replacing the ester group with iodide, which is amenable to catalytic hydrogenation. For these ester displacements, the sugar may be in either the cyclic or the acyclic (usually dithioacetal) form. Sulfonic esters at nonterminal positions do not undergo these replacements.

$$R-CH_2OH \longrightarrow R-CH_2OSO_2R' \longrightarrow R-CH_3$$

$$R-CH_2I$$

A promising alternative procedure has also been used, in which the primary alcohol is converted directly into the iodide by using the triphenylphosphite complex with methyl iodide, iodine, or bromine.[91,107] The iodide is then reduced catalytically to the deoxy sugar. The method is not specific for primary alcohols (see under 4-deoxy sugars above), in contrast to the sulfonic ester route, although sodium cyanoborohydride has subsequently been reported[109] to afford selective reduction. During the reaction with the complex, acetal groups can migrate, and in this way 1,2:5,6-di-*O*-isopropylidene-α-D-glucofuranose yields 6-deoxy-6-iodo-1,2:3,5-di-*O*-isopropylidene-α-D-glucofuranose. An example of halogenolysis of a partially benzoylated pyranoside to give a 6-bromo-6-deoxyfuranose (and subsequent reduction) has been reported.[109a]

Action of acetylsalicyloyl chloride on partially protected sugar derivatives converts unsubstituted, primary hydroxyl groups into chloro substituents and acetylates secondary hydroxyl groups. 1,2-*O*-Isopropylidene-α-D-glucofuranose was thus converted into the 3,5-di-*O*-acetyl-6-chloro-6-deoxy derivative; *p*-toluenesulfonylation or *N*-phenylcarbamoylation of O-3 did not alter the reaction at O-5 and O-6, but acetylation of O-6 precluded the isolation of chlorodeoxy products.[109b]

Where benzylidene acetals that include the terminal sugar hydroxyl group are available, the terminal deoxy sugar is very conveniently prepared by treatment of the derivative with *N*-bromosuccinimide[109] or *N*,*N*-dibromobenzenesulfonamide[110] and reduction of the resulting bromodeoxy sugar.

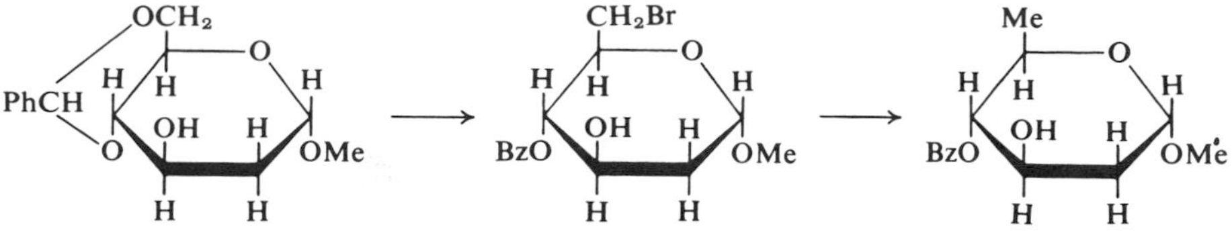

46

Thus, methyl 4,6-*O*-benzylidene-2-deoxy-α-D-*arabino*-hexopyranoside may readily be converted into the 4-*O*-benzoylglycoside[111] (**46**) of 2,6-dideoxy-D-*arabino*-hexose (*chromose C*). This has become a standard method for preparation of terminal deoxy sugars. One-step conversion of benzylidene acetals into deoxy sugars occurs by action of di(*tert*-butyl) peroxide, but the reaction is apparently not always positionally specific.[112]

Other methods include treatment of 3-*O*-benzyl-1,2-*O*-isopropylidene-α-D-*xylo*-pentodialdo-1,4-furanose with methylmagnesium iodide to furnish the 6-deoxy-L-idofuranose isomer (**47**) stereospecifically[113]; further procedures are the degradation of 6-deoxyhexoses to 5-deoxypentoses, and the isomerization of terminal deoxy sugars, via sulfonic ester displacements with sodium benzoate in *N,N*-dimethylformamide—for example, the 5-deoxy-D-xylose derivative **48** leading[114] to the 5-deoxy-D-ribose isomer **49**. Another route for

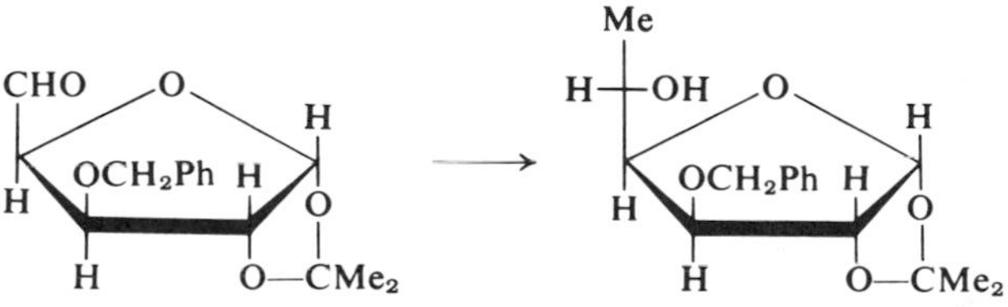

47

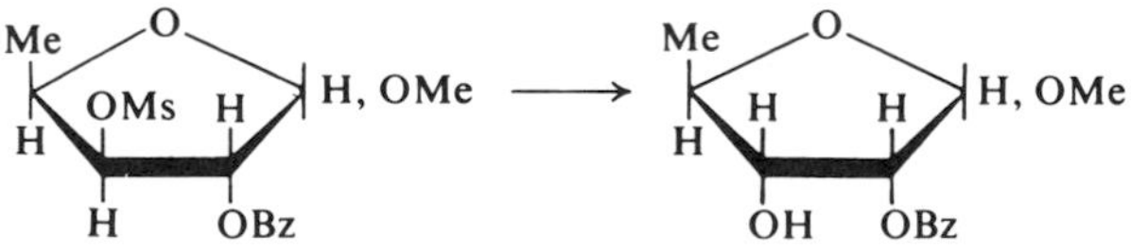

48 **49**

isomerization is via the aldosulose formed by oxidation of a suitable deoxy sugar derivative, which may be stereospecifically reduced to the isomeric deoxy sugar in certain cases, usually when the isomerized position initially has an equatorial hydroxyl group, as in the L-rhamnose derivative **50**, which is readily oxidized to the hexosulose **51**, which may then be reduced catalytically or with lithium aluminum hydride to give predominantly the 6-deoxy-L-taloside **52** having the C-4 hydroxyl group axial in the favored conformation.[115] Both of these isomerization reactions are suited to general application.

References start on p. 791.

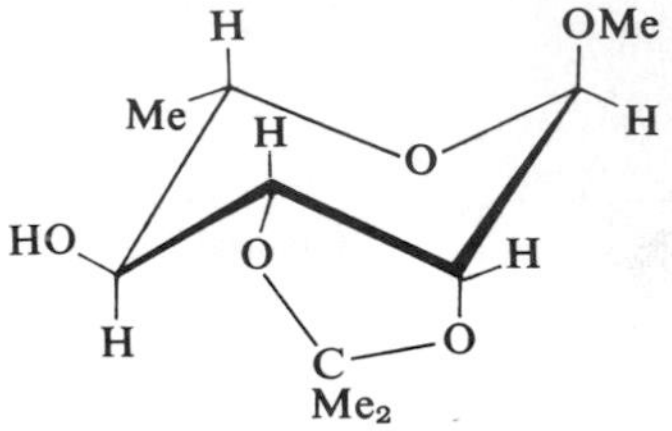

50

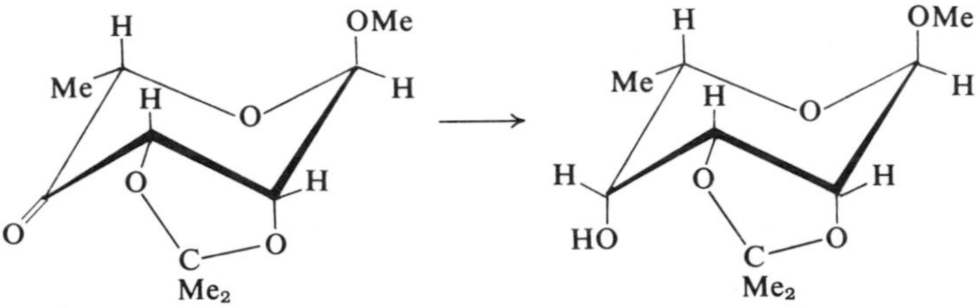

51 → **52**

6. *Dideoxy and Trideoxy Sugars*

These sugars have been synthesized by direct application of the methods developed for monodeoxy sugars. Which deoxy group is introduced first is a matter of choice, frequently governed by the availability of suitable mono-deoxy sugars. In most cases the terminal deoxy group has been introduced last, although in a few cases the deoxy groups are formed at the same stage, as in the synthesis of methyl D-digitoxoside (**53**) from the 2,3-anhydro-6-*O*-*p*-tolylsulfonyl-D-alloside **54** by using lithium aluminum hydride.[116] (See also the synthesis of 4-deoxy sugars above.)

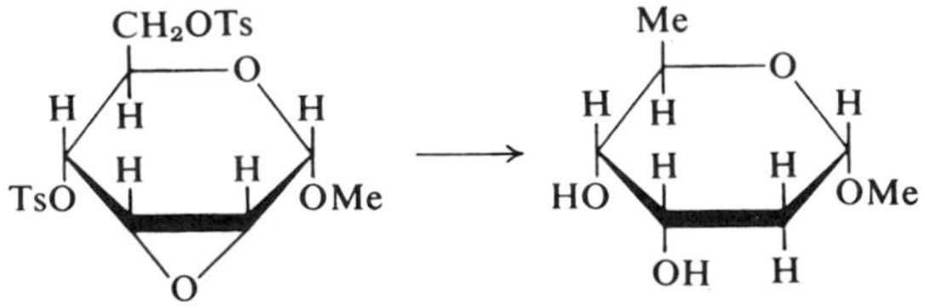

54 → **53**

The 2,3,6-trideoxyhexoses have been synthesized by a special method involving the hydrogenation of the ethyl hex-2-enoside **55** (see Chapter 19) to the 2,3-dideoxy sugar **56**, finally introducing the 6-deoxy group in the usual way.[117] The resulting ethyl 2,3,6-trideoxy-α-D-*erythro*-hexopyranoside (ethyl amicetoside) (**57**) was then isomerized by benzoate exchange with the

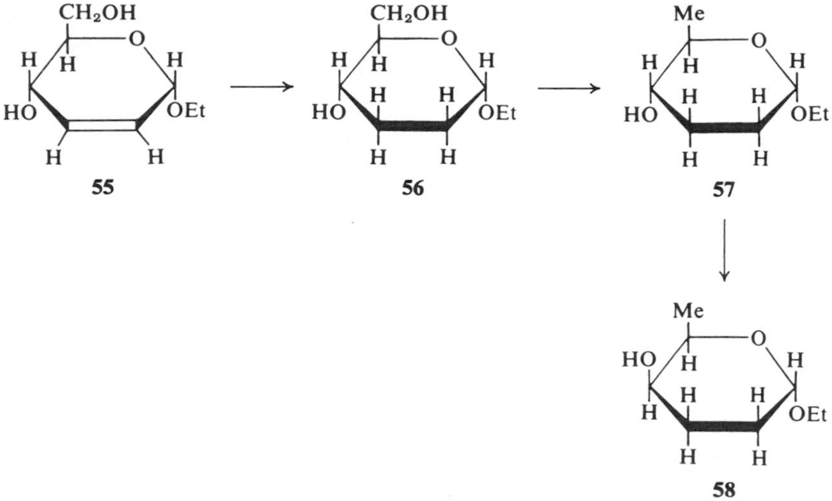

4-methanesulfonate to the D-*threo* isomer (**58**), the free sugar from which was shown to be enantiomorphic with natural rhodinose.

Alternative syntheses starting from L-rhamnose[118] and *aldehydo*-L-xylose tetraacetate[119] also proceeded by hydrogenation of a 2,3-unsaturated group. A synthesis of 3-amino-2,3,6-trideoxy-L-*lyxo*-hexose (*daunosamine*) from D-mannose employed base-induced cleavage of a benzylidene acetal and hydrogenation of a 5,6-double bond to generate the deoxy positions.[120] The 1,6-anhydro-di-[121] and -tri-deoxy-β-D-hexopyranoses[122] have been prepared by standard methods.

Reduction of methyl 2,3,4,6-tetra-*O*-*p*-tolylsulfonyl-α-D-glucopyranoside by lithium aluminum hydride affords[123] a mixture of products from which methyl 3,6-dideoxy-α-D-*ribo*-hexopyranoside was isolated in 16% yield.

C. PROPERTIES

There are few reactions that clearly distinguish deoxy sugars from normal sugars, which they otherwise closely resemble. Entirely predictable differences arise where an essential hydroxyl group is lacking in the deoxy sugar; thus, 2-deoxy sugars cannot form osazones, 4-deoxy sugars furanose derivatives, nor 5-deoxy sugars pyranose derivatives.

2-Deoxy sugars are the most clearly distinguished as a class, their most notable difference from normal sugars being the enhanced reactivity of the aldehydic carbon atom. This is reflected most obviously in the much readier hydrolysis of their glycosides, their pyranosides being of comparable lability to normal furanosides.[3] The difference may be attributed (see Vol. IA,

Chapter 9) to the absence of the electron-attracting oxygen atom adjacent to the glycosidic center, which in normal sugars discourages reactions that involve formation of a positive charge at that center. 3-Deoxy sugars show a similar, but much less marked, rate enhancement.[124] 2-Deoxy sugars also show a number of special color reactions useful for their detection and determination,[125] notably the Keller–Kiliani and the Dische reaction, the cysteine–sulfuric acid test, and the xanthydrol test.[10] Other color reactions have been developed for 3-deoxy and 6-deoxy sugars.[126]

III. BRANCHED-CHAIN SUGARS

A. OCCURRENCE

Table II lists principal, naturally occurring branched-chain sugars together with their sources. The first two, having hydroxymethyl side chains, occur in plants. Whereas the distribution of hamamelose is principally in the hamamelitannin of witch hazel and a few other plants, apiose is now known to occur widely in a variety of plants, either in glycosides or in polysaccharides,[134,135] a rich source being *Lemna* (duckweed).[135] The remaining sugars arise in antibiotic compounds derived from microorganisms, mainly various strains of *Streptomyces*.[7] So far, methyl, formyl, hydroxymethyl, 1-hydroxyethyl, acetyl, and 2-hydroxyacetyl side chains have been found in these sugars, and almost all of them are terminal deoxy sugars as well. With the exception of blastmycinone[137] [2-*C*-butyl-2,5-dideoxy-3-*O*-(3-methylbutanoyl)-DL-arabinono-1,4-lactone], these branched-chain sugars found in Nature have a polar substituent at the branching carbon-atom; tertiary alcohols are commonest, but arcanose, cladinose, and nogalose have a methyl ether group, L-chromose B has an acetate group, aldgarose has a cyclic carbonate group (bridging to the hydroxyethyl side-chain), and vancosamine (Chapter 16) and evernitrose have an amino and a nitro group, respectively, at the branch point. The deoxy sugar cordycepose was formerly regarded as the 3-deoxy analogue of apiose, and as such the first exception, but this sugar was subsequently shown to be the straight-chain deoxy sugar 3-deoxy-D-*erythro*-pentose by spectroscopic[12,168] and chemical-degradative[169] methods.

The configuration of aldgarose was determined during the course of its synthesis, principally by comparison of n.m.r. data measured for isomeric pairs of intermediates,[170] whereas X-ray crystallography of appropriate derivatives has provided an indisputable method of stereochemical assignment in such favorable[171] examples as nogalose[143] and streptose.[172] Pillarose, which was at first thought[167a] to be a 2,3,6-trideoxy-2-*C*-glycolylhexos-4-ulose, has been shown to be 2,3,6-trideoxy-4-*C*-glycolyl-L-*threo*-hexose by

TABLE II

PRINCIPAL BRANCHED-CHAIN SUGARS OF NATURAL OCCURRENCE

Sugar	Trivial name	Source	Reference
3-C-(Hydroxymethyl)-D-glycero-tetrose	Apiose	Parsley (including apiin)	127
		Celery (graveobiosides A and B)	128
		Dalbergia lanceolaria (lanceolarin)	129
		Posidonia australis	130
		Taraxacum kok-saghyz	131
		Hevea brasiliensis	132
		Zostera marina	133, 134
		Lemna spp.	135
		Platycodon grandiflora (platycodin D)	135a
2-C-(Hydroxymethyl)-D-ribose	Hamamelose	Hamamelis virginiana (hamamelitannin)	136
3,4-Anhydro-2-C-methyl-L-erythrose		Cotylenins A and C	136a
2-C-Methyl-L-erythrose		Cotylenin D	136a
4-Chloro-4-deoxy-2-C-methyl-L-erythrose		Cotylenin B	136a
2-C-Methyl-D-erythritol		Convolvulus glomeratus	136b
2-C-(2-Methyl-1-propyl)-D-erythronic acid		Loroglossine	136c
2-C-Butyl-2,5-dideoxy-3-O-(3-methylbutanoyl)-DL-arabinono-1,4-lactone	Blastmycinone	Blastmycin	137
3-C-Formyl-L-lyxose	Hydroxystreptose	Hydroxystreptomycin	138
5-Deoxy-3-C-formyl-L-lyxose	Streptose	Streptomycin	139
5-Deoxy-3-C-hydroxymethyl-L-lyxose	Dihydrostreptose	Bluensomycin	140
		Glebomycin	141
6-Deoxy-3-C-Methyl-D-mannose	D-Evalose	Everninomycin B	18a
		Everninomycin C	18b
		Everninomycin D	18c
		Everninomycin-2	18d
6-Deoxy-3-C-methyl-L-mannose		Flambamycin	17
		Flambeurekanose	19
2,3,4-trimethyl ether 6-Deoxy-3-C-methyl-L-talose	Nogalose	Nogalamycin	142, 143
2-methyl ether 6-Deoxy-5-C-methyl-L-lyxo-hexose	Vinelose	Acetobacter vinelandii	144
4-methyl ether	Noviose	Novobiocin	145

(Continued)

References start on p. 791.

TABLE II—*Continued*

Sugar	Trivial name	Source	Reference
2,6-Dideoxy-3-*C*-methyl-D-*arabino*-hexose	D-Evermicose	Everninomycin	146, 147
		Everninomycins C;D,-2	18b–d
2,6-Dideoxy-3-*C*-methyl-L-*arabino*-hexose	Olivomycose	Olivomycin	148
	Deacetylchromose Epimycarose	Chromomycin A$_3$	27, 32
3-acetate	L-Chromose B	Chromomycin A$_3$	149
2,6-Dideoxy-3-*C*-methyl-L-*ribo*-hexose	Mycarose	Magnamycin	150
		Carbomycin	151
		Spiramycin	152
		Formacidin	153
		Tylosin	14, 154
		Angolamycin	155
		Leucomycin	156
		Erythromycin C	157
		Erythromycin D	157a
		Chromocyclomycin	29
		Josamycin	158
		Kitsamycin	158
		Plateomycin(s)	159
		Maridomycin	159a
3-methyl ether	Cladinose	Erythromycin	160
2,6-Dideoxy-3-*C*-methyl-L-*xylo*-hexose	Axenose	Axenomycins A,B,D	161
3-methyl ether	Arcanose	Lankamycin	162
2,6-Dideoxy-4-*C*-[1(*S*)-hydroxyethyl]-L-*xylo*-hexose		Quinocycline A	163
		Isoquinocycline B	163, 164
4-*C*-Acetyl-2,6-dideoxy-L-*xylo*-hexose		Quinocycline B	163
		Isoquinocycline B	163
4-*C*-Acetyl-6-deoxyhexonic acid		Flambamycin	17b
		Flambeurekanose	19a
4,6-Dideoxy-3-*C*-[1(*S*)-hydroxyethyl]-D-*ribo*-hexose			
3,1′-carbonate	Aldgarose	Aldgamycin E	165
2,3,6-Trideoxy-3-*C*-methyl-3-nitro-L-*arabino*-hexose			
4-methyl ether	Evernitrose	Everninomycins B; C, and D	166
2,3,6-Trideoxy-4-*C*-glycolyl-L-*threo*-hexose	Pillarose	Pillaromycin A	167b–d

X-ray crystallographic analysis of pillaromycin[167b] and by synthesis of both of the proposed structures.[167c,d] X-ray crystallography also permitted assignment of the configuration[166] of evernitrose.[166a]

B. SYNTHESIS

Two principal approaches have been used in the synthesis of branched-chain sugars. The first route utilizes a suitable carbonyl derivative of a sugar, usually a dicarbonyl sugar or a sugar acid, which is treated with diazomethane, an organometallic, or a Wittig reagent to introduce the branching substituent; other reagents have been used in the specific case of apiose. This route has the advantage of giving a known, fixed configuration at all of the asymmetric centers other than the branching position, its limitation being the availability of suitable sugar precursors. Alternatively, the sugar has been built up from simple noncarbohydrate compounds, usually involving the stereospecific hydroxylation of unsaturated intermediates. The drawback here is that the products are racemates and furnish the natural material only after resolution.

1. Addition Reactions of Carbohydrate Carbonyl Derivatives

Aldgarose,[170,173] apiose,[133,174,177] D-arcanose,[178] axenose,[179] D-[180] and L-cladinose,[181] D-evermicose,[181a] dihydrostreptose,[182] hamamelose,[183,184] D-[180] and L-mycarose,[181] nogalose,[185] noviose,[186,187] olivomycose,[188] streptose,[182,184] and vinelose[189,190] have been prepared by sequences involving addition reactions to suitable carbohydrate derivatives having free carbonyl groups.

Streptose was synthesized[182] from 5-deoxy-1,2-O-isopropylidene-α-L-arabinofuranose (59) by oxidation to the L-threo-pentos-3-ulose derivative 60 with the Pfitzner–Moffatt reagent, followed by treatment with vinylmagnesium bromide to give 5-deoxy-1,2-O-isopropylidene-3-C-vinyl-α-L-lyxofuranose (61) stereospecifically. Ozonolysis of 61 gave the 3-C-formyl derivative 62 [later[184] prepared by reaction of 60 with 2-(1,3-dithianyl)lithium and subsequent demercaptalation], which, on mild acid hydrolysis, furnished streptose (63). Borohydride reduction of 62 followed by acid hydrolysis gave dihydrostreptose (64). Addition reactions to carbonyl groups have also been accomplished using allyl or propargyl bromide and zinc.[190a]

By a similar sequence, methyl 4,6-O-benzylidene-2-deoxy-α-D-arabino-hexopyranoside (65) was oxidized with chromium trioxide in pyridine, or ruthenium tetraoxide, to the hexos-3-ulose (66), which reacted with methylmagnesium iodide to yield the 3-C-methyl-D-ribo-hexose isomer (67). After

---

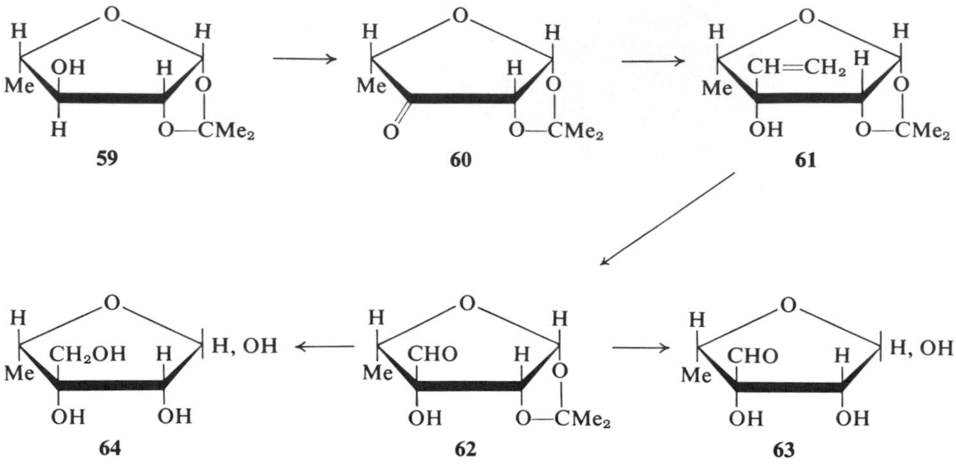
59
60
61
64
62
63

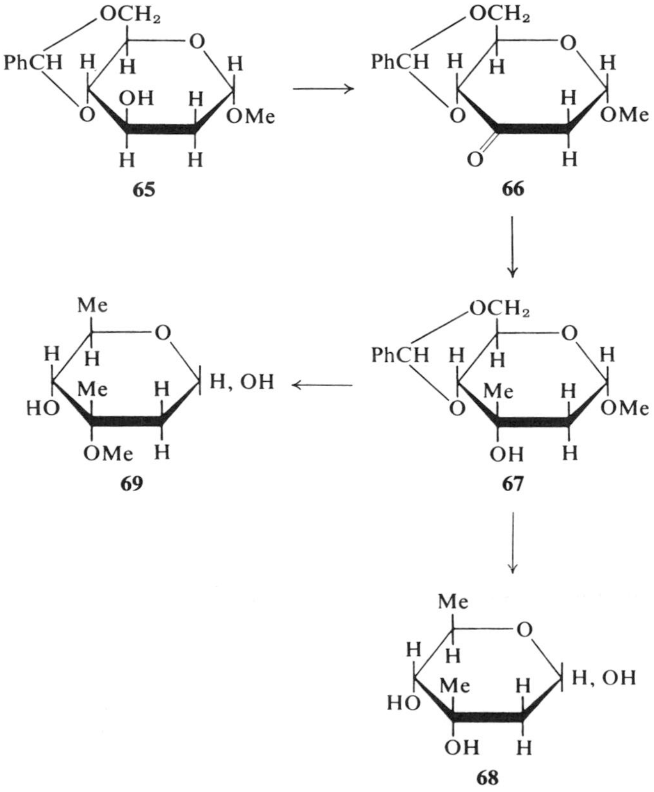
65
66
69
67
68

removal of the benzylidene group, the 6-deoxy group was introduced in the usual way to furnish[180] the D-enantiomorph (68) of mycarose. Methylation of 67 followed by the same transformations gave D-cladinose (69). Methyl 3-O-benzoyl-4,6-O-benzylidene-α-D-*arabino*-hexopyranosid-2-ulose likewise exhibits strong stereoselectivity in reacting with C-nucleophiles, methylmagnesium iodide giving the *gluco* adduct, and diazomethane or nitromethane giving *manno* products.[190b]

The L-isomer of 66 was likewise converted into the natural sugars.[181] Alternatively, the hexos-3-ulose 70 reacted with methylenetriphenylphosphorane to yield the unsaturated sugar 71, which was hydrated to yield the 3-C-methyl sugar 72 epimeric with 67 at C-3, as addition to the unsaturated group takes place from the less-hindered side of both 66 and 71. Treatment of 72 by Hanessian's procedure[109] gives olivomycose[188] (73).

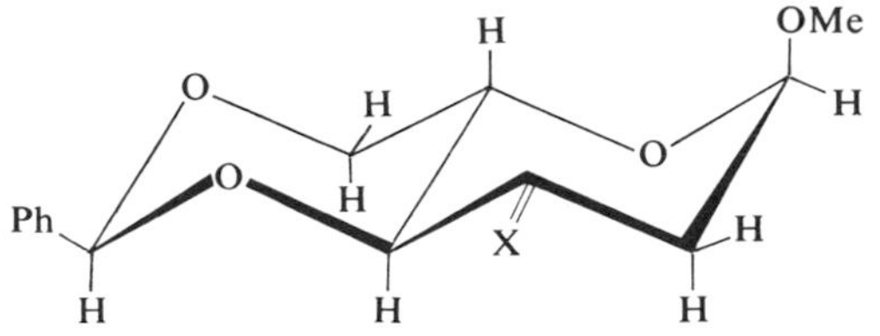

70 X = O
71 X = CH₂

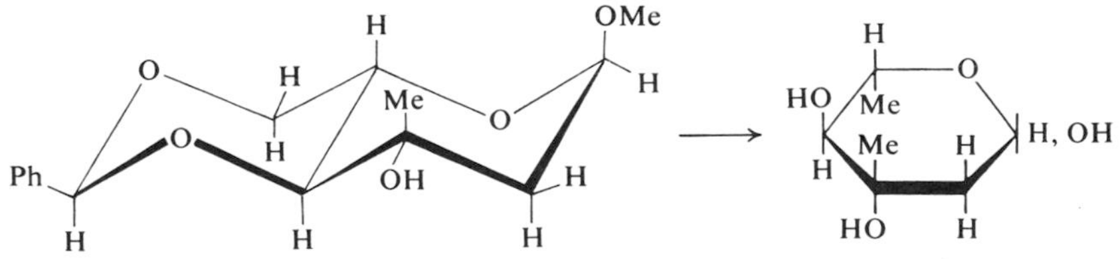

72   73

Noviose has been prepared[186,187] from the acid lactone 74 by using methylmagnesium iodide to obtain the 1,1-di-C-methyl-D-galactitol derivative 75. Sequential benzoylation, debenzylation, and oxidation with lead tetraacetate furnished the 2-O-benzoyl derivative (76) of noviose which, on alkaline hydrolysis, gave noviose. These workers used the intermediate 76 to prepare the 3-O-carbamoyl derivative of noviose necessary for their synthesis of novobiocin.

References start on p. 791.

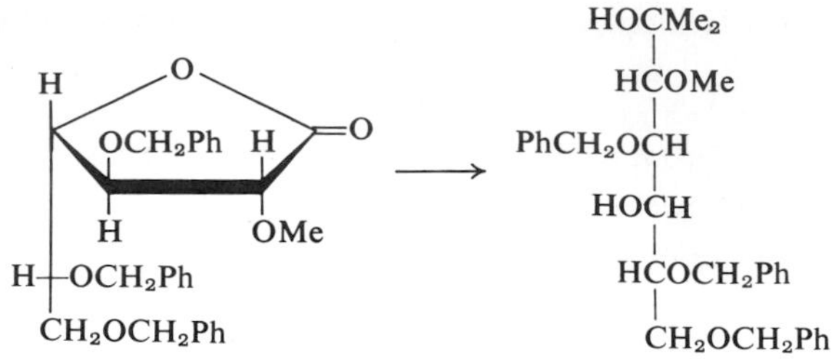

HOCMe$_2$
HCOMe
PhCH$_2$OCH
HOCH
HCOCH$_2$Ph
CH$_2$OCH$_2$Ph

74 → **75**

HOCMe$_2$
HCOMe
HOCH
BzOCH
CHO

$\equiv$

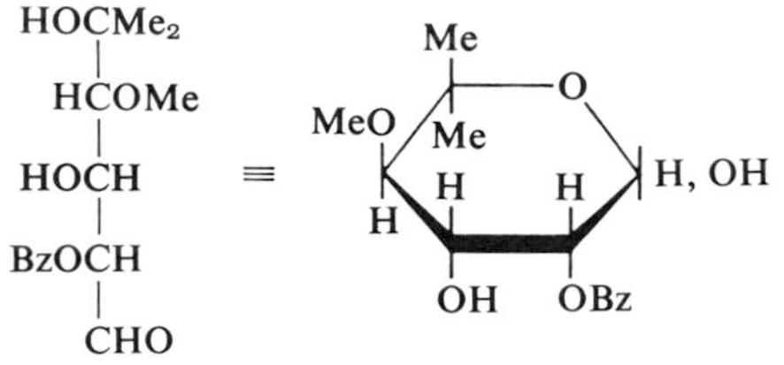

76

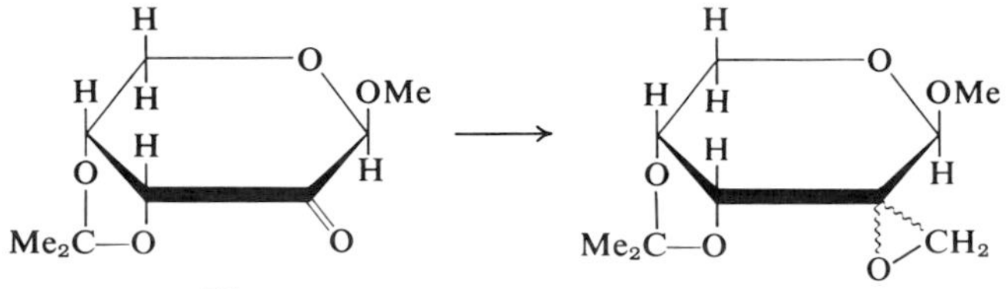

77 → **78**

↓

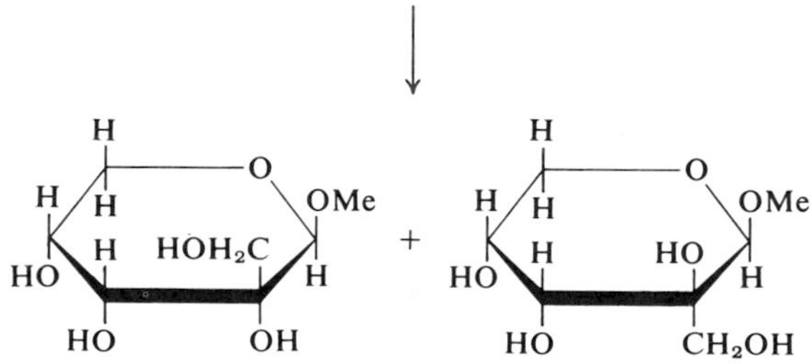

79 + **80**

Treatment of the oxo sugar **77** (readily obtained from D-arabinose) with diazomethane furnished a mixture of epoxides (**78**). Alkaline hydrolysis of the epoxide ring followed by deacetonation gave a mixture of the 2-*C*-(hydroxymethyl)-D-ribo- and arabino-pyranosides (**79** and **80**, respectively), the former being the major product, easily separable by crystallization. Acid hydrolysis yielded hamamelose and epihamamelose, respectively.[183] Condensation of methyl 3,4-*O*-isopropylidene-α-L-*erythro*-hexopyranosid-2-ulose, with 2-(1,3-dithianyl)lithium, followed by demercaptalation, reduction of the 2-*C*-formyl group thus liberated, and mild acidic hydrolysis afforded methyl α-hamameloside.[184]

Two independent syntheses[189,190] of vinelose proceeded from isopropylidene acetals of 3-*C*-methyl-D-allofuranose along the same line, namely inversion of C-5 by epoxidation, reductive opening of the oxirane, and exposure of O-2 for methylation. 1,2:5,6-Di-*O*-isopropylidene-3-*C*-formyl-D-allofuranose, prepared by condensation of the corresponding glycos-3-ulose with nitromethane[191] or 2-(1,3-dithianyl)lithium[192] and oxidation or acidic cleavage, respectively, was sequentially reduced, selectively deprotected at O-5 and O-6, cleaved with periodate, and oxidized, to afford 3-*C*-(hydroxymethyl)-D-riburonic acid, which was not identical to a constituent of a bilirubin glycoside that had tentatively been assigned the same structure.

The synthesis of aldgarose as the methyl β-glycoside was accomplished by rearrangement of methyl 2-*O*-benzyl-6-deoxy-3,4-di-*O*-*p*-tolylsulfonyl-β-D-galactopyranoside in the presence of hydrazine to afford the glycos-3-ulose **81**. Addition of 2-(2-methyl-1,3-dithianyl)lithium gave a separable mixture of epimers, and sequential demercaptalation, hydrogenation, and treatment with phosgene converted the D-*ribo* adduct into a separable mixture of 1'(*R*) and 1'(*S*) cyclic carbonate derivatives (whose configurations were assigned by comparison of spectroscopic data), the latter being identical to the natural product (**82**).[192a] A second route[173] to **82** involved addition of vinylmagnesium bromide to the glycos-3-ulose **83**, protection of the 2,3-diol by acetonation, epoxidation and subsequent reductive opening of the side-chain, treatment with 2,2,2-trichloroethoxycarbonyl chloride to form the 1'(*S*),3-cyclic carbonate, deprotection, and glycosidation. A model approach[193] was based on addition of ethynylmagnesium bromide to 1,2:5,6-di-*O*-isopropylidene-α-D-*ribo*-hexofuranose-3-ulose and mercury(II)-catalyzed hydration of the triple bond to afford a 3-*C*-acetyl intermediate that was reduced and carbonated to form a mixture of 1',3-cyclic carbonate derivatives of 3-*C*-(hydroxymethyl)-D-allose.

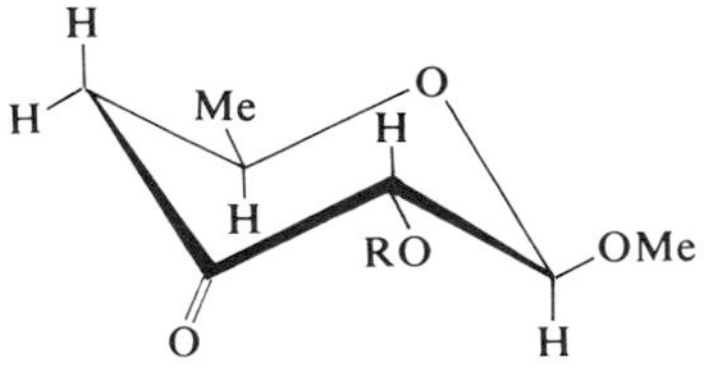

81 R = CH₂Ph
83 R = Bz

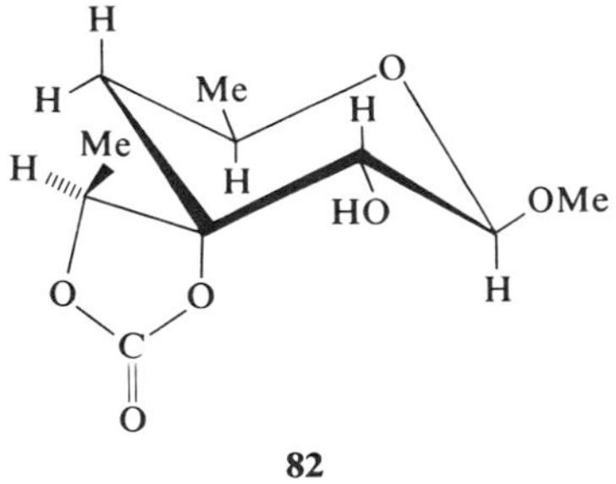

82

Several methods have been proposed for the synthesis of apiose. Two start from D-fructose derivatives and introduce the branching group either via diazomethane as above,[174] or via the cyanohydrin formed by addition of hydrogen cyanide to the free keto group.[175] The fructose chain is subsequently cleaved to furnish the apiose skeleton by the action of periodate or lead tetraacetate. Two other methods introduce the branch hydroxymethyl group by treating a formyl sugar derivative with alkaline formaldehyde, furnishing compounds having the dihydroxymethyl system of apiose, and leading to L-[176] and D-[133] apiose; the latter method[133] is recommended as a satisfactory procedure. The *aldehydo*-arabinose derivative **84** is converted into the bis(hydroxymethyl) compound **85**, which is hydrolyzed selectively to the mono-isopropylidene acetal **86**. The latter, on periodate oxidation, yields 2,3-*O*-isopropylidene-D-apiose (**87**), of fixed stereochemistry at C-3, which may be readily hydrolyzed to the free sugar.

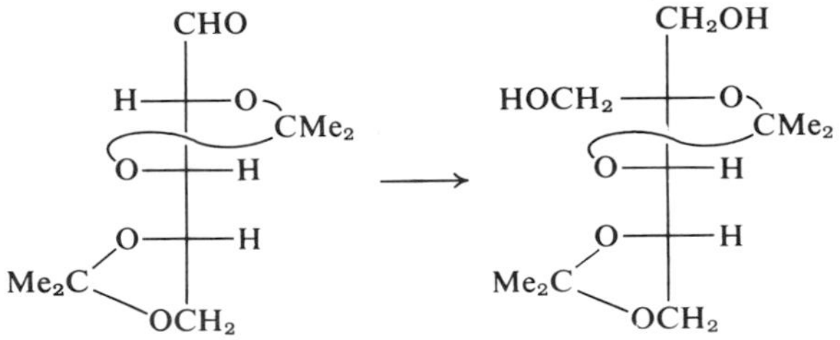

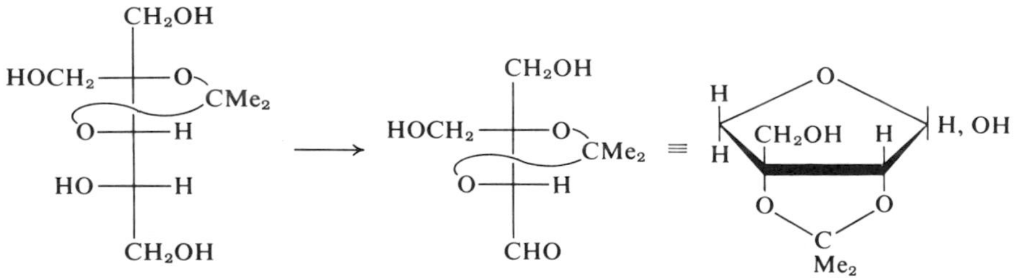

L-Dendroketose [4-*C*-(hydroxymethyl)-L-*glycero*-pentulose], the ketose homologue of apiose, was selectively metabolized[194] by a microorganism from a racemic mixture isolated from base-catalyzed autocondensation[194,195] of 1,3-dihydroxy-2-propanone to afford the D-isomer. Stereospecific synthesis of the L-isomer from 2,3:4,5-di-*O*-isopropylidene-D-arabinose was effected[196] by aldol condensation with formaldehyde (and spontaneous reduction by a crossed Cannizzaro process), hydrolysis of the primary acetal, protection of O-1, oxidation of O-2, and deprotection. Oxidation of 2-*C*-(hydroxymethyl)-D-*erythro*-pentitol by *Acetobacter suboxydans* also gives L-dendroketose.[196a]

2,6-Dideoxy-4-C-[1(S)-hydroxyethyl]-L-xylo-hexose (**104**, a constituent of the quinocycline complex) and its 4-epimer were prepared[196b] by selective addition of vinyllithium to C-4 of methyl 2,3-anhydro-6-deoxy-α-L-ribo-hexopyranosid-4-ulose; subsequent epoxidation of the newly added 4-C-vinyl group, and simultaneous reduction of both epoxide groups with lithium aluminum hydride. These being methylated homologues of apiose, C-4 is not asymmetric, and hydrolysis of both produces the common, bicyclic anhydride **105**. Addition of 2-(1,3 dithianyl)lithium to the starting glycosid-4-ulose was followed, in the 3,4-*trans* product, by attack of the O-4 anion at C-3 to give a 3,4-epoxide.[196b]

The preparation of blastmycinone[137] and its stereoisomers was accomplished by opening of methyl 2,3-anhydro-4,6-O-benzylidene-α-D-mannopyranoside with butylmagnesium bromide to afford a 3-C-butyl-3-deoxy-D-*altro* product. This type of oxirane ring-opening provides a general route to derivatives having a hydrogen atom at the branch point.[197] Glycol cleavage of the 1,2-bond produced the 2-C-butyl-2-deoxypentonate skeleton, and conventional reactions were used to deoxygenate at C-5 and invert or retain stereochemistry at C-2 and C-4. 2-C-Butyl-2,5-dideoxy-L-arabinono-1,4-lactone was found to correspond to the (racemic) natural product.[137] Other routes to derivatives of this class include addition of carbon electrophiles to glycals,[197a] displacement of acyloxyl groups vicinal to nitro groups by carbon nucleophiles,[197b] and condensation of aldehydes with 1,3-dioxaphospholens.[197c]

2. Preparations from Noncarbohydrate Precursors

Apiose,[198,199] mycarose,[200-202] cladinose,[201] and epimycarose[200-202] have been synthesized by this approach. Raphael and Roxburgh reported the first synthesis of a naturally occurring branched-chain sugar, preparing DL-apiose in seven steps from the intermediate **88** obtained from reaction of diethyl malonate and 2-bromoacetaldehyde diethyl acetal.[198] (A similar route gave the sugar thought at the time to be cordycepose.) The ester-acid chloride derivative (**89**) of L-threaric acid may be converted into the keto intermediate **90**, which with diazomethane and subsequent hydrolysis furnished the branched-chain sugar acid **91**, which undergoes Ruff degradation to yield L-apiose.[199]

Several syntheses of mycarose (and epimycarose) have been reported, all based on the stereospecific hydroxylation of a suitable alkene. In one,[200] the unsaturated lactone (**92**) prepared from 2-hydroxybutanal and ketene was converted into the unsaturated sugar (**93**) by reduction with lithium aluminum hydride and subsequent acetylation. On *cis*-hydroxylation from the less-hindered side of the double bond, by action of iodine and silver acetate,

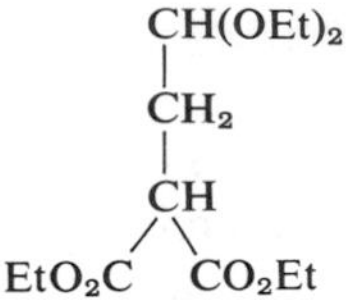

$$CH(OEt)_2$$
$$CH_2$$
$$CH$$
$$EtO_2C \quad CO_2Et$$

88

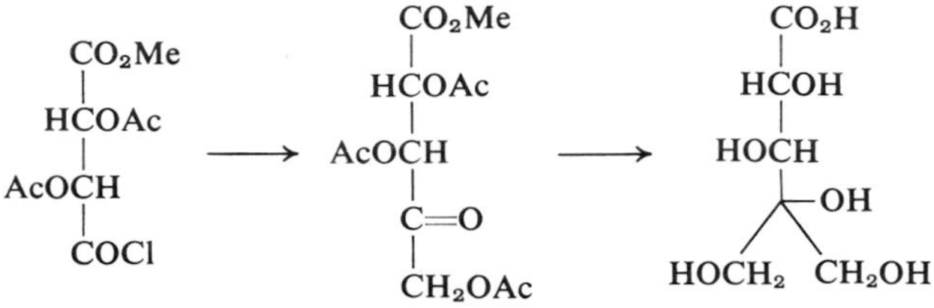

89   **90**   **91**

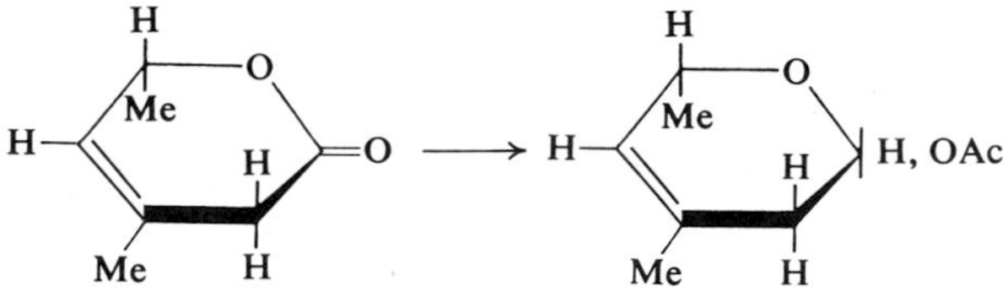

92 (racemic)   **93**

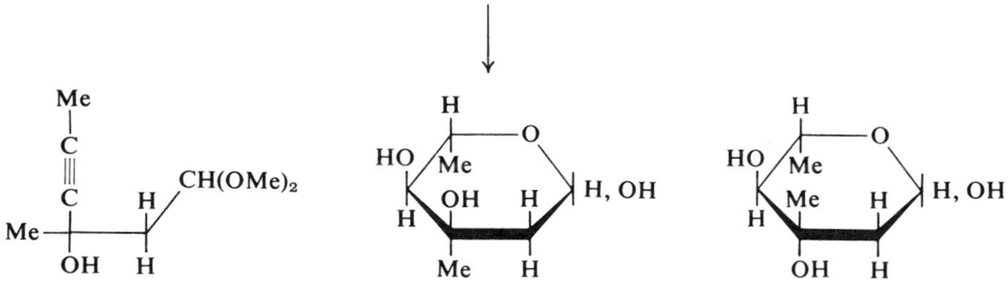

96   **94**   **95**

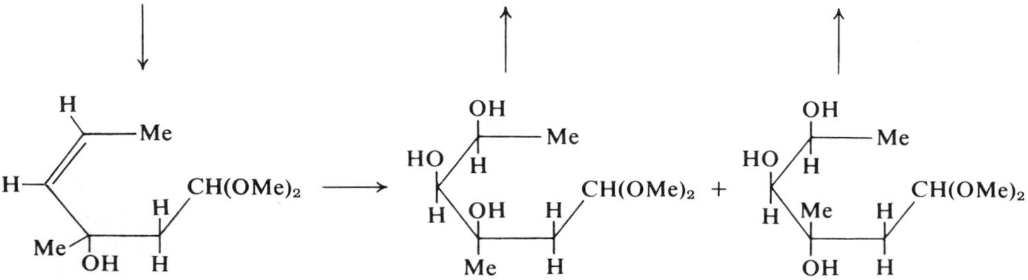

97

followed by alkaline hydrolysis, the 3,4-diol **94** was obtained, which proved to be DL-mycarose. Formation of the *trans*-diol from the olefins via epoxidation with a peroxyacid and acidic hydrolysis led to DL-epimycarose (**95**). In an alternative procedure,[201] the acetylenic alcohol **96** was stereospecifically reduced to the *cis*-olefinic alcohol **97**, which was then *cis*-hydroxylated with alkaline permanganate to give a mixture of DL-mycarose and DL-epimycarose. In the latter synthesis, the mixture was separated by using the 4-*O*-*p*-tolyl-sulfonyl derivative of mycarose, which could be crystallized, and the 3,5-*O*-isopropylidene furanoside derivative **98**, which only epimycarose can form, thus demonstrating its configuration as a 2,6-dideoxy-*arabino*-hexose. The DL-mycarose was resolved via the 4-*O*-(+)-bornanol-10-sulfonate of the methyl glycoside, furnishing the natural L-mycarose.[201] These workers also converted mycarose into its 3-methyl ether (*cladinose*).

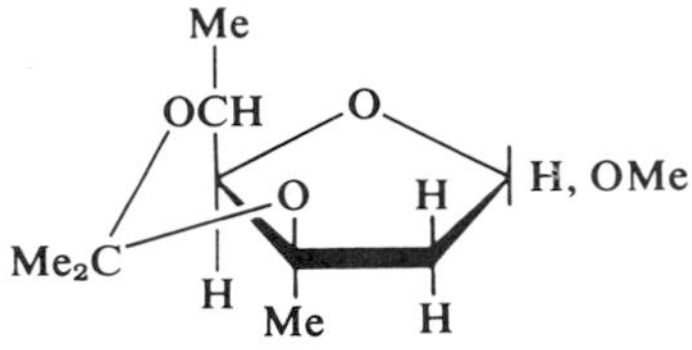

98

In a somewhat similar synthesis, DL-mycarose and epimycarose have been prepared from the unsaturated hydroxy ester **99** by epoxidation and acidic hydrolysis to the lactone **100**, obtained as a mixture of isomers. This mixture, on alkylborane reduction, gave a mixture of mycarose and epimycarose, which was separated by paper chromatography.[202]

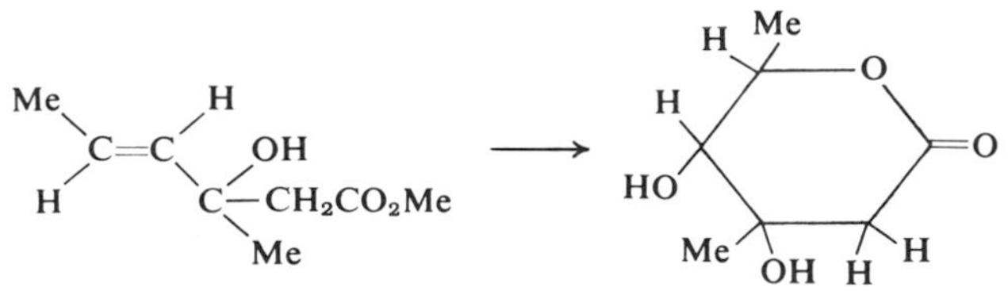

99   **100**

C. PROPERTIES

The chemistry of branched-chain sugars has not been extensively studied. In general, they exhibit properties similar to those of simple unbranched sugars, but, depending on the nature of the branching group and, more particularly, its location relative to the glycosidic carbon atom, some interesting differences may arise.

Sugars branching at C-2 cannot form osazones, and these sugars are also much more resistant to decomposition by both acid and alkali, presumably because they cannot tautomerize to an enediol intermediate.

The tertiary hydroxyl group at the branching carbon atom is usually found to resist esterification by the usual procedures with acid chloride or anhydride in pyridine, although unstable p-toluenesulfonates have been formed under forcing conditions with the sodium derivative and the acid chloride in ether solvents. By contrast, acid-catalyzed acetylation has been found to proceed readily.[203] The tertiary hydroxyl group has also been found to be less reactive than a secondary one as a member of a *cis*-diol system in forming acetals, and the failure of methyl mycaroside (**101**) to form an isopropylidene acetal

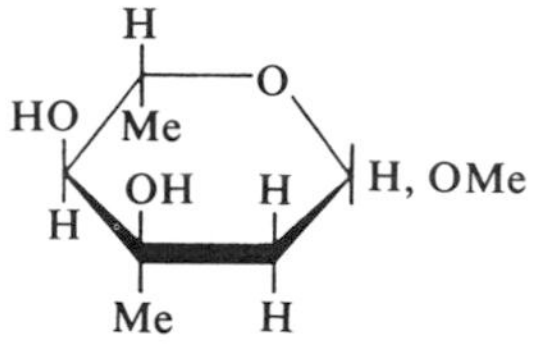

101

initially suggested that the hydroxyl groups at C-3 and C-4 were *trans*, not *cis*, related.[204] The reason for this difference would appear to be the added steric strain of forming a bicyclic system involving the tertiary carbon atom, since the tertiary alcohol at C-5 in noviose readily participates in the intramolecular hemiacetal link with the aldehyde group, and more generally it forms ethers without undue difficulty. Reaction of tertiary alcohols with thionyl chloride in pyridine causes dehydration. This reaction may be useful in assigning configurations to such centers, as endocyclic, *trans*-elimination is reported[205] to occur preferentially, exocyclic unsaturation arising only in those examples in which no vicinal proton is present in a *trans* relationship to the tertiary hydroxyl group. The branched-chain sugar, aldgarose[165] (**82**) is very interesting in having a cyclic-carbonate ester group which includes the tertiary hydroxyl group.

The side-chain hydroxyl group in sugars containing hydroxyalkyl chain branches at C-3 or C-4 can participate in the formation of acetals with the aldehyde group as readily as the main-chain hydroxyl groups. In the case of apiose, two diastereoisomeric furanose derivatives are therefore possible, **102**

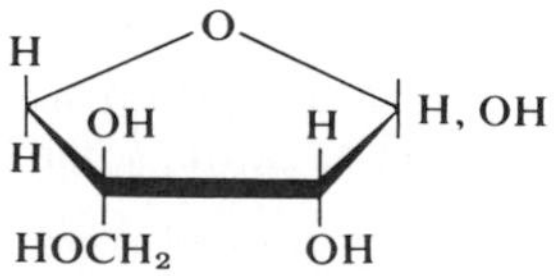

102 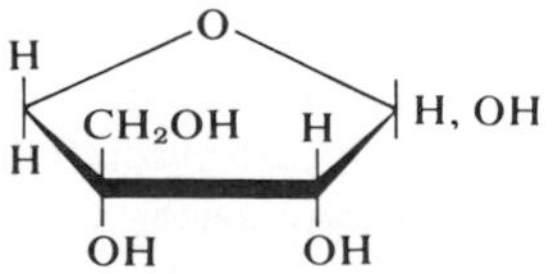   **103**

and **103**, the branching carbon atom becoming asymmetric on ring formation, and this complicates its chemistry; for example, acetonation of apiose leads to the formation of two isomeric 1,2:3,1'-di-*O*-isopropylidene derivatives, both of which can be isolated.[133,206] The 2,6-dideoxy-4-*C*-(1'-hydroxyethyl)-hexose **104** occurs as a glycoside in the intact antibiotic from *Streptomyces aureofaciens*, but during isolation involving acidic hydrolysis, the free sugar forms the interesting bicyclic acetal **105**, which is nonreducing, being a stable bis(pyranoside).[164]

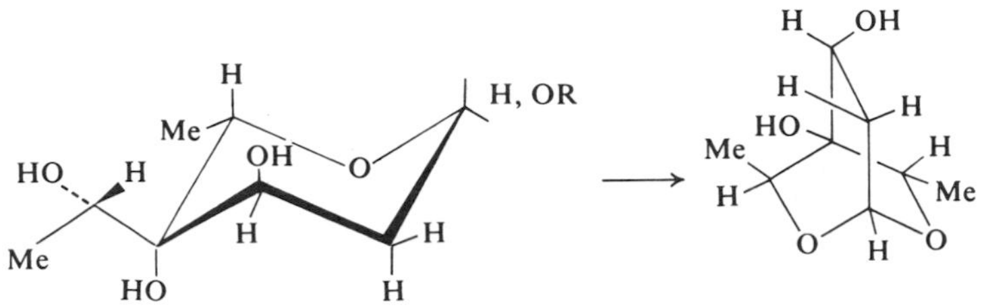

104   **105**

REFERENCES

1. J. Staněk, M. Černý, J. Kocourek, and J. Pacák, "The Monosaccharides," Academic Press, New York, 1963, pp. 83, 401.
2. L. Hough and A. C. Richardson, "Rodd's Chemistry of Carbon Compounds," 2nd Edition (S. Coffey, ed.), Vol. 1F, Elsevier, Amsterdam, 1967, pp. 500, 529.
3. W. G. Overend and M. Stacey, *Advan. Carbohyd. Chem.*, **8**, 45 (1953).
4. S. Hanessian, *Advan. Carbohyd. Chem.*, **21**, 143 (1966); R. F. Butterworth and S. Hanessian, *Advan. Carbohyd. Chem. Biochem.*, **26**, 279 (1971).
5. F. Shafizadeh, *Advan. Carbohyd. Chem.*, **11**, 263 (1956).
6. J. S. Brimacombe, *Angew. Chem.*, **83**, 261 (1971).
7. H. Grisebach and R. Schmid, *Angew. Chem.*, **84**, 192 (1972); see also H. Grisebach, *Advan. Carbohyd. Chem. Biochem.*, **35**, 81 (1978) for a discussion of the biosynthesis of these sugars.
8. H. Paulsen, *Staerke*, **25**, 389 (1973).
9. O. Westphal and O. Lüderitz, *Angew. Chem.*, **72**, 881 (1960).
10. T. Reichstein and E. Weiss, *Advan. Carbohyd. Chem.*, **17**, 65 (1962).
11. H. R. Bentley, K. G. Cunningham, and F. S. Spring, *J. Chem. Soc.*, 2301 (1951).
12. E. A. Kaczka, N. R. Trenner, B. Arison, R. W. Walker, and K. Folkers, *Biochem. Biophys. Res. Commun.*, **14**, 456 (1964).
13. J. Keleti, H. Mayer, I. Fromme, and O. Lüderitz, *Eur. J. Biochem.*, **16**, 284 (1970).
14. H. Achenbach, W. Regel, and W. Karl, *Ber.*, **108**, 2481 (1975).
15. A. M. Spichtig and A. Vasella, *Helv. Chim. Acta*, **54**, 1191 (1971).
16. J. S. Brimacombe, I. Da'Aboul, and L. C. N. Tucker, *J. Chem. Soc.* (*C*), 3762 (1971).
17. (a) W. D. Ollis, C. Smith, and D. E. Wright, *Chem. Commun.*, 881, 882 (1974); (b) W. D. Ollis, C. Smith, I. O. Sutherland, and D. E. Wright, *ibid.*, 350 (1976).
18. (a) A. K. Ganguly and A. K. Saksema, *Chem. Commun.*, 531 (1973); *J. Antibiot.*, **28**, 707 (1975); (b) A. K. Ganguly and S. Szmulewicz, *ibid.*, **28**, 710 (1975); (c) A. K. Ganguly, O. Z. Sarre, D. Greeves, and J. Morton, *J. Amer. Chem. Soc.*, **97**, 1982

(1975); (d) A. K. Ganguly, S. Szmulewicz, O. Z. Sarre, and V. M. Girijavallabhan, *Chem. Commun.*, 609 (1976).

19. W. D. Ollis, C. Smith, and D. E. Wright, *Chem. Commun.*, 348 (1976).
20. E. Percival and M. Young, *Carbohyd. Res.*, **32**, 195 (1974).
21. H. Björndal, B. Lindberg, and W. Nimmich, *Acta Chem. Scand.*, **24**, 3414 (1970).
22. J. Weckesser, H. Mayer, and G. Drews, *Eur. J. Biochem.*, **16**, 158 (1970).
23. C. Villé and M. Gastambide-Odier, *Carbohyd. Res.*, **12**, 97 (1970).
24. J. Weckesser, H. Mayer, and I. Fromme, *Biochem. J.*, **135**, 293 (1973).
25. H. B. Borén, K. Eklind, P. J. Garegg, B. Lindberg, and Å. Pilotti, *Acta Chem. Scand.*, **26**, 4143 (1972).
26. P. Studer, S. K. Pavanaram, C. R. Gavilanes, H. Linde, and K. Meyer, *Helv. Chim. Acta*, **46**, 23 (1963).
27. M. Miyamoto, Y. Kawamatsu, M. Shinohara, K. Nakanishi, Y. Nakadaira, and N. S. Bhacca, *Tetrahedron Lett.*, 2371 (1964).
28. Yu. A. Berlin, S. E. Esipov, M. N. Kolosov, M. M. Shemyakin, and M. G. Brazhnikova, *Tetrahedron Lett.*, 1323, 3513 (1964).
29. Yu. A. Berlin, M. N. Kosolov, and I. V. Yartseva, *Khim. Prir. Soedin.*, **9**, 539 (1973); *Chem. Abstr.*, **80**, 27, 439 (1974).
29a. F. W. Lichtenthaler and T. Kulikowski, *J. Org. Chem.*, **41**, 600 (1976).
30. R. N. Tursunova, V. A. Maslennikova, and N. K. Abubakirov, *Khim. Prir. Soedin.*, **11**, 171 (1975); *Chem. Abstr.*, **83**, 114, 803 (1975).
30a. R. N. Tursunova, V. A. Maslennikova, and N. K. Abubakirov, *Tezisy Dokl.— Vses. Simp. Bioorg. Khim.*, 9 (1975); *Chem. Abstr.*, **85**, 108, 933 (1976).
31. H. Els, W. D. Celmer, and K. Murai, *J. Amer. Chem. Soc.*, **80**, 3777 (1958).
32. M. Miyamoto, Y. Kawamatsu, M. Shinohara, Y. Asahi, Y. Nakadaira, H. Kakisawa, K. Nakanishi, and N. S. Bhacca, *Tetrahedron Lett.*, 693 (1963); see O. J. Varela, A. Fernandez Cirelli, and R. M. de Lederkremer, *Carbohyd. Res.*, **70**, 27 (1979) for a synthesis of ascarylose.
33. H. Brockmann, B. Scheffer, and C. Stein, *Tetrahedron Lett.*, 3699 (1973).
34. H. Brockmann and H. Greve, *Tetrahedron Lett.*, 831 (1975).
35. A. Zeeck, *Ann.*, 2079 (1975).
35a. N. P. Maksyutina, *Khim. Prir. Soedin.*, 603 (1975); *Chem. Abstr.*, **84**, 74, 570 (1976).
36. H. Takai, H. Taki, and K. Takiura, *Tetrahedron Lett.*, 3647 (1975).
36a. V. A. Maslennikova and N. K. Abubakirov, *Tezisy Dokl.—Vses. Simp. Bioorg. Khim.*, 10 (1975); *Chem. Abstr.*, **85**, 108, 934 (1976).
37. (a) C. Fouquey, J. Polonsky, and E. Lederer, *Bull. Soc. Chim. Biol.*, **39**, 101 (1957); (b) E. Lederer, *ibid.*, **42**, 1367 (1960); (c) Zh. M. Putneva, L. G. Mzhel'skaya, T. T. Gorovits, E. S. Kondratenko, and N. K. Abubakirov, *Tezisy Dokl.—Vses. Simp. Bioorg. Khim.*, 11 (1975); *Chem. Abstr.*, **85**, 108, 935 (1976).
38. (a) P. W. K. Woo, H. W. Dion, and Q. R. Bartz, *J. Amer. Chem. Soc.*, **83**, 3352 (1961); (b) P. W. K. Woo, H. W. Dion, and L. F. Johnson, *ibid.*, **84**, 1066 (1962).
39. E. Gäumann, R. Hütter, W. Keller-Schierlein, L. Neipp, V. Prelog, and H. Zähner, *Helv. Chim. Acta*, **43**, 601 (1960); W. Keller-Schierlein and G. Roncari, *ibid.*, **45**, 138 (1962).
40. C. L. Stevens, K. Nagarajan, and T. H. Haskell, *J. Org. Chem.*, **27**, 2991 (1962).
41. E. J. Czerwinski and L. K. Steinraug, *Biochem. Biophys. Res. Commun.*, **45**, 1284 (1971).
42. T. J. Petcher and H.-P. Weber, *Chem. Commun.*, 697 (1974).
43. N. D. Jones, M. O. Chaney, J. W. Chamberlin, R. L. Hamill, and S. Chen, *J. Amer. Chem. Soc.*, **95**, 3399 (1973).

43a. J. W. Chamberlin, *U.S. Publ. Pat. Appl. B*, 524,179 (1976); *Chem. Abstr.*, **84**, 150,931 (1976).

44. J. F. Blount, R. H. Evans, Jr., C. M. Lin, T. Hermann, and J. W. Westley, *Chem. Commun.*, 853 (1975).

45. H. Brockmann and T. Waehneldt, *Naturwissenschaften*, **50**, 43 (1963); J. K. Rinehart, Jr., and D. B. Borders, *J. Amer. Chem. Soc.*, **85**, 4037 (1963).

46. E. Fischer, M. Bergmann, and H. Schotte, *Ber.*, **53**, 509 (1920).

47. R. E. Deriaz, W. G. Overend, M. Stacey, and L. F. Wiggins, *J. Chem. Soc.*, 2836 (1949).

48. L. Vargha and J. Kuszmann, *Ber.*, **96**, 2016 (1963).

49. L. Vargha and J. Kuszmann, *Ber.*, **96**, 411 (1963).

50. J. Staněk and V. Schwarz, *Collect. Czech. Chem. Commun.*, **20**, 42 (1955).

51. R. U. Lemieux and S. Levine, *Can. J. Chem.*, **42**, 1473 (1964).

52. G. R. Inglis, J. C. P. Schwarz, and L. McLaren, *J. Chem. Soc.*, 1014 (1962).

53. P. T. Manolopoulos, M. Mednick, and N. N. Lichtin, *J. Amer. Chem. Soc.*, **84**, 2203 (1962).

54. S. Honda, K. Izumi, and K. Takiura, *Carbohyd. Res.*, **23**, 427 (1972).

55. D. Horton, J. M. Tarelli, and J. D. Wander, *Carbohyd. Res.*, **23**, 440 (1972).

55a. I. Lundt and C. Pedersen, *Acta Chem. Scand.*, *Ser. B*, **30**, 680 (1976).

56. M. Gut, D. A. Prins, and T. Reichstein, *Helv. Chim. Acta*, **30**, 743 (1947).

57. D. A. Prins, *J. Amer. Chem. Soc.*, **70**, 3955 (1948).

58. G. Casini and L. Goodman, *J. Amer. Chem. Soc.*, **85**, 235 (1963).

59. L. Goodman, *J. Amer. Chem. Soc.*, **86**, 4167 (1964).

60. For a review, see H. H. Baer, *Advan. Carbohyd. Chem. Biochem.*, **24**, 67 (1969).

61. J. C. Sowden and H. O. L. Fischer, *J. Amer. Chem. Soc.*, **69**, 1047 (1947).

62. W. W. Zorbach and A. P. Ollapally, *J. Org. Chem.*, **29**, 1790 (1964).

63. For a review, see Yu. A. Zhdanov, Yu. E. Alexseev, and V. G. Alexeeva, *Advan. Carbohyd. Chem. Biochem.*, **27**, 227 (1972).

64. Yu. A. Zhdanov and V. G. Alexeeva, *Carbohyd. Res.*, **10**, 184 (1969); J. M. J. Tronchet, E. Doelker, and B. Baehler, *Helv. Chim. Acta*, **52**, 308 (1969); M. F. Shostakovskii, N. N. Aseeva, and A. I. Polyakov, *Izv. Akad. Nauk SSSR, Ser. Khim.*, 892 (1970).

65. J. M. J. Tronchet, S. Jaccard-Thorndahl, and B. Baehler, *Helv. Chim. Acta*, **52**, 817 (1969).

65a. H. Zinner and J. Weber, *J. Prakt. Chem.*, **316**, 13 (1974).

66. A. M. Sepulchre, G. Vass, and S. D. Gero, *Tetrahedron Lett.*, 3619 (1973).

67. R. J. Bayly and J. C. Turner, *J. Chem. Soc. (C)*, 705 (1966).

68. D. C. C. Smith, *Chem. Ind.* (London), 92 (1955). See also E. Hardegger, *Methods Carbohyd. Chem.*, **1**, 177 (1962).

69. J. C. Sowden, *J. Amer. Chem. Soc.*, **76**, 3541 (1954); H. W. Diehl and H. G. Fletcher, Jr., *Arch. Biochem. Biophys.*, **78**, 386 (1958).

70. R. M. de Lederkremer and L. F. Sola, *Carbohyd. Res.*, **40**, 385 (1975); R. M. de Lederkremer, M. I. Litter, and L. F. Sola, *ibid.*, **36**, 185 (1974).

71. Y. Matsushima, *Sci. Papers Osaka Univ.* No. 35, 13 (1951); *Chem. Abstr.*, **46**, 7053 (1952).

72. M. L. Wolfrom and A. B. Foster, *J. Amer. Chem. Soc.*, **78**, 1399 (1956).

73. P. A. J. Gorin and J. K. N. Jones, *Nature*, **172**, 1051 (1953).

74. G. Rembarz, *Ber.*, **95**, 1565 (1962).

75. F. Weygand and H. Wolz, *Ber.*, **85**, 256 (1952).

76. H. Zinner, G. Wulf, and R. Heinatz, *Ber.*, **97**, 3536 (1964).

77. W. A. Szarek and A. Dmytraczenko, *Synthesis*, 579 (1974).
78. S. Jacobsen and C. Pedersen, *Acta Chem. Scand.*, **27**, 3111 (1973).
79. A. Klemer and G. Rodemeyer, *Ber.*, **107**, 2612 (1974).
79a. D. Horton and M. Sakata, *Carbohyd. Res.*, **48**, 41 (1976).
79b. V. B. Mochalin, A. N. Kornilov, I. S. Varpakhorskaya, and A. N. Vulf'son, *Zh. Org. Khim.*, **12**, 58 (1976).
80. S. Mukherjee and A. R. Todd, *J. Chem. Soc.*, 969 (1947).
81. P. W. Kent, M. Stacey, and L. F. Wiggins, *J. Chem. Soc.*, 1232 (1949).
82. E. Walton, R. F. Nutt, S. R. Jenkins, and F. W. Holly, *J. Amer. Chem. Soc.*, **86**, 2952 (1964).
83. M. Černý and J. Pacák, *Collect. Czech. Chem. Commun.*, **21**, 1003 (1956).
84. M. Černý, J. Pacák, and V. Jina, *Monatsh. Chem.*, **94**, 632 (1963).
85. R. H. Bell, D. Horton, and D. M. Williams, *Chem. Commun.*, **323** (1968); R. H. Bell, D. Horton, D. M. Williams, and E. Winter-Mihaly, *Carbohyd. Res.*, **58**, 109 (1977).
86. H. B. Wood and H. G. Fletcher, Jr., *J. Org. Chem.*, **26**, 1969 (1961).
87. D. H. Murray and J. Prokop, *J. Pharm. Sci.*, **54**, 1637 (1965).
87a. T. Okutani, T. Asako, K. Yoshioka, K. Hiraga, and M. Kida, *J. Amer. Chem. Soc.*, **99**, 1278 (1977).
88. E. J. Hedgley, W. G. Overend, and R. Rennie, *J. Chem. Soc.*, 4701 (1963).
89. M. Černý, J. Pacák, and J. Staněk, *Chem. Ind.* (London), 945 (1961).
90. P. W. Kent and P. F. V. Ward, *J. Chem. Soc.*, 416 (1953).
91. N. K. Kochetkov and A. I. Usov, *Tetrahedron Lett.* 519 (1963); *Izv. Akad. Nauk SSSR, Ser. Khim.*, 492 (1965).
92. B. T. Lawton, D. J. Ward, W. A. Szarek, and J. K. N. Jones, *Can. J. Chem.*, **47**, 2899 (1969).
93. B. T. Lawton, W. A. Szarek, and J. K. N. Jones, *Carbohyd. Res.*, **14**, 255 (1970).
94. H. Arita, K. Fukukawa, and E. Y. Matsushima, *Bull. Chem. Soc. Japan*, **45**, 3614 (1972).
95. R. G. Edwards, L. Hough, A. C. Richardson, and E. Tarelli, *Carbohyd. Res.*, **35**, 111 (1974).
96. J. Hill, L. Hough, and A. C. Richardson, *Proc. Chem. Soc.*, 314 (1963).
97. M. L. Wolfrom, K. Matsuda, F. Komitsky, Jr., and T. E. Whiteley, *J. Org. Chem.*, **28**, 3551 (1963); see also W. A. Szarek, R. G. S. Ritchie, and D. M. Vyas, *Carbohyd. Res.*, **62**, 89 (1978) for the use of iodine trifluoroacetate in net hydration of the 5,6-double bond in a derivative of **41**.
98. R. E. Gramera, T. R. Ingle, and R. L. Whistler, *J. Org. Chem.*, **29**, 2074 (1964).
99. E. J. Hedgley, O. Meresz, W. G. Overend, and R. Rennie, *Chem. Ind.* (London), 938 (1960).
100. E. J. Hedgley, O. Meresz, and W. G. Overend, *J. Chem. Soc.* (*C*), 888 (1967).
101. E. J. Reist, R. R. Spencer, and B. R. Baker, *J. Org. Chem.*, **23**, 1757 (1958). See ref. 102, footnote 11.
102. K. J. Ryan, H. Arzoumanian, E. M. Acton, and L. Goodman, *J. Amer. Chem. Soc.*, **86**, 2503 (1964).
103. J. A. Montgomery and K. Hewson, *J. Org. Chem.*, **29**, 3436 (1964).
104. A. M. Sepulchre, G. Lukacs, G. Vass, and S. D. Gero, *Angew. Chem.*, **84**, 111 (1972).
105. K. Eklind, P. J. Garegg, B. Lindberg, and Å. Pilotti, *Acta Chem. Scand.*, *Ser. B*, **28**, 260 (1974).
106. Yu. A. Zhdanov, Yu. E. Alekseev, and Kh. A. Kurdanov, *Zh. Obshch. Khim.*, **40**, 943 (1970).
107. N. K. Kochetkov and A. I. Usov, *Tetrahedron*, **19**, 973 (1963).

108. H. Kuzuhara, K. Sato, and S. Emoto, *Carbohyd. Res.*, **43**, 293 (1975).
109. S. Hanessian, *Carbohyd. Res.*, **2**, 86 (1966).
109a. A. Fogh, I. Lundt, and C. Pedersen, *Acta Chem. Scand.*, *Ser. B*, **30**, 624 (1976).
109b. A. A. Akhrem, G. V. Zaitseva, and I. A. Mihkailopulo, *Carbohyd. Res.*, **50**, 143 (1976).
110. Y. Kamiya and S. Takemura, *Chem. Pharm. Bull.* (Tokyo), **22**, 201 (1974).
111. S. Hanessian and N. R. Plessas, *J. Org. Chem.*, **34**, 1035 (1969).
112. L. M. Jeppssen, I. Lundt, and C. Pedersen, *Acta Chem. Scand.*, **27**, 3579 (1973).
113. M. L. Wolfrom and S. Hanessian, *J. Org. Chem.*, **27**, 1800 (1962).
114. K. J. Ryan, H. Arzoumanian, E. M. Acton, and L. Goodman, *J. Amer. Chem. Soc.*, **86**, 2497 (1964).
115. P. M. Collins and W. G. Overend, *J. Chem. Soc.*, 1912 (1965).
116. H. R. Bolliger and P. Ulrich, *Helv. Chim. Acta*, **35**, 93 (1952).
117. C. L. Stevens, P. Blumbergs, and D. L. Wood, *J. Amer. Chem. Soc.*, **85**, 3592 (1964).
118. A. H. Haines, *Carbohyd. Res.*, **21**, 99 (1972).
119. R. Knollmann and I. Dyong, *Ber.*, **108**, 2021 (1975).
120. D. Horton and W. Weckerle, *Carbohyd. Res.*, **44**, 227 (1975).
121. J. Pecka, J. Staněk, Jr., and M. Černý, *Collect. Czech. Chem. Commun.*, **39**, 1192 (1974).
122. J. Pecka and M. Černý, *Collect. Czech. Chem. Commun.*, **38**, 132 (1973).
123. G. Ekborg and S. Svensson, *Acta Chem. Scand.*, **27**, 1437 (1973).
124. G. N. Richards, *Chem. Ind.* (London), 228 (1955).
125. Z. Dische, *Methods Carbohyd. Chem.*, **1**, 503 (1962).
126. Z. Dische, *Methods Carbohyd. Chem.*, **1**, 501 (1962).
127. E. Vongerichten, *Ann. Chem.*, **318**, 121 (1901); C. G. Nordstrom, T. Swain, and A. J. Hamblin, *Chem. Ind.* (London), 85 (1953). See also C. S. Hudson, *Advan. Carbohyd. Chem.*, **4**, 57 (1949).
128. M. O. Farooq, S. R. Gupta, M. Kiamuddin, W. Rahman, and T. R. Seshadri, *J. Sci. Ind. Res.* (India), **12B**, 400 (1953).
129. A. Malhotra, V. V. S. Murti, and T. R. Seshadri, *Tetrahedron Lett.*, 3191 (1965).
130. D. J. Bell, F. A. Isherwood, and N. E. Hardwick, *J. Chem. Soc.*, 3702 (1954).
131. J. Crastil, *Chem. Listy*, **50**, 163 (1956); *Chem. Abstr.*, **50**, 4455 (1956).
132. A. D. Patrick, *Nature*, **178**, 216 (1956).
133. D. T. Williams and J. K. N. Jones, *Can. J. Chem.*, **42**, 69 (1964).
134. J. S. D. Bacon, *Biochem. J.*, **89**, 103P (1963).
135. R. B. Duff and K. M. Knight, *Biochem. J.*, **88**, 33P (1963).
135a. A. Tada, Y. Kaneiwa, J. Shoji, and S. Shibata, *Chem. Pharm. Bull.* (Tokyo), **23**, 2965 (1975).
136. E. Fischer and K. Freudenberg, *Ber.*, **45**, 2709 (1912); W. Mayer, W. Kanz, and F. Loebich, *Ann.*, **688**, 232 (1965); H. Gilck, A. Thainbichler, J. Sellmair, and E. Beck, *Carbohyd. Res.*, **39**, 161 (1975). See also ref. 5.
136a. T. Sassa, M. Togashi, and T. Kitaguchi, *Agr. Biol. Chem.*, **39**, 1735 (1975).
136b. T. Anthonsen, S. Hagen, M. A. Kazi, S. W. Shah, and S. Tagar, *Acta Chem. Scand.*, *Ser. B*, **30**, 91 (1976).
136c. D. Behr, J. Dahmen, and K. Leander, *Acta Chem. Scand.*, *Ser. B*, **30**, 309 (1976).
137. M. Kinoshita, S. Aburako, and N. Konishi, *Asahi Garasu Kogyo Gijutsu Shoreikai Kenkya Hokoku*, **25**, 103 (1974); *Chem. Abstr.*, **84**, 5268 (1975).
138. F. H. Stodola, O. L. Shotwell, A. M. Borud, R. G. Benedict, and A. C. Riley, Jr., *J. Amer. Chem. Soc.*, **73**, 2290 (1951). See also ref. 37b.
139. R. U. Lemieux and M. L. Wolfrom, *Advan. Carbohyd. Chem.*, **3**, 337 (1948); See also ref. 37b.
140. B. Bannister and A. D. Argoudelis, *J. Amer. Chem. Soc.*, **85**, 234 (1963).

796   NEIL R. WILLIAMS AND JOSEPH D. WANDER

141. T. Naito, *Peniskirin Sono Ta Koseibusshitsu*, **15**, 373 (1962).
142. P. F. Wiley, F. A. MacKellar, E. L. Caron, and R. B. Kelly, *Tetrahedron Lett.*, 663 (1968); P. F. Wiley, R. B. Kelly, E. L. Caron, V. H. Wiley, J. H. Johnson, F. A. MacKellar, and S. A. Mizsak, *J. Amer. Chem. Soc.*, **99**, 542 (1977).
143. P. F. Wiley, D. J. Duchamp, V. Hsiung, and C. C. Chidester, *J. Org. Chem.*, **36**, 2670 (1971).
144. D. Okuda, N. Suzuki, and S. Suzuki, *J. Biol. Chem.*, **242**, 958 (1967).
145. (a) J. W. Hinman, H. Hoeksema, E. L. Caron, and W. G. Jackson, *J. Amer. Chem. Soc.*, **78**, 1072 (1956); (b) C. H. Shunk, C. H. Stammer, E. A. Kaczka, E. Walton, C. F. Spenser, A. N. Wilson, J. W. Richter, F. W. Holly, and K. Folkers, *ibid.*, **78**, 1770 (1956); (c) E. Walton, J. O. Rodin, C. H. Stammer, F. W. Holly, and K. Folkers, *ibid.*, **80**, 5168 (1958).
146. A. K. Ganguly and O. Z. Sarre, *Chem. Commun.*, 1149 (1969).
147. A. K. Ganguly, O. Z. Sarre, and S. Szmulewicz, *Chem. Commun.*, 924 (1971).
148. Yu. A. Berlin, S. E. Esipov, M. N. Kolosov, M. M. Shemyakin, and M. G. Brazhnikova, *Tetrahedron Lett.*, 1323, 3513 (1964).
149. H. Umezawa, "Recent Advances in Chemistry and Biochemistry of Antibiotics," Microbial Chemistry Research Foundation, Tokyo, 1964, p. 18.
150. (a) P. P. Regna, F. A. Hochstein, R. L. Wagner, and R. B. Woodward, *J. Amer. Chem. Soc.*, **75**, 4625 (1953); (b) F. A. Hochstein and K. Murai, *ibid.*, **76**, 5080 (1954); (c) see R. B. Woodward, *Angew. Chem.*, **69**, 50 (1957).
151. F. W. Tanner, A. R. English, T. M. Lees, and J. B. Routien, *Antibiot. Chemother.*, **2**, 441 (1952).
152. R. Paul and S. Tchelitcheff, *Bull. Soc. Chim. Fr.*, 443 (1957), 734; 150 (1960).
153. R. Corbaz, L. Ettlinger, E. Gäumann, W. Keller-Schierlein, F. Kradolfer, F. Kyburz, L. Neipp, V. Prelog, A. Wettstein, and H. Zähner, *Helv. Chim. Acta*, **39**, 304 (1956).
154. R. L. Hamill, M. E. Haney, Jr., P. F. Wiley, and M. C. Stamper, *Antibiot. Chemother.*, **11**, 328 (1961).
155. R. Hütter, W. Keller-Schierlein, and H. Zähner, *Arch. Mikrobiol.*, **39**, 158 (1961).
156. T. Watanabe, N. Nishida, and K. Satake, *Bull. Chem. Soc. Japan*, **34**, 1285 (1961); *J. Biochem.* (Tokyo), **50**, 197 (1961).
157. L. W. Hofheinz and H. Grisebach, *Z. Naturforsch.*, **B17**, 852 (1962).
157a. J. Majer, J. R. Martin, R. S. Egan, and J. W. Corcoran, *J. Amer. Chem. Soc.*, **99**, 1620 (1977).
158. K. Kawahara, T. Yoshida, T. Watanabe, K. Miyauchi, B. Nomiga, S. Tada, and S. Kuwahara, *Chemother.* (Tokyo), **20**, 633 (1972).
159. A. Kinamaki, I. Takamori, Y. Sugawara, Y. Seki, M. Suzuki, and T. Okuda, *J. Antibiot.* **27**, 117 (1974); A. Kinumaki, I. Takamori, Y. Sugawara, M. Suzuki, and T. Okuda, *ibid.*, **27**, 107 (1974).
159a. M. Muroi, M. Izawa, and T. Kishi, *Chem. Pharm. Bull.* (Tokyo), **24**, 450,463 (1976).
160. P. F. Wiley and O. Weaver, *J. Amer. Chem. Soc.*, **77**, 3422 (1955).
161. F. Arcamone, G. Barbieri, G. Franceschi, S. Penco, and A. Vigevani, *J. Amer. Chem. Soc.*, **95**, 2008 (1973).
162. G. Roncari and W. Keller-Schierlein, *Helv. Chim. Acta*, **45**, 138 (1962); **47**, 78 (1964); **49**, 705 (1966).
163. U. Matern, H. Grisebach, W. Karl, and H. Achenbach, *Eur. J. Biochem.*, **29**, 1 (1972).
164. (a) J. S. Webb, R. W. Broschard, D. B. Cosulich, J. H. Mowat, and J. E. Lancaster, *J. Amer. Chem. Soc.*, **84**, 3183 (1962); (b) D. B. Cosulich, J. H. Mowat, R. W. Broschard, J. B. Patrick, and W. E. Meyer, *Tetrahedron Lett.*, 453 (1963); A. Tulinsky, *J. Amer. Chem. Soc.*, **86**, 5368 (1964).

165. M. P. Kunstmann, L. A. Mitscher, and N. Bohonos, *Tetrahedron Lett.*, 839 (1966); J. E. Lancaster, L. A. Mitscher, and G. Morton, *Tetrahedron*, **23**, 3893 (1967).

166. A. K. Ganguly, O. Z. Sarre, A. T. McPhail, and K. O. Onán, *Chem. Commun.*, 313 (1977).

166a. A. K. Ganguly, O. Z. Sarre, and H. Reimann, *J. Amer. Chem. Soc.*, **90**, 7129 (1968).

167. (a) M. Asai, *Chem. Pharm. Bull.* (Tokyo), **18**, 1713 (1970); (b) J. D. Pezzanite, J. Clardy, P. Y. Lau, G. Wood, D. L. Walker, and B. Fraser-Reid, *J. Amer. Chem. Soc.*, **97**, 6250 (1975); (c) D. L. Walker and B. Fraser-Reid, *ibid.*, **97**, 6251 (1975); (d) H. Paulsen, K. Roden, V. Sinnwell, and W. Koebernick, *Angew. Chem.*, **88**, 477 (1976); H. Paulsen and W. Koebernick, *Carbohyd. Res.*, **56**, 53 (1977).

168. S. Fredericksen, H. Malling, and K. Klenow, *Biochim. Biophys. Acta*, **95**, 189 (1965).

169. R. J. Suhadolnik and J. G. Cory, *Biochim. Biophys. Acta*, **91**, 661 (1964).

170. H. Paulsen and H. Redlich, *Ber.*, **107**, 2992 (1974).

171. See: J. D. Wander and D. Horton, *Advan. Carbohyd. Chem. Biochem.*, **33**, 15 (1976).

172. W. Depmeier, O. Jarchow, P. Stadler, V. Sinnwell, and H. Paulsen, *Carbohyd. Res.*, **34**, 219 (1974).

173. J. S. Brimacombe, C. W. Smith, and J. Minshall, *Tetrahedron Lett.*, 2997 (1974); J. S. Brimacombe, J. Minshall, and C. W. Smith, *J. Chem. Soc. Perkin I*, 682 (1975).

174. A. Khalique, *J. Chem. Soc.*, 2515 (1962).

175. P. A. J. Gorin and A. S. Perlin, *Can. J. Chem.*, **36**, 480 (1958).

176. R. Schaffer, *J. Amer. Chem. Soc.*, **81**, 5454 (1959).

177. A. D. Ezekiel, W. G. Overend, and N. R. Williams, *Tetrahedron Lett.*, 1635 (1969).

178. G. B. Howarth, W. A. Szarek, and J. K. N. Jones, *Carbohyd. Res.*, **7**, 284 (1968).

179. P. J. Garegg and T. Norberg, *Acta Chem. Scand.*, *Ser. B*, **29**, 507 (1975).

180. B. Flaherty, W. G. Overend, and N. R. Williams, *J. Chem. Soc.* (*C*), 398 (1966).

181. G. B. Howarth and J. K. N. Jones, *Can. J. Chem.*, **45**, 2253 (1967).

181a. M. Funabashi, N. Hong, H. Kodama, and J. Yoshimura, *Carbohyd. Res.*, **67**, 139 (1978); compare I. Dyong and D. Glittenberg, *Ber.*, **110**, 2721 (1977).

182. J. R. Dyer, W. E. McGonigal, and K. C. Rice, *J. Amer. Chem. Soc.*, **87**, 655 (1965).

183. W. G. Overend and N. R. Williams, *J. Chem. Soc.*, 3446 (1965).

184. H. Paulsen, V. Sinnwell, and P. Stadler, *Angew. Chem.*, **84**, 1100 (1972); *Ber.*, **105**, 1978 (1972).

185. J. S. Brimacombe and A. J. Rollins, *J. Chem. Soc. Perkin I*, 1568 (1974).

186. B. P. Vaterlaus, K. Foebel, J. Kiss, A. I. Rachlin, and H. Spiegelberg, *Experientia*, **19**, 383 (1963).

187. B. P. Vaterlaus, J. Kiss, and H. Spiegelberg, *Helv. Chim. Acta*, **47**, 381 (1964).

188. E. H. Williams, W. A. Szarek, and J. K. N. Jones, *Can. J. Chem.*, **47**, 4467 (1969).

189. M. Funabashi, S. Yamazaki, and J. Yoshimura, *Tetrahedron Lett.*, 4331 (1974); *Carbohyd. Res.*, **44**, 275 (1975).

190. J. S. Brimacombe, S. Mahmood, and A. J. Rollins, *J. Chem. Soc. Perkin I*, 1292 (1975).

190a. Yu. A. Zhdanov, Yu. E. Alexeev, and V. A. Tyumenev, *Dokl. Akad. Nauk SSSR*, **226**, 1334 (1976).

190b. J. Yoshimura, K. Mikami, K. Sato, and C. Shin, *Bull. Chem. Soc. Japan*, **49**, 1686 (1976).

191. W. P. Blackstock, C. C. Kuenzle, and C. H. Eugsker, *Helv. Chim. Acta*, **57**, 1003 (1974).

192. H. Paulsen and W. Stenzel, *Tetrahedron Lett.*, 25 (1974); *Ber.*, **107**, 3020 (1974).

192a. H. Paulsen and H. Redlich, *Ber.*, **107**, 2992 (1974).

193. D. C. Baker, D. K. Brown, D. Horton, and R. G. Nickol, *Carbohyd. Res.*, **32**, 299 (1974).

194. L. M. Utkin, *Dokl. Akad. Nauk SSSR*, **67**, 301 (1949).

195. J. Konigstein, D. Anderle, and F. Janecek, *Chem. Zvesti*, **28**, 701 (1974).

196. E. B. Rathbone and G. R. Woolard, *Carbohyd. Res.*, **46**, 183 (1976); H. C. Jarrell, W. A. Szarek, and J. K. N. Jones, *ibid.*, **64**, 283 (1978).

196a. W. A. Szarek, G. W. Schnarr, H. C. Jarrell, and J. K. N. Jones, *Carbohyd. Res.*, **53**, 101 (1977).

196b. H. Paulsen and V. Sinnwell, *Angew. Chem.*, **88**, 476 (1976).

197. See N. R. Williams, *Advan. Carbohyd. Chem. Biochem.*, **25**, 109 (1970).

197a. K. Heyns, R. Hohlweg, J. I. Park, and J. Thiem, *Tetrahedron Lett.*, 1481 (1976).

197b. T. Sakakibara, M. Yamada, and R. Sudoh, *J. Org. Chem.*, **41**, 737 (1976).

197c. S. David, M.-C. Lepine, G. Aranda, and G. Vass, *Chem. Commun.*, 747 (1976).

198. R. A. Raphael and C. M. Roxburgh, *J. Chem. Soc.*, 3405 (1955).

199. F. Weygand and R. Schmiechen, *Ber.*, **92**, 535 (1959).

200. F. Korte, U. Claussen, and K. Goehring, *Tetrahedron*, **18**, 1257 (1962).

201. D. M. Lemal, P. D. Pacht, and R. B. Woodward, *Tetrahedron*, **18**, 1275 (1962).

202. H. Grisebach, W. Hofheinz, and N. Doerr, *Ber.*, **96**, 1823 (1963).

203. B. R. Baker and D. H. Buss, *J. Org. Chem.*, **31**, 217 (1966).

204. A. B. Foster, T. D. Inch, J. Lehmann, L. F. Thomas, J. M. Webber, and J. A. Wyer, *Proc. Chem. Soc.*, 254 (1962).

205. T. D. Inch, G. J. Lewis, and N. E. Williams, *Carbohyd. Res.*, **19**, 17 (1971).

206. D. H. Ball, F. A. Carey, K. L. Klundt, and L. Long, Jr., *Carbohyd. Res.*, **10**, 121 (1969).

18. THIO SUGARS AND DERIVATIVES

DEREK HORTON AND JOSEPH D. WANDER*

I. INTRODUCTION

Carbohydrate derivatives in which one or more of the oxygen atoms bonded directly to the carbon skeleton have been replaced by sulfur are termed *thio sugars*; other group VIA elements may be substituted likewise, and examples of synthetic seleno sugars are known. A number of reviews have appeared describing aspects of this general subject. Classical studies (up to 1945) were summarized by Raymond[1] and, later, by Reid[2]; Horton and Hutson[3] analyzed developments in this area between 1945 and 1963 in the light of modern principles. The monograph by Staněk *et al.* includes discussions of thioglycosides, dithioacetals, and other thio sugars[4]; the first two topics are discussed in Vol. IA, Chapters 9,VI and 10,II. General preparative methods have been outlined for 1-thioaldoses,[5] 1-thioaldosides,[6] 4-thioaldofuranoses,[7] and 5-thioaldopyranoses.[8]

These group VIA homologues are isoelectronic (in their valence configuration) with the oxygenated sugars and exhibit similar reactivity under many different reaction conditions; replacement of oxygen by sulfur generally

* The authors thank Professor J. M. J. Tronchet for preliminary work on this chapter.

alters only slightly the conformational behavior of acyclic[9] or cyclic[10] derivatives. In the second-period and heavier elements, however, low-lying, vacant d-orbitals are available to afford stabilization to intermediates (especially carbanions) and to products in which the coordination number of the group VIA element exceeds 2; these factors, coupled with the general instability of multiple bonds formed by overlap between the small p-orbitals of carbon and the larger p-orbitals of elements beyond the first period, lead to many properties that are especially characteristic of thio sugars.

II. 1-THIO SUGARS

A. 1-THIOGLYCOSIDES OF ALDOSES

1. *General Considerations*

Most of the thio sugars found in Nature consist of a noncarbohydrate, organic molecule having a thiol group bearing a glycosyl substituent. Known examples of 1-thioglycosides are limited to those of D-glucose and D-glucuronic acid, and the sugar component appears to function as a protecting group to stabilize the (reactive) thiol compound, either for storage (via S-D-glucosides) or for detoxification (as the S-D-glucosyluronic acid derivatives).[10a]

Laboratory syntheses of 1-thioaldose derivatives have been accomplished in many configurational series. Those derivatives thereof having an alkyl or aryl group as the S-substituent are termed 1-thioglycosides; they do not normally undergo reactions involving cleavage of the sulfur–aglycon bond. Other 1-thioaldose derivatives, in which the S–aglycon bond is labile to scission, are discussed in Section II,B.

2. *Naturally Occurring 1-Thioglycosides*

Complex glycosides of 1-thio-D-glucose occur most commonly in the multitudinous[11] mustard-oil glucosides (*glucosinolates*) of the Cruciferae, Capparidaceae, and Resedaceae plant families, and examples have been known[12] since 1831. These compounds have been reviewed by Raymond,[1] Zinner,[13] Kjær,[11] Horton and Hutson,[3] and VanEtten *et al.*[14] The D-glucoside *sinigrin*, isolated from the seed of black mustard [*Brassica nigra* (L.) Koch], is typical of this group of substances; it is the potassium salt of an acid (myronic acid), and it undergoes hydrolysis[15] by the enzyme myrosinase[16] (EC 3.2.3.1), a thio-D-glucosidase present in special cells in the seed, to give D-glucose, hydrogensulfate ion, and an isothiocyanate (in this case, allyl isothiocyanate). On the basis of this and related evidence, Gadamer[15] proposed structure **1** (R = CH_2=$CHCH_2$—) for sinigrin; the isolation of 1-thio-D-glucose by treatment of sinigrin with potassium methoxide[17] supported this assignment.

Glucotropaeolin reacts similarly with myrosinase to give benzyl isothiocyanate and was thus formulated as **1** (R = $PhCH_2$—); gluconasturtiin was considered to have structure **1** (R = $PhCH_2CH_2$—) by similar reasoning. Gadamer's formulation (**1**) was, however, inconsistent with evidence from direct chemical degradation,[18] which gave not the amines (RNH_2) expected on the basis of formula **1**, but nitriles (RCN) or carboxylic acids (RCO_2H) having the same number of carbon atoms as the isothiocyanate (RNCS) produced enzymically. Finally, hydrogenolysis of sinigrin in the presence of Raney nickel afforded butylamine (RCH_2NH_2), and, in 1956, Ettlinger and Lundeen[19] proposed the currently accepted, general formulation **2** for mustard-oil D-glucosides.

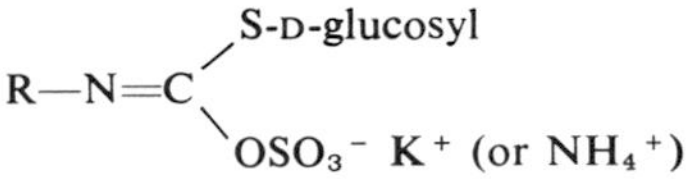

$$R-N=C\begin{array}{c} \text{S-D-glucosyl} \\ \diagup \\ \diagdown \\ OSO_3^- \ K^+ \ (\text{or } NH_4^+) \end{array}$$

1

$$R-C\begin{array}{c} \text{S-}(\beta\text{-D-glucopyranosyl}) \\ \diagup \\ \diagdown \\ N-OSO_3^- \end{array}$$

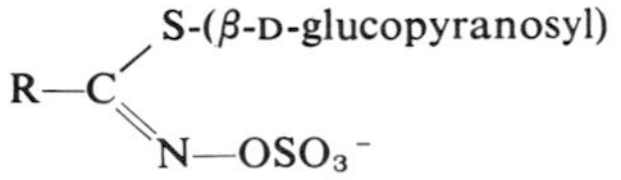

2

The same authors verified the general structure **2** by direct condensation of phenylacetothiohydroxamic acid (**3**) with tetra-*O*-acetyl-α-D-glucopyranosyl bromide to afford, after sulfation with pyridine–sulfur trioxide and subsequent saponification, glucotropaeolin (**4**), identical with the natural product.[20]

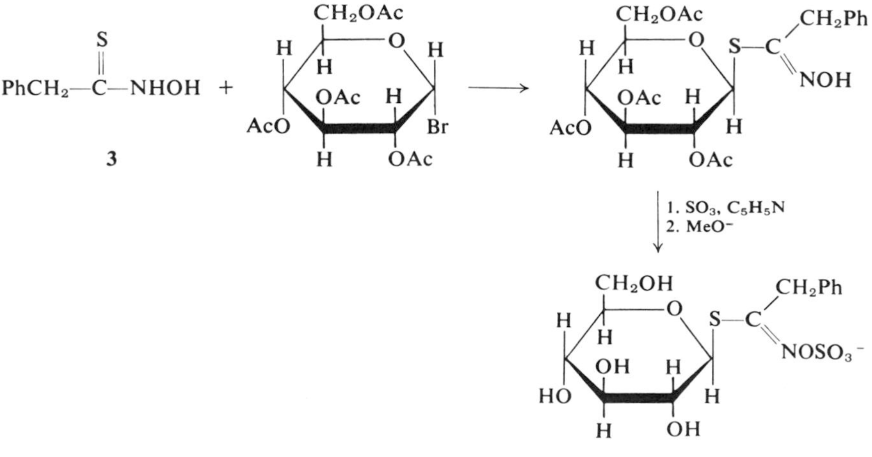

A second synthesis, reported by Jensen and Kjær,[21] coupled the racemic chloro-oxime (**5**) with 2,3,4,6-tetra-*O*-acetyl-1-thio-β-D-glucopyranose (**6**) to afford, after saponification and sulfation, a mixture of progoitrin (**7**, found in rutabaga,[22] kohlrabi,[23] and turnips[22]) and epiprogoitrin (**8**, found[24] in *Crambe abyssinica*). Benn and Yelland[25] prepared glucocochlearin [**2**, R = (*S*)-2-butyl] by nucleophilic addition of **6** to (*S*)-2-methylbutanonitrile oxide, obtained from L-isoleucine (via decarboxylation and oximation); the latter is presumed to be the biogenetic precursor also.

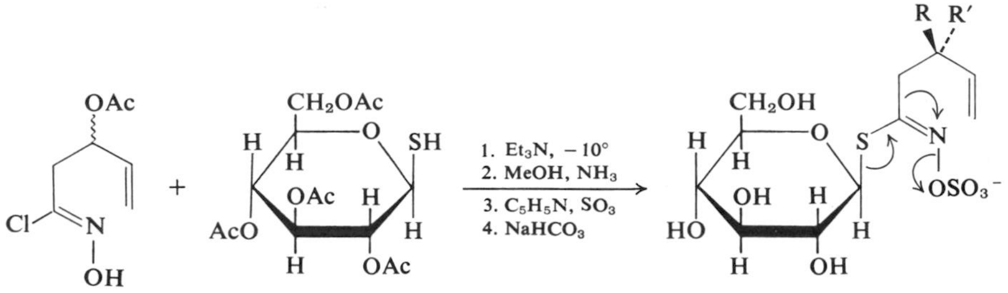

7 R = H, R' = OH
8 R = OH, R' = H

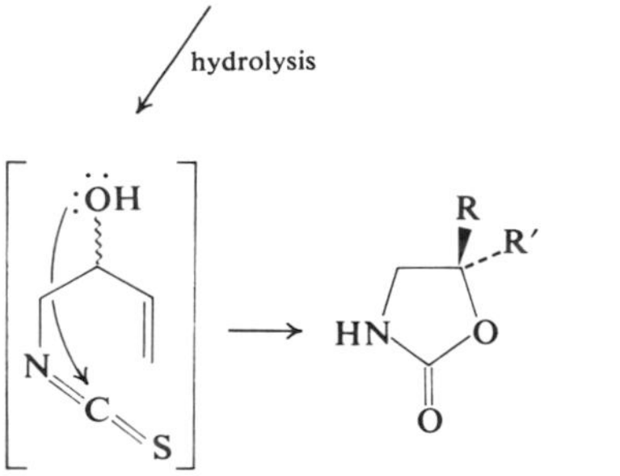

9

10 R = —CH=CH₂, R' = H
11 R = H, R' = —CH=CH₂

Confusion in the early formulation of the general structure arose from isolation of products formed in the Lossen type of rearrangement[26] accompanying enzymic or mild chemical cleavage of the glycosidic bond. Further reactions, such as the cyclization of the hydroxyisothiocyanate (**9**) to the oxazolidinethiones (**10, 11**), complicate the identification of aglycons having additional, nucleophilic substituents; the latter conversion has agricultural significance as well, because the products, goitrin (**10**, formed from **7**) and epigoitrin (**11**, from **8**), (as well as more-toxic nitriles[14]) are formed during storage, processing, or digestion of several plants of the Cruciferae family. Studies on laboratory animals and human subjects have demonstrated[27] an irreversible inhibition of iodine absorption by the thyroid gland in the presence of **10** or **11** (which are equally potent), so that the possible utilization of these protein-rich seed meals as feedstocks (after extraction of the oils) awaits implementation either of suitable extraction or deactivation procedures or of botanical modifications.

The results of a study[28] of the thioglucosidase of the seed of yellow mustard (*Sinapis alba*) indicated that there are at least two enzymes in the extract, both catalyzing the same reaction; one enzyme (myrosinase) is indifferent to L-ascorbic acid, whereas the other requires L-ascorbic acid as a coenzyme.[29] It was shown by the ascorbate activation that this enzyme is a specific thioglucosidase, and that the Lossen rearrangement is nonenzymic. Reese and co-workers[30] screened 300 microorganisms for thioglucosidase (sinigrinase) activity and found it to be of very rare occurrence.[31] One fungal source, *Aspergillus sydowi* QM31c, was found to produce a high sinigrinase activity, the properties of which are similar to those of myrosinase. Intestinal microorganisms, such as *Paracolobacterium*,[27,32] liberate **10** (or **11**) from **7** (or **8**). These thioglycosidases are without effect on simple alkyl or aryl 1-thioglycosides, although myrosinase will cleave[33] 6-(β-D-glucopyranosylthio)-9*H*-purine and a number of related analogues.[34] Derivatives of this type are also cleaved by a thioglucosidase found[33] widely distributed in the tissues of mammalian species, although this mammalian thioglucosidase is unable to cleave simple alkyl or aryl 1-thioglycosides. The β-D-glucosidase from sweet almonds is likewise unable to effect hydrolysis of phenyl 1-thio-β-D-glucopyranoside.[34a] Although methyl 1-thio-β-D-galactopyranoside is not hydrolyzed by the β-D-galactosidase of *Escherichia coli* ML, it acts to induce[34b] mobilization of this enzyme; this phenomenon was later interpreted in the context of a fundamental model advanced by Monod and co-workers[34c] to describe the general phenomenon of allosterism in enzymes.

1-Thio-β-D-glucopyranosiduronic acids have been isolated from human and mouse urine subsequent to the administration of thiophenol,[35] benzothiazole-2-thiol,[36] 9-(alkylmercapto)purines,[37] and *N,N*-diethyldithiocarbamic acid[38,39]; the same reaction is observed in plant[40] and animal tissue[39] preparations treated with *N,N*-diethyldithiocarbamic acid. Studies[41] made on intact mouse-liver tissue and on homogenates have established that substituted thiophenols and *N,N*-diethyldithiocarbamic acid are acted upon by an enzyme that transfers a D-glucopyranosyluronic acid residue from uridine 5'-(α-D-glucopyranosyluronic acid pyrophosphate) (UDP-D-glucuronate) to the thiol; an analogous transfer reaction from uridine 5'-(α-D-glucopyranosyl pyrophosphate) (UDP-D-glucose) to convert thiophenol or the carbamate into 1-thio-β-D-glucopyranosides was demonstrated in digestive-tissue homo-

References start on p. 833.

genates and in intact cells of the mollusc *Arion ater*. A β-D-glucopyran-osiduronase from rat preputial gland hydrolyzed the *p*-nitro- and *o*-aminoaryl thioglycosides, but exerted no effect on phenyl 1-thio-β-D-glucopyran-osiduronic acid; the two thioglucosides were not affected by almond emulsin β-D-glucosidase.[41]

3. *Chemical Synthesis of 1-Thioglycosides*

Methods of preparation of 1-thioglycosides (as well as 2-thioglycosides of 2-ketoses) are outlined in refs. 1, 3, 4, and 6. The common methods—namely, direct, acid-catalyzed condensation of a sugar or derivative with a thiol, and partial demercaptalation of a dithioacetal—are discussed in Vol. IA, Chapter 9, Sections VI,A,1 and 2.

The first preparation of a 1-thioglycoside, phenyl 1-thio-β-D-glucopyran-oside, was accomplished (according to principles outlined in Vol. IA, Chapter 7, Section II,C) through replacement of the halide in tetra-*O*-acetyl-α-D-glucopyranosyl bromide by the thiophenoxide anion.[42] Similar condensations have been effected with methyl 2,3,4-tri-*O*-acetyl-1-bromo-1-deoxy-α-D-glucopyranuronate, as in the glycosidation of an antineoplastic nitrogen mustard, 4-[bis(2-chloroethyl)amino]thiophenol (which was undertaken in an effort to moderate the toxicity of the thiol[43]), and with 2-amino-3,4,6-tri-*O*-benzoyl-2-deoxy-α-D-glucopyranosyl bromide[44] and the corresponding 3,4,6-triacetate,[45] which undergo halide replacement by neutral thiols with net inversion of configuration.

Bicyclic 1,2-acetoxonium ions, postulated as intermediates in reactions of *trans*-2-acylglycosyl halides[46,47] and in the Lewis acid-catalyzed conversion of glycopyranose peracetates into the corresponding 1,2-*trans*-thioglycoside peracetates,[48] are presumably also involved in displacement reactions of glycosyl halides with thiols or thioxides conducted in ionizing solvents, as the products isolated generally have the 1,2-*trans* configuration.[3] In poorly ionizing solvents, however, kinetic studies have shown that such strong nucleophiles as thiophenoxide ion react bimolecularly with 2,3,4,6-tetra-*O*-methyl-α-D-glucopyranosyl chloride to afford the 1-thio-β-D-glucoside, the product of inversion at C-1, whereas the corresponding α-D-*manno*-halide still reacts by a unimolecular process and the product is the 1-thio-α-D-glycoside.[49] A subsequent study of the reaction of a number of per-*O*-acetylaldopyranosyl bromides with the same anion in 1-pentanol–toluene likewise revealed second-order kinetics, except in the reaction of 2,3,4,6-tetra-*O*-acetyl-α-D-manno-pyranosyl bromide.[50] These observations were rationalized in terms of a transient, intermediate structure, **12**, in which approach of the nucleophile completes the separation of the axially disposed halide from C-1, thus effecting inversion. The different behavior of the *manno* derivatives was attributed to steric crowding in the approach of the nucleophile, because of the axially attached substituent on C-2 in structure **13**, which would require virtually complete separation of the halide ion (presumably inviting participation by the 2-acetoxyl group) before the nucleophile could gain access to the anomeric carbon atom.

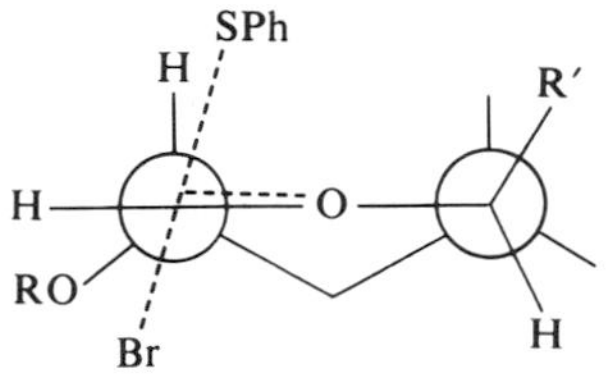

12 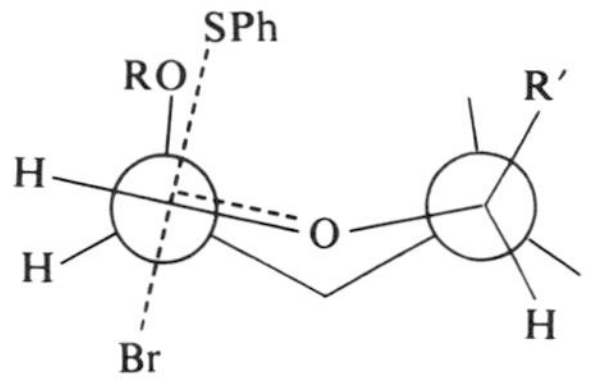   **13**

3,5-Di-O-benzoyl-α,β-D-arabinofuranosyl chloride reacts with the ethane-thioxide ion to afford only the α-thioglycoside,[51] in accordance with the *trans* rule. This exclusive product could arise by direct displacement from the 1,2-*cis* (β) isomer and by participation of the free 2-hydroxyl group of the 1,2-*trans* (α) chloride, with subsequent opening of the epoxide ring by the thiol to effect net retention of configuration. Boron trifluoride catalyzes anomerization of alkyl 1-thio-β-D-glucopyranoside peracetates.[51a] Reaction of acetohalogen sugars with p-thiocresol in nitromethane containing 2,4,6-trimethylpyridine affords a thio ortho ester,[51b] whence an alkyl ortho ester[51c] or the 1,2-*trans* p-tolyl 1-thioglycoside[51d] are obtained by treatment with Raney nickel in an alcohol solvent or with aged catalyst in the presence p-thiocresol, respectively; this suggests an initial, ionic dissociation of the C–S bond in at least some reductive desulfurization reactions.

The nonsymmetrical, sulfur-bridged disaccharide 6-S-β-D-glucopyranosyl-6-thio-D-glucose was prepared by the base-catalyzed reaction of methyl 6-thio-α-D-glucopyranoside with tetra-O-acetyl-α-D-glucopyranosyl bromide and subsequent removal of the protecting groups.[52] 2,3,4,6-Tetra-O-acetyl-1-thio-β-D-glucopyranose reacts with methyl 6-O-p-tolylsulfonyl-α-D-glucopyranoside in the presence of sodium carbonate in refluxing aqueous acetone to afford, after deprotection, the same disaccharide, although in inferior yield.[52] The use of anomeric thiols as nucleophiles is discussed in Vol. IA, Chapter 9, Section VI,A,4.

Electrophilic addition of glycosylsulfenyl bromides (Gl—S—Br) to alkene systems or to activated aromatic molecules—for example, the addition of 2,3,4,6-tetra-O-acetyl-β-D-glucopyranosylsulfenyl bromide (**15**) to cyclo-hexene or N,N-dimethylaniline to afford *trans*-2-bromocyclohexyl and 4-(dimethylamino)phenyl 1-thio-β-D-glucopyranoside (**14** and **16**, respectively)—has been demonstrated.[53] Alkanesulfenyl bromides are generated during the conversion of protected dithioacetals into 1-bromo-1-thioalditols by direct

---

bromination,[54] and cyclohexene, which may be added to the reaction mixture to decompose the excess of bromine, readily undergoes addition to form a racemic *trans*-2-bromocyclohexyl alkyl sulfide; the latter may compete with the 1-bromo-1-thioalditol in further reactions.[55]

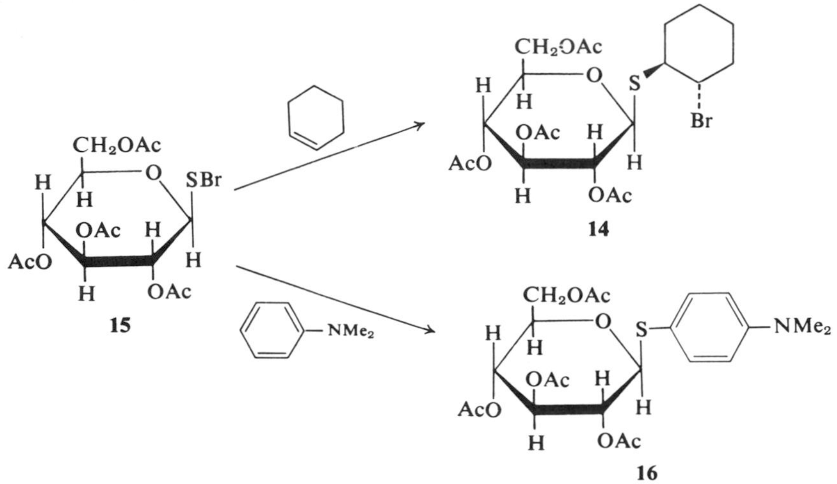

B. 1-THIOALDOSES

4. *Reactions of 1-Thioglycosides*

The conventional reactions of 1-thioglycosides (acid-catalyzed hydrolysis and replacement reactions, oxidation of the alkylthio group to the corresponding sulfone, and reductive desulfurization) are outlined in Vol. IA, Chapter 9,VI,B,1–4, and discussed at length by Horton and Hutson.[3] Methyl (dichloromethyl) ether was subsequently demonstrated[55a] to be a suitable, alternative reagent for effecting halogenolysis of anomeric thioalkyl groups in per-*O*-acyl derivatives to afford the corresponding acetochloro sugars.

Collins and Whitton subjected tetra-*O*-acetyl-D-glucofuranosyl-[55b] and -pyranosyl phenyl sulfone[55c] to photolysis in benzene, obtaining the corresponding (acylated) anhydroalditols, bis(anhydro-1-deoxyalditol-1-yl) derivatives, and 1-*C*-arylanhydroalditols, whence they concluded that the reaction proceeds by homolysis of carbon–sulfur bonds and subsequent recombination of the glycosyl (and aryl) free radicals.

1. *General Considerations*

The free 1-thioaldoses exhibit behavior typical both of reducing sugars [mutarotation (albeit very slowly[55d]) in hydroxylic solvents, reduction of

Fehling solution] and of thiols (oxidation to the disulfide by a limited amount of hydrogen peroxide[56] or iodine[57,58]). The color of unreacted iodine provides an indicator for the titrimetric determination of the free thiol content. Additionally, the disulfide bridge serves as a convenient protecting group because the thiol may be regenerated by reductive cleavage at any point in a synthetic sequence—for example, through the action of aluminum amalgam in acetic acid–ethanol[58-60]; the corresponding reduction by zinc in acetic acid–acetic anhydride affords the thiol protected as the S-acetyl derivative.[59]

1-Thio sugars readily form stable metal derivatives[56] by virtue of the weak acidity of the thiol group, and these derivatives are convenient for stabilization of the 1-thio sugars, which, in the free state, exhibit poor crystallizing properties.

The high nucleophilicity of the thiol group makes this group the favored point of attack in alkylation reactions.[56,59,61] Alkyl halides readily convert 1-thio sugars into 1-thioglycosides (see Vol. IA, Chapter 9, Section VI,A,4). Condensation of tetra-O-acetyl-α-D-glucopyranosyl bromide with 2,3,4,6-tetra-O-acetyl-1-thio-β-D-glucopyranose produces bis(tetra-O-acetyl-β-D-glucopyranosyl) sulfide.[61]

The reductive desulfurization procedure employing Raney nickel[62] can be used for replacing a C-1 thiol group by hydrogen[60]; the reaction is analogous to the desulfurization of 1-thioglycosides. The S-esterified 1-thio sugar derivatives react analogously; for example, 2,3,4,6-tetra-O-acetyl-β-D-glucopyranosyl ethylxanthate gives 2,3,4,6-tetra-O-acetyl-1,5-anhydro-D-glucitol (polygalitol tetraacetate).[63] Migration (S → N) of acyl groups[64,65] may complicate attempted hydrogenolysis of 1-S-acyl-2-amino sugars, as in the conversion of 2-(3,4,6-tri-O-acetyl-2-amino-2-deoxy-β-D-glucopyranosyl)-2-thiopseudourea hydrobromide into a product presumed to be 3,4,6-tri-O-acetyl-1,5-anhydro-2-deoxy-2-guanidino-D-glucitol by the action of Raney nickel.[66] S → O migration of an acetyl group has been reported[66a] to occur during chromatography on silica gel.

β-D-Mannopyranosyl ethylxanthate and β-D-xylopyranosyl ethylxanthate have been found[67] to be nontoxic inhibitors of Ehrlich ascites carcinoma cells.

2. Synthesis and Reactions

General routes for the preparation of 1-thioaldoses were discussed by Raymond,[1] and methods for the preparation of 1-thio-D-glucose have been reviewed.[5] Most of the synthetic procedures for 1-thioaldoses involve condensation of a poly-O-acylglycosyl halide with a suitable thiol derivative, to yield a fully substituted product from which the S-protecting group can be removed. The first such synthesis recorded[68] employed o-ethylbenzyl thio-imidocarbonate and tetra-O-acetyl-α-D-glucopyranosyl bromide; saponification of the initial product yielded 1-thio-D-glucose, isolated as the silver salt. Potassium disulfide,[59,69] potassium thiolacetate,[70] potassium ethyl-xanthate,[56,60] and thiourea[71] undergo similar condensation with protected glycosyl halides, usually affording products of the general structural types **17–20**, respectively; in acetone, however, potassium methyl- or benzyl-xanthate reacts[72] with tetra-O-acetyl-α-D-glucopyranosyl bromide to form the bis(tetra-O-acetyl-β-D-glucopyranosyl) sulfide in high yield.

$$2\ R\!-\!Hal + K_2S_2 \longrightarrow R\!-\!S\!-\!S\!-\!R \qquad \textbf{17}$$

$$R\!-\!Hal + KSAc \longrightarrow R\!-\!SAc \qquad \textbf{18}$$

$$R\!-\!Hal + KSC(\!=\!S)OEt \longrightarrow R\!-\!SC(\!=\!S)OEt \qquad \textbf{19}$$

$$R\!-\!Hal + (NH_2)_2C\!=\!S \longrightarrow R\!-\!S\!-\!C(NH_2)_2{}^+ \qquad \textbf{20}$$

Attack by either of the sulfur atoms in the bidentate nucleophile generates equivalent products in the formation of **17** and **19**, whereas the exceptionally high nucleophilicity of the sole sulfur atom militates against attack by the other heteroatoms in the preparation of **18** and **20**. The 1-thioaldose is obtained from **17** by reductive cleavage[58–60] of the S–S bond, and **18–20** undergo ready saponification to liberate the thiol group. Direct conversions of the thiol group are also possible, as in (*a*) the sodium iodide-catalyzed elimination of carbonyl sulfide from alkylxanthates to afford the corresponding thioglycosides,[73] (*b*) the condensation of 2-(2,3,4,6-tetra-O-acetyl-β-D-glucopyranosyl)-2-thio(or seleno)pseudourea hydrobromide with chloro-dimethylarsine to form the corresponding S- (or Se-)dimethylarsino deriva-tive,[74] and (*c*) the conversion of the former into trialkylphosphinatogold 1-thio-D-glucose derivatives by condensation with trialkylphosphinatogold(I) chloride.[75]

The amidino group in the salt (**20**) is particularly labile to bases, and it may be cleaved with potassium carbonate (or merely by heating a solution of the hydrogencarbonate salt[76]) without scission of the O-acyl groups. 2,3,4,6-Tetra-O-acetyl-1-thio-β-D-glucose[59] is readily prepared[77] by this method from 2-(tetra-O-acetyl-β-D-glucopyranosyl)-2-thiopseudourea hydrogen car-bonate. Hydrolysis with sodium pyrosulfite in a water–carbon tetrachloride system has been recommended[78] for large-scale reactions; as fast as the 2,3,4,6-tetra-O-acetyl-1-thio-D-glucose is formed, it is extracted into the organic layer, and subsequent side-reactions are thus minimized. Saponifica-tion of the resulting tetra-O-acetyl derivative with sodium methoxide gives 1-thio-D-glucose (in 90% yield) as its stable crystalline, sodium salt dihydrate.[56,78]

1-Thio-β-D-glucose and related analogues have been used for stabilization of aqueous solutions of L-ascorbic acid,[79] thiamine,[80] and streptomycin[81] to light and heat; 1-thio-D-glucose is reported[82] to degrade insulin. Radioactive-labeling studies have established that 1-thio-D-glucose is not an intermediate in the biosynthesis of glucosinolates (see Section II,A,2).[83]

Gold(I) 1-thio-D-glucopyranose[84] (*Solganol, aurothioglucose*), which can be prepared by treatment of the sodium salt with an equivalent of aurous iodide,[76] is physiologically active, and it has been examined as a prophylaxis against sunburn[85] and as an antiarthritic agent[75,86]; animal and tissue studies, mainly with rodents, have revealed severe toxic effects in the brain,[87] primarily in the ventromedial region of the hypothalamus,[88] which are associated with local loss of lactic acid dehydrogenase activity[89] and with development of obesity in the animal. The latter effect lends support to the theory that the affected region of the brain is concerned with the mechanism that restricts intake of food. Injection of the gold thiolate also induces progressive, histologic changes in the liver,[90] hemorrhagic ulcers of the stomach and duodenum,[91] degeneration of the myocardium,[92] and necrosis and atrophy of the kidney[93]; in human subjects, enhanced secretion of adrenocortical hormone is reportedly[94] stimulated by administration of the salt. Gold 1-thio-D-glucose has been examined as a potential stabilizing agent for photographic emulsions.[95]

Peracylated alkyl 1-thioglycopyranosides generally react with bromine or chlorine in inert solvents to give the corresponding acylated glycosyl halides.[3,96] In contrast, 1-thio-β-D-glucopyranose pentaacetate (21, R = Ac) reacts with bromine to give[97,98] the crystalline, unstable 2,3,4,6-tetra-*O*-acetyl-β-D-glucopyranosylsulfenyl bromide (15); this same product is also obtained when *tert*-butyl 2,3,4,6-tetra-*O*-acetyl-1-thio-β-D-glucopyranoside[98] is used. These differences in the course of reaction suggested that a bromo-sulfonium ion (22) is formed as an intermediate and that this subsequently decomposes to displace the more-stable cation and form either the alkyl (R = Me, Et, and so on) or glycosyl (R = Ac, CMe₃) sulfenyl bromide.[97,98] The sulfenyl bromide (15) reacts as an *S*-electrophile (see Section II,A,3) and decomposes in the presence of an excess of bromine or protonic solvents.[98]

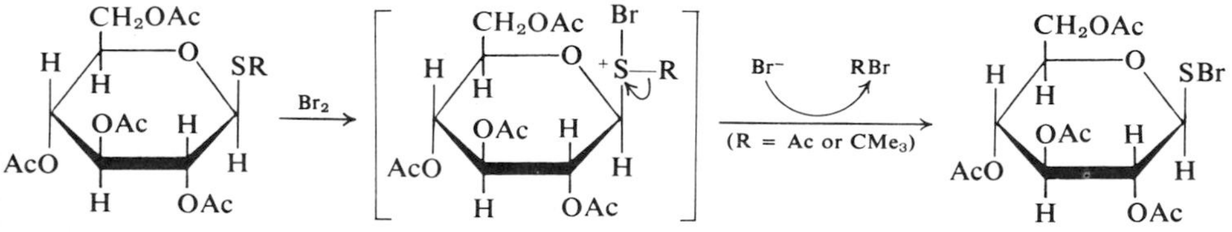

An exceptional example is ethyl 3,5,6-tri-O-acetyl-2-S-ethyl-1,2-dithio-α-D-mannopyranoside, which undergoes both displacement and further chlorination by an excess of chlorine to generate a 3,5,6-tri-O-acetyl-2-chloro-2-S-ethyl-2-thio-D-manno(or gluco)furanosyl chloride; treatment with bromine, however, produced 3,5,6-tri-O-acetyl-2-S-ethyl-2-thio-D-mannofuranosyl bromide.[99]

Glycosyl thiocyanates (for example, 21, R = CN) result[100] from treatment of per-O-acylglycosyl halides with potassium thiocyanate. They undergo saponification by sodium methoxide to afford 1-thioaldoses; silver thiocyanate produces the corresponding isothiocyanate[101] (Gl—N=C=S), which is also formed by thermal rearrangement of the thiocyanate.

Direct thiolation of D-glucose by hydrogen sulfide proceeds in meager yield in the presence of pyridine,[102] and not at all with N,N-dimethylformamide as the solvent[103]; however, the so-called Pictet anhydride (supposed[103] to be 1,2-anhydro-α-D-glucopyranose) and its 3,4,6-triacetate (*Brigl's anhydride*) combine with hydrogen sulfide at slightly elevated temperatures and pressures, and satisfactory yields of 1-thio-D-glucose can be obtained.[103]

1,2-*trans* Derivatives of 1-thio-D-glucuronic acid result from condensation of methyl 2,3,4-tri-O-acetyl-1-bromo-1-deoxy-α-D-glucopyranuronate with potassium ethylxanthate[104] or with potassium thiocyanate[105]; saponification of the product (23) of the former reaction causes complete deprotection and spontaneous cyclization to 1-thio-β-D-glucopyranurono-1,6-lactone, which reopens on treatment with diazomethane and acetylation to yield[104] methyl (methyl 2,3,4-tri-O-acetyl-1-thio-β-D-glucopyranosid)uronate[106] (24).

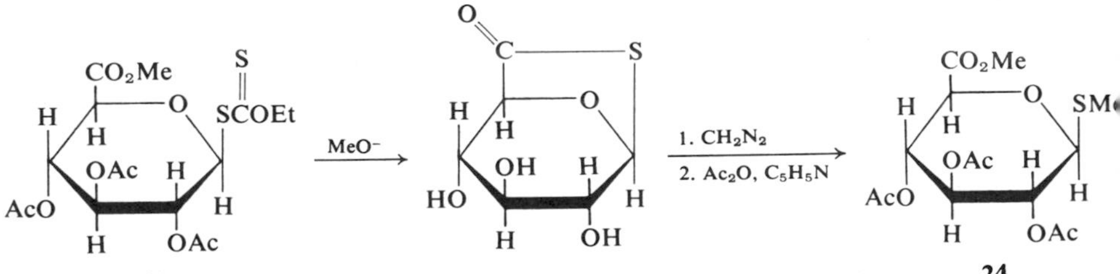

3. 2-Deoxy- and 2-Amino-2-deoxy-1-thio Sugar Derivatives

Certain aliphatic thiol derivatives [for example, 2-(2-aminoethyl)-2-thiopseudourea] having amino or guanidino groups one, or two, carbon atoms removed from the thiol group act to protect biological systems against the damaging effects of ionizing radiation[107]; however, their intrinsic toxicity limits their utility at high dose levels. Condensation of N-protected 3,4,6-tri-O-acetyl-2-amino-2-deoxy-α-D-glucopyranosyl halides with potassium thiolacetate,[108,109] potassium ethylxanthate,[64,109] or thiourea[109] generates N-substituted 2-amino-2-deoxy analogues of **18–20**. Removal of protecting groups from the Schiff base (**26**, R = Ac) was accomplished by O,S-de-acetylation with sodium methoxide, and removal of the p-methoxybenzylidene group in ammonium hydroxide, to afford the unstable aminothiol (**25**) as a crude sodium salt[64]; attempted deacetylation of **26** in the presence of an unprotected amino group, as in **27**, results in O (or S)→N acyl migration[64] to produce only the stable 2-acylamido derivative (**28**, R = Ac), isolated as the crystalline sodium salt.[108]

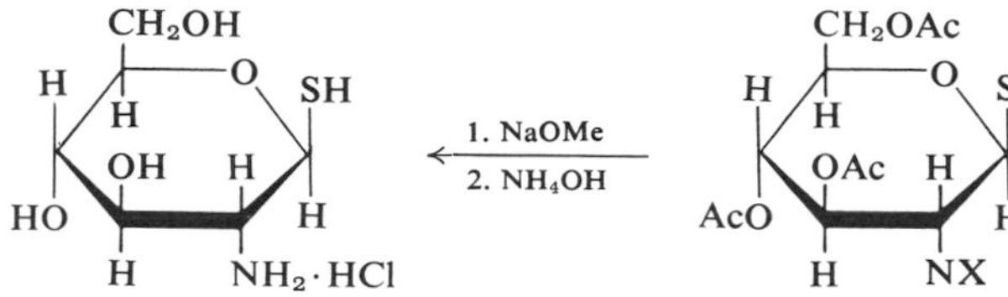

3,4,6-Tri-O-acetyl-2-deoxy-2-guanidino-1-thio-β-D-glucopyranose hydro-bromide[66] [**28**, R = C(NH$_2$)$_2$Br, R′ = Ac] and its β-D-*galacto* isomer[110] have been prepared from the corresponding pseudothiourea derivatives {**26** [X = H$_3$Br, R = C(NH$_2$)$_2$Br] and its 4-epimer} by a base-catalyzed re-arrangement that occurs spontaneously on dissolution in aqueous phosphate buffer. Several 2-amino-2-deoxyglycosyl alkylxanthates and thiolacetates have been converted into glycosyl halides by the action of chlorine or bromine.[97] Compound **26** [R = (C=S)OEt, X = Ac or 2,4-dinitrophenyl] was converted into the corresponding halide, whereas **26** [R = (C=S)OEt, X = =CHC$_6$H$_4$OMe-p] underwent concurrent cleavage of the X-group; related 2-glycosyl-2-thiopseudourea salts [such as **26**, R = C(NH$_2$)$_2$Br, X = Ac] resisted halogenolysis, presumably owing to the electron-withdraw-ing character of the S-substituent.

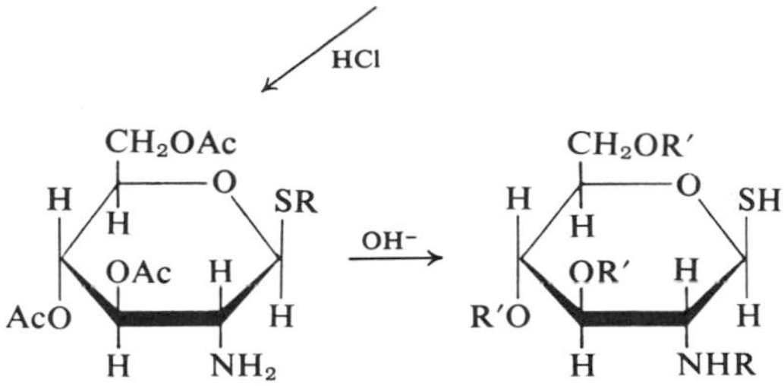

Physical methods of analysis are of value with these derivatives, as the infrared absorption frequencies of acetate groups are characteristic (N—Ac 6.1, S—Ac 5.95, O—Ac 5.75 μm) of the heteroatom[109]; differentiation is also possible by n.m.r. spectroscopy (see Chapter 27, Section I). Glycosyl ethylxanthates[109,111] and derivatives exhibit optically active absorption bands in the u.v. region that give rise to strong Cotton effects in their optical rotatory dispersion spectra. Raman spectroscopy is useful for anomeric assignment.[111a]

3,4,6-Tri-O-acetyl-2-deoxy-β-D-*arabino*-hexopyranosyl ethylxanthate was prepared from 3,4,6-tri-O-acetyl-1,5-anhydro-2-deoxy-D-*arabino*-hex-1-enitol (*tri-O-acetyl-D-glucal*, 29) by hydrobromination to give the glycosyl bromide and subsequent displacement of the anomeric bromine atom; direct treatment of the glycal (29) with thiolacetic acid afforded[112] the 2,3-unsaturated sugar derivative (30) in yields exceeding 50%. Bromination of 29 and subsequent treatment with potassium ethylxanthate produced a 2-bromo-2-deoxyhexosyl ethylxanthate.[113]Allylic ethylxanthates in sugar systems readily undergo allylic rearrangements, with net $\alpha \rightarrow \gamma$ migration of the substituent.[114]

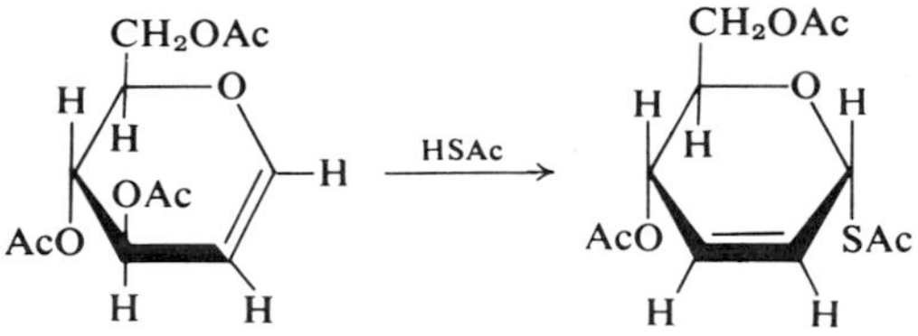

III. OTHER THIO SUGARS

The chemistry of thio sugars having a sulfur atom elsewhere than at the anomeric center had been very little explored at the time of Raymond's review[1] in 1945; since then, however, much attention has been focused on these derivatives, both as an end in themselves and as intermediates in the synthesis of deoxy sugars (see Chapter 17). 1,3-Dithian-2-yl derivatives, which provide a useful route to branched-chain sugars, are discussed in Chapter 17, and Wander and Horton[113a] have reviewed the chemistry of dithioacetals of sugars in detail.

A. THIOLATION BY NUCLEOPHILIC DISPLACEMENT REACTIONS

1. *Displacements of Halides and Sulfonates*

a. Simple, Bimolecular Reactions.—The first nonglycosidic thio sugar derivative reported was 6,6'-thiobis(methyl 2,3,4-tri-O-acetyl-6-deoxy-β-D-glucopyranoside), which was prepared by Wrede[114] by the action of potassium hydrogensulfide on methyl 2,3,4-tri-O-acetyl-6-bromo-6-deoxy-β-D-glucopyranoside. Subsequently,[115] the displacement of primary *p*-tolylsulfonyloxy groups from ω-sulfonylated 1,2-*O*-isopropylidene-α-D-xylo- and -D-gluco-furanoses by ethanethioxide was used to obtain ethyl thioethers. Sulfonyloxy groups[116] (see also Vol. IA, Chapter 6, Section IX) are generally more convenient than halides as leaving groups in such displacement reactions, and their synthetic uses have been discussed in detail by Ball and Parrish.[117] Similarly, 6-*O*-*p*-tolylsulfonyl-D-galactose diethyl dithioacetal,[118] 1,2:3,4-di-*O*-isopropylidene-6-*O*-*p*-tolylsulfonyl-α-D-galactopyranose,[119] and 2,3:4,5-di-*O*-isopropylidene-1-*O*-*p*-tolylsulfonyl-β-D-fructopyranose[119] react with ethanethioxide ion to afford the respective ethyl thioethers; the rates of these conversions are governed[119] by the steric accessibility of the primary carbon atom undergoing attack.

Thiocyanate ion in acetone converts 1,2,3,4-tetra-*O*-acetyl-6-*O*-*p*-tolyl-sulfonyl-β-D-glucopyranose at 130° into the corresponding 6-thiocyanato derivative,[100b] which may be hydrogenolyzed[120] to the 6-deoxy derivative or transformed into the 6-thiocyanato glycosyl bromide[100b] and thence into glycosides. The reaction of polysaccharide *p*-toluenesulfonic esters with sodium thiocyanate in 2,5-hexanedione[116] at 110° has been used[121] as a method for the quantitative determination of the content of primary hydroxyl group, as secondary *p*-toluenesulfonic ester groups give no apparent reaction under these conditions. The results are in close agreement with those obtained by the (more common) method of iodide displacement.[116] 6-Thioamylose having a degree of substitution (d.s.) 0.8 has been prepared[122] from amylose by a sequence of reactions involving tritylation, carbamoylation to protect O-2 and O-3, detritylation, *p*-toluenesulfonylation, displacement by potassium thiocyanate, and removal of all substituents by reduction with lithium aluminum hydride; a similarly modified polysaccharide has been obtained by the action of thiocyanate on 6-*O*-*p*-tolylsulfonylamylose. 6-*O*-*p*-Tolylsulfonyl-cellulose (in water)[123] and the 2,3-diacetate (in aqueous 1-propanol)[124] react with sodium thiosulfate to form a 6-thiosulfonic derivative (*Bunte salt*). A crystalline, monomeric analogue, 1,2,3,4-tetra-*O*-acetyl-6-deoxy-6-thiosulfo-β-D-glucopyranose, was converted,[123] by sequential reaction with iodine, mild acid, and acetylating reagents, into 6,6′-thiobis(1,2,3,4-tetra-*O*-acetyl-6-deoxy-α-D-glucopyranose). The selenocyanate ion has been used successfully in preparing 5-deoxy-1,2-*O*-isopropylidene-5-selenocyanato-α-D-xylofuranose from the corresponding 5-*p*-toluenesulfonate; however, saponification of the selenocyanato group afforded several products.[125] Decomposition of a 6-isoselenouronium salt (prepared by action of *N*,*N*-dimethylselenourea on 6-deoxy-6-iodo-D-glucose gave[125a] bis(6-deoxy-D-glucosyl) diselenide, whence 6-*Se*-dimethylarsino-6-seleno-D-glucose was prepared.

References start on p. 833.

The thiolacetate ion has been shown, largely by the work of Owen and co-workers,[126–129] to be a suitable nucleophile for displacement of primary *p*-tolylsulfonyloxy groups, particularly when the reaction is performed in hot *N,N*-dimethylformamide solution.[126,130] Deprotection may be accomplished under deacetylating conditions, and 6-thio-[127] and 1,6-dithio-D-mannitol[128] and 6-thio-D-glucitol[129] have been thus prepared. Potassium thiolbenzoate has been similarly employed[131] in a double displacement with 2,3,4-tri-*O*-acetyl-6-deoxy-6-iodo-α-D-glucopyranosyl bromide to yield 2,3,4-tri-*O*-acetyl-1,6-di-*S*-benzoyl-1,6-dithio-β-D-glucopyranose; 1,6-dithioalditols result from simultaneous displacement of leaving groups from both termini of substituted alditols by acylthiolate ions,[131a,b] and 1,6-anhydro-1-thioalditols from reaction of alditol 1,6-disulfonates with sodium sulfide,[131c] although stepwise displacements using protecting groups afford superior yields of the latter.[131b]

Action of sodium ethylxanthate and subsequent saponification converts sterically accessible, primary sulfonic esters into the corresponding thiols; 1,2:3,4-di-*O*-isopropylidene-6-*O*-*p*-tolylsulfonyl-α-D-galactopyranose reacts with the dry salt at 145° to give the corresponding 6,6′-disulfide.[131d]

Reaction of 3,4,5,6-tetra-*O*-acetyl-1-chloro-1-deoxy-D-psicose with thiourea initially proceeds by displacement of the halide by sulfur, but subsequent attack of one of the newly introduced nitrogen atoms at C-2 causes spontaneous cyclization to afford 2-amino-4-(1,2,3,4-tetra-*O*-acetyl-D-*ribo*-tetritol-1-yl)-6*H*-1,3-thiazole.[131e]

The action of sulfur nucleophiles on secondary sulfonates is much less readily effected, often requiring forcing conditions in polar, aprotic solvents.[116,126,130,132] 1,2:5,6-Di-*O*-isopropylidene-3-*O*-*p*-tolylsulfonyl-α-D-glucofuranose reacts very slowly with nucleophiles; conversion into the 3-*S*-benzoyl-3-thio-D-allose analogue by action of thiolbenzoate ion in refluxing *N,N*-dimethylformamide requires 4 hours to afford an 8% yield; the use of thiolacetate and an extended time of reaction raised[132a] the yield of (3-*S*-acetyl) product to 53%. Attack of thiolbenzoate occurs from the sterically more-accessible side of the corresponding 3-sulfonate having D-*allo* stereochemistry, effecting its conversion into 1,2:5,6-di-*O*-isopropylidene-3-*S*-benzoyl-3-thio-α-D-glucofuranose at five times the rate (although in poor yield) under similar[132] conditions. The D-*gluco* isomer appears to be inert[133] to alkanethioxide nucleophiles. 4-Thio-D-glucopyranosyl phosphate was prepared[134] by a sequence involving the reaction of methyl 2,3,6-tri-*O*-benzoyl-4-*O*-(methylsulfonyl)-α-D-galactopyranoside with potassium thiolbenzoate in *N,N*-dimethylformamide at 140°, followed by conversion into the phosphate by way of an intermediate per-*O*-acylglycosyl bromide. The thiolacetate ion was used under similar conditions to convert methyl 2,3-*O*-isopropylidene-4-*O*-*p*-tolylsulfonyl-α-L-lyxopyranoside into methyl 4-*S*-acetyl-2,3-*O*-isopropylidene-4-thio-β-D-ribopyranoside.[7] The 2,5-di-*S*-acetyl derivative formed by the action of potassium thiolacetate on 1,4:3,6-dian-hydro-2,5-di-O-(methylsulfonyl)-D-mannitol (31) was assigned[128] the D-*manno* stereochemistry by the original authors, although it has been suggested[3] that an L-*ido* configuration (32) would be more in accord with the stereochemical course generally[117,135] observed with bimolecular, nucleophilic displacement reactions of secondary sulfonic esters by thiolate ions.[136]

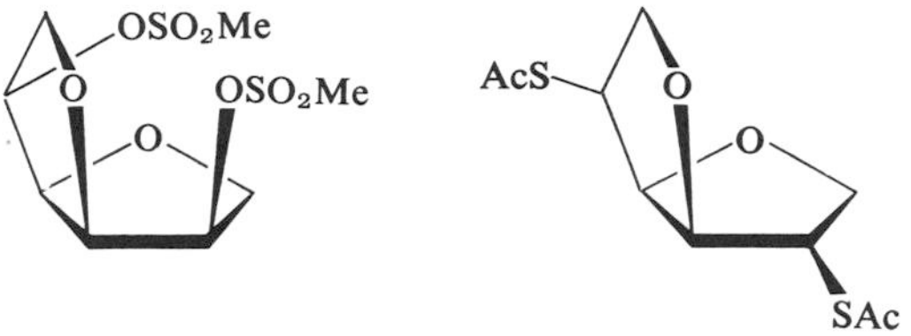

31   32

Secondary halides have also been displaced by the thiolacetate ion, as in the conversion[112] of a 3,4,6-tri-O-acetyl-2-bromo-2-deoxy-D-hexopyranosyl ethylxanthate derived from tri-O-acetyl-D-glucal (29) into a 1,2-dithiohexo-pyranose pentaacetate; this reaction may, however, involve initial participa-tion at C-2 by the sulfur atom of the thiocarbonyl group.

b. Displacements Assisted by Intramolecular Participation.—Considerable enhancement of reactivity is associated with processes involving cyclization; this subject has been reviewed in detail by Goodman.[137] Freudenberg and Wolf[138] noted that the 3-O-(methylthio)thiocarbonyl derivative (33) of 1,2:5,6-di-O-isopropylidene-α-D-glucofuranose undergoes pyrolysis to form not the expected alkene but an isomer (34), having the thio substituent at C-3, of the starting material. The D-*gluco* configuration is retained in this reaction,[139] which probably occurs with an internal, nucleophilic displacement by front-side attack, as illustrated. Similar rearrangements are reported[140] for 2-O-[(alkylthio)thiocarbonyl] derivatives of methyl 3,4-O-isopropylidene-β-D(and L)-arabinopyranoside and methyl 4,6-O-benzylidene-3-O-methyl-α-D-altropyranoside under the conditions used for the Chugaev elimination. The

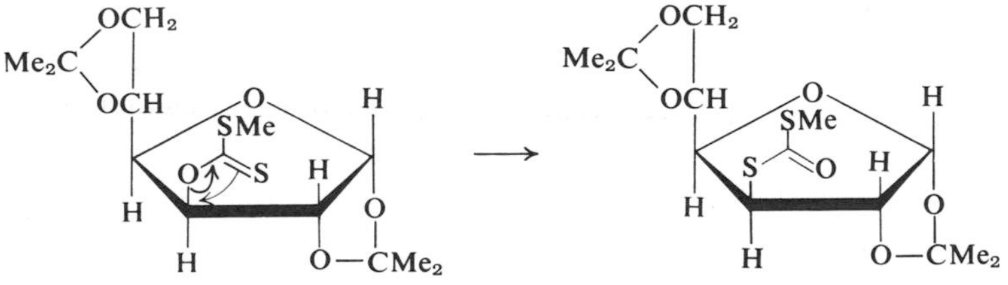

33   34

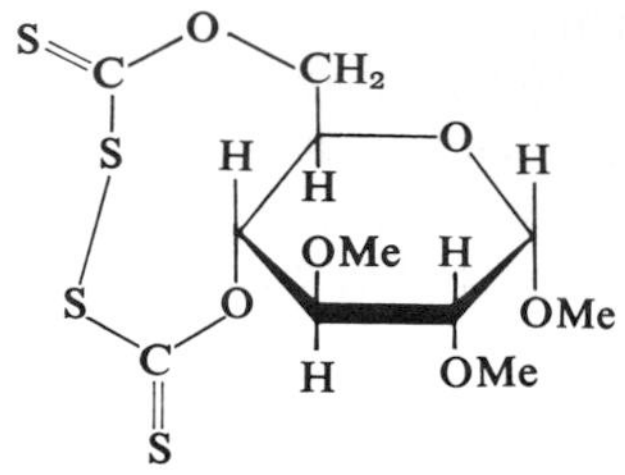

35

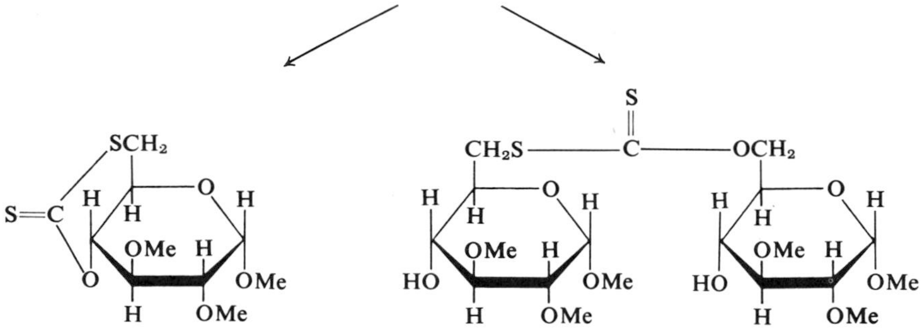

35a **35b**

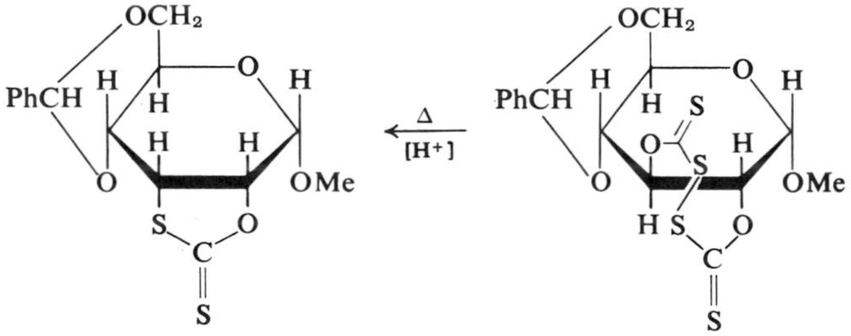

36 **37**

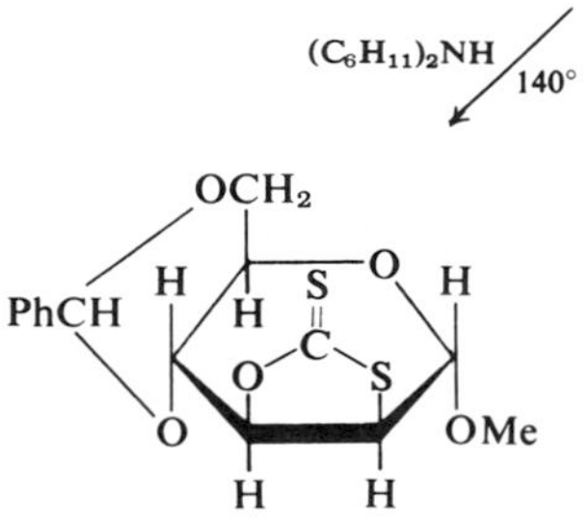

38

drastic reaction conditions and attendant low yields of product tend to limit the usefulness of this approach to thio sugars.[140a]

Doane and co-workers[141] have devised synthetically practicable rearrangements of cyclic thiocarbonyl derivatives in several model systems having no unprotected hydroxyl groups. 3-O-Acetyl-1,2-O-isopropylidene-α-D-glucofuranose 5,6-thionocarbonate in acetonitrile containing potassium iodide was found to rearrange during 15 hours at 125° to afford 3-O-acetyl-1,2-O-isopropylidene-6-thio-α-D-glucofuranose 5,6-cyclic carbonate in 84% yield[141a]; 1,2-O-isopropylidene-3-O-methyl-6-thio-α-D-glucofuranose 5,6-cyclic thionocarbonate formed spontaneously after dissolution of 1,2-O-isopropylidene-3-O-methyl-α-D-glucofuranose 5,6-cyclic dithiobis(thioformate) in acetone,[141b] and thermolysis of 1,2-O-isopropylidene-3-O-methyl-5,6-bis[(methylthio)thiocarbonyl]-α-D-glucofuranose at 245° gave the same product in approximately 70% yield.[142] Methyl 2,3-di-O-methyl-α-D-glucopyranoside 4,6-cyclic dithiobis(thioformate) (35) underwent base-catalyzed rearrangement and degradation, to form the corresponding 6-thio 4,6-cyclic thionocarbonate derivative (35a) [and, on extended treatment, an unsymmetrical 6,6'-bis(hexosyl)xanthate derivative (35b)][141c] Methyl 4,6-O-benzylidene-α-D-glucopyranoside 2,3-cyclic dithiobis(thioformate) (37) is reported[141d] to undergo inversion of configuration at a secondary position during pyrolysis; in the presence of a trace of residual acidity, the product (in 20% yield) is the 3-thio-D-altrose derivative (36), whereas dicyclohexylamine catalyzes the formation (in 47% yield) of the 2-thio-D-mannose isomer (38). Similar treatment of the D-*manno* (2,3-*cis*) isomer of 37 merely restores the *cis*-2,3-diol group. 2'-Thiocytidine 2',3'-cyclic phosphate results from internal displacement of a 2,2'-anydro grouping by a 3'-O-phosphorothioyl group.[142a]

Nongeminal displacement processes have been more commonly encountered, and have been exploited effectively for enhancing the nucleophilicity of sulfur reagents. Methyl 2-O-thionobenzoyl-3,5-di-O-p-tolylsulfonyl-α-D-xylofuranoside reacts in the presence of sodium benzoate by internal displacement of the 3-C-substituent to form a presumed oxathiolane intermediate that undergoes decomposition by water to form[142b] the 2,3-*cis*-hydroxythiol—namely, methyl 2,5-di-O-benzoyl-3-thio-α-D-ribofuranoside.

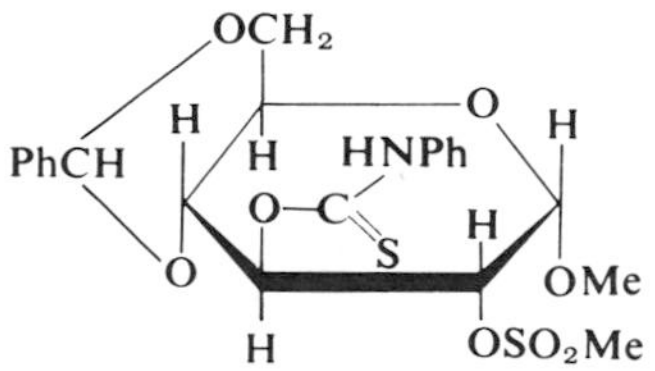

39

Likewise, base-catalyzed cyclization of the thiocarbamate (39) led, in part, to the 2-thio-D-mannoside, although, in this instance, competitive displacement by the nitrogen atom gave 2-amino-2-deoxy-D-mannoside products also.[143]

Methyl 4,6-O-benzylidene-3-deoxy-3-[(methylthio)thiocarbonyl]amino-2-O-(methylsulfonyl)-α-D-altropyranoside rearranges similarly in boiling pyridine to form a 2-(methylthio)thiazolidine derivative; partial reduction and hydrolysis then afford[144] crystalline 3-amino-1,6-anhydro-3-deoxy-2-thio-β-D-allopyranose hydrochloride. Similar preparations have been recorded for derivatives of 2-amino-2-deoxy-3-thio-D-allose,[145] methyl 3-amino-3-deoxy-2-thio-α-D-ribofuranoside,[146] and 3'-amino-3'-deoxy-2'-thiouridine.[147]

Vicinal cis-1,2-dithiol derivatives have been obtained by internal displacement reactions with 3,4,6-tri-O-acetyl-2-O-(methylsulfonyl)-β-D-glucopyranosyl ethylxanthate (acetate ion-catalyzed) and N,N-dimethyl(or ethyl)dithiocarbamate (thiolacetate ion-catalyzed) to form the D-*manno* 1,2-cyclic dithiocarbonate[148] and the corresponding 2-S-acetyl-2-thio-β-D-mannopyranosyl N,N-dialkyldithiocarbamate,[149] respectively. Under acetate ion catalysis, the dithiocarbamate group migrates with elimination of the sulfonate group to form the 2-thioglycal derivative (40); this result verifies[149] the presumed participation of the thiocarbonyl group.

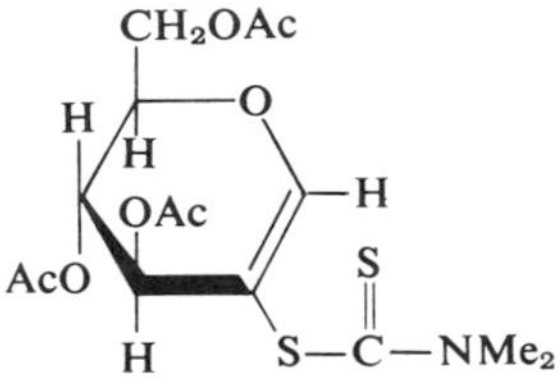

40

Methyl 6-deoxy-2,3-O-isopropylidene-4-O-p-tolylsulfonyl-α-L-mannopyranoside (41) reacts with the thiolbenzoate ion[150] to form methyl 5-S-benzoyl-6-deoxy-2,3-O-isopropylidene-5-thio-α-L-talofuranoside[151] (42). Owen correlated this skeletal rearrangement with the antiparallel arrangement of the C-4–O-sulfonyl and C-5–O-5 bonds, postulating an initial, cyclic rearrangement to cause inversion at C-4 and C-5, followed by thiolate displacement of the new 5-sulfonate to effect net inversion at C-4 only.[152]

Formation and reopening of an intermediate 2,3-oxonium ion has been postulated[153] in the conversion[115] of methyl 2-bromo-2-deoxy-β-D-glucopyranoside into methyl 3-S-methyl-3-thio-β-D-altropyranoside by the action of methanethioxide ion (see also Vol. IIA, Chapter 29, Sections IV,B and IV,C).

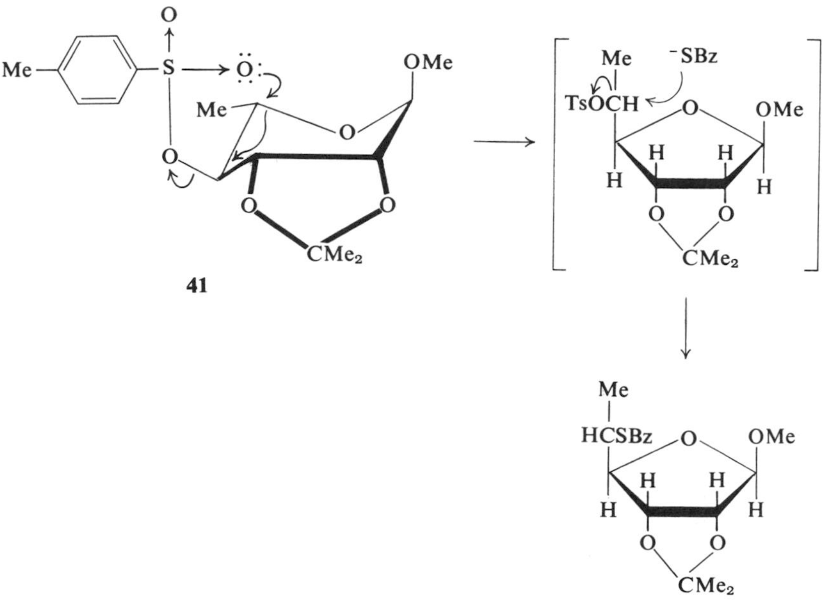

41

42

2. Nucleophilic Opening of Epoxides and Episulfides

a. Terminal.—Ohle and Mertens[154] prepared 1,2-*O*-isopropylidene-6-thio-α-D-glucofuranose by the action of hydrogen sulfide on the corresponding 5,6-anhydro derivative; although displacement occurs at the primary center only, the 5-thio analogue (**45**) is accessible[8] by double inversion of **43** with thiourea to form the episulfide (**44**), displacement by potassium acetate, and deprotection.

Primary episulfides may also conveniently be formed in two steps from suitable disulfonate derivatives. Thus, 1,2:3,4-di-*O*-isopropylidene-5,6-di-*O*-*p*-tolylsulfonyl-D-mannitol is converted into the 6-thiolacetate by the action of thiolacetate ion; saponification of the thioester liberates a 6-thioxide ion that cyclizes spontaneously by displacement of the remaining sulfonate residue to afford 1,2:3,4-di-*O*-isopropylidene-5,6-thioanhydro-L-gulitol.[127] By this procedure, 1,2-*O*-isopropylidene-α-D-glucofuranose was converted[155] into 1,2-*O*-isopropylidene-5,6-thioanhydro-β-L-idofuranose in 90% yield; saponification of the 3,5,6-triacetate of 1,2-*O*-isopropylidene-6-thio-α-D-glucofuranose produces the same episulfide, but in only 10% yield.[127,155] The

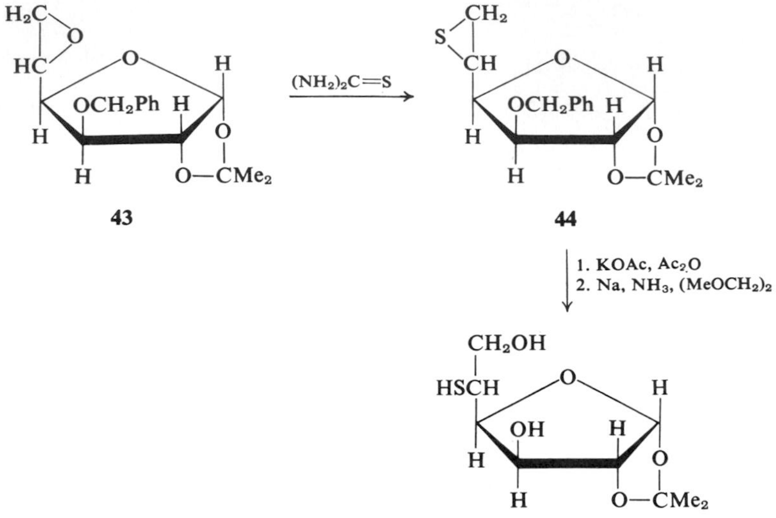

43   **44**

1. KOAc, Ac$_2$O
2. Na, NH$_3$, (MeOCH$_2$)$_2$

45

terminal episulfide reacts with lithium aluminum hydride to form an ω-deoxy thiol,[127] with potassium methylxanthate to form a cyclic trithiocarbonate[127,155] (presumably[3] by initial displacement of the primary C–S bond and subsequent return of the 5-thiolate ion to displace the alkoxide substituent), and with itself[155] in alkali to form a poly(thioether) by a presumably similar, intermolecular mechanism; the action of lithium aluminum hydride on cyclic trithiocarbonates yields the parent dithiols.[156]

b. Secondary.—Epoxides fused to hexopyranosyl rings generally react with sulfur nucleophiles to form products of *trans*-diaxial opening (*Fürst–Plattner rule*). Methyl 2,3-anhydro-4,6-*O*-benzylidene-α-D-allopyranoside (**46**, X = O) is attacked at C-2 to form[157] the 2-(methylthio)-D-*altro* product (**47**), whereas the isomeric epoxide (**48**, X = O) undergoes displacement at C-3 to form a 3-methylthio[158] (**49**, R = Me) or 3-thiocyanato[159] (**49**, R = CN) product having the same configuration. Similarly, the 4-epimer of **46** (X = O) reacts with methanethioxide ion to form[133a] the 4-epimer of **47**.

1,5-Di-*O*-acetyl-2,7:3,4-dianhydro-β-D-*manno*-heptulopyranose[160] and 1,6:3,4-dianhydro-β-D-galactopyranose[161] react with thiolate anions to form, respectively, 3- and 4-thio derivatives, having the all-axial *gluco* configuration; however, 1,6:2,3-dianhydro-β-D-mannopyranose, a lower homologue of the former, mainly undergoes[162] opening at C-3 to form 1,6-anhydro-3-*S*-benzyl-3-thio-β-D-altropyranose, the diequatorial product. Methyl 3,4-anhydro-β-D-galactopyranoside is attacked by methanethioxide ion to give methyl

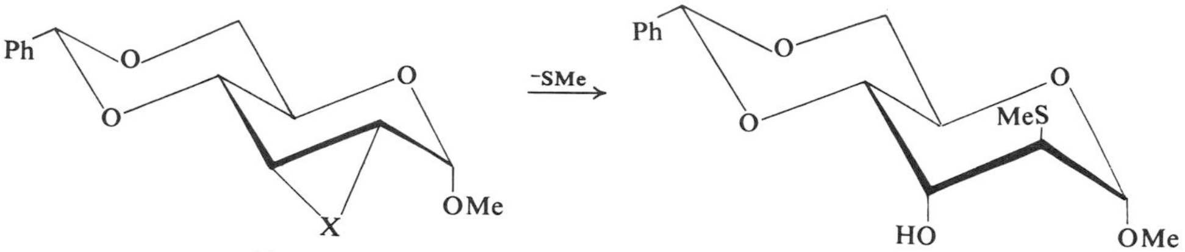

3-S-methyl-3-thio-β-D-gulopyranoside[163]; uncertainty about the conformation of the initial compound makes impossible the application of the principle of diaxial opening to this example. The 6,6-*cis*-fused epoxides methyl 2,3-anhydro-4,6-O-benzylidene-α-D-altro- and -D-gulopyranoside react with sodium methylxanthate to afford 2,3-*trans* trithiocarbonates[163a] having the D-*ido* and D-*galacto* configuration, respectively, inversion occurring at C-3 in both instances.

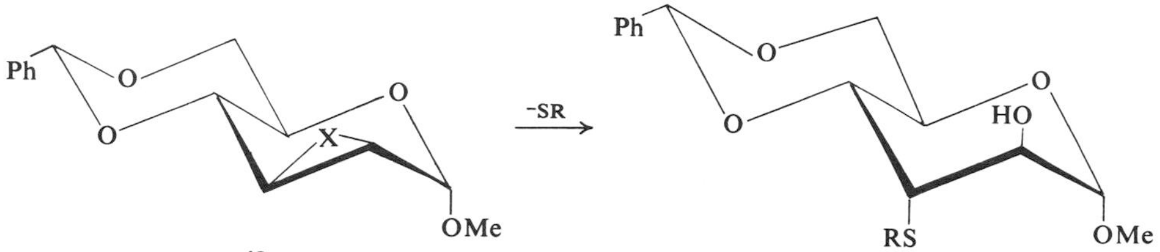

2,3-Anhydropentopyranose derivatives generally[3] undergo favored attack by nucleophiles at C-3. Methyl 2,3-anhydro-β-L-ribopyranoside reacts with methanethioxide ion to yield[164] methyl 3-S-methyl-3-thio-β-L-xylopyranoside; however, ethyl 3,4-anydro-β-L-ribopyranoside is attacked by benzylthioxide ion at C-4, affording ethyl 4-S-benzyl-4-thio-α-D-lyxopyranoside to the virtual exclusion[165] of the 3-thio-L-*xylo* isomer.

Methyl 4,6-O-benzylidene-2,3-thioanhydro-α-D-allopyranoside (**46**, X = S) is formed directly from the D-*manno* epoxide (**48**, X = O) by the action of thiocyanate ion[166]; the yield is meager, evidently because of conformational restraints in the intermediate 2-imino-oxathiolane structure. The analogous conversion of the *allo* epoxide (**46**, X = O) into the *manno* episulfide (**48**, X = S) occurs[167] in 6% yield, the major product being a poly(ethylene sulfide) derivative. The episulfide **46** (X = S) is available[166] by a multistep sequence wherein the *manno* epoxide (**48**, X = O) reacts with ammonium thiocyanate to produce a 3-thiocyanato derivative whose 2-O-(methylsulfonyl) analogue reacts with base to effect hydrolysis of the thiocyanato group and

References start on p. 833.

cyclization of the product to give **46** (X = S). Triethyl phosphite efficiently[159b] converts the episulfide **46** (X = S) into methyl 4,6-*O*-benzylidene-2,3-dideoxy-α-D-*erythro*-hex-2-enopyranoside (**50**). Direct conversion of methyl 2,3-anhydro-4,6-*O*-benzylidene-α-D-altropyranoside into the corresponding D-*gulo* 2,3-episulfide has also been accomplished[167a] by treatment of the former with thiocyanate ion.

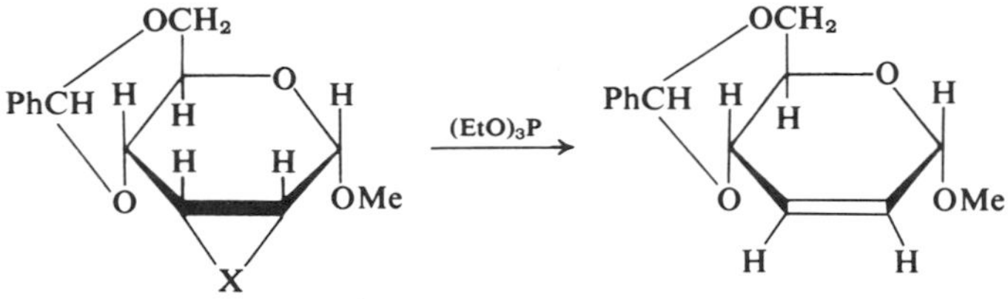

The obvious implications for nucleoside chemistry lend particular interest to the 2,3-anhydroaldofuranosyl series. Reaction of 7-(2,3-anhydro-α-D-lyxofuranosyl)theophylline and its D-*ribo* isomer with ethanethioxide ion at C-3′ is favored to generate the 3-*S*-ethyl-3-thio-D-*arabino* and 3-*S*-ethyl-3-thio-D-*xylo* nucleosides as the respective major[168] products; similar results were observed for methyl 2,3-anhydro-β-D-ribofuranoside,[169] 9-(2,3-anhydro-β-D-ribofuranosyl)purine,[170] and methyl 2,3-anhydro-α-D-lyxofuranoside.[171] Methyl 2,3-anhydro-β-D-lyxofuranoside was found[172] to favor attack at C-2 by benzylthioxide ion, producing a 3:2 mixture of methyl 2-*S*-benzyl-2-thio-β-D-xylofuranoside and methyl 3-*S*-benzyl-3-thio-β-D-arabinofuranoside. 9-(2,3-Anhydro-5-deoxy-β-D-ribofuranosyl)adenine and its β-D-*lyxo* isomer were subsequently treated with a series of nucleophiles; the former gave the 3-*S*-benzyl-3-thio product almost exclusively (although nucleophiles smaller than benzylthioxide afforded slightly greater proportions of 2-substitution products), whereas the latter gave approximately equal amounts of 2-substituted D-*xylo* and 3-substituted D-*arabino* products, irrespective of the nucleophile, indicating at least some degree of steric control over the product distributions.[173]

Methyl 2,3-thioanhydro-β-D-ribofuranoside and methyl 2,3-thioanhydro-α-D-lyxofuranoside were prepared from the corresponding β-D-*lyxo* and α-D-*ribo* epoxides by addition of thiocyanate, methanesulfonylation of the newly formed hydroxyl group, and saponification of the thioester.[174] Methyl 2-*S*-benzoyl-2-thio-3-*O*-*p*-tolylsulfonyl-5-*O*-trityl-β-D-xylofuranoside (**51**) and the isomeric glycoside **53** (both isolated from the products of opening of the D-*lyxo* epoxide by thiolbenzoate) were each treated with sodium benzoate in *N,N*-dimethylformamide to determine which heteroatom, if either, of the thiolbenzoate group would participate in displacement of the sulfonyloxy group; clear evidence for sulfur participation (through an intermediate such as **54**) was found in the observation[175] that the sole product of both reactions was methyl 2,3-thioanhydro-5-*O*-trityl-β-D-ribofuranoside (**52**).

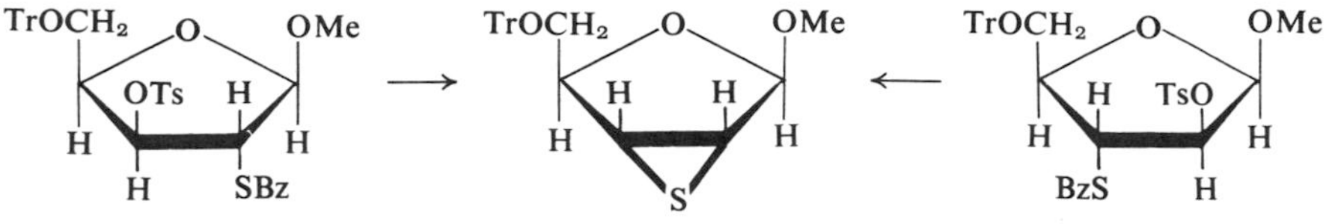

3. *Addition of Thiol Reagents to Unsaturated Carbon Atoms*

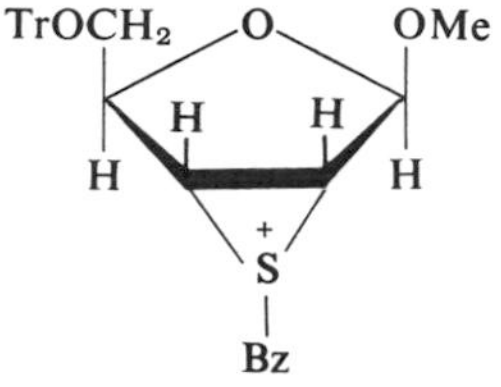

Replacement of the iodine atom of methyl 6-deoxy-6-iodo-α-D-gluco-pyranoside by certain sulfur nucleophiles was found by Lehmann and Benson[176] to proceed by initial elimination to afford a 5,6-alkene. Subsequent addition of the sulfite radical anion to C-6 afforded methyl 6-deoxy-α-D-glucopyranoside 6-sulfonic acid. Lehmann and Weckerle[177] employed conformational interaction parameters to deduce that the sulfite addition reaction occurs under product-development control.

5,6-Dideoxy-1,2-*O*-isopropylidene-α-D-*xylo*-hex-5-enofuranose adds thiolacetic acid under ultraviolet irradiation, forming[178] the "anti-Markovnikov" product—namely, 6-*S*-acetyl-5-deoxy-1,2-*O*-isopropylidene-6-thio-α-D-*xylo*-hexofuranose; α-toluenethiol was similarly added to 5-deoxy-1,2-*O*-isopropyl-idene-3-*O*-methyl-β-L-*threo*-pent-4-enofuranose to give 5-*S*-benzyl-1,2-*O*-isopropylidene-3-*O*-methyl-5-thio-α-D-xylofuranose.[179] Tri-*O*-acetyl-D-glucal (**29**) adds thiolacetic acid under free-radical catalysis by cumene hydroperoxide, the product being a 3:1 mixture of 3,4,6-tri-*O*-acetyl-2-*S*-acetyl-1,5-anhydro-2-thio-D-mannitol and -glucitol.[180]

(*E*)-3,4,5,6-Tetra-*O*-acetyl-1,2-dideoxy-1-nitro-D-*arabino*-hex-1-enitol was converted into a 4:1 mixture of the diastereoisomeric Markovnikov addition

---

products, 3,4,5,6-tetra-O-acetyl-2-S-acetyl(benzoyl or benzyl)-1-deoxy-1-nitro-2-thio-D-mannitol and -glucitol, by treatment with the appropriate thiol in pyridine or piperidine solution[181]; the anionic intermediate in the addition is stabilized by the nitro group, and the 4-acetoxyl group hinders approach of the incoming nucleophile. Subsequent treatment with strong mineral acid (the Nef reaction) afforded 2-thio-D-mannose derivatives.[181]

Ingles[182] prepared substituted-ammonium salts of the hydrogen sulfite adducts of D-glucose and of *aldehydo*-D-glucose pentaacetate, both in the free form and as the 1-O-acetyl derivatives. Paulsen and Steinert[182a] earlier used such derivatives, prepared by the action of aqueous sulfur dioxide, as intermediates in the generation of derivatives having a nitrogen atom in the pyranose ring.

4. Reactions Involving Bridged Sulfonium Intermediates

Intramolecular participation by sulfur atoms, as in the conversion of the vicinal benzoylthio *p*-toluenesulfonates (51 and 53) into an episulfide (52) (Section III,A,2,b), is now recognized as a common phenomenon in displacement reactions of thio sugars.

a. Acid-catalyzed Reactions.—Brigl et al.[183] found that ethanethiol in the presence of hydrogen chloride "overmercaptalates" 3,4,5,6-tetra-O-benzoyl-D-glucose diethyl dithioacetal (55) to yield a tetra-O-benzoyl-2-S-ethyl-2-thiohexose dithioacetal (56), later shown[184,185] to have the D-*manno* configuration; the analogous reaction of 3,4,5-tri-O-benzoyl-D-xylose diethyl dithioacetal affords[186] a 2-S-ethyl-2-thio derivative that probably has the D-*lyxo* configuration. The intermediacy of an episulfonium ion (56) was demonstrated[9a,185] by the acid-catalyzed addition of thiophenol to 55, which produced 3,4,5,6-tetra-O-benzoyl-2-S-ethyl-2-thio-D-mannose ethyl phenyl dithioacetal, evidently by attack of the nucleophile at C-1 of 56.

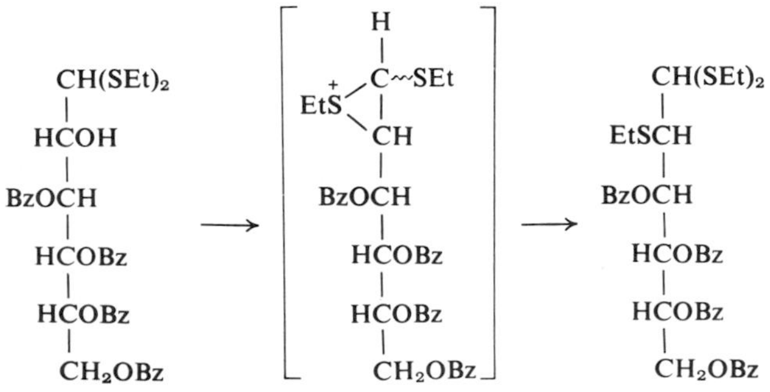

CH(SEt)₂ | HCOH | BzOCH | HCOBz | HCOBz | CH₂OBz   **55**

56

CH(SEt)₂ | EtSCH | BzOCH | HCOBz | HCOBz | CH₂OBz   **57**

Deacylation followed by mercuric chloride-catalyzed hydrolysis of **57** (or its methyl phenyl analogue) yields[187] 2-*S*-ethyl-2-thio-D-mannose,[185] which epimerizes to the D-*gluco* isomer under very mild conditions (pH 8). The D-*gluco* product has also been prepared by diazotization[188] of 2-amino-2-deoxy-D-glucose diethyl dithioacetal at pH 0; this product probably arises by way of the D-*manno* derivative. At pH 6, the deamination reaction[189] gives, as the major product,[190] ethyl 2-*S*-ethyl-1,2-dithio-α-D-mannofuranoside, which was also prepared[185] by deacylation followed by partial demercaptalation of **57**. Opening of the intermediate episulfonium ion (the parent tetrol of **56**) evidently occurs[9a] by attack of a solvent (water) molecule at acid pH and by intramolecular attack of O-4 near neutral pH.

Such partially acylated aldose derivatives as **55** are commonly disposed to undergo "overmercaptalation." The use of Lewis acid catalysts in the mercaptalation of aldose peracetates may also lead to the introduction of additional sulfur atoms. β-D-Glucopyranose pentaacetate reacts[191] with ethanethiol in the presence of zinc chloride to give, after deacetylation, 2-*S*-ethyl-2-thio-D-mannose diethyl dithioacetal,[185] whereas α-L-arabinopyranose tetraacetate reacts with ethanethiol in the presence of boron trifluoride to give, after deacetylation, 5-*S*-ethyl-5-thio-L-arabinose diethyl dithioacetal.[192] The suspected intermediate in the latter reaction—namely, 2,3,4-tri-*O*-acetyl-L-arabinose diethyl dithioacetal—reacts with methanethiol in the presence of zinc chloride to form 5-*S*-ethyl-5-thio-L-arabinose ethyl methyl thioacetal; this result supports the concept[193] of an intermediate 1,5-bridged ethyl-sulfonium ion. β-D-Ribopyranose tetraacetate reacts[194] with methanethiol in the presence of zinc chloride to afford a mixture of 4-*S*-methyl-4-thio-L-lyxose dimethyl dithioacetal, methyl 1,5-dithio-β-D-ribopyranoside, methyl 1-thio-β-D-ribopyranoside, and a trace of 5-*S*-methyl-5-thio-D-ribose dimethyl dithioacetal. Hughes and co-workers[194] proposed a sequence of reactions involving transient formation of a 2,3,4-tri-*O*-acetyl dialkyl dithioacetal that reacts via the 4,5-acetoxonium ion (**58**), followed by displacement at O-4 or O-5 to introduce a 4- or 5-thio substituent, respectively. Ethanethiolysis of 3,5,6-tri-*O*-benzoyl-1,2-*O*-isopropylidene-α-D-glucofuranose affords a tetra-thio product[187] that was shown by Bethell and Ferrier[195] to be 4,5,6-tri-*O*-benzoyl-2,3-di-*S*-ethyl-2,3-dithio-D-allose diethyl dithioacetal; the authors[195] proposed a similar mechanism involving sequential displacement of benzoyl-oxonium ions by adjacent ethylthio groups, as illustrated for attack of the final thiol molecule on the intermediate **59**. Acid-catalyzed ethanethiolysis of 3-*O*-benzoyl-1,2:5,6-di-*O*-isopropylidene-α-D-glucofuranose affords ethyl 4-*O*-benzoyl-2,3,6-tri-*S*-ethyl-1,2,3,6-tetrathio-α-D-mannopyranoside.[195a] Re-investigation of the zinc chloride-catalyzed reaction of *keto*-D-fructose

---

pentaacetate and *keto*-L-sorbose pentaacetate with ethanethiol revealed that, although the diethyl dithioacetal pentaacetate (Vol. IA, Chapter 9, Section II,A) is the major product (37%), a substantial proportion (19%) of the corresponding 3,4,5,6-tetra-*O*-acetyl-1-*S*-ethyl-1-thiohexulose diethyl dithioacetal[195b] is formed, together with minor proportions of more-extensively thiolated products.

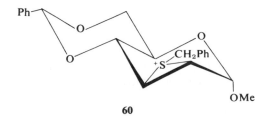

58   59

Concentrated hydrochloric acid converts 2-deoxy-4,5-*O*-isopropylidene-3-*O*-methyl-D-*erythro*(or D-*threo*)-pent-1-enose diphenyl dithioacetal into a complex mixture containing ~15% of 5-*O*-acetyl-2-deoxy-3-*S*-phenyl-3-thio-D-*threo*-pentono-1,4-lactone and ~10% of the D-*erythro* diastereoisomer.[196]

Acidic methanolysis of 5-*Se*-benzyl-1,2-*O*-isopropylidene-5-seleno-α-D-xylofuranose yields a mixture of products, including 4-*Se*-benzyl-2,3-*O*-isopropylidene-5-*O*-methyl-4-seleno-D-xylose dimethyl acetal, which is presumed[196a] to arise via an intermediate, 4,5-bridged, cationic selenium species.

b. Reactions under Basic or Neutral Conditions—Ammonolysis of the episulfide (**46**, X = S) results only in polymerization; however, methyl 2-*S*-benzyl-4,6-*O*-benzylidene-3-*O*-(methylsulfonyl)-2-thio-α-D-altropyranoside reacts in the presence of sodium azide by intramolecular displacement of the sulfonate group to form an intermediate episulfonium ion (**60**) that is attacked (axially) by the azide ion, yielding 3-azido-2-*S*-benzyl-4,6-*O*-benzylidene-3-deoxy-2-thio-α-D-altropyranoside.[159b] Analogous conversions in the aldofuranose series offer a route to 2-thioaldofuranose derivatives; thus after *p*-toluenesulfonylation, methyl 3-*S*-ethyl-3-thio-β-D-xylofuranoside

60

reacts[169] with the acetate ion to form methyl 3,5-di-O-acetyl-2-S-ethyl-2-thio-β-D-arabinofuranoside as the major product, the 3-thio-D-$xylo$ isomer being formed to the extent of 20%. Benzyl 3,4-O-cyclohexylidene-2-O-(methylsulfonyl)-1-thio-β-D-arabinopyranoside and benzyl 3,5-di-O-benzoyl-2-O-(methylsulfonyl)-1-thio-α-D-arabinofuranoside react with sodium hydrogen-carbonate and silver carbonate in dry methanol to form[197] the respective methyl 2-S-benzyl-2-thio-D-ribosides in excellent yields.

Dutcher and co-workers[198] interpreted the partial hydrogenolysis of the 2,3,4,5-tetra-O-benzoyl derivative of 3-amino-3,6-dideoxy-D-mannose (*mycosamine*) diethyl dithioacetal to give a 3-amino-1,2,3,6-tetradeoxy product as proceeding by internal displacement of the 2-benzoyloxy group and hydrogenolysis of the resulting episulfonium ion. Van Es[199] reported that *p*-toluenesulfonylation of 2-O-methyl- and 2,3-di-O-methyl-D-xylose diethyl dithioacetal (**61**, R' = H; R = H, Me) produces ethyl 2-S-ethyl-5-O-methyl- and -3,5-di-O-methyl-2-thio-α-D-lyxofuranoside, respectively, whereas similar reaction of the 2,3,4-tri-O-methyl analogue (**61**, R = R' = Me) causes formation of 1,2-di-S-ethyl-3,4,5-tri-O-methyl-1,2-dithio-D-*threo*-pent-1-enitol; it was speculated that (*a*) formation of a 1,2-episulfonium ion frees the methoxyl group on C-2 so that it can replace the primary sulfonate group, and (*b*) the product (**62**) of step (*a*) undergoes internal attack if a 4-hydroxyl group is available (R = H), or elimination if it is not (R = Me).

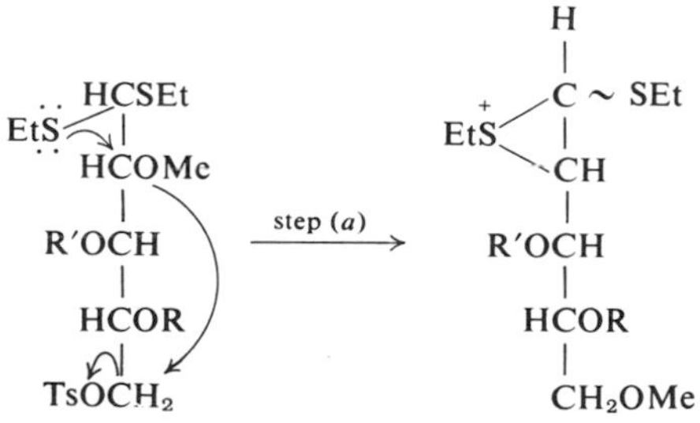

61   62

Acid-catalyzed methanolysis of methyl 1-thio-6-O-*p*-tolylsulfonyl-β-D-glucopyranoside affords methyl 6-S-methyl-6-thio-α-D-glucopyranoside, evidently via a 1,6-bridged sulfonium ion; acetolysis of the corresponding triacetate affords 1,2,3,4-tetra-O-acetyl-6-S-methyl-6-thio-β-D-glucopyranose, which indicates occurrence of a second intramolecular displacement (to form a 1,2-acetoxonium ion) prior to attack by the solvent.[200] 5-O-*p*-Tolylsulfonyl-L-arabinose diethyl dithioacetal undergoes thermolysis in aqueous solution via a sulfonium ion (**64**, R = Et) related to 5-thio-L-arabinopyranose, ultimately yielding ethyl 5-S-ethyl-5-thio-α,β-L-arabinofuranoside (**63**)[201];

References start on p. 833.

replacement by the dibenzyl dithioacetal in a later experiment (performed in acetone containing sodium iodide) produced an intermediate ion (64, R = PhCH$_2$) from which benzyl 1,5-dithio-α,β-L-arabinopyranoside (65) was liberated by nucleophilic attack at the benzyl group.[202]

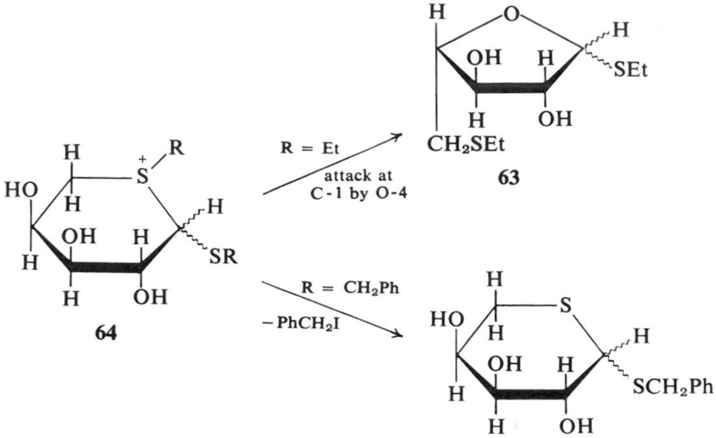

Whereas 2,3,4-tri-O-methyl-5-O-p-tolylsulfonyl-D-arabinose dibenzyl dithioacetal is converted[202a] into the tri-O-methyl derivative of 65 by treatment with sodium iodide in hot acetone, the same conditions convert the D-*ribo* and D-*xylo* isomers into 2,5-anhydro-3,4-di-O-methyl dithioacetals. Action of sodium iodide and zinc in hot N,N-dimethylformamide converts 2,5-anhydro-3,4-di-O-p-tolylsulfonyl-D-xylose ethylene dithioacetal into a D-*glycero* 3,4-alkene (see Chapter 19, Section IV,A,2), but the corresponding diisobutyl dithioacetal reacts under similar conditions to give a mixture of 2-(isobutyl-thiomethyl)furan and its 3- and 4-isobutylthio derivatives.[202b] Azidolysis of 8-deoxy-1,2:3,4-di-O-isopropylidene-6-O-(methylsulfonyl)-β-L-*glycero*-D-*galacto*-octo-1,5-pyranos-7-ulose 7-(trimethylene dithioacetal) (65a) proceeds[202c] by way of presumed 6,7-episulfonium ions to afford the epimeric 7-azido-6,7-dithio products 65b and 65c; the major (28% yield) product, the 6,7-dithio-7-ene 65d, presumably arises by deprotonation of an intermediate 7-sulfonium ion.

2,3,5-Tri-O-methyl-4-O-p-tolylsulfonyl dibenzyl dithioacetals of D-ribose, D-xylose, and D-lyxose rearrange in sodium iodide–acetone[202b] or hot pyridine[202a] to form benzyl 1,4-dithiofuranosides with inversion of configuration at C-4; reaction of similarly substituted diethyl dithioacetals with pyridine gives the corresponding, acyclic 4-S-ethyl-4-thiopentose derivatives, presumably because the 1,4-episulfonium ion postulated to form by intra-

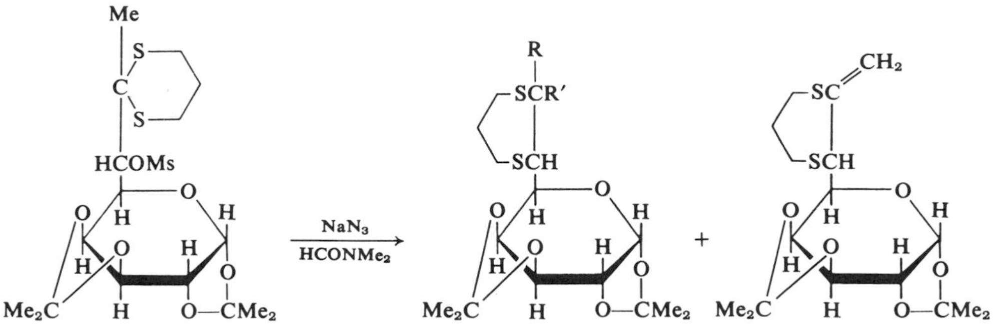

65a

65b R = Me, R′ = N₃
65c R = N₃, R′ = Me

65d

molecular attack at C-4 is less stabilized by an ethyl than by a benzyl substituent. A variety of substituents act upon 1,2-O-isopropylidene-3,4-di-O-(methylsulfonyl)-5-thio-α-D-xylopyranose to cause ring contraction by initial intramolecular displacement, the final products being 5-substituted 1,2-O-isopropylidene-3-O-(methylsulfonyl)-4-thiopentofuranose derivatives.[202d]

S-Alkylation of 6-S-benzyl-1,2-O-isopropylidene-3-O-methyl-6-thio-α-D-xylo-hexofuranos-5-ulose with methyl fluorosulfonate and subsequent deprotonation of the resulting 6-sulfonium salt produced the corresponding 6-ylid, which was added to the alkenic bond of acrolein to form a disubstituted cyclopropane.[202e]

B. Sugars Having Sulfur in the Ring

This class of synthetic sugars is the subject of a detailed article by Paulsen and Todt.[203] Table I lists a number of reported examples containing this structural feature.

Simple aliphatic 4- and 5-thioaldehydes cyclize spontaneously in solution, whereas 6-thioaldehydes appear to favor the acyclic tautomer[220]; 4-thioaldofuranose and 5-thioaldopyranose derivatives are formed by liberation of the nucleophilic thiol group from appropriate precursors under hydrolytic conditions. Thus, 5-thio-D-xylopyranose (67), the first example[211,212] of this series, was generated[211] by acid hydrolysis of 66, and 4-thio-D-ribofuranose (as the tetraacetate) resulted[7] from saponification of methyl 4-S-acetyl-2,3-O-isopropylidene-4-thio-β-D-ribopyranoside and acetolysis. 5-Seleno-D-xylopyranose[125] has been prepared under mild conditions; the same sequence of reactions was unsuccessful in the D-ribo series.[221] The 6-thioseptanoside

TABLE I

REFERENCES TO CONFIGURATIONAL SERIES IN WHICH S-RING SUGAR DERIVATIVES
HAVE BEEN PREPARED

	References		
Parent Sugar	4-Thio	5-Thio	6-Thio
D-Ribose	7,204 (L)	205–207	
D-Arabinose	208	209 (L)	
D-Xylose	210	211, 212	
2-Deoxy-D-*erythro*-pentose	213	206	
D-Glucose	150, 214	8, 215	
6-Deoxy-L-mannose		215a	
L-Idose		216	
6-Deoxy-D-idose	216a	916 (L)	
D-Galactose			217, 218
6-Deoxy-L-talose		151	
D-Fructose		219	219
1-Deoxy-D-fructose			219b
2-Acetamido-2-deoxy-D-glucose		219a	

(**68**, R = R′ = Ac) was isolated in low yield by acetolysis of 3,5-di-*O*-methyl-6-thio-D-galactofuranose, whereas aqueous hydrolysis afforded 3,5-di-*O*-methyl-1,6-thioanhydro-β-D-galactofuranose (**69**). 2,3,4,5-Tetra-*O*-methyl-6-thio-D-galactose ethylene acetal undergoes acid hydrolysis, with the formation[216] of the protected septanose (**68** R = Me, R′ = H), and acetolysis of 6-thio-*aldehydo*-D-galactose pentaacetate is reported[218] to generate 6-thio-α,β-D-galactoseptanose pentaacetate.

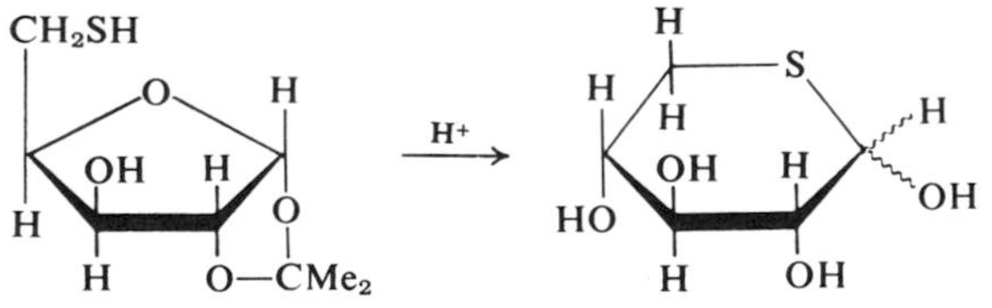

66   67 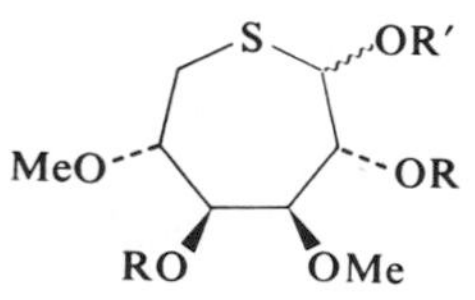   68 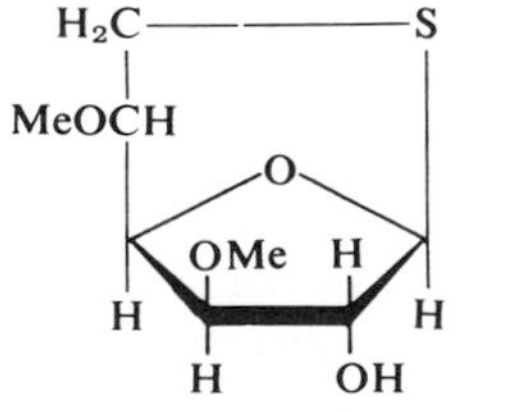   69

The S-ring structures are quite stable, and undergo mutarotation[55a] very slowly, if at all, because of the decrease in basicity wrought by introduction of the sulfur atom. The 5-thioaldopyranose ring is unaffected by iodine, whereas 4-thioaldofuranoses equilibrate with the 4-thiopyranose tautomer and are converted into disulfides[204b] by the halogen. 4-Thio-α-D-xylopyranose undergoes the Amadori rearrangement and then the browning reaction in the presence of glycine, but only to a fraction of the extent[222] of that for D-xylopyranose. 5-Thio-D-xylopyranose[222] and 4-thio-D-ribofuranose[223] have been converted into acetylated glycosyl halides, and these coupled with a variety of nucleophiles—for instance, with 2,4-diethoxy-5-methylpyrimidine to form, after deprotection, 1-(5-thio-β-D-xylopyranosyl- and -4-thio-α,β-D-ribofuranosyl)thymine, respectively. 2,3,4,6-Tetra-O-acetyl-5-thio-α-D-glucopyranosyl bromide reacts with silver diphenyl phosphate to afford, after removal of the phenyl and acetyl groups, 5-thio-D-glucopyranosyl phosphate.[224] 4-Thio-D-ribofuranose tetraacetate quite readily undergoes base-catalyzed elimination to yield[225] 3-acetoxy-5-(acetoxymethyl)thiophene.

5-Thio-D-xylopyranose acts as an inhibitor of β-D-xylosidase and xylobiase,[226] and 5-thio-D-glucopyranose inhibits release of insulin,[82] transport of glucose,[82,227] and phosphorylation of D-mannose by ox-heart particulate hexokinase.[228] 5-Thio-D-glucopyranose also causes *reversible* inhibition of sperm-cell development in mice, without displaying acute toxicity.[228a]

The acid-lability of 5-thioaldopyranosides[203,207] exceeds that of the oxygenated analogues; however, S-oxidation to the sulfoxide or sulfone reverses this trend, the latter compounds being inert[229] to $0.5M$ hydrochloric acid at 75°.

The sulfoxide of 2,3-di-O-methyl-1,4-thioanhydro-DL-threitol reacts with 1 equivalent of acetic anhydride in refluxing benzene[230] to form 1,2,3-tri-O-acetyl-4-thio-DL-threofuranose. Foster and co-workers[231,232] have studied 1,4-oxathian-2-ol derivatives as precursors of optically active sulfoxides. $(2R,6S)$-2-Methoxy-6-methyl-1,4-oxathiane results[231] from methyl α-L-rhamnopyranoside after sequential treatment with periodate, borohydride, p-toluenesulfonyl chloride, and sodium sulfide; the bicyclic analogue methyl 2,6-thioanhydro-β-D-talopyranoside (**70**) is formed by saponification[232] of

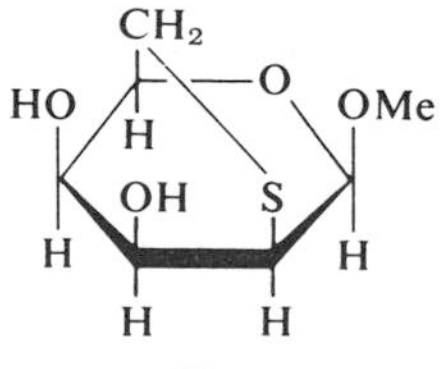

70 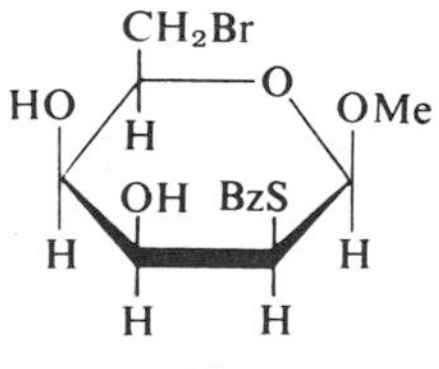   71

a 2-benzoylthio-6-bromo-6-deoxy precursor (71). Sulfoxides of the latter oxathiolane were prepared by periodate oxidation of the cyclic carbonate derivative, and their chirality at sulfur was ascertained on the basis of the n.m.r. shifts induced by the S–O bond.[232]

C. Naturally Occurring Terminal-Thio Sugar Derivatives

9-(5-S-Methyl-5-thio-β-D-ribofuranosyl)adenine (72) was identified[233] as a component[234] of a yeast nucleoside. The sulfonium ion (73) is an intermediate ("active methionine")[235] in enzymic transmethylation reactions in biological systems, operating reversibly with S-(5'-deoxyadenosine-5'-yl)-homocysteine (74) as a carrier of the one-carbon fragment in biochemical methylation. Compound 74 has been isolated from natural sources,[236] prepared enzymically,[237] and synthesized[238]; although 72 invariably contaminates the product isolated from biological sources,[239] pure 73 has been synthesized by the action of methyl iodide[240] on 74. Dilute mineral acid[237b] or enzymes present in crude extracts of rat liver[241] cleave the glycosidic bond, and hydrolysis with a strong acid liberates D-ribose and L-homocysteine.[237b]

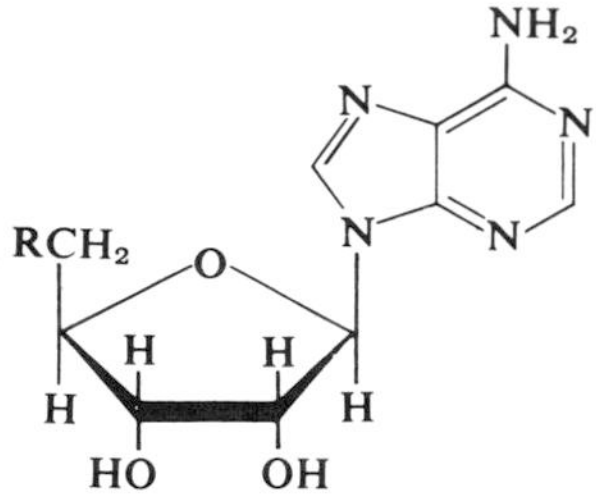

72 R = MeS—
73 R = $^-O_2CCH(NH_2)CH_2CH_2{}^+S(CH_3)$— (L)
74 R = $HO_2CCH(NH_2)CH_2CH_2S$—(L)

A 1-deoxyglycerol-1-yl glycoside of 6-deoxy-α-D-glucopyranose 6-sulfonic acid was identified[242] by synthesis of the aldose derivative as a component in the sulfolipids isolated from chloroplasts; other syntheses of 6-deoxy-D-glucose 6-sulfonic acid are discussed in Section III,A,3, and its biological aspects are discussed in Vol. IIB, Chapter 44, Section IV,B,2. Simpler C-sulfonic acids arise as artifacts of sulfite-pulping processes (see Chapter 23, Section II,D,3). A 2-amino-2,6-dideoxyhexose 6-sulfonic acid was identified in hydrolyzates of *Halococcus* sp. cell walls.[242a]

D. Thioalditols and Related Acyclic Derivatives

Partial hydrogenolysis[243] or photolysis[244] converts dithioacetals into the corresponding 1-thioalditol derivatives; displacement reactions (Section III,A) offer an alternative source, and borohydride reduction of 6-S-ethyl-6-thio-D-galactose was used to produce[118] 1-S-ethyl-1-thio-L-galactitol. Aqueous solutions of D-glucose react with hydrogen sulfide under elevated pressure at 150° to form[245] 1-thio-D-glucitol.

Protected 1-halo-1-thioalditols (see Vol. IA, Chapter 10, Section V,B) undergo nucleophilic displacements to afford a variety of 1-C-substituted 1-thioalditol derivatives; 2,3,4,5,6-penta-O-acetyl-1-bromo-1-S-alkyl(methyl or ethyl)-1-thio-D-galactitol reacts with wet acetone, with thiourea (and subsequently with sodium pyrosulfite), and with 2,4-bis(trimethylsilyl)-thymine, forming the 2,3,4,5,6-pentaacetates of a monothiohemiacetal,[246] a dithiohemiacetal,[247] and the acyclic nucleoside analogue 1-S-ethyl-1-thio-1-(thymin-1-yl)-D-galactitol,[248,248a] respectively.

In a successful attempt to minimize the toxicity of 1,2-dithioglycerol (*British Anti-Lewisite*) without loss of its capacity to detoxify heavy-metal ions, 3-O-(β-D-glucopyranosyl)-1,2-dithio-DL-glycerol was prepared[249] by bromination of allyl β-D-glucopyranoside and subsequent displacement of bromine by thiolacetate ion followed by saponification, whereas attempted saponification of 1,3-bis(acetylthio)-2-propyl-β-D-glucopyranoside (as the tetraacetate) resulted in formation[249] of the anhydride **75**. The corresponding (deacetylated) 1,3-bis(benzylthio)propyl glycoside underwent debenzylation by sodium in liquid ammonia to yield[250] a dithiol.

Acid hydrolysis of 1,2:5,6-di-O-isopropylidene-D-talitol 3,4-(trithio-carbonate) yielded 3,4-S-isopropylidene-3,4-dithio-D-talitol; an external-return mechanism was proposed.[251] 3,4-Di-O-(methylsulfonyl)-1,2:5,6-bis(thioanhydro)-L-iditol is reported[252] to react with carbon disulfide plus sodium methoxide to form the hexathiohexitol derivative (**76**).

75 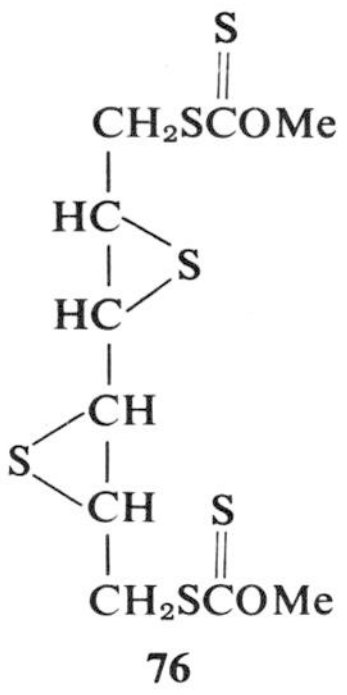   76

REFERENCES

1. A. L. Raymond, *Advan. Carbohyd. Chem.*, **1**, 129 (1945).
2. E. E. Reid, "Organic Chemistry of Bivalent Sulfur," Vol. I, Chemical Publishing Co., New York, 1958, pp. 391–394; Vol. II, 1960, pp. 221–222; Vol. III, 1960, pp. 349–361.

3. D. Horton and D. H. Hutson, *Advan. Carbohyd. Chem.*, **18**, 123 (1963).

4. J. Staněk, M. Černý, J. Kocourek, and J. Pacák, "The Monosaccharides," Academic Press, New York, 1963.

5. D. Horton, *Methods Carbohyd. Chem.*, **2**, 433 (1963).

6. D. Horton, *Methods Carbohyd. Chem.*, **2**, 368 (1963).

7. M. Bobek and R. L. Whistler, *Methods Carbohyd. Chem.*, **6**, 292 (1972).

8. R. L. Whistler and W. C. Lake, *Methods Carbohyd. Chem.*, **6**, 286 (1972).

9. (a) D. Horton, *Pure Appl. Chem.*, **42**, 301 (1975); (b) D. Horton and J. D. Wander, *J. Org. Chem.*, **39**, 1859 (1974).

10. P. L. Durette and D. Horton, *Chem. Commun.*, 1608 (1970); *Carbohyd. Res.*, **18**, 419 (1971); R. L. Girling and G. A. Jeffrey, *ibid.*, **27**, 257 (1973).

10a. D. Keglević, *Advan. Carbohyd. Chem. Biochem.*, **36**, 57 (1979).

11. A. Kjær, *Fortschr. Chem. Org. Naturst.*, **18b**, 122 (1960); M. G. Ettlinger and A. Kjær, *Recent Advan. Phytochem.*, **1**, 58 (1968); A. Kjær and P. O. Larsen, *Biosynthesis*, **4**, 179 (1976).

12. P. J. Robiquet and F. Boutron, *J. Pharm.*, **17**, 279 (1831).

13. G. Zinner, *Deut. Apoth. Ztg.*, **98**, 335 (1958).

14. C. H. VanEtten, M. E. Daxenbichler, and I. A. Wolff, *J. Agr. Food Chem.*, **17**, 483 (1969).

15. J. Gadamer, *Arch. Pharm.*, **235**, 44, 83 (1897); *Ber.*, **30**, 2322, 2327 (1897).

16. A. Bussy, *J. Pharm.*, **26**, 39 (1840); *Ann.*, **34**, 223 (1840); F. Boutron and E. Frémy, *Ann.*, **34**, 230 (1840).

17. W. Schneider and F. Wrede, *Ber.*, **47**, 225 (1914).

18. H. Will and W. Körner, *Ann.*, **125**, 257 (1863); H. Will and A. Laubenheimer, *Ann.*, **199**, 150 (1879); J. Gadamer, *Arch. Pharm.*, **237**, 111, 507 (1899); H. Schmid and P. Karrer, *Helv. Chim. Acta*, **31**, 1017, 1087 (1948); O.-E. Schultz and R. Gmelin, *Arch. Pharm.*, **287**, 342 (1954).

19. M. G. Ettlinger and A. J. Lundeen, *J. Amer. Chem. Soc.*, **78**, 4172 (1956).

20. M. G. Ettlinger and A. J. Lundeen, *J. Amer. Chem. Soc.*, **79**, 1764 (1957).

21. S. R. Jensen and A. Kjær, *Acta Chem. Scand.*, **25**, 3891 (1971).

22. E. B. Astwood and M. A. Greer, *J. Biol. Chem.*, **181**, 121 (1949).

23. K. A. Jensen, J. Conti, and A. Kjær, *Acta Chem. Scand.*, **7**, 1267 (1953).

24. M. E. Daxenbichler, C. H. VanEtten, and I. A. Wolff, *Biochemistry*, **4**, 318 (1965).

25. M. H. Benn and L. Yelland, *Can. J. Chem.*, **45**, 1595 (1967).

26. R. D. Bright and C. R. Hauser, *J. Amer. Chem. Soc.*, **61**, 618 (1939).

27. A detailed discussion of this topic has been presented by M. A. Greer, *Recent Progr. Hormone Res.*, **18**, 187 (1962).

28. M. G. Ettlinger, G. P. Dateo, B. W. Harrison, T. J. Mabry, and C. P. Thompson, *Proc. Nat. Acad. Sci. U.S.*, **47**, 1875 (1961).

29. Z. Nagashima and M. Uchiyama, *Bull. Agr. Chem. Soc. Jap.*, **23**, 555 (1959).

30. E. T. Reese, R. C. Clapp, and M. Mandels, *Arch. Biochem. Biophys.*, **75**, 228 (1958).

31. M. A. Jermyn, *Aust. J. Biol. Sci.*, **8**, 577 (1965).

32. E. L. Oginsky, A. E. Stein, and M. A. Greer, *Proc. Soc. Exp. Biol. Med.*, **119**, 360 (1965).

33. I. Goodman, J. R. Fouts, E. Bresnick, R. Menegas, and G. H. Hitchings, *Science*, **130**, 450 (1959).

34. Burroughs Wellcome & Co. (U.S.) Inc., British Patent 913,348 (1962); *Chem. Abstr.*, **59**, 738 (1963).

34a. W. W. Pigman, *J. Res. Nat. Bur. Stand.*, **26**, 197 (1941); see also ref. 42.

34b. J. Monod, in "Enzymes: Units of Biological Structure and Function" (O. Gaebler, ed.), Academic Press, New York, 1955, pp. 7–28; *Chem. Abstr.*, **50**, 12137 (1956).

34c. J. Monod, J. P. Changeaux, and F. Jacob, *J. Mol. Biol.*, **6**, 306 (1963); J. Monod, J. W. Wyman, and J. P. Changeaux, *ibid.*, **12**, 88 (1965); see E. Zeffren and P. L. Hall, "The Study of Enzyme Mechanisms," Wiley (Interscience), New York, 1973, pp. 259–264.

35. D. V. Parke, Ph.D. Dissertation, London, 1952; cited by R. T. Williams in "Detoxication Mechanisms," 2nd ed., Chapman and Hall, London, 1959, p. 492.

36. J. W. Clapp, *J. Biol. Chem.*, **223**, 207 (1956).

37. H. J. Hansen, W. G. Giles, and S. B. Nadler, *Proc. Soc. Exp. Biol. Med.*, **113**, 163 (1963).

38. J. Kaslander, *Biochim. Biophys. Acta*, **71**, 730 (1963).

39. J. H. Strömme, *Biochem. Pharmacol.*, **14**, 393 (1965).

40. J. Kaslander, A. Kaars Sijpestein, and G. J. M. Van der Kerk, *Biochim. Biophys. Acta*, **52**, 396 (1961).

41. G. J. Dutton and H. P. A. Illing, *Biochem. J.*, **129**, 539 (1972).

42. E. Fischer and K. Delbrück, *Ber.*, **42**, 1476 (1909).

43. M. H. Benn, L. N. Owen, and A. M. Creighton, *J. Chem. Soc.*, 2800 (1958).

44. H. Weidmann, H. K. Zimmerman, Jr., and J. R. Monk, *Ann.*, **628**, 255 (1959).

45. M. L. Wolfrom, W. A. Cramp, and D. Horton, *J. Org. Chem.*, **30**, 3056 (1965).

46. H. S. Isbell, *Annu. Rev. Biochem.*, **9**, 65 (1940).

47. R. U. Lemieux, *Advan. Carbohyd. Chem.*, **9**, 1 (1954).

48. R. U. Lemieux, *Can. J. Chem.*, **29**, 1079 (1951).

49. A. J. Rhind-Tutt and C. A. Vernon, *J. Chem. Soc.*, 4637 (1960).

50. B. Capon, P. M. Collins, A. A. Levy, and W. G. Overend, *J. Chem. Soc.*, 3242 (1964); compare M. Blanc-Muesser, J. Defoye, and H. Driguez, *Carbohyd. Res.*, **67**, 305 (1978) for synthetic applications in stereoselective syntheses of 1-thio-aldosides.

51. E. J. Reist, P. A. Hart, L. Goodman, and B. R. Baker, *J. Amer. Chem. Soc.*, **81**, 5176 (1959).

51a. B. Erbing and B. Lindberg, *Acta Chem. Scand.*, *Ser. B*, **330**, 611 (1976).

51b. G. Magnusson, *J. Org. Chem.*, **41**, 4110 (1976).

51c. G. Magnusson, *Carbohyd. Res.*, **56**, 188 (1977).

51d. G. Magnusson, *J. Org. Chem.*, **42**, 913 (1977).

52. D. H. Hutson, *J. Chem. Soc. (C)*, 442 (1967).

53. R. H. Bell, D. Horton, and M. J. Miller, *Carbohyd. Res.*, **9**, 201 (1969).

54. F. Weygand, H. Ziemann, and H. J. Bestmann, *Ber.*, **91**, 2534 (1958).

55. J. Defaye, D. Horton, S. S. Kokrady, and Z. Machon, *Carbohyd Res.*, **43**, 265 (1975).

55a. I. Farkaš, R. Bognár, M. M. Meynhart, A. K. Tarnai, M. Bihari, and J. Tamas, *Acta Chim. Acad. Sci. Hung.*, **84**, 325 (1975); *Chem. Abstr.*, **83**, 10, 707 (1975).

55b. P. M. Collins and B. R. Whitton, *Carbohyd. Res.*, **36**, 293 (1974).

55c. P. M. Collins and B. R. Whitton, *J. Chem. Soc. Perkin I*, 1069 (1974).

55d. H. S. Isbell and W. W. Pigman, *Advan. Carbohyd. Chem. Biochem.*, **24**, 13 (1969).

56. W. Schneider, R. Gille, and K. Eisfeld, *Ber.*, **61**, 1244 (1928).

57. W. Schneider and H. Leonhardt, *Ber.*, **62**, 1384 (1929).

58. W. Schneider and A. Bansa, *Ber.*, **64**, 1321 (1931).

59. F. Wrede, *Z. Physiol. Chem.*, **119**, 46 (1922).

60. N. K. Richtmyer, C. J. Carr, and C. S. Hudson, *J. Amer. Chem. Soc.*, **65**, 1477 (1943).

61. M. Černý and J. Pacák, *Chem. Listy*, **52**, 2090 (1958); *Collect. Czech. Chem. Commun.*, **24**, 2566 (1959).
62. J. Bougalt, E. Cattelain, and P. Chabrier, *C. R. Acad. Sci.*, **208**, 657 (1939); *Bull. Soc. Chim. Fr.*, 780 (1940); M. L. Wolfrom and J. V. Karabinos, *J. Amer. Chem. Soc.*, **66**, 909 (1944); H. G. Fletcher, Jr., and N. K. Richtmyer, *Advan. Carbohyd. Chem.*, **5**, 1 (1950).
63. H. G. Fletcher, Jr., *J. Amer. Chem. Soc.*, **69**, 706 (1947).
64. W. Meyer zu Reckendorf and W. A. Bonner, *J. Org. Chem.*, **26**, 4596 (1961).
65. T. Ito, *Agr. Biol. Chem.* (Tokyo), **26**, 831 (1962).
66. M. L. Wolfrom, D. Horton, and D. H. Hutson, *J. Org. Chem.*, **28**, 845 (1963).
66a. R. L. Whistler, A. K. M. Anisuzzaman, and J. C. Kim, *Carbohyd. Res.*, **31**, 237 (1973).
67. M. Akagi, S. Tejima, M. Haga, Y. Hirokawa, M. Yamada, M. Ishiguro, and D. Mizuno, *Yakugaku Zasshi*, **87**, 287 (1967).
68. W. Schneider, D. Clibbens, G. Hüllweck, and W. Steibelt, *Ber.*, **47**, 1258 (1914).
69. F. Wrede, *Ber.*, **52**, 1756 (1919).
70. M. Gehrke and W. Kohler, *Ber.*, **64**, 2696 (1931); Schering–Kahlbaum A.-G., German Patent 557,247 (1930); *Chem. Abstr.*, **27**, 374 (1933); British Patent 373,755 (1932); *Chem. Zentralbl.*, **103**, II, 2992 (1932).
71. W. Schneider and K. Eisfeld, *Ber.*, **61**, 1260 (1928); W. A. Bonner and J. E. Kahn, *J. Amer. Chem. Soc.*, **73**, 2241 (1951).
72. M. Akagi, S. Tejima, M. Haga, and M. Sakata, *Chem. Pharm. Bull.* (Tokyo), **11**, 1081 (1963).
73. M. Sakata, M. Haga, and S. Tejima, *Carbohyd. Res.*, **13**, 379 (1970).
74. R. A. Zingaro and J. K. Thomson, *Carbohyd. Res.*, **29**, 147 (1973); compare J. R. Daniel and R. A. Zingaro, *ibid.*, **64**, 69 (1978).
75. E. R. McGusty and B. M. Sutton, German Patent 2,051,495 (1971); *Chem. Abstr.*, **75**, 77223 (1971).
76. V. Horák, *Chem. Listy*, **48**, 414 (1954); *Collect. Czech. Chem. Commun.*, **19**, 1238 (1954).
77. M. Černý, J. Vrkoć, and J. Staněk, *Chem. Listy*, **52**, 311 (1958); *Collect. Czech. Chem. Commun.*, **24**, 64 (1959).
78. M. Černý and J. Pacák, *Collect. Czech. Chem. Commun.*, **26**, 2084 (1961).
79. J. Opplt, U.S. Patent 2,585,580 (1952); *Chem. Abstr.*, **46**, 5270 (1952).
80. M. D. Bray, U.S. Patent 2,498,200 (1950); *Chem. Abstr.*, **44**, 4641 (1950).
81. E. J. Hanus, U.S. Patent 2,719,812 (1955); *Chem. Abstr.*, **50**, 1269 (1956).
82. B. Hellmann, Å. Lernmark, J. Sehlin, I. B. Taljedal, and R. L. Whistler, *Biochem. Pharmacol.*, **22**, 29 (1973); *Chem. Abstr.*, **78**, 69506 (1973).
83. E. W. Underhill and L. R. Wetter, *Plant Physiol.*, **44**, 589 (1969); *Chem. Abstr.*, **71**, 10361 (1969).
84. Schering–Kahlbaum A.-G., German Patent 573,629 (1933); *Chem. Abstr.*, **27**, 4349 (1933); British Patent 386,562 (1933); *Chem. Abstr.*, **27**, 4350 (1933); H. A. Swartz, F. N. Andrews, and J. E. Christian, *Can. Pharm. J.*, **93**, 64 (1960).
85. M. M. Tinao and A. F. Zubiri-Vidal, *Arch. Inst. Farmacol. Exp.* (Madrid), **4**, 87 (1952); *Chem. Abstr.*, **47**, 5636 (1953); M. M. Tinao, *An. Inst. Farmacol. Espan.*, **2** 345 (1953); *Chem. Abstr.*, **48**, 8947 (1954).
86. B. M. Sutton, E. (R.) McGusty, D. T. Salz, and M. J. DiMartino, *J. Med. Chem.*, **15**, 1095 (1972).
87. T. Matsumura, *Yonago Igaku Zasshi*, **21**, 508 (1970); *Chem. Abstr.*, **77**, 552 (1972).
88. C. J. V. Smith, *Physiol. Behav.*, **9**, 391 (1972); *Chem. Abstr.*, **78**, 122,264 (1973); D. Simkova and K. Boda, *Vet. Med.* (Prague), **16**, 586 (1971); *Chem. Abstr.*, **76**,

108,698 (1972); J. A. Owen, Jr., W. Parson, and K. R. Crispell, *Metabolism*, **2**, 362 (1953); J. Mayer and N. B. Marshall, *Nature*, **178**, 1399 (1956); *C. R. Acad. Sci.*, **242**, 169 (1956).

89. J. Markanik and R. Skarda, *Vet. Med.* (Prague), **17**, 27 (1972); *Chem. Abstr.*, **77**, 32,297 (1972).

90. G. F. Gray, R. A. Liebelt, and A. G. Liebelt, *Cancer Res.*, **20**, 1101 (1960); G. Levai, J. Laczko, and L. Muszbek, *Acta Histochem.*, *Suppl.*, 193 (1971); *Chem. Abstr.*, **76**, 21,586 (1972).

91. T. J. Luparello, *J. Psychosom. Res.*, **13**, 113 (1969); *Chem. Abstr.*, **71**, 11183 (1969).

92. K. Ei, *Showa Igakku Zasshi*, **32**, 318 (1972); *Chem. Abstr.*, **78**, 52,734 (1973).

93. K. Takaraji, *Showa Igakku Zasshi*, **31**, 353 (1971); *Chem. Abstr.*, **77**, 70,139 (1972).

94. R. Nakade, *Nippon Naibumpi Gakkai Zasshi*, **34**, 572 (1958); *Chem. Abstr.*, **53**, 3,465 (1959).

95. R. E. Damschroeder, U.S. Patent 2,597,856 (1952); *Chem. Abstr.*, **47**, 435 (1953).

96. C. D. Hurd and W. A. Bonner, *J. Amer. Chem. Soc.*, **67**, 1764 (1945).

97. D. Horton, M. L. Wolfrom, and H. G. Garg, *J. Org. Chem.*, **28**, 2992 (1963).

98. R. H. Bell and D. Horton, *Carbohyd. Res.*, **9**, 187 (1969).

99. D. Horton and M. Sakata, *Carbohyd. Res.*, **39**, 67 (1975).

100 (a) E. Fischer, B. Helferich, and P. Ostman, *Ber.*, **53**, 873 (1920); (b) A. Müller and A. Wilhelms, *Ber.*, **74**, 698 (1941).

101. E. Fischer, *Ber.*, **47**, 1378 (1914).

102. W. Schneider, *Ber.*, **49**, 1638 (1916); W. Schneider and O. Stiehler, *Ber.*, **52**, 2131 (1919); W. Schneider and A. Beuther, *Ber.*, **52**, 2135 (1919).

103. V. Prey and F. Grundschober, *Monatsh. Chem.*, **91**, 358 (1960).

104. M. Akagi, S. Tejima, and M. Haga, *Chem. Pharm. Bull.* (Tokyo), **8**, 1114 (1960).

105. Y. Nitta, M. Kuranari, and T. Kondo, *Yakugaku Zasshi*, **81**, 1166 (1961); *Chem. Abstr.*, **56**, 6075 (1962).

106. B. Helferich, D. Türk, and F. Stoeber, *Ber.*, **89**, 2220 (1956).

107. D. R. Kalkwarf, *Nucleonics*, **18** (5), 76 (1960); D. G. Doherty, in "Radiation Protection and Recovery" (A. Hollaender, ed.), Pergamon Press, London, 1960, p. 45.

108. M. Akagi, S. Tejima, and M. Haga, *Chem. Pharm. Bull.* (Tokyo), **9**, 360 (1961); W. Meyer zu Reckendorf and W. A. Bonner, *Ber.*, **94**, 2431 (1961).

109. D. Horton and M. L. Wolfrom, *J. Org. Chem.*, **27**, 1794 (1962).

110. M. L. Wolfrom, W. A. Cramp, and D. Horton, *J. Org. Chem.*, **29**, 2302 (1964).

111. Y. Tsuzuki, K. Tanabe, M. Akagi, and S. Tejima, *Bull. Chem. Soc. Jap.*, **37**, 162 (1964); **40**, 628 (1967).

111a. A. T. Tu, J. Lee, and Y. C. Lee, *Carbohyd. Res.*, **67**, 295 (1978).

112. T. Maki, H. Nakamura, S. Tejima, and M. Akagi, *Chem. Pharm. Bull.* (Tokyo), **13**, 764 (1965); H. Nakamura, S. Tejima, and M. Akagi, *ibid.*, **13**, 1478 (1965).

113. R. J. Ferrier and N. Vethaviyasar, *Carbohyd. Res.*, **58**, 481 (1977).

113a. J. D. Wander and D. Horton, *Advan. Carbohyd. Chem. Biochem.*, **32**, 15 (1976).

114. F. Wrede, *Z. Physiol. Chem.*, **115**, 284 (1921).

115. A. L. Raymond, *J. Biol. Chem.*, **107**, 85 (1934).

116. R. S. Tipson, *Advan. Carbohyd. Chem.*, **8**, 107 (1953).

117. D. H. Ball and F. W. Parrish, *Advan. Carbohyd. Chem. Biochem.*, **24**, 139 (1969).

118. S. B. Baker, *Can. J. Chem.*, **33**, 1102 (1955).

119. S. B. Baker, *Can. J. Chem.*, **33**, 1459 (1955).

120. J. Fernández-Bolaños and R. Guzmán de Fernández-Bolaños, *An. Real Soc. Espan. Fis. Quim.*, **55B**, 693 (1959).

121. J. F. Carson and W. D. Maclay, *J. Amer. Chem. Soc.*, **70**, 2220 (1948).

122. R. L. Whistler and D. G. Medcalf, *Abstr. Pap. Amer. Chem. Soc. Meeting*, **142**, 15D (1962).
123. G. Hebblethwaite, R. F. Schwenker, and E. Pacsu, *Proc. Cellul. Conf., 2nd (Syracuse)*, 214 (1959); L. Lifland and E. Pacsu, *Text. Res. J.*, **32**, 170 (1962).
124. E. F. Izard and P. W. Morgan, *Ind. Eng. Chem.*, **41**, 617 (1949).
125. T. van Es and R. L. Whistler, *Tetrahedron*, **23**, 2849 (1967).
125a. G. C. Chen, R. A. Zingaro, and C. R. Thompson, *Carbohyd. Res.*, **39**, 61 (1975).
126. T. J. Adley and L. N. Owen, *Proc. Chem. Soc.*, 418 (1961).
127. A. M. Creighton and L. N. Owen, *J. Chem. Soc.*, 1024 (1960).
128. P. Bladon and L. N. Owen, *J. Chem. Soc.*, 585 (1950).
129. J. H. Chapman and L. N. Owen, *J. Chem. Soc.*, 579 (1950).
130. C. J. Clayton and N. A. Hughes, *Chem. Ind.* (London), 1795 (1962).
131. J. Kocourek, M. Tichá, and V. Kiraček, *Angew. Chem.*, **76**, 50 (1964).
131a. G. E. McCasland, A. B. Zanlungo, and L. J. Durham, *J. Org. Chem.*, **39**, 1462 (1974); *ibid.*, **41**, 1125 (1976).
131b. J. Kuszmann and P. Sohár, *Carbohyd. Res.*, **48**, 23 (1976).
131c. M. Jarman and L. J. Griggs, *Carbohyd. Res.*, **44**, 317 (1975).
131d. D. Trimnell, E. I. Stout, W. M. Doane, C. R. Russell, V. Beringer, M. Saul, and G. Van Gessel, *J. Org. Chem.*, **40**, 1337 (1975).
131e. M. Fuertes, M. T. García-Lopez, G. García Muñoz, and R. Madroñero, *J. Carbohyd., Nucleosides, Nucleotides*, **2**, 277 (1975).
132. J. M. Heap and L. N. Owen, *J. Chem. Soc. (C)*, 712 (1970).
132a. U. G. Nayak and R. L. Whistler, *J. Org. Chem.*, **34**, 3819 (1969).
133 (a) A. C. Maehly and T. Reichstein, *Helv. Chim. Acta*, **30**, 496 (1947); (b) N. C. Jamieson and R. K. Brown, *Can. J. Chem.*, **39**, 1765 (1961).
134. N. K. Kochetkov, V. N. Shibaev, Yu. Yu. Kusov, and M. F. Troitskiĭ, *Izv. Akad. Nauk SSSR, Ser. Khim.*, 425 (1973).
135. J. A. Mills, *Advan. Carbohyd. Chem.*, **10**, 1 (1955).
136. J. Hill, L. Hough, and A. C. Richardson, *Proc. Chem. Soc.*, 346 (1963).
137. L. Goodman, *Advan. Carbohyd. Chem.*, **22**, 109 (1967).
138. K. Freudenburg and A. Wolf, *Ber.*, **60**, 232 (1927); see also A. K. Sanyal and C. B. Purves, *Can. J. Chem.*, **34**, 426 (1956); M. Černý, J. Pacák, and V. Jina, *Monatsh. Chem.*, **94**, 632 (1963).
139. D. Horton and H. S. Prihar, *Carbohyd. Res.*, **4**, 115 (1967).
140. M. L. Wolfrom and A. B. Foster, *J. Amer. Chem. Soc.*, **78**, 1399 (1956).
140a. Such dithiocarbamates as **33** are, however, useful as precursors for deoxy sugars by action of free-radical reagents; see R. H. Bell, D. Horton, D. M. Williams, and E. Winter-Mihaly, *Carbohydr. Res.*, **58**, 109 (1977), and specifically deuterated deoxy sugars are also accessible; J. J. Patroni and R. V. Stick, *Chem. Commun.*, 449 (1978).
141. (a) D. Trimnell, W. M. Doane, C. R. Russell, and C. E. Rist, *Carbohyd. Res.*, **17**, 319 (1971); D. Trimnell and W. M. Doane, *Methods Carbohyd. Chem.*, **7**, 41 (1976); (b) B. S. Shasha and W. M. Doane, *Carbohyd. Res.*, **34**, 370 (1974); (c) D. Trimnell, W. M. Doane, and C. R. Russell, *ibid.*, **22**, 351 (1972); (d) B. S. Shasha, D. Trimnell, and W. M. Doane, *ibid.*, **32**, 349 (1974); B. S. Shasha, D. Trimnell, E. I. Stout, and W. M. Doane, *Methods Carbohyd. Chem.*, **7**, 36 (1976).
142. G. Descotes and A. Faure, *Synthesis*, 449 (1976).
142a. E. Bradbury and J. Nagyvary, *Nucleic Acid Res.*, **3**, 2437 (1976).
142b. E. M. Acton, K. J. Ryan, and L. Goodman, *J. Amer. Chem. Soc.*, **89**, 467 (1967).
143. B. R. Baker, K. Hewson, L. Goodman, and A. Benitez, *J. Amer. Chem. Soc.*, **80**, 6577 (1958).

144. L. Goodman and J. E. Christensen, *J. Amer. Chem. Soc.*, **83**, 3823 (1961).
145. W. Meyer zu Reckendorf and W. A. Bonner, *Proc. Chem. Soc.*, 429 (1961).
146. L. Goodman and J. E. Christensen, *J. Org. Chem.*, **28**, 2610 (1963).
147. T. Sekiya and T. Ukita, *Chem. Pharm. Bull.* (Tokyo), **15**, 542, 1503 (1967).
148. K. Araki and S. Tejima, *Chem. Pharm. Bull.* (Tokyo), **14**, 1303 (1966).
149. S. Ishiguro and S. Tejima, *Chem. Pharm. Bull.* (Tokyo), **15**, 1478 (1967).
150. L. N. Owen and P. L. Ragg, *J. Chem. Soc.*, 1291 (1966).
151. C. L. Stevens, R. P. Glinski, G. E. Gutowski, and J. P. Dickerson, *Tetrahedron Lett.*, 649 (1967).
152. L. N. Owen, *Chem. Commun.*, 526 (1967).
153. T. van Es, *Carbohyd. Res.*, **11**, 282 (1969).
154. H. Ohle and W. Mertens, *Ber.*, **68**, 2176 (1935).
155. L. D. Hall, L. Hough, and R. A. Pritchard, *J. Chem. Soc.*, 1537 (1961).
156. S. M. Iqbal and L. N. Owen, *J. Chem. Soc.*, 1030 (1960).
157. R. W. Jeanloz, D. A. Prins, and T. Reichstein, *Helv. Chim. Acta*, **29**, 371 (1946).
158. H. R. Bolliger and D. A. Prins, *Helv. Chim. Acta*, **29**, 1061 (1946).
159. (a) J. E. Christensen and L. Goodman, *J. Amer. Chem. Soc.*, **82**, 4738 (1960); (b) **83**, 3827 (1961).
160. E. Zissis, *J. Org. Chem.*, **32**, 660 (1967).
161. L. Vegh and E. Hardegger, *Helv. Chim. Acta*, **56**, 2020 (1973).
162. E. Hardegger and W. Schüep, *Helv. Chim. Acta*, **53**, 951 (1970).
163. M. Dahlgard, B. H. Chastain, and R.-J. Lee Han, *J. Org. Chem.*, **27**, 932 (1962).
163a. M. V. Jesudason and L. N. Owen, *J. Chem. Soc. Perkin I*, 2024 (1974).
164. S. Mukherjee and A. R. Todd, *J. Chem. Soc.*, 969 (1947).
165. J. P. H. Verheyden and J. G. Moffatt, *J. Org. Chem.*, **34**, 2643 (1969).
166. R. D. Guthrie, *Chem. Ind.* (London), 2121 (1962).
167. M. Kojima, M. Watanabe, and T. Taguchi, *Tetrahedron Lett.*, 839 (1968).
167a. M. V. Jesudason and L. N. Owen, *J. Chem. Soc. Perkin I*, 2019 (1974).
168. J. Davoll, B. Lythgoe, and S. Trippett, *J. Chem. Soc.*, 2230 (1951).
169. C. D. Anderson, L. Goodman, and B. R. Baker, *J. Amer. Chem. Soc.*, **81**, 898 (1959).
170. C. D. Anderson, L. Goodman, and B. R. Baker, *J. Amer. Chem. Soc.*, **80**, 6453 (1958); **81**, 3967 (1959).
171. J. E. Christensen and L. Goodman, *J. Org. Chem.*, **28**, 2995 (1963).
172. G. Casini and L. Goodman, *J. Amer. Chem. Soc.*, **85**, 235 (1963); **86**, 1427 (1964).
173. E. J. Reist, D. F. Calkins, and L. Goodman, *J. Org. Chem.*, **32**, 2538 (1967).
174. L. Goodman, *Chem. Commun.*, 219 (1968).
175. K. J. Ryan, E. M. Acton, and L. Goodman, *J. Org. Chem.*, **33**, 3727 (1968).
176. J. Lehmann and A. A. Benson, *J. Amer. Chem. Soc.*, **86**, 4469 (1964).
177. J. Lehmann and W. Weckerle, *Carbohyd. Res.*, **22**, 23, 317 (1972).
178. D. Horton and W. N. Turner, *Carbohyd. Res.*, **1**, 444 (1966).
179. S. Inokawa, H. Yoshida, C.-C. Wang, and R. L. Whistler, *Bull. Chem. Soc. Jap.*, **41**, 1472 (1968).
180. K. Igarashi and T. Honma, *Tetrahedron Lett.*, 751 (1968); *J. Org. Chem.*, **35**, 606 (1970); compare J. Borowiecka and M. Michalska, *Carbohyd. Res.*, **68**, C8 (1979), who showed exclusive thiation of the glycal at C-1 in the uncatalyzed addition of *O,O*-dialkylphosphorodithioic acids.
181. P. Wirz and E. Hardegger, *Helv. Chim. Acta*, **54**, 2017 (1971).
182. D. L. Ingles, *Chem. Ind.* (London), 50 (1969); *Aust. J. Chem.*, **22**, 1789 (1969).
182a. H. Paulsen and G. Steinert, *Ber.*, **100**, 2467 (1967); **103**, 1834 (1970).
183. P. Brigl, H. Mühlschlegel, and R. Schinle, *Ber.*, **64**, 2921 (1931).

184. A. Ducruix, C. Pascard-Billy, D. Horton, and J. D. Wander, *Carbohyd. Res.*, **29** 276 (1973).
185. B. Berrang and D. Horton, *Chem. Commun.*, 1038 (1970).
186. M. L. Wolfrom and W. von Bebenburg, *J. Amer. Chem. Soc.*, **82**, 2817 (1960).
187. P. Brigl and R. Schinle, *Ber.*, **65**, 1890 (1932).
188. A. E. El Ashmawy, D. Horton, L. G. Magbanua, and J. M. J. Tronchet, *Carbohyd. Res.*, **6**, 299 (1968).
189. J. Defaye, *Bull. Soc. Chim. Fr.*, 1101 (1967).
190. J. Defaye, T. Nakamura, D. Horton, and K. D. Philips, *Carbohyd. Res.*, **16**, 133 (1971); J. Defaye, A. Ducruix, and C. Pascard-Billy, *Bull. Soc. Chim. Fr.*, 4514 (1970).
191. R. U. Lemieux, *Can. J. Chem.*, **29**, 1079 (1951).
192. M. L. Wolfrom and T. E. Whiteley, *J. Org. Chem.*, **27**, 2109 (1962).
193. N. A. Hughes and R. Robson, *Chem. Commun.*, 1383 (1968).
194. N. A. Hughes, R. Robson, and S. A. Saeed, *Chem. Commun.*, 1381 (1968).
195. G. S. Bethell and R. J. Ferrier, *J. Chem. Soc. Perkin I*, 1033, 2873 (1972).
195a. G. S. Bethell and R. J. Ferrier, *J. Chem. Soc. Perkin I*, 1400 (1973).
195b. G. S. Bethell and R. J. Ferrier, *Carbohyd. Res.*, **34**, 194 (1974).
196. B. Berrang, D. Horton, and J. D. Wander, *J. Org. Chem.*, **38**, 187 (1973).
196a. T. van Es and J. (J.) Rabelo, *Carbohyd. Res.*, **36**, 408 (1974).
197. K. J. Ryan, E. M. Acton, and L. Goodman, *J. Org. Chem.*, **36**, 2646 (1971).
198. M. von Saltza, J. D. Dutcher, J. Reid, and O. Wintersteiner, *J. Org. Chem.*, **28**, 999 (1963).
199. T. van Es, *Carbohyd. Res.*, **37**, 373 (1974).
200. E. V. E. Roberts, J. C. P. Schwarz, and C. A. McNab, *Carbohyd. Res.*, **7**, 311 (1968).
201. N. A. Hughes and R. Robson, *J. Chem. Soc. (C)*, 2366 (1966).
202. J. Harness and N. A. Hughes, *Chem. Commun.*, 811 (1971).
202a. T. van Es, *Carbohyd. Res.*, **46**, 237 (1976).
202b. P. Angibeaud, J. Defaye, H. Franconie, and M. Blanc-Muesser, *Carbohyd. Res.*, **49**, 209 (1976).
202c. A. Gateau-Olesker, S. D. Gero, C. Pascard-Billy, C. Riche, A. M. Sepulchre, G. Vass, and N. A. Hughes, *Chem. Commun.*, 811 (1974).
202d. W. Clegg and N. A. Hughes, *Chem. Commun.*, 300 (1975).
202e. J. M. J. Tronchet and H. Eder, *Helv. Chim. Acta*, **58**, 1799 (1975).
203. H. Paulsen and K. Todt, *Advan. Carbohyd. Chem.*, **23**, 115 (1968).
204 (a) E. J. Reist, D. E. Gueffroy, and L. Goodman, *J. Amer. Chem. Soc.*, **85**, 3715 (1963); (b) **86**, 5658 (1964).
205. C. J. Clayton and N. A. Hughes, *Chem. Ind.* (London), 1795 (1962).
206. D. C. Ingles and R. L. Whistler, *J. Org. Chem.*, **27**, 3896 (1962).
207. C. J. Clayton and N. A. Hughes, *Carbohyd. Res.*, **4**, 32 (1967); C. J. Clayton, N. A. Hughes, and T. D. Inch, *ibid.*, **45**, 55 (1975).
208. R. L. Whistler, U. G. Nayak, and A. W. Perkins, Jr., *J. Org. Chem.*, **35**, 519 (1970).
209. R. L. Whistler and R. M. Rowell, *J. Org. Chem.*, **29**, 1259 (1964).
210. E. J. Reist, L. V. Fisher, and L. Goodman, *J. Org. Chem.*, **33**, 189 (1968).
211. J. C. P. Schwarz and K. C. Yule, *Proc. Chem. Soc.*, 417 (1961); T. J. Adley and L. N. Owen, *ibid.*, 418 (1961).
212. R. L. Whistler, M. S. Feather, and D. L. Ingles, *J. Amer. Chem. Soc.*, **84**, 122 (1962); R. L. Whistler and D. L. Ingles, *Abstr. Pap. Amer. Chem Soc. Meeting*, **141**, 10D (1962).

213. U. G. Nayak and R. L. Whistler, *Chem. Commun.*, 434 (1969); *Ann.*, **741**, 131 (1970).

214. L. N. Owen, *Chem. Commun.*, 526 (1967); see also, L. Vegh and E. Hardegger, *Helv. Chim. Acta*, **56**, 2020 (1973).

215. R. M. Rowell and R. L. Whistler, *J. Org. Chem.*, **31**, 1514 (1966); R. L. Whistler, T. J. Luttnegger, and R. M. Rowell, *ibid.*, **33**, 396 (1968).

215a. A. K. M. Anisuzzaman and R. L. Whistler, *Carbohyd. Res.*, **55**, 205 (1977).

216. T. J. Adley and L. N. Owen, *J. Chem. Soc.* (C), 1287 (1966).

216a. B. Gross and F.-X. Oriez, *Carbohyd. Res.*, **36**, 385 (1974).

217. J. M. Cox and L. N. Owen, *J. Chem. Soc.* (C), 1121 (1967).

218. R. L. Whistler and C. S. Campbell, *J. Org. Chem.*, **31**, 816 (1966).

219. M. S. Feather and R. L. Whistler, *J. Org. Chem.*, **28**, 1567 (1963); M. Chmielewski and R. L. Whistler, *ibid.*, **40**, 639 (1975); M. Chmielewski, M.-S. Chen, and R. L. Whistler, *Carbohyd. Res.*, **49**, 479 (1976); M. Chmielewski and R. L. Whistler, *ibid.*, **69**, 259 (1979).

219a. A. Hasegawa, Y. Kawai, H. Kasugai, and M. Kiso, *Carbohyd. Res.*, **63**, 131 (1978).

219b. T. van Es and M. S. Feather, *Carbohyd. Res.*, **22**, 420 (1972).

220. J. M. Cox and L. N. Owen, *J. Chem. Soc.* (C), 1130 (1967).

221. J. J. Rabelo and T. van Es, *Carbohyd. Res.*, **30**, 381 (1973).

222. D. L. Ingles, *Chem. Ind.* (London), 1901 (1963).

223. B. Urbas and R. L. Whistler, *J. Org. Chem.*, **31**, 813 (1966).

224. R. L. Whistler and J. H. Stark, *Carbohyd. Res.*, **13**, 15 (1970).

225. R. L. Whistler and D. J. Hoffman, *Carbohyd. Res.*, **11**, 137 (1969).

226. M. Claeyssens and C. K. De Bruyne, *Naturwissenschaften*, **52**, 515 (1965).

227. R. L. Whistler and W. C. Lake, *Biochem. J.*, **130**, 919 (1972).

228. S. V. Paranjpe and V. Jagunathan, *Indian J. Biochem. Biophys.*, **8**, 227 (1971); *Chem. Abstr.*, **76**, 109,531 (1972).

228a. J. R. Zysk, R. L. Whistler, and A. A. Bushway, *Abstr. Pap. Amer. Chem. Soc. Meeting*, **168**, CARB-22 (1974).

229. R. M. Rowell and R. L. Whistler, *Carbohyd. Res.*, **5**, 337 (1967).

230. J. E. McCormick and R. S. McElhinney, *Chem. Commun.*, 171 (1969).

231. K. W. Buck, F. A. Fahim, A. B. Foster, A. R. Perry, M. N. Qadir, and J. M. Webber, *Carbohyd. Res.*, **2**, 14 (1966).

232. A. B. Foster, J. M. Duxbury, T. D. Inch, and J. M. Webber, *Chem. Commun.*, 881 (1967).

233. F. Weygand, O. Trauth, and R. Löwenfeld, *Ber.*, **83**, 563 (1950); K. Sato, *J. Biochem.* (Tokyo), **40**, 485, 557, 563 (1953); R. Falconer and J. M. Gulland, *J. Chem. Soc.*, 765 (1936); 1912 (1937).

234. U. Suzuki, S. Odake, and T. Mori, *Biochem. Z.*, **154**, 278 (1924); P. A. Levene and H. Sobotka, *J. Biol. Chem.*, **65**, 551 (1925); G. Wendt, *Z. Physiol. Chem.*, **272**, 152 (1942).

235. S. K. Shapiro and F. Schlenk, *Advan. Enzymol.*, **22**, 237 (1960).

236. G. L. Cantoni and E. Scarano, *J. Amer. Chem. Soc.*, **76**, 4744 (1954).

237 (a) G. de la Haba and G. L. Cantoni, *J. Biol. Chem.*, **234**, 603 (1959); (b) J. A. Duerre, *Arch. Biochem. Biophys.*, **96**, 70 (1962).

238. J. Baddiley and G. A. Jamieson, *J. Chem. Soc.*, 1085 (1955); R. T. Borchardt, J. A. Huber, and Y. S. Wu, *J. Org. Chem.*, **41**, 565 (1976).

239. G. L. Cantoni, *J. Biol. Chem.*, **204**, 403 (1953).

240. W. Sakami and A. Stevens, *Bull. Soc. Chim. Biol.*, **40**, 1787 (1958); A. M. Yurkevich,

A. A. Amagaeva, I. P. Rudakova, and N. A. Preobrazhenskiĭ, *Zh. Obshch. Khim.*, **39**, 434 (1969).

241. J. A. Duerre and C. H. Miller, *J. Label. Compounds*, **4**, 171 (1968).

242. H. Daniel, M. Miyano, R. O. Mumma, T. Yagi, M. Lepage, I. Shibuya, and A. A. Benson, *J. Amer. Chem. Soc.*, **83**, 1765 (1961).

242a. R. Reistad, *Carbohyd. Res.* **54**, 308 (1977).

243. J. K. N. Jones and D. L. Mitchell, *Can. J. Chem.*, **36**, 206 (1958); B. Lindberg and L. Nordén, *Acta Chem. Scand.*, **15**, 958 (1961).

244. D. Horton and J. S. Jewell, *J. Org. Chem.*, **31**, 509 (1966); K. Matsuura, Y. Araki, and Y. Ishido, *Bull. Chem. Soc. Jap.*, **46**, 2261 (1973).

245. A. R. Procter and R. H. Wiekenkamp, *Carbohyd. Res.*, **10**, 459 (1969).

246. H. Zinner and M. Schlutt, *J. Prakt. Chem.*, **313**, 1181 (1971).

247. H. Zinner, R. Kleeschätzky, and M. Schlutt, *J. Prakt. Chem.*, **313**, 855 (1971).

248. M. L. Wolfrom, H. B. Bhat, P. McWain, and D. Horton, *Carbohyd. Res.*, **23**, 289 (1972).

248a. Compare D. Horton, D. C. Baker, and S. S. Kokrady, *Ann. N.Y. Acad. Sci.*, **255**, 131 (1975).

249. N. S. Johary and L. N. Owen, *J. Chem. Soc.*, 1299 (1955).

250. N. S. Johary and L. N. Owen, *J. Chem. Soc.*, 1302 (1955).

251. G. E. McCasland and A. B. Zanlungo, *Carbohyd. Res.*, **17**, 475 (1971).

252. J. Kuszmann and L. Vargha, *Carbohyd. Res.*, **11**, 165 (1969).

19. UNSATURATED SUGARS

R. J. FERRIER

I. INTRODUCTION

Many stable, discrete carbohydrate derivatives possessing a double bond in the carbon skeleton are now known, and these alone will be considered in this section; unstable, unsaturated reaction intermediates, enols, enediols, enones, dienes, unsaturated cyclitols, and carbohydrates bearing unsaturated substituents, all of which have considerable significance, will not be treated. Despite these restrictions, a highly diversified group of compounds, which can vary in the position and geometry of the double bond, in the substituents on the sp^2-hybridized carbon atoms, in the nature of the carbon chain, and in stereochemical features, remains for discussion.

The 1,2-unsaturated derivatives of cyclic aldoses have long been known as "glycals," and this terminology is still frequently employed. A more-general system that is now widely utilized involves first naming the saturated parent-structure (see Vol. IIB, Chapter 46) and then inserting the infix "-*m*-en" (or "-*m*-yn") after the stem name (such as "hex" or "pent") to denote a double (or triple) bond at position number *m*.

Fuller accounts of this field have been presented in *Advances in Carbohydrate Chemistry and Biochemistry*.[1]

II. GLYCALS

The glycals, cyclic compounds having a double bond between C-1 and C-2, are vinyl ethers and consequently can take part in a wide variety of selective addition reactions. Pyranoid and furanoid members are known, and the esters of each also undergo rearrangements to give 2,3-unsaturated products. The anomalous name used for this series originated from the aldehydic characteristics shown by impure preparations obtained in early work. (See Section II,B.)

A. PREPARATION

By a reaction first reported in 1913 by Fischer and Zach, per-*O*-acylglyco-pyranosyl halides [usually bromides, such as tetra-*O*-acetyl-α-D-glucopyranosyl bromide (**1**)] are reduced by zinc and acetic acid to the corresponding acetylated glycals [in this case tri-*O*-acetyl-D-glucal (**2**)], and by utilization of a

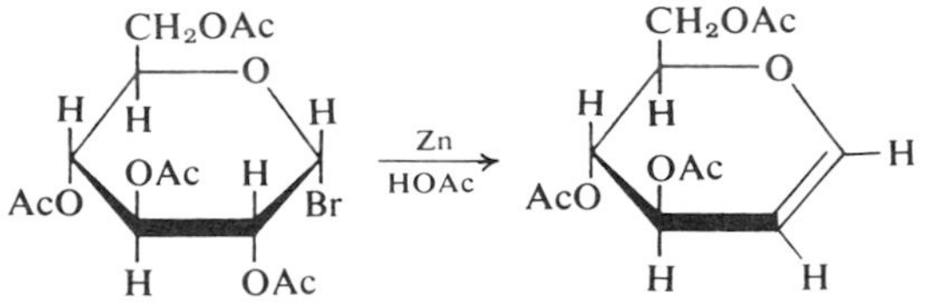

direct method for preparing glycosyl halides, the unsaturated products can be obtained satisfactorily from free sugars without isolation of any intermediates.[2] Platinum and copper salts catalyze the reduction step, which is thought to proceed by addition of two electrons from the metal to the carboxonium ion at C-1 formed by initial ionization of the halide. Elimination of the acetoxy ion at C-2 from the resulting carbanion gives the acylated glycals, which can be de-esterified smoothly with traces of sodium methoxide in methanol or with methanolic ammonia.

A less-predictable synthesis of 4,6-*O*-benzylidene-D-allal (**3**) involves treatment of an ethereal solution of methyl 2,3-anhydro-4,6-*O*-benzylidene-α-D-allopyranoside with methyllithium and lithium iodide. This reaction initially affords methyl 4,6-O-benzylidene-2-deoxy-2-iodo-α-D-altropyranoside, which then undergoes elimination to give the glycal derivative **3**. The same conditions convert methyl 4,6-O-benzylidene-2-deoxy-2-iodo-α-D-idopyranoside into 4,6-O-benzylidene-D-gulal. When the *allo*-epoxide is treated with halide-free methyllithium, the three-membered ring is opened by the carbon nucleophile, subsequent elimination affording[3] the 2-methylglycal derivative **4**.

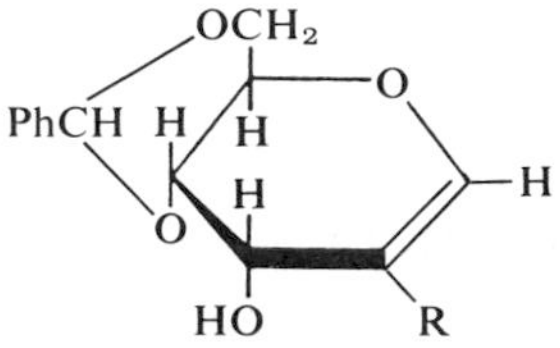

3 R = H
4 R = Me

Addition of chlorosulfonyl isocyanide to 4,6-di-O-acetyl-3-O-benzyl-D-glucal gives a 2-cyano-2-deoxyglycosyl halide that is activated for triethylamine-catalyzed elimination of hydrogen halide, giving[4] the branched-chain glycal **5**. Intramolecular processes are involved in the conversion of ethyl 4,6-di-O-acetyl-2,3-dideoxy-D-*erythro*-hex-2-enopyranoside into 4,6-di-O-acetyl-3-deoxy-3-ethoxycarbonylamino-D-glucal[5] (**6**) by action of chlorosulfonyl isocyanate, and in the reduction by lithium aluminum hydride of similar precursors to afford deoxyglycals, as in the conversion of methyl 4,6-O-benzylidene-2,3-dideoxy-α-D-*erythro*-hex-2-enopyranoside into 1,5-anhydro-4,6-O-benzylidene-2,3-dideoxy-D-*erythro*-hex-1-enitol. Studies using deuterium labels showed that the hydrogen atom enters *cis* to the departing oxygen atom,[6] an example being methyl 3,4-dideoxy-6-O-trityl-α-D-*threo*-hex-3-enopyranoside, which reacts with lithium aluminum deuteride in refluxing 1,4-dioxane during 19 hours to afford a 2:1 mixture of methyl 2,3,4-trideoxy-4-C-deuterio-6-O-trityl-α-D-*threo*-hexopyranoside and the glycal **7** formed by two, stereospecific, nucleophilic displacements involving allylic rearrangements.

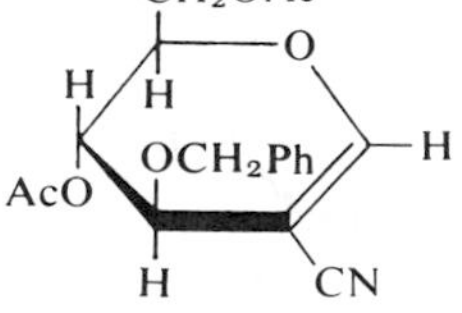

5

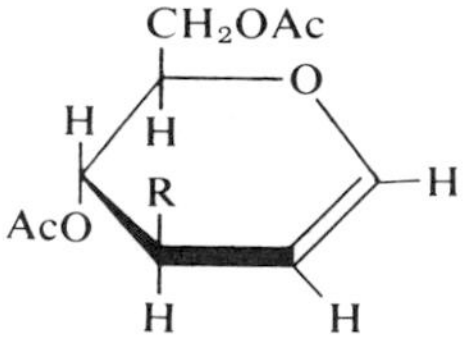

6 R = NHCO₂Et

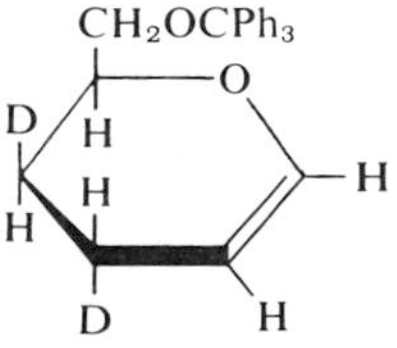

7

The reactivity of the furanoid analogs (Section II,B) precludes their synthesis by the zinc–acetic acid method. Instead, compounds having readily displaceable groups at C-2 have been employed; 3,5-di-O-benzoyl-2-O-(p-nitrophenylsulfonyl)-β-D-ribofuranosyl bromide (8) on treatment with sodium iodide in acetone solution gives[7] the benzoylated, furanoid glycal 9. The nucleoside 10 was prepared[8] from a fully protected derivative of 9-(2'-deoxy-2'-iodo-β-D-arabinofuranosyl)adenine by action of 1,5-diazabicyclo[4.3.0]non-5-ene and subsequent deprotection.

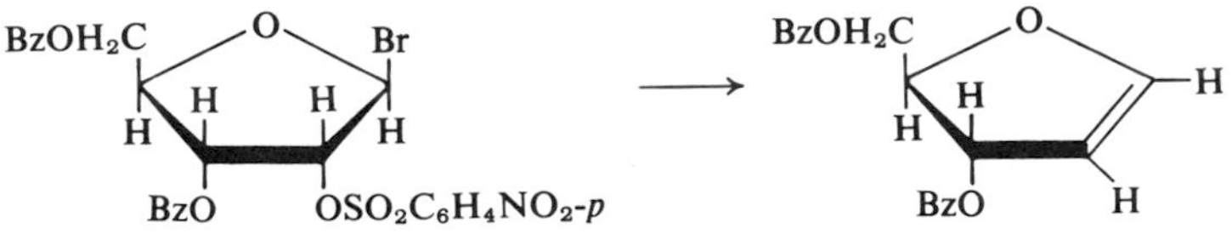

B. Reactions

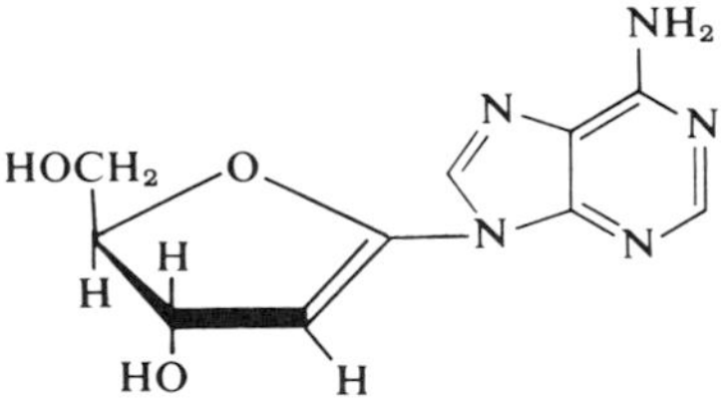

The addition and rearrangement reactions undergone by glycal derivatives make them among the most useful compounds in carbohydrate synthesis.

1. Addition Reactions

Hydrogenations proceed readily, although hydrogenolysis of allylic ester groups occasionally occurs[9] as a competing process, and polar additions to the double bond take place with high electrospecificity; the mesomeric influence of the ring-oxygen atom ensures that the electrophile enters at position 2. Many additions—for example, of water, alcohols, phenols, and

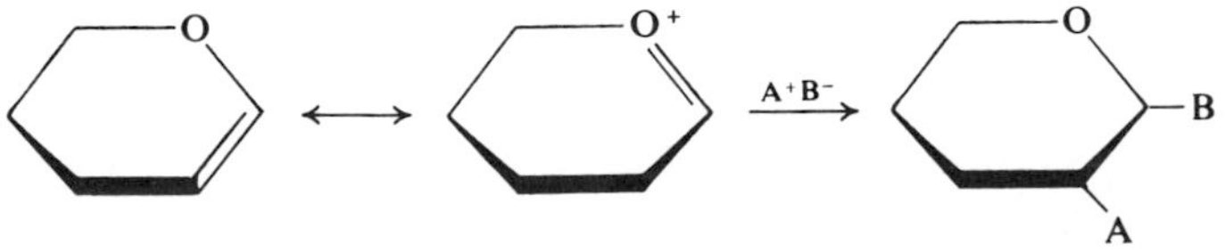

carboxylic acids—require an acid catalyst and lead exclusively to 2-deoxy-aldoses, -aldosides, and -aldosyl esters. 2-Deoxy-pentoses, -hexoses, and

-disaccharides,[10] methyl 2-deoxy-α-D-*arabino*-hexopyranoside,[11] phenyl 3,4,6-tri-O-acetyl-2-deoxy-α-D-*lyxo*-hexopyranoside,[12] and 1,3,4,6-tetra-O-acetyl-2-deoxy-α-D-*lyxo*-hexopyranose[13] have, for example, all been prepared in this way from glycals or their acetates. Since anomerization occurs in the presence of the acid, the method gives predominantly the thermodynamically stable products (usually α-D). In a reaction analogous to these, 6-chloropurine and 3,4-di-O-acetyl-D-arabinal, on fusion in the presence of *p*-toluenesulfonic acid, give[14] the anomeric 6-chloro-9-(3,4-di-O-acetyl-2-deoxy-D-*erythro*-pento-pyranosyl)purines (**11**). A prototype synthesis of sucrose was based on addition of 1,3,4,6-tetra-O-acetyl-D-fructofuranose to a conjugated glycal, whence the disaccharide was prepared[15] in several more steps. Hydrogenation of the glycal **10** affords[8] predominantly 2′-deoxyadenosine plus a smaller amount of the α-anomer. Such enzymes as emulsin catalyze hydration of D-glucal.[15a]

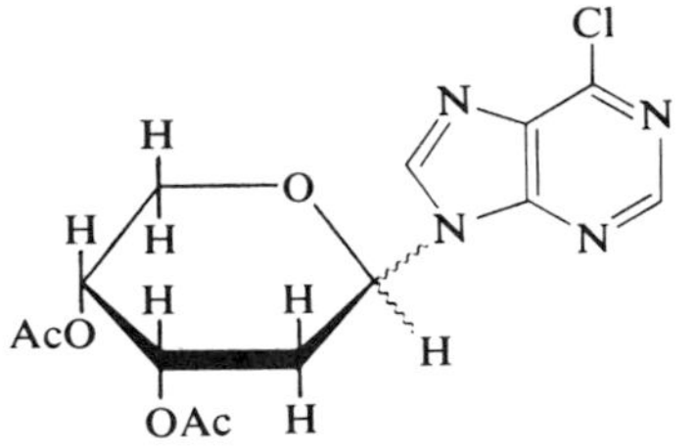

11

Methoxymercuration of glycals and their acetates, brought about by treatment with mercuric acetate in methanol, gives methyl 2-deoxyglycosides (or their acetates) having carbon–mercury bonds at C-2. Tri-O-acetyl-D-glucal, on such treatment with subsequent replacement of the ionic acetate by chloride, gives methyl 3,4,6-tri-O-acetyl-2-(chloromercuri)-2-deoxy-β-D-glucopyranoside (**12**) together with the α-D-*manno* isomer. Both of these can

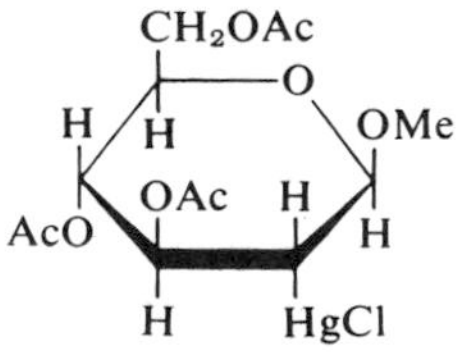

12

be isolated in satisfactory yield and converted, by reductive cleavage of the carbon–metal bond and deacetylation, into the respective anomeric methyl 2-deoxyglycosides. Direct methoxymercuration of D-glucal affords methyl

---

References start on p. 875.

2-(acetoxymercuri)-2-deoxy-α-D-mannopyranoside in high yield.[16] Photolysis of these organomercurials also gives[17] the corresponding 2-deoxy glycosides.

Direct halogenation of glycals produces mixtures of adducts which, as glycosyl halide derivatives, are glycosylating agents. Thus tri-*O*-acetyl-D-glucal with bromine affords 60% and 30% of the α-D-*gluco* (13) and α-D-*manno* (14) products, respectively,[18] and these on treatment with methanol in the presence of silver carbonate give high yields of β-D-glycoside derivatives (15, 16) from which methyl 2-deoxy-β-D-*arabino*-hexopyranoside (17) is obtainable by

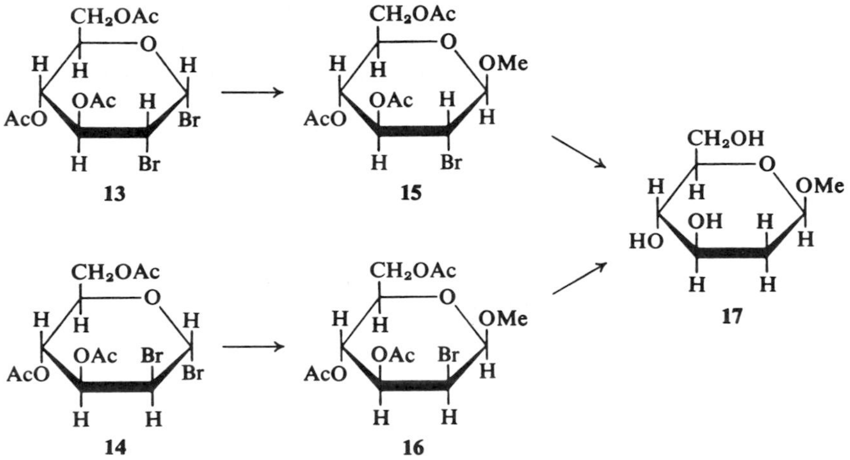

reduction and deacetylation. Alternatively, when halogenation of tri-*O*-acetyl-D-glucal is effected in alcoholic solutions and in the presence of silver salts, α-D-glycoside derivatives are formed predominantly, and from these 2-deoxy-α-D-*arabino*-hexopyranosides may be obtained.[18] Either method can clearly be applied to the synthesis of 2-deoxyaldoses; 2-deoxy-D-*erythro*-pentose ("*2-deoxy-D-ribose*"), for example, has been synthesized from 3,4-di-*O*-acetyl-D-arabinal by way of the methyl 3,4-di-*O*-acetyl-2-chloro-2-deoxy-D-*erythro*-pentopyranosides.[19] Treatment of the di-*O*-acetyl-D-arabinal–chlorine adduct with theophyllinesilver gives products from which the anomeric (2-deoxy-D-*erythro*-pentopyranosyl)theophyllines (18) have been obtained.[20] The dihalogeno adducts are convertible into 2-halogenoglycals by dehydrohalogenation,[21] and on hydrolysis give 2-deoxy-2-halogenoaldoses, which generally form phenylosazones on treatment with phenylhydrazine and, with lead oxide, lose the elements of hydrogen halide to give 2-deoxyaldonolactones.[22] Later studies showed[23] that the orientation of addition of chlorine is solvent-dependent, nonpolar media favoring *cis*-addition, and polar, *trans*.

Chlorination of di-*O*-acetyl-D-xylal in methanol gives[24] a mixture of acetylated (1,2-*trans*) methyl 2-chloro-2-deoxypentosides.

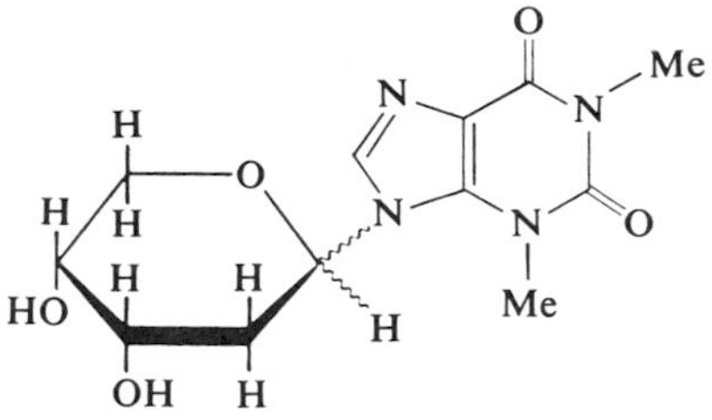

18

N-Bromosuccinimide and hydrogen fluoride in ether solution at $-60°$ react with glycals to produce mainly 1,2-*trans*-2-bromo-2-deoxyglycosyl fluorides,[25] whereas, with hydrogen fluoride in dichloromethane in the presence of lead tetraacetate, 2,5-anhydro-1-deoxy-1,1-difluoroalditols are obtained instead of the anticipated 1,2-difluoro derivatives.[26] Presumably, in the latter case, the ring-oxygen atom attacks C-2 in an intermediate containing

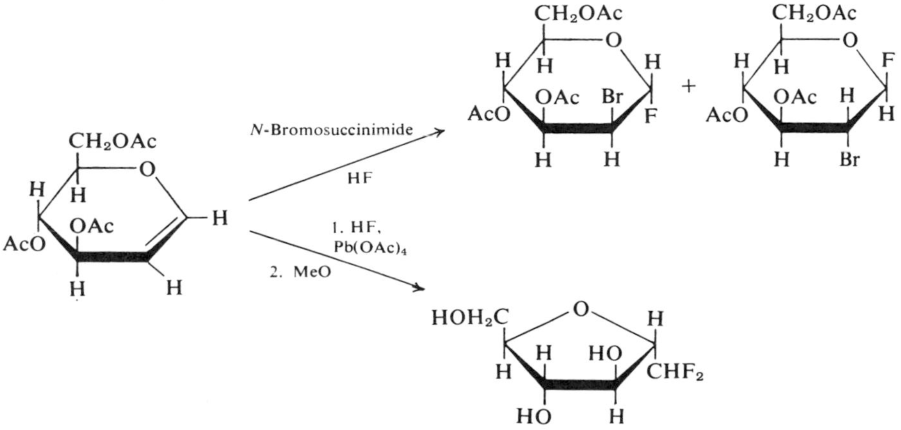

a readily displaceable substituent at this site, in a reaction similar to that occurring on treatment of 2-amino-2-deoxyaldoses with nitrous acid (see Chapter 16).

Hydrogen halides generally add to glycals to give 2-deoxyglycosyl halides which, as glycosylating agents, can be employed in the synthesis of 2-deoxy-glycopyranoses, -pyranosides, or -pyranosyl purines or pyrimidines. However, there are exceptions to the generalization. Hydrogen bromide plus tri-*O*-acetyl-D-glucal in acetic acid gives, in addition to the expected products,

4,6-di-*O*-acetyl-3-bromo-2,3-dideoxy-α-D-*arabino*-hexose,[27] and the action of hydrogen fluoride in benzene solution affords[28] 4,6-di-*O*-acetyl-2,3-dideoxy-D-*erythro*-hex-2-enopyranosyl fluoride.

Hydroxylation offers a means of reverting to aldoses. Oxidation of unsubstituted glycals with peroxybenzoic acid affords mainly products having the *cis*-2,3-diol arrangement, since the hydroxyl group at C-3 stabilizes, by hydrogen bonding, the transition state leading to the *cis*-epoxide, which then opens by cleavage of the C-1–oxygen bond. Alternatively, in 3-*O*-substituted glycals, this stabilization is not possible, and, conversely, the C-3 group shields one side of the double bond, causing the preferential formation of the alternative epoxides, and hence products having the 2,3-*trans*-diol structure. A method is thus available for obtaining, for example, D-mannose compounds from D-glucose derivatives (provided the hydroxyl group at C-3 is free), and also D-glucoses unsubstituted at C-2; 3,4,6-tri-*O*-methyl-D-glucose has been prepared in this way from 3,4,6-tri-*O*-methyl-D-glucal.[29] Hydroxylation may also be carried out with hydrogen peroxide in *tert*-butyl alcohol in the presence of osmium tetraoxide, but under these conditions both D-glucal and its tri-acetate give predominantly D-*gluco* products. Good yields of 2,3-*cis* diols are available[30] by the reaction of glycals with hydrogen peroxide in the presence of molybdenum trioxide.

An additional means of hydroxylating the double bond in glycals is provided by the adducts formed on heating with phenanthrenequinone in boiling benzene under ultraviolet light. From tri-*O*-acetyl-D-glucal the *gluco* adduct (**19**) is obtainable in 50% yield, and this, on ozonolysis, followed by de-esterification, gives diphenic acid and D-glucose.[2] Condensation of the product of deacetylation of **19** with tetra-*O*-acetyl-α-D-glucopyranosyl bromide gives mainly the 6-linked disaccharide derivative, from which gentiobiose can be prepared in satisfactory yield.[2]

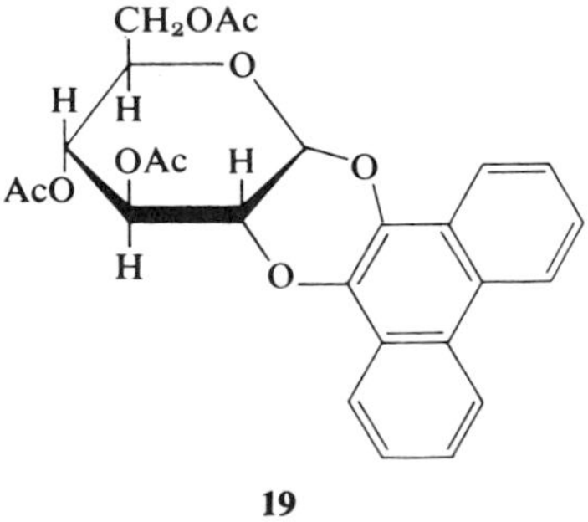

19 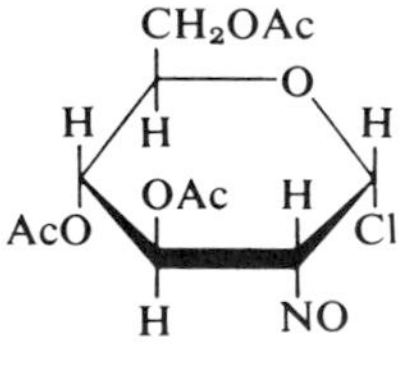   **20**

Compounds of considerable interest for the synthesis of 2-amino-2-deoxy-aldose derivatives are produced on addition of nitrosyl chloride to acetylated glycals. Tri-O-acetyl-D-glucal affords, for example, tri-O-acetyl-2-deoxy-2-nitroso-α-D-glucopyranosyl chloride (20), which exists in a dimeric form, and which, on treatment with silver acetate in acetic acid, gives 2-deoxy-2-nitroso-D-glucose tetraacetate.[31] When reduced in acetic acid with copper–zinc, or when treated with alcohols, compound 20 yields 1,3,4,6-tetra-O-acetyl-2-amino-2-deoxy-D-glucose and alkyl 3,4,6-tri-O-acetyl-2-oximino-α-D-arabino-hexosides, respectively. The oximino compounds have been applied in highly satisfactory preparations of alkyl 2-amino-2-deoxy-D-glucosides[32] (see Chapter 16,II,G).

In contrast with alcohols (Section II,B,2), thiolacetic acid[33] and other thiols[34] undergo photocatalyzed addition across the double bond of tri-O-acetylglucal to afford an epimeric mixture of 1,5-anhydro-2-thioalditols. Photoaddition of 1,3-dioxolane to the same glycal was less specific, giving a mixture of several different products[35]; however, photoaddition of acetone shows a predictable[36] dependence upon conditions, ninefold and greater excesses of acetone giving the adduct 21, whereas a limited (fivefold or less) excess leads to formation of the oxetane 22.

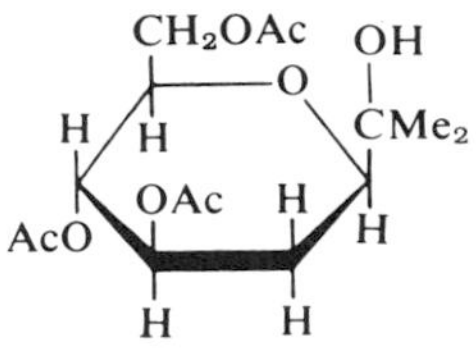

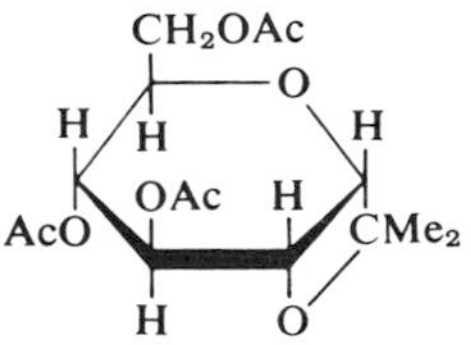

Hydroformylation, the reaction undergone by olefins on heating under pressure with hydrogen and carbon monoxide in the presence of cobalt octacarbonyl, gives 2,6-anhydro-3-deoxyalditols.[37] From di-O-acetyl-D-arabinal (23), 1,5-anhydro-4-deoxy-D-lyxo- and -L-ribo-hexitol (24 and 25) (that is, 2,6-anhydro-3-deoxy-D-arabino- and -D-ribo-hexitols) are obtained by hydroformylation, deacetylation, and reduction of the formyl compounds formed initially.[38]

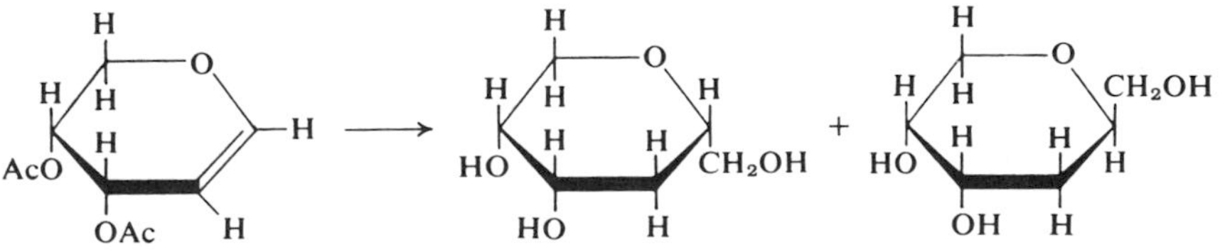

2. *Rearrangement Reactions*

Water, alcohols, and phenols do not add to the double bonds of glycals or their esters in the absence of acidic catalysts, but with the esters cause nucleophilic displacement of the allylic acyloxy group and migration of the double bond to the 2,3 position.[39] With water, the reaction occurs at 100°, and, for example, tri-*O*-acetyl-D-glucal gives 4,6-di-*O*-acetyl-2,3-dideoxy-D-*erythro*-hex-2-enose ("*diacetylpseudoglucal*"; **26**). High temperatures are required for

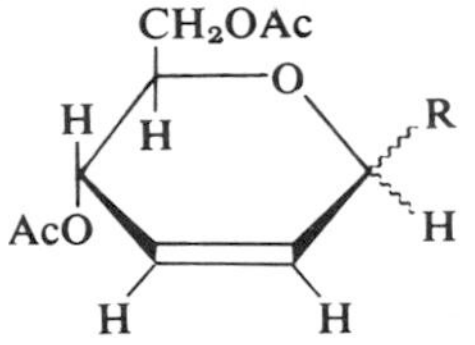

26 R = OH
27 R = OEt, α anomer

uncatalyzed reaction with alcohols, and mixtures of anomers are formed. Nevertheless, this procedure can be used to prepare ethyl 4,6-di-*O*-acetyl-2,3-dideoxy-α-D-*erythro*-hex-2-enopyranoside[39] (**27**) and methyl 4,6-di-*O*-acetyl-2,3-dideoxy-α-D-*threo*-hex-2-enopyranoside,[13] since these isomers crystallize from the mixed products. Aryl 2,3-unsaturated glycosides may similarly be obtained from glycal esters.[40] Other nucleophilic reagents that have been shown to take part in this reaction are thiolacetic acid[41] and the purine theophylline.[14]

Boron trifluoride catalyzes this reaction of tri-*O*-acetyl-D-glucal with alcohols and also causes anomerization of the products. Glycosides consequently predominate, and the reaction can be applied to the efficient synthesis of simple and complex 2,3-unsaturated glycosides.[42] Reaction of tri-*O*-acetyl-D-glucal with lactononitrile under a low-pressure mercury lamp gave[43] the corresponding 2-cyanoethyl 4,6-di-*O*-acetyl-2,3-dideoxy-D-*erythro*-hex-2-enopyranosides in an analogous reaction which, it was concluded, was dependent on the presence of catalytic acid.

From the observation that tri-*O*-acetyl-D-glucal or tri-*O*-acetyl-D-allal undergo boron trifluoride-catalyzed reaction with dimethyl phosphonite to afford the same distribution of dimethyl (hex-2-enopyranosyl)phosphonates, Paulsen and Thiem[44] interpreted the failure of the C-3 substituent to influence the course of the reaction to indicate a unimolecular ionization-process giving an allylic oxonium ion as an intermediate.

A striking feature of the benzoylated furanoid glycal **9** is the ease with which it takes part in this type of reaction. While pyranoid glycal esters react with water and alcohols only at high temperatures, the five-membered cyclic compound loses benzoic acid at room temperature in both aqueous alcohol and 3:1 methanol–dichloromethane to give, in the latter case, 2-(benzoyloxy-methyl)-2,5-dihydro-5-methoxyfuran.[7] Such relative reactivities are consistent with the ease of allylic solvolytic displacements from cyclohexene and cyclopentene rings.

Further rearrangements occur on deacetylation of "pseudoglycal" esters, so that unsubstituted sugars having 2,3-double bonds are not known. From 4,6-di-O-acetylpseudoglucal (**26**) "isoglucal" and "protoglucal" are readily obtained, the former being 3,6-anhydro-2-deoxy-D-*arabino*-hexose, which can also be prepared by treatment of 2-deoxy-D-*arabino*-hexose with alkali.[45] Apparently, in this last reaction, an elimination β to the carbonyl group occurs to give "pseudoglucal," which is then attacked at C-3 by the C-6 nucleophile. "Protoglucal" has not been characterized fully, but Isbell[46] has suggested, very reasonably, that it is the enone **28**. Other structures that have been proposed for these compounds may be considered to be wrong.

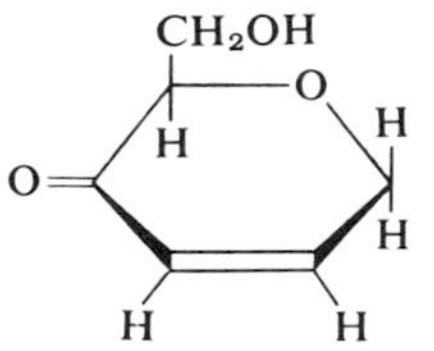

28

Tri-O-acetyl-D-glucal is isomerized to 1,4,6-tri-O-acetyl-2,3-dideoxy-D-*erythro*-hex-2-enopyranose in boiling nitrobenzene or on treatment with traces of boron trifluoride etherate; higher concentrations of the Lewis acid cause electrophilic addition of a presumed allylic oxonium ion to the rearranged product, giving[47] the dimeric product **29**.

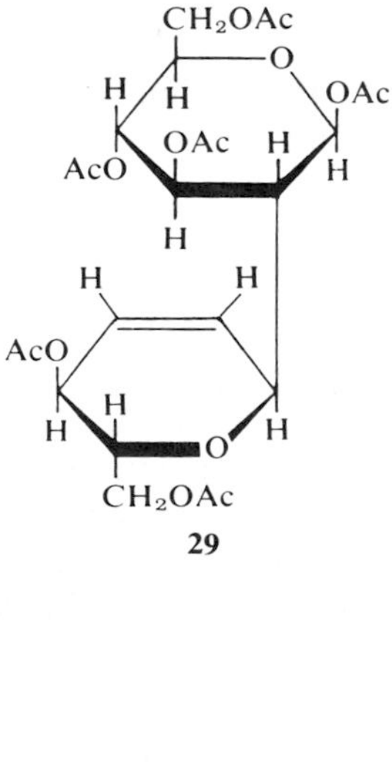

29

A further, historically significant reaction of tri-O-acetyl-D-glucal was elucidated by the recognition[48] that, treated with refluxing aqueous 1,4-dioxane, it gave some of the *trans*-isomer of the 2-enose **26**—that is, a free aldehyde that may have been present in impure samples of the glycal and led to Emil Fischer's selection of the anomalous "glycal" name. Further, more-efficient syntheses of the α,β-unsaturated aldehyde have appeared: (*a*) by treatment of 2-acetoxymercuri-3,4,6-tri-O-acetyl-2-deoxy-D-glucose, formed by hydroxymercuration of tri-O-acetyl-D-glucal, with limited quantities of sodium borohydride,[48a] and (*b*) by treatment of tri-O-acetyl-D-glucal with catalytic quantities of mercuric ion in dilute sulfuric acid.[48b] A comprehensive review of these rearrangements has been published.[48c]

III. 2-HYDROXYGLYCALS

Elimination of the elements of hydrogen halide from acylated glycosyl halides gives acylated 2-hydroxyglycals (*1,2-glycosenes*)—for example, 2-acetoxy-3,4,6-tri-O-acetyl-D-glucal (**30**)—which on de-esterification do not give discrete products. The esters, however, are stable, crystalline compounds, best obtained by treatment of per-O-acetylated glycosyl bromides in acetonitrile with diethylamine in the presence of tetrabutylammonium bromide,[49]

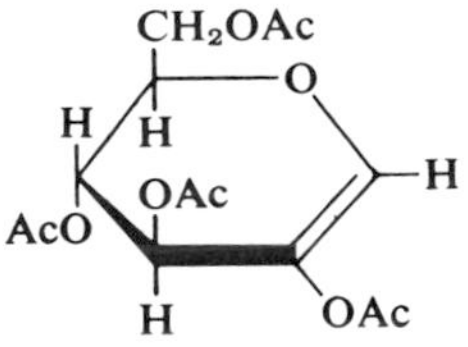

30

or by conversion of bromide into iodide prior to treatment with diethylamine,[50] or, better, by treatment of the halide with 1,5-diazabicyclo[5.4.0]-undec-5-ene.[51]

Hydrogenation of acylated 2-hydroxyglycals gives mixtures of 1,5-anhydro-alditol derivatives, although hydrogenolytic cleavage of ester groups often complicates[9] this reaction, and chlorination gives 1,2-dichloro compounds. The latter very readily undergo hydrolysis to give 1,2-dihydroxy derivatives closely related to aldosuloses. These products, in the hexose series, on treatment with acetic anhydride in pyridine give diacetylkojic acid (**31**) instead of direct products of acetylation.[52] Dichloro adducts formed from benzoylated analogues are readily converted into 4-deoxyglyc-3-ene-2-ulose derivatives.[53] Oxidation with permanganate gives esters of aldonic acids formed by loss of C-1.

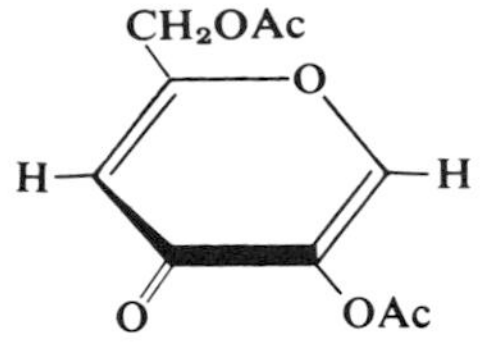

CH$_2$OAc

31

Phenylhydrazine reacts with 2-acetoxy-3,4,6-tri-*O*-acetyl-D-glucal to give a 2,3-diulose phenylosazone **32**. In contrast, the product of deacetylation of the glycal derivative, when treated with this reagent, undergoes an intermediate oxidation in place of an elimination step and forms[54] the osazone **33**. Photocatalyzed addition of 1,3-dioxolane to **30** gives 2-(tetra-*O*-acetyl-D-glucopyranosyl)-1,3-dioxolane,[35] whereas photoaddition of thiols gives 1-thioglycosides.[34]

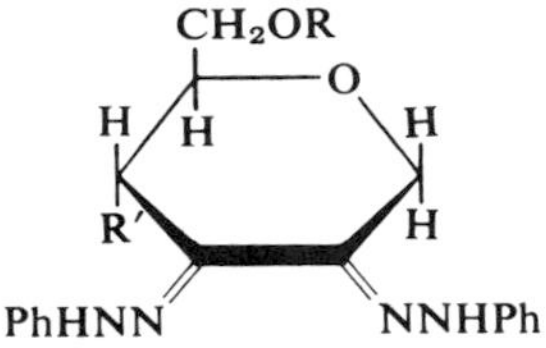

CH$_2$OR

32 R = Ac, R′ = H
33 R = H, R′ = OH

Rearrangement reactions analogous to those occurring in the glycal series (Section II,B) also take place with 2-hydroxyglycal esters. Thus, on treatment with boiling acetic acid, 2-acetoxy-3,4,6-tri-*O*-acetyl-D-glucal rearranges completely to the anomeric tetra-*O*-acetyl-3-deoxy-D-*erythro*-hex-2-enopyranoses (**33a**). The latter are separable by fractional crystallization

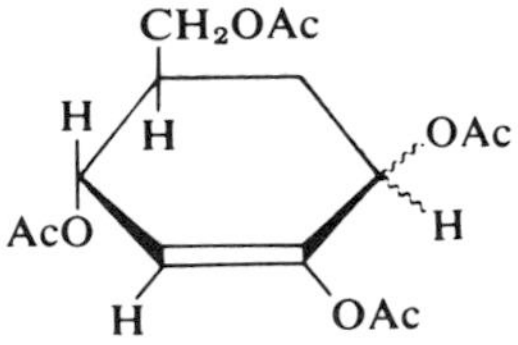

CH$_2$OAc

33a

and give, on hydrogenation and deacetylation, mixtures of 3-deoxy-D-*ribo*- and -*arabino*-hexose together with products formed by hydrogenolysis.[55] Such rearrangements may be effected at room temperature by using acetic anhydride with zinc chloride as catalyst, but, under these conditions, a further, slower isomerization occurs, due to epimerization at the C-4 allylic center, and the α-D-*threo*-2,3-unsaturated compound can be isolated from the products.[56] This last isomer is also formed, almost exclusively, when 2-acetoxy-3,4,6-tri-*O*-acetyl-D-galactal is heated in acetic acid, and several other hydroxyglycal esters have been shown to take part in this rearrangement.[50] Both methods for isomerizing 2-acetoxy-3,4,6-tri-*O*-acetyl-D-glucal favor the thermodynamically more stable α-D product; the β-D product is isolated only with difficulty, but on heating the glycal derivative in an inactive, high-boiling solvent (nitrobenzene, for example), stereospecific allylic migration of an acetoxy group from C-3 to C-1 occurs and affords a convenient means of synthesizing the less-stable isomer.[57]

Like glycal esters, 2-hydroxyglycal esters react in inert solvents with alcohols in the presence of boron trifluoride etherate. In this way, for example, methyl 2,4,6-tri-*O*-benzoyl-3-deoxy-α-D-*erythro*-hex-2-enopyranoside (**34**) can readily be obtained from 2-benzoyloxy-tri-*O*-benzoyl-D-glucal. Heating of the latter compound in benzene solution with trichloroacetic acid affords the glycosyl trichloroacetate **35**, which, on solvolytic displacement of the trichloroacetyl group, gives[58] the unsaturated β-glycoside **36**. 2-(*N*-Acetylacetamido)-tri-*O*-acetyl-D-glucal gives analogous α-glycosides and α-glycosyl esters by reaction with phenols or acids.[59] 3-Deoxyglyc-2-enopyransoyl halides are formed as initial products by the action of hydrogen halides on various 2-hydroxyglycals, and silver(I) benzoate causes[60] formation from them of the corresponding 1-*O*-benzoyl derivatives.

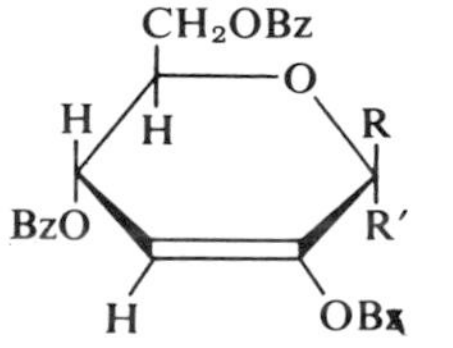

34 R = H, R′ = OMe
35 R = H, R′ = OCOCCl₃
36 R = OMe, R′ = H

2-Hydroxyglycal derivatives having a five-membered ring have been prepared from 1,3,4,6-tetra-*O*-benzyl-D-fructofuranosyl chloride[61] and from 2,3,5-tri-*O*-benzoyl-α-L-arabinofuranosyl bromide[62]; however, the product from the latter halide is unstable, spontaneously rearranging at room temperature to a mixture of glyc-2-enofuranose and furan derivatives.

IV. METHODS OF SYNTHESIS OF OTHER UNSATURATED SUGARS

Other unsaturated sugar derivatives represent a broad class of compounds of potential value in synthetic work, and they have also been invoked as biosynthetic intermediates. Interest in the investigation of their preparation and reactions has been stimulated by the finding in Nature of the antibiotic nucleosides cytosinine[63] (37) and decoyinine[64] (38), and of 2,6-diamino-2,3,4,6-tetradeoxy-D-*glycero*-hex-4-enopyranose and 2,3,6-trideoxy-L-*erythro*-hex-2-enono-1,5-lactone as constituents, respectively, of sisomycin, an antibiotic produced[65] by *Micromonospora inyoensis*, and osmundalin, produced[66] by the ferns *Osmunda japonica* and *O. regalis*.

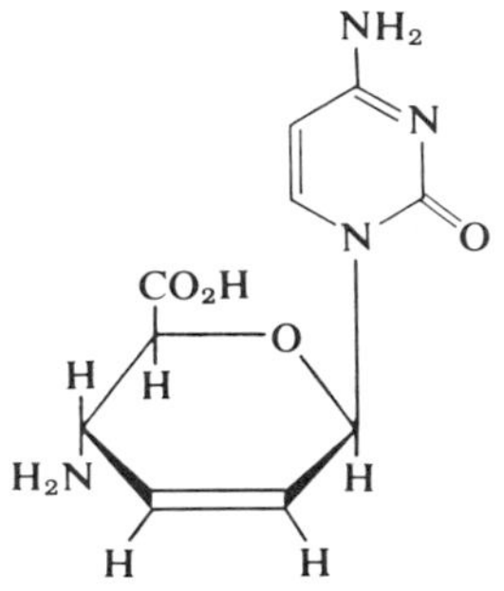

37 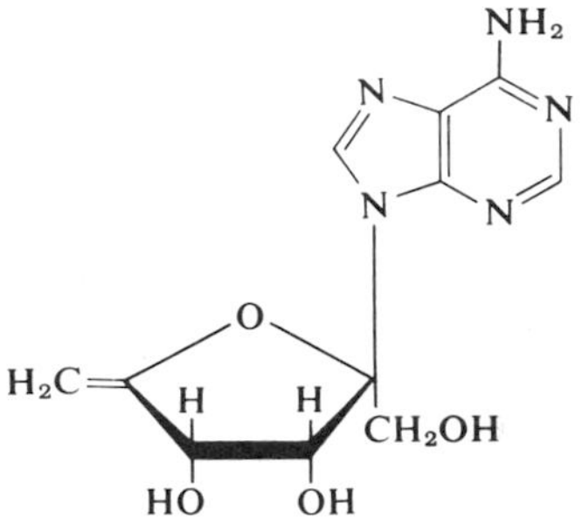   38

Syntheses based on direct elimination of groups having good leaving properties, on elimination induced by strongly electron-withdrawing groups, on allylic rearrangement of compounds already containing double bonds, and on the addition of unsaturated extensions to the carbon chains of sugars have all been employed in the preparation of various unsaturated carbohydrates.

A. DIRECT ELIMINATIONS

Many methods for direct introduction of double bonds are known, and together they represent the most useful means of preparing unsaturated sugars.

1. *Methods from Monohydroxy Derivatives*

Base-catalyzed elimination of sulfonyloxy groups has been used most frequently in the synthesis of olefinic and enolic derivatives from monohydroxy carbohydrates. Heating the derived sulfonic esters under diminished pressure with soda lime or treatment with sodium methoxide or potassium *tert*-butoxide in methyl sulfoxide has been employed to effect efficient elimination. Examples are illustrated:

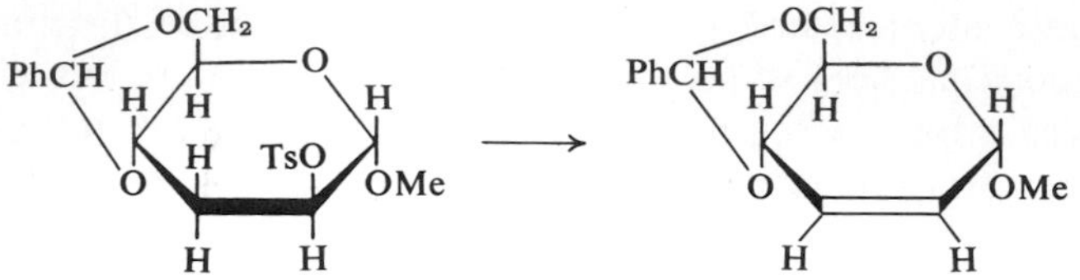

(Ref. 67)

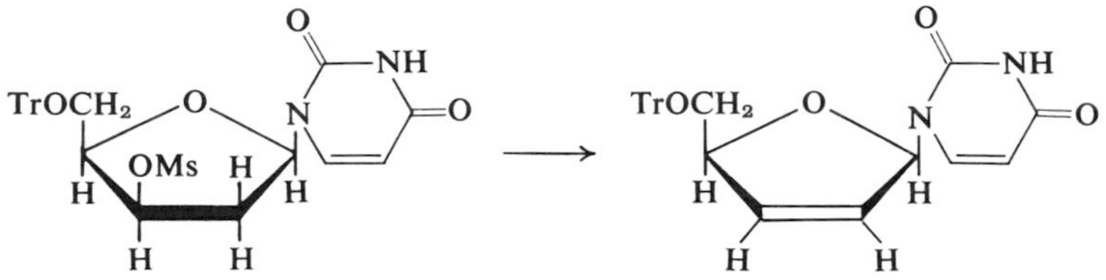

(Ref. 68)

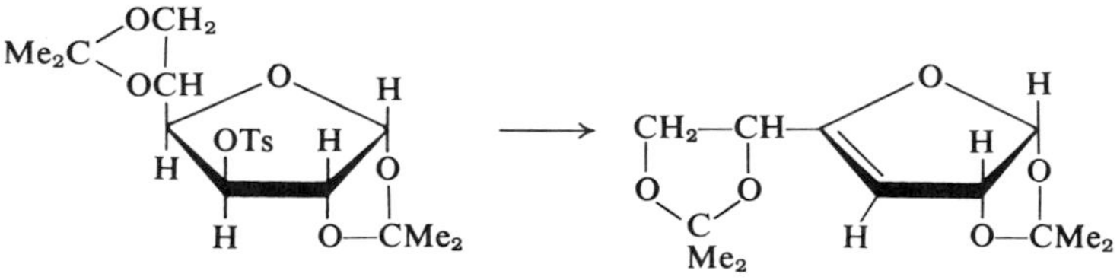

(Ref. 69)

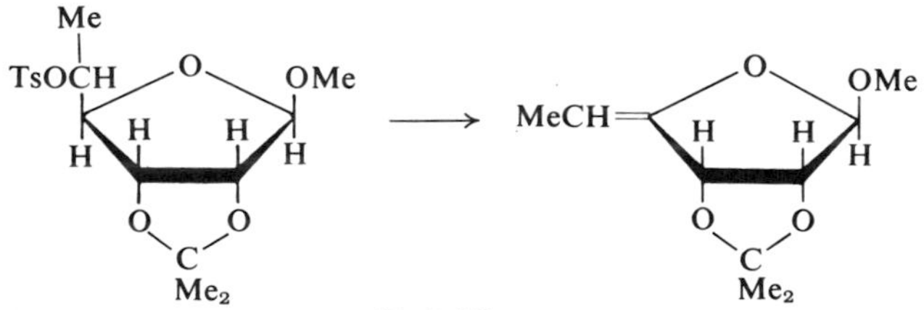

(Ref. 70)

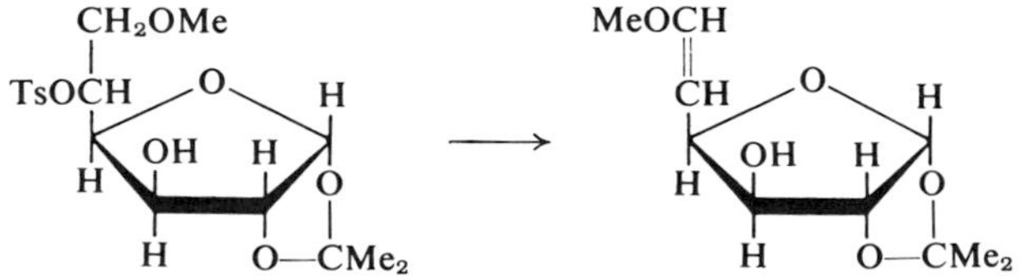

(Ref. 71)

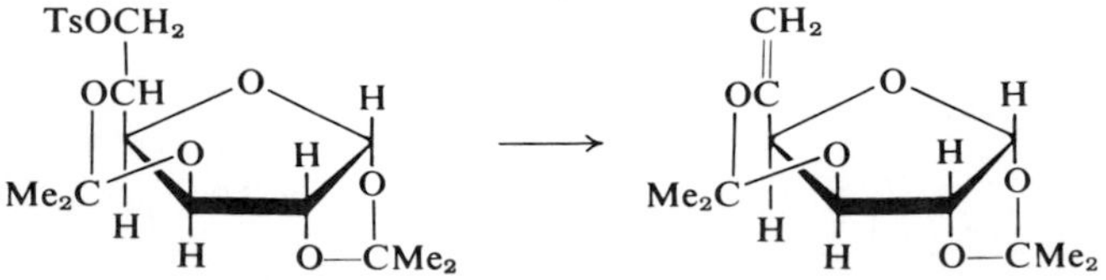

(Ref. 72)

Halogenated derivatives obtained by halide displacement of sulfonyloxy groups can also be employed; for example, 6-deoxy-6-iodo or -bromo derivatives of suitably protected aldohexoses or hexosides, on treatment with silver fluoride in pyridine, give compounds having 5,6-double bonds—for instance, **39** (ref. 73); related halides derived from opening of 4,6-O-benzylidene derivatives by N-bromosuccinimide likewise afford[74] 5,6-enes. 1,6-Dichloro-1,6-dideoxy-2,4:3,5-di-O-methylene-D-mannitol loses hydrogen chloride by action of alcoholic potassium hydroxide to form[75] the dienolic compound **40**.

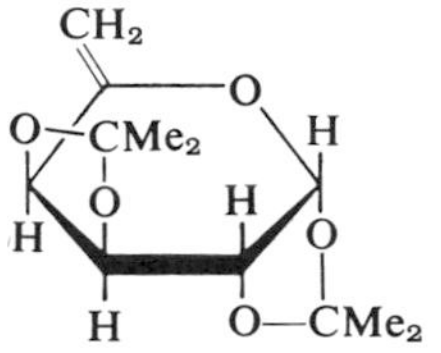

39 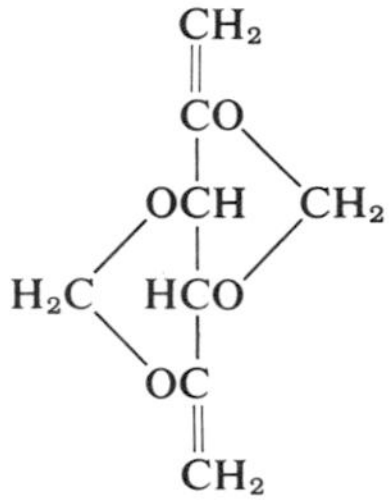   **40**

Pyrolysis of xanthate esters provides a useful means of preparing olefins from alcohols (Chugaev reaction), but with carbohydrate esters a rearrangement to the corresponding (alkylthio)thiocarbonyl derivatives [—O(C=S)SR → —S(C=O)SR] normally occurs,[76] and this has been applied in the synthesis of specific deoxy derivatives. However, in those cases that have been investigated in which a methyl or methylene group adjoins the ester function, straightforward elimination does occur: compounds **41** and **42**

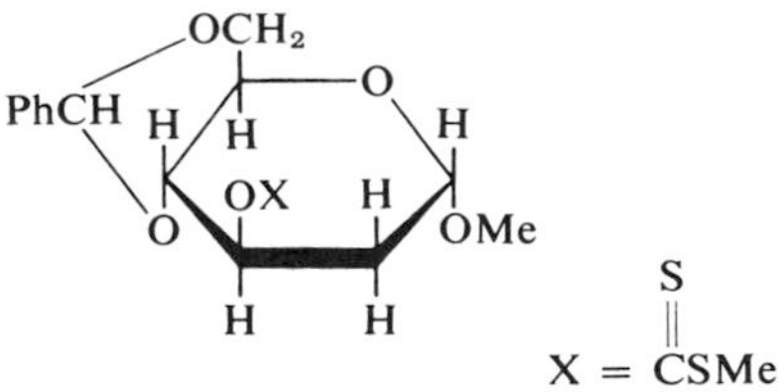

41 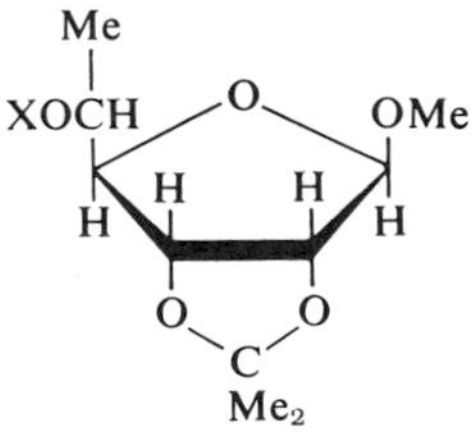   **42**

give products[39,77] having 2,3- and 5,6- double bonds, respectively. Branched-chain tertiary alcohols (Chapter 17) undergo dehydration by action of thionyl chloride in pyridine, and this has found synthetic use.[78] Butyllithium-induced eliminations from 4,6-O-benzylidene-2,3-di-O-methylhexopyranosides, a 2,3-benzylidene acetal, and methyl 3,4-O-isopropylidene-2-O-methyl-β-D-arabinopyranoside afforded a 3-deoxy-2-enol ether,[79] a 2-deoxyglycos-3-ulose,[80].

and a 4-deoxypent-4-enopyranoside,[80a] respectively, the second product presumably forming by way of an enol.

Oxidation of a secondary hydroxyl group to the corresponding ketone offers an additional means of introducing a double bond, since derived hydrazones of the ketones can be caused to undergo elimination; methyl 4,6-*O*-benzylidene-2,3-dideoxy-α-D-*glycero*-hex-3-enopyranoside (**43**) is produced on base-catalyzed elimination from methyl 4,6-*O*-benzylidene-2-deoxy-α-D-*erythro*-hexopyranosid-3-ulose *p*-tolylsulfonylhydrazone[39] (**44**).

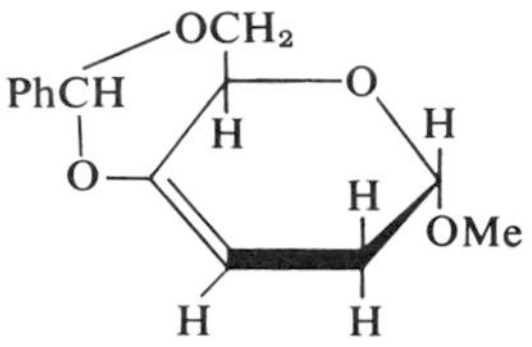

43 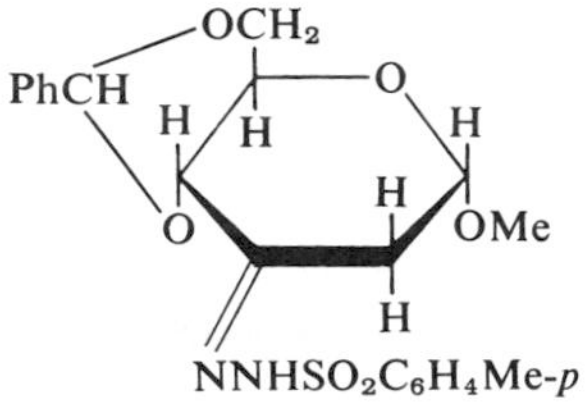   **44**

2. *Methods from α-Diols*

The Corey and Winter procedure,[81] by which α-diols may be converted into olefins (by heating their derived thionocarbonates in trimethyl phosphite), offers a suitable means of introducing an olefinic group into a sugar, provided, in the case of cyclic compounds, the hydroxyl groups have the *cis* orientation. Olefins **45–47** have been prepared in this way from the appropriate derivatives of D-mannitol,[82] L-arabinose,[82] and D-glucofuranose,[83] respectively. Alternatively, olefins have been produced from *trans*-diols by treating the derived disulfonic esters with potassium ethylxanthate in boiling butyl alcohol. Methyl 4,6-*O*-benzylidene-2,3-di-*O*-(methylsulfonyl)-α-D-glucopyranoside is converted satisfactorily, for instance, into the 2,3-unsaturated glycoside derivative by this procedure.[83] Terminal disulfonic esters undergo elimination readily on treatment with sodium iodide in acetone solution, and in this way several unsaturated alditol and hexofuranosyl derivatives have been synthesized. Similarly, nonterminal, vicinal disulfonates will react with iodide in

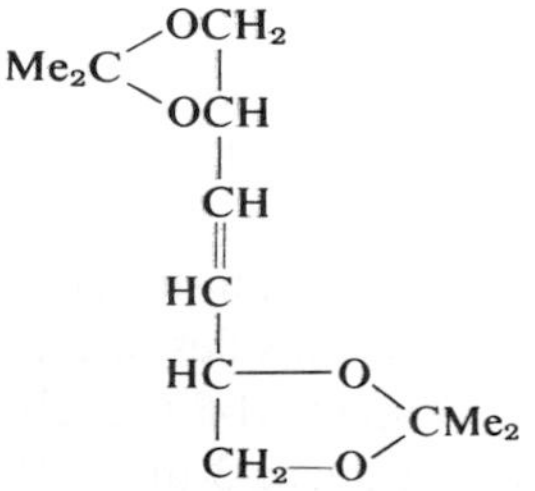

45 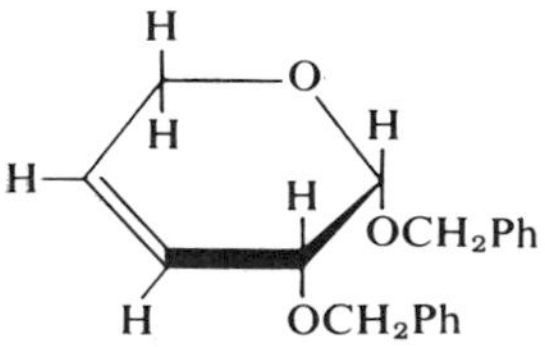   **46** 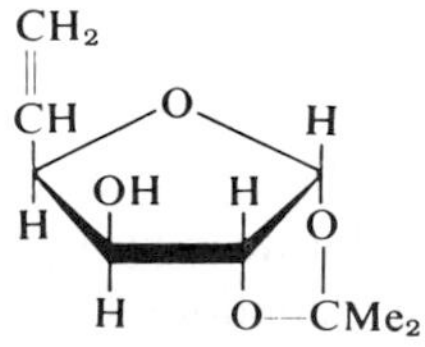   **47**

N,*N*-dimethylformamide to give isolable olefinic products provided zinc is present to remove the molecular iodine that is also produced in the reaction.[83–85] A report[86] claiming that this reaction is not generally applicable has since been corrected,[85] and it has been useful both in the preparation of naturally occurring vicinal, dideoxy sugars (for example amicetose[87] and rhodinose[87a]) and in the synthetic modification of antibiotics (such as in conversion of kanamycin into 3′,4′-dideoxykanamycin).[88]

3. *Methods from Epoxides and Episulfides*

As epoxides and episulfides are generally prepared from diols, this section is closely related to the last, but nevertheless it may conveniently be considered separately.

Treatment of epoxides with methylmagnesium iodide in ether, or with sodium iodide in acetone and acetic acid, gives mainly iodohydrins (—CHI—CHOH—), which on sulfonylation afford derivatives that undergo ready elimination when treated with sodium iodide. Methyl 4,6-*O*-benzylidene-2,3-dideoxy-α-D-*erythro*-hex-2-enopyranoside[86,89] and the unsaturated nucleosides **48** (ref. 90) and **49** (ref. 68) have, for example, been prepared by this method. More simply, the first of these has been prepared directly from methyl 2,3-anhydro-4,6-*O*-benzylidene-α-D-allopyranoside by heating in butyl alcohol with potassium ethylxanthate.[83]

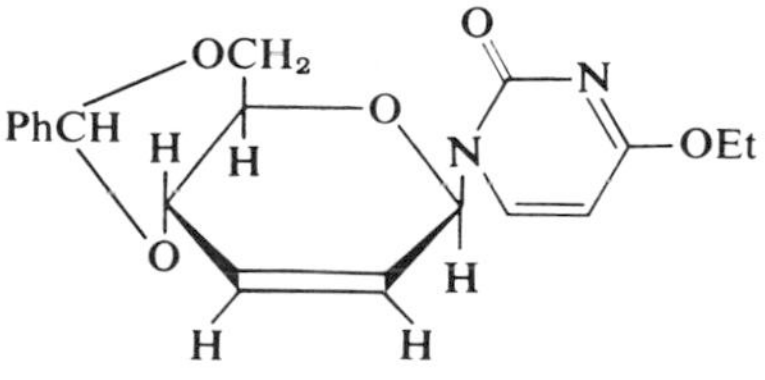

48 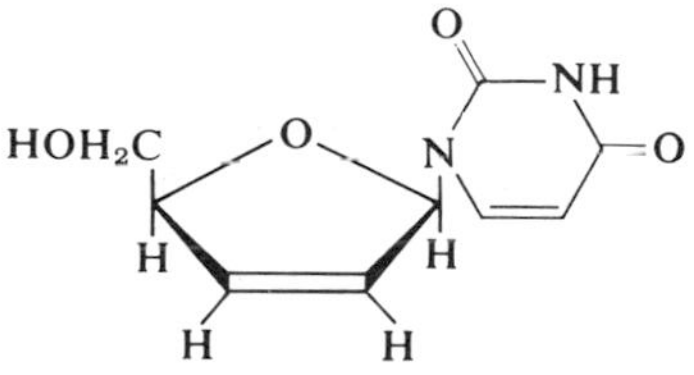   **49**

Although some noncarbohydrate epoxides have been converted into olefins by treatment with hot trimethyl phosphite, the above alloside derivative was inert under these conditions. However, the epithio analogue undergoes elimination smoothly,[91] suggesting another general route to carbohydrate derivatives possessing an olefinic group. Metal alkyls may react with some epoxides to give olefins; both 5,6-anhydro-3-*O*-benzyl-1,2-*O*-isopropylidene-α-D-glucofuranose and the β-L-*ido* isomer give the corresponding 5,6-olefin when treated with methyllithium in ether.[92] Ammonolysis of epoxides and subsequent *N*-methylation of the resulting amino sugars affords dimethyl-amino or trimethylammonium derivatives, which may more generally be

---

References start on p. 875.

converted into olefins by the Cope[93,94] or Hoffmann[94] eliminations. Deamination (see Chapter 16,III,B) of a sugar spiro-aziridine gave[95] the corresponding exocyclic alkene. Lastly, heating epoxides with potassium selenocyanate in aqueous 2-methoxyethanol may afford an efficient means of introducing unsaturation into carbohydrate derivatives.[96]

B. ELIMINATIONS INDUCED BY ELECTRON-WITHDRAWING GROUPS

1. *Cyclic Compounds*

Aldoses having alkali-stable substituents or a deoxy group at C-2 undergo β elimination in alkali to give 2,3-unsaturated products, which may react further [2-deoxy-D-*arabino*-hexose finally affords the 3,6-anhydro derivative (Section II,B)]; 2,3,4,6-tetra-O-methyl-D-glucose, for example, gives 3-deoxy-2,4,6-tri-O-methyl-D-*erythro*-hex-2-enose[97] (**50**).

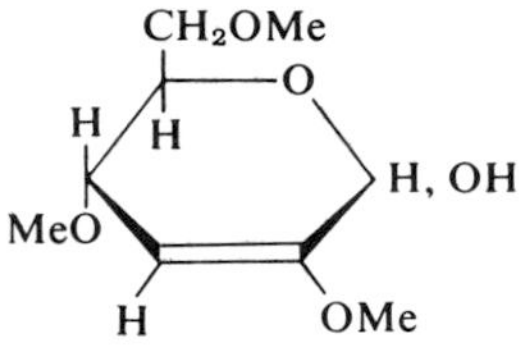

50

Conversely, a 2,3 double bond may be introduced as a result of the influence of an activating group at C-3; methyl 4,6-O-benzylidene-2,3-dideoxy-3-C-nitro-β-D-*erythro*-hex-2-enopyranoside (**51**) arises by a facile, base-catalyzed elimination from the corresponding 2-O-acetyl-3-deoxy-3-C-nitro sugar.[98]

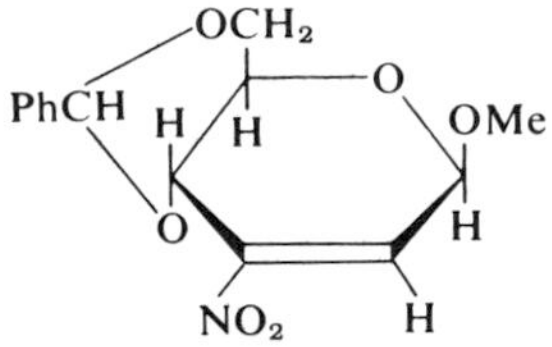

51

In similar fashion, 2-substituted aldonic acids can give 2,3-unsaturated products, which may exist as lactones; alternatively, hexopyranosiduronic esters produce compounds having 4,5-double bonds. Thus methyl (methyl α-D-galactopyranosid)uronate (**52**), on treatment with sodium methoxide, gives methyl (methyl 4-deoxy-β-L-*threo*-hex-4-enopyranosid)uronate[99] (**53**) in a reaction that resembles the one occurring in certain enzymic eliminative degradations of glycuronans from which disaccharides having 4,5-double bonds in the nonreducing moiety are produced; similar disaccharides have been obtained[99a] by methylation of birch xylan according to the Hakomori procedure. Extended exposure to conditions of acylation of aldono-1,5-lactones having O-3 unsubstituted may lead to very efficient β elimination.[100]

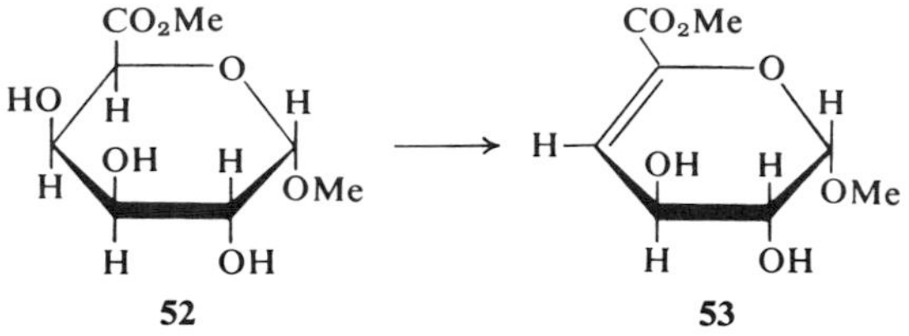

2. *Acyclic Compounds*

Under the influence of basic catalysts, acyclic carboxylic acid derivatives having deoxy or alkoxy groups at the α positions give α,β-unsaturated products as do hydrazones of 2-deoxyaldoses (particularly when the hydroxyl group at C-3 is substituted), but 1,2-unsaturated products are obtained by elimination of acetic acid, as shown, from acetylated aldose phenylhydrazones.[101] α,β-Unsaturated aldonate esters may also be prepared by treatment of acyclic

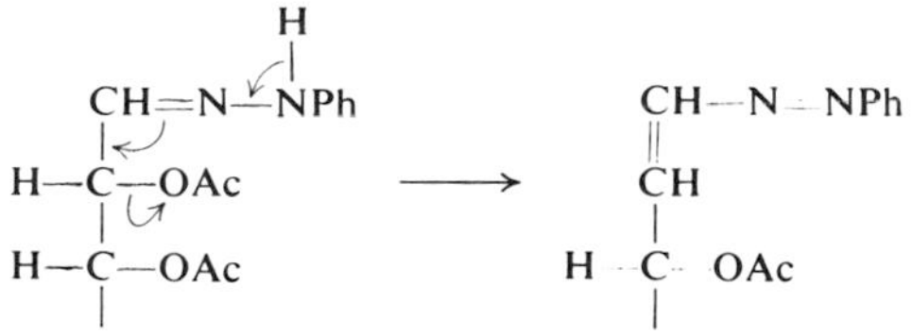

aldoses with Knoevenagel[102,103] or Wittig[104] reagents; in the latter case either acyclic acetates or free aldoses react with phosphoranes (such as $Ph_3P = CHCO_2Et$) to give compounds from which derivatives of higher sugars may be prepared after hydroxylation.

Two other activating groups in this context are sulfonyl[105] and nitro,[106] which, when introduced into terminal positions in acyclic compounds, will induce 1,2 eliminations, particularly when good leaving groups, such as esters, are present at C-2. Fully acetalated aldose dithioacetals react with strong bases[107] to afford 1,2-unsaturated derivatives that are formally ketenic dithioacetals. 2,3:4,5-Di-O-isopropylidene-*aldehydo*-D-arabinose has been converted into the (Z) isomer of the enol acetate[107a] by treatment with hot sodium acetate–acetic anhydride; photoisomerization of the product gave (E)-1-O-acetyl-2,3:4,5-di-O-isopropylidene-D-*erythro*-pent-1-enitol.

References start on p. 875.

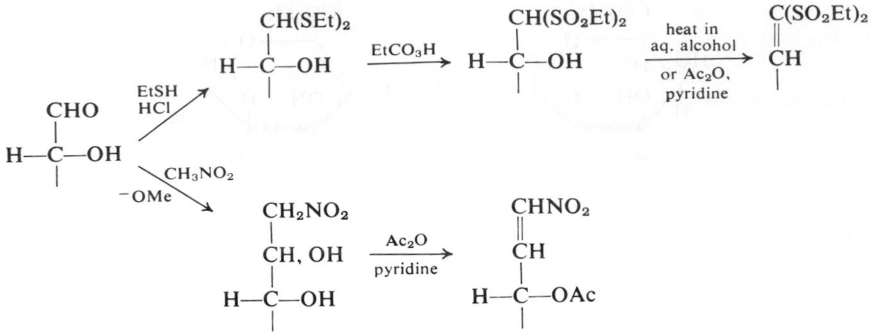

$$\text{CHO} \atop \text{H—C—OH}$$

via EtSH/HCl:

$$\text{CH(SEt)}_2 \atop \text{H—C—OH}$$ $\xrightarrow{\text{EtCO}_3\text{H}}$ $$\text{CH(SO}_2\text{Et)}_2 \atop \text{H—C—OH}$$ $\xrightarrow[\substack{\text{aq. alcohol} \\ \text{or Ac}_2\text{O,} \\ \text{pyridine}}]{\text{heat in}}$ $$\text{C(SO}_2\text{Et)}_2 \atop \text{CH}$$

via CH₃NO₂/−OMe:

$$\text{CH}_2\text{NO}_2 \atop \substack{\text{CH, OH} \\ \text{H—C—OH}}$$ $\xrightarrow[\text{pyridine}]{\text{Ac}_2\text{O}}$ $$\text{CHNO}_2 \atop \substack{\text{CH} \\ \text{H—C—OAc}}$$

C. METHODS BASED ON ALLYLIC DISPLACEMENTS

Although of considerable potential importance, methods based on allylic displacements have been little exploited. The elimination with double-bond migration, which occurs on reduction of methyl 6-deoxy-2,3-O-isopropylidene-4-O-(methylsulfonyl)-α-D-*lyxo*-hex-5-enopyranoside[108] (**54**) with lithium aluminum hydride, illustrates the value of such reactions; however, not all

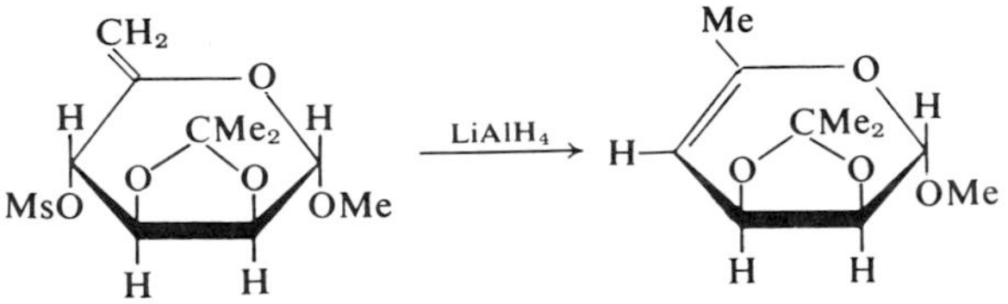

54

allylic sulfonyloxy-group displacements proceed in this fashion. Treatment of methyl 2,3-dideoxy-4,6-di-O-(methylsulfonyl)-α-D-*erythro*-hex-2-enopyranoside (**55**), for example, in hot N,N-dimethylformamide with sodium benzoate causes direct displacement at C-4, followed by displacement at the primary position.[109] The *threo* analog of **55** behaves in similar fashion, and this procedure can be employed for specific inversions at allylic sites.

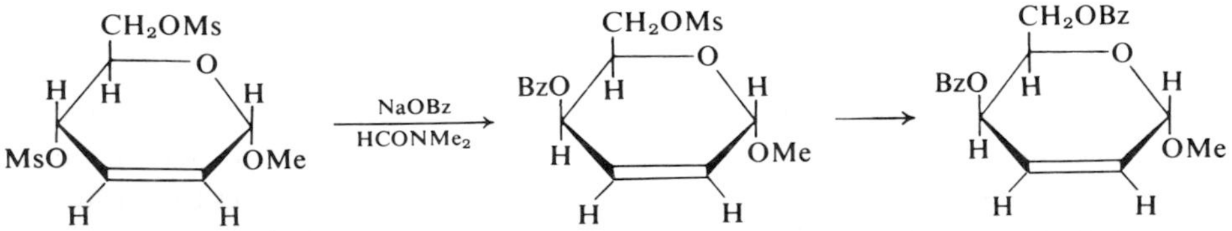

55

Other allylic displacements that have been effected, which offer a means of relocating double bonds in unsaturated compounds, include the formation of compound **43** by treatment of methyl 4,6-*O*-benzylidene-2,3-dideoxy-α-D-*erythro*-hex-2-enopyranoside with potassium *tert*-butoxide in methyl sulfoxide,[110] the thermal rearrangement[111] of the 2,3-unsaturated thiocyanate **56** to the 3,4-unsaturated isothiocyanate **57**, and the Claisen rearrangement[112] of **58** to the branched-chain dialdose **59**.

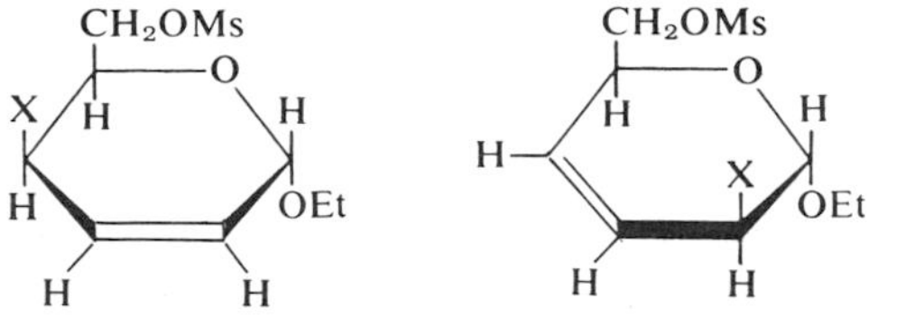

56 X = —SCN
58 X = —OCH=CH₂

57 X = —NCS
59 X = —CH₂CHO

D. Extension of Sugar Chains by Addition of Unsaturated Groups

1. *Addition of Grignard Reagents*

Reactions of unsaturated Grignard reagents with carbohydrates containing a free aldehydic group offer alternative means of synthesizing sugars possessing either vinylic or acetylenic groups at the end of the carbon chains. Compounds **60** and **61** have, for example, been synthesized from the aldehydes obtained by oxidation of 1,2:3,4-di-*O*-isopropylidene-α-D-galactopyranose and from 2,3-*O*-isopropylidene-*aldehydo*-D-glyceraldehyde, respectively, by use of ethynylmagnesium bromide, and are of potential interest as synthetic precursors of vinylic analogues and of a wide variety of saturated compounds.[113] During hydroboronation of derived esters of these acetylenes, elimination of acid occurs and α,β-unsaturated aldehydes result.

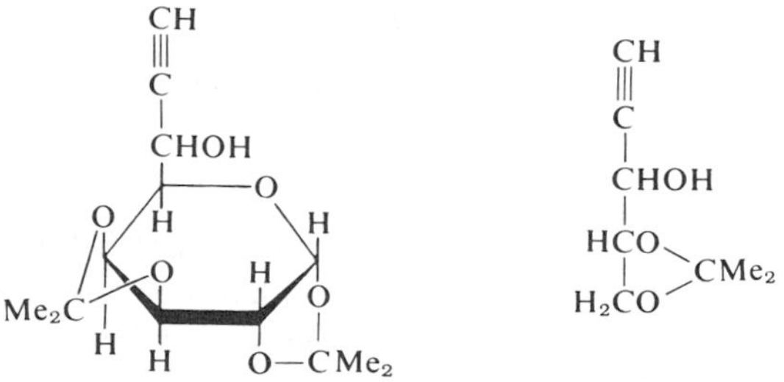

60

61

Grignard reagents (and organolithium compounds) add to sugars having exposed ketonic centers to give branched-chain products, and a number of examples are described in Chapter 17. Michael-type addition of vinyl-magnesium bromide to a 6-deoxy-6-nitrohex-5-enose derivative gave a 5,6,7-trideoxy-5-C-(nitromethyl)hept-6-enose.[114] Alkyl 2,3,6-trideoxy-L-*glycero*-hex-2-enopyranosid-4-uloses have been prepared by addition of 3,3-dialkoxypropyn-1-yl magnesium halides to 2-O-acyl-L-lactyl halides and subsequent reduction and deprotection.[114a]

2. Condensation of Carbonyl Sugars with Wittig Reagents

Methylene ylids condense with aldehydo and keto sugars to afford the corresponding alkenes, which are versatile intermediates for syntheses of branched-chain sugars. By this reaction, 1,2:5,6-di-O-isopropylidene-α-D-*ribo*-hexofuranos-3-ulose (62) has been converted into 63–70, although some

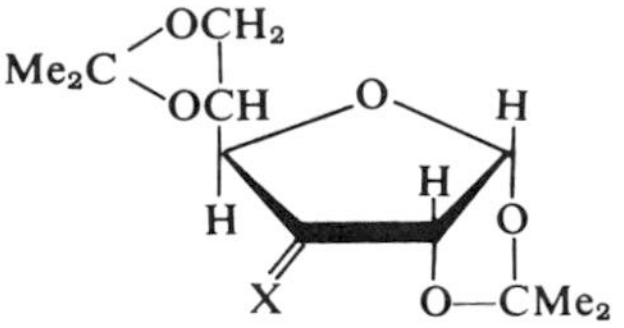

62 X = O	67 X = CHCOMe (ref. 119)
63 X = CH₂ (ref. 115)	68 X = CH=CHCHO (ref. 119)
64 X = CHSMe (ref. 116)	69 X = CHCO₂Me (ref. 120)
65 X = CH(Cl, Br, or I) (ref. 117)	70 X = CHCN (ref. 121, 122)
66 X = CF₂ (ref. 118)	71 X = CHP(=O)(OEt)₂ (ref. 123)

72 X = (EtO₂C)CNHCHO (ref. 124)

73 X = 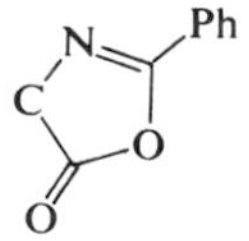   (ref. 125)

74 X = 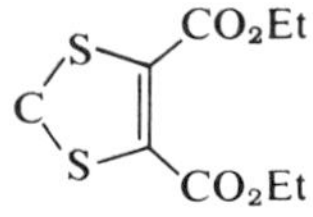   (ref. 126)

inversion of stereochemistry at C-4 of 62 may occur[116] during the course of the Wittig condensation. A Grignard-related condensation of 62 with dichloromethane and magnesium has been shown[126a] to afford a usable, alternative route to 63. Quaternization of the trigonal ring-atom generally favors formation of the product having the opposite stereochemistry to that formed by direct nucleophilic addition, as by a Grignard reagent or other organometallic addends; thus hydrogenation,[122–124] hydroboronation,[126b] hydroxylation by action of permanganate,[120] azidolysis,[127] and addition of nitryl iodide[128] proceed from the less-hindered side to afford mainly products having the branched substituent *cis* to O-2.

Products **71–73** are formed by base-catalyzed condensation of active methylene compounds with **62**, and **74** is prepared by treatment of **62** with carbon disulfide and tributylphosphine and subsequent addition of ethyl butynedioate. Addition of active methylene compounds to aldehydo sugars provides a useful route to various furan derivatives.[128a]

V. REACTIONS OF OTHER UNSATURATED SUGARS

A. ADDITIONS

1. *To Isolated Double Bonds*

Hydrogenation of vinylic bonds affords a means of synthesizing dideoxy sugars. For example, the antibiotic sugar amicetose (2,3,6-trideoxy-D-*erythro*-hexose) has been prepared from ethyl 2,3-dideoxy-α-D-*erythro*-hex-2-eno-pyranoside by hydrogenation followed by specific removal of oxygen from the primary site.[129] Enolic systems usually undergo hydrogenation to give two products, one of which is frequently strongly favored, so that highly specific syntheses may be effected, but in some cases both products of reduction may

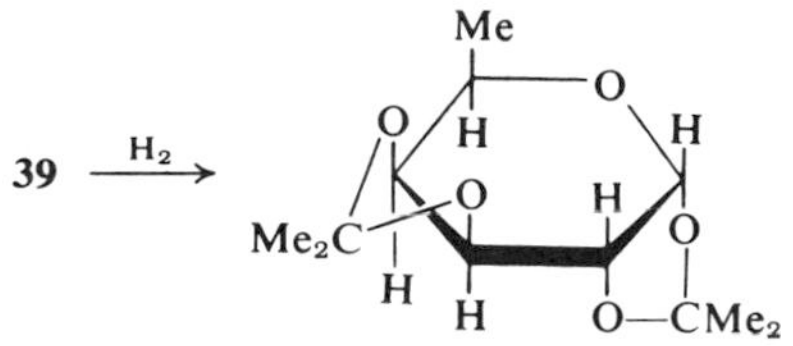

(Ref. 130)

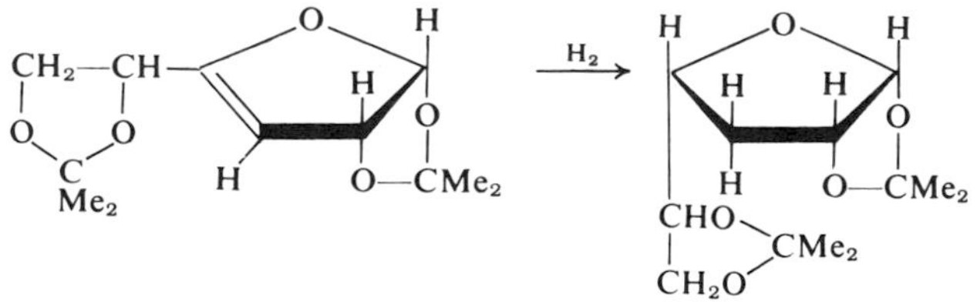

(Ref. 69)

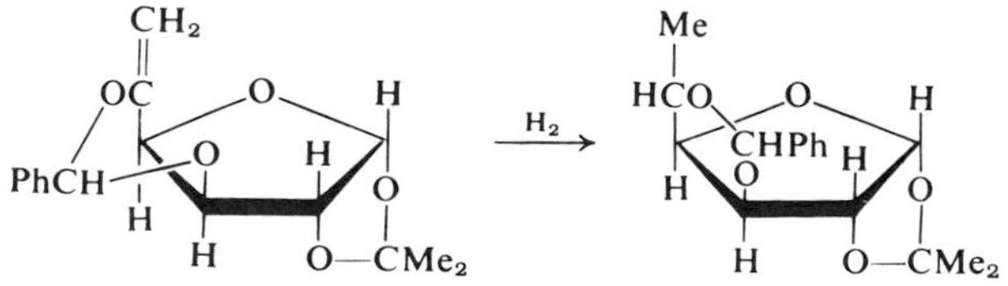

(Ref. 131)

be isolated. For example, from the 4,5-unsaturated uronate **53**, methyl 4-deoxy-α-D-*xylo*- and β-L-*arabino*-hexoside have been obtained.[132] Frequently, hydrogenolysis of allylic groups accompanies hydrogenation and may interfere with the preparation of simple deoxy compounds. Thus, in the treatment of 2,3-unsaturated derivatives with hydrogen in the presence of palladium or, especially, platinum catalysts, acetoxy, alkoxy, or aryloxy groups at C-1 may be lost.

A few halogen adducts have been prepared, but they tend to be unstable (bromides and iodides in particular), and little is known of their structures; however, bromination of methyl 4,6-*O*-benzylidene-2,3-dideoxy-α-D-*erythro*-hex-2-enopyranoside in methanol in the presence of barium carbonate gives crystalline methyl 4,6-*O*-benzylidene-2,3-dibromo-2,3-dideoxy-α-D-altro-pyranoside in 70% yield. On treatment with potassium *tert*-butoxide in refluxing xylene, this product undergoes elimination of the elements of hydrogen bromide and gives methyl 4,6-*O*-benzylidene-2-bromo-2,3-dideoxy-α-D-*threo*-hex-3-enopyranoside in 90% yield.[133] In the same way, hydroxylation reactions have not been examined fully, but their potential value in synthetic work has been illustrated:

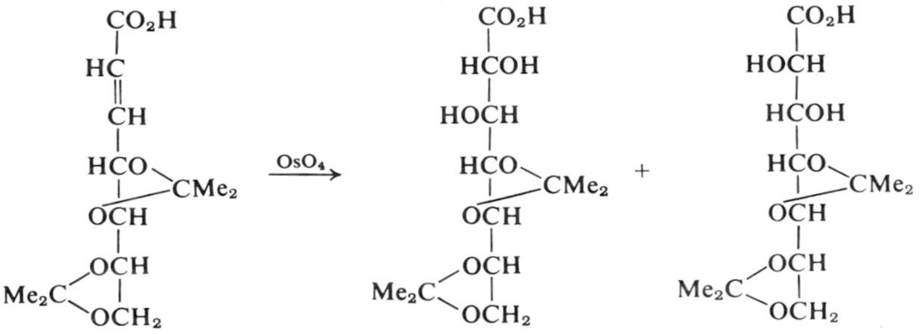

(Ref. 134)

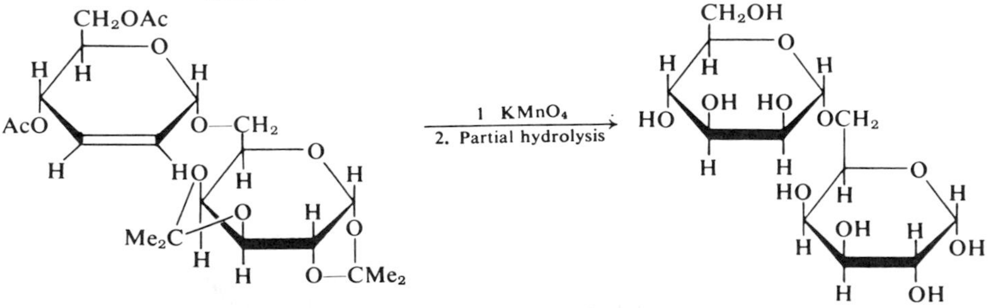

(Ref. 135)

cis-Hydroxylation of methyl glyc-2-enofuranosides by permanganate gave the 1,2-*trans*-disposed adducts.[136] Acetoxylation and epoxidation have been

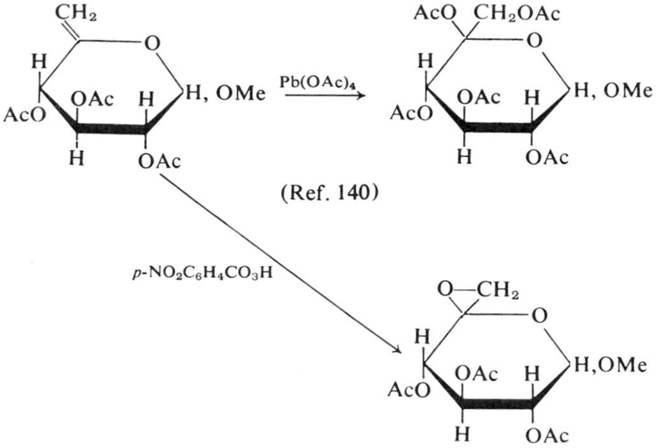

(Ref. 140)

(Ref. 141)

effected in a limited number of cases, and treatment of alditol derivatives having terminal double bonds with hypobromous acid followed with sodium acetate has been shown to be a method by which hexitols can be interconverted.[137] Stereospecific conversion of (E)-3,4-dideoxy-1,2:5,6-di-O-isopropylidene-D-threo-hex-3-enitol into its cis (Z) isomer was effected[138] by epoxidation and base-catalyzed opening to give the 3,4-diol, followed by regeneration of the 3,4-double bond by the Corey–Winter procedure. Epoxidation of 2-methoxyglyc-2-enoses has been shown[139] to afford 1,2-orthoperoxy esters in addition to the expected 2,3-anhydro-2-methoxy sugars.

Hydroboronation, by which the elements of water can be added in the anti-Markownikov sense to double bonds, has been used to synthesize deoxy sugars, and may be employed for specific labeling; and addition of thiolacetic acid, which also affords deoxy derivatives, has been carried out under the influence of ultraviolet light:

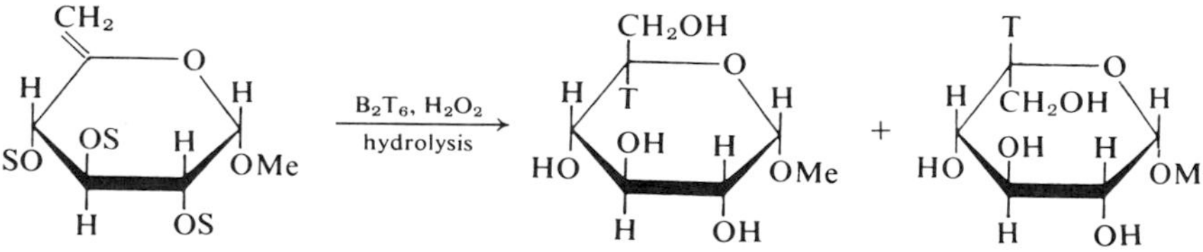

S = trimethylsilyl; T = tritium
(Ref. 108)

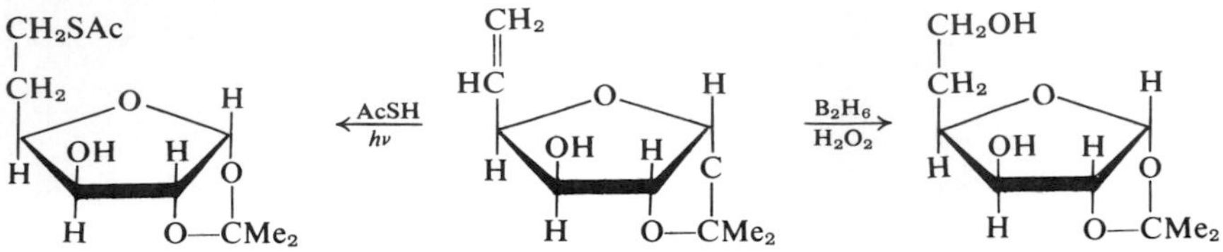

(Ref. 143)   (Ref. 142)

Branched-chain adducts have been reported to result from photoaddition of 1,3-dioxolane,[144] 2-propanol,[145] and formamide.[146] Hydroformylation[37,147] also offers a means of chain alteration, and addition of diiodomethane in the presence of zinc–copper alloys converts unsaturated sugars into the corresponding cyclopropane[133,148] derivatives. The cyclopropane **75** is presumably formed by tautomerization of an enolate anion.[149]

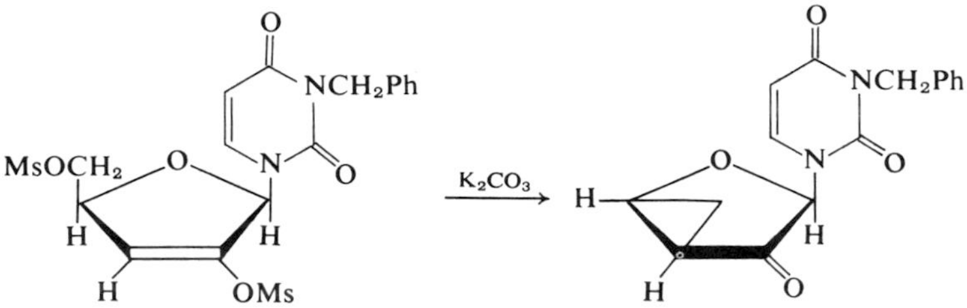

75

2. *To Double Bonds Conjugated with Electron-Withdrawing Groups*

Those compounds resulting from eliminations induced by electron-withdrawing groups (Section IV,B) or from allylic oxidation,[150] undergo addition reactions which can be applied with profit despite the production of diastereoisomers:

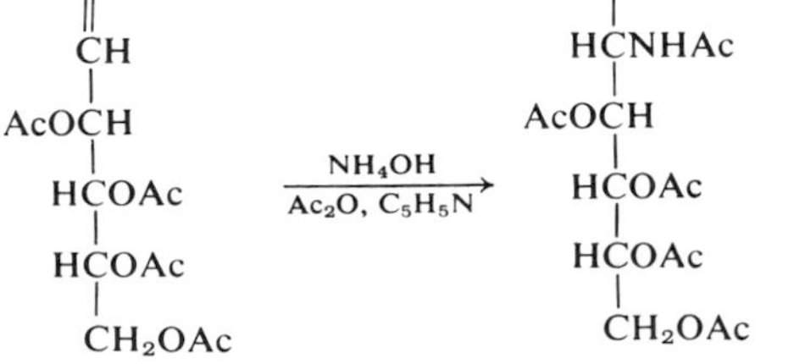

(Ref. 105)

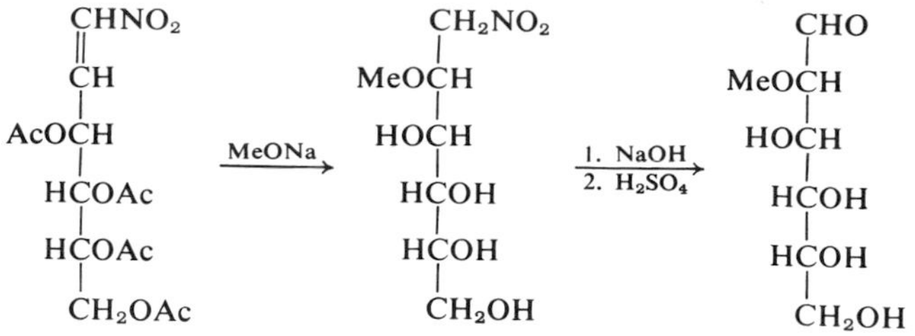

(Ref. 151)

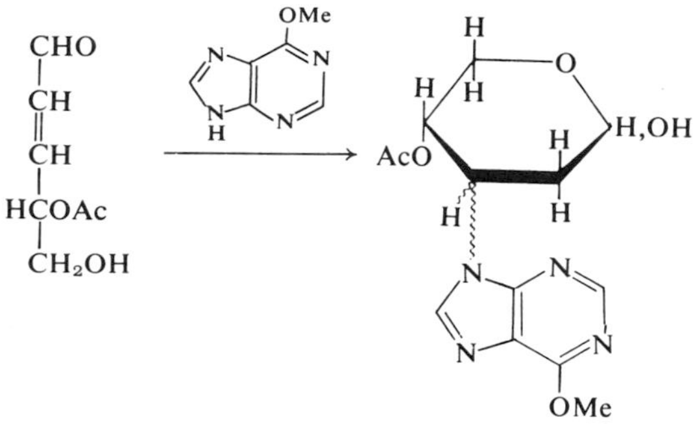

(Ref. 152)

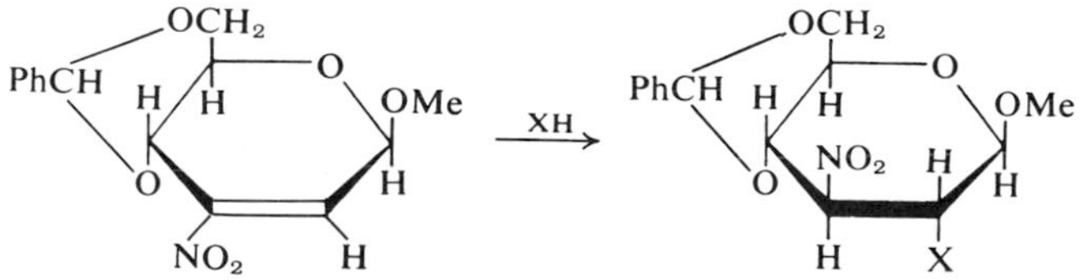

(Ref. 98) X = OCOR
(Ref. 153) X = OR, NHR, SR
(Ref. 154) X = CN

Strictly analogous are the intramolecular additions which occur during deacetylation of diacetylpseudoglucal (Section II,B), and on heating the bis(ethylsulfonyl) derivatives of hexoses or 1-deoxy-1-nitroalditols in inert solvents.

Nucleophilic addition of azide ion[156] and photocatalyzed addition of alcohols[157] to conjugated hexenopyranosulose derivatives affords products of

---

References start on p. 875.

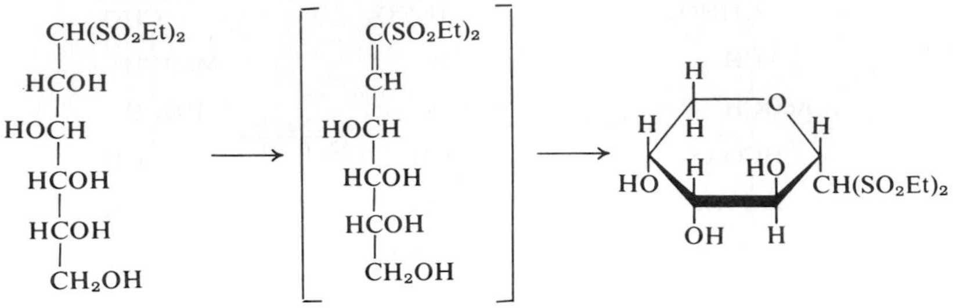

(Ref. 155)

Michael-type addition, and 2,3-dideoxy-2-phenylazopyranosides,[158] furanosides,[159] and the corresponding 3-phenylazofuranosides[159] have been demonstrated to undergo 1,4-addition of a wide range of nucleophiles to afford α-substituted phenylhydrazones; the latter reactions occur faster in β glycosides than α and in furanosides than in pyranosides.

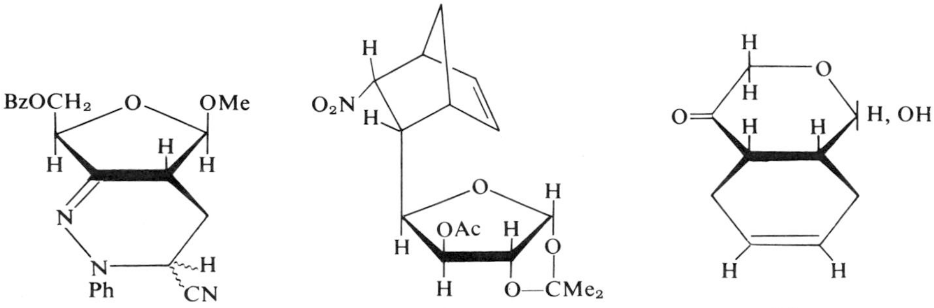

Diels-Alder adducts (76–78) have been prepared by reaction of methyl 5-O-benzoyl-2,3-dideoxy-3-phenylazo-β-D-glycero-pentofuranoside[160] with acrylonitrile, 3-O-acetyl-5,6-dideoxy-1,2-O-isopropylidene-6-C-nitro-α-D-xylo-hexofuranose[161] with cyclopentadiene, and 2,3-dideoxypentopyranos-4-ulose[162] with 1,3-butadiene, respectively.

The conjugated diene 1,5-anhydro-4,6-O-benzylidene-2,3-dideoxy-3-C-methylene-D-erythro-hex-1-enitol[163] has been prepared by several methods. The extensive investigations of reactions of unsaturated and conjugated sugar derivatives have been summarized by Fraser-Reid.[164]

B. OZONOLYSIS

Oxidative cleavage of double bonds offers means of determining the positions of the unsaturated centers and also for synthesizing carbonyl derivatives:

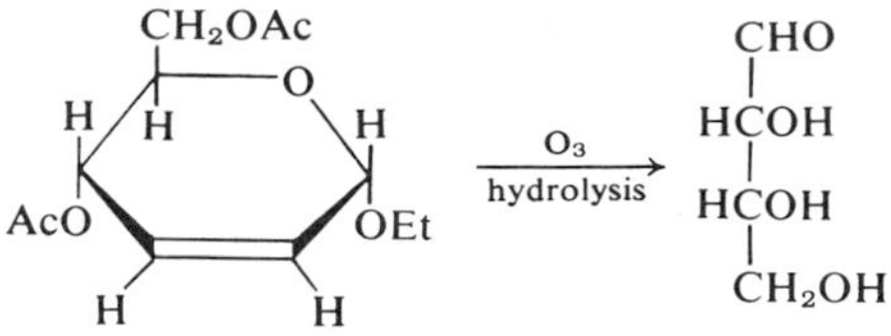

(Ref. 165)

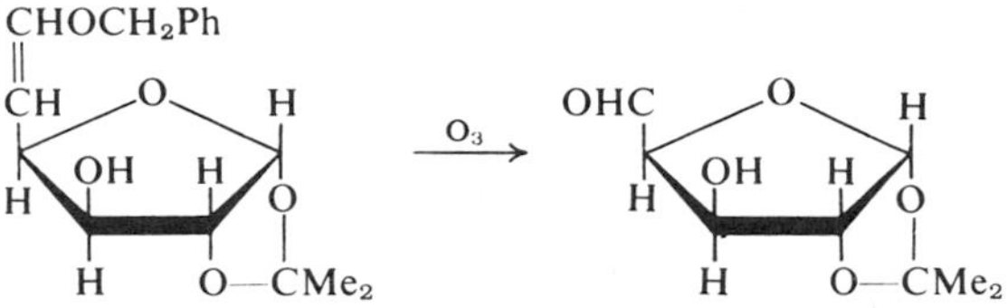

(Ref. 166)

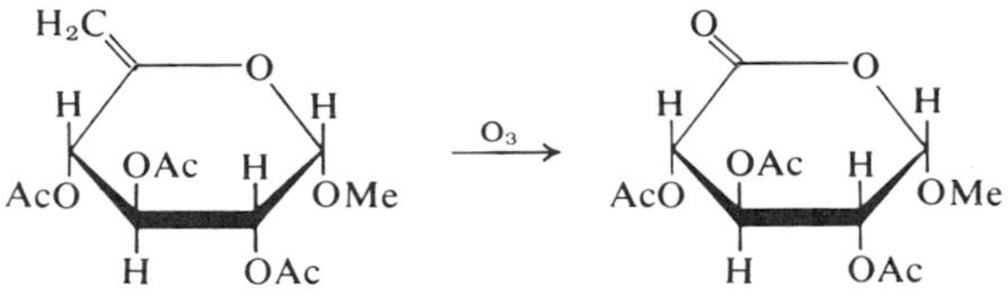

(Ref. 167)

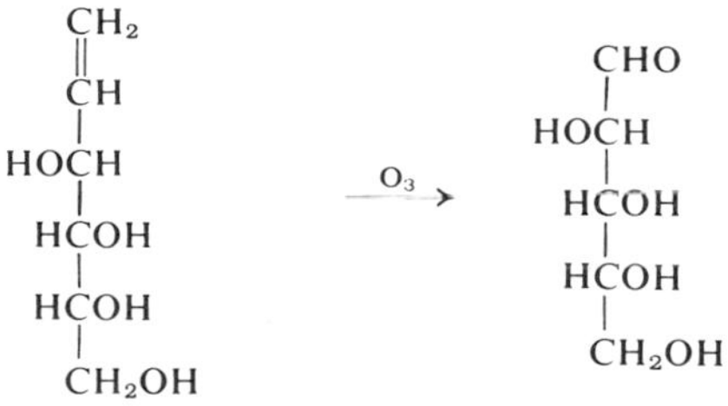

(Ref. 168)

C. HYDROLYSIS AND OXIDATION OF ENOLIC COMPOUNDS

Enolic ethers and esters derived from free sugars or glycosides are hydrolyzed readily in acidic media and provide means of obtaining deoxydicarbonyl sugars, which may undergo facile elimination and rearrangement. Oxidation with peroxyacids or with lead tetraacetate, on the other hand, causes, initially, hydroxylation of the double bond and, after hydrolysis, the formation of new carbonyl groups.

References start on p. 875.

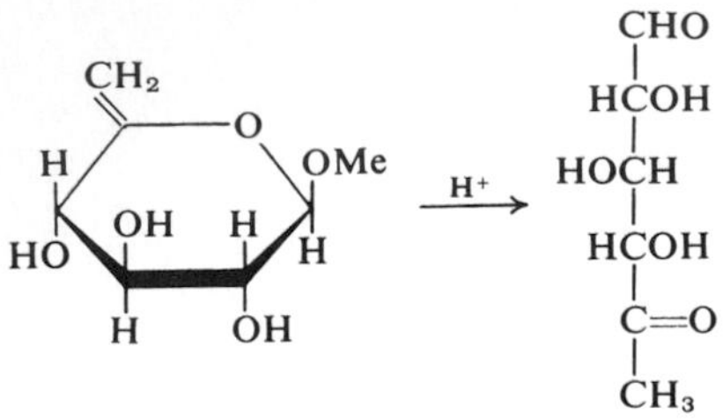

(Ref. 73)

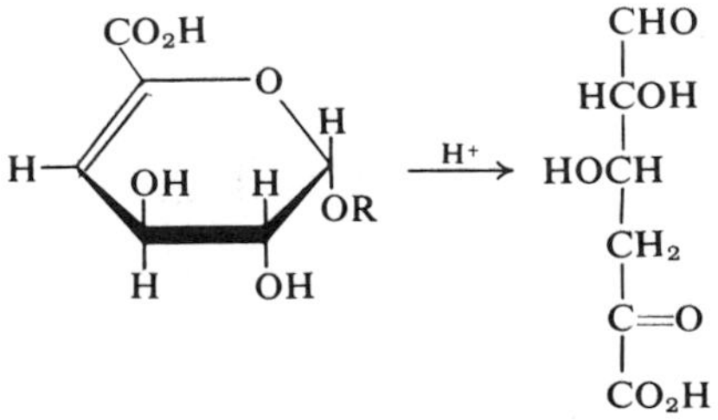

(Ref. 169)

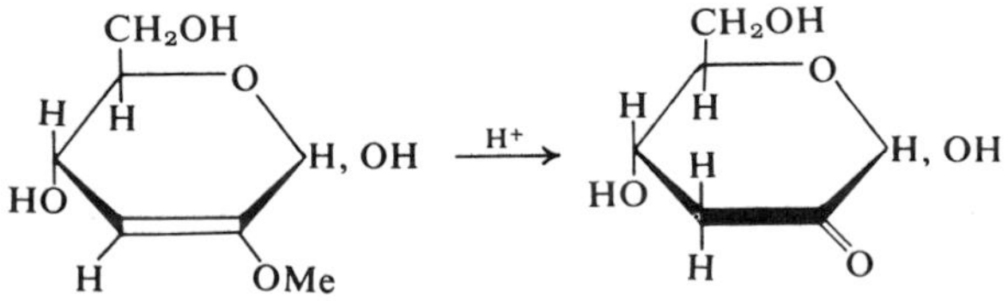

(Ref. 97)

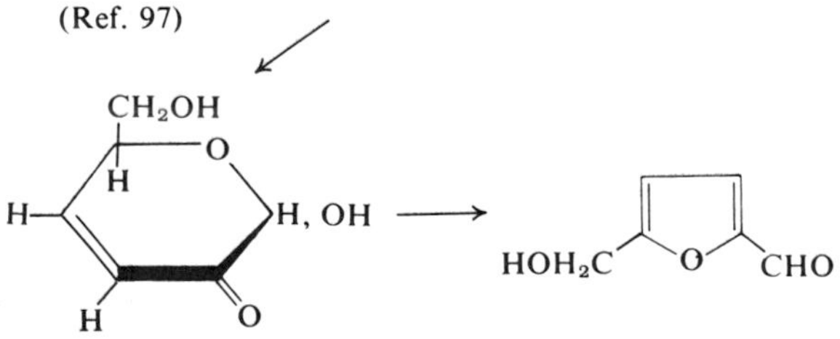

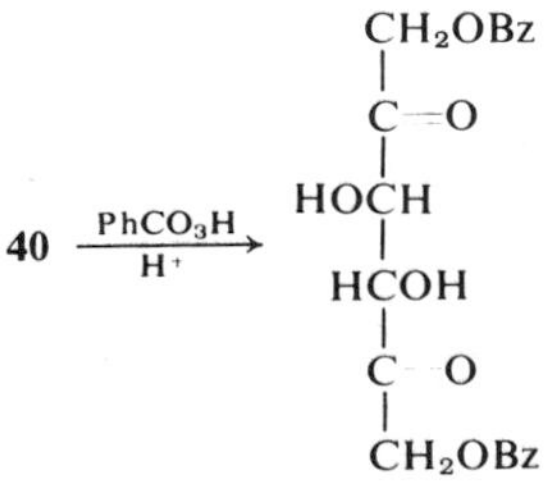

(Ref. 170)

REFERENCES

1. R. J. Ferrier, *Advan. Carbohyd. Chem.*, **20**, 67 (1965); *Advan. Carbohyd. Chem. Biochem.*, **24**, 199 (1969).
2. B. Helferich, E. N. Mulcahy, and H. Zeigler, *Ber.*, **87**, 233 (1954).
3. A. A. J. Feast, W. G. Overend, and N. R. Williams, *J. Chem. Soc.*, 7378 (1965); M. Sharma and R. K. Brown, *Can. J. Chem.*, **44**, 2825 (1966); see also ref. 86.
4. R. H. Hall and A. Jordaan, *J. Chem. Soc. Perkin I*, 1059 (1973).
5. R. H. Hall, A. Jordaan, and G. J. Lourens, *J. Chem. Soc. Perkin I*, 38 (1973); R. H. Hall, A. Jordaan, and O. G. DeVilliers, *ibid.*, 626 (1975).
6. S. Y.-K. Tam and B. Fraser-Reid, *Tetrahedron Lett.*, 4897 (1973).
7. R. K. Ness and H. G. Fletcher, Jr., *J. Org. Chem.*, **28**, 435 (1963); M. Haga and R. K. Ness, *ibid.*, **30**, 158 (1965).
8. M. J. Robins and R. A. Jones, *J. Org. Chem.*, **39**, 113 (1974).
9. G. R. Gray and R. Barker, *J. Org. Chem.*, **32**, 2764 (1967).
10. J. Staněk, M. Černý, J. Kocourek, and J. Pacák, "The Monosaccharides," Academic Press, New York, 1963.
11. F. Shafizadeh and M. Stacey, *J. Chem. Soc.*, 3608 (1952).
12. K. Wallenfels and J. Lehmann, *Ann.*, **635**, 166 (1960).
13. D. M. Ciment and R. J. Ferrier, *J. Chem. Soc. (C)*, 441 (1966).
14. W. A. Bowles and R. K. Robins, *J. Amer. Chem. Soc.*, **86**, 1252 (1964).
15. D. E. Iley and B. Fraser-Reid, *J. Amer. Chem. Soc.*, **97**, 2563 (1975).
15a. See J. Lehmann and E. Schröter, *Carbohyd. Res.* **58**, 65 (1977) and earlier papers.
16. G. R. Inglis, J. C. P. Schwarz, and L. McLaren, *J. Chem. Soc.*, 1014 (1962); K. Takiura and S. Honda, *Carbohyd. Res.*, **21**, 379 (1972); S. Honda, K. Kaheki, H. Takai, and K. Takiura, *ibid.*, **29**, 477 (1973).
17. D. Horton, J. M. Tarelli, and J. D. Wander, *Carbohyd. Res.*, **23**, 440 (1972).
18. R. U. Lemieux and B. Fraser-Reid, *Can. J. Chem.*, **42**, 532 (1964); **43**, 1460 (1965); see ref. 1 for a fuller treatment of this subject.
19. L. Vargha and J. Kuszmann, *Ber.*, **96**, 411 (1963).
20. J. Davoll and B. Lythgoe, *J. Chem. Soc.*, 2526 (1949).
21. C. D. Hurd and H. Jenkins, *Carbohyd. Res.*, **2**, 240 (1966); P. R. Bradley and E. Buncel, *Can. J. Chem.*, **46**, 3001 (1968).
22. A. M. Gakhokidze, *J. Gen. Chem. USSR* (Eng. Transl.), **16**, 1914 (1946).
23. K. Igarashi, T. Honma, and T. Imagawa, *J. Org. Chem.*, **35**, 610 (1970).
24. T. Van Es, *J. S. Afr. Chem. Inst.*, **26**, 152 (1973).
25. J. C. Campbell, R. A. Dwek, P. W. Kent, and C. K. Prout, *Carbohyd. Res.*, **10**, 71 (1969); L. D. Hall and J. F. Manville, *Can. J. Chem.*, **47**, 361 (1969).
26. P. W. Kent, J. E. G. Barnett, and K. R. Wood, *Tetrahedron Lett.*, 1345 (1963); P. W. Kent and J. E. G. Barnett, *Tetrahedron*, Suppl. **7**, 69 (1966); K. R. Wood and P. W. Kent, *J. Chem. Soc. (C)*, 2422 (1967).
27. T. Maki and S. Tejima, *Chem. Pharm. Bull.* (Tokyo), **15**, 1069 (1967).
28. I. Lundt and C. Pedersen, *Acta Chem. Scand.*, **20**, 1369 (1966).
29. R. Kuhn, I. Löw, and H. Trischmann, *Ber.*, **90**, 203 (1957).
30. V. Bílik and Š. Kučár, *Carbohyd. Res.*, **13**, 311 (1970); also see ref. 60b.
31. W. J. Serfontein, J. H. Jordaan, and J. White, *Tetrahedron Lett.*, 1069 (1964); R. U. Lemieux, T. L. Nagabhushan, and I. K. O'Neill, *ibid.*, 1909 (1964); *Can. J. Chem.*, **46**, 413 (1968).
32. R. U. Lemieux, S. W. Gunner, and T. L. Nagabhushan, *Tetrahedron Lett.*, 2149 (1965); *Can. J. Chem.*, **46**, 397, 405 (1968).

33. K. Igarashi and T. Honma, *J. Org. Chem.*, **35**, 606 (1970).
34. Y. Araki, K. Matsuura, Y. Ishido, and K. Kushida, *Chem. Lett.*, 383 (1973).
35. K. Matsuura, S. Maeda, Y. Araki, Y. Ishido, and T. Murai, *Tetrahedron Lett.*, 2869 (1970).
36. K. Matsuura, Y. Araki, Y. Ishido, A. Murai, and K. Kushida, *Carbohyd. Res.*, **29**, 459 (1973); compare Matsuura *et al.*, *ibid.*, **64**, 109 (1978).
37. A. Rosenthal, *Advan. Carbohyd. Chem.*, **23**, 59 (1968).
38. A. Rosenthal and D. Abson, *J. Amer. Chem. Soc.*, **86**, 5356 (1964).
39. R. J. Ferrier, *J. Chem. Soc.*, 5443 (1964); see also G. Grynkiewicz, W. Priebe, and A. Zamojski, *Carbohyd. Res.* **68**, 33 (1979).
40. R. J. Ferrier, W. G. Overend, and A. E. Ryan, *J. Chem. Soc.*, 3667 (1962).
41. T. Maki, H. Nakamura, S. Tejima, and M. Agaki, *Chem. Pharm. Bull.* (Tokyo), **13**, 764 (1966).
42. R. J. Ferrier and N. Prasad, *J. Chem. Soc.* (*C*), 570 (1969).
43. K. Matsuura, K. Senna, Y. Araki, and Y. Ishido, *Bull. Chem. Soc. Japan*, **47**, 1197 (1974).
44. H. Paulsen and J. Thiem, *Ber.*, **106**, 3850 (1973).
45. R. J. Ferrier, W. G. Overend, and A. E. Ryan, *J. Chem. Soc.*, 1488 (1962).
46. H. S. Isbell, *J. Res. Nat. Bur. Stand.*, **32**, 45 (1944).
47. R. J. Ferrier and N. Prasad, *J. Chem. Soc.* (*C*), 581 (1969).
48. B. Fraser-Reid and B. Radatus, *J. Amer. Chem. Soc.*, **92**, 5288 (1970).
48a. K. Takiura and S. Honda, *Carbohyd. Res.*, **23**, 369 (1972).
48b. F. Gonzalez, S. Lesage, and A. S. Perlin, *Carbohyd. Res.*, **42**, 267 (1975).
48c. R. J. Ferrier in "International Review of Science," Organic Chemistry, Series 2, (G. O. Aspinall, Ed.) Vol. 7, Butterworths, London, 1976, p. 35.
49. R. U. Lemieux and D. R. Lineback, *Can. J. Chem.*, **43**, 94 (1965).
50. R. J. Ferrier and G. H. Sankey, *J. Chem. Soc.* (*C*), 2339 (1966).
51. D. R. Rao and L. M. Lerner, *Carbohyd. Res.*, **19**, 133 (1971); **22**, 345 (1972); N. A. Hughes, *ibid.*, **25**, 242 (1972); also see ref. 62.
52. M. G. Blair, *Advan. Carbohyd. Chem.*, **9**, 97 (1954).
53. F. W. Lichtenthaler and E. Fischer, *Angew. Chem.*, **86**, 591 (1974); E. Fischer and F. W. Lichtenthaler, *ibid.*, **86**, 590 (1974).
54. W. M. Corbett, *J. Chem. Soc.*, 3213 (1959).
55. R. J. Ferrier, W. G. Overend, and G. H. Sankey, *J. Chem. Soc.*, 2830 (1965).
56. R. U. Lemieux, D. R. Lineback, M. L. Wolfrom, F. B. Moody, E. G. Wallace, and F. Komitsky, Jr., *J. Org. Chem.*, **30**, 1092 (1965).
57. R. J. Ferrier, N. Prasad, and G. H. Sankey, *J. Chem. Soc.* (*C*), 974 (1968).
58. R. J. Ferrier, N. Prasad, and G. H. Sankey, *J. Chem. Soc.* (*C*), 587 (1969).
59. N. Pravdić, B. Zidoveć, and H. G. Fletcher, Jr., *Croat. Chim. Acta*, **42**, 523 (1970).
60. (a) K. Bock and C. Pedersen, *Acta Chem. Scand.*, **24**, 2465 (1970); (b) *ibid.*, **25**, 1021 (1971).
61. E. Zissis, R. K. Ness, and H. G. Fletcher, Jr., *Carbohyd. Res.*, **20**, 9 (1971).
62. R. J. Ferrier and J. R. Hurford, *Carbohyd. Res.*, **38**, 125 (1974).
63. J. J. Fox and K. A. Watanabe, *Tetrahedron Lett.*, 897 (1966).
64. H. Hoeksema, G. Slomp, and E. E. van Tamelen, *Tetrahedron Lett.*, 1787 (1964).
65. H. Riemann, D. J. Cooper, A. K. Mallams, R. S. Jaret, A. Yehaskel, M. Kugelman, H. F. Vernay, and D. Schumacher, *J. Org. Chem.*, **39**, 1451 (1974).
66. K. H. Hollenbeak and M. E. Kuehne, *Tetrahedron*, **30**, 2307 (1974).
67. H. R. Bolliger and D. A. Prins, *Helv. Chim. Acta*, **29**, 1061 (1946).
68. J. P. Horwitz, J. Chua, I. L. Klundt, M. A. DaRooge, and M. Noel, *J. Amer. Chem. Soc.*, **86**, 1896 (1964).

69. F. Weygand and H. Wolz, *Ber.*, **85**, 256 (1952).
70. H. Arzoumanian, E. M. Acton, and L. Goodman, *J. Amer. Chem. Soc.*, **86**, 74 (1964).
71. R. E. Gramera, T. R. Ingle, and R. L. Whistler, *J. Org. Chem.*, **29**, 1083 (1964).
72. H. Ohle and R. Deplanque, *Ber.*, **66**, 12 (1933).
73. B. Helferich and E. Himmen, *Ber.*, **62**, 2136 (1929).
74. D. Horton and W. Weckerle, *Carbohyd. Res.*, **44**, 227 (1975).
75. W. N. Haworth, R. L. Heath, and L. F. Wiggins, *J. Chem. Soc.*, 155 (1944).
76. D. Horton and D. H. Hutson, *Advan. Carbohyd. Chem.*, **18**, 123 (1963).
77. K. J. Ryan, H. Arzoumanian, E. M. Acton, and L. Goodman, *J. Amer. Chem. Soc.*, **86**, 2503 (1964).
78. A. Rosenthal and C. M. Richards, *Carbohyd. Res.*, **31**, 331 (1973).
79. A. Klemer and G. Rodemeyer, *Ber.*, **108**, 1896 (1975).
80. A. Klemer and G. Rodemeyer, *Ber.*, **107**, 2612 (1974); D. M. Clode, D. Horton, and W. Weckerle, *Carbohyd. Res.*, **49**, 305 (1976).
80a. A. Klemer, G. Rodemeyer, and F. J. Linnenbaum, *Ber.*, **109**, 2849 (1976).
81. E. J. Corey and R. A. E. Winter, *J. Amer. Chem. Soc.*, **85**, 2677 (1963).
82. A. H. Haines, *Carbohyd. Res.*, **1**, 214 (1965).
83. D. Horton and W. N. Turner, *Tetrahedron Lett.*, 2531 (1964); E. L. Albano, D. Horton, and T. Tsuchiya, *Carbohyd. Res.*, **2**, 349 (1966); D. Horton, J. K. Thomson, and C. G. Tindall, Jr., *Methods Carbohyd. Chem.*, **6**, 297 (1972).
84. R. S. Tipson and A. Cohen, *Carbohyd. Res.*, **1**, 338 (1965).
85. T. Yamazaki, H. Sugiyama, N. Yamaoka, K. Matsuda, and S. Seto, *Carbohyd. Res.*, **50**, 279 (1976).
86. R. U. Lemieux, E. Fraga, and K. A. Watanabe, *Can. J. Chem.*, **46**, 61 (1968).
87. E. L. Albano and D. Horton, *J. Org. Chem.*, **34**, 3519 (1969).
87a. A. H. Haines, *Carbohyd. Res.*, **21**, 99 (1972).
88. S. Umezawa, Y. Okazaki, and T. Tsuchiya, *Bull. Chem. Soc. Japan*, **45**, 3619 (1972); T. Miyake, T. Tsuchiya, S. Umezawa, and H. Umezawa, *Carbohyd. Res.*, **49**, 141 (1976).
89. F. H. Newth, *J. Chem. Soc.*, 471 (1956).
90. C. L. Stevens, N. A. Nielsen, and P. Blumbergs, *J. Amer. Chem. Soc.*, **86**, 1894 (1964).
91. J. E. Christensen and L. Goodman, *J. Amer. Chem. Soc.*, **83**, 3827 (1961).
92. J. English and M. F. Levy, *J. Amer. Chem. Soc.*, **78**, 2846 (1956).
93. A. Banaszek and A. Zamojski, *Carbohyd. Res.*, **25**, 453 (1972).
94. C. L. Stevens and D. Chitharanjan, *J. Org. Chem.*, **40**, 2474 (1975).
95. J.-M. Bourgeois, *Helv. Chim. Acta*, **57**, 2553 (1974).
96. T. van Es, *Carbohyd. Res.*, **5**, 282 (1967).
97. A. Klemer, H. Lukowski, and F. Zerhusen, *Ber.*, **96**, 1515 (1963); E. F. L. J. Anet, *Chem. Ind.* (London), 1035 (1963).
98. H. H. Baer and T. Neilson, *Can. J. Chem.*, **43**, 840 (1965); H. H. Baer, F. Kienzle, and T. Neilson, *ibid.*, **43**, 1829 (1965).
99. P. Heim and H. Neukom, *Helv. Chim. Acta*, **45**, 1735, 1737 (1962).
99a. K. Shimizu, *Mokuzai Gakkaishi*, **22**, 51 (1976); *Chem. Abstr.*, **84**, 180, 495 (1976).
100. R. M. de Lederkremer, M. I. Litter, and L. F. Sala, *Carbohyd. Res.*, **36**, 185 (1974); G. Deák, K. Gall-Istok, and P. Sohár, *Acta Chim. Acad. Sci. Hung.*, **75**, 189 (1973).
101. M. L. Wolfrom, A. Thompson, and D. R. Lineback, *J. Org. Chem.*, **27**, 2563 (1962); M. L. Wolfrom, G. Fraenkel, D. R. Lineback, and F. Komitsky, Jr., *J. Org. Chem.*, **29**, 457 (1964).
102. H. Zinner, E. Wittenburg, and G. Rembarz, *Ber.*, **92**, 1614 (1959).
103. N. K. Kochetkov and B. A. Dmitriev, *Chem. Ind.* (London), 864 (1963).

104. For a review, see Yu. A. Zhdanov, Yu. E. Alexeev, and V. G. Alexeeva, *Advan. Carbohyd. Chem. Biochem.*, **27**, 227 (1972).
105. D. L. MacDonald and H. O. L. Fischer, *J. Amer. Chem. Soc.*, **74**, 2087 (1952).
106. J. C. Sowden, *J. Amer. Chem. Soc.*, **72**, 808 (1950).
107. D. Horton and J. D. Wander, *Carbohyd. Res.*, **13**, 33 (1970); B. Berrang, D. Horton, and J. D. Wander, *J. Org. Chem.*, **38**, 187 (1973).
107a. A. Ducruix, C. Pascard-Billy, S. J. Eitelman, and D. Horton, *J. Org. Chem.*, **41**, 2652 (1976).
108. J. Lehmann, *Angew. Chem. (Intern. Ed.)*, **4**, 874 (1965).
109. D. M. Ciment, R. J. Ferrier, and W. G. Overend, *J. Chem. Soc. (C)*, 446 (1966).
110. S. Hanessian and N. R. Plessas, *Chem. Commun.*, 706 (1968).
111. R. J. Ferrier and N. Vethaviyasar, *Chem. Commun.*, 1385 (1970); *J. Chem. Soc. (C)*, 1907 (1971).
112. R. J. Ferrier and N. Vethaviyasar, *J. Chem. Soc. Perkin I*, 1791 (1973).
113. D. Horton, J. B. Hughes, and J. M. J. Tronchet, *Chem. Commun.*, 481 (1965); D. Horton and J. M. J. Tronchet, *Carbohyd. Res.*, **2**, 315 (1966); D. Horton, J. B. Hughes, and J. K. Thomson, *J. Org. Chem.*, **33**, 728 (1968); D. Horton, A. Liav, and S. E. Walker, *Carbohyd. Res.*, **28**, 201 (1973).
114. T. Iida, M. Funabashi, and J. Yoshimura, *Bull. Chem. Soc. Japan*, **46**, 3203 (1973).
114a. S. Yamada, K. Koga, and M. Yo, *Japan Kokai*, **75**, 106, 913 (1975); *Chem. Abstr.*, **84**, 31, 374 (1976).
115. A. Rosenthal and M. Sprinzl, *Can. J. Chem.*, **47**, 3941 (1969).
116. J. M. J. Tronchet and J.-M. Bourgeois, *Helv. Chim. Acta*, **53**, 1463 (1970).
117. J. M. J. Tronchet and D. Schwarzenbach, *Carbohyd. Res.*, **30**, 395 (1973).
118. J. M. J. Tronchet, D. Schwarzenbach, and F. Barbalat-Rey, *Carbohyd. Res.*, **46**, 9 (1976).
119. J. M. J. Tronchet and B. Gentile, *Carbohyd. Res.*, **44**, 23 (1975)
120. A. Rosenthal, C. M. Richards, and K. Shudo, *Carbohyd. Res.*, **27**, 353 (1973); A. Rosenthal and L. Nguyen (Benzing), *Tetrahedron Lett.*, 2393 (1967).
121. J. M. J. Tronchet and J.-M. Bourgeois, *Helv. Chim. Acta*, **54**, 1718 (1971).
122. A. Rosenthal and D. A. Baker, *J. Org. Chem.*, **38**, 193, 198 (1973).
123. H. P. Albrecht, G. H. Jones, and J. G. Moffatt, *J. Amer. Chem. Soc.*, **92**, 5511 (1970).
124. A. J. Brink and A. Jordaan, *Carbohyd. Res.*, **34**, 1 (1974).
125. A. Rosenthal and K. Dooley, *J. Carbohyd., Nucleosides, Nucleotides*, **1**, 61 (1974).
126. J. M. J. Tronchet, T. Nguyen-Xuan, and M. Rouiller, *Carbohyd. Res.*, **36**, 404 (1974).
126a. J. Yoshimura, K. Sato, H. Wakai, and M. Funabashi, *Bull. Chem. Soc. Japan*, **49**, 1169 (1976).
126b. A. Rosenthal and M. Sprinzl, *Carbohyd. Res.*, **16**, 337 (1971).
127. J. S. Brimacombe, J. A. Miller, and U. Zakir, *Carbohyd. Res.*, **49**, 233 (1976).
128. W. A. Szarek, J. S. Jewell, I. Szczerek, and J. K. N. Jones, *Can. J. Chem.*, **47**, 4473 (1969).
128a. F. J. Lopez Aparicio, F. J. Lopez Herrera, and J. Sanchez Ballesteros, *Carbohyd. Res.*, **69**, 55 (1979).
129. C. L. Stevens, P. Blumbergs, and D. L. Wood, *J. Amer. Chem. Soc.*, **86**, 3592 (1964); see also ref. 87.
130. K. Freudenberg and K. Raschig, *Ber.*, **62**, 373 (1929).
131. A. S. Meyer and T. Reichstein, *Helv. Chim. Acta*, **29**, 139 (1946).
132. A. F. Cook and W. G. Overend, *J. Chem. Soc.*, 1549 (1966); see also H. W. H. Schmidt and H. Neukom, *Tetrahedron Lett.*, 2063 (1964).

133. E. L. Albano, D. Horton, and J. H. Lauterbach, *Carbohyd. Res.*, **9**, 149 (1969).
134. N. K. Kochetkov and B. A. Dmitriev, *Chem. Ind.* (London), 115 (1963).
135. R. J. Ferrier and N. Prasad, *J. Chem. Soc.* (*C*), 575 (1969).
136. V. Bilik, *Collect. Czech. Chem. Commun.*, **39**, 1621 (1974).
137. P. Bladon and L. N. Owen, *J. Chem. Soc.*, 598 (1950).
138. G. O. Aspinall, N. W. H. Cheetham, J. Furdová, and S. C. Tam, *Carbohyd. Res.*, **36**, 257 (1974).
139. G. O. Aspinall, R. R. King, and Z. Pawlák, *Can. J. Chem.*, **51**, 388 (1973); G. O. Aspinall and R. R. King, *ibid.*, **51**, 394 (1973).
140. B. Helferich and N. M. Bigelow, *Z. Physiol. Chem.*, **200**, 263 (1931).
141. J. Defaye, *C. R. Acad. Sci.*, **255**, 795 (1962).
142. M. L. Wolfrom, K. Matsuda, F. Komitsky, and T. E. Whiteley, *J. Org. Chem.*, **28**, 3551 (1963).
143. D. Horton and W. N. Turner, *Carbohyd. Res.*, **1**, 444 (1966).
144. J. S. Jewell and W. A. Szarek, *Tetrahedron Lett.*, 43 (1969).
145. K. Matsuura, Y. Araki, Y. Ishido, and S. Satoh, *Chem. Lett.*, 849 (1972); K. Matsuura, Y. Araki, Y. Ishido, and M. Kainosho, *ibid.*, 853 (1972).
146. A. Rosenthal and M. Ratcliffe, *Carbohyd. Res.*, **39**, 79 (1975).
147. A. Rosenthal and G. Kan, *J. Org. Chem.*, **36**, 592 (1971).
148. D. Horton and C. G. Tindall, Jr., *Carbohyd. Res.*, **15**, 215 (1970).
149. T. Sasaki, K. Minamoto, and H. Suzuki, *J. Org. Chem.*, **38**, 598 (1973).
150. N. L. Holder and B. Fraser-Reid, *Can. J. Chem.*, **51**, 3357 (1973); B. Fraser-Reid, D. L. Walker, S. Y.-K. Tam, and N. Holder, *ibid.*, **51**, 3950 (1973).
151. J. C. Sowden, M. L. Oftedahl, and A. Kirkland, *J. Org. Chem.*, **27**, 1791 (1962).
152. J. A. Carbon, *J. Amer. Chem. Soc.*, **86**, 720 (1964).
153. H. H. Baer, T. Neilson, and W. Rank, *Can. J. Chem.*, **45**, 991 (1967).
154. H. Paulsen and W. Greve, *Ber.*, **107**, 3013 (1974).
155. R. Barker and D. L. MacDonald, *J. Amer. Chem. Soc.*, **82**, 2297 (1960).
156. J. Cléophax, S. D. Gero, J. Leboul, and A. Forchioni, *Chem. Commun.*, 710 (1973).
157. D. L. Walker, B. Fraser-Reid, and J. K. Saunders, *Chem. Commun.*, 319 (1974); B. Fraser-Reid, D. R. Hicks, D. L. Walker, D. E. Iley, M. B. Yunker, S. Y.-K. Tam, R. C. Anderson, and J. Saunders, *Tetrahedron Lett.*, 297 (1975).
158. P. M. Collins, W. G. Overend, and V. M. Racz, *Carbohyd. Res.*, **45**, 127 (1975).
159. P. M. Collins, J. R. Hurford, and W. G. Overend, *J. Chem. Soc. Perkin I*, 2163 (1975).
160. P. M. Collins, J. R. Hurford, and W. G. Overend, *J. Chem. Soc. Perkin I*, 2178 (1975).
161. W. A. Szarek and J. S. Jewell, *Can. J. Chem.*, **48**, 1030 (1970).
162. G. Jones, *Tetrahedron Lett.*, 2231 (1974).
163. S. Y. K. Tam, D. E. Iley, N. L. Holder, D. R. Hicks, and B. Fraser-Reid, *Can. J. Chem.*, **51**, 3150 (1973).
164. B. Fraser-Reid, *Accts. Chem. Research*, **8**, 192 (1975).
165. A. B. Foster, R. Harrison, J. Lehmann, and J. M. Webber, *J. Chem. Soc.*, 4471 (1963).
166. R. E. Gramera, T. R. Ingle, and R. L. Whistler, *J. Org. Chem.*, **29**, 878 (1964).
167. B. Helferich and E. Himmen, *Ber.*, **61**, 1825 (1928).
168. Y. Hamamura, *Bull. Agr. Chem. Soc. Japan*, **18**, 36, 37, 49 (1942).
169. A. Linker, P. Hoffman, K. Meyer, P. Sampson, and E. D. Korn, *J. Biol. Chem.*, **235**, 3061 (1960).
170. F. Micheel and K. Horn, *Ann.*, **515**, 1 (1934).

20. GLYCOSYLAMINES*

H. Paulsen and K.-W. Pflughaupt

I. INTRODUCTION

Cyclic sugars that possess a free reducing group are able to react with primary and secondary amines, as shown in Scheme I with D-glucose, to give glycosylamines, derivatives in which the glycosidic hydroxyl group is replaced by an

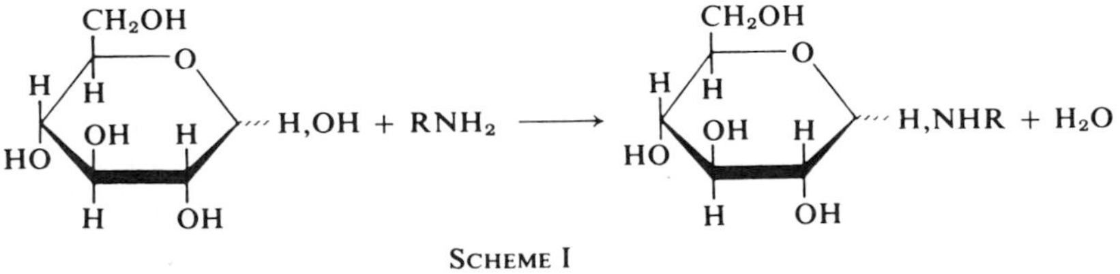

Scheme I

* Translated from the German by Ernst K. Just and D. Horton.

881

amino group. The reaction can be effected with alkyl- and aryl-amines, ammonia, esters of amino acids, and urea derivatives. Amides do not react because they are insufficiently basic. Authentic Schiff bases are obtained only from suitably protected *aldehydo*-aldoses (see Vol. IB, Chapter 21). Glycosylamines in solution frequently exhibit mutarotation. This may be caused by one or more of several factors: (1) hydrolytic cleavage into sugar and amine, (2) isomerization to various ring forms and open-chain structures, or (3) rearrangement of the aldosylamine into an acid-stable 1-amino-1-deoxyketose.

Amino derivatives react on heating with sugars to give brown products. Irreversible degradation of the carbohydrates takes place to give reactive intermediates that interact with amines to give dark, polymeric pigments. These reactions occur in foodstuffs on heating, and are of considerable importance in food chemistry.

The glycosylamines that are of most biochemical interest are the nucleosides, which are treated separately in Vol. IIA, Chapter 29. Pharmaceutical products having free amino groups can be condensed with sugars to give glycosylamines. The pharmaceutical activity of these products is generally the same as that of the free base owing to the ease with which the glycosylamine is cleaved, but the water solubility of the product is increased. *N*-Glycosyl derivatives of long-chain aliphatic amines have been proposed as wetting agents and textile softeners, and *N*-arylglycosylamines as antioxidants for rubber.

II. GLYCOSYLAMINES[1,1a]

A. Preparation

Aldohexoses[2-5] and aldopentoses[5,6] form glycosylamines when treated in the cold with concentrated alcoholic solutions of ammonia, and the products generally crystallize when the solutions are kept for a prolonged period. D-Glucose gives β-D-glucopyranosylamine[3] (**1**). Disaccharides such as cellobiose[7] react similarly, but require more vigorous conditions.

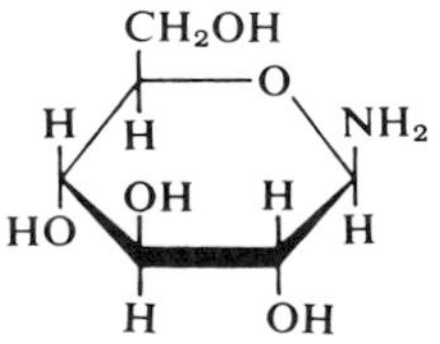

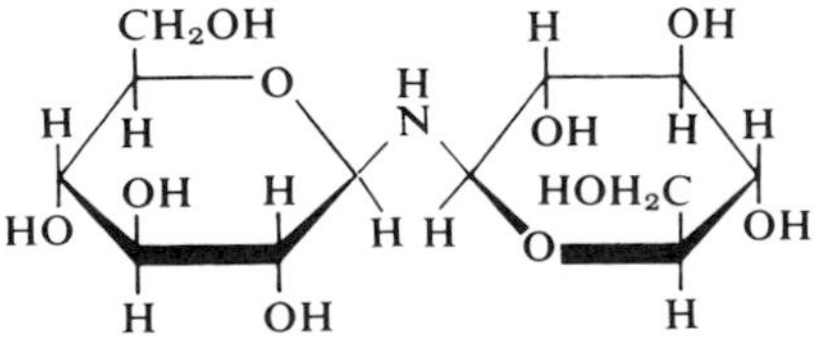

The glycosylamines so obtained are frequently contaminated with diglycosylamines of the type **2**. According to Isbell and Frush,[5] substances **1** and **2** are present in equilibrium in alcoholic solution, and **2** is formed by transglycosylation from two molecules of **1** by splitting out of ammonia. High concentrations of ammonia decrease the proportion of **2** that is formed, whereas addition of ammonium chloride causes increased formation of **2**. Liquid ammonia converts D-glucose into a mixture[3] of **1** and **2**. The yield of **2** can be raised to 65% by adding anhydrous calcium sulfate as well as ammonium chloride. Chromatography of sugars with ammoniacal eluents frequently gives rise to ninhydrin-positive substances because of reversible formation of glycosylamines.[8]

N-Alkylaldosylamines are easily prepared by direct interaction of aldoses with primary and secondary aliphatic amines[9-15]; for example, *N*-butyl-D-glucopyranosylamine[9] is obtained from D-glucose and butylamine, and *N*-D-glucopyranosylpiperidine[10] is similarly prepared. *N*-Alkylglycosylamines are reactive compounds and undergo hydrolysis rapidly. They are frequently difficult to obtain crystalline if dark-colored side products are present in the reaction mixture. Well crystallized *N*-alkylaldosylamines are obtained by treatment of aliphatic amines with 4,6-*O*-benzylidene-D-glucopyranose,[16,17] and the products are easily converted into the corresponding 1-amino-1-deoxyketoses through the Amadori rearrangement (see Section II,A).

Aldosylamines having amino acids as the nitrogen-containing components can be obtained only when the aldose is treated with the amino acid in the form of a sodium salt,[18-20] a metal complex,[20] or as an ester[21] or peptide[22] of the amino acid. The condensation products are likewise easily cleaved. If aldoses are condensed directly with the free amino acids in alcoholic solution, the Amadori rearrangement (see Section II,A) of the initially formed *N*-aldosyl-(amino acid) occurs immediately.

N-Arylaldosylamines have been known for a long time.[23] They are best prepared by the method of Sorokin,[24] by heating the sugar with the appropriate aromatic amine in alcoholic[25-29] or aqueous[30,31] solution. The products usually crystallize well and are far more resistant to hydrolysis than are the *N*-alkylglycosylamines. A large number of *N*-arylglycosylamines, with various substituents on the aromatic ring,[32,33] have been prepared. With weakly basic amines, addition of the amine hydrochloride or ammonium chloride as catalyst is necessary for the condensation. Addition of molecular sieves as a desiccant is reported[33a] to facilitate the reaction of aldoses with *p*-toluidine. Condensation of 5-*O*-methyl-D-ribofuranose[33b] and D-glucose[33c] with substituted 2,3-dihydroindoles in hot, absolute alcohol affords anomeric mixtures of the corresponding heterocyclic glycosylamines. Direct condensation of D-glucose with 2-aminobenzenethiol gives a 1-deoxyalditol-1-ylbenzothiazoline,

---

whereas addition of mercury(II) acetate to the mixture causes reversible formation[33d] of N-(2-thiophenyl)-D-glucofuranosylamines.

Ketoses react with methanolic ammonia to give the corresponding ketosylamines, of which L-sorbosylamine and di-L-sorbosylamine have been isolated crystalline.[34] Ketosylamines tend to undergo the Heyns rearrangement spontaneously on standing (see Section II,B). N-Alkylketosylamines have been prepared from D-fructose[35-37] and L-sorbose[38] by direct reaction in the cold with such amines as benzylamine and butylamine. They are easily hydrolyzed. Ketoses react with aromatic amines with considerably more difficulty.[29,39-41] D-Fructose is converted into N-p-tolyl- or N-phenyl-D-fructosylamine by prolonged heating with the hydrochloride of p-toluidine[41] or aniline.[39]

Glycosylamines of amino sugars are also known. If 2-acetamido-2-deoxy-D-glucose is heated with p-toluidine in methanol, the corresponding N-p-tolyl-aldosylamine is obtained.[42] 1-Deoxy-1-p-toluidino-D-fructose is converted into the ketosylamine (3) when treated with benzylamine in the cold.[43]

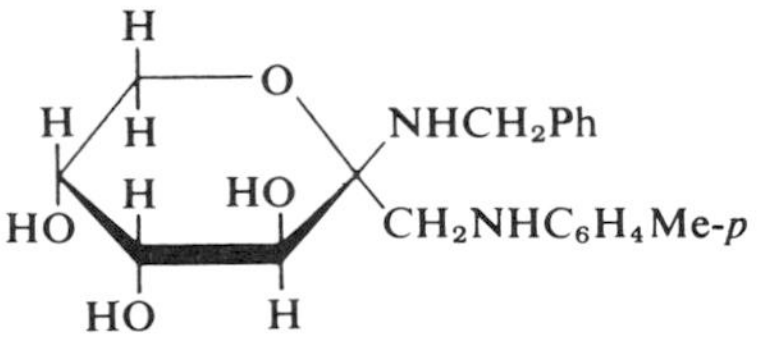

5-Thio-α-D-xylopyranose, in which the ring oxygen atom is replaced by sulfur, reacts readily with aromatic amines,[44,45] as, for example, with p-toluidine, which forms N-p-tolyl-5-thio-D-xylopyranosylamine.[44]

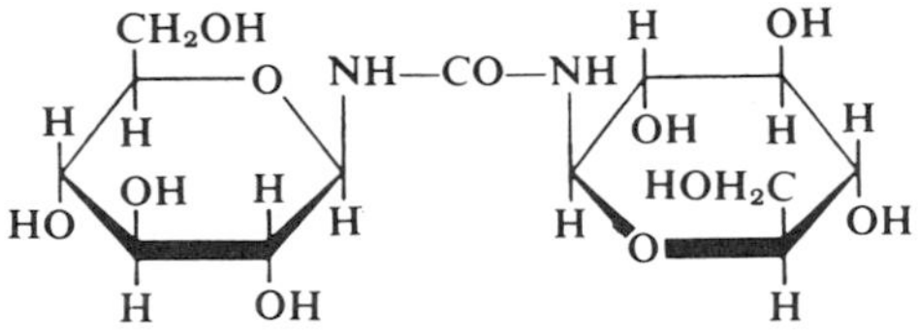

Aldoses undergo acid-catalyzed condensation with urea and thiourea.[46-49] If equimolar quantities of D-glucose and urea in 5% aqueous sulfuric acid are heated to 70°, N-β-D-glucopyranosylurea is preferentially formed.[49] If the molar concentration of D-glucose is doubled, N,N'-bis(β-D-glucosyl)urea (4) is isolated as the principal product.[49] Urea derivatives of D-ribose have been examined in detail.[49,50]

Peracetylated aldoses react with aromatic[51] and aliphatic[40] amines to form acetylated glycosylamines. In this way, penta-O-acetyl-β-D-glucopyranose is converted in alcohol containing acetic acid with p-toluidine into 2,3,4,6-tetra-O-acetyl-N-p-tolyl-D-glucosylamine. A mixture of the α-D and β-D anomers results.[28,51,52] Under similar conditions, piperidine reacts to give the C-2 unsubstituted product, N-(3,4,6-tri-O-acetyl-D-glucosyl)piperidine,[11] and the latter has been used as a starting material for the preparation of 2-O-methyl-D-glucose.[11] Indazole reacts with D-ribofuranose tetraacetate and D-glucopyranose pentaacetate to afford[52a] the respective 1-(β-D-glycosyl)indazoles. Derivatives of aldoses in which only the anomeric hydroxyl groups are unprotected react with phthalimide in the presence of triphenylphosphine and ethyl azodicarboxylate to afford anomeric mixtures of the corresponding O-acylglycosyl phthalimides in fair yield.[52b]

The reaction of O-acetylated glycosyl halides with amines is a general method for preparation of glycosylamines which is particularly useful with amines of low reactivity. The anomeric configuration of the product corresponds to that expected on the basis of the *trans* rule (see Vol. IIA, Chapter 28), if subsequent anomerization is avoided through careful isolation of the reaction product. For example, tetra-O-acetyl-α-D-glucopyranosyl bromide (**5**) reacts with p-toluidine to form 2,3,4,6-tetra-O-acetyl-N-p-tolyl-β-D-glucopyranosyl-amine,[52] but with ammonia the diglucosylamine (**2**) results.[53] Aliphatic secondary amines such as diethylamine cause elimination of hydrogen bromide from **5** to give the corresponding 1,2-unsaturated compound, 2,3,4,6-tetra-O-acetyl-1,5-anhydro-D-*arabino*-hex-1-enitol[54] (see Chapter 19). In contrast, glycosylamines of the type **6** are obtained with p-substituted benzylmethyl-amines.[55] Tertiary amines such as triethylamine[56] and pyridine[57] yield crystal-line quaternary glycosylammonium salts (**7**) with **5**, which can be converted in good yield into 1,6-anhydro-D-glucopyranose (see Vol. IA, Chapter 13) by treatment with barium hydroxide.[56] Tri-O-acetyl-β-D-ribopyranosyl bromide can be converted with nicotinamide into the corresponding β-D-ribosylpyri-dinium bromide, which is related to nicotinamide adenine dinucleotide.[53] The reaction of acetylated glycosyl halides with mercury salts of purine and other bases present in nucleic acids offers a most valuable method for the synthesis of nucleosides and their analogues (see Vol. IIA, Chapter 29). The nucleoside bases do not condense directly with sugars. Mercury(II) appears to direct substitution to the heterocyclic nitrogen atom, as 1-phenyltetrazole-5-thione reacts with **5** to give a thioglycoside, whereas addition of mercury(II) to the mixture causes formation[57a] of a 4-(tetra-O-acetyl-β-D-glucopyranosyl)-1-phenyltetrazole-5-thione.

With sodium azide, the glycosyl bromide (**5**) yields a glycosyl azide[58,59] (**8**), which is readily reduced to 2,3,4,6-tetra-O-acetyl-β-D-glucopyranosylamine[60] (**9**), and this product gives N-acyl derivatives (**10a**) with acid anhydrides or acid chlorides.[32] N'-Substituted D-glycosylureas[32] (**10b**) or D-glycosylthioureas (**11**) are obtained by treating the acetylated glycosylamine (**9**) with isocyanates or isothiocyanates. If suitable substituents (R) are chosen, ring closure to give heterocycles is possible. For example, tetra-O-acetyl-β-D-glucopyranosylurea (**10d**) (R $=$ NH$_2$) can be converted with cyanoacetic acid by way of the inter-mediate **12** (R $=$ H) into the pyrimidine nucleoside **13** through cyclization.[61] This synthetic scheme is also suitable for the preparation of other nucleosides (see Vol. IIA, Chapter 29). A further, simple route to glycosyl azides is by the

References start on p. 921.

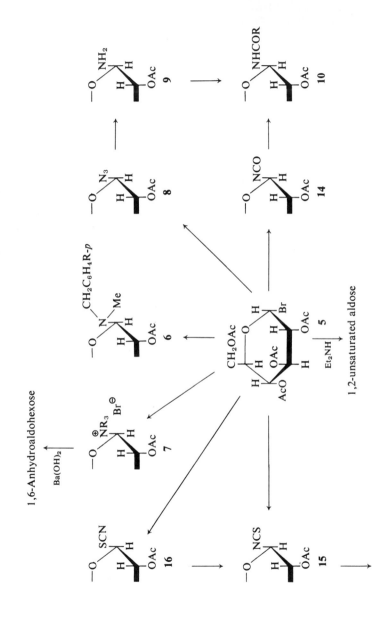

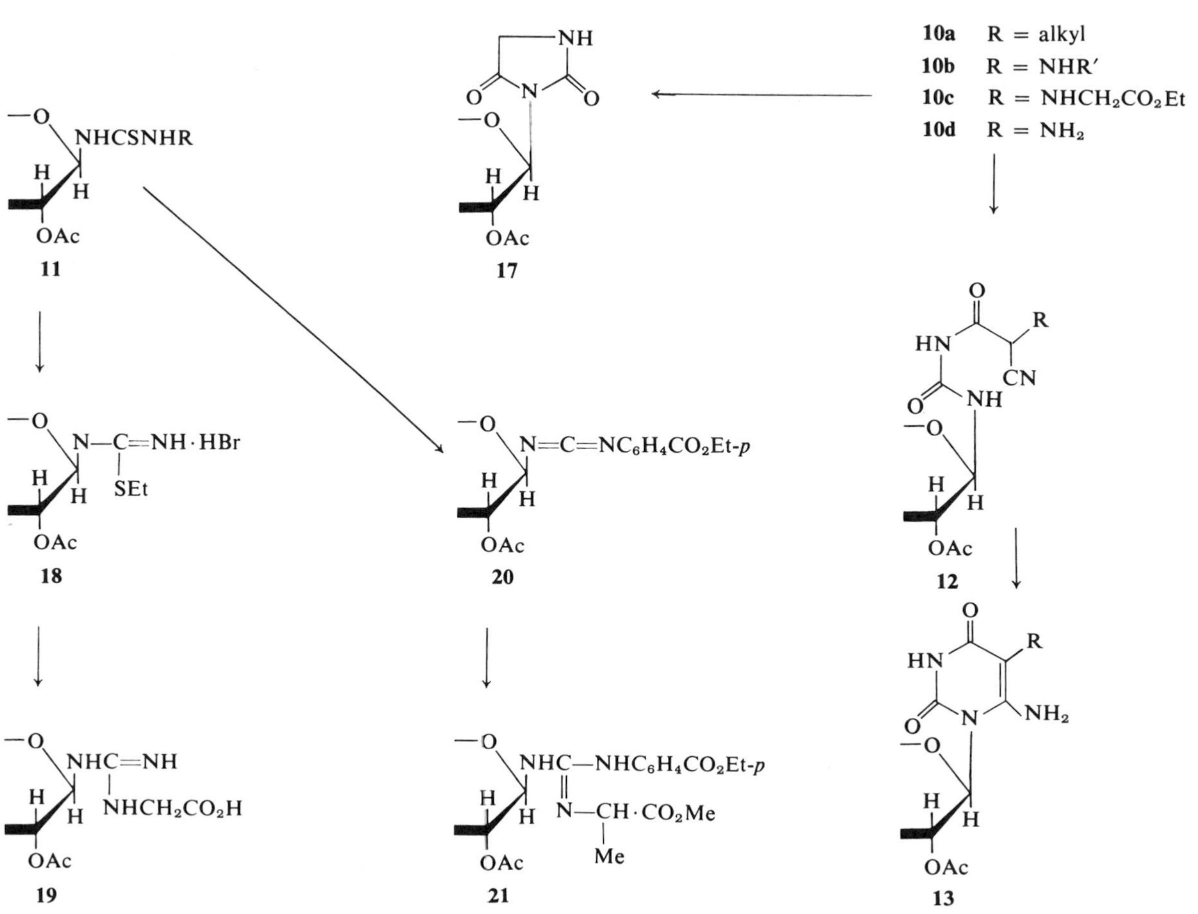
10a R = alkyl
10b R = NHR′
10c R = NHCH₂CO₂Et
10d R = NH₂
NHCSNHR
OAc
11
17
N—C=NH·HBr
SEt
18
N=C=NC₆H₄CO₂Et-p
20
12
NH₂
13
NHC=NH
NHCH₂CO₂H
19
NHC—NHC₆H₄CO₂Et-p
N—CH·CO₂Me
Me
21

References start on p. 921.

reaction of peracetylated aldoses with trimethylsilyl azide in the presence of stannic chloride.[61a]

Tetra-O-acetyl-α-D-glucopyranosyl bromide (5) reacts with silver cyanate to give the acetylated glycosyl isocyanate[62] (14), and with silver thiocyanate to form the acetylated glycosyl isothiocyanate[62] (15). Treatment of the bromide (5) with potassium thiocyanate gives the acetylated glycosyl thiocyanate (16); the latter can be converted thermally into the isothiocyanate[63] (15). The reactive compounds 14 and 15 are very useful starting materials for the synthesis of D-glucosylureas (10b) and D-glucosylthioureas (11). Various amines and amino acid esters can be condensed with 14 and 15 in this way and give N'-substituted N-glycosylurea derivatives.[46] The product (10c) formed with a glycine ester can, after saponification,[64] be cyclized under acid catalysis to the N-D-glucosylhydantoin 17. N-(Tetra-O-acetyl-D-glucopyranosyl)thiourea (11) (R = H) can be converted with ethyl bromide into 2-S-ethyl-1-N-(D-glucopyranosyl)-2-thiopseudourea hydrobromide[65] (18), which affords the N-(D-glucosyl)-guanidine derivative (19) on treatment with amino acids in the presence of alkali.[65] By desulfurization of the N-(D-glucosyl)thiourea derivative (11) (R = $C_6H_4CO_2Et$-p) with mercuric oxide, a carbodiimide derivative (20) is obtained which can be converted[66] with amino acids into an N-(D-glucopyranosyl)guanidine derivative of the type 21. Compounds formed between carbohydrates and amino acids are of interest because bonding of the type shown in 22 occurs in Nature in glycoproteins.[67,68] Corresponding syntheses

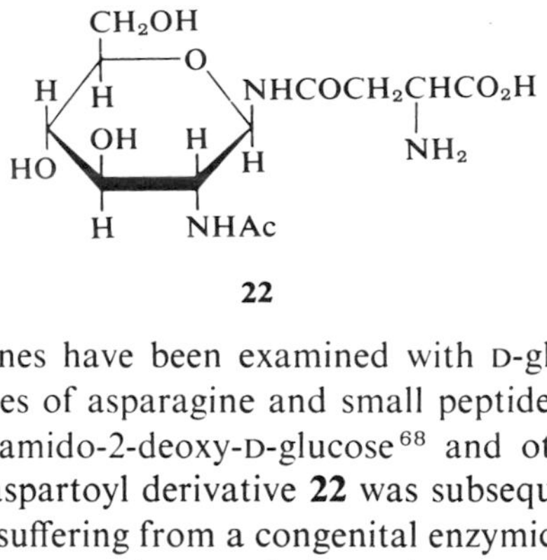

22

of some glycosylamines have been examined with D-glucose and L-asparagine.[67] Glycosylamines of asparagine and small peptides[68a,b] were also synthesized from 2-acetamido-2-deoxy-D-glucose[68] and other saccharides.[68c-e] A 2-acetamido-1-N-aspartoyl derivative 22 was subsequently identified[68f] in the urine of patients suffering from a congenital enzymic defect.

Transglycosylation, an exchange of glycosyl groups by another aglycon, is possible with O-glycosides only with difficulty, but occurs easily with glycosylamines.[69] The reaction, which is catalyzed by hydrochloric acid, acetic acid, or ammonium chloride, leads to the establishment of an equilibrium. Bognár and co-workers[70] have described a series of reaction types. In the simplest case, the amino group of the glycosylamine is replaced by another amine according to Scheme II. A glycosylamine, A, can transfer its amino residue to

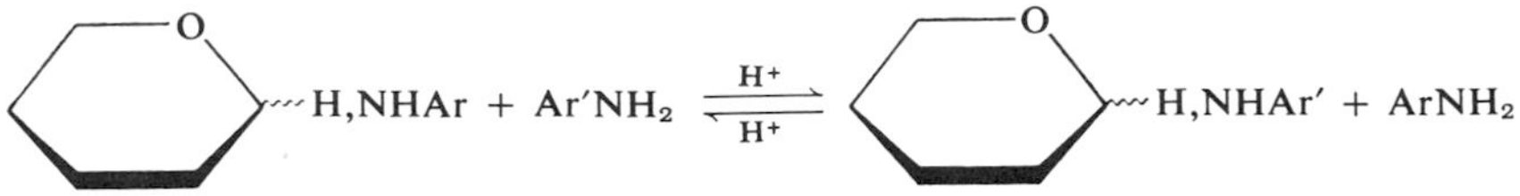

SCHEME II

the aldose, B, as shown in Scheme III. Additionally, on heating two different glycosylamines, an equilibration with exchange of the amino residues (double transglycosylation), is observed. Also, acetylated glycosylamines are susceptible to transglycosylation in the same manner.[70] Action of urea converts N-(p-nitrophenyl)-D-glucosylamine into N-D-glucosylurea[70c] in 65% yield.

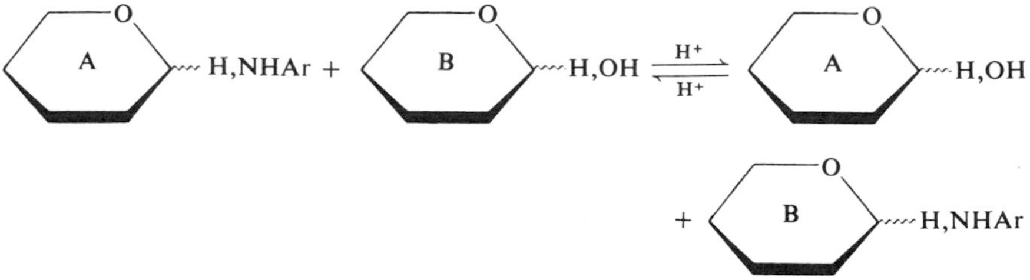

SCHEME III

The transglycosylation of an O-glycoside to give a glycosylamine was discussed by Lee and co-workers[71] in the conversion of methyl α-D-glucofuranoside (23) with p-nitroaniline in the presence of a catalytic amount of hydrochloric acid. The acyclic aldehyde-ammonia (24) could be isolated as an intermediate product, which was converted on heating into N-p-nitrophenyl-β-D-glucopyranosylamine (25). A further $O \rightarrow N$ transglycosylation was found in the glycoside (26) of 2-hydroxypyridine, which on heating with mercuric bromide changed into the glucosylamine[72] (27); the reverse ($N \rightarrow O$) process is catalyzed[72a] by p-toluenesulfonamide. Analogous $S \rightarrow N$ migration occurs upon treatment of 5-(1-phenyltetrazolyl) 1-thioglycosides with mercury(II) salts.[57a]

1,1-Diacetamido-1-deoxyalditols were first obtained[73] through the Wohl degradation of acetylated cyanohydrins with ammonia according to Scheme IV. By treating acetylated or benzoylated hexoses and pentoses with ammonia, 1,1-diacetamido- or 1,1-dibenzamido-1-deoxyalditols, respectively, are also obtained.[74] Peracetylated pyranoses and furanoses afford the same reaction products as the open-chain aldose derivatives having free aldehyde groups.[74,75]

References start on p. 921.

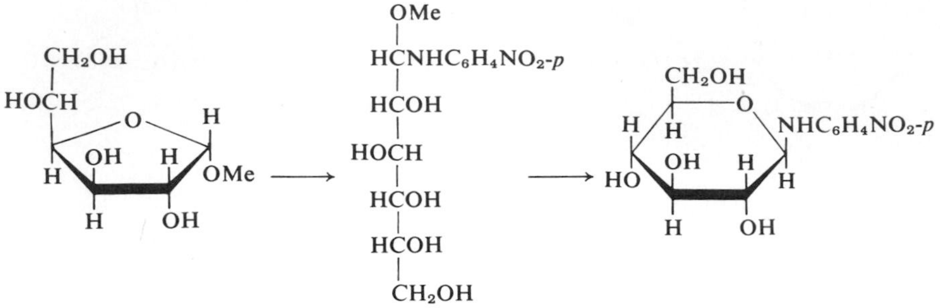

23   **24**   **25**

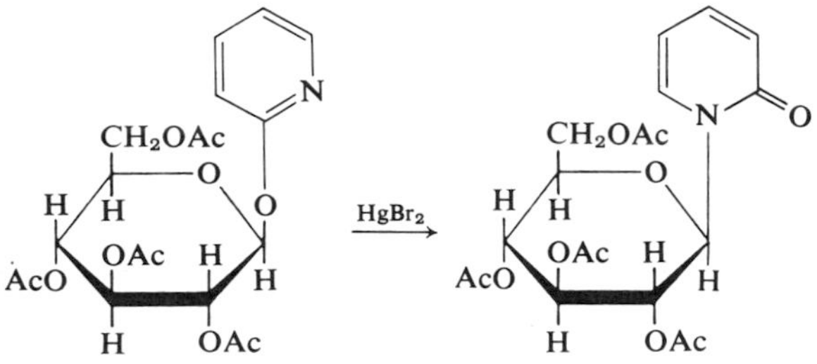

26   **27**

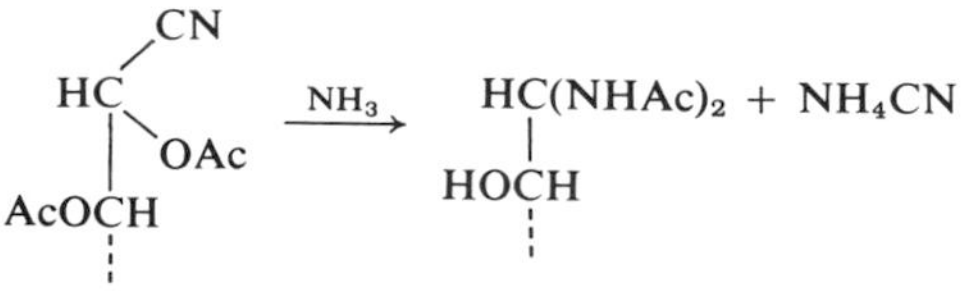

SCHEME IV

Acetylated disaccharides react to form the corresponding bis(acylamido) compounds, but in lower yields.[76]

A mechanism for the ammonolysis reaction has been proposed by Isbell,[77] whose findings agree with the experimental results obtained by using isotopically labeled compounds.[78,79] According to this scheme, a molecule of ammonia reacts, as Scheme V shows, with the released carbonyl group to form Vb. Through intramolecular acyl migration by way of a cyclic orthoester (Vc), the N-acylglycosylamine derivative (Vd) is obtained. Most probably, ammonia then adds to an acyl group to form Ve and, through the intermediacy of Vf, the diacylamido derivative (Vg) is formed. By experiments with labeled sugar benzoates, it was found[79] that the acyl groups migrate to C-1 preferentially from C-3 and C-4 and less frequently from C-2. Hence, compounds that

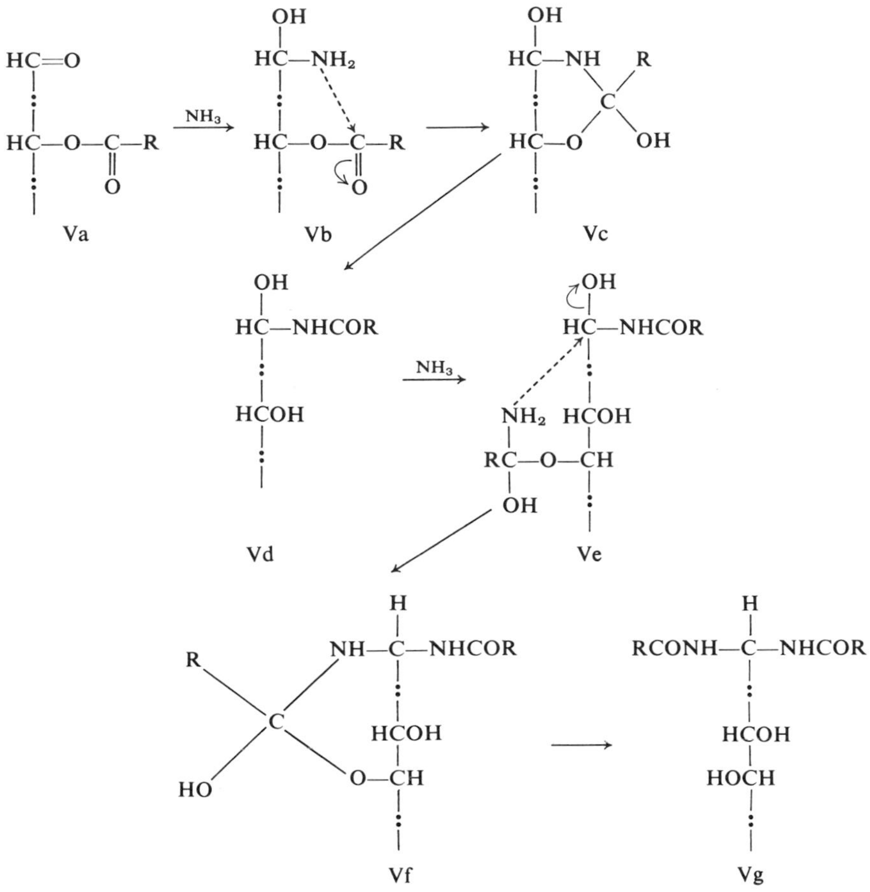

$$\text{HC}{=}\text{O} \quad \xrightarrow{\text{NH}_3} \quad \text{HC}{-}\text{NH}_2 \quad \longrightarrow \quad \text{HC}{-}\text{NH} \quad R$$

Va   Vb   Vc

$$\xrightarrow{\text{NH}_3}$$

Vd   Ve

Vf   Vg

SCHEME V

are not acetylated at C-2 also react similarly. If 3,5,6-tri-*O*-benzoyl-D-glucofuranose and 3,4,5,6-tetra-*O*-benzoyl-*aldehydo*-D-glucose are subjected to ammonolysis, the same product, 1,1-dibenzamido-1-deoxy-D-glucitol, is produced.[75] Ammonolysis of 2,3,4,6-tetra-*O*-benzoyl-D-glucopyranosylamine produces[79a] mainly D-glucopyranosylamine, plus some D-glucose and only small amounts of *N*-benzoylated products, from which it appears that $O \rightarrow N$ acyl migration to glycosylamines is not favored under conditions of ammonolysis.

Heterocyclic glycosylamines have also been prepared by addition and cyclization reactions of simpler glycosylamines. *N*-Glycosylimidazoles result from reaction of glycosylamines with ethyl (ethoxymethylideneamino)cyano-acetate,[79b,c] and from reaction of ethyl formamidate[79b] and 2-methyloxazol-ine[79d] derivatives with ethyl 2-aminocyanoacetate. Casieniroedine,[79e] isolated from the Mexican fruit *Zapote blanco*, was shown to be a β-D-glucopyran-osylimidazole, and 2-(β-D-glucopyranosyl)isoxazolidones have been identi-fied[79f] as constituents of seedlings of several species.

B. STRUCTURE

The absence of a $C = N$ band in the infrared spectra of glycosylamines indicates that they have the cyclic structure already formulated, and are not Schiff bases (see Vol. IB, Chapter 21, for a discussion of derivatives having the Schiff-base type of structure). The pyranose ring form was demonstrated by methylation of N-p-tolyl-β-D-glucosylamine (**28**), whereby 2,3,4,6-tetra-O-methyl-N-p-tolyl-α-D-glucopyranosylamine[80,81] (**29**) was obtained, hydrolysis of which gave the known 2,3,4,6-tetra-O-methyl-D-glucose (**30**). Conversely,

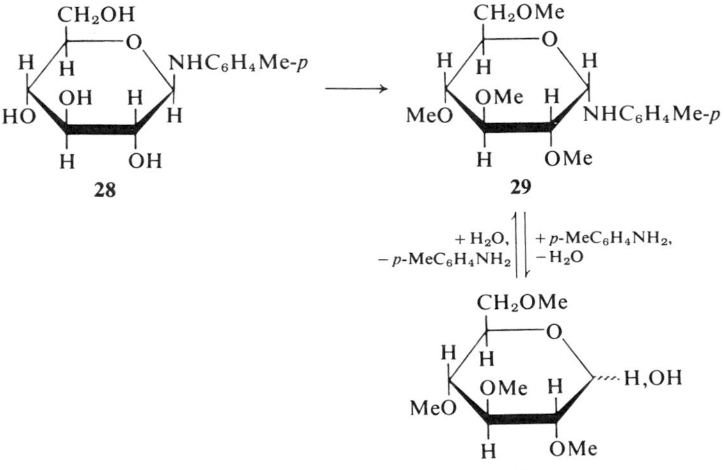

29 may be prepared from **30** with p-toluidine.[80] Methylation of **28** gives the α-D anomer (**29**). The β-D anomer may be produced by acid-catalyzed anomeri-zation in acetone.[81]

N-Acetyl-β-D-glucosylamine, which can be obtained by treating β-D-glucosylamine with ketene, has been subjected to periodate cleavage.[82] Under the conditions used, the –NHAc group is not affected by periodate, and the cleavage follows the route given in Scheme VI. Oxidation of the product (VIb) with bromine–water gave D-glyceric acid (VIc), thus proving that the glycosyl-amine existed in the pyranose form (VIa), since cleavage and oxidation of a furanosylamine would have given 2-hydroxymalonic acid (glyceraric acid).[82]

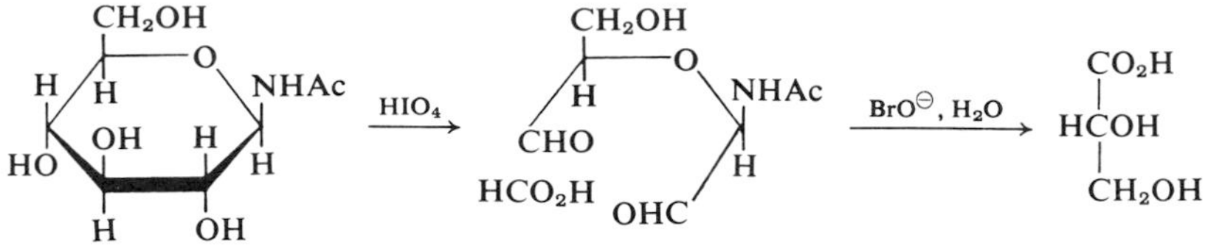

VIa VIb VIc

SCHEME VI

N-Acetyl-L-arabinosylamine[6] and *N*-acetyl-D-galactosylamine were similarly shown to have a pyranoid ring.

Generally, aldosyl- and ketosyl-amines favor the pyranoid structure. Well crystallized per-*O*-acetyl derivatives of *N*-substituted glycopyranosylamines can be obtained by treatment with acetic anhydride in pyridine. Such acetates and the analogous benzoates readily lose the amine moiety on mild hydrolysis to give the sugar acetate (or benzoate) having the reducing group free.[10,40,80,83] Partial hydrolysis of the tetraacetate of *N*-*p*-tolyl-D-fructosylamine yields the known 1,3,4,5-tetra-*O*-acetyl-D-fructopyranose, thus establishing the six-membered ring structure of the D-fructosylamine derivative.[39] The assumed furanoid structures of the *o*-nitroaniline derivatives[26] of D-ribose and L-arabinose, and also an *N*-phenyl-D-ribosylamine,[84,85] have not been firmly verified. An undoubtedly authentic D-glucofuranosylamine results from treatment of 5,6-di-*O*-methyl-D-glucofuranose with *p*-toluidine.[86]

Authentic Schiff bases are normally obtainable only from acylated *aldehydo* sugars. 3,4,5,6-Tetra-*O*-benzoyl-*aldehydo*-D-glucose reacts with *p*-toluidine to yield the 1,1-bis(*p*-toluidino) derivative (**31**), which, on heating in ether, loses one mole of amine to give the Schiff base[87] **32**. Addition of *p*-toluidine to the latter regenerates the bis(*p*-toluidino) derivative (**31**). Treatment of **32**

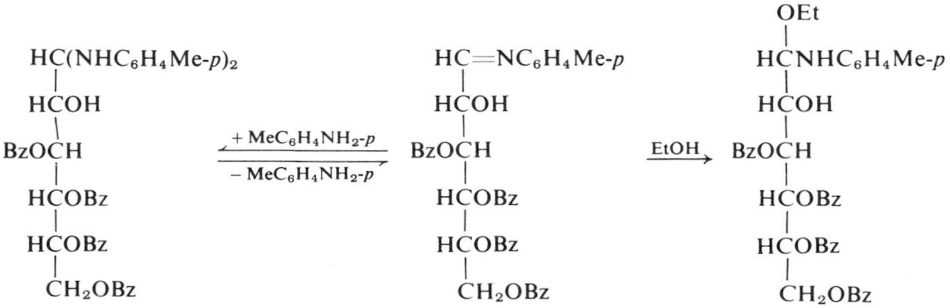

31 **32** **33**

References start on p. 921.

with ethanol gives the 1-ethoxy-1-p-toluidino derivative[87] (33). Corresponding compounds are obtained from 2,3,4,5-tetra-O-acetyl-$aldehydo$-D-ribose.[1] Earlier reports concerning the behavior of acylic, unblocked glycosylamines[88] could not be confirmed with N-o-tolyl-D-glucosylamine.[85] The infrared spectra of cyclic glycosylamines have been studied in detail[85,86] (see also Chapter 27, Section V).

The reaction of carbohydrates with amines frequently leads to anomeric mixtures of glycosylamines, in proportions strongly dependent on the reaction conditions and the type of the amine. In the reaction with aliphatic amines, the anomer having the amino group equatorial is strongly favored; in the $^4C_1(D)$ conformation this is the β anomer.[86a] As a rule, this anomer is generally the more stable, the reason being that the anomeric effect of an amino group is either very small or absent, whence steric reasons give preference to the equatorial position. This conclusion was reached from a study of the conformational equilibria of various 2,3,4-tri-O-acetylpentopyranosylamines.[86a] Introduction of a phosphinamido group increases the electron density of the nitrogen atom linked to C-1, considerably increasing the anomeric effect and shifting the conformational equilibrium accordingly; however, increasing the positive charge of the nitrogen atom linked to C-1 by incorporating it into an imidazolinium group produces an inverse anomeric effect, which additionally stabilizes the substituent in the equatorial position.[86a,86b] In the reaction of D-glucose with ammonia or an aliphatic amine, the β-D form generally crystallizes preferentially. This is shown by the small or negative optical rotation and from the n.m.r. spectra of the compounds obtained.[1] With D-glucosylamine and N-butyl-D-glucosylamine, a large 1,2-diaxial coupling (8.0 Hz), characteristic of the β-D configuration, is observed.[89] Since free N-arylglycosylamines readily mutarotate, it is expedient in isolating the α-D and β-D forms to convert them into the more-stable tetra-O-acetyl-N-arylglycosylamines, which are separable into the anomeric forms by fractional crystallization.[28,51,52,90,91] Tetra-O-acetyl-N-arylglycosylamines also undergo anomerization with hydrochloric or acetic acids as catalyst. This anomerization is of use if the minor form produced in the reaction must be enriched for purposes of isolation.[91,92]

The pure, free anomers of N-arylglycosylamines can be obtained by carefully controlled saponification of the corresponding peracetates with a catalytic amount of sodium methoxide.[91,92] In this way pure α and β anomers of N-phenyl-, N-p-tolyl-, N-p-bromophenyl-, and N-p-nitrophenyl-D-glucosylamine and N-phenyl-D-galactosylamine have been obtained.[92]

Capon and Connett[93] have examined the anomeric N-arylglycosylamines by n.m.r. spectroscopy. In pyridine solution, the signals of the anomeric protons are observed at relatively low field strengths. Products having negative optical rotations consistently show a large diaxial coupling ($J_{1,2} = 7.0$ to 8.0 Hz), as expected for a β-D form. Those compounds having positive rotations show $J_{1,2} = 4.0$ Hz, which indicates that an equatorial–axial orientation is present in the α-D form. Therefore, with simple glycosylamines, Hudson's rule of isorotation[94] is applicable, according to which the α-D form has a more positive optical rotation than the β-D form. Deviations from the rule occur if the amine possesses a chromophore that exerts a vicinal effect. This is the case with a series of nucleosides.[95]

Substituted *N*-arylglycosylamines exhibit quantitative divergence from Hudson's rules, the extent increasing with the polarizability of the aglycon.[95a] Rotation about the C-1–N bond of *N*-glycosylimides was shown[95b] to be less hindered in 5-membered than in 6-membered ring imides.

C. MUTAROTATION AND HYDROLYSIS[1a]

Glycosylamines display simple mutarotation in solution. Figure 1 shows the mutarotation of *N-p*-tolyl-α- and β-D-glucosylamine in methanol.[92] A kinetic investigation of the mutarotation of *N-p*-tolylglycosylamines in aqueous sodium hydroxide at pH 12.99 revealed that the reaction is first-order.[96,96a] The mechanism of the mutarotation is considered in Scheme VII. Isbell and Frush[5] assumed initial protonation of the ring-oxygen atom. As an intermediate step they postulated the open-chain immonium structure (VIIb), which, through intramolecular reaction of a hydroxide group with the positive carbon atom at C-1, then recloses to give the hemiacetal ring (VIIc). Mutarotation of *N*-aryl-D-fructopyranosylamines is reported[96b] to proceed by rearrangement to D-fructofuranosylamines.

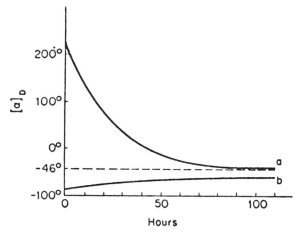

FIG. 1. Mutarotation of *N-p*-tolyl-α-D-glucosylamine (*a*) and *N-p*-tolyl-β-D-glucosylamine (*b*) in methanol (--- equilibrium mixture). Taken from ref. 92.

References start on p. 921.

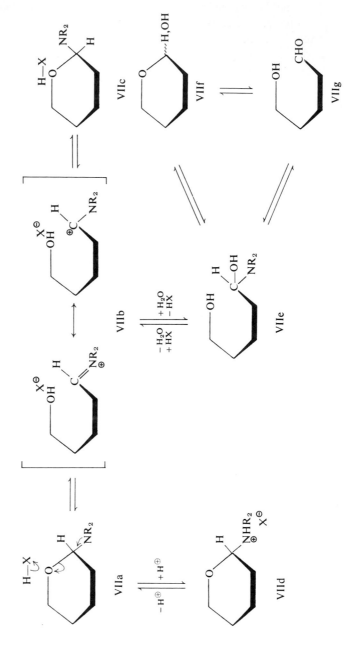

Scheme VII

In aqueous solution, hydrolysis of the glycosylamine occurs concurrently with mutarotation, and the favored outcome of the reaction is strongly dependent on the pH of the solution. With L-arabinosylamine, mutarotation unaccompanied by hydrolysis is observed[6] only at pH values above 8.0. In acid solution, concurrent hydrolysis occurs, with a maximum rate of cleavage at pH 5. In very strongly acidic solution, the rate of hydrolysis again decreases. Consequently L-arabinosylamine is stable in strongly acid (pH < 1.5) and strongly alkaline (pH > 9) solutions.[6]

Simon and co-workers,[97] with the aid of radioisotope dilution analysis, followed the rate of hydrolysis as a function of the pH for the N-D-glucosyl derivatives of p-nitroaniline, p-toluidine, benzylamine, and piperidine. The hydrolysis is susceptible to general acid catalysis. For each of the four glycosylamines, a specific, pH-dependent maximum for the rate of hydrolysis was found. Figure 2 presents the dependence of reaction rate as a function of the acidity function, H_0. It can be seen that the maximum rate of hydrolysis is shifted to higher pH values the larger the pK value of the amine. The rate of acid-catalyzed mutarotation has been used as a measure of the dissociation constants[97a] of a series of substituted benzoic acids.

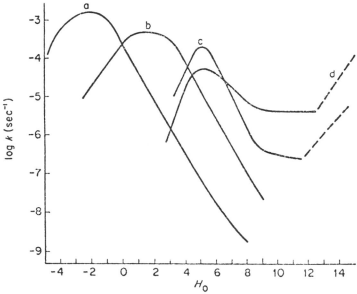

FIG. 2. Dependence of hydrolytic rate constant on the acidity function, H_0, for N-β-D-glucopyranosyl derivatives of p-nitroaniline (a), p-toluidine (b), benzylamine (c), and piperidine (d). Taken from ref. 97.

References start on p. 921.

The kinetic data can be interpreted in accord with the mechanism of hydrolysis given in Scheme VII. Thus, protonation of the oxygen atom and ring opening to the immonium ion VIIb occurs initially. For the cleavage of the amine from VIIe it was postulated[97] that intramolecular displacement by a hydroxyl group, which leads directly to the pyranose derivative VIIf, occurs. Capon and co-workers[96] considered that cleavage of the amino group to form structure VIIg, followed by a secondary ring closure to give VIIf, is more probable. In strongly acidic solution, glycosylamines form ammonium ions (VIId), which are stable toward hydrolysis. β-D-Glucosylpiperidine, for example, is completely stable in 2N hydrochloric acid for 17 hours at 0°. The concentration of the species VIIa through which hydrolysis occurs is so small under these conditions that practically no cleavage occurs.[97] The stability of glycosylamines toward hydrolytic cleavage is determined largely by the nature of the amine. The nature of the sugar is of minor importance, but follows in general the trends observed with glycosides.[98] The stability of glycosylamines toward acid increases according to the amine, in the following series[9,99]: aliphatic amines $\simeq$ amino acids and their esters < aromatic amines < acid amides $\simeq$ ureas, thioureas.

The ease of ring opening of the glycosylamine (VIIa) to the immonium ion (VIIb) is determined by the availability of electrons from the nitrogen atom. With aliphatic amines, the free pair of electrons on the nitrogen atom is readily available, so that formation of VIIb and subsequent hydrolysis is favored. In N-arylglycosylamines, the nonbonded pair of electrons on the nitrogen atom overlaps with the π electrons of the aromatic ring, so that formation of VIIb and its subsequent hydrolysis are hindered.[5,99] Acid hydrolysis of **22** to 2-amino-2-deoxy-D-glucose appears to proceed independently at each amide function, as no evidence was found[68] for $N \rightarrow N$ acyl migration.

The great stability of N-glycosylureas and glycosylguanidines toward acid hydrolysis can be attributed to the involvement of the electron pair of the nitrogen atom at C-1 in the resonance of the amide or guanidine group, which impedes the formation of VIIb. Similar reasons may be given for the particular stability toward hydrolytic cleavage of many nucleosides. Here the nitrogen atom at C-1 is a constituent of the resonance system of a heterocycle.[99] Glycosylamines of tertiary amines of the type **7** are an exception. They are not readily hydrolyzed, but instead form stable glycosylammonium salts.[56]

Acetolysis of a glycosylamine causes drastic alterations in the molecule. N-p-Tolyl-β-D-hexopyranosylamines thus treated afford[99a] N-acetyl-p-toluidine in excellent yield, whereas the 2-acetamido-2-deoxy-D-glucose analogue is converted[99b] into 4-(D-*arabino*-tetraacetoxybutyl)-2-methyl-1-p-tolylimidazole. N-(2,4-Dinitroanilino)-tetra-O-acetyl-β-D-glucopyranosylamine undergoes acetolysis to afford[99c] a mixture of D-glucose 2,4-dinitrophenylhydrazone pentaacetate and the corresponding imino derivative of D-glucono-1,4-lactone tetraacetate.

Reaction of β-D-glucopyranosylamine pentaacetate (**10a**, R = Me) with dinitrogen tetraoxide affords an *N*-nitroso derivative, which reacts with acetic acid or methanol in dichloromethane–pyridine to afford very high yields of α,β-D-glucopyranose pentaacetate or methyl α,β-D-glucopyranoside tetra-acetate, respectively.[99d]

III. REARRANGEMENT REACTIONS OF GLYCOSYLAMINES

A. THE AMADORI REARRANGEMENT[100]

Amadori[101] fused D-glucose with various aromatic amines, such as *p*-toluidine. In each case, he obtained two products, which were termed the "labile" and "stable" isomers. The "labile" product was shown to be a D-glucosylamine (**34**). Kuhn and Weygand[25] were able to prove that the "stable" product is a 1-arylamino-1-deoxy-D-fructose (**35** or **36**). The latter products

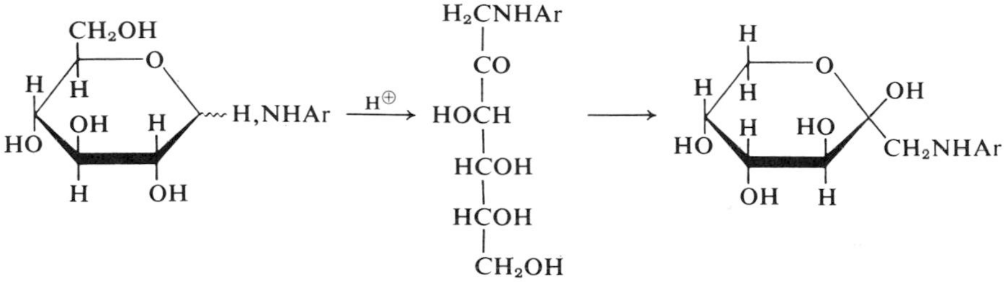

are formed by rearrangement of the glycosylamine (**34**) in the presence of a catalytic amount of acid. Compounds of the type **35** (or **36**) are known as "Amadori compounds." Cleavage of the amino group from **35** (or **36**) by acid hydrolysis is not possible.

In a standard route[30] for the preparation of the Amadori compounds, 1 mole of aldose is heated for 30 minutes with 1.1 to 1.4 moles of an aromatic amine, 2.5 to 3.0 moles of water, and 0.002 to 0.02 mole of acid (acetic acid or hydrochloric acid). The products usually crystallize from ethanol in 50 to 80% yield. 1-Arylamino-1-deoxy-D-fructoses (**36**) undergo hydrogenation much more easily over platinum or Raney nickel than glycosylamines or aldoses, and the reaction proceeds stereoselectively to give derivatives of 1-amino-1-deoxy-D-mannitol.[25,27,102] Occurrence of the Amadori rearrangement on treating D-galactose, D-xylose, and D-arabinose with arylamines can be demonstrated by hydrogenation of the products and identification of the corresponding substituted 1-amino-1-deoxy-hexitols or -pentitols.[30,102] Amadori rearrangement of N-phenyl-D-glucopyranosylamine has been demonstrated[102a] to be a first-order process, independent of the concentration of acid.

Solutions of Amadori compounds in dilute sodium hydroxide, when treated dropwise with a solution of o-dinitrobenzene, at once produce a deep-violet color. This color reaction is also given by aldoses and ketoses, but only after a delay. Amadori compounds show characteristic reducing behavior with methylene blue,[25] potassium ferricyanide,[103] and 2,6-dichlorophenol-indophenol[104]; the latter can be used for quantitative determination. The enhanced reducing ability is due to the ready enolization in alkaline solution of the acyclic form **35** to form the 1,2-enaminol, which behaves as a reductone.[105] The enolization is readily reversible.[106] The catalytic oxidation of 1-deoxy-1-p-toluidino-D-fructose in ammoniacal solution results in a cleavage of the enaminol to give D-arabinonic acid.[107]

Amadori compounds react with phenylhydrazine to form phenylosazones more easily and in higher yields (72–80%) than do the aldoses.[108] The phenylhydrazones can be obtained free from osazones only under carefully controlled conditions.[11,109] Crystalline oximes have been prepared from 1-deoxy-1-p-toluidino- and 1-deoxy-1-piperidino-D-fructose.[11] 1-Deoxy-1-arylamino-ketoses couple with diazonium salts to form[110] well crystallized diazoamino derivatives (**37**). These products undergo cleavage by acid to regenerate the Amadori compound and diazonium salt. Because N-arylglycosylamines do not undergo this coupling reaction, the 1-amino-1-deoxyketoses may be separated from the former by this method.[110]

Micheel and co-workers[33] examined the Amadori rearrangement of substituted arylamines as a function of the position and nature of the substituents. Substituents that decrease the electron density at the glycosidic nitrogen atom were found to prevent or impede the rearrangement. Whereas o- and p-toluidine and o- and p-anisidine react readily, m-toluidine, m-anisidine, and amines of lower basicity, such as m- and p-chloroaniline, m- and p-nitroaniline, and the o-, m-, and p-aminobenzoic acids, do not cause rearrangement.

Hodge and Rist[10,11] showed that the rearrangement is possible with secondary aliphatic amines such as piperidine, morpholine, and dibenzylamine, when the reaction is catalyzed by 2,4-pentanedione or ethyl malonate. Better yields are obtained when the components are brought into reaction in a mixture of acetic acid and triethylamine.[111] Micheel and co-workers[12] obtained the rearrangement products of D-glucose with such aliphatic amines as n-propylamine, n-butylamine, and benzylamine by treating the N-alkylglucosylamines

or the original reactants with oxalic acid in *p*-dioxane under rigorously anhydrous conditions. The Amadori compounds can be readily precipitated as
their oxalates, and the latter can be converted into the free bases with sodium
methoxide.[12] Huber and co-workers[112,113] treated D-glucose with alkyl-
benzylamines to give 1-alkylbenzylamino-1-deoxy-D-fructoses, from which
1-alkylamino-1-deoxy-D-fructoses are accessible by hydrogenolysis. D-Glucose
and such long-chain amines as octadecylamine[14,15] also give Amadori
compounds.

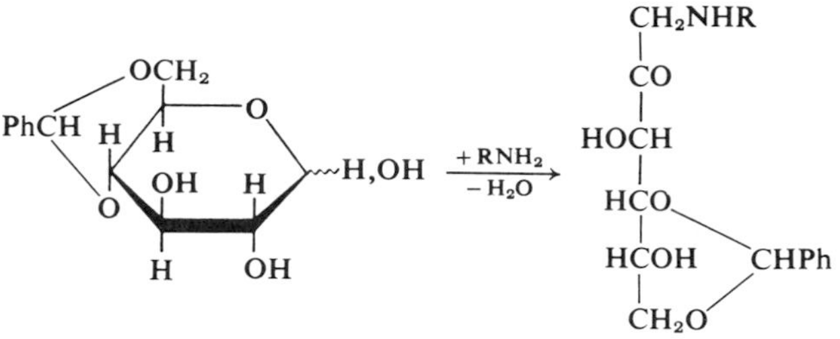

38   **39**

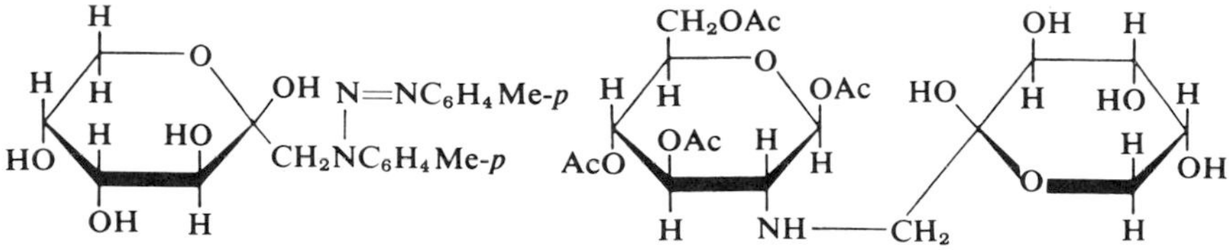

37   **40**

4,6-*O*-Benzylidene-D-glucopyranose (**38**) is an example for which the Amadori rearrangement is particularly facile.[16,114] Aromatic amines often react
without acid catalysis; for example, *m*-anisidine[16] and *N*-methyl-*p*-toluidine,[115]
which do not rearrange with D-glucose, do so with **38**. However, weakly basic
amines (*m*-nitroaniline and *p*-aminobenzoic acid), do not give the rearrangement with **38**. The Amadori compounds are formulated in the open-chain
form (**39**) on the basis of the carbonyl absorption band in their infrared
spectra.[16] D-Glucosylamines formed from **38** and aliphatic amines,[16] amino
acid esters,[21] or taurine[115a] react under mild conditions in the presence of
oxalic acid to give oxalates of the Amadori products (**39**). A glycosylamine of
38 reacts with 1,3,4,6-tetra-*O*-acetyl-2-amino-2-deoxy-β-D-glucose in triethylamine–acetic acid to afford,[116] after removal of the benzylidene residue, a
"disaccharide" of the type **40**. An explanation for the enhanced reactivity of

4,6-O-benzylidene-D-glucose has not been given. 4,6-Di-O-methyl-D-gluco-pyranose reacts readily with aliphatic amines, but not aromatic amines,[12] to form Amadori compounds.

D-Glucuronic acid undergoes the Amadori rearrangements more readily than D-glucose, indicating that the C-6 substituent exerts an influence. The relative ease of the rearrangement is: D-glucuronic acid > D-glucose > 6-deoxy-D-glucose.[117] Amadori rearrangement products of D-glucuronic acid with aromatic[118] and aliphatic amines[119] and amino acids[120,121] have been prepared. D-Glucuronic acid 6,3-lactone reacts with aliphatic amines with rearrangement to yield 1-alkylamino-1-deoxy-D-*arabino*-hexulosuronamides.[119]

Amino acids readily react with aldoses on heating the components in methanol or methyl sulfoxide[103,122,123] or by keeping[124] syrupy, water-containing mixtures at 30°. Without additional acid catalysis it is possible to obtain rearranged compounds such as 1-(carboxymethylamino)-1-deoxy-D-fructose (**42**), which have been termed "fructose–amino acids." The intermediate N-(D-glucosyl)amino acids (**41**) are not isolable. Such "fructose–amino acids" may be crystallized after purification by column chromatography[123] and have been prepared from various amino acids.[103,122–126]

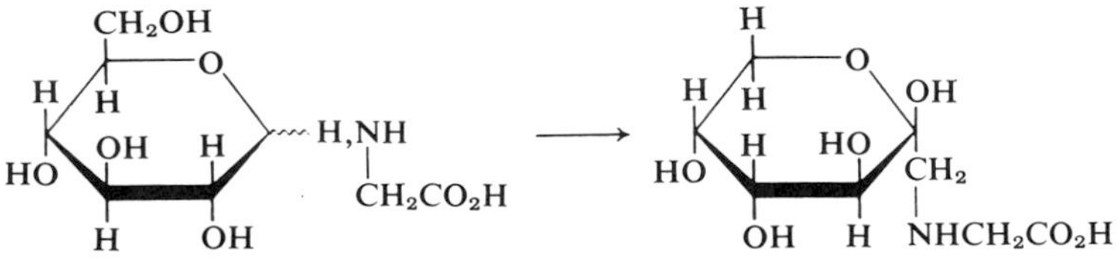

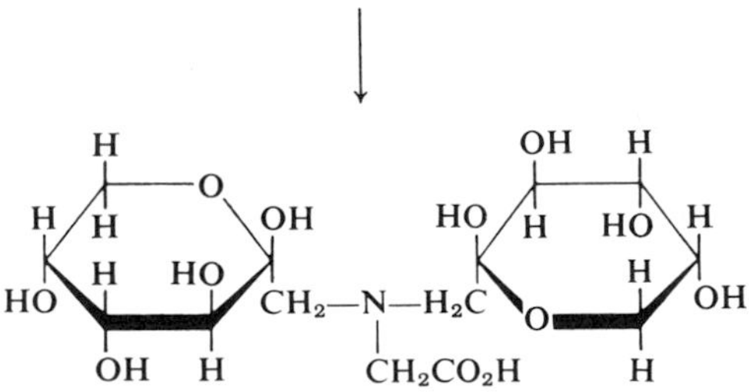

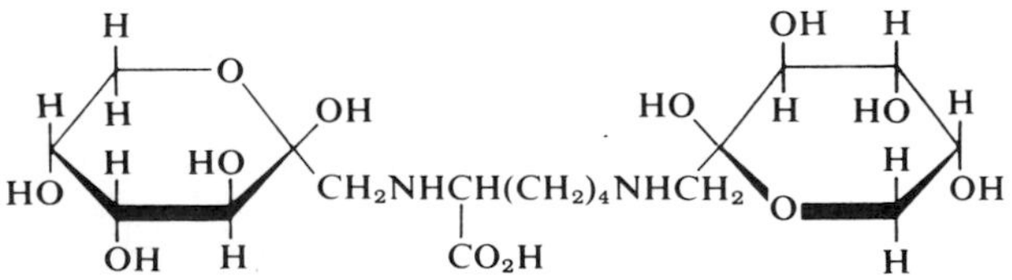

"D-Fructose–glycine" (**42**) reacts with an excess of D-glucose to undergo further rearrangement forming "di-D-fructose–glycine"[127] (**43**). L-Lysine hydrochloride reacts[128] initially with D-glucose to form a mixture of "ε-D-fructose–L-lysine" and "α-D-fructose–L-lysine" in a 2.5:1 ratio. Both compounds are rapidly converted into "α,ε-di-D-fructose–L-lysine" (**44**) by further reaction.[125,128]

"Fructose–amino acids" have been detected in liver extracts.[103,129] Their biochemical significance is questionable, since they were not detectable in liver tissue shortly after death.[123] The biosynthesis of tryptophan[130,131] in Enterobacteriaceae is connected with 1-(o-carboxyanilino)-1-deoxy-D-*erythro*-pentulose ("D-*erythro*-pentulose–anthranilic acid"), a substance also obtained synthetically.[132]

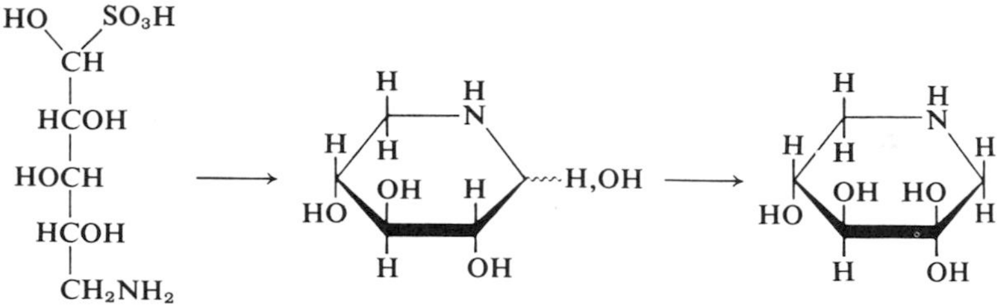

45   46   47

5-Amino-5-deoxy-D-xylopyranose (**46**) has been shown[133] to undergo a cyclic Amadori rearrangement. Alkaline cleavage of the 5-amino-5-deoxy-D-xylose bisulfite adduct (**45**) gives a sugar having a nitrogen-containing ring[134] (**46**). Amadori rearrangement of **46** to the cyclic piperidin-3-one hydrate **47** follows directly upon acidification. A threefold elimination of water with aromatization to give 3-hydroxypyridine occurs concurrently.[135] The hydrochloride of **47** crystallizes as a stable hydrate.

Amadori products generally favor the pyranoid ring structure. The ring size of the 1-*N*,2-*O*-benzylidene derivative (VIIIb) of 1-deoxy-1-*p*-toluidino-D-fructose (VIIIa) is demonstrated[136] by the reaction sequence given in Scheme VIII. After methylation and removal of protecting groups, the product (VIIId) was converted into the known 1,3,4,5-tetra-*O*-methyl-β-D-fructopyranose (VIIIf) by deamination followed by methylation.[136] "Fructose–glycine" (**42**) was assigned the β-D configuration on the basis of its optical rotation.[124]

Micheel and co-workers[137] proposed an infrared absorption at 3570 cm^{-1} as characteristic for Amadori compounds that have the pyranoid ring and possess a secondary nitrogen atom. Exceptions were found for "tagatose–glycine"[43] and the furanosidic 1-alkylamino-1-deoxy-D-*arabino*-hexulosuronic

---

References start on p. 921.

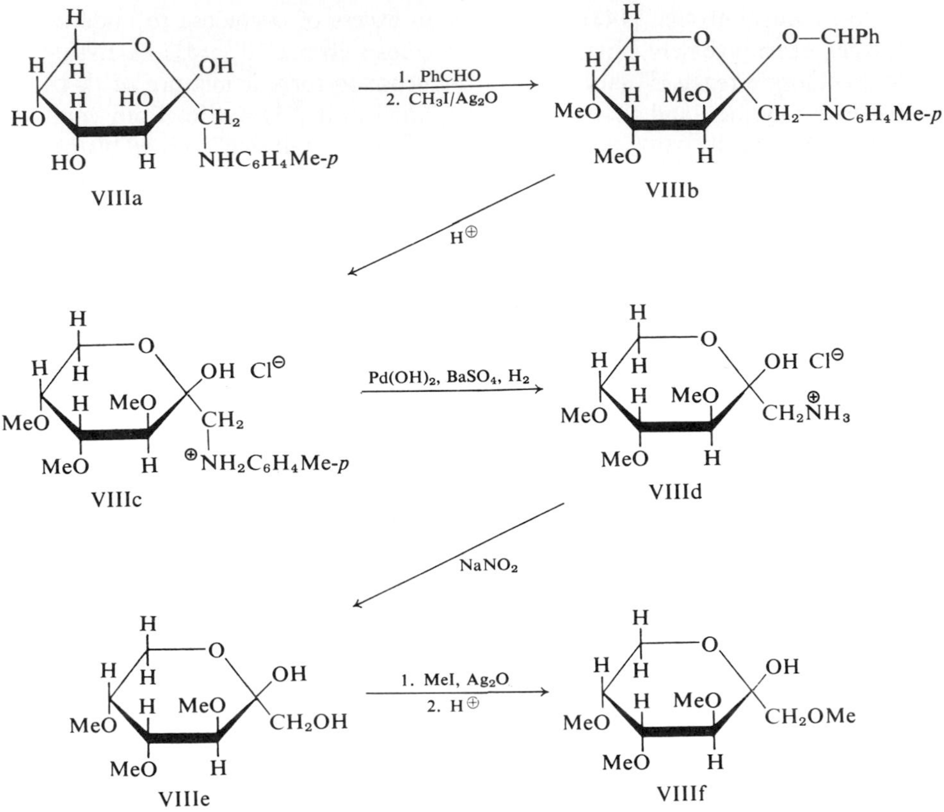

SCHEME VIII

acids.[119] Examples of acyclic Amadori compounds are compounds **48** (ref. 115), **49, 50** (ref. 118), and products obtained by rearrangement from 4,6-O-benzylidene-D-glucose (**38**) and various amines.[16]

A plausible reaction mechanism for the Amadori rearrangement is given in Scheme IX. Weygand[30] first proposed that initial protonation of the glycosylamine at the nitrogen atom, forming IXb, leads to IXd on ring opening. Isbell and Frush[5] subsequently developed the currently accepted[1a] idea that protonation of the ring-oxygen atom to form IXc is more probable because of the analogy between mutarotation and hydrolysis of glycosylamines. The electrophilic carbon atom (C-1) of the immonium structure (IXd) withdraws electrons from C-2, weakening the H–C bond at C-2 so that a proton can be abstracted by base to form the enaminol (IXe). This intermediate rearranges to the amino-ketone (IXf) and thence to the corresponding ring form (IXg).

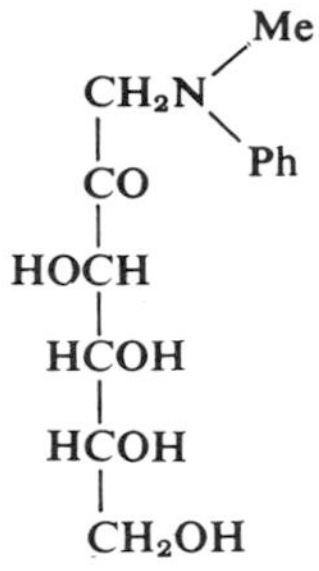

$$
\begin{array}{c}
\text{CH}_2\text{N}\!\stackrel{\text{Me}}{\underset{\text{Ph}}{\diagdown}} \\
\mid \\
\text{CO} \\
\mid \\
\text{HOCH} \\
\mid \\
\text{HCOH} \\
\mid \\
\text{HCOH} \\
\mid \\
\text{CH}_2\text{OH} \\
\mathbf{48}
\end{array}
\qquad
\begin{array}{c}
\text{CH}_2\text{NHCH}_2\text{CO}_2\text{H} \\
\mid \\
\text{CO} \\
\mid \\
\text{HCOH} \\
\mid \\
\text{HCOH} \\
\mid \\
\text{CH}_2\text{OH} \\
\mathbf{49}
\end{array}
\qquad
\begin{array}{c}
\text{CH}_2\text{NHC}_6\text{H}_4\text{Me-}p \\
\mid \\
\text{CO} \\
\mid \\
\text{HOCH} \\
\mid \\
\text{HCOH} \\
\mid \\
\text{HCOH} \\
\mid \\
\text{CO}_2\text{H} \\
\mathbf{50}
\end{array}
$$

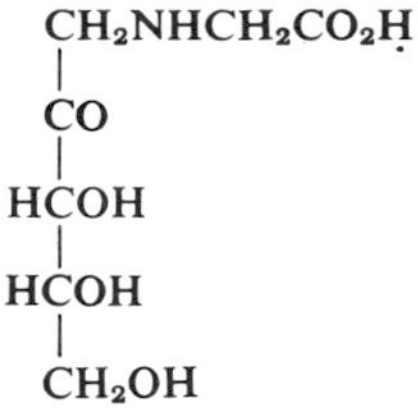

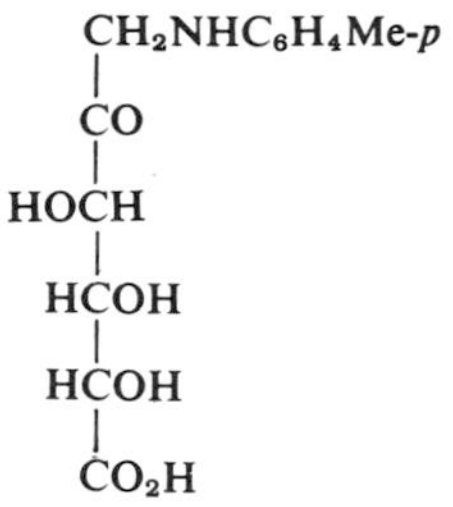

The Amadori rearrangement of D-[2-³H]glucose with *p*-toluidine is accompanied by complete removal of the tritium label,[138] in agreement with the step IXd → IXe. The loss of hydrogen at C-2 in IXd is considered to be the rate-determining step.[138] The Amadori rearrangement requires a carefully balanced acid–base catalysis. The ring is opened by acid to give IXd, and the base acts

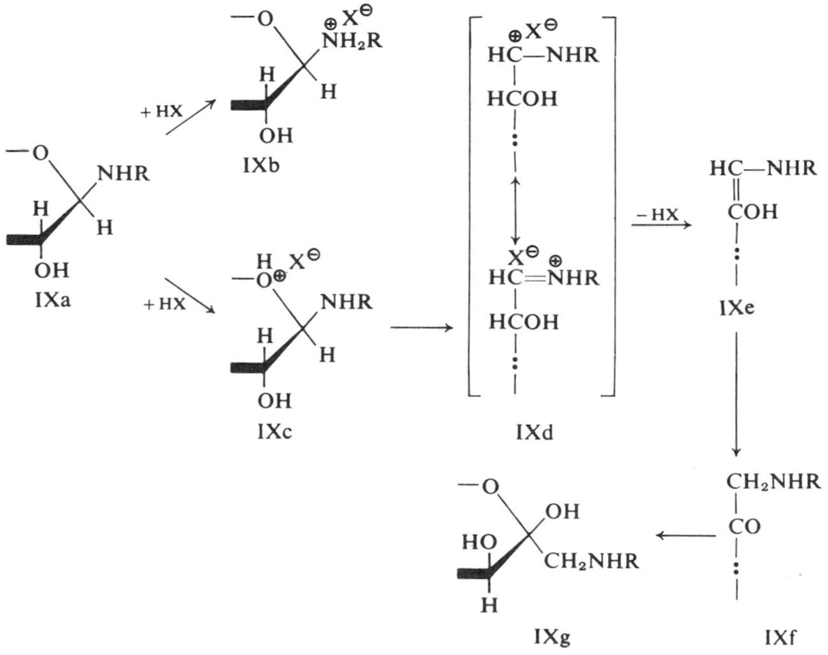

SCHEME IX

as an acceptor for the proton in the step IXd to IXe. Formation of the immonium structure (IXd) is precluded with glycosylamines prepared from sugars with nitroanilines, acetamide, or urea derivatives, because of resonance participation of the nitrogen atom, and the Amadori rearrangement does not occur.

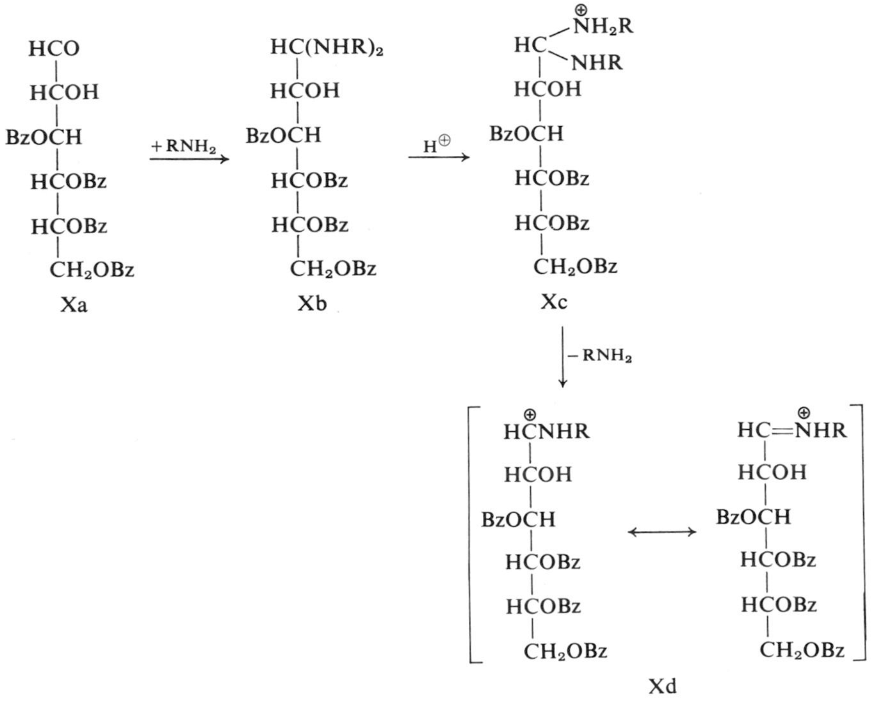

$$
\begin{array}{ccc}
\text{HCO} & \text{HC(NHR)}_2 & \overset{\oplus}{\text{NH}_2\text{R}} \\
| & | & \text{HC}\overset{}{\underset{\text{NHR}}{\diagdown}} \\
\text{HCOH} & \text{HCOH} & \text{HCOH} \\
| & | & | \\
\text{BzOCH} \xrightarrow{+\text{RNH}_2} & \text{BzOCH} \xrightarrow{\text{H}^{\oplus}} & \text{BzOCH} \\
| & | & | \\
\text{HCOBz} & \text{HCOBz} & \text{HCOBz} \\
| & | & | \\
\text{HCOBz} & \text{HCOBz} & \text{HCOBz} \\
| & | & | \\
\text{CH}_2\text{OBz} & \text{CH}_2\text{OBz} & \text{CH}_2\text{OBz} \\
\text{Xa} & \text{Xb} & \text{Xc}
\end{array}
$$

$$\downarrow -\text{RNH}_2$$

$$
\left[
\begin{array}{ccc}
\overset{\oplus}{\text{HCNHR}} & & \text{HC}=\overset{\oplus}{\text{NHR}} \\
| & & | \\
\text{HCOH} & & \text{HCOH} \\
| & & | \\
\text{BzOCH} & \longleftrightarrow & \text{BzOCH} \\
| & & | \\
\text{HCOBz} & & \text{HCOBz} \\
| & & | \\
\text{HCOBz} & & \text{HCOBz} \\
| & & | \\
\text{CH}_2\text{OBz} & & \text{CH}_2\text{OBz}
\end{array}
\right]
$$

Xd

SCHEME X

Micheel and co-workers[87] proposed a special mechanism (Scheme X) for the Amadori rearrangement in the case of the open-chained 3,4,5,6-tetra-*O*-benzoyl-*aldehydo*-D-glucose (Xa). With *p*-toluidine, Xa gives the 1,1-diamino derivative (Xb). Protonation of Xb to give Xc is followed by cleavage of a molecule of amine yielding the immonium structure (Xd). The latter reacts further according to the general route of Scheme IXd–g. The mechanism is supported by the observation that the authentic Schiff base of Xa, obtained by heating Xb in ether, does not rearrange to the Amadori product by acid catalysis, whereas Xb readily reacts. Isotope-labeling experiments proved that the Amadori rearrangement of unsubstituted, cyclic glycosylamines (such as *N-p*-tolyl-D-glucopyranosylamine) does not proceed[139] by way of a 1,1-diamino derivative of the type Xb.

1-Amino-1-deoxy-D-fructose (*"isoglucosamine"*) cannot be satisfactorily obtained from D-glucosylamine by the Amadori rearrangement.[3] It can be prepared by hydrogenolysis of D-*arabino*-hexulose phenylosazone,[140,141] although hydrogenolysis of 1-deoxy-1-*p*-toluidino-D-fructopyranose[142] or 1-deoxy-1-(dibenzyl)amino-D-fructopyranose[112] is preferable. 1-Amino-1-deoxy-D-*arabino*-hexulosuronic acid is also accessible by this route.[118]

B. THE HEYNS REARRANGEMENT

Ketosylamines undergo a rearrangement to 2-amino-2-deoxyaldoses. This rearrangement bears a formal resemblance to the Amadori rearrangement and has been termed the "Heyns rearrangement." Since a new asymmetric center is produced at C-2, two isomeric rearrangement products can be expected; for example rearrangement of a D-fructosylamine (51) leads to derivatives of 2-amino-2-deoxy-D-glucose and -D-mannose (52 and 53).

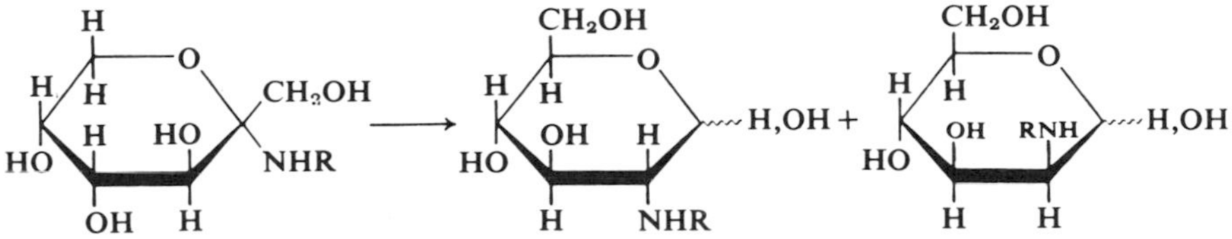

51   52   53

51a, 52a, 53a, R = —CHR'CO$_2$H

D-Fructose forms a D-fructosylamine in liquid ammonia or (better) methanolic ammonia which rearranges stereoselectively to 2-amino-2-deoxy-D-glucose by acid catalysis.[34,143,144] D-Tagatosylamine spontaneously rearranges under the same conditions to 2-amino-2-deoxy-D-galactose.[34] The stable, crystalline L-sorbosylamine may be converted into 2-amino-2-deoxy-L-gulose and -L-idose in equal parts on heating with oxalic acid in methanol.

D-Fructose reacts spontaneously with cyclohexyl-,[35] isopropyl-,[35] and butylamine[36] in the cold to give the rearranged products (2-alkylamino-2-deoxy-D-glucoses). Crystalline *N*-alkyl-D-fructosylamines from ethyl-[36] and benzylamine[37] rearrange easily in 15 to 55% yield with acetic acid. 1-*N*-Benzyl-2-benzylamino-2-deoxy-D-glucosylamine was obtained by prolonged treatment of D-fructose with benzylamine and benzylamine hydrochloride.[38] The configurational assignment of the rearrangement products to the D-*gluco* series was made by optical rotatory comparisons and by hydrogenation of 2-benzyl-

References start on p. 921.

amino-2-deoxy-D-glucose to give 2-amino-2-deoxy-D-glucose.[37,38] The Heyns rearrangement has also been observed [144a] with N-(D-fructosyl)pyrrolidine and various N-alkyl-D-threo-pentulosylamines. A stepwise epimerization of D-fructose to D-psicose was observed when D-fructose was treated with piperidine, and a rearranged product was not isolated.[145] The rearrangement of ketosylamines of D-fructose with such aromatic amines as aniline, p-toluidine, and p-anisidine, which traditionally presented difficulties, was finally accomplished by heating in 1,4-dioxane–acetic acid.[145a]

Free amino acids react readily with an excess of D-fructose by way of the Heyns rearrangement when the reactants are heated in methanol or methyl sulfoxide.[146,147] The N-D-fructosyl(amino acids) (51a), which are not isolable as intermediate products, rearrange to a mixture of 2-(1-carboxyalkyl)amino-2-deoxy-D-glucose ("D-glucose–amino acids," 52a) and 2-(1-carboxyalkyl)-amino-2-deoxy-D-mannose ("D-mannose–amino acids," 53a). A large number of amino acids, and peptides such as gelatin, have been brought into reaction with D-fructose in this manner.[124,125,145–148]

In the reaction mixture there can be found, together with 52a and 53a, "D-fructose–amino acids" of the type 42. The latter are normally formed only by Amadori rearrangement after reaction of aldoses with amino acids, and arise by a secondary reaction of "D-glucose- or D-mannose–amino acids" with excess amino acid in a fashion similar to the Amadori rearrangement.[128,146] Isolation and separation of "hexose–amino acids" can be accomplished by ion-exchange chromatography. Heating 52a and 53a in dilute acetic acid may again regenerate D-fructose and the amino acids. The Heyns rearrangement is therefore reversible.[146]

Quantitative analysis of the products from rearrangement of D-fructose with different amino acids indicates [128] that "D-mannose–amino acid," having the highest initial rate of formation, generally preponderates in the early stages of reaction, whereas more-extended times of reaction allow formation of "D-glucose–amino acid" and "D-fructose–amino acid" to increase to final, equilibrium values. The relative concentrations at equilibrium depend on the reaction conditions.

The mechanism of the Heyns rearrangement may be represented in a manner similar to the Amadori rearrangement, as shown in Scheme XI. Cleavage of hydrogen as a proton (step XIb–c), was proved by performing the rearrangement in CH_3OD and observing exchange of the hydrogen atom at C-1 by deuterium.[149]

The Heyns rearrangement is related to the biosynthesis of 2-amino-2-deoxy-D-glucose. 2-Amino-2-deoxy-D-glucose 6-phosphate is biochemically formed from D-fructose 6-phosphate and glutamine by a route similar to the Heyns rearrangement, whereby the amide group of glutamine is transferred to the sugar.[150]

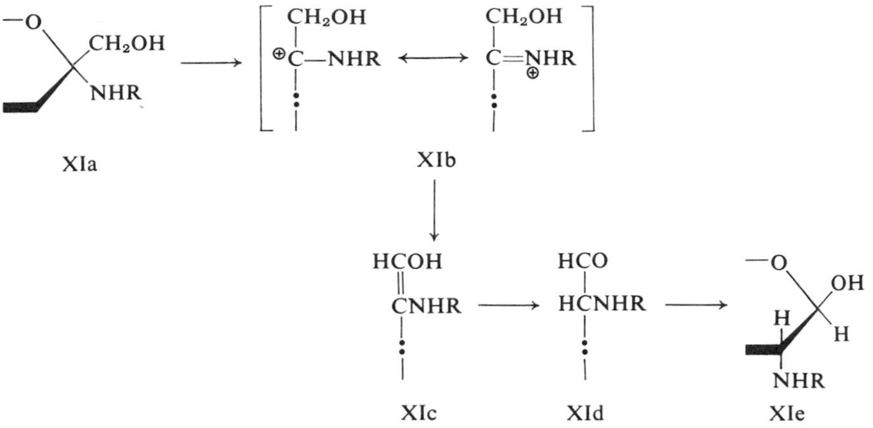

$$\text{XIa} \longrightarrow \left[\begin{array}{ccc} \text{CH}_2\text{OH} & & \text{CH}_2\text{OH} \\ \overset{\oplus}{\text{C}}\text{—NHR} & \longleftrightarrow & \text{C}\overset{}{=}\underset{\oplus}{\text{NHR}} \\ \vdots & & \vdots \end{array} \right]$$

XIa   XIb

$$\begin{array}{ccccc} \text{HCOH} & & \text{HCO} & & \\ \parallel & & | & & \\ \text{CNHR} & \longrightarrow & \text{HCNHR} & \longrightarrow & \\ \vdots & & \vdots & & \\ | & & | & & \\ \text{XIc} & & \text{XId} & & \text{XIe} \end{array}$$

SCHEME XI

C. 1-AMINO-1-DEOXYALDITOLS [151]

Alditols in which a CH_2OH group is replaced by CH_2NH_2 or CH_2NHR are named systematically as 1-amino-1-deoxyalditols; the trivial name "glyca-mine" has also been used. They can be obtained [10,152] by hydrogenation of

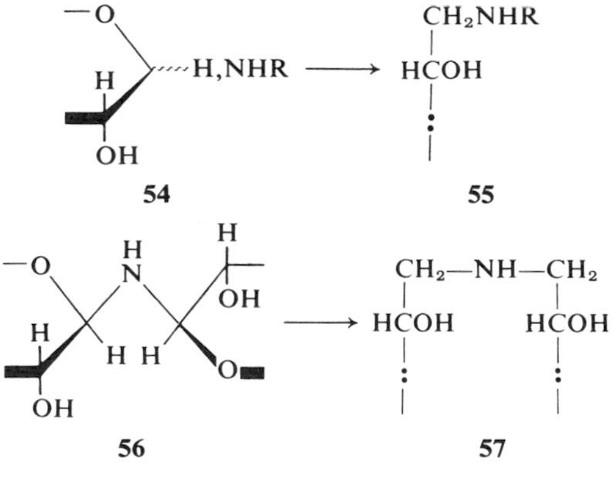

$$54 \xrightarrow{\text{H,NHR}} \begin{array}{c} \text{CH}_2\text{NHR} \\ | \\ \text{HCOH} \\ \vdots \\ | \end{array}$$

54   55

$$56 \longrightarrow \begin{array}{ccc} \text{CH}_2\text{—NH—CH}_2 \\ | \qquad\qquad | \\ \text{HCOH} \quad \text{HCOH} \\ \vdots \qquad\quad \vdots \\ | \qquad\qquad | \end{array}$$

56   57

glycosylamines ($54 \rightarrow 55$) or mixtures of aldoses with ammonia or amines. Aldoses in the presence of excess amine can be hydrogenated to give 1-amino-1-deoxyalditols with hydrogen at ~160 atmospheres pressure at 40° to 120° in the presence of Raney nickel or platinum. [153] Hydrogenation of D-glucose and D-galactose in methanolic ammonia, or preferably in liquid ammonia, yields

1-amino-1-deoxy-D-glucitol or -D-galactitol, isolated as the benzylidene or salicylidene Schiff bases.[154,155] Hydrogenation of di-D-glucosylamine (56) leads to bis(1-deoxy-D-glucit-1-yl)amine[3] (57). Mixtures of D-glucose (or other hexoses and pentoses), with ammonium chloride in methanolic ammonia, are likewise hydrogenated to bis(1-deoxyaldit-1-yl)amines ("*dialditylamines*").[3]

1-Alkylamino-1-deoxyalditols can be prepared by hydrogenation of aldoses in the presence of aliphatic amines.[156,157] 1-Benzylamino-1-deoxyalditols are cleaved by hydrogenolysis to give the free 1-amino-1-deoxyalditols. This method is advantageous for the preparation of 1-amino-1-deoxy-L-arabinitol, -D-ribitol, and -D-xylitol.[156] 1-Deoxy-1-methylaminoalditols[157] and their nitroso analogues,[158] prepared from several hexoses and pentoses, are known. N-(4,5-Dimethyl-2-nitrophenyl)-D-ribosylamine is converted by hydrogenation into 1-(2-amino-4,5-dimethylanilino)-1-deoxy-D-ribitol, which can be condensed with alloxan to form vitamin B_2 (lactoflavin).[26]

$$
\begin{array}{ccccc}
\text{CH}_2\text{NHR} & & \text{CH}_2\text{NHR} & & \text{CH}_2\text{NHR} \\
| & & | & & | \\
\text{HOCH} & \xleftarrow{\ \text{H}_2\ } & \text{CO} & \xrightarrow{\ \text{H}_2\ } & \text{HCOH} \\
| & & | & & | \\
\vdots & & \vdots & & \vdots \\
| & & | & & | \\
\mathbf{58} & & \mathbf{59} & & \mathbf{60}
\end{array}
$$

Amadori products (1-amino-1-deoxyketoses) (59) are easily hydrogenated at normal pressure and room temperature, in contrast to aldosylamines.[25,31] The 1-amino-1-deoxyalditols (58 and 60) epimeric at C-2 are obtained in this way. Hydrogenation of 1-arylamino-1-deoxy-D-fructoses in alkaline solution leads preferentially to 1-arylamino-1-deoxy-D-mannitols,[30,102] whereas 1-deoxy-1-piperidino-D-fructose affords preponderantly the D-*gluco* analogue.[11] The course of hydrogenation of other Amadori products is dependent on the reaction conditions and the catalyst.[102] In acid solution 1-deoxy-1-(3,4-dimethylanilino)-L-*erythro*-pentulose yields the corresponding L-ribitol derivative, whereas in alkaline solution the L-*arabino* derivative is obtained.

$$
\begin{array}{ccccccc}
\text{HC=NNHPh} & & \text{H}_2\text{CNH}_2 & & \text{CONH}_2 & & \text{H}_2\text{CNH}_2 \\
| & & | & & | & & | \\
\text{HCNH}_2 & \longrightarrow & \text{HCNH}_2 & & \text{HCOR} & \longrightarrow & \text{HCOR} \\
| & & | & & | & & | \\
\vdots & & \vdots & & \vdots & & \vdots \\
| & & | & & | & & | \\
\mathbf{61} & & \mathbf{62} & & \mathbf{63} & & \mathbf{64}
\end{array}
$$

Oximes and phenylhydrazones of aldoses have been hydrogenated to the 1-amino-1-deoxy derivatives of D-arabinitol, D-galactitol, D-glucitol, and D-xylitol.[159,160] The phenylhydrazone of 2-amino-2-deoxy-D-glucose (61) is converted by hydrogenation into 1,2-diamino-1,2-dideoxy-D-glucitol[160] (62).

Ether-soluble protected derivatives of aldonamides (63), such as 3,4:5,6-di-O-isopropylidene-D-gluconamide, can be reduced by lithium aluminum hydride to the corresponding 1-amino-1-deoxyalditols[161] (64).

1-Amino-1-deoxyalditols can be obtained by the nitromethane synthesis of Sowden and Fischer.[162] Thus, 2,4-O-benzylidene-L-xylose yields 2,4-O-benzylidene-6-deoxy-6-nitro-D-glucitol, which can be reduced to the corresponding 6-amino derivative.[163] 4,6-O-Benzylidene-D-glucose can similarly be converted into 1-amino-1-deoxy-D-*glycero*-D-*gulo*-heptitol.[164]

1-Amino-1-deoxyalditols are strong bases. They are deaminated by nitrous acid to give anhydroalditols[165]; for example, 1-amino-1-deoxy-D-glucitol reacts with sodium nitrite in acid solution to form 1,4-anhydro-D-glucitol[166] (see Vol. IA, Chapter 13). 1-Deoxy-1-methylamino-D-glucitol forms complexes with ions of heavy metals.[167,168] It is used as a hydrophilic component in pharmacologic preparations, particularly to obtain water-soluble salts of such X-ray contrast materials as 2,4,6-triiodobenzoic acid.

IV. BROWNING REACTIONS BETWEEN SUGARS AND AMINO DERIVATIVES[169-172]

A. REACTION WITH AMINO ACIDS (MAILLARD REACTION)[171]

The Maillard reaction is the process whereby mixtures of sugars with amino acids, peptides, or proteins lead on heating to very dark-colored products (melanoidins). According to Maillard,[173] a coloration of reddish brown to brownish black occurs when D-glucose (1 mole) and glycine (1.67 moles) are heated with an equal quantity of water to 100°. During the course of the browning reaction, the viscosity rises and carbon dioxide is evolved. In the final phase a dark, polymeric pigment, which is insoluble in water and acid, is obtained in 56% yield. It is soluble in strong alkali and is reprecipitated by acid. The empirical formula $C_{5.0}H_{4.7}N_{0.8}O_{1.5}[CH_2(CO_2H)]_{0.3}$ indicates the net loss of 0.5 mole of CO_2 and 4 moles of water.[174] A pigment obtained in the primary phase of the reaction is water-soluble and acidic.[169] The structure of the polymeric melanoidins has not been elucidated. Infrared spectra give evidence for OH, $C=O$, and conjugated —$C=C$— groups.[169]

Numerous variations of the Maillard reaction have been examined. The sugar and amino acid have been varied, and the influence of concentration, temperature, pH, buffer, metal catalysis, and acid studied. The reaction is extremely complex. Various reaction sequences that are far from fully elucidated occur simultaneously. Reactions significant in the initial phase of the Maillard reaction that have been clarified by work on model systems are discussed below.

The formation of *N*-(D-glucosyl)amino acids is assumed to be the initial step of the Maillard reaction, and these are converted into "D-fructose–amino acids" by the Amadori rearrangement (see Section II,A). D-Fructose forms the corresponding "D-glucose–amino acids" by the Heyns rearrangement (see Section II,B). Anet and Reynolds[175] have isolated numerous "D-fructose–amino acids" from freeze-dried apricot and peach pulps that had been strongly "browned." The Amadori rearrangement to "D-fructose–amino acids" need not be the only path leading to browning products; other reaction paths may occur simultaneously.

Anet[167a,176,177] found that "di-D-fructose–glycine" (65) (see Section II,A) is split in weakly acidic solution (pH 5.5, 100°) to give "D-fructose–glycine" and 3-deoxy-D-*erythro*-hexosulose (72) in a few minutes. Small quantities of the unsaturated hexosulose 73 (*trans* form) and 74 (*cis* form) were also isolated.[178] The "D-fructose–glycine" released may then react with D-glucose by way of the Amadori rearrangement to form "di-D-fructose–glycine"[127] (65), thus establishing a cycle whereby the conversion of an unlimited quantity of D-glucose into the 3-deoxyhexosulose (72) (and the unsaturated products 73 and 74) is theoretically possible. The extremely reactive dicarbonyl compounds 72–74 can readily react further with amines to give brown, polymeric products.[170] Acids may convert 72–74 quantitatively into 5-(hydroxymethyl)-2-furaldehyde.[179] The 3-deoxyhexosulose (72) can also be formed from "D-fructose–glycine" (42), although the degradation evidently takes place more slowly than it does by way of "di-D-fructose–glycine"[180] (65).

The release of the 3-deoxyhexosulose from 65 may be traced back to the facile enolization of the "diketose–amino acid" (65) to give 66, which occurs at pH 5.5. Protonation and dehydration of 66 yield the cation 67, which, by splitting out a molecule of "D-fructose–glycine," is converted into the enol 68, most of which is stabilized[177,180a] by forming the hexosulose 72. A small proportion of 68 reacts by elimination to form 69. The *trans* structure 70 yields[178] the *trans*-hexosulose 73, whereas the *cis* structure 71 forms the *cis*-hexosulose 74.

Kato,[181] studying the Amadori rearrangement of *N*-butyl-D-glucosylamine with acetic acid in methanol, isolated 3.8% of the 3-deoxyhexosulose 72, presumably formed by the same general pathway as formulated through the intermediate steps 66 and 68. The question of whether the enol (66) reacts as the actual intermediate of the rearrangement, or whether 66 is formed secondarily from the corresponding Amadori product (65), remains open. Furthermore, the 3-deoxyhexosulose 72 was detected in the browning mixtures formed from D-glucose and glycine[182] in soy sauce and "miso."[183]

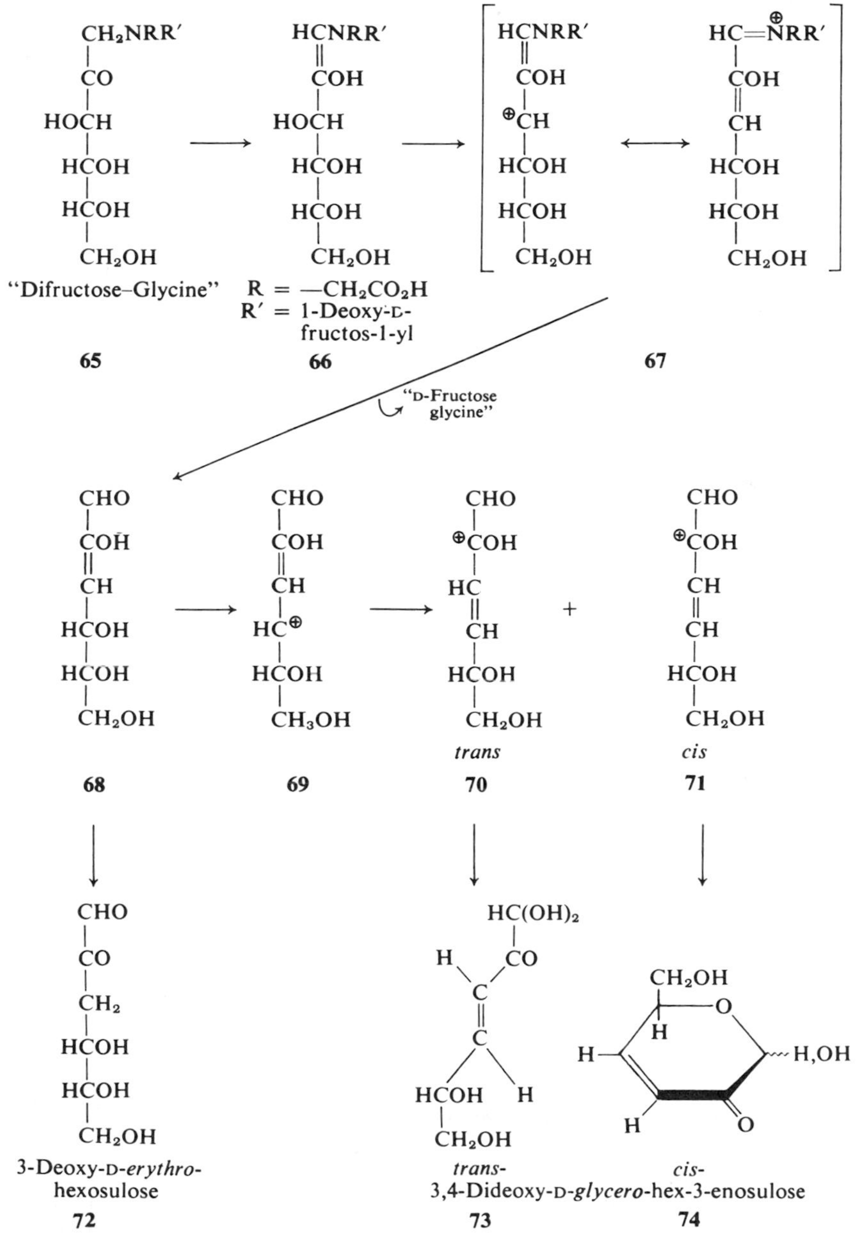

$$\begin{array}{c}
\text{CH}_2\text{NRR}' \\
| \\
\text{CO} \\
| \\
\text{HOCH} \\
| \\
\text{HCOH} \\
| \\
\text{HCOH} \\
| \\
\text{CH}_2\text{OH}
\end{array}
\longrightarrow
\begin{array}{c}
\text{HCNRR}' \\
\| \\
\text{COH} \\
| \\
\text{HOCH} \\
| \\
\text{HCOH} \\
| \\
\text{HCOH} \\
| \\
\text{CH}_2\text{OH}
\end{array}
\longrightarrow
\left[
\begin{array}{c}
\text{HCNRR}' \\
\| \\
\text{COH} \\
| \\
\overset{\oplus}{\text{CH}} \\
| \\
\text{HCOH} \\
| \\
\text{HCOH} \\
| \\
\text{CH}_2\text{OH}
\end{array}
\longleftrightarrow
\begin{array}{c}
\text{HC}{=}\overset{\oplus}{\text{N}}\text{RR}' \\
| \\
\text{COH} \\
| \\
\text{CH} \\
| \\
\text{HCOH} \\
| \\
\text{HCOH} \\
| \\
\text{CH}_2\text{OH}
\end{array}
\right]$$

"Difructose–Glycine" $\quad$ R = —CH$_2$CO$_2$H
$\qquad$ R' = 1-Deoxy-D-
$\qquad\qquad$ fructos-1-yl

65 $\qquad\qquad$ **66** $\qquad\qquad\qquad\qquad\qquad\qquad$ **67**

"D-Fructose
glycine"

$$\begin{array}{c}
\text{CHO} \\
| \\
\text{COH} \\
\| \\
\text{CH} \\
| \\
\text{HCOH} \\
| \\
\text{HCOH} \\
| \\
\text{CH}_2\text{OH}
\end{array}
\longrightarrow
\begin{array}{c}
\text{CHO} \\
| \\
\text{COH} \\
| \\
\text{CH} \\
| \\
\text{HC}^{\oplus} \\
| \\
\text{HCOH} \\
| \\
\text{CH}_3\text{OH}
\end{array}
\longrightarrow
\begin{array}{c}
\text{CHO} \\
| \\
\overset{\oplus}{\text{COH}} \\
| \\
\text{HC} \\
\| \\
\text{CH} \\
| \\
\text{HCOH} \\
| \\
\text{CH}_2\text{OH} \\
trans
\end{array}
+
\begin{array}{c}
\text{CHO} \\
| \\
\overset{\oplus}{\text{COH}} \\
| \\
\text{CH} \\
\| \\
\text{CH} \\
| \\
\text{HCOH} \\
| \\
\text{CH}_2\text{OH} \\
cis
\end{array}$$

68 $\qquad\qquad$ **69** $\qquad\qquad$ **70** $\qquad\qquad$ **71**

$$\begin{array}{c}
\text{CHO} \\
| \\
\text{CO} \\
| \\
\text{CH}_2 \\
| \\
\text{HCOH} \\
| \\
\text{HCOH} \\
| \\
\text{CH}_2\text{OH}
\end{array}$$

3-Deoxy-D-*erythro*-
hexosulose

72

$$\begin{array}{c}
\text{HC(OH)}_2 \\
\text{H} \quad \text{CO} \\
\diagdown \diagup \\
\text{C} \\
\| \\
\text{C} \\
\diagup \diagdown \\
\text{HCOH} \quad \text{H} \\
| \\
\text{CH}_2\text{OH}
\end{array}$$

trans-

cis-
3,4-Dideoxy-D-*glycero*-hex-3-enosulose

73 $\qquad\qquad$ **74**

Maillard's hypothesis[173] that the carbon dioxide released in the browning reaction is formed from the carboxyl group of the added amino acid was confirmed by isotopic-labeling experiments. By reaction of unlabeled D-glucose with glycine labeled at the carboxyl group, and labeled D-glucose with unlabeled glycine, it was demonstrated that 90% of the carbon dioxide originates from the carboxyl group of glycine.[184]

The decarboxylation of the amino acids proceeds by a Strecker type of degradation.[185] Dicarbonyl derivatives having the general formula $-CO[-C=C]_n-CO-$, including 3-deoxy-D-erythro-hexosulose (72), react according to Scheme XII. The aldehyde (XIIf) and the amino enol (XIIe) may undergo polymerization in the presence of amino derivatives.

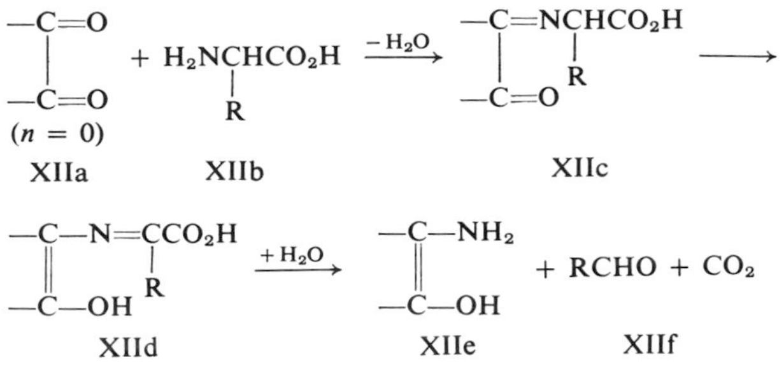

$$
\begin{array}{ccc}
\underset{\substack{| \\ -C=O \\ (n=0) \\ \text{XIIa}}}{-C=O} + \underset{\substack{| \\ R \\ \text{XIIb}}}{H_2NCHCO_2H} & \xrightarrow{-H_2O} & \underset{\substack{| \\ -C=O \\ \text{XIIc}}}{-C=NCHCO_2H \atop |\ R} \longrightarrow
\end{array}
$$

$$
\begin{array}{ccc}
\underset{\substack{\| \\ -C-OH \\ \text{XIId}}}{-C-N=CCO_2H \atop |\ R} & \xrightarrow{+H_2O} & \underset{\substack{\| \\ -C-OH \\ \text{XIIe}}}{-C-NH_2} + RCHO + CO_2 \\
& & \quad\quad\quad \text{XIIf}
\end{array}
$$

SCHEME XII

3-Deoxy-D-erythro-hexosulose (72) is considered[170] to be an important pigment-forming intermediate, which in either the keto or enol (68) form can react rapidly with amines. The 3-deoxyhexosulose (72), released by heating "di-D-fructose–glycine" (65) with glycine at pH 5.5, reacts with glycine twice as fast as 5-(hydroxymethyl)-2-furaldehyde to give a pigment.[176] The proportion of 5-(hydroxymethyl)-2-furaldehyde incorporated into the pigment, in the reaction[170,185a] of "di-D-fructose–glycine" (65) with glycine at pH 3.6, is less than 20%.

B. DEGRADATION OF SUGARS WITH ALIPHATIC SECONDARY AMINES (FORMATION OF REDUCTONES)[186]

By pyrolysis of 1-deoxy-1-piperidino-D-fructose, or by heating D-glucose with piperidine and acetic acid in ethanol, Hodge[11] and co-workers obtained a product of sugar decomposition in 27% yield, which they called "piperidino-hexose reductone." The product[11] resembled reductone itself,[105] and reduced silver nitrate in the cold and 2,6-dichlorophenol-indophenol in acid solution. Titration with iodine solution required 2 equivalents of iodine per mole of aminohexose reductone. The compound possesses the structure **75**. Through splitting out of water, the reductone (**75**) is converted into an anhydro-piperi-dino-hexose reductone (**76**), which can be hydrogenated to the dihydro-anhydro-piperidino-hexose reductone (**77**). The structure of **77** was confirmed by independent synthesis, through treatment of **78** with piperidine.[187]

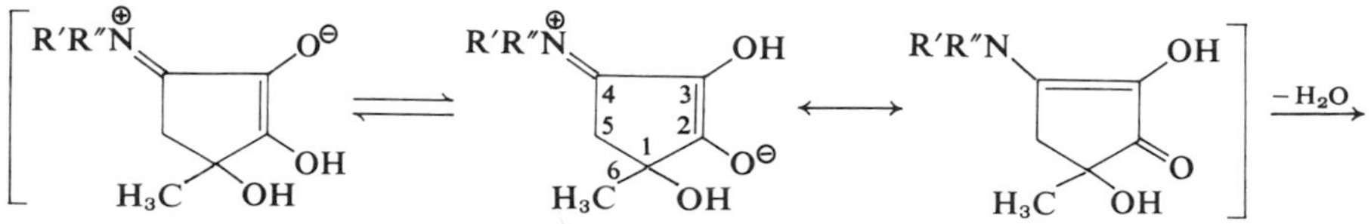

75

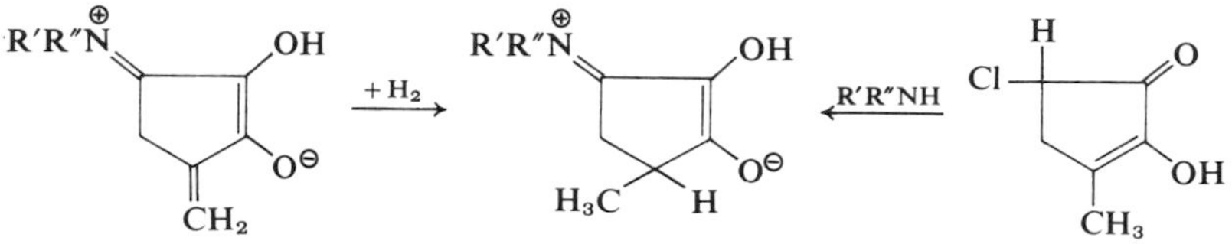

76   **77**   **78**

R', R" = —(CH₂)₅—

The same reductone **75** was obtained by treatment of D-mannose, D-galactose, D-fructose, or L-sorbose with piperidine.[188] D-Glucose reacts with morpholine, dimethylamine, diallylamine, and dibutylamine to form the corresponding N-substituted reductones.[188] Hodge and co-workers[186] showed that the reaction of benzylmethylammonium acetate with D-glucose gave in 35% yield a reductone of the type **75**, together with a 5% yield of a product shown to be a four-carbon reductone. The structure **79** was assigned to the latter and was proved[189] by synthesis from 1-bromobutane-2,3-dione (**81**) by way of the aminodiketone **80**.

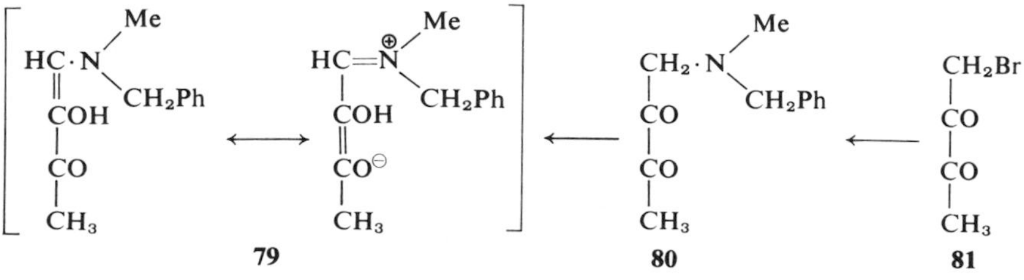

79   **80**   **81**

To explain the mechanism of formation of reductones, Simon[190,191] prepared the reductone 75 from D-[1-^{14}C]-, -[2-^{14}C]-, and -[6-^{14}C]-glucose, and also from 1-deoxy-1-piperidino-D-[1-^{14}C]fructose, and examined the distribution of labeling. He found that 72% of the C-6 atoms and 28% of the C-1 atoms of the D-glucose were incorporated in the methyl group (C-6) of the reductone (75). The remaining 72% of atoms from C-1 and 28% of those from C-6 of D-glucose were found in the C-5 atom of 75, whereas no incorporation of ^{14}C at C-1, 2, 3, and 4 was detected[190] by use of D-glucose labeled at C-1 or C-6. The retention of all of the activity in only two carbon atoms of the reductone (75) shows that the sugar chain in 75 is maintained intact, since recondensation from smaller fragments to give 75 would have led to a broader distribution of activity. In the four-carbon reductone 79, 80% of the C-6 of D-glucose is found in the methyl group (C-4 of 79), and 20% of the C-6 of D-glucose is found in C-1 of 79. When D-glucose labeled at C-1 is used, the reductone (79) obtained is devoid of activity,[191] indicating that C-1 and C-2 of D-glucose are not involved in the formation of 79.

Simon[191] put forward a mechanism for formation of reductones that explains the results observed. The initial intermediate is the Amadori derivative (82), which subsequently undergoes an irreversible 2,3-enolization to the enol (83); this process was postulated as the rate-determining step.[191]

By the use of model compounds it was proved that allylically positioned amino groups are eliminated more easily than hydroxyl groups positioned allylically. Accordingly, the amine residue is easily cleaved from 83 to give the enol 84, which ketonizes to give 85. The reaction of D-glucose with hot, ethanolic piperidine also affords 3,5-dihydroxy-2-methyl-5,6-dihydro-4H-pyran-4-one (106) in 15% yield. This compound forms (via 82) by cyclization[191a] of the intermediate 86.

The 1-deoxydiulose 85 occupies a key role, since, according to Hodge,[186] it is also involved in the formation of maltol and isomaltol. If the OH group at C-4 in 85 is unsubstituted, the dicarbonyl compound 85 may undergo numerous enolizations and ketonizations to yield isomerized intermediates. One of these intermediates (86) may undergo elimination to form "diacetylformoin" (89) by way of the intermediates 87 and 88. "Diacetylformoin" (89) then cyclizes with the participation of an amine to form the reductone 75 in two ways, either by reaction of C-1 with C-5 or by reaction of C-6 with C-2. The isotopic distribution observed indicates that the first route (starred in the formula scheme) is favored[186,191] by 72%. The four-carbon reductone 79 is formed from the deoxydiulose 86 by way of steps 90 → 93 and subsequent cleavage of 93, whereby the activity at C-6 in the original sugar appears in the methyl group in the reductone (79), and that from C-1 appears in the acetic acid that is split off.[191]

$$
\begin{array}{ccccc}
\text{H}_2\text{CN} & \text{H}_2\text{CN} & \text{CH}_2 & \text{CH}_3 \\
| & | & \parallel & | \\
\text{CO} & \text{COH} & \text{COH} & \text{CO} \\
| & \parallel & | & | \\
\text{HOCH} & \text{COH} & \text{CO} & \text{CO} \\
| & | & | & | \\
\text{HCOH} & \text{HCOH} & \text{HCOH} & \text{HCOH} \\
| & | & | & | \\
\text{HCOH} & \text{HCOH} & \text{HCOH} & \text{HCOH} \\
| & | & | & | \\
\text{CH}_2\text{OH} & \text{CH}_2\text{OH} & \text{CH}_2\text{OH} & \text{CH}_2\text{OH} \\
\textbf{82} & \textbf{83} & \textbf{84} & \textbf{85}
\end{array}
\longrightarrow
$$

$$
\begin{array}{ccccc}
\text{CH}_3 & \text{CH}_3 & \text{CH}_3 & {}^{*}\text{CH}_3 \\
| & | & | & | \\
\text{CO} & \text{CO} & \text{CO} & \text{CO} \\
| & | & | & | \\
\text{HCOH} & \text{HCOH} & \text{HCOH} & \text{HCOH} \\
| & | & | & | \\
\text{CO} & \text{COH} & \text{CO} & \text{CO} \\
| & \parallel & | & | \\
\text{HCOH} & \text{COH} & \text{COH} & \text{CO} \\
| & | & \parallel & | \\
\text{CH}_2\text{OH} & \text{CH}_2\text{OH} & \text{CH}_2 & {}^{**}\text{CH}_3 \\
\textbf{86} & \textbf{87} & \textbf{88} & \textbf{89}
\end{array}
$$

$$
\begin{array}{c}
\text{NC}=\!=\text{COH} \\
\text{H}_2^{*}\text{C} \quad \text{CO} \\
\text{C} \\
{}^{**}\text{CH}_3 \quad \text{OH} \\
\textbf{75}
\end{array}
$$

$$
\begin{array}{ccccc}
\text{CH}_3 & \text{CH}_3 & \text{CH}_3 & {}^{*}\text{CH}_3 & \text{CH}_3^{*}\text{COOH} \\
| & | & | & | & \\
\text{CO} & \text{CO} & \text{CO} & \text{CO} & + \\
| & | & | & | & \\
\text{HCN} & \text{HCN} & \text{HCN} & \text{HCN} & \text{HCN} \\
| & | & | & | & \parallel \\
\text{CO} & \text{COH} & \text{CO} & \text{CO} & \text{COH} \\
| & \parallel & | & | & | \\
\text{HCOH} & \text{COH} & \text{COH} & \text{CO} & \text{CO} \\
| & | & \parallel & | & | \\
\text{CH}_2\text{OH} & \text{CH}_2\text{OH} & \text{CH}_2 & {}^{**}\text{CH}_3 & {}^{**}\text{CH}_3 \\
\textbf{90} & \textbf{91} & \textbf{92} & \textbf{93} & \textbf{79}
\end{array}
$$

Formation of piperidino-hexose reductone and four-carbon reductones

Hodge[192] obtained maltol (**101**) in 2% yield by heating maltose with piperidine phosphate, and under the same conditions obtained β-D-galactosylisomaltol (**96**) from lactose in 35% yield. The formation of both of these compounds is in accord with the mechanism proposed for formation of reductones.[186] The reaction of both disaccharides follows the general route by way of an Amadori product (**82**) to give the deoxydiulose (**85**). Since the OH group at C-4 of maltose and lactose is substituted, rearrangement to the 2,4-diulose structure (**86**) cannot occur. The product (**94**) arising from maltose reacts

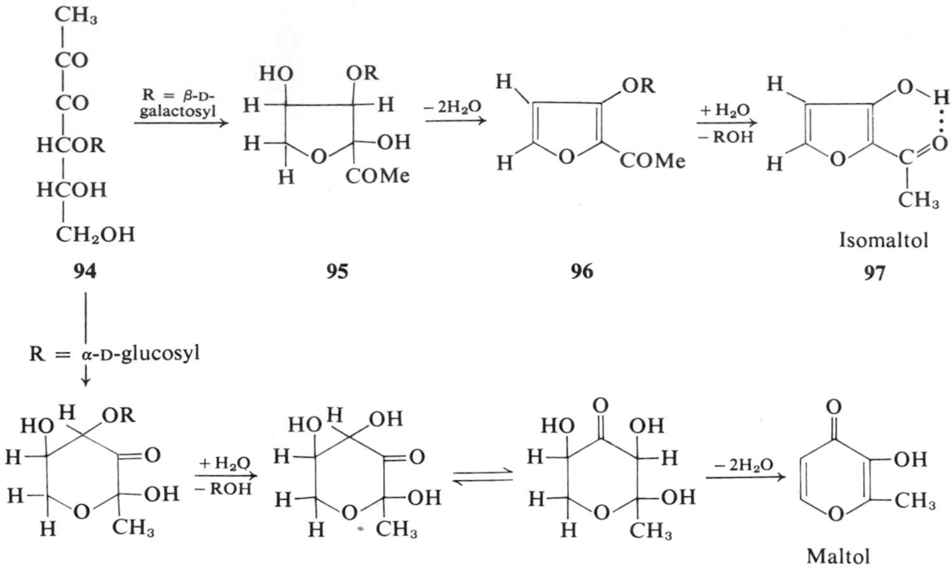

CH$_3$
|
CO
|
CO   R = β-D-
| galactosyl
HCOR ⟶
|
HCOH
|
CH$_2$OH

94   **95**   **96**   Isomaltol **97**

|
R = α-D-glucosyl
↓

98   **99**   **100**   Maltol **101**

further in the pyranoid form (**98**). The D-glucosyl moiety is α-D-linked and is relatively easily cleaved to give **99**, which isomerizes to **100** and forms maltol (**101**) by loss of water.[186] The corresponding diulose (**94**) from lactose, on the other hand, reacts by way of the furanoid form (**95**) with loss of water, to give β-D-galactosylisomaltol (**96**), from which isomaltol (**97**) is obtained by hydrolysis.[186,193]

The behavior of 6-deoxy-L-galactose on heating with piperidine acetate to form 2,5-dimethyl-4-hydroxy-2,3-dihydro-3-furanone (**104**) is also explained by the mechanism for formation of reductones. Here, the reaction proceeds through the Amadori product to an analogue of **86** in which further elimination to **88** or **92** is prevented. The 6-deoxy derivative (**102**) corresponding to **86** loses water from the furanoid form (**103**) to give[186] **104**. As a pseudo-reductone, the ring of compound **104** can easily be opened in acid with addition of the

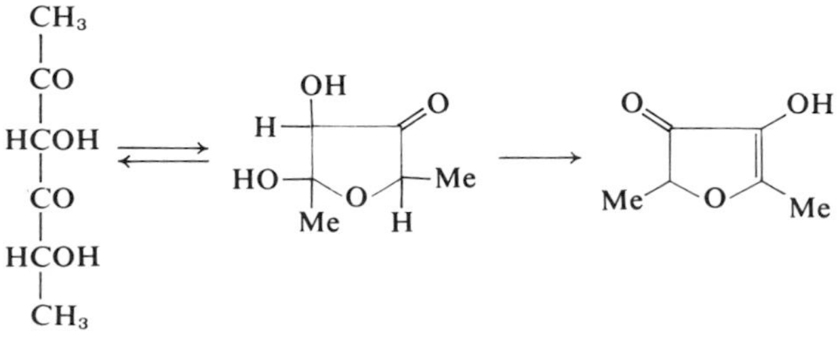

CH$_3$
|
CO
|
HCOH
|
CO
|
HCOH
|
CH$_3$

102   **103**   **104**

elements of water. Maltol and isomaltol are important flavoring agents occurring in numerous browned foodstuffs, such as bread.

Considerable research has been done on the volatile flavor compounds[194] that develop during heating of 1-deoxy-1-piperidino-D-fructose. In addition to amides of piperidine, there were found **75, 79, 104**, and the reductone **105**. The latter may result from cleavage between C-4 and C-5 of **93**; formation of **104** is explained by reductive cyclization of **89**. Heating of !-deoxy-1-(L-prolino)-D-fructose or a mixture of D-glucose with methylamine afforded[195–197] the pyranone **106**.

Pyrrole compounds have not been detected among products of decomposition of Amadori compounds of hexoses. By heating *N*-alkyl-D-xylosylamines or mixtures of D-xylose with alkylamines or amino acids in water–acetic acid, Kato[198] obtained the corresponding *N*-substituted pyrrol-2-aldehydes as degradation products. Rearrangement to a 3,4-dideoxypent-3-enosulose is assumed[198] by analogy to the reaction **65–70**; reaction of the enosulose with amines and subsequent cyclization yields the pyrrole. Similar products arise from reaction of hexoses with primary amines.[196,199]

Amadori compounds heated in $6N$ HCl[200] show a behavior different from the reaction sequences **65–72** or **82–85**; 2,3-enolization of **108** gives **109**, and subsequent elimination occurs at C-4, whereas the amino group at C-1 is not split off. Cyclization of the assumed intermediate **110** affords *N*-(2-furoylmethyl)amine compounds (**111**). Thus "fructose–ϵ-lysine" undergoes[201] acid-

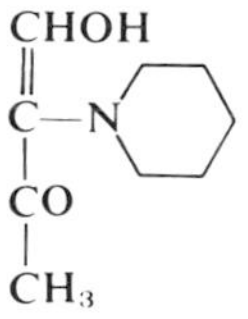

105

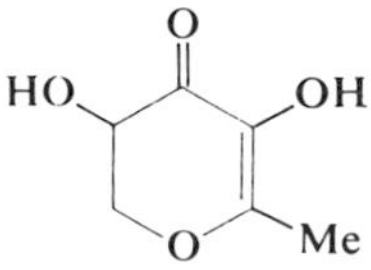

106

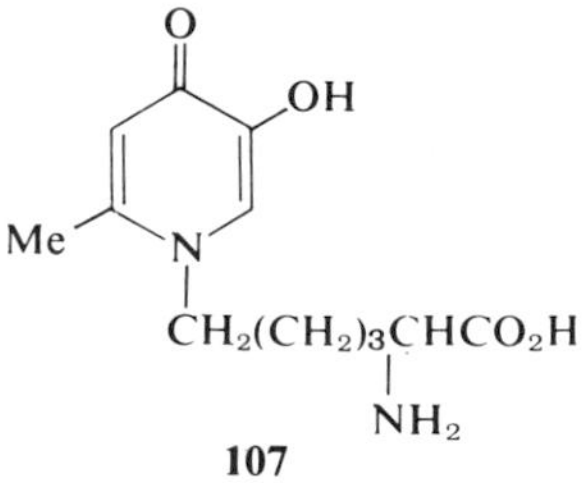

107

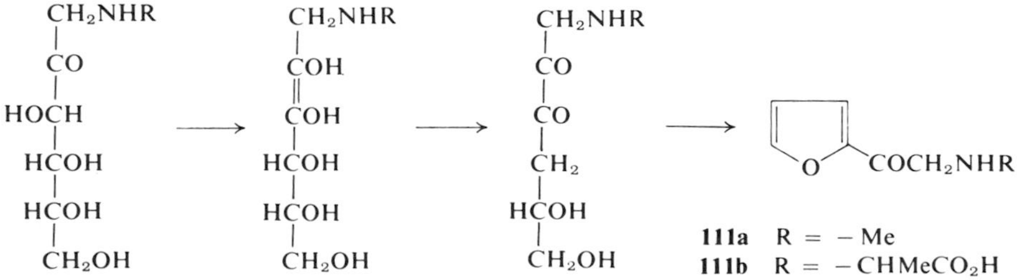

108   **109**   **110**

111a R = − Me
111b R = − CHMeCO$_2$H
111c R = − (CH$_2$)$_4$CHCO$_2$H
                              |
                              NH$_2$

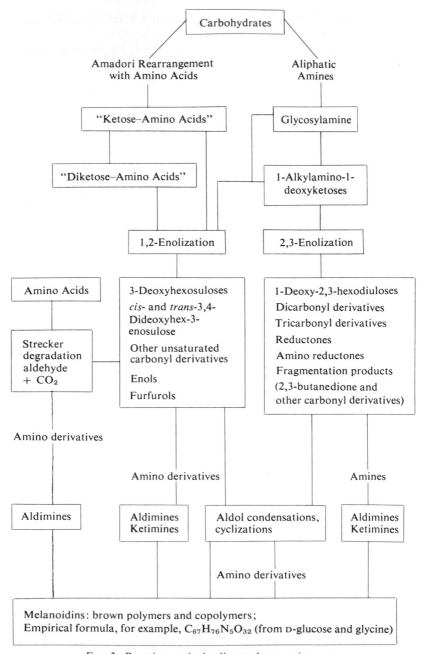

FIG. 3. Reaction paths leading to brown pigments.

catalyzed conversion into **111c**, together with minor amounts of **107**. Under conditions fostering autoxidation, *N*-aryl-D-glucopyranosylamines are rearranged and degraded to afford isatin derivatives.[202]

C. SYNOPSIS OF THE REACTION PATHS OF THE BROWNING REACTION

A synopsis of the reaction paths involved in the reaction of sugars with amino acids and aliphatic amines, leading to brown pigments, is presented in Fig. 3. The left branch indicates the reaction with amino acids, wherein the 1,2-enol of the Amadori compound is regarded as the key substance from which the 3-deoxyhexosuloses are released. Additionally, other unsaturated keto derivatives are formed, and furfurols are produced in acid solution. The Strecker decomposition of amino acids occurs simultaneously, affording carbon dioxide. The mixture of keto derivatives undergoes condensation and cyclization to give molecules of intermediate molecular weight, which subsequently react with amines to give the melanoidin pigment. However, it is conceivable that direct reaction of carbonyl derivatives with amines, to form polymers by way of ketimines, can also occur.

The right branch of Fig. 3 deals with reactions involving aliphatic amines. A 1,2-enolization from the Amadori compound or the glycosylamine can also occur here. The rearrangement of consequence is a 2,3-enolization. It leads to the 1-deoxy-2,3-hexodiulose, which has not yet been isolated but which appears in the mechanism of Simon[191] and Hodge[186] as the most important intermediate product. Further reaction leads to the reductones, aminoreductones, and polycarbonyl derivatives. These extremely reactive products are able to condense with themselves or with amines to form polymers. The scheme gives a survey of the entire reaction sequences as presently known. Additional research will be necessary to establish further reaction paths and to obtain detailed understanding of the individual steps.

REFERENCES

1. G. P. Ellis and J. Honeyman, *Advan. Carbohyd. Chem.*, **10**, 95 (1955).
1a. H. Simon and A. Kraus, *Fortschr. Chem. Forsch.*, **14**, 430 (1970).
2. C. A. Lobry de Bruyn and A. P. N. Franchimont, *Rec. Trav. Chim.* (Pays-Bas), **12**, 286 (1893).
3. J. E. Hodge and B. F. Moy, *J. Org. Chem.*, **28**, 2784 (1963).
4. H. L. Frush and H. S. Isbell, *J. Res. Nat. Bur. Stand.*, **47**, 239 (1951).
5. H. S. Isbell and H. L. Frush, *J. Org. Chem.*, **23**, 1309 (1958).
6. H. S. Isbell and H. L. Frush, *J. Res. Nat. Bur. Stand.*, **46**, 132 (1951).
7. F. Micheel, R. Frier, E. Plate, and A. Hiller, *Ber.*, **85**, 1092 (1952).
8. D. Raacke-Fels, *Arch. Biochem. Biophys.*, **43**, 289 (1953).
9. E. Mitts and R. M. Hixon, *J. Amer. Chem. Soc.*, **66**, 483 (1944).
10. J. E. Hodge and C. E. Rist, *J. Amer. Chem. Soc.*, **74**, 1494 (1952).
11. J. E. Hodge and C. E. Rist, *J. Amer. Chem. Soc.*, **75**, 316 (1953).
12. F. Micheel and G. Hagemann, *Ber.*, **92**, 2836 (1959); **93**, 2381 (1960).

13. B. N. Stepanenko and R. D. Greshnykh, *Dokl. Akad. Nauk SSSR*, **170**, 121 (1966).
14. G. R. Ames and T. A. King, *J. Org. Chem.*, **27**, 390 (1962).
15. J. G. Erickson, *J. Amer. Chem. Soc.*, **77**, 2839 (1955).
16. F. Micheel and A. Frowein, *Ber.*, **90**, 1599 (1957).
17. H. E. Zaugg, *J. Org. Chem.*, **26**, 603 (1961).
18. F. Micheel and A. Klemer, *Ber.*, **84**, 212 (1951).
19. F. Micheel and A. Klemer, *Ber.*, **85**, 1083 (1952).
20. G. Weitzel, H.-U. Geyer, and A.-M. Fretzdorff, *Ber.*, **90**, 1153 (1957).
21. M. L. Wolfrom, R. D. Schuetz, and L. F. Cavalieri, *J. Amer. Chem. Soc.*, **71**, 3518 (1949); F. Micheel and A. Frowein, *Ber.*, **92**, 304 (1959).
22. H. B. F. Dixon, *Biochem. J.*, **129**, 203 (1972).
23. H. Schiff, *Ann.*, **154**, 30 (1870).
24. B. Sorokin, *Ber.*, **19**, 513 (1886); *J. Chem. Soc.*, **54**, 807 (1888); *J. Prakt. Chem.*, [2], **37**, 291 (1888).
25. R. Kuhn and F. Weygand, *Ber.*, **70**, 769 (1937).
26. R. Kuhn and R. Ströbele, *Ber.*, **70**, 773 (1937).
27. R. Kuhn and L. Birkofer, *Ber.*, **71**. 621 (1938).
28. J. Honeyman and A. R. Tatchel, *J. Chem. Soc.*, 967 (1950); G. P. Ellis and J. Honeyman, *ibid.*, 1490 (1952).
29. J. Sykulski, *Rocz. Chem.*, **40**, 1495 (1966).
30. F. Weygand, *Ber.*, **72**, 1663 (1939); **73**, 1259 (1940).
31. B. N. Stepanenko, E. S. Volkova, and M. G. Chentsova, *Dokl. Akad. Nauk SSSR*, **177**, 607 (1967); S. Adachi, *Carbohyd. Res.*, **10**, 165 (1969). Hydroxyamination of glycals is also a feasible route; see I. Dyong, G. Schulte, Q. Lam-Chi, and H. Friege, *Carbohyd. Res.*, **68**, 257 (1979).
32. B. Helferich and A. Mitrowsky, *Ber.*, **85**, 1 (1952).
33. F. Micheel and B. Schleppinghoff, *Ber.*, **89**, 1702 (1956).
33a. M. B. Kozikowski and G. Kuperszewski, *Rocz. Chem.*, **47**, 1899 (1973).
33b. V. I. Mukhanov, M. N. Preobrazhenskaya, N. P. Kosyuchenko, T. Ya. Filipenko, and N. N. Suvorov, *Zh. Org. Khim.*, **10**, 587 (1974).
33c. M. N. Preobrazhenskaya, T. D. Miniker, V. S. Martynov, L. N. Yakhontov, N. P. Kostyuchenko, and D. M. Krasnokutskaya, *Zh. Org. Khim.*, **10**, 745 (1974).
33d. D. S. Boolieris, R. J. Ferrier, and L. A. Branda, *Carbohyd. Res.*, **35**, 131 (1974).
34. K. Heyns, H. Paulsen, R. Eichstedt, and M. Rolle, *Ber.*, **90**, 2039 (1957).
35. J. F. Carson, *J. Amer. Chem. Soc.*, **77**, 1881 (1955).
36. J. F. Carson, *J. Amer. Chem. Soc.*, **77**, 5957 (1955).
37. J. F. Carson, *J. Amer. Chem. Soc.*, **78**, 3728 (1956).
38. K. Heyns, R. Eichstedt, and K.-H. Meinecke, *Ber.*, **88**, 1551 (1955).
39. C. P. Barry and J. Honeyman, *J. Chem. Soc.*, 4147 (1952).
40. B. Helferich and W. Portz, *Ber.*, **86**, 604 (1953).
41. F. Knotz, *Monatsh. Chem.*, **88**, 703 (1957).
42. Y. Inouye, K. Onodera, and S. Kitaoka, *J. Agr. Chem. Soc. Japan*, **29**, 139 (1955).
43. K. Onodera, T. Uehara, and S. Kitaoka, *Bull. Chem. Soc. Japan*, **24**, 703 (1960).
44. T. van Es and R. L. Whistler, *J. Org. Chem.*, **29**, 1087 (1964).
45. D. L. Ingles, *Chem. Ind.* (London), 1901 (1963).
46. I. Goodman, *Advan. Carbohyd. Chem.*, **13**, 215 (1958).
47. M. N. Schoorl, *Rec. Trav. Chim.* (Pays-Bas), **22**, 31 (1903).
48. B. Helferich and W. Kosche, *Ber.*, **59**, 69 (1926).
49. M. H. Benn and A. S. Jones, *J. Chem. Soc.*, 3337 (1960); A. S. Jones and G. W. Ross, *Tetrahedron*, **18**, 189 (1962); W. E. Jensen, A. S. Jones, and G. W. Ross, *J. Chem. Soc.*, 2463 (1965).

50. E. A. M. Badani, A. S. Jones, and M. Stacey, *Tetrahedron*, Suppl. No. 7, 281 (1966).

51. M. Frèrejacque, *C. R. Acad. Sci.*, **202**, 1190 (1936); **207**, 638 (1938).

52. R. Bognár and P. Nánási, *J. Chem. Soc.*, 185 (1955).

52a. I. A. Korbulch, F. F. Blanco, and M. N. Preobrazhenskaya, *Tetrahedron Lett.*, 4619 (1973).

52b. J. Jurczak, G. Grynkiewicz, and A. Zamojski, *Carbohyd. Res.*, **39**, 147 (1975).

53. M. Viscontini, R. Hochreuter, and P. Karrer, *Helv. Chim. Acta*, **36**, 1778 (1953); M. Viscontini, M. Marti, and P. Karrer, *ibid.*, **37**, 1374 (1954); M. Viscontini, D. Hoch, M. Marti, and P. Karrer, *ibid.*, **38**, 646 (1955).

54. M. G. Blair, *Advan. Carbohyd. Chem.*, **9**, 97 (1954).

55. J. W. Baker, *J. Chem. Soc.*, 1205 (1929).

56. P. Karrer and A. P. Smirnoff, *Helv. Chim. Acta*, **4**, 817 (1921).

57. E. Fischer and K. Raske, *Ber.*, **43**, 1750 (1910).

57a. G. Wagner, G. Walz, B. Dietech, and G. Fischer, *Pharmazie*, **29**, 90 (1974).

58. A. Bertho, *Ber.*, **63**, 836 (1930).

59. F. Micheel and A. Klemer, *Advan. Carbohyd. Chem.*, **16**, 95 (1961).

60. A. Bertho and J. Maier, *Ann.*, **498**, 50 (1932).

61. I. Goodman, *Federation Proc.*, **15**, 264 (1956).

61a. H. Paulsen, Z. Györgydeák, and M. Friedmann, *Ber.*, **107**, 1568 (1974).

62. E. Fischer, *Ber.*, **47**, 1377 (1914).

63. A. Müller and A. Wilhelms, *Ber.*, **74**, 698 (1941).

64. K. M. Harin and T. B. Johnson, *J. Amer. Chem. Soc.*, **55**, 395 (1933).

65. F. Micheel, W. Berlenbach, and K. Weichbrodt, *Ber.*, **85**, 189 (1952).

66. F. Micheel and W. Brunkhorst, *Ber.*, **88**, 481 (1955).

67. R. D. Marshall and A. Neuberger, *Biochemistry*, **3**, 1596 (1964); A. Gottschalk, "Glycoproteins," Elsevier, Amsterdam, 1966.

68. D. E. Cowley, L. Hough, and M. Y. Khan, *Carbohyd. Res.*, **19**, 242 (1971); compare B. Paul and W. Korytnyk, *ibid.*, **67**, 457 (1978).

68a. H. G. Garg and R. W. Jeanloz, *Carbohyd. Res.*, **32**, 145 (1974).

68b. T. Y. Shen, J. P. Li, C. P. Dorn, D. Ebel, R. Bugianesi, and R. Fecher, *Carbohyd. Res.*, **23**, 87 (1972).

68c. D. Dunstan and L. Hough, *Carbohyd. Res.*, **23**, 17 (1972); **25**, 246 (1972).

68d. E. Walker and R. W. Jeanloz, *Carbohyd. Res.*, **32**, 37 (1974).

68e. M. Tanaka and J. Yamashina, *Carbohyd. Res.*, **27**, 175 (1973).

68f. R. J. Pollitt and K. M. Pretty, *Biochem. J.*, **141**, 141 (1974).

69. R. Bognár and P. Nánási, *J. Chem. Soc.*, 189, 193 (1955).

70. R. Bognár and P. Nánási, *Tetrahedron*, **14**, 175 (1961).

70a. V. A. Afanas'ev and Zh. A. Dzhananbaev, *Khim. Prir. Soedin.*, **10**, 176 (1974); *Chem. Abstr.*, **81**, 25, 897 (1974).

71. J. B. Lee and M. M. el Sawi, *Tetrahedron*, **6**, 91 (1959).

72. G. Wagner and H. Pischel, *Arch. Pharm.*, **295**, 373 (1962); **296**, 699 (1963).

72a. M. Yamada, S. Inaba, T. Yoshino, and Y. Ishido, *Carbohyd. Res.*, **31**, 151 (1973).

73. V. Deulofeu, *Advan. Carbohyd. Chem.*, **4**, 119 (1949).

74. V. Deulofeu and J. O. Deferrari, *J. Org. Chem.*, **17**, 1087, 1093, 1097 (1952); **22**, 802 (1957); **24**, 183 (1959).

75. P. Brigl, H. Mühlschlegel, and R. Schinle, *Ber.*, **64**, 2921 (1931).

76. J. O. Deferrari and R. A. Cadenas, *J. Org. Chem.*, **28**, 1070, 1072, 2613 (1953); **30**, 2007 (1965).

77. H. S. Isbell and H. L. Frush, *J. Amer. Chem. Soc.*, **71**, 1579 (1949).

78. R. C. Hockett, V. Deulofeu, and J. O. Deferrari, *J. Amer. Chem. Soc.*, **82**, 1840 (1950).

79. E. G. Gros, M. A. Ondetti, J. O. Sproviero, V. Deulofeu, and J. O. Deferrari, *J. Org. Chem.*, **27**, 924 (1962).

79a. J. F. Sproviero, A. Salinas, and E. S. Bertiche, *Carbohyd. Res.*, **19**, 81 (1971).

79b. N. J. Cusack, D. H. Robinson, P. W. Rugy, G. Shaw, and R. Lofthouse, *J. Chem. Soc. Perkin I*, 73 (1974).

79c. V. V. Alenin and V. D. Domkin, *Khim. Geterotsikl. Soedin.*, 275 (1974); *Chem. Abstr.*, **81**, 4179 (1974).

79d. D. H. Robinson and G. Shaw, *J. Chem. Soc. Perkin I*, 774 (1974).

79e. R. P. Panzica and L. B. Townsend, *J. Amer. Chem. Soc.*, **95**, 8737 (1973).

79f. L. Van Rompay, F. Lambein, R. De Gussen, and R. Van Parijs, *Biochem. Biophys. Res. Commun.*, **56**, 199 (1974); F. Lambein and R. Van Parijs, *ibid.*, **40**, 557 (1970).

80. G. P. Ellis and J. Honeyman, *J. Chem. Soc.*, 2053 (1952).

81. R. Bognár, P. Nánási, and A. Lipták, *Acta Chim. Acad. Sci. Hung.*, **45**, 47 (1965).

82. C. Niemann and J. T. Hays, *J. Amer. Chem. Soc.*, **62**, 2960 (1940).

83. J. G. Douglas and J. Honeyman, *J. Chem. Soc.*, 3674 (1955).

84. L. Berger and J. Lee, *J. Org. Chem.*, **11**, 75 (1946).

85. G. P. Ellis, *J. Chem. Soc. (B)*, 572 (1966).

86. P. Nánási, E. Nemes-Nánási, and P. Cerletti, *Gazz. Chim. Ital.*, **95**, 966 (1965).

86a. H. Paulsen, Z. Györgydeák, and M. Friedmann, *Ber.*, **107**, 1590 (1974).

86b. W. A. Szarek and D. Horton (eds.), Anomeric Effect: Origin and Consequences, *Amer. Chem. Soc. Symp. Ser.*, **87** (1979).

87. F. Micheel and J. Dijong, *Ann.*, **658**, 120 (1962); V. D. Shcherbubin, B. N. Stepanenko, and E. S. Volkova, *Dokl. Akad. Nauk SSSR*, **174**, 775 (1967); R. S. Tipson, A. S. Cerezo, V. Deulofeu, and A. Cohen, *J. Res. Nat. Bur. Stand.*, **71**, 73 (1967).

88. F. Legay, *C. R. Acad. Sci.*, **234**, 1612 (1952).

89. H. Paulsen and F. Leupold, unpublished results (1966).

90. W. Pigman and K. C. Johnson, *J. Amer. Chem. Soc.*, **75**, 3464 (1953).

91. P. Nánási and R. Bognár, *J. Chem. Soc.*, 323 (1961).

92. P. Nánási and R. Bognár, *Magyar Kem. Foly.*, **67**, 32 (1962).

93. B. Capon and B. E. Connett, *J. Chem. Soc.*, 4492 (1965).

94. C. S. Hudson, *J. Amer. Chem. Soc.*, **31**, 66 (1909).

95. T. L. V. Ulbricht, T. R. Emerson, and R. J. Swan, *Tetrahedron Lett.*, 1561 (1966).

95a. A. Lipták and R. Bognár, *Acta Chim. Acad. Sci. Hung.*, **72**, 309 (1972).

95b. H. von Voethenberg, A. Skrzekewski, J. C. Jochims, and W. Pfleiderer, *Tetrahedron Lett.*, 4063 (1974).

96. B. Capon and B. E. Connett, *J. Chem. Soc.*, 4497 (1965).

96a. T. Jalinski and K. Smiataczowa, *Z. Phys. Chem.*, **235**, 49 (1967); T. Jalinski, K. Smiataczowa, and J. Sokolowski, *Rocz. Chem.*, **42**, 107 (1968).

96b. K. Heyns and W. Beilfuß, *Ber.*, **106**, 2680 (1973).

97. H. Simon and D. Palm, *Ber.*, **98**, 433 (1965); compare H. Simon and G. Philipp, *Carbohyd. Res.*, **8**, 424 (1968).

97a. N. Galicka, K. Smiataczowa, and T. Jasinski, *Rocz. Chem.*, **44**, 411 (1970).

98. F. Shafizadeh, *Advan. Carbohyd. Chem.*, **13**, 24 (1958).

99. F. Micheel and A. Heesing, *Ber.*, **94**, 1814 (1961).

99a. S. Hirano and R. Yamasaki, *Nippon Nogei Kagaku Zasshi*, **39**, 995 (1975).

99b. S. Hirano and R. Yamasaki, *Carbohyd. Res.*, **43**, 377 (1975).

99c. Á. Gerecs, *Acta Chim. Acad. Sci. Hung.*, **84**, 467 (1975).

99d. O. Larm, *Carbohyd. Res.*, **43**, 192 (1975).

100. J. E. Hodge, *Advan. Carbohyd. Chem.*, **10**, 169 (1955).

101. M. Amadori, *Atti Real. Acad. Naz. Lincei*, [6] **2**, 337 (1925); [6] **9**, 68 (1929); [6] **9**, 226 (1929); [6] **13**, 72 (1931).

102. F. Weygand, *Ber.*, **73**, 1278 (1940).
102a. S. Kolka and J. Sokolowski, *Rocz. Chem.*, **46**, 147 (1972).
103. A. Abrams, P. H. Lowy, and H. Borsook, *J. Amer. Chem. Soc.*, **77**, 4794 (1955).
104. L. Rosen, J. W. Woods, and W. Pigman, *Ber.*, **90**, 1038 (1957).
105. H. v. Euler and B. Eistert, "Chemie und Biochemie der Reduktone und Reduktonate," F. Enke Verlag, Stuttgart, 1957.
106. R. Kuhn and A. Dansi, *Ber.*, **69**, 1745 (1936).
107. F. Weygand and A. Bergmann, *Ber.*, **80**, 261 (1947).
108. F. Weygand, *Ber.*, **73**, 1284 (1940).
109. J. Dijong and F. Micheel, *Ann.*, **684**, 216 (1965).
110. R. Kuhn, G. Krüger, and A. Seeliger, *Ann.*, **628**, 240 (1959).
111. J. E. Hodge and B. E. Fischer, *Methods Carbohyd. Chem.*, **2**, 99 (1963).
112. J. Druey and G. Huber, *Helv. Chim. Acta*, **40**, 342 (1957).
113. G. Huber, O. Schier, and J. Druey, *Helv. Chim. Acta*, **43**, 713 (1960).
114. B. Helferich and A. Porck, *Ann.*, **582**, 233 (1953).
115. F. Weygand, H. Simon, and R. von Ardenne, *Ber.*, **92**, 3117 (1959).
115a. K. Heyns, H. Behre, and H. Paulsen, *Carbohyd. Res.*, **5**, 225 (1967).
116. F. Micheel, K. H. Heinemann, K. H. Schwieger, and A. Frowein, *Tetrahedron Lett.*, 3769 (1965).
117. K. Heyns, T. Chiemprasert, and W. Baltes, *Ber.*, **103**, 2877 (1970).
118. K. Heyns and W. Baltes, *Ber.*, **91**, 622 (1958).
119. K. Heyns and W. Baltes, *Ber.*, **93**, 1616 (1960).
120. K. Heyns and W. Schulz, *Ber.*, **93**, 128 (1960).
121. K. Heyns and W. Schulz, *Ber.*, **95**, 709 (1962).
122. A. Gottschalk, *Biochem. J.*, **52**, 455 (1952).
123. K. Heyns and H. Paulsen, *Ann.*, **622**, 160 (1959).
124. E. F. L. J. Anet, *Aust. J. Chem.*, **10**, 193 (1957).
125. K. Heyns and H. Noack, *Ber.*, **95**, 720 (1962); compare P. A. Finot and J. Mauron, *Helv. Chim. Acta*, **52**, 1488 (1969).
126. K. Heyns and H. Noack, *Ber.*, **97**, 415 (1964).
127. E. F. L. J. Anet, *Aust. J. Chem.*, **12**, 280 (1959).
128. K. Heyns, G. Müller, and H. Paulsen, *Ann.*, **703**, 202 (1967).
129. H. Borsook, A. Abrams, and P. H. Lowy, *J. Biol. Chem.*, **215**, 111 (1955).
130. C. H. Doy and F. Gibson, *Biochem. J.*, **72**, 586 (1959).
131. F. Lingens and W. Lück, *Hoppe-Seyler's Z. Physiol. Chem.*, **333**, 190 (1963).
132. F. Lingens and E. Schraven, *Ann.*, **655**, 167 (1962).
133. H. Paulsen, *Ann.*, **683**, 187 (1965); compare H. Paulsen and F. Leupold, *Ber.*, **102**, 2804, 2822 (1969).
134. H. Paulsen, *Angew. Chem.*, **78**, 501 (1966); *Angew. Chem. Intern. Ed. Engl.*, **5**, 495 (1966).
135. H. Paulsen, *Ann.*, **665**, 166 (1963).
136. R. Kuhn and G. Krüger, *Ann.*, **618**, 82 (1958).
137. F. Micheel and V. Hühne, *Ber.*, **93**, 2383 (1960).
138. D. Palm and H. Simon, *Z. Naturforsch.*, **18b**, 419 (1963).
139. D. Palm and H. Simon, *Z. Naturforsch.*, **20b**, 32 (1965).
140. E. Fischer, *Ber.*, **19**, 1920 (1886).
141. R. Kuhn and W. Kirschenlohr, *Ber.*, **87**, 1547 (1954).
142. R. Kuhn and H. J. Haas, *Ann.*, **600**, 148 (1956).
143. K. Heyns and K.-H. Meinecke, *Ber.*, **86**, 1453 (1953).
144. K. Heyns and W. Koch, *Z. Naturforsch.*, **7b**, 486 (1952).

144a. K. Heyns, K.-W. Pflughaupt, and H. Paulsen, *Ber.*, **101**, 2800 (1968); K. Heyns, K.-W. Pflughaupt, and D. Müller, *ibid.*, **101**, 2807 (1968); K. Heyns and W. Beilfuß, *ibid.*, **103**, 2873 (1970).

145. K. Heyns, H. Paulsen, and H. Schroeder, *Tetrahedron*, **13**, 247 (1961).

145a. K. Heyns and W. Beilfuß, *Ber.*, **106**, 2693 (1973).

146. K. Heyns, H. Breuer, and H. Paulsen, *Ber.*, **90**, 1374 (1957).

147. K. Heyns and H. Breuer, *Ber.*, **91**, 2750 (1958).

148. K. Heyns and M. Rolle, *Ber.*, **92**, 2439, 2451 (1959).

149. R. Schumacher, *Dissertation* Hamburg *1961*.

150. E. A. Davidson, *in* "The Amino Sugars" (E. A. Balazs and R. W. Jeanloz, ed.), Vol. IIB, Academic Press, New York, 1966, p. 10.

151. J. E. Hodge, *Advan. Carbohyd. Chem.*, **10**, 191 (1955).

152. R. Kuhn and L. Birkofer, *Ber.*, **71**, 6216 (1938).

153. E. I. du Pont de Nemours and Co., U.S. Patent 2,016,962, *Chem. Abstr.*, **29**, 8007 (1935); VEB Fahlberg-List, German Patent (DDR) 13746; *Chem. Abstr.*, **53**, 11261 (1959).

154. W. Wayne and H. Adkins, *J. Amer. Chem. Soc.*, **62**, 3314 (1940).

155. F. W. Holly, E. W. Peel, R. Mozingo, and K. Folkers, *J. Amer. Chem. Soc.*, **72**, 5416 (1950).

156. F. Kagan, M. A. Rebenstorf, and R. V. Heinzelmann, *J. Amer. Chem. Soc.*, **79**, 3541 (1957).

157. H. Dorn, H. Welfle, and R. Liebig, *Ber.*, **99**, 812 (1966).

158. H. Dorn, G. Bacigalupo, and H. Wand, *Ber.*, **99**, 1208 (1966).

159. L. Maquenne and E. Roux, *C. R. Acad. Sci.*, **132**, 980 (1901); E. Roux, *ibid.*, **134**, 291 (1902); **135**, 691 (1902).

160. M. L. Wolfrom, F. Shafizadeh, J. O. Wehrmüller, and R. K. Armstrong, *J. Org. Chem.*, **23**, 571 (1958).

161. J. W. W. Morgan and M. L. Wolfrom, *J. Amer. Chem. Soc.*, **78**, 2496 (1956).

162. J. C. Sowden, *Advan. Carbohyd. Chem.*, **6**, 291 (1951).

163. J. C. Sowden and H. O. L. Fischer, *J. Amer. Chem. Soc.*, **67**, 1713 (1945).

164. J. C. Sowden and H. O. L. Fischer, *J. Amer. Chem. Soc.*, **68**, 1511 (1946).

165. R. Barker, *J. Org. Chem.*, **29**, 869 (1964); D. D. Heard, B. G. Hudson, and R. Barker, *ibid.*, **35**, 464 (1970); compare M. Blanc-Muesser, J. Defaye, and D. Horton, *Carbohyd. Res.*, **68**, 175 (1979).

166. V. G. Bashford and L. F. Wiggins, *J. Chem. Soc.*, 299 (1948).

167. R. S. Juvet, Jr., *J. Amer. Chem. Soc.*, **81**, 1796 (1959).

168. J. E. Hodge, E. C. Nelson, and B. F. Moy, *J. Agr. Food Chem.*, **11**, 126 (1963).

169. T. M. Reynolds, *Advan. Food Res.*, **12**, 1 (1963).

170. T. M. Reynolds, *Advan. Food Res.*, **14**, 167 (1965).

171. G. P. Ellis, *Advan. Carbohyd. Chem.*, **14**, 63 (1959).

172. J. E. Hodge, *J. Agr. Food Chem.*, **1**, 928 (1953).

173. L. C. Maillard, *C. R. Acad. Sci.*, **154**, 66 (1912); **155**, 1554 (1912); *Ann. Chim.* (Paris), [11] **5**, 258 (1916); **7**, 113 (1917).

174. M. L. Wolfrom, R. C. Schlicht, A. W. Langer, Jr., and C. S. Rooney, *J. Amer. Chem. Soc.*, **75**, 1013 (1953).

175. E. F. L. J. Anet and T. M. Reynolds, *Aust. J. Chem.*, **10**, 182 (1957); **11**, 575 (1958).

176. E. F. L. J. Anet, *Aust. J. Chem.*, **12**, 491 (1959).

176a. E. F. L. J. Anet, *Advan. Carbohyd. Chem.*, **19**, 181 (1964).

177. E. F. L. J. Anet, *Aust. J. Chem.*, **13**, 393 (1960).

178. E. F. L. J. Anet, *Aust. J. Chem.*, **15**, 503 (1962); **18**, 240 (1965); compare M. S. Feather and K. R. Russell, *J. Org. Chem.*, **34**, 2650 (1969).
179. E. F. L. J. Anet, *Aust. J. Chem.*, **14**, 295 (1961).
180. E. F. L. J. Anet, *J. Amer. Chem. Soc.*, **82**, 1502 (1960).
180a. G. Fodor and J. P. Sachetto, *Tetrahedron Lett.*, 401 (1968), proposed a 1,2-hydride shift mechanism; compare critical remarks by E. F. L. J. Anet, *ibid.*, 3525 (1968).
181. H. Kato, *Agr. Biol. Chem.* (Tokyo), **26**, 187 (1962); compare H. Kato, M. Yamamoto, and M. Fujimaki, *ibid.*, **33**, 939 (1969).
182. H. Kato, *Agr. Biol. Chem.* (Tokyo), **27**, 461 (1963).
183. H. Kato and Y. Sakurai, *Agr. Biol. Chem.* (Tokyo), **27**, A 33 (1963); *J. Agr. Chem. Soc. Japan*, **37**, 423 (1963).
184. F. H. Stadtman, C. O. Chichester, and G. Mackinney, *J. Amer. Chem. Soc.*, **74**, 3194 (1952); M. L. Wolfrom, R. C. Schlicht, A. W. Langer, Jr., and C. S. Rooney, *ibid.*, **75**, 1013 (1953).
185. A. Schönberg and R. Moubacher, *Chem. Rev.*, **50**, 261 (1953).
185a. P. S. Song, C. O. Chichester, and F. H. Stadtman, *J. Food Sci.*, **31**, 906 (1966); P. S. Song and C. O. Chichester, *ibid.*, **31**, 914 (1966).
186. J. E. Hodge, B. E. Fisher, and E. C. Nelson, *Amer. Soc. Brewing Chemists Proc.*, 84 (1963).
187. F. Weygand, H. Simon, W. Bitterlich, J. E. Hodge, and B. E. Fisher, *Tetrahedron*, **6**, 123 (1959).
188. J. E. Hodge, U.S. Patent 2,936,308 (1960); *Chem. Abstr.*, **54**, 17281 (1960).
189. H. Simon, G. Heubach, W. Bitterlich, and H. Gleinig, *Ber.*, **98**, 3692 (1965).
190. H. Simon, *Ber.*, **95**, 1003 (1962).
191. H. Simon and G. Heubach, *Ber.*, **98**, 3703 (1965).
191a. G. A. M. van den Ouweland and H. G. Peer, *Rec. Trav. Chim.* (Pays-Bas), **89**, 750 (1970).
192. J. E. Hodge and E. C. Nelson, *Cereal Chem.*, **38**, 207 (1961).
193. B. E. Fisher and J. E. Hodge, *J. Org. Chem.*, **29**, 776 (1964).
194. F. D. Mills, B. G. Baker, and J. E. Hodge, *J. Agr. Food Chem.*, **17**, 723 (1969); *Carbohyd. Res.*, **15**, 205 (1970).
195. F. D. Mills, D. Weisleder, and J. E. Hodge, *Tetrahedron Lett.*, 1243 (1970).
196. G. R. Jurch, Jr., and J. H. Tatum, *Carbohyd. Res.*, **15**, 233 (1970).
197. P. E. Shaw, J. H. Tatum, and R. E. Berry, *Carbohyd. Res.*, **16**, 207 (1971).
198. H. Kato and M. Fujimaki, *J. Food Sci.*, **33**, 445 (1968); H. Kato, *Agr. Biol. Chem.* (Tokyo), **31**, 1086 (1967).
199. H. Kato, H. Shigematsu, T. Kurata, and M. Fujimaki, *Agr. Biol. Chem.* (Tokyo), **36**, 1639 (1972); H. Kato and M. Fujimaki, *Lebensm.-Wiss. Technol.*, **5**, 172 (1972).
200. K. Heyns, J. Heukeshoven, and K.-H. Brose, *Angew. Chem.*, **80**, 627 (1968); *Angew. Chem. Internat. Ed.*, **7**, 628 (1968).
201. P. A. Finot, J. Bricout, R. Viani, and J. Mauron, *Experientia*, **25**, 134 (1969).
202. T. Ozawa and N. Kinae, *Yakugaku Zasshi*, **90**, 665 (1970).

21. HYDRAZINE DERIVATIVES AND RELATED COMPOUNDS

L. Mester and H. S. El Khadem

I. INTRODUCTION

Glycoses react with hydrazine derivatives to give a variety of products, the nature of which depends on the particular hydrazine and sugar derivative used (and on their ratios), on the nature of the medium, and on the time of reaction. This complexity results because initial nucleophilic attack by nitrogen on the potential carbonyl group of the sugar, to give a Schiff-base type of product, is often followed by rearrangement. Reaction with a second hydrazine molecule is also possible—for example, in the formation of osazones. This process may proceed even further and involve the total length of the glycose chain, leading to alkazones [poly(hydrazones)]. A logical classification of the products of glycose–hydrazine reactions is according to the number of hydrazine residues that are ultimately combined to the sugar residue. These types include: (1) azines, which are formed by the combination of one hydrazine molecule with two sugar molecules; (2) hydrazones, obtained when one hydrazine molecule combines with one sugar molecule; (3) bis(hydrazones), which have two hydrazone residues; and (4) poly(hydrazones), which have more than two hydrazone residues linked to each sugar residue.

The type of product formed is strongly dependent on the type of hydrazine used. Table I shows the derivatives obtained from various types of hydrazine and sugar. Semicarbazide and related compounds have been included with acylhydrazines.

In addition to the condensation products listed in Table I, phenylhydrazine is capable of degrading periodate-oxidized polysaccharides to give hydrazine derivatives of the monomeric fragments; these products are discussed in Section VI together with the hydrazones of periodate-oxidized monosaccharides.

Additionally there are the oximes, a group of compounds closely related to hydrazones and obtained by condensation of reducing sugars with hydroxylamine. Also included are the diazo sugars, which can be regarded as dehydro analogue of sugar hydrazones, and which have been applied in the synthesis of ketoses; and the azido derivatives, useful precursors of amino, imino, and aldehydo derivatives of sugars.

TABLE I

HYDRAZINE DERIVATIVES OF SUGARS

	Hydrazine derivative			
Product	Unsubstituted	Alkyl and acyl	Aroyl	Aryl
	Type of sugar precursor [a]			
Azine	A, B	—	—	—
Hydrazone	A	A	A, B, C	A, B, C
Bis(hydrazone)	—	C[b]	C	A, B, C
Poly(hydrazone)	—	—	D	A, D

[a] A, aldose; B. ketose; C, aldosulose; D, diulose.
[b] Only acylhydrazines.

II. AZINES

Aldoses, ketoses, and disaccharides react with hydrazine in methanol to give colorless azines, which on prolonged treatment with excess hydrazine give the hydrazones.[1,2]

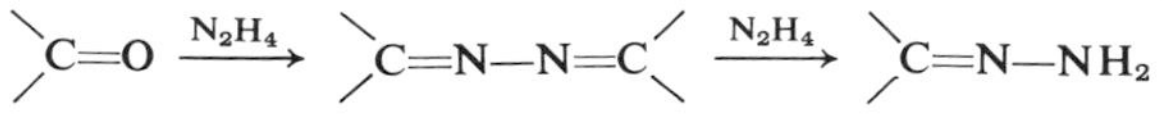

$$\text{>C=O} \xrightarrow{\text{N}_2\text{H}_4} \text{>C=N-N=C<} \xrightarrow{\text{N}_2\text{H}_4} \text{>C=N-NH}_2$$

Azine   Hydrazone

In the case of aldosulose 1-(N-methyl-N-phenylhydrazones), hydrazine alone yields the hydrazone, from which the azine may be obtained, whereas hydrazine and acetic acid yield an azine directly.[3]

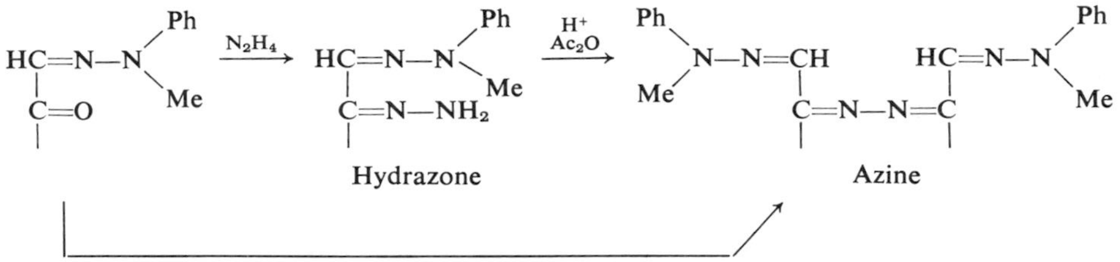

Hydrazone   Azine

On acetylation, the azines yield acetates[2] that have O-acetyl but no N-acetyl groups, indicating that they have acyclic structures. On the other hand, the unacetylated azines probably exist as equilibrated mixtures of cyclic and acyclic forms; a strong absorption band at 260 nm and a C=N stretching band at 1620 cm^{-1} are indicative that the acyclic form is preponderant. Photolysis of the azine of D-galactose in methanol leads to D-galactose with concomitant chain degradation to give also D-lyxose.[3a]

III. HYDRAZONES

Saccharide hydrazones can be obtained by interaction of unsubstituted hydrazine,[2,4] or monosubstituted hydrazines having alkyl,[5] acyl,[6] aroyl,[7] sulfonyl,[8,9] aryl,[10–13] or heterocyclic[14] groups, or disubstituted N,N-dialkyl-[5] or N,N-diaryl-[15] or N-alkyl-N-arylhydrazines,[9,16–18] with aldoses, ketoses, aldosuloses, or disaccharides.[19,20]

Semicarbazones[6] and thiosemicarbazones[21] are included with hydrazones, but hydrazones of periodate-oxidized monosaccharide derivatives will be discussed in Chapter 25 and on p. 969, together with other hydrazine derivatives obtained by the hydrazinolysis of periodate-oxidized polysaccharides.

A. FORMATION

Unsubstituted saccharide hydrazones are prepared by prolonged heating of sugars with a large excess of methanolic hydrazine. The excess reagent is used to obviate the formation of azines.[2,4] Substituted hydrazones are prepared by heating equimolar amounts of the saccharide and the substituted hydrazine as the free base,[8,9,22] or by treating cold, aqueous solutions of the sugar with a weakly acidic solution of the acetate salt of the substituted hydrazine (from the hydrochloride salt and sodium acetate). Aldosulose monohydrazones are prepared either by treating aldosuloses with one mole of hydrazine[23] or by removing one hydrazone residue from osazones by using nitrous acid[24] or copper(II) sulfate.[25] Both reactions must be performed under mild conditions to prevent the formation of the bis(hydrazone) in the first case and the aldosulose in the second.

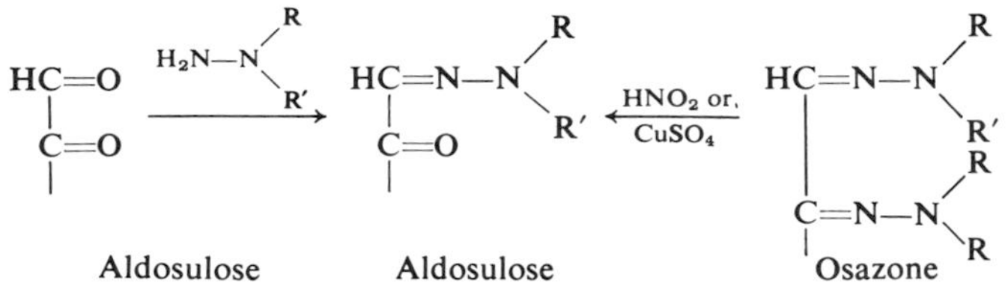

Aldosulose   Aldosulose   Osazone
       monohydrazone

Hydrazones are formed most rapidly at pH 4 to 5 in the presence of high concentrations of a buffer. Phosphate ions are reported to have a greater catalytic effect than acetate ions. Hydrochloric acid catalyzes formation of hydrazones but not of osazones, particularly in the absence of air.[22,26,27] The reaction presumably proceeds (pathway A) by nucleophilic attack of the hydrazine at the hemiacetal carbon atom of a protonated form of the sugar, giving hydrazonium intermediates which lead ultimately to the hydrazone. An alternative possible route[28] (pathway B) would involve the acyclic form of the sugar. Weakly basic hydrazines are better suited for this condensation than strongly basic ones, since they form salts less readily and their addition products can split out a proton more easily.[8,9,17,28]

The reaction has been investigated kinetically. One report suggested a second-order reaction,[22] but it seems that the reaction is pseudo-unimolecular[29] when a solution of hydrazine hydrochloride buffered with acetate ions is used. The rates of formation of hydrazone for *aldehydo*-D-galactose penta-acetate and two isomeric cyclic tetraacetates of D-galactose having the C-1 hydroxyl group free are quite different, being highest for the acyclic acetate, denoting that the rate-determining step is either the opening of the rings (for the cyclic acetates) to form the acyclic derivatives, or the direct reaction of the original cyclic forms with the substituted hydrazine.[29]

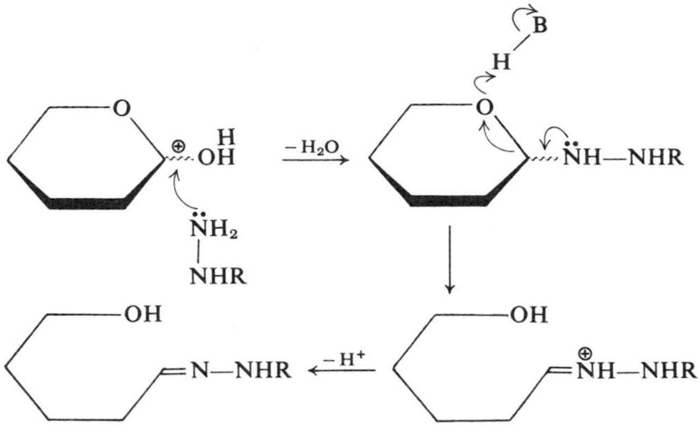

Pathway A

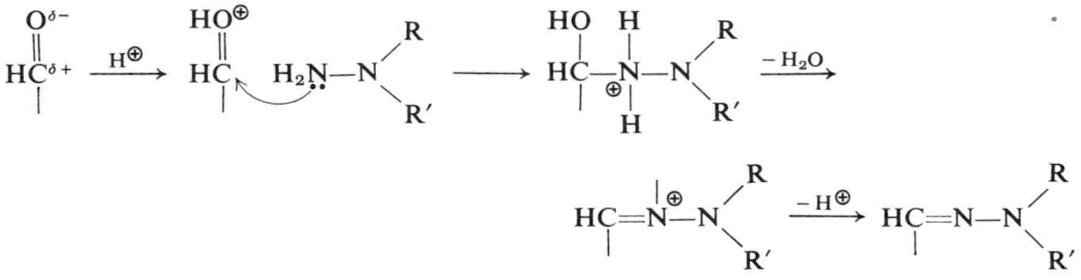

Pathway B
Formation of Hydrazones

Detailed studies on the mechanism of formation of various hydrazones[8,9,17,28,30] revealed, among other things, that when a given sugar reacts with various substituted hydrazines the rate of formation of hydrazone depends on the basicity of the hydrazine, the rate decreasing with increasing basicity. For a given hydrazine reacting with different sugars, the rate of formation of hydrazone follows roughly the rate of ring opening of the pyranose forms of these sugars.

B. STRUCTURE

Hydrazones of sugars are capable of existing in various tautomeric forms; evidence for these was early recognized from the complex mutarotation curves, which usually pass through maxima or minima, for such hydrazones in solution.[23,24,31,32] The failure of the mutarotational behavior to follow the first-order equation indicates that more than two substances take part in the equilibrium.

The principal structures encountered in hydrazones of sugars are the acyclic Schiff-base type and the cyclic hydrazino forms. In some cases three forms

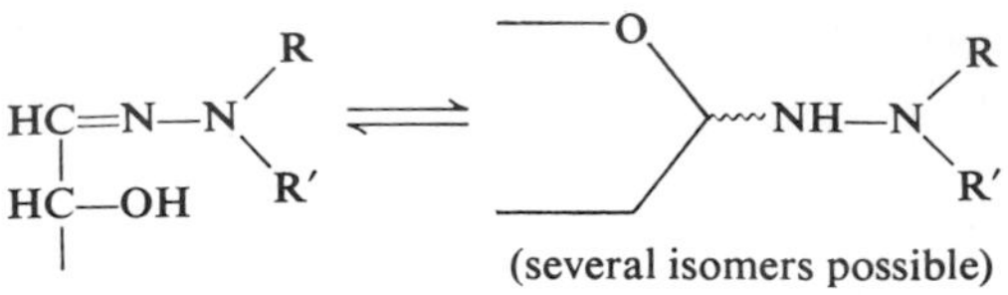

$$HC{=}N{-}N\overset{R}{\underset{R'}{\diagdown}} \rightleftharpoons \overset{\overline{O}}{\diagup}{\diagdown}\hspace{-0.3em}\sim\hspace{-0.3em}NH{-}N\overset{R}{\underset{R'}{\diagdown}}$$

(several isomers possible)

Tautomerism of Hydrazones

have been isolated because of the possibility for anomeric and ring-size isomerization with the cyclic forms. The best known example where three forms are involved is the case of the three isomeric phenylhydrazones of D-glucose, designated "α," "β," and "γ" forms.[33] Schiff-base derivatives are suitable for polarographic determination, and this method was used to estimate the percentages of the acyclic forms present in various oximes and semicarbazones of sugars (Table II).[34] The values accord with those obtained for the corresponding phenylhydrazones by the formazan method (see p. 940).

Ultraviolet and infrared spectroscopy (see Vol. IB, Chapter 27) have been used to detect acyclic hydrazones of sugars by making use[2,4] of the absorption in the u.v. of the $C{=}N$ chromophore at 260 nm, and of the infrared $C{=}N$ stretching vibrations at[2,4] 1620 cm^{-1}. These absorptions are applicable with the unsubstituted hydrazones, and with hydrazones having no phenyl rings. Aryl groups give rise to similar absorptions in the u.v., and their $C{=}C$ vibrations may be mistaken for $C{=}N$ ones.[35] Nuclear magnetic resonance spectroscopy is also useful, since the cyclic hydrazones invariably have one imino proton more than the acyclic hydrazones.[36,37]

Another method of establishing the cyclic or acyclic structures depends on the reaction of benzenediazonium chloride with acyclic phenylhydrazones to

TABLE II

POLAROGRAPHICALLY ESTIMATED PERCENT OF ACYCLIC FORM
OF SUGAR DERIVATIVE PRESENT IN SOLUTION

Sugar	Acyclic form (%)	
	Oxime	Semicarbazone
Glucose	47	12
Galactose	71	36
Mannose	100	53
Xylose	100	100
Arabinose	100	100
Lyxose	100	100
Ribose	100	100

form diphenylformazans; no well-defined product is obtained from the cyclic isomers.[38] Thus, for example, "β"-D-glucose phenylhydrazone yields D-glucose diphenylformazan, whereas no reaction is observed with the other two (cyclic) isomers. The reaction can also be used for estimating the proportions of acyclic forms during mutarotation of sugar phenylhydrazones.[12,36,38]

Chemical evidence for the open-chain structure of D-galactose hydrazones was provided by the reaction of N-methyl-N-phenylhydrazine with 2,3,4,6-tetra-O-acetyl-D-galactopyranose (1), 2,3,5,6-tetra-O-acetyl-D-galactofuranose (5), and penta-O-acetyl-aldehydo-D-galactose (4). The hydrazones (2 and 6) formed from 1 and 5 could be converted by acetylation into the same penta-O-acetyl-D-galactose N-methyl-N-phenylhydrazone (3) as obtained from 4, and so it was concluded that the hydrazones 2 and 6 have open-chain structures.[29]

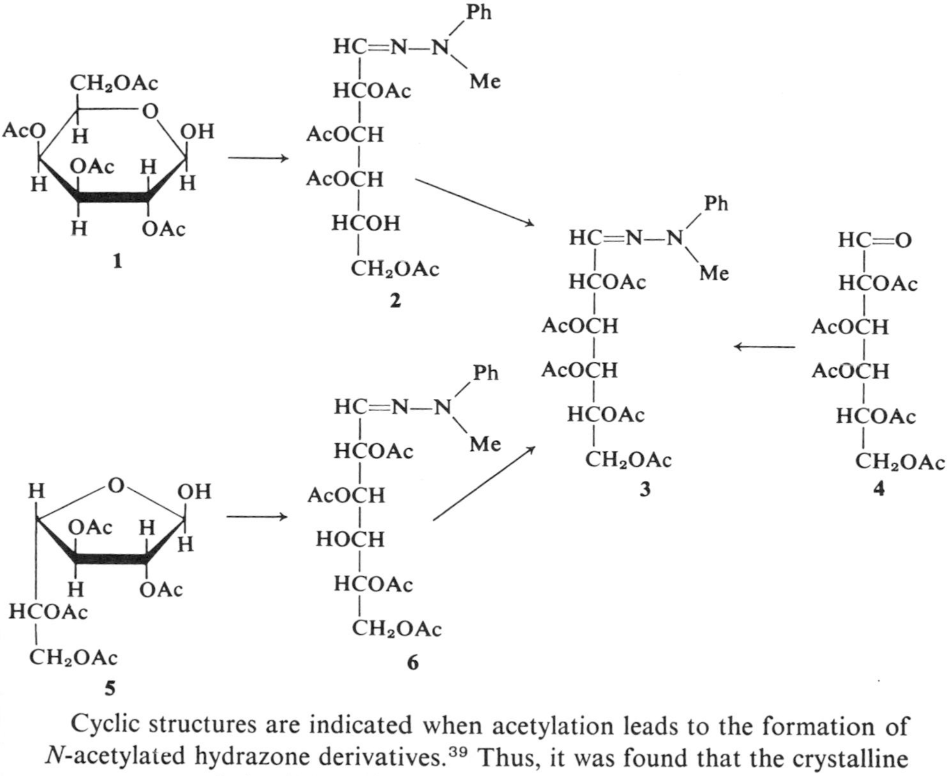

Cyclic structures are indicated when acetylation leads to the formation of N-acetylated hydrazone derivatives.[39] Thus, it was found that the crystalline pentaacetate of the "α"-D-glucose phenylhydrazone was a cyclic isomer, because removal of the phenylhydrazone group gave rise to N-acetyl-N-phenylhydrazine, indicating that one acetyl group was attached to a nitrogen atom. The acetylated "β" isomer on similar treatment gives phenylhydrazine, suggesting that it is acyclic. The distinction between OAc and NAc groups by chemical means is quite tedious, but the groups are readily differentiated by spectroscopy (OAc ν = 1740, NAc ν = 1690 cm^{-1}).

The Schiff-base structure of the acetylated semicarbazones of D-glucose and D-mannose has been confirmed by preparing the same acetylated semi-carbazone **8**, either from the 2,3,4,5,6-penta-*O*-acetylaldose (**9**) and semi-carbazide, or from the semicarbazone **7** by acetylation.[40,44] In contrast, acetylation of the semicarbazones of maltose and cellobiose leads to cyclic derivatives.

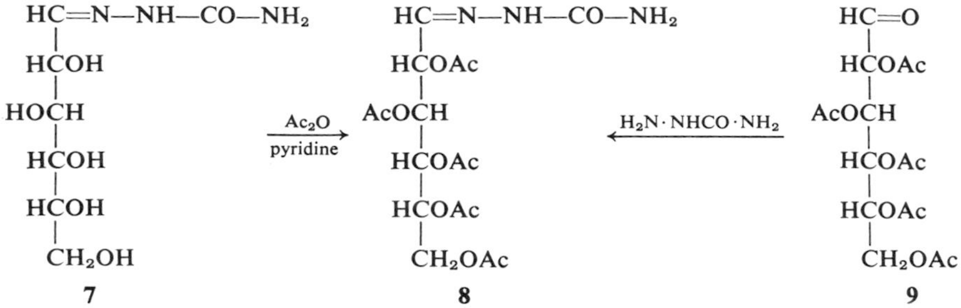

HC=N—NH—CO—NH$_2$ HC=N—NH—CO—NH$_2$ HC=O

HCOH		HCOAc	HCOAc
HOCH	$\xrightarrow[\text{pyridine}]{\text{Ac}_2\text{O}}$ AcOCH	$\xleftarrow{\text{H}_2\text{N·NHCO·NH}_2}$	AcOCH
HCOH	HCOAc		HCOAc
HCOH	HCOAc		HCOAc
CH$_2$OH	CH$_2$OAc		CH$_2$OAc
7	**8**		**9**

Isonicotinoylhydrazine reacts with sugars to give cyclic[45] or acyclic[46] sugar isonicotinoylhydrazones, which equilibrate in solution. The acyclic, acetylated isonicotinoylhydrazones of sugars having confirmed Schiff-base structures are obtained from acetylated *aldehydo*-aldoses.[47]

Aldosuloses react with substituted hydrazines to give mono- and bis-hydrazones. The monohydrazones have the hydrazone residues attached to C-1. This position of substitution was established conclusively with D-*arabino*-hexosulose mono-(*N*-methyl-*N*-phenyl)hydrazone (**10**) by hydrogenating it to give D-mannose *N*-methyl-*N*-phenylhydrazone[48] (**11**).

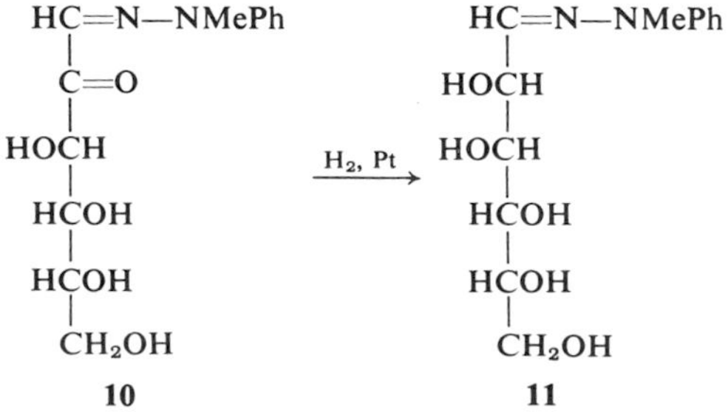

HC=N—NMePh   HC=N—NMePh

C=O		HOCH
HOCH	$\xrightarrow{\text{H}_2,\ \text{Pt}}$	HOCH
HCOH		HCOH
HCOH		HCOH
CH$_2$OH		CH$_2$OH
10		**11**

The "benzyl rule" correlates the configuration of the C-2 hydroxyl group with the sign of rotation of certain aldose hydrazones. If, in the Fischer

projection formula, the C-2 hydroxyl group is to the right, the corresponding
N-benzyl-N-phenylhydrazone is levorotatory; if the group is to the left, the
derivative is dextrorotatory.[49-52]

C. REACTIONS

1. Cleavage of the Hydrazone Residue

This reaction is useful for the purification of sugars from other contaminants.
A suitable hydrazone is prepared, purified by crystallization, and decomposed
to regenerate the starting sugar pure.[20] Useful reagents for removal of
hydrazone residues are: hydrochloric acid, benzaldehyde, copper sulfate, and
nitrous acid. Many hydrazones, however, are too water-soluble to be useful

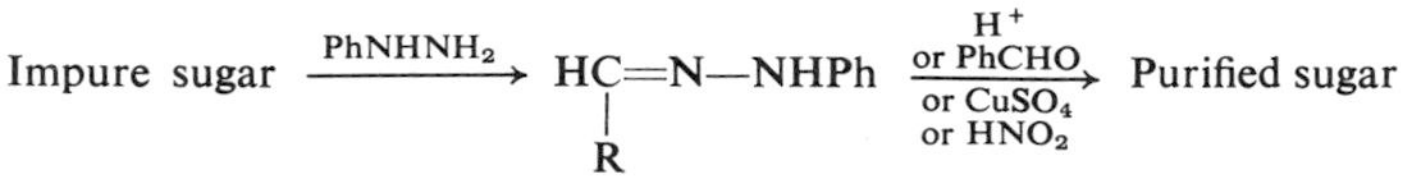

$$\text{Impure sugar} \xrightarrow{\text{PhNHNH}_2} \underset{\underset{\text{R}}{|}}{\text{HC}}=\text{N—NHPh} \xrightarrow[\substack{\text{or CuSO}_4 \\ \text{or HNO}_2}]{\substack{\text{H}^+ \\ \text{or PhCHO}}} \text{Purified sugar}$$

Purification of Sugars by Way of Hydrazones

in this regard. Some of the less-soluble ones are very suitable for the characterization of sugars.[53,54] Another application of this reaction is for the
preparation of acetylated *aldehydo*-aldoses by using, for example, semicarbazone acetates and removing the semicarbazone residues with sodium
nitrite.[40-44]

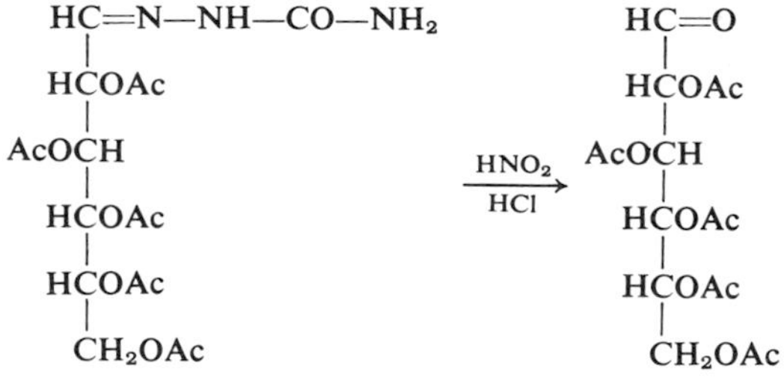

$$
\begin{array}{ccc}
\text{HC}=\text{N—NH—CO—NH}_2 & & \text{HC}=\text{O} \\
| & & | \\
\text{HCOAc} & & \text{HCOAc} \\
| & & | \\
\text{AcOCH} & \xrightarrow[\text{HCl}]{\text{HNO}_2} & \text{AcOCH} \\
| & & | \\
\text{HCOAc} & & \text{HCOAc} \\
| & & | \\
\text{HCOAc} & & \text{HCOAc} \\
| & & | \\
\text{CH}_2\text{OAc} & & \text{CH}_2\text{OAc}
\end{array}
$$

Preparation of Acetylated Aldehyde Sugars

2. Action of Hydrazines

Arylhydrazines react in hot solution with aldose and ketose hydrazones in
the presence of acidic acetate buffers to give simple or mixed osazones. This
reaction is not suitable for the formation of mixed osazones because transhydrazonation may lead to partial and sometimes to complete exchange of

References start on p. 981.

hydrazone residues. To prepare mixed osazones, aldosulose monohydrazones are treated in the cold with substituted hydrazines.[26,27,55]

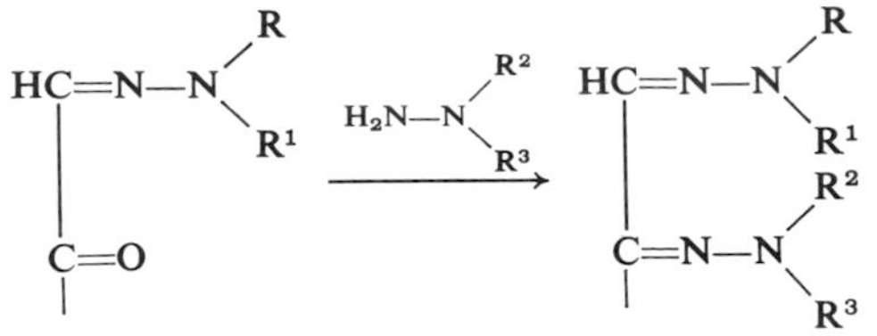

Preparation of Mixed Osazones

3. Action of Bases

Bases cause degradative fission of the saccharide chain adjacent to the hydrazone residue, and thus alcoholic potassium hydroxide gives with phenylhydrazones a cyclization product, 1-phenylpyrazole[56] (12). Hot pyridine leads, instead, to forming a dimerization product, glyoxal bis(phenylhydrazone)[57] (13). Hydrogen–tritium exchange experiments have revealed that in alkaline medium, hydrazones undergo a series of rapid tautomerizations, which may lead to these degradation products.[57a]

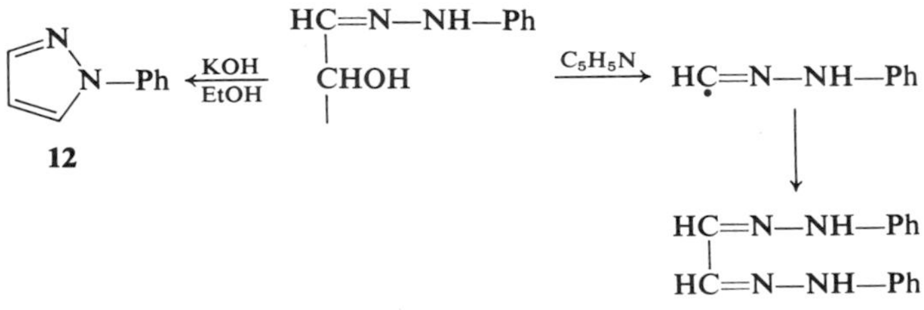

4. Reduction

Reduction of saccharide hydrazones can be made to afford the corresponding 1-amino-1-deoxyalditols. The phenylhydrazones have been found useful in this reaction, especially when catalytic hydrogenation is used.[58]

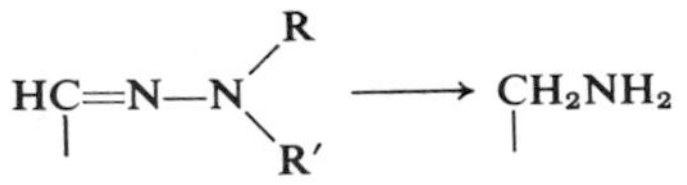

Reduction of Hydrazones

5. Derivatives

Acetylation of saccharide hydrazones with acetic anhydride in pyridine affords the per-O-acetyl derivatives with acyclic hydrazines, and the N-acetyl-O-acetyl derivatives with the cyclic ones.[7,29,39,57] Under these conditions an NH group attached to the aryl moiety is not acetylated, but an NH group attached to the sugar moiety is. The O-acetyl groups are split off by base more readily than the N-acetyl groups,[59] and methods for selective saponification have been devised. With acetyl chloride in N,N-dimethylaniline the NH group attached to the aryl residue becomes acylated and the acyclic hydrazones give N-acetyl-O-acetyl derivatives[57]; similarly, benzoylation with benzoyl chloride in pyridine affords N-benzoyl-O-benzoyl derivatives.[60]

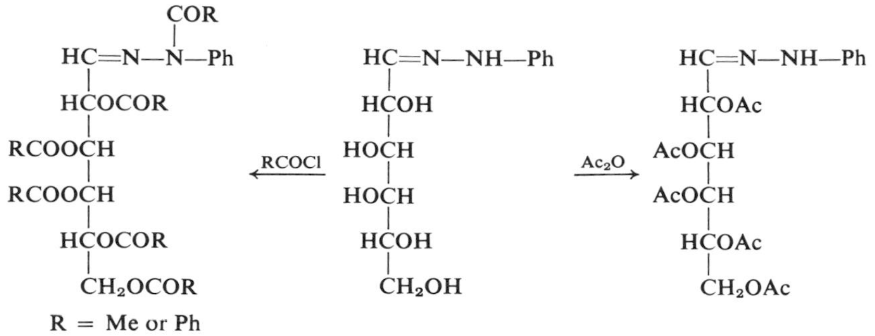

$$
\begin{array}{ccc}
& & \overset{\displaystyle \text{COR}}{\underset{\displaystyle |}{\;}} \\
\text{HC}{=}\text{N}{-}\text{N}{-}\text{Ph} & \text{HC}{=}\text{N}{-}\text{NH}{-}\text{Ph} & \text{HC}{=}\text{N}{-}\text{NH}{-}\text{Ph} \\
| & | & | \\
\text{HCOCOR} & \text{HCOH} & \text{HCOAc} \\
| & | & | \\
\text{RCOOCH} \xleftarrow{\text{RCOCl}} & \text{HOCH} \xrightarrow{\text{Ac}_2\text{O}} & \text{AcOCH} \\
| & | & | \\
\text{RCOOCH} & \text{HOCH} & \text{AcOCH} \\
| & | & | \\
\text{HCOCOR} & \text{HCOH} & \text{HCOAc} \\
| & | & | \\
\text{CH}_2\text{OCOR} & \text{CH}_2\text{OH} & \text{CH}_2\text{OAc} \\
\text{R} = \text{Me or Ph}
\end{array}
$$

Acylation of Arylhydrazones

6. Formation of Azoethylenes

The penta-O-acetyl derivatives of D-galactose and D-mannose aryl-hydrazones, when warmed for a short time with small amounts of pyridine, lose the elements of acetic acid and afford 3,4,5,6-tetra-O-acetyl-1-phenylazo-*trans*-1-hexenes. The structures of these compounds have been confirmed by n.m.r. spectroscopy.[57,61–63]

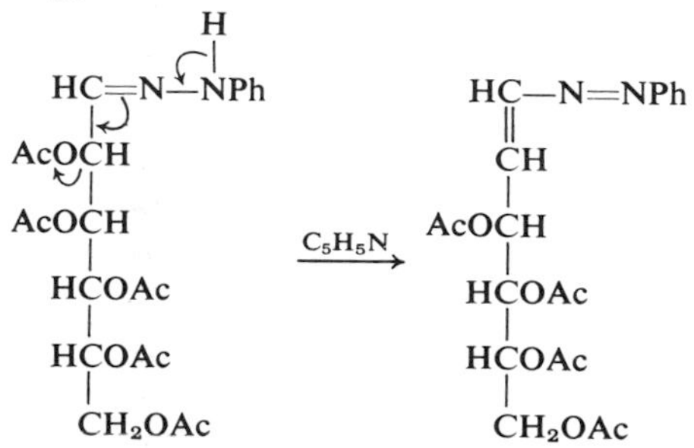

$$
\begin{array}{cc}
\overset{\displaystyle \text{H}}{\underset{\displaystyle |}{\;}} & \\
\text{HC}{=}\text{N}{-}\text{NPh} & \text{HC}{-}\text{N}{=}\text{NPh} \\
| & \| \\
\text{AcOCH} & \text{CH} \\
| & | \\
\text{AcOCH} \xrightarrow{\text{C}_5\text{H}_5\text{N}} & \text{AcOCH} \\
| & | \\
\text{HCOAc} & \text{HCOAc} \\
| & | \\
\text{HCOAc} & \text{HCOAc} \\
| & | \\
\text{CH}_2\text{OAc} & \text{CH}_2\text{OAc} \\
\text{Arylhydrazone} & \text{Azoethylene derivative}
\end{array}
$$

References start on p. 981.

D. Formazans

1. Formation of Formazans[64]

Aldose phenylhydrazones react in pyridine with cold solutions of aryldiazonium salts to give brilliant red, crystalline formazan derivatives of the sugars.[38,64a] Two structural features are required for formation of formazans:

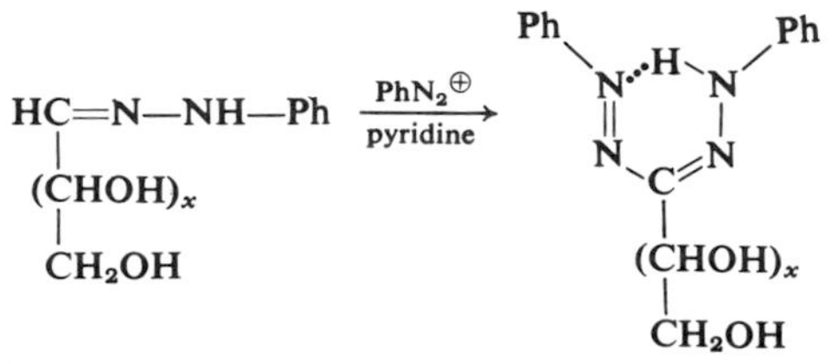

$$\text{HC}=\text{N}-\text{NH}-\text{Ph} \xrightarrow[\text{pyridine}]{\text{PhN}_2{}^{\oplus}}$$

$$\underset{\displaystyle \overset{|}{\text{CH}_2\text{OH}}}{\overset{|}{(\text{CHOH})_x}}$$

Aldose phenylhydrazone   Aldose diphenylformazan

(1) the presence of an aldehyde arylhydrazone (Schiff-base) type of structure, and (2) the presence of a free methine hydrogen atom in the arylhydrazone group.

Thus, aldose phenylhydrazones in the cyclic hemiacetal forms, and ketose phenylhydrazones, do not give the formazan reaction. N,N-Disubstituted hydrazones, such as aldose N-methyl-N-phenylhydrazones, also fail to yield formazans.

2. Structure

Formazans are known to have a chelated structure, which permits the equilibration of monosubstituted isomers through fast tautomerization.[65] For example, attempts to prepare two isomeric N-phenyl-N'-p-bromophenyl-formazans of D-galactose by the routes shown (p. 941) gave, in fact, the same product.[66]

Investigation of sugar formazans and their acetates by circular dichroism may permit establishment of the configuration at the C-2 atom and the conformation of the sugar chain.[66a]

3. Reactions

Formazans react with salts of heavy metals to give stable complexes[67]; they are also oxidized by N-bromosuccinimide, or (after acetylation) by lead tetraacetate to yield tetrazolium salts.[68–70] The latter give back the formazans on reduction. (See illustration on p. 942.)

Reductive decomposition of sugar formazans with hydrogen sulfide leads to aldothionic acid phenylhydrazines.[71,72] The mechanism that has been proposed for the reaction[72a] is depicted on p. 942.

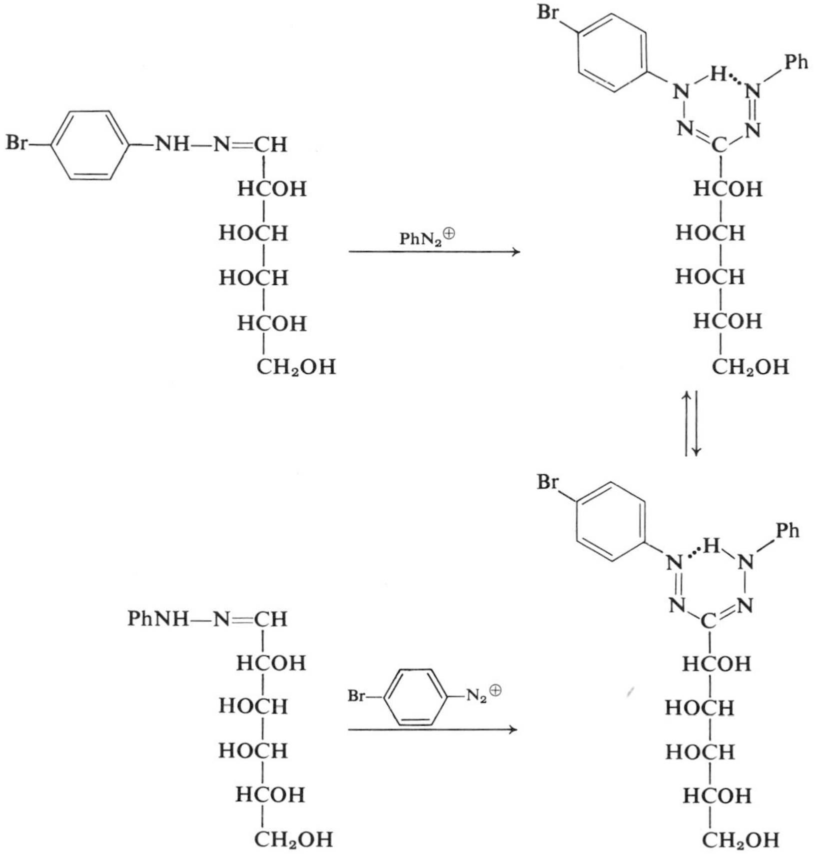

Formation of Mixed Formazans

Thioaldonic acid phenylhydrazides, treated for one or two minutes with benzaldehyde in the presence of hydrochloric acid as catalyst, yield sugar thiadiazoline derivatives.[71]

Because of their sharp melting points and characteristic crystalline forms, sugar formazans can be used to identify aldoses[73] (ketoses do not give formazans).

To identify aldoses, it is usually not necessary to isolate the hydrazone; the sugar is treated with phenylhydrazine, and the mixture is allowed to react with a diazonium salt in the presence of pyridine to give the formazan.

References start on p. 981.

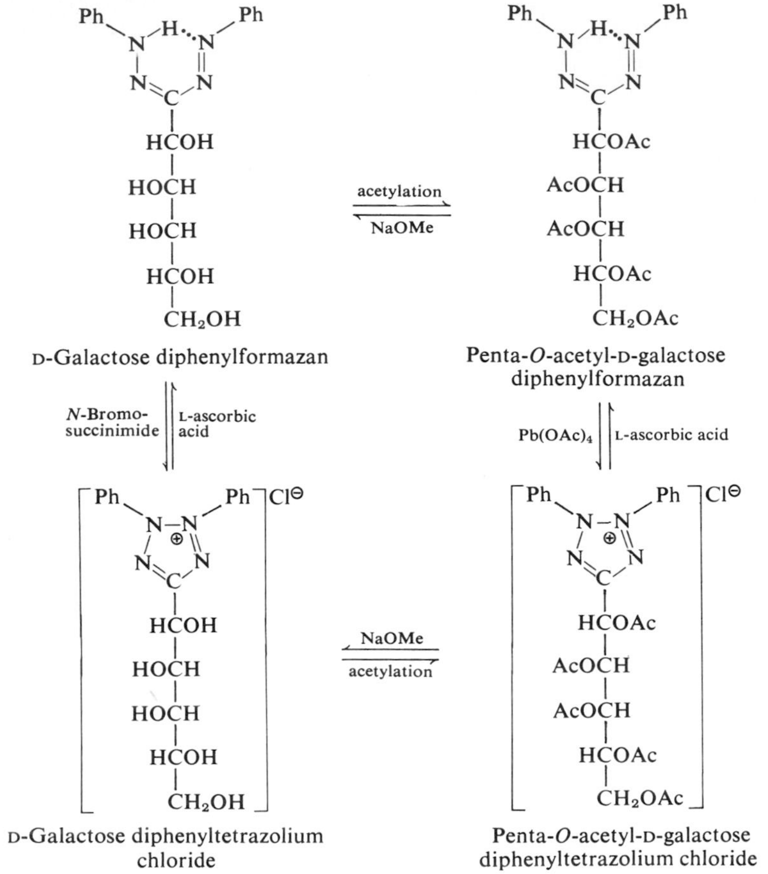

D-Galactose diphenylformazan

Penta-O-acetyl-D-galactose diphenylformazan

acetylation / NaOMe

N-Bromo-succinimide / L-ascorbic acid

Pb(OAc)₄ / L-ascorbic acid

D-Galactose diphenyltetrazolium chloride

Penta-O-acetyl-D-galactose diphenyltetrazolium chloride

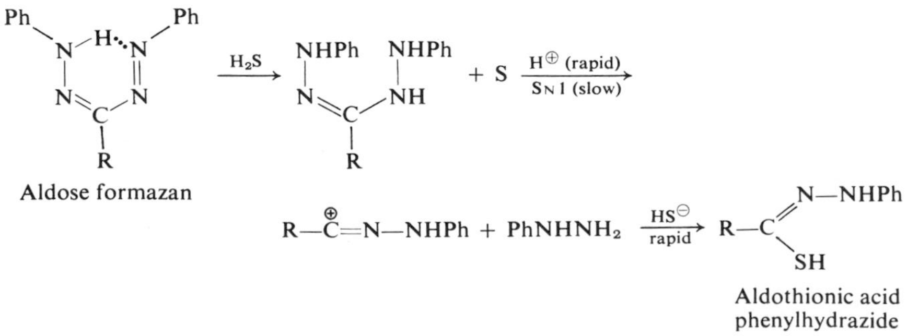

NaOMe / acetylation

Aldose formazan

H_2S

$+ S$ $\xrightarrow{H^{\oplus} \text{ (rapid)}}_{S_N I \text{ (slow)}}$

$R-\overset{\oplus}{C}=N-NHPh + PhNHNH_2 \xrightarrow[\text{rapid}]{HS^{\ominus}} R-C$

Aldothionic acid phenylhydrazide

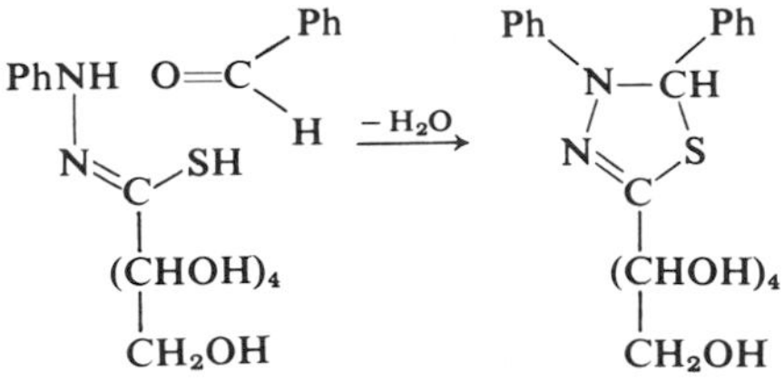

Formation of Sugar Thiadiazoline

TABLE III

Melting Point and Crystal Forms of the Diphenylformazans of Some Sugars

Diphenylformazan of	M.p. (°C)	Crystal form
D-Glucose	177–178	Red needles, frequent rosettes
D-Galactose	167–168	Bronze-red tablets
D-Mannose	174–175	Bunches of microscopic russet needles
L-Arabinose	173–174	Bunches of fine, bright-red needles
L-Rhamnose	175–176	Brilliant red needles
D-Xylose	123–124	Lanceolate red needles

Aldoses can be recovered from the formazans via the thioaldonic acid phenylhydrazides, through reduction of the corresponding aldonolactones.[74]

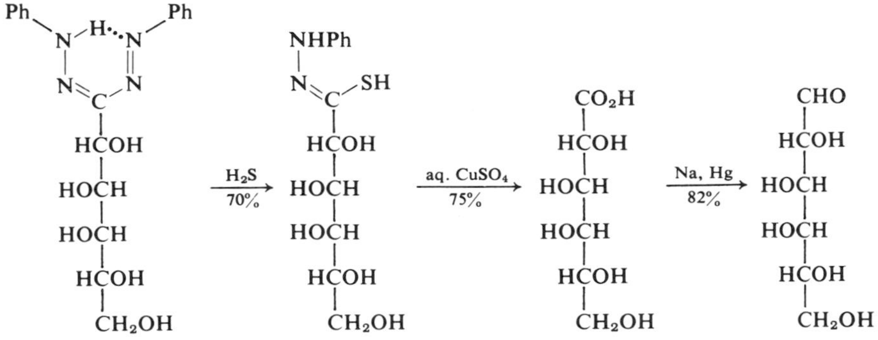

Recovery of Aldoses from Formazans

Sugar diphenylformazans produce highly colored, resonance-stabilized cations in the presence of strong acids such as perchloric acid in acetic acid.[73a] Fluoroboric acid in acetic anhydride transforms penta-*O*-acetyl-D-galactose diphenylformazan into mono-*N*-acetyl-penta-*O*-acetyl-D-galactose diphenylformazan. On saponification, a 2,6-anhydro-D-galactose diphenylformazan is obtained.[73b]

References start on p. 981.

IV. 1,2-BIS(HYDRAZONES) INCLUDING OSAZONES

A. FORMATION

1,2-Bis(hydrazones) are formally derivatives of aldosuloses. However, the most familiar examples are the arylosazones,[75] which are usually prepared by treating monocarbonyl sugars, such as aldoses, ketoses, and disaccharides with aryl-, N-alkyl-N-aryl-, and N,N-diarylhydrazines. Alkyl- or acyl-hydrazines generally form monohydrazones from monocarbonyl sugars and require aldosuloses as starting materials to form bis(hydrazones) in reasonable yield. Mixed bis(hydrazones) can be obtained by interacting aldosulose mono-hydrazones with differently substituted hydrazines.

Aldoses and ketoses react with aroylhydrazines in the presence of aromatic amines to give first the bis(aroylhydrazones)[74a] and, on prolonged treatment, the 3-deoxyaldos-2-ulose bis(aroylhydrazones). The latter can be converted with benzaldehyde into the 3-deoxyaldos-2-uloses.[74b]

The reaction between aldosuloses or aldosulose monohydrazones and hydrazines is usually effected with equimolar quantities in the cold and in the presence of acetic acid. On the other hand, for forming osazones from aldoses, ketoses, and disaccharides, an excess of arylhydrazine is used at 100° in the presence of acetate buffers at pH 4 to 6; in more acid solutions, and with the free base[75-79] without buffer, the monohydrazone is formed.[23,24] Hydrazines having electron-withdrawing groups, such as p-nitrophenylhydrazine, favor formation of osazones, whereas electron-releasing groups inhibit formation of osazones; methylhydrazine fails to give such derivatives.[80]

The nomenclature of bis(hydrazones) has long been nondefinitive; thus the names glucosazone, glucose phenylosazone, fructose phenylosazone, and mannose phenylosazone have been used for the same compound. The name from the corresponding ketose[81] contains no redundant stereochemical information, and so D-*arabino*-hexulose phenylosazone is the accepted semi-trivial name for this compound. The fully systematic name is that of a bis(hydrazone) of the parent aldosulose, as, in this case, D-*arabino*-hexosulose bis(phenylhydrazone) (see also Vol. IIB, Chapter 47).

B. MECHANISM OF OSAZONE FORMATION

During formation of an osazone, three moles of hydrazine are consumed per mole of sugar, one mole being reduced during the reaction to yield one mole each of aniline and ammonia.[82,83] To explain this reaction, Emil Fischer proposed the following mechanism.[83]

$$
\begin{array}{c}
\underset{|}{\text{HC}=\text{O}} \\
\underset{|}{\text{HCOH}}
\end{array}
\xrightarrow{\text{PhNHNH}_2}
\begin{array}{c}
\underset{|}{\text{HC}=\text{N}-\text{NHPh}} \\
\underset{|}{\text{HCOH}}
\end{array}
\xrightarrow{\text{PhNHNH}_2}
\begin{array}{c}
\underset{|}{\text{HC}=\text{N}-\text{NHPh}} \\
\underset{|}{\text{C}=\text{O}}
\end{array}
\;+\;\text{NH}_3\;+\;\text{PhNH}_2
$$

$$\big\downarrow \text{PhNHNH}_2$$

$$
\begin{array}{c}
\underset{|}{\text{HC}=\text{N}-\text{NHPh}} \\
\underset{|}{\text{C}=\text{N}-\text{NHPh}}
\end{array}
$$

Fischer Mechanism for Formation of Osazones

However it seems unlikely that a reducing agent as mild as the secondary hydroxyl group (on C-2) could reduce the phenylhydrazine, especially since titanium trichloride does not.

To overcome this anomaly, it was suggested that the phenylhydrazonium cation, and not phenylhydrazine itself, is responsible for the oxidation.[84] This hypothesis was supported by the fact that substituted hydrazine salts undergo reductive decomposition by heat more readily than do the corresponding free hydrazines.

In agreement with this supposition is the fact that primary and secondary alcohols containing an aromatic nucleus, or two or more double bonds in conjugation, are oxidized by the action of 2,4-dinitrophenylhydrazine to give the corresponding hydrazones.[85] In the sugar series, the formation of the intermediate aldosulose monophenylhydrazone may be represented as follows:

$$
\begin{array}{c}
\underset{|}{\text{HC}=\text{N}-\text{NHPh}} \\
\underset{|}{\text{CHOH}} \\
\text{R}
\end{array}
\xrightarrow{\text{PhNHNH}_3^{\oplus}\text{Cl}}
\begin{array}{c}
\text{HC}=\text{N}-\text{NHPh} \\
\underset{|}{\text{R}-\text{C}} \overset{\text{H}}{\underset{\text{O}}{\diagdown}} \overset{\text{NH}_3^{\oplus}}{\underset{\text{H}}{\diagup}} \text{NHPh}
\end{array}
\longrightarrow
\begin{array}{c}
\underset{|}{\text{HC}=\text{N}-\text{NHPh}} \\
\underset{|}{\text{C}=\text{O}} + \text{NH}_4^{\oplus} + \text{PhNH}_2 \\
\text{R}
\end{array}
$$

In 1940, Weygand[86] established that 1-arylamino-1-deoxy-D-fructoses, obtained by Amadori rearrangement (see Chapter 20) from N-aryl-D-glucosylamines, yield osazones more rapidly, and in higher yield, than do the free sugars by the usual procedure. Moreover, hydrazine and methylhydrazine are able to convert 1-deoxy-1-arylaminoketoses into the aldosuloses. Later[87] Weygand proposed two mechanisms, designated scheme A and scheme B, which involve Amadori rearrangements.

A similar mechanism for formation of osazones, utilizing the intermediate (14) of scheme B, involves the formation of a cyclic transition state (15) between the intermediate (14) and a molecule of phenylhydrazine hydrochloride, which is then cleaved reductively to yield the osazone directly and at the same time give ammonia and aniline.[88]

References start on p. 981.

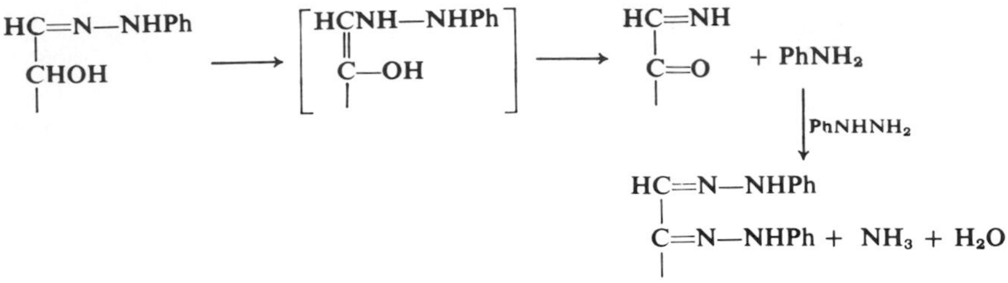

$$
\begin{array}{c}
\underset{|}{HC}{=}N{-}NHPh \\
\underset{|}{CHOH}
\end{array}
\longrightarrow
\left[
\begin{array}{c}
\underset{\parallel}{HCNH}{-}NHPh \\
\underset{|}{C}{-}OH
\end{array}
\right]
\longrightarrow
\begin{array}{c}
\underset{|}{HC}{=}NH \\
\underset{|}{C}{=}O
\end{array}
\; + \; PhNH_2
$$

$$\Big\downarrow PhNHNH_2$$

$$
\begin{array}{c}
\underset{|}{HC}{=}N{-}NHPh \\
\underset{|}{C}{=}N{-}NHPh
\end{array}
\; + \; NH_3 \; + \; H_2O
$$

Weygand's Scheme A

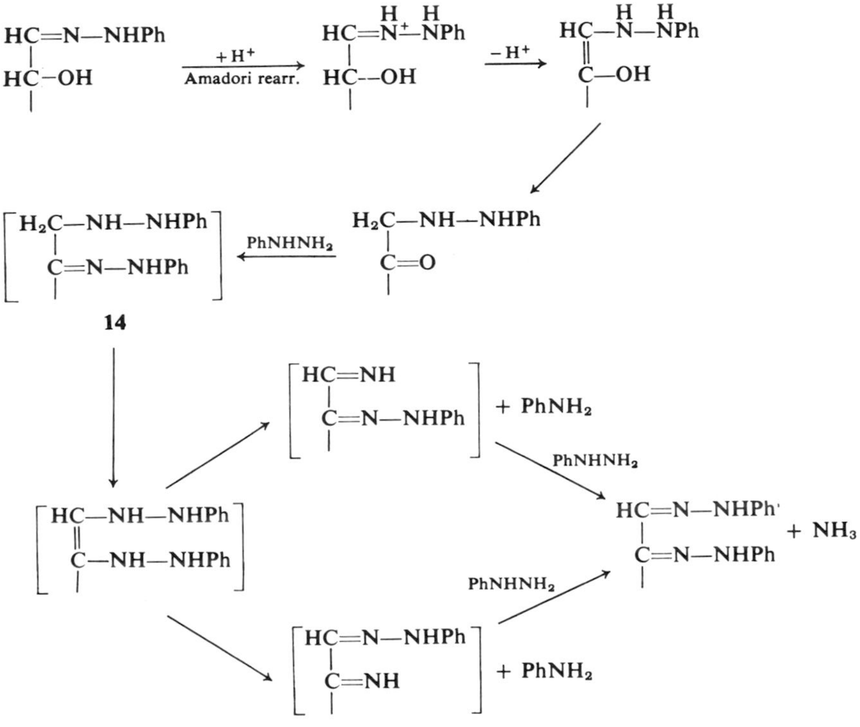

Weygand's Scheme B

Since 1,2-disubstituted hydrazines are very readily oxidized to the corresponding hydrazones, it was suggested[89] that Weygand's 1-hydrazino-2-ketose may be oxidized directly by air or by the hydrazonium cation to the corresponding aldosulose 1-phenylhydrazone.

Subsequent investigations[90] on the formation of osazones, starting from 4,6-O-benzylidene-D-glucose (16) have led to postulation of a new mechanism involving intermolecular oxidation-reduction. According to this scheme, phenylhydrazine is decomposed to give benzene, ammonia, and aniline.

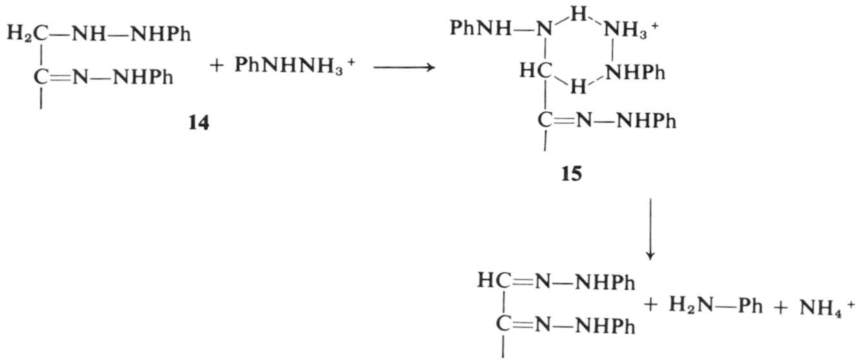

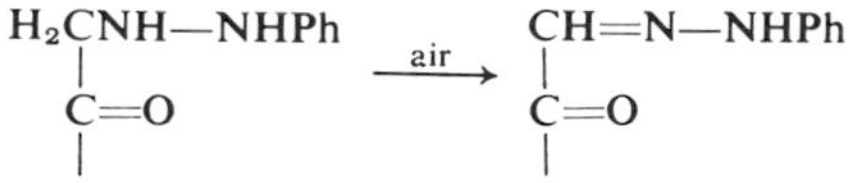

$$H_2C-NH-NHPh$$
$$\underset{|}{\overset{|}{C}}=N-NHPh \quad + \; PhNHNH_3{}^+ \longrightarrow$$

14

$$PhNH-N\overset{H}{\diagdown}NH_3{}^+$$
$$HC\diagdown_{H}\diagup NHPh$$
$$\underset{|}{\overset{|}{C}}=N-NHPh$$

15

$$\downarrow$$

$$HC=N-NHPh$$
$$\underset{|}{\overset{|}{C}}=N-NHPh \quad + \; H_2N-Ph \; + \; NH_4{}^+$$

$$H_2CNH-NHPh \qquad CH=N-NHPh$$
$$\underset{|}{\overset{|}{C}}=O \quad \xrightarrow{\;air\;} \quad \underset{|}{\overset{|}{C}}=O$$

Transamination of the initial phenylhydrazone (**17**) by the aniline yields 4,6-O-benzylidene-N-phenyl-D-glucosylamine (**18**), which undergoes an Amadori rearrangement to give the 1-anilino-1-deoxyhexulose derivative (**20**). The corresponding phenylhydrazone (**19**) then formed is reduced to the osazone by the third molecule of phenylhydrazine as shown. This mechanism seems, however, to apply only to this specific example.

Many laborious and often contradictory investigations have been carried out with labeled compounds in attempts to decide which of the proposed mechanisms is correct.

Labeled D-mannose p-bromophenylhydrazone was treated with unlabeled p-bromophenylhydrazine, but rapid transhydrazonation occurred and prevented definite conclusions from being reached.[91]

Attempts have also been made[92-94] to differentiate between Fischer's mechanism and Weygand's two schemes by the action of unlabeled p-nitrophenylhydrazine on p-nitrophenylhydrazones labeled with ^{15}N. The results[95] seem to favor Weygand's scheme A.

The expected distribution of ^{15}N in the ammonia and osazone is given in Table IV.

D-[1-^{3}H]Glucose has been converted with phenylhydrazine into the phenylosazone without loss of tritium, in accord with a Fischer type of mechanism.[96] This behavior was later attributed to the isotope effect between tritium and hydrogen, since, when D-glucose labeled with both tritium and deuterium on C-1 was used, the results favored scheme B for about one-third of the reacting

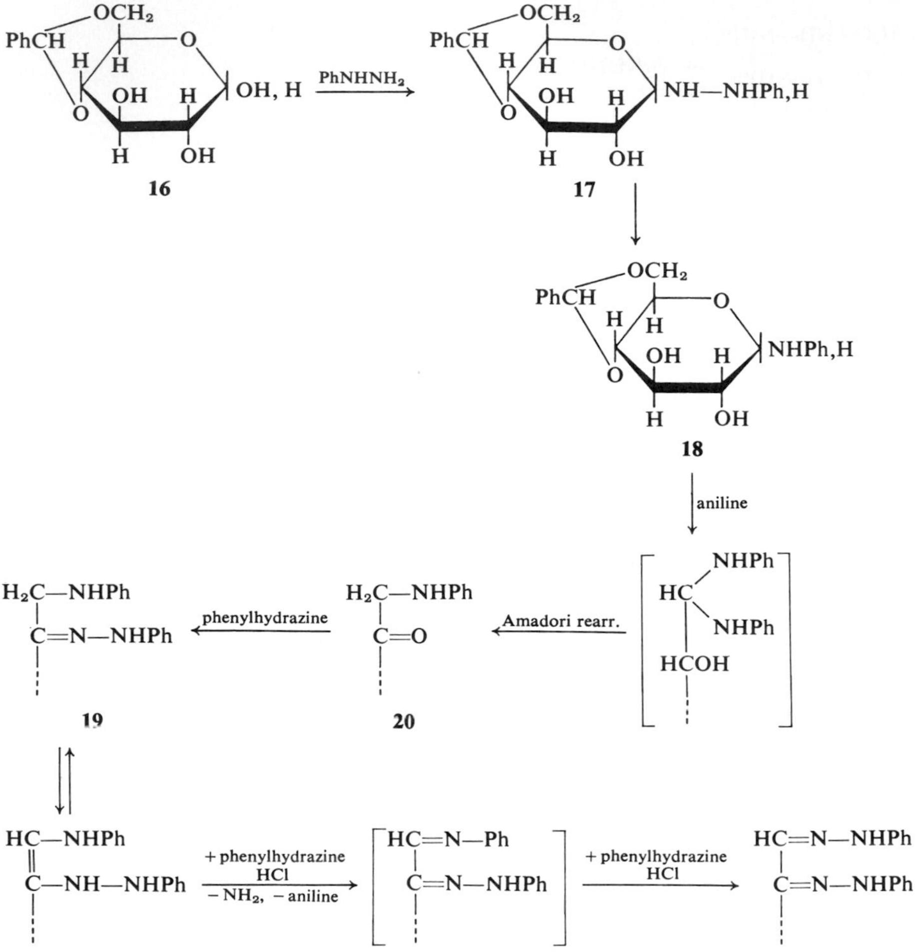

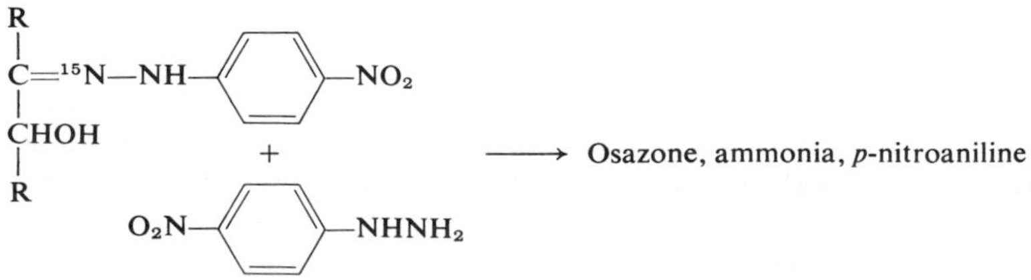

16 $\xrightarrow{\text{PhNHNH}_2}$ **17**

18

aniline

$$\text{H}_2\text{C—NHPh} \quad \xleftarrow{\text{phenylhydrazine}} \quad \text{H}_2\text{C—NHPh} \quad \xleftarrow{\text{Amadori rearr.}} \quad \left[\begin{array}{c} \text{HC} \stackrel{\text{NHPh}}{\diagdown_{\text{NHPh}}} \\ \text{HCOH} \end{array}\right]$$

$$\overset{|}{\underset{|}{\text{C}}}=\text{N—NHPh} \qquad\qquad \overset{|}{\underset{|}{\text{C}}}=\text{O}$$

19   **20**

$$\text{HC—NHPh} \xrightarrow[\;-\text{NH}_2,\ -\text{aniline}\;]{\substack{+\text{phenylhydrazine}\\ \text{HCl}}} \left[\text{HC}=\text{N—Ph}\right] \xrightarrow{\substack{+\text{phenylhydrazine}\\ \text{HCl}}} \text{HC}=\text{N—NHPh}$$

$$\text{C—NH—NHPh} \qquad\qquad \text{C}=\text{N—NHPh} \qquad\qquad \text{C}=\text{N—NHPh}$$

$$\begin{array}{c} \text{R} \\ | \\ \text{C}=^{15}\text{N—NH}\!-\!\!\!\!\bigcirc\!\!\!\!-\text{NO}_2 \\ | \\ \text{CHOH} \\ | \\ \text{R} \end{array} \quad + \quad \text{O}_2\text{N}\!-\!\!\!\!\bigcirc\!\!\!\!-\text{NHNH}_2 \qquad \longrightarrow \qquad \text{Osazone, ammonia, } p\text{-nitroaniline}$$

TABLE IV

EXPECTED DISTRIBUTION OF ^{15}N IN AMMONIA AND OSAZONE ACCORDING TO
THE DIFFERENT MECHANISMS PROPOSED

Mechanism	Ammonia ^{15}N (%)	Osazone ^{15}N (%)
Fischer	0	100
Weygand scheme A	100	0
Weygand scheme B	50	50

molecules, but with participation of another mechanism, most probably scheme A, for the other two-thirds.[97,98]

An investigation[98a] of the rate of formation of 2-hydroxycyclohexanone phenylhydrazone and 1-*N*-methyl-*N*-phenylhydrazone, together with the corresponding 2-methoxy, 2-acetoxy, and 2-chloro derivatives, suggested 1,4 eliminations by two competing mechanisms as follows:

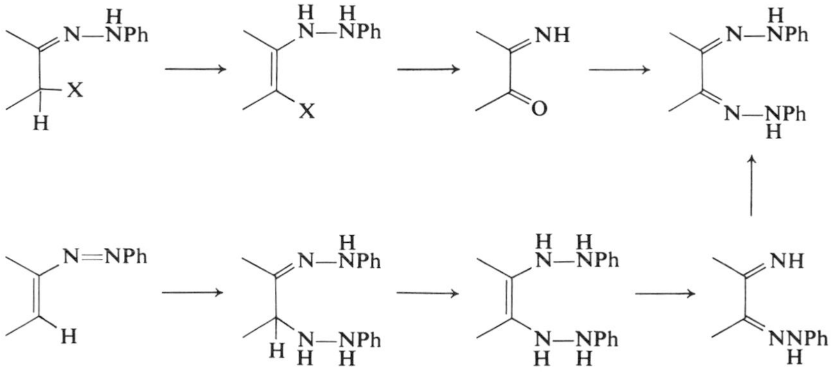

An interesting intermediate in formation of osazones has been isolated from the reaction of D-glucose with phenylhydrazine in acetic or hydrochloric acid. This is 3-(D-*arabino*-tetrahydroxybutyl)cinnoline, which is possibly formed by cyclization of D-*arabino*-hexosulose 2-phenylhydrazone or of the aldimino-2-phenylhydrazone derived through Weygand's scheme B.[79]

C. STRUCTURE

The most extensively studied compounds of this group are the osazones. When Emil Fischer first prepared the sugar phenylosazones,[83] he represented

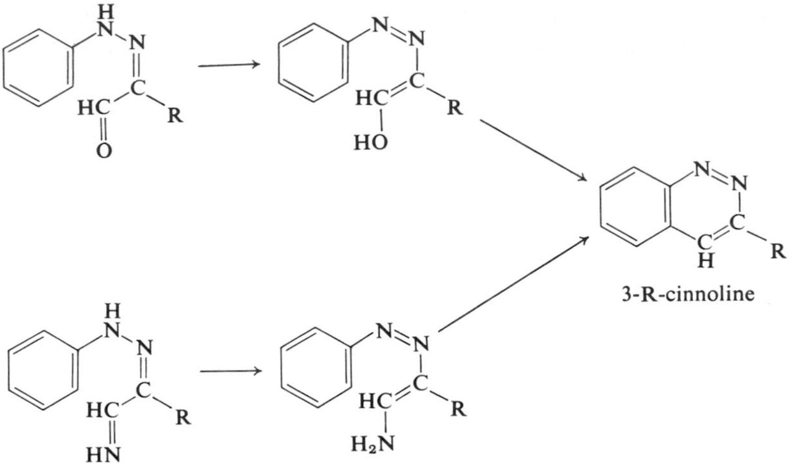

3-R-cinnoline

R = D-*arabino*-tetrahydroxybutyl

them as open-chain compounds, such as shown for D-*arabino*-hexulose phenyl-osazone (**21**), and this formulation was later confirmed by the following evidence.

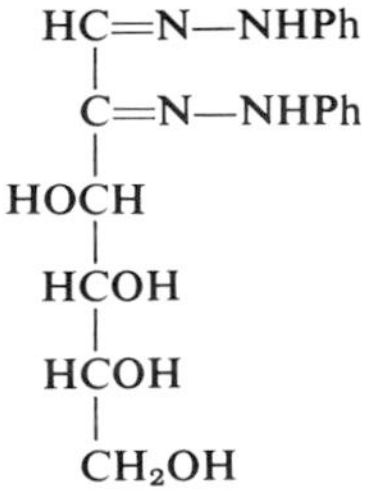

$$
\begin{array}{l}
HC{=}N{-}NHPh \\
| \\
C{=}N{-}NHPh \\
| \\
HOCH \\
| \\
HCOH \\
| \\
HCOH \\
| \\
CH_2OH
\end{array}
$$

21

Many of the reactions involving the polyhydroxyalkyl chain of hexulose and pentulose osazones are consistent with the idea that all the carbon atoms beyond the bis(hydrazone) group are hydroxylated, and the possibility of a ring between the sugar chain and one of the hydrazone residues is thus

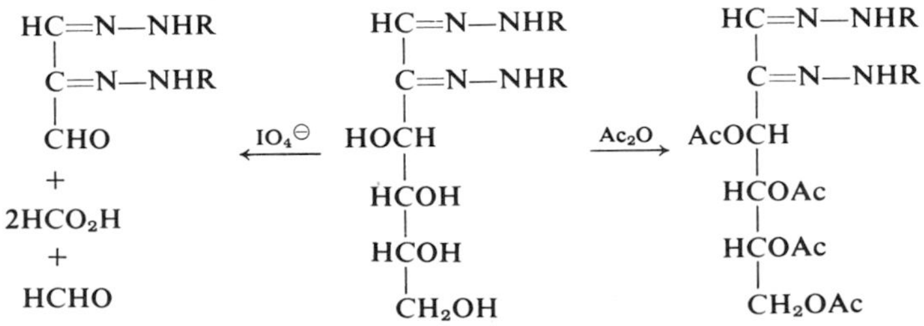

$$
\begin{array}{lll}
HC{=}N{-}NHR & HC{=}N{-}NHR & HC{=}N{-}NHR \\
| & | & | \\
C{=}N{-}NHR & C{=}N{-}NHR & C{=}N{-}NHR \\
| & | & | \\
CHO & \xleftarrow{IO_4{}^{\ominus}} \;\; HOCH & \xrightarrow{Ac_2O} \;\; AcOCH \\
+ & | & | \\
2HCO_2H & HCOH & HCOAc \\
+ & | & | \\
HCHO & HCOH & HCOAc \\
& | & | \\
& CH_2OH & CH_2OAc
\end{array}
$$

Reactions of Hexulose Osazones   R = Ph or Bz

excluded. Irrespective of the substituent attached to the hydrazone residues, hexulose osazones and hexosulose bis(hydrazones) consume three moles of periodate[99] and form acyl derivatives possessing four groups[59,100-103] attached to their oxygen atoms. Methylation experiments also provide evidence in favor of the acyclic structure of sugar osazones (see page 967).

Fischer's open-chain structure fails, however, to explain certain properties of osazones such as:

(1) Why osazone formation stops at the second carbon atom instead of continuing to the third one and ultimately to the end of the chain.

(2) Why the two phenylhydrazone groups in the sugar osazones behave differently; only one can be methylated, acetylated, or benzoylated, the other cannot.

(3) Why there is a distinct difference between the phenylosazones and the 1-(N-methyl-N-phenyl)-2-phenylosazones, on the one hand, and the bis-(N-methyl-N-phenyl)osazones, on the other.

1. *Chelated Structures*

Fieser and Fieser[104] proposed, on purely theoretical grounds, two stabilized, chelated structures (22 and 24) for sugar osazones, in tautomeric equilibrium with the forms 23 and 25. They considered that the formation of a chelate ring between C-1 and C-2 prevented the further electron displacement necessary for formation of a polyhydrazone, and thus explained why the reaction stops at C-2.

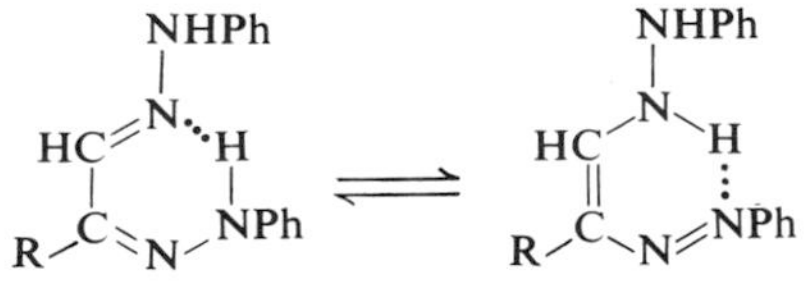

The evidence for the chelated structure of osazones may be summarized as follows:

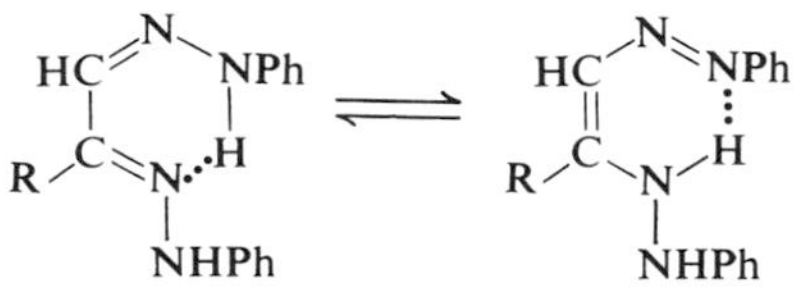

a. Formazan Reaction.[64]—Sugar osazones react with diazotized aniline to

References start on p. 981.

give osazone formazans. The reaction proceeds in ethanolic potassium hydroxide but, unlike the hydrazones, not in pyridine. The function of the strong alkali appears to be to break the chelated ring of the osazones.[105] [Non-sugar bis(hydrazones), such as phenylglyoxal bis(phenylhydrazone), do not yield formazans but instead undergo phenylation at C-1.[106]]

$$
\begin{array}{ccc}
\underset{\substack{| \\ \text{NHPh}}}{\text{H}\diagdown\underset{\text{C}}{}\diagup^{\text{N}}\diagdown\text{H}} & & \\
\overset{|}{\underset{\diagup\text{C}}{}}\diagup\diagdown\underset{\text{Ph}}{\overset{\text{N}}{}} & \xrightarrow{\text{NaOH}} & \begin{array}{c}\text{HC}=\text{N}-\text{NHPh} \\ | \\ \text{C}=\text{N}-\text{NHPh} \\ |\end{array} & \xrightarrow{\text{PhN}_2^{\oplus}} & \begin{array}{c}\text{N}=\text{NPh} \\ | \\ \text{C}=\text{N}-\text{NHPh} \\ | \\ \text{C}=\text{N}-\text{NHPh} \\ |\end{array}
\end{array}
$$

Formation of Osazone Formazan

b. *Polarographic Behavior.*—The chelated structure of sugar osazones and mixed osazones has also been confirmed with the aid of polarography.[107] D-*arabino*-Hexulose phenylosazone, 1-(N-methyl-N-phenyl)-2-phenylosazone, and bis(N-methyl-N-phenyl)osazone exhibit a four-electron reduction wave, which for the third compound is 200 millivolts more positive than for the first two.

The difference between D-*arabino*-hexulose phenylosazone and 1-(N-methyl-N-phenyl)-2-phenylosazone, on the one hand, and the bis(N-methyl-N-phenyl)osazone, on the other, is a further manifestation of the structural difference that characterizes these two groups of compounds. The fact that the first two are less readily reduced (their reduction potential is more negative) is explained by the increased stability conferred by their chelate structures, in contrast to the bis(N-methyl-N-phenyl)hydrazone structure, which cannot be stabilized by chelation.

c. *Ultraviolet Spectra.*—Comparison of the spectra of D-*arabino*-hexulose phenylosazone, 1-(N-methyl-N-phenyl)-2-phenylosazone, bis(N-methyl-N-phenyl)osazone, and bis(benzyl phenylhydrazone) reveals that the spectra of the first two compounds are very similar and differ from the spectra of the bis(N-methyl-N-phenyl)osazone and of the non-sugar bis(hydrazone).[108] The electronic spectra of sugar osazones cannot be explained either by the presence of a classical bis(hydrazone) nor of an azoethylene chromophore.[65a,105,108]

d. *X-Ray Crystallography.*—A very important contribution to knowledge of the structure of sugar osazones has been the X-ray crystallographic analysis of the p-bromophenylosazones from D-*arabino*-hexulose, D-*threo*-pentulose, and D-*erythro*-pentulose, all of which have the same general crystal structure.[109,110]

The electron-density projection in the direction of the b axis of these crystals (see Fig. 1) confirmed the open-chain structure of the sugar osazones. Hydrogen

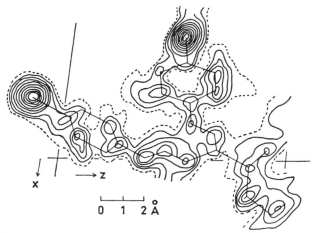

FIG. 1. Electron-density projection in the direction of the *b* axis of D-*threo*-pentulose *p*-bromophenylosazone. The distances between lines of equal density are arbitrary.

bonding between the hydroxyl group on C-3 and the phenylhydrazone group on C-2 was excluded because of the distance separating both groups. The six-membered, chelate ring is approximately planar (its bond angles are close to 120°) and essentially coplanar with the two benzene rings.

e. Nuclear Magnetic Resonance Spectra.—Confirmation of the structure of sugar osazones in solution has been obtained[101,111–114] by n.m.r. spectroscopy.

The chemical shifts of the signals of the protons on C-1 and on the nitrogen atoms in sugar osazones, dehydro-osazones, and a number of related compounds[111,112] are given in Table V. In the spectra of compounds **26–30**, a signal corresponding to a chelated N—H proton is observed at very low field (between $\tau - 2$ and -3), but is absent in the spectrum of compound **31**. The structure of compound **31** has been proved to be a true aldehyde phenylhydrazone. The signal for a chelated N—H proton is also absent from the spectrum of the symmetrical bis(phenylhydrazone) (**32**).

A signal for the nonchelated, imino proton of **31** is observed at $\tau - 1$ in pyridine-d_5 (at $\tau - 0.12$ in methyl sulfoxide) is also observed in the spectra of compounds **26–30**. A two-proton signal in the same region was present in the spectrum of **32**. Furthermore, a signal between τ 2 and 3, corresponding to the C-1 proton (H-1), is present in the spectra of compounds **28–30** but is absent in the spectra of the cyclic derivatives **26** and **27**.

References start on p. 981.

TABLE V

NUCLEAR MAGNETIC RESONANCE SPECTRAL SIGNALS FOR THE PROTONS ON C-1 AND ON THE NITROGEN ATOMS OF SUGAR OSAZONES, DEHYDRO-OSAZONES, AND RELATED COMPOUNDS [a]

		In methyl sulfoxide			In pyridine-d_5		
	Compound	N—H (chelated)	N—H	H-1	N—H (chelated)	N—H	H-1
(26)	Dehydro-D-*arabino*-hexulose phenylosazone	−2.45	0.65	—	−2.95	0.39	—
(27)	Tetra-*O*-acetyl-dehydro-D-*arabino*-hexulose phenylosazone	−2.48	0.45	—	−2.90	−0.05	—
(28)	D-*arabino*-Hexulose phenylosazone	−2.20	−0.66	2.11	−2.65	−1.20	1.32
(29)	Tetra-*O*-acetyl-D-*lyxo*-hexulose phenylosazone	−2.20	−0.80	2.27	−2.86	−1.49	1.95
(30)	D-Triose phenylosazone	−2.05	−0.70	2.21	−2.63	−1.27	1.77
(31)	Tetra-*O*-acetyl-D-*lyxo*-hexulose phenylhydrazone	—	−0.12	[b]	—	−1.00	2.42
(32)	Glyoxal bis(phenylhydrazone)	—	−0.33	2.31	—	−1.10	1.88

[a] The data are chemical shifts (τ) relative to the signal of tetramethylsilane (internal standard).

[b] Signal included in the benzene peaks.

Further evidence for these chelated structures is obtained by study of the deuterium-exchange of the hydrogen atoms bonded to nitrogen.[111–114] In chloroform-*d* the n.m.r. signal of the free imino proton is observed between τ 2 and 3 (Table VI). It disappears very rapidly when deuterium oxide is added and the mixture is shaken. The signal at τ − 2 to −3, corresponding to the proton involved in the chelation, disappears considerably less rapidly, indicating that it is more firmly held in the molecule. The signal of the C-1 proton is located at τ 2.44 in compounds **29** and **34**, but in the spectra of compounds **30** and **33** it is overlapped by the signals of the protons on the aromatic rings. In compound **27** there is no C-1 proton.

The fact that replacement of the α-imino hydrogen in the phenylhydrazone residue on C-1 by a methyl group (compound **33**) has practically no influence on the position of the signal of the chelated NH group, whereas the signal of the nonchelated NH group disappears, proves that the position of the nonchelated proton in the sugar osazones is the same as the position of the methyl group in the 1-*N*-methyl-*N*-phenyl-2-phenylosazones. This position is

TABLE VI

Nuclear Magnetic Resonance Spectral Signals of the Protons on C-1 and on the
Nitrogen Atoms of Some Osazones in Chloroform-d[a]

	Compound	N—H (chelated)	N—H	H-1
27	Tetra-O-acetyl-dehydro-D-*arabino*-hexulose phenylosazone	−2.50	2.03	—
29	Tetra-O-acetyl-D-*lyxo*-hexulose phenylosazone	−2.34	2.11	2.46
30	D-Triose phenylosazone	−2.16	2.22	[b]
33	Tetra-O-acetyl-D-*arabino*-hexulose 1-(N-methyl-N-phenyl)-2-phenylosazone	−2.54	—	[b]
34	Tri-O-acetyl-D-*erythro*-pentulose phenylosazone	−2.42	2.12	2.44

[a] The data are chemical shifts (τ) relative to the signal of tetramethylsilane (internal standard).

[b] Signal included in the benzene peaks.

chemically well defined.[48,105] Thus the sugar osazones must have the structures[101] 22 and 23 rather than 24 and 25. The long-range spin–spin coupling observed between H-1 and the nonchelated N—H proton constitutes a cogent argument for the suggested geometry of the molecule.[112]

The limited equilibration of sugar osazones during mutarotation through the action of a basic solvent[115] can also be ascribed to the stability of the chelate ring.[116]

On the basis of these considerations[105,111,112,116–119] (including increased stability and chemical properties approaching those of aromatic compounds), Mester proposed a strongly conjugated, chelated structure (35) having an acyclic π-electron system (a so-called "quasi-aromatic" structure[117,118]) for the sugar osazones. Dorman[119] has attributed a "nonclassical aromatic"

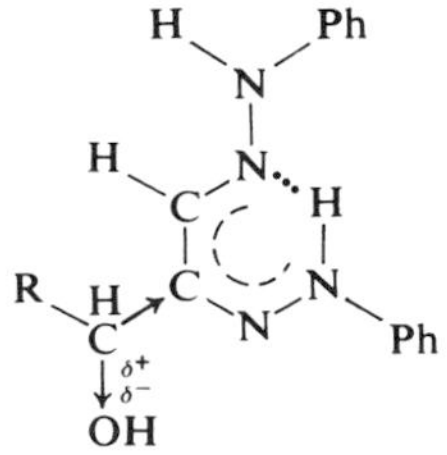

35

structure to the osazones, having π-electron interaction through the hydrogen bond. Chapman[119] has refuted this later concept: the chelated proton cannot provide a p-orbital orthogonal to the ring that would allow formation of a cyclic π-orbital system. He has argued in favor of the classical bis(phenyl-hydrazone) structure. However, the results obtained by analysis of the n.m.r. spectra of ^{15}N-labeled sugar osazones are incompatible with a classical bis(hydrazone) structure.[120] A detailed, molecular-orbital treatment of these structures seems to be necessary in order to permit formulation of definitive structures for the sugar osazones.

D. Mutarotation

Mutarotation of sugar osazones was first observed in 1909,[121] and since then various authors[122-127] have suggested a multitude of structures for the initial and final forms involved in this transformation. From the foregoing section it is evident that the initial form must be a chelated structure (**22**); the presently accepted structure[126] for the final form being **37** or its iso-mer[113] **36**. These formulations are consistent with the following observations regarding mutarotation.[125]

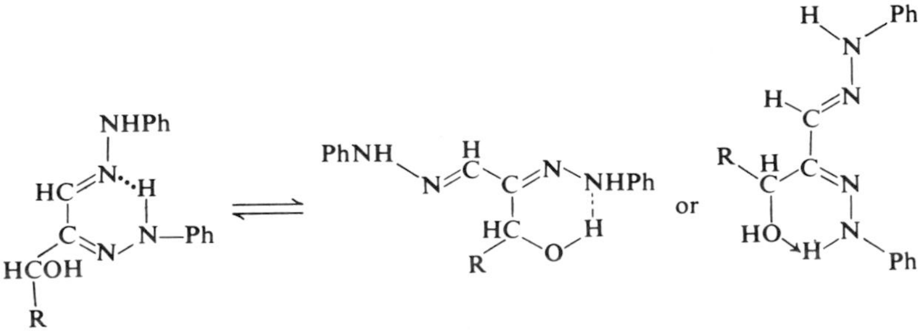

Mutarotation of Ozazones

(1) Sugar osazones mutarotate only in basic solvents such as pyridine, by a first-order process. p-Dioxane is sufficiently basic to cause mutarotation.

(2) Removal of the solvent after mutarotation regenerates the initial osazone as the sole product.

(3) Mutarotation of sugar osazones is accompanied by a shift of the long-wavelength ultraviolet absorption maximum at 395 nm to shorter wavelengths and by a gradual decrease in intensity at the absorption minimum near 345 nm; the process is reversible, and the material recovered from solution behaves in the same way.

(4) As mutarotation progresses, the ability of D-*arabino*-hexulose phenyl-osazone to give the formazan decreases.

Base-catalyzed mutarotation is shown only by sugar osazones having a chelated structure. Bis(alkylphenyl)hydrazones do not undergo this change; however, the optical rotations of these compounds are greatly influenced by temperature. This interesting optical property is termed "thermomutarotation."[3,128]

The n.m.r. spectra of osazones undergoing mutarotation exhibit a series of changes which suggest[113,115] the formation of O-chelated structures (36 or 37). These changes are:

(1) The appearance of two new N—H signals in the region characteristic of solvent-bonded N—H protons.

(2) Appearance of a new C—H signal at slightly higher field than that of the C—H signal of the initial aldimine.

(3) Appearance of a new signal for the 3-hydroxyl proton, shifted slightly downfield.

Addition of water to a solution of the osazones undergoing mutarotation in p-dioxane leads to a downfield shift of the signal of the nonchelated N—H proton by about 1 ppm, because of the solvation by water of the N—H group.[115] Chelated imino protons do not show this phenomenon. These observations establish the presence of a chelated imino group, as well as a nonchelated one, in the final form of sugar osazones after mutarotation. The N-chelated N—H proton signal is shifted to lower field, suggesting that the chelated imino proton in the final form of sugar osazones is O- rather than N-chelated.

The long-range spin–spin coupling[111,115] between the C-1 proton and the nonchelated N—H proton in the initial form of sugar osazones is observed also in the final form. This evidence indicates that the geometrical disposition of these protons in the final form of sugar osazones after mutarotation is very similar to that present in the initial form (22). A *rule of mutarotation* has been formulated.[116] The sign of the initial value of the optical rotation in pyridine–ethanol solution is *positive* if the C-4 hydroxyl group of the osazone is *at the left* in the standard Fischer projection of the compound, and it is *negative* if the same hydroxyl group is *at the right*. The progression of the specific rotation during mutarotation toward more-positive or more-negative values may be predicted from the configuration of sugars, and *vice versa*.

It has been also reported that the chromophore in sugar osazones produces circular-dichroism spectra exhibiting multiple Cotton effects that conform to the following rule.[116] A *positive* Cotton effect in the 250–270 nm region will result if the C-3 hydroxyl group is *at the right* in the Fischer projection of the

---

References start on p. 981.

compound, and it will be *negative*, if the same hydroxyl group is *at the left*. Similar spectra were reported for osazone derivatives of cyclic sugars (anhydro-, dianhydro-, and dehydro-osazones).[129]

E. REACTIONS

1. *Cleavage of Hydrazone Residues*

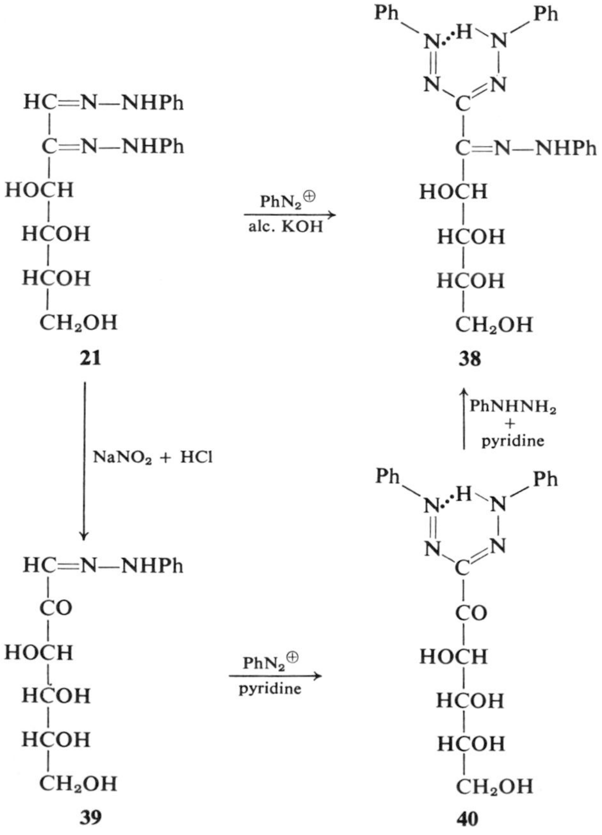

Fischer[23] was the first to report that sugar osazones are hydrolyzed by the action of hydrochloric acid, with the formation of the corresponding dicarbonyl derivatives (aldosuloses), which he termed "osones." The nitrogen-containing residues may also be removed by the action of benzaldehyde,[130,131] pyruvic acid,[132] or nitrous acid.[133] Partial hydrolysis of sugar osazones by nitrous acid leads to the aldosulose 1-phenylhydrazones. This type of compound has been systematically investigated by Henseke and collaborators.[128,134-136] Aldosulose 1-(N-methyl-N-phenyl)hydrazones are obtained with special ease. It has been reported that D-*arabino*-hexosulose 1-phenylhydrazone (**39**) reacts with diazotized aniline to yield a crystalline formazan, 1-phenylazo-D-*arabino*-hexulose phenylhydrazone (**40**).[134,137] This compound can be converted by the action of phenylhydrazine into the corresponding "osazone formazan," identical with the 1-phenylazo-D-*arabino*-hexulose phenylosazone (**38**) obtained by the direct action of diazotized aniline on the osazone **21**.[105]

Aldosuloses[138] and aldosulose hydrazones[134,139] react readily with o-phenylenediamine to form quinoxaline derivatives (**41**) of sugars. The hydroxyl group

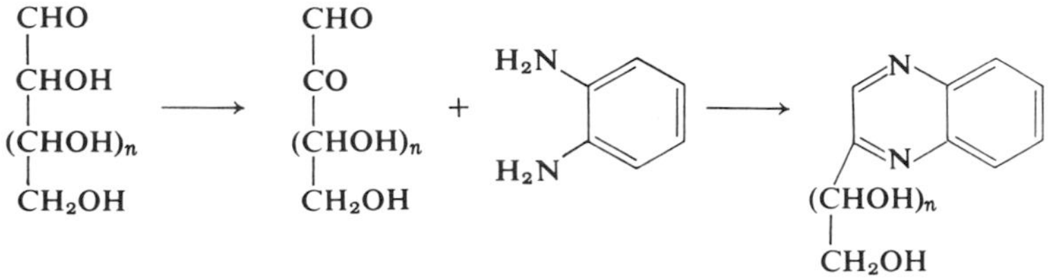

quinoxaline

41

at C-3 undergoes a reaction analogous to that of osazone formation[140]: it is first converted by action of phenylhydrazine into a keto group, which then condenses with a second molecule of phenylhydrazine to give the product **42**. In this particular case, the reaction proceeds further, yielding the tricyclic derivative (**43**) by the action of a third molecule of phenylhydrazine.

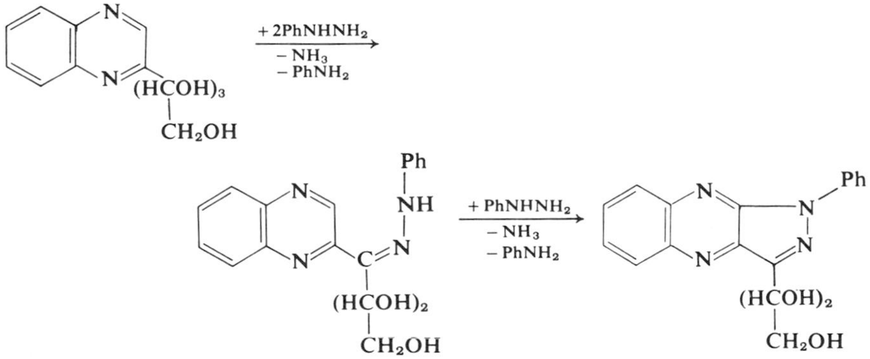

42   **43**

2. Action of Bases

The bis(hydrazones), are stable even to cold concentrated alkalies, but are progressively degraded. Thus phenylosazones give rise to glyoxal bis(phenylhydrazone).[56,141,142]

$$
\begin{array}{l}
\text{HC=N—NHPh} \\
\;\;| \\
\text{C=N—NHPh} \\
\;\;| \\
\text{(CHOH)}_n \\
\;\;| \\
\text{CH}_2\text{OH}
\end{array}
\quad\xrightarrow{\text{OH}^\ominus}\quad
\begin{array}{l}
\text{HC=N—NHPh} \\
\;\;| \\
\text{HC=N—NHPh}
\end{array}
$$

Glyoxal bis(phenylhydrazone)

3. Reduction

By reduction of D-*arabino*-hexulose phenylosazone (**21**) with zinc and acetic acid, Fischer[143] obtained 1-amino-1-deoxy-D-fructose (**44**, "*iso*-D-*glucos-amine*"). The yield has been increased by conducting the reduction in acetic acid over a palladium-on-carbon catalyst.[100,144] To open the chelated ring and facilitate the reduction, D-*arabino*-hexulose phenylosazone can be reduced over Raney nickel in 2 N alcoholic potassium hydroxide; the reaction gives 1,2-diamino-1,2-dideoxy-D-mannitol and -D-glucitol (**45** and **46**).[58,145]

$$
\begin{array}{c}
\text{CH}_2\text{NH}_2 \\
| \\
\text{C=O} \\
| \\
\text{HOCH} \\
| \\
\text{HCOH} \\
| \\
\text{HCOH} \\
| \\
\text{CH}_2\text{OH} \\
\mathbf{44}
\end{array}
\xleftarrow[\text{HOAc}]{\text{Zn}}
\begin{array}{c}
\text{HC=N—NHPh} \\
| \\
\text{C=N—NHPh} \\
| \\
\text{HOCH} \\
| \\
\text{HCOH} \\
| \\
\text{HCOH} \\
| \\
\text{CH}_2\text{OH} \\
\mathbf{21}
\end{array}
\xrightarrow[\text{H}_2,\,\text{Ni}]{\text{KOH}}
\begin{array}{c}
\text{CH}_2\text{NH}_2 \\
| \\
\text{H}_2\text{NCH} \\
| \\
\text{HOCH} \\
| \\
\text{HCOH} \\
| \\
\text{HCOH} \\
| \\
\text{CH}_2\text{OH} \\
\mathbf{45}
\end{array}
+
\begin{array}{c}
\text{CH}_2\text{NH}_2 \\
| \\
\text{HCNH}_2 \\
| \\
\text{HOCH} \\
| \\
\text{HCOH} \\
| \\
\text{HCOH} \\
| \\
\text{CH}_2\text{OH} \\
\mathbf{46}
\end{array}
$$

This reaction has also been applied with disaccharide phenylosazones[146,147] and dehydroascorbic acid osazone.[148,149]

4. Action of Oxidizing Agents

Vigorous oxidizing agents completely rupture the poly(hydroxyalkyl) chain of bis(hydrazones) and give a number of products in low yield. Mild oxidizing agents, however, give rise to high yields of cyclic products of considerable interest.[150]

a. *Dehydro-osazones.*—These yellow compounds, very similar in appearance to the parent phenylosazones, are obtained by oxidation of phenylosazones in alkaline media with atmospheric oxygen.[151] Structural investigations[152,153] have shown that the dehydro derivative obtained from D-*arabino*-hexulose phenylosazone has a pyranoid ring because it consumed one mole of periodate and gave a dialdehyde which was characterized by phenylhydrazinolysis (see p. 970). The general structure of a 1,5-anhydro-1,2-dideoxy-2-phenylazo-1-phenylhydrazinohex-1-enitol has been assigned to the dehydro-osazones

formed from hexose and hexulose precursors. The dehydro-osazone from
D-glucose gave no formazan, and the n.m.r. spectrum of its tri-O-acetyl
derivative suggested that the hydroxyl group on C-3 was not equatorial, as
would have been expected for the D-*arabino* configuration, but was axial
(D-*ribo* configuration).

In the D-hexose series, four dehydro-osazones (derived from D-*arabino*-
hexulose, D-*lyxo*-hexulose, D-*ribo*-hexulose, and D-*xylo*-hexulose) would be

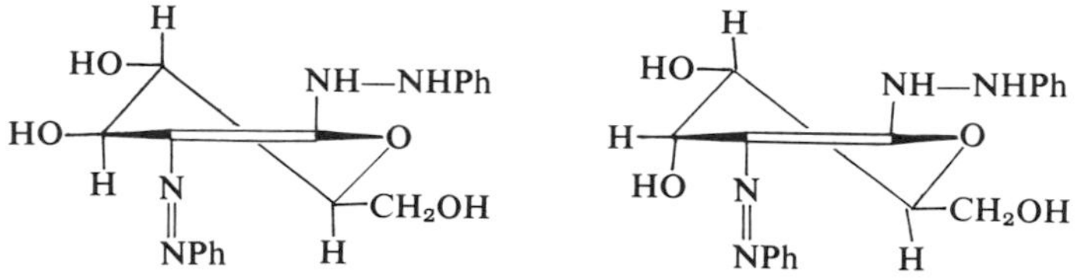

D-*arabino* (not observed)   D-*ribo* (observed)

Dehydro-osazone Derived from D-Glucose

expected if no inversion of the 3-OH group occurs. However, only two derivatives are obtained. D-Glucose (or D-*arabino*-hexulose) gives the same dehydro-
osazone as D-allose (or D-*ribo*-hexulose), and D-galactose (or D-*lyxo*-hexulose)
gives the same derivative as D-gulose (or D-*xylo*-hexulose), thus confirming
that inversion at C-3 occurs in one of each pair of examples.

b. *Saccharide 1,2,3-Triazoles.*—Two types of 1,2,3-triazoles are obtained from
bis(hydrazones). The first and best known are the 2-aryl-4-(hydroxyalkyl)-
1,2,3-triazoles, better known as osotriazoles; the others are 1-arylamino-4-
(hydroxyalkyl)-1,2,3-triazoles obtained from saccharide bis(aroylhydrazones).
Other types of triazoles can be prepared by the reaction of acetylenic[154]
derivatives with azides (see Section X).

Saccharide osotriazoles. This type of triazole has been by far the most
extensively studied. The first examples were prepared by Hudson and co-
workers,[155] who refluxed arylosazones with aqueous copper(II) sulfate. Subsequently, numerous osotriazoles have been prepared to characterize the
osazones of monosaccharides,[155–165] disaccharides,[166–172] and anhydro-
osazones.[173–177] The conversion of arylosazones into the corresponding
osotriazoles necessitates the presence of an oxidizing agent, and it is obvious
that the process does not involve a simple removal of aniline from the osazone,
as the equation might suggest. Apart from copper(II) sulfate, which is the
reagent most commonly used, other oxidizing heavy-metal salts, such as
mercuric acetate[178] and ferric sulfate and chloride,[179] have been used.
Halogens[180] and nitrososulfonates have also been used.[181] The acetylated

osazones are converted by nitrous acid into osotriazoles[182]; this reagent decomposes the unacetylated osazones to give the aldosuloses.[183] The structure of the osotriazoles was confirmed by oxidation with periodic acid, which leads to 2-phenyl-1,2,3-triazole-4-carboxaldehyde,[155] and with permanganate. which gives 2-phenyl-1,2,3-triazole-4-carboxylic acid.[179,180]

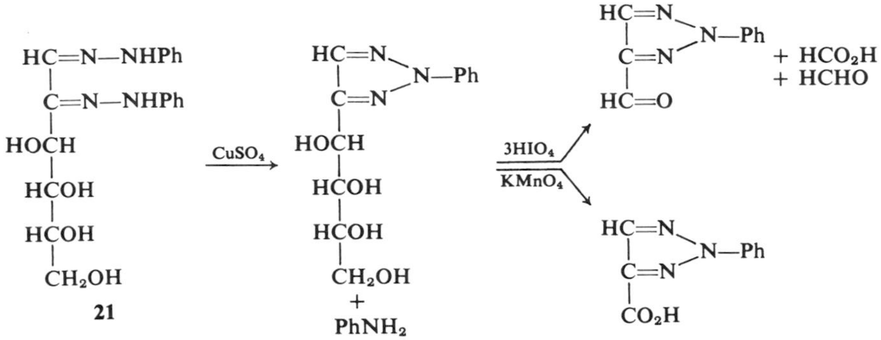

D-*arabino*-Hexulose phenylosotriazole

The formation of the former compound by the oxidation of labeled osotriazoles has been used for determining the position of ^{14}C in aldoses.[184]

To ascertain from which hydrazone residue the aniline is removed during formation of osotriazoles, Weygand[91] prepared ^{82}Br-labeled D-*arabino*-hexulose p-bromophenylosazone (47) labeled principally at the C-1 substituent. The distribution of label was demonstrated by treating the osazone with periodic acid, and reducing the resultant pyrazolinone (48) with stannous chloride. The p-bromoaniline split off was derived from the hydrazone residue at C-2 of the sugar and had 23% of the label, denoting that most of the label was located in the C-1 phenylhydrazone residue. When the unequally labeled osazone (47) was converted into the osotriazole (49), only 18% of the label remained in the osotriazole, indicating that the aniline eliminated had come from the C-1 phenylhydrazone residue. Similar observations have been made[183] by using mixed arylosazones. Similarly, during conversion of osazone formazans into the corresponding 1-phenylazo-osotriazoles, elimination of aniline takes place from the C-1 substituent.[185]

By using formazan labeled with ^{14}C, it was found that in osazone formazans from nonsugar precursors, such as the phenylosazone formazan (50) obtained by coupling pyruvaldehyde phenylhydrazone with diazotized ^{14}C-aniline and condensing the product with nonradioactive phenylhydrazine, exactly half the radioactivity is lost with the aniline in the osotriazole-forming reactions.

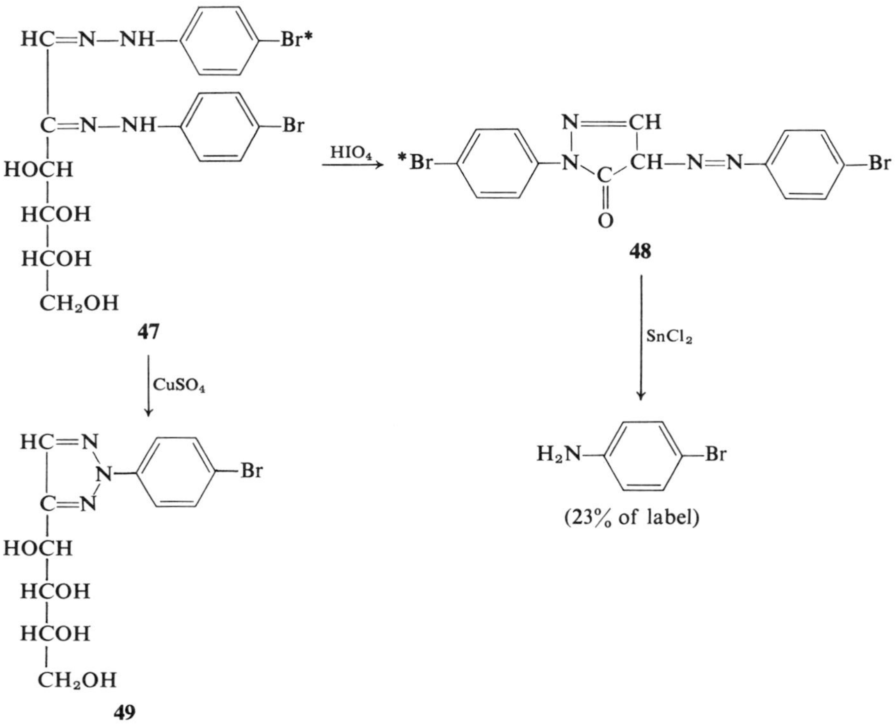

47

48

SnCl₂

49
(18% of label)

H₂N—⬡—Br

(23% of label)

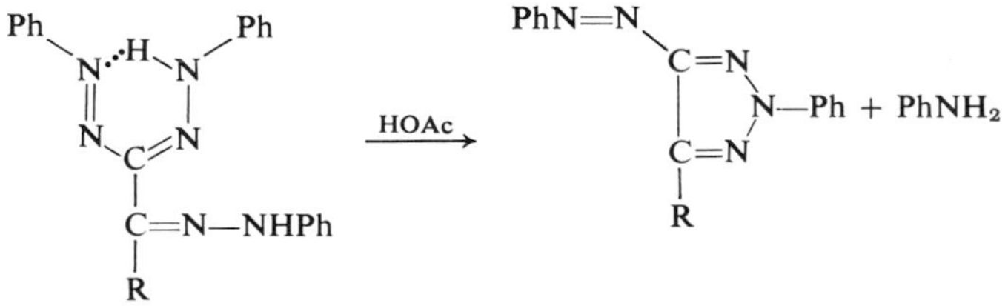

Osazone formazan

1-Phenylazo-osotriazole

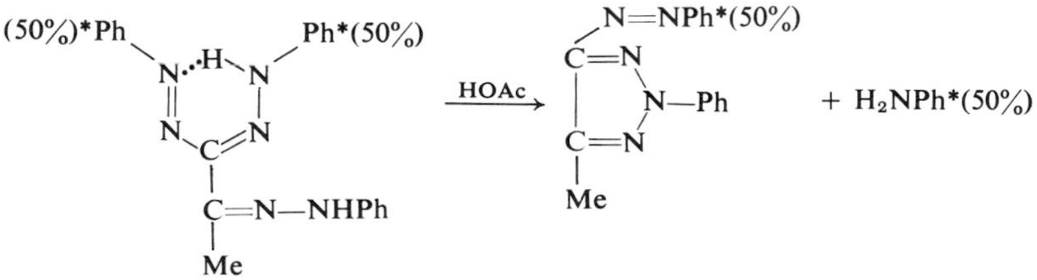

50

In contrast, if a saccharide osazone is coupled with diazotized ^{14}C-aniline and the osazone formazan obtained is converted into the phenylazo-osotriazole (osotriazole formazan), 58% of the total radioactivity is found in the phenylazo-osotriazole and only 42% in the aniline liberated. This observation is a supplementary argument in favor of the chelated structure of sugar osazones.

Osazone formazan

Phenylazo-osotriazole

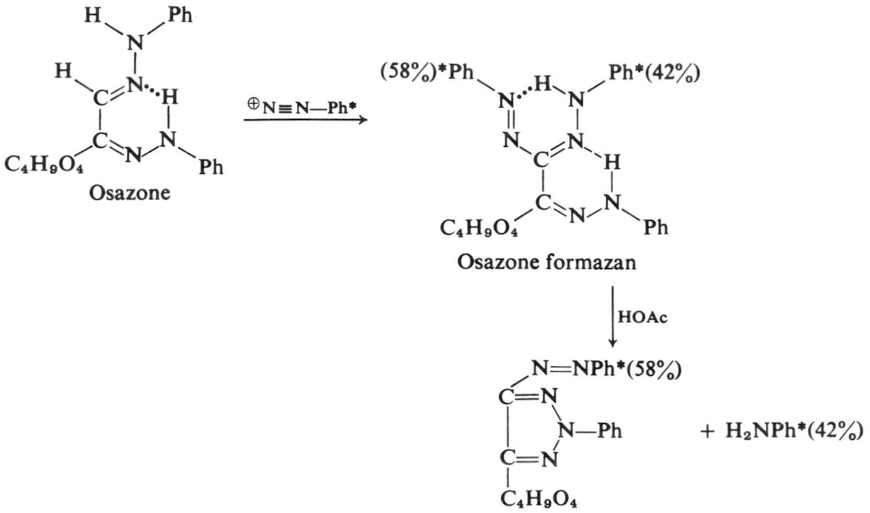

Correlations have been reported between the configuration of the hydroxyl group attached to the C-3 of the sugar osotriazoles and the sign of their rotations[186,187] and the sign of their Cotton effects.[188,189] Comparative n.m.r. studies of osotriazoles as a function of configuration of the side chain demonstrated that the chain adopts a "sickle" conformation if the planar zig-zag arrangement of carbon atoms would have given rise to a parallel 1,3-interaction between hydroxyl[190] or acetoxyl groups[191] (see also Vol. IA, Chapter 5).

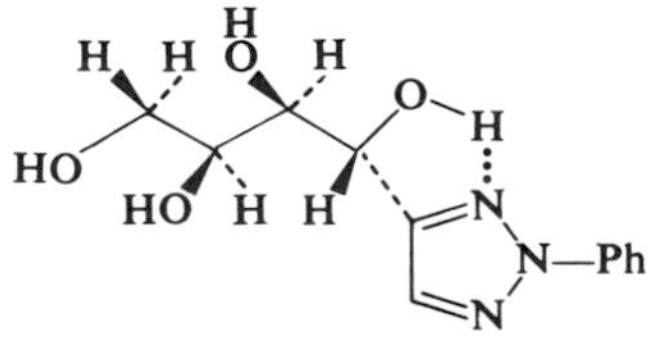

The Favored ("Sickle") Conformation of L-*xylo*-Hexulose Phenylosotriazole, as Revealed by N.M.R. Spectroscopy[190]

1-Arylamidotriazoles. Compounds of this type have been obtained by the action of iodine on bis(aroylhydrazone) acetates. Their structure was determined by periodate oxidation of their deacetylated analogues, and also by n.m.r. spectroscopy.[192,193]

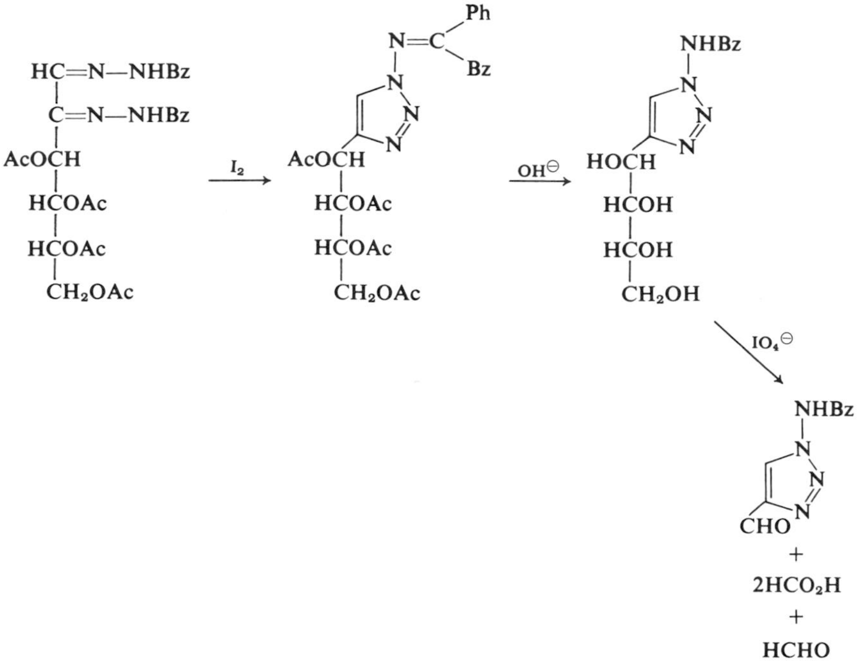

Formation and Characterization of 1-Arylamidotriazoles

5. Anhydro-osazones

Formation of an anhydro-osazone was first reported by Fischer.[83,194] Similar compounds were obtained by refluxing osazones in methanol containing some sulfuric acid as catalyst.[195] Their structure was studied by many investigators,[141,196] and eventually they were shown to be 3,6-anhydro-hexulose phenylosazones.[197,198] During the formation of anhydro-osazones, inversion at C-3 sometimes takes place, especially when the hydroxyl groups at C-3 and C-4 in the parent osazone are "*trans*" to each other, as in the *arabino*- and *xylo*-hexulose derivatives. When these hydroxyl groups are *cis*-oriented, as in the *ribo* and *lyxo* derivatives, no inversion takes place.[174] 3,6-Anhydro-osazones are also formed when osazones are refluxed with acetic anhydride[199];

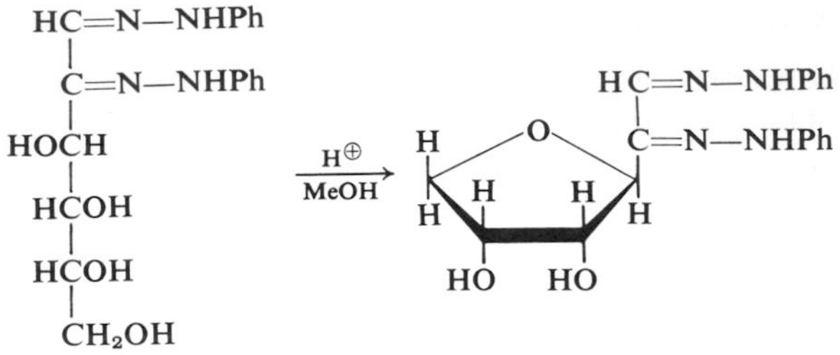

Formation of a 3,6-Anhydrohexulose Phenylosazone

these products are accompanied by dianhydro-osazones having pyrazole rings —for example, **51**. The structure of the pyrazole derivatives was established by degradation, including oxidation to a known pyrazoledicarboxylic acid (**52**), and also by n.m.r. spectroscopy.[60,102,199,200] An analogue of pyrazole **51** was obtained by heating 3,4,5-tri-O-acetylpentulose phenylosazones with pyridine.[201] Another type of dianhydro-osazone has been obtained by

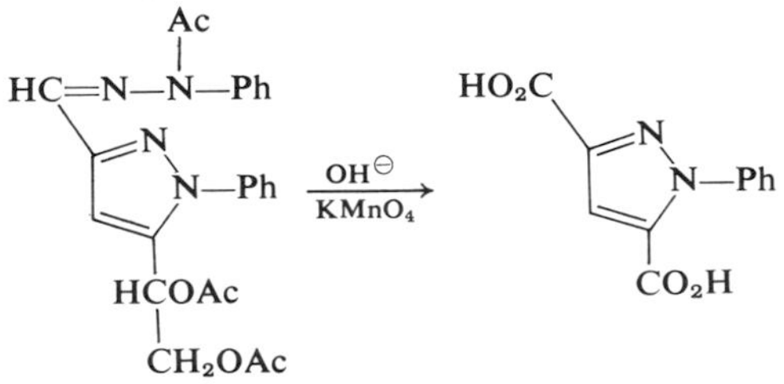

deacetylation of acetylated hexulose phenylosazones with sodium hydroxide. Two enantiomorphic compounds are obtained from hexose precursors, one from the D-hexoses and the other from the L-hexoses.[202,203] Tricyclic structures were first assigned to these compounds, but these were later revised[204] to the bicyclic chelated structures **53** and **54**. Subsequently, n.m.r. data provided evidence for structure **53**, which possesses one imino proton.[205]

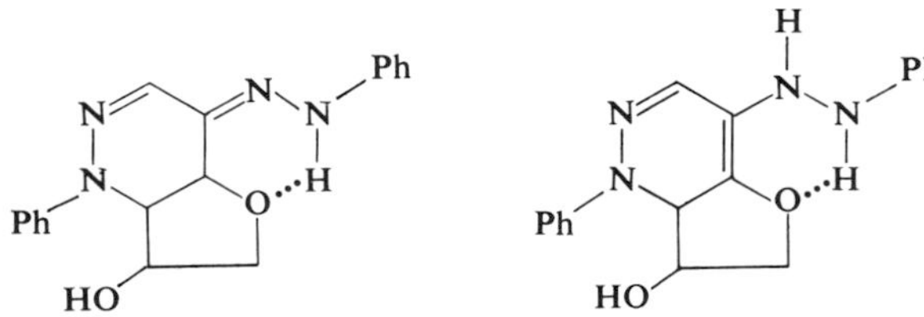

6. Derivatives

a. Esters.—Mild acetylation of D-*arabino*- and D-*lyxo*-hexulose phenyl-osazones leads to tetra-*O*-acetyl derivatives.[59,100,103] Stronger acetylating agents, such as acetyl chloride in *N,N*-dimethylaniline, yield the 1-*N*-acetyl-tetra-*O*-acetyl derivative.[103]

```
                                                                  Ac
                                                                  |
     HC=N—NHPh                 HC=N—NHPh                 HC=N—N—Ph
       |                          |                          |
     C=N—NHPh                  C=N—NHPh                  C=N—NH—Ph
       |            Ac₂O         |            AcCl          |
     AcOCH        ←——————      HOCH         ——————→       AcOCH
       |                          |                          |
     HCOAc                      HCOH                       HCOAc
       |                          |                          |
     HCOAc                      HCOH                       HCOAc
       |                          |                          |
     CH₂OAc                     CH₂OH                      CH₂OAc
```

Acylation of Osazones

Benzoylation of hexulose osazones with benzoyl chloride in pyridine affords crystalline pentabenzoates.[102,114] These contain one *N*-benzoyl and four *O*-benzoyl groups. The fully acetylated and benzoylated sugar phenylosazones thus have acyclic structures.

b. Ethers.—Methylation of D-*arabino*-hexulose phenylosazone (**21**) under mild conditions has been stated to give rise preferentially to the 5-methyl ether.[206,206a] Repeated methylation leads to a mixture of methylated products.

When treated with acetone in the presence of an acidic catalyst, the osazone (**21**) yields a 5,6-isopropylidene acetal (**55**), which is converted into the *O*-isopropylidene-*N*-methyl-di-*O*-methyl derivative (**56**) on methylation with methyl sulfate.[207] These reactions also provide evidence in favor of the acyclic structure of the sugar osazones.

```
                                                                    Me
                                                                    |
     HC=N—NHPh                 HC=N—NHPh                   HC=N—N—Ph
       |                          |                           |
     C=N—NHPh                  C=N—NHPh                    C=N—NHPh
       |           Me₂CO          |           Me₂SO₄          |
     HOCH        ——————→        HOCH        ——————→        MeOCH
       |            H⊕            |            NaOH           |
     HCOH                       HCOH                        HCOMe
       |                          |                           |
     HCOH                       HCO                         HCO
       |                          |  \CMe₂                    |  \CMe₂
     CH₂OH                      H₂CO/                       H₂CO/
       21                         55                          56
```

Mono (*N*-methyl)phenylosazones are obtained either by direct methyla-tion[207,208] or by transhydrazonation of bis(*N*-methyl-*N*-phenyl)osazones with phenylhydrazine.[209] Two mono-*N*-methyl derivatives of D-*arabino*-hexulose phenylosazone, designated mixed osazones "A" and "B," have been described in the literature. The so-called mixed osazone "A" has been identified as 1-(*N*-methyl-*N*-phenyl)-2-phenylosazone, whereas mixed osazone "B" proved to be a mixture of compound "A" and D-*arabino*-hexulose phenyl-osazone.[105,115] No method has yet been proposed that leads to bis-*N*-methylated osazones by direct methylation.

7. *Identification and Isolation of Sugars as Their Osazones*

The phenylosazones of sugars, because of their insolubility, are of con-siderable value for the identification of sugars. Since the asymmetry of carbon 2 is lost during their preparation, the same osazone is formed from three related sugars, the two epimeric aldoses and the corresponding ketose.

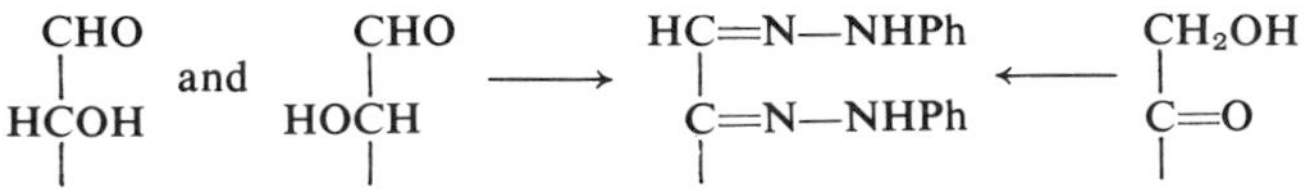

$$
\begin{array}{cccc}
\text{CHO} & \text{CHO} & \text{HC}=\text{N}-\text{NHPh} & \text{CH}_2\text{OH} \\
| & \text{and} \quad | & \longrightarrow \quad | & \longleftarrow \quad | \\
\text{HCOH} & \text{HOCH} & \text{C}=\text{N}-\text{NHPh} & \text{C}=\text{O} \\
| & | & | & |
\end{array}
$$

There are four D- and four (enantiomorphous) L-hexulose phenylosazones, and two D- and two L-pentulose analogues. Preparation of the osazone of an unknown sugar is therefore utilized only for the preliminary identification of the unknown product as one of a group of sugars. Final identification has to be made on the basis of other derivatives, such as the dithioacetals, the hydrazones, or the formazans, which are characteristic of the individual sugar. Confirmation of the identity of the osazone can be achieved by conversion into the corresponding osotriazole, or by chromatographic separation on paper,[210,211] thin layers,[212–214] or columns.[174,215–218] Photomicrographs of many phenylosazones are of considerable value for identification purposes.[219]

V. POLY(HYDRAZONES)

If the assumption is correct that the chelated ring in osazones stops the osazone-forming reaction from proceeding beyond C-2, it would be expected that osazones incapable of forming such chelated rings would undergo extended reaction, ultimately involving the whole poly(hydroxyalkyl) chain. This expectation was verified in the case of the *N*-methyl-*N*-phenylosazones, which are incapable of forming chelated rings. Treating trioses, tetroses, and pentoses with *N*-methyl-*N*-phenylhydrazine afforded tris-, tetrakis-, and pentakis-hydrazones. These compounds were given the generic name of alkazones.[220]

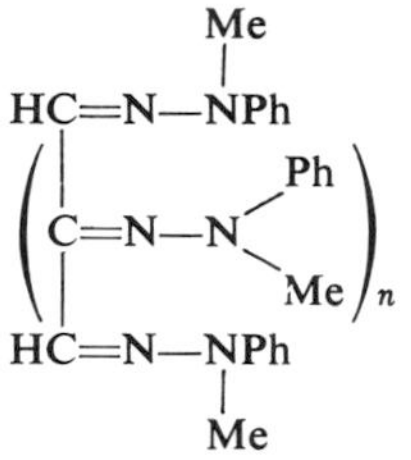

An Alkazone

Triose tris(hydrazones) can also be obtained from periodate-oxidized bis(hydrazones) by treatment with hydrazines. Mesoxalaldehyde tris(benzoyl-hydrazone) (57) was found to react with iodine in a manner similar to that of hexosulose bis(benzoylhydrazones) (see p. 965) to give a triazole derivative[193] (58).

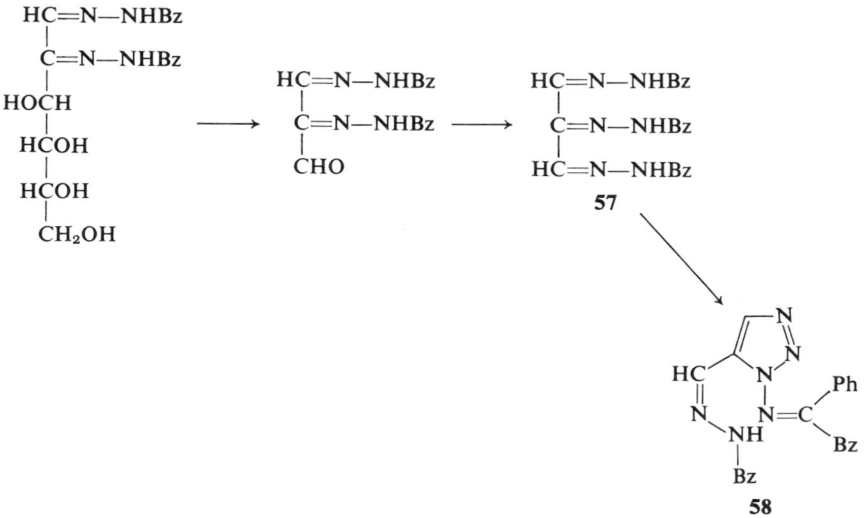

VI. PHENYLHYDRAZONES OF PERIODATE-OXIDIZED SUGAR DERIVATIVES, AND THEIR HYDRAZINOLYSIS

The structure of hydrazine derivatives from periodate-oxidized sugars remained controversial for a long time.[221] Structures based on parent di-aldehydes, hemiacetals, and aldehydo-hemiacetals have been proposed[222-225]

References start on p. 981.

for the phenylhydrazones obtained by treatment of the oxidation products with ice-cold solutions of phenylhydrazine.

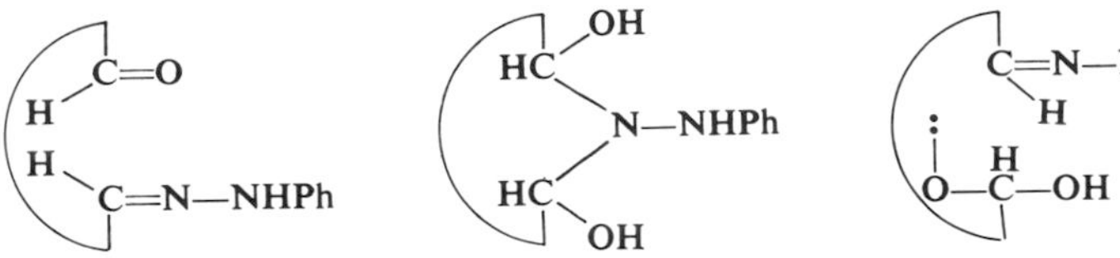

In the special cases of methyl 4,6-*O*-benzylidene-α-D-glucopyranoside and of methyl 4,6-*O*-benzylidene-β-D-galactopyranoside, reaction of the "dialdehyde" with phenylhydrazine at higher temperatures leads to 3-phenylazo-3-deoxyaldohexose derivatives such as **59**, through reconstitution of the C-2–C-3 linkage.[221,226–228]

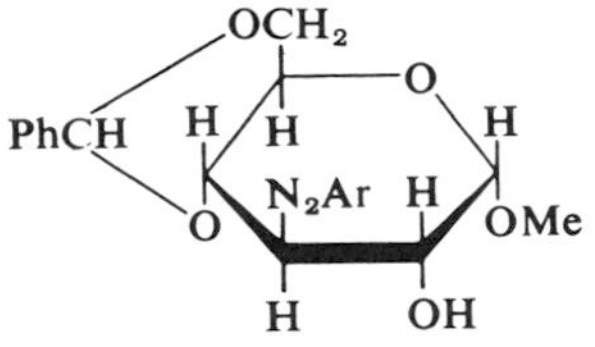

59

By catalytic reduction of the phenylazo sugars, new 3-amino-3-deoxy-hexoses were obtained.[229] Through isomerization of the phenylazo sugars to phenylhydrazones, and consecutive cleavage with benzaldehyde, the corresponding hexosid-3-ulose derivatives are obtained.[230]

Most of the monophenylhydrazones formed from periodate-oxidized sugars react with diazotized aniline to give formazans, thus supporting the aldehyde–hemiacetal structures assigned.[231–234] For example, the 5-phenylhydrazone (**60**) of 1,2-*O*-isopropylidene-D-*xylo*-pentodialdo-1,4-furanose and the 1,5-bis(phenylhydrazone) (**62**) of *xylo*-pentodialdose afford a monoformazan (**61**) and a bis(formazan) (**63**), respectively.[232]

A monoformazan derivative (**67**) has been obtained from the phenylhydrazone (**66**) of periodate-oxidized methyl α-glucopyranoside (**64** ≡ **65**), and supports the aldehyde–hemiacetal structure proposed for this oxidation product.[233]

Similarly, periodate-oxidized oligosaccharides and polysaccharides give red formazans containing a formazan group for each sugar residue oxidized.[231,234]

Periodate-oxidized polysaccharides, when heated with an excess of phenylhydrazine in acetic acid solution, are completely degraded to osazones. This reaction, known as the Barry degradation,[224,235–236] has often been used for

60

61

62

63

determining the structure of polysaccharides. The osazones obtained can be separated by crystallization or on circular paper chromatograms. In this way a periodate-oxidized D-xylan containing (1 → 3) and (1 → 4) links gave glyoxal bis(phenylhydrazone), triosulose bis(phenylhydrazone), and D-*threo*-pentulose phenylosazone. The reaction is especially useful for determining the structure of polysaccharides having (1 → 3) linkages. The nonreducing end group is first oxidized by periodate and then removed by treatment with phenylhydrazine, to give glyoxal bis(phenylhydrazone) and a new polysaccharide having one sugar residue fewer.

References start on p. 981.

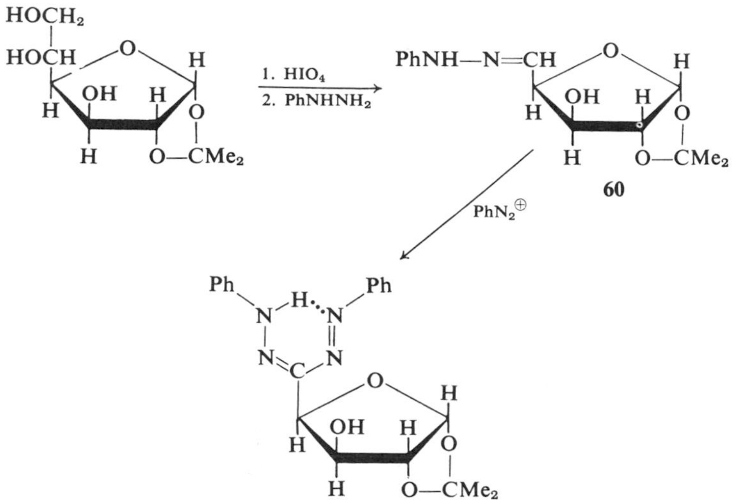

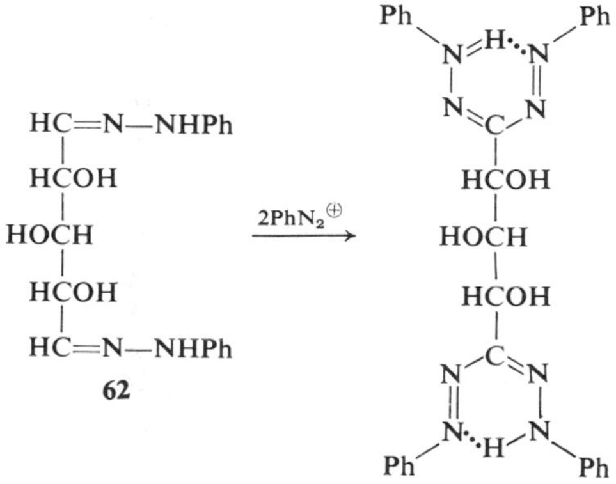

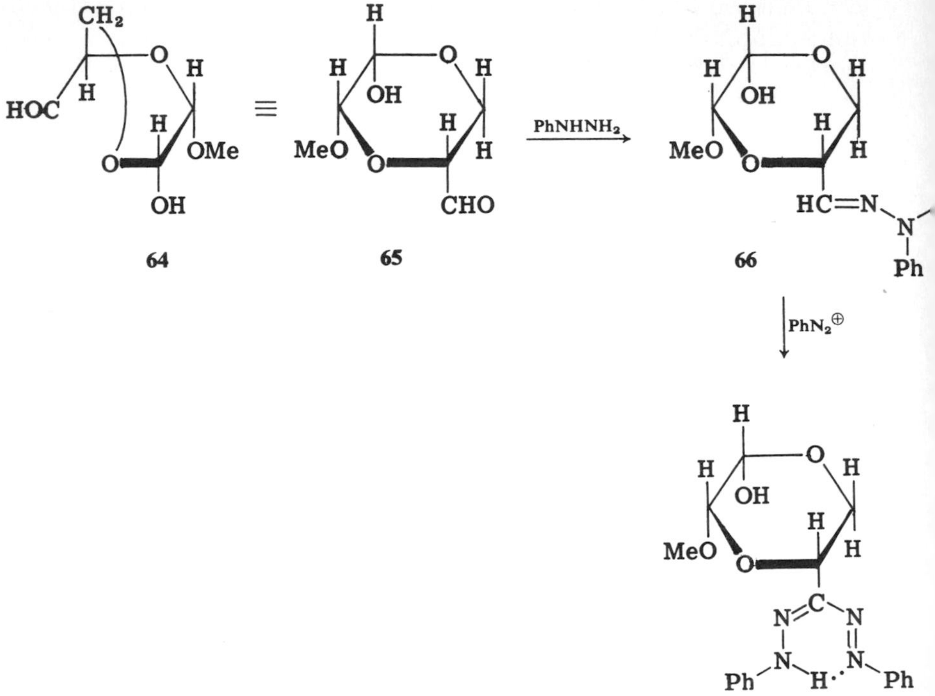

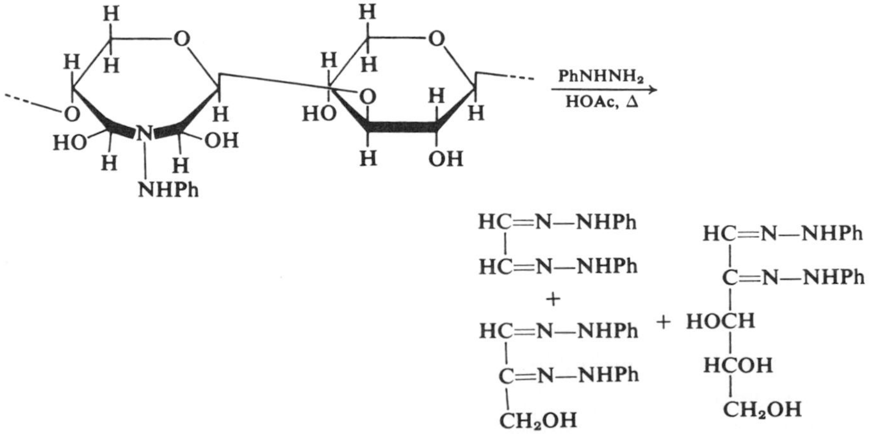

Barry Degradation of a β-D-Xylan Containing $(1 \rightarrow 3)$ and $(1 \rightarrow 4)$ Linkages

It was also found[237] that, during the degradation of the polysaccharides, a small proportion of a red product is always formed; it was identified as rubazonic acid [1-phenyl-4-(1-phenyl-5-oxopyrazolin-4-ylideneamino)pyrazol-5-one]. Rubazonic acid is formed not only from triosulose bis(phenylhydrazone) during the Barry degradation, but also from D-*arabino*-hexosulose 1-phenylhydrazone through the action of 1-methyl-1-phenylhydrazine.

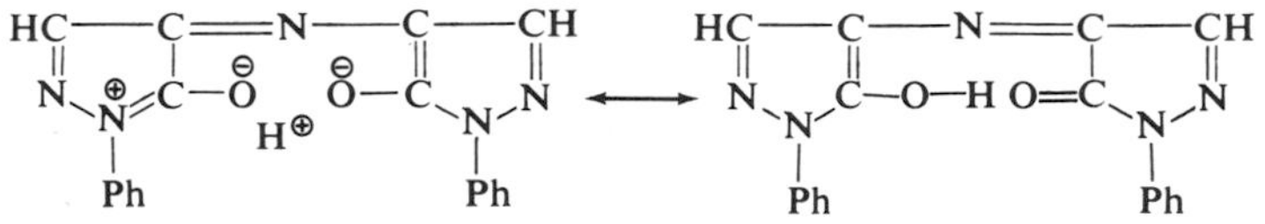

Rubazonic acid

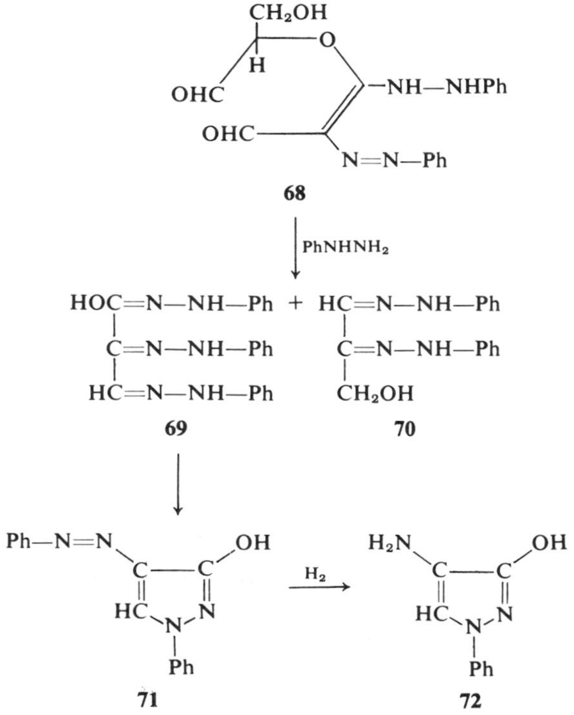

Another type of phenylhydrazinolysis occurs with periodate-oxidized dehydro-D-hexulose phenylosazone.[152] Degradation of the oxidation product (**68**) by refluxing with phenylhydrazine gives triosulose bis(phenylhydrazone) (**70**) and the phenylhydrazide (**69**) of formylglyoxylic acid bis(phenylhydrazone). Acid hydrolysis of the latter compound causes elimination of

References start on p. 981.

phenylhydrazine to give 3-hydroxy-1-phenyl-4-(phenylazo)pyrazole (71), which on catalytic hydrogenation gives 4-amino-3-hydroxy-1-phenylpyrazole (72).

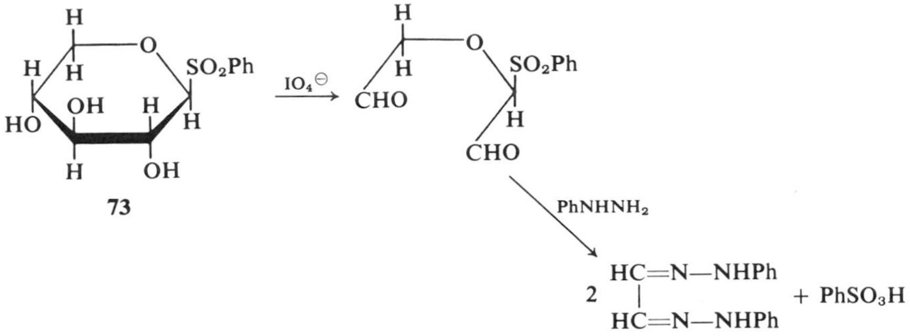

Periodate oxidation of phenylsulfonyl α-D-xylopyranoside (73) followed by degradation with phenylhydrazine leads to benzenesulfonic acid and glyoxal bis(phenylhydrazone).[238] It should be noted, however, that phenylsulfonyl glucosides may give osazones without periodate oxidation.[239]

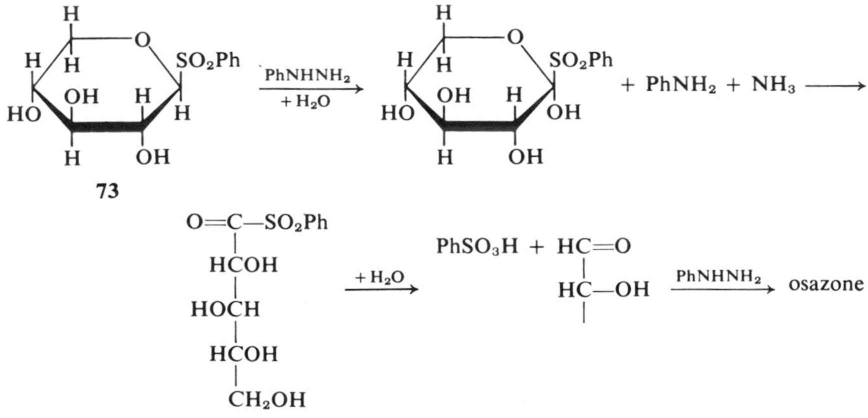

The phenylhydrazinolysis of periodate-oxidized sugar phosphates,[240] nucleosides,[241] phospholipids,[242] and glycerol teichoic acids[243] has been reported. As an example, the degradation of periodate-oxidized glycerol phosphates may be represented as follows[244]:

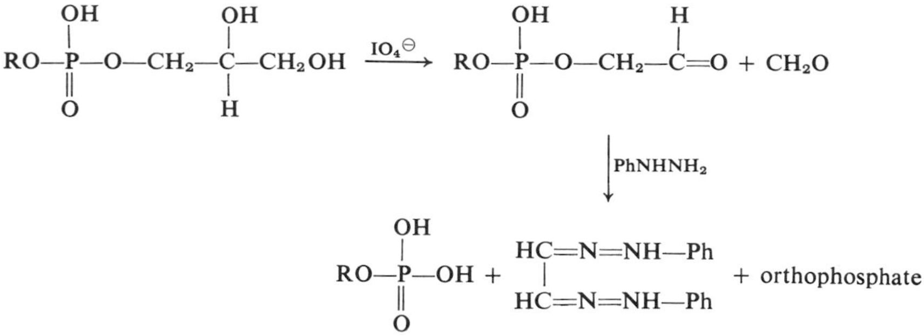

$$RO-\overset{OH}{\underset{\overset{\|}{O}}{P}}-O-CH_2-\overset{OH}{\underset{\overset{|}{H}}{C}}-CH_2OH \xrightarrow{IO_4^{\ominus}} RO-\overset{OH}{\underset{\overset{\|}{O}}{P}}-O-CH_2-\overset{H}{\underset{}{C}}{=}O + CH_2O$$

$$\Big\downarrow PhNHNH_2$$

$$RO-\overset{OH}{\underset{\overset{\|}{O}}{P}}-OH + \underset{HC{=}N{=}NH-Ph}{\overset{HC{=}N{=}NH-Ph}{|}} + \text{orthophosphate}$$

Attempts have also been made to effect similar degradations of glycosaminoglycans.[245–252]

VII. OXIMES

Reducing sugars react with hydroxylamine to give oximes which may be represented in the acyclic or cyclic forms (**74** and **75**). Owing to their solubilities

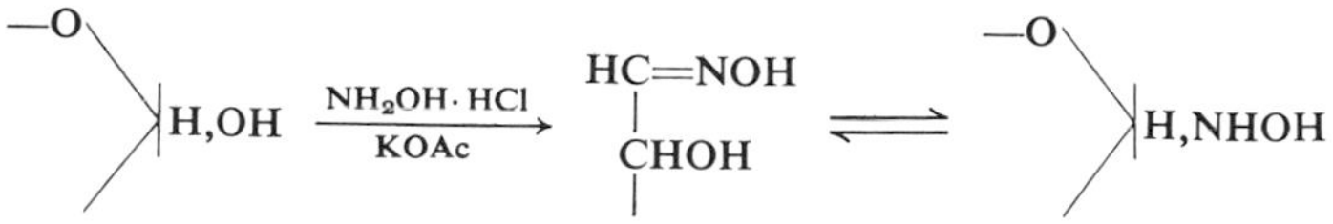

$$\underset{}{\overset{-O}{\diagdown}}\!\!\!\big\rangle H,OH \xrightarrow[\text{KOAc}]{NH_2OH\cdot HCl} \underset{\overset{|}{CHOH}}{\overset{HC{=}NOH}{|}} \rightleftharpoons \underset{}{\overset{-O}{\diagdown}}\!\!\!\big\rangle H,NHOH$$

74　　　　　　　**75**

Tautomeric oximes of aldoses

in water and ethanol, oximes are not convenient derivatives for the identification of sugars, but have been used as intermediates in the Wohl degradation and in the preparation of acyclic sugar derivatives (see Vol. IA, Chapter 10). Like hydrazones, the oximes of monosaccharides exhibit mutarotation,[253] which has been attributed to the establishment of an equilibrium between the cyclic and acyclic forms. It is possible, however, that *syn* and *anti* forms might coexist in solution; in the solid state only one isomer is isolated. Two crystalline hexaacetates of D-glucose oxime have been isolated[254,255]: an acyclic derivative prepared from penta-O-acetyl-*aldehydo*-D-glucose by acetylation of its oxime, and a cyclic form prepared by acetylation of D-glucose oxime at low temperature.[256,257] The acyclic acetate was found to possess six O-acetyl groups, whereas the cyclic form (**79**) possessed only five O-acetyl groups; one group

(N-acetyl) resisted mild alkaline hydrolysis.[59] Upon complete deacetylation both acetates gave the same D-glucose oxime, which suggests that the cyclic and acyclic forms of the oxime are interconvertible.

When D-glucose oxime is acetylated at higher temperatures, the nitrile pentaacetate (77) is the main product. Since at the higher temperatures the acyclic hexaacetate (78) of D-glucose oxime also gives the nitrile pentaacetate (77) in good yields, it was concluded that the acyclic oxime (76) is probably an intermediate in the formation of the nitrile (77), and that both cyclic and acyclic acetates (78 and 79) were formed during acetylation.

The products obtained upon acetylation of the sugar oximes depend not only on the temperature but also on the configuration of the sugar involved.[258,260] Nevertheless, use of controlled conditions provides via the acetylated aldononitriles a useful quantitative micro-analytical method for sugars.[260a]

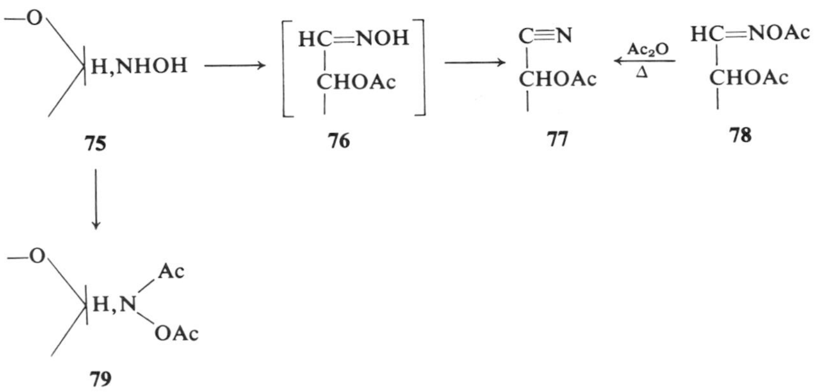

The oximes of arabinose, rhamnose, xylose, and 2-amino-2-deoxy-D-glucose, on low-temperature acetylation, yield only the nitriles. D-Glucose gives the cyclic hexaacetate, D-mannose and L-fucose give the acyclic hexa-acetates, and D-galactose yields a mixture of all three types. At higher temperatures, the proportion of the nitrile in the reaction mixtures increases considerably. Detailed studies by ^{13}C-n.m.r. and g.l.c. have been made on *syn* and *anti* forms of O-methyl oximes of monosaccharides.[260b]

The Wohl method for shortening the carbon chain of sugars utilizes the acetylated nitrile prepared by the above procedure and is described in more detail in Vol. IA, Chapter 3.

Reduction of monosaccharide oximes by Winestock and Plaut,[261] with hydrogen over platinum oxide in acetic acid, yields 1-amino-1-deoxyalditols, thus resembling the hydrazones. The oximes of keto derivatives of sugars have been utilized extensively in synthesis of amino sugars (see Chapter 16).

VIII. HYDRAZINO DERIVATIVES

Derivatives in which one of the hydroxyl groups of the sugar is replaced by a hydrazino group were first reported by Freudenberg and co-workers[261a]; since that time, numerous examples of such products have been prepared by hydrazinolysis of various sulfonic esters of sugars, and these have served, by subsequent reduction, as precursors for the corresponding amino sugars. (See Chapter 16, Section II, D, 2.) These procedures have also been applied with derivatives of amylose[261b] and cellulose.[262]

IX. DIAZO DERIVATIVES

Three types of diazo derivatives of sugars are known, of which the most extensively studied are the diazoketones, formed by interaction of aldonic acid chlorides with diazomethane or diazoethane. Diazoketones are involved in the Arndt–Eistert synthesis,[263] which was first applied to carbohydrates by

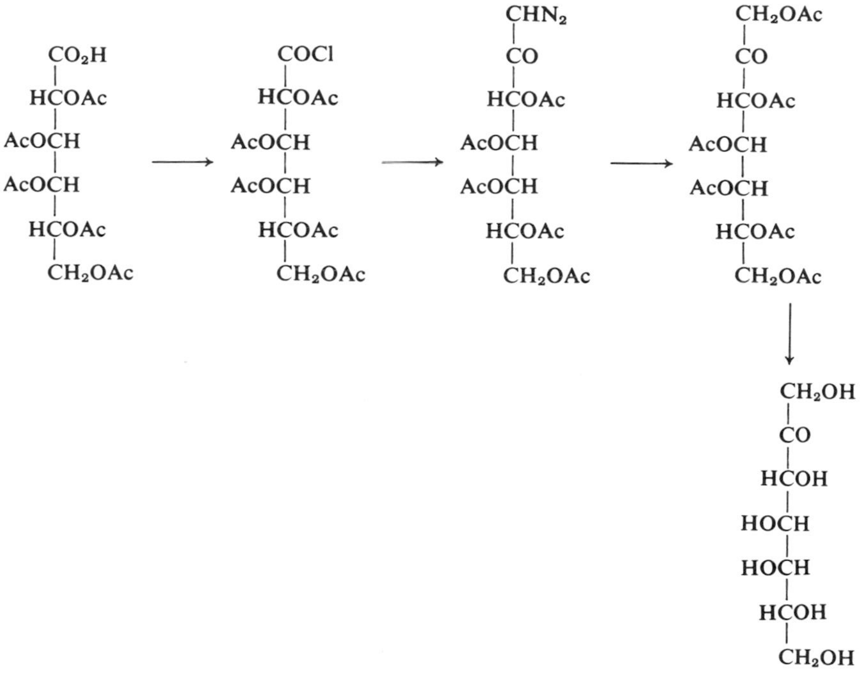

$$
\begin{array}{ccccc}
& & & \text{CHN}_2 & \text{CH}_2\text{OAc} \\
& & & | & | \\
\text{CO}_2\text{H} & \text{COCl} & \text{CO} & \text{CO} \\
| & | & | & | \\
\text{HCOAc} & \text{HCOAc} & \text{HCOAc} & \text{HCOAc} \\
| & | & | & | \\
\text{AcOCH} & \text{AcOCH} & \text{AcOCH} & \text{AcOCH} \\
| \quad \longrightarrow & | \quad \longrightarrow & | \quad \longrightarrow & | \\
\text{AcOCH} & \text{AcOCH} & \text{AcOCH} & \text{AcOCH} \\
| & | & | & | \\
\text{HCOAc} & \text{HCOAc} & \text{HCOAc} & \text{HCOAc} \\
| & | & | & | \\
\text{CH}_2\text{OAc} & \text{CH}_2\text{OAc} & \text{CH}_2\text{OAc} & \text{CH}_2\text{OAc}
\end{array}
$$

$$
\begin{array}{c}
\downarrow \\
\text{CH}_2\text{OH} \\
| \\
\text{CO} \\
| \\
\text{HCOH} \\
| \\
\text{HOCH} \\
| \\
\text{HOCH} \\
| \\
\text{HCOH} \\
| \\
\text{CH}_2\text{OH} \\
\textbf{80}
\end{array}
$$

The Diazoketone Route to Higher Ketoses

Reichstein and Gätzi[263] to convert L-xylose into L-sorbose. Later, Wolfrom[265] isolated several diazoketones and used these yellow crystalline compounds for the synthesis of pentuloses such as D-*erythro*-pentulose,[266] the hexuloses D-sorbose[267] and D-psicose,[81] the heptuloses having the D-*gluco*-[268,269] D-*manno*,[270] D-*altro*,[271] and D-*galacto*[272] configurations, together with octuloses[273,274] and nonuloses.[275] The reaction may be exemplified by the conversion of D-galactonic acid into D-*galacto*-heptulose (**80**).

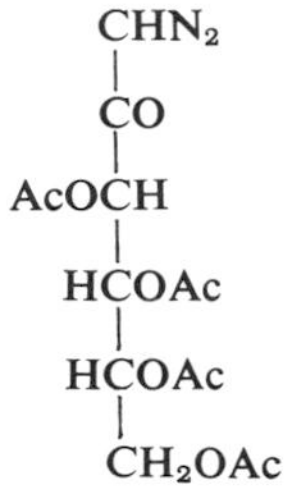

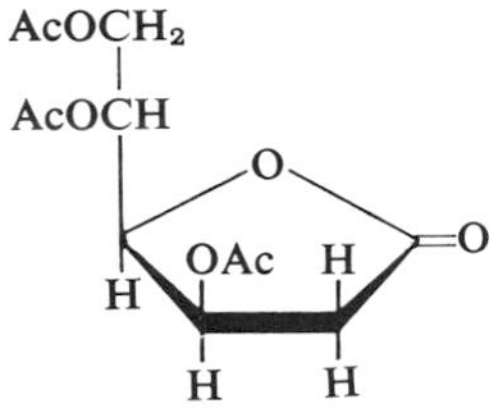

Diazoketones such as **81** undergo the Wolff rearrangement in water in the presence of catalytic amounts of silver oxide,[276] to give the 2-deoxyaldonic acid, isolated as the acetylated lactone (**82**). Diazoketones are readily converted into the corresponding 1-deoxy-1-halogenoketoses (**83**) with hydrogen halides[276,277] and yield[278] 1-deoxyketoses (**84**) on vigorous reduction with aluminum amalgam. Milder reduction with hydrogen sulfide affords[278] the aldosulose 1-hydrazone (**85**).

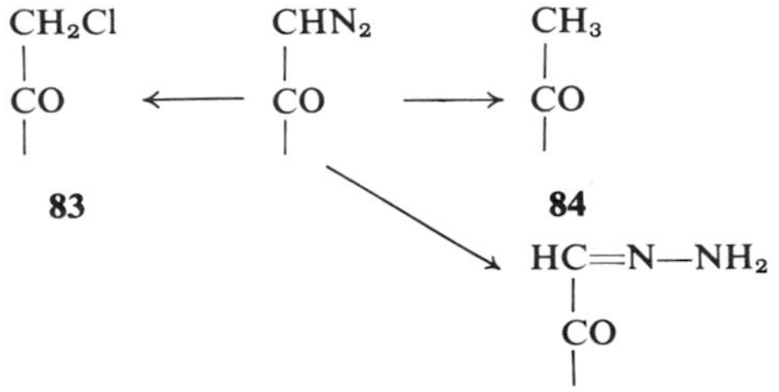

A nonterminal diazo derivative was encountered by Levene,[279] who treated ethyl 2-amino-2-deoxy-D-gluconate (**86**) with nitrous acid and obtained a yellow crystalline compound that gave correct analyses for the 2-diazo derivative (**87**). The diazo structure of this compound and a series of homologues has been established[280] by n.m.r. and mass spectroscopy.

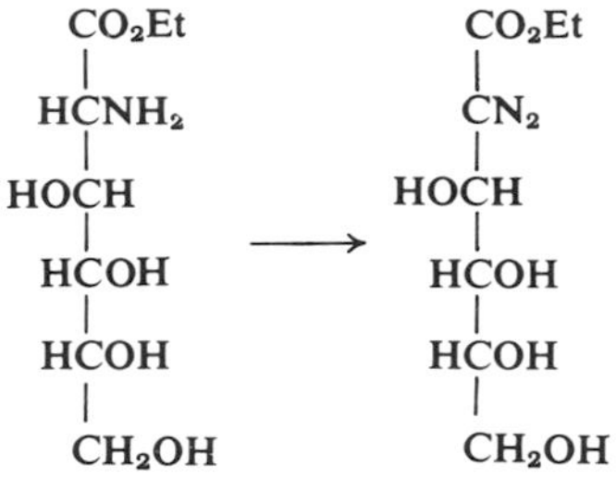

$$
\begin{array}{ccc}
\text{CO}_2\text{Et} & & \text{CO}_2\text{Et} \\
| & & | \\
\text{HCNH}_2 & & \text{CN}_2 \\
| & & | \\
\text{HOCH} & & \text{HOCH} \\
| & \longrightarrow & | \\
\text{HCOH} & & \text{HCOH} \\
| & & | \\
\text{HCOH} & & \text{HCOH} \\
| & & | \\
\text{CH}_2\text{OH} & & \text{CH}_2\text{OH} \\
\mathbf{86} & & \mathbf{87}
\end{array}
$$

Nonterminal diazo derivatives in cyclic systems can readily be prepared[281] from aldosulose derivatives by the Bamford–Stevens[282] route, through pyrolysis of the sodium salts of appropriate sulfonylhydrazones. For example, pyrolysis of the sodium salt of the 2,4,5-trichlorobenzenesulfonylhydrazone (**88**) obtained from 1,6-anhydro-2,3-*O*-isopropylidene-*β*-D-*lyxo*-hexopyranos-4-ulose yields the 4-diazo derivative (**89**) as a stable, yellow crystalline compound.

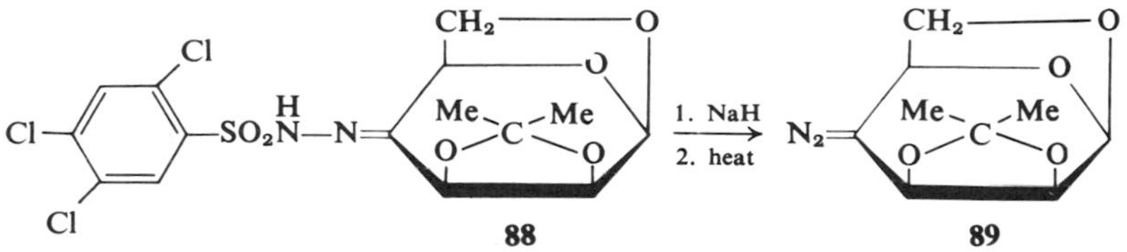

X. AZIDO DERIVATIVES

Azido derivatives of sugars are commonly prepared by displacing a halogen atom or a sulfonyl group by azide ion.[283,283a-c] Like other organic azides they are characterized by a strong absorption in the infrared at ν 2200 cm^{-1}, which is quite useful for diagnostic purposes. Azides of sugars are widely utilized in synthesis of amino sugars (see Chapter 16).

Azides can also be used as starting materials for synthesis of a multitude of other nitrogenous derivatives, such as hydrazines, imines, nitrogen heterocycles, and also carbonyl derivatives. Thus, on reduction they first yield hydrazino derivatives and then amines. If flanked by a suitable leaving group in a *trans*-diaxial orientation, they yield aziridines.[283d]

References start on p. 981.

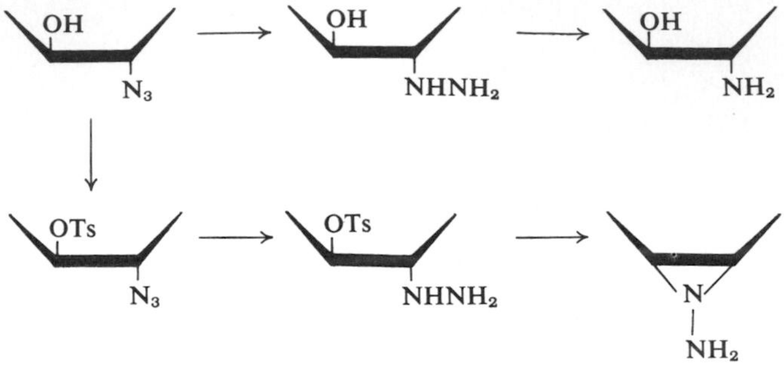

Some Reactions of Azido Sugars

Photolysis of terminal azides yields polymeric imines, convertible in high yields into aldehydes. Thus, photolysis of methyl 6-azido-6-deoxy-α-D-glucopyranoside (90) or its triacetate afforded[284] the polymeric imine (91), which on hydrolysis gives in high yield the 6-aldehyde (92). Secondary azides

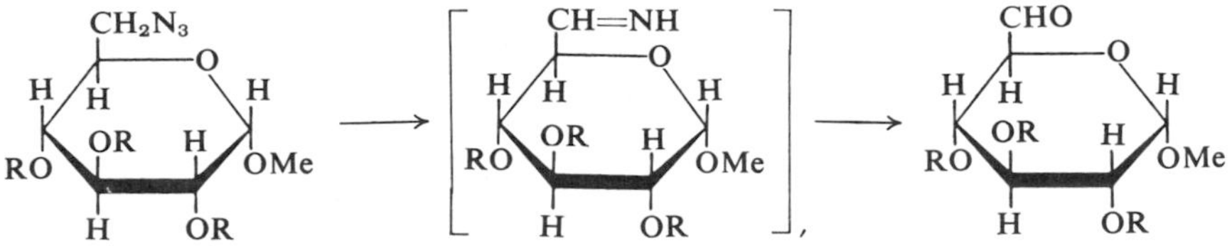

R = H or Ac

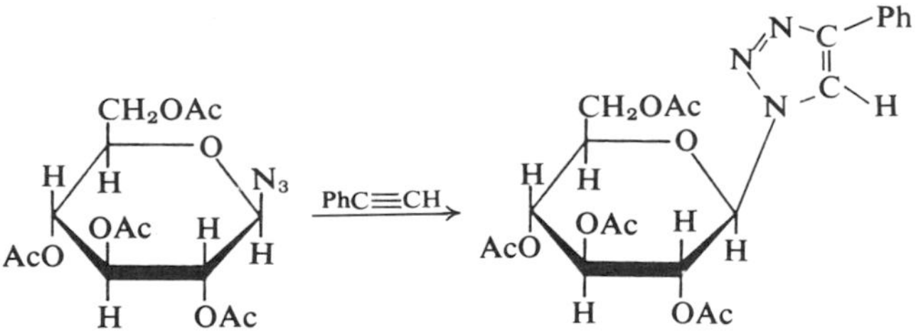

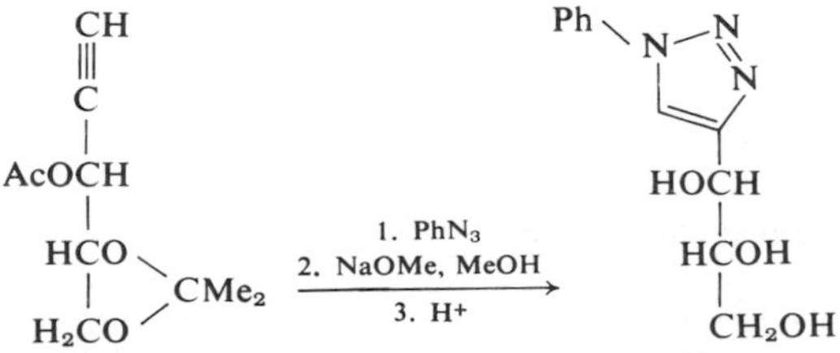

likewise give ketones, but the yields are not sufficiently high for preparative utility.[285] A synthetic route to glycosyl triazoles such as **94** is provided by condensation of glycosyl azides (such as **93**) with phenylacetylene.[286,287]

Related triazoles, positional isomers of the well-known osotriazoles, are obtained by 1,3-dipolar cycloaddition of phenyl azide with acetylenic sugar derivatives. Thus, addition of phenyl azide to the 5-carbon acetylenic sugar derivative **95**, followed by removal of the protecting groups, yields 4-(D-*threo*-trihydroxypropyl)-1-phenyl-1,2,3-triazole (**96**), the 1-phenyl isomer of D-*threo*-hexulose phenylosotriazole.[287] Similar reduction of epimeric hept-1-ynitols gave small amounts of 5-(hydroxyalkyl)-1-phenyl-1,2,3-triazoles in addition to the 4-substituted (major) product.[288]

REFERENCES

1. E. Davidis, *Ber.*, **29**, 2308 (1896).
2. H.-H. Stroh, A. Arnold, and H.-G. Scharnow, *Ber.*, **98**, 1404 (1965).
3. G. Henseke and H. J. Binte, *Ber.*, **88**, 1167 (1955).
3a. R. W. Binkley and W. W. Binkley, *Carbohyd. Res.*, **13**, 163 (1970).
4. R. S. Tipson, *J. Org. Chem.*, **27**, 2272 (1962).
5. H.-H. Stroh and H.-G. Scharnow, *Ber.*, **98**, 1588 (1965).
6. B. Helferich and H. Schirp, *Ber.*, **84**, 469 (1951); **86**, 547 (1953); L. Maquenne and W. Goodwin, *Bull. Soc. Chim. Fr.*, 1075 (1964).
7. H. Wolff, *Ber.*, **28**, 160 (1895); E. L. Hirst, J. K. N. Jones, and E. A. Woods, *J. Chem. Soc.*, 1048 (1947); B. Holmberg, *Arkiv. Kemi*, 7, 501 (1954); H. Zinner, J. Brock, B. Peter, and H. Schaukellis, *J. Prakt. Chem.*, [4] 29, 101 (1965); H.-H. Stroh and H. Tengler, *Ber.*, **101**, 751 (1968).
8. K. Freudenberg and F. Blummel, *Ann.*, **440**, 45 (1924).
9. H. Zinner, H. Brenken, W. Braun, I. Falk, E. Fechtner, and E. Hahner, *Ann.*, **622**, 133 (1959); L. Hough and J. K. N. Jones, *J. Chem. Soc.*, 1122, 3191 (1951); H. Zinner and W. Rehpenning, *Carbohyd. Res.*, **5**, 176 (1967).
10. H.-H. Stroh and E. Repte, *Ber.*, **93**, 1148 (1960).
11. H.-H. Stroh and H. Lamprecht, *Ber.*, **96**, 651 (1963).
12. H.-H. Stroh and B. Ihlo, *Ber.*, **96**, 658 (1963).
13. H.-H. Stroh and K. Milde, *Ber.*, **98**, 941 (1965).
14. H.-H. Stroh and R. Apel, *Ber.*, **98**, 2500 (1965).
15. H.-H. Stroh, W. Kegel, and G. Lehmann, *Ber.*, **98**, 1956 (1965).
16. H.-H. Stroh, *Ber.*, **90**, 352 (1957).
17. H.-H. Stroh and H. E. Nikolajewski, *Ber.*, **95**, 562 (1962).
18. H.-H. Stroh, *Ber.*, **91**, 2657 (1958).
19. L. Mester and A. Messmer, *Methods Carbohyd. Chem.*, **2**, 117 (1963).
20. For early literature see A. W. van der Haar, "Anleitung zum Nachweis, zur Trennung und Bestimmung der Monosaccharide und Aldehydsäuren," Borntraeger, Berlin, 1920.
21. C. Neuberg and W. Neimann, *Ber.*, **35**, 2055 (1902).
22. E. G. R. Ardagh and F. C. Rutherford, *J. Amer. Chem. Soc.*, **57**, 1085 (1935).
23. E. Fischer, *Ber.*, **22**, 87 (1889).
24. G. Henseke and W. Liebenow, *Ber.*, **87**, 1068 (1954).

25. H. El Khadem, *J. Chem. Soc.*, 3452 (1953).
26. G. H. Stempel, Jr., *J. Amer. Chem. Soc.*, **56**, 1351 (1934).
27. A. Orning and G. H. Stempel, Jr., *J. Org. Chem.*, **4**, 410 (1939).
28. H.-H. Stroh, *Ber.*, **91**, 2645 (1958).
29. J. Compton and M. L. Wolfrom, *J. Amer. Chem. Soc.*, **56**, 1, 1157 (1934).
30. H.-H. Stroh and G. Westphal, *Ber.*, **96**, 184 (1963).
31. H. Jacobi, *Ann.*, **272**, 170 (1892).
32. C. L. Butler and L. H. Cretcher, *J. Amer. Chem. Soc.*, **53**, 4358 (1931).
33. Z. H. Skraup, *Monatsh. Chem.*, **10**, 401 (1889).
34. J. W. Haas, Jr., J. D. Strorey, and C. C. Lynch, *Anal. Chem.*, **34**, 145 (1962).
35. H. S. Blair and G. P. Roberts, *J. Chem. Soc. (C)*, 2357 (1969).
36. L. Mester and G. Vass, *Tetrahedron Lett.*, 5191 (1968).
37. J. R. Holker, *Chem. Ind.* (London), 546 (1964).
38. L. Mester and A. Major, *J. Amer. Chem. Soc.*, **77**, 4297 (1955).
39. R. Behrend and W. Reinsberg, *Ann.*, **377**, 189 (1910).
40. M. L. Wolfrom, *J. Amer. Chem. Soc.*, **51**, 2188 (1929).
41. M. L. Wolfrom, *J. Amer. Chem. Soc.*, **52**, 2464 (1930).
42. M. L. Wolfrom, L. W. Georges, and S. Soltzberg, *J. Amer. Chem. Soc.*, **56**, 1794 (1934).
43. M. L. Wolfrom and L. W. Georges, *J. Amer. Chem. Soc.*, **58**, 1781 (1936).
44. M. L. Wolfrom and S. Soltzberg, *J. Amer. Chem. Soc.*, **58**, 1783 (1936).
45. H. H. Fox and J. T. Gibas, *J. Org. Chem.*, **17**, 1653 (1952).
46. H. L. Yale, K. Losee, J. Martins, M. Holsing, F. M. Perry, and J. Bernstein, *J. Amer. Chem. Soc.*, **75**, 1933 (1953).
47. H. Zinner and W. Bock, *Ber.*, **89**, 1124 (1956).
48. G. Henseke and H. Hantschel, *Ber.*, **87**, 477 (1954).
49. C. S. Hudson, *J. Amer. Chem. Soc.*, **39**, 462 (1917).
50. E. Votoček, F. Valentin, and O. Leminger, *Coll. Trav. Chim. Tchecosl.*, **3**, 250 (1931).
51. E. Votoček and Z. Allan, *Coll. Trav. Chim. Tchecosl.*, **8**, 313 (1936).
52. E. Votoček and O. Wichterle, *Coll. Trav. Chim. Tchecosl.*, **8**, 322 (1936).
53. E. A. Lloyd and D. G. Doherty, *J. Amer. Chem. Soc.*, **74**, 1214 (1952); see also L. M. White and G. E. Secor, *Anal. Chem.*, **27**, 1016 (1955).
54. F. L. Humoller, S. J. Kuman, and F. H. Snyder, *J. Amer. Chem. Soc.*, **61**, 3370 (1939).
55. G. Henseke and E. Brose, *Ber.*, **91**, 2273 (1958).
56. H. Simon and W. Moldenhauer, *Ber.*, **100**, 3121 (1967).
57. H. El Khadem, M. L. Wolfrom, Z. M. El-Shafei, and S. H. El Ashry, *Carbohyd. Res.*, **4**, 225 (1967).
57a. H. Simon and W. Moldenhauer, *Ber.*, **101**, 2124 (1968).
58. M. L. Wolfrom, F. Shafizadeh, J. O. Wehrmüller, and R. K. Armstrong, *J. Org. Chem.*, **23**, 571 (1958).
59. M. L. Wolfrom, M. Konigsberg, and S. Soltzberg, *J. Amer. Chem. Soc.*, **58**, 490 (1936).
60. H. El Khadem, *J. Org. Chem.*, **29**, 2073 (1964).
61. M. L. Wolfrom and M. G. Blair, *J. Amer. Chem. Soc.*, **68**, 2110 (1946).
62. M. L. Wolfrom, A. Thompson, and D. R. Lineback, *J. Org. Chem.*, **27**, 2563 (1962).
63. M. L. Wolfrom, G. Fraenkel, D. R. Lineback, and F. Komitsky, Jr., *J. Org. Chem.*, **29**, 457 (1964).
64. L. Mester, *Advan. Carbohyd. Chem.*, **13**, 105 (1958).
64a. L. Mester and A. Messmer, *Methods Carbohyd. Chem.*, **2**, 119 (1963).
65. G. V. D. Tiers, S. Ploven, and S. Searles, *J. Org. Chem.*, **25**, 285 (1960); H. S. Isbell

and A. J. Fatiadi, *Carbohyd. Res.*, **2**, 204 (1966); L. Mester, A. Stephen, and J. Parello, *Tetrahedron Lett.*, 4119 (1968); P. B. Fischer, B. L. Kaul, and H. Zollinger, *Helv. Chim. Acta*, **51**, 1449 (1968).

65a. A. J. Fatiadi and H. S. Isbell, *Carbohyd. Res.*, **5**, 302 (1967).

66. G. Zemplén, L. Mester, A. Messmer, and E. Eckhart, *Acta Chim. Acad. Sci. Hung.*, **2**, 25 (1952).

66a. L. Mester, G. Vass, and M. Mester, *Z. Chem.*, **10**, 395 (1970).

67. L. Mester, *J. Polymer Sci.*, **30**, 329 (1958).

68. G. Zemplén, L. Mester, and E. Eckhart, *Ber.*, **86**, 472 (1953).

69. L. Mester and A. Messmer, *J. Chem. Soc.*, 3802 (1957).

70. L. Mester and A. Messmer, *Methods Carbohyd. Chem.*, **2**, 123 (1963).

71. G. Zemplén, L. Mester, and A. Messmer, *Ber.*, **86**, 697 (1953).

72. L. Mester and A. Messmer, *Methods Carbohyd. Chem.*, **2**, 126 (1963).

72a. A. Messmer and L. Mester, *Chem. Ind.* (London), 423 (1957).

73. L. Mester and A. Major, *J. Amer. Chem. Soc.*, **78**, 1403 (1956).

73a. H. S. Isbell and A. J. Fatiadi, *Carbohyd. Res.*, **11**, 303 (1969).

73b. F. M. Soliman, I. Pinter, and A. Messmer, *Acta Chim. Acad. Sci. Hung.*, **65**, 397 (1970); compare V. Zsoldos, A. Messmer, I. Pinter, and A. Neszmélyi, *Carbohyd. Res.* **62**, 105 (1978).

74. L. Mester, *Ber.*, **93**, 1684 (1961).

74a. H. El Khadem, G. H. Labib, and M. A. Nashed, *Carbohyd. Res.*, **3**, 509 (1967).

74b. H. El Khadem, D. Horton, M. H. Meshreki, and M. A. Nashed, *Carbohyd. Res.*, **13**, 317 (1970); **17**, 183 (1971).

75. The chemistry of osazones has been reviewed: H. El Khadem, *Advan. Carbohyd. Chem.*, **20**, 139 (1965); L. Mester, *Angew. Chem., Intern. Ed. Engl.*, **4**, 574 (1965); L. Mester, ed., "Dérivés hydraziniques des glucides," Hermann, Paris, 1967; H. Simon and A. Kraus, *Fortschr. Chem. Forsch.*, **14**, 430 (1970).

76. E. Fischer, *Ber.*, **17**, 579 (1884).

77. D. D. Garard and H. C. Sherman, *J. Amer. Chem. Soc.*, **40**, 955 (1918).

78. G. J. Bloink and K. H. Pausacker, *J. Chem. Soc.*, 622 (1951).

79. H. J. Haas and A. Seeliger, *Ber.*, **96**, 2427 (1963).

80. J. Ashmore and A. E. Renold, *J. Amer. Chem. Soc.*, **76**, 6189 (1954).

81. M. L. Wolfrom, A. Thompson, and E. F. Evans, *J. Amer. Chem. Soc.*, **67**, 1793 (1945).

82. E. Knecht and F. P. Thompson, *J. Chem. Soc.*, **125**, 222 (1924).

83. E. Fischer, *Ber.*, **20**, 821 (1887).

84. J. Kenner and E. C. Knight, *Ber.*, **69**, 341 (1936).

85. E. A. Braude and W. F. Forbes, *J. Chem. Soc.*, 1762 (1951).

86. F. Weygand, *Ber.*, **73**, 1284 (1940).

87. F. Weygand and M. Reckhaus, *Ber.*, **82**, 438 (1949).

88. G. J. Bloink and K. H. Pausacker, *J. Chem. Soc.*, 661 (1952).

89. V. C. Barry and P. W. D. Mitchell, *Nature*, **175**, 220 (1955).

90. F. Micheel and I. Dijong, *Ann.*, **669**, 136 (1963); I. Dijong and F. Micheel, *Ann.*, **684**, 216 (1965).

91. F. Weygand, H. Grisebach, K. D. Kirchner, and M. Haselhorst, *Ber.*, **88**, 487 (1955).

92. M. M. Shemyakin and V. I. Maimind, *Dokl. Akad. Nauk SSSR*, **102**, 1147 (1955).

93. E. M. Bandas, K. M. Ermolaev, V. I. Maimind, and M. M. Shemyakin, *Chem. Ind.* (London), 1195 (1959).

94. M. M. Shemyakin, V. I. Maimind, K. E. Ermolaev, and E. M. Bandas, *Dokl. Akad. Nauk SSSR*, **128**, 564 (1959).

95. M. M. Shemyakin, V. I. Maimind, K. M. Ermolaev, and E. M. Bandas, *Tetrahedron*, **21**, 2771 (1965).
96. F. Friedberg and L. Kaplan, *J. Amer. Chem. Soc.*, **79**, 2600 (1957).
97. F. Weygand, H. Simon, and J. Klebe, *Ber.*, **91**, 1567 (1958).
98. H. Simon, K. D. Keil, and F. Weygand, *Ber.*, **95**, 17 (1962).
98b. H. Simon and W. Moldenhauer, *Ber.*, **102**, 1191 (1969).
99. E. Chargaff and B. Magasanik, *J. Amer. Chem. Soc.*, **69**, 1959 (1947).
100. K. Maurer and B. Schiedt, *Ber.*, **68**, 2187 (1935).
101. L. Mester, H. El Khadem, and D. Horton, *J. Chem. Soc.* (*C*) 2567 (1970).
102. H. El Khadem, Z. M. El-Shafei, and M. A. A. Rahman, *Carbohyd. Res.*, **1**, 31 (1965).
103. H. El Khadem and M. A. A. Rahman, *Carbohyd. Res.*, **6**, 470 (1968).
104. L. F. Fieser and M. Fieser, "Organic Chemistry," D. C. Heath and Co., Boston, Mass., 1944, p. 351.
105. L. Mester, *J. Amer. Chem. Soc.*, **77**, 4301 (1955); G. Zemplén, L. Mester, A. Messmer, and A. Major, *Acta Chim. Acad. Sci. Hung.*, **7**, 455 (1955).
106. W. Reid and K. Sommer, *Ann.*, **611**, 108 (1958).
107. B. Jambor and L. Mester, *Acta Chim. Acad. Sci. Hung.*, **9**, 485 (1956).
108. G. Henseke and H. J. Binte, *Chimia*, **12**, 103 (1958).
109. K. Bjamer, S. Dahm, S. Furberg, and C. S. Petersen, *Acta Chem. Scand.*, **17**, 559 (1963).
110. S. Furberg, *Svensk. Kem. Tidskr.*, **77**, 4 (1965).
111. L. Mester, E. Moczar, and J. Parello, *Tetrahedron Lett.*, 3223 (1964).
112. L. Mester, E. Moczar, and J. Parello, *J. Amer. Chem. Soc.*, **87**, 596 (1965).
113. O. L. Chapman, R. W. King, W. J. Welstead, Jr., and T. J. Murphy, *J. Amer. Chem. Soc.*, **86**, 4968 (1964).
114. H. El Khadem, M. L. Wolfrom, and D. Horton, *J. Org. Chem.*, **30**, 838 (1965).
115. L. Mester, E. Moczar, G. Vass, and A. Schimpl, *Carbohyd. Res.*, **5**, 406 (1967).
116. L. Mester, *Chimia*, **23**, 133 (1969).
117. D. M. G. Lloyd and D. R. Marshall, *Chem. Ind.* (London), 1760, (1964); *Jerusalem Symp. Quantum Chem. Biochem.*, III, 85 (1971).
118. L. C. Dorman, *Tetrahedron Lett.*, 459 (1966).
119. O. L. Chapman, *Tetrahedron Lett.*, 2599 (1966).
120. L. Mester, G. Vass, A. Stephan, and J. Parello, *Tetrahedron Lett.*, 4053 (1968).
121. P. A. Levene and W. A. Jacobs, *Ber.*, **42**, 3249 (1909).
122. E. Zerner and R. Waltuch, *Monatsh. Chem.*, **35**, 1025 (1914).
123. W. A. Haworth, "The Constitution of Sugars," Edward Arnold & Co., London, 1929, p. 7.
124. L. L. Engel, *J. Amer. Chem. Soc.*, **57**, 2419 (1935).
125. L. Mester and A. Major, *J. Amer. Chem. Soc.*, **79**, 3232 (1957).
126. L. Mester, "Carbohydrate Chemistry of Substances of Biological Interest," 4th Intern. Congr. of Biochem., 1958, Pergamon Press, Vol. I, p. 173.
127. G. Henseke and H. Köhler, *Ann.*, **614**, 105 (1958).
128. G. Henseke, *Acta Chim. Acad. Sci. Hung.*, **12**, 173 (1957).
129. L. Mester, H. El Khadem, and G. Vass, *Tetrahedron Lett.*, 4135 (1969).
130. E. Fischer and E. F. Armstrong, *Ber.*, **35**, 3141 (1902).
131. S. Bayne, *Methods Carbohyd. Chem.*, **2**, 421 (1963).
132. L. Brüll, *Ann. Chim. Appl.*, **26**, 415 (1936).
133. H. Ohle, G. Henseke, and Z. Czyzewski, *Ber.*, **86**, 316 (1953).
134. G. Henseke and M. Winter, *Ber.*, **89**, 956 (1956).

135. G. Henseke and G. Badicke, *Ber.*, **89**, 2910 (1956).
136. G. Henseke and M. Winter, *Ber.*, **92**, 3156 (1959).
137. L. Mester and A. Major, *J. Chem. Soc.*, 3227 (1956).
138. H. Ohle, *Ber.*, **67**, 155 (1934).
139. G. Henseke and C. Bauer, *Ber.*, **92**, 501 (1959).
140. F. Weygand and A. Bergmann, *Ber.*, **80**, 255 (1947).
141. O. Diels, R. Meyer, and O. Onnen, *Ann.*, **525**, 94 (1936).
142. H. Simon and A. Kraus, in "Synthetic Methods for Carbohydrates," H. El Khadem, ed., *ACS Symp. Ser.*, **39**, 188 (1976).
143. E. Fischer, *Ber.*, **19**, 1920 (1886).
144. T. Neilson and H. C. S. Wood, *J. Chem. Soc.*, **44** (1962).
145. M. L. Wolfrom and J. L. Minor, *J. Org. Chem.*, **30**, 841 (1965).
146. R. Kuhn and W. Kirschenlohr, *Ber.*, **87**, 1547 (1954).
147. M. L. Wolfrom, H. El Khadem, and J. Vercellotti, *Chem. Ind.* (London), 545 (1964); *J. Org. Chem.*, **29**, 3284 (1964).
148. F. Micheel, G. Bode, and R. Siebert, *Ber.*, **70**, 1862 (1937).
149. B. Gross, M. El Sekily, S. Mancy, and H. El Khadem, *Carbohyd. Res.*, **37**, 384 (1974).
150. For nitrogen heterocycles obtained by this method see H. El Khadem, *Advan. Carbohyd. Chem. Biochem.*, **25**, 351 (1970).
151. O. Diels, E. Cluss, H. J. Stephan, and R. König, *Ber.*, **71**, 1189 (1938).
152. L. Mester and E. Moczar, *Chem. Ind.* (London), 554 (1962).
153. L. Mester and E. Moczar, *J. Org. Chem.*, **29**, 247 (1964).
154. The chemistry of osotriazoles has been reviewed by H. El Khadem, *Advan. Carbohyd. Chem.*, **18**, 99 (1963).
155. R. M. Hann and C. S. Hudson, *J. Amer. Chem. Soc.*, **66**, 735 (1944).
156. W. T. Haskins, R. M. Hann, and C. S. Hudson, *J. Amer. Chem. Soc.*, **67**, 939 (1945); **68**, 1766 (1946); **69**, 1050, 1461 (1947); **70**, 2288 (1948).
157. D. A. Rosenfeld, N. K. Richtmyer, and C. S. Hudson, *J. Amer. Chem. Soc.*, **73**, 4907 (1951).
158. L. C. Stewart, N. K. Richtmyer, and C. S. Hudson, *J. Amer. Chem. Soc.*, **74**, 2206 (1952).
159. J. W. Pratt, N. K. Richtmyer, and C. S. Hudson, *J. Amer. Chem. Soc.*, **74**, 2210 (1952).
160. J. V. Karabinos, R. M. Hann, and C. S. Hudson, *J. Amer. Chem. Soc.*, **75**, 4320 (1953).
161. N. K. Richtmyer and T. S. Bodenheimer, *J. Org. Chem.*, **27**, 1892 (1962).
162. P. P. Regna, *J. Amer. Chem. Soc.*, **69**, 246 (1947).
163. V. Ettel and J. Liebster, *Collect. Czech. Chem. Commun.*, **14**, 80 (1949).
164. E. Hardegger and H. El Khadem, *Helv. Chim. Acta*, **30**, 900, 1478 (1947).
165. E. Hardegger, H. El Khadem, and E. Schreier, *Helv. Chim. Acta*, **34**, 253 (1951).
166. J. F. Carson, *J. Amer. Chem. Soc.*, **77**, 1881 (1955).
167. W. Z. Hassid, M. Doudoroff, and H. A. Barker, *Arch. Biochem.*, **14**, 29 (1947).
168. W. Z. Hassid, M. Doudoroff, A. L. Potter, and H. A. Barker, *J. Amer. Chem. Soc.*, **70**, 306 (1948).
169. F. H. Stodola, H. J. Koepsel, and E. S. Sharpe, *J. Amer. Chem. Soc.*, **74**, 3202 (1952).
170. A. Thompson and M. L. Wolfrom, *J. Amer. Chem. Soc.*, **76**, 5173 (1954).
171. F. H. Stodola, E. S. Sharpe, and H. J. Keopsel, *J. Amer. Chem. Soc.*, **78**, 2514 (1956).
172. D. Rutherford and N. K. Richtmyer, *Carbohyd. Res.*, **11**, 341 (1969).
173. E. Hardegger and E. Schreier, *Helv. Chim. Acta*, **35**, 623 (1952).

174. H. El Khadem, E. Schreier, G. Stohr, and E. Hardegger, *Helv. Chim. Acta*, 35, 993 (1952).
175. S. Bayne, *J. Chem. Soc.*, 4993 (1952).
176. E. Schreier, G. Stohr, and E. Hardegger, *Helv. Chim. Acta*, 37, 574 (1954).
177. G. Hanisch and G. Henseke, *Ber.*, 100, 3225 (1967); 101, 2074 (1968).
178. A. J. Fatiadi, *Chem. Ind.* (London), 617 (1969).
179. H. El Khadem and Z. M. El-Shafei, *J. Chem. Soc.*, 3117 (1958); 1655 (1959).
180. H. El Khadem, Z. M. El-Shafei, and M. H. Meshreki, *J. Chem. Soc.*, 2957 (1961); 1524 (1965); H. El Khadem, Z. M. El-Shafei, and Y. S. Mohammed, *ibid.*, 3993 (1960); H. El Khadem, A. M. Kolkaila, and M. H. Meshreki, *ibid.*, 2531 (1963).
181. H. J. Tauber and G. Jellinek, *Ber.*, 85, 95 (1952).
182. M. L. Wolfrom, H. El Khadem, and H. Alfes, *J. Org. Chem.*, 29, 2072 (1964).
183. G. Henseke and M. Winter, *Ber.*, 93, 45 (1960).
184. H. Simon and J. Steffen, *Ber.*, 95, 358 (1962).
185. L. Mester and F. Weygand, *Bull. Soc. Chim. Fr.*, 350 (1960).
186. H. El Khadem, *J. Org. Chem.*, 28, 2478 (1963).
187. J. A. Mills, *Aust. J. Chem.*, 17, 277 (1964).
188. W. S. Chilton and R. C. Krahn, *J. Amer. Chem. Soc.*, 89, 4129 (1967).
189. G. G. Lyle and M. J. Piazza, *J. Org. Chem.*, 33, 2478 (1968).
190. H. El Khadem, D. Horton, and J. D. Wander, *J. Org. Chem.*, 37, 1630 (1972).
191. H. El Khadem, D. Horton, and T. F. Page, Jr., *J. Org. Chem.*, 33, 734 (1968).
192. H. El Khadem and M. A. E. Shaban, *Carbohyd. Res.*, 2, 178 (1966); *J. Chem. Soc.* (*C*), 519 (1967).
193. H. El Khadem, M. A. M. Nassr, and M. A. E. Shaban, *J. Chem. Soc.* (*C*), 1465 (1968); 1416 (1969).
194. E. Fischer and K. Zach, *Ber.*, 45, 456 (1912).
195. O. Diels and R. Meyer, *Ann.*, 519, 157 (1935).
196. E. G. V. Percival, *J. Chem. Soc.*, 783 (1945).
197. E. Hardegger and E. Schreier, *Helv. Chim. Acta*, 35, 232 (1952).
198. L. Mester and A. Major, *J. Amer. Chem. Soc.*, 77, 4305 (1955).
199. H. El Khadem, Z. M. El-Shafei, and M. M. Mohammed-Ali, *J. Org. Chem.*, 29, 1565 (1964).
200. H. El Khadem and M. M. Mohammed-Ali, *J. Chem. Soc.*, 4929 (1963).
201. A. Kraus and H. Simon, *Ber.*, 105, 954 (1972).
202. E. G. V. Percival, *J. Chem. Soc.*, 1770 (1936).
203. E. G. V. Percival, *J. Chem. Soc.*, 1384 (1938).
204. G. Henseke, V. Müller, and G. Badicke, *Ber.*, 91, 2270 (1958).
205. H. El Khadem and M. A. A. Rahman, *Tetrahedron Lett.*, 579 (1966); *J. Org. Chem.*, 31, 1178 (1966).
206. E. E. Percival and E. G. V. Percival, *J. Chem. Soc.*, 137, 1398 (1935).
206a. O. T. Schmidt, G. Zinke-Allmang, and V. Holzach, *Ber.*, 90, 1331 (1957).
207. S. Akiya and S. Tejima, *Yakugaku Zasshi*, 72, 894, 1574, 1577, 1580 (1952).
208. E. E. Percival and E. G. V. Percival, *J. Chem. Soc.*, 750 (1941).
209. E. Votoček and R. Vondraček, *Ber.*, 37, 3848 (1904).
210. S. Haq and W. J. Whelan, *Nature*, 178, 1222 (1956).
211. B. Arreguin, *Anal. Chem.*, 31, 1371 (1959).
212. D. A. Applegarth, G. G. S. Dutton, and Y. Tanaka, *Can. J. Chem.*, 40, 2177 (1962).
213. H. J. Haas and A. Seeliger, *J. Chromatog.*, 13, 573 (1964).
214. P. T. Jossang, *Anal. Biochem.*, 7, 123 (1964).
215. H. F. Schaffer, *J. Chem. Educ.*, 25, 20 (1948).

216. D. E. Kidder and W. D. Robertson, *School Sci. Rev.*, **21**, 346 (1950).
217. P. F. Jorgensen, *Dansk. Tidsskr. Farm.*, **24**, 1 (1962).
218. E. F. L. J. Anet, *J. Chromatog.*, **9**, 291 (1962).
219. W. Z. Hassid and R. M. McCready, *Ind. Eng. Chem., Anal. Ed.*, **14**, 683 (1942).
220. O. L. Chapman, W. J. Welstead, Jr., T. J. Murphy, and R. W. King, *J. Amer. Chem. Soc.*, **86**, 732 (1964).
221. R. D. Guthrie, *Advan. Carbohyd. Chem.*, **16**, 105 (1961).
222. E. L. Jackson and C. S. Hudson, *J. Amer. Chem. Soc.*, **59**, 2049 (1937).
223. G. Jayme and M. Sätre, *Ber.*, **77**, 242, 248 (1944).
224. V. C. Barry and P. W. D. Mitchell, *J. Chem. Soc.*, 3631 (1953).
225. S. Akiya, S. Okui, and S. Suzuki, *Yakugaku Zasshi*, **72**, 785 (1952).
226. R. D. Guthrie, *Proc. Chem. Soc.*, 387 (1960).
227. R. D. Guthrie and L. F. Johnson, *J. Chem. Soc.*, 4166 (1961).
228. G. J. F. Chittenden and R. D. Guthrie, *J. Chem. Soc.*, 1045 (1964).
229. G. J. F. Chittenden and R. D. Guthrie, *J. Chem. Soc.*, 2358 (1963).
230. G. J. F. Chittenden and R. D. Guthrie, *Proc. Chem. Soc.*, 289 (1964).
231. L. Mester, *J. Amer. Chem. Soc.*, **77**, 5452 (1955).
232. L. Mester and E. Moczar, *J. Chem. Soc.*, 3228 (1956).
233. L. Mester and E. Moczar, *Chem. Ind.* (London), 761 (1957).
234. L. Mester and E. Moczar, *Chem. Ind.* (London), 764 (1957).
235. V. C. Barry, J. E. McCormick, and P. W. D. Mitchell, *J. Chem. Soc.*, 3692 (1954).
236. V. C. Barry and P. W. D. Mitchell, *J. Chem. Soc.*, 4020 (1954).
237. V. C. Barry and J. E. McCormick, *Ann.*, **648**, 96 (1961).
238. W. A. Bonner and R. W. Drisko, *J. Amer. Chem. Soc.*, **73**, 3701 (1951).
239. E. Blanchfield and T. Dillon, *J. Amer. Chem. Soc.*, **75**, 647 (1953).
240. J.-E. Courtois, *Bull. Soc. Chim. Biol.*, **23**, 133 (1941).
241. J. X. Khym and W. E. Cohn, *J. Amer. Chem. Soc.*, **82**, 380 (1960).
242. D. M. Brown, G. E. Hall, and R. Letters, *J. Chem. Soc.*, 3547 (1959).
243. J. Baddiley, personal communication.
244. P. Fleury, J.-E. Courtois, and A. Desjobert, *Bull. Soc. Chim. Fr.*, 458 (1952).
245. M. L. Wolfrom and B. O. Juliano, *J. Amer. Chem. Soc.*, **82**, 2588 (1960).
246. Z. Yosizawa, *Biochim. Biophys. Acta*, **52**, 588 (1961).
247. Z. Yosizawa and T. Sato, *Biochim. Biophys. Acta*, **52**, 591 (1961).
248. Z. Yosizawa, *J. Biochem.* (Tokyo), **51**, 145 (1962).
249. Z. Yosizawa and T. Sato, *Tohoku J. Exptl. Med.*, **76**, 100 (1961).
250. Z. Yosizawa and T. Sato, *Tohoku J. Exptl. Med.*, **77**, 213 (1962).
251. Z. Yosizawa, *J. Biochem.* (Tokyo), **51**, 1 (1962).
252. Z. Yosizawa, K. Ino, and Y. Fujisaza, *J. Biochem.* (Tokyo), **51**, 162 (1962).
253. H. Jacobi, *Ber.*, **24**, 696 (1891).
254. M. L. Wolfrom and A. Thompson, *J. Amer. Chem. Soc.*, **53**, 622 (1931).
255. A. Wohl, *Ber.*, **26**, 730 (1893).
256. R. Behrend, *Ann.*, **353**, 106 (1907).
257. J. C. Irvine and R. Gilmour, *J. Chem. Soc.*, **93**, 1429 (1908).
258. R. M. Hann and C. S. Hudson, *J. Amer. Chem. Soc.*, **59**, 1898 (1937).
259. E. Restelli de Labriola and V. Deulofeu, *J. Amer. Chem. Soc.*, **62**, 1611 (1940).
260. V. Deulofeu, *Advan. Carbohyd. Chem.*, **4**, 119 (1949).
260a. B. W. Li, T. W. Cochran, and J. R. Vercellotti, *Carbohyd. Res.*, **59**, 567 (1977); F. R. Seymour, E. C. M. Chen, and S. H. Bishop, *ibid.*, **73**, 19 (1979); and references cited therein.
260b. W. Funcke and C. von Sonntag, *Carbohyd. Res.*, **69**, 247 (1979).

261. C. H. Winestock and W. E. Plaut, *J. Org. Chem.*, **26**, 4456 (1961).
261a. K. Freudenberg and O. Brauns, *Ber.*, **55**, 3233 (1922).
261b. M. L. Wolfrom, M. I. Taha, and D. Horton, *J. Org. Chem.*, **28**, 3553 (1963).
262. R. L. Whistler and B. Shasha, *J. Org. Chem.*, **29**, 880 (1964).
263. F. Arndt and B. Eistert, *Ber.*, **68**, 200 (1935).
264. T. Reichstein and K. Gätzi, *Helv. Chim. Acta*, **21**, 186 (1938).
265. M. L. Wolfrom, D. I. Weisblat, W. H. Zophy, and S. W. Waisbrot, *J. Amer. Chem. Soc.*, **63**, 201 (1941).
266. D. L. MacDonald, J. D. Crum, and R. Barker, *J. Amer. Chem. Soc.*, **80**, 3379 (1958).
267. M. L. Wolfrom, S. M. Olin, and E. F. Evans, *J. Amer. Chem. Soc.*, **66**, 204 (1944).
268. M. L. Wolfrom, D. I. Weisblat, E. F. Evans, and J. B. Miller, *J. Amer. Chem. Soc.*, **79**, 6454 (1957).
269. M. L. Wolfrom, J. D. Crum, J. B. Miller, and D. I. Weisblat, *J. Amer. Chem. Soc.*, **81**, 243 (1959).
270. M. L. Wolfrom and H. B. Wood, Jr., *J. Amer. Chem. Soc.*, **73**, 730 (1951).
271. M. L. Wolfrom, J. M. Berkebile, and A. Thompson, *J. Amer. Chem. Soc.*, **74**, 2197 (1952).
272. M. L. Wolfrom, R. L. Brown, and E. F. Evans, *J. Amer. Chem. Soc.*, **65**, 1021 (1943).
273. M. L. Wolfrom, S. W. Waisbrot, and R. L. Brown, *J. Amer. Chem. Soc.*, **64**, 2329 (1942).
274. M. L. Wolfrom and P. W. Cooper, *J. Amer. Chem. Soc.*, **71**, 2068 (1949).
275. M. L. Wolfrom and H. B. Wood, Jr., *J. Amer. Chem. Soc.*, **77**, 3096 (1955).
276. M. L. Wolfrom, S. W. Waisbrot, and R. L. Brown, *J. Amer. Chem. Soc.*, **64**, 1701 (1942).
277. M. L. Wolfrom and R. L. Brown, *J. Amer. Chem. Soc.*, **65**, 1516 (1943).
278. M. L. Wolfrom and J. B. Miller, *J. Amer. Chem. Soc.*, **80**, 1678 (1958).
279. P. A. Levene, *J. Biol. Chem.*, **36**, 89 (1918).
280. D. Horton and K. D. Philips, *Carbohyd. Res.*, **22**, 151 (1972); Y. Gelas-Mialhe and D. Horton, *J. Org. Chem.*, **43**, 2307 (1978).
281. D. Horton and E. K. Just, *Chem. Commun.*, 116 (1969); compare M. Alexander and D. Horton, *Abstr. Papers Amer. Chem. Soc. Meeting*, **174**, CARB-38 (1977).
282. W. R. Bamford and T. S. Stevens, *J. Chem. Soc.*, 4735 (1952).
283. For a review see F. Micheel and A. Klemer, *Advan. Carbohyd. Chem.*, **16**, 95 (1961).
283a. A. Bertho, *Ber.*, **63**, 836 (1930).
283b. B. Helferich and A. Mitrowsky, *Ber.*, **85**, 1 (1952).
283c. A. Bertho and M. Bentler, *Ann.*, **562**, 229 (1949).
283d. D. H. Buss, L. Hough, and A. C. Richardson, *J. Chem. Soc.*, 5295 (1963).
284. D. Horton, A. E. Luetzow, and J. C. Wease, *Carbohyd. Res.*, **8**, 366 (1968).
285. D. M. Clode and D. Horton, *Carbohyd. Res.*, **14**, 405 (1970).
286. F. Micheel and G. Baum, *Ber.*, **90**, 1595 (1957).
287. H. El Khadem, D. Horton, and M. H. Meshreki, *Carbohyd. Res.*, **16**, 409 (1971); compare D. Horton and J.-H. Tsai, *ibid.*, **67**, 357 (1978).
288. D. Horton and A. Liav, *Carbohyd. Res.*, **47**, 81 (1976).

22. REDUCTION OF CARBOHYDRATES

John W. Green

I. INTRODUCTION

The term *reduction* describes either the addition of the elements of two protons plus two electrons across an unsaturated bond or the effective displacement of an electronegative substituent by a hydride ion; the latter process is generally termed *hydrogenolysis* or *reductive cleavage*. In the carbohydrates, the former process is often, but not exclusively applied to polarized unsaturations such as those present in aldehydes, ketones, nitriles, and carboxyl derivatives, whereas hydrogenolysis is more commonly applied to functional groups having a single bond between carbon and a heteroatom; as the hydroxyl function is a poor leaving group, it must normally be replaced by a better leaving group, such as a sulfonic ester or a halide atom, prior to hydrogenolysis.

II. TYPES OF REDUCTANTS

Numerous different reagents and processes have been employed in reduction reactions. The most generally useful types include (1) inorganic molecules containing hydrogen atoms bonded directly to electropositive atoms, commonly aluminum or boron, which function essentially as a source of hydride (H^-) ions; (2) active metals in the presence of proton sources—for example, sodium in liquid ammonia or zinc in hydrochloric acid; (3) metal catalysts, such as palladium, platinum, or nickel, acting in conjunction with molecular hydrogen; and (4) electrolytic or photolytic methods, in which the hydrogen atoms are derived indirectly. In all these methods, replacement of the proton source by an equivalent deuterium (or tritium) source provides one means whereby isotopically labeled atoms[1] may be introduced into a molecule.

The two principal hydride reagents in common use are lithium aluminum hydride and sodium borohydride.[2] They are nucleophilic, and they effectively add hydrogen (as a "hydride ion") to the carbon atom of a polarized unsaturated function such as a C=O group, the C=N bond of an imine, or the C≡N bond of a nitrile.

$$\begin{array}{c}\diagup\\ \diagdown\end{array}C{=}O + [AlH_4]^{\ominus} \longrightarrow \begin{array}{c}\diagup\\ \diagdown\end{array}C\begin{array}{c}H\\ \diagdown\\ OAlH_3{}^{\ominus}\end{array} \xrightarrow{H_2O} \begin{array}{c}\diagup\\ \diagdown\end{array}C\begin{array}{c}H\\ \diagdown\\ O^{\ominus}\end{array} + Al(OH)_3 \qquad (!)$$

$$\begin{array}{c}\diagup\\ \diagdown\end{array}C{=}O + [BH_4]^{\ominus} \longrightarrow \begin{array}{c}\diagup\\ \diagdown\end{array}C\begin{array}{c}H\\ \diagdown\\ OBH_3{}^{\ominus}\end{array} \longrightarrow \begin{array}{c}\diagup\\ \diagdown\end{array}C\begin{array}{c}H\\ \diagdown\\ O^{\ominus}\end{array} + H_3BO_3 \qquad (2)$$

$$R'CO_2R \longrightarrow R'CH_2OH + ROH \qquad (3)$$

Reactions with lithium aluminum hydride (Eq. 1) are usually performed in benzene, ether, or tetrahydrofuran; aqueous solutions cannot be used, as the reagent reacts violently with water. Sugar substrates are generally derivatized to afford solubility in organic solvents; acetylated derivatives are de-esterified (Eq. 3) by this reagent with the formation of ethanol. Excess reagent is readily converted into the insoluble aluminum hydroxide by addition of excess ethyl acetate.

In contrast, sodium borohydride is relatively stable in water at pH values of 7 and above, so that it can be used in aqueous solution; reactions in alcohols or other organic solvents generally proceed more slowly. Boric acid is usually removed from reaction mixtures as the volatile methyl borate because borate–sugar complexes may be retained by anion exchangers. The extent of reaction of a substrate with borohydride can be controlled by regulating the pH. Carboxylic acids cannot be reduced directly, however, because the alkaline reagent generates the carboxylate anion, and reaction of this negative species with the borohydride anion is prohibitively slow. In contrast, lithium aluminum hydride reduces free acids because the nonpolar solvents used do not promote ionization.

Variants of these two reagents have been prepared through reaction with alcohols to give products, such as LiAlH(OEt)$_3$, containing less "hydride" hydrogen. Other cations—for example, potassium and lithium—have been used with the borohydride ion, and reaction of borohydride with very strong acids (such as boron trifluoride) leads to the formation of diborane (B$_2$H$_6$), itself a useful reductant.

Active metals are often used as reducing agents; common examples include sodium amalgam in ethanol, iron and zinc dust in acetic acid, and sodium in liquid ammonia or ethanol. These metals have electron-rich surfaces and react with protons in a proton-donating solvent (Brønsted acid) to form hydrogen atoms (Eq. 4).[3] These hydrogen atoms may combine to form molecular hydrogen or else react with a radical ion of the substrate; the latter is also formed by reaction with the metal (Eq. 5). In both instances, atoms of the metal are oxidized and go into solution as cations, giving up

$$M \cdot + H^{\oplus} \longrightarrow M^{\oplus} + H \cdot \tag{4}$$

$$M \cdot + R_2C{=}O \longrightarrow M^{\oplus} + R_2\overset{\displaystyle O^-}{\underset{\cdot}{C}} \tag{5}$$

electrons to either the proton or the substrate. The hydrogen radical and the substrate radical ion are probably bound by adsorption to the electron-rich surface of the remaining metal. When the system is a homogeneous solution, such as sodium in liquid ammonia,[4] a soluble complex has been suggested. Combination of the hydrogen radical with the radical ion completes the reduction process (Eq. 6); the resulting alkoxide ion can then accept a proton from the solution. In the ammonia system, a little alcohol must be present.

$$M{:}H + M{:}CR_2O^{\ominus} \longrightarrow 2M \cdot + R_2CHO^{\ominus} \tag{6}$$

In electrochemical and polarographic reductions,[4,5] electrons are supplied by an electric current, and the anions thus formed react with protons available from the solvent or from acids in the solution.[6] Such reductions can be accomplished in a variety of ways, and reductions may occur in either one-electron or two-electron steps. The reactions of the substrate with electrons

$$R^{\oplus} + 2e^{\ominus} \longrightarrow R^{\ominus}; \qquad R^{\ominus} + H^{\oplus} \longrightarrow RH \tag{7}$$

and with protons occur separately; protons apparently do not react with electrons to form hydride ions. Electrochemical processes can be controlled very closely by careful variations in electrode surface, applied voltage, or other experimental conditions. In polarographic reactions, a dropping-mercury electrode constantly provides a fresh electrode surface.

References start on p. 1007.

Whereas the preceding methods based on metal reduction provide electron-rich surfaces that are constantly being renewed, catalytic hydrogenation employs fixed surfaces of such metals as nickel, palladium, or platinum. These surfaces attract electron-deficient species, such as the double bonds of carbonyl groups, and also split the hydrogen molecule into hydrogen atoms. The resulting radicals are held to the surface of the metal by a relatively weak binding (about 30 kcal mole^{-1}) that has been termed chemisorption.[7] Reductions employing such catalysts can be conducted at high temperatures and pressures of hydrogen, to the extent that carbon–carbon bonds may be broken. Steric effects are more pronounced at such catalytic surfaces than in reductions with hydride reagents or with active metals.

Both hydride ions and radicals are electron-rich and readily donate electrons to a suitable substrate. The hydride ions can be considered to be nucleophiles, whereas radicals are more neutral in their reaction. The reagent diborane, B_2H_6, formed by the reaction of borohydride ion and a strong acid (Eq. 8) is, however, the dimer of the electron-deficient species BH_3; as an electrophile it reacts readily with free carboxyl groups (Eq. 9).[8]

$$2[BH_4]^{\ominus} + 2H^{\oplus} \longrightarrow B_2H_6 + 2H_2 \qquad (8)$$

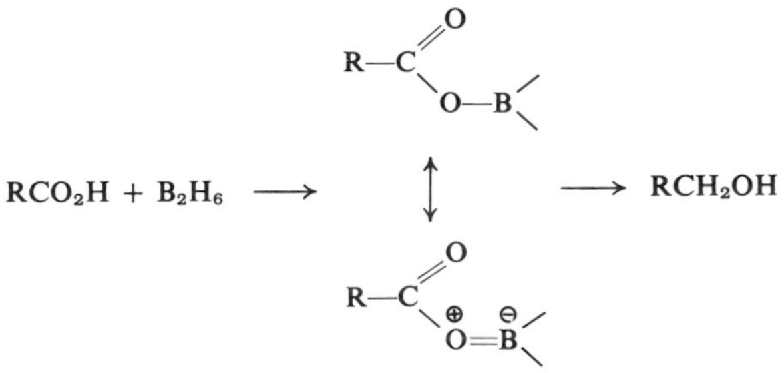

$$RCO_2H + B_2H_6 \longrightarrow \quad\updownarrow\quad \longrightarrow RCH_2OH \qquad (9)$$

The "carbonyl" character of the carboxyl group in the intermediate is stabilized by resonance, and the group is, therefore, readily reduced; in contrast, ester groups are reduced rather slowly because the "carbonyl" character of the intermediate (Eq. 10) is lessened by resonance. There is also

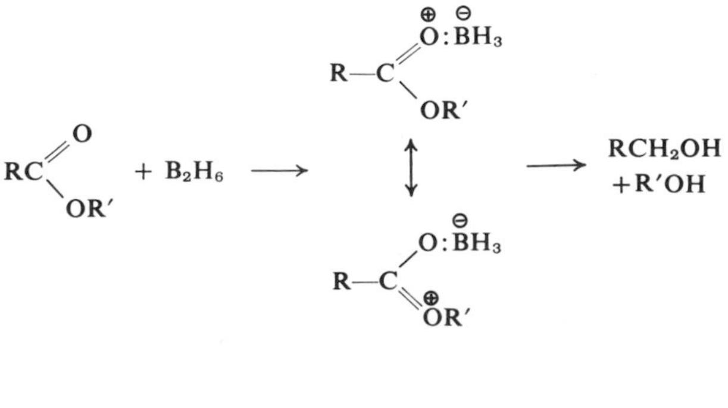

$$RC{\overset{O}{\underset{OR'}{}}} + B_2H_6 \longrightarrow \quad\updownarrow\quad \longrightarrow \begin{array}{l} RCH_2OH \\ +R'OH \end{array} \qquad (10)$$

$$RCO_2R' + B_2H_6 \longrightarrow RCH_2OR' \tag{11}$$

a tendency for diborane to reduce ester groups directly without cleavage; ethers are thus formed (Eq. 11). Little is known of the mechanism for this last reaction.[9]

III. REDUCTION OF LACTONES AND ALDOSES

D-Glucono-1,5-lactone (1) is reduced by sodium borohydride in very weakly acidic solution (pH 5) to D-glucose (2); the reaction proceeds no further at this pH, because the hydroxyl group at C-1 of the sugar in its hemiacetal form is not readily displaced by hydride. If the solution is then

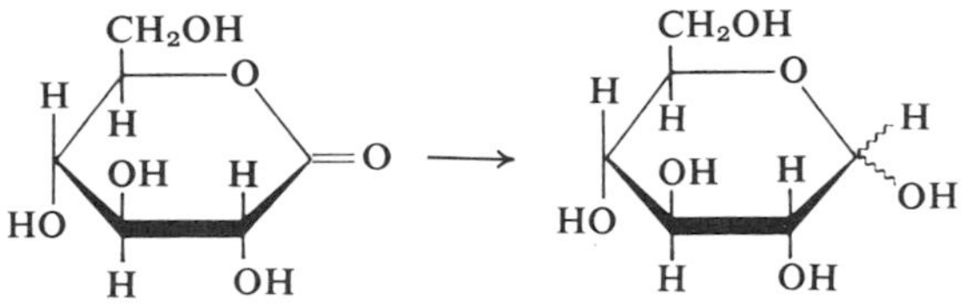

made alkaline, the ring of 2 is opened more readily, and the sugar is then reduced through the free carbonyl form, giving D-glucitol (3) as the product. Wolfrom and Wood[10] thus obtained good yields of various aldoses and alditols from the appropriate lactones and aldoses. Frush and Isbell[11] converted 1 into 3 by a two-stage reduction, first with an acid buffer and then

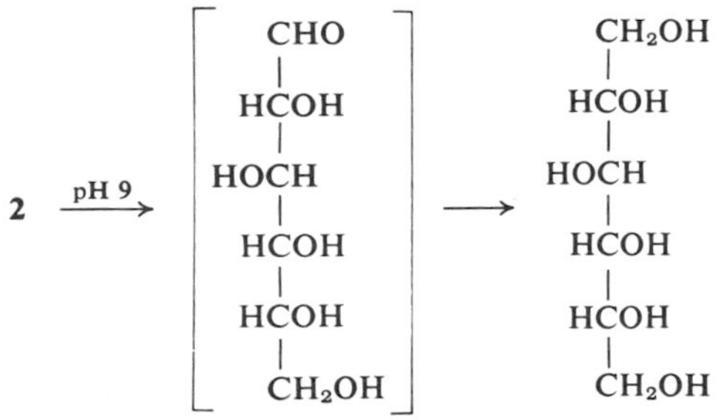

$$2 \xrightarrow{\text{pH 9}} \begin{bmatrix} \text{CHO} \\ | \\ \text{HCOH} \\ | \\ \text{HOCH} \\ | \\ \text{HCOH} \\ | \\ \text{HCOH} \\ | \\ \text{CH}_2\text{OH} \end{bmatrix} \longrightarrow \begin{array}{c} \text{CH}_2\text{OH} \\ | \\ \text{HCOH} \\ | \\ \text{HOCH} \\ | \\ \text{HCOH} \\ | \\ \text{HCOH} \\ | \\ \text{CH}_2\text{OH} \\ \mathbf{3} \end{array} \tag{11}$$

at pH 9, and achieved an overall yield exceeding 90%. This type of reduction can also be effected with the 1,4-lactones; the furanoid hemiacetals of the sugars are obtained as the initial products in weakly acid solution.

Alternatively, lithium aluminum hydride converts ether-soluble derivatives of lactones into the alditol derivatives in one step. Ness et al.[12] obtained D-glucitol in 47% yield from 2,3,5,6-tetra-O-acetyl-D-glucono-1,4-lactone by this route; the acetyl groups were reductively cleaved (Eq. 3) by the action of the reagent. Roseman[13] similarly converted 1,2-O-isopropylidene-D-glucurono-6,3-lactone into 1,2-O-isopropylidene-α-D-glucofuranose; here the protecting acetal groups were not removed. It should be pointed out that the reduction of the lactone linkage, to either an aldose or the alditol, is a reductive cleavage. The reaction sequence: (a) protection of hydroxyl groups by formation of tetrahydropyran-2-yl ethers, (b) reduction with diborane in tetrahydrofuran or with lithium aluminum hydride–aluminum chloride in ether, and (c) deprotection has been recommended[13a] for microscale preparations of aldoses from aldonolactones.

Reductions of lactones to aldoses have been traditionally[14] carried out with sodium amalgam in acid solution (pH 3). L-Galactose has thus been prepared from L-galactono-1,4-lactone,[15] and D-gulose from D-gulono-1,4-lactone.[14] These reactions have employed buffers such as sodium hydrogenoxalate, sulfuric acid,[16] and ion-exchange resins.[17]

Reduction of 1,4- or 1,5-lactones does not proceed readily in alkaline solution because formation of the sodium salt of the aldonic acid impedes reaction with the nucleophilic reductant. Similarly, the free acids are not reduced by borohydride, even in acidic solution. Esters of the acids can, however, be reduced. Methyl D-arabinonate can be converted into D-arabinose[18] by sodium amalgam, and the methyl esters of methyl glycosiduronic acids have been successfully reduced to the corresponding glycosides.[19]

The susceptibility of lactones to reduction and the contrasting inertness of free acid groups have been exploited for preparation of uronic acids by differential reduction of the monolactones of aldaric acids. The reduction of uronic acids, or their sodium salts, to aldonic acids is also possible because of the difference in reactivity between carbonyl groups and carboxyl groups. These procedures have been conducted by electrochemical generation of sodium amalgam, which serves to reduce aldaric acid monolactones in two steps (Eq. 12) to aldonic acids by way of the uronic acids.[20]

Mild reduction of D-glucurono-6,3-lactone (4), either with borohydride[10] at pH 5 and low temperatures or with sodium amalgam[21] at pH 3.4, gives

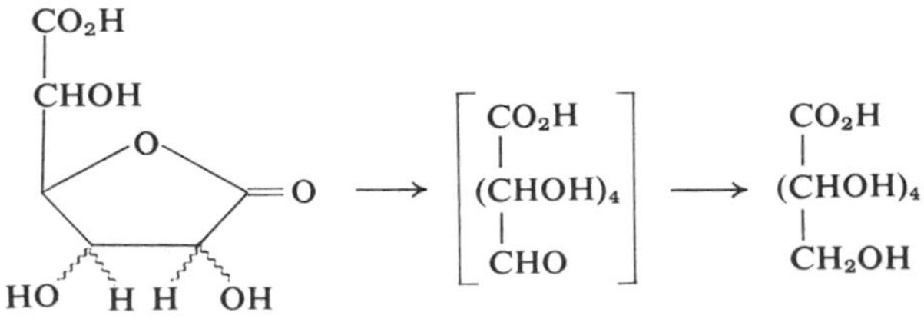

$$(12)$$

about 15% of D-*gluco*-hexodialdose (**5**); small proportions of D-glucose and L-gulose are also formed when the latter reductant is used.

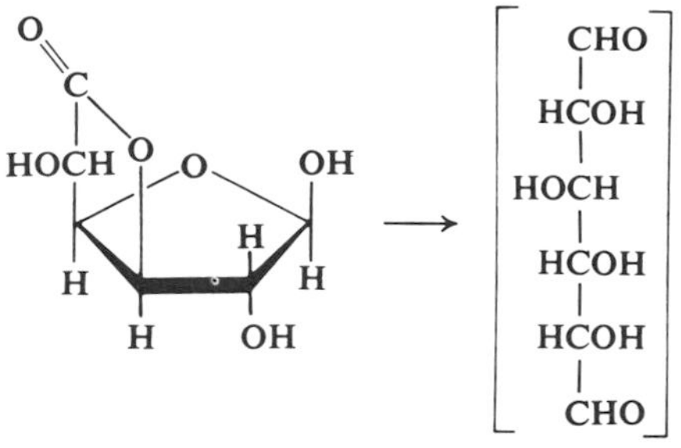

4   **5**

The carboxyl group is more readily reduced by lithium aluminum hydride in nonpolar solvents than by borohydride or sodium amalgam in water. With polysaccharides especially, however, complications may arise because of partial de-esterification of acylated derivatives and consequent precipitation of the unconverted substrate. Methylated derivatives are often used to avoid this complication.

Diborane, an electrophilic reductant, was first suggested by Smith and Stephen[22] for application to carbohydrates. This gaseous reagent can be generated *in situ* from borohydride by the addition of boron trifluoride; alternatively, it can be prepared from the same reagents in a separate flask and distilled into the reaction vessel. 2,2′-Dimethoxydiethyl ether ($MeOCH_2CH_2OCH_2CH_2OMe$, "*diglyme*"), is generally used as a solvent. Methyl α-D-galactopyranosiduronic acid and tetra-*O*-acetylgalactaric acid were converted into methyl α-D-galactopyranoside and galactitol, respectively, by reduction with diborane. This reagent also reduced 82% of the uronic acid residues in a propionylated alginic acid. Many such applications have been made in the polysaccharide field.[23]

Although diborane is a very effective reductant, side reactions, such as depolymerization of the polysaccharide,[24] reductive cleavage of acyl groups, and conversion of acyl groups into ether groups,[24,25] may detract from its utility. Lithium borohydride in hot tetrahydrofuran has been found[26] to be quite effective in reducing methyl esters of some alginates; the reagent causes minimal depolymerization or formation of ether groups.[24]

Kohn *et al.*[27] demonstrated that diisoamylborane, which does not cleave ester groups as does diborane, can be utilized to reduce acetylated aldono-1,4-lactones to the corresponding acetylated aldofuranoses; as the lactone group is reduced but the acetates remain, this reaction provides a valuable route to aldofuranose esters and such derivatives as aldofuranosyl nucleosides. The use of freshly prepared reductant is recommended[28] for this reaction.

The preparation of alditols from aldoses has been accomplished with many reductants, catalytic hydrogenation being mainly used for large-scale applications. Raney nickel is a very effective catalyst in giving a pure product; such reductions are often conducted in the presence of calcium carbonate to prevent the formation of acids.[29] Electrochemical reductions have been performed in alkaline solution to favor ring-opening; epimerization sometimes occurs, however, as in the reduction of D-glucose to give D-mannitol as well as D-glucitol.[30] Sodium borohydride (or sodium borodeuteride) is commonly used in converting aldoses or oligosaccharides into alditol derivatives for examination by gas–liquid chromatography–mass spectrometry.[31]

The rate of reduction of sugars is very much lower in neutral or slightly acidic solutions than in alkaline solutions; one example of this is the reduction of sedoheptulose with sodium amalgam.[32] Overend et al.[33] have correlated the rate of polarographic reduction for various sugars with the rate of conversion of the ring form into the open-chain form. Such rates have been shown to be pH-dependent for D-ribose and 2-deoxy-D-erythro-pentose, attaining a minimum value at pH 3.5 to 6. In the reduction of 2,4:3,5-di-O-ethylidene-aldehydo-L-xylose, no such dependence was shown; it was concluded that the carbonyl group of this compound is hydrated (compare ref. 33a), and that the rate-determining step is dehydration of this nonreducible species to the free carbonyl form. It has been suggested that L-xylo-hexulosonic acid exists as a cyclic hemiacetal, which is nonreducing in a polarographic system.[34] The different rates of reduction of aldoses versus ketoses and of pentoses versus hexoses, with sodium amalgam in buffered acid solution, have been measured; trends were rationalized in terms of ring-opening.[35]

The open-chain form of an aldose or ketose may be formulated in a linear or zigzag conformation; a substituent at the C-3 position could thus be positioned quite close to the carbonyl group at C-1 and might exert a steric effect on the approach of a reducing agent. Bragg and Hough[36] found that 3-O-methyl-D-glucose was reduced more slowly than D-glucose by borohydride at pH 10.4 (Eq. 13). Laminarabiose, having a bulky hexosyl group at C-3, was reduced at an even lower rate.

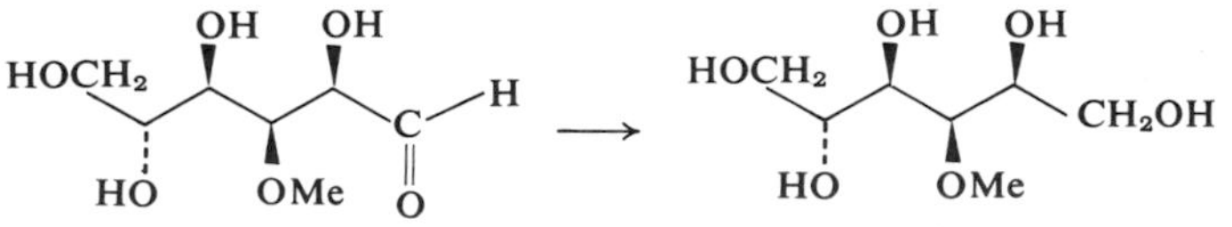

$$\text{(13)}$$

IV. REDUCTION OF KETO GROUPS

The principle that reduction of a keto group to a secondary alcohol group generally produces two epimers provides a method of locating the position of the keto group in a compound of unknown structure. Thus D-*erythro*-pentulose reacts with borohydride to give both D-arabinitol and ribitol,[37] showing that the carbonyl group is at C-2. Likewise, Raney nickel acts upon D-*xylo*-hex-5-ulosonic acid to give both D-gluconic and L-idonic acids.[38]

Theander[39] has shown that the axial alcohol is the preponderant product in the reduction of several methyl D-hexopyranosiduloses (see Table I). This effect is more pronounced for catalytic hydrogenation than for reduction either by borohydride or by sodium amalgam. This demonstration of the von Auwers–Skita rule[40] evidently indicates approach of the reductant from the less-hindered side of the carbonyl group to form an equatorially disposed carbon–hydrogen bond (Eq. 14). A similar effect has been shown in the

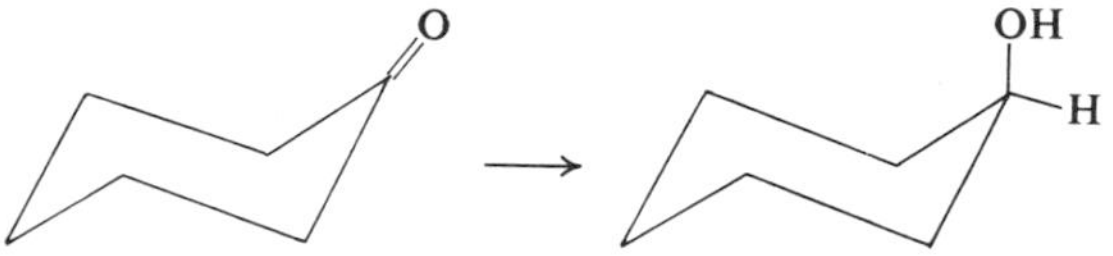

$$(14)$$

reduction of the 3-ulose derivative[41] (**6**) formed by oxidation of 1,2:5,6-di-*O*-isopropylidene-α-D-glucofuranose. Treatment of **6** with potassium borohydride rapidly produces the D-*allo* isomer in over 98% yield.[41,42] Selective oxidation followed by reduction, as in this sequence, is a useful means for effecting epimerization when the stereochemistry is favorable.

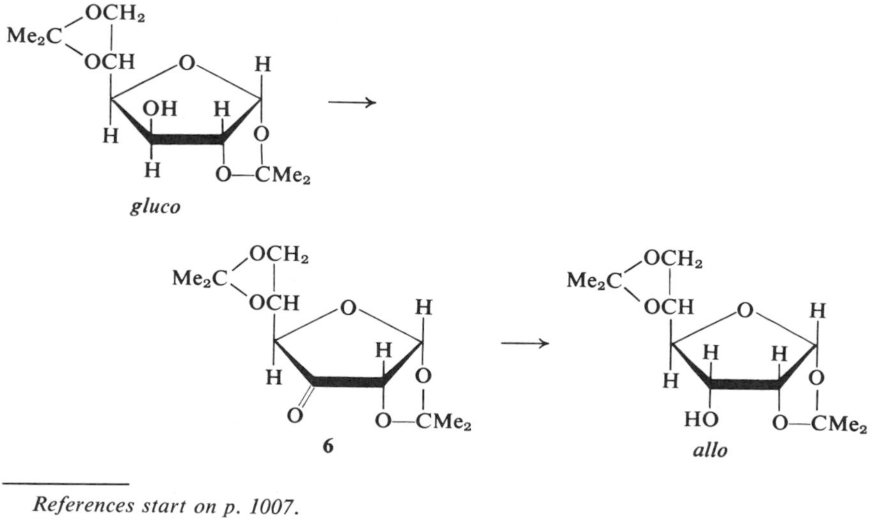

gluco

6   *allo*

TABLE 1

PREFERENTIAL REDUCTION OF KETO GROUPS IN METHYL HEXOPYRANOSIDULOSES TO AXIAL
ALCOHOL GROUPS[39]

| | | | Percent axial/percent equatorial alcohol formed | | |
| | | | | | |
Reducing agent	pH	Temperature (degrees)	β-D-arabino 2-ulose	β-D-xylo-3-ulose	α-D-xylo-3-ulose
Hydrogen and platinum	6	20	94/6	77/23	86/14
Raney nickel	6	78	74/26	62/38	91/9
Sodium amalgam	6	20	53/47	66/34	63/37
Sodium borohydride	9.5	20	65/35	49/51	96/4

An unexpected, stereospecific side reaction, enolic exchange of H-2 for a deuterium atom, was found[42a] to compete with reduction of methyl 2-*O*-acetyl-4,6-*O*-benzylidene-α-D-*ribo*-hexopyranosid-3-ulose by sodium boro-deuteride in anhydrous, alcoholic solvents; however, reduction to the 3-*C*-deuterio-D-allose derivative occurs conventionally in moist alcohols.

Reduction to give the axial alcohol preferentially, and catalytic oxidation, which favors attack at axial alcohol groups (see Chapter 24), are closely related; the determining factor in both reactions is the steric accessibility of the carbon–hydrogen bond that is to be formed or broken.

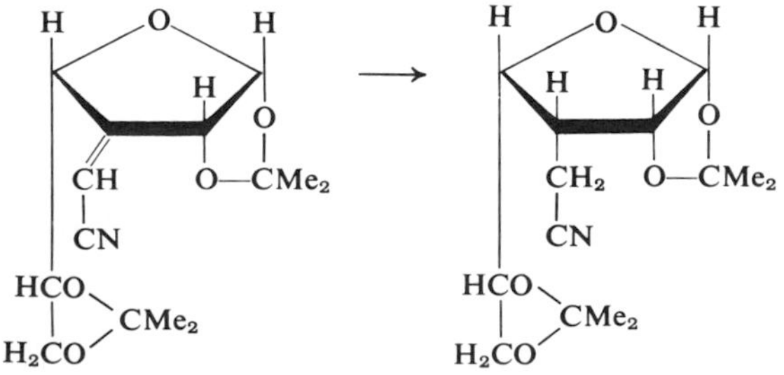

Reduction of keto groups generally stops with formation of the secondary alcohol; Lindberg and Theander,[43] however, were able to reduce methyl β-D-*ribo*-hexopyranosid-3-ulose with hydrogen in the presence of platinum in acid solution to afford the 3-deoxy derivative in 59% yield. This conversion of a ketone into the "hydrocarbon" in acid solution is similar to the Clemmensen reduction[44]; the acid medium facilitates removal of the oxygen atom from the carbonyl group by favoring protonation. An adsorbed, methylene diradical has been suggested as an intermediate. Such a reduction to the deoxy derivative proceeds only from a carbonyl group; adjacent alcohol groups are not reduced.

A related type of reduction, leading to branched-chain derivatives, can be accomplished following addition of a Wittig reagent to the ketone. Thus, catalytic hydrogenation of the Wittig adduct 3-C-cyanomethylene-3-deoxy-1,2:5,6-di-O-isopropylidene-α-D-xylo-hexofuranose (7) with fresh palladium black gives the 3-deoxy-D-gulose derivative (8) stereospecifically.[45]

V. REDUCTIVE CLEAVAGE (HYDROGENOLYSIS) OF HYDROXYL GROUPS

Tetra-O-acetyl-α-D-glucopyranosyl bromide (9) can be readily converted by lithium aluminum hydride into 1,5-anhydro-D-glucitol (10)[46]; in contrast, D-glucose is reduced only via the open-chain form, yielding D-glucitol. Displacement of a hydroxyl group by a hydrogen atom is very difficult, whereas the bromide group, which gives rise to a more-stable anion,[47] is

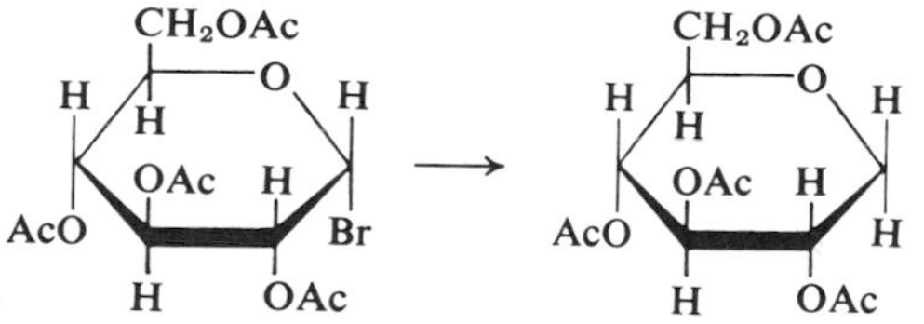

more readily displaced by the hydride group. 1,5-Anhydroalditols can also be made by reduction of the appropriate acetylated glycosyl nitrates with borohydride,[48] or by the action of Raney nickel on aryl thioglycoside acetates[49]; as with the bromide group, the nitrate and arylthio groups are easily displaced as the corresponding anions.

Dialkyl dithioacetals of sugars can also be reduced, with displacement of the alkylthio groups. Wolfrom and Karabinos[50] converted D-galactose diethyl dithioacetal pentaacetate with hydrogen-saturated Raney nickel into 1-deoxy-D-galactitol pentaacetate. Lindberg and Nordén[51] monitored this reduction by paper chromatography and isolated intermediate 1-S-ethyl-1-thiohexitols resulting from partial hydrogenolysis of this dithioacetal and of the D-mannose analogue; a similar reaction can be achieved by photolysis in methanolic solution.[52]

References start on p. 1007.

Reduction of a primary hydroxyl group is usually conducted through the sequence:

ω-O-p-Tolylsulfonyl $\longrightarrow$ ω-deoxy-ω-halo $\longrightarrow$ ω-deoxy derivative

Thus, reduction of 6-deoxy-6-iodo-1,2:3,4-di-O-isopropylidene-α-D-galacto-pyranose with Raney nickel gives 6-deoxy-1,2:3,4-di-O-isopropylidene-α-D-galactopyranose in 90% yield.[53] Kochetkov et al.[54] simultaneously reduced off the iodo and both benzylthio groups from 6-deoxy-6-iodo-D-galactose dibenzyl (or diethyl) dithioacetal by the action of Raney nickel and obtained 1,6-dideoxygalactitol directly. Action of sodium cyanoborohydride in hexa-methyl phosphoric triamide is reported[54a] to effect selective hydrogenolysis of primary halides in the presence of azide and benzyloxyl substituents; this reductant acts only slowly on primary sulfonates.

In favorable circumstances, however, the intermediate halide step can be omitted; thus, the action of lithium aluminum hydride on 6-O-p-tolylsulfonyl-D-mannose diethyl dithioacetal yields D-rhamnose diethyl dithioacetal directly,[55] because, in contrast to their reactivity toward Raney nickel, alkylthio groups are inert to hydride reductants. Hydrogenolysis of 6-O-benzoyl-1,2-O-isopropylidene-5-O-p-tolylsulfonyl-α-D-glucofuranose and its 3-deoxy homologue using lithium aluminum hydride gives mainly the corre-sponding 6-deprotected 5-deoxy sugar acetal.[55a]

Whereas halide groups can be introduced readily by replacement of primary alcohol groups, such reactions with secondary hydroxyl groups are more difficult. Kochetkov et al.[54,56] utilized triphenoxyphosphonium halides as reagents to substitute primary or secondary alcohol groups by halide groups. 1,2:5,6-Di-O-isopropylidene-α-D-glucofuranose was converted with rearrange-ment into 6-bromo-6-deoxy-1,2:3,5-di-O-isopropylidene-α-D-glucofuranose by the action of triphenoxyphosphonium dibromide; the location of the halogen substituent was established by hydrolysis and hydrogenolysis of the product to give 6-deoxy-D-glucose (quinovose).[54] Tributyltin hydride, in the presence of a trace of a free-radical initiator, hydrogenolyzes secondary chlorides or bromides at 60°; removal of primary halide groups occurs satis-factorily at 120°, thus allowing selective reductions.[57] Selective introduction of a primary bromo group in hexopyranosides is readily accomplished through oxidation of a 4,6-benzylidene acetal group by N-bromosuccinimide, which produces a 4-O-benzoyl-6-bromo-6-deoxy derivative.[58] Condensation of a sugar derivative possessing a free, primary hydroxyl group with 2-pyrimidinethiol, and hydrogenolysis of the resultant thioether, are reported[59] to produce ω-deoxy sugars in two steps without need of protecting groups.

Dimethylthiocarbamates of secondary alcohols may be converted by photolysis into the corresponding deoxy sugars[59a]; the same net reaction takes place when secondary dithiocarbonates are treated with tributyltin hydride.[59b]

The carbon–oxygen bond in epoxy derivatives is readily broken by reducing agents. Methyl 2,3-anhydro-4,6-O-benzylidene-α-D-allopyranoside reacts with lithium aluminum hydride to give the 2-deoxy product in 58% yield[60]; reduction of this epoxide with Raney nickel, however, gives the 3-deoxy isomer as the major product, in similar yield. Likewise, methyl 3,4-anhydro-D-galactopyranoside has been reported[61] to react with lithium aluminum hydride, giving a 73% yield of the 3-deoxy product, but to produce a 30% yield of the 4-deoxy isomer on reduction with Raney nickel; methyl 2,3-anhydro-α-D-mannopyranoside, however, gives only the 3-deoxy product under both conditions of reduction.[60] Catalytic hydrogenation of 1,6:3,4-dianhydro-2-O-p-tolylsulfonyl-β-D-galactopyranose[62] gave 65% of the 4-deoxy product.

Prior to the introduction of lithium aluminum hydride, deoxy derivatives were frequently obtained in two steps from epoxides by first opening the ring with sodium methanethioxide and subsequently reducing the thioether with Raney nickel.[63] Anderson et al.[64] explored the utility of episulfonium intermediates from thioethers of glycofuranosides for effecting configurational alterations through neighboring-group participation; subsequent reduction with Raney nickel was employed in syntheses of various deoxynucleosides.

Primary and secondary alkoxy groups adjacent to a free carbonyl group in a sugar derivative may be removed directly by photolysis, through the operation of a Norrish type II photochemical process.[65] Indirect replacement of the oxygen atom adjacent to the anomeric center of an aldose may also be accomplished by conversion of the sugar into a glycal with subsequent oxymercuration to form a 2-deoxymercuric derivative, which can be hydrogenolyzed with sodium borohydride[66] or photolyzed[67] to yield a deoxy sugar; a similar sequence, with iodonium collidine perchlorate as an iodinating agent, proceeds through a deoxyiodo intermediate to yield likewise a 2-deoxy sugar.[68]

VI. REDUCTIVE CLEAVAGE (HYDROGENOLYSIS) OF ESTER AND OTHER PROTECTING GROUPS

Whereas replacement of the groups X in RX or OX in ROX by hydrogen gives the deoxy derivative RH as a product, in many instances the group ROX can also be cleaved by hydrogenolysis to produce ROH, the free hydroxyl derivative, together with the reduced form of the original substituent group. The cleavage of carboxylic acid esters is noted in Eq. (11); other examples of reductive cleavage are given as follows (Eqs. 15–19). These are of two types: (1) those in which the central atom, N or S, is changed in valence, and (2) those in which the displaced group—for example, the benzyl or benzylidene group—is stabilized, as a radical or ion, by the adjacent aromatic ring.

$$RONO_2 \longrightarrow ROH + HNO_2 \qquad (15)$$
$$ROSO_2Ar \longrightarrow ROH + ArSO_2H \qquad (16)$$
$$ROSO_3H \longrightarrow ROH + H_2SO_3 \qquad (17)$$
$$ROCH_2Ph \longrightarrow ROH + PhCH_3 \qquad (18)$$
$$RSCH_2Ph \longrightarrow RSH + PhCH_3 \qquad (19)$$

The nitric ester group ($-ONO_2$) can readily be displaced (with N–O cleavage) from anomeric centers by a hydride ion or a hydrogen atom; this ester group is not displaced[48] by borohydride from secondary alcoholic groups or from primary carbon atoms, but it is labile to hydrogenolysis by other reagents. Ansell and Honeyman[69] denitrated methyl 4,6-O-isopropylidene-α-D-glucopyranoside 2,3-dinitrate with lithium aluminum hydride; some 3-nitrate was also obtained. Primary nitrates have been hydrogenolyzed with iron in hot acetic acid[70]; the reductant did not remove iodo groups.[71] Other reductants used for denitration have been sodium sulfide, sodium amalgam, and hydrogen at high pressures in the presence of palladium.[71] The action of the last-named catalyst has been used by Wright and Hayward for the almost quantitative conversion of various pentitol nitrates into the respective pentitols.[72] Hoffman et al.[73] have denitrated sugar nitrates in acetic anhydride by the action of zinc dust and anhydrous hydrogen chloride; the sugar acetates were obtained as products.

Schmid and Karrer[74] found that 1,2:5,6-di-O-isopropylidene-3-O-p-tolylsulfonyl-α-D-glucofuranose was detosylated slowly by the action of lithium aluminum hydride; similar treatment removed the 1-substituent from 1-O-p-tolylsulfonyl-2,3:4,5-di-O-isopropylidene-β-D-fructopyranose.[74] Detosylation of 1,2-O-isopropylidene-5-O-p-tolylsulfonyl-β-L-arabinofuranose was accomplished with sodium amalgam in 80% ethanol.[75] p-Toluenesulfinic acid was obtained as a product in all instances. Raney nickel[75] and sodium naphthalenide[76] have also been used to remove sulfonic ester groups, as has photolysis in methanolic solution.[77]

Sulfuric ester groups can likewise be removed by lithium aluminum hydride; Grant[78] de-esterified 1,2:5,6-di-O-isopropylidene-α-D-glucofuranose 3-sulfate by treatment with this reductant in 1,4-dioxane.

Triphenylmethyl ethers are commonly removed hydrolytically, although reduction with sodium in liquid ammonia is a mild alternative that can be utilized to allow several different types of protecting groups to be removed at the same time; thus, methyl 4-O-acetyl-3-O-benzyl-2-O-methyl-6-O-triphenylmethyl-α-D-glucopyranoside is converted directly into methyl 2-O-methyl-α-D-glucopyranoside by this reagent.[79]

The ready reductive cleavage of O-benzyl and O-benzylidene groups depends on the presence of the aromatic ring adjacent to the —CH— grouping; formation of a radical can be stabilized by the electrons in the aromatic ring (Eq. 20). Such hydrogenolyses are usually effected with hydrogen

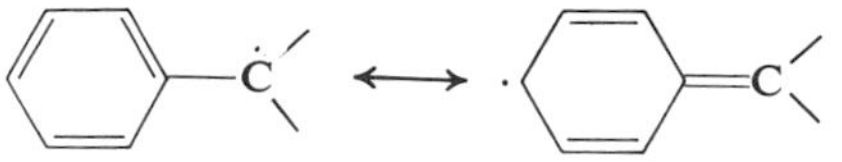

$$
\text{(structure)} \quad (20)
$$

and palladium black at room temperature; thus benzyl 2-O-(methylsulfonyl)-β-D-arabinopyranoside is converted into 2-O-(methylsulfonyl)-D-arabinose in 96% yield.[80] Benzyl 2-O-benzoyl-3,4-O-isopropylidene-β-D-arabinopyranoside in 1,4-dioxane is reduced quantitatively to the substituted pentoside; although no reductive cleavage of the benzoic ester occurs, hydrogenolysis of allylic O-isopropylidene groups[81] and 2-enosyl aglycones[82] has been recorded. Acyl-group migration from O-1 to O-2 has been reported[83] to accompany the platinum black-catalyzed hydrogenolysis of several N-protected 1-O-(aminoacyl)-2,3,4,6-tetra-O-benzyl-α-D-glucopyranoses, giving the 2-O-(aminoacyl)-D-glucose as the major product; the hydrogenolysis proceeds normally, however, for the β anomers.

Electrochemical reduction[83a] of benzylidene acetals also liberates the hydroxyl groups, whereas borane in tetrahydrofuran or lithium aluminum hydride–aluminum chloride effect conversion[83b] into hydroxy ethers. Action of di-*tert*-butyl peroxide hydrogenolyzes one of the hydroxyl groups involved in benzylidene acetals, giving,[83c] after removal of protecting groups, a deoxy sugar.

Reist et al.[84] have used sodium in liquid ammonia as an O-debenzylating agent, and the same reagent was employed by Whistler et al.[85] to convert 5-S-benzyl-5-thio-D-xylose into the 5-thio sugar (compare Eq. 19). Although the benzylthio group gives a stable anion, this group is not displaced from the sugar derivative by hydrogen because the extreme stabilization of the benzyl radical favors the breaking of the sulfur–methylene bond over C-5–sulfur scission.

O-Benzylidene groups have been removed by similar hydrogenolysis. Methyl 2,3-di-O-benzoyl-4,6-O-benzylidene-α-D-galactopyranoside is converted quantitatively by hydrogen and palladium in hot ethanol into methyl 2,3-di-O-benzoyl-α-D-galactopyranoside[86]; Raney nickel in hot ethanol can also be used.[87] Partial hydrogenolysis of acetals to form hydroxy ethers occurs during treatment with a 1:1 complex of aluminum chloride and lithium aluminum hydride; treatment of an ortho ester with this complex causes successive conversion into an acetal and thence into a dihydroxy ether.[88]

VII. REDUCTION OF NITROGEN-CONTAINING DERIVATIVES

Various carbohydrate derivatives containing nitrogen–carbon multiple bonds can be reduced to the aminodeoxy derivatives. In most cases the C–N bond is retained; the amino group resembles the hydroxyl group in its resistance to replacement by hydrogen. Azido derivatives are hydrogenated in the presence of platinum,[89] palladium,[90] or Raney nickel,[91,92] or by hydrazine and air[93] to the amino derivatives; hydrazino[91] and phenylazo[94] groups are similarly reduced. In all instances the nitrogen–nitrogen bond is broken but not the carbon–nitrogen bond. Reduction of 3,4:5,6-di-O-isopropylidene-D-gluconamide by lithium aluminum hydride gave 1-amino-1-deoxy-3,4:5,6-di-O-isopropylidene-D-glucitol[95]; in this reaction the C=O group is reduced, whereas the amino group is left intact.

Schiff bases and glycosylamines are also reduced to aminodeoxyalditols. Hydrogenation of N-benzyl-D-galactopyranosylamine in the presence of platinum at 40° to 50° and 2 to 3 atm gave 1-benzylamino-1-deoxy-D-galactitol.[96] A mixture of D-glucose and methylamine, reduced with hydrogen and Raney nickel, gives 1-deoxy-1-methylamino-D-glucitol.[97] Similar hydrogenation of D-galactose in liquid ammonia containing Raney nickel at 35 atm produces 1-amino-1-deoxy-D-galactitol.[96]

Oximes are reduced to the amino derivatives; catalytic reduction of oximino groups in pyranoid rings favors formation of the axial amines, in keeping with the preferential reduction of keto derivatives to the axial alcohols[98]; reduction with sodium amalgam, however, leads to a substantial proportion of the equatorial amines. Roth et al.[99] reduced D-fructose oxime with hydrogen and platinum in acid solution to afford 2-amino-2-deoxy-D-mannitol in preference to the D-glucitol epimer. The *manno* isomer may arise through favored generation of the amino group in a *trans* relationship to the C-4 hydroxyl group, thereby averting an eclipsed interaction (illustrated in Eq. 13) that would otherwise destabilize the molecule in an extended, zigzag conformation (see also Vol. IA, Chapter 5). α,β Elimination of a polar group has resulted[99a] from treatment of a glyculose O-benzyloxime with lithium aluminum hydride.

Several examples have been reported in which carbon–nitrogen scission occurs. Hydrogenation of D-fructose cyanohydrin in the presence of platinum gave 2-C-(hydroxymethyl)-D-glucose plus some 1-amino-1-deoxy-2-C-(hydroxymethyl)-D-glucitol.[100] Similar catalytic hydrogenations of various 2-benzamido-2-deoxyhexononitriles[101] and aldononitriles[102] have given the respective aldoses as products. This partial reduction may go through the sequence $RCN \rightarrow RCH=NH \rightarrow RCH_2NH_2$, with competing hydrolysis of the intermediate Schiff base to give the free sugar.

VIII. MISCELLANEOUS REDUCTIONS

Wolfrom and Usdin[103] reduced the bis(acid chloride) of galactaric acid to give *galacto*-hexodialdose. This example of the Rosenmund reduction[104] was performed with hydrogen in the presence of palladium; under the conditions used, the aldehyde product is not reduced. 2,3:4,5-Di-*O*-isopropylidene-L-*manno*-hexodialdose[105] showed a resistance to hydrogenation catalyzed by platinum or palladium; reduction was achieved satisfactorily, however, with aluminum isopropoxide, a hydride-transfer reagent.[106]

Angyal and McHugh[107] reduced hexahydroxybenzene in the presence of platinum and obtained, as the initial product, a good yield of *cis*-inositol (Eq. 21). This result indicates that the benzene ring becomes oriented on the platinum surface, and that the approach of hydrogen occurs from only one side of the ring, to generate alternating axial and equatorial hydroxyl groups in the resulting cyclohexane ring.

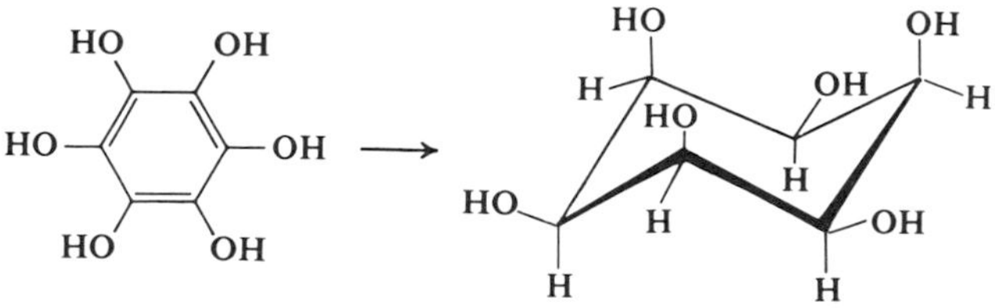

$$\text{(21)}$$

Acetylenic sugar derivatives are reduced to *trans*-alkenes[108] by lithium aluminum hydride, whereas propargylic acetate groups are displaced by diisoamylborane[109]; hydrogen peroxide converts the initial boron adduct into an α,β-unsaturated aldehyde.

1,2-Glycals are usually prepared by reduction of acylated glycosyl halides, involving removal of the halide from C-1 and the acetate group from C-2; the reducing agent, zinc and acetic acid, is generally catalyzed by addition of a trace of platinic chloride.[110] This reaction is illustrated for the conversion of tetra-*O*-acetyl-α-D-glucopyranosyl bromide (**9**) into 3,4,6-tri-*O*-acetyl-1,5-anhydro-2-deoxy-D-*arabino*-hex-1-enitol (**11**, "*tri-O-acetyl-D-glucal*").

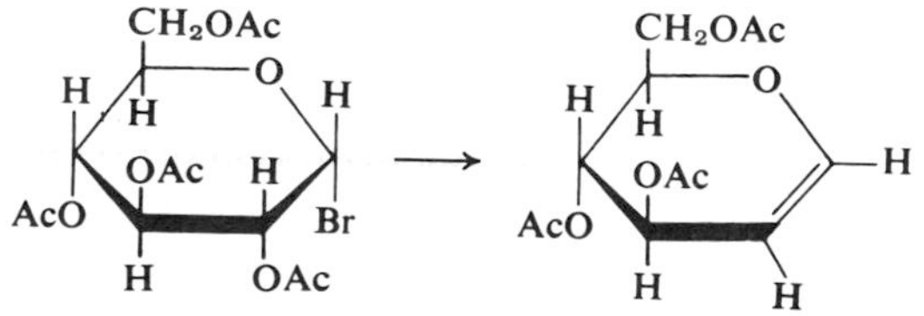

The mechanism suggested by Prins[111] involves three steps: ionization of **9** to form Br⁻ and a glycosyl cation (**12**), reduction of **12** to form the (presumably unstable) carbanion (**13**), and, finally, elimination of the acetate group from C-2; the acetate group is lost too rapidly to allow a proton to add at C-1 and form the anhydroalditol. Derivatives methylated at O-2 do not give this reaction, presumably because the methoxyl group is a poorer leaving group.

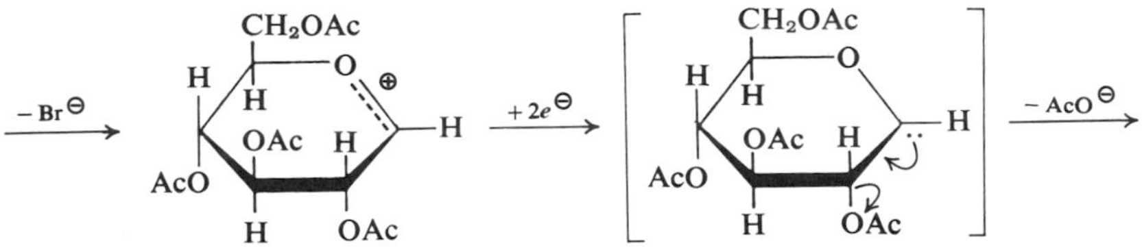

Palladium-catalyzed hydrogenation of the double bond in the tetraacetate of "2-hydroxyglucal" (1,5-anhydro-D-*arabino*-hex-1-enitol) gives 1,5-anhydro-D-mannitol tetraacetate, the product having O-2 axial; however, direct hydrogenation of **9** in the presence of triethylamine gives only the tetraacetate of 1,5-anhydro-D-glucitol.[112]

Sugar derivatives containing the phosphonic ester [C—P(O)(OR)₂] group react with hydrides of aluminum to yield phosphines[113] and phosphine oxides,[114] which can be used to prepare cyclic sugars having phosphorus as the ring heteroatom.

IX. DRASTIC HYDROGENOLYSIS

The energy required to break a carbon–carbon bond by hydrogenation is much greater than that for other bonds. Values for C–C cleavage vary from 18 to 48 kcal mole⁻¹, in comparison with 17 kcal mole⁻¹ for the carbon–oxygen bond, and 5–10 kcal mole⁻¹ for the double bond in the carbonyl group.[115] D-Glucitol has been converted in the presence of nickel at 215° into glycerol in 40% yield; erythritol and xylitol were also obtained.[116] Xylitol has been hydrogenolyzed[117] at 200° to 245° to give mixtures of glycerol, erythritol, and ethylene glycol. Breaking of carbon–oxygen bonds has been observed in the nickel-catalyzed hydrogenolysis of unhydrolyzed larch gum at 100° to 160°, and galactitol and L-arabinitol were found.[118] Copper–chromium oxide-catalyzed hydrogenation of sucrose in ethanol gives a variety of products,[119] including 2,6-anhydro-β-D-fructofuranose, erythritol, ethyl α- and β-D-fructofuranoside, D-fructose, D-glucitol, glycerol, 5-(hydroxymethyl)-2-furaldehyde, and D-mannitol. Interestingly, D-glucitol was the only derivative of D-glucose isolated.

Isomerization at asymmetric centers has been demonstrated under vigorous conditions of reduction. Heating D-glucitol in hydrogen in the presence of nickel at 170° gave an equilibrium mixture of 41% of D-glucitol, 31% of D-mannitol, and 26% of L-iditol.[120] It was found that both a hydrogen atmosphere and an alkaline solution are necessary for this epimerization; acid inhibits it. As a mechanism, dehydrogenation at C-1 or C-6 was suggested, followed by alkaline isomerization through the enediol and hydrogenation back to a hexitol. Perlin et al.[121] converted methyl β-L-arabinopyranoside at 180° under similar conditions into the glycosides of D-ribose, D-xylose, and D- and L-lyxose; no anomerization was observed. Gorin[122] heated various methyl glycopyranosides in hydrogen in the presence of a copper chromite catalyst at 180° and observed greater reactivity toward epimerization for pentosides than for hexosides, and for α-glycosides than for β-glycosides; the D-glucosides and D-talosides were the most stable of the glycosides investigated.

Hot deuterium oxide containing Raney nickel effects deuterium exchange in methyl glycosides and alditols at carbon atoms bound to free hydroxyl groups; configuration is retained except under conditions of prolonged exposure to the reagent and the reaction thus provides a valuable route to ^{2}H-labeled sugars.[123]

REFERENCES

1. J. E. G. Barnett and D. L. Corina, *Advan. Carbohyd. Chem. Biochem.*, **27**, 127 (1972).
2. N. G. Gaylord, "Reduction with Complex Metal Hydrides," Wiley (Interscience), New York, 1956.
3. C. R. Noller, "Chemistry of Organic Compounds," 3rd ed., W. B. Saunders Co., Philadelphia, Pennsylvania, 1957, pp. 217–219.
4. M. Fedoroňko, *Advan. Carbohyd. Chem. Biochem.*, **29**, 103 (1974).
5. A. J. Birch, *Quart. Rev. Chem. Soc.*, **4**, 69 (1950).
6. J. E. Page, *Quart. Rev. Chem. Soc.*, **6**, 262 (1952).
7. G. C. Bond, *Quart. Rev. Chem. Soc.*, **8**, 279 (1954); P. M. Gundry and F. C. Tompkins, *ibid.*, **14**, 257 (1960).
8. H. C. Brown, *Tetrahedron*, **12**, 117 (1961); H. C. Brown and B. C. Subba Rao, *J. Amer. Chem. Soc.*, **82**, 681 (1960); R. B. Wetherill, H. C. Brown, and B. C. Subba Rao, *J. Org. Chem.*, **22**, 1136 (1957).
9. G. R. Pettit and T. R. Kasturi, *J. Org. Chem.*, **25**, 875 (1960); **26**, 4553 (1961); G. R. Pettit, U. R. Ghatak, B. Green, T. R. Kasturi, and D. M. Piatak, *ibid.*, **26**, 1685 (1961); G. R. Pettit and D. M. Piatak, *ibid.*, **27**, 2127 (1962).
10. M. L. Wolfrom and H. B. Wood, Jr., *J. Amer. Chem. Soc.*, **73**, 2933 (1951).
11. H. L. Frush and H. S. Isbell, *J. Amer. Chem. Soc.*, **78**, 2844 (1956).
12. R. K. Ness, H. G. Fletcher, Jr., and C. S. Hudson, *J. Amer. Chem. Soc.*, **73**, 4759 (1951).
13. S. Roseman, *J. Amer. Chem. Soc.*, **74**, 4467 (1952).
13a. S. S. Bhattacharjee, J. A. Schwarcz, and A. S. Perlin, *Carbohyd. Res.*, **42**, 259 (1975).

14. E. Fischer and R. Stahel, *Ber.*, **24**, 528 (1891); H. S. Isbell, *Methods Carbohyd. Chem.*, **1**, 135 (1962).

15. H. L. Frush and H. S. Isbell, *Methods Carbohyd. Chem.*, **1**, 127 (1962); U.S. Patent 2,632,005 (1953); *Chem. Abstr.*, **48**, 1434 (1954).

16. F. L. Humoller, *Methods Carbohyd. Chem.*, **1**, 105 (1962).

17. N. K. Kochetkov and B. A. Dmitriev, *Izv. Akad. Nauk SSSR, Ser. Khim.*, 2095 (1964).

18. N. Sperber, H. E. Zaugg, and W. M. Sandstrom, *J. Amer. Chem. Soc.*, **69**, 915 (1947).

19. M. L. Wolfrom and K. Anno, *J. Amer. Chem. Soc.*, **74**, 5583 (1952).

20. L. A. Mai, USSR Patent 102,627 (1956); *Chem. Abstr.*, **52**, 4682i (1958).

21. F G. Fischer and H. Schmidt, *Ber.*, **93**, 658 (1961).

22. F. Smith and A. M. Stephen, *Tetrahedron Lett.*, 17 (1960).

23. M. L. Wolfrom, J. R. Vercellotti, and G. H. S. Thomas, *J. Org. Chem.*, **26**, 2160 (1961); G. O. Aspinall, A. J. Charlson, E. L. Hirst, and R. Young, *J. Chem. Soc.*, 1696 (1963); S. C. McKee and E. E. Dickey, *J. Org. Chem.*, **28**, 1561 (1963).

24. J. H. Manning and J. W. Green, *J. Chem. Soc.* (*C*), 2357 (1967).

25. E. L. Hirst, E. (E.) Percival, and J. K. Wold, *J. Chem. Soc.*, 1493 (1964); R. J. Ross and N. S. Thompson, *Tappi*, **48**, 376 (1965); E. F. Walker, *ibid.*, **48**, 298 (1965).

26. D. A. Rees and J. W. B. Samuel, *Chem. Ind.* (London), 2008 (1965).

27. P. Kohn, R. H. Samaritano, and L. M. Lerner, *J. Amer. Chem. Soc.*, **87**, 5475 (1965).

28. L. M. Lerner, *Methods Carbohyd. Chem.*, **6**, 131 (1972).

29. Atlas Chemical Industries Inc., French Patent 1,377,972 (1964); *Chem. Abstr.*, **62**, 13222b (1965).

30. M. L. Wolfrom, M. Konigsberg, F. B. Moody, and R. M. Goepp, Jr., *J. Amer. Chem. Soc.*, **68**, 122 (1946).

31. J. Lönngren and S. Svensson, *Advan. Carbohyd. Chem. Biochem.*, **29**, 41 (1974).

32. A. T. Merrill, W. T. Haskins, R. M. Hann, and C. S. Hudson, *J. Amer. Chem. Soc.*, **69**, 70 (1947).

33. W. G. Overend, A. R. Peacocke, and J. B. Smith, *J. Chem. Soc.*, 3487 (1961).

33a. D. Horton and J. D. Wander, *Carbohyd. Res.*, **16**, 477 (1971).

34. J. M. Los and N. J. Gasper, *Rec. Trav. Chim.* (Pays-Bas), **79**, 112 (1960).

35. H. H. Stroh, G. Göldenitz, and H. Patzig, *Z. Chem.*, **5**, 459 (1965).

36. P. D. Bragg and L. Hough, *J. Chem. Soc.*, 4347 (1957).

37. D. H. Rammler and D. L. MacDonald, *Arch. Biochem. Biophys.*, **78**, 359 (1958).

38. Daichi Seiyaku Co. Ltd., Japanese Patent 20,016 (1961); *Chem. Abstr.*, **57**, 15221e (1962); R. M. Alieva, *Vestr. Leningrad. Univ.*, **18** (21), *Ser. Biol. No. 4*, 148 (1963); *Chem. Abstr.*, **60**, 9346c (1964).

39. O. Theander, *Acta Chem. Scand.*, **12**, 1883 (1958).

40. E. L. Eliel, "Stereochemistry of Organic Compounds," McGraw-Hill, New York. 1962, p. 243; D. H. R. Barton, *J. Chem. Soc.*, 1027 (1953), footnote 23.

41. O. Theander, *Acta Chem. Scand.*, **18**, 2209 (1964).

42. D. C. Baker, D. Horton, and C. G. Tindall, Jr., *Carbohyd. Res.*, **24**, 192 (1972); *Methods Carbohyd. Chem.*, **7**, 3 (1976).

42a. D. C. Baker, C. Boeder, J. Defaye, A. Gadelle, and D. Horton, *J. Org. Chem.*, **41**, 3834 (1976).

43. B. Lindberg and O. Theander, *Acta Chem. Scand.*, **13**, 1226 (1959).

44. E. L. Martin, *Org. React.*, **1**, 155 (1952).

45. A. Rosenthal and D. A. Baker, *Carbohyd. Res.*, **26**, 163 (1973); compare J. M. J. Tronchet and J. Tronchet, *ibid.*, **33**, 237 (1974).

46. R. K. Ness, H. G. Fletcher, Jr., and C. S. Hudson, *J. Amer. Chem. Soc.*, **72**, 4547 (1950).

47. E. S. Gould, "Mechanism and Structure in Organic Chemistry," Holt, Rinehart, and Winston, New York, 1959, p. 258.

48. F. A. H. Rice and M. Inatome, *J. Amer. Chem. Soc.*, **80**, 4709 (1958).

49. H. G. Fletcher, Jr., and C. S. Hudson, *J. Amer. Chem. Soc.*, **69**, 1672 (1947).

50. M. L. Wolfrom and J. V. Karabinos, *J. Amer. Chem. Soc.*, **66**, 909 (1944).

51. B. Lindberg and L. Nordén, *Acta Chem. Scand.*, **15**, 958 (1961).

52. D. Horton and J. S. Jewell, *J. Org. Chem.*, **31**, 509 (1966).

53. O. T. Schmidt, *Methods Carbohyd. Chem.*, **1**, 191 (1962).

54. N. K. Kochetkov, L. I. Kudryashov, and A. I. Usov, *Dokl. Akad. Nauk SSSR*, **133**, 1091 (1960).

54a. H. Kuzuhara, K. Sato, and S. Emoto, *Carbohyd. Res.*, **43**, 293 (1975).

55. W. W. Zorbach and C. O. Tio, *J. Org. Chem.*, **26**, 3543 (1961).

55a. A. Zobačová, V. Hermanková, Z. Kefurtová, and J. Jarý, *Collect. Czech. Chem. Commun.*, **40**, 3505 (1975).

56. N. K. Kochetkov and A. I. Usov, *Tetrahedron*, **19**, 973 (1963).

57. H. Arita, N. Ueda, and Y. Matsushima, *Bull. Chem. Soc. Jap.*, **45**, 567 (1972).

58. S. Hanessian, *Carbohyd. Res.*, **2**, 86 (1966); D. L. Failla, T. L. Hullar, and S. B. Siskin, *Chem. Commun.*, 716 (1966); see also W. A. Szarek, *Advan. Carbohyd. Chem. Biochem.*, **28**, 225 (1973).

59. A. Holý, *Tetrahedron Lett.*, 585 (1972).

59a. R. H. Bell, D. Horton, D. M. Williams, and E. Winter-Mihaly, *Carbohyd. Res.*, **58**, 109 (1977).

59b. D. H. R. Barton and S. W. McCombie. *J. Chem. Soc. Perkin I*, 1574 (1975).

60. D. A. Prins, *J. Amer. Chem. Soc.*, **70**, 3955 (1948).

61. M. Dahlgard, B. H. Chastain, and R. L. Han, *J. Org. Chem.*, **27**, 929 (1962).

62. M. Černý and J. Pacák, *Collect. Czech. Chem. Commun.*, **27**, 94 (1962).

63. H. G. Fletcher, Jr., and N. K. Richtmyer, *Advan. Carbohyd. Chem.*, **5**, 19 (1950).

64. C. D. Anderson, L. Goodman, and B. R. Baker, *J. Amer. Chem. Soc.*, **80**, 5247 (1958); **81**, 898, 3967 (1962).

65. P. M. Collins, P. Gupta, and R. Iyer, *J. Chem. Soc. Perkin I*, 1670 (1972).

66. K. Takiura and S. Honda, *Carbohyd. Res.*, **21**, 379 (1972).

67. D. Horton, J. M. Tarelli, and J. D. Wander, *Carbohyd. Res.*, **23**, 440 (1972).

68. S. Honda, K. Kaheki, and K. Takiura, *Carbohyd. Res.*, **29**, 488 (1973).

69. E. G. Ansell and J. Honeyman, *J. Chem. Soc.*, 2778 (1952).

70. J. W. H. Oldham, *J. Chem. Soc.*, **127**, 2840 (1925).

71. J. Honeyman and J. W. W. Morgan, *Advan. Carbohyd. Chem.*, **12**, 117 (1957).

72. I. G. Wright and L. D. Hayward, *Can. J. Chem.*, **38**, 316 (1960); see also D. J. Bell, *Methods Carbohyd. Chem.*, **1**, 260 (1962).

73. D. O. Hoffman, R. S. Bower, and M. L. Wolfrom, *J. Amer. Chem. Soc.*, **69**, 249 (1947).

74. L. Schmid and P. Karrer, *Helv. Chim. Acta*, **32**, 1371 (1949).

75. R. S. Tipson, *Methods Carbohyd. Chem.*, **2**, 250 (1963).

76. H. C. Jarrell, R. G. S. Ritchie, W. A. Szarek, and J. K. N. Jones, *Can. J. Chem.*, **51**, 1767 (1973).

77. S. Zen, S. Tashima, and S. Koto, *Bull Chem. Soc. Jap.*, **41**, 3025 (1969).

78. D. Grant, *Chem. Ind.* (London), 1492 (1959).

79. P. Kováč and Š. Bauer, *Tetrahedron Lett.*, 2349 (1972).

80. H. B. Wood, Jr., and H. G. Fletcher, Jr., *J. Amer. Chem. Soc.* **80**, 5242 (1958); see also H. G. Fletcher, Jr., *Methods Carbohyd. Chem.*, **2**, 386 (1964).

81. W. Meyer zu Reckendorf and J. C. Jochims, *Ber.*, **102**, 4199 (1969).

82. B. Fraser-Reid and B. Radatus, *J. Amer. Chem. Soc.*, **92**, 6661 (1970); S. Y.-K. Tam and B. Fraser-Reid, *Tetrahedron Lett.*, 4897 (1973).

83. D. Keglević, Š. Valenteković, G. Riglić, D. Goleš, and F. Plavšić, *Carbohyd. Res.*, **29**, 25 (1973).

83a. V. G. Mairanovskii, N. F. Logovina, A. M. Ponomarev, and A. Ya. Veinberg, *Elektrokhimya*, **10**, 172 (1974); *Chem. Abstr.*, **80**, 108, 770 (1974).

83b. S. S. Bhattacharjee and P. A. J. Gorin, *Can. J. Chem.*, **47**, 1207 (1969); B. Fleming and H. I. Bolker, *ibid.*, **52**, 888 (1974); A. Lipták, I. Jodál, and P. Nánási, *Carbohyd. Res.*, **44**, 1 (1975).

83c. L. M. Jeppssen, I. Lundt, and C. Pedersen, *Acta Chem. Scand.*, **27**, 3579 (1973).

84. E. J. Reist, V. J. Bartuska, and L. Goodman, *J. Org. Chem.*, **29**, 3725 (1964).

85. R. L. Whistler, M. S. Feather, and D. L. Ingles, *J. Amer. Chem. Soc.*, **84**, 122 (1962).

86. E. J. Reist, R. R. Spencer, and B. R. Baker, *J. Org. Chem.*, **24**, 1618 (1959).

87. G. Rembarz, *J. Prakt. Chem.*, [4], **19**, 315 (1963).

88. S. S. Bhattacharjee and P. A. J. Gorin, *Can. J. Chem.*, **47**, 1195 (1969).

89. E. J. Reist, R. R. Spencer, B. R. Baker, and L. Goodman, *Chem. Ind.* (London), 1794 (1962).

90. S. Hanessian, *Chem. Ind.* (London), 1296 (1965).

91. M. L. Wolfrom, J. Bernsmann, and D. Horton, *J. Org. Chem.*, **27**, 4505 (1962).

92. M. S. Patel and J. K. N. Jones, *Can. J. Chem.*, **43**, 3105 (1965).

93. R. D. Guthrie and R. D. Wells, *Carbohyd. Res.*, **24**, 11 (1972).

94. R. D. Guthrie, British Patent 1,007,001 (1965); *Chem. Abstr.*, **64**, 8290f (1966).

95. J. W. W. Morgan and M. L. Wolfrom, *J. Amer. Chem. Soc.*, **78**, 2496 (1956).

96. F. Kagen, M. A. Rebenstorf, and R. V. Heinzelman, *J. Amer. Chem. Soc.*, **79**, 3541 (1962); see also J. W. Long and G. N. Bollenback, *Methods Carbohyd. Chem.*, **2**, 81 (1963).

97. G. I. Vishnevskaya, L. M. Yagupol'skii, and A. D. Gorbunova, USSR Patent 148,800 (1962); *Chem. Abstr.*, **58**, 12662h (1963).

98. O. Theander, *Advan. Carbohyd. Chem.*, **17**, 223 (1962).

99. W. Roth, W. W. Pigman, and I. Danishevsky, *Tetrahedron*, **20**, 1675 (1964).

99a. J. Plenkiewicz, W. A. Szarek, P. A. Sipos, and M. K. Phibbs, *Synthesis*, 56 (1974).

100. R. Kuhn and H. Grassner, *Ann.*, **612**, 55 (1957).

101. R. Kuhn and W. Kirschenlohr, *Angew. Chem.*, **67**, 786 (1955).

102. R. Kuhn and P. Klesse, *Ber.*, **91**, 1989 (1958).

103. M. L. Wolfrom and E. Usdin, *J. Amer. Chem. Soc.*, **75**, 4318 (1953).

104. E. Mosettig and R. Mozingo, *Org. React.*, **4**, 362 (1948).

105. S. J. Angyal, C. G. MacDonald, and N. K. Matheson, *J. Chem. Soc.*, 3321 (1953).

106. A. L. Wilds, *Org. React.*, **2**, 178 (1944); T. Bersin, in "Newer Methods of Preparative Organic Chemistry," (Wiley: Interscience), New York, 1948, p. 125.

107. S. J. Angyal and D. J. McHugh, *J. Chem. Soc.*, 3682 (1957).

108. R. Hems, D. Horton, J. B. Hughes, and M. Nakadate, *Carbohyd. Res.*, **25**, 205 (1972); D. C. Baker, D. K. Brown, D. Horton, and R. G. Nickol, *ibid.*, **32**, 299 (1974).

109. D. Horton, A. Liav, and S. E. Walker, *Carbohyd. Res.*, **28**, 201 (1973).

110. E. Fischer and K. Zach, *Sitzungsber. Kgl. Preuss. Akad. Wiss.*, **16**, 311 (1913); see also W. Roth and W. W. Pigman, *Methods Carbohyd. Chem.*, **2**, 405 (1963).

111. W. G. Overend and M. Stacey, *Advan. Carbohyd. Chem.*, **8**, 71 (1953); R. W. Jeanloz and D. A. Prins, *Annu. Rev. Biochem.*, **17**, 67 (1948).
112. L. Zervas, *Ber.*, **63**, 1689 (1930); see also N. K. Richtmyer, C. J. Carr, and C. S. Hudson, *J. Amer. Chem. Soc.*, **65**, 1477 (1943).
113. R. L. Whistler and C.-C. Wang, *J. Org. Chem.*, **33**, 4455 (1968).
114. S. Inokawa, H. Kitigawa, K. Seo, H. Yoshida, and T. Ogata, *Carbohyd. Res.*, **30**, 127 (1973).
115. A. A. Balandin and N. A. Vasyunina, *Dokl. Akad. Nauk SSSR*, **117**, 84 (1957); P. H. Brahme, M. U. Pai, and G. Narsimhan, *Brit. Chem. Eng.*, **9**, 684 (1964).
116. I. T. Clark, *Ind. Eng. Chem.*, **50**, 1125 (1958).
117. N. A. Vasyunina, A. A. Balandin, and Yu. Mamatov, *Kinet. Katal.*, **4**, 156 (1963); *Chem. Abstr.*, **59**, 731c (1963).
118. A. F. Zaitseva and L. A. Zaitseva, *Izv. Vyssh. Ucheb. Zaved. Les. Zh.*, **3**, 153 (1960); *Chem. Abstr.*, **55**, 15358h (1961).
119. H. R. Goldschmid and A. S. Perlin, *Can. J. Chem.*, **38**, 2178 (1960).
120. L. Wright and L. Hartmann, *J. Org. Chem.*, **26**, 1588 (1961).
121. A. S. Perlin, E. v. Rudloff, and A. P. Tulloch, *Can. J. Chem.*, **36**, 921 (1958).
122. P. A. J. Gorin, *Can. J. Chem.*, **38**, 641 (1960).
123. H. J. Koch and R. S. Stuart, *Carbohyd. Res.*, **67**, 341 (1978).

23. ACIDS AND OTHER OXIDATION PRODUCTS

OLOF THEANDER

I. INTRODUCTION

This chapter deals with the low-molecular-weight carbohydrates that can be formally considered as oxidation products of mono- or oligo-saccharides in which an aldehyde group and/or one or more hydroxyl groups have been oxidized to carbonyl and/or carboxyl groups. They are divided into two principal categories—namely, acids (and lactones) and neutral compounds.

II. ACIDS

The main classes of acidic carbohydrates, which are considered separately in Sections II,A–C, are:

$$
\begin{array}{ccc}
CO_2H & CHO & CO_2H \\
| & | & | \\
(CHOH)_n & (CHOH)_n & (CHOH)_n \\
| & | & | \\
CO_2H & CO_2H & CO_2H \\
\text{Aldonic acids} & \text{Glycuronic acids} & \text{Aldaric acids}
\end{array}
$$

1013

Aldonic acids having other substituents are discussed in Section II,D, and the ascorbic acids are treated in a separate section (II,E). A comprehensive collection of literature references prior to 1964, together with tabulations of physical constants for the acids and many of their derivatives, can be found in the monographs[1,2] by Staněk *et al.*

Some acids—for instance, L-ascorbic, glycuronic, and aldonic acids of various types—occur naturally as such. They have also been found (or suggested to occur) both as metabolic intermediates in animals, plants, and microorganisms (discussed in Vol. IIA, Chapters 33 and 34) and as products of the action of microorganisms on carbohydrates. Others are constituents of naturally occurring, acidic polysaccharides or in such polysaccharides as cellulose and starch (see Vol. IIA, Chapter 36; Vol. IIB, Chapter 38) that have been technically modified.

Acids of great commercial importance are, particularly, the ascorbic and gluconic acids. However, there are others that are either important pharmaceutical intermediates (such as D-ribonic and D-glucuronic acids) or have interesting potentialities.

A. Aldonic Acids

The aldonic acids (or their lactones) are the initial oxidation products of aldoses and can readily be produced by the action of a number of common oxidants upon the corresponding aldose:

$$HOCH_2(CHOH)_nCHO \xrightarrow{\text{oxidation}} HOCH_2(CHOH)_nCO_2H$$

As the aldonic acids readily form crystalline salts, lactones, hydrazides, and other crystalline carboxylic derivatives, they have been valuable for characterization of sugars. A reducing disaccharide containing two different aldoses can be converted into the corresponding aldobionic acid and, after subsequent hydrolysis and identification of the products, the position of the reducing group in the original disaccharide can thus be proved.

The 2-ketoses generally need more energetic oxidation for reaction, and thereby undergo oxidation at a primary hydroxyl group (see Section II,B) or chain-splitting (sometimes to a lower aldonic acid). This difference in rate of oxidation can be used analytically for the determination of aldoses in the presence of ketoses.

D-Gluconic acid occurs as a metabolic intermediate in animals, plants, and microorganisms, and L-gulonic acid is almost certainly an intermediate in the biosynthesis of L-ascorbic acid. Numerous bacteria and fungi produce D-gluconic acid and other aldonic acids from their corresponding sugars. L-Arabinonic acid has been isolated from *Austrocedrus chilensis* (Cupressales).[3]

D-Gluconic acid is of great commercial importance. The main methods for its industrial production are various microbial processes, electrolytic oxidation, and catalyzed air oxidation. (For references to the many patents concerning the preparation of the acid and its salts, and their uses, see ref. 1 and references in articles by Prescott *et al.*[4] and by Mehltretter.[5]) The commercial product is mainly a 50% aqueous solution of calcium, sodium, or other salts but the free acid and the 1,4- and 1,5-lactones are known in crystalline form. D-Gluconic acid has achieved increased prominence as a sequestering agent for calcium, iron, and other metals because it prevents the precipitation of hydroxides or such salts as oxalates by converting the divalent and trivalent ions into chelated, water-soluble forms. A number of these salts are valuable pharmaceutical chemicals; for instance, ferrous D-gluconate is used in the treatment of iron-deficiency anemia, and the calcium salt is an excellent agent for the treatment of calcium deficiency in man. In the textile industry, both the free acid and the 1,5-lactone are used as acid catalysts, and they, as well as various salts, have numerous other applications.

1. Preparation

Aldonic acids can be prepared by: (*a*) oxidation of the parent sugar, (*b*) methods involving addition or degradative removal of asymmetric carbon atoms, (*c*) epimerization of other aldonic acids, or (*d*) reduction of carbohydrates in a higher oxidation state.

a. Oxidation of Aldoses to Aldonic Acids. —A more-detailed discussion of the formation of aldonic acids (and lactones) from aldoses, glycosides, and polysaccharides by various oxidants, and also the mechanisms involved, is given in Chapter 24. The methods given here will be the more stoichiometric and preparative ones.

Hypoiodite[6,7] and chlorous acid,[8] which can afford stoichiometric yields of aldonic acids under controlled conditions, have been used for the analysis of aldoses. Moore and Link[9] prepared various aldonic acids in good yield by the oxidation of aldoses with potassium hypoiodite in methanol. This reagent has been recommended for the oxidation of acid-sensitive compounds—for instance, 2,3-*O*-isopropylidene-α-D-lyxofuranose.[10]

The most widely used chemical method, which is not quite so stoichiometric as those just discussed, is oxidation with bromine in aqueous solution. It was first used by Hlasiwetz[11] in 1861. The hydrobromic acid formed as a by-product lowers the rate of oxidation; this effect is, however, minimized in buffered solutions (pH 5 to 6), and for this purpose the addition of barium carbonate or barium benzoate is convenient.[12] This method has been used for the preparation of a number of aldonic acids from both mono- and oligo-saccharides.[1] Essentially the same reaction occurs in the electrolytic oxidation procedure of Isbell and co-workers,[13] in which bromine is generated by the oxidation of catalytic amounts of calcium bromide in a solution containing the sugar and calcium carbonate. The procedure has also been adapted to a convenient laboratory scale for the preparation of aldonates of mono- and di-saccharides.[14] It was shown that oxidation by chlorine is partly a radical and partly an ionic process. When the former is retarded by chlorine dioxide, the main products from oxidation of methyl β-D-glucopyranoside are D-gluconic and D-arabinonic acids.[14a]

Another type of halogen oxidant, namely N-bromo compounds, has also been used. For example N-bromocarbamide was recommended by Kiss[15] as a selective and convenient compound for oxidizing benzylated sugars to their corresponding aldonolactones in yields exceeding 90%.

Morrison and Perry[16] reported a method for the quantitative determination and characterization of aldoses by oxidation with hypoiodite or hypobromite to the corresponding aldonic acids. After conversion into per-O-(trimethylsilyl)-1,4-lactones, they were analyzed by gas–liquid partition chromatography.

Catalytic oxidation of aldoses to the corresponding aldonic acids is readily conducted with a platinum-on-carbon catalyst in the presence of an equivalent of alkali, as demonstrated by Heyns and Paulsen.[17] Oxidation of aldoses with mercuric acetate in aqueous solution has been reported[18] to give yields of aldonic acids up to about 50%.

Numerous organisms are able to transform aldoses into aldonic acids (or lactones), and many oxidizing enzymes from various, natural sources have been isolated (see Chapter 34). A dehydrogenase enzyme, called notatin, isolated from various molds, especially *Aspergillus niger* and *Penicillium glaucum*,[19] has been studied by many workers. Pazur and Kleppe[20] examined its purification and studied the rates of oxidation of D-aldoses, monodeoxy-D-hexoses, and O-methyl-D-glucoses. As the rate of oxidation of β-D-glucose is much higher than that of other aldoses, this process has been widely used for determination of D-glucose.

D-Gluconic acid can be produced in good yield by many species of bacteria and molds, and microbial oxidation of other monosaccharides to their aldonic acids has also been reported (see Chapter 24). Lactose undergoes stoichiometric oxidation to lactobionic acid by *Pseudomonas graveolens*.[21]

b. Synthesis from Lower Aldoses or by Oxidative Degradation.—By the Kiliani cyanohydrin synthesis, aldonic acids may be prepared from an aldose having one carbon atom fewer (see Chapter 3). A new asymmetric center is created, and many methods have been used for separating the two epimeric acids formed. In the form of lactones or esters, they can be reduced to aldoses.

An alternative, general method involving elongation of the carbon chain by the two carbon atoms has been developed by Kochetkov and Dmitriev[22] (see Chapter 3); it is based on transformation of an *aldehydo*-aldose into the homologous *trans*-2,3-dideoxyald-2-enonic acid or its ester and subsequent hydroxylation of the double bond to afford the two diastereoisomeric aldonic acids having the D-*threo* and L-*threo* stereochemistry, respectively, at the newly created asymmetric centers, and the same configuration as the original aldose in the rest of the molecule:

$$\text{RCHO} \longrightarrow \text{RCH=CHCO}_2\text{R}' \longrightarrow \text{RCHOHCHOHCO}_2\text{R}'$$

R = monosaccharide residue; R' = H or Et

The chain-extension step is the condensation of the protected aldehydo derivative with malonic acid (Knoevenagel–Doebner condensation) or with (ethoxycarbonylmethylene)triphenylphosphorane (Wittig reaction), followed by hydroxylation with osmium tetraoxide in the presence of a suitable oxidant; the optimal oxidant differs for various types of unsaturated intermediates. Many examples of preparations of heptonic and octonic acids in good yields have been reported.

Controlled, degradative, alkaline oxidation of aldoses affords aldonic acids containing one fewer carbon atom; accordingly, the same acid can be formed from several sugars. Spengler and Pfannenstiel[23] oxidized D-fructose or D-glucose with oxygen in alkaline solution to give sodium D-arabinonate in 70% yield. By the same method, D-ribose and L-arabinose have been degraded to D- and L-erythronic acid, respectively, and D-xylose to D-threonic acid.[24] Isbell *et al.*[24a] elucidated the mechanism of this process (see Chapter 24). During industrial alkaline oxidation of polysaccharides, the reducing end-groups are transformed into aldonic acid residues. See ref. 24b and Section III,B,1,*b*.

The method of Reichstein *et al.*[25] for the preparation of L-threono-1,4-lactone by permanganate oxidation of 5,6-*O*-isopropylidene-L-ascorbic acid was improved by Perel and Dayton[26] to afford the crystalline lactone in 65% yield.

2-Glyculosonic ("2-*ketoaldonic*") acids can be converted by the Ruff degradation into aldonic acids having one carbon atom fewer; thus, L-xylonic acid is formed from L-*xylo*-hexulosonic acid by the action of iron(III) and hydrogen peroxide.[27] 2-Glyculosonic acids are presumed to be intermediates in the Ruff degradation of aldoses (see Chapter 3, Section IV,A).

References start on p. 1085.

c. Configurational Interchange.—By alkaline treatment at elevated temperatures, the aldonic acids can undergo epimerization, causing equilibration of configuration about C-2 (see Section II,A,2); similar interconversions may also occur upon heating. This method is of particular interest for the production of D-ribonic acid from D-arabinonic acid by epimerization in pyridine or other media.[28] D-Ribonic acid is reduced to D-ribose, and this sugar or its tetraacetate is used in the important production of riboflavin, which is nowadays consumed in considerable quantities. It is notable that the D-arabinonic acid used is prepared from D-glucose by oxidative degradation.

d. Reduction of More-Highly Oxidized Acid Derivatives.—Glyculosonic acids can be reduced to aldonic acids; thus, L-idonic acid has been prepared from D-*xylo*-5-hexulosonic acid by reduction in the presence of Raney nickel.[29] Bestmann and Schmiechen[30] synthesized L-threono-1,4-lactone from L-threaric acid via reduction of methyl 2,3-di-*O*-acetyl-L-threaroyl chloride.

2. Properties and Reactions

a. Lactonization.—In aqueous solutions of aldonic acids, an equilibrium between the acid and lactone forms will exist if it is not prevented by substitution. For example, the carboxyl group of D-gluconic acid (**2**) may lactonize by condensation with the hydroxyl group on C-4 or C-5 to generate the 1,4-(γ, **3**) or 1,5-(δ, **1**) lactone, respectively; the proportions of acid and lactone depend on the configuration of the asymmetric carbon atoms. There

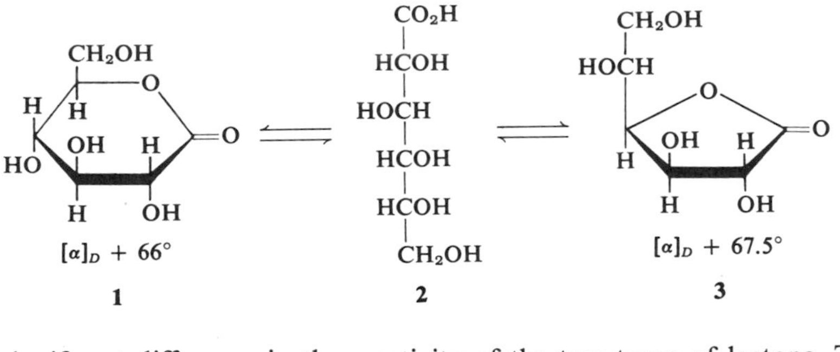

is a significant difference in the reactivity of the two types of lactone. The 1,5-lactones usually hydrolyze readily and may exhibit rapid mutarotation in aqueous solution, whereas the 1,4-lactones are more stable. This is also true for the methylated lactones, and the distinction between the two types of lactone is very clearly shown by the investigations of Haworth.[31] (Fig. 1) The rates of hydrolysis of the lactones depend on the structure of the molecules; for instance, the lactones of D-mannonic acid are more stable than those of D-gluconic acid, and the 1,4-lactones of 2-deoxyaldonic acids are

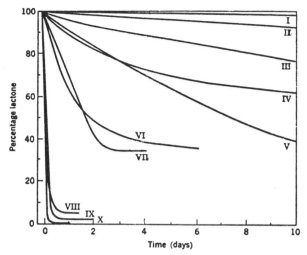

FIG. 1. Mutarotation of methylated lactones. (After Haworth.[31]) I. Tetra-*O*-meth-ylmannono-1,4-lactone; II. Tetra-*O*-methylgalactono-1,4-lactone; III. Tri-*O*-meth-ylxylono-1,4-lactone; IV. Tri-*O*-methylarabinono-1,4-lactone; V. Tetra-*O*-methyl-glucono-1,4-lactone; VI. Tetra-*O*-methylmannono-1,5-lactone; VII. Tri-*O*-methyl-xylono-1,5-lactone; VIII. Tetra-*O*-methylglucono-1,5-lactone; IX. Tetra-*O*-meth-ylgalactono-1,5-lactone; X. Tri-*O*-methylarabinono-1,5-lactone.

more stable than the corresponding aldonolactones. The final attainment of equilibrium between free aldonic acids and their lactones is reached only after many days at room temperature; it is, however, accelerated by the presence of strong acids. A more detailed discussion of the formation and hydrolysis of aldonolactones is available in a review by Shafizadeh,[32] and the conformations and stabilities of aldonolactones have been discussed by Lemieux.[33] Detailed analyses of D-glucono-1,5-lactone and other lactones have been reported.[33a]

On paper chromatograms, the equilibrium mixture of acid and lactones often gives a slow-moving, elongated spot for the acid and one or two faster-moving spots corresponding to the lactones. On electrophoretograms, a nonmigrating spot appears for the lactones, whereas the spot for the acid migrates. The lactones give a violet coloration when the paper is treated with the hydro-xamic reagent.[34]

The equilibrium proportions of the constituents vary with temperature, concentration, and solvent and are characteristic for each individual aldonic acid. Dehydration *in vacuo*, or by evaporation from suitable solvents, promotes lactone formation. By suitable choice of conditions, many aldonic acids and both types of lactone have been obtained crystalline[1]; considerations have been

References start on p. 1085.

detailed by Isbell and Frush.[35] The carbonyl frequency in the infrared spectrum provides a fairly characteristic method for differentiating between γ and δ lactones of aldonic acids. With few exceptions, the absorptions are in the range 1790 to 1765 cm^{-1} and 1760 to 1725 cm^{-1}, respectively.[36] Proton n.m.r. spectra of D-gluconic acid and lactones have also been studied.[37]

The relationship between structure and paper-chromatographic mobility of aldono-1,4-lactones has been discussed.[38] Butanone, saturated with water, is suitable for the preparative separation of lactones on cellulose columns; free acids and reducing sugars move very slowly in this system. For example, the separation of homologous aldono-1,4- and other lactones, obtained by oxidative and/or alkaline degradations of hexoses and glycosides, can be effected.[39] Samuelson and Simonson[40] have developed a method for separation of different aldonic acids on an anion-exchange resin in the borate form, by using aqueous sodium tetraborate as the eluant. The method can be automated and can also be applied to the separation of such complex mixtures as pulping waste liquors, which contain many other types of compounds. Separation of aldono-1,4-lactones[16] and of aldonic acids[41] as the O-(trimethylsilyl) derivatives has been accomplished. The mass spectra of such derivatives are characteristic, so that it is possible to identify aldonic acids[41a] and lactones[41b] by means of a combined gas chromatograph–mass spectrometer.

Under dehydrating conditions, it is probable that, in addition to lactonization, intermolecular esterification can occur, yielding esters of aldonic acids and polymers.

b. Optical Rotatory Relationships.—A number of empirical relationships have been derived from the optical rotations of acids, lactones, salts, and various derivatives (see also Chapter 27, Section III). These relationships have been important in proving the configurations of new acids, in particular of the epimers produced in the cyanohydrin synthesis. They have also been useful in establishing the ring size of lactones.

Because of the conformational restraints imposed by ring formation, the lactones have considerably higher rotations than the free acids. Hudson[42] correlated the optical rotations of a number of lactones and found that the configurations of the hydroxyl groups on C-4 and C-5 exert a profound influence on the rotations. His "lactone rule" stipulates that a lactone is more dextrorotatory than the corresponding free acid if the hydroxyl group involved in lactone formation lies on the right-hand side in the Fischer projection formula; conversely, the lactone is more levorotatory if the hydroxyl group is on the left side. Thus, 1,4- and 1,5-lactones of both D-gluconic acid and D-mannonic acid are dextrorotatory, whereas D-galactonic acid and D-gulonic acid form levorotatory 1,4-lactones and dextrorotatory 1,5-lactones.

The rotations of acyclic derivatives of the aldonic acids are influenced most strongly by the configuration of the adjacent carbon atom (C-2). Those amides[43] are dextrorotatory in which the hydroxyl group at C-2 lies to the right in the Fischer formula; for example, the amide of D-gluconic acid has $[\alpha]_D + 31°$, whereas the amide of D-mannonic acid is levorotatory, $[\alpha]_D - 17°$. A similar rule also applies for phenylhydrazides[44,45] and benzimidazoles.[46] More details about these and related rules, with a discussion of exceptions and supplementations, are found in ref. 1 and in a review by Klyne.[47]

Optical rotatory dispersion studies on lactones of aldonic acids and other carbohydrate acids were made by Hirst et al.[48] and later by Okuda et al.[49] The latter workers related the absolute configuration at C-2 of aldono-1,4-lactones to the sign of the Cotton effect.

c. *Other Properties and Reactions.*—Aldonic acids possess the properties of carboxylic acids, and their aqueous solutions have a pH of 2 to 3. They form salts, some of which are sparingly soluble (for example, lead and cadmium salts), and these can be used for separation of epimeric acids. The lactones (in particular the 1,4-lactones) require an excess of alkali or elevated temperatures for conversion into salts in a reasonable period of time. Aldonic acids and lactones are stable toward acids. However, by treatment in alkaline solution, or by heating, the configuration at C-2 of aldonic acids can be altered. An example is the classical epimerization of D-mannonic acid performed by Emil Fischer[50] in 1890. A mixture of D-gluconic and D-mannonic acid was prepared by heating a solution of the latter in pyridine.

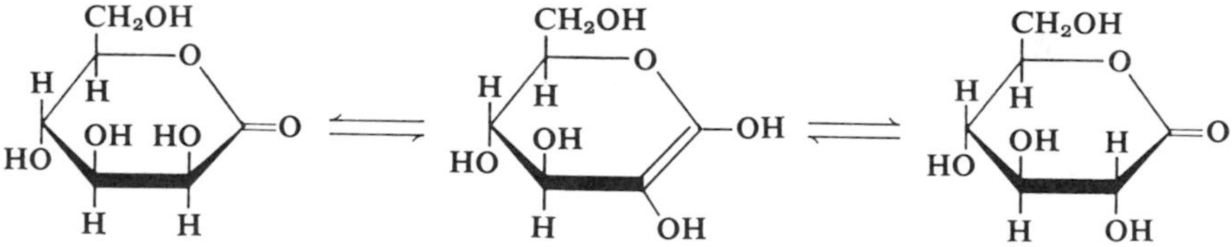

(1)

A second example is the manufacture of D-ribose, via epimerization of D-arabinonic into D-ribonic acid.[28] Epimerization may also occur when the hydroxyl group at C-2 is methylated; thus the tetra-*O*-methyl derivatives of both D-glucono-1,4- and -1,5-lactones can be converted into the corresponding tetra-*O*-methyl-D-mannonolactones.[51] The epimerization may take place, as with the sugars, through an intermediate enediol (see Eq. 1); this mechanism also accounts for the reaction of methylated derivatives, as the methoxyl group on C-2 is present merely as a substituent (see Chapter 4). Later investigations showed that marked epimerization of aldonic acids also occurs[52] in water at pH 8 to 7 at 60 to 100°. Two useful preparations of lower sugars by degradation of aldonic acids are discussed in Chapters 3 and 24. These are the traditional oxidation of salts of aldonic acids with hydrogen peroxide in the presence of ferric acetate (Ruff degradation) and the related oxidation of aldonic acids by hypochlorite, which was developed by Whistler and Schweiger.[53]

Aldonic acids form crystalline esters, amides, hydrazides, and benzimidazoles. Many of these derivatives, as well as the lactones, have been valuable in the characterization and identification not only of aldonic acids but also of sugars and methyl ethers after their conversion into aldonic acids.

Esters are prepared by reaction of aldonic acids or lactones with an alcohol in the presence of hydrogen chloride[54]; the reaction is slowest with 1,4-lactones. Amides are readily formed from lactones by reaction with liquid ammonia followed by evaporation of the solvent,[55] or by treatment with concentrated ammonium hydroxide and subsequent precipitation with ethanol.[56] Aldonamides having five or more carbon atoms in the sugar chain are hydrolyzed in aqueous solution with mutarotation[57]; those of tetronic acids are, however, stable.[58] The preparation of lower sugars from aldonamides by the Weerman degradation is discussed in Chapter 3. Phenylhydrazides, prepared by reaction of the lactone or ester with phenylhydrazine in ethanol–benzene, (or ether), can be converted into lactones by the action of nitrous acid.[59] D-Gluconic hydrazide reacts with chlorine in the presence of hydrogen chloride to afford a 1,2-bis(hydrazide) derivative.[60] The benzimidazoles are useful and readily prepared derivatives of all classes of carbohydrate-derived acids (see Chapter 21 and a survey by Richtmyer[61]).

Fully acetylated aldonic acids can be prepared by oxidation of *aldehydo*-sugar acetates.[62] Direct acetylation of certain aldonic acid salts is possible, and cadmium salts in particular afford high yields of acetates.[63] The synthesis can also be accomplished by deamination of the readily prepared, acetylated amides with nitrous acid[64] or nitrosyl chloride.[65] Examples of the various methods are given in ref. 66.

Fully acetylated aldonyl chlorides have been prepared by treatment of acetylated aldonic acids with phosphorus pentachloride[67] or thionyl chloride.[68] Dichloromethyl methyl ether was later shown to afford this derivative of D-gluconic acid in almost quantitative yield.[69] The action of diazomethane converts acetylated aldonyl chlorides into 1-diazoketose derivatives having one carbon atom more in the carbon chain; treatment with acids permits replacement of the diazo group by any one of a number of different nucleophiles. This versatile synthesis of ketoses, widely applied by Wolfrom and Thompson,[70] is discussed in Chapter 3.

Treatment with silver oxide in water causes these α-diazoketones to undergo the Wolff rearrangement, yielding 2-deoxyaldonic acids.[71]

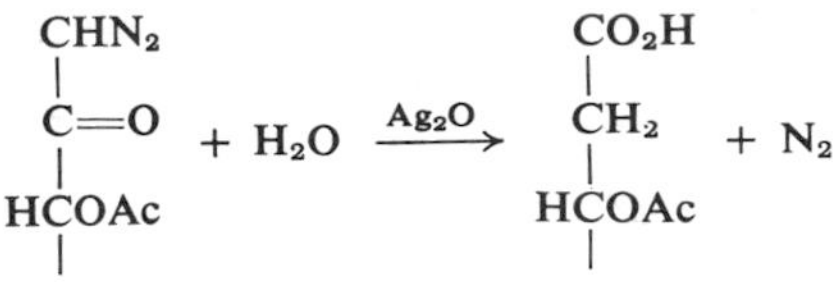

$$
\begin{array}{ccccc}
CHN_2 & & & CO_2H & \\
| & & & | & \\
C{=}O & + H_2O & \xrightarrow{Ag_2O} & CH_2 & + N_2 \\
| & & & | & \\
HCOAc & & & HCOAc & \\
| & & & | &
\end{array}
\tag{2}
$$

Bognár et al. prepared a series of acetylated aldonic acid chlorides[72] and bromides[73] by the action of dihalomethyl methyl ethers and other reagents. Crystalline azides, anilides, and 4-aminosulfonylanilides were prepared from acetylated aldonyl chlorides in good yields; crystalline products also resulted on saponification of the acyl protecting groups.

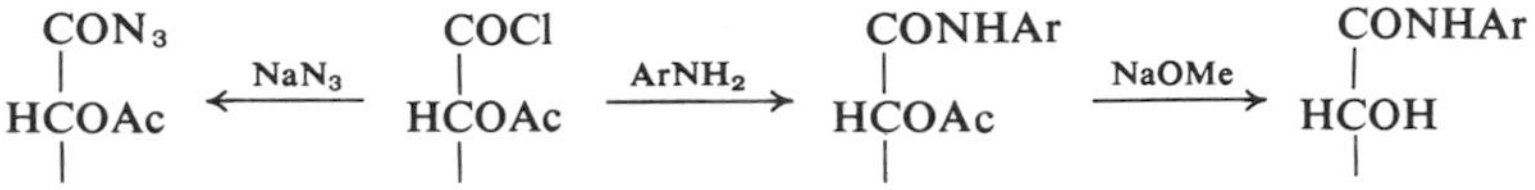

$$
\begin{array}{ccccccc}
CON_3 & & COCl & & CONHAr & & CONHAr \\
| & \xleftarrow{\;NaN_3\;} & | & \xrightarrow{\;ArNH_2\;} & | & \xrightarrow{\;NaOMe\;} & | \\
HCOAc & & HCOAc & & HCOAc & & HCOH \\
| & & | & & | & & |
\end{array}
\tag{3}
$$

Degradation of aldonic acids to lower sugars via azides was also studied.[74] For instance, penta-O-acetyl-D-galactonyl azide rearranges to 1-acetamido-1,2,3,4,5-penta-O-acetyl-D-lyxitol in acetic anhydride; subsequent reaction with ammonia gives 1-deoxy-1,1-bis(acetamido)-D-lyxitol, which may be converted into D-lyxose by treatment with hydrogen chloride.

Acetylated aldonyl chlorides[75] and thioesters[76] can be reduced catalytically to open-chain derivatives of aldoses (see Chapter 3). The important reduction of aldonic esters and lactones to sugars and alditols is discussed in Chapter 22; esters are treated in Chapter 8.

2,3:5,6-Di-O-isopropylidene-D-mannonic acid and its 1,4-lactone are known,[77] as are isopropylidene acetals of D-glycero-D-gulo-heptono-1,4-lactone[78] and benzylidene acetals of L-gulono-1,4-lactone.[79] The preparation and synthetic utilization of dibenzylidene acetals of pentonic acids, esters, amides, and hydrazides have been reported by Zinner et al.[80]

B. GLYCURONIC ACIDS

The glycuronic acids can be regarded as aldoses in which the primary alcoholic group has been oxidized to a carboxyl group or, alternatively, as aldos-ω-onic acids. Like aldoses, they can exist in both furanose and pyranose forms, and, like aldonic acids, they can lactonize. Their structures are illustrated by the three most common, naturally occurring examples, which are depicted in the pyranose forms: D-glucuronic acid (4), D-galacturonic acid (5), and D-mannuronic acid (6).

When combined in natural products, the uronic acids always seem to occur as glycopyranosiduronates.

References start on p. 1085.

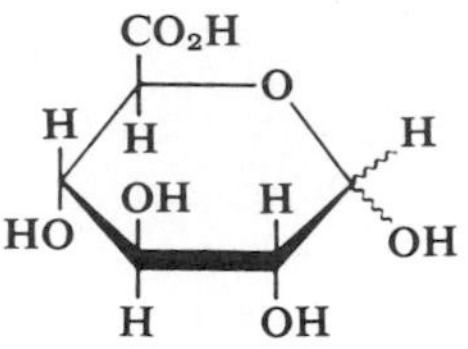

$$\begin{array}{ccc} \mathbf{4} & \mathbf{5} & \mathbf{6} \end{array}$$

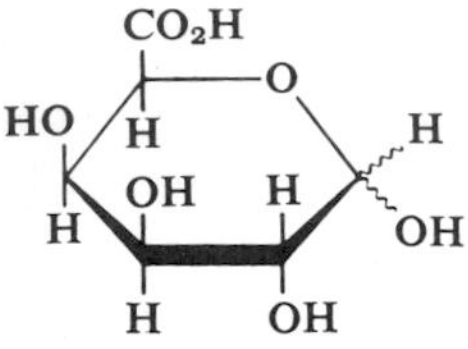

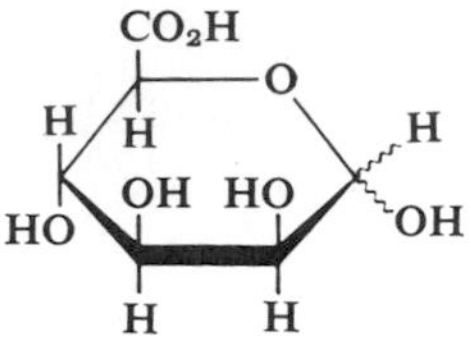

1. Occurrence

The uronic acids are very important biologically, occurring as constituents in many animal polysaccharides,[81,81a] microbial polysaccharides,[82] and plant polysaccharides[83,83a] (see also Vols. IIA and IIB). D-Glucuronic acid also occurs in many metabolic products of low molecular weight. The monograph "Glucuronic Acid" by Dutton[84] considers various aspects of the chemistry of this acid, including polysaccharides and conjugates containing the acid.

D-Glucuronic acid (**4**) serves as a detoxifying agent in mammals, and some poisonous compounds in metabolic products are excreted in the urine and bile as alkyl or aryl D-glucosiduronic acids.[84,85] The acid was first isolated[86] from such a conjugate by Jaffé in 1878, and was prepared biosynthetically for many years, for example, by feeding borneol or menthol to animals. The acid also occurs in many flavanol glycosides of plants. The important role of D-glucuronic acid and derivatives in the biosynthesis of polysaccharides and in the metabolism of carbohydrates is treated in other chapters. 4-O-(L-1-Carboxyethyl)-D-glucuronic acid has been identified[86a] as a constituent of the capsular polysaccharide from *Klebsiella* type 37.

Noteworthy polysaccharides containing D-glucuronic acid include heparin, chondroitin 4- and 6-sulfates, hyaluronic acid, *Pneumococcus* polysaccharides, and a number of plant polysaccharides, hemicelluloses, and gums (for instance, the commercially important gum arabic). Further details may be found in Vol. II. In this connection, it should also be mentioned that the 4-methyl ether of D-glucuronic acid is a very common constituent of plant polysaccharides.

D-Galacturonic acid (**5**) is also a common polysaccharide component in the plant kingdom. It forms the building unit of pectic acid (in which some of the carboxyl groups may be esterified with methanol), and it occurs, together with other components, in many plant gums, animal polysaccharides, and bacterial polysaccharides.

D-Mannuronic acid (**6**) residues constitute the major component of alginic acid, the main polysaccharide of the brown algae, in which L-guluronic acid is also found. D-Glucuronic acid has been found in another complex polysaccharide from brown algae.[87]

Dermatan sulfate, a polysaccharide structurally related to the chondroitin sulfates, contains L-iduronic acid residues. Aminohexuronic acids have been found in some bacterial polysaccharides, such as 2-amino-2-deoxy-D-galacturonic acid, which was isolated from the Vi-antigen of *Escherichia coli* and other organisms.[88,89] The presence of 2-amino-2-deoxy-D-mannuronic acid[90] and 2-amino-2-deoxy-D-glucuronic acid[91] has been demonstrated in other polymers. 5-Amino-5-deoxy-D-alluronic acid has been identified as the sugar moiety common to the polyoxins[91a] (Vol. IIA, Chapter 31).

The isolation of uronic acids from polysaccharides is usually difficult, and yields of monomeric uronic acids are usually low. The reason is that the glycosidic linkages of uronic acids are more stable than glycosidic linkages of neutral constituents, whereas the free uronic acids are more readily degraded than neutral sugars. The resistance of the glycosiduronic linkages to scission can sometimes be used for the identification of glycosidic linkages by separation and identification of the accumulated aldobiouronic (disaccharidic) fragments. Enzymic hydrolysis of pectin has been used for the isolation of D-galacturonic acid in good yield.

2. Preparation

There are three principal methods for the chemical synthesis of glycuronic acids: (*a*) reduction of the monolactones of aldaric acids, (*b*) oxidation of the primary alcoholic group of aldoses or derivatives, and (*c*) various oxidative degradations of higher acids.

Although used for the first chemical synthesis of D-glucuronic acid (in low yield from D-glucaro-1,4-lactone),[92] the reductive approach has not been developed, owing to the difficulty of obtaining pure 1,4-lactones for reduction.

Considerably more attention has been paid to oxidative methods, which require either protection of the aldehyde group and of the secondary hydroxyl groups or else the use of selective oxidants. The main interest has centered around the synthesis of D-glucuronic acid,[93] because of its possible therapeutic value—for instance, in the treatment of rheumatic diseases.[94] It is, in fact, used to some extent in a number of therapeutic compounds.[5]

The most suitable protecting group used during oxidation has been the isopropylidene group, which is stable to neutral and alkaline conditions but can readily be removed by acid hydrolysis after oxidation. The use of oxygen with platinum–carbon catalysts affords good yields of product; thus, Mehltretter *et al.*[95] prepared D-glucuronic acid by catalytic oxidation of 1,2-*O*-isopropylidene-α-D-glucofuranose (7) at pH 8 to 9. The intermediate 1,2-*O*-isopropylidene-α-D-glucofuranuronic acid, obtained in 50 to 60%

---

yield, was hydrolyzed almost quantitatively to D-glucuronic acid, which was isolated as the crystalline lactone (**8**). At pH 3 to 4.5, oxidation of the 5-hydroxyl group also occurs, giving in good yield the 5-keto analogue (**8a**),[95a] an intermediate in a synthesis of vitamin C (see Section II,G). Many glycosiduronic acids have been prepared in good yield by similar catalytic oxidation methods.[17]

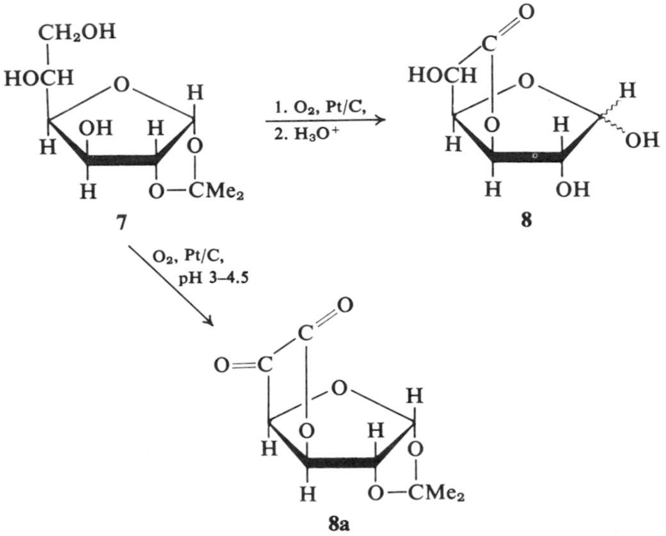

In aqueous alkaline solution or in acetic acid, potassium permanganate is a nonspecific oxidant that has been much used in the preparation of uronic acids and derivatives; it was the reagent of choice for such transformations before use of the catalytic oxidation procedures became general.

The protecting groups most commonly used have been isopropylidene acetals and acetates; the former are stable under neutral or alkaline conditions, and the latter may be employed under acidic conditions. Thus, good yields of D-galacturonic acid are obtained by oxidation of 1,2:3,4-di-O-isopropylidene-α-D-galactopyranose with alkaline potassium permanganate.[96] Many other oxidants have been used (see Chapter 24); D-glucuronic acid can be prepared from starch by nitric acid oxidation.[84]

D-Xyluronic acid is prepared by oxidative degradation of 1,2-O-isopropylidene-α-D-glucofuranose with neutral permanganate,[97] and D-lyxuronic acid results from degradation of D-galactaric acid monoamide with hydrogen peroxide and iron salts.[98] Some glycuronic acids have been prepared by glycol cleavage of higher acids. α-D-Glucofuranurono-6,3-lactone (**8**) was treated with an equimolar proportion of lead tetraacetate and, after hydrolysis of the resulting formic ester group, gave a good yield of D-arabinurono-5,2-lactone, together with some of the free acid. D-Galactofuranuronic acid and potassium D-glucofuranuronate have likewise been degraded to D-threuronic and D-erythruronic acids, respectively.[99] Oxidation of D-galactono-1,4-lactone by one molar equivalent of periodate affords L-lyxuronic acid, which may be isolated as methyl L-lyxuronate.[100]

Glycuronic acids epimerize much more readily than aldoses in heated aqueous solutions,[100a] although the preparative utility of this reaction is limited by the complexity of the product mixtures.

The route to ω-aldehydo derivatives of unprotected hexopyranosides via photolysis of a primary azide (see Section III,A,1 and also refs. 100b–d) is also a potentially useful route to glycuronic acids and their glycosides. Thus, the primary hydroxyl groups in cellulose were selectively converted into carboxyl groups by chlorite oxidation of aldehydo groups (see ref. 100d).

3. Oligoglycuronic Acids

There are three main categories of disaccharide that contain glycuronic acids: (a) *aldobiouronic acids*, in which the nonreducing component is a uronic acid glycosidically linked to an aldose (or ketose); (b) *pseudoaldobiouronic acids*, in which an aldose (or ketose) is glycosidically linked to a uronic acid; and (c) *diglycuronic acids*, in which a glycosyluronic group is glycosidically linked to a uronic acid residue. The three types of biouronic acid are here exemplified by acids 9–11; the methyl β-glycosides of 9–11 and the free aldobiouronic acid 4-*O*-(α-D-glucopyranosyluronic acid)-D-glucose (9) were isolated by Abbott and Weigel[101] after catalytic oxidation of methyl β-maltoside:

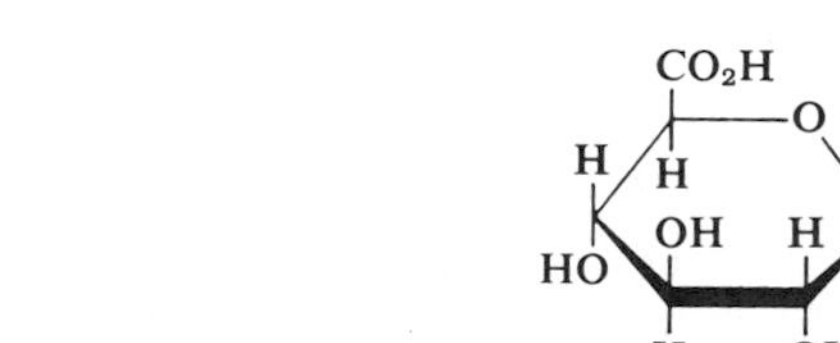

Maltobiouronic acid (9)

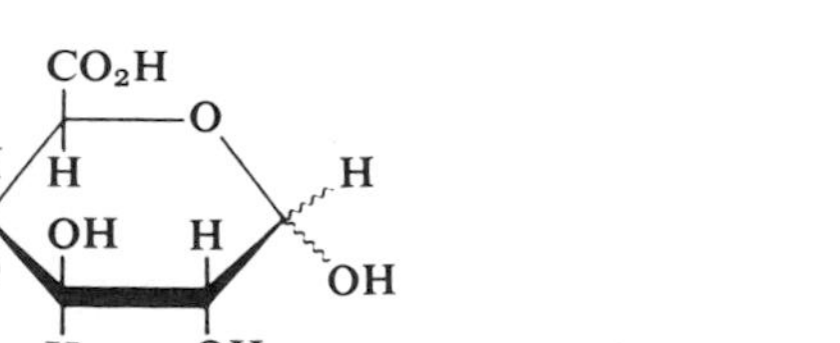

Pseudomaltobiouronic acid (10)

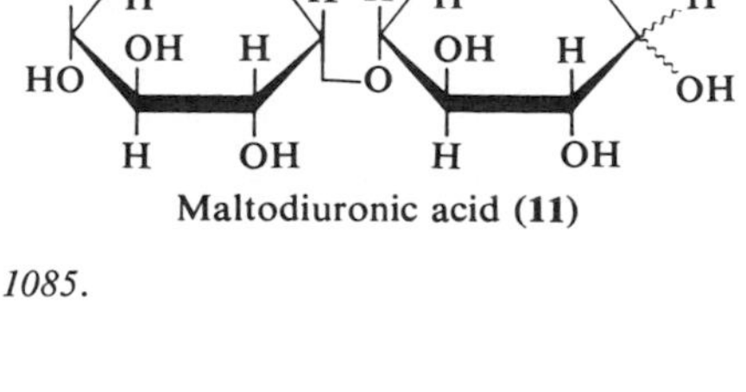

Maltodiuronic acid (11)

Most of the oligoglycuronic acids known are aldobiouronic acids. Because of the stability of the glycosidic linkage of aldobiouronic acids toward acid hydrolysis, they are readily isolated after vigorous hydrolysis of polysaccharides that contain uronic acid residues. Many aldobiouronic acids are known; they have been obtained from wood hemicelluloses, plant mucilages, gums, and bacterial and animal polysaccharides; in addition, several synthetic examples are known. (See ref. 2 for tables of aldobiouronic acids and other oligoglycuronic acids, and more-complete references.)

A common example that has been isolated from many plant polysaccharides is 2-O-(4-O-methyl-α-D-glucopyranosyluronic acid)-α-D-xylose (12). Hyalobiouronic acid [2-amino-2-deoxy-3-O-(β-D-glucopyranosyluronic acid)-D-glucose (13)], whose N-acetyl derivative is the repeating unit of the glycosaminoglycan hyaluronic acid, has been synthesized by Takanashi et al.[102] and by Flowers and Jeanloz.[103]

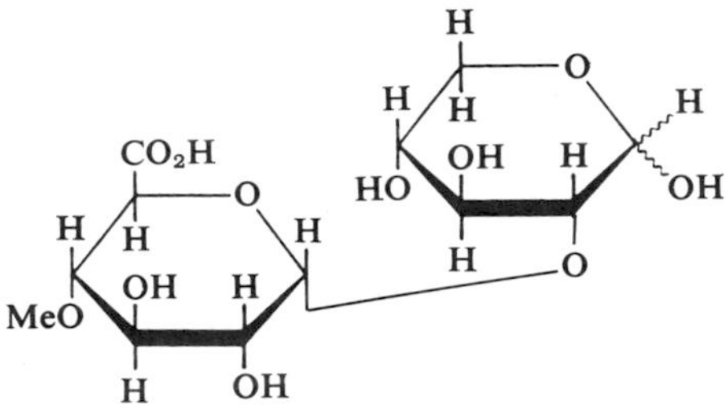

12 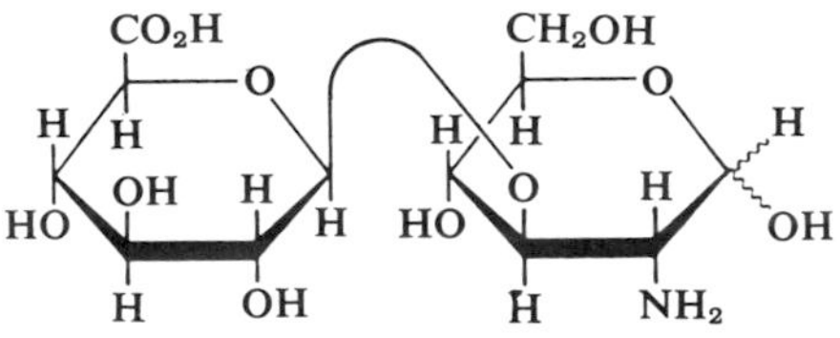   13

Some capsular *Pneumococcus* polysaccharides,[84] for example, S III, initially examined by Heidelberger and Goebel,[104] contain cellobiouronic acid [4-O-(β-D-glucopyranosyluronic acid-D-glucose] residues. Jayme and Demmig[105] synthesized this acid by platinum-on-carbon–catalyzed oxidation of benzyl β-cellobioside with oxygen. It is notable that only the primary alcoholic group of the D-glucose residue remote from the benzyl group is oxidized. Hydrogenolysis of the benzyl group liberated the crystalline acid. Lindberg and Selleby[106] independently prepared the acid by permanganate oxidation of the single, unprotected hydroxyl group of 2,3,2',3',4'-penta-O-acetyl-1,6-anhydrocellobiose. Maltobiouronic acid (9) has been prepared by similar methods.[101,107,108]

Aldobiouronic acids have also been prepared by the Koenigs–Knorr reaction.[2] Bishop[109] synthesized 3-O-(β-D-glucopyranosyluronic acid)-D-xylose, starting with the condensation of an ester of 2,3,4-tri-O-acetyl-1-bromo-1-deoxy-α-D-glucopyranuronic acid with 1,2-O-isopropylidene-5-O-trityl-α-D-xylofuranose.

Barker et al.[110] reported the enzymic synthesis of a pseudoaldobiouronic acid, 2-O-α-D-glucopyranosyl-D-glucuronic acid. Synthetic 4-O-β-D-gluco-pyranosyl-D-glucuronic acid (14) was prepared by Johansson et al.[111] by preparation and subsequent permanganate oxidation of 1,2,3,2′,3′,4′,6′-hepta-O-acetyl-β-cellobiose; 4-O-α-D-glucopyranosyl-D-glucuronic acid was similarly prepared.[112]

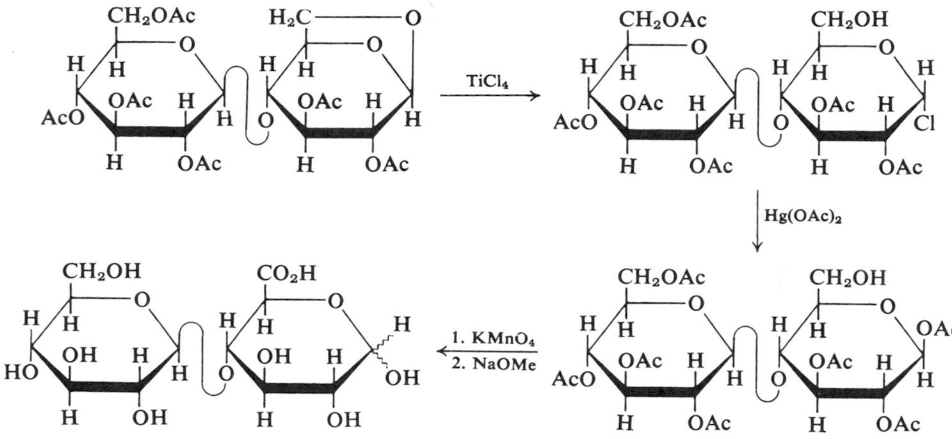

Di-D-galacturonic acids and higher homologues, isolated from enzymic or acidic hydrolyzates of pectin, have been studied by many workers.[2] A di-D-mannuronic acid was prepared[113] from alginic acid, and other diglycuronic acids or derivatives (such as 11)[101] have also been synthesized.[2]

4. Modified Glycuronic Acids

The development of suitable conditions for catalytic oxidation has made possible the preparation of the naturally occurring aminoglycuronic acids in good yield.[17] Weidmann and Zimmerman[114] prepared 2-amino-2-deoxy-D-glucuronic acid by oxidation of benzyl 2-(benzyloxycarbonyl)amino-2-deoxy-α-D-glucopyranoside with oxygen–platinum dioxide; the crystalline amino acid was liberated by simultaneous hydrogenolysis of the benzyl and the N-(benzyloxycarbonyl) groups in the oxidation product.

Derivatives of 2-deoxyglycuronic acids have been obtained by catalytic oxidation[115] of suitably protected alcohols, and by hydroformylation[115a] of terminal epoxide groups in aldonic acids. 6-Deoxyhepturonic acids or derivatives have been prepared by cyanide displacement of a 6-sulfonyloxy group from derivatives of hexoses and hydrolysis of the resulting nitriles.[116,117] 5-Deoxy-D-xylo-hexuronic acid was prepared in good yield via the reaction of

1,2-*O*-cyclohexylidene-5-*O*-{methylsulfonyl)-α-D-glucofuranurono-1,4-lactone with hydrazine.[118]

The formation of 4,5-unsaturated uronic acid residues by enzymic or alkaline degradation of polysaccharides containing uronic acid residues has been the object of many investigations.[119] Hasegawa and Nagel[120] obtained the unsaturated diglycuronic acid 4-*O*-(4-deoxy-β-L-*threo*-hex-4-enopyranosyl-uronic acid)-D-galacturonic acid (**15**), after hydrolysis of pectic acid by a pectolytic enzyme from *Bacillus polymyxa*. Preiss and Ashwell[121] degraded a D-galacturonan by action of an extract of a pseudomonad and isolated D-galacturonic acid and 4-deoxy-L-*threo*-hexos-5-ulosuronic acid (**16**), obviously products of continued enzymolysis of **15**; acid **16** had been obtained by Linker *et al.*[121a] from enzymic degradation of hyaluronic acid. Pectin (the partial methyl ester of pectic acid) is alkali-labile[122] and undergoes the well known[122a] β-elimination reaction. Methyl (methyl α-D-galactopyranosid)uronate was used by Heim and Neukom[123] as a model compound for studying this reaction, and alkaline treatment of it afforded the expected methyl (methyl 4-deoxy-β-L-*threo*-hex-4-enopyranosid)uronate (**17**). Subsequent study revealed that even poor leaving groups (for instance, isopropylidene acetals) undergo facile β elimination under similar conditions.[123a] (Compare the β elimination of hexodialdo-1,5-pyranosides, Section III,A,2.) Timpe *et al.*[133b] studied the elimination–reduction reaction converting D-hexofuranosidurono-6,3-lactones into 3-deoxyhex-2-enono-1,4-lactones by treatment with sodium borohydride in hexamethylphosphoric triamide.

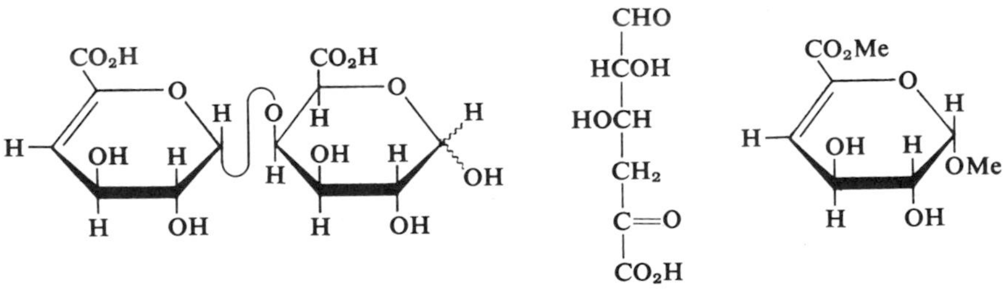

15   16   17

The 3-keto derivative of D-glucuronic acid, D-*ribo*-3-hexosulosuronic acid, has been prepared by chromic acid oxidation of methyl 5-*O*-acetyl-1,2-*O*-isopropylidene-α-D-glucofuranuronate[124]; the 5-keto derivative of 1,2-*O*-isopropylidene-α-D-glucofuranurono-6,3-lactone (**8a**) has been prepared by oxidation with manganese dioxide,[125] chromium trioxide,[126] or oxygen in the presence of platinum.[95a] 3-*C*-(Hydroxymethyl)-D-riburonic acid has been synthesized for comparison with a naturally occurring bilirubin conjugate.[126a]

5. Properties and Reactions of Glycuronic Acids

The uronic acids and their derivatives can exist as α and β anomers and in pyranose (4–6) and furanose forms; where stereochemical conditions are favorable, there is a tendency for lactone formation as well. Thus, whereas the crystalline glycofurano-6,3-lactones of D-glucuronic and D-mannuronic acids ["*glucurone*" (8) and "*mannurone*," respectively] are known, D-galactofuranuronic acid, both anomers of which are crystalline, is unable to lactonize, because O-3 and C-5 are in a *trans* relationship with respect to the furanose ring; however, such derivatives as 2,4,5-tri-O-acetyl-D-galactofuranurono-6,3-lactone dialkyl dithioacetals and the corresponding demercaptalated aldehydo derivatives are known.[127] The binding of sodium ions to D-glucuronic acid has been studied crystallographically.[127a]

Separation of glycuronic acids from their lactones has been accomplished by paper chromatography.[84] Electrophoretic separations are based on differential stabilities of complexes of the carbonyl groups present with hydrogen sulfite[128] or of the carboxyl groups with cations (for example, with a borax–calcium chloride electrolyte).[129] Glycuronic and aldonic acids have been separated by chromatography on an anion-exchange resin, with acetic acid as eluant and a three-channel analyzer for determination.[130] Perry and Hulyalkar[131] have reported a method for the characterization and determination of hexuronic acids by quantitative reduction to aldonic acids and subsequent separation of the aldono-1,4-lactones by gas–liquid chromatography as the per(trimethylsilyl) ethers. Petersson[131a] described the use of hydroxylamine instead of reducing agents to transform the carbonyl group before trimethylsilylation and analysis by g.l.c. and mass spectrometry. Mass-spectrometric analysis may be applied directly to pure samples of underivatized glycuronic acids and lactones.[131b] Although drastic hydrolytic conditions, accompanied by considerable decomposition, are needed in order to afford free uronic acids from uronic acid-containing polymers, individual uronic acids can normally be identified by the methods already discussed. For identification and a more-quantitative determination of individual uronic acid residues, they may be esterified and the esters reduced to the corresponding sugar residues before hydrolysis, which then proceeds without difficulty. Thus, 4-O-methyl-D-glucuronic acid was identified as a residue in a xylan, by esterification with propylene oxide followed by reduction with borohydride.[131c] The methyl ester groups in permethylated uronic acid residues can be reduced by lithium aluminum deuteride, and the methyl ethers derived from uronic acid residues can be distinguished from those derived from sugar residues by their mass spectra.[131d] Diborane has also been used for reduction of glycosyluronic

---

residues in methylated polysaccharides.[131e] A method for degradation of uronic acid-containing polymers to permit identification of the component residues is conveniently accomplished by the sequence: methylation of hydroxyl and carboxyl groups, alkali-catalyzed β elimination, mild hydrolysis with acid, methylation with CD_3 groups, hydrolysis, and acetylation, followed by g.l.c.–m.s. analysis.[131f]

One of the most important reactions of hexuronic acids and glycuronans is the decarboxylation caused by treating with strong acids (usually $\sim 12\%$ hydrochloric acid). The rapid and stoichiometric evolution of one mole of carbon dioxide per mole, which was first observed by Lefèvre and Tollens,[132] has been developed as an analytical method by many workers[84] —in particular, for the determination of the uronic acid content of polysaccharides; it has also been applied in analysis of conjugates and free acids. When a sufficient amount of material is available (corresponding to ~ 200 mg of uronic acid), the gravimetric method of Whistler et al.[133] is very accurate, provided that correction is made for the slow evolution of carbon dioxide from nonglycuronic carbohydrates. Adaptation of this method to smaller samples is accomplished by trapping evolved carbon dioxide as barium carbonate, converting the barium salt into barium iodate, and titrating the anion iodometrically.[134] In a semi-micro method, the sample is decarboxylated with concentrated hydriodic acid, and the reaction is monitored continuously by recording the change of conductivity during the absorption of carbon dioxide into a barium hydroxide solution.[134a] The method is rapid, and seems especially suitable for routine testing. Methods based on the formation of colored phenolic compounds by uronic acids in strongly acidic media are discussed in Chapter 45 and ref. 84.

The mechanism of acid-catalyzed decarboxylation of hexuronic acids is still not fully elucidated, although it has been the subject of many investigations,[135,136] from the results of which, various mechanisms have been suggested.[135–138] The formation of carbon dioxide (see Eq. 4) is accompanied by the formation of 2-furaldehyde, $C_5H_4O_2$ (18) as the main product, in addition to considerable amounts of "humins"; however, both 5-formyl-2-furoic acid[139] (19) and reductic acid[140] (20) have been isolated as end products from strong acid treatment of hexuronic acids. This field has been reviewed by Feather and Harris,[140a] and these authors came to the conclusion that the

$$C_6H_{10}O_7 \longrightarrow C_5H_4O_2 + CO_2 + 3H_2O \qquad (4)$$

18   19   20

experimental facts are in favor of Isbell's original scheme[137] for the formation of 18 and 20 (compare also Chapter 4, Section II,C).

Hexulosonic acids and L-ascorbic acid likewise undergo acid-catalyzed decomposition to produce carbon dioxide and 2-furaldehyde,[136] and the latter is formed in a higher yield than from D-galacturonic acid.[140a] Feather and Harris[141] used ^{14}C-labeled uronic acids in a study of the relationships of these compounds to the decarboxylation products formed in 12% hydrochloric acid during 4 hours under reflux. The 2-furaldehyde arising from the 1-^{14}C-labeled uronic acids contained over 99% of the activity in the aldehyde group, whereas 20 appeared to be formed via two different mechanisms. D-Galacturonic acid, D-glucurono-6,3-lactone, alginic acid, D-xylo-5-hexulosonic acid, and D-arabino-hexulosonic acid all produced 20, but only D-galacturonic acid and D-xylo-5-hexulosonic acid produced compound 19. Pentoses gave no (or less than 1% of) 20, and no pentoses could be detected during the decomposition of compounds that were sources of 20.

Zweifel and Deuel[142] demonstrated that glycuronic acids are decarboxylated under relatively mild conditions in the presence of heavy metals; thus, L-arabinose was isolated from D-galacturonic acid. This decarboxylation, which appears to follow a different mechanism from the one already mentioned, was practically complete in pyridine at 100° with nickel acetate as the catalyst but was slower in an aqueous medium. On treatment with N,N-dimethylformamide dineopentyl acetal at elevated temperatures,[142a] methyl 2,3-di-O-benzyl-β-D-glucopyranoside (21) undergoes decarboxylative β elimination to generate methyl 2,3-di-O-benzyl-4-deoxy-β-L-threo-pent-4-enopyranoside (22).

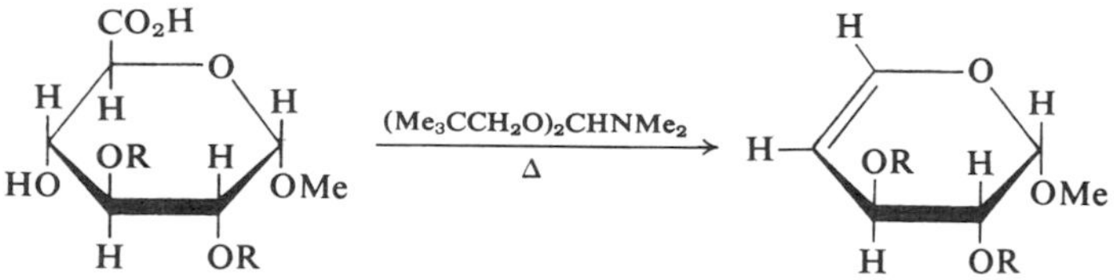

On treatment of D-glucurono-6,3-lactone (8) in aqueous sodium hydroxide at room temperature, Siddiqui and Purves[143] obtained 8, D-mannuronic acid, and an unidentified keto acid; from D-galacturonic acid, they isolated D-taluronic acid. α-D-Glucofuranurono-6,3-lactone reacts, in a way presumably thermodynamically controlled, with aqueous calcium hydroxide at room temperature to afford an optimum yield (70%) of D-lyxo-5-hexulosonic acid, which precipitated from the reaction mixture after 11 days.[144] On heating D-glucuronic acid in aqueous solution[144a] at pH 7, the following products were identified: D-*lyxo*-5-hexulosonic acid (main product; previously[144b] obviously mistaken for L-iduronic acid, having similar chromatographic properties), L-*ribo*-5-hexulosonic acid, D-mannuronic, D-altruronic, D-alluronic, and, later,[144c] L-*ribo*-4-hexulosonic acid. L-Iduronic acid was not detected, but a glycosidically linked 4-*O*-methyl-D-glucuronic acid residue in a xylan was shown to be isomerized to some extent to the corresponding L-iduronic acid residue by treatment at high alkalinity and temperature (conditions for kraft pulping).[144d] Similar treatment at pH 7 of D-galacturonic acid yielded a proportion of isomerization products that was smaller than from D-glucuronic acid and, again, the C-2 epimer, D-*arabino*-5-hexulosonic acid, was the main product, identified together with three other hexulosonic acids, D-taluronic, D-guluronic, and D-iduronic acids.[144e] It was also shown that, at 100° and pH 7, the rate of isomerization increases in the order: D-galactose < D-glucose < D-galacturonic acid < D-glucuronic acid < 4-*O*-methyl-D-glucuronic acid. By treatment of **8** with potassium hydroxide in a nonaqueous solvent, Ishidate *et al.*[145] isolated an interesting, crystalline, strongly reducing compound, which, from its properties and elemental analysis, was tentatively identified as (*E*)-4,5-dideoxy-D-*glycero*-hex-4-enos-3-ulosuronic acid.

It is obvious that such conditions as pH, temperature, solvent, cations present, and the nature of the glycuronic acid studied strongly influence the transformation and degradation of uronic acids and the proportions of products formed. Aso[145a] reported the isolation of small amounts of 3,8-dihydroxy-2-methylchromone (**22g**) after treating alginic acid and other glycuronans in water at 160°. Popoff and Theander[146] isolated and identified **18**, **20**, 2-furoic acid, and a series of phenolic compounds (**22a–e, 22g**), and, later,[146a] also the chromones **22i** and **22j**, on heating D-glucuronic acid in aqueous solution at 96° and pH 3.5 or 4.5. Under these slightly acidic conditions, the glycuronic acid was also isomerized to the same acids as at pH 7 (as already described) but at a lower rate, and also small proportions of pentoses and pentuloses were detected. The nonaromatic precursors of **22e** and **22g** were also isolated[146]—namely, **22f** and **22h**. D-Galacturonic acid yielded the same compounds under these conditions, but in lower yields, and some of the phenolic compounds from treatment of D-xylose and L-arabinose were also identified. It was considered likely[146] that many of these very labile compounds are intermediates in color formation during heating, aging, and humification of cellulosic and plant material containing polysaccharides possessing uronic acids and pentoses as structural units. It is notable that, under neutral conditions, glycuronic acids give a quite different product pattern; no chromones, but a series of monocyclic benzenoid compounds, were isolated; also, D-glucose and D-fructose under slightly acidic conditions yield phenols in a much lower proportion and in another product pattern than do pentoses and hexuronic acids.[146b]

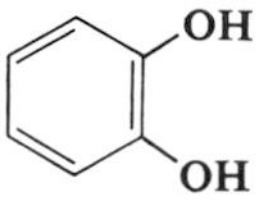

22a 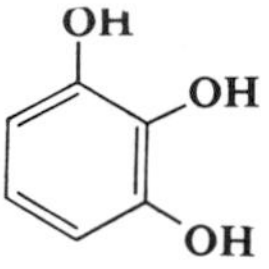   22b 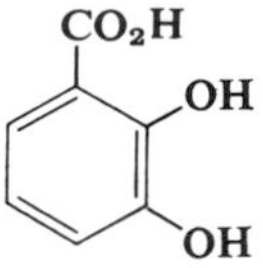   22c 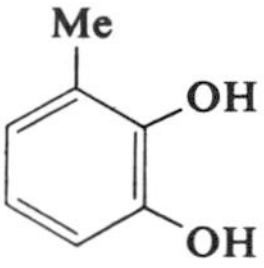   22d

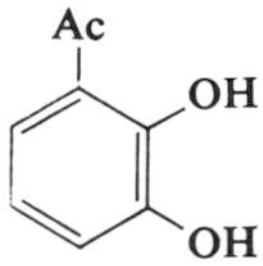

22e 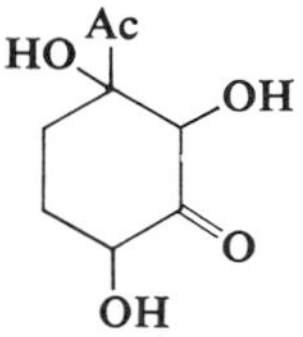   22f 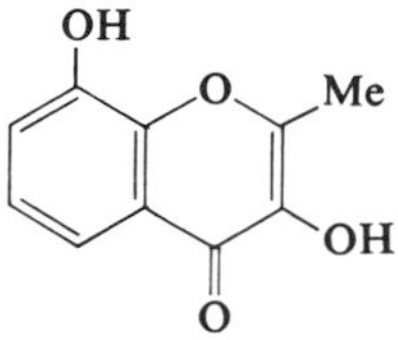   22g

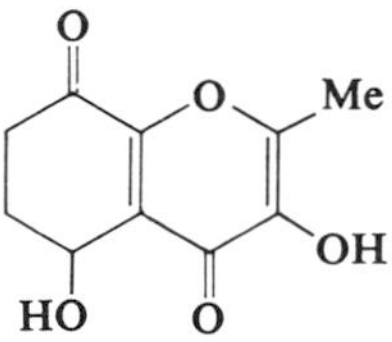

22h 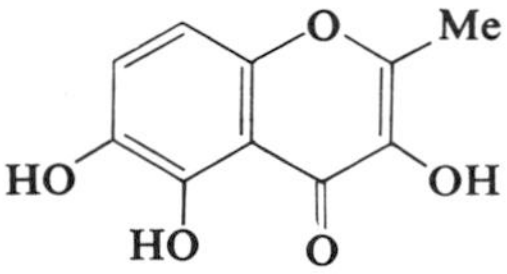   22i 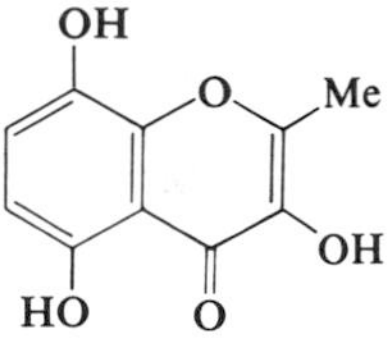   22j

The reducing group in uronic acids can be oxidized by nitric acid or bromine yielding aldaric acids. Oxidations that give rise to chain shortening are discussed in Section II,B,2.

Uronic acids, as salts, can be quantitatively reduced to the corresponding aldonic acids by borohydride.[131] Lactones are readily reduced, as in the controlled conversion of D-glucurono-6,3-lactone into D-*gluco*-hexodialdose (see Section III,A,1). The carboxyl group of glycuronic acids may likewise be reduced to a primary alcoholic group. (See Chapter 22.)

The presence of both aldehyde and carboxylic acid groups in uronic acids allows the formation of numerous types of derivatives. Phenylhydrazine may combine with uronic acid derivatives to form hydrazides, hydrazones, or osazones. Amides and benzimidazoles are useful as characterizing derivatives. Esterification and glycosidation are competing reactions when uronic acids are treated with an alcohol under acidic conditions. D-Galacturonic acid reacts with methanolic hydrogen chloride to undergo esterification 25 to 55 times as fast as glycoside formation.[147] Methyl and ethyl D-glucuronate are conveniently prepared from D-glucurono-6,3-lactone by base-catalyzed esterification,[148,149] which proceeds smoothly at room temperature. Diazomethane is the reagent of choice for methylation of the carboxyl group in glycosiduronic acids. Morgenlie[149a] obtained methyl (1,2-*O*-isopropylidene-α-D-xylofuran)-uronate in good yield by treatment of 1,2-*O*-isopropylidene-α-D-glucofuranose with silver carbonate–Celite in refluxing methanol.

The acetylated glycuronic acids, and the per-*O*-acetylglycosyluronic halides of their methyl esters, are important precursors for the synthesis of glycosiduronic acids. D-Glucofuranurono-6,3-lactone (**8**) reacts with acetic anhydride in pyridine to form mainly 1,2,5-tri-*O*-acetyl-α-D-glucofuranurono-6,3-lactone, whereas zinc chloride catalyzes formation of the β anomer (**23**) as the main product[150]; a very good yield of crystalline **23** was obtained by boron trifluoride-catalyzed acetylation[151] of **8**. Acetylation of methyl D-glucopyranuronate with acetic anhydride in the presence of various catalysts gives mixtures of the α and β anomers of methyl 1,2,3,4-tetra-*O*-acetyl-D-glucopyranuronate,[148] which can be converted into methyl 2,3,4-tri-*O*-acetyl-1-deoxy-1-halo-D-glucopyranuronates. A number of β-D-glucopyranosiduronic acids have been prepared from methyl 2,3,4-tri-*O*-acetyl-1-bromo-1-deoxy-β-D-glucopyranuronate (**24**), initially prepared by Goebel and Babers[152] by the action of hydrogen bromide in acetic acid solution on methyl tetra-*O*-acetyl-α- or -β-D-glucopyranuronates.

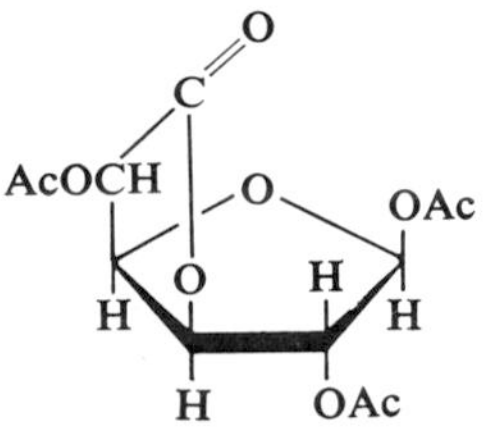

23 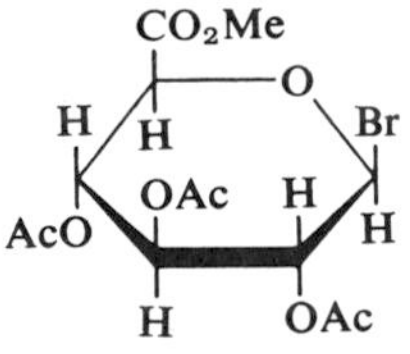   **24**

Both anomers of D-glucopyranosyluronic acid phosphate have been synthesized (see Chapter 8); this has made possible the synthesis of uridine 5'-(α-D-glucopyranosyluronic acid pyrophosphate).[84]

Numerous glycosiduronic acid derivatives are known, most of which are synthetic or biosynthetic D-glucopyranosiduronic acids (conjugates) in the form of glycosides (alkyl, aryl, or steroidal), 1-*O*-acyl derivatives, glycosylamines (including ureides and sulfonamides), or thioglycosides. This field was reviewed by Marsh[84] (see also refs. 1 and 85), who also discussed the acid hydrolysis of D-glucosiduronic acids.

The two principal routes for the synthesis of glycosiduronic acids differ in the sequence in which glycosidation and oxidation are performed. The first, oxidation of glycosides as their derivatives, is discussed in Section II,B,2; permanganate and catalytic oxidation[17] are methods of choice. The second, condensation of a derivative of the desired aglycon with a suitable glycuronic acid derivative, employs methods similar to those used for the preparation of glycosides (see Chapter 9).

Methyl α-D-glucofuranosidurono-6,3-lactone and its β anomer result from Fischer glycosidation of the lactone (8) at room temperature with methanol in the presence of hydrogen chloride[153] or a strong cation-exchange resin.[154] Extended treatment at a higher temperature converts the furanosides (25) into the pyranosides (26) in which the carboxyl group has been esterified.[153] By treating D-galacturonic acid with acidified methanol, dimethyl acetals of D-galacturono-6,3-lactone and of methyl D-galacturonate were detected, in addition to the four methyl glycosides.[154a]

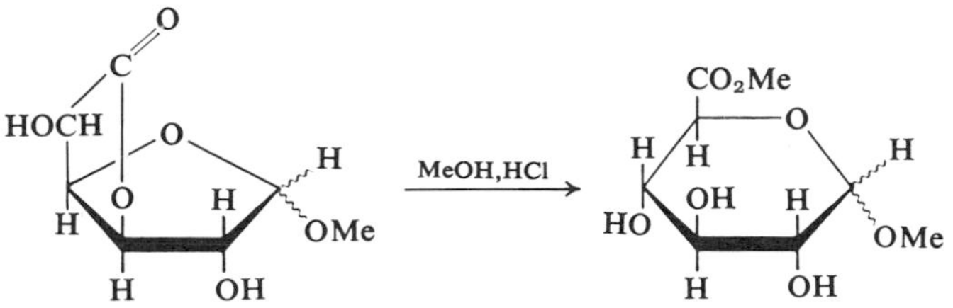

Many aryl D-glucopyranosiduronic acids have been prepared by the Helferich method; for instance, compound 23 has been condensed with 2-naphthol,[151] and Bollenback et al.[148] used methyl tetra-O-acetyl-β-D-glucopyranuronate in the preparation of many aryl glycosiduronic acids.

A large number of glycosiduronic acids and esters have been prepared from esters of acetylated glycosyluronic halides by condensation with alcohols, phenols, and other nucleophiles; silver oxide or carbonate (Koenigs–Knorr reaction), mercury(II) salts, and inorganic or organic bases have been used as hydrogen halide acceptors,[155] and the method has been used almost exclusively under conditions that permit Walden inversion at C-1. Thus Wolfrom and McWain[156] synthesized the crystalline nucleoside derivative adenin-9-yl β-D-glucopyranosiduronamide starting from methyl tri-O-acetyl-1-bromo-1-deoxy-α-D-glucopyranuronate (24).

The acid hydrolysis of glycosiduronic acids, which has been the subject of many investigations, is discussed in Chapter 9 (see also ref. 84). It has been suggested that the stability of aldobiouronic acids under the conditions for acid hydrolysis results from the inductive effect of the carboxylic group. By the same argument, the presence of glycuronic acids in a polysaccharide should weaken a glycosidic–glycosyluronic acid linkage. In an investigation of a 4-O-methyl-D-glucuronoxylan, however, McKee and Dickey[157] obtained evidence suggesting that a conformational effect determines this reactivity.

The findings of Johansson *et al.*[111] that the rates of hydrolysis of pseudo-cellobiouronic acid (**14**) and cellobiose are almost identical, whereas the rate of hydrolysis of cellobiouronic acid is considerably lower, also militate against the inductive-effect hypothesis. An intensive investigation by Timell *et al.*[158] compared the rates of hydrolysis of different alkyl and aryl glycosiduronic acids in dilute acids with the rates of hydrolysis for the corresponding glycosides. Methyl and other D-glucosiduronic acids were only slightly more stable than the D-glucoside analogues, and many glycosiduronic acids were hydrolyzed even faster than the corresponding glycosides. The results indicated that both polar and conformational effects may be operative, and that the two classes of compounds may be hydrolyzed by different mechanisms. Other investigations have assayed the influence of pH (the effect of ionized and un-ionized forms) on the hydrolysis of glycosiduronic linkages.[159,160]

Many methyl ethers of hexuronic acids have been isolated and synthesized, in connection with structural studies on polysaccharides,[161] The naturally occurring 4-*O*-methyl-D-glucuronic acid has been synthesized via catalytic oxidation starting with benzyl 2,3-di-*O*-benzyl-β-D-glucopyranoside.[162] A number of dithioacetals of uronic acids have been prepared (see Chapter 10).

C. Aldaric Acids

Aldaric acids (previously called saccharic acids) are polyhydroxy dicarboxylic acids, $HO_2C(CHOH)_nCO_2H$, that are formally produced by the oxidation of aldoses at both termini; they are primarily obtained from sugars, aldonic acids, and oligo- or poly-saccharides by reaction with strong oxidizing agents. They are named systematically by replacing "ose" in the name of the corresponding aldose with "aric acid," although there are also many established trivial names. End-to-end symmetry allows internal compensation, so that (optically inactive) *meso* forms may exist, in addition to the pure active forms and racemic modifications; for an example, see the tetraric acids in Chapter 1. Different aldoses may afford identical aldaric acids; for instance, D-glucose and L-gulose give the same acid: D-glucaric acid ≡ L-gularic acid. The former name, appearing first in the alphabet, is preferred; D-glucaric acid is also occasionally described by the obsolete trivial name, saccharic acid. Many characteristic, difficultly soluble acidic salts and many aldaro-lactones are known.

1. *Glyceraric and Tetraric Acids*

Glyceraric (commonly called tartronic acid or hydroxymalonic acid), $HO_2CCH(OH)CO_2H$, and the tetraric (tartaric) acids, classical compounds in organic stereochemistry, have for many years been known as products of the oxidative degradation of carbohydrates (refs. 163 and 164, respectively; a later example from the degradation of D-galacturonic acid appears in ref. 165). These acids can also be prepared synthetically, but L-threaric acid [known as "ordinary" or (+)-tartaric acid] and its 3-deoxy derivative, malic acid, are best known as widely distributed plant products. Malic acid occurs in fruits and berries, whereas L-threaric acid is found principally in the juice of grapes, from which it deposits as the potassium hydrogen salt after fermentation. The acid and many of its salts have commercial and therapeutic value.

When L-threaric acid is heated in water at 150° to 170° or in alkaline solution at lower temperatures, various proportions of erythraric acid (*meso*-tartaric acid) are produced, depending on the conditions, together with the racemate (DL-threaric acid), which can be isolated from the mixture.[166] Pasteur originally resolved this form by mechanical separation of crystals of the ammonium sodium salt, and Marckwald later resolved D-threaric acid by his classical procedure using the diastereoisomeric cinchonine salts.[167] Practically quantitative resolution of the D-acid has been accomplished in one crystallization[168] by the method of Haskins and Hudson[169] by using 2-(1-deoxy-D-*glycero*-D-*gulo*-hexitol-1-yl)benzimidazole as the resolving agent.

Erythraric (*meso*-tartaric) acid can be prepared by prolonged heating of a solution of L-threaric acid,[166] by oxidation of erythritol with nitric acid, or by *cis*-hydroxylation of maleic acid.

2. *Pentaric and Hexaric Acids*

There are four pentaric acids and ten hexaric acids:

Pentaric (sometimes known as hydroxyglutaric) *acids*: ribaric (*meso*) formerly *ribo*-trihydroxyglutaric, xylaric (*meso*) formerly *xylo*-trihydroxyglutaric, and D- and L-arabinaric (D- and L-lyxaric) formerly D- and L-*arabino*-trihydroxyglutaric.

Hexaric acids: allaric (*meso*) formerly *allo*-mucic, galactaric (*meso*) formerly mucic, D- and L-glucaric (L- and D-gularic) formerly D- and L-*gluco*-saccharic, D- and L-mannaric formerly D- and L-*manno*-saccharic, D- and L-idaric formerly D- and L-*ido*-saccharic, and D- and L-altraric (D- and L-talaric) formerly D- and L-*talo*-mucic acid.

The pentaric acids and their methyl ethers often appear in the structural proofs of many compounds. They can be prepared by oxidation of the corresponding aldopentoses with nitric acid.

Several of the hexaric acids are of especial interest. Galactaric acid was first isolated by Scheele in 1780, by the oxidation of lactose. Its low solubility in water is characteristic, and it has been utilized in the quantitative determination of D- and L-galactose in oligo- and poly-saccharides by oxidation with nitric acid, and isolation of galactaric acid from the products. (See the review by Lewis *et al.*,[170] which also details the preparation of various derivatives.) Interestingly, the peracetate of galactaric acid is more soluble in water than is galactaric acid. The acid lactonizes with difficulty, whereas 2-deoxy-D-*lyxo*-hexaric acid, prepared from 2-deoxy-D-*lyxo*-hexose by oxidation with nitrogen dioxide,[171] forms a monolactone.

D-Glucaric acid, which is reported to occur as the magnesium salt in the sap of *Ficus elastica*,[172] is readily prepared by oxidation of D-glucose or starch with nitric acid, as described by Bose *et al.*[173] (and many previous authors), who also developed a procedure for the isolation of the 1,4- and 6,3-lactones. The acid is generally characterized as the potassium hydrogen salt. The 1,4-lactone, which has been shown by Levvy[174] to have a specific anti-(β-D-glucosiduronase) activity, has wide application as a biochemical reagent. It is also of therapeutic interest, and it is reported to be effective in the treatment of bladder cancer.[175] It may be obtained in 50–55% yield from D-glucose or D-gluconic acid by the action of oxygen–platinum.[175a]

D-Glucaric and D-mannaric acids also yield dilactones, which reduce Fehling solution, and the same behavior is shown by the ester lactones of D-glucaric acid. Smith *et al.*[176] attributed the reducing character to the opening of a lactone ring by the alkaline reagent, with subsequent formation of an "enol of dehydration," as exemplified by the conversion of the 1,4:3,6-dilactone (**27**) of D-glucaric acid into **28**. In alkaline solution, ozone attacks the double bond, affording oxalic acid and L-threuronic acid. Reaction with diazomethane affords an esterified enol ether.

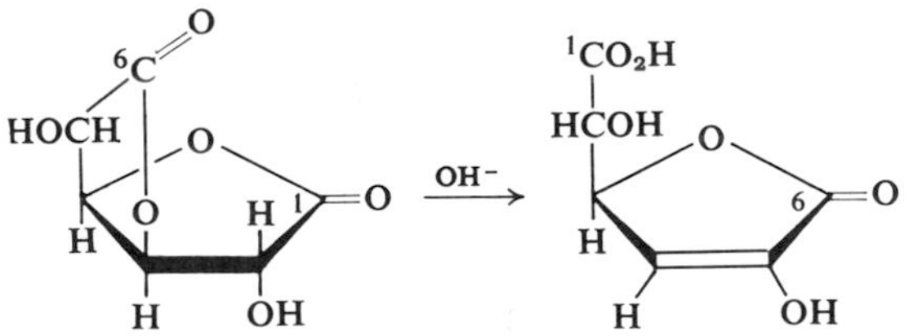

Morgan and Wolfrom[177] isolated an α-pyrone, probably 4-acetoxy-6-(ethoxycarbonyl)-2*H*-2-pyrone (**29**), from the reaction of ethyl hydrogen DL-galactarate or its monolactone with acetic anhydride and sodium acetate at 100°; the authors postulated the intermediate formation of an enolic lactone in this transformation.

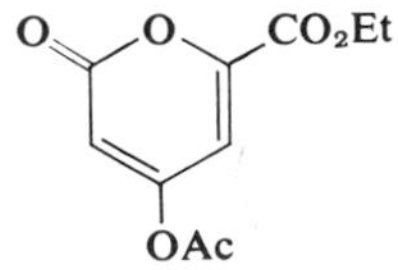

Aldaric acids or their derivatives undergo ready epimerization in pyridine; treatment with mineral acids then affords furan-2,5-dicarboxylic acid. Monolactones of aldaric acids can be reduced to glycuronic acids. Majs[178] studied the formation and hydrolysis of various aldaric lactones, and Bird *et al.*[179] examined acetals of dimethyl D-glucarate.

2,5-Anhydroaldaric acids are formed in the deamination and oxidation of 2-amino-2-deoxyaldoses (see Chapter 13), and branched-chain aldaric acids have been obtained by periodate oxidation of 2-C-(hydroxymethyl)-D-gluconic acid[180] and alkaline degradation of 4-O-methyl-D-glucuronic acid.[181] The two, diastereoisomeric 3-deoxy-2-C-(hydroxymethyl)pentaric acids obtained as products from the latter acid have, as previously postulated,[181a] also been obtained by treatment of alginates with alkali.[181b] Alkaline treatment of 4-O-methyl-D-glucuronic acid in the presence of air affords 4-O-methyl-D-glucaric and -D-mannaric acids as well as lower O-methylaldaric acids.[181] Evidence has been obtained for the formation of 3-deoxy-D-*erythro*-2-hexulosaric acid and 3-deoxy-L-*threo*-2-hexulosaric acid from D-glucaric acid and of the latter acid from galactaric acid in metabolism by *Escherichia coli.*[182] Attempted acyloin cyclization of dimethyl tetrakis-O-(trimethylsilyl)galactarate led to ring closure, but uncontrolled elimination reactions precluded the isolation of an inosose[182a]; instead, a catechol derivative was obtained.

D. Modified Aldonic Acids

This section deals with less-general groups of acids, which are formally modified aldonic acids, such as keto, sulfo, deoxy, and branched-chain acids (including the so-called *saccharinic acids* as the most important members). The aminoaldonic acids, which are oxidation products of amino sugars, and, in particular, the important nonulosaminic acids and muramic acid, are discussed in Chapter 16. The formation of saccharinic acids by the treatment of sugars with alkali, and the mechanisms involved, are discussed in Chapter 4; saccharinic acids,[183] the four-carbon acids[184] and their formation by alkaline degradation of polysaccharides,[185] and the preparation of the most common six-carbon saccharinic acids[186] have been reviewed.

1. *Glyculosonic Acids* ("*Ketoaldonic Acids*")

The glyculosonic acids fall mainly into two categories: the 2-keto acids, which have also been called osonic acids because of their preparation by the oxidation of glycos-2-uloses ("*osones*"), and the 5-keto acids of the hexose

References start on p. 1085.

series (or 4-keto acids in the pentose series), which have also been termed *keturonic acids* (glycuronic acids related to ketoses). Some other keto and ketodeoxy acids are known. The glyculosonic acids are prepared by chemical or enzymic oxidation of hexoses or aldonic acids, or by isomerization of glycuronic acids. They show great similarity to glycuronic acids in their color reactions and in the property of decarboxylation (see Section II,B,5). Their analytical determination is discussed in Chapter 45.

A number of chemical methods are available for the synthesis of 2-aldulosonic acids,[1] including oxidation of glycosuloses or osazones with bromine water, direct oxidation of ketoses with various oxidants, oxidation of aldonic acids with chlorates in the presence of vanadium pentaoxide, or oxidation of ketose diisopropylidene acetals having the primary hydroxyl group at C-1 free. Thus L-*xylo*-2-hexulosonic acid ("*2-keto-*L-*gulonic acid*"), an important intermediate in the synthesis of L-ascorbic acid (see Section II,E), is obtained by the oxidation of 2,3:4,6-di-*O*-isopropylidene-L-sorbose with potassium permanganate or one of the other patented oxidants.

A number of papers described the preparation of these acids by bacteria. The genus *Pseudomonas* is reported to be particularly efficient for the preparation of D-*arabino*-2-hexulosonic acid (**30**). Lockwood[187] described the preparation of this acid in good yield by such a strain. 2-Aldulosonic acids have been investigated primarily as intermediates in the synthesis of L-ascorbic acid, to which they are readily enolized. D-*arabino*-2-Hexulosonic acid (**30**) was isolated from the fungus *Cyttaria harioti* by careful hydrolysis and chromatography.[187a] 2-Glyculosonic acids can be degraded to lower aldonic acids by the Ruff procedure (see Section II,A,1*b*). Catalytic decarboxylation of 2-aldulosonic acids to pentuloses and pentoses has been reported by Matsui *et al.*[188]

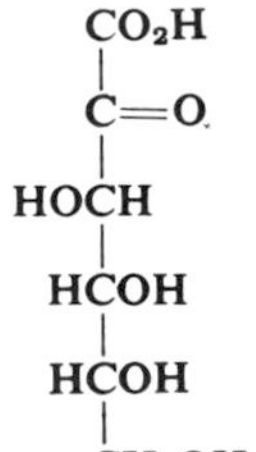

$$
\begin{array}{ll}
CO_2H & CO_2H \\
| & | \\
C\!=\!O & HCOH \\
| & | \\
HOCH & HOCH \\
| & | \\
HCOH & HCOH \\
| & | \\
HCOH & C\!=\!O \\
| & | \\
CH_2OH & CH_2OH \\
\mathbf{30} & \mathbf{31}
\end{array}
$$

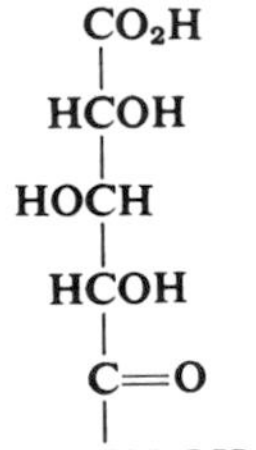

5-Aldulosonic acids are formed by direct oxidation of suitable aldonic acids with different oxidants—for example, bromine[189] or chlorine[190] in water. L-*arabino*-5-Hexulosonic acid was prepared via oxidation of 1,2:3,4-di-*O*-isopropylidene-D-tagatopyranose.[191] Acylated methyl β-D-hexopyranosides and the corresponding α-or β-D-furanosides are oxidized by chromium trioxide to methyl 5-hex- and 4-hex-ulosonates, respectively.[191a] D-*xylo*-5-Hexulosonic acid (31), is, however, best prepared by the action of *Acetobacter suboxydans* on D-glucose.[187] The same organism also dehydrogenates D-arabinonic acid to the 4-aldulosonic acid.[192] D-*lyxo*-5-Hexulosonic acid was prepared in 70% yield by tautomerization of D-glucurono-6,3-lactone in aqueous calcium hydroxide solution.[144] Reduction of 5-aldulosonic acids has been used for the preparation of aldonic acids of the L series (see Section II,A,1,*d*).

The rather unstable D-*threo*-2,5-hexodiulosonic acid was obtained by the oxidation of D-glucose with *Acetobacter melanogenum*[193,194] or *Gluconobacter liquefaciens*[195]; this acid was shown to be further oxidized to three 4-pyrone derivatives by the latter organism,[195] and to be degraded to oxalic acid and L-*glycero*-tetrulose in aqueous solutions of calcium hydroxide.[195a] 2,3-Diulosonic acids are formed by oxidation of L-ascorbic acids (see Section II,E), and their formation by oxidation with hydrogen peroxide is discussed in Chapter 24.

2. Deoxyketoaldonic Acids

The 3-deoxy-2-glyculosonic acids, a comparatively new class of biochemically important deoxy sugars, have been reviewed.[195b] The first example, 3-deoxy-D-*erythro*-2-hexulosonic acid 6-phosphate, was isolated by MacGee and Doudoroff,[196] as the crystalline sodium salt, from the action of *Pseudomonas saccharophila* on D-gluconic acid 6-phosphate. 3-Deoxy-D-*glycero*-2-pentulosonic acid[197] and 3-deoxy-D-*threo*-2-hexulosonic acid[198] were also isolated from the oxidative action of the same organism on D-arabinose and D-galactose, respectively. A 3-deoxy-*threo*-2-hexulosonic acid is a component in the extracellular polysaccharide of *Azobacter vinelandii*,[198a] and 3-deoxy-L-*glycero*-2-pentulosonic acid has been found[198b] as a component of the capsular polysaccharide of *Klebsiella* type 38. A 3-deoxy-2-heptulosonic acid[199] has been isolated from extracts of *Escherichia coli*, and 3-deoxy-D-*manno*-2-octulosonic acid occurs as a cell-wall lipopolysaccharide constituent of that organism[200] and of *Xanthomonas sinensis*[200a]; it is readily transformed into anhydro derivatives by treatment with acid. An enzyme from *Pseudomonas aeruginosa* combines D-arabinose 5-phosphate and enolpyruvic acid phosphate to form a 3-deoxy-2-octulosonic acid 8-phosphate.[201]

The synthesis of 3-deoxy-2-glyculosonic acids by oxidation with chlorate–vanadium pentaoxide[196,198,199] of the hydroxyl group on C-2 of 3-deoxy-aldonic acids ("*metasaccharinic acids*") involves large-scale chromatographic separations if pure compounds are needed. A more convenient method, developed by Kuhn et al.,[202] converts aldoses into iminolactones of enamines (for example, 33), which then undergo stepwise hydrolysis to afford 3-deoxy-2-glyculosonic acids. Paerels[203] used this method in preparing the first crystalline members of this group, namely 3-deoxy-D-*erythro*-2-hexulo-2,6-pyranosonic acid (34), and the L isomer starting from D-ribose (32) and L-arabinose, respectively. The synthesis of the former is illustrated.

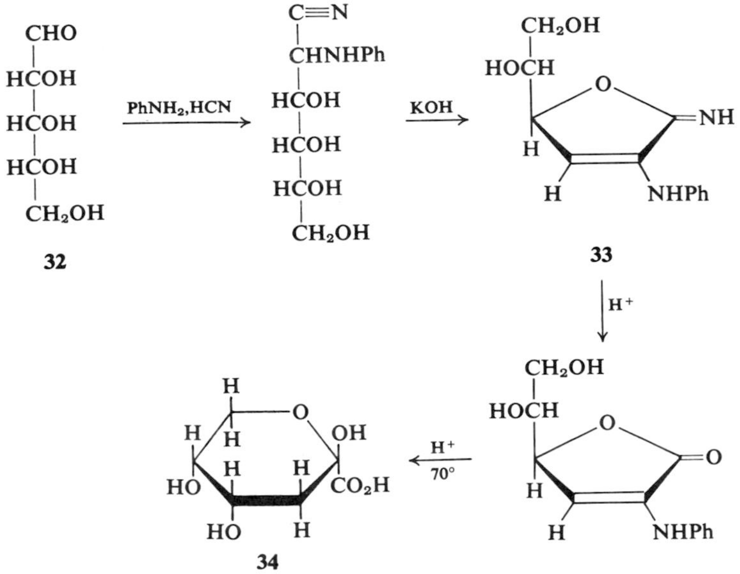

3-Deoxy-D-*arabino*-2-heptulosonic acid, which crystallizes as a lactone in which the ketone group has enolized, was later prepared from D-glucose.[204] 3,7-Dideoxy-D-*threo*-2,6-heptodiulosonic acid,[205] which has been postulated to be an intermediate in the enzymic formation of 5-dehydroquinate, was synthesized by the same route. Another general synthesis is the (reversible[205a]) aldol reaction of oxobutanoic acid with *aldehydo*-aldose derivatives; 3-deoxy-5-O-methyl-D-*manno*-2-octulosonic acid[205b] and 3,6-dideoxy-D-*erythro*- and -D-*threo*-2-hexulosonic acids[205c] were prepared by this method, starting from 2-O-methyl-D-arabinose and 3-deoxy-D-glyceraldehyde, respectively. These and related compounds have been used in investigations[205b,d] of the mechanism of the Warren color test (see Chapter 45) for 3-deoxy-2-glyculosonates.[205e]

The enzymic production of 3-deoxy-D-*glycero*-2,5-hexodiulosonic acid from D-galacturonan[206] and a di(D-galacturonic[207] acid) has been demonstrated. A derivative of 2-deoxy-D-*manno*-3-octulosonic acid was prepared[207a] by the Reformatsky reaction of ethyl bromoacetate with 2,3:5,6-di-O-cyclo-hexylidene-D-mannono-1,4-lactone. 3,4-Dideoxy-2-glyculosonic acids occur as intermediates in the synthesis of 3,4-dideoxyaldos-2-uloses by chain extension of aldehydo derivatives using the Wittig reagent 3-(2-carboxy-2-oxoethyli-dene)triphenyl phosphorane.[207b]

The 3-deoxy-2-glyculosonic acids are readily decomposed in acid solution.[208]

3. Sulfoaldonic Acids

The few known members of this group have been identified in connection with studies on the chemistry of the sulfite pulping process and of nonenzymic browning reactions.

Hägglund et al. found that D-glucose is oxidized to D-gluconic acid[209] under the conditions of acid sulfite pulping, but that, at pH 6 to 6.5 and 130° a more complex reaction ensues, from which carbohydrate-derived sulfonic acids could be isolated.[210] Adler[211] later isolated two acids from this mixture, one of which also contained a carbonyl group.

Lindberg and Theander[212] treated D-mannitol with aqueous sulfite under various pulping conditions and showed that oxidation to monosaccharides occurs at pH 1.7 to 6.0. At pH 4.4 and 6.0, and 160 to 180°, reactions occur that result in the formation of "saccharinic acids" and acidic fragmentation products (similar to those produced in alkaline solutions at room tempera-ture), together with considerable proportions of sulfonic acids. It was sug-gested that the saccharinic acids and sulfonic acids are formed from common dicarbonyl intermediates—for instance, 3-deoxy-D-erythro-hexos-2-ulose (36), an intermediate in the formation of the D-glucometasaccharinic acids (35). The isolation of a compound tentatively assigned as a 3,4-dideoxy-4-sulfo-D-hexos-2-ulose (37) by Ingles,[213] afforded support for this mechanism. He proposed that this compound is formed either from 36 (or its enol) or from the unsaturated aldos-2-ulose (38). The structure of 37 was confirmed by Lindberg et al.,[214] who also showed that it was readily formed from 36 by action of aqueous sulfite at pH 6.5. Anet and Ingles[215] converted 38 into 37 by the action of sulfur dioxide at room temperature, under conditions that inhibit nonenzymic browning. It was subsequently shown[214] that the acids (39 and 40) first isolated by Adler[211] are formed from 37 via an oxidation process and a benzilic acid type of rearrangement, respectively, both of which have analogies in the reactions of D-mannitol.[212] The sulfonic acids discussed are most probably mixtures of isomers, but it is probable that the configura-tion at C-5 is retained.

The C_5 homologue of 39 was isolated after treatment of D-xylose[216,217] or L-arabinose[217] with aqueous sulfite at pH 6.5 and 135°. (A mistaken

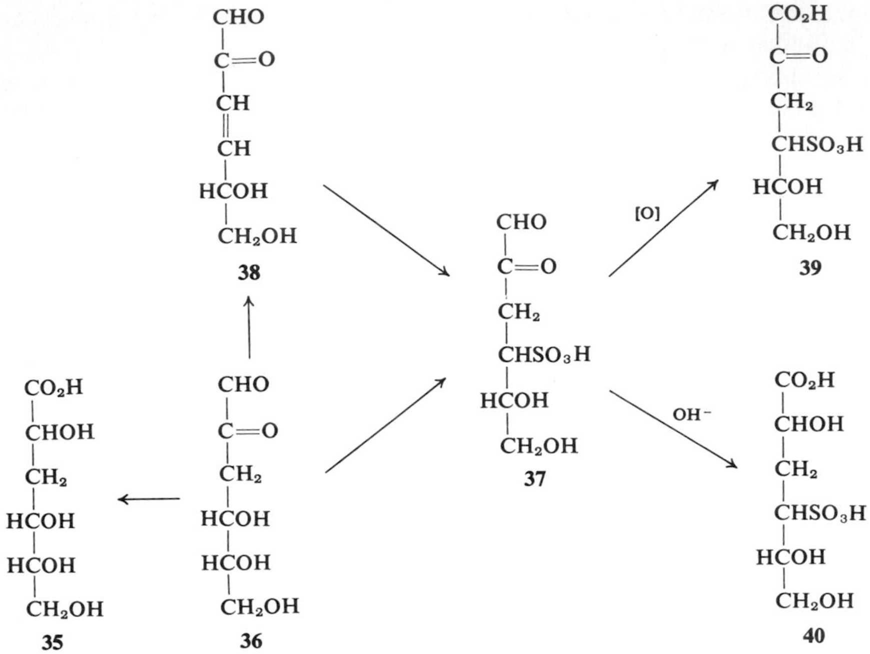

$$
\textbf{38}:\quad
\begin{array}{l}
\text{CHO} \\
\text{C}=\text{O} \\
\text{CH} \\
\text{CH} \\
\text{HCOH} \\
\text{CH}_2\text{OH}
\end{array}
\qquad
\textbf{37}:\quad
\begin{array}{l}
\text{CHO} \\
\text{C}=\text{O} \\
\text{CH}_2 \\
\text{CHSO}_3\text{H} \\
\text{HCOH} \\
\text{CH}_2\text{OH}
\end{array}
\xrightarrow{[\text{O}]}
\textbf{39}:\quad
\begin{array}{l}
\text{CO}_2\text{H} \\
\text{C}=\text{O} \\
\text{CH}_2 \\
\text{CHSO}_3\text{H} \\
\text{HCOH} \\
\text{CH}_2\text{OH}
\end{array}
$$

$$
\textbf{35}:\quad
\begin{array}{l}
\text{CO}_2\text{H} \\
\text{CHOH} \\
\text{CH}_2 \\
\text{HCOH} \\
\text{HCOH} \\
\text{CH}_2\text{OH}
\end{array}
\;\leftarrow\;
\textbf{36}:\quad
\begin{array}{l}
\text{CHO} \\
\text{C}=\text{O} \\
\text{CH}_2 \\
\text{HCOH} \\
\text{HCOH} \\
\text{CH}_2\text{OH}
\end{array}
\qquad\xrightarrow{\text{OH}^-}\qquad
\textbf{40}:\quad
\begin{array}{l}
\text{CO}_2\text{H} \\
\text{CHOH} \\
\text{CH}_2 \\
\text{CHSO}_3\text{H} \\
\text{HCOH} \\
\text{CH}_2\text{OH}
\end{array}
$$

identification of the compound as the C_5 homologue of **40** was later revised.[218]) The C_5 homologue of **40** was also later isolated, but in lower yield, from D-xylose by this method.[218] Similar treatment of D-erythrose afforded 3-sulfopropionic acid and the C_4 homologue of **40**, $HO_3SCH_2CHOHCO_2H$, which was synthesized independently from acrolein via addition of bisulfite followed by a cyanohydrin synthesis.[218] The C_5 homologue of **40** is also one of the main acids found in the liquor when cellulose is treated with sulfite at pH 7 at 180°.

The sulfonic acids are strong acids that have been obtained only in the amorphous state; they have been isolated from mixtures, both by fractional recrystallization of barium and brucine salts and by chromatographic separation of their ammonium salts on cellulose. Paper electrophoresis in an aqueous pyridine–acetic acid buffer of pH 3 is a useful means for identification of these acids in mixtures, particularly because some lactonizable acids yield two migrating spots.[214] Acids containing a keto group have been separated as the difficultly soluble barium salts of hydrazones.

These acids seem to be very stable; thus, no color formation was observed when wood polysaccharides were treated with sulfite at pH 2 to 7 at elevated temperature (conditions for sulfite pulping), most probably because degradation products are transformed into sulfonic acids instead of undergoing further reaction to colored material[218a] Similar treatments with buffer solutions without sulfite, or with kraft pulping liquor, gave considerable browning.

4. Deoxyaldonic Acids

Many deoxyaldonic acids (mainly 2-deoxy) and derivatives have been prepared[219]; oxidation of deoxyaldoses is the usual source, and bromine water, permanganate, or barium hypoiodite (introduced by Goebel[220]) have commonly been used as oxidants. 2-Deoxy-D-*arabino*-hexonic acid (**41**) or its lactone can be obtained crystalline in three steps from tri-*O*-acetyl-D-glucal[221]; the 6-phosphate of **41** has also been prepared.[222] Acetylated diazomethyl ketoses undergo the so-called Wolff rearrangement to give 2-deoxyaldonic acids[71] (see Section II,A,2,*c*). L-Arabinitol has been converted into 2-deoxy-L-*lyxo*-hexonic acid via displacement of a 5-*p*-tolylsulfonyloxy group to form a deoxyhexononitrile.[223] 2-Deoxyaldonolactones are typical aldonolactones in that the rates of hydrolysis of the 1,5-lactones far exceed those of the 1,4-lactones. It is significant, however, that 2-deoxyaldono-1,4-lactones are considerably more stable than the corresponding aldono-1,4-lactones.[32]

The 3-deoxyaldonic acids ("*metasaccharinic acids*") arise by alkaline treatment of monosaccharides,[183] especially those substituted at C-3. They have also been obtained from (1 → 3)-linked oligo- and poly-saccharides. The most common ones are 3-deoxy-D-*ribo*-hexonic acid ("α"-D-*gluco-metasaccharinic acid*, **42**) and 3-deoxy-D-*arabino*-hexonic acid ("β"-D-*glucometasaccharinic acid*, **43**), which can most conveniently be prepared by treatment of 3-*O*-methyl-D-glucose or laminaran with lime-water.[186]

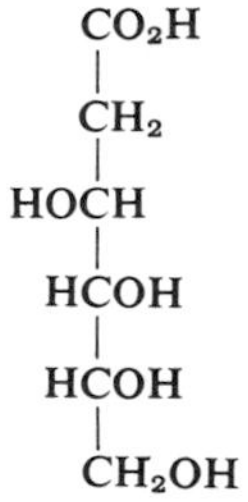

CO_2H	CO_2H	CO_2H
CH_2	HCOH	HOCH
HOCH	CH_2	CH_2
HCOH	HCOH	HCOH
HCOH	HCOH	HCOH
CH_2OH	CH_2OH	CH_2OH
41	**42**	**43**

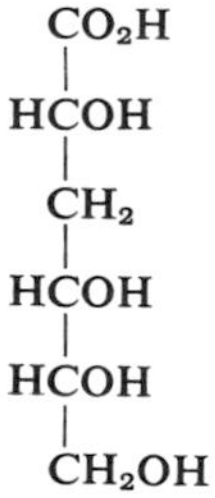

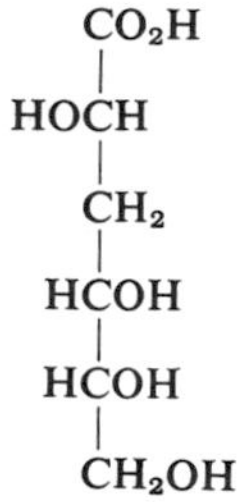

Compounds **42** and **43** are obtained together, but they can be isolated in pure form by fractional recrystallization of their calcium salts. The 1,4-lactones of **42** and **43** are also crystalline. Reversible epimerization of the

---

lactones results from alkaline treatment or heating.[212,224] The conversion of 2-deoxy-D-*erythro*-pentose into **42** and **43** via the cyanohydrin synthesis and the reduction of the intermediate 3-deoxyaldono-1,4-lactones to the corresponding 3-deoxyaldoses was described by Wood and Fletcher.[225] Conversely, the biologically significant sugar 2-deoxy-D-*erythro*-pentose is available by Ruff degradation of **42** or **43**; the original method of Nef[226] has been modified repeatedly, and the procedure of Diehl and Fletcher[227] utilizes, without isolation, the intermediate metasaccharinic acids produced by the action of lime-water on D-glucose.

3-Deoxy-D-*xylo*-hexonic acid ("α"-D-*galactometasaccharinic acid*) can be prepared from D-galactose[186]; the "β" isomer can be isolated in smaller amounts from the mother liquors. Examples of 3-deoxy-tetrònic and -pentonic acids are known.[183,184] Both the 5-phosphates and the 6-phosphates of mixtures of **42** and **43** have been prepared[228] by Lewak and Szabó. Brucine salts, anilides, and phenylhydrazides are useful crystalline derivatives of meta (and other) saccharinic acids. Dideoxyaldonic acids have been prepared, both by hydrogenation of derivatives of the intermediate (*E*)-2,3-dideoxyald-2-enonic acids in the aldonic acid synthesis of Kochetkov and Dmitriev[22] (see Section II,A,1,*b*) and by hydroxylation of various unsaturated acids as reported by Lukeš *et al.* in a series of papers.[229]

5. *Branched-Chain Aldonic Acids*

This group includes saccharinic acids, which are products of the alkaline degradation of sugars, together with synthetic acids prepared by condensation reactions or by oxidation of branched-chain sugars.

2-*C*-Methyl-D-ribonic acid ("α"-D-*glucosaccharinic acid*) is readily prepared by the action of lime-water on D-fructose or "inverted" sucrose. It was isolated in the mid-nineteenth century as the crystalline calcium salt,[183] and its 1,4-lactone[186] (**44**) is also well known. In 1883, Kiliani[230] identified it as a 2-*C*-methylaldonic acid by reducing **44** with hydrogen iodide in the presence of red phosphorus (a classical technique used in many structural studies) to the known 2-methylglutaric acid. It was not until much later, however, that the configuration of the acid was determined by Sowden and Strobach[231] and Foster *et al.*[232] The latter group of workers correlated the configuration of **44** with that of the branched-chain sugar hamamelose,

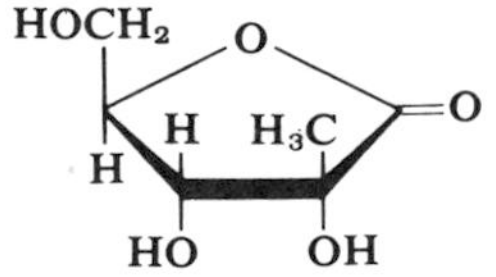

44

2-C-(hydroxymethyl)-D-ribose, by converting each into 5-deoxy-2-C-methyl-D-ribitol.

Feast et al.[233] demonstrated that only 2-C-methyl-D-ribonic acid and no C-2 ("β") isomer results from the action of aqueous calcium hydroxide on D-fructose or its 1-benzyl ether; in connection with this study, 2-C-methyl-L-arabinono-1,4-lactone was synthesized by oxidation of 2-C-methyl-L-arabinose with bromine. In a subsequent investigation,[234] 1-deoxy-D-erythro-2,3-hexodiulose, a predicted dicarbonyl intermediate in the formation of the D-glucosaccharinic acids, was synthesized and found to rearrange almost exclusively to the "α" acid in aqueous calcium hydroxide, and to afford only low yields of the "β" acid (2-C-methyl-D-arabinonic acid) from the action of aqueous sodium hydroxide. The competing fragmentation affording D-erythronic acid was considerable, particularly for the latter reaction.

Ishizu et al.[235] found that D-xylose and D-fructose react with aqueous calcium hydroxide to produce thirteen lactonizable saccharinic and other acids, some of which were obviously products of recombination. These were identified after separation by cellulose column and gas–liquid chromatography, and the previously unknown C_5-saccharinic acids 2-C-methyl-D-threonic acid (**45a**) and 2-C-methyl-D-erythronic acid (**45b**) were among those isolated. Ishizu et al.[235a] later reported that L-sorbose reacts similarly, to generate fourteen lactones, including the final two saccharinic acids—namely, 2-C-methyl-L-xylono-1,4-lactone (**46a**) and 2-C-methyl-L-lyxono-1,4-lactone (**46b**) —which were also prepared from 1-deoxy-L-threo-pentulose via the cyanohydrin reaction.

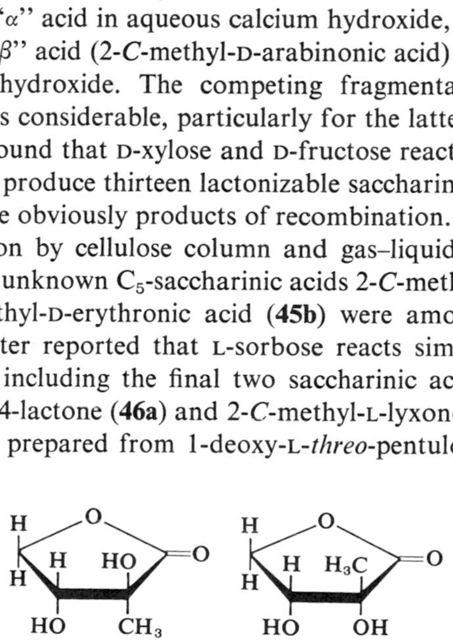

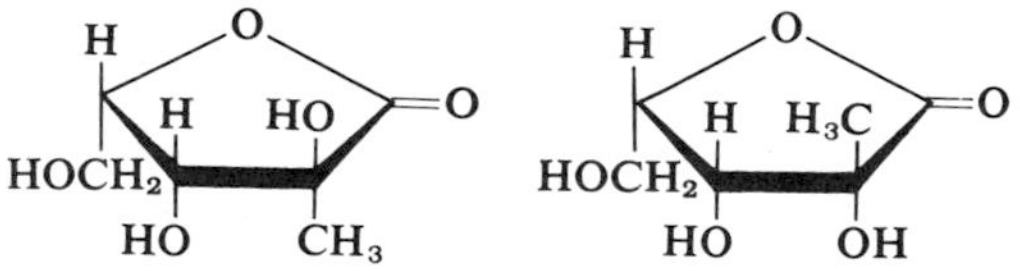

3-Deoxy-2-C-(hydroxymethyl)-D-erythro-pentonic acid (**47**, "α"-D-gluco-isosaccharinic acid) has also been known for a long time.[183] It is most conveniently prepared by treatment of 4-O-substituted D-glucose derivatives, such as lactose, maltose, and cellobiose with lime-water,[186] although it is also found among the numerous products of alkaline decomposition of D-glucose, D-mannose, and D-fructose. The acid can be isolated as the crystalline calcium salt, or the 1,4-lactone. The configuration of C-2 was finally settled by X-ray crystallography.[236] For a long time there had been indications in the literature that an epimeric "β" acid is also formed,[186] before this *threo* isomer was ultimately isolated from the mother liquors of the α isomer and characterized as the crystalline tri-*O*-benzoyl-1,4-lactone.[233] The two isomers were found to be formed in approximately equal amounts on treatment of various 4-*O*-methyl-D-glucose derivatives with aqueous calcium hydroxide. Support for the mechanism of base-catalyzed saccharinic acid formation, proposed by Nef and refined by Isbell,[137] was found in the pattern of isotope incorporation in the products of the reaction between maltose, turanose, and inulin with barium hydroxide in deuterated[236a] or tritiated[236b] water; the radioisotopic labels incorporated into the smaller fragments in the latter study facilitated identification. 2-*C*-(2,3-Dihydroxypropyl)glyceraric acid (**48a**) was obtained from oxygen and alkaline solutions of cellulose[236c] and cellobiose,[236d] and from alkaline treatment of L-ascorbic acid[236d]; action of oxygen and alkali on xylan gave the corresponding 2-*C*-(2-hydroxyethyl)tartronic acid.[236e]

$$
\begin{array}{ccc}
\mathrm{CO_2H} & \mathrm{CO_2H} & \mathrm{CO_2H} \\
| & |\;\diagup^{\mathrm{CH_2OH}} & |\;\diagup^{\mathrm{CO_2H}} \\
\mathrm{HOCH_2COH} & \mathrm{C}\diagdown_{\mathrm{OH}} & \mathrm{C}\diagdown_{\mathrm{OH}} \\
| & | & | \\
\mathrm{CH_2} & \mathrm{CH_2} & \mathrm{CH_2} \\
| & | & | \\
\mathrm{HCOH} & \mathrm{CH_2OH} & \mathrm{CHOH} \\
| & & | \\
\mathrm{CH_2OH} & & \mathrm{CH_2OH} \\
\mathbf{47} & \mathbf{48} & \mathbf{48a}
\end{array}
$$

The five-carbon isosaccharinic acid 3-deoxy-2-*C*-(hydroxymethyl)-DL-*glycero*-tetronic acid (**48**) was obtained from alkali-degraded oligo- and poly-saccharides containing β-(1 → 4)-linked D-xylose residues[237,238] (and also in low yield from D-xylose and D-fructose[235]).

Barker *et al.*[239] reported a micro-method for quantitation of each of the three types of saccharinic acid—namely, D-glucosaccharinic, D-glucoiso-saccharinic, and D-glucometasaccharinic acid ("α" and "β" isomers determined together)—in a mixture, which was later improved into an automated method[239a] for determination of meta- and iso-saccharinic acids. The method utilizes a color reaction between the periodate-oxidized acids and barbituric acid (see Chapter 45). The 1,4-lactones of the six acids have been separated as their trimethylsilyl ethers by gas–liquid chromatography.[233]

In 1885, Kiliani[240] first attempted the synthesis of a branched-chain mono-saccharide by applying the cyanohydrin reaction to D-fructose; the 2-epimeric acids formed have been isolated. Schmidt and Weber-Molster[241] assigned configuration to these two acids, 2-C-(hydroxymethyl)-D-gluconic acid ("α-D-*fructoheptonic acid*") and 2-C-(hydroxymethyl)-D-mannonic acid ("β-D-*fructoheptonic acid*"), by application of the isorotation rules. Wood and Neish[180] confirmed the configuration of the latter acid by converting it into a series of branched-chain compounds. For instance, one mole of the lactone (**49**) of 2-C-(hydroxymethyl)-D-gluconic acid was degraded by one mole of periodic acid to crystalline 4-C-(hydroxymethyl)-L-xyluronic acid (**50**), which was further oxidized to 2-C-(hydroxymethyl)-D-xylaric acid (**51**) by nitric acid or bromine, and also hydrogenated in the presence of Raney nickel to crystalline 1,4-lactone (**52**) of 2-C-(hydroxymethyl)-D-xylonic acid.

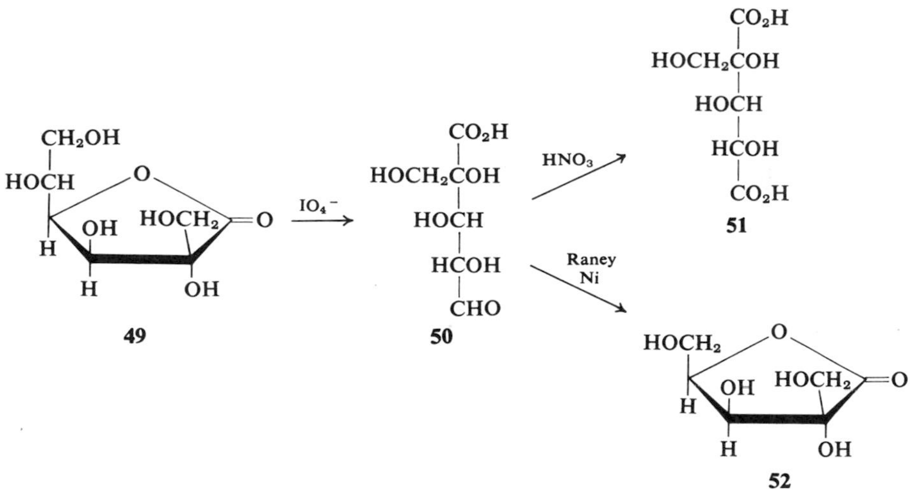

Compound **52** was also reduced by sodium amalgam to the branched-chain sugar 2-C-(hydroxymethyl)-D-xylose. Lactones of glucosaccharinic and glucoisosaccharinic acids can similarly be reduced to branched-chain sugars by treatment with sodium amalgam or borohydride.[233]

Ferrier[242] prepared the 2-C-(hydroxymethyl)-L-*ribo*- and -L-*arabino*-, and -D-*xylo*- and -D-*lyxo*-pentono-1,4-lactones via the cyanohydrin reaction, starting from L-*erythro*-pentulose, and D-*threo*-pentulose respectively. Several branched-chain aldonic acids have been obtained as oxidation products of branched-chain aldoses (see Chapter 2).

E. L-ASCORBIC ACIDS

L-Ascorbic acids may be considered to be reductones (**53**), which are enolic lactones of 2- and 3-glyculosonic acids. They are illustrated by the formula of the most important member of the group, L-*threo*-hex-2-enono-1,4-lactone (**54**), also known as vitamin C (see ref. 243 for a controversial discussion of its function and dietary significance), "L-*xylo*"-ascorbic acid, or simply L-ascorbic acid; the common nomenclature of the L-ascorbic acids is based on the configuration of the glyculosonic acid actually or hypothetically used in its preparation. Smith reviewed the L-ascorbic acids, including their deoxy and amino derivatives,[244] and more-recent aspects have been treated by Crawford and Crawford.[244a]

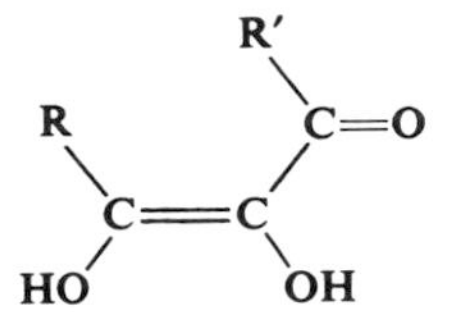

53 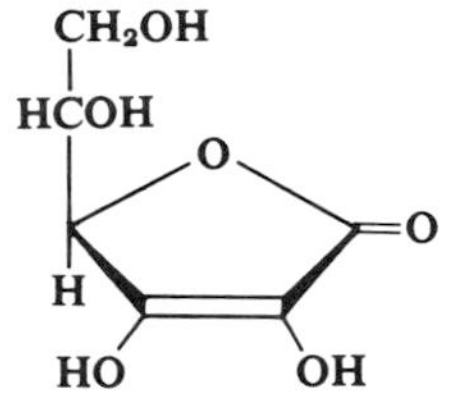   54

1. General Properties and Reactions

Of the many analogues of vitamin C that have been synthesized, only those having the lactone ring on the right of the formula when written according to the Fischer convention display antiscorbutic activity. Thus, the 6-deoxy analogue of **54** has about one-third of the activity of the natural vitamin, but a 2-amino-2,6-dideoxy analogue and a seven-carbon analogue of **54** (L-*rhamnoascorbic acid*) exhibit rather high activities; however, the C-5 epimer of **54** (D-*arabinoascorbic acid*) has a much lower activity.

The L-ascorbic acids have generally been synthesized by four main routes: (*a*) cyanohydrin synthesis from glycos-2-uloses, (*b*) enolization and lactonization of glyculosonic acids or esters, (*c*) condensation of hydroxy aldehydes with ethyl glyoxylate or mesoxalate, and (*d*) condensation of esters of hydroxy acids.

The addition of cyanide to a glycos-2-ulose has been used for the preparation of many ascorbic acids, including the first synthesis of **54**, by Reichstein *et al.*[245] and Ault *et al.*,[246] starting from L-*threo*-pentosulose (**55**). In the latter synthesis (see formulas), the nitrile **56** undergoes immediate enolization and ring closure, to form the iminolactone **57**, which can be isolated crystalline; **57** is readily converted by dilute acid into **54**. Yields are dependent on the purity of the starting material.

Simultaneous enolization and lactonization of 2-glyculosonic acids or esters is generally the method of choice for the preparation of L-ascorbic acids (illustrated for the preparation of **54**), provided that the requisite 2-glyculosonic acid is a practicably accessible starting material; Maurer and Schiedt[247] and

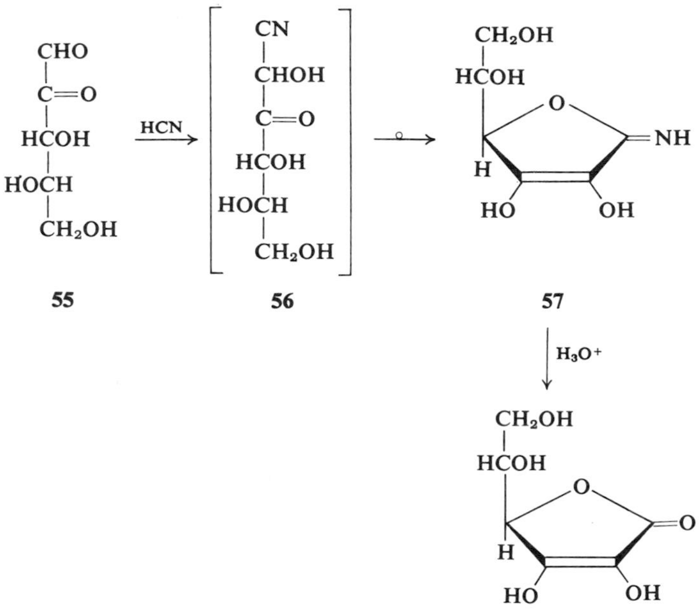

$$
\begin{array}{c}
\text{CHO} \\
|\\
\text{C}=\text{O} \\
|\\
\text{HCOH} \\
|\\
\text{HOCH} \\
|\\
\text{CH}_2\text{OH}
\end{array}
\xrightarrow{\text{HCN}}
\left[
\begin{array}{c}
\text{CN} \\
|\\
\text{CHOH} \\
|\\
\text{C}=\text{O} \\
|\\
\text{HCOH} \\
|\\
\text{HOCH} \\
|\\
\text{CH}_2\text{OH}
\end{array}
\right]
\longrightarrow
$$

55   **56**   **57**

(structure **57**: CH₂OH–HCOH–furanose ring with =NH, HO, OH)

$$\downarrow \text{H}_3\text{O}^+$$

(structure **54**: CH₂OH–HCOH–furanose lactone ring with =O, HO, OH)

54

Ohle *et al.*[248] first accomplished this (almost quantitative) double transformation by the action of sodium methoxide on methyl 2-glyculosonates. A substantial literature and a large number of patents[249] attest that acids can also catalyze the same transformation of 2-glyculosonic acids, their esters, or their diisopropylidene acetals, and the use of ion exchange resins as catalysts has been reported in patents. L-*erythro*-2-Pentulosonic acid was not isolated from the acid hydrolysis of the methyl 3,4-*O*-isopropylidene-L-*erythro*-pentofuranosid-2-ulosonic acid, because of the rapid transformation of the former into the unsaturated 1,4-lactone.[250]

The third method bears analogy to the acyloin condensation; thus, D-*arabino*-hept-2-enono-1,4-lactone was prepared by condensation of D-glucose with ethyl glyoxylate.[251] In the fourth method, a crossed-Claisen condensation, L-ascorbic acids result from condensation of ethyl benzoyloxyacetate with perbenzoates of polyhydroxy esters.[252]

The enediol grouping in these acids causes their acidity, reducing properties, and instability in alkaline solution. Ascorbic acids reduce Fehling solution in the cold, and react with ferric chloride to produce the violet color typical of enolic compounds. They are very readily oxidized to their reversible, primary oxidation products, 2,3-glycodiulosono-1,4-lactones (commonly known

---

References start on p. 1085.

as *dehydroascorbic acids*), by such mild oxidizing agents as aqueous iodine; titration with such oxidants constitutes a quantitative method for distinguishing ascorbic acids from 2-glyculosonic acids.

Brenner *et al.*[253] studied the isomerization of ascorbic acids. Epimerization at C-4 occurs in refluxing 50% aqueous methanol containing potassium hydroxide, and an approximately equal mixture of epimers was obtained after 16 to 24 hours. Thus, the rare L-*erythro*- and D-*threo* isomers of ascorbic acid were isolated as solids by fractional recrystallization of epimerized mixtures afforded by the more readily available L-*threo*- (**54**) and D-*erythro*-hex-2-enono-1,4-lactones.

The action of boiling 12% hydrogen chloride converts ascorbic acids into 2-furaldehyde in high yield[140a]; L-*xylo*-hexulosonic acid[253a] and the 3,6-anhydro derivative[253b] of **54** have been proposed as intermediates in different mechanisms for this transformation of **54**.

2. Vitamin C (L-threo-*Hex-2-enono-1,4-lactone*, **54**)[249,254]

Vitamin C was first isolated as a strongly reducing crystalline substance, called "hexuronic acid," by Szent-Györgyi[255] in 1928 from adrenal glands, oranges, and cabbages, and it was later shown to possess the antiscorbutic activity long known in lemon juice, from which **54** was later isolated.[256] The vitamin is widely distributed in Nature; in addition to citrus fruits, other particularly rich sources include black currants, paprika, and rose hips.

The structure of **54** was elucidated, shortly after its isolation, in extensive contributions by many groups. (Early work on the isolation and structural elucidation of **54** was discussed by Haworth and Hirst.[257]) Several important discoveries relating to its constitution were reported in 1933, the same year in which the first successful synthesis was described.[245,246] Thus, Herbert *et al.*[258] found that the primary oxidation product of **54**, the 2,3-diulosono-lactone **58** (*dehydroascorbic acid*), could be quantitatively oxidized by sodium hypoiodite to generate oxalic acid (**60**) and L-threonic acid (**61**), which was identified as crystalline tri-*O*-methyl-L-threonamide. These results showed that dehydroascorbic acid is converted into L-*threo*-2,3-hexodiulosonic acid (**59**) and also established the stereochemical relationship between **54** and L-gulonic (and L-idonic) acid. Karrer *et al.*[259] and Micheel and Kraft[260] showed the presence of a double bond and four hydroxyl groups in L-ascorbic acid; two of these groups were identified as enolic because of their facile methylation by diazomethane. Micheel and Kraft[261] also established the presence of the furanoid ring and showed that the acid contains a double bond, which they hydrogenated to produce L-idonic acid. X-ray studies[262] verified the elucidated structure of **54**.

The equilibrium between **54** and the reversible oxidation product **58** is vital

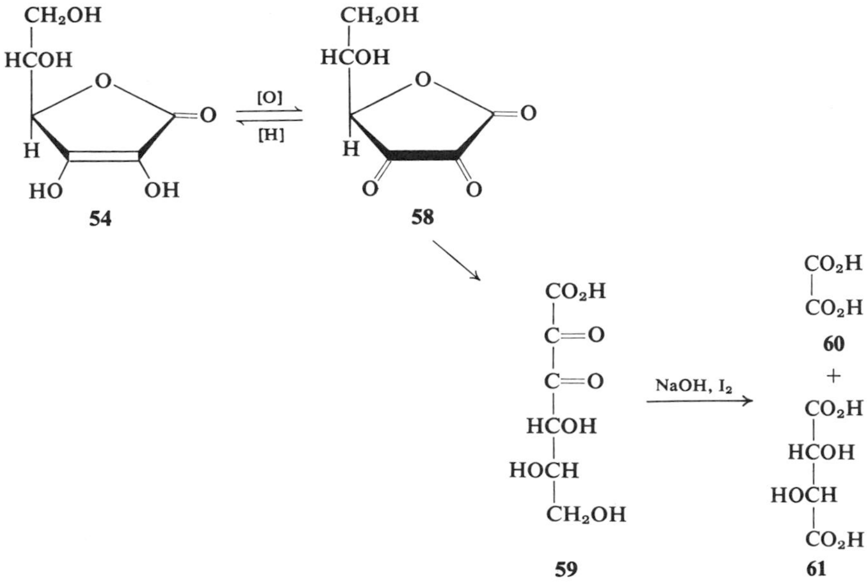

CH$_2$OH
|
HCOH
[structure with O-ring, =O]
H
HO   OH
54

$\xrightleftharpoons[\text{[H]}]{\text{[O]}}$

CH$_2$OH
|
HCOH
[structure with O-ring, =O]
H
O   O
58

CO$_2$H
|
C=O
|
C=O
|
HCOH
|
HOCH
|
CH$_2$OH
59

$\xrightarrow{\text{NaOH, I}_2}$

CO$_2$H
|
CO$_2$H
60

+

CO$_2$H
|
HCOH
|
HOCH
|
CO$_2$H
61

to plant and animal life[254]; the **54–58** couple apparently functions to mediate the transfer of hydrogen atoms. Compound **58** hydrates spontaneously in water,[263] affording **59**; **58** may, however, be stored as a methanol complex, which is far more stable than the parent substance.[264] Compound **59** has no antiscorbutic properties; although the lactone **58** regenerates **54** by the action of hydrogen sulfide, **59** is not similarly reduced. Solutions of **54**, which is readily oxidized, are more sensitive to alkalies than to acids, and the rates of oxidation are lowest under slightly acid conditions. The stability of **54** is very important in the food industry. Oxidation in milk is accelerated by copper and by sunlight. Enzymes in plant cells, including a copper-containing L-ascorbic acid oxidase, degrade vitamin C when plant tissues are damaged.

Most plants and animals (except man, monkey, and guinea pig) seem able to synthesize the vitamin. For the various metabolisms involved, see refs. 84 and 243 and monographs on biochemistry. The most satisfactory chemical methods for determination of L-ascorbic acid are based on the rapid reduction of 2,6-dichlorophenol-indophenol to its leuco form.

The commercial production of **54** largely follows the method of Reichstein and Grüssner,[265] utilizing L-sorbose (prepared from D-glucose). Oxidation of 2,3:4,6-di-*O*-isopropylidene-α-L-sorbofuranose (**62**) with permanganate or other oxidants gives the 2-glyculosonic acid derivative **63**, which is cyclized to

54 after hydrolysis of the isopropylidene protecting groups. Many modifications of the method have been investigated, including direct oxidation of D-sorbose to the glyculosonic acid, which is conveniently monitored by g.l.c.–m.s. analysis of per(trimethylsilyl)ated aliquots[265a] and the use of other 2-glyculosonic acid intermediates; there are many patents.[249]

A simple synthesis[95a,265b] of **54** in few steps proceeds from 1,2-O-isopropylidene-α-D-glucofuranose by platinum-catalyzed oxidation to a glycos-5-ulosurono-6,3-lactone derivative (**8a**); in one acidic step the isopropylidene group is removed, accompanied by enolization and generation of the free aldehyde group, which can be reduced by borohydride, affording **54**.

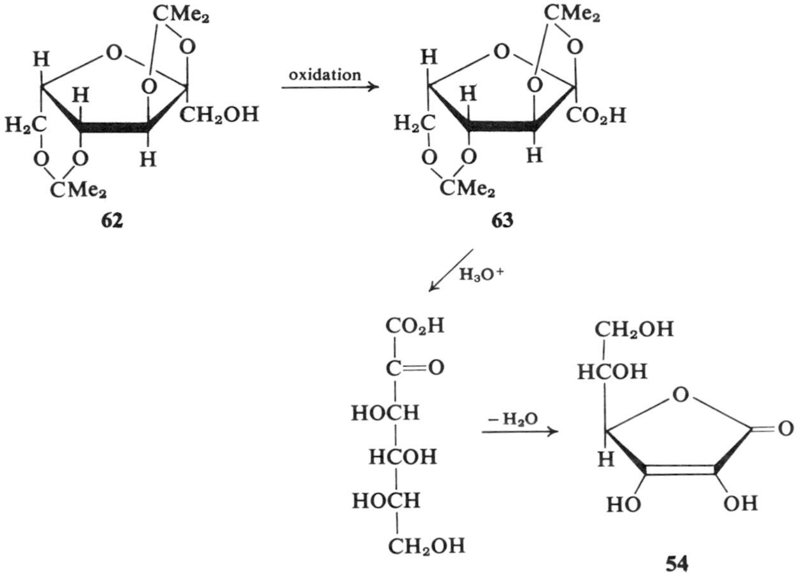

Vargha[266] prepared the 5,6-isopropylidene acetal of **54**, and von Schuching and Frye[267] prepared the corresponding cyclohexylidene acetal. These were found to be more resistant than **54** toward oxidation, and **54** could readily be regenerated by acid hydrolysis. The derivative was used in the synthesis of ^{14}C-labeled vitamin C.

O-Sulfation of **54** by pyridine–sulfur trioxide occurs[267a] exclusively at O-2,[267b] whereas phosphorylation affords several products.[267c]

Jackson and Jones[268] reported the alkylation of L-ascorbic acid with benzyl chloride, which produced a 2-C-benzyl-3-hexulosonolactone and 3-O-benzyl-L-*threo*-hex-2-enono-1,4-lactone.

III. NEUTRAL OXIDATION PRODUCTS

This section deals with those neutral carbohydrates having two carbonyl groups, either free or substituted, present in the same carbon atom chain. Some of these compounds are prepared by splitting exocyclic or inositol α-glycol groups; the formation of dialdehydes by oxidative cleavage between endocyclic α-glycol groups is considered in Chapter 25. The neutral dicarbonyl compounds are divided into three categories: (A) dialdehydes (dialdoses), (B) ketoaldehydes (glycosuloses), and (C) diketo compounds (glycodiuloses). The review by Theander[269] contains more-detailed information about the various dicarbonyl carbohydrates known up to 1962. Methods for quantitative analysis of the content of free carbonyl groups in starch[269a] and cellulose[269a,b] have been established.

A. DIALDOSES

Several compounds of this type, formally derived from aldoses by oxidation of the terminal —CH$_2$OH group to —CHO, have been prepared. Dialdoses arise as intermediates in structural studies, but they are also valuable starting materials for synthetic conversions. A naturally occurring branched-chain dialdose, streptose, occurs as a component of the antibiotic streptomycin.

1. Preparation

erythro-Tetrodialdose has been synthesized from acetylene[270] and from furan.[271] With a few exceptions, the general approaches to the preparation of dialdoses and their derivatives have employed (a) reduction of one, or two carboxyl group(s), (b) controlled oxidation of one, or two, primary alcoholic group(s), (c) oxidative cleavage of α-glycols, or (d) ozonolysis of double bonds to generate aldehyde groups.

1,2-O-Isopropylidene-α-D-xylo-pentodialdo-1,4-furanose (64), which has been used for many syntheses, was prepared by Iwadare[272] through oxidation of 1,2-O-isopropylidene-α-D-glucofuranose with lead tetraacetate. Periodate was shown by Schaffer and Isbell[273] to be a superior oxidant for this transformation; they isolated and identified[274] the crystalline dimeric compound 65; reaction of 65 with benzylamine results in replacement of the oxygen atom connecting C-5 and C-5' by a benzylimino group, and the crystalline hydrogen sulfate adduct of monomeric 64 was also prepared.[274a] The free dialdose is obtained by acidic hydrolysis. Exchange of O-5 in H$_2$^{18}O converts 65 into the [5-^{18}O] analogue, whence D-[5-^{18}O]xylose was prepared.[274b]

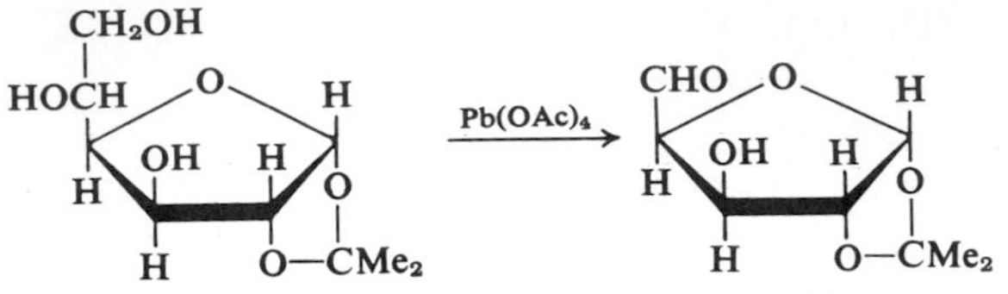

64

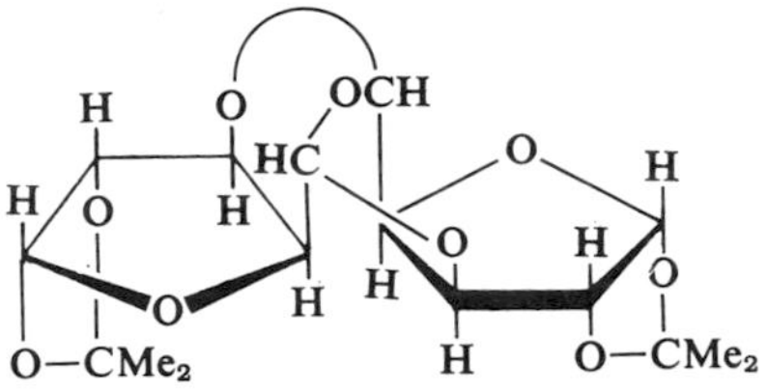

65

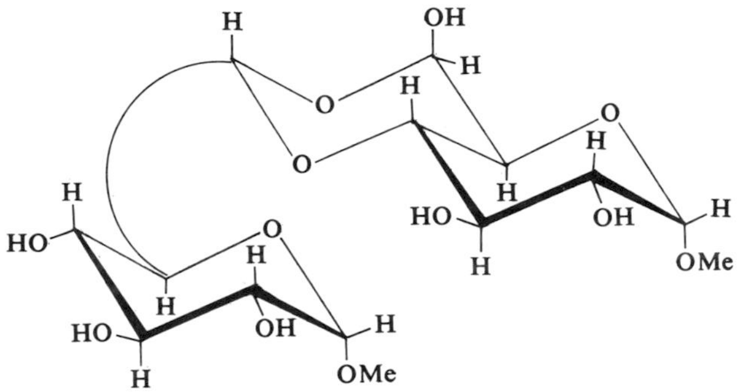

66

Unsymmetrical dimers are formed by condensation of 6-*aldehydo*-hexo-dialdo-1,5-pyranosides that have a free hydroxyl group at C-4. A dimer (66) of methyl α-D-*gluco*-hexodialdo-1,5-pyranoside has been prepared,[274c] and Maradufu and Perlin[274d] reported a β-D-*galacto* analogue.

Hexodialdoses have been obtained by α-glycol cleavage of suitably substituted cyclitols. In the elegant structural elucidation by Dangschat and Fischer[275] of the naturally occurring cyclitol conduritol (67), the sequence of reactions illustrated was employed to convert 67 into 2,3,4,5-tetra-*O*-acetyl-*galacto*-hexodialdose (68), a compound also afforded by Rosenmund reduction of the dichloride of tetra-*O*-acetylgalactaric acid.[276,277]

D-*gluco*-Hexodialdose has been prepared by controlled, catalytic hydrogenation of D-glucurono-6,3-lactone (see Chapter 22), and 6-deoxy-1,2-*O*-isopropylidene-α-D-*gluco*-heptodialdo-1,4-furanose-3,7-pyranose (70) was prepared by Meyer zu Reckendorf[278] by reduction of the corresponding deoxy-lactone derivative with diisobutylaluminum hydride. This reducing agent was also used to prepare 1,2-*O*-isopropylidene-α-D-*gluco*-hexodialdo-1,4-furanose from 1,2-*O*-isopropylidene-α-D-glucofuranurono-6,3-lactone. Compound **70** has also been prepared in 78% yield by hydroformylation[279] of 5,6-anhydro-1,2-*O*-isopropylidene-α-D-glucofuranose (**69**). Careful hydrolysis of **70** yielded a mixture containing amorphous 6-deoxy-D-*gluco*-heptodialdose and a crystalline trioxane derivative (**71**) related to adamantane.[278]

Crystalline 1,2-O-isopropylidene-α-D-*gluco*-hexodialdo-1,4:6,3-difuranose (**72**) was obtained by Henseke and Hanisch[280] via methyl sulfoxide oxidation of **69**, and by Theander[281] via chromic acid oxidation of 1,2-O-isopropylidene-α-D-glucofuranose; **72** was converted into the crystalline tetracyclic compound **73** by the action of acidic acetone.[282] (Note that **72** and **73** have the same configuration, and that the Fischer representation of **72** is misleading about actual stereochemical relationships.)

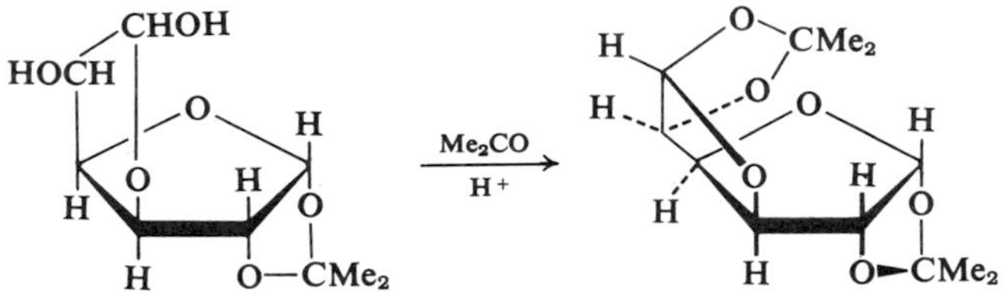

Horton *et al.*[283] used methyl sulfoxide oxidation (Pfitzner–Moffatt reagent) for the preparation of 1,2:3,4-di-O-isopropylidene-α-D-*galacto*-hexodialdo-1,5-pyranose from the corresponding D-galactose derivative. Direct oxidation of the primary hydroxyl group in methyl hexopyranosides having unprotected secondary hydroxyl groups generally gives only low yields of ω-aldehydo compounds.[269]

Methyl pentodialdofuranosides[284] and methyl hexodialdopyranosides,[285,286] were prepared in quite good yields by equimolar periodate oxidation of the exocyclic α-glycol grouping of the next higher homologous methyl glycoside. The galactose oxidase (EC1.1.3.9) that catalyzes the oxidation of D-galactose (and derivatives) to D-*galacto*-hexodialdose (and derivatives) is useful for analytical determinations[287] but provides a somewhat unsatisfactory method for large-scale preparation of methyl D-*galacto*-hexodialdo-1,5-pyranosides.[274c,286]

Horton *et al.*[100b] have shown that photolysis of methyl 2,3,4-tri-O-acetyl-6-azido-6-deoxy-α-D-glucopyranoside affords methyl 2,3,4-tri-O-acetyl-α-D-*gluco*-hexodialdo-1,5-pyranoside in good yield. Subsequent investigations into the generality of this reaction have revealed that 5′-azido-5′-deoxy-nucleosides[100c] and 6-azido-6-deoxycellulose[100d] are likewise converted readily into ω-aldehydo derivatives by photolysis. Gibson *et al.*[287a] subsequently prepared ω-aldehydo derivatives by oxidative deamination of 6-amino-6-deoxycyclohexaamylose and 6-amino-6-deoxy-1,2:3,4-di-O-isopropylidene-α-D-galactopyranose using ninhydrin.

The structure of the branched dialdose streptose [5-deoxy-3-C-formyl-L-lyxose (**75**)], a constituent of the antibiotic streptomycin, was elucidated after extensive investigations of its derivatives and transformation products.[288]

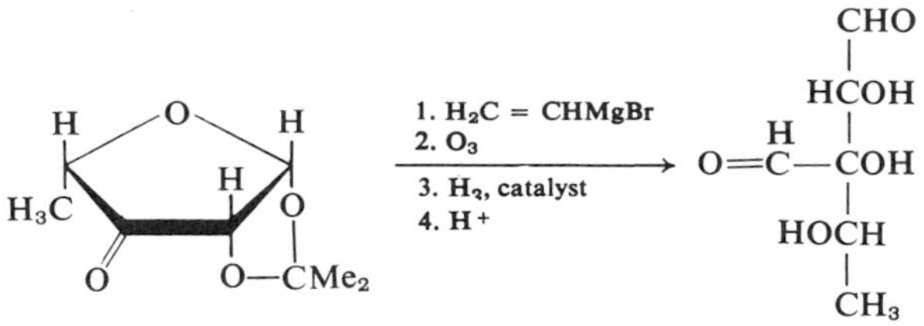

$$\text{74} \qquad \xrightarrow[\substack{\text{3. H}_2\text{, catalyst} \\ \text{4. H}^+}]{\substack{\text{1. H}_2\text{C} = \text{CHMgBr} \\ \text{2. O}_3}} \qquad \begin{array}{c} \text{CHO} \\ | \\ \text{HCOH} \\ \text{H} \quad | \\ \text{O}=\text{C}-\text{COH} \\ | \\ \text{HOCH} \\ | \\ \text{CH}_3 \end{array}$$

74   **75**

Dyer *et al.*[289] succeeded in synthesizing this compound from 5-deoxy-1,2-*O*-isopropylidene-β-L-*threo*-pentofuranos-3-ulose (**74**), according to the sequence of conversions illustrated. (The use of glycosulose derivatives in the preparation of rare sugars is discussed in Section III,B,2.) 3-Deoxy-3-*C*-formyl- and 2-deoxy-2-*C*-formylpentofuranosides have been identified among the products of deamination of methyl 3-amino-3-deoxy-α-D-glucopyranoside[289a] and of methyl 2-amino-2-deoxy-α- and -β-D-glucopyranosides,[289b] respectively.

2. Properties and Reactions

Dialdoses and their derivatives that contain one or two unsubstituted carbonyl group(s) generally appear as multiple spots on chromatograms and undergo mutarotation in aqueous solution[269]; these phenomena reveal that an equilibrium is established between different cyclic and acyclic forms of dialdoses. The rates at which these equilibria are established approximate those for conversions between lactones and the free acid form of carbohydrate acids. The different components of the equilibrium can be obtained as well separated fractions by column chromatography on cellulose or carbon; however, each reverts to the same equilibrium mixture after a few days. Many forms are possible, depending, among other things, on the number of hydroxyl groups available for ring formation; crystalline dialdose derivatives spontaneously adopt many different (dimeric, hemiacetal, and polycyclic) forms, and evidence has also been found for free aldehyde forms, hydrates, and hemialdals.[269]

Theander[290] investigated the hydrolysis of some methyl β-D-*gluco*-hexo-dialdo-1,5-pyranosides in dilute sulfuric acid and found that they are hydrolyzed considerably faster than the corresponding D-glucosides. A plausible explanation, at least in part, for this effect is that aldehydo derivatives (such as **76**, R, R' = Me) have a tendency to form cyclic hemiacetals in aqueous solution. The formation of **77**, requires, however, that the molecule undergo

---

conformational inversion into the $^1C_4(D)$ form, in which all of the bulky substituents occupy axial positions; accordingly, the factors that act to accelerate the hydrolysis of 3,6-anhydro-β-D-glucopyranosides are also present in **77**. As **76** (R, R′ = poly[(1→4)-α-D-glucopyranosyl] chains linked through the 4 and 1 positions, respectively) may also be a component of cellulose, it follows that oxidation of primary alcoholic groups in polysaccharides to aldehyde groups is a potential source of "weak linkages" toward acid hydrolysis (compare Section II,B,5); this was later confirmed,[290a] when cellulose containing ω-aldehyde groups became available.[100d] It was further found that such a modified cellulose, as well as low-molecular methyl hexodialdopyranosides (including methyl 6,6′-dialdehydo-β-cellobioside[290b]), readily undergo β elimination at C-4 at pH 3 and up.[290c]

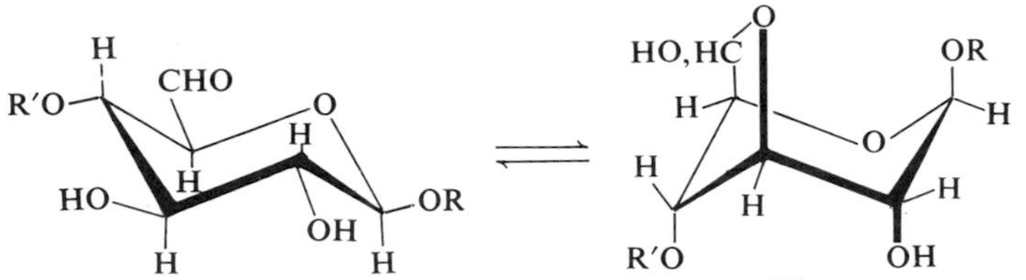

76 **77**

The unsaturated hexodialdopyranosides formed are very readily transformed into brown, polymerized products, by either alkaline or acid treatment. From acid treatment of **77a**, the β-elimination product of methyl α-D-*gluco*-hexodialdo-1,5-pyranoside, the reactive intermediates **77b** and **77c** were isolated.[290c] The 2,3-diacetate of **77a** has previously been obtained by Perlin *et al.* by an oxidation–β-elimination reaction.[290d]

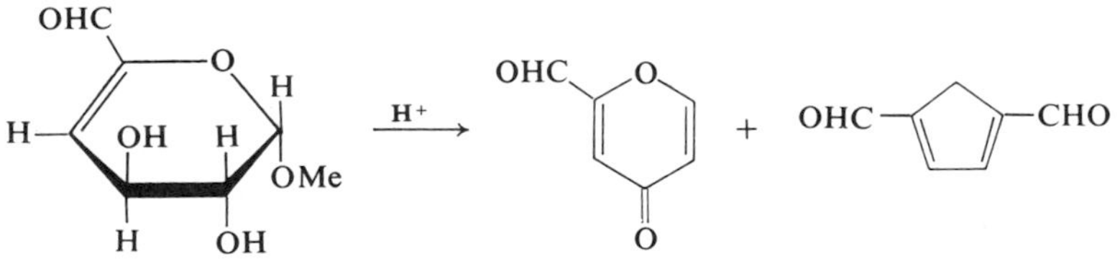

77a **77b** **77c**

Acyclic dialdoses, and derivatives having one free aldehyde group, generally exhibit properties typical of acetylated aldehyde sugars and aliphatic aldehydes rather than of monosaccharides; thus, they give a positive Schiff test and exhibit strong electrophoretic migration in the presence of hydrogen sulfite (see Table I).

Ballou and Fischer[291] treated D-*manno*-hexodialdose and its 2,3:4,5-di-O-isopropylidene derivative with methanol under the Fischer glycosidation conditions, obtaining not only glycosides but also some dimethyl acetals;

a crystalline difuranoside (probably the α,α anomer) was obtained in 20%
yield. Crystalline bis(diethyl dithioacetal) derivatives are afforded by reaction
of D-*manno*-hexodialdose[292] and D-*gluco*-hexodialdose[293,294] with a stronger
nucleophile, ethanethiol, under acidic conditions. Grosheintz and Fischer[295]
prepared 6-deoxy-6-nitro derivatives of D-glucose and L-idose by condensation
of nitromethane with 1,2-*O*-isopropylidene-α-D-*xylo*-pentodialdo-1,4-furanose
(**65**), as intermediates in a synthesis of some nitrocyclitol derivatives.
Lichtenthaler[296] described a synthesis of ^{14}C-labeled *myo*-inositol starting
from *xylo*-pentodialdose. D-[6-^{14}C]Glucose, L-[6-^{14}C]idose, and the corresponding labeled glycuronic acids have been prepared from **65** via the
cyanohydrin reaction, a method that has been improved upon through the
extensive studies of Schaffer and Isbell.[297]

Schaffer[298] described a synthesis of L-apiose starting from the reaction of
65 with formaldehyde, and Horton *et al.*[273b] developed a method for extension
of sugar chains through ethynylation of **65** and other aldehydo compounds by
the Grignard reaction.

Stereoselective addition to suitably protected dialdose derivatives has been
recommended as a starting reaction for the preparation of asymmetrically
substituted, noncarbohydrate molecules.[298a]

B. GLYCOSULOSES

These compounds, formally derived from aldoses by oxidation of a secondary hydroxyl group to a ketone group, will be considered in two sections, the
first being a discussion of glycosuloses and derivatives that have both carbonyl
groups free, and the second being a discussion of derivatives in which at least
one carbonyl group is not free.

1. *Unsubstituted Glycosuloses*

The well-known aldos-2-uloses (usually termed simply *aldosuloses*, as the
-2- is implicit by convention; or *osones* in former usage) have long been known
in the form of their bis(hydrazone) derivatives, the osazones, which are discussed in Chapter 21. Deoxyaldosuloses have been implicated as intermediates
in a variety of degradation reactions. Aldos-3-, -4-, and -5-uloses have been
prepared, principally as intermediates for synthesis. In Nature, a few sugars
are found as components of various antibiotics that belong to this group (see
Chapter 31); thus, 6-deoxy-D-*arabino*-hexos-5-ulose and 6-deoxy-L-*lyxo*-
hexosulose are components in hygromycin[299] and angustmycin A,[300]
respectively. Evidence has been obtained that the carbohydrate component

---

References start on p. 1085.

in the cardiac glycoside gomphoside is derived from a 4,6-dideoxy-hexosulose.[301]

The early chemistry of the aldos-2-uloses has been reviewed by Bayne and Fewster.[302]

a. Preparation.—Aldos-2-uloses are prepared by acid hydrolysis of osazones or by transfer of the phenylhydrazine residues to simpler carbonyl compounds. In 1888, Fischer obtained D-*arabino*-hexos-2-ulose (**78**, "D-*glucosone*") by treating the phenylosazone of D-glucose with hydrochloric acid; however, nitrous acid[303] is a more suitable hydrolytic agent, and transhydrazonation with benzaldehyde is a common procedure.[304] Purity of the osazone is important in these treatments.

A more direct method of preparation is oxidation of aldoses, and optimal yields[305] are afforded by the action of cupric acetate in methanol or ethanol, according to the procedure of Weidenhagen.[306] The method is suitable for large-scale preparation of intermediates; however, a pure product is obtained only by chromatographic separation from the unreacted sugar by-products. The chromium trioxide·(pyridine)$_2$ complex affords fair yields of dicarbonyl compounds by oxidation of suitably protected precursors; 2,3:4,5-di-*O*-isopropylidene-β-D-*arabino*-hexos-2-ulose-2,6-pyranose was thus prepared in 53% yield from 2,3:4,5-di-*O*-isopropylidene-β-D-fructopyranose[306a] (see Chapter 24). Oxidative deamination of 1-amino-1-deoxy-D-fructose by 3,5-bis(*tert*-butyl)-1,2-benzoquinone in oxygen-free pyridine solution is reported[306b] to afford **78**. 3,4,6-Tri-*O*-acetyl-D-*arabino*-hexos-2-ulose has been prepared from 2,3,4,6-tetra-*O*-acetyl-1,5-anhydro-D-*arabino*-hexitol ("2-hydroxy-D-glucal tetraacetate") by the method of Maurer.[307] Collins *et al.*[307a] reported that the sugar ring in methyl 4,6-*O*-benzylidene-3-deoxy-α-D-*erythro*-hexopyranosid-2-ulose 2-phenylhydrazone opened in pyridine to afford an aldehydo derivative that was subsequently transhydrazonated to afford the 4,6-benzylidene acetal of the acyclic aldosulose. Such partially substituted derivatives have been extensively studied.[302]

Passage of aqueous solutions of lactose through cation-exchange resins in the borate form results, rather unexpectedly, in a low degree (1 to 10%) of conversion into 4-*O*-β-D-galactopyranosyl-D-*arabino*-hexos-2-ulose.[307b]

The 3-deoxyhexos-2-uloses, which had long been hypothesized as intermediates in many important degradation reactions of carbohydrates, have been isolated and studied, in particular by Anet, who presented a review[308] of the subject. Examples of these reactions include the formation of metasaccharinic acids (see Chapter 4 and Section II,D,4) by the action of alkali on reducing sugars or derivatives, the degradation of carbohydrates by acids to 2-furaldehydes (see Chapter 4), the degradation of carbohydrates to the Maillard products by the action of amines (see Chapter 20), the decomposition of sugars to form sulfonic acids (see Section II,D,3) by the action of aqueous sulfite ion, and the pyrolytic degradation of carbohydrates.[308a]

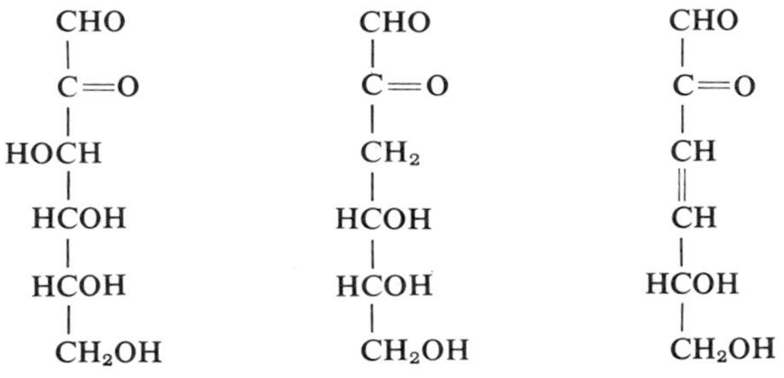

$$
\begin{array}{ccc}
\text{CHO} & \text{CHO} & \text{CHO} \\
| & | & | \\
\text{C}=\text{O} & \text{C}=\text{O} & \text{C}=\text{O} \\
| & | & | \\
\text{HOCH} & \text{CH}_2 & \text{CH} \\
| & | & \| \\
\text{HCOH} & \text{HCOH} & \text{CH} \\
| & | & | \\
\text{HCOH} & \text{HCOH} & \text{HCOH} \\
| & | & | \\
\text{CH}_2\text{OH} & \text{CH}_2\text{OH} & \text{CH}_2\text{OH} \\
\mathbf{78} & \mathbf{79} & \mathbf{80}
\end{array}
$$

Kato[309] reported the isolation of 3-deoxy-D-*erythro*-hexos-2-ulose (**79**) from the reaction of *N*-butyl-D-glucosylamine with acetic acid. Anet[310] prepared **79** starting from "di-D-fructose–glycine" and isolated, as by-products, the *trans* and *cis* forms of 3,4-dideoxy-D-*glycero*-hex-3-enos-2-ulose (**80**).[311] Machell and Richards[312] isolated **79** on treatment of 3-*O*-benzyl-D-glucose with aqueous sodium hydroxide. The best general synthesis of 3-deoxy-aldosuloses is the *p*-toluidine-catalyzed rearrangement of the corresponding aldose in the presence of benzoylhydrazine; transhydrazonation affords the product in fair to good yields.[312a]

Compound **79** has been isolated as a component in pathological mammalian liver tissue after treatment with arsenic(III) oxide[312b]; however, it has been reported[312c] that **79** is formed from the common hexoses by the action of arsenic(III) oxide under conditions similar to those employed in early tissue experiments. Other 3-deoxyaldos-2-uloses are also known,[269,308] as well as terminally phosphorylated derivatives generated by condensation of enol-pyruvate phosphate with aldose ω-phosphates in the presence of an extract from *Pseudomonas aeruginosa*[312c]; methyl ethers derived from the enolic form of **79** arise during alkaline treatment of methylated sugars.[308]

Fukui and Hochster[313] prepared D-*ribo*-hexos-3-ulose (**85a**) by enzymic oxidation and hydrolysis of sucrose.

A few hexos-5-uloses are known; Helferich and Bigelow[314] synthesized D-*xylo*-hexos-5-ulose (**82**) from methyl 2,3,4-tri-*O*-acetyl-6-deoxy-β-D-*xylo*-hex-5-enopyranoside (**81**) by oxidative acetoxylation of the double bond with lead tetraacetate, followed by hydrolysis of the protecting groups to afford **82**. (As the ring forms in which aldosuloses exist have yet to be elucidated, an acyclic representation is employed in illustrations for this section.) Kiely and

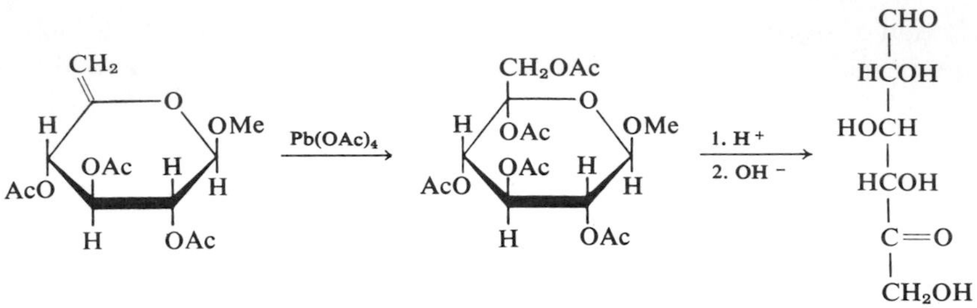

$$81 \qquad\qquad\qquad\qquad\qquad\qquad 82$$

Fletcher[314a] cyclized **82** to *myo*-inositol by an aldol reaction, thereby support-ing the intermediate formation of **82** (or a phosphate thereof) in the bio-synthesis of inositols. Helferich and Himmen[314b] obtained 6-deoxy-D-*xylo*-hexos-5-ulose in good yield by acid hydrolysis of deacetylated **81**; it was later isolated from an acid hydrolyzate of 5,6-unsaturated amylose.[314c]

b. Properties and Reactions.—The most characteristic and common proper-ties of aldos-2-uloses and deoxyaldos-2-uloses (all isolated in the amorphous state), are their reduction with Fehling solution in the cold, and their ready formation of osazones. The occurrence of the 3-deoxyaldos-2-uloses as labile intermediates in various carbohydrate reactions (see Section III,B,1,*a*) is discussed under the respective reactions.

The principal synthetic application of the aldos-2-uloses has been as intermediates in the preparation of L-ascorbic acids (see Section II,E,1 for an example). Aldos-2-uloses can be oxidized to 2-glyculosonic acids. They are more stable to acids than to alkalies, which catalyze enolization reactions; D-*arabino*-hexosulose (**78**) was converted into kojic acid by the action of alkali.[302] In connection with studies on the possible role of a D-*arabino*-hexos-2-ulose (**78**) group in the formation of acidic end units in cellulose during alkaline, oxidative, technical processes, compound **78** was treated with alkali; pentoses, 2-pentuloses, 3-deoxypentonic acids, and a series of aldonic acids were detected.[314d] Fischer reduced **78** to D-fructose by the action of zinc dust in aqueous acetic acid, and the method was later used by his group for other syntheses.[315] Henseke and Liebenow[316] prepared various 1-hydrazones by the reaction of aldos-2-uloses with disubstituted hydrazines under neutral or acidic conditions; from this observation, they decided that the (*intrinsically more reactive*) aldehyde group is present as the free carbonyl form.

The structure of aldos-2-uloses, as syrups and in solution, has been a sub-ject of great interest and discussion.[302] Studies by ultraviolet spectroscopy indicated that there is no appreciable proportion of the enolic form present in acidic or neutral solution. These compounds exhibit slight mutarotation, and appear as either multiple or tailed spots on paper chromatograms (compare Section III,A,2). The latter physical evidence accords with an equilibrium involving a number of isomeric ring forms, whereas the selective reactivity of the aldehyde group may indicate that it exists in the free form to a greater extent than does the ketone group, or that the former is more readily regenerated than the latter from their respective hydrated or hemiacetal forms.

Corroboration of this interpretation was presented by Assarsson and Theander,[317] who found that **78** reacts with methanolic hydrogen chloride to afford D-*arabino*-hexos-2-ulose 1,1-dimethyl acetal (**83**) and crystalline methyl β-D-*arabino*-hexos-2-ulose-2,5-furanoside 1,1-dimethyl acetal (**84**) as the main components, to the exclusion of *keto*-glycosides or 2,2-acetals. Further support was found in the observation that **78** migrates rapidly on hydrogen sulfite electrophoretograms (see Table I), as do the aldehydo sugar derivatives, dialdoses, and *aldehydo*-glycosides; conversely, **83**, like the keto sugar derivatives and monosaccharides, migrates only slightly, if at all.[128,269]

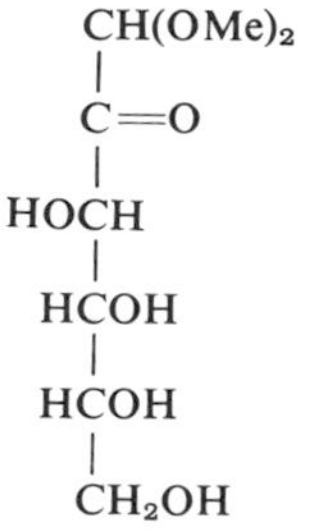

83 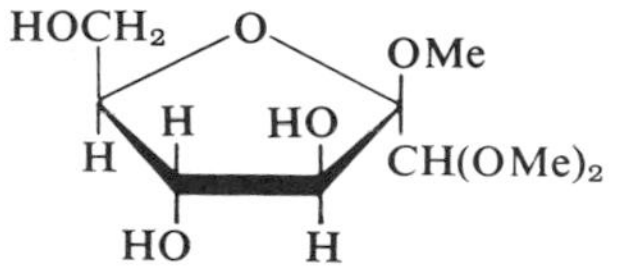   **84** 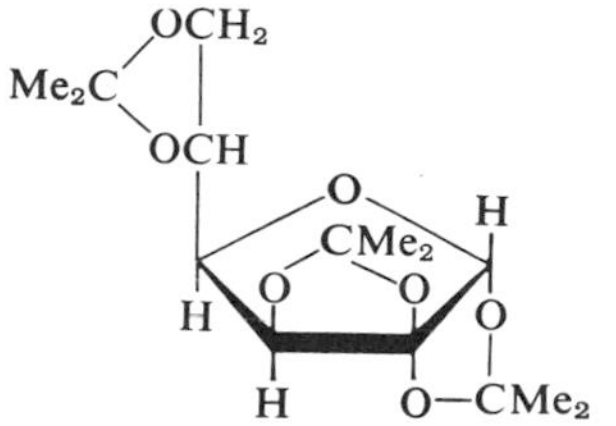   **85**

Crystalline 1,2:2,3:5,6-tri-*O*-isopropylidene-2(*R*)-α-D-*arabino*-hexofuranos--2-ulose (**85**), which Bayne *et al.*[318] prepared by acetonation of **78**, is an example in which the aldehyde group is engaged in a lactol ring, whereas the ketone group formally is not. Compound **85** is useful for the purification of **78**, as it can be partially hydrolyzed, affording the 1,2:2,3-di-*O*-isopropylidene derivative, or completely, regenerating **78**. Aldos-2-ulose 1,1-dithioacetals are also crystalline derivatives.[319,320]

D-*ribo*-Hexos-3-ulose (**85a**), treated with methanolic hydrogen chloride, afforded **85b** and **85c** as main products.[320a] They were converted into the glycoside **85d** by mild treatment with acid; this product was further converted into D-*ribo*-3-hexulose (**85e**), as shown in the scheme. The preparation of **85e** by partial reduction of **85a** was not practicable.

TABLE I

M_v Values[a] of Some Carbonyl Carbohydrates in Paper Electrophoresis in Hydrogensulfite (pH 4.7, 0.1M, at 50°)[128,269]

Compound	M_v
D-Glucose	0.05
D-Fructose	0
D-Xylose	0.32
aldehydo-D-Glucose pentaacetate	1.18
D-arabino-Hexosulose (78)	1.24
D-arabino-Hexosulose 1,1-dimethyl acetal (83)	0
Methyl β-D-gluco-hexodialdo-1,5-pyranoside (76, R = Me, R¹ = H)	1.25
Methyl 4-O-methyl-β-D-gluco-hexodialdo-1,5-pyranoside (76, R = R¹ = Me)	1.22
Methyl β-D-arabino-hexopyranosid-2-ulose (86)	0.79
Methyl β-D-ribo-hexopyranosid-3-ulose (87)	0.56
Methyl α-D-ribo-hexopyranosid-3-ulose	0.07
Methyl 4-O-methyl-β-D-ribo-hexopyranosid-3-ulose	0.38
Methyl β-D-xylo-hexopyranosid-4-ulose (88)	0.81
Methyl β-D-erythro-pentopyranosid-3-ulose[330]	0.99
Methyl α-D-erythro-pentopyranosid-3-ulose[330]	0.26
1,2-O-Isopropylidene-α-D-gluco-hexodialdo-1,4-furanose[281] (72)	1.11
1,2-O-Isopropylidene-α-D-ribo-hexofuranos-3-ulose[281]	0.12
1,2-O-Isopropylidene-α-D-xylo-hexofuranos-5-ulose[281]	0.17
1,2-O-Isopropylidene-α-D-xylo-pentodialdo-1,4-furanose[281] (64)	1.36
1,5-Anhydro-D-erythro-hexo-2,3-diulose	1.62

[a] The M_v value is defined as the true distance of migration of the carbonyl compound compared with that of vanillin.

When **85a** was treated with alkali, the main end-products were 3-deoxy-D-threo- and 3-deoxy-D-erythro-pentonic acids, and the anticipated 3-deoxy-D-glycero-hexos-2-ulose was isolated as an intermediate.[320b]

2. Glycosuloses Having Substituted Aldehyde Groups

The known members of this group are mainly keto-glycosides (including some keto-oligosaccharides produced by microorganisms) and compounds in which the aldehyde group participates in the formation of an isopropylidene acetal ring; 1,1-acetals and dithioacetals are oxidized by Acetobacter suboxydans[320c] to ketones at the penultimate carbon atom.

Aldosiduloses were first isolated and later studied in connection with model studies on the oxidation of cellulose[321]; thus, Lindberg and Theander[322] isolated methyl β-D-ribo-hexopyranosid-3-ulose (together with a dialdose

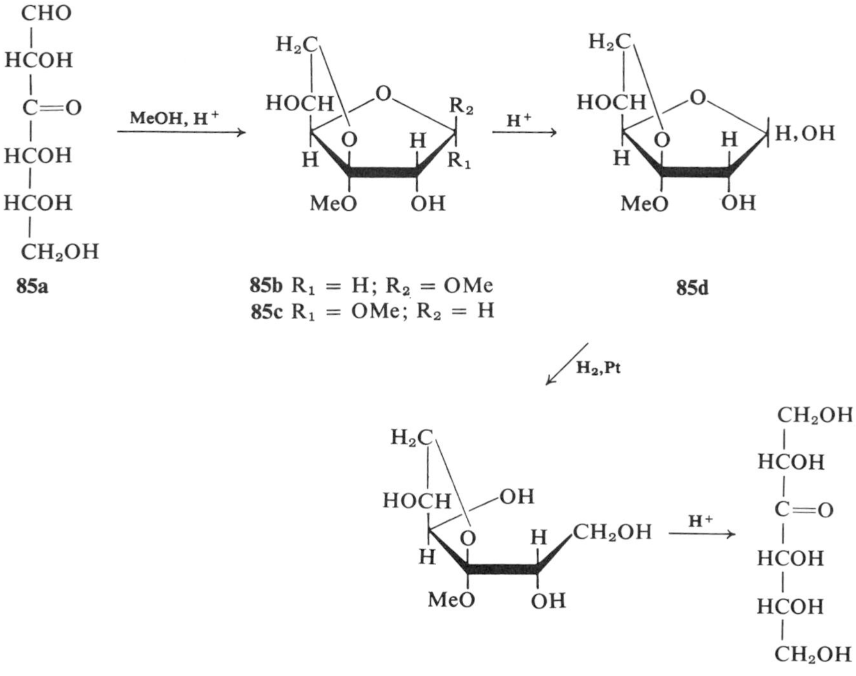

CHO
|
HCOH
|
C=O
|
HCOH
|
HCOH
|
CH₂OH

85a

$\xrightarrow{\text{MeOH, H}^+}$

(structure with H₂C, HOCH, O, R₂, MeO, OH, H, R₁)

$\xrightarrow{\text{H}^+}$

(structure with H₂C, HOCH, O, MeO, OH, H, H,OH)

85b R₁ = H; R₂ = OMe
85c R₁ = OMe; R₂ = H

85d

$\swarrow$ H₂, Pt

(structure with H₂C, HOCH, OH, O, H, CH₂OH, MeO, OH)

$\xrightarrow{\text{H}^+}$

CH₂OH
|
HCOH
|
C=O
|
HCOH
|
HCOH
|
CH₂OH

85e

derivative) from oxidation of methyl β-D-glucopyranoside with dichromate. Interest in the preparation of these derivatives of glycosuloses increased considerably after their versatility in the synthesis of rare sugars possessing amino, deoxy, and branched-chain functions (see Chapter 2, 16, and 17) had been established; Brimacombe has reviewed[323] this topic. The probable role of aldos-4-ulose derivatives as intermediates in the biosynthesis of many different sugars has also been discussed.[324] (See Chapter 34.)

a. Preparation.—Except for a few special examples, the substituted glycosuloses have been prepared by direct oxidation of a secondary alcoholic group to a ketone group; methods for this purpose include preparative catalytic oxidation, and oxidation with chromic acid, methyl sulfoxide–acid anhydride, or ruthenium tetraoxide. A more detailed discussion of oxidative conditions is presented in Chapter 24.

Preparations of unsubstituted aldopyranosiduloses are more difficult, necessitating chromatographic separations, and give lower yields than oxidations of sugar derivatives in which all but one of the hydroxyl groups are protected during the oxidation; properly selected protecting groups need not be disturbed until the ketone has been converted into a more-stable product.

Many unsubstituted pento- and hexo-pyranosiduloses have been prepared, mainly in connection with model studies on the properties of oxidized poly-saccharides. Methods used include the rather nonspecific chromic acid oxidation (in pyridine or acetone) of glycosides[269] having none, one, or two of the hydroxyl groups protected (sometimes oxidation is accompanied by elimination reactions[324a]), and catalytic oxidation[17,269,324b] of axially attached hydroxyl groups. As an example, chromic acid oxidation[325] and subsequent detritylation of methyl 6-O-trityl-β-D-glucopyranoside affords three mono-keto derivatives—methyl β-D-*arabino*-hexopyranosid-2-ulose (**86**), methyl β-D-*ribo*-hexopyranosid-3-ulose (**87**), and methyl β-D-*xylo*-hexopyranosid-4-ulose (**88**). Similar oxidation of methyl 4-O-methyl-β-D-glucopyranoside yields[326] the three 4-O-methyl monocarbonyl derivatives. 6-O-Tritylamylose is, however, oxidized to a glycos-2-ulose polymer by treatment with methyl sulfoxide–acetic anhydride.[327] Other examples of glycosulose residues in polysaccharides are given in reviews.[269,327a]

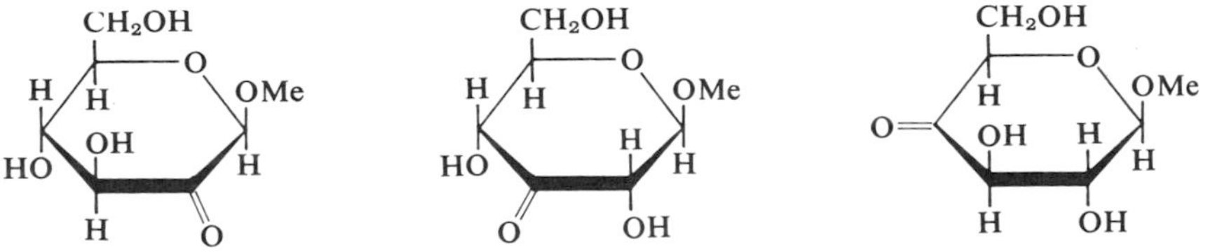

The product mixtures may contain numerous other products arising from further oxidation, rearrangement, attack on the aglycon, and cleavage of the glycosidic linkage; considerable losses of the sensitive keto compounds may also result from the removal of protecting groups after oxidation, so that net yields of glycosiduloses thus obtained are often quite low.

Simpler reaction products and higher yields of glycosiduloses have been obtained by catalytic oxidation; however, preparation of a suitable starting material having an axially attached hydroxyl group properly situated and no unprotected primary hydroxyl groups present may present an insurmountable obstacle. The method has been used by Heyns et al.[328] and Brimacombe et al.[329] for the preparation of some 6-deoxyhexopyranosiduloses.

Bromine oxidation (pH 7)[329a] of unprotected methyl glycosides has been found promising. Methyl α-D-glucopyranoside yielded mainly the 2-keto and 4-keto derivatives (~30% together); methyl α-D-galacto- and -D-manno-pyranoside, having axial hydroxyl groups in the 4 and 2 positions, respectively, gave mainly the 4-keto and 2-keto derivatives. This indicates that oxidation at a ring carbon atom having an axially disposed hydrogen atom is hindered when bulky substituents are present in a *syn*-diaxial relationship.

Lindberg and Slessor[330] reported the preparation of the crystalline methyl D-*erythro*-pentopyranosid-3-uloses in 50 to 60% yield by using a 2,4-phenyl-boronate protecting group during oxidation of the respective methyl D-xylopyranosides with methyl sulfoxide–acetic anhydride. As this protecting group is removed under very mild conditions, its introduction is a valuable improvement in the preparation of those unsubstituted glycopyranosiduloses having hydroxyl groups so arranged as to complex with phenylboronic acid.

Chromatography on carbon–Celite columns[325,326] and on anion-exchange resins in the bisulfite form,[328,329,330] often followed by chromatography on cellulose, has been a valuable tool in the separation and purification of unsubstituted glycopyranosiduloses. The quantitative conversion of mixtures of methyl glycopyranosiduloses into the more stable *O*-methyloxime derivatives simplifies the fractionation and purification by chromatography. The parent glycosidulose can then be regenerated in good yield by treatment with weak acid.[329a]

The oxidation of glycoside derivatives having only one free hydroxyl group generally affords the desired glycosidulose as a major product, often in very good yield; exacting chromatographic separations are seldom required, because the crude products can generally be purified by crystallization or distillation.

In connection with configurational studies of the aldose moiety of the antibiotic novobiocin, Walton *et al.*[331] prepared the exocyclic keto compound methyl 6-deoxy-2,3-*O*-isopropylidene-L-*lyxo*-hexofuranosid-5-ulose, in 73% yield, by oxidation of methyl 2,3-*O*-isopropylidene-L-rhamnofuranoside with chromic acid–pyridine. Burton *et al.*[332] used the same oxidant for the preparation of the anomers of methyl 3,4-*O*-isopropylidene-D-*erythro*-pento-pyranosidulose from the corresponding methyl D-arabinopyranoside derivatives. Baker and Buss[333] used the Pfitzner–Moffatt procedure for the preparation of methyl 3-benzamido-4,6-*O*-benzylidene-3-deoxy-α-D-*arabino*-hexopyranosid-2-ulose and methyl 2-benzamido-4,6-*O*-benzylidene-2-deoxy-α-D-*ribo*-hexopyranosid-3-ulose in yields of ∼90%. Beynon *et al.*[334] suggested the use of ruthenium tetraoxide as an oxidant. They,[334] and Parikh and Jones,[335] synthesized many glycopyranosidulose derivatives that had previously been prepared by chromic acid oxidation; for several of these reactions, better yields were obtained with ruthenium tetraoxide, which may conveniently be generated *in situ* (see also Chapter 24) by converting catalytic amounts of ruthenium dioxide into the tetraoxide with periodate.[335] Glycosyl linkages

---

are reported to be unaffected by this oxidant,[334] including those of glycosyl chlorides.[335a]

Heyns *et al.* obtained 1,6-anhydro-β-D-*xylo*-hexopyranos-3-ulose (**89**)[336] and, later,[336a] six other 1,6-anhydro-β-D-hexopyranosuloses by oxygen–platinum oxidation of seven isomeric 1,6-anhydro-β-D-hexopyranoses. Horton *et al.* reported preparation of 1,6-anhydro-2,3-*O*-isopropylidene-β-D-*lyxo*-hexopyranos-4-ulose[337] (**90**) from a D-*manno* precursor and of 1,6-anhydro-3,4-*O*-isopropylidene-β-D-*lyxo*-hexopyranosulose[337a] (**90a**) from a D-*galacto* precursor in high yield using various oxidation conditions. Methyl anhydro-glycosiduloses containing an α,β-epoxy carbonyl grouping have been prepared and found to be very versatile in the synthesis of deoxy, amino, and branched-chain sugars.[337b,c]

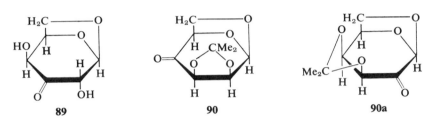

89   90   90a

Chittenden and Guthrie[338] rearranged methyl 4,6-*O*-benzylidene-3-deoxy-3-phenylazoaldopyranosides to afford glycopyranosid-3-uloses. A synthesis of α-D-glucopyranosides, developed by Lemieux *et al.*,[338a] involves preparation of tri-*O*-acetyl-2-deoxy-2-nitroso-α-D-glycopyranosyl chlorides (by addition of nitrosyl chloride to tri-*O*-acetylglycals), conversion into the 2-oximino compounds, deoximation, and liberation of the keto derivatives. The action of hydrazine on vicinal *cis*-disulfonate derivatives[339] is reported to yield deoxyglycopyranosiduloses. Lemieux and Fraser-Reid[340] treated methyl 3,4,6-tri-*O*-acetyl-2-deoxy-2-iodo-α-D-mannopyranoside with silver acetate and bromine in acetic acid, and isolated methyl 3,3,4,6-tetra-*O*-acetyl-2-deoxy-2-iodo-α-D-*arabino*-hexopyranosid-3-ulose in good yield; hydrolysis of the aldehydrol acetate groups produced a 2-deoxy-2-iodohexopyranosid-3-ulose derivative that underwent reductive dehalogenation in the presence of zinc in acetic acid to afford methyl 4,6-di-*O*-acetyl-2-deoxy-α-D-*erythro*-hexopyranosid-3-ulose. Jennings and Jones[341] noted that treatment of methyl 4,6-*O*-ethylidene-α-D-mannopyranoside with sulfuryl chloride afforded a crystalline product that rearranged spontaneously in cold pyridine to afford methyl 2-deoxy-4,6-*O*-ethylidene-α-D-*erythro*-hexopyranosid-3-ulose in 10% overall yield. Lichtenthaler *et al.*[342] have prepared and studied the properties of acylated 4-deoxyhex-3-enopyranosiduloses, and Paulsen *et al.*[343] have reported addition reactions on various enopyranosiduloses.

Some microorganisms effect the oxidation of oligosaccharides to glycosid-3-uloses; this conversion has been reported for maltose, lactose, and their respective bionic acids by Bernaerts and De Ley,[344] and for sucrose by Feingold et al.[345] The 3-keto derivatives of sucrose[346] and of trehalose, maltose, and lactose[347] were obtained crystalline by Fukui et al. by the action of cultures of Agrobacterium tumefaciens. Gabriel and Ashwell[348] prepared tritium-labeled sucrose and α-D-[3-³H]allopyranosyl β-D-fructofuranoside by reduction of α-D-ribo-hexopyranosyl-3-ulose β-D-fructofuranoside (91) with tritiated borohydride; from the action of sucrose phosphorylase on the labeled sucrose they isolated α-D-[3-³H]glucopyranosyl phosphate and thymidine 5'-D-([3-³H]glucopyranosyl pyrophosphate). Aldos-4-ulose derivatives have been postulated as intermediates in the biosynthesis of many sugars (see Chapter 34); thymidine 5'-(6-deoxy-D-xylo-hexopyranosyl-4-ulose pyrophosphate) was isolated by Okazaki et al.[349] as an intermediate in the synthesis of thymidine 5'-(L-rhamnosyl pyrophosphate) from thymidine 5'-(α-D-glucopyranosyl pyrophosphate) by Escherichia coli.

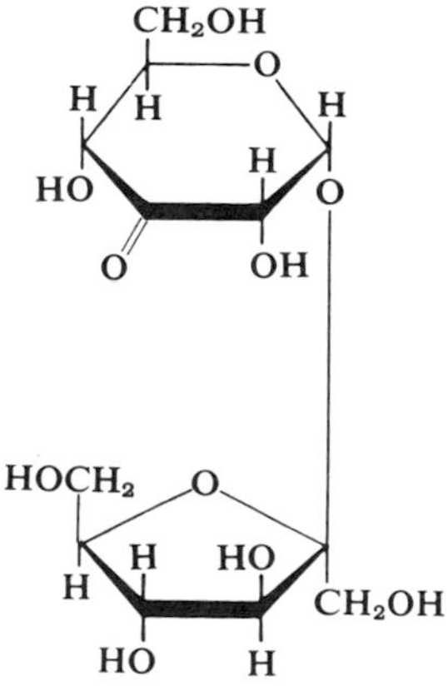

91

Ohle et al. prepared epimeric glycosulose derivatives by equimolar addition of a Grignard reagent to 1,2:3,4-di-O-isopropylidene-α-D-galactopyranuronic acid,[350] and 6-deoxy-1,2-O-isopropylidene-α-D-xylo-hexofuranos-5-ulose by mild hydrolysis of 6-deoxy-1,2:3,5-di-O-isopropylidene-α-D-xylo-hex-5-eno-furanose with acid.[351]

Methyl 6-deoxy-2,3-O-isopropylidene-α-L-lyxo-hexopyranosid-4-ulose is reported[352] to epimerize to the corresponding β-D-ribo derivative in hot, aqueous pyridine; although, in this example, the equilibrium is unfavorable, the approach is promising.

---

b. Properties and Reactions.—The unsubstituted glycopyranosiduloses generally give sharp, round spots on paper chromatograms.[269] This suggests that these derivatives adopt only one form; Collins and Overend, [353] however, reported evidence of equilibria between different forms of some methyl *O*-isopropylideneglycopyranosiduloses in alcoholic or aqueous 1,4-dioxane, as revealed by chromatographic behavior, mutarotation, and changes of the carbonyl absorption in the ultraviolet spectrum. 1,2-*O*-Isopropylidene-α-D-*xylo*-hexofuranos-3-ulose crystallizes as a hemiacetal and mutarotates in solution.[281] There are also examples of other crystalline structures; methyl β-D-*ribo*-hexopyranosid-3-ulose (**87**) and its 4-methyl ether,[269] and the disaccharide[346] **91**, crystallize as free ketones, whereas 1,2:5,6-di-*O*-isopropylidene-α-D-*ribo*-hexofuranos-3-ulose crystallizes as the hydrate.[282,334] Horton and Just[354] isolated the aldol dimer **92** by the action of acetic anhydride and triethylamine on **90** at room temperature. Heyns *et al.*[355] found that unsymmetrical, *spiro*, dimeric hemiacetals are readily formed from 1,6-anhydrohexopyranosuloses having an axially attached hydroxyl group adjacent to the carbonyl group; thus, 1,6-anhydro-β-D-*ribo*-hexopyranosulose undergoes[356] acid-catalyzed conversion into the dimer **93**.

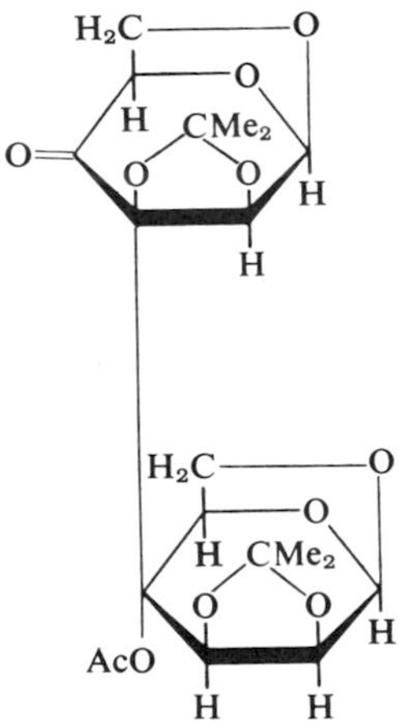

92 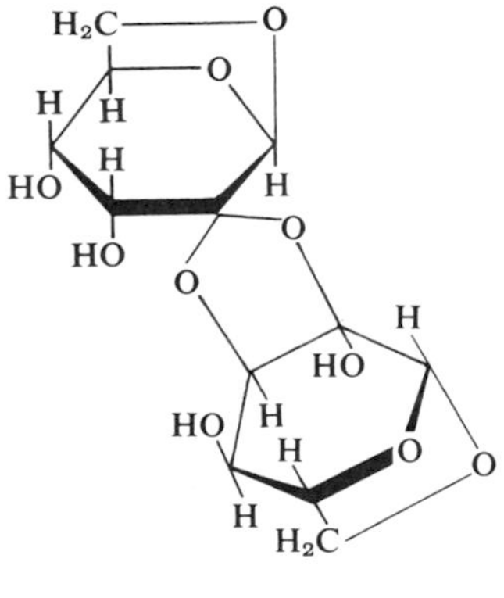   93

Glycosulose derivatives can conveniently be studied by paper[269] and thin-layer[281] chromatography; paper electrophoresis in hydrogensulfite[128,269] has also been useful in the characterization of both glycosulose derivatives and other reducing carbohydrates (see examples in Table I). The last method can also provide structural information, identifying, for example, ring forms and such steric effects as 1,3-diaxial interactions; interaction between the bulky, axial groups at C-1 and the sulfite complex at C-3 (either OH or SO_3^- in the axial orientation) in the α-D-glycopyranosid-3-uloses destabilizes the hydrogensulfite adducts and, therefore, decreases M_v values relative to those of the corresponding β-D anomers. Conversion into the (alkali-stable) O-methyloxime and subsequent per(trimethylsilyl)ation renders methyl glycopyranosiduloses suitable for separation and quantitative analysis by gas–liquid chromatography.[329a]

In general, the aldosuloses readily reduce alkaline copper and silver reagents. The formation of glycosiduloses by oxidation of glycosides with various oxidants is often accompanied by an attack at the glycosidic linkage, resulting in nonhydrolytic cleavage.[269,356a] Also, the ketone groups formed initiate cleavage of carbon–carbon bonds, yielding acids (see Chapter 24). These reactions are very dependent on the conditions; Heyns et al.[357] obtained considerable proportions of dicarboxylic acids and only small yields of glycosiduloses from the catalytic oxidation of benzyl glycopyranosides at pH 7.3.

Reduction of substituted glycosuloses has been valuable in synthetic schemes; identification of the two epimers obtained by reduction of a ketone has been used to locate the position of the ketone group (see Chapter 22). Kashimura et al.[357a] suggested a method for locating linkage positions in polysaccharides, consisting in permethylation, hydrolysis, oxidation of liberated hydroxyl groups, and identification of the carbonyl sugars as (2,4-dinitrophenyl)hydrazones. Steric factors influence various reducing reagents differently; therefore, one epimer can sometimes be an almost exclusive product.[357b] A good example of stereochemistry is found in the reduction of 1,2:5,6-di-O-isopropylidene-α-D-*ribo*-hexofuranos-3-ulose (**94**), which was

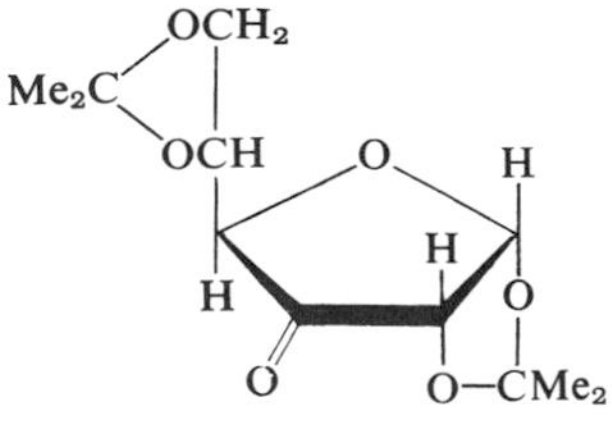

94

isolated in low yield by Theander[282] from the oxidation of 1,2:5,6-di-O-isopropylidene-α-glucofuranose with a chromic acid reagent, by Beynon et al.[334] employing ruthenium tetraoxide, and later by other workers[335,358,359] using conventional oxidants; borohydride converts **94** into 1,2:5,6-di-O-isopropylidene-α-D-allofuranose[282] as virtually the only product of the reaction, affording an inexpensive and convenient[359] route to the rare sugar D-allose. Lindberg and Theander[360] found the catalytic reduction of methyl β-D-*ribo*-hexopyranosid-3-ulose (**87**) to be strongly dependent upon conditions; in acid solution they obtained primarily the 3-deoxy derivative **95**, whereas under neutral conditions they isolated a mixture of the expected epimers—namely, methyl β-D-glucopyranoside (**96**) and methyl β-D-allopyranoside (**97**).

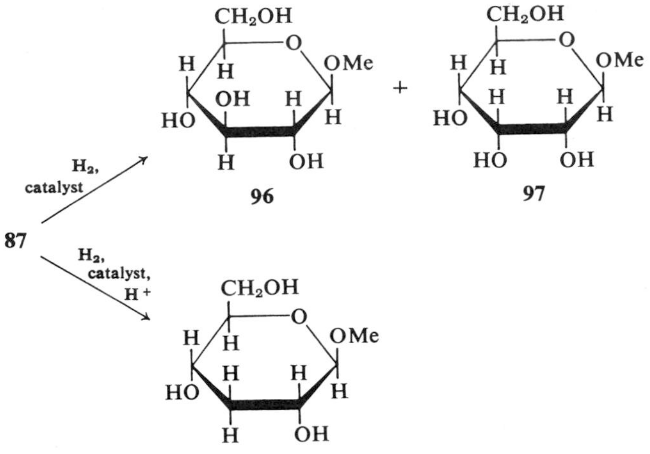

Collins[361] reported studies on *syn-* and *anti-*oximes of aldopyranosiduloses. Reduction of glycopyranosidulose oximes provides a synthetic route to amino sugars, Lindberg and Theander[360] prepared the 3-amino-3-deoxy analogues of **96** and **97** by reduction of the oxime of **87**. Taking advantage of the improvements in yields and stability brought about by protecting groups, Collins and Overend[353,362] prepared methyl 2-amino-2-deoxy-3,4-*O*-isopropylidene-β-L-ribo- and -arabino-pyranoside by reduction of methyl 2,3-*O*-isopropylidene-β-L-*erythro*-pentopyranosidulose oxime; reduction of the corresponding hydrazone was shown to be preparatively inferior. Brimacombe and Cook[363] converted methyl 6-deoxy-α-L-*arabino*-hexopyranosidulose oxime into methyl 2-amino-2,6-dideoxy-α-L-mannopyranoside. The oxime route was used similarly to afford good yields of 2-amino-1,6-anhydro-2-deoxy-β-D-talopyranose and 4-amino-1,6-anhydro-4-deoxy-β-D-talopyranose.[364] In separate, stereoselective reactions, Rosenthal *et al.* prepared methyl 3-*C*-(2-aminoethyl)-4,6-*O*-benzylidene-2-deoxy-α-D-*ribo*-[365] and -arabino-[366] hexopyranoside by hydrogenation of cyanomethylene or cyanomethyl groups introduced into methyl 4,6-*O*-benzylidene-2-deoxy-α-D-*erythro*-hexopyranosid-3-ulose via a Wittig reaction or a modified aldol reaction with acetonitrile, respectively.

Walton *et al.*[331] prepared methyl 2,3-*O*-isopropylidene-5-*C*-methyl-L-*lyxo*-hexofuranoside by addition of a methyl Grignard reagent to the free carbonyl group of the corresponding glycos-5-ulose. Wolfrom and Hanessian[367] synthesized the related, branched glycoside, 3-*O*-benzyl-6-deoxy-1,2-*O*-isopropylidene-5-*C*-methyl-α-D-*xylo*-hexofuranose, from 3-*O*-benzyl-1,2-*O*-isopropylidene-α-D-*xylo*-pentodialdo-1,4-furanose by addition of one equivalent of methylmagnesium iodide, oxidation of the resulting 6-deoxy-L-idofuranose derivative to the corresponding glycos-5-ulose, and addition of a second equivalent of the Grignard reagent. Overend *et al.* reported numerous syntheses of sugars having methyl, hydroxymethyl, and formyl groups attached to carbon atoms of the sugar by the use of Grignard reagents,[332] organolithium compounds,[368] and diazomethane.[369] (See ref. 369a for an early review of branched sugars; see also Chapter 17.) 2-*C*-(Hydroxymethyl)-L-arabinose (L-*hamamelose*) was thus prepared[332] by addition of 2-styrylmagnesium iodide to methyl 3,4-*O*-isopropylidene-β-L-*erythro*-pentopyranosidulose, and subsequent ozonolysis, reduction, and hydrolysis; the (natural) D isomer (**100**) was prepared[369] by the action of diazomethane on the enantiomorph (**98**) of the preceding glyculose to form the epoxide **99**; the anhydro ring of **99** was opened by treatment with aqueous base, the product deacetonated in aqueous acid, and the product hydrolyzed to **100** by the action of a cation-exchange resin in hot water.

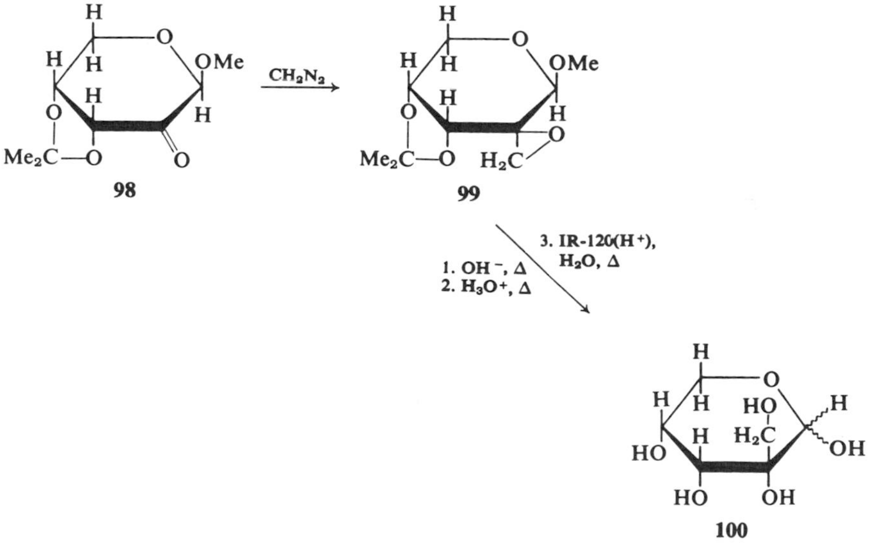

1,3-Dithian-2-yllithium is a superior nucleophile for the introduction of oxygenated, chain-branching features by addition to carbonyl[370] (and epoxide[371]) groups. Paulsen *et al.* reported the addition of this reagent to several glycosuloses, including **98** (affording **100** in good yield after demer-captalation, reduction, and hydrolysis), the 1,6-anhydrohexopyranosuloses **90** and **90a** (affording branched products having the D-*talo* configuration), **94** (to give a product having the D-*allo* configuration), and 5-deoxy-1,3-*O*-isopropylidene-α-L-*erythro*-pentofuranos-3-ulose [affording a protected derivative of streptose (**74**; Section III,A,1) directly]. Methyl 4,6-dideoxy-3-*C*-[1(*S*)-hydroxyethyl]-β-D-*ribo*-hexopyranoside 3,3¹-carbonate (**101**) was synthesized by Paulsen and Redlich[372] according to the scheme illustrated, in which both carbon atoms of the branched chain derive from 2-methyl-1,3-dithian-2-yllithium; of the four epimeric products thus formed, the $3^1(S)$-D-*ribo* isomer (**101**) was shown to be identical with the methyl glycoside of natural aldgarose.

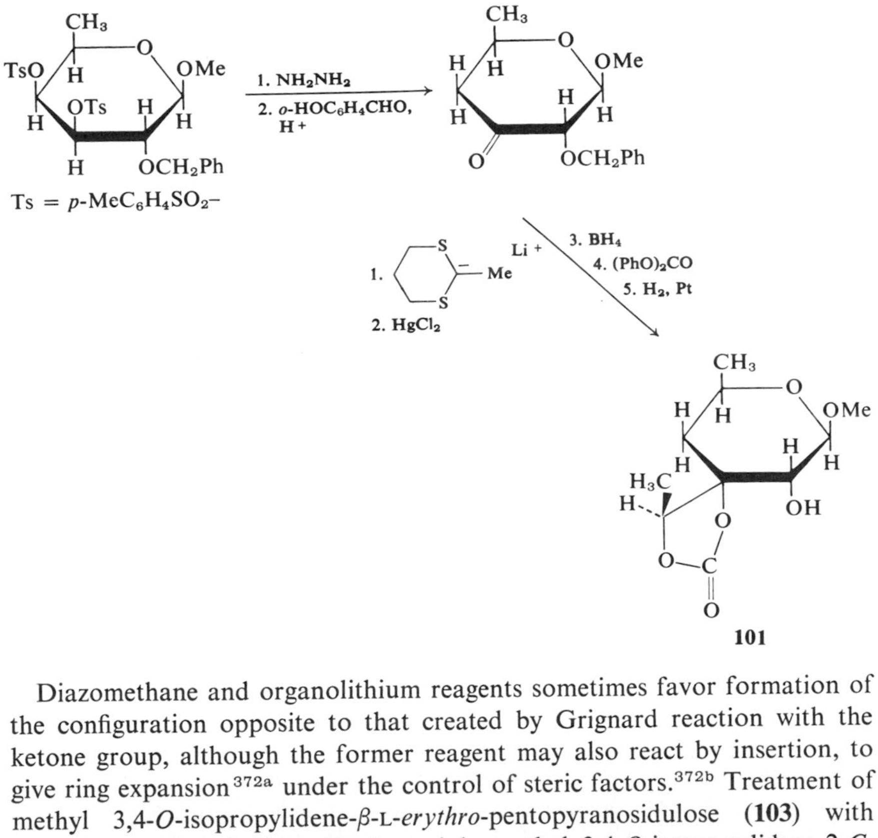

101

Diazomethane and organolithium reagents sometimes favor formation of the configuration opposite to that created by Grignard reaction with the ketone group, although the former reagent may also react by insertion, to give ring expansion[372a] under the control of steric factors.[372b] Treatment of methyl 3,4-*O*-isopropylidene-β-L-*erythro*-pentopyranosidulose (**103**) with methylmagnesium iodide affords mainly methyl 3,4-*O*-isopropylidene-2-*C*-methyl-β-L-arabinopyranoside (**102**), whereas alkylation with methyllithium generates mainly[368] the 2-epimer (**104**); however, **94** is reported to yield a preponderance of the D-*allo* product on addition of either 1,3-dithian-2-yllithium[370] or ethynylmagnesium bromide.[373]

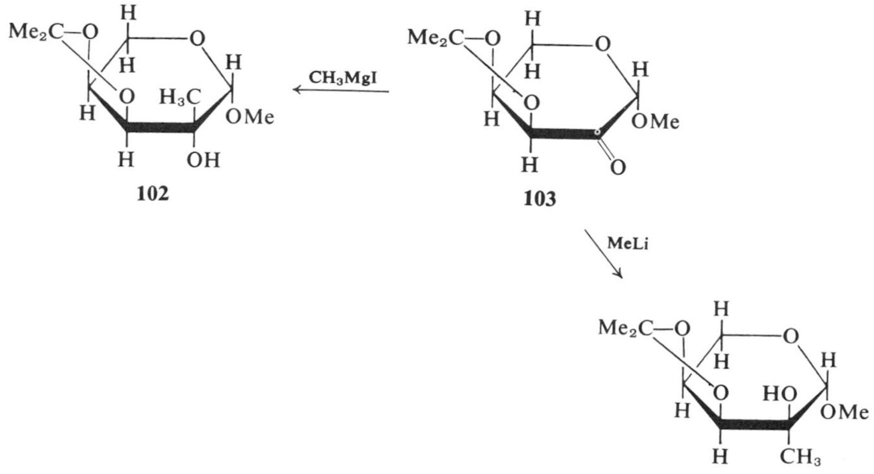

Attachment of a carbon substituent in the opposite configuration at C-3 of **94** was first accomplished by Rosenthal and Nguyen[374] by the action of methyl (dimethylphosphino)acetate on **94** to form an $(E),(Z)$ mixture of the 3-*C*-(methoxycarbonylmethylene) derivative **105** and subsequent catalytic hydrogenation of the double bond from the favored side to form the branched ester (**106**). The preparation of both epimers of a branched aminodeoxy sugar via Grignard[366] and Wittig[365] processes was discussed earlier in this Section.

Use of the Wittig reagents as a means of chain branching (and extension) was exploited by Tronchet *et al.*,[375,376] who reported the use of permanganate to convert 3-*C*-(cyanomethylene)-1,2-*O*-isopropylidene-α-D-*glycero*-tetro-furanose, which had been prepared from the glycos-3-ulose precursor **108** by condensation with (cyanomethylene)triphenylphosphorane, into a 3-*C*-substituted D-threose derivative that underwent reduction by borohydride to afford 3-*C*-(hydroxymethyl)-1,2-*O*-isopropylidene-β-L-threofuranose (**107**); the diastereoisomer, 3-*C*-(hydroxymethyl)-1,2-*O*-isopropylidene-α-D-erythro-furanose (**109**), was earlier obtained from **108** by addition of an unsaturated Grignard reagent, ozonolysis, and reduction.[377] Successful addition of a Grignard reagent to a 2-benzamido-2-deoxy-D-*ribo*-hexopyranosid-3-ulose derivative has been reported.[378] Dimethyl phosphite has been added to glycosulose derivatives (for example, **94**) as a means of preparing *C*-phosphono

---

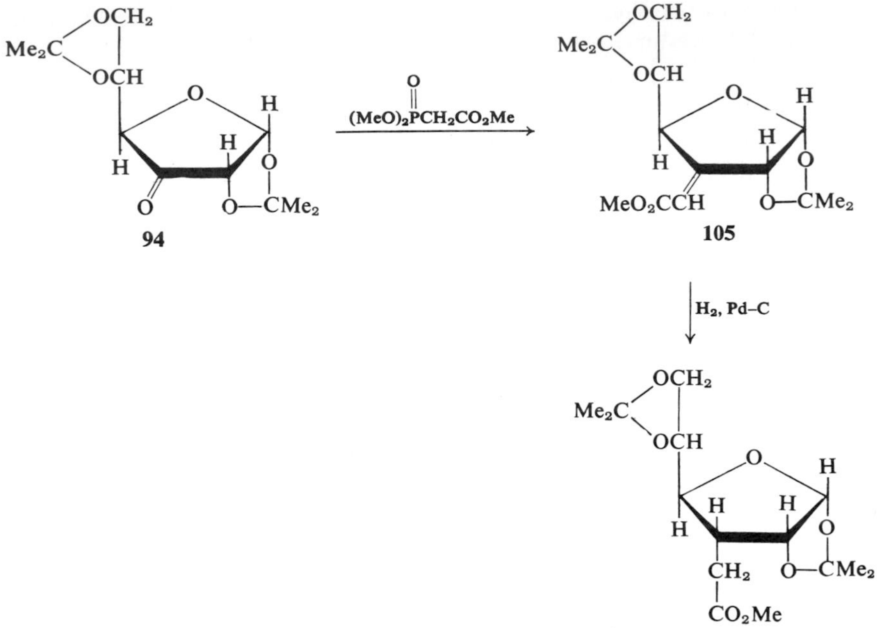

94

$(MeO)_2\overset{O}{\overset{\|}{P}}CH_2CO_2Me$

105

H_2, Pd–C

106

sugars.[379] Methyl 2,3,6-tri-*O*-benzoyl-α-D-*ribo*-hexopyranosid-4-ulose reacts with 1,2-ethanedithiol to form the corresponding 4,4-ethylene dithioacetal.[379a]

Collins *et al.* examined the photochemical reactions of aldosulose derivatives. Photolysis of **103** in *tert*-butyl alcohol results in the expulsion of the

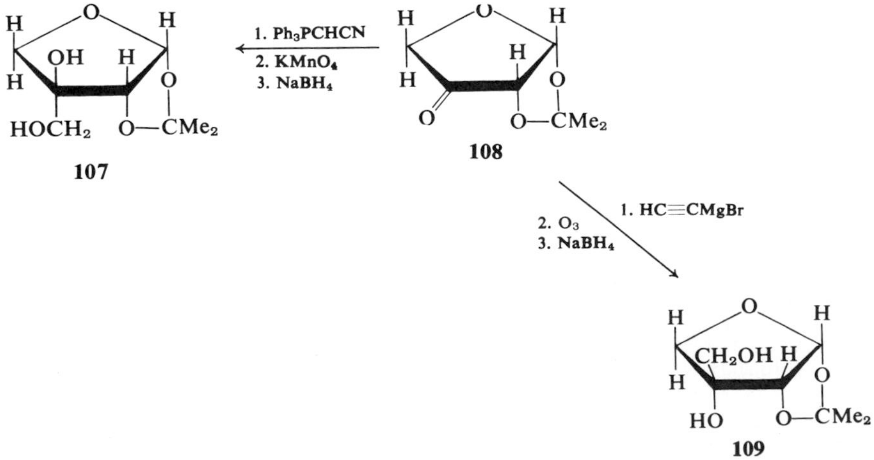

107

1. Ph$_3$PCHCN
2. KMnO$_4$
3. NaBH$_4$

108

2. O$_3$
3. NaBH$_4$

1. HC≡CMgBr

109

elements of formaldehyde by a Norrish type II process, to generate 1,5-anhydro-3,4-O-isopropylidene-L-*erythro*-pentulose.[380] Molecules that do not have an alkoxyl substituent adjacent to the carbonyl group undergo ring contraction by loss of carbon monoxide as a consequence of photolysis; methyl 6-deoxy-2,3-O-isopropylidene-β-D-*ribo*- and -α-L-*lyxo*-hexopyranosid-4-ulose undergo photolytic decarbonylation to a mixture of methyl 5-deoxy-2,3-O-isopropylidene-β-D-ribo- and -α-L-lyxo-furanosides.[381] Photolysis of the α-β-unsaturated ketone ethyl 2,3-dideoxy-α-D-*glycero*-hex-2-enopyranosid-4-ulose in methanol solution results in 1,4 addition of methanol to afford a branched glycoside—namely, ethyl 2,3-dideoxy-2-C-(hydroxymethyl)-α-D-*erythro*-hexopyranosid-4-ulose.[382]

Enolization processes, which occur during several of the photochemical transformations, are also seen in chemically mediated, synthetic transformations. Enol acetals were prepared from 1,2:5,6-di-O-isopropylidene-α-D-*ribo*-hexofuranos-3-ulose[383,384] (**94**) and other di-O-isopropylidene-D-hexofuranos-3-uloses,[384] and from 1,6-anhydro-2,3-O-isopropylidene-β-D-*lyxo*-hexopyranos-4-ulose[354] (**90**); 1,6-anhydro-β-D-hexopyranosuloses having an axially attached hydroxyl group form enediol acetates by enolization processes involving the other hydroxyl group; 1,6-anhydro-β-D-*lyxo*-hexopyranos-4-ulose (**110**) rearranges in acetic anhydride–pyridine to afford 2,4-di-O-acetyl-1,6-anhydro-β-D-*lyxo*-hexopyranos-3-ulose[385] (**111**).

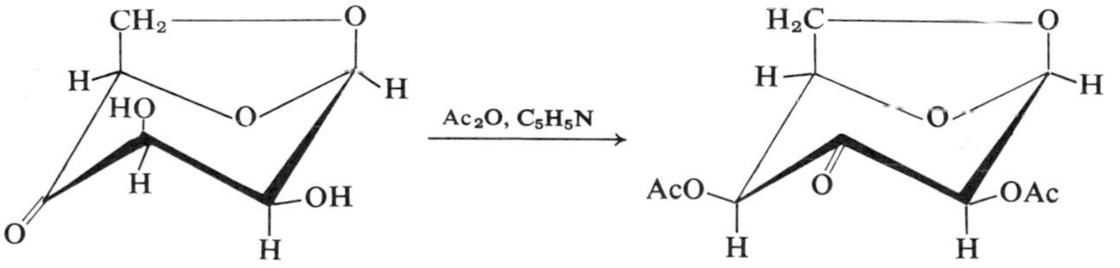

Reactions of enolate anionic species formed by basic catalysis have been used to effect chain branching of **90** by condensation with formaldehyde,[386] and to prepare specifically deuterated analogues of isopropylidenated anhydro sugars for spectroscopic studies[337a,387]; steric constraints within the isopropylidene acetals **90** (and the 3,4-acetalated 2-glyculose, **90a**) prohibit exchange of H-5 (or H-1), because of the bridgehead location of C-1 and C-5, whereas the greater strain inherent in *trans*-fused 6,5-bicyclic rings militates against epimerization at C-3, so that the enolate reactions of **90** and **90a** are positionally and stereochemically specific.

The substituted enediol **113** was postulated[388] as an intermediate in the conversion of methyl 2,3-*O*-isopropylidene-6-*O*-*p*-tolylsulfonyl-α-D-*lyxo*-hexofuranosid-5-ulose (**112**) into the L-*erythro*-hexos-4,5-diulose derivative **114**.

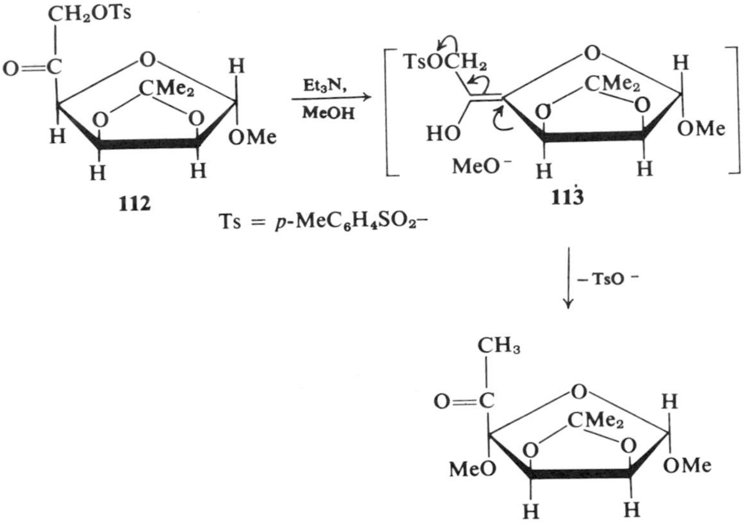

Studies by Theander[389,390] showed that the glycopyranosiduloses **86** and **87**, which were investigated as models for oxidized cellulose and starch, are extremely alkali-labile; for example, the half-life of **87** in $0.02M$ lime-water at 25° is *ca.* 1 minute. Alkaline degradation of **87** (and its α anomer) and **86** afforded mainly acidic end products and took place mainly by initial elimination of the aglycon (see Chapter 4); however, degradation of the 4-methyl ether (**117**), occurred[321] primarily via elimination at C-4. The labile diulose intermediates **116** (which is common for the unsubstituted glycosiduloses,[389]) and **118** (which derives[321] from **117**) were isolated and identified.

The same type of degradation occurs at pH 4 to 7 at elevated temperatures[390] and is probably an important side reaction in many oxidations. It was evident that the hexopyranosidulose derivatives **86** and **87** afforded mainly a 2,3-enediol intermediate (**115**), which, under mild alkaline conditions at lower temperatures, or under neutral or slightly acidic conditions at higher temperatures, isomerized to produce principally **87** and stereoisomers. Heyns *et al.*[357] presented evidence indicating that some pentopyranosid-3-uloses exist mainly in a 3,4-enediol form under similar conditions; they discussed conformational effects on the enolization of glycopyranosid-3-uloses. In connection with the development of specific degradation methods for polysaccharides,[396a] Kenne *et al.* studied the alkaline degradation of methyl ethers of methyl pyranosiduloses; methyl 4-*O*-ethyl-3-*O*-methyl-β-D-*threo*-pento-pyranosid-2-ulose yielded ethanol and methyl 4-deoxy-3-*O*-methylpent-3-eno-pyranosid-2-ulose. (Mild acid hydrolysis released the substituents at C-1 and C-3.)[390b] Treatment of methyl 2,4,6-tri-*O*-methyl-α- and -β-D-*ribo*-hexosid-3-ulose with sodium ethoxide in ethanol yielded a 1,5-anhydro-2,4,6-tri-*O*-methyl-D-hex-1-en-3-ulose and several isomeric ethyl 2,3,6-tri-*O*-methyl-hexopyranosid-3-uloses.[390c]

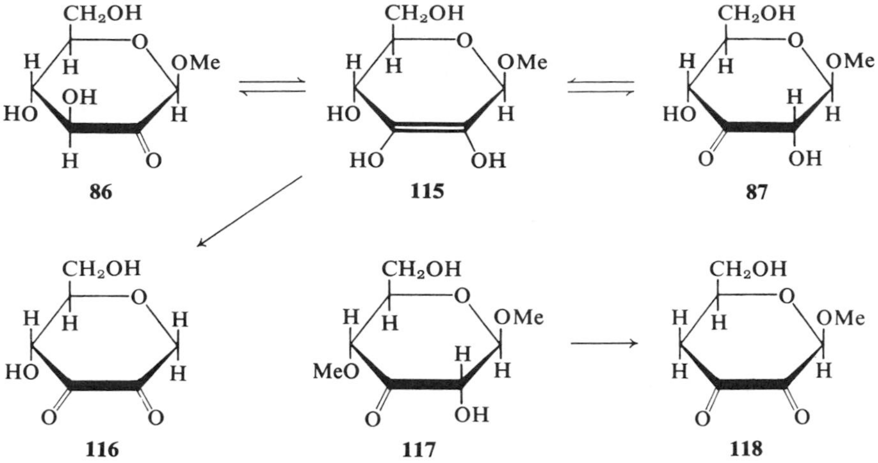

Other hexopyranosidulose derivatives have also been shown to undergo β elimination during their preparation and purification.[334] Rearrangements similar to those discussed for glycosiduloses were suggested by Corbett and Kidd[391] in studies on acetylated 2-hydroxyglycals (see Chapter 19).

On treatment with dilute sulfuric acid, **86** and **87** yield very complex mixtures. The expected D-*arabino*-hexos-2-ulose (**78**) was isolated from acidic hydrolyzates of **86**, whereas D-*ribo*-hexos-3-ulose was not detected among the products formed by acid hydrolysis of **87**; reductic acid (**20**) was, however, one of the main products present in the hydrolyzates of both ketones.[390]

C. GLYCODIULOSES

This group thus far contains only a few members, obtained by synthesis or by microbial oxidation; a few examples have been isolated as intermediates in the degradation of carbohydrates.

Micheel and Horn[392] synthesized crystalline 1,6-dideoxy-D-*threo*-2,5-hexodiulose (**119**) by acid hydrolysis of 1,6-dideoxy-2,4:3,5-di-*O*-methylene-D-*threo*-hexa-1,5-dienitol (**120**); treatment of **120** with peroxybenzoic acid (or lead tetraacetate), and subsequent hydrolysis, afforded the diester **121** (or its 1,2,2,5,5,6-hexa-*O*-acetyl analogue), from which D-*threo*-2,5-hexodiulose **(122)** was obtained by careful saponification. This glycodiulose was later obtained crystalline by Tereda *et al.*[393] by oxidation of D-fructose with many species of *Acetobacter*. X-ray crystallographic analysis of **122** later showed that it crystallizes as an unsymmetrical dimer.[393a]

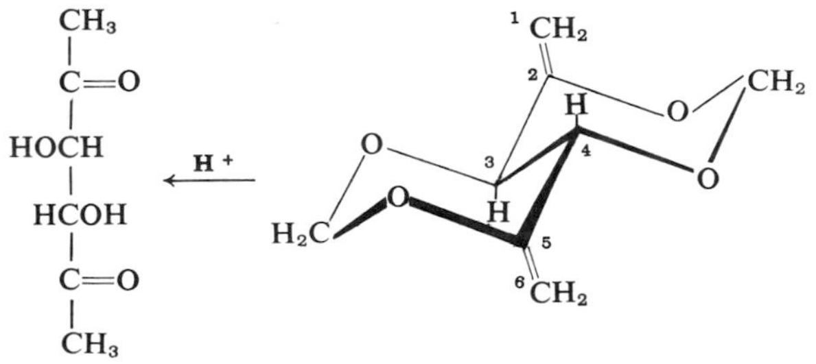

Heyns *et al.* prepared[394] 1,4:3,6-dianhydro-D-*threo*-2,5-hexodiulose by catalytic oxidation of 1,4:3,6-dianhydro-D-mannitol; the same group[395] synthesized 2,7-anhydro-4,5-*O*-isopropylidene-1-*O*-trityl-β-D-*ribo*-2,3-hepto-diulo-2,6-pyranose by the action of ruthenium tetraoxide on the corresponding 2,7-anhydro-β-D-*altro*-heptulopyranose (*sedoheptulosan*) derivative. Pettersson and Theander[286] reported a diketo derivative obtained in low yield from oxidation of methyl 3-*O*-methyl-β-D-glucopyranoside. Collins *et al.*[396] attempted the preparation of methyl 4,6-*O*-benzylidene-α-D-*erythro*-hexo-pyranosid-2,3-diulose, but were able to isolate a stable product only after conversion of the glycopyranosid-2,3-diulose into a quinoxaline derivative by condensation with *o*-phenylenediamine.

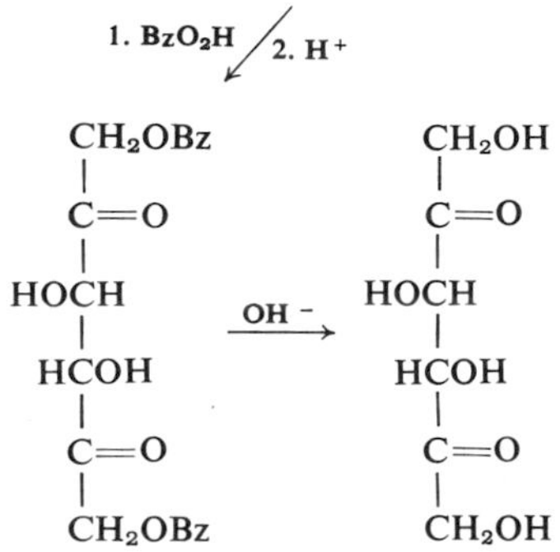

Ishizu *et al.*[234] synthesized 1-deoxy-D-*erythro*-hexo-2,3-diulose **(123)** as a possible intermediate in the production of saccharinic and other acids (see

Section II,D,3–5). Glycodiuloses (116) and glycopyranosid-2,3-diuloses (118) have been documented[389,390] as being intermediates in the alkaline degradation of carbohydrates. A related compound, 1,5-anhydro-6-deoxy-L-*erythro*-2,3-hexodiulose, was obtained by pyrolysis of a cardiac glycoside of *Calotropis procera* by Crout *et al.*[397] 4-Deoxy-D-*glycero*-2,3-hexodiulose (124) was isolated by Machell and Richards[398] and by Whistler and BeMiller[399] as an intermediate in the formation of "D-glucoisosaccharinic" acids in the degradation of 4-*O*-substituted D-glucose derivatives. Harris and Feather[400] discussed the possible role of 123 and 1,5-dideoxyhex-4-eno-2,3-diulose (125) as intermediates in the acid-catalyzed degradation of D-glucose and D-fructose. Methyl β-D-*erythro*-hexopyranosid-2,3-diulose[401] (later also the corresponding 3,4-diulose[402]) and 4-deoxy-D-*glycero*-hexos-2,3-diulose[320b] (126) have been proposed as intermediates in alkaline decompositions of methyl β-D-glucopyranoside (in the presence of oxygen or peroxide) and D-*ribo*-3-hexulose, respectively, to afford acids.

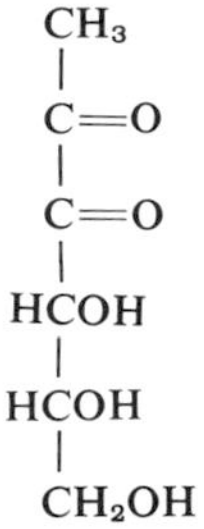

$$
\begin{array}{cccc}
CH_3 & CH_2OH & CH_3 & CHO \\
| & | & | & | \\
C{=}O & C{=}O & C{=}O & C{=}O \\
| & | & | & | \\
C{=}O & C{=}O & C{=}O & C{=}O \\
| & | & \| & | \\
HCOH & CH_2 & COH & CH_2 \\
| & | & \| & | \\
HCOH & HCOH & CH & HCOH \\
| & | & | & | \\
CH_2OH & CH_2OH & CH_2OH & CH_2OH \\
\mathbf{123} & \mathbf{124} & \mathbf{125} & \mathbf{126}
\end{array}
$$

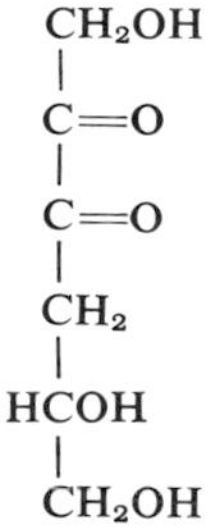

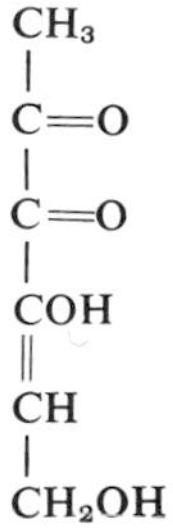

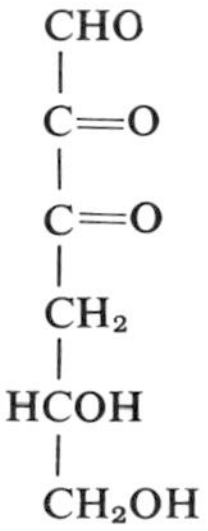

Photolysis of 1,2:4,5-di-*O*-isopropylidene-β-D-*erythro*-2,3-hexodiulo-2,6-pyranose causes expulsion of carbon monoxide, and ring contraction, yielding 1,2:3,4-di-*O*-isopropylidene-β-D-*erythro*-pentofuranos-2-ulose.[402]

REFERENCES

1. J. Staněk, M. Černy, J. Kocourek, and J. Pacák, "The Monosaccharides," Academic Press, New York and London, 1963.
2. J. Staněk, M. Černy, and J. Pacák, "The Oligosaccharides," Academic Press, New York and London, 1965.
3. A. Assarsson, B. Lindberg, and H. Vorbrüggen, *Acta Chem. Scand.*, 13, 1395 (1959).
4. F. J. Prescott, J. K. Shaw, J. P. Bilello, and G. O. Cragwall, *Ind. Eng. Chem.*, 45, 338 (1953).
5. C. L. Mehltretter, *Staerke*, 15, 313 (1963).
6. R. Willstätter and G. Schudel, *Ber.*, 51, 780 (1918).
7. G. M. Kline and S. F. Acree, *Bur. Stand. J. Res.*, 5, 1963 (1930); *Ind. Eng. Chem., Anal. Ed.*, 2, 413 (1930).
8. A. Jeanes and H. S. Isbell, *J. Res. Nat. Bur. Stand.*, 27, 125 (1941).

9. S. Moore and K. P. Link, *J. Biol. Chem.*, **133**, 293 (1940).
10. R. Schaffer and H. S. Isbell, *Methods Carbohyd. Chem.*, **2**, 11 (1963).
11. H. Hlasiwetz, *Ann.*, **119**, 281 (1861).
12. C. S. Hudson and H. S. Isbell, *J. Amer. Chem. Soc.*, **51**, 2225 (1929); H. S. Isbell, *Methods Carbohyd. Chem.*, **2**, 13 (1963).
13. H. S. Isbell, H. L. Frush, and F. J. Bates, *Ind. Eng. Chem.*, **24**, 375 (1932); H. S. Isbell, U.S. Pat. 1,976,731 (1934).
14. H. L. Frush and H. S. Isbell, *Methods Carbohyd. Chem.*, **2**, 14 (1963).
14a. P. S. Fredericks, B. O. Lindgren, and O. Theander, *Cellulose Chem. Technol.*, **4**, 533 (1970).
15. J. Kiss, *Chem. Ind.* (London), 73 (1964).
16. I. M. Morrison and M. B. Perry, *Can. J. Biochem.*, **44**, 1115 (1966).
17. K. Heyns and H. Paulsen, *Advan. Carbohyd. Chem.*, **17**, 169 (1962).
18. G. N. Dorofeenko and E. I. Antsupova, *Nauchn. Zap. Luganskogo Sel'skokhoz. Inst.*, **7**, 208 (1960); *Chem. Abstr.*, **58**, 4633g (1963).
19. D. Müller, *Biochem. Z.*, **199**, 136 (1928); *Enzymologia*, **10**, 40 (1941).
20. J. H. Pazur and K. Kleppe, *Biochemistry*, **3**, 578 (1964).
21. Y. Nishizuka and O. Hayaishi, *J. Biol. Chem.*, **237**, 2721 (1962).
22. N. K. Kochetkov and B. A. Dmitriev, *Tetrahedron*, **21**, 803 (1965); see also, *Methods Carbohyd. Chem.*, **6**, 150 (1972).
23. O. Spengler and A. Pfannenstiel, *Z. Wirtschaftsgruppe Zuckerind.*, **85**, 546 (1935); *Chem. Abstr.*, **30**, 4470 (1936).
24. E. Hardegger, K. Kreis, and H. El Khadem, *Helv. Chim. Acta*, **34**, 2343 (1951).
24a. H. S. Isbell and H. L. Frush, *Carbohyd. Res.*, **28**, 295 (1973); H. S. Isbell, H. L. Frush, and E. T. Martin, *ibid.*, **26**, 287 (1973).
24b. P. Ahlgren, A. Ishizu, I. Szabó, and O. Theander, *Sv. Papperstidn.*, **71**, 355 (1968); O. Samuelson, and L. Thede, *Tappi*, **52**, 99 (1969); O. Samuelson, and L. Stolpe, *ibid.*, **52**, 1709 (1969); H. Kolmodin, and O. Samuelson, *Sv. Papperstidn.*, **76**, 71 (1973).
25. T. Reichstein, A. Grüssner, and W. Bosshard, *Helv. Chim. Acta*, **18**, 602 (1935).
26. J. M. Perel and P. G. Dayton, *J. Org. Chem.*, **25**, 2044 (1960).
27. T. S. Gardner and E. Wenis, *J. Amer. Chem. Soc.*, **73**, 1855 (1951).
28. L. H. Sternbach, U. S. Pat. 2,438,881 (1948); H. Spupsack, Ger. Pat. 1,148,991 (1963); Z. Perina and V. Sapara, Czech. Pat. 107,676 (1963), and others cited in ref. 1.
29. T. Miki, T. Hasegawa, and Y. Sakashi, *J. Vitaminol.*, **6**, 205 (1960); *Chem. Abstr.*, **55**, 18607b (1961).
30. H. J. Bestmann and R. Schmiechen, *Ber.*, **94**, 751 (1961).
31. W. N. Haworth, "The Constitution of Sugars," Arnold, London, 1929, p. 24.
32. F. Shafizadeh, *Advan. Carbohyd. Chem.*, **13**, 10 (1958).
33. R. U. Lemieux, in "Molecular Rearrangements" (P. de Mayo, ed.), Part 2, p. 709, Interscience, New York, 1964.
33a. Y. Pocker and E. Green, *J. Amer. Chem. Soc.*, **95**, 113 (1973); D. Horton and Z. Walaszek, *Abstr. Papers Amer. Chem. Soc. Meeting*, **170**, CARB-10 (1975); C. R. Nelson, *Carbohyd. Res.*, **68**, 55 (1979).
34. M. Abdel-Akher and F. Smith, *J. Amer. Chem. Soc.*, **73**, 5859 (1951).
35. H. S. Isbell and H. L. Frush, *Methods Carbohyd. Chem.*, **2**, 16 (1963).
36. S. A. Barker, E. J. Bourne, R. M. Pinkard, and D. H. Whiffen, *Chem. Ind.* (London), 658 (1958).

37. D. T. Sawyer and J. R. Brannan, *Anal. Chem.*, **38**, 192 (1966).
38. R. J. Ferrier, *J. Chromatogr.*, **9**, 251 (1962).
39. P. H. Abenius, A. Ishizu, B. Lindberg, and O. Theander, *Sv. Papperstidn.*, **70**, 612 (1967).
40. O. Samuelson and R. Simonson, *Sv. Papperstidn.*, **65**, 363 (1962), and later papers by O. Samuelson *et al.*
41. L. A. T. Verhaar and H. G. J. deWilt, *J. Chromatogr.*, **41**, 168 (1969); I. Matsunaga, T. Imanari, and Z. Tanura, *Chem. Pharm. Bull.* (Tokyo), **18**, 2535 (1970).
41a. G. Petersson, *Tetrahedron*, **26**, 3413 (1970).
41b. G. Petersson, O. Samuelson, K. Anjou, and E. von Sydow, *Acta Chem. Scand.*, **21**, 1251 (1967).
42. C. S. Hudson, *J. Amer. Chem. Soc.*, **32**, 338 (1910).
43. C. S. Hudson, *J. Amer. Chem. Soc.*, **40**, 813 (1918).
44. C. S. Hudson, *J. Amer. Chem. Soc.*, **39**, 462 (1917).
45. P. A. Levene and G. M. Meyer, *J. Biol. Chem.*, **31**, 623 (1917).
46. N. K. Richtmyer and C. S. Hudson, *J. Amer. Chem. Soc.*, **64**, 1612 (1942).
47. W. Klyne, in "Determination of Organic Structures by Physical Methods" (E. A. Braude and F. C. Nachod, eds.), Academic Press, New York, 1955, p. 94.
48. T. L. Harris, E. L. Hirst, and C. E. Wood, *J. Chem. Soc.*, 1825 (1934).
49. T. Okuda, S. Harigaya, and A. Koyomoto, *Chem. Pharm. Bull.* (Tokyo), **12**, 504 (1964).
50. E. Fischer, *Ber.*, **23**, 799 (1890).
51. W. N. Haworth and C. W. Long, *J. Chem. Soc.*, 345 (1929).
52. U. B. Larsson, I. Norstedt, and O. Samuelson, *J. Chromatogr.*, **22**, 102 (1966).
53. R. L. Whistler and R. Schweiger, *J. Amer. Chem. Soc.*, **81**, 5190 (1959), and later papers by these authors.
54. O. F. Hedenberg, *J. Amer. Chem. Soc.*, **37**, 345 (1915).
55. J. W. E. Glattfeld and D. Macmillan, *J. Amer. Chem. Soc.*, **56**, 2481 (1934).
56. J. W. W. Morgan and M. L. Wolfrom, *J. Amer. Chem. Soc.*, **78**, 1897 (1956).
57. M. L. Wolfrom, R. B. Bennett, and J. D. Crum, *J. Amer. Chem. Soc.*, **80**, 944 (1958).
58. M. L. Wolfrom and R. B. Bennett, *J. Org. Chem.*, **30**, 458 (1965).
59. A. Thompson and M. L. Wolfrom, *J. Amer. Chem. Soc.*, **68**, 1509 (1946).
60. Z. Wypych, *Rocz. Chem.*, **38**, 1255 (1964); *Chem. Abstr.*, **63**, 3024g (1965).
61. N. K. Richtmyer, *Advan. Carbohyd. Chem.*, **6**, 175 (1951).
62. R. T. Major and E. W. Cook, *J. Amer. Chem. Soc.*, **58**, 2474 (1936); see ref. 75.
63. K. Ladenburg, M. Tishler, J. W. Wellman, and R. D. Babson, *J. Amer. Chem. Soc.*, **66**, 1217 (1944).
64. C. D. Hurd and J. C. Sowden, *J. Amer. Chem. Soc.*, **60**, 235 (1938).
65. M. L. Wolfrom, M. Konigsberg, and D. I. Weisblat, *J. Amer. Chem. Soc.*, **61**, 574 (1939).
66. M. L. Wolfrom and A. Thompson, *Methods Carbohyd. Chem.*, **2**, 21 (1963).
67. R. T. Major and E. W. Cook, *J. Amer. Chem. Soc.*, **58**, 2477 (1936).
68. J. W. E. Glattfeld and B. D. Kribben, *J. Amer. Chem. Soc.*, **61**, 1720 (1939).
69. H. Gross and I. Farkaš, *Ber.*, **93**, 95 (1960).
70. M. L. Wolfrom and A. Thompson, *J. Amer. Chem. Soc.*, **68**, 791 (1946).
71. M. L. Wolfrom, S. W. Waisbrot, and R. L. Brown, *J. Amer. Chem. Soc.*, **64**, 1701 (1942).
72. R. Bognár, I. Farkaš, I. F. Szabó, and G. D. Szabó, *Ber.*, **96**, 689 (1963).
73. I. Farkaš, M. Menyhart, R. Bognár, and H. Gross, *Ber.*, **98**, 1419 (1965).

74. R. Bognár, I. Farkaš, and I. Szabó, *Magy. Kem. Folyoirat*, **71**, 382 (1965); *Chem. Abstr.*, **64**, 8283g (1966).
75. E. W. Cook and R. T. Major, *J. Amer. Chem. Soc.*, **58**, 2410 (1936).
76. M. L. Wolfrom and J. V. Karabinos, *J. Amer. Chem. Soc.*, **68**, 1455 (1946).
77. E. H. Goodyear and W. N. Haworth, *J. Chem. Soc.*, 3136 (1927).
78. J. S. Brimacombe and L. C. N. Tucker, *Carbohyd. Res.*, **2**, 341 (1966).
79. M. Matsui, M. Okada, and M. Ishidate, *Yakugaku Zasshi*, **86**, 110 (1966); *Chem. Abstr.*, **64**, 15967a (1966).
80. H. Zinner, H. Voight, and J. Voight, *Carbohyd. Res.*, **7**, 38 (1968).
81. M. Stacey and S. A. Barker, "Carbohydrates of Living Tissues," Van Nostrand, London, 1962.
81a. J. S. Brimacombe and J. M. Webber, "Mucopolysaccharides," Elsevier, Amsterdam, 1964.
82. M. Stacey and S. A. Barker, "Polysaccharides of Micro-organisms," Oxford University Press, London, 1960.
83. F. Smith and R. Montgomery, "The Chemistry of Plant Gums and Mucilages," Reinhold, New York, 1959.
83a. G. O. Aspinall, *Advan. Carbohyd. Chem. Biochem.*, **24**, 333 (1969).
84. G. J. Dutton, "Glucuronic Acid," Academic Press, New York and London, 1966.
85. R. S. Teague, *Advan. Carbohyd. Chem.*, **9**, 185 (1954); D. Keglević, *Advan. Carbohyd. Chem. Biochem.*, **36**, 57 (1979).
86. M. Jaffé, *Z. Physiol. Chem.*, **2**, 47 (1878).
86a. B. Lindberg, B. Lindqvist, and J. Lönngren, *Carbohyd. Res.*, **49**, 411 (1976).
87. B. Larsen, A. Haug, and T. J. Painter, *Acta Chem. Scand.*, **20**, 219 (1966).
88. W. R. Clark, J. McLaughlin, and M. E. Webster, *J. Biol. Chem.*, **230**, 81 (1958).
89. K. Heyns, G. Kiessling, W. Lindenberg, H. Paulsen, and M. E. Webster, *Ber.*, **92**, 2435 (1959).
90. H. R. Perkins, *Biochem. J.*, **86**, 475 (1963).
91. S. Hanessian and T. H. Haskell, *J. Biol. Chem.*, **239**, 2758 (1964).
91a. K. Isono, K. Asahi, and S. Suzuki, *J. Amer. Chem. Soc.*, **91**, 7490 (1969).
92. E. Fischer, *Ber.*, **22**, 2204 (1889).
93. C. L. Mehltretter, *Advan. Carbohyd. Chem.*, **8**, 231 (1953).
94. E. A. Peterman, *Lancet*, **67**, 451 (1947).
95. C. L. Mehltretter, B. H. Alexander, R. L. Mellies, and C. E. Rist, *J. Amer. Chem. Soc.*, **73**, 2424 (1951); C. L. Mehltretter, U.S. Pat. 2,559,652 (1951).
95a. O. Theander, Swed. Pat. 325,409 (1970); J. Bakke and O. Theander, *Chem. Commun.*, 175 (1971).
96. C. Niemann and K. P. Link, *J. Biol. Chem.*, **104**, 195 (1934).
97. S. Akiya and T. Watanabe, *Yakugaku Zasshi*, **64**, 37 (1944).
98. M. Bergmann, *Ber.*, **54**, 1362 (1921).
99. P. A. J. Gorin and A. S. Perlin, *Can. J. Chem.*, **34**, 693 (1956).
100. R. K. Hulyalkar and M. B. Perry, *Can. J. Chem.*, **43**, 3241 (1965).
100a. B. Carlsson, O. Samuelson, T. Popoff, and O. Theander, *Acta Chem. Scand.*, **23**, 261 (1969); B. Carlsson and O. Samuelson, *ibid.*, **23**, 318 (1969); K. Larsson and O. Samuelson, *Carbohyd. Res.*, **31**, 81 (1973).
100b. D. Horton, A. E. Luetzow, and J. C. Wease, *Carbohyd. Res.*, **8**, 366 (1968).
100c. D. C. Baker and D. Horton, *Carbohyd. Res.*, **21**, 393 (1972).
100d. D. Horton, A. E. Luetzow, and O. Theander, *Carbohyd. Res.*, **26**, 1 (1973).
101. D. Abbott and H. Weigel, *J. Chem. Soc.*, 5157 (1965).
102. S. Takanashi, Y. Hirasaka, M. Kawada, and M. Ishidate, *J. Amer. Chem. Soc.*, **84**, 3029 (1962).

103. H. M. Flowers and R. W. Jeanloz, *Biochemistry*, **3**, 123 (1964).
104. M. Heidelberger and W. F. Goebel, *J. Biol. Chem.*, **74**, 613 (1927); W. F. Goebel, *ibid.*, **110**, 391 (1935).
105. G. Jayme and W. Demmig, *Ber.*, **93**, 356 (1960).
106. B. Lindberg and L. Selleby, *Acta Chem. Scand.*, **14**, 1051 (1960).
107. Y. Hirasaka, *Yakugaku Zasshi*, **83**, 960 (1963).
108. G. G. S. Dutton and K. N. Slessor, *Can. J. Chem.*, **42**, 1110 (1964).
109. C. T. Bishop, *Can. J. Chem.*, **31**, 134 (1953).
110. S. A. Barker, A. Gómez Sánchez, and M. Stacey, *J. Chem. Soc.*, 3264 (1959).
111. I. Johansson, B. Lindberg, and O. Theander, *Acta Chem. Scand.*, **17**, 2019 (1963).
112. Y. Hirasaka and I. Matsunaga, *Chem. Pharm. Bull.* (Tokyo), **13**, 176 (1965).
113. G. Jayme and K. Kringstad, *Ber.*, **93**, 2263 (1960).
114. H. Weidmann and H. K. Zimmerman, Jr., *Ann.*, **639**, 198 (1961).
115. W. G. Overend, F. Shafizadeh, M. Stacey, and G. Vaughan, *J. Chem. Soc.*, 3633 (1954).
115a. A. Rosenthal and J. N. C. Whyte, *Can. J. Chem.*, **46**, 2239 (1968).
116. R. Grewe and G. Rockstroh, *Ber.*, **86**, 536 (1953).
117. V. Prey and O. Szaboles, *Monatsh. Chem.*, **89**, 350 (1958).
118. H. Paulsen and D. Stoye, *Ber.*, **99**, 908 (1966).
119. R. J. Ferrier, *Advan. Carbohyd. Chem.*, **20**, 67 (1965).
120. S. Hasegawa and C. W. Nagel, *J. Biol. Chem.*, **237**, 619 (1962).
121. J. Preiss and G. Ashwell, *J. Biol. Chem.*, **238**, 1571 (1963).
121a. A. Linkder, P. Hoffman, K. Meyer, P. Sampson, and E. D. Korn, *J. Biol. Chem.*, **235**, 3061 (1960).
122. H. Neukom and H. Deuel, *Chem. Ind.* (London), 683 (1958).
122a. J. Kiss, *Advan. Carbohyd. Chem. Biochem.*, **29**, 229 (1974).
123. P. Heim and H. Neukom, *Helv. Chim. Acta*, **45**, 1735 (1962).
123a. P. Kováč, J. Hirsch, and V. Kováčik, *Carbohyd. Res.*, **32**, 360 (1974).
123b. W. Timpe, K. Dax, N. Wolf, and H. Weidmann, *Carbohyd. Res.*, **39**, 53 (1975).
124. M. Ishidate, Z. Tamura, and T. Kinoshita, *Chem. Pharm. Bull.* (Tokyo), **10**, 1258 (1962); **13**, 99 (1965).
125. H. Weidmann and G. Olbrich, *Tetrahedron Lett.*, 725 (1965).
126. W. Mackie and A. S. Perlin, *Can. J. Chem.*, **43**, 2921 (1965).
126a. W. P. Blackstock, C. C. Kuenzle, and C. H. Eugster, *Helv. Chim. Acta*, **57**, 1003 (1974).
127. H. Zinner, W. Thielebeule, and G. Rembarz, *Ber.*, **91**, 1006 (1958).
127a. L. J. DeLucas, G. L. Gartland, and C. E. Bugg, *Carbohyd. Res.*, **62**, 213 (1978).
128. O. Theander, *Acta Chem. Scand.*, **11**, 717 (1957).
129. A. Haug and B. Larsen, *Acta Chem. Scand.*, **15**, 1395 (1961).
130. B. Carlsson, T. Isaksson, and O. Samuelson, *Anal. Chim. Acta*, **43**, 47 (1968).
131. M. B. Perry and R. K. Hulyalkar, *Can. J. Biochem.*, **43**, 573 (1965).
131a. G. Petersson, *Carbohyd. Res.*, **33**, 47 (1974).
131b. V. Kováčik and Š. Bauer, *Collect. Czech. Chem. Commun.*, **34**, 326 (1969); V. Kováčik, Š. Bauer, and P. Šipoš, *ibid.*, **34**, 2409 (1969).
131c. E. Sjöström, S. Juslin, and E. Seppälä, *Acta Chem. Scand.*, **23**, 3610 (1969).
131d. H. Björndal and B. Lindberg, *Carbohyd. Res.*, **12**, 29 (1970).
131e. K. W. Talmadge, K. Keegstra, W. D. Bauer, and P. Albersheim, *Plant Physiol.*, **51**, 158 (1973).
131f. B. Lindberg, J. Lönngren, and J. K. Thomson, *Carbohyd. Res.*, **28**, 351 (1973).
132. K. U. Lefèvre and B. Tollens, *Ber.*, **40**, 4513 (1907).

133. R. L. Whistler, A. R. Martin, and M. Harris, *J. Res. Nat. Bur. Stand.*, **24**, 13 (1940); R. L. Whistler and M. S. Feather, *Methods Carbohyd. Chem.*, **1**, 404 (1962).
134. A. Johansson, B. Lindberg, and O. Theander, *Sv. Papperstidn.*, **57**, 41 (1954).
134a. M. Bylund and A. Donetzhuber, *Sv. Papperstidn.*, **71**, 505 (1968).
135. E. Stutz and H. Deuel, *Helv. Chim. Acta*, **41**, 1722 (1958) and preceding papers.
136. D. M. W. Anderson and S. Garbutt, *J. Chem. Soc.*, 3204 (1963) and preceding papers; see also M. A. Madson and M. S. Feather, *Carbohyd. Res.*, **70**, 307 (1979).
137. H. S. Isbell, *J. Res. Nat. Bur. Stand.*, **32**, 45 (1944).
138. V. P. Kiseleva, A. A. Konkin, and Z. A. Rogovin, *Zh. Prikl. Khim.*, **27**, 1133 (1954).
139. E. Stutz and H. Deuel, *Helv. Chim. Acta*, **39**, 2126 (1956).
140. T. Reichstein and R.·Oppenauer, *Helv. Chim. Acta*, **16**, 988 (1933); **17**, 390 (1934).
140a. M. S. Feather and J. F. Harris, *Advan. Carbohyd. Chem. Biochem.*, **28**, 161 (1973).
141. M. S. Feather and J. F. Harris, *J. Org. Chem.*, **31**, 4018 (1966).
142. G. Zweifel and H. Deuel, *Helv. Chim. Acta*, **39**, 662 (1956).
142a. K. D. Philips, J. Žemlička, and J. P. Horowitz, *Carbohyd. Res.*, **30**, 281 (1974).
143. I. R. Siddiqui and C. B. Purves, *Can. J. Chem.*, **41**, 382 (1963).
144. E. R. Nelson and P. F. Nelson, *Aust. J. Chem.*, **21**, 2323 (1968).
144a. B. Carlsson, O. Samuelson, T. Popoff, and O. Theander, *Acta Chem. Scand.*, **23**, 261 (1969).
144b. F. G. Fischer and H. Schmidt, *Ber.*, **92**, 2184 (1959).
144c. B. Carlsson and O. Samuelson, *Acta Chem. Scand.*, **23**, 318 (1969).
144d. P. F. Nelson and O. Samuelson, *Sv. Papperstidn.*, **71**, 325 (1968).
144e. B. Carlsson and O. Samuelson, *Carbohyd. Res.*, **11**, 347 (1969).
145. M. Ishidate, M. Kimura, and M. Kawata, *Chem. Pharm. Bull.* (Tokyo), **11**, 1083 (1963).
145a. K. Aso, *J. Agr. Chem. Soc. Japan*, **10**, 1189 (1934); **11**, 1083 (1936).
146. T. Popoff and O. Theander, *Carbohyd. Res.*, **22**, 135 (1972).
146a. T. Popoff and O. Theander, *Acta Chem. Scand.*, Ser. B, **30**, 705 (1976).
146b. T. Popoff and O. Theander, *Acta Chem. Scand.*, Ser. B, **30**, 397 (1976).
147. E. F. Jansen and R. Jang, *J. Amer. Chem. Soc.*, **68**, 1475 (1946).
148. G. N. Bollenback, J. W. Long, D. G. Benjamin, and J. A. Lindquist, *J. Amer. Chem. Soc.*, **77**, 3310 (1955).
149. Y. Hirasaka and K. Umemoto, *Yakugaku Zasshi*, **82**, 1676 (1962).
149a. S. Morgenlie, *Acta Chem. Scand.*, **27**, 2217 (1973).
150. W. F. Goebel and F. H. Babers, *J. Biol. Chem.*, **100**, 743 (1933).
151. K. C. Tsou and A. M. Seligman, *J. Amer. Chem. Soc.*, **74**, 5605 (1952).
152. W. F. Goebel and F. H. Babers, *J. Biol. Chem.*, **111**, 347 (1935).
153. L. N. Owen, S. Peat, and W. J. G. Jones, *J. Chem. Soc.*, 339 (1941).
154. E. M. Osman, K. C. Hobbs, and W. E. Walston, *J. Amer. Chem. Soc.*, **73**, 2726 (1951).
154a. K. Larsson and G. Petersson, *Carbohyd. Res.*, **34**, 323 (1974).
155. J. Conchie, G. A. Levvy, and C. A. Marsh, *Advan. Carbohyd. Chem.*, **12**, 157 (1957).
156. M. L. Wolfrom and P. McWain, *J. Org. Chem.*, **30**, 1099 (1965).
157. S. C. McKee and E. E. Dickey, *J. Org. Chem.*, **28**, 1561 (1963).
158. T. E. Timell, W. Enterman, F. Spencer, and E. J. Soltes, *Can. J. Chem.*, **43**, 2296 (1965).
159. B. Capon and B. C. Ghosh, *Chem. Commun.*, 586 (1965).
160. O. Smidsrød, A. Haug, and B. Larsen, *Acta Chem. Scand.*, **20**, 1027 (1966).
161. G. O. Aspinall, *Advan. Carbohyd. Chem.*, **9**, 131 (1954).
162. A. Wacek, F. Leitinger, and P. Hochbahn, *Monatsh. Chem.*, **90**, 562 (1959).

163. C. F. Cross, E. J. Bevan, and C. Smith, *J. Chem. Soc.*, **73**, 459 (1898).
164. A. Kekulé and R. Anschütz, *Ber.*, **13**, 2150 (1880).
165. R. H. Zienius and C. B. Purves, *Can. J. Chem.*, **37**, 1820 (1959).
166. A. F. Holleman, *Org. Syn. Coll. Vol.*, **1**, 497 (1941).
167. W. Marckwald, *Ber.*, **29**, 42 (1896).
168. N. K. Richtmyer, *Methods Carbohyd. Chem.*, **1**, 167 (1962).
169. W. T. Haskins and C. S. Hudson, *J. Amer. Chem. Soc.*, **61**, 1266 (1939).
170. B. A. Lewis, F. Smith, and A. M. Stephen, *Methods Carbohyd. Chem.*, **1**, 469 (1962).
171. W. G. Overend, F. Shafizadeh, M. Stacey, and G. Vaughan, *J. Chem. Soc.*, 3633 (1954).
172. M. K. Gorter, *Rec. Trav. Chim.* (Pays-Bas), **31**, 281 (1912).
173. R. J. Bose, T. L. Hullar, B. A. Lewis, and F. Smith, *J. Org. Chem.*, **26**, 1300 (1961).
174. G. A. Levvy, *Biochem. J.*, **52**, 464 (1952); G. A. Levvy and J. Conchie, in ref. 84, p. 333.
175. E. Boyland and D. C. Williams, *Biochem. J.*, **64**, 578 (1956).
175a. J. M. H. Dirkx, H. S. van der Baan, and J. M. A. J. J. van den Broek, *Carbohyd. Res.*, **59**, 63 (1977).
176. See F. Smith, *Advan. Carbohyd. Chem.*, **20**, 101 (1946).
177. J. W. W. Morgan and M. L. Wolfrom, *J. Amer. Chem. Soc.*, **78**, 1897 (1956).
178. L. Majs, *Zh. Obshch. Khim.*, **31**, 2635 (1961), and preceding papers.
179. T. P. Bird, W. A. P. Black, E. T. Dewar, and H. W. T. Rintoul, *J. Chem. Soc.*, 4512 (1964).
180. R. J. Wood and A. C. Neish, *Can. J. Chem.*, **31**, 471 (1953).
181. L. Löwendahl, G. Petersson, and O. Samuelson, *Carbohyd. Res.*, **43**, 355 (1975).
181a. R. L. Whistler and G. N. Richards, *J. Amer. Chem. Soc.*, **80**, 4888 (1958).
181b. R. L. Whistler and J. N. BeMiller, *J. Amer. Chem. Soc.*, **82**, 457 (1960).
182. See H. J. Blumenthal and T. Jepson, *Biochem. Biophys. Res. Commun.*, **17**, 282 (1964).
182a. D. Detert and B. Lindberg, *Acta Chem. Scand.*, **23**, 690 (1969).
183. J. C. Sowden, *Advan. Carbohyd. Chem.*, **12**, 35 (1957).
184. J. D. Crum, *Advan. Carbohyd. Chem.*, **13**, 169 (1958).
185. R. L. Whistler and J. N. BeMiller, *Advan. Carbohyd. Chem.*, **13**, 289 (1958).
186. R. L. Whistler and J. N. BeMiller, *Methods Carbohyd. Chem.*, **2**, 477, 483, 484 (1963); W. M. Corbett, *ibid.*, **2**, 480 (1963).
187. L. B. Lockwood, *Methods Carbohyd. Chem.*, **2**, 51 (1963).
187a. A. F. Cirelli and R. M. de Lederkremer, *Chem. Ind.* (London), 1139 (1971).
188. M. Matsui, M. Uchiyama, and C.-E. Liau, *Agr. Biol. Chem.* (Tokyo), **27**, 180 (1963).
189. J. P. Hart and M. R. Everett, *J. Amer. Chem. Soc.*, **61**, 1822 (1939).
190. A. Dyfverman, B. Lindberg, and D. Wood, *Acta Chem. Scand.*, **5**, 253 (1951).
191. T. Reichstein and W. Bosshard, *Helv. Chim. Acta*, **17**, 753 (1934).
191a. S. J. Angyal and K. James, *Aust. J. Chem.*, **23**, 1209 (1970).
192. J. Liebster, M. Kulhánek, and M. Tadra, *Chem. Listy*, **47**, 1075 (1953); *Chem. Abstr.*, **48**, 4044a (1954).
193. H. Katznelson, S. W. Tanenbaum, and E. L. Tatum, *J. Biol. Chem.*, **204**, 43 (1953).
194. K. Kondo, M. Ameyama, and T. Yamaguchi, *J. Agr. Chem. Soc. Japan*, **30**, 419 (1956).
195. K. Aida, M. Fujii, and T. Asai, *Proc. Japan Acad.*, **32**, 595 (1956).
195a. K. Imada, K. Inoue, and M. Sato, *Carbohyd. Res.*, **34**, Cl (1974).
195b. B. A. Dmitriev and L. V. Backinovskiǐ, *Usp. Biol. Khim.*, **9**, 182 (1968); *Chem. Abstr.*, **70**, 78261a (1969).

196. J. MacGee and M. Doudoroff, *J. Biol. Chem.*, **210**, 617 (1954).
197. N. J. Palleroni and M. Doudoroff, *J. Biol. Chem.*, **223**, 499 (1956).
198. J. de Ley and M. Doudoroff, *J. Biol. Chem.*, **227**, 745 (1957).
198a. D. Claus, *Biochem. Biophys. Res. Commun.*, **20**, 745 (1965).
198b. B. Lindberg, K. Samuelsson, and W. Nimmich, *Carbohyd. Res.*, **30**, 63 (1973).
199. A. Weissbach and J. Hurwitz, *J. Biol. Chem.*, **234**, 705 (1959).
200. E. C. Heath and M. A. Ghalambor, *Biochem. Biophys. Res. Commun.*, **10**, 340 (1963).
200a. W. A. Volk, N. L. Salomonsky, and D. Hunt, *J. Biol. Chem.*, **247**, 3881 (1972).
201. D. H. Levin and E. Racker, *J. Biol. Chem.*, **234**, 2532 (1959).
202. R. Kuhn, D. Weiser, and H. Fischer, *Ann.*, **628**, 207 (1959).
203. G. B. Paerels, *Rec. Trav. Chim.* (Pays-Bas), **80**, 985 (1961).
204. G. B. Paerels and H. W. Geluk, *Nature*, **197**, 379 (1963).
205. M. Adlersberg and D. B. Sprinson, *Biochemistry*, **3**, 1855 (1964).
205a. B. A. Dmitriev, L. V. Backinovskiĭ, and N. K. Kochetkov, *Dokl. Akad. Nauk SSSR*, **193**, 1304 (1970).
205b. D. Charon and L. Szabó, *Eur. J. Biochem.*, **29**, 184 (1972).
205c. D. Portsmouth, *Carbohyd. Res.*, **8**, 193 (1968); C. Hershberger and S. B. Binkley, *J. Biol. Chem.*, **243**, 1578 (1968).
205d. D. Charon and L. Szabó, *Carbohyd. Res.*, **34**, 271 (1974).
205e. L. Warren, *J. Biol. Chem.*, **234**, 1971 (1959).
206. J. Preiss and G. Ashwell, *J. Biol. Chem.*, **238**, 1577 (1963).
207. K. Okamoto, C. Hatanaka, J. Ozawa, *Agr. Biol. Chem.* (Tokyo), **27**, 596 (1963); *Chem. Abstr.*, **59**, 15626 (1963).
207a. Yu. A. Zhdanov, Yu. E. Alexseev, and Ch. A. Khurdanov, *Carbohyd. Res.*, **14**, 422 (1970).
207b. B. A. Dmitriev, N. E. Bairamova, L. V. Backinovskiĭ, and N. K. Kochetkov, *Dokl. Akad. Nauk SSSR*, **73**, 350 (1967).
208. T. Kurata and P. Sakurai, *Agr. Biol. Chem.* (Tokyo), **31**, 177 (1967).
209. E. Hägglund and T. Johnson, *Sv. Kem. Tidskr.*, **41**, 8, 55 (1929); E. Hägglund, *Ber.*, **62**, 437 (1929).
210. E. Hägglund, T. Johnson, and H. Urban, *Ber.*, **63**, 1387 (1930).
211. E. Adler, *Sv. Papperstidn.*, **49**, 339 (1946).
212. B. Lindberg and O. Theander, *Sv. Papperstidn.*, **65**, 509 (1962).
213. D. L. Ingles, *Aust. J. Chem.*, **15**, 342 (1962).
214. B. Lindberg, J. Tanaka, and O. Theander, *Acta Chem. Scand.*, **18**, 1164 (1964).
215. E. F. L. J. Anet and D. L. Ingles, *Chem. Ind.* (London), 1319 (1964).
216. S. Yllner, *Acta Chem. Scand.*, **10**, 1251 (1956).
217. R. H. Cordingly, *Tappi*, **42**, 654 (1959).
218. H.-L. Hardell and O. Theander, *Acta Chem. Scand.*, **25**, 877 (1971).
218a. O. Theander, unpublished results.
219. W. G. Overend and M. Stacey, *Advan. Carbohyd. Chem.*, **8**, 45 (1953).
220. W. F. Goebel, *J. Biol. Chem.*, **72**, 801 (1927).
221. W. M. Corbett, *Methods Carbohyd. Chem.*, **2**, 18 (1963).
222. M. L. Wolfrom and N. E. Franks, *J. Org. Chem.*, **29**, 3645 (1964).
223. R. Grewe and H. Pachaly, *Ber.*, **87**, 46 (1954) (compare ref. 116).
224. B. Alfredsson and O. Samuelson, *Sv. Papperstidn.*, **65**, 690 (1962).
225. H. B. Wood, Jr., and H. G. Fletcher, Jr., *J. Org. Chem.*, **26**, 1969 (1961).
226. J. U. Nef, *Ann.*, **376**, 1 (1910).
227. H. W. Diehl and H. G. Fletcher, Jr., *Arch. Biochem. Biophys.*, **78**, 386 (1958);

Chem. Ind. (London), 1087 (1958); G. N. Richards, *Methods Carbohyd. Chem.*, **1**, 180 (1962).

228. S. Lewak and L. Szabó, *J. Chem. Soc.*, 3975 (1963).
229. R. Lukeš, J. Jarý, and J. Němec, *Collect. Czech. Chem. Commun.*, **27**, 735 (1962), and preceding papers.
230. H. Kiliani, *Ann.*, **218**, 361 (1883).
231. J. C. Sowden and D. R. Strobach, *J. Amer. Chem. Soc.*, **82**, 3707 (1960).
232. A. B. Foster, T. D. Inch, J. Lehmann, and J. M. Webber, *J. Chem. Soc.*, 948 (1964).
233. A. A. J. Feast, B. Lindberg, and O. Theander, *Acta Chem. Scand.*, **19**, 1127 (1965).
234. A. Ishizu, B. Lindberg, and O. Theander, *Carbohyd. Res.*, **5**, 329 (1967).
235. A. Ishizu, B. Lindberg, and O. Theander, *Acta Chem. Scand.*, **21**, 424 (1967).
235a. A. Ishizu, K. Yoshida, and N. Yamazaki, *Carbohyd. Res.*, **23**, 23 (1972).
236. M. von Glehn, P. Kirkegaard, R. Norrestam, O. Rönnquist, and P. E. Werner, *Chem. Commun.*, 291 (1967).
236a. R. F. Burns and P. J. Somers, *Carbohyd. Res.*, **31**, 289 (1973).
236b. R. F. Burns and P. J. Somers, *Carbohyd. Res.*, **31**, 301 (1973).
236c. L. Löwendahl and O. Samuelson, *Sv. Papperstidn.*, **77**, 593 (1974).
236d. L. Löwendahl, G. Petersson, and O. Samuelson, *Acta Chem. Scand., Ser. B*, **29**, 975 (1975).
236e. L. Löwendahl, G. Petersson, and O. Samuelson, *Acta Chem. Scand., Ser. B*, **29**, 526 (1975).
237. G. O. Aspinall, M. E. Carter, and M. Los, *J. Chem. Soc.*, 4807 (1956).
238. R. L. Whistler and W. M. Corbett, *J. Amer. Chem. Soc.*, **78**, 1003 (1956).
239. S. A. Barker, A. R. Law, P. J. Somers, and M. Stacey, *Carbohyd. Res.*, **3**, 435 (1967).
239a. A. M. Y. Ko and P. J. Somers, *Carbohyd. Res.*, **33**, 281 (1974).
240. H. Kiliani, *Ber.*, **18**, 3066 (1885).
241. O. T. Schmidt and C. C. Weber-Molster, *Ann.*, **515**, 43 (1934).
242. R. J. Ferrier, *J. Chem. Soc.*, 3544 (1962).
243. I. Stone, *Amer. Lab.*, **6** (4), 32 (1974).
244. F. Smith, *Advan. Carbohyd. Chem.*, **2**, 79 (1946).
244a. T. C. Crawford and S. A. Crawford, *Advan. Carbohyd. Chem. Biochem.*, **37**, 79 (1979).
245. T. Reichstein, A. Grüssner, and R. Oppenauer, *Helv. Chim. Acta*, **16**, 561, 1019 (1933); **17**, 510 (1934).
246. R. G. Ault, D. K. Baird, H. C. Carrington, W. N. Haworth, R. Herbert, E. L. Hirst, E. G. V. Percival, F. Smith, and M. Stacey, *J. Chem. Soc.*, 1419 (1933).
247. K. Maurer and B. Schiedt, *Ber.*, **66**, 1054 (1933).
248. H. Ohle, H. Erlbach, and H. Carls, *Ber.*, **67**, 324 (1934).
249. See J. J. Burns, in "Encyclopedia of Chemical Technology" (R. E. Kirk and D. F. Othmer, eds.), Interscience, New York, 1963, Vol. 2, p. 747.
250. T. Reichstein, *Helv. Chim. Acta*, **17**, 1003 (1934).
251. B. Helferich and O. Peters, *Ber.*, **70**, 465 (1937).
252. F. Micheel and H. Haarkoff, *Ann.*, **545**, 28 (1940).
253. G. S. Brenner, D. F. Hinkley, L. M. Perkins, and S. Weber, *J. Org. Chem.*, **29**, 2389 (1964).
253a. N. Goshima, N. Maezono, and K. Tokuyama, *Carbohyd. Res.*, **17**, 245 (1971).
253b. K. Tokuyama, K. Goshima, N. Maezono, and T. Maeda, *Tetrahedron Lett.*, 2503 (1971).
254. B. B. Lloyd and H. M. Sinclair, in "Biochemistry and Physiology of Nutrition" (G. H. Bourne and G. W. Kidder, eds.), Academic Press, New York, 1953, Vol.

I, p. 369; F. Smith, in "The Vitamins" (W. H. Sebrell and R. S. Harris, eds.), Academic Press, New York, 1954, Vol. I, p. 180; B. Åberg, in "Encyclopedia of Plant Physiology" (W. Ruhland, ed.), Springer-Verlag, Berlin, 1958, Vol. VI, p. 479.

255. A. Szent-Györgyi, *Biochem. J.*, **22**, 1387 (1928).

256. C. G. King and W. A. Waugh, *Science*, **75**, 357 (1932); W. A. Waugh and C. G. King, *J. Biol. Chem.*, **97**, 325 (1932).

257. W. N. Haworth and E. L. Hirst, *Ergeb. Vitamin- u. Hormonforsch.*, **2**, 160 (1939); E. L. Hirst, *Fortschr. Chem. Org. Naturstoffe*, **2**, 132 (1939).

258. R. W. Herbert, E. L. Hirst, E. G. V. Percival, R. J. W. Reynolds, and F. Smith, *J. Chem. Soc.*, 1270 (1933).

259. P. Karrer, H. Salomon, K. Schöpp, and R. Morf, *Biochem. Z.*, **258**, 4 (1933); P. Karrer, G. Schwarzenbach, and K. Schöpp, *Helv. Chim. Acta*, **16**, 302 (1933).

260. F. Micheel and K. Kraft, *Z. Physiol. Chem.*, **215**, 215 (1933).

261. F. Micheel and K. Kraft, *Z. Physiol. Chem.*, **218**, 280 (1933); **222**, 235 (1933).

262. E. G. Cox, *Nature*, **130**, 205 (1932); E. G. Cox and E. L. Hirst, *ibid.*, **131**, 402 (1933); E. G. Cox and T. H. Goodwin, *J. Chem. Soc.*, 769 (1936).

263. J. R. Penney and S. S. Zilva, *Biochem. J.*, **39**, 1 (1945).

264. B. Pecherer, *J. Amer. Chem. Soc.*, **73**, 3827 (1951).

265. T. Reichstein and A. Grüssner, *Helv. Chim. Acta*, **17**, 311 (1934).

265a. H. G. J. De Wilt, *J. Chromatogr.*, **63**, 379 (1971).

265b. O. Theander, Swed. Pat. 351,638 (1972).

266. L. von Vargha, *Nature*, **130**, 847 (1932).

267. S. L. von Schuching and G. H. Frye, *Biochem. J.*, **98**, 652 (1966).

267a. R. O. Mumma, A. J. Verlangieri, and W. W. Weber, II, *Carbohyd. Res.*, **19**, 127 (1971).

267b. S. F. Quadri, P. A. Seib, and C. W. Deyoe, *Carbohyd. Res.*, **29**, 259 (1973).

267c. H. Nomura, T. Ishiguro, and S. Morimoto, *Chem. Pharm. Bull.* (Tokyo), **17**, 381, 387 (1969).

268. K. G. A. Jackson and J. K. N. Jones, *Can. J. Chem.*, **43**, 450 (1965).

269. O. Theander, *Advan. Carbohyd. Chem.*, **17**, 223 (1962).

269a. M. Levin, *Methods Carbohyd. Chem.*, **6**, 76 (1972).

269b. M. Pašteka, *Methods Carbohyd. Chem.*, **6**, 81 (1972).

270. A. Wohl and B. Mylo, *Ber.*, **45**, 322 (1912).

271. N. Clauson-Kaas and J. Fakstorp, *Acta Chem. Scand.*, **1**, 216 (1947).

272. K. Iwadare, *Bull. Chem. Soc. Jap.*, **16**, 40 (1941).

273. R. Schaffer and H. S. Isbell, *J. Res. Nat. Bur. Stand.*, **56**, 191 (1956); D. Horton and J.-H. Tsai, *Carbohyd. Res.*, **58**, 89 (1977), and earlier papers in this series.

274. R. Schaffer and H. S. Isbell, *J. Amer. Chem. Soc.*, **79**, 3864 (1957); see also, T. D. Inch, *Carbohyd. Res.*, **5**, 53 (1967).

274a. H. Paulsen and K. Todt, *Ber.*, **101**, 3358 (1968).

274b. W. D. Hitz, D. C. Wright, P. A. Seib, M. K. Hoffman, and R. M. Caprioli, *Carbohyd. Res.*, **46**, 195 (1976).

274c. A. E. Luetzow, Ph.D. dissertation (D. Horton, preceptor), The Ohio State University, 1971.

274d. A. Maradufu and A. S. Perlin, *Carbohyd. Res.*, **32**, 127 (1974).

275. G. Dangschat and H. O. L. Fischer, *Naturwissenschaften*, **27**, 756 (1939).

276. W. Uenzelmann, Dissertation, Göttingen, Ger., 1930.

277. M. L. Wolfrom and E. Usdin, *J. Amer. Chem. Soc.*, **75**, 4318 (1953).

278. W. Meyer zu Reckendorf, *Angew. Chem.*, **78**, 673 (1966).
279. A. Rosenthal and G. Kan, *Tetrahedon Lett.*, 477 (1967); *Carbohyd. Res.*, **19**, 145 (1971).
280. G. Henseke and G. Hanisch, *Angew. Chem.*, **75**, 420 (1963).
281. O. Theander, *Acta Chem. Scand.*, **17**, 1751 (1963).
282. O. Theander, *Acta Chem. Scand.*, **18**, 2209 (1964).
283. D. Horton, J. B. Hughes, and J. M. J. Tronchet, *Chem. Commun.*, 481 (1965); D. Horton, M. Nakadate, and J. M. J. Tronchet, *Carbohyd. Res.*, **7**, 56 (1968).
284. O. Kjølberg, *Acta Chem. Scand.*, **14**, 1118 (1960).
285. N. J. Antia and M. B. Perry, *Can. J. Chem.*, **38**, 1917 (1960).
286. B. Pettersson and O. Theander, *Acta Chem. Scand.*, *Ser. B*, **25**, 29 (1974).
287. G. Avigad, D. Amaral, C. Asensio, and B. L. Horecker, *J. Biol. Chem.*, **237**, 2736 (1962).
287a. A. R. Gibson, L. D. Melton, and K. N. Slessor, *Can. J. Chem.*, **52**, 3905 (1974).
288. R. U. Lemieux and M. L. Wolfrom, *Advan. Carbohyd. Chem.*, **3**, 338 (1948); F. Shafizadeh, *ibid.*, **11**, 263 (1956).
289. J. R. Dyer, W. E. McGonigal, and K. C. Rice, *J. Amer. Chem. Soc.*, **87**, 654 (1965).
289a. S. Inoue and H. Ogawa, *Chem. Pharm. Bull.* (Tokyo), **8**, 79 (1960).
289b. C. Erbing, B. Lindberg, and S. Svensson, *Acta Chem. Scand.*, **27**, 2699 (1973).
290. O. Theander, *Acta Chem. Scand.*, **18**, 1297 (1964).
290a. A. E. Luetzow and O. Theander, *Sv. Papperstidn.*, **77**, 312 (1974).
290b. H. F. G. Beving, A. E. Luetzow, and O. Theander, *Carbohyd. Res.*, **41**, 105 (1975).
290c. H. F. G. Beving and O. Theander, *Acta Chem. Scand.*, *Ser. B*, **29**, 577, 641 (1975).
290d. A. S. Perlin, D. M. Mackie, and C. P. Dietrich, *Carbohyd. Res.*, **18**, 185 (1971).
291. C. E. Ballou and H. O. L. Fischer, *J. Amer. Chem. Soc.*, **75**, 4695 (1953).
292. C. E. Ballou and H. O. L. Fischer, *J. Amer. Chem. Soc.*, **75**, 3673 (1953).
293. D. L. MacDonald and H. O. L. Fischer, *J. Amer. Chem. Soc.*, **78**, 5025 (1956).
294. F. G. Fischer and H. Schmidt, *Ber.*, **93**, 658 (1960).
295. J. M. Grosheintz and H. O. L. Fischer, *J. Amer. Chem. Soc.*, **70**, 1476, 1479 (1948).
296. F. W. Lichtenthaler, *Angew. Chem.*, **75**, 93 (1963).
297. See R. Schaffer and H. S. Isbell, *Methods Carbohyd. Chem.*, **1**, 281 (1962).
298. R. Schaffer, *J. Amer. Chem. Soc.*, **81**, 5452 (1959).
298a. T. D. Inch, *Advan. Carbohyd. Chem. Biochem.*, **27**, 191 (1972).
299. R. L. Mann and D. O. Woolf, *J. Amer. Chem. Soc.*, **79**, 120 (1957).
300. H. Yüntsen, *J. Antibiotics* (Tokyo) *Ser. A*, **10**, 41 (1958); **11**, 233 (1958).
301. R. G. Coombe and T. R. Watson, *Proc. Chem. Soc.* (London), 214 (1962).
302. S. Bayne and J. A. Fewster, *Advan. Carbohyd. Chem.*, **11**, 44 (1956).
303. H. Ohle, G. Henseke, and A. Czyzewski, *Ber.*, **86**, 316 (1953); G. Henseke and M. Winter, *ibid.*, **89**, 956 (1956).
304. See S. Bayne, *Methods Carbohyd. Chem.*, **2**, 421 (1963).
305. J. K. Hamilton and F. Smith, *J. Amer. Chem. Soc.*, **74**, 5162 (1952).
306. R. Weidenhagen, *Z. Wirtschaftsgruppe Zuckerind.*, **87**, 711 (1937).
306a. R. E. Arrick, D. C. Baker, and D. Horton, *Carbohyd. Res.*, **26**, 441 (1973).
306b. H. J. Haas and P. Schlimmer, *Ann.*, **759**, 208 (1972).
307. K. Maurer, *Ber.*, **62**, 332 (1929).
307a. P. M. Collins, S. Kumar, and W. G. Overend, *Carbohyd. Res.*, **22**, 187 (1972).
307b. B. N. While and R. Carabelli, *Carbohyd. Res.*, **33**, 366 (1974).
308. E. F. L. J. Anet, *Advan. Carbohyd. Chem.*, **19**, 181 (1964).

308a. F. Shafizadeh, *Advan. Carbohyd. Chem.*, **23**, 419 (1968); see also F. Shafizadeh and Y. L. Fu, *Carbohyd. Res.*, **29**, 113 (1973).

309. H. Kato, *Bull. Agr. Chem. Soc. Japan*, **24**, 1 (1960); *Agr. Biol. Chem.* (Tokyo), **26**, 187 (1962).

310. E. F. L. J. Anet, *J. Amer. Chem. Soc.*, **82**, 1502 (1960); *Aust. J. Chem.*, **13**, 396 (1960).

311. E. F. L. J. Anet, *Chem. Ind.* (London), 345 (1961).

312. G. Machell and G. N. Richards, *J. Chem. Soc.*, 1938 (1960).

312a. H. El. Khadem, D. Horton, M. H. Meshreki, and M. A. Nashed, *Carbohyd. Res.*, **13**, 317 (1970); **17**, 183 (1971); **22**, 381 (1972).

312b. L. G. Együd [presented as Eguyd], *Carbohyd. Res.*, **23**, 307 (1972).

312c. H. Otsuka and L. G. Együd. *Biochim. Biophys. Acta*, **165**, 172 (1968); E. Jellum, *ibid.*, **170**, 430 (1968).

312d. D. H. Levine and E. Racker, *Arch. Biochem. Biophys.*, **79**, 396 (1959); *J. Biol. Chem.*, **234**, 2532 (1959).

313. S. Fukui and R. M. Hochster, *J. Amer. Chem. Soc.*, **85**, 1697 (1963).

314. B. Helferich and N. M. Bigelow, *Z. Physiol. Chem.*, **200**, 263 (1931).

314a. D. E. Kiely and H. G. Fletcher, Jr., *J. Amer. Chem. Soc.*, **90**, 3289 (1968).

314b. B. Helferich and E. Himmen, *Ber.* **62**, 2136 (1929).

314c. S. H. El Ashry and D. Horton, *Abstr. Papers Intern. Symp. Carbohyd. Chem.*, *VIIth*, *Madison, Wis.*, (1972).

314d. B. Lindberg, and O. Theander, *Acta Chem. Scand.*, **22**, 1782 (1968); B. Ericsson, B. O. Lindgren, and O. Theander, *Cellulose Chem. Technol.*, **7**, 581 (1973); see also L.-Å. Lindström and O. Samuelson, *Carbohyd. Res.*, **64**, 57 (1978).

315. See E. Fischer, *Ber.*, **23**, 2114 (1890).

316. G. Henseke and W. Liebenow, *Ber.*, **87**, 1069 (1954).

317. A. Assarsson and O. Theander, *Acta Chem. Scand.*, **17**, 47 (1963).

318. S. Bayne, G. A. Collie, and J. A. Fewster, *J. Chem. Soc.*, 2766 (1952).

319. F. Weygand, E. Klieger, and H. J. Bestmann, *Ber.*, **90**, 645 (1957).

320. S. Bayne, *Proc. Chem. Soc.*, 170 (1958).

320a. H. P. Humphries and O. Theander, *Carbohyd. Res.*, **16**, 317 (1971).

320b. H. P. Humphries and O. Theander, *Acta Chem. Scand.*, **25**, 883 (1971).

320c. D. T. Williams, J. K. N. Jones, N. J. Dennis, R. J. Ferrier, and W. G. Overend, *Can. J. Chem.*, **43**, 955 (1965); D. T. Williams and J. K. N. Jones, *ibid.*, **45**, 741 (1967).

321. For a summary, see O. Theander, *Tappi*, **48**, 105 (1965).

322. B. Lindberg and O. Theander, *Acta Chem. Scand.*, **8**, 1870 (1954).

323. J. S. Brimacombe, *Chem. Brit.*, **2**, 99 (1966); *Angew. Chem.*, **81**, 415 (1969).

324. L. Glaser, in *Carbohydrates*, *MTP International Review of Science*, (G. O. Aspinall, ed.), Butterworth, London, 1973, p. 191.

324a. D. M. Mackie and A. S. Perlin, *Carbohyd. Res.*, **24**, 67 (1972).

324b. K. Heyns and P. Köll, *Methods Carbohyd. Chem.*, **6**, 342 (1972).

325. O. Theander, *Acta Chem. Scand.*, **11**, 1557 (1957).

326. A. Assarsson and O. Theander, *Acta Chem. Scand.*, **18**, 727 (1964).

327. M. L. Wolfrom and P. Y. Wang, *Carbohyd. Res.*, **12**, 109 (1970); D. Horton and E. K. Just, *ibid.*, **30**, 349 (1973).

327a. J. F. Kennedy, *Advan. Carbohyd. Chem. Biochem.*, **29**, 305 (1974).

328. K. Heyns, A. L. Baron, and H. Paulsen, *Ber.*, **97**, 921 (1964).

329. J. S. Brimacombe, M. C. Cook, and L. C. N. Tucker, *J. Chem. Soc.*, 2292 (1965).

329a. O. Larm, E. Scholander, and O. Theander, *Carbohyd. Res.*, **49**, 69 (1976).

330. B. Lindberg and K. N. Slessor, *Carbohyd. Res.*, **1**, 492 (1966); *Acta Chem. Scand.*, **21**, 910 (1967).

331. E. Walton, J. O. Rodin, C. H. Stammer, F. W. Holly, and K. Folkers, *J. Amer. Chem. Soc.*, **80**, 5168 (1958).
332. J. S. Burton, W. G. Overend, and N. R. Williams, *Chem. Ind.* (London), 175 (1961); *J. Chem. Soc.*, 3433 (1965).
333. B. R. Baker and D. H. Buss, *J. Org. Chem.*, **30**, 2308 (1965).
334. P. J. Beynon, P. M. Collins, and W. G. Overend, *Proc. Chem. Soc.*, 342 (1964); P. J. Beynon, P. M. Collins, P. T. Doganges, and W. G. Overend, *J. Chem. Soc.* (*C*), 1131 (1966).
335. V. M. Parikh and J. K. N. Jones, *Can. J. Chem.*, **43**, 3452 (1965).
335a. P. M. Collins, W. G. Overend, and B. A. Rayner, *Carbohyd. Res.*, **31**, 1 (1973).
336. K. Heyns, J. Weyer, and H. Paulsen, *Ber.*, **98**, 327 (1965).
336a. K. Heyns, J. Weyer, and H. Paulsen, *Ber.*, **100**, 2317 (1967).
337. D. Horton and J. S. Jewell, *Carbohyd. Res.*, **2**, 251 (1966).
337a. D. Horton, J. S. Jewell, E. K. Just, J. D. Wander, and R. L. Foltz, *Biomed. Mass Spectrom.*, **1**, 145 (1974), and earlier papers.
337b. H. Paulsen, K. Eberstein, and W. Koebernick, *Tetrahedron Lett.*, 4377 (1974); H. Paulsen and K. Eberstein, *ibid.*, 1495 (1975).
337c. G. S. Hajivarnava, W. G. Overend, and N. R. Williams, *Carbohyd. Res.*, **49**, 93 (1976).
338. G. J. F. Chittenden and R. D. Guthrie, *Proc. Chem. Soc.*, 289 (1964); *J. Chem. Soc.* (*C*), 695 (1966).
338a. R. U. Lemieux, R. Svemitsu, and S. W. Gunner, *Can. J. Chem.*, **46**, 1040 (1968).
339. H. Paulsen and D. Stoye, *Ber.*, **102**, 834 (1969).
340. R. U. Lemieux and B. Fraser-Reid, *Can. J. Chem.*, **42**, 539 (1964).
341. H. J. Jennings and J. K. N. Jones, *Can. J. Chem.*, **41**, 1151 (1963).
342. F. W. Lichtenthaler, K. Strobel, and G. Reidel, *Carbohyd. Res.*, **49**, 57 (1976).
343. H. Paulsen, W. Koebernick, and H. Koebernick, *Tetrahedron Lett.*, 2297 (1976).
344. M. J. Bernaerts and J. De Ley, *Biochim. Biophys. Acta*, **30**, 661 (1958); *J. Gen. Microbiol.*, **22**, 129, 137 (1960).
345. D. S. Feingold, R. Durbin, and E. E. Grebner, *Abstr. Papers Amer. Chem. Soc. Meeting*, **140**, 3D (1961).
346. S. Fukui, R. M. Hochster, R. Durbin, E. E. Grebner, and D. S. Feingold, *Bull. Res. Counc. Israel, Sect. A*, **11**, 262 (1963).
347. S. Fukui and R. M. Hochster, *Can. J. Biochem. Physiol.*, **41**, 2363 (1963).
348. O. Gabriel and G. Ashwell, *J. Biol. Chem.*, **240**, 4123 (1965).
349. R. Okazaki, T. Okazaki, J. L. Strominger, and A. M. Michelson, *J. Biol. Chem.*, **237**, 3014 (1962).
350. H. Ohle and C. Dambergis, *Ann.*, **481**, 255 (1930).
351. H. Ohle and R. Deplanque, *Ber.*, **66**, 12 (1933).
352. C. L. Stevens and K. K. Balasubramanian, *Carbohyd. Res.*, **21**, 166 (1972).
353. P. M. Collins and W. G. Overend, *J. Chem. Soc.*, 3448 (1965).
354. D. Horton and E. K. Just, *Carbohyd. Res.*, **9**, 129 (1969).
355. K. Heyns, P. Köll, and H. Paulsen, *Ber.*, **104**, 2553 (1971); K. Heyns and P. Köll, *ibid.*, **104**, 3835 (1971).
356. K. Heyns and P. Köll, *Ber.*, **106**, 611 (1973).
356a. P. S. Fredricks, B. O. Lindgren, and O. Theander, *Sv. Papperstidn.*, **74**, 597 (1971).
357. K. Heyns, J. Lenz, and H. Paulsen, *Ber.*, **95**, 2964 (1962).
357a. N. Kashimura, K. Yoshida, and K. Onodera, *Carbohyd. Res.*, **25**, 264 (1972).
357b. O. Theander, *Acta Chem. Scand.*, **12**, 1883 (1958).
358. W. Sowa and G. H. S. Thomas, *Can. J. Chem.*, **44**, 836 (1966).

359. D. C. Baker, D. Horton, and C. G. Tindall, Jr., *Carbohyd. Res.*, **24**, 192 (1972); *Methods Carbohyd. Chem.*, **7**, 3 (1976).

360. B. Lindberg and O. Theander, *Acta Chem. Scand.*, **13**, 1226 (1959).

361. P. M. Collins, *Chem. Commun.*, 164 (1966).

362. P. M. Collins and W. G. Overend, *Chem. Ind.* (London), 375 (1963).

363. J. S. Brimacombe and M. C. Cook, *Chem. Ind.* (London), 1281 (1963); *J. Chem. Soc.*, 2663 (1964).

364. A. K. Chatterjee, D. Horton, J. S. Jewell, and K. D. Philips, *Carbohyd. Res.*, **7**, 173 (1968).

365. A. Rosenthal and C. M. Richards, *Carbohyd. Res.*, **32**, 53 (1974).

366. A. Rosenthal and G. Schöllnhammer, *Can. J. Chem.*, **52**, 51 (1974).

367. M. L. Wolfrom and S. Hanessian, *J. Org. Chem.*, **27**, 1800, 2107 (1962).

368. A. A. J. Feast, W. G. Overend, and N. R. Williams, *J. Chem. Soc.* (*C*), 303 (1966).

369. W. G. Overend and N. R. Williams, *J. Chem. Soc.*, 3446 (1965).

369a. W. G. Overend, *Chem. Ind.* (London), 342 (1963).

370. H. Paulsen, V. Sinnwell, and P. Stadler, *Angew. Chem.*, **84**, 112 (1972).

371. A.-M. Sepulchre, G. Lukacs, G. Vass, and S. D. Gero, *Angew. Chem.*, **84**, 111 (1972).

372. H. Paulsen and H. Redlich, *Angew. Chem.*, **84**, 1100 (1972); *Ber.*, **107**, 2992 (1974).

372a. B. Flaherty, W. G. Overend, and N. R. Williams, *Chem. Commun.*, 434 (1966).

372b. T. D. Inch, G. J. Lewis, and N. E. Williams, *Carbohyd. Res.*, **19**, 17 (1971); T. D. Inch, G. J. Lewis, and R. P. Peel, *ibid.*, **19**, 29 (1971).

373. D. C. Baker, D. K. Brown, D. Horton, and R. G. Nickol, *Carbohyd. Res.*, **32**, 299 (1974).

374. A. Rosenthal and L. Nguyen (Benzing), *Tetrahedron Lett.*, 2393 (1967).

375. J. M. J. Tronchet, J.-M. Bourgeois, R. Graf, and J. Tronchet, *C.R. Acad. Sci., Ser. C*, **269**, 420 (1969).

376. J. M. J. Tronchet and J. Tronchet, *Carbohyd. Res.*, **33**, 237 (1974).

377. J. M. J. Tronchet and J. Tronchet, *Helv. Chim. Acta*, **54**, 1466 (1971).

378. B. R. Baker and D. H. Buss, *J. Org. Chem.*, **31**, 217 (1966).

379. H. Paulsen, W. Bartsch, and J. Theim, *Ber.*, **104**, 2545 (1971); L. Evelyn, L. D. Hall, L. Lynn, P. R. Steiner, and D. H. Stokes, *Carbohyd. Res.*, **27**, 21 (1973).

379a. J. F. Batey, C. Bullock, J. Hall, and J. M. Williams, *Carbohyd. Res.*, **40**, 275 (1975).

380. P. M. Collins, P. Gupta, and R. Iyer, *J. Chem. Soc. Perkin I*, 1670 (1972).

381. P. M. Collins, *J. Chem. Soc.* (*C*), 1960 (1971); P. M. Collins and P. Gupta, *ibid.*, 1965 (1971).

382. T. D. Inch and G. J. Lewis, *Carbohyd. Res.*, **22**, 91 (1972).

383. W. Meyer zu Reckendorf, *Ber.*, **102**, 1071 (1969).

384. K. N. Slessor and A. S. Tracey, *Can. J. Chem.*, **48**, 2900 (1970).

385. K. Heyns, P. Köll, and H. Paulsen, *Ber.*, **104**, 3096 (1971).

386. D. Horton and E. K. Just, *Carbohyd. Res.*, **18**, 81 (1971).

387. D. Horton, J. S. Jewell, E. K. Just, and J. D. Wander, *Carbohyd. Res.*, **18**, 49 (1971).

388. A. Dmytraczenko, W. A. Szarek, and J. K. N. Jones, *Carbohyd. Res.*, **26**, 297 (1973).

389. O. Theander, *Acta Chem. Scand.*, **12**, 1887 (1958).

390. O. Theander, *Acta Chem. Scand.*, **12**, 1897 (1958).

390a. B. Lindberg, J. Lönngren, and S. Svensson, *Advan. Carbohyd. Chem. Biochem.*, **31**, 185 (1975).

390b. L. Kenne and S. Svensson, *Acta Chem. Scand.*, **26**, 2144 (1972).

390c. L. Kenne, O. Larm, and S. Svensson, *Acta Chem. Scand.*, **26**, 2473 (1972), **27**, 2797 (1973).

391. W. M. Corbett and J. Kidd, *J. Chem. Soc.*, 1594 (1959); W. M. Corbett, *ibid.*, 3213 (1959).
392. F. Micheel, *Ann.*, **496**, 77 (1932); F. Micheel and K. Horn, *ibid.*, **515**, 1 (1934).
393. O. Terada, K. Tomizawa, S. Suzuki, and S. Kinoshita, *Bull. Agr. Chem. Soc. Jap.*, **24**, 535 (1960).
393a. L. K. Hansen, A. Hordvik, and R. Hove, *Chem. Commun.*, 572 (1976).
394. K. Heyns, W.-P. Trautwein, and H. Paulsen, *Ber.*, **96**, 3195 (1963).
395. K. Heyns, W.-D. Soldat, and P. Köll, *Ber.*, **106**, 1677 (1973).
396. P. M. Collins, D. Gardiner, and W. G. Overend, *Carbohyd. Res.*, **32**, 203 (1974).
397. D. H. G. Crout, R. F. Curtis, C. H. Hassall, and T. L. Jones, *Tetrahedron Lett.*, 63 (1963).
398. G. Machell and G. N. Richards, *J. Chem. Soc.*, 1932 (1960).
399. R. L. Whistler and J. N. BeMiller, *J. Amer. Chem. Soc.*, **82**, 3705 (1960).
400. D. W. Harris and M. S. Feather, *Carbohyd. Res.*, **30**, 359 (1973).
401. B. Ericsson, B. O. Lindgren, O. Theander, and G. Petersson, *Carbohyd. Res.*, **23**, 323 (1972).
402. B. Ericsson, B. O. Lindgren, and O. Theander, *Cellulose Chem. Technol.*, **8**, 363 (1974).
403. P. M. Collins and P. Gupta, *Chem. Commun.*, 1288 (1969).

24. OXIDATIVE REACTIONS AND DEGRADATIONS

JOHN W. GREEN

I. INTRODUCTION

The oxidation of an organic compound is accomplished by the transfer of electrons from the substrate to the oxidant, a process that is usually accompanied by the breaking of carbon–hydrogen or carbon–carbon bonds. Such a reaction, which can formally be regarded as occurring between a nucleophile (the substrate) and an electrophile (the oxidant), is greatly affected by factors that alter the nucleophilicity or electrophilicity of the respective reactants; steric effects are often important also. A brief discussion of these factors is given first, and more-specific examples are presented in Sections II–XIV. For detailed background, the reader is referred to certain reviews on oxidation,[1-4] to an article outlining various mechanisms,[5] and to two books discussing nucleophilicity and electrophilicity.[6] Synthetic aspects of these reactions have been reviewed[7]; these are also discussed in Chapter 23.

A. HETEROLYTIC OXIDATIONS

The two-electron oxidation of a secondary alcohol group can be regarded as an elimination of two hydrogen atoms and the formation of a double bond between the carbon atom and the oxygen atom. Equation 1 illustrates breakage of a C–H bond with elimination of a hydride ion: this anion is a very

$$\text{Ox} + \text{H} - \text{CR}_2 - \text{O} - \text{H} \longrightarrow \text{Ox}^{2-} + \text{H}^+ + \text{CR}_2{=}\text{O} + \text{H}^+ \tag{1}$$

poor leaving group, but its removal is aided by the oxidant, which captures the pair of electrons and converts the hydride ion into a proton. The O–H bond dissociates with the liberation of a proton; this bond-breaking is readily accomplished with the aid of a base, usually the solvent. Thus, the breaking of the C–H bond is generally the slow (rate-determining) step of the reaction; for example, the oxidation of β-D-glucopyranose by bromine at $0°$ is much faster ($k_H/k_T = 4$) than that of β-D-[1-^{3}H]glucopyranose.[8]

In principle, the flow of electrons shown in Eq. 1 can also occur in the reverse direction, with elimination of the hydrogen atom from the hydroxyl group as a hydride ion. Swain et al.[9] studied the oxidation of ethanol by bromine, and concluded that the isotope effect wrought by substitution in the C–H bond is consistent with hydride removal as the mode of breakage of this bond. Vaska and DiLuzio[10] isolated an iridium hydride complex that was formed by the oxidation of ethanol; studies using deuterated compounds showed that the hydride ion came from the methylene group.

The oxidant may aid the elimination in a concerted or E2 type of mechanism as illustrated in Eq. 1; for such examples, the oxidant is not bonded to the substrate, except possibly in the transition state. Other oxidants—for example, chromic acid—have been shown to form intermediate esters, which subsequently decompose by a related, bimolecular elimination (Eq. 2); here the leaving group is the reduced form of the oxidant, and the C–H bond must necessarily break with the liberation of a proton. As in Eq. 1, the capture of electrons by the oxidant is the driving force of the reaction, so that the breaking of the C–H bond occurs simultaneously in the rate-determining step.

$$B\!:\; +\; H\!-\!CR_2\!-\!O\!-\!Ox \longrightarrow BH + CR_2\!=\!O + Ox^- \tag{2}$$

Variations of the foregoing two reactions are often encountered. Alkoxide ions eliminate a hydride ion (Eq. 3) more readily than does the neutral alcohol, because the negative charge acts as a strong driving force. Such alkoxide anions, being relatively strong nucleophiles, can be considered to be reducing agents.[11] Ethanol vapor reacts with sodium hydroxide at 230° to 250° to generate acetaldehyde (initially) and molecular hydrogen; the hydrogen is probably formed by a secondary reaction of hydride ions with protons in the system. This formation of molecular hydrogen has also been observed in the preparation of oxalic acid by caustic fusion of cellulose, and in similar alkaline treatments of D-glucose and D-fructose[12]; under more stringent conditions, these reactions produce sodium carbonate.

$$H\!-\!CR_2\!-\!O^- \longrightarrow H^- + R_2C\!=\!O \tag{3}$$

Glycols are more acidic than monohydric alcohols,[13] and the C-1 group on an aldopyranoid compound is even more acidic, owing to the inductive effect of the ring-oxygen atom; therefore, sugars and glycosides are more readily oxidized than ordinary alcohols. For the free sugars, however, oxidations at higher pH are accompanied by competing processes of epimerization and of degradation to saccharinic acids; this factor will be discussed in detail in Section II.

B. STERIC EFFECTS

The β anomers of sugars are generally oxidized more rapidly than the α anomers[14]; a similar pattern is seen in the faster oxidation of the β-glycosides. Derevitskaya et al.[15] showed that the 2-hydroxyl group of methyl β-D-glucopyranoside is more acidic than the corresponding group in the α-D anomer. Although the disparate rates of oxidation of β-D-glucopyranose and the α-D anomer have been attributed to the equatorial orientation of the anomeric hydroxyl group in the $^4C_1$ conformation of the former, the formation of the anion may also prove to be a factor in the oxidation of aldosides.

The oxidation of a secondary hydroxyl group of glycopyranosides involves the conversion of a tetrahedral atom into a trigonal atom, and there may be some resistance to the introduction of this constraint into the ring. Baker and Haines[16] pointed out the greater ease of oxidation of acyclic than of cyclic sugar derivatives; however, more-powerful oxidants have given good yields of glycosiduloses.[17]

C. Homolytic Oxidations

As termolecular reactions are very rare, the transfer of two electrons must necessarily occur in two successive steps in reactions of one-electron oxidants. The first step is the formation of a radical (Eq. 4); the oxidant facilitates removal of the hydrogen atom, converting it into a proton. Resonance stabilization (Eq. 5) of the resulting radical is regarded as more important than inductive effects. For such stabilization to occur, an adjacent atom (such as oxygen) having unshared pairs of electrons is necessary; therefore, initial attack generally occurs at a C–H bond adjacent to an oxygen atom, rather than at an O–H bond adjacent to a saturated carbon atom, which is incapable of participating in resonance stabilization. This is illustrated by the favored, catalytic oxidation of glycosides having axially attached hydroxyl groups, because the attack is initially upon the equatorial C–H bonds, which are sterically more available.

$$Ox + H{-}CR_2{-}O{-}H \longrightarrow Ox^- + H^+ + \cdot CR_2{-}O{-}H \qquad (4)$$

$$R_2\dot{C}{-}\ddot{O}{-}H \longleftrightarrow R_2\ddot{C}{-}\dot{O}^+{-}H \qquad (5)$$

$$R_2\dot{C}{-}O{-}H + Ox \longrightarrow R_2C{=}O + H^+ + Ox^- \qquad (6)$$

The final step is homolysis of the O–H bond in the radical to afford the carbonyl product plus a second hydrogen radical, which is converted by the oxidant into a proton (Eq. 6). Equation 4 is the slow or rate-determining step; the removal of the second hydrogen atom is rapid.

D. Electrophilic Nature of the Oxidant

Although the effect of increasing pH, already mentioned, is to increase the susceptibility of the substrate to oxidation, this increase generally decreases the effectiveness of the electrophilic oxidant at the same time. Permanganic acid is more effective than the permanganate anion as an oxidant.[18] Wet combustions, the complete oxidation of organic compounds under strongly acidic conditions to carbon dioxide and water, may occur because of the extreme electrophilicity of the protonated oxidant. Thus, the two extremes of pH—extremely strong acid and extremely strong alkali—favor complete oxidation.

The rate of oxidation often reaches a maximum near pH 7; this is a common observation in halogen oxidations. Although this effect has been attributed to increasing concentration of the effective oxidant at this pH, it could also be explained by the mutual interaction between diametrically opposed effects of pH on the effectiveness of oxidant and substrate.

The effectiveness of an oxidant as an electrophile is roughly proportional to its "acidity" or lack of basicity. Phenylhydrazine, a weak base, is a very weak oxidant; however, substitution of nitro groups on the aromatic ring decreases the basicity and increases the potency of the hydrazine as an oxidant.

Another factor is the leaving-group principle,[19] which is of importance in oxidations by peroxides and halogens. Such oxidants, reacting with a substrate S, tend to form an HO^+ or X^+ cation in the transition state and lose the rest of the molecule as an anion; the more effective is X^- as a leaving group, the better is HOX or X_2 as an oxidant. Hydrogen peroxide, which must displace a strongly basic OH^- group, is a poor oxidant, whereas peroxyacetic acid, which reacts to displace the resonance-delocalized acetate anion, is a good oxidant.

$$S + HO\!-\!X \longrightarrow S\text{---}HO^+\text{---}X^- \qquad (7)$$

Oxidants often tend to disproportionate by interaction of the acidic or undissociated oxidant with the anion of the same oxidant[19]; this is a nucleophilic displacement, by the OX^- anion, of X^- from the oxygen atom of the electrophilic HOX (Eq. 8). Two points should be emphasized here. First, the maximum decomposition occurs when HOX and OX^- are in equal concentrations—that is, when the pH of the reaction is equal to the pK_a of the oxidant (Eq. 9). Second, since this is a displacement, the reaction goes more readily when the group X^- is easily displaced; thus formation of iodate from hypoiodite is more extensive than chlorate formation from hypochlorite.

$$2OX^- + HOX \longrightarrow 2X^- + HXO_3 \qquad (8)$$

$$K_a = [H^+]\frac{[OX^-]}{[HOX]} = [H^+]; \; pK_a = pH \qquad (9)$$

Although examples in Eqs. 1–7 are given as secondary alcoholic groups, primary alcoholic groups can be treated similarly. Oxidation of the anomeric carbon atom in reducing sugars ($H\!-\!CR_2\!-\!OH$ grouping) proceeds much as for a secondary hydroxyl group. Oxidation of an $H\!-\!CR_2\!-\!OMe$ grouping (in, for example, an aldoside) is, however, quite different, because the readily

---

displaceable hydrogen atom has been replaced by a methyl group, which is eliminated with great difficulty. Thus, oxidative attack on aldosides often takes place on available secondary or primary alcoholic groups, rather than on C-1.

II. HALOGEN OXIDATIONS

Bromine and hypoiodite oxidations are particularly useful in the preparation of aldonic acids from aldoses and of aldaric acids from glycuronic acids. Primary alcohols also undergo oxidation by these reagents, although this conversion is of less value; glycosides can thus be converted into glycosiduronic acids, and alditols into aldoses and aldonic acids.

Secondary alcoholic groups are slowly oxidized to ketone groups, and 2-glyculosonic and 5-glyculosonic acids are formed in this way. More-extended oxidation results in the cleavage of carbon–carbon bonds and the production of chain-shortened acids.

The halogens and their oxy acids, particularly chlorine and hypochlorous acid, are widely used as bleaching agents. The chemistry of bleaching and oxidizing agents, with emphasis on the variations of redox potentials, has been reviewed.[20] Mixed bromine–chlorine systems have also been studied.[21]

A. HALOGENS AND HYPOHALITES

The use of halogens and hypohalites as oxidizing agents is complicated by changes in the nature of the oxidation as the conditions of temperature, acidity, and concentration are changed. The halogens show considerable differences in the positions of the various equilibria and the speed with which the equilibria are attained (see Table I). In acidic solution, the equilibrium between free halogen and hypohalous acid (Eq. 10) lies far to the left, and the concentration of hypohalous acid is very low. When alkali is added to the system, the concentration of hypohalite ion increases.

$$X_2 + H_2O \rightleftharpoons HOX + HX \tag{10}$$

$$X_2 + 2OH^- \rightleftharpoons OX^- + X^- + H_2O \tag{11}$$

Hence, the concentrations of free halogen, halic acid, and hypohalite vary greatly with the acidity. For $0.02M$ chlorine solutions at room temperature, for example, Ridge and Little[28] have shown that, at pH 1, 82% of the total chlorine present exists as free chlorine and 18% as hypochlorous acid. At pH 4, only 0.4% is free chlorine and 99.6% is hypochlorous acid; at pH 8, 21%

TABLE I

SELECTED PROPERTIES OF HALOGENS, HALIDES, AND HYPOHALOUS ACIDS

	$X =$			
	Cl	Br	I	Ref.
Solubility of X_2, g/100 ml of water, at 20°	1.85	3.58	0.28	22
K for $X_2 + H_2O \leftrightarrows HOX + HX$	4.5×10^{-4}	2.4×10^{-8}	3.6×10^{-13}	23
k, in sec^{-1}, for $X_2 + H_2O \rightarrow$ HOX + HX	11.0	110	3.0	24
K_a, for HOX $\leftrightarrows H^+ + OX^-$	2.95×10^{-8}	2.06×10^{-9}	2.3×10^{-11}	25
E_0, in volts, for $2X^- \leftrightarrows X_2 + 2e^-$	-1.3583	-1.087	-0.5345	26
Leaving-group tendency, k_X/k_{Br} (average)	0.02	1.0	3	27

exists as hypochlorous acid and 79% as hypochlorite. Obviously, the concentration of the oxidant and, probably, the nature of the oxidation are influenced greatly by the acidity.

Hypohalites are converted into halates according to the following equations:

$$HOX + OX^- \longrightarrow HXO_2 + X^- \tag{12}$$

$$HXO_2 + OX^- \longrightarrow HXO_3 + X^- \tag{13}$$

As mentioned in Section I, disproportionation occurs most rapidly at a pH corresponding to the pK_a value (Eq. 9); for hypochlorous acid (X = Cl),[29] minimum and maximum stability are observed at pH 6.75 and at pH 13, respectively. Whistler and Schweiger[30] noted a 29% conversion into chlorate at pH 7. As predicted from the K_a values in Table I, the values for minimum stability shift to higher pH regions for hypobromite and hypoiodite. The velocity of formation of halate increases greatly in the order $ClO_3^- < BrO_3^- < IO_3^-$; this order correlates directly with the leaving-group tendencies of the respective halide ion (see Table I).

1. Oxidation in Acidic Solutions

In acidic solution, the active oxidant is normally the free halogen molecule (for chlorine or bromine systems), and the hypohalous acid is less effective;

this may derive from the relative stabilities of the anionic products of the oxidation, as the halogen, X_2, has two potentially good leaving groups (see Eq. 7), whereas the acid HOX has only one such group. The order of effectiveness as an oxidant, however, is in the order $Br_2 > Cl_2 > I_2$, the last being very ineffective. This order corresponds neither to the oxidation potentials (Table I) nor to the order of leaving groups; it does, however, correspond to the rate of hydrolysis of the free halogen by water (k in Table I), and also to the solubility in water. This order has not yet been explained, especially the ineffectiveness of free iodine as an oxidant. It is possible that either the leaving tendency of the group X^- in X_2 is unrelated to the behavior of the same group as a substituent on carbon (as given in Table I), or the various X^+ cations differ greatly in their ability to capture electrons from the substrate as they are formed in the transition state; the Br^+ ion would be the strongest of the three in this respect.

The relative proportions of free halogen and hypohalous acid vary with the acidity of the solution and the nature of the halogen; however, unless a buffer or neutralizing compound is present, the solution becomes strongly acidic as a result of the formation of hydrohalic acid.

$$RCHO + Br_2 + H_2O \longrightarrow RCO_2H + 2HBr \tag{14}$$

$$RCHO + HOBr \longrightarrow RCO_2H + HBr \tag{15}$$

Hlasiwetz first used two of the halogens for the oxidation of sugars.[31] Lactose was treated with bromine, and D-glucose with chlorine. D-Gluconic acid was formed from D-glucose and isolated as the calcium salt. Kiliani found that sugars are readily oxidized by bromine at room temperature, and he obtained yields of 50 to 70% of various aldonic acids.[32]

The accumulation of hydrogen bromide during oxidations by bromine profoundly inhibits the rate of further oxidation. This effect results from more than the simple increase in acidity, for, although other strong acids also inhibit the rate, the effect is largest for hydrogen bromide and chloride.[33] With bromine, part of this effect may be due to complexation of free bromine as Br_3^-, which is ineffective as an oxidant.

To minimize[34] this inhibiting influence, the reaction may be conducted in the presence of a solid buffer, such as barium carbonate or calcium benzoate. In general, the presence of a buffer increases the yield of aldonic acid, and, in addition, it precludes the hydrolysis of disaccharides. Yields of 96% of D-gluconic acid and of 90% of D-xylonic acid (as salts) have been obtained from oxidation of the respective aldoses in buffered solutions.

When the oxidation period is extended, particularly under unbuffered conditions, keto acids may be formed in small yields. Thus, L-rhamnose and hexoses afford 6-deoxy-L-*lyxo*-5-hexulosono-1,4-lactone[35] and the hex-5-ulosonic acids,[36] respectively. Under more-drastic conditions, carbon–carbon bonds are cleaved, to yield chain-shortened acids.

A variation of the bromine oxidation process that seems to be particularly feasible for the commercial production of aldonic acids involves the electrolysis, between carbon electrodes, of solutions containing sugars, a small proportion of a bromide, and a solid buffer, such as calcium carbonate.[37] Presumably, the reaction occurs because of the formation of free bromine at the anode; the bromine oxidizes the aldose to the aldonic acid and is itself reduced to bromide. Yields are almost theoretical in many cases. If the electrolytic method is not well controlled, aldaric acids and 2- and 5-glyculosonic acids may be produced.[38] Whereas the electrolytic oxidation is normally conducted with direct current, a yield of 55% of D-gluconic acid was obtained by using alternating current and platinum electrodes[39]; a very low efficiency was observed with graphite electrodes. Electrolytic reactions of sugars have been reviewed.[39a]

The ketoses are generally resistant to the action of bromine[40]; bromine oxidation is sometimes used to remove aldoses from such mixtures as invert sugar. By extending the period of oxidation and employing high temperatures, Kiliani obtained oxalic acid, bromoform, and glycolic acid from 2-ketoses.[41] Milder conditions give keto acids, such as D-*lyxo*-5-hexulosonic acid from D-fructose and D-*xylo*-5-hexulosonic acid from L-sorbose.[42] Calcium D-*arabino*-2-hexulosonate yielded 65% of calcium D-arabinonate by electrolytic oxidation with bromine.[43]

More-drastic conditions are needed for bromine oxidations of polyols than for aldoses. The oxidation products of D-glucitol give two osazones derived from D-*arabino*-hexos-2-ulose and L-*xylo*-hexos-2-ulose.[44] In the cyclitol series, a 12% yield of D-*myo*-inosose-1 was isolated (as the phenylosazone) after the action of bromine in a sodium acetate buffer on L-inositol.[45]

2. Mechanism of Bromine Oxidation of Aldoses

The mechanism of the oxidation of aldoses by bromine in the presence of barium carbonate and bromides (pH about 5.4) was studied by Isbell and Pigman.[14] Under these conditions the active oxidant is free bromine, not hypobromous acid. Molecular chlorine was found[46] to be the active oxidant in the oxidation of D-glucose by buffered chlorine-water at pH 2.2 and 3.

Interestingly, the cyclic forms of an aldose, not free aldehyde, are oxidized directly under these conditions.[47] Pyranoses afford 1,5-lactones, and furanoses, 1,4-lactones, directly, in high yield. The direct formation of the 1,5-lactone

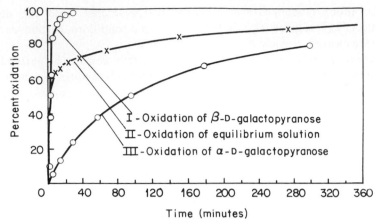

FIG. 1. Rate of oxidation of D-galactose by bromine ($\sim 0°$, pH = 5.4, buffered). (After Isbell and Pigman.[47a])

from an aldose provides strong evidence that the crystalline sugars are, in general, pyranoses (see Chapter 1).

α-D-Glucopyranose   +   Br$_2$ $\longrightarrow$ D-Glucono-1,5-lactone   +   2HBr

The equilibrium solutions are oxidized at rates intermediate between those for the individual anomers (see Figs. 1 and 2), and the oxidation curve is composed of a rapid phase followed by a slow phase. Extrapolation of the slow portion (on a semilogarithmic plot) to zero time gives the amount of the two anomers in the equilibrium solution. The composition of equilibrium solutions of several sugars as determined in this way agrees with that obtained by studies of optical rotation (see Table I, Chapter 4).

One form of D-mannose (D-mannose·CaCl$_2$·4H$_2$O, Fig. 2) exhibits an oxidation curve intermediate between those for the α- and β-pyranose anomers, and a considerable proportion of D-mannono-1,4-lactone is eventually found in the solution. Consequently, it appears that, in this compound, the modification is a D-mannofuranose.

Overend *et al.*[48] suggested that the rates of bromine oxidation observed for α-D anomers are actually an expression of the rates of mutarotation into

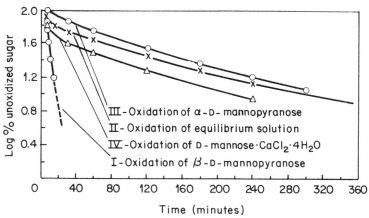

FIG. 2. Rate of oxidation of D-mannose by bromine (∼0°, pH = 5.4, buffered). (After Isbell and Pigman.[47a])

the β-D anomers, and that the true rates of oxidation of the α-D anomers are much lower. As a test of this hypothesis, they plotted the rate constants of mutarotation for the respective anomers against the rate constants for bromine oxidation of nine aldoses and the reducing disaccharide lactose (Fig. 3); the random scatter of data points found for the β-D anomers showed no correlation, whereas the rather good, straight-line relationship observed for the α-D anomers indicated a definite correlation between the observed rate of oxidation by bromine and the rate of mutarotation for the sugars studied. Hence, the rate-determining step in the oxidation of the α-D anomers is anomerization to generate the faster-reacting β-D anomer. For D-glucose, the true rate of oxidation of the α anomer was found to be about 1/250th that of the β anomer.

It has been suggested[48] that the higher rate of oxidation of β-D-glucopyranose is due to the equatorial orientation of the hydroxyl group on C-1 (see Fig. 4); this group is favorably oriented to react with the bromine molecule, and neighboring-group participation by the ring-oxygen atom is possible. A slow, irreversible formation of the conjugate acid (1) of the hypobromite ester is followed by reversible loss of a proton. A free hypobromous ester (2) then undergoes an E2 elimination of hydrobromic acid to form D-glucono-1,5-lactone. The formation of the ester is enhanced by higher pH, and the axial disposition of H-1 allows the *trans* elimination to proceed from a relatively uncrowded position. In contrast, similar elimination from β-D-glucopyranosyl hypobromite would occur from a conformation destabilized

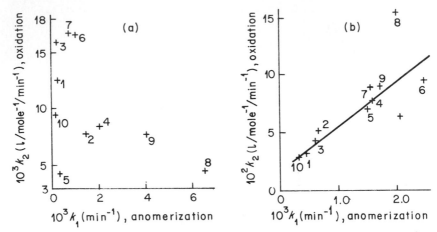

FIG. 3. Variation of oxidation rate constants with anomerization rate constants for a series of (a) β-D anomers and (b) α-D anomers, based on measurements of Isbell and co-workers,[47] reproduced by permission of the Chemical Society and the senior author, from ref. 48. 1, Glucose; 2, mannose; 3, galactose; 4, talose; 5, gulose; 6, arabinose; 7, xylose; 8, lyxose; 9, rhamnose; 10, lactose.

by crowding between the bromine molecule and the protons axially attached to C-3 and C-5.

Isbell[49] proposed an alternative explanation based on the difference in free energy between respective ground and (presumed) transition states of the two anomeric anions derived from D-glucopyranose (see Fig. 5). The transition states are assumed to be similar in structure for both anomers; however, the transition state for the β-D anomer suffers less destabilization by nonbonded interactions. Lewin and Albeck[50] then advanced a similar mechanism for the oxidation of cellulose by bromine.

Perlmutter-Hayman and Persky[51] had postulated that the increase in the rate of oxidation of aldoses with increase in pH from 1.25 to 4.5 arises owing to the progressively enhanced extent of participation by the anion of the aldose, which undergoes oxidation faster than the neutral aldose; however, this implies a difference in reactivity of about 10^{11}, and Overend et al.[48] argued that such an extreme difference in rate could not be explained by a difference in acidity only.* Furthermore, relative rates of oxidation by bromine decrease in the order: β-L-arabinose > β-D-galactose > β-D-glucose

---

* Stewart et al.[18] noted a difference in reactivity of $\sim 10^3$ for the oxidation of fluoral dihydrate and of the mono-anion by permanganate; in this case, the oxidant is the MnO_4^- anion. These authors considered that the negative charge on the substrate is very helpful in aiding the reaction, and that the interaction with the negative charge on the oxidant does not slow the reaction appreciably.

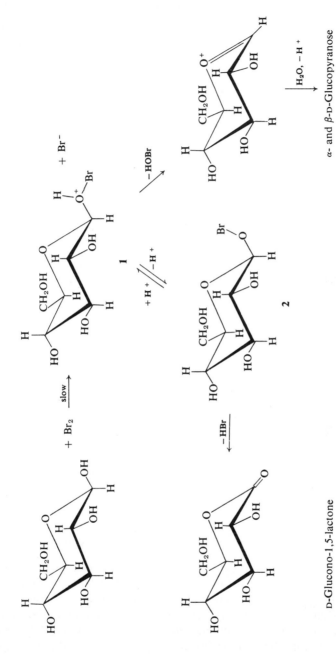

D-Glucono-1,5-lactone

Fig. 4. Mechanism for the oxidation of β-D-glucopyranose by bromine, as proposed by Barker et al.[48]

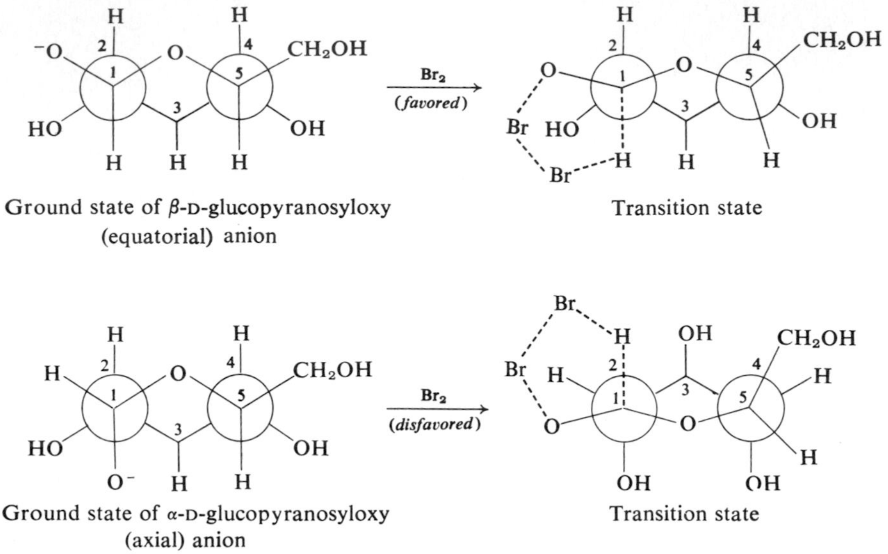

Ground state of β-D-glucopyranosyloxy (equatorial) anion

Transition state

Ground state of α-D-glucopyranosyloxy (axial) anion

Transition state

FIG. 5. Ground states and transition states proposed[49] for the reaction of the D-glucopyranoses with bromine.

> β-D-mannose, which precisely reverses the order of the extents of dissociation of these aldoses.

Whereas the foregoing reactions of aldoses involve a free hydroxyl group at C-1, alkyl β-D-glucosides are also more readily oxidized by acidic chlorine, and also by alkaline hypochlorite, than the α-D-glucosides. No hypohalite ester can be formed directly at the anomeric carbon atom, and a direct relationship of oxidizability of the sugar to the acidity of the anion[15] may be relevant. The difference in rates for the anomeric aldosides is, however, minor in comparison to that for the free aldoses, and oxidation occurs[51a] at carbon atoms bearing axial hydroxyl groups.

3. Oxidation of Aldopyranosides with Acidic Chlorine Systems

Lindberg et al.[52-54] carried out several investigations of the oxidation of methyl D-aldopyranosides with saturated chlorine-water (unbuffered, pH ~ 1 initially). The main product was D-gluconic acid, which was oxidized further, to D-xylo-2-hexulosonic acid and D-glucaric acid. Similar studies were made on methyl D-galactosides, D-mannosides, D-xylosides, and cellobiosides. For all of them, the β-D-glycopyranoside, in which the aglycon adopts an equatorial orientation in the $^4C_1(D)$ conformation, is oxidized at a higher rate (two to ten times) than the α anomer. The difference in rate between the two D-xylosides was not so great as that for the hexosides. Bentley[55] noted a similar difference for the methyl L-arabinopyranosides; the α-L anomer, which, in the $^1C_4(L)$ conformation, has an equatorially attached group at C-1, is oxidized more rapidly.

In the oxidation of methyl β-cellobioside,[54] cellobionic acid was formed initially and oxidized further to D-gluconic acid, indicating that the methyl–D-glucosyloxy linkage is cleaved more readily than the D-glucosyl–D-glucosyloxy linkage.

Such reactions with chlorine are very slow; treatment for 14 days at room temperature was needed for conversion of methyl β-D-glucopyranoside into D-gluconic acid (in 50% yield).[52] The D-glucoside is, apparently, oxidized directly to D-gluconic acid, perhaps via a cyclic orthoester chloride intermediate; the possibility of an initial hydrolysis to liberate D-glucose is precluded.

The oxidation of methyl β-D-glucoside at a controlled pH of 4.5 was studied by Henderson.[56] D-Glucose, D-arabinose, oxalic acid, and carbon dioxide were the main products; traces of the 2-glyculosonic and glyco-2,5-diulosonic acids were detected.

Theander[57] investigated the oxidation of methyl β-D-glucopyranoside at several pH values. The major products formed at pH 4 were D-gluconic, D-erythronic, and glyoxylic acids; minor products were D-glucose, D-arabinose, and hexopyranosiduloses, the latter being possible intermediates in the formation of D-erythronic and glyoxylic acids. Hypochlorous acid was considered to be the oxidant, based on the observations that the concentration of this species is at a maximum at pH 4 and that the yield of neutral products from the oxidation was also at a maximum at this pH value.

The formation of aldonic acids from the aldosides suggests an initial attack at C-1. Staudinger[58] proposed the formation of carbonic ester linkages in oxidized cellulose, based on such an attack. Kaverzneva et al.[59] investigated the product of oxidation of cellulose by aqueous chlorine, buffered at pH 4.5; treatment with dilute sodium hydroxide caused the evolution of carbon dioxide. A carbonic ester was postulated, formed by oxidative cleavage of the C-1–C-2 bond, that had the original oxygen linkages to C-1 still intact. Daniel[60] prepared an oxidized cellulose, labeled at C-1 and C-6 with ^{14}C, and determined that less than 20% of the carbon dioxide evolved by the action of alkali was derived from C-1. Similarly, a small proportion of the carbon dioxide evolved from the oxidized cellulose was formed from C-6 by oxidation to D-glucuronic acid intermediates, on the basis of data from the study of the liberation of carbon dioxide; oxidative attack on cellulose appears to occur at random.

References start on p. 1156.

4. *Oxidation with Hypohalites in Alkaline Solution*

In alkaline solution, the halogens are converted into hypohalous acid and hypohalite ions. Hypohalite oxidation is likely to be more drastic than action of the free halogens. Thus, whereas free iodine does not act as an oxidant, hypoiodite is a powerful oxidizing agent. Hypobromite and hypochlorite, in particular, are prone to oxidize primary and secondary alcoholic groups, and to cause cleavage of carbon–carbon bonds; these processes are complicated by the tendency of hypohalites to disproportionate to halate ions (see Eqs. 12 and 13).

The action of chlorine in alkaline media is much slower than that of bromine. Lewin[21] reported that the rate of oxidation of D-glucose at pH 9.8 by hypobromite is 1360 times that by hypochlorite at the same pH. For cellulose, the ratio is much smaller (33 to 1). The complexity of the latter system is, however, revealed by the variability of this ratio over the pH range of 8 to 13; at pH 6 to 7, the action of hypochlorite is actually slightly faster than that of hypobromite.

Maximal rates of oxidation have been observed near neutrality; Mehrotya and Grover[61] found such a maximum for the bromine oxidation of D-glucose at pH 7.8, and Theander[57] noted a maximum at pH 7 for the chlorine oxidation of methyl β-D-glucopyranoside (see Fig. 6). This phenomenon was first observed in the oxidation of cellulose with chlorine, in the classical work of Clibbens and Ridge[62] (Fig. 6); it has also been noted for such oxidation of amylopectin and alginic acid.

The attack of hypochlorite on amylopectin[30] and alginic acid[63] has been shown to occur at C-2 and C-3; methyl 2-*O*-methyl-α-D-glucopyranoside[64] undergoes attack at C-3 and C-4. From studies utilizing ^{14}C-labeled methyl β-D-glucopyranoside, Crossman and Green[65] were able to determine that oxidative attack of aqueous chlorine at pH 7 occurs primarily on the D-glucosyloxy group, releasing methanol; only 5.7% of the oxidation occurred by attack on the methyl group. Oxidation of D-glucose by hypochlorite at pH 11, followed by a second oxidation at pH 5, afforded D-arabinose[66]; the intermediate, formed in the alkaline step, is D-arabinonic acid. This method has also been applied to disaccharides.

The identity of the active oxidant in alkaline, aqueous solutions of halogens is uncertain. Epstein and Lewin[67] suggested a free radical mechanism for the chlorine oxidation of cellulose, whereas Whistler *et al.*[30,63] stated that hypochlorous acid is the oxidant in this reaction; Wolfrom and Lucke,[68] working at pH 9 and 50.5°, found the reaction to be zero order in substrate and second order in hypochlorite, and they suggested that the intermediate is either $Cl_2O_2^-$ or $Cl_2O_2^{2-}$. Mehrotya and Grover[61] concluded that hypobromous acid is the oxidant in aqueous systems at pH ~ 7, whereas Lewin[21]

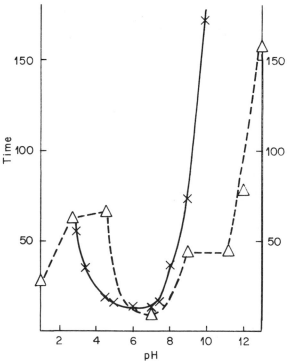

FIG. 6. Rate of consumption of chlorine–hypochlorite oxidant by methyl β-D-glucopyranoside and by cotton, as a function of pH. The vertical axis represents that time (minutes) needed for consumption of 0.05 equivalents of oxidant per mole of methyl β-D-glucopyranoside[57] (——— × ———) or the time (hours) needed for consumption of half the total amount of oxidant reacting with cotton[62] (---△---).

considered the reactive species to be Br_2OOH. Theander's group interpreted kinetic data from oxidations of methyl β-D-glucopyranoside as signifying the simultaneous operation of ionic and radical mechanisms.[68a]

Methyl 4-*O*-methyl-β-D-glucopyranoside reacts[69] more rapidly than the α anomer with hypochlorite at pH 9 and 11. The oxidation of β-D-glucopyranose by hypoiodous acid at pH 9.8 is, initially, at least twenty-five times as fast as that of the α anomer.[70] As the oxidation progresses, the simultaneous mutarotation tends to equalize the two rates. This difference in rates has been observed in the pH range of 7.6 to 11.8.

Alkaline hypoiodite has been proposed as a reagent for the quantitative determination of aldehyde groups.[71] With careful control of conditions, aldoses are converted almost quantitatively into aldonic acids (see Chapter

23). Measurement of the iodine consumed permits quantitation of the amount of aldose originally present.

$$RCHO + I_2 + 3\ NaOH \longrightarrow RCO_2Na + 2\ NaI + 2\ H_2O \qquad (16)$$

In this reaction, the rate of iodate formation should be lower than the oxidation of the aldose. The reaction is slowed by the presence of such buffers as borax.[72] Colbran and Nevell[73] have studied the effect of iodide ion in suppressing iodate formation; too much iodide ion, however, causes over-oxidation by maintaining a higher concentration of oxidant (hypoiodous acid) during the reaction.

Isbell et al.[74] showed a large, primary isotope effect in the oxidation of D-[1-^{3}H,6-^{14}C]glucose by alkaline iodine; the reaction was analyzed by monitoring the ratio of the two isotopes.

Hypoiodites are used for preparatory as well as analytical purposes. Goebel used barium hypobromite in the preparation of calcium D-gluconate and maltobionate[75]; high yields of aldonic acids are obtained in methanol solution.[76] Wolfrom and Frank[77] used a modified barium hypoiodite method. Hypoiodite has been suggested as the reagent of choice for oxidizing fragments from periodate-oxidized polysaccharides to aldaric and lower acids for separation and identification.[77a]

Ketoses are essentially inert to the action of hypoiodites under the conditions used for the determination of aldoses, although, for accurate work, small corrections may be necessary. The action of an excessive proportion of alkali at slightly elevated temperatures affords oxalic acid.[78]

More-drastic oxidation of aldoses with hypoiodite leads to glyculosonic acids and finally to cleavage of carbon–carbon bonds. Hönig and Tempus[79] claimed to have oxidized D-glucose stepwise to D-gluconic acid, D-*arabino*-2-hexulosonic acid, and D-arabinonic acid; however, other workers[80] claimed that the main product is D-*xylo*-5-hexulosonic acid.

Aldosides are converted by hypoiodite or hypobromite into glycosiduronic acids in rather low yield.[81] Jackson and Hudson[82] isolated the brucine salt of methyl α-D-mannopyranosiduronic acid (4) after oxidation of methyl α-D-mannopyranoside (3); however, cleavage of carbon–carbon bonds also occurred.

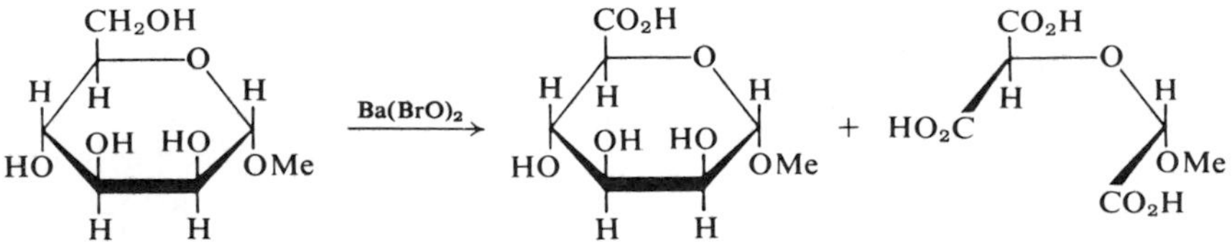

Alditols are oxidized by alkaline solutions of some halogens. Fischer and Tafel[83] obtained 20% yields of 3-hydroxypropanonal 1,2-bis(phenylhydrazone) from glycerol by the action of bromine and sodium carbonate and subsequent treatment with phenylhydrazine. Similar reaction of galactitol yielded an osazone that appeared to be DL-*lyxo*-2-hexulose phenylosazone. Presumably, the oxidation by halogen takes place mainly at the primary alcoholic group.

Aldonamides having free hydroxyl groups at C-2 are degraded to the aldose with one carbon atom less by treatment with hypochlorites. This is the basis of the Weerman method of degrading aldoses, discussed in Chapter 3.

B. HALIC ACIDS (HXO₃)

Chloric acid in conjunction with catalysts, particularly vanadium pentaoxide,[84] has as its principal use the oxidation of aldonic acids or lactones to the 2-glyculosonic acids, intermediates in the preparation of L-ascorbic acid and analogues as discussed in Chapter 23.

D-Glucono-1,4-lactone (**5**) and potassium D-galactonate in methanol, in the presence of phosphoric acid and vanadium pentaoxide, are oxidized by chloric acid to methyl D-*arabino*-2-hexulosonate (**6**) and methyl D-*lyxo*-2-hexulosonate, respectively.[85]

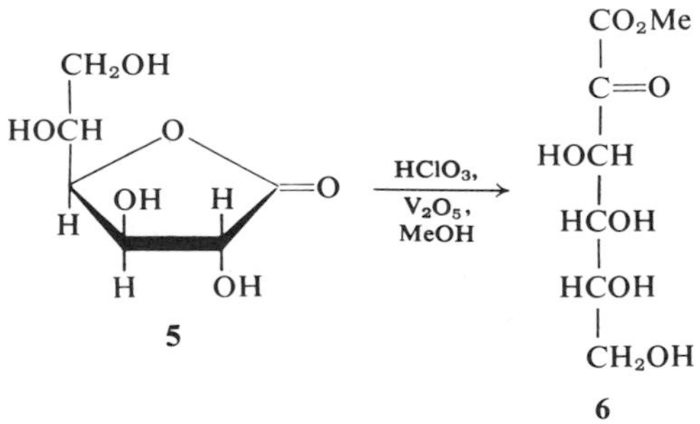

Iodic acid in concentrated sulfuric acid at 100° is reported[86] to show rather remarkable specificity; ketoses, sucrose, and aldopentoses are oxidized, but aldohexoses and lactose are not attacked. At still higher temperatures, hexoses are quantitatively oxidized to carbon dioxide and water.[87]

At moderate temperatures in the absence of a catalyst, aldoses, ketoses, and sucrose are inert to the action of chlorates over several weeks' time[88]; bromates in alkaline solution also exert no oxidative action.[89]

References start on p. 1156.

C. Chlorous Acid ($HClO_2$)

Chlorous acid is of particular interest, because of its use for the removal of lignin and other noncarbohydrate components from woody tissue without appreciable action on the carbohydrates (see Chapter 37). It also is reported to be an effective bleaching agent.

Jeanes and Isbell[88] found that, under mild conditions, aldoses are oxidized to aldonic acids, but that nonreducing carbohydrates and ketoses are oxidized only slowly. The rate of oxidation decreases in the order: pentoses > hexoses > disaccharides; however, in contrast to other oxidants, chlorous acid oxidizes α-hexoses more rapidly than the β anomers. The yields of aldonic acids are, however, less than from bromine oxidations.[90] The equation for the oxidation in acidic solution was expressed as:

$$RCHO + 3\,HClO_2 \longrightarrow RCO_2H + HCl + 2\,ClO_2 + H_2O \qquad (17)$$

The quantitative stoichiometry of the D-glucose–chlorous acid reaction has been studied in detail; the reagent used was sodium chlorite in a phosphoric acid–phosphate buffer at pH 2.4 to 3.4. The molar ratio of oxidant consumed to D-glucose consumed was 3:1; no overoxidation occurred over extended periods of time. The decomposition of the reagent was monitored throughout the oxidation, and the rate was found to be proportional to the geometric mean of the chlorite concentration. The method is recommended for the determination of aldehyde groups in carbohydrates, especially alkali-sensitive carbohydrates.

D-Glucose in admixture with D-fructose can be determined quantitatively by oxidation with sodium chlorite at pH 4.0; the chlorine dioxide evolved in the reaction is measured.[91]

Zienius and Purves[92] compared the effect of various oxidants on D-galacturonic acid. The order of effectiveness was: sodium chlorite at pH 2.8 and 75° > sodium hypochlorite at pH 10 and 23° > chlorine dioxide at pH 1.3 or 5 at 75° > chlorine at pH 5 and 0°; the action of hydrogen peroxide at pH 10 was negligible.

Becker et al.[93] studied the action of chlorine dioxide at pH 3 and 56° on cellotetraose and the corresponding methyl β-glycoside and alditol. The reaction was found to be zero order in oxidant; at higher concentrations of chlorine dioxide, two successive zero-order reactions were discerned. This kinetic order agrees with that found by Somsen[94] for C_2, C_3, and C_4 alcohols and carbonyl derivatives. The major products from the tetrasaccharides were D-gluconic, D-arabinonic, D-erythronic, and glyoxylic acids; some oxalic acid was formed later in the reaction. Identification of D-mannose and D-allose among the products obtained by reduction of the oxidation mixture, followed by hydrolysis, implicated C-2 and C-3, respectively, as sites of attack.

D. MISCELLANEOUS HALOGEN OXIDANTS

Whistler and co-workers[94a] used chlorine in nonaqueous solvents (acetic acid or carbon tetrachloride) to effect chlorinolysis of glycosides. The initial products of this reaction are a glycosyl chloride and a hypochlorous ester (Eq. 18).

$$GlOCH_2R + Cl_2 \longrightarrow GlCl + RCH_2OCl \tag{18}$$

The presence of the glycosyl chloride was demonstrated by its conversion by ethanol in the presence of a silver salt into an ethyl glycoside. The hypochlorite ester formed by chlorinolysis of a methyl glycoside underwent elimination of hydrogen chloride to give formaldehyde. A similar reaction with amylopectin oxidized the secondary alcohol group on C-4 to a ketone group; the position of oxidation was proved by reduction and hydrolysis of the product, to generate derivatives of both D-glucose and D-galactose.

N-Bromosuccinimide was used by Hanessian[95] to effect oxidative opening of the benzylidene acetal ring in methyl 4,6-O-benzylidene-α-D-glucopyranoside; the product was methyl 4-O-benzoyl-6-bromo-6-deoxy-α-D-glucopyranoside. A free-radical mechanism was proposed. Whereas a halogen atom is also incorporated into this product, the same reagent (also, N-bromoacetamide or N-bromocarbamide) in methanol was shown by Kiss to oxidize benzylated sugars to the aldonolactones.[96] The first two reactions in this section are exceptions to the generalization that halogens as oxidants do not normally introduce halogen atoms into the compound.

Meiners and Morris[97] utilized u.v. light in the oxidation of starch by chlorine. The reaction was very much faster under illumination than in the dark, and three times as fast at pH 4 as at pH 0.5. The product isolated in 91% yield from reaction at pH 4 had a high carbonyl content (25 mole%). Photolysis of primary azidodeoxy derivatives has been demonstrated as a convenient, general means of converting the primary alcohol into an aldehyde group in such diverse molecules as those of nucleosides,[97a] starch,[97b] cellulose,[97c] and methyl β-cellobioside.[97d] Binkley[97e] has demonstrated a two-step oxidation of free primary or secondary hydroxyl groups by initial conversion into a pyruvyl ester and subsequent photolysis.

III. ORGANIC PEROXY ACIDS

Peroxy acids, RCO_2OH, are comparable with the hypohalous acids, HOX, in that they possess a good leaving group, RCO_2^-, and so can develop an electrophilic center, HO^+ in the transition state (Eq. 7, $X^- = RCO_2^-$). Such oxidants are either used as such, or prepared, as needed, by the direct addition of hydrogen peroxide to the organic acid in the reaction mixture.

Peroxy acids as an alternative to permanganate for the oxidation of aldose diethyl dithioacetals to disulfones were first employed by MacDonald and Fischer.[98] Thus, D-glucose diethyl dithioacetal pentaacetate (7) was oxidized by peroxyphthalic acid in ether to 3,4,5,6-tetra-O-acetyl-1,2-dideoxy-1,1-bis(ethylsulfonyl)-D-*arabino*-hex-1-enitol (8); this unsaturated disulfone may be cleaved at the double bond by hydrazine to afford D-arabinose and bis(ethylsulfonyl)methane. D-Lyxose was similarly prepared. Unacetylated aldose dithioacetals can be converted by the action of aqueous peroxypropionic acid into disulfone derivatives, which are readily degraded in aqueous ammonia to the corresponding aldose having one carbon atom fewer.[99] D-Erythrose, D-threose, and D-arabinose were obtained in high yields from the D-arabinose, D-xylose, and D-mannose dithioacetals. The ring in *scyllo-myo*-inosose was broken by oxidation of the dithioacetal and subsequent ammonolysis of both carbon–carbon bonds to the sulfonylated center, affording *xylo*-pentodialdose. D-Fructose diethyl dithioacetal reacts with peroxypropionic acid in 1,4-dioxane to afford a (presumably unstable) disulfone that decomposes spontaneously, yielding D-erythrose in a single step.

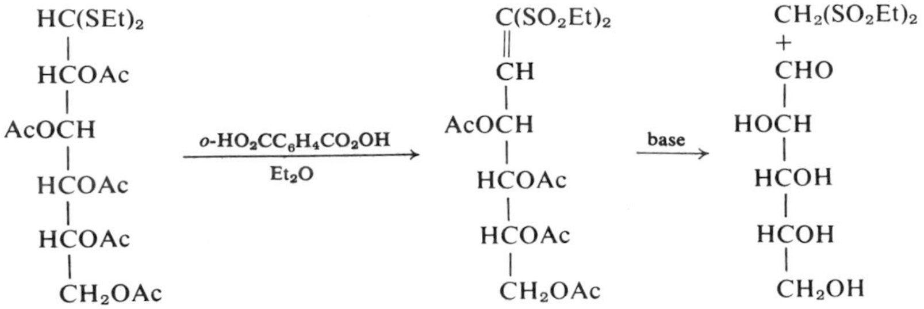

$$
\begin{array}{ccc}
HC(SEt)_2 & C(SO_2Et)_2 & CH_2(SO_2Et)_2 \\
| & \parallel & + \\
HCOAc & CH & CHO \\
| & | & | \\
AcOCH & AcOCH & HOCH \\
| \xrightarrow[Et_2O]{o\text{-}HO_2CC_6H_4CO_2OH} & | \xrightarrow{\text{base}} & | \\
HCOAc & HCOAc & HCOH \\
| & | & | \\
HCOAc & HCOAc & HCOH \\
| & | & | \\
CH_2OAc & CH_2OAc & CH_2OH \\
\mathbf{7} & \mathbf{8} &
\end{array}
$$

Hough and Richardson[100] explored the reaction of unacetylated hexose dithioacetals with peroxypropionic acid; the sulfone 9 initially formed is only marginally stable, and dehydrates to mixtures of the unsaturated disulfone (10) and the 2,6-anhydro-1-deoxy-1,1-bis(ethylsulfonyl)alditol (11). All three products are degraded by aqueous ammonia to the next lower sugar plus bis(ethylsulfonyl)methane. This degradation is much slower for 10 than for 9, and it was concluded that hydration of 10 to 9 precedes disproportionation. Displacement of the bis(ethylsulfonyl)methyl carbanion is favored by resonance delocalization of the negative charge into the sulfone groups. Addition to the double bond had been demonstrated by MacDonald and Fischer[98] in the reaction of **8** with ammonia to give a 2-acetamido-1,2-dideoxy-1,1-bis(ethylsulfonyl)hexitol.

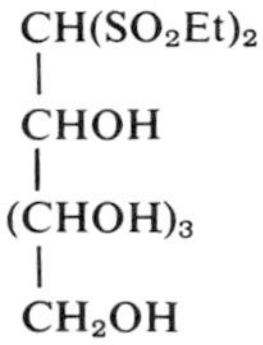

$CH(SO_2Et)_2$
|
CHOH
|
$(CHOH)_3$
|
CH_2OH

9

$C(SO_2Et)_2$
||
CH
|
$(CHOH)_3$
|
CH_2OH

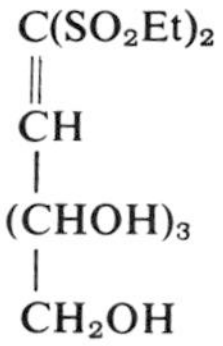

10

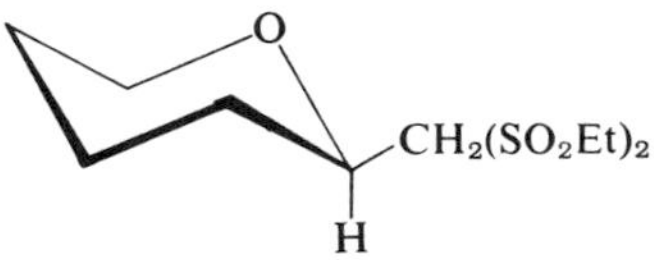

11

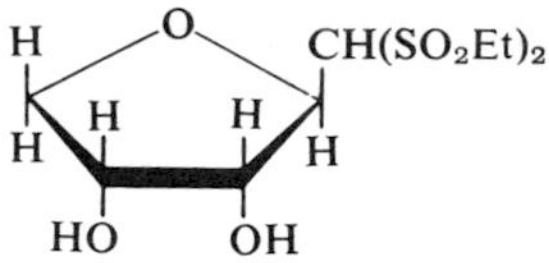

12

The cyclodehydration of **10** to the anhydroalditol derivative **11** has been explained as an attack by the 6-hydroxyl group on the double bond. Studies by Hall *et al.*[101] of acyclic sulfones formed from two heptose dithioacetals epimeric at C-2 revealed their ready conversion in hot water into a common 2,6-anhydride, which was shown by n.m.r. analysis to have the bulky bis-(ethylsulfonyl)methyl group in the "β" (equatorial) position. The sulfone prepared from D-arabinose diethyl dithioacetal cyclizes to 2,5-anhydro-1-deoxy-1,1-bis(ethylsulfonyl)-D-ribitol (**12**).[101a]

The kinetics of the nucleophilic hydrolysis of **11** to the lower sugar plus bis(ethylsulfonyl)methane are complicated; an S_N1 mechanism is favored, with breaking of the carbon–carbon bond occurring in the rate-determining step. Participation by the adjacent hydroxyl group (O-3 in the 2,6-anhydro ring) is not needed for occurrence of the reaction, which is essentially unaltered in the presence of a 3-*O*-methyl substituent. The initial nucleophilic attack is considered to take place on the ring-oxygen atom.

Although peroxy acids as reagents in sugar chemistry have been confined largely to oxidations of dithioacetals, they have also been used in the hydroxylation of alkenes to give the dihydroxy derivatives. Such hydroxylation is normally *trans*; the initial epoxide, formed by *cis* addition to the double bond, suffers inversion of stereochemistry in the course of hydrolysis, to generate a *trans*-glycol grouping ultimately.

Griesebach *et al.*[102] used monoperoxyphthalic acid to convert methyl 3-hydroxy-3-methyl-4-hexenoate into the 3-epimeric mycaric lactones.

References start on p. 1156.

Peroxybenzoic acid was used similarly with *trans*-1-acetoxy-4-ethoxy-2,4-pentadiene (EtOCH=CH—CH=CHCH$_2$OAc) in a preparation of DL-arabinose and DL-ribose by Iwai and Tomita.[103] Aspinall and King[103a] used *m*-chloroperoxybenzoic acid to epoxidize 1,4,6-tri-*O*-acetyl-3-deoxy-2-*O*-methyl-α-D-*erythro*-hex-2-enopyranose (**14**, R = Ac); exposure to water caused hydrolysis of the 2,3-epoxide group of **13**, to afford mainly 1,4,6-tri-*O*-acetyl-α-D-*ribo*-hexopyranosulose, which was degraded to D-ribonic acid derivatives by the action of the same peroxy acid in trifluoroacetic acid. 1-*O*-Acetyl-3-deoxy-2,4,6-tri-*O*-methyl-α-D-*erythro*-hex-2-enopyranose (**14**, R = Me) reacts with *m*-chloroperoxybenzoic acid to give the mixed orthoperoxyanhydride **15**, which decomposes spontaneously, in the presence of trifluoroacetic acid, to D-ribonolactone derivatives.[103b]

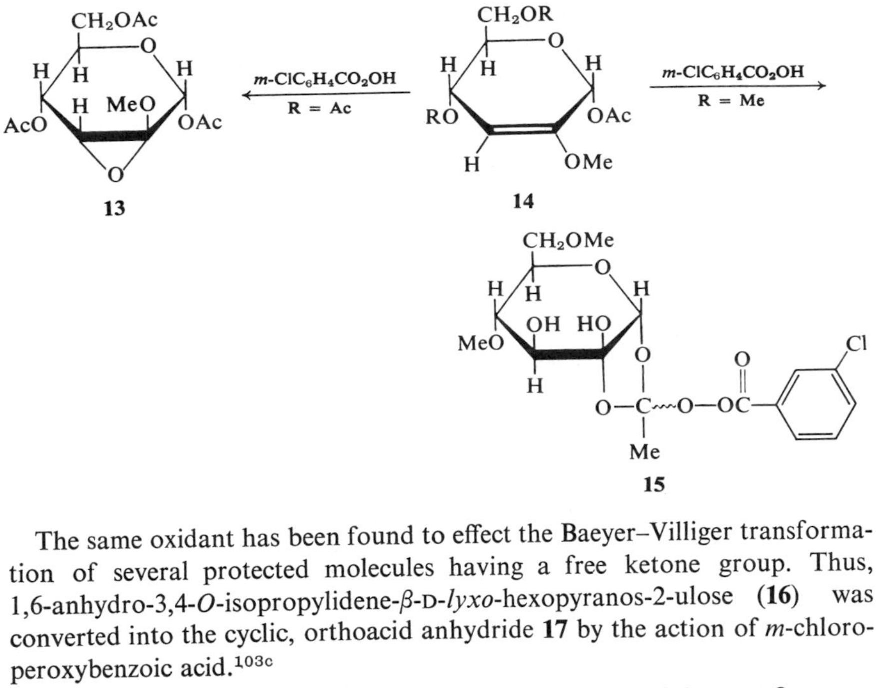

The same oxidant has been found to effect the Baeyer–Villiger transformation of several protected molecules having a free ketone group. Thus, 1,6-anhydro-3,4-*O*-isopropylidene-β-D-*lyxo*-hexopyranos-2-ulose (**16**) was converted into the cyclic, orthoacid anhydride **17** by the action of *m*-chloroperoxybenzoic acid.[103c]

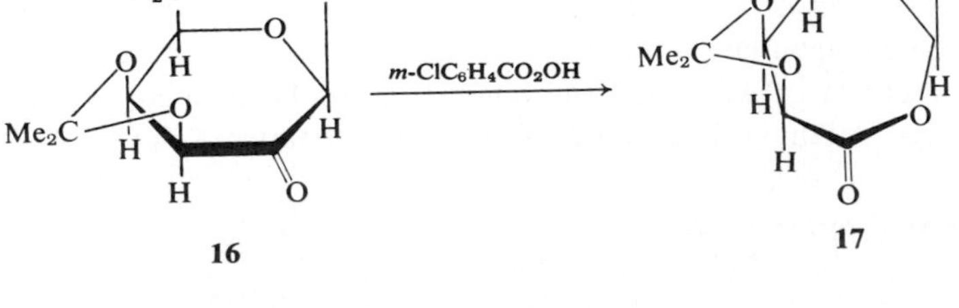

IV. PHENYLHYDRAZINE AS AN OXIDANT

Under controlled, acidic conditions, an excess of phenylhydrazine acts specifically to convert the (usually terminal) —CO—CHOH— grouping in aldoses or ketoses into a bis(phenylhydrazone) residue, which undergoes ready hydrolysis to liberate a —CO—CO— grouping (see Chapter 23); the bis(phenylhydrazones) are also termed phenylosazones (see Chapter 21). Simultaneously, reductive cleavage of the nitrogen–nitrogen bond of the oxidant gives aniline and ammonia as products.

In some respects, hydrazine might be regarded as a nitrogen analogue of hydrogen peroxide; however, hydrazine is a base and a reducing agent, not an oxidant (for example, in the Huang–Minlon modification of the Wolff–Kishner reduction).[103d] Free phenylhydrazine is a weaker base than hydrazine, but it is not an oxidant; the conjugate acid ($PhNH—NH_3^+$) is, however, an electrophile and an effective oxidant. Thus, phenylhydrazine in neutral solution reacts with sugars to afford only the corresponding phenylhydrazones, whereas oxidation with formation of the phenylosazones, occurs in acid solution.[104] (Nitrophenyl)osazone formation is faster, as the electron-withdrawing nitro group decreases the basicity of the substituted phenylhydrazine, whereas 1-methyl-1-phenylhydrazine is a predictably less effective oxidant, because the electron-releasing methyl group enhances the basicity of the 1-methyl-1-phenylhydrazine.

Phenylosazone formation from D-fructose is faster than from D-glucose. The primary alcoholic group adjacent to the ketone, having two C–H bonds, is statistically favored for undergoing oxidation (relative to a secondary alcoholic group). More important, the primary group is isolated, in contrast to a secondary alcohol group, which is surrounded by polar groups; the inductive effects of polar substituents decrease the electron density on the carbon atom affected, and disfavor the removal of a hydride ion (or the transfer of electrons). The weaker oxidizing ability of 1-methyl-1-phenylhydrazine is apparent from the observation that, under normal conditions, only ketoses (and not aldoses) are oxidized by this reagent to form 1-methyl-1-phenylosazones.

Chapman et al.[105] accomplished the complete oxidation of all the alcohol groups of several sugars with an excess of 1-methyl-1-phenylhydrazine; the products are trivially known as *alkazones*, and the same product is obtained from aldoses, such as D-arabinose and D-xylose, that have the same structure from C-3 to C-ω. Pentoses and smaller sugars react at similar rates, whereas hexoses react only slowly. (See also Chapter 21.)

Bamdas et al.[106] studied osazone formation with D-fructose and specifically

---

labeled (*p*-nitrophenyl)hydrazine; the isotopic composition of the ammonia evolved was determined. The results of this experiment were in accord with the mechanism proposed by Weygand[107] (Eq. 19), which proceeds via a 1,4 elimination; the less basic is the $ArNH_2$ molecule, the more readily is it eliminated as a leaving group. Simon *et al.*[108] used tritiated D-fructose to study the formation of phenylosazones, and established that breaking of the carbon–hydrogen bond on C-1 of this 2-ketose is the rate-determining step; the slow step is somewhat stereoselective, because the two hydrogen atoms on C-1 are not equivalent (see Chapter 21).

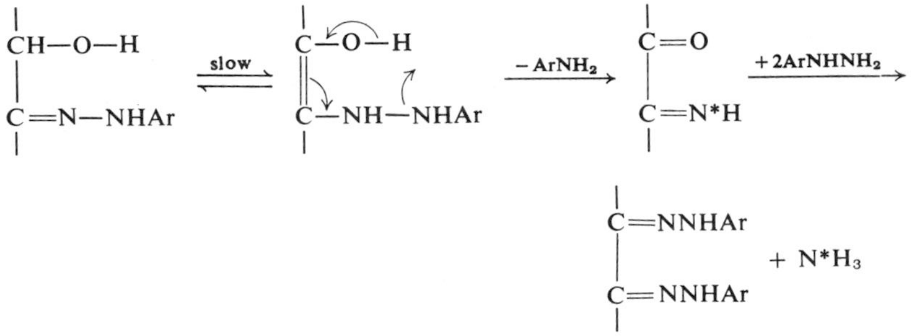

$$ (19) $$

V. OXYGEN IN ALKALINE AND NEUTRAL SOLUTION

Study of the action of molecular oxygen on sugars is of considerable interest insofar as it relates to the mechanism of the *in vivo* oxidation of sugars; the influence of configurational and conformational features on the course of catalytic oxidations is a related area of study. Molecular oxygen has synthetic utility for the degradation of sugars to acids having shorter chains (see Chapter 3).

A. OXIDATION IN ALKALINE SOLUTION

In alkaline solution, oxygen degrades aldoses to aldonic acids having one carbon atom fewer. Air or oxygen may be used, and relatively high yields of acids are obtained.[109] For example, potassium D-arabinonate was prepared from D-glucose (in 60 to 75% yield) and from a series of cello-oligosaccharides.[109a,b] Ketoses act similarly; oxidation of L-sorbose affords L-*xylo*-2-hexulosonic acid plus L-xylonic acid in good yield.[110] Formation of the

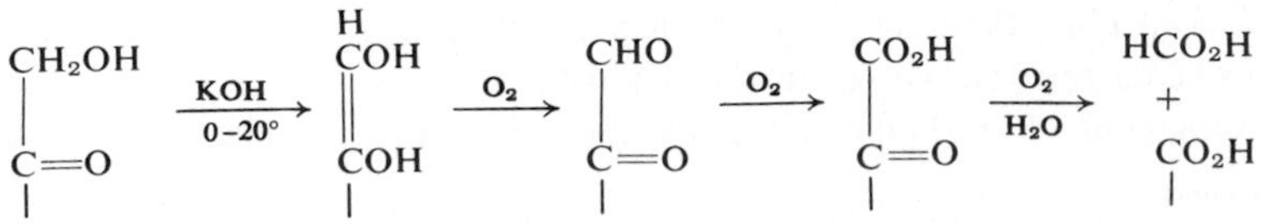

$$ (20) $$

glyculosonic acid indicates that an aldosulose, which is probably formed from the enediol, occurs as an intermediate in the reaction.

3-O-(α-D-Glucopyranosyl)-D-arabinonic acid and its β anomer were prepared by oxidation of maltose and cellobiose, respectively[111]; 3-O-(β-D-galactopyranosyl)-D-arabinonic acid is formed similarly from lactose. Quantitative examination of the products formed by the action of oxygen on D-glucose in aqueous potassium hydroxide solution showed that the main products are D-arabinonic acid and formic acid; lactic, oxalic, and carbonic acids are formed in minor proportions.[112]

In its ground state, the oxygen molecule exists as a diradical; it is a stable molecule in isolation, and its reactivity is greatly enhanced in the presence of catalysts. In alkali-catalyzed auto-oxidation, fragmentation of carbon–hydrogen bonds is either caused (Eq. 29) or greatly facilitated (Eq. 21) by alkali; subsequent transfers of electrons are relatively rapid and easy processes.

The reaction of oxygen with cellulose in alkali (auto-oxidation or alkaline aging) was interpreted by Entwistle *et al.*[113] as being a free-radical process; the initiation steps are: loss, from an activated molecule (for instance, an enol), of a labile hydrogen atom (Eq. 23), capture of oxygen by the radical to generate a hydroperoxy radical (Eq. 24), and abstraction of a hydrogen atom from the substrate (RH) to afford the radical R· (Eq. 25).

$$R^*H + O_2 \xrightarrow{\text{slow}} R^*_\cdot + HOO\cdot \tag{21}$$

$$R^*_\cdot + O_2 \longrightarrow R^*OO\cdot \tag{22}$$

$$R^*OO\cdot + RH \longrightarrow R^*OOH + R\cdot \tag{23}$$

$$R\cdot + O_2 \longrightarrow ROO\cdot \tag{24}$$

$$ROO\cdot + RH \longrightarrow ROOH + R\cdot \tag{25}$$

Once it is formed, the radical R· is constantly regenerated by a chain process (Eqs. 26 and 27). The hydroperoxides formed according to Eqs. 23 and 25 can decompose in various ways to give an oxidation product (Eq. 26), or to produce oxy radicals (Eq. 27) that can react with molecules of RH (Eq. 28), thus accelerating its oxidation.

$$ROOH + H_2O \longrightarrow ROH + HOOH \tag{26}$$

$$ROOH \longrightarrow RO\cdot + \cdot OH \tag{27}$$

$$RO\cdot + RH \longrightarrow ROH + R\cdot \tag{28}$$

References start on p. 1156.

Gersman *et al.*[114] proposed another mode of initiation, which proceeds by deprotonation of RH to a carbanion (Eq. 29) that transfers one electron to an oxygen molecule (Eq. 30); capture of oxygen (Eq. 24) produces a peroxy radical that oxidizes another molecule of the initial carbanion (Eq. 31), the last two steps constituting a chain-propagation process.

$$RH + OH^- \longrightarrow R^- + H_2O \tag{29}$$

$$R^- + O_2 \longrightarrow R\cdot + O_2{}^{\overline{\cdot}} \tag{30}$$

$$ROO\cdot + R^- \longrightarrow ROO^- + R\cdot \tag{31}$$

Direct combination of the carbanion with oxygen is considered to be a highly improbable initiation process; Russell[115] pointed out that such a step would require bond creation and a change in multiplicity.

$$R^-: + \uparrow \ddot{O}{-}\ddot{O}: \longrightarrow R:\ddot{O}{-}\ddot{O}:^- \tag{32}$$

Mattor[116] proposed an ionic mechanism to account for early failures to detect appreciable amounts of free-radical intermediates. The mechanism, which is based on hydride transfer (Eq. 33) from a doubly ionized aldehydrol group and subsequent attack by the hydroperoxyl anion thus produced on the cellulose molecule, incorporates a step similar to the one (Eq. 32) disputed by Russell.[115]

$$RCH\begin{smallmatrix} O^- \\ \\ O^- \end{smallmatrix} + O_2 \longrightarrow RCO_2^- + HOO^- \tag{33}$$

Later studies verified the intermediate formation of free radicals. Arthur and Hinojosa[116a] reported e.s.r. spectra obtained during auto-oxidation of cellulose, and hydrogen peroxide was identified[116b] as a by-product in the auto-oxidation of D-glucitol. Similar oxidations of cellulose in the presence of alkenic monomers afforded graft copolymers. Auto-oxidation of cellulose and of the cello-oligosaccharides is more extensive in the presence of transition-metal cations,[116c] and is suppressed by magnesium(II)[109a,116b,d] or iodide ions.[116b] Treatment with alkali also causes nonoxidative transformations in cellulose prior, and in addition, to auto-oxidation.[116e] Reviews have been published.[116f] (See also Chapter 26.)

B. CATALYTIC OXIDATION

In the presence of platinum catalysts in neutral solution, the main process effected on sugars by oxygen is dehydrogenation.[117] Isotopic oxygen is not incorporated into the organic product, and the dehydrogenation is not reversible in the presence of $H_2{}^{18}O$. The oxidation is a radical process, and initial attack is on a C–H bond (Eq. 4); the stability of the resulting radical is the important factor (Eq. 5). As it occurs on a catalytic surface, the reaction is profoundly influenced by steric effects: (1) generally, primarily alcoholic groups are oxidized rather than secondary alcoholic groups; (2) when only secondary hydroxyl groups on a pyranoid ring are available, the favored attack is on a carbon atom bearing an axially attached hydroxyl group; (3) in the case of rigid, bicyclic molecules, the attack favored is on the carbon atom bearing an *endo* hydroxyl group.

Heyns and Beck[118] obtained both L-gulose and D-glucose from platinum-catalyzed oxidation of D-glucitol. The specificity of this reaction is probably due to both statistical and steric factors. A related example is the conversion of L-*xylo*-2-hexulosonic acid in 62% yield[119,120]; Russian workers claimed an 88% yield.[121] Similar oxidation of 1,2-*O*-isopropylidene-α-D-glucofuranose[120] and 1,2-*O*-cyclohexylidene-α-D-xylofuranose[122] afforded the respective glycuronic acids in 53% and 40% yields, whereas, under more drastic conditions, the former acetal was converted into 1,2-*O*-isopropylidene-α-D-*xylo*-hexo-1,4-furanos-5-ulosono-6,3-lactone, a synthetic precursor of ascorbic acids.[122a]

This method of oxidation has been extensively used in the preparation of alkyl and aryl glycopyranosiduronic acids. Barker *et al.*[123] so prepared methyl α-D-glucopyranosiduronic acid and its β anomer, and numerous other examples have been recorded.[124] This approach has also been applied to the oxidation of other glycopyranosides, including aldopentofuranosides.[125] Disaccharides are suitable for this reaction; benzyl β-maltoside was oxidized to the biouronic acid,[126] whereas the analogous reaction of methyl β-maltoside yielded all three possible glycuronate disaccharides, indicating random attack at both primary hydroxyl groups.[127] Aspinall *et al.*[128] oxidized an arabino-galactan in order to convert the arabinofuranosyl residues into arabino-furanosiduronic acid residues and thereby increase the resistance of the polymer to acid hydrolysis.[128]

Favored oxidation of axially attached secondary alcoholic groups occurs when primary alcoholic groups are either absent or protected. Thus, Heyns and Lenz[129] converted benzyl β-D-arabinopyranoside into a 4-glyculoside, and β-D-ribopyranoside into a 3-glyculoside, correlating with attack on the equatorially attached hydrogen atoms of the (axial) alcohol groups; benzyl β-D-xylopyranoside, which has no equatorial hydrogen atoms available, is oxidized very slowly. Lindberg *et al.*[130] obtained an 82% yield of the 2-keto derivative from similar oxidation of methyl 4,6-*O*-ethylidene-α-D-manno-pyranoside. Brimacombe *et al.*[131] obtained 2-, 3-, and 4-glycosidulose deriva-tives from 6-deoxy-manno-, -allo-, and -galactopyranosides, respectively; the

---

6-deoxyglucopyranoside was virtually inert to oxidation. Both benzyl and methyl D-xylopyranoside have been shown to react to a certain extent; for the pentoside, which has no bulky hydroxymethyl group stabilizing the $^4C_1(D)$ conformation (18), ring inversion into the $^1C_4(D)$ conformation (19) presumably precedes attack by the oxidant. Oxidation of a series of partially protected acetals of ketoses revealed that primary hydroxyl groups react faster than axial secondary hydroxyl groups, that proximity to the ketogenic center slows the reaction of the former more than the latter, and that equatorial secondary hydroxyl groups are nearly inert.[131a]

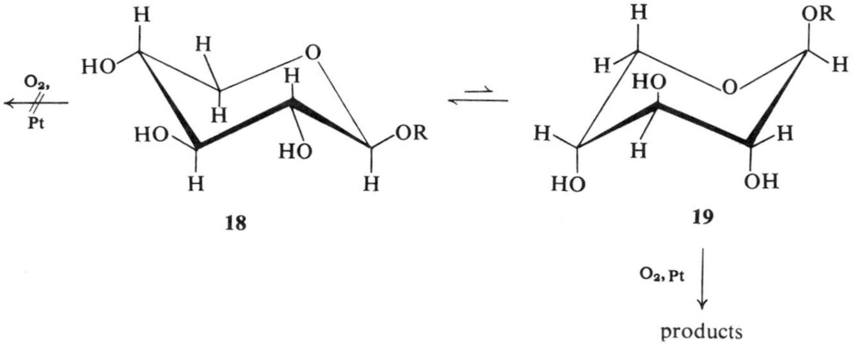

Favored attack was demonstrated for anhydro sugars. The *endo* alcohol groups in 1,4:3,6-dianhydro-D-mannitol are oxidized to afford 1,4:3,6-dianhydro-D-*threo*-hexo-2,5-diulose as the product.[132] 1,6-Anhydro-β-D-hexopyranoses undergo favored oxidation at hydroxyl groups in the order[132a]: $3ax > 4ax > 2ax > 4eq > 2eq > 3eq$. This procedure has been recommended for the preparation of 1,6-anhydro-β-D-*xylo*-hexopyranos-3-ulose from 1,6-anhydro-β-D-galactopyranose.[132b] 1,4-Anhydrohexitols exhibit selective oxidation of hydroxyl groups in the order: primary > quasi-axial > quasi-equatorial ~ side chain.[132c] Antonakis and Leclerq reported[132d] that 1,2-O-isopropylidene-6-O-p-tolylsulfonyl-β-D-glucofuranose reacts with oxygen in the presence of a platinum catalyst to produce only a 3-hexulofuranose derivative, owing to the effect of the sulfonyl group.

C. Oxidation under Neutral Conditions

Under conditions simulating biological processes (neutral, aqueous solutions at a temperature of 37.5°), oxygen attacks D-glucose, glyceraldehyde, glycerol, and related polyhydric alcohols.[133] One mole of carbon dioxide is formed per mole of D-glucose; sodium ferropyrophosphate has been used as the catalyst. D-Fructose is much more sensitive than D-glucose in the presence of a phosphate or arsenate as the catalyst, and the rate of oxidation depends on the concentration of salt present, but not[134] on the pH.

Alkaline solutions of aldonic acids and alditols are relatively stable to oxygen in the absence of a catalyst; however, oxygen is consumed in the presence of salts of cobalt, copper, iron, or nickel.[135] Carbon dioxide and formic acid are among the oxidation products.

From the standpoint of the conditions encountered during the manufacture of sucrose, the action of oxygen on sucrose solutions in the presence and absence of lime is important. Carbon dioxide is liberated from hot, neutral solutions, and acids are formed. The increased acidity then results in inversion of the sucrose and decomposition of the resulting hexoses, whereas increased alkalinity enhances oxidation.[136]

The reaction of cellulose with oxygen or air at 1 atmosphere at 170° has been investigated[137]; a later study[137a] fitted kinetic data from the latter reaction to a rate equation that was calculated to account for the simultaneous operation of two different mechanisms. Two studies were made with oxygen at pressures of ~ 50 atmospheres; in one, crystalline methyl β-D-glucopyranoside was oxidized[138] at 100° to 150°, and in the other, an aqueous solution of uniformly labeled D-[^{14}C]glucose underwent oxidation[139] at 120° to 140°. Both reactions were found to fit rate expressions to the order 1.5 in carbohydrate (Eq. 34):

$$\text{Rate} = k[\text{RH}]^{1.5}[\text{O}_2]^{0.5} \tag{34}$$

This rate order is considered typical of a nonautocatalytic, free-radical mechanism that is initiated by attack by molecular oxygen.[140] The initial product in the oxidation of D-glucose was D-gluconic acid, which was slowly converted into D-arabinose and D-arabinonic acid; this decarboxylation is similar to that encountered in the Ruff degradation (see Section VI,A). The initial reaction was considered to be the formation of a hydroperoxyl radical (Eq. 35).

$$\text{RH} + \text{O}_2 \longrightarrow \text{R·} + \text{HOO·} \tag{35}$$

VI. HYDROGEN PEROXIDE

Hydrogen peroxide is a very ineffective oxidant in a polar solvent at acidic or neutral pH; formation of a hydroxide ion in the transition state (Eq. 7) is quite difficult. Use of this oxidant in the presence of organic acids generally involves intermediate conversion into peroxy acids, which have better leaving groups for displacement (see Section III). In alkaline solutions, hydrogen peroxide has been used as a bleaching agent; its initial action on cellulose[141] and on amylopectin[142] is depolymerization. Oxidation seems to occur mainly on reducing end-units and other partially oxidized positions; the (strongly nucleophilic) hydroperoxide anion is more likely to attack polarized carbonyl than alcohol groups. Isbell *et al.* showed that alkaline peroxide degrades aldoses,[142a] ketoses,[142b] and reducing disaccharides[143c] sequentially to formic acid, by a series of enolization and oxidation processes. With glycuronic and glyculosonic acids, oxalic acid as well as formic acid is a product.[142d] Hydroperoxy intermediates have been proposed[142e] in the oxidation of methyl β-D-glucopyranoside.

Use of catalysts, generally ferric or ferrous salts, promotes radical reactions with hydrogen peroxide; the oxidizing action produced by ferrous ions is the more vigorous. Molybdenum(III) chloride was used to catalyze the oxidation of 1,2:5,6-di-O-isopropylidene-α-D-glucofuranose to the 3-glycosulose by 2-(2-methylbutyl)hydroperoxide.[142f]

A. Hydrogen Peroxide and Ferric Ions

Ferric ion catalyzes the formation of the hydroperoxyl radical (Eq. 36), a radical that appears to constitute the oxidant in the Ruff method (see Chapter 3) of degrading aldonic acids to the next lower aldoses.[143] The hydroperoxyl radical, which is not so effective an oxidant as the hydroxyl radical, does not attack aliphatic alcohols; accordingly, an appreciable yield (about 50%) of the aldose is obtained from the higher aldonic acid. In the presence of an excess of hydrogen peroxide, however, the accumulation of ferrous ions in solution catalyzes the production of hydroxyl radicals and lowers the yield of aldose. (See Eq. 40.)

$$Fe^{3+} + HOOH \longrightarrow Fe^{2+} + H^+ + \cdot OOH \tag{36}$$

The mechanism of this reaction has not yet been investigated. (A possible sequence of steps in the decarboxylation is given in Eqs. 37–39.)

$$RCHOHCO_2H + \cdot OOH \longrightarrow RCHOHCO_2\cdot + HOOH \tag{37}$$

$$RCHOH-\overset{O\cdot}{\underset{O}{C}} \longrightarrow R\dot{C}HOH + CO_2 \tag{38}$$

$$R\dot{C} \underset{H}{\overset{|}{-}} O-H \longrightarrow RCHO + H\cdot \tag{39}$$

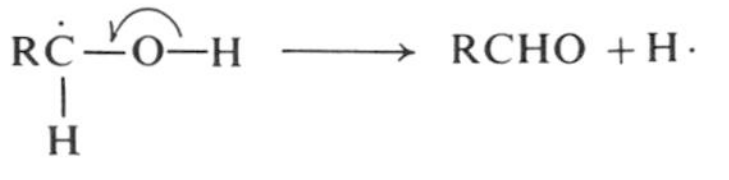

Hydrogen peroxide in acetonitrile has been shown to convert a pento-furanosidulose oxime into the corresponding nitro derivative.[143a]

B. HYDROGEN PEROXIDE AND FERROUS IONS

A mixture of hydrogen peroxide and a ferrous salt is known as the Fenton reagent.[144] According to Haber and Weiss,[145] the ferrous salts bring about the formation of hydroxyl radicals (Eq. 40). Hydroxyl radicals are very effective in abstracting hydrogen atoms, in contrast to the weaker action ascribed

$$Fe^{2+} + HOOH \longrightarrow Fe^{3+} + HO\cdot + OH^- \qquad (40)$$

(Section VI,A) to hydroperoxyl radicals. Waters[146] suggested that neutral hydroxyl radicals are the oxidizing species in ferrous-ion-catalyzed oxidations of α-hydroxy acids to 2-glyculosonic acids:

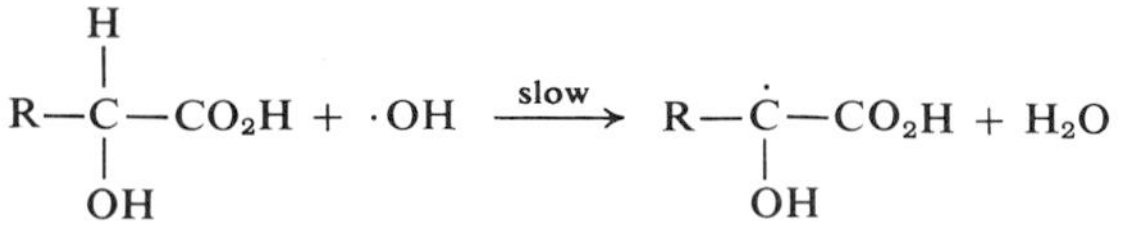

$$\underset{\underset{OH}{|}}{\overset{\overset{H}{|}}{R-C-CO_2H}} + \cdot OH \xrightarrow{\text{slow}} \underset{\underset{OH}{|}}{R-\overset{\cdot}{C}-CO_2H} + H_2O \qquad (41)$$

$$\underset{\underset{OH}{|}}{R-\overset{\cdot}{C}-CO_2H} + HOOH \xrightarrow{-HO\cdot} \underset{\underset{OH}{|}}{\overset{\overset{OH}{|}}{R-C-CO_2H}} \xrightarrow{-H_2O} \underset{R}{\overset{O}{\underset{}{\overset{||}{C}}}}\diagdown_{CO_2H} \qquad (42)$$

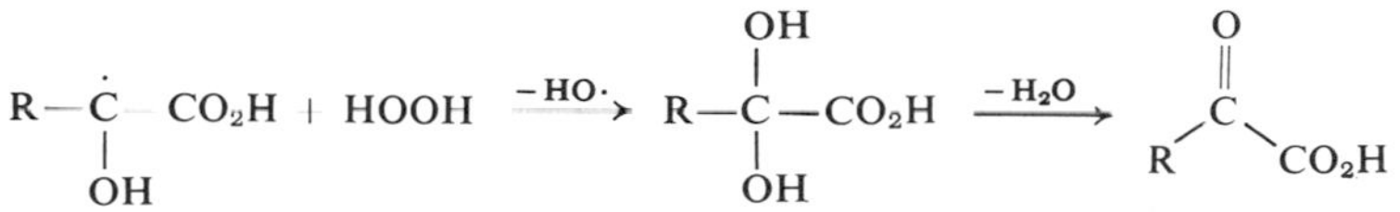

The slow step (Eq. 41) involves abstraction of the carbon-bonded hydrogen atom by a hydroxyl radical, which is converted into water; regeneration of the hydroxyl radical in the next step maintains the chain reaction.

At low temperatures, D-glucose (20) and D-fructose in the presence of ferrous sulfate are converted into D-*arabino*-hexos-2-ulose (21), which can be degraded by further oxidation to glycolic acid, glyoxylic acid, and D-erythronic acid.[147] Cold, uncatalyzed oxidation is very slow, but, at high temperatures,[148] the main product is carbon dioxide together with some formic acid.

The nature of the products formed under various conditions and the mechanism of the reaction were investigated by Küchlin.[149] In dilute solution in the presence of ferrous sulfate at low temperature, compound 21, D-*arabino*-2-hexulosonic acid (22), and D-*erythro*-hexo-2,3-diulosonic acid (23) were formed from D-glucose and identified as derivatives; in concentrated

---

References start on p. 1156.

solutions, formaldehyde was also found. The formation of these products at low temperature was ascribed to the following series of reactions:

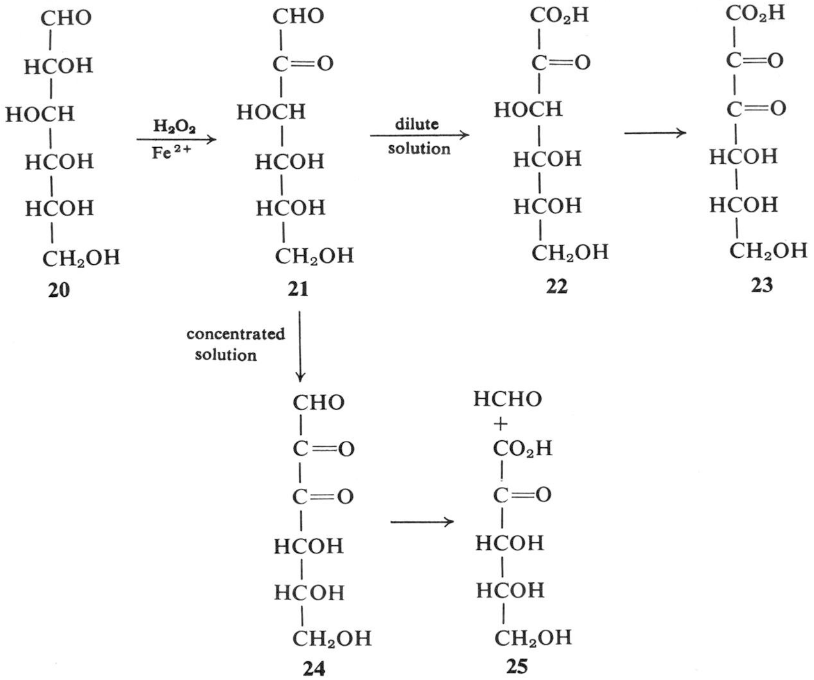

$$
\begin{array}{ccccc}
\text{CHO} & \text{CHO} & \text{CO}_2\text{H} & \text{CO}_2\text{H} \\
| & | & | & | \\
\text{HCOH} & \text{C}=\text{O} & \text{C}=\text{O} & \text{C}=\text{O} \\
| & \xrightarrow[\text{Fe}^{2+}]{\text{H}_2\text{O}_2} & | & \xrightarrow[\text{solution}]{\text{dilute}} & | & \longrightarrow & | \\
\text{HOCH} & \text{HOCH} & \text{HOCH} & \text{C}=\text{O} \\
| & | & | & | \\
\text{HCOH} & \text{HCOH} & \text{HCOH} & \text{HCOH} \\
| & | & | & | \\
\text{HCOH} & \text{HCOH} & \text{HCOH} & \text{HCOH} \\
| & | & | & | \\
\text{CH}_2\text{OH} & \text{CH}_2\text{OH} & \text{CH}_2\text{OH} & \text{CH}_2\text{OH} \\
\textbf{20} & \textbf{21} & \textbf{22} & \textbf{23}
\end{array}
$$

concentrated solution

$$
\begin{array}{ccc}
\text{CHO} & & \text{HCHO} \\
| & & + \\
\text{C}=\text{O} & & \text{CO}_2\text{H} \\
| & & | \\
\text{C}=\text{O} & \longrightarrow & \text{C}=\text{O} \\
| & & | \\
\text{HCOH} & & \text{HCOH} \\
| & & | \\
\text{HCOH} & & \text{HCOH} \\
| & & | \\
\text{CH}_2\text{OH} & & \text{CH}_2\text{OH} \\
\textbf{24} & & \textbf{25}
\end{array}
$$

At higher temperatures, carbon dioxide, formic acid, oxalic acid, glycolic acid, hydroxymalonic acid, glyceric acid, and other acids were shown to be formed. The formation of carbon dioxide is ascribed to decarboxylation of **23**, and oxalic acid and D-erythronic acid arise from cleavage of the C-2–C-3 bond; compound **24** is cleaved to glyoxylic acid plus D-erythronic acid. Compound **25** is oxidized further to D-*glycero*-2,3-pentodiulosonic acid and is subsequently cleaved to oxalic and glyceric acids.

The effect of variations in the conditions of the reaction has been studied by Küchlin.[150] Ferrous sulfate-catalyzed oxidation of D-fructose attains the maximum rate of reaction between pH 3.2 and 5.4. The effects of an increase in temperature on the reaction is small in strongly acid solution, but increases as the solution becomes more alkaline. The initial reaction velocity is proportional to the concentrations of catalyst and of hydrogen peroxide, whereas the rate is independent of the proportion of *ferric* salts.

Primary alcoholic groups are oxidized to aldehydes by the action of peroxide and ferrous ions. D-Mannose (as the phenylhydrazone) has been synthesized[151] from D-mannitol in yields of ~40%. In the absence of ferrous ions, even with ferric ions present, no reaction occurs. Presumably, the proportion of ferrous ions present is critical, for it would be expected that with sufficient catalyst the reaction would proceed further, as for **20**.

The action of the Fenton reagent on 3,4-di-O-methyl-D-mannitol gives not only 3,4-di-O-methyl-D-mannose but also 3-O-methyl-D-mannose, D-mannitol, and 3-O-methyl-D-mannitol[152]; the symmetry (C_2) of the starting alditol limits the number of isomeric products. This demethylation shows that the hydroxyl radical attacks the C—O—Me grouping as well as primary alcoholic carbon atoms; the alkoxymethyl radical formed by abstraction of hydrogen from the methoxyl group can be stabilized by resonance interaction with the adjoining oxygen atom (as in Eq. 5),

Oxidation of 1 mole of methyl β-D-glucopyranoside with 1.5 moles of hydrogen peroxide in the presence of ferrous ion gives small proportions of pyranosiduloses and D-glucose; undetermined amounts of D-erythronic and D-arabinonic lactones were also formed.[153]

In the absence of a catalyst, the oxidation may proceed by quite a different mechanism. D-Glucuronic acid has been prepared[154] in low yield by uncatalyzed oxidation of D-glucose (**20**) with hydrogen peroxide at 37°. Küchlin[149] provided evidence that the main products at moderately high temperatures are formic acid and tartronic acid. He suggested that the oxidation proceeds through initial formation of glycuronic acid and subsequent, serial, oxidative cleavage of the terminal bond to generate formic acid. The formation of carbon dioxide was considered to be a secondary reaction. Initial oxidation at the primary alcoholic group is, presumably, a general reaction.

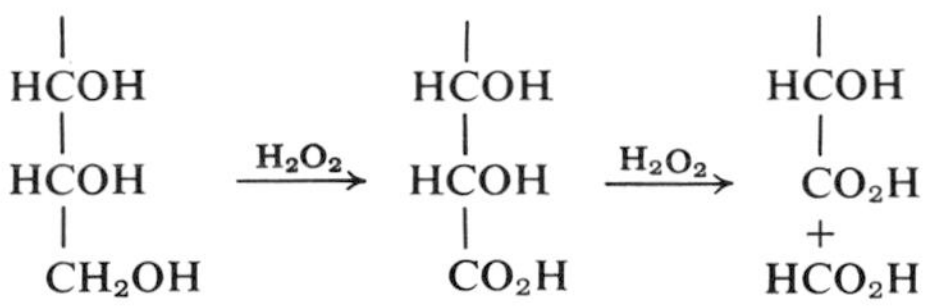

$$
\begin{array}{ccccc}
| & & | & & | \\
HCOH & & HCOH & & HCOH \\
| & & | & & | \\
HCOH & \xrightarrow{H_2O_2} & HCOH & \xrightarrow{H_2O_2} & CO_2H \\
| & & | & & + \\
CH_2OH & & CO_2H & & HCO_2H \\
\mathbf{20} & & & &
\end{array}
$$

The use of hydrogen peroxide with a platinum catalyst has been reported briefly.[155] The action is similar to that of oxygen, affording crude methyl α-D-galactopyranosiduronic acid in $\sim 40\%$ yield from methyl α-D-galactopyranoside at 20°.

VII. NITRIC ACID AND NITROGEN DIOXIDE

Nitric acid is a strong acid; it is a potent oxidant, but rather unreactive as its salts. Under strongly acidic conditions, it converts primary alcoholic and aldehydic groups into carboxylic acid groups. Frequently, however, cleavage of carbon–carbon bonds occurs. Conversion of D-galactose (26) into (insoluble) galactaric ("*mucic*") acid (28) occurs to an extent > 70%, and the reaction is used for the quantitative determination of this sugar.[156,157] Interruption of this oxidation before completion allows the isolation of L-galacturonic acid (27) in 10% yield.[158] Smaller yields of D-glucaric acid were obtained from D-glucose, together with considerable proportions of oxalic acid and some tetraric acid.[159] Among the products of the oxidation of D-fructose are formic acid, oxalic acid, erythraric acid, and glycolic acid, but the reaction seems to require more severe conditions than for D-glucose; at low temperature, the ketoses are not attacked by 32% acid.

$$
\begin{array}{ccc}
\text{CHO} & \text{CO}_2\text{H} & \text{CO}_2\text{H} \\
| & | & | \\
\text{HCOH} & \text{HCOH} & \text{HCOH} \\
| & | & | \\
\text{HOCH} & \text{HOCH} & \text{HOCH} \\
| & | & | \\
\text{HOCH} & \text{HOCH} & \text{HOCH} \\
| & | & | \\
\text{HCOH} & \text{HCOH} & \text{HCOH} \\
| & | & | \\
\text{CH}_2\text{OH} & \text{CHO} & \text{CO}_2\text{H} \\
\mathbf{26} & \mathbf{27} & \mathbf{28}
\end{array}
$$

Oxidation of methylated sugars with nitric acid was used extensively by early workers for finding the position of unsubstituted hydroxyl groups.[159a]

Cleavage of carbon–carbon bonds appears to be facilitated by the presence of such catalysts as vanadium salts; the latter promote formation of tetraric and oxalic acids at the expense of aldaric acids.[160] As hot nitric acid acts as a hydrolyzing agent, as well as an oxidant, oligo- and poly-saccharides may be used directly.

Kiliani[161,162] made an extensive study of the oxidation of carbohydrates with nitric acid. He found that aldoses are oxidized to aldonic and aldaric acids or their lactones. D-Glucose, for example, is oxidized by nitric acid to D-gluconic acid and D-glucaric acid. Alditols are oxidized to aldonic acids; oxidation of glycerol gives DL-glyceric acid. Aldonic acids are oxidized to 2-glyculosonic acids, aldaric acids, and glycuronic acids,[158,162] The formation of these products suggests that the oxidation, by nitric acid, of aldoses without cleavage of carbon–carbon bonds proceeds through the following sequence of transformations:

$$
\begin{array}{cccccc}
\text{CHO} & & \text{CO}_2\text{H} & & \text{CO}_2\text{H} & & \text{CO}_2\text{H} \\
| & & | & & | & & | \\
(\text{CHOH})_n & \longrightarrow & (\text{CHOH})_n & \longrightarrow & (\text{CHOH})_n & \longrightarrow & (\text{CHOH})_n \\
| & & | & & | & & | \\
\text{CH}_2\text{OH} & & \text{CH}_2\text{OH} & & \text{CHO} & & \text{CO}_2\text{H}
\end{array}
$$

The reaction probably proceeds via the cyclic forms of the sugars and lactones, but this refinement of the mechanism has not yet been clarified; under the strongly acidic conditions of these oxidations, however, equilibria between the various cyclic forms and the acyclic form are, presumably, established quickly. In this connection, it should be noted that, whereas D-galactose undergoes oxidation to (acyclic) galactaric acid, similar reaction of D-mannonic acid affords a dilactone.

The isolation of 2- and 5-hexulosonic acids constitutes presumptive evidence that cleavage of carbon–carbon bonds results from further oxidation of such intermediates. Vanadium salts appear to promote cleavage reactions[162a]:

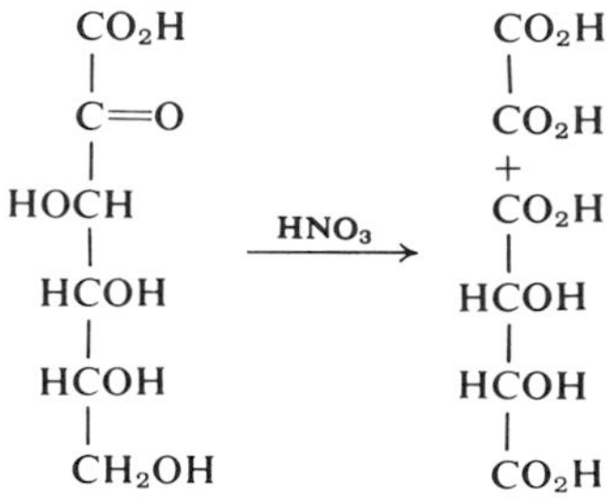

$$
\begin{array}{ccc}
\text{CO}_2\text{H} & & \text{CO}_2\text{H} \\
| & & | \\
\text{C}{=}\text{O} & & \text{CO}_2\text{H} \\
| & & + \\
\text{HOCH} & \xrightarrow{\text{HNO}_3} & \text{CO}_2\text{H} \\
| & & | \\
\text{HCOH} & & \text{HCOH} \\
| & & | \\
\text{HCOH} & & \text{HCOH} \\
| & & | \\
\text{CH}_2\text{OH} & & \text{CO}_2\text{H} \\
\end{array}
$$

22

The specificity of the oxidation may be enhanced by the use of nitrogen dioxide (NO_2) instead of nitric acid.[163] In gaseous form, or in nonaqueous solution, this reagent exhibits a marked specificity for the oxidation of primary alcoholic (and aldehydic) groups (Eq. 43); it converts glycosides into glycuronic acids, and cellulose into a D-glucuronan.

$$
\begin{array}{ccc}
\text{CH}_2\text{OH} & \xrightarrow{\text{NO}_2} & \text{CO}_2\text{H} \\
| & & |
\end{array}
\tag{43}
$$

The oxidation and subsequent hydrolysis of starch to α-D-glucofuranurono-6,3-lactone has been studied[164]; nitric acid alone, or in combination with nitrites, and nitrogen dioxide, have been examined as oxidants, and 14–16N nitric acid in the presence of formic acid was recommended for this oxidation. The solvent of choice for reactions with nitrogen dioxide is carbon tetrachloride, and better results are obtained with high concentrations of this oxidant.

Assarsson and Theander[165] treated methyl β-D-glucopyranoside with liquid nitrogen dioxide for 5 hours at 12°; the presence of free oxygen did not affect the results. About 60% of the glycoside was converted into acidic products, mostly D-glucaric acid; three isomeric hexopyranosiduloses accounted for an additional 6% of the products (see Chapter 23).

References start on p. 1156.

The nature of the oxidant in reactions with nitric acid and nitric oxide has been investigated.[157] Concentrated nitric acid encounters an initial induction period as an oxidizing agent, and it exerts no oxidizing action in the presence of urea, which removes nitrous acid. The induction period is eliminated by the addition of fuming nitric acid, oxides of nitrogen, nitrous acid, or various other compounds.[157,166] Nitrogen dioxide requires the presence of water for its oxidizing action. These observations indicate that the true oxidant is not nitric acid, but that the effective agent is nitrous acid, which, in the presence of nitric oxide, establishes an equilibrium with nitric acid according to the equation[167]:

$$HNO_3 + 2NO + H_2O \rightleftharpoons 3\,HNO_2 \tag{44}$$

The catalytic effect of oxides of nitrogen and of sodium nitrite appears to operate by the establishment of equilibria similar to those in Eq. 44. The oxidizing species from nitrogen dioxide may be similar, as, in the presence of water, a similar equilibrium condition is reached.

$$2\,NO_2 + H_2O \rightleftharpoons HNO_2 + HNO_3 \tag{45}$$

Under conditions simultaneously favorable to the establishment of these equilibria and unfavorable for carbon–carbon bond cleavage, the specificity of the reaction is greatly increased. For example, catalyzed oxidation of D-galactose at low temperatures affords galactaric acid in 90% yield,[157] whereas, with hot nitric acid, the usual yield is only $\sim 75\%$.

Oxidation of primary alcohol groups appears to proceed through the intermediate formation of an ester of nitric (or nitrous) acid.[168] In the initial stages of oxidation by these reagents, cellulose incorporates nitrogen, the

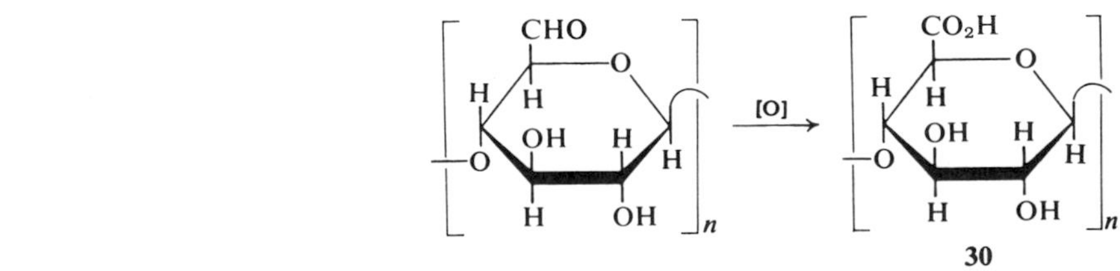

29   30

proportion of which increases and then slowly decreases. Nitric acid appears to act as a catalyst for the de-esterification.

Kinetic studies on the reaction between cellulose and nitrogen dioxide revealed the initial formation[168a] of a nitrous ester (29) that undergoes reaction with an additional molecule of nitrogen dioxide to give dinitrogen trioxide (N_2O_3) and an alkoxyl free radical. The latter reacts optimally with nitrogen containing 7 to 10% of nitric oxide to produce[169] an aldehyde which is subsequently oxidized to the acid 30. Formation of 29 was inhibited by the addition of nitrogen trioxide.

The facile oxidation of primary alcohol groups in D-glucosides and polysaccharides was attributed by Lenshina et al.[170] to (undefined) factors associated with the six-membered ring; this conclusion was based on the slow oxidation of the acyclic "dialcohol" prepared by reduction of periodate-oxidized cellulose with borohydride.

VIII. CHROMIC ACID

Chromic acid, like nitric acid, is a powerful oxidant.[171] The mechanism of its reactions with alcohols has been investigated by Westheimer[172]; chromic esters have been shown to be intermediates in the oxidations. Isopropyl chromate has been isolated and converted into acetone by pyridine-catalyzed elimination in benzene solution (Eq. 2).

Chromic acid (chromium trioxide) is a more generally useful oxidant than either nitric acid or permanganate because it is stable in organic solvents. Oxidation of alcohols by chromic acid is much faster in acetic acid than in mineral acids of the same concentration; this higher rate has been ascribed either to the decrease in dielectric constant, which favors ester formation, or to formation of the more-reactive acetic chromic anhydride.[173] tert-Butyl chromate has been used as an oxidant in acetic acid or in acetic acid–benzene; it acts by rapid transesterification to oxidize primary and secondary alcohols. Mixtures and complexes of chromium trioxide with pyridine have been used. As a reaction solvent, acetone inhibits further oxidation of carbonyl-containing products and is, therefore, extremely useful.

Reagents prepared from chromic acid function most efficiently in the oxidation of isolated, alcoholic groups in partially protected sugar derivatives.[174] Oxidation of methyl glycosides having several unprotected hydroxyl groups with chromium trioxide–acetone gave each of the several keto derivatives possible, but in yields not exceeding 5%; this reactant acted upon methyl β-D-xylopyranoside to afford the pentopyranosid-2-, -3-, and -4-ulose products in 4.3% yield.

A two-phase system has been used to protect products somewhat from further oxidation by isolating the oxidation products in an oxidant-free organic layer.[175] Theander oxidized 1,2-*O*-isopropylidene-α-D-glucofuranose with chromic acid in aqueous butanone; the 15% yield of products contained mostly 1,2-*O*-isopropylidene-α-D-*gluco*-hexo- and -D-*xylo*-pentodialdo-1,4-furanose, together with trace amounts of several aldos-3- and -5-uloses.[176]

Jones *et al.*[177] oxidized the unprotected hydroxyl group of 6-deoxy-1,3:2,5-di-*O*-methylene-L-mannitol to the ketone in 67% yield by the action of chromic acid–acetic acid; Sera and Goto[178] likewise converted 1,5-di-*O*-benzoyl-2,4-*O*-methylenexylitol into the corresponding *erythro*-3-pentulose derivative. However, Angyal and Evans[179] reported that chromic acid in acetic acid–acetic anhydride causes five-membered-ring methylene acetals to undergo oxidation to cyclic carbonic esters, and effects oxidative removal of larger-ring methylene acetal groups.

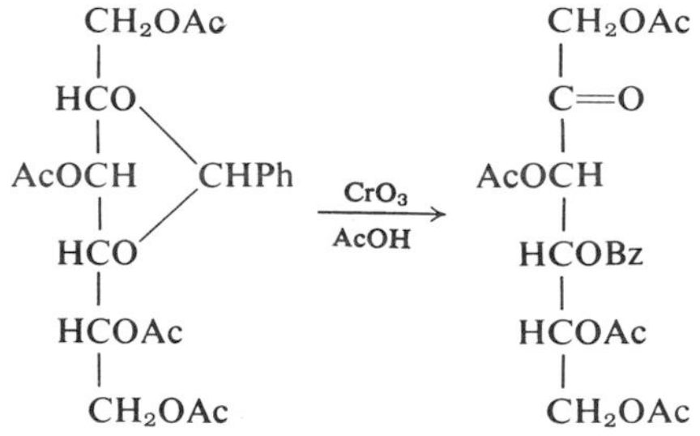

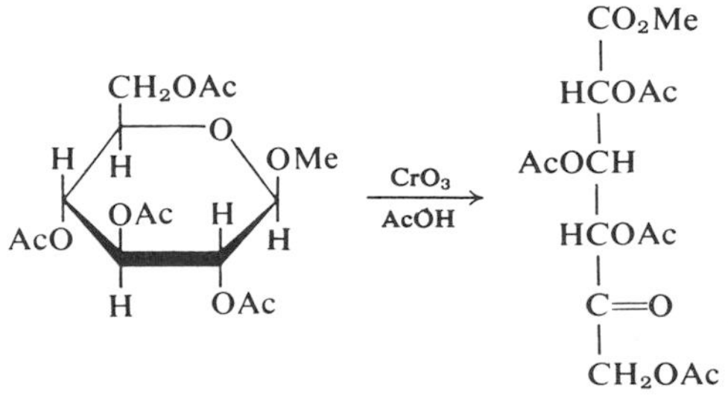

The oxidizability of acetals was investigated by Angyal and James,[180] who found that acetylated methyl glycosides react with chromium trioxide–acetic acid to give acetylated methyl aldonates having a keto group in place of the oxygen atom formerly involved in the furanoid or pyranoid ring; methyl β-D-glucopyranoside tetraacetate (**31**) thus afforded methyl 2,3,4,6-tetra-*O*-acetyl-D-*xylo*-5-hexulosonate (**32**). Primary alkylidene acetals are opened to esters of ketoses by the same reagent[181,182]; 1,3,5,6-tetra-O-acetyl-2,4-O-benzylidene-D-glucitol (33) is thus oxidized to 1,3,5,6-tetra-O-acetyl-4-O-benzoyl-*keto*-D-fructose (34) and a trace of the corresponding 5-O-benzoyl-*keto*-L-*xylo*-3-hexulose.[181] Primary alkyl groups as aglycons[183] or ether substituents[184] are similarly oxidized to acyloxy substituents.

Oxidation of cyclohexanol derivatives, including steroids, is known[185,186] to proceed more rapidly and in higher yield for axial and other hindered alcohols than for their respective epimers. Theander[187] noted the same effect in the oxidation of 1,2:5,6-di-O-isopropylidene-α-D-glucofuranose and its 3-epimer; both undergo oxidation by chromic acid–pyridine at the 3-hydroxyl group to form 1,2:5,6-di-O-isopropylidene-α-D-*ribo*-hexofuranos-3-ulose, the former in 6% yield and the latter in > 50% yield. Platinum-catalyzed oxidation of these isopropylidene acetals was even more strongly affected, the yields being 0% and 80%, respectively, from the foregoing 3-alcohols.

Chromic acid oxidation of β-D-hexopyranosides is much faster than for their anomers,[180] and Hoffman *et al.*[188] have suggested the use of this reagent to cleave β-D-linkages in heteropolysaccharides selectively; ozone in acetic anhydride–sodium acetate exhibits similar specificity.[189] Two rationalizations have been advanced to account for this more facile oxidation of the alcohol having the *endo* hydroxyl group: (*a*) relief of steric strain in the transition state as the trigonal carbonyl group is being formed; and (*b*) unimolecular decomposition of the intermediate chromic ester through a cyclic transition state that is favored by the decrease in rotational freedom of the hindered ester.[185] In some cases, however, ester formation is the rate-determining step, and there is no isotope effect in the breaking of the C–H bond.[186] Wu and Sugihara[190] noted that, whereas 1-O-benzoyl-2,3:5,6-di-O-isopropylidene-DL-galactitol (35) undergoes ready oxidation at C-4 by chromic acid to afford 6-O-benzoyl-1,2:4,5-di-O-isopropylidene-DL-*xylo*-3-hexulose, the stereoisomeric *manno* (36), *gulo*, and *allo* derivatives do not react with this reagent (or with methyl sulfoxide–phosphoric anhydride). They argued that

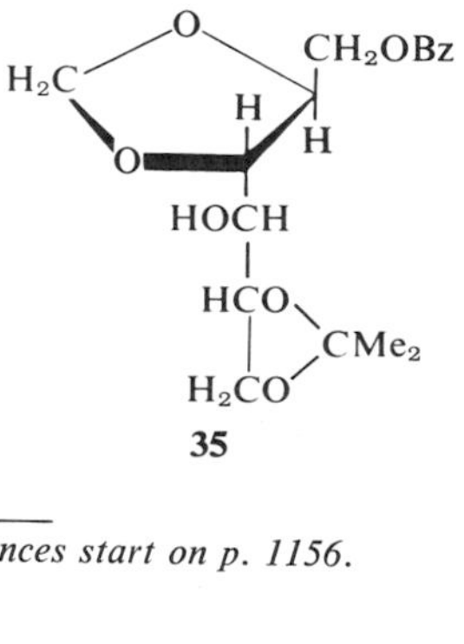

35

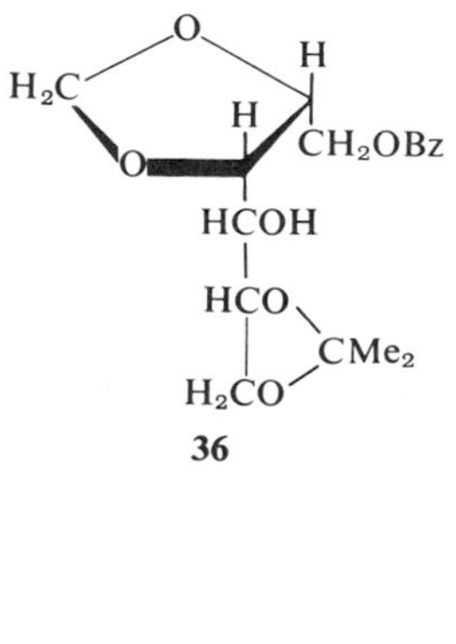

36

the *cis* arrangement of the 1-*O*-benzoyl and 4-hydroxyl groups (as in **36**) is too crowded for initial attack by the oxidant to occur.

Mizoiri *et al.*[191] used a chromic acid–pyridine complex to oxidize 6-*O*-benzoyl-1,2:3,4-di-*O*-isopropylidene-D-glucitol to 1-*O*-benzoyl-3,4:5,6-di-*O*-isopropylidene-*keto*-L-sorbose in 49% yield; methyl 6-deoxy-3,4-*O*-isopropylidene-α-L-galactopyranoside was likewise converted into the corresponding L-*xylo*-hexopyranosid-4-ulose in 35% yield.[192] Horton *et al.*[193] reported good to fair yields for chromic acid–pyridine oxidations of a number of derivatives having an unprotected hydroxyl group in a primary or exocyclic secondary position, but generally poor reactions of endocyclic alcohol groups. The basicity of pyridine can alter the stereochemistry in the carbonyl product, as in the spontaneous formation of methyl 4,6-*O*-benzylidene-3-*O*-methyl-α-D-*arabino*-hexopyranosid-2-ulose through (presumably enolic) epimerization of the D-*ribo* epimer transiently formed by oxidation of methyl 4,6-*O*-benzylidene-3-*O*-methyl-α-D-altropyranoside with chromium trioxide–pyridine.[194]

Investigation of a cellulose that had been oxidized by aqueous chromic acid in the presence of sulfuric acid showed that most of the attack had occurred at C-3, as reduction and subsequent hydrolysis gave some D-allose, but no D-mannose.[195] The action of chromic acid on the 2-amino-2-deoxy analogue (chitosan) as the perchlorate (salt) was, however, directed almost exclusively of C-6 to afford an aminodeoxyglycopyranuronan.[196] The rate of perchloric acid-catalyzed oxidation of D-glucose by chromic acid exhibits first-order dependence on the concentration of the oxidant and the substrate, and varies as the second power of the concentration of acid.[197]

IX. METHYL SULFOXIDE

Pfitzner and Moffatt reported the use of methyl sulfoxide–*N*,*N*′-dicyclohexylcarbodiimide as an oxidizing reagent[198]; it was subsequently shown that other anhydrides, and also boron trifluoride[199] and, more commonly, phosphorus pentaoxide[200] and acetic anhydride,[201] in combination with methyl sulfoxide, effect oxidation of isolated hydroxyl groups. Lyophilization is a general means (see ref. 216) of removing the excess of the reagents. A review[202] of this oxidation method in organic chemistry has been written, and Jones and co-workers[203] presented a detailed comparison of the relative suitability of this and other oxidation methods for the oxidation of 1,2:3,4-di-*O*-isopropylidene-α-D-galactopyranose (**38**) to the corresponding dialdose derivative **37**. Primary hydroxyl groups tend to react faster[204] in such unprotected derivatives as 1,3-*O*-benzylidene-D-arabinitol; however, acetic anhydride promotes formation of (methylthio)methyl ethers (such as **39**) instead of oxidation at primary hydroxyl groups,[205] whereas N,N'-dicyclo-hexylcarbodiimide–phosphorus pentaoxide[200] afford the aldehyde. This effect also occurs occasionally in oxidations of secondary alcoholic groups.[206]

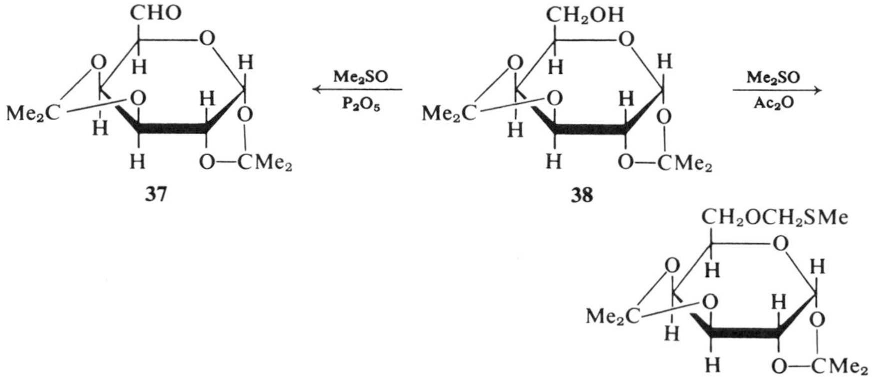

Other side reactions have been reported. Partially acylated derivatives commonly undergo elimination of a molecule of acid to generate an unsaturation conjugated with the newly formed carbonyl group[207,208]; methyl 3,4,6-tri-O-benzoyl-α-D-glucopyranoside is converted into methyl 3,6-di-O-benzoyl-4-deoxy-α-D-$glycero$-hex-3-enopyranosid-2-ulose by the action of methyl sulfoxide–acetic anhydride.[207] The potent Lewis acid boron trifluoride has been reported to effect oxidative opening of epoxide groups in the presence of methyl sulfoxide. 5,6-Anhydro-1,2-O-isopropylidene-α-D-glucofuranose is converted into 1,2-O-isopropylidene-α-D-$gluco$-hexodialdo-1,4-furanose,[209] and methyl 2,3-anhydro-4,6-O-benzylidene-α-D-mannopyranoside (40) is similarly converted[210] into 1,5:4,6-dianhydro-D-$threo$-2,3-hexodiulose (41). An example of phosphorus pentaoxide-catalyzed oxidation has also been shown to be accompanied by a side reaction; methyl 4,6-O-benzylidene-α-D-glucopyranoside reacts with the oxidant to afford[211] a mixture containing the 3-glyculoside and methyl 4,6-O-benzylidene-2,3-O-(methylenoxymethylene)-α-D-glucopyranoside (42). Methyl 4,6-O-benzylidene-α-D-galactopyranoside reacts with methyl sulfoxide–acetic anhydride to form a 2-O-acetylglycosid-3-ulose.[211a]

Protected, reducing derivatives of aldoses have been converted into aldonolactones by this method of oxidation[212]; 2-acetamido-3,4,6,-tri-O-benzoyl-2-deoxy-D-glucopyranose thus affords the corresponding D-glucono-1,5-lactone derivative,[213] whereas 2-azido-3,4,6-tri-O-benzyl-2-deoxy-D-altro-pyranose undergoes epimerization concurrent with, or subsequent to, oxidation, yielding a D-allono-1,5-lactone derivative.[214]

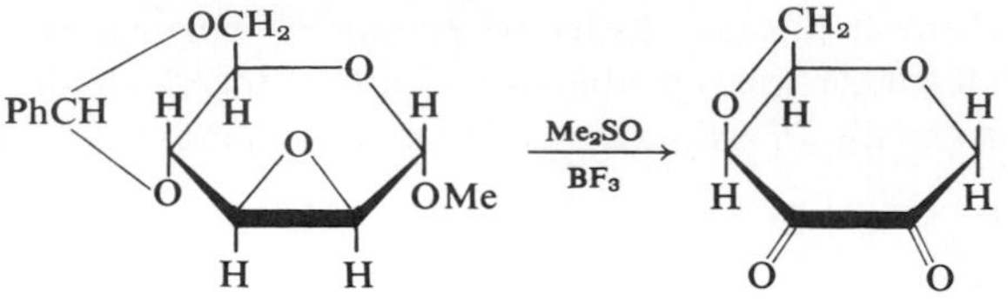

40   41

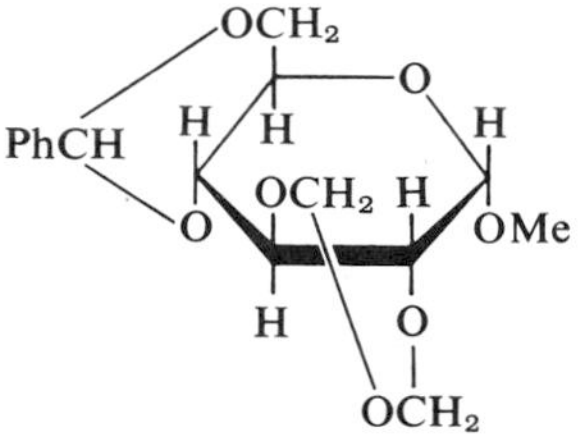

42

Oxidation of isolated, secondary alcohol groups in otherwise protected molecules generally proceeds in good yield, whereas complicating side reactions are sometimes associated[210,211] with similar reactions in the presence of more than one unprotected hydroxyl group. 1,6-Anhydro-2,3-O-isopropylidene-β-D-mannopyranose was oxidized in 77% yield[212,215] to the corresponding D-*lyxo*-hexos-4-ulose derivative by the action of methyl sulfoxide–acetic anhydride; similar good yields were obtained in the analogous preparation of 1,6-anhydro-3,4-O-isopropylidene-β-D-*lyxo*-hexopyranos-2-ulose from a D-*galacto* precursor.[216,217] 1,2:5,6-Di-O-isopropylidene-α-D-glucofuranose is converted into the corresponding glycose-3-ulose derivative by methyl sulfoxide–acetic anhydride[218] or, better, methyl sulfoxide–phosphorus pentaoxide.[219] The vicissitudes of oxidizing unprotected methyl α- and β-D-xylopyranoside by this method are neatly avoided by the temporary introduction of a 2,4-cyclic phenylboronate protecting group[220] prior to the oxidation; by this method,[221] methyl β-D-xylopyranoside (**43**) was converted into methyl β-D-*erythro*-pentopyranosid-3-ulose (**44**) in 45% overall yield.[215]

Successful examples of selective oxidation of only one of several free hydroxyl groups have been reported. Grindley et al.[222] prepared and oxidized 1,6-di-O-benzyl-2,5-O-methylene-D-mannitol, which possesses C_2 symmetry and, accordingly, affords only a single mono-oxidation product. 6-O-Tritylamylose was oxidized to afford a substantial content of glucopyranosyl-2-ulose residues,[223] and similar selectivity appears to operate in the oxidation of 6-O-tritylcellulose.[224] Kashimura et al.[225] suggested permethylation, methanolysis, and methyl sulfoxide oxidation of polysaccharides as a means of preparing characterizing derivatives [dicarbonyl sugar (2,4-dinitrophenyl)-hydrazones] from which linkage positions in the polymer may be inferred.

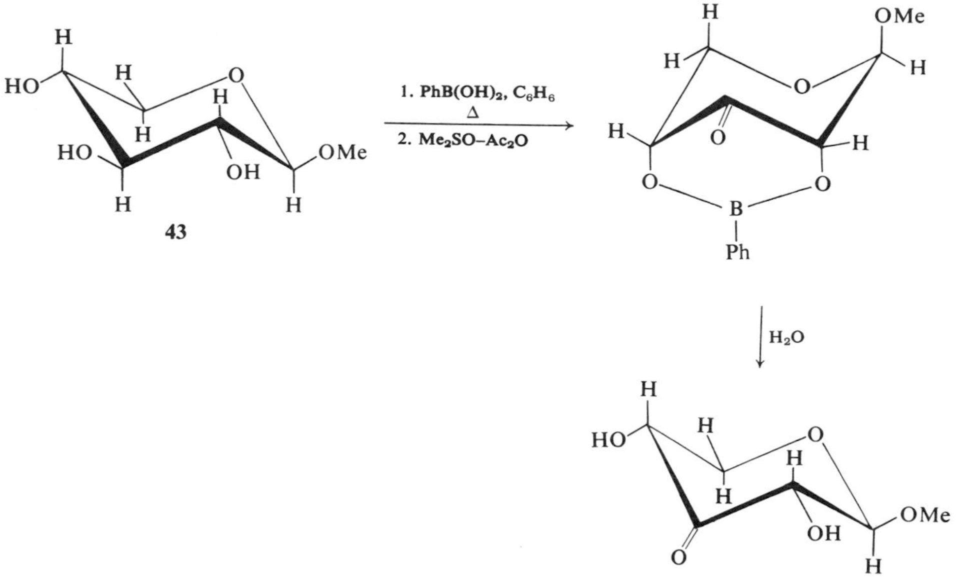

The mechanism of this oxidation may[226,227] proceed through formation, and deprotonation, of an alkoxydimethylsulfonium ion (**45**) to form a methylene ylide that decomposes in a final, concerted elimination step; crystalline 1-(2-methylpropyl)oxydimethylsulfonium (**45**, R = Me_2CH–) tetraphenylboronate was shown[226] to afford 2-methylpropanal under the conditions of the oxidation.

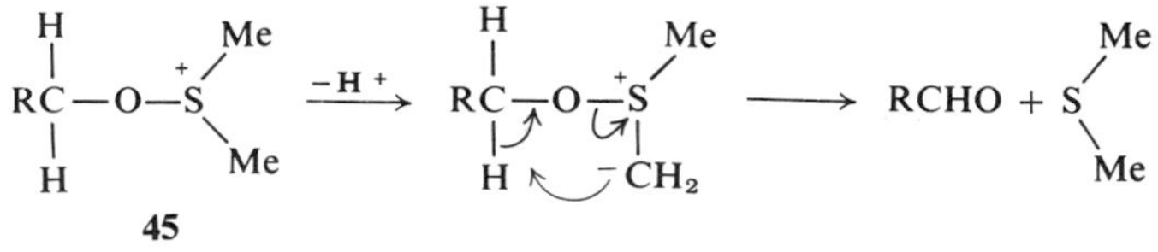

X. RUTHENIUM TETRAOXIDE

Ruthenium tetraoxide is an enormously powerful oxidant; it is more reactive than osmium tetraoxide, and combines explosively with ether or benzene,[228,229] so that it is generally used as a dilute solution in carbon tetrachloride. Beynon *et al.*[230] first demonstrated the usefulness of this

References start on p. 1156.

reagent in carbohydrate chemistry by converting methyl 4,6-*O*-benzylidene-2-deoxy-α-D-*ribo*- and -D-*arabino*-hexopyranoside into methyl 4,6-*O*-benzylidene-2-deoxy-α-D-*erythro*-hexopyranosid-3-ulose.

The oxidant may be prepared before the reaction in the stoichiometric amount needed, or generated *in situ* by reaction of a catalytic amount of ruthenium dioxide with periodate[231,231a] or hypochlorite.[232] The catalytic method serves to minimize further oxidations, such as a Baeyer–Villiger reaction (see Section III); the latter reaction has been demonstrated to occur when the period of oxidation is extended. Thus, 6-*O*-benzoyl-1,2-*O*-isopropylidene-α-D-glucofuranose undergoes initial oxidation by an excess of ruthenium tetraoxide to afford the 3-ulose derivative **46**, which is subsequently overoxidized[233] to the lactone diacetal **47**.

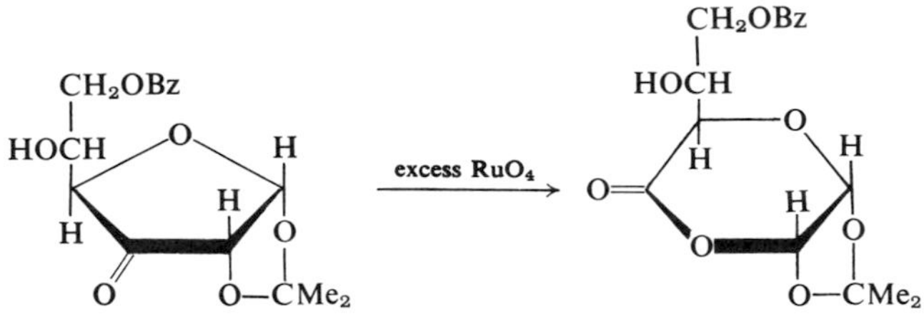

Ruthenium tetraoxide selectively oxidizes primary hydroxyl groups to carboxylic acids; 1,2-*O*-isopropylidene-α-D-xylofuranose is converted into the corresponding D-xyluronic acid derivative by this reagent.[235] Isopropylidene acetals and acetic and benzoic esters are inert to ruthenium tetraoxide, benzylidene acetals are attacked slowly, and benzyl ethers are oxidized as rapidly as primary hydroxyl groups. Collins *et al.*[236] prepared 3,4,6-tri-*O*-acetyl-1-chloro-1-deoxy-α-D-*arabino*- (and α-D-*lyxo*-)hexopyranos-2-ulose by direct oxidation of 3,4,6-tri-*O*-acetyl-α-D-gluco- and -D-galacto-pyranosyl chloride, respectively.

The use of ruthenium tetraoxide as an oxidant in organic chemistry has been reviewed.[237]

XI. PERMANGANATE AND MANGANESE OXIDES

Primary and secondary alcohols are rapidly oxidized by alkaline permanganate; the reaction is slower in neutral and mildly acid solutions.[238] The rapid oxidation in alkaline solution has been attributed to the formation of a nucleophilic alkoxide anion.[239] The choice of solvent is dictated by the type of protecting group used: acetates are oxidized in acetic acid, and isopropylidene acetals in alkaline solution. Unprotected glycol groupings and double bonds are usually attacked by alkaline permanganate, often resulting in carbon–carbon bond cleavage. Under controlled conditions permanganate[239a] (or osmium tetraoxide[239b]) effects *cis*-hydroxylation of double bonds.

1,2:3,4-Di-*O*-isopropylidene-α-D-galactopyranose was converted into the corresponding glycuronic acid by the action of permanganate[240]; however, the unprotected glycol grouping in 1,2-*O*-isopropylidene-3-*O*-methyl-α-D-glucofuranose is attacked to afford a D-xyluronic acid derivative.[241] The C-2–C-3 bond is broken in the conversion of 5,6-*O*-isopropylidene-L-*threo*-hex-2-enono-1,4-lactone into 3,4-*O*-isopropylidene-L-threonic acid.[242] Low yields of tetraric acids have been obtained[243] by the action of alkaline permanganate on D-glucaric acid at 0°.

Partially acetylated derivatives of maltose[244] and cellobiose[245] were treated with potassium permanganate in acetic acid to afford the corresponding biouronic and pseudobiouronic acids.

In neutral or slightly acid solution at room temperature, the reactivity[246] of a number of sugars toward permanganate is in the order: maltose > fructose > arabinose > galactose > mannose > glucose > lactose. For D-fructose, the maximal rate of oxidation is at pH 3.5 to 4.5.

β-D-Glucopyranose undergoes oxidation at 1.7 times the rate for the α anomer at pH 2.3; this easier oxidation of the anomer having the equatorially attached, anomeric hydroxyl group is similar to that observed with other oxidants, although not so great as that encountered with bromine oxidation. Stroh *et al.*[247] interpreted the observation that ketoses are oxidized faster than aldoses by this reagent as a function of the relative stability of cyclic forms and acyclic tautomers and their rate of interconversion.

Manganese dioxide was used by Foster *et al.*[248] to effect degradation of aldoses to the next lower aldoses. Reducing disaccharides are converted into *O*-hexosylpentoses; further degradation to the *O*-hexosyltetrose is apparently concealed by oxidation of the tetrose residue to acidic products; small yields of tetroses have been obtained from pentoses. Such reactions in aqueous solution gave lower yields of acidic products at 50° than at 100°. In aqueous suspension, manganese dioxide is slightly alkaline (pH 10), and the carbon–carbon bond cleavage may involve the corresponding enediol as an intermediate. Manganese dioxide has also been used for the oxidation of isolated alcohol groups; Weidenhagen thus oxidized 1,2-*O*-isopropylidene-α-D-glucofuranurono-6,3-lactone to the glycos-5-ulose in acetone.[249] Attempted oxidation of the cyanohydrin prepared from (*E*)-4,5,6-tri-*O*-acetyl-*aldehydo*-D-*erythro*-hex-2-enose by this reagent resulted in a series of allylic elimination

---

References start on p. 1156.

reactions; the ultimate product was methyl (E,E)-6-hydroxy-2,4-hexadieno-ate.[250]

Very little is known of the mechanism of oxidations with this oxidant in heterogeneous systems,[251] although there is a profound dependence upon steric effects.[252]

XII. MISCELLANEOUS OXIDANTS

A. FERRICYANIDE

Singh et al.[253] studied the oxidation of various sugars with alkaline ferricyanide; the 1,2-enediol is an intermediate, and the rate of oxidation observed is actually the rate of formation of the enediol. The rate of disappearance of the oxidant is directly proportional to the concentrations of both the sugar and the hydroxide ion, but is independent of the concentration of the oxidant. D-Fructose reacts faster than D-glucose or D-galactose; this is in agreement with its known tendency to enolize more rapidly. The oxidation of D-xylose is faster than that of L-arabinose. Ferricyanide oxidations conducted in air show a short inhibition period, owing to re-oxidation of the reduced oxidant; Kasper monitored the oxidation of $0.1M$ D-glucose potentiometrically and found that the decrease in oxidant and sodium hydroxide are first-order reactions, whereas the decrease in the concentration of D-glucose is second-order.[254]

B. CUPRIC OXIDE

Studies of oxidation by cupric oxide in the presence of alkaline citrate and tartrate as chelating agents[255] revealed that, as with ferricyanide, the rate of reaction is zero-order in oxidant and directly proportional to the concentration of sugar and hydroxide ion; Marshall and Waters[256] obtained similar results for the oxidation of D-glucose with cupric ion complexed as the picolinate. Use of acetoin as a model system demonstrated that oxidation proceeds via a chelated, cuprous complex of the enediol; this complex is readily attacked by cupric ions, and its presence is necessary to the overall

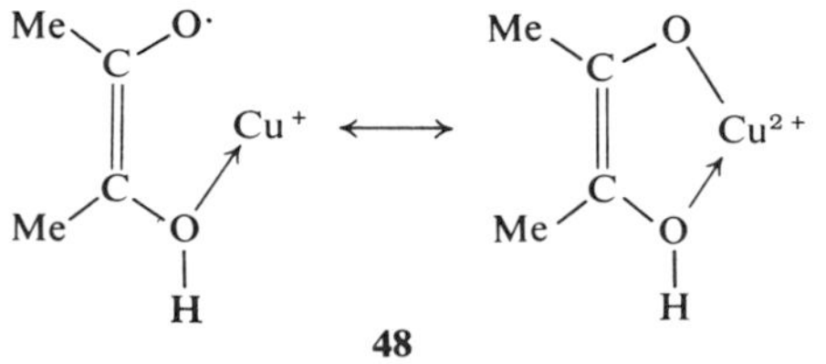

48

reaction. A resonance hybrid (**48**) of cupric ion complexed with the enediol has been proposed as a source of cuprous ion for initiating the reaction. The accumulation of **48** is minimal, owing to continuous precipitation of cuprous oxide from the system.

The most important methods for the quantitative determination of reducing sugars are based on oxidation with hot, alkaline solutions of copper salts (see Chapter 45). The composition of the oxidation products has been investigated; in general, monobasic acids having one to six carbon atoms are formed, accompanied by oxalic acid, carbon dioxide, and lactic acid.

Copper(II) sulfate in sodium carbonate solution (the Soldaini reagent) oxidizes D-glucose during 8 hours at 100° to a mixture of acids, more than 60% of which are nonvolatile; the latter include D-gluconic, D-mannonic, D-arabinonic, D-erythronic, D-threonic, D-glyceric, and glycolic acids.[257] The same products are formed by the action of the (more alkaline) Fehling solution on D-glucose.[258] From 199 g of D-fructose, Nef reported the isolation of carbon dioxide (2.4 g), formic acid (13.8 g), glycolic acid (22 g), glyceric acid (18 g), tetronic acids (35 g), and hexonic acids (30 g). According to Nef,[259] the oxidation with copper(II) acetate in neutral solution proceeds differently; much more oxygen is consumed, greater proportions of carbon dioxide are produced, and erythronic acid seems to be the main oxidation product.

In ammoniacal solutions of copper salts, the oxidation products are likely to contain nitrogen[260]; D-glucose, D-fructose, and D-mannose give oxalic acid, imidazoles, hydrogen cyanide, and urea.

In the oxidation of methyl α-D-glucopyranoside by ammoniacal copper(II) ion, cleavage occurs[261] between C-2 and C-3.

C. SILVER OXIDE

Aldohexoses, arabinose, erythritol, D-fructose, D-galactonolactone, D-glucaric acid, and glyceraldehyde are oxidized by silver oxide at 50° (in water or potassium hydroxide) to carbon dioxide, oxalic acid, formic acid, and glycolic acid.[262]

Alkaline silver solutions have been used for detection of spots on paper chromatograms[263]; premature precipitation of silver oxide can be avoided either by complexation of silver(I) with ammonia or by application of the silver salt and subsequently of the base to the paper. The reaction with reducing sugars is fast, whereas that with alditols and glycosides is slow. Reaction of alkyl β-D-glucopyranosides is faster than that of the α anomers[264]; methyl 3-O-methyl-α-D-glucopyranoside, which contains no free glycol grouping, does not react.[265] Complexation of the silver ion with the acidic glycol grouping is, presumably, involved in the oxidation.

Fétizon and Golfier[266] first reported the use of silver carbonate suspended on Celite (diatomaceous earth) as a mild oxidant. Tronchet *et al.*[267] used this reagent to oxidize 1,5-anhydro-2-deoxy-D-*arabino*-hex-1-enitol (D-glucal, **49**) to the 3-glyculose **50**. In a series of papers, Morgenlie demonstrated the use of this oxidant in the conversion of aldoses into aldonolactones,[268] and of aldoses and ketoses into lower sugars by cleavage of the endocyclic, carbon–carbon bond next to the anomeric center[268,269]; molecules having an alkyl group protecting the hydroxyl group adjacent to the anomeric center,[268,270] or too few carbon atoms to permit ring closure,[271] undergo oxidation without degradation.

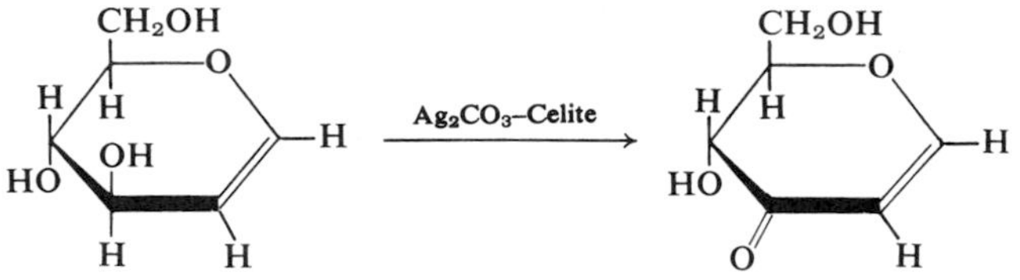

Lee and Clarke[272] used silver(II) picolinate as an oxidant for isolated hydroxyl groups; with this reagent, they converted 2,3:5,6-di-O-isopropyl-idene-D-mannofuranose into the corresponding D-mannono-1,4-lactone derivative, and methyl 6-deoxy-2,3-O-isopropylidene-L-mannofuranoside into an L-*lyxo*-hexofuranosid-5-ulose derivative.

D. SULFITE PULPING

The sulfite process for the delignification of wood in the paper industry involves heating with acidic sulfite solutions at elevated temperatures under pressure. Under such conditions, oxidation occurs to a certain extent and, presumably, the sulfite is partly reduced to sulfide. Nord and co-workers[273] isolated aldonic acids from cooking liquors after 14 hours at 156° and pH 4. Such liquors contain mostly aldohexoses and aldopentoses formed by the hydrolysis of cellulose and hemicelluloses. The aldonic acids isolated were mainly D-xylonic, D-mannonic, and D-galactonic acids, together with a smaller proportion of D-gluconic acid.

Lindberg and Theander[274] treated D-mannitol with sulfite solution at pH 1.7 to 6, in a temperature range of 130 to 180°; both D-mannose and D-fructose were obtained as products. Hydrogen sulfide was also formed. They suggested that the hexitol is first converted into the two sugars and also into the presumed D-*arabino*-3-hexulose; these could be oxidized further, or else react with sulfite to afford compounds having carboxylic and sulfonic acid groups. Theander[275] reported the formation of alkanesulfonic acids from similar treatment of methyl β-D-glucopyranoside (see Chapter 23).

E. WET COMBUSTIONS

Oxidations conducted under strongly acidic or strongly alkaline conditions are quite drastic, often completely degrading the substrate to carbon dioxide and water. Extremes of pH enhance either (a) the effectiveness of the oxidant by converting it into a stronger Lewis acid (at low pH), or (b) the nucleophilicity of the organic substrate by deprotonating it to an anion (or carbanion) (at high pH); the high electron density of the latter, negatively charged species facilitates the removal of either a hydride ion or a hydrogen atom, and expedites electron transfer to the oxidant.

The Van Slyke reagent,[276] a mixture of potassium iodate and dichromate in concentrated sulfuric and phosphoric acids, is used for the total combustion of ^{14}C-labeled compounds, as a means of determining the content of ^{14}C. Kleinert[277] used iodate in concentrated sulfuric acid to determine the degree of oxidation of oxidized cellulose; the ratio of carbon dioxide formed to reagent consumed was used as a measure of the oxidation.

Hot solutions of chromic acid and acid solutions of ceric sulfate oxidize carbohydrate materials to carbon dioxide, formic acid, and formaldehyde. The quantity of oxidant consumed is a fairly precise measure of the amount of carbohydrate present.[278]

The equations for the oxidation of several sugars by ceric sulfate are as follows:

$$2\,C_6H_{12}O_6\;(\text{D-glucose}) + 13\,O \longrightarrow H_2O + CO_2 + 11\,HCO_2H \qquad (46)$$

$$2\,C_6H_{12}O_6\;(\text{D-fructose}) + 13\,O \longrightarrow 3\,H_2O + 7\,HCO_2H + 2\,HCHO \quad (47)$$

$$C_{12}H_{22}O_{11}\;(\text{sucrose}) + 13\,O \longrightarrow 2\,CO_2 + 9\,HCO_2H + H_2O \qquad (48)$$

For determination of sucrose, an accuracy to within $\pm 0.3\%$ is claimed.

Chromic acid acts similarly, and it has been used for the volumetric determination of cellulosic materials.[279] Under the same conditions, acid permanganate would presumably cause similar reactions.

$$C_6H_{10}O_5 + 4\,Cr_2O_7^{2-} + 32\,H^+ \longrightarrow 6\,CO_2 + 21\,H_2O + 8\,Cr^{3+} \qquad (49)$$

A summary[280] of the effect of four oxidants on a number of organic compounds is given in Table II. Most of the determinations with permanga-

TABLE II

EXTENT[a] OF OXIDATION OF VARIOUS ORGANIC COMPOUNDS BY
ALKALINE PERMANGANATE, AND PERIODIC, SULFATOCERIC,
AND CHROMIC ACIDS[b]

Compound	Molar uptake of oxidant per mole			
	MnO_4^-	IO_4^-	Ce^{4+}	CrO_3
Acetic Acid	X[c]	X	X	X
Oxalic acid			2	
Formic acid	2	X	X	
Glycolic acid	6	X	3.95	
Malonic acid			6.66	
Threaric acid	10	2	7.20	
Succinic acid			X	
Malic acid	12	X	9.25	
Citric acid	18	X	15.85	
Pyruvic acid			2	
Ethylene glycol	10[d]	2		
Glycerol	14	4	8	14
Erythritol	18	6		
Arabinitol		8		
Mannitol	26			
Phenol	28			
Salicyclic acid	28			
Gallic acid	24			
Formaldehyde	4	X		
Glucose	24	10		
Fructose	24	8		
Sucrose	48			
Ethanol		X		

[a] Figures given are the number of moles of oxidant consumed
per mole; whole numbers are for stoichiometric reactions,
and nonintegers were determined from empirical reactions.
[b] From ref. 280. [c] X indicates that no oxidation occurs.
[d] Italicized values were obtained by the improved method of
Stamm.[281]

nate employed the method of Stamm[281]; ceric sulfate oxidations were performed by the procedure of Willard and Young[282]; perchlorate–ceric acid
[$H_2Ce(ClO_4)_6$] oxidations were performed according to Smith and Duke.[283]

D-Glucose may be oxidized completely to carbon dioxide and water by hot,
weakly alkaline solutions of potassium permanganate[284]; as the alkalinity
increases above 0.03N, oxalic acid is produced, and, in 1.8N potassium
hydroxide, yields of 42% of oxalic acid are obtained.[285] Similar results are

obtained from other hexoses, pentoses, and glyceraldehyde.[286] The ratio of carbon dioxide to oxalic acid differs for various sugars, but at high temperatures, the differences become small. The alditols are oxidized to the same products as the corresponding sugars, and the effect of alkalinity is the same as for the corresponding sugars.[287] Hence, it would appear that the sugars are intermediate products in the oxidation. The equivalents of oxidant required per mole are given in Table II for several carbohydrates.

Beer[288] pointed out that 17 to 20 equivalents of silver(I) are consumed in the oxidation of hexoses and alditols with ammoniacal silver nitrate on paper chromatograms; it is interesting that this extensive oxidation occurs on a cellulose support.

Higher temperatures also favor complete oxidation; mercuric cyanide has been used at 500° to 550° as an oxidant to convert all of the ^{18}O in ^{18}O-labeled organic substrates into carbon dioxide.[289] Vanadium pentaoxide effects complete oxidation of aldoses and related compounds[290]; the authors[290] speculated that the degradation proceeds in a stepwise manner, removing individual carbon atoms starting at the carbonyl group. The kinetics and mechanism of oxidation of aldoses by vanadium (V) in strong acids has been studied.[290a]

Hyperoxidized transition metals, such as copper(III)[291] and silver(III),[292] effect complete combustion of aldoses and ketoses, (but the latter reagent requires a temperature of 80° for complete reaction); silver(III) in the cold[292] and cobalt(III)[293] effect only partial oxidation of sugars.

XIII. ENZYMIC OXIDATIONS

Oxidation of carbohydrates is effected *in vivo* and *in vitro* by numerous enzymes that are present in living tissues; these enzymically induced changes of oxidation state, which are usually dehydrogenations or hydrogenations, are necessary for carbohydrate metabolism.

An enzyme, originally called "glucose oxidase," and later *notatin* or *penicillin B*, catalyzes the oxidation of D-glucose by molecular oxygen to D-gluconic acid. It has been isolated from various molds, especially *Aspergillus niger* and *Penicillium glaucum*.[294] The enzyme is highly specific for β-D-glucopyranose.[295] As the initial reaction product is D-glucono-1,5-lactone, the pyranose ring is unbroken, as with bromine oxidation; hydrogen peroxide is also formed.

This enzyme had been called an oxidase, but experiments with ordinary molecular oxygen and water enriched with ^{18}O showed that the action is that of a dehydrogenase; the enzyme catalyzes the transfer of two hydrogen atoms from D-glucose to gaseous oxygen.[296] The enzyme is commonly used for determining D-glucose, and has been used in a method for determining the aeration capacity of fermentation tanks.[297]

A purified D-glucose dehydrogenase, isolated from animal liver, catalyzes the oxidation of D-glucose to D-gluconic acid.[298] A coenzyme (either nicotinamide adenine dinucleotide or its 2'-phosphate) is necessary in these reactions. The reaction is reversible, and specific for β-D-glucopyranose; it does not proceed with the α-D anomer, and it occurs only slowly with D-xylose. In this type of reaction the pyranose ring of D-glucose 6-phosphate is unbroken, and D-glucono-1,5-lactone 6-phosphate is the product.[299] Other enzymes have been reported that give the corresponding 1,4-lactone as a product.[300]

D-Galactose oxidase (EC 1.1.3.9) removes the *pro-S* hydrogen atom[301] from C-6 of D-galactopyranose derivatives to produce the D-*galacto*-hexodialdo-1,5-pyranose analogues initially.[302] Rogers and Thompson[303] used the action of this enzyme, and subsequent bromine oxidation, to effect the conversion of a galactomannan into a galacturonomannan; a similar sequence has been used with glycoproteins as a method for linkage-analysis.[303a]

The energy released by enzymic oxidation of carbohydrates has been suggested[304] as a potential source of power for fuel cells.

XIV. MICROBIAL OXIDATIONS[305]

Fermentative processes are of considerable value for the production of useful organic chemicals from carbohydrate precursors. Large amounts of acetic acid, acetone, butanol, citric acid, ethanol, D-gluconic acid, lactic acid, and L-sorbose are made industrially by fermentation. In fermentative processes, oxidizing as well as reducing conditions may be employed, depending on the product desired. Laboratory and industrial preparation of many other substances, such as glycols, have been carried out by using enzymes.

Micro-organisms exhibit a marked specificity in their choice of substrates, and in the reaction products. This property is useful for the qualitative and quantitative determination of sugars, as well as for the identification of micro-organisms. The formation of glycuronic, aldonic, and glyculosonic acids, and glycosuloses is considered in Chapter 23.

D-Gluconic acid is produced by the action of many species of bacteria and molds on D-glucose[306]; Boutroux found that D-gluconic acid is a metabolic product of acetic acid bacteria. D-Glucose dehydrogenases from molds, bacteria, and liver cause this oxidation,[307] and the process is used for the commercial production of D-gluconic acid and its lactones and salts. Molds of the *Aspergillus* and *Penicillium* species are particularly suitable for large-scale production. By use of *Aspergillus niger* in fermentations under air pressure in rotating drums, yields of 90 to 99% of D-gluconic acid are obtained.[308] When calcium carbonate is present, calcium D-gluconate crystallizes directly from the medium.

Some strains of *A. niger* oxidize D-mannose to D-mannonic acid, and D-galactose to D-galactonic acid.[309] Many species of *Pseudomonas* and *Acetobacter xylinum* oxidize pentoses to the corresponding pentonic acids.[310] The yields are not always high, but could probably be increased by improvements in the strains and in the conditions of culture.

Acetobacter suboxydans can oxidize D-glucose to D-*xylo*-5-hexulosonic acid[311]; D-gluconic acid is initially formed and is subsequently converted into the 5-hexulosonic acid. D-*arabino*-2-Hexulosonic acid is formed by some *Acetobacter* species, and *Pseudomonas* species afford yields of over 80% after 25 hours when the solution is strongly aerated in a rotating drum.[312]

Bacterial oxidation has been used for the preparation of L-*xylo*-2-hexulosonic acid from L-idonic acid.[313] D-*threo*-2,5-Hexodiulosonic acid was reported from the oxidation of D-glucose or D-gluconic acid with *P. albosesamae*.[314] A wild strain of *A. suboxydans* converted D-fructose into D-*lyxo*-hexos-5-ulose[315]; strains of *A. suboxydans* oxidized L-sorbose to L-*threo*-2,5-hexodiulose,[316] and acyclic aldose derivatives to pentos-4-ulose derivatives.[317,318] Thus, D-galactose diethyl dithioacetal was oxidized to L-*lyxo*-hexos-5-ulose 1,1-(diethyl dithioacetal).[318] *Pseudomonas putida* was found[319] to oxidize D-galactose and 3-deoxy-D-*xylo*-hexose to D-*lyxo*-hexos-2-ulosonic and 3-deoxy-D-*threo*-hexos-2-ulosonic acids, respectively.

D-Mannitol (**51**) is readily oxidized by *A. suboxydans* to D-fructose (**54**); the oxidation may occur on either C-2 or C-5 because the two atoms are related by symmetry. Isbell *et al.*[320] used isotopically labeled D-mannitol to prove that dissociation of the carbon–hydrogen bond at C-2 (or C-5) is the rate-determining step. Use of D-[2-³H]mannitol (**52**) revealed a large isotope

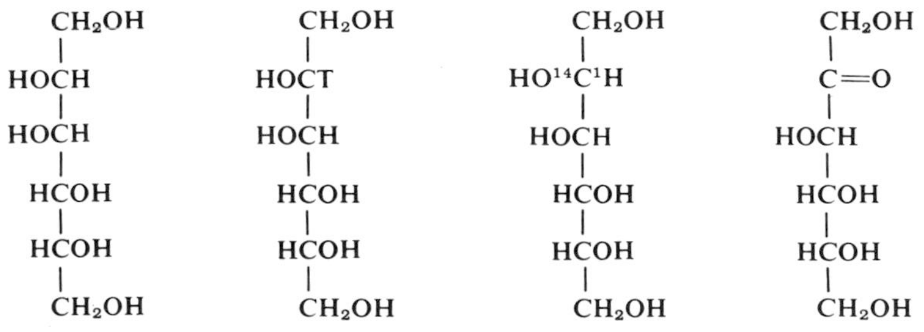

CH₂OH	CH₂OH	CH₂OH	CH₂OH
HOCH	HOCT	HO¹⁴C¹H	C=O
HOCH	HOCH	HOCH	HOCH
HCOH	HCOH	HCOH	HCOH
HCOH	HCOH	HCOH	HCOH
CH₂OH	CH₂OH	CH₂OH	CH₂OH
51	**52**	**53**	**54**

effect, $k_3/k_1 = 0.23$. The isotope effect was smaller with D-[2-^{14}C]mannitol (**53**) but still appreciable, considering the smaller differences in mass; $k_{14}/k_{12} = 0.93$. D-[3-^{3}H]Mannitol undergoes oxidation at a rate ($k_3/k_1 = 0.70$) intermediate between those for **52** and **53**; the secondary isotope effect was attributed to lessened hyperconjugation in the tritiated analogue, which lessens the electron density on C-2 and thus inhibits the oxidation; however, labeling at C-1 revealed the unexpected absence of any isotope effect.

Agrobacterium tumefaciens possesses a D-glucoside 3-dehydrogenase enzyme that has been shown to convert sucrose into 1-*O*-β-D-fructofuranosyl-α-D-*ribo*-hexopyranos-3-ulose[321] (**55**), and cellobiose into 4-*O*-(β-D-*ribo*-hexopyranosyl-3-ulose)-D-glucose[322] (**56**).

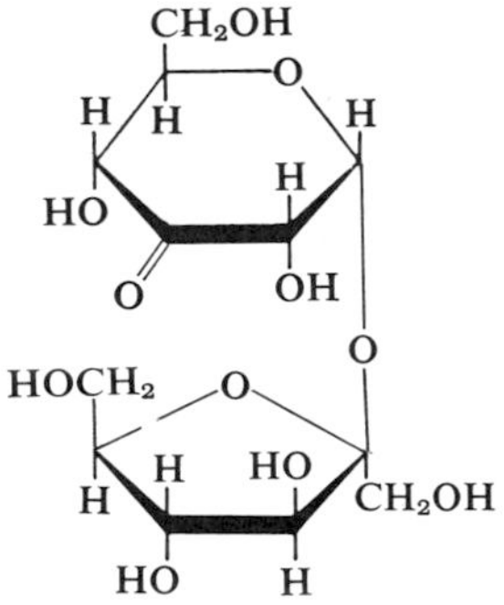

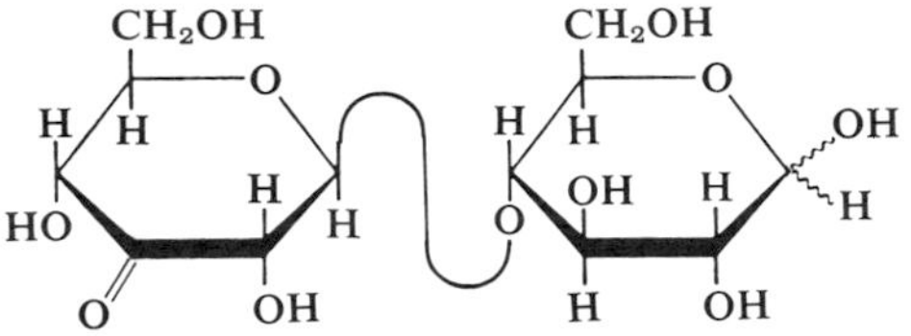

REFERENCES

1. R. Stewart, "Oxidation Mechanisms," Benjamin, New York, 1964.
2. W. A. Waters, "Mechanisms of Oxidation of Organic Compounds," Wiley, New York, 1964; Methuen and Co. Ltd., London, 1960.
3. "Oxidation in Organic Chemistry" (K. B. Wiberg, ed.), Academic Press, New York, 1965.
4. "Selected Oxidation Processes," E. K. Fields, Symposium Chairman, *Advan. Chem. Ser.*, **51** (1965).
5. J. S. Littler, *J. Chem. Soc.*, 2190 (1962).
6. E. S. Gould, "Mechanism and Structure in Organic Chemistry," Holt, Rinehart, and Winston, New York, 1959; J. Hine, "Physical Organic Chemistry," McGraw-Hill, New York, 1962.
7. R. F. Butterworth and S. Hanessian, *Synthesis*, 70 (1971); J. S. Brimacombe, *Angew. Chem.*, **81**, 415 (1969).
8. F. Friedberg and L. Kaplan, *Abstr. Papers Amer. Chem. Soc. Meeting*, **131**, 86-O (1957).
9. C. G. Swain, R. A. Wiles, and R. F. W. Bader, *J. Amer. Chem. Soc.*, **83**, 1945 (1961).
10. L. Vaska and J. W. DiLuzio, *J. Amer. Chem. Soc.*, **84**, 4990 (1962).
11. E. S. Lewis and M. C. R. Symons, *Quart. Rev. Chem. Soc.*, **12**, 232 (1958).
12. H. S. Fry and E. Otto, *J. Amer. Chem. Soc.*, **50**, 1138 (1928).
13. P. Ballinger and F. A. Long, *J. Amer. Chem. Soc.*, **82**, 795 (1960).

14. H. S. Isbell and W. W. Pigman, *Bur. Stand. J. Res.*, **10**, 337 (1933).
15. V. A. Derevitskaya, G. S. Smirnova, and Z. A. Rogovin, *Dokl. Akad. Nauk SSSR*, **141**, 1090 (1961).
16. B. R. Baker and A. H. Haines, *J. Org. Chem.*, **28**, 438 (1963).
17. B. Lindberg and K. N. Slessor, *Carbohyd. Res.*, **1**, 492 (1966).
18. R. Stewart and M. M. Mocek, *Can. J. Chem.*, **41**, 1160 (1963); see also, ref. 1, pp. 67–70.
19. J. O. Edwards, "Inorganic Reaction Mechanisms," Benjamin, New York, 1964, pp. 82–87.
20. G. Holst, *Chem. Rev.*, **54**, 169 (1954).
21. M. Lewin, *Tappi*, **48**, 333 (1965).
22. A. Seidell, "Solubilities of Inorganic and Metal Organic Compounds," Vol. 1, Van Nostrand, New York, 1940.
23. J. W. Mellor, "A Comprehensive Treatise on Inorganic and Theoretical Chemistry," Vol. 2, Longmans Green, New York, 1927.
24. W. M. Eigen and K. Kustin, *J. Amer. Chem. Soc.*, **84**, 1355 (1962).
25. "Handbook of Chemistry and Physics," 46th ed., Chemical Rubber Co., Cleveland, 1964, p. D-79.
26. W. M. Latimer, "The Oxidation States of the Elements and Their Potentials in Aqueous Solutions," Prentice Hall, Englewood Cliffs, N.J., 1938, p. 45.
27. R. Breslow, "Organic Reaction Mechanisms," Benjamin, New York, 1966, p. 79.
28. B. P. Ridge and A. H. Little, *J. Text. Inst.*, **33**, T33 (1942).
29. R. M. Chapin, *J. Amer. Chem. Soc.*, **56**, 2211 (1934).
30. R. L. Whistler and R. Schweiger, *J. Amer. Chem. Soc.*, **79**, 6460 (1957).
31. H. Hlasiwetz, *Ann.*, **119**, 281 (1861); H. Hlasiwetz and J. Habermann, *ibid.*, **155**, 120 (1870).
32. H. Kiliani and S. Kleeman, *Ber.*, **17**, 1296 (1884).
33. H. H. Bunzel and A. P. Mathews, *J. Amer. Chem. Soc.*, **31**, 464 (1909).
34. G. H. A. Clowes and B. Tollens, *Ann.*, **310**, 164 (1899); C. S. Hudson and H. S. Isbell, *J. Amer. Chem. Soc.*, **51**, 2225 (1929); *Bur. Stand. J. Res.*, **3**, 57 (1929).
35. E. Votoček and S. Malachata, *An. Real Soc. Españ. Fís. Quím.*, **27**, 494 (1929); *Chem. Abstr.*, **24**, 341 (1930).
36. J. P. Hart and M. R. Everett, *J. Amer. Chem. Soc.*, **61**, 1822 (1939).
37. H. S. Isbell and H. L. Frush, *Bur. Stand. J. Res.*, **6**, 1145 (1931); H. S. Isbell, U.S. Pat. 1,976,731 (Oct. 16, 1934); E. L. Helwig, U.S. Pat. 1,895,414 (Jan. 24, 1933).
38. R. Pasternack and P. P. Regna, U.S. Pat. 2,222,155 (Nov. 19, 1940); E. W. Cook and R. T. Major, *J. Amer. Chem. Soc.*, **57**, 773 (1935).
39. A. N. Kappanna and K. M. Joshi, *J. Indian Chem. Soc.*, **29**, 69 (1952); *Chem. Abstr.*, **47**, 56c (1953).
39a. M. Fedoroňko, *Advan. Carbohyd. Chem. Biochem.*, **29**, 107 (1974).
40. R. L. Kiliani and C. Scheibler, *Ber.*, **21**, 3276 (1888).
41. H. Kiliani, *Ann.*, **205**, 145 (1880).
42. M. R. Everett and F. Sheppard, "Oxidation of Carbohydrates; Keturonic Acids; Salt Catalysis," University of Oklahoma Medical School, Norman, 1944.
43. C. L. Mehltretter, W. Dvonch, and C. E. Rist, *J. Amer. Chem. Soc.*, **72**, 2294 (1950); C. L. Mehltretter and W. Dvonch, U.S. Pat. 2,502,472 (1950).
44. C. Vincent and A. B. Delachanal, *C. R. Acad. Sci.*, **111**, 51 (1890); E. Fischer, *Ber.*, **23**, 3684 (1890); H. W. Talen, *Rec. Trav. Chim. (Pays-Bas)*, **44**, 891 (1925).
45. A. J. Fatiadi, *Carbohyd. Res.*, **8**, 135 (1968).
46. W. N. Lichtin and M. H. Saxe, *J. Amer. Chem. Soc.*, **77**, 1875 (1955).

46a. H. S. Isbell and W. W. Pigman, *Bur. Stand. J. Res.*, **18**, 141 (1937).
47. H. S. Isbell, *Bur. Stand. J. Res.*, **8**, 615 (1932).
47a. H. S. Isbell and C. S. Hudson, *Bur. Stand. J. Res.*, **8**, 327 (1932).
48. I. R. L. Barker, W. G. Overend, and C. W. Rees, *J. Chem. Soc.*, 3254 (1964); *Chem. Ind.* (London), 558 (1961).
49. H. S. Isbell, *J. Res. Nat. Bur. Stand.*, **66A**, 233 (1962).
50. M. Lewin and M. Albeck, cited in ref. 21, p. 339 and ref. 83 therein.
51. B. Perlmutter-Hayman and A. Persky, *J. Amer. Chem. Soc.*, **82**, 276 (1960).
51a. O. Larm, E. Scholander, and O. Theander, *Carbohyd. Res.*, **49**, 69 (1976).
52. A. Dyfverman, B. Lindberg, and D. Wood. *Acta Chem. Scand.*, **5**, 253 (1951).
53. B. Lindberg and D. Wood, *Acta Chem. Scand.*, **6**, 791 (1952).
54. A. Dyfverman, *Acta Chem. Scand.*, **7**, 280 (1953).
55. R. Bentley, *J. Amer. Chem. Soc.*, **79**, 1720 (1957).
56. J. T. Henderson, *J. Amer. Chem. Soc.*, **79**, 5304 (1957).
57. O. Theander, *Sv. Papperstidn.*, **61**, 581 (1958).
58. H. Staudinger and A. W. Sohn, *Ber.*, **72**, 1709 (1939); *J. Prakt. Chem.*, **155**, 177 (1940).
59. E. D. Kaverzneva, *Izv. Akad. Nauk SSSR*, 6, 791 (1951); E. D. Kaverzneva, V. I. Ivanov, and A. S. Salova, *ibid.*, 185, 751 (1952); *Bumazh. Prom.*, **28**, 6 (1953); *Chem. Abstr.*, **47**, 12,799 (1953).
60. J. W. Daniel, *Tappi*, **42**, 534 (1959).
61. R. C. Mehrotya and K. C. Grover, *Vijnana Parishad Anusandhan Patrika*, **4**, 255 (1961); *Chem. Abstr.*, **59**, 6493e (1963).
62. D. A. Clibbens and B. P. Ridge, *J. Text. Inst.*, **18**, T133 (1927).
63. R. L. Whistler and R. Schweiger, *J. Amer. Chem. Soc.*, **80**, 5701 (1958).
64. R. L. Whistler and S. J. Kazeniac, *J. Org. Chem.*, **21**, 468 (1956).
65. J. K. Crossman and J. W. Green, *Tappi*, **49**, 131 (1966).
66. R. L. Whistler and R. Schweiger, *J. Amer. Chem. Soc.*, **81**, 5190 (1959); R. L. Whistler and K. Yagi, *J. Org. Chem.*, **26**, 1050 (1961).
67. J. A. Epstein and M. Lewin, *Textil-Rundschau*, **16**, 494 (1961).
68. M. L. Wolfrom and W. E. Lucke, *Tappi*, **47**, 189 (1964).
68a. P. S. Fredricks, B. O. Lindgren, and O. Theander, *Sv. Papperstidn.*, **74**, 597 (1971).
69. R. L. Whistler, E. G. Linke, and S. J. Kazeniac, *J. Amer. Chem. Soc.*, **78**, 4704 (1956).
70. K. D. Reeve, *J. Chem. Soc.*, 172 (1951).
71. G. Romijn, *Z. Anal. Chem.*, **36**, 349 (1897).
72. K. Myrbäck and E. Gyllensvard, *Sv. Kem. Tidskr.*, **54**, 17 (1942).
73. R. L. Colbran and T. P. Nevell, *J. Chem. Soc.*, 2427 (1957).
74. H. S. Isbell, L. T. Sniegoski, and H. L. Frush, *Anal. Chem.*, **34**, 982 (1962).
75. W. F. Goebel, *J. Biol. Chem.*, **72**, 809 (1927).
76. S. Moore and K. P. Link, *J. Biol. Chem.*, **133**, 293 (1940).
77. M. L. Wolfrom and N. E. Franks, *J. Org. Chem.*, **29**, 3645 (1964).
77a. P. V. Peplow and P. J. Somers, *Carbohyd. Res.*, **22**, 281 (1972).
78. K. Bailey and R. H. Hopkins, *Biochem. J.*, **27**, 1965 (1933).
79. M. Hönig and F. Tempus, *Ber.*, **57**, 787 (1924).
80. T. Reichstein and O. Neracher, *Helv. Chim. Acta*, **18**, 892 (1935); W. Ruzicka, *Z. Zuckerind. Böhmen-Mähren*, **64**, 219 (1941).
81. M. Bergmann and W. W. Wolff, *Ber.*, **56**, 1060 (1923); K. Smolenski, *Rocz. Chem.*, **3**, 153 (1924); *Chem. Abstr.*, **19**, 41 (1925).
82. E. L. Jackson and C. S. Hudson, *J. Amer. Chem. Soc.*, **59**, 994 (1937).

83. E. Fischer and J. Tafel, *Ber.*, **20**, 3384 (1887); **22**, 106 (1889).
84. R. Pasternack and P. P. Regna, U.S. Pat. 2,203,923 (June 11, 1940); U.S. Pat. 2,207,991 (July 16, 1940); U.S. Pat. 2,188,777 (Jan. 30, 1940).
85. P. P. Regna and B. P. Caldwell, *J. Amer. Chem. Soc.*, **66**, 243 (1944); H. S. Isbell, *J. Res. Nat. Bur. Stand.*, **33**, 45 (1944).
86. R. J. Williams and M. Woods, *J. Amer. Chem. Soc.*, **59**, 1408 (1937).
87. W. Hurka, *Mikrochemie ver. Mikrochim. Acta*, **30**, 259 (1942).
88. A. Jeanes and H. S. Isbell, *J. Res. Nat. Bur. Stand.*, **27**, 125 (1941).
89. P. Van Fossen and E. Pacsu, *Text. Res. J.*, **16**, 163 (1946).
90. H. F. Launer and Y. Tomimatsu, *Anal. Chem.*, **26**, 382 (1954); *J. Amer. Chem. Soc.*, **76**, 2591 (1954).
91. F. Stitt, S. Friedlander, H. J. Lewis, and F. E. Young, *Anal. Chem.*, **26**, 1478 (1954).
92. R. H. Zienius and C. B. Purves, *Can. J. Chem.*, **37**, 1820 (1959).
93. E. S. Becker, J. K. Hamilton, and W. E. Lucke, *Tappi*, **48**, 60 (1965).
94. R. A. Somsen, *Tappi*, **43**, 154, 157 (1960).
94a. R. L. Whistler, T. W. Mittag, and T. R. Ingle, *J. Amer. Chem. Soc.*, **87**, 4218 (1965).
95. S. Hanessian, *Carbohyd. Res.*, **2**, 86 (1966); see also, D. L. Failla, T. L. Hullar, and S. B. Siskin, *Chem. Commun.*, 716 (1966).
96. J. Kiss, *Chem. Ind.* (London), 73 (1964).
97. A. F. Meiners and F. V. Morriss, *J. Org. Chem.*, **29**, 449 (1964).
97a. D. C. Baker and D. Horton, *Carbohyd. Res.*, **21**, 393 (1972).
97b. D. M. Clode and D. Horton, *Carbohyd. Res.*, **17**, 365 (1971).
97c. D. Horton, A. E. Luetzow, and O. Theander, *Carbohyd. Res.*, **26**, 1 (1973).
97d. H. F. G. Beving, A. E. Luetzow, and O. Theander, *Carbohyd. Res.*, **41**, 105 (1975).
97e. R. W. Binkley, *Carbohyd. Res.*, **48**, C1 (1976); *ibid.*, **58**, C10 (1977).
98. D. L. MacDonald and H. O. L. Fischer, *J. Amer. Chem. Soc.*, **74**, 2087 (1952); *Biochim. Biophys. Acta*, **12**, 203 (1953); R. Barker and D. L. MacDonald, *J. Amer. Chem. Soc.*, **82**, 2297 (1960).
99. L. Hough and T. J. Taylor, *J. Chem. Soc.*, 1212, 3544 (1955); 970 (1956).
100. L. Hough and A. C. Richardson, *J. Chem. Soc.*, 5561 (1961); 1019, 1024 (1962).
101. L. D. Hall, L. Hough, S. H. Shute, and T. J. Taylor *J. Chem. Soc.*, 1154 (1965).
101a. A. Farrington and L. Hough, *Carbohyd. Res.*, **16**, 59 (1971).
102. H. Griesebach, W. Hofheinz, and N. Doerr, *Ber.*, **96**, 1823 (1963).
103. I. Iwai and K. Tomita, *Chem. Pharm. Bull.* (Tokyo), **11**, 184 (1963); *Chem. Abstr.*, **59**, 7622 (1963).
103a. G. O. Aspinall and R. R. King, *Can. J. Chem.*, **51**, 394 (1973).
103b. G. O. Aspinall and Z. Pawlák, *Can. J. Chem.*, **51**, 388 (1973).
103c. P. Köll, R. Dürrfeld, U. Wolfmeier, and K. Heyns, *Tetrahedron Lett.*, 5081 (1972).
103d. D. Horton, W. Weckerle, L. Odier, and R. J. Sorenson, *J. Chem. Soc. Perkin I*, 1564 (1977).
104. E. G. V. Percival, *Advan. Carbohyd. Chem.*, **3**, 23 (1948); H. El Khadem, *ibid.*, **20**, 139 (1965).
105. O. L. Chapman, W. J. Welstead, Jr., T. J. Murphy, and R. W. King, *J. Amer. Chem. Soc.*, **86**, 732 (1964).
106. E. M. Bamdas, K. M. Ermolaev, V. J. Maimind, and M. M. Shemyakin, *Chem. Ind.* (London), 1195 (1959).
107. F. Weygand, *Ber.*, **73**, 1284 (1940).
108. H. Simon, J. D. Dorrer, and A. Trebst, *Ber.*, **96**, 1285 (1963).
109. J. U. Nef, *Ann.*, **403**, 204 (1914); O. Spengler and A. Pfannenstiehl, *Z. Wirtschafts-gruppe Zuckerind.*, **85**, 546 (1935); *Chem. Abstr.*, **30**, 4470[4] (1936).

109a. R. Malinen and E. Sjöström, *Pap. Puu*, **54**, 451 (1972).

109b. R. Malinen and E. Sjöström, *Pap. Puu*, **55**, 547 (1973).

110. O. Dalmers and K. Heyns, U.S. Pat. 2,190,377 (Feb. 13, 1939); H. S. Isbell, *J. Res. Nat. Bur. Stand.*, **29**, 227 (1942).

111. E. Hardegger, K. Kreis, and H. El Khadem, *Helv. Chim. Acta*, **35**, 618 (1952).

112. B. Warshowsky and W. M. Sandstrom, *Arch. Biochem. Biophys.*, **37**, 46 (1952).

113. D. Entwistle, E. H. Cole, and N. S. Wooding, *Text. Res. J.*, **19**, 527, 609 (1949).

114. H. R. Gersmann, H. J. W. Nieuwenhuis, and A. F. Bickel, *Tetrahedron Lett.*, 1383 (1963).

115. G. A. Russell, E. G. Janzen, A. G. Bemis, E. J. Geels, A. J. Moye, and E. T. Strom, in ref. 4, pp. 113–114.

116. J. A. Mattor, *Tappi*, **46**, 586 (1963).

116a. J. C. Arthur, Jr., and O. Hinojosa, *J. Polym. Sci. C*, **36**, 53 (1971).

116b. J. L. Minor and N. Sanyer, *J. Polym. Sci. C*, **36**, 73 (1971).

116c. B. Ericsson, B. O. Lindgren, and O. Theander, *Sv. Papperstidn.*, **74**, 757 (1971); *Chem. Abstr.*, **76**, 73968 (1972).

116d. O. Samuelson and L. Stolpe, *Sv. Papperstidn.*, **74**, 545 (1971); *Chem. Abstr.*, **76**, 26614 (1972).

116e. A. Palma, J. Jovanović, and G. V. Schulz, *Makromol. Chem.*, **169**, 219 (1973).

116f. J. E. Charles, C. de Choudens, and P. Monzie, *ATIP*, **27** 139 (1973); A. Ishizu, *Kam Pa Gikyoshi*, **27**, 371 (1973); *Chem. Abstr.*, **79**, 93591 (1973).

117. K. Heyns and H. Paulsen, *Advan. Carbohyd. Chem.*, **17**, 169 (1962).

118. K. Heyns and M. Beck, *Ber.*, **91**, 1720 (1958).

119. Knoll, A.-G. Chemische Fabriken, Ger. Pat. 935,968, *Chem. Abstr.*, **52**,19,972 (1958).

120. K. Heyns and H. Paulsen, *Angew. Chem.*, **69**, 600 (1957).

121. L. O. Schnaidman and I. N. Kushchinskaya, *Tr. Vses. Nauchn.-Issled Vitamin Inst.*, **8**, 13 (1961); *Chem. Abstr.*, **58**, 2495c (1963).

122. K. Heyns and J. Lenz, *Ber.*, **94**, 384 (1961).

122a. J. Bakke and O. Theander, *Chem. Commun.*, 175 (1971).

123. S. A. Barker, E. J. Bourne, J. G. Fleetwood, and M. Stacey, *J. Chem. Soc.*, 4128 (1958).

124. Ref. 117, Table I, pp. 177–178.

125. Ref. 117, Table III, p. 190.

126. Y. Hirasaka, *Yakugaku Zasshi*, **83**, 960, 971 (1963); *Chem. Abstr.*, **60**, 4232c, 4233d (1964); G. G. S. Dutton and K. N. Slessor, *Can. J. Chem.*, **42**, 1110 (1964).

127. D. Abbott and H. Weigel, *J. Chem. Soc.*, 5157 (1965).

128. G. O. Aspinall and A. Nicolson, *J. Chem. Soc.*, 2503 (1960); G. O. Aspinall and I. M. Cairncross, *ibid.*, 3877, 3998 (1960).

129. K. Heyns and J. Lenz, *Angew. Chem.*, 73, 299 (1961).

130. B. Lindberg, I. Svensson, and O. Theander, *Acta Chem. Scand.*, **17**, 930 (1963); J. S. Brimacombe and M. C. Cook, *J. Chem. Soc.*, 2663 (1964).

131. J. S. Brimacombe, M. C. Cook, and L. C. N. Tucker, *J. Chem. Soc.*, 2292 (1965).

131a. K. Heyns, W. D. Soldat, and P. Köll, *Ber.*, **108**, 3619 (1975).

132. K. Heyns, W. P. Trautwein, and H. Paulsen, *Ber.*, **96**, 3195 (1963); Atlas Chemical Industries Inc., Fr. Pat. 1,426,204 (Jan. 28, 1966); *Chem. Abstr.*, **65**, 15,490g (1967).

132a. K. Heyns, J. Weyer, and H. Paulsen, *Ber.*, **100**, 2317 (1967).

132b. K. Heyns and P. Köll, *Methods Carbohyd. Chem.*, **6**, 342 (1972).

132c. K. Heyns, E: Alpers, and J. Weyer, *Ber.*, **101**, 4199, 4209 (1968).

132d. K. Antonakis and F. Leclerq, *C. R. Acad. Sci.*, Ser. C, **265**, 1004 (1967).

133. H. A. Spoehr and H. W. Milner, *J. Amer. Chem. Soc.*, **56**, 2068 (1934).

134. M. Clinton, Jr., and R. S. Hubbard, *J. Biol. Chem.*, **119**, 467 (1937).
135. W. Traube and F. Kuhbier, *Ber.*, **69**, 2664 (1936).
136. M. Garino, M. Parodi, and V. Vignolo, *Gazz. Chim. Ital.*, **65**, 132 (1935).
137. W. D. Major, *Tappi*, **41**, 530 (1958).
137a. W. J. Harrison, F. H. Jackson, and I. K. Waller, *N. Z. J. Sci.*, **16**, 199 (1973).
138. J. A. Church, *Tappi*, **48**, 185 (1965).
139. R. E. Olson and J. W. Green, unpublished data.
140. H. R. Cooper and H. W. Melville, *J. Chem. Soc.*, 1984 (1951).
141. J. F. Haskins and M. J. Hogsed, *J. Org. Chem.*, **15**, 1264 (1950).
142. R. L. Whistler and R. Schweiger, *J. Amer. Chem. Soc.*, **81**, 3136 (1959).
142a. H. S. Isbell, H. L. Frush, and E. T. Martin, *Carbohyd. Res.*, **26**, 287 (1973).
142b. H. S. Isbell and H. L. Frush, *Carbohyd. Res.*, **28**, 295 (1973).
142c. H. S. Isbell and R. G. Naves, *Carbohyd. Res.*, **36**, C1 (1974).
142d. H. S. Isbell, H. L. Frush, and Z. Orhanovic, *Carbohyd. Res.*, **36**, 283 (1974); **43**, 93 (1975); see also H. S. Isbell, *ibid.*, **39**, C4 (1975).
142e. J. W. Weaver, L. R. Schroeder, and N. S. Thompson, *Carbohyd. Res.*, **48**, C5 (1976).
142f. G. A. Tolstikov, U. M. Gemilev, and V. P. Yurjev, *Zh. Obsch. Khim.*, **42**, 1611 (1972).
143. O. Ruff, *Ber.*, **31**, 1573 (1898); **34**, 1362 (1901); O. Ruff and G. Ollendorf, *ibid.*, **33**, 1798 (1900); H. G. Fletcher, Jr., H. W. Diehl, and C. S. Hudson, *J. Amer. Chem. Soc.*, **72**, 4546 (1950).
143a. T. Takamoto, R. Sudoh, and T. Nakagawa, *Carbohyd. Res.*, **27**, 135 (1973).
144. H. J. H. Fenton and H. O. Jones, *J. Chem. Soc.*, **77**, 69 (1900).
145. F. Haber and J. Weiss, *Proc. Roy. Soc.* London, *Ser. A*, **147**, 332 (1934); J. H. Baxendale, M. G. Evans, and G. S. Park, *Trans. Faraday Soc.*, **42**, 155 (1946).
146. See W. A. Waters, *Annu. Rep. Progr. Chem.*, **42**, 145 (1945).
147. C. F. Cross, E. J. Bevan, and C. Smith, *J. Chem. Soc.*, **73**, 463 (1898); R. S. Morrell and J. M. Crofts, *ibid.*, **75**, 786 (1899); H. A. Spoehr, *J. Amer. Chem. Soc.*, **43**, 227 (1910).
148. J. H. Payne and L. Foster, *J. Amer. Chem. Soc.*, **67**, 1654 (1945); A. A. Kultyugin and L. H. Sokoliva, *Arch. Sci. Biol. USSR*, **41**, 145 (1936).
149. A. T. Küchlin, *Rec. Trav. Chim.* (Pays-Bas), **51**, 887 (1932), and earlier papers.
150. A. T. Küchlin, *Biochem. Z.*, **261**, 411 (1933).
151. H. J. H. Fenton and H. Jackson, *J. Chem. Soc.*, **75**, 1 (1899).
152. B. Fraser-Reid, J. K. N. Jones, and M. B. Perry, *Can. J. Chem.*, **39**, 555 (1961).
153. A. N. DeBelder, B. Lindberg, and O. Theander, *Acta Chem. Scand.*, **17**, 1012 (1963).
154. A. Jollès, *Biochem. Z.*, **34**, 242 (1911).
155. M. Stacey, *J. Chem. Soc.*, 1529 (1939); H. Ohle and G. Berend, *Ber.*, **58**, 2585 (1925); R. G. Ault, W. N. Haworth, and E. L. Hirst, *J. Chem. Soc.*, 517 (1935); M. Stacey and P. I. Wilson, *ibid.*, 587 (1944).
156. See C. A. Brown and F. W. Zerban, "Sugar Analysis," Wiley, New York, 1941, pp. 691 and 728; A. W. Van der Haar, "Monosaccharide und Aldehydsäuren," Borntraeger, Berlin, 1920.
157. W. W. Pigman, B. L. Browning, W. H. McPherson, C. R. Calkins, and R. L. Leaf, Jr., *J. Amer. Chem. Soc.*, **71**, 2200 (1949).
158. W. E. Militzer and R. Angier, *Arch. Biochem. Biophys.*, **10**, 291 (1946).
159. H. Kiliani, *Ann.*, **205**, 163, 172 (1880); *Ber.*, **54**, 463 (1921); W. E. Stokes and W. E. Barch, U.S. Pat. 2,257,284 (Sept. 30, 1941).
159a. See R. A. Laidlaw and E. G. V. Percival, *Advan. Carbohyd. Chem.*, **7**, 1 (1952).

160. J. K. Dale and W. F. Rice, Jr., *J. Amer. Chem. Soc.*, **55**, 4984 (1933); S. Soltzberg, U.S. Pat. 2,380,196 (July 10, 1945); A. F. Odell, U.S. Pat. 1,425,605 (Aug. 15, 1922).
161. H. Kiliani, *Ber.*, **55**, 2817 (1922); **56**, 2016 (1923); **58**, 2344 (1925).
162. H. Kiliani, *Ber.*, **54**, 456 (1921); **55**, 75 (1922).
163. K. Maurer and G. Drehfahl, *Ber.*, **75**, 1489 (1942); K. Maurer and G. Reiff, *J. Makromol. Chem.*, **1**, 27 (1943); E. C. Yackel and W. O. Kenyon, *J. Amer. Chem. Soc.*, **64**, 121 (1942); C. C. Unruh and W. O. Kenyon, *ibid.*, **64**, 127 (1942).
164. Y. Nakajima and N. Ando, Jap. Pat. 8272 (1956); *Chem. Abstr.*, **52**, 12,899b (1958); Y. Nitta, H. Nakajima, J. Shihota, A. Shirase, and N. Ando, Jap. Pat. 9247 (1957); *Chem. Abstr.*, **52**, 15,572g (1958); M. Hiramoto and M. Susai, Jap. Pat. 5830 (1958), *Chem. Abstr.*, **53**, 18,882d (1959); H. Nakajima (Chugai Drug Co.) Jap. Pat. 3309 (1959); *Chem. Abstr.*, **54**, 14,144f (1960); Yodogawa Pharm. Products Co., Jap. Pat. 12,114 (1961); *Chem. Abstr.*, **58**, 1408b (1963); T. Ishikawa, *Nippon Kagaku Zasshi*, **82**, 1536 (1961); *Chem. Abstr.*, **59**, 4,018h (1963); B. G. Yasnitskiĭ, E. B. Dol'berg, and V. A. Oridoroga, *Ukr. Khim. Zh.*, **38**, 76 (1972); *Chem. Abstr.*, **76**, 129,055 (1972); A. Takahashi and S. Takahashi, *Sen'i Gakkaishi*, **29**, 280 (1973); *Chem. Abstr.*, **79**, 106254 (1973).
165. A. Assarsson and O. Theander, *Acta Chem. Scand.*, **18**, 553 (1964).
166. H. Kiliani, *Ber.*, **54**, 456 (1921); G. D. Hiatt, U.S. Pat. 2,256,391 (Sept. 16, 1941); J. G. Bremmer, R. H. Stanley, D. G. Jones, and A. W. C. Taylor, U.S. Pat. 2,389,950 (Nov. 27, 1945).
167. V. H. Veley, *Phil. Trans. Roy. Soc.* (London), **52**, 27 (1893); *J. Chem. Soc. (Abstr.)* **64**, II, 413 (1893).
168. P. A. McGee, W. F. Fowler, Jr., E. W. Taylor, C. C. Unruh, and W. O. Kenyon, *J. Amer. Chem. Soc.*, **69**, 355 (1947).
168a. M. M. Pavlyuchenko, I. N. Ermolenko, and F. N. Kaputskiĭ, *Zh. Priklad. Khim.*, **33**, 1385 (1957).
169. E. D. Kaverzneva and A. S. Salova, *Izv. Akad. Nauk SSSR*, 344 (1959).
170. N. Ya. Lenshina, V. S. Ivanova, and V. I. Ivanov, *Izv. Akad. Nauk SSSR*, 519 (1961).
171. K. B. Wiberg, in ref. 3, p. 69.
172. F. H. Westheimer and A. Novick, *J. Chem. Phys.*, **11**, 506 (1943); M. Cohen and F. H. Westheimer, *J. Amer. Chem. Soc.*, **74**, 4387 (1952).
173. Ref. 1, p. 38.
174. O. Theander, *Acta Chem. Scand.*, **11**, 1557 (1957); A. Assarsson and O. Theander, *ibid.*, **12**, 1507 (1958); E. Brimacombe, J. S. Brimacombe, and B. Lindberg, *ibid.*, **14**, 2236 (1960); E. Brimacombe, J. S. Brimacombe, B. Lindberg, and O. Theander, *ibid.*, **15**, 437 (1961).
175. H. C. Brown and C. P. Garg, *J. Amer. Chem. Soc.*, **83**, 2952 (1961).
176. O. Theander, *Acta Chem. Scand.*, **17**, 1751 (1963).
177. J. W. Bird and J. K. N. Jones, *Can. J. Chem.*, **41**, 1877 (1963).
178. A. Sera and R. Goto, *Bull. Chem. Soc. Japan*, **38**, 1415 (1965).
179. S. J. Angyal and M. E. Evans, *Aust. J. Chem.*, **25**, 1513 (1972).
180. S. J. Angyal and K. James, *Aust. J. Chem.*, **23**, 1209 (1970).
181. S. J. Angyal and K. James, *Aust. J. Chem.*, **24**, 1219 (1971).
182. S. J. Angyal and M. E. Evans, *Aust. J. Chem.*, **25**, 1495 (1972).
183. M. Bertolini and C. P. J. Glaudemans, *Carbohyd. Res.*, **15**, 263 (1970).
184. S. J. Angyal and K. James, *Carbohyd. Res.*, **12**, 147 (1970).
185. Ref. 1, p. 45.
186. Ref. 2, p. 63.

187. O. Theander, *Acta Chem. Scand.*, **18**, 2209 (1964).
188. J. Hoffman, B. Lindberg, and S. Svensson, *Acta Chem. Scand.*, **26**, 661 (1972).
189. P. Deslongchamps and C. Moreau, *Can. J. Chem.*, **49**, 2465 (1971).
190. G. Y. Wu and J. M. Sugihara, *Carbohyd. Res.*, **13**, 89 (1970).
191. H. Mizoiri, A. Sera, and R. Goto, *Bull. Chem. Soc. Japan*, **37**, 1023 (1964).
192. P. M. Collins and W. G. Overend, *Chem. Ind.* (London), 375 (1963).
193. R. E. Arrick, D. C. Baker, and D. Horton, *Carbohyd. Res.*, **26**, 441 (1973).
194. A. F. Krasso, E. K. Weiss, and T. Reichstein, *Helv. Chim. Acta*, **46**, 2538 (1963).
195. B. Lindberg and O. Theander, *Acta Chem. Scand.*, **11**, 1355 (1957).
196. D. Horton and E. K. Just, *Carbohyd. Res.*, **29**, 173 (1973).
197. S. Chandra and R. K. Mittal, *Carbohyd. Res.*, **19**, 123 (1971); K. K. S. Gupta, S. Gupta, and S. N. Basu, *ibid.*, **71**, 75 (1979).
198. K. E. Pfitzner and J. G. Moffatt, *J. Amer. Chem. Soc.*, **85**, 3027 (1963).
199. G. Henseke and G. Hanisch, *Angew. Chem.*, **75**, 420 (1963).
200. K. Onodera, S. Hirano, and N. Kashimura, *J. Amer. Chem. Soc.*, **87**, 4651 (1965); *Carbohyd. Res.*, **6**, 276 (1968).
201. J. D. Albright and L. Goldman, *J. Amer. Chem. Soc.*, **87**, 4214 (1965).
202. W. W. Epstein and F. W. Sweat, *Chem. Rev.*, **67**, 247 (1967).
203. D. J. Ward, W. A. Szarek, and J. K. N. Jones, *Carbohyd. Res.*, **21**, 305 (1972).
204. T. A. Guidici and J. J. Griffin, *Carbohyd. Res.*, **33**, 287 (1974).
205. J. L. Godman and D. Horton, *Carbohyd. Res.*, **6**, 229 (1968); D. Horton, M. Nakadate, and J. M. J. Tronchet, *ibid.*, **7**, 56 (1968); B. A. Dmitriev, A. A. Kost, and N. K. Kochetkov, *Izv. Akad. Nauk SSSR*, 697 (1969); R. F. Butterworth and S. Hanessian, *Can. J. Chem.*, **49**, 2755 (1971).
206. Z. Pawlak, *Rocz. Chem.*, **46**, 737 (1972).
207. F. W. Lichtenthaler, *Methods Carbohyd. Chem.*, **6**, 348 (1972).
208. D. M. Mackie and A. S. Perlin, *Carbohyd. Res.*, **24**, 67 (1972).
209. G. Hanisch and G. Henseke, *Ber.*, **101**, 2074 (1968).
210. G. Hanisch and G. Henseke, *Ber.*, **101**, 4170 (1968).
211. Y. Kondo and F. Takao, *Can. J. Chem.*, **51**, 1476 (1973).
211a. J. Defaye and A. Gadelle, *Carbohyd. Res.*, **42**, 373 (1975).
212. D. Horton and J. S. Jewell, *Carbohyd. Res.*, **2**, 251 (1966).
213. N. Pravdić and H. G. Fletcher, Jr., *Carbohyd. Res.*, **19**, 353 (1971).
214. N. Kuzuhara, N. Oguchi, H. Ohrui, and S. Emoto, *Carbohyd. Res.*, **23**, 217 (1972).
215. B. Lindberg, *Methods Carbohyd. Chem.*, **6**, 323 (1972).
216. D. Horton and J. S. Jewell, *Carbohyd. Res.*, **5**, 149 (1967).
217. G. H. Jones and J. G. Moffatt, *Methods Carbohyd. Chem.*, **6**, 315 (1972).
218. W. Sowa and G. H. S. Thomas, *Can. J. Chem.*, **44**, 836 (1966).
219. K. Onodera and N. Kashimura, *Methods Carbohyd. Chem.*, **6**, 331 (1972).
220. R. J. Ferrier, *Methods Carbohyd. Chem.*, **6**, 419 (1972).
221. B. Lindberg and K. N. Slessor, *Carbohyd. Res.*, **1**, 492 (1966); *Acta Chem. Scand.*, **21**, 910 (1967).
222. T. B. Grindley, J. W. Bird, W. A. Szarek, and J. K. N. Jones, *Carbohyd. Res.*, **24**, 212 (1972).
223. M. L. Wolfrom and P. Y. Wang, *Carbohyd. Res.*, **12**, 109 (1970); D. Horton and E. K. Just, *ibid.*, **30**, 349 (1973).
224. K. Bredereck, *Tetrahedron Lett.*, 695 (1967).
225. N. Kashimura, K. Yoshida, and K. Onodera, *Carbohyd. Res.*, **25**, 264 (1972).
226. K. Torssell, *Tetrahedron Lett.*, 4445 (1966).
227. A. H. Fenselau and J. G. Moffatt, *J. Amer. Chem. Soc.*, **88**, 1762 (1966).

228. C. Djerassi and R. R. Engle, *J. Amer. Chem. Soc.*, **75**, 3838 (1953).

229. L. M. Berkowitz and P. N. Rylander, *J. Amer. Chem. Soc.*, **80**, 6682 (1958).

230. P. J. Beynon, P. M. Collins, and W. G. Overend, *Proc. Chem. Soc.*, 342 (1964); P. J. Beynon, P. M. Collins, D. Gardiner, and W. G. Overend, *Carbohyd. Res.*, **6**, 431 (1968).

231. V. M. Parikh and J. K. N. Jones, *Can. J. Chem.*, **43**, 3452 (1965); compare R. H. Hall and K. Bischofberger, *Carbohyd. Res.*, **65**, 139 (1978).

231a. D. C. Baker, D. Horton, and C. G. Tindall, Jr., *Methods Carbohyd. Chem.*, **7**, 3 (1976).

232. C. L. Stevens and C. P. Bryant, *Methods Carbohyd. Chem.*, **6**, 337 (1972).

233. R. F. Nutt, B. Arison, F. W. Holly, and E. Walton, *J. Amer. Chem. Soc.*, **87**, 3273 (1965).

234. J. M. J. Tronchet and J. Tronchet, *Helv. Chim. Acta*, **53**, 1174 (1970).

235. J. Šmejkal and L. Kalvoda, *Collect. Czech. Chem. Commun.*, **38**, 1981 (1973).

236. P. M. Collins, W. G. Overend, and B. A. Rayner, *Carbohyd. Res.*, **31**, 1 (1973).

237. J. L. Courtney and K. F. Swansborough, *Rev. Pure Appl. Chem.*, **22**, 47 (1972).

238. R. Stewart, in ref. 3, pp. 2–68; ref. 2, pp. 82–83; and ref. 1, pp. 63–75; see also A. J. Fatiadi, *Synthesis*, **6**, 229 (1974).

239. R. Stewart, *J. Amer. Chem. Soc.*, **79**, 3057 (1957).

239a. J. M. J. Tronchet and J. Tronchet, *Carbohyd. Res.*, **33**, 237 (1974).

239b. N. K. Kochetkov and B. A. Dmitriev, *Methods Carbohyd. Chem.*, **6**, 150 (1972).

240. C. Niemann and K. P. Link, *J. Biol. Chem.*, **104**, 195, 743 (1934).

241. W. Bosshard, *Helv. Chim. Acta*, **18**, 956 (1935).

242. J. M. Perel and P. G. Dayton, *J. Org. Chem.*, **25**, 2044 (1960).

243. E. Fischer and A. W. Crossley, *Ber.*, **27**, 394 (1894).

244. Y. Hirasaka, *Yakugaku Zasshi*, **83**, 971 (1963); *Chem. Abstr.*, **60**, 4233d (1964); Y. Hirasaka and I. Matsunaga, *Chem. Pharm. Bull.* (Tokyo), **13**, 176 (1965); *Chem. Abstr.*, **62**, 14,799a (1965).

245. I. Johansson, B. Lindberg, and O. Theander, *Acta Chem. Scand.*, **17**, 2019 (1963).

246. R. Kuhn and T. Wagner-Jauregg, *Ber.*, **58**, 1441 (1925).

247. H. H. Stroh, G. Geoldenitz, and H. Patzig, *Z. Chem.*, **5**, 459 (1965); *Chem. Abstr.*, **65**, 7256e (1966).

248. J. L. Bose, A. B. Foster, M. Stacey, and J. M. Webber, *Nature*, **184**, 1301 (1959); J. L. Bose, A. B. Foster, N. Salim, M. Stacey, and J. M. Webber, *Tetrahedron*, **14**, 201 (1961).

249. H. Weidmann and G. Olbrich, *Tetrahedron Lett.*, 725 (1965).

250. S. Y.-K. Tam and B. Fraser-Reid, *Tetrahedron Lett.*, 3151 (1972).

251. R. M. Evans, *Quart. Rev. Chem. Soc.*, **13**, 61 (1959).

252. B. Fraser-Reid, D. L. Walker, S. Y.-K. Tam, and N. L. Holder, *Can. J. Chem.*, **51**, 3950 (1973); P. M. Collins, *J. Chem. Soc.*, (*C*), 1670 (1972).

253. M. P. Singh, *Vijnana Parishad Anusandhan Patrika*, **3**, 97 (1960); *Chem. Abstr.*, **59**, 7625c (1963); N. Nath and M. P. Singh, *Z. Physik. Chem.*, **224**, 419 (1963); *J. Phys. Chem.*, **69**, 2038 (1965); R. K. Srivastava, N. Nath, and M. P. Singh, *Bull. Chem. Soc. Japan*, **39**, 833 (1965); *Tetrahedron*, **23**, 1189 (1967).

254. E. Kasper, *Z. Physik. Chem.*, **224**, 427 (1963); *Chem. Abstr.*, **60**, 14579c (1964).

255. M. P. Singh and S. Ghosh, *Z. Physik. Chem.*, **204**, 1 (1955); M. P. Singh, B. Krishna and S. Ghosh, *Proc. Nat. Acad. Sci. India, Sect. A*, **28**, 21 (1959); *Chem. Abstr.*, **54**, 17015d (1960); M. P. Singh, *Z. Physik. Chem.*, **216**, 13, 17 (1961); M. P. Singh, *Vijnana Parishad Anusandhan Patrika*, **4**, 277 (1961); *Chem. Abstr.*, **59**, 8843c (1963).

256. B. A. Marshall and W. A. Waters, *J. Chem. Soc.*, 2392 (1960).

257. F. W. Jensen and F. W. Upson, *J. Amer. Chem. Soc.*, **47**, 3019 (1925).

258. E. Anderson, *Amer. Chem. J.*, **42**, 401 (1909); J. U. Nef, *Ann.*, **357**, 214 (1907); see also, J. Habermann and M. Hönig, *Monatsh. Chem.*, **3**, 651 (1882); **5**, 208 (1884).
259. J. U. Nef, *Ann.*, **335**, 247 (1904); **357**, 259 (1907).
260. P. Girard and J. Parrod, *C.R. Acad. Sci.*, **190**, 328 (1930); J. Parrod, *ibid.*, **192**, 1136 (1931); **200**, 1884 (1935); **212**, 610 (1941).
261. V. I. Ivanov and K. M. Sokova, *Dokl. Akad. Nauk SSSR*, **42**, 179 (1944); *Chem. Abstr.*, **39**, 281^2 (1945).
262. H. Kiliani, *Ann.*, **205**, 145 (1880); K. Dreyer, *ibid.*, **416**, 203 (1918); K. G. A. Busch, J. W. Clark, L. B. Genung, E. F. Schroeder, and W. L. Evans, *J. Org. Chem.*, **1**, 1 (1936).
263. W. E. Trevelyan, D. P. Procter, and J. S. Harrison, *Nature*, **166**, 444 (1950).
264. L. R. Schroeder, unpublished data.
265. G. F. Bayer, unpublished data.
266. M. Fétizon and M. Golfier, *C.R. Acad. Sci.*, *Ser. C*, **267**, 900 (1968); *Chem. Commun.*, 1102, 1118 (1969).
267. J. M. J. Tronchet, J. Tronchet, and A. Birkhaeuser, *Helv. Chim. Acta*, **53**, 1489 (1970).
268. S. Morgenlie, *Acta Chem. Scand.*, **25**, 1155 (1971); **26**, 2518 (1972).
269. S. Morgenlie, *Acta Chem. Scand.*, **26**, 1709 (1972); **27**, 1557, 2607 (1973).
270. S. Morgenlie, *Acta Chem. Scand.*, **25**, 2773 (1971); *Carbohyd. Res.*, **73**, 315 (1979); Compare *ibid.*, **59**, 73 (1977).
271. S. Morgenlie, *Acta Chem. Scand.*, **27**, 3009 (1973).
272. J. B. Lee and T. G. Clarke, *Tetrahedron Lett.*, 415 (1967).
273. S. Nord, O. Samuelson, and R. Simonson, *Sv. Papperstidn.*, **65**, 767 (1962).
274. B. Lindberg and O. Theander, *Sv. Papperstidn.*, **65**, 509 (1962); see also S. Yllner, *Acta Chem. Scand.*, **10**, 1251 (1956); R. Cordingly, *Tappi*, **42**, 654 (1959).
275. O. Theander, *Tappi*, **48**, 105 (1965).
276. D. D. Van Slyke, J. Plazin, and J. R. Weisiger, *J. Biol. Chem.*, **191**, 299 (1951); see also M. Gibbs, P. K. Kindel, and M. Busse, *Methods Carbohyd. Chem.*, **2**, 496 (1963).
277. T. N. Kleinert and W. Wincor, *Holzforschung*, **10**, 80 (1956); see also, T. N. Kleinert and R. Dloughy, *Tappi*, **40**, 827 (1957).
278. G. Birstein and M. Blumenthal, *Bull. Soc. Chim. Fr.*, [5] **11**, 573 (1944).
279. H. F. Launer, *J. Res. Nat. Bur. Stand.*, **20**, 87 (1938); **18**, 333 (1937).
280. G. F. Smith, "Cerate Oxidimetry," G. F. Smith Co., Columbus, Ohio, 1942.
281. H. Stamm, *Angew. Chem.*, **47**, 791 (1934).
282. H. H. Willard and P. Young, *J. Amer. Chem. Soc.*, **50**, 1322 (1928).
283. G. F. Smith and F. R. Duke, *Ind. Eng. Chem.*, *Anal. Ed.*, **13**, 558 (1941).
284. A. Smolka, *Monatsh. Chem.*, **8**, 1 (1887).
285. E. J. Witzemann, *J. Amer. Chem. Soc.*, **38**, 150 (1916).
286. W. L. Evans, C. A. Buehler, C. D. Looker, R. A. Crawford, and C. W. Holl, *J. Amer. Chem. Soc.*, **47**, 3085 (1925); W. L. Evans and C. A. Buehler, *ibid.*, **47**, 3098 (1925).
287. W. L. Evans and C. W. Holl, *J. Amer. Chem. Soc.*, **47**, 3102 (1925).
288. J. Z. Beer, *Carbohyd. Res.*, **1**, 297 (1965).
289. J. S. Lee, *Anal. Chem.*, **34**, 835 (1962).
290. P. Malangeau and M. Guernet, *Carbohyd. Res.*, **24**, 499 (1972).
290a. K. M. Haldorsen, *Carbohyd. Res.*, **63**, 61 (1978); K. K. S. Gupta and S. N. Basu, *ibid.*, **72**, 139 (1979).
291. P. K. Jaiswal, *Microchem. J.*, **15**, 434 (1970); S. Chandra and K. L. Yadava, *ibid.*, **17**, 4 (1972).
292. P. K. Jaiswal, *Microchem. J.*, **15**, 122 (1970).

293. M. A. Ali Butt, R. U. Malik, and J. Zýka, *Microchem. J.*, **17**, 612 (1972).

294. D. Müller, *Biochem. Z.*, **199**, 136 (1928); *Enzymologia*, **10**, 40 (1941).

295. D. Keilin and E. F. Hartree, *Biochem. J.*, **50**, 331 (1952).

296. R. Bentley and A. Neuberger, *Biochem. J.*, **45**, 584 (1949).

297. K. Linek and M. Sobotka. *Collect. Czech. Chem. Commun.*, **38**, 2819 (1973).

298. H. J. Strecker and S. Korkes, *Nature*, **168**, 913 (1951); *J. Biol. Chem.*, **196**, 769 (1952).

299. B. L. Horecker, P. Z. Smyrniotis, and J. E. Seegmiller, *J. Biol. Chem.*, **193**, 383 (1951); J. E. Seegmiller and B. L. Horecker, *ibid.*, **194**, 261 (1952).

300. R. Weimberg, *Biochim. Biophys. Acta*, **67**, 359 (1963).

301. A. Maradufu, G. M. Cree, and A. S. Perlin, *Can. J. Chem.*, **49**, 3429 (1971).

302. A. Maradufu and A. S. Perlin, *Carbohyd. Res.*, **32**, 127 (1974).

303. J. K. Rogers and N. S. Thompson, *Carbohyd. Res.*, **7**, 66 (1968).

303a. J. S. Myers and O. Gabriel, *Carbohyd. Res.*, **67**, 223 (1978).

304. M. Shaw, *Proc. Ann. Power Sources Conf.*, **17**, 59 (1963); *Chem. Abstr.*, **60**, 5069d (1964).

305. For a general summary of the subject, see J. R. Porter, "Bacterial Chemistry and Physiology," Wiley, New York, 1946, pp. 896–1030.

306. L. Boutroux, *C. R. Acad. Sci.*, **91**, 236 (1880); A. J. Brown, *J. Chem. Soc.*, **49**, 172, 432 (1886).

307. W. Franke and F. Lorenz, *Ann.*, **532**, 1 (1937).

308. A. J. Moyer, P. A. Wells, J. J. Stubbs, H. T. Herrick, and O. E. May, *Ind. Eng. Chem.*, **29**, 777 (1937); E. A. Gastrock, N. Porges, P. A. Wells, and A. J. Moyer, *ibid.*, **30**, 782 (1938); N. Porges, T. F. Clark, and S. I. Aronovsky, *ibid.*, **33**, 1065 (1941).

309. H. Knobloch and H. Mayer, *Biochem. Z.*, **307**, 285 (1941).

310. L. B. Lockwood and G. E. Nelson, *J. Bacteriol.*, **52**, 581 (1946); G. Bertrand, *C.R. Acad. Sci.*, **127**, 124, 728 (1898).

311. J. J. Stubbs, L. B. Lockwood, E. T. Roe, B. Tabenkin, and G. E. Ward, *Ind. Eng. Chem.*, **32**, 1626 (1940); L. B. Lockwood, *Methods Carbohyd. Chem.*, **2**, 51 (1963); A. J. Kluyver and A. G. J. Boezaardt, *Rec. Trav. Chim.* (Pays-Bas), **57**, 609 (1938); K. R. Butlin and W. H. D. Wince, *J. Soc. Chem. Ind.*, **58**, 363 (1939).

312. L. B. Lockwood, B. Tabenkin, and G. E. Ward, *J. Bacteriol.*, **42**, 51 (1941); see also ref. 256.

313. M. Yamasaki, Jap. Pat. 5331 (1955), *Chem. Abstr.*, **52**, 2899c (1958); Chas. Pfizer and Co. Inc., Brit. Pat. 800,634 (Aug. 27, 1958); *Chem. Abstr.*, **53**, 8009h (1959).

314. Y. Wakisaka, *Agr. Biol. Chem.*, (Tokyo), **28**, 819 (1964); *Chem. Abstr.*, **62**, 14795b (1965).

315. R. Weidenhagen and G. Bernsee, *Angew. Chem.*, **72**, 109 (1960).

316. K. Sato, Y. Yamada, K. Aida, and T. Uemura, *Agr. Biol. Chem.* (Tokyo), **31**, 877 (1967).

317. D. T. Williams, J. K. N. Jones, N. J. Dennis, R. J. Ferrier, and W. G. Overend, *Can. J. Chem.*, **43**, 955 (1965).

318. D. T. Williams and J. K. N. Jones, *Can. J. Chem.*, **45**, 741 (1967).

319. H. W. Schiwara and G. F. Domagk, *Z. Physiol. Chem.*, **349**, 297, 1321 (1968).

320. L. T. Sniegoski, H. L. Frush, and H. S. Isbell, *J. Res. Nat. Bur. Stand.*, **65A**, 441 (1961).

321. E. E. Grebner and D. S. Feingold, *Biochem. Biophys. Res. Commun.*, **19**, 37 (1965); H. Hayano and S. Fukui, *J. Biol. Chem.*, **242**, 3665 (1967).

322. H. Hayano and S. Fukui, *J. Biochem.* (Tokyo), **64**, 901 (1969); *Chem. Abstr.*, **70**, 65465 (1969).

25. GLYCOL-CLEAVAGE OXIDATION

A. S. PERLIN

I. INTRODUCTION

The carbohydrates provide a profusion of compounds that contain hydroxyl groups on two or more adjacent carbon atoms, and the fact that this type of carbon–carbon bond generally undergoes oxidative scission selectively and quantitatively has been a major factor contributing to the current status of carbohydrate chemistry. The cleavage reaction was discovered by Malaprade,[1,2] who observed that polyols are rapidly oxidized by periodate ion. Criegee[3] subsequently found that lead tetraacetate cleaves 1,2-diols, and Fleury and Lange[4] reported that the success of Malaprade's reaction depends on the presence of contiguous hydroxyl groups in the compound.

For many purposes, these oxidants are interchangeable; however, the fact that periodate functions best in water, and lead tetraacetate in organic solvents, makes glycol-cleavage oxidation possible with all types of carbohydrates and derivatives.[5,6] For favorable examples, the behavior of the two reagents toward a given compound is sufficiently different to provide complementary information.

Oxidative glycol cleavage, one of the most widely used methods for determining constitution, has also been a highly effective means for deducing stereochemistry. In addition, it furnishes a general, degradative method for the preparation of compounds in the aldotriose, aldotetrose, and aldopentose series, and in the synthesis of a wide range of sugars and derivatives, including isotopically labeled compounds.

Several reviews of glycol-cleavage oxidations have appeared.[5-7]

II. CHARACTERISTICS OF PERIODATE AND LEAD TETRAACETATE OXIDATIONS

A. MECHANISMS OF GLYCOL CLEAVAGE[7]

The high selectivity of periodate and lead tetraacetate as glycol-cleaving oxidants is attributed mainly to the ability of the central atom of the reagent to complex with a 1,2-diol and effect a two-electron transfer.[8] That an intermediate complex is formed has been shown for periodate by pH[1,9] and ultraviolet spectral[10] changes, and, for both oxidants, less directly by consideration of the reaction kinetics.[7,11]

Based on the original proposals of Criegee et al.,[11] it is generally considered that the cleavage reaction involves reversible formation of a cyclic complex or intermediate (**3** or **4**), via an acyclic ester (**1** or **2**), and that the intermediate decomposes via a cyclic transition state to the products.

Criegee et al. also suggested that the planar conformation of the cyclic complex is optimal for cleavage, because the most reactive compounds are those in which the adjacent hydroxyl groups are virtually eclipsed, as in 1,2-acenaphthanediol,[11] methyl 2,6-anhydro-α-D-altropyranoside,[12] and cis-1,2-camphanediol.[13] Slightly less reactive are diols that may be able to attain the eclipsed condition with little strain—for instance, cis-1,2-cyclopentanediol[14] and cis-2,3-tetrahydrofurandiol.[15] Equally consistent with the mechanism of Criegee and co-workers are (a) the inert behavior of antiparallel trans-1,2-diols,[13,14,16,17] which clearly cannot form a complex of type **3** or **4**, and (b) the intermediate oxidation rates found for six-membered-ring and acyclic 1,2-diols, which would be expected to form a puckered rather than a planar complex. Generally, the cis isomers of these cyclic diols are more

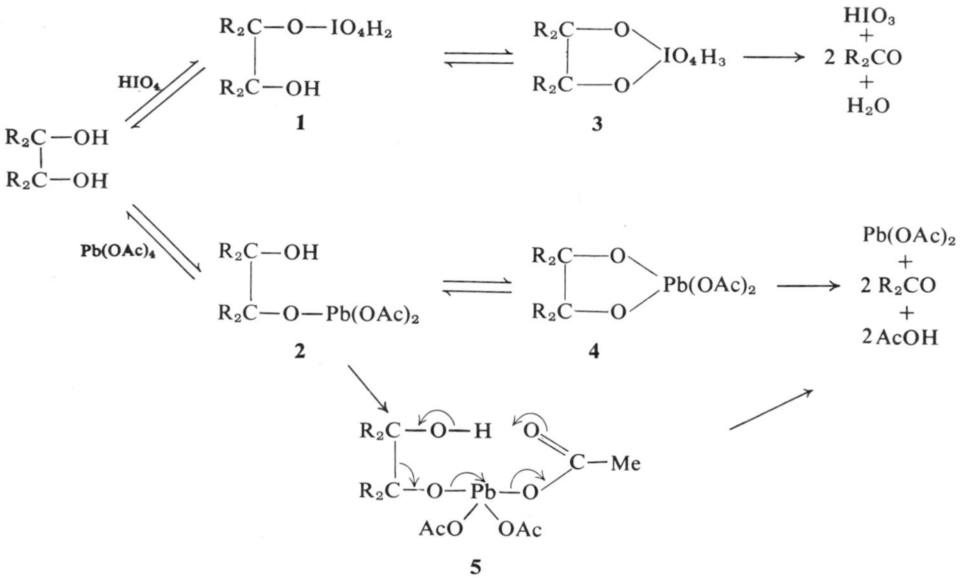

$$R_2C-O-IO_4H_2$$

$$R_2C-OH$$

1

$$\begin{array}{c} R_2C-O \\ | \quad\quad\;\; \rangle IO_4H_3 \\ R_2C-O \end{array}$$

3

$$\xrightarrow{} \begin{array}{c} HIO_3 \\ + \\ 2\,R_2CO \\ + \\ H_2O \end{array}$$

$$R_2C-OH$$
$$R_2C-OH$$

$$\xrightarrow{HIO_4}$$

$$\xrightarrow{Pb(OAc)_4}$$

$$R_2C-OH$$

$$R_2C-O-Pb(OAc)_2$$

2

$$\begin{array}{c} R_2C-O \\ | \quad\quad\;\; \rangle Pb(OAc)_2 \\ R_2C-O \end{array}$$

4

$$\xrightarrow{} \begin{array}{c} Pb(OAc)_2 \\ + \\ 2\,R_2CO \\ + \\ 2\,AcOH \end{array}$$

$$\begin{array}{c} R_2C-O-H \quad O \\ \quad\quad\quad\quad \backslash \\ \quad\quad\quad\quad\quad C-Me \\ R_2C-O-Pb-O \\ \quad AcO \quad OAc \end{array}$$

5

reactive than the *trans*,[11,17-19] presumably because there is lessened internal strain and a closer approach to coplanarity when the *a,e* hydroxyl groups of the former isomer coordinate with the oxidant.[11,18,20] Similarly, *threo*-1,2-diols are usually oxidized faster than the *erythro* isomers,[17,21,22] and the former also appear better able to accommodate a cyclic complex.[19,22]

Direct, kinetic evidence for the formation of cyclic intermediates is available in only a limited number of instances, largely through studies by Bunton and colleagues,[7,9,10,19] who have shown that the singly charged anions **6** and **7** are transient intermediates in periodate cleavage. The concentrations of **6** and **7** are at the maxima at pH 4 to 5, a pH range that is also maximal for the rate of oxidation. Most probably the entity that actually decomposes to products is the dehydrated form **6**. The doubly charged intermediate **8** preponderates at higher pH values, but it should be inert, because it cannot likewise suffer dehydration; hence, the oxidation rate is low in alkaline media.

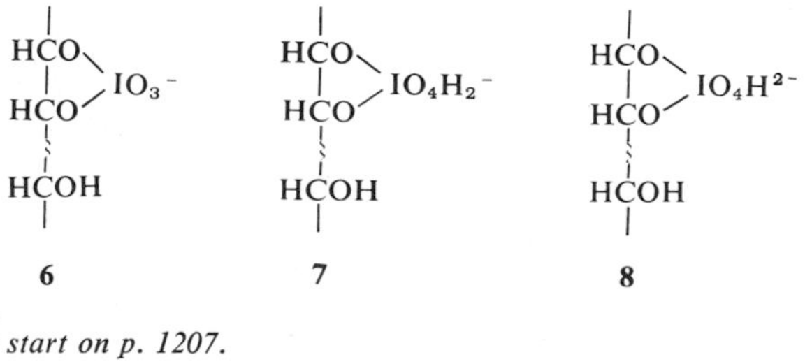

$$\begin{array}{c} HCO \\ | \quad\;\; \rangle IO_3{}^- \\ HCO \\ \; \\ HCOH \end{array}$$

6

$$\begin{array}{c} HCO \\ | \quad\;\; \rangle IO_4H_2{}^- \\ HCO \\ \; \\ HCOH \end{array}$$

7

$$\begin{array}{c} HCO \\ | \quad\;\; \rangle IO_4H^{2-} \\ HCO \\ \; \\ HCOH \end{array}$$

8

$$
\begin{array}{cc}
\begin{array}{l}
|\\
\text{HCO} \diagdown \\
|\\
\text{HCO}-\text{IO}_3\text{H}^- \\
|\\
\text{HCO} \diagup \\
|
\end{array}
&
\begin{array}{l}
|\\
\text{HCO} \diagdown \\
|\\
\text{HCO}-\text{IO}_3^{2-} \\
|\\
\text{HCO} \diagup \\
|
\end{array}
\\
\mathbf{9} & \mathbf{10}
\end{array}
$$

The detection of stable periodate complexes having a tridentate structure also supports the concept of the cyclic intermediate. These complexes are formed at neutral and high pH by α-D-ribopyranose and other *cis,cis*-1,2,3-triols,[23] and by such compounds as 1,2-*O*-isopropylidene-α-D-glucofuranose, O-3, O-5, and O-6 of which coordinate with the iodine atom of the reagent.[24] There is evidence for two types of tridentate complex (**9** and **10**), corresponding to **7** and **8**, respectively, and their stability appears to be consistent with Bunton's findings, because neither can be converted into a dehydrated form.[24]

The ease of complex formation is not the sole factor governing oxidation rates. Reactions proceeding through a planar, cyclic complex should favor concerted electron transfer and, therefore, concerted bond-breaking, which would accelerate the decomposition of the complex to products.[7] A clear demonstration of the relative importance of the *decomposition* step is found in the reaction between the periodate dianion ($H_3IO_6^{2-}$) and the isomeric 1,2-cyclohexanediols or 2,3-butanediols; in these examples, strong, non-bonded interactions cause the rate of decomposition of the intermediate to products to be a more decisive factor than the equilibrium constant for its formation.[19]

The glycol-cleaving action of periodate is generally more consistent with the concept of a five-membered ring intermediate than is the action of lead tetraacetate, because the latter reagent can oxidize a number of diols incapable of forming a type **4** complex[14,17] and can effect oxidation under one set of conditions but not another.[25,26] Such seeming inconsistencies as these were rationalized by Criegee's suggestion[17] that other pathways are possible within the general framework. For example, a concerted displacement of electrons (as in **5**) can be envisaged for the lead complex **2**. In addition, **2** may decompose by proton transfer to a Lewis base, as suggested by the fact that lead tetraacetate oxidations are base-catalyzed.[27] A wider selection of pathways open to the lead tetraacetate reagent is thus consistent with the fact that it promotes a number of oxidations in addition to glycol cleavage.[28]

Rates of periodate oxidation have also been shown to be solvent-dependent; solvents having unhindered, basic oxygen atoms,[28a] such as *N,N*-dimethylformamide, slow down cleavage reactions[28b] by altering the conformation of the diol, or by competitive complexation with the oxidant, whereas 1,4-dioxane enhances the rate of oxidation of simple diols.

B. Some General Properties of Cleavage Reactions; Experimental Methods[29]

In addition to *vic*-diols, other 1,2-dioxygenated groups—2-hydroxy aldehydes,[1,3] 1,2-dicarbonyl compounds,[30] α-hydroxy and α-keto acids[31,32]— and α-amino alcohols[33] are oxidatively cleaved both by periodate and by lead tetraacetate; however, lead tetraacetate oxidizes α-hydroxy acids much more readily than does periodate,[32,34,35] and both reagents attack 2-hydroxy aldehydes and 1,2-dicarbonyl compounds relatively slowly.[32,34,36–38a] Mechanistically, these reactions appear to resemble glycol scission.[7]

A terminal 1,2-diol yields formaldehyde,[1-3,39] whereas the carboxylic derivatives yield carbon dioxide.[31,40] Several methods are available for detecting and determining these important products in reaction mixtures: formaldehyde may be determined as a sparingly soluble methone,[41] by its highly specific color reaction with chromotropic acid,[42] or polarographically[42a]; conventional, manometric techniques are utilized for measurement of the carbon dioxide[40,43] from carboxylic acids.

Formic acid is produced from a 1,2,3-triol grouping. As this acid is stable toward periodate, it is readily determined by volumetric[42a,44] or potentiometric[45] titration of the reaction mixture, by highly sensitive spectrophotometric methods,[45a] or manometrically.[46] Lead tetraacetate slowly oxidizes formic acid to carbon dioxide, which may complicate the stoichiometry of the glycol-cleavage reaction.[31,36] Corrected values for uptake of oxidant may be obtained[46] by introducing potassium acetate to catalyze this secondary oxidation (and the rate of glycol cleavage as well)[27,46] and measuring the liberated carbon dioxide manometrically. Cleavage of an aldehydic α-hydroxy hemiacetal group leads to a formic ester, whereas a ketonic α-hydroxy hemiacetal group may yield an ester of either glycolic or glyoxylic acid (see Section V,B).

Nuclear magnetic resonance spectroscopy may provide a convenient measure of the formaldehyde[47] or formic acid[48] produced in some periodate oxidations, and may also permit differentiation between free formic acid and that bound as formic esters[48]; deuterium oxide is a convenient solvent for these determinations. Mechanical, automated techniques for determining the amounts of various oxidation product have been developed.[48a]

Most of the common types of substituent group—esters, acetals, ethers— are stable to conditions normally used in glycol-cleavage oxidations. Notable exceptions are thio derivatives (see Section II,C). Phenolic constituents of *C*-glycosyl derivatives[49] and other naturally occurring saccharides[50] may also undergo oxidation; however, the phenolic groups may be adequately stabilized by alkylation.[49] With partially acylated sugars, the danger exists that ester

---

References start on p. 1207.

migration might occur in the oxidizing medium and lead to inconclusive results. Such a possibility was specifically considered in regard to certain cleavage reactions, but no evidence for migration was obtained,[38,51–53] nor has an authentic example of acyl migration accompanying oxidation been reported; however, the general need for caution with potentially labile substituents is illustrated by the observation that the acetal group of 1,4-anhydro-2,3-O-benzylidene-D-mannitol migrates to O-5,O-6 as the compound dissolves in acetic acid, so that treatment with lead tetraacetate effects glycol cleavage in the latter anhydro acetal.[54] Trityl groups may also be hydrolyzed slowly under the same conditions.[55,56] The oxidation of conduritol oxide to 2,3-epoxysuccinaldehyde[57] indicates that an oxirane ring is stable. Nevertheless, there is evidence that such a ring can be opened, and recyclized through an adjacent position, by lead tetraacetate. Thus, Criegee and Fiedler[58] explained the unexpectedly high rate of oxidation of trans-3,4-epoxy-1,2-cyclobutanediol 11 by the reaction sequence involving 12 and 13.

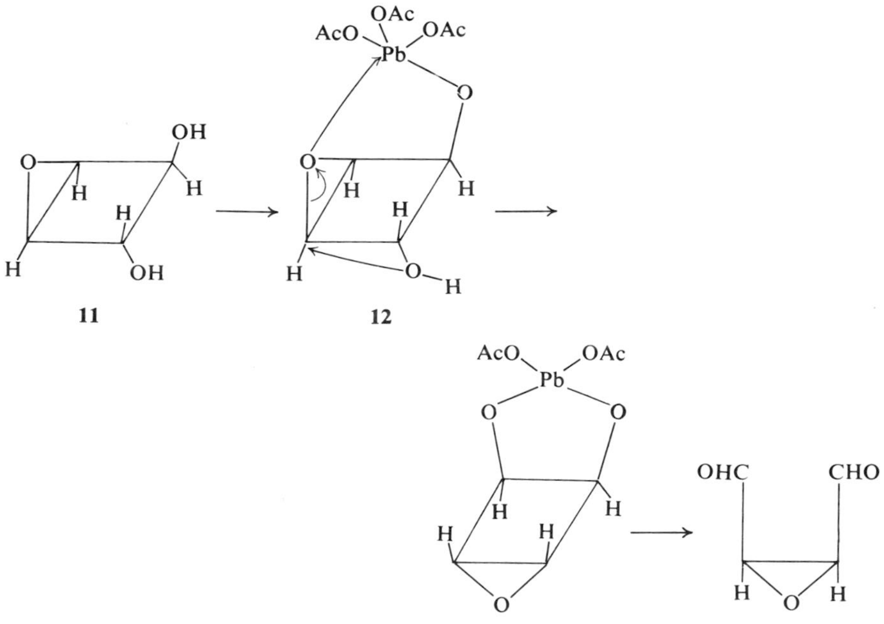

The consumption of oxidant is most generally determined by volumetric methods—periodate by titration with arsenite,[59] and both oxidants by iodimetry.[60,61] Spectrophotometric methods are also frequently used for determining[61a] periodate[62–64] and lead tetraacetate,[65,66] and are especially useful for micro-oxidations or highly dilute reaction mixtures.[66] Spectro-photometric determination of the extent of conversion of the violet dye tris[2,4,6-tris(2-pyridyl)-1,3,5-triazino]iron(II) into its colorless ferric state reportedly[66a] provides a means of quantitating nanomole amounts of un-reacted periodate. Polarographic analysis has also been shown to be a practical[66b] method for measurement of concentrations of periodate (and iodate).

Periodate and lead tetraacetate both decompose quite rapidly at elevated temperatures and, therefore, are employed at room (or lower) temperature; in addition, periodate oxidations are best conducted in the absence of light.[67]

C. OXIDATIONS OTHER THAN GLYCOL CLEAVAGE; "OVEROXIDATION"

Although periodate and lead tetraacetate are among the most highly specific of oxidants, they nevertheless promote a number of oxidations in the carbohydrate series besides glycol cleavage.

A problem frequently encountered is "overoxidation" or "non-Mala-pradian" oxidation. Most examples of this kind of behavior involve the for-mation, as a product of the cleavage reaction, of tartronaldehyde and related compounds containing an active hydrogen atom. Glycuronic acids and some other carboxylic derivatives are subject to extensive overoxidation[68–71] attributable to instability[71a] of the product of normal scission—for instance, **14**—or to dehydrogenation at C-5. Similarly, substituted tartronaldehydes (**17**), which are formed during oxidation of hexofuranosides,[36,39] partially substituted sugars,[72–74] oligosaccharides (see Section V,C), and polysac-charides (see Section VI,A), are readily degraded oxidatively.

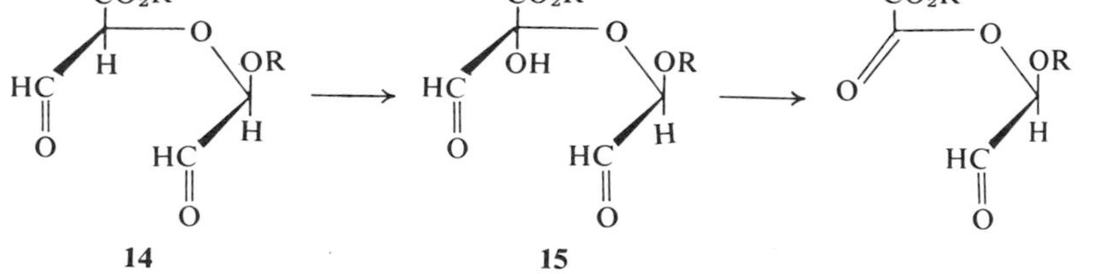

$$
\begin{array}{ccc}
\text{HC}{=}\text{O} & \text{HC}{=}\text{O} & \text{HC}{=}\text{O} & \text{HC}{=}\text{O} & \text{HC}{=}\text{O} \\
| & | & | & | & | \\
\text{HOCOR} \leftarrow & \text{HCOR} \rightleftharpoons & \text{COR} \rightarrow & \text{HOCOR} \rightarrow & \text{CO}_2\text{R} \\
| & | & \| & | & \textbf{20} \\
\text{HC}{=}\text{O} & \text{HC}{=}\text{O} & \text{HCOH} & \text{HOCOH} & + \\
& & & \text{H} & \\
\textbf{16} & \textbf{17} & \textbf{18} & \textbf{19} & \text{HCO}_2\text{H}
\end{array}
$$

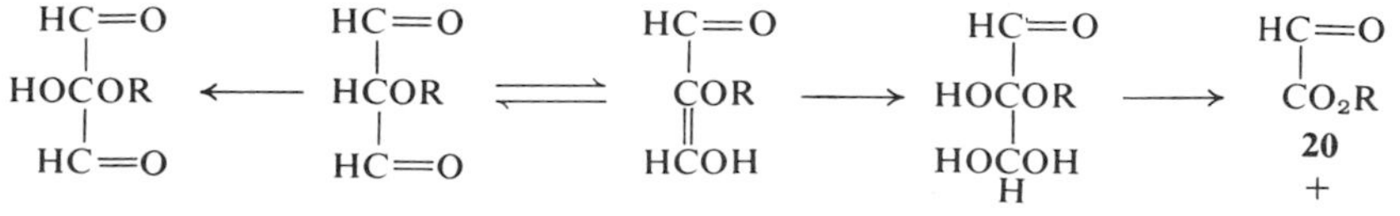

Overoxidation by periodate may involve direct hydroxylation as the first step[69-71a,75] (as in the oxidation of hydrocarbons),[75a] with the resulting hydroxy aldehyde (15 or 16) being cleaved in the conventional way; however, hydroxylation of the enolic form 18 is the more probable pathway.[76-79] Direct evidence for the latter possibility is provided by the finding[79] that the trialdehyde (22) produced from 1,4-anhydro-D-allitol (21) enolizes to yield 25. The latter, which has been isolated as a crystalline compound, is oxidized to the same products as are obtained without interrupting the reaction,[79] probably via 24 or 23 and 26.

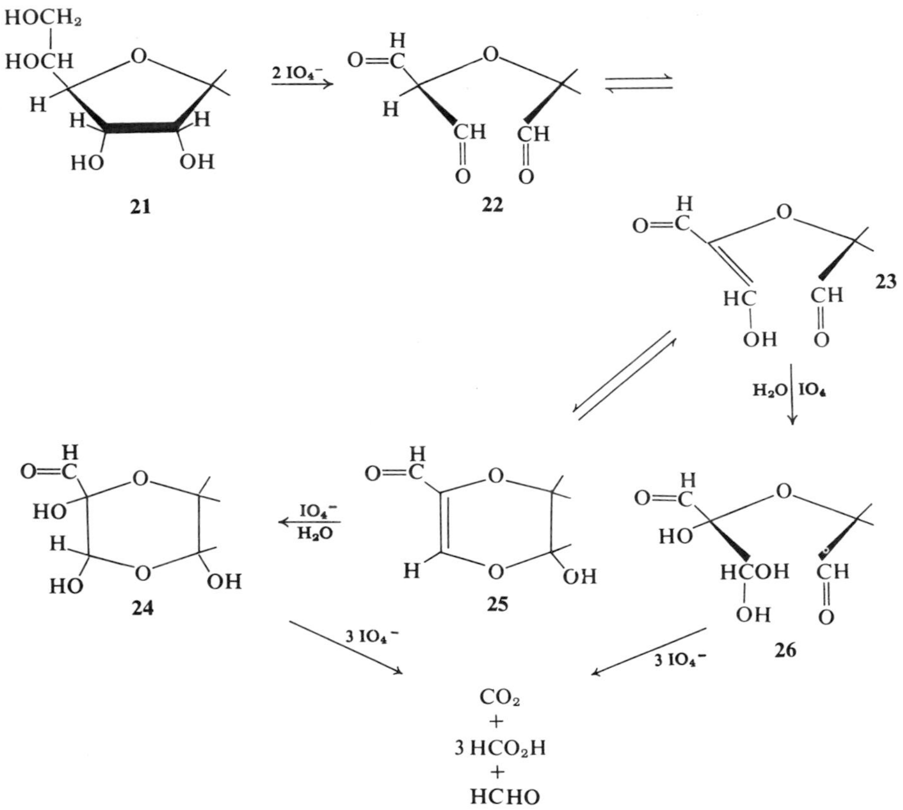

A similar route has been suggested to account for the formation of glyoxylic acid (20, R = H) and methyl glyoxylate (20, R = Me) during the periodate oxidation of inositols[76] and O-methylinositols,[78] respectively. Formation of these carboxy derivatives was depicted[76,78] by sequences involving enolization to such reductones as 18, which are hydroxylated to 19 and then cleaved oxidatively.

Direct hydroxylation is a probable step in the overoxidation phase observed during the periodate oxidation of glycals; attack presumably occurs at the allylic position adjacent to the carbonyl group of the initial product formed by cleavage of the 3,4-diol grouping.[80]

Products formed in the overoxidation of such compounds with lead tetraacetate have not been so well characterized, although acetoxylation might be expected to occur.[36,61,81] Some compounds examined as models of these activated methine groups, or as possible products of glycol cleavage—for instance, 2,4-pentanedione and formic, glycolic, and oxalic acids—are readily oxidized by lead tetraacetate[11,31,35,36,46] but not by periodate.[5,82]

Effective O-demethylation has been observed to occur in the periodate oxidation of 3-deoxy-4-O-methylaldosulosonic acids, which produce 2-oxobutanedioic acid and give a positive test with thiobarbituric acid.[82a]

N-Substituted derivatives in the amino sugar series show a variety of oxidative characteristics. Whereas N-acetylation or N-benzoylation prevents cleavage of a 1,2-amino alcohol,[83-88] an N-ethoxycarbonyl[89] or N-p-tolyl-sulfonyl group[84] may permit oxidation to the imine (R—CH=N—CO$_2$Et, or R—CH=N—SO$_2$C$_7$H$_7$-p, respectively). Nonspecific oxidation has been observed with some N-methyl derivatives, the extent of the "anomalous" reaction being pH-dependent.[90] The dimethyl ether of methyl amosaminide is probably N-demethylated, because one mole of it reduces 2.3 moles of periodate to yield one mole of formaldehyde.[90] Cleavage of the α-hydroxy-dimethylamino group of desosamine has also been reported,[91] whereas methyl mycaminoside is resistant[92] to periodate at pH 4.5. "Anomalous" results were also obtained with 3-amino-3-deoxyribofuranosides, which consume two moles (not the expected one mole) of periodate per mole.[93] An interesting steric effect is encountered in the periodate oxidation of methyl 2-amino-4,6-O-benzylidene-2-deoxy-α-D-altropyranoside at pH 6.9 (but not at pH 4), as the corresponding 2,3-diol is not attacked.[94]

Re-examination of the conditions of reaction of the conformationally fixed 2-amino-2-deoxy-D-altropyranoside derivative and the 3-amino-3-deoxy isomer revealed that they slowly undergo C-2–C-3 cleavage, whereas the 2,3-diol is unreactive[94a]; the latter amino sugars are subject to overoxidation, whereas stereoisomeric 2-amino-4,6-O-benzylidene-2-deoxyaldopyranosides, which are oxidized much faster, consume the stoichiometric amount of oxidant.[94b] Several 1,2-O-isopropylidenefuranos-3-ulose 3-p-nitrophenyl-hydrazones have been converted into the corresponding geminal azo alcohols in which O-2 is *trans* to O-3 by treatment with lead tetraacetate.[94c]

The ability of periodate and lead tetraacetate to oxidize sulfides may produce unexpected results when the effects of these reagents are used to examine thio

---

sugars and derivatives. Some alkyl and aryl 1-thioglycosides show complex stoichiometry,[95-98] sometimes accompanied by the release of iodine, whereas the corresponding sulfones undergo normal glycol cleavage under suitable conditions.[95] The sulfur atoms of sugar dithioacetals are relatively stable in a solvent of low dielectric constant (commonly, benzene), but S-oxidation is rapid in a more-polar solvent (such as acetic acid) and may take precedence over glycol cleavage.[99,100] By contrast, one mole of 1,2-O-isopropylidene-α-D-glucofuranose 5,6-thionocarbonate consumes only 0.5 mole of lead tetraacetate in acetic acid, but over two moles in pyridine.[101] Cyclic thioethers appear to be less reactive; one mole of methyl 5-thio-α-D-ribopyranoside reportedly[101a] reduces three moles of periodate.

Overoxidation accompanied by release of inorganic phosphate or sulfate has been observed in the periodate oxidation of some hexose monophosphates[102] or monosulfates,[103,104] although the ester hydrolysis is largely obviated by use of dilute reaction mixtures. The products of the periodate oxidation of D-glucose 2-sulfate and D-galactose 2-sulfate suggest that the acyclic form of each is oxidized; 3-sulfates are, however, oxidized normally.[104a]

Instances of *apparent* oxidation sometimes occur that are actually attributable to analytical difficulties. Thus, one mole of methyl 2-deoxy-α-D-*xylo*-hexopyranoside appears to consume two moles of periodate, not one, as measured by the arsenite method, because the methylene group of the oxidation product is iodinated during the back-titration.[105] Also, some tridentate, periodate complexes are decomposed only slowly by arsenite, resulting in spurious uptake values.[23,24] A 3,6-anhydro-4,5-O-isopropylidene-octitol of unspecified configuration at C-2 and C-7 was reported to undergo stoichiometric oxidation by periodate to afford 2,5-anhydro-3,4-isopropylidene-DL-allose directly[105a]; however, re-examination of this reaction revealed that the actual precursor in the reported oxidation is a heptitol, and that oxidation of the octitol gives the (expected) 2,5-anhydroheptoseptanose.[105b]

D. Vicinal Diols Resistant to Glycol Cleavage

Periodate and lead tetraacetate are routinely used to test for the presence or absence of a 1,2-diol grouping. Nevertheless, a relatively large number of examples of marked resistance to cleavage is known. This consideration attaches some uncertainty to the absolute validity of a negative oxidation test, particularly for a complex compound.

1,6-Anhydro-α-D-glucofuranose[16] was mentioned (in Section II,A) as an α-glycol that is resistant to cleavage, in accord with Criegee's cyclic intermediate mechanism. The D-*galacto* isomer behaves similarly,[106,107] and related inert compounds are 2,7-anhydroheptulofuranoses that contain a 3,4-*trans*-diol grouping.[108] Other examples are methyl 4,6-O-benzylidene-α-D-altropyranoside,[94]   2,6-anhydro-β-D-fructofuranose,[25]   and   1,4-anhydro-*epi*-inositol[109]; the last two are sterically analogous to *trans*-2,3-camphanediol.[13] Each of these is a bicyclic, fused- or bridged-ring compound in which the inactive *trans*-diol grouping appears to be held rigidly, so that the minimum dihedral angle between the two C–O bonds probably exceeds 100°; however, these compounds are oxidized by lead tetraacetate in pyridine solution,[25] (a particularly vigorous form of the oxidant[56]) perhaps by way of an acylic mechanism (see Section II,A). In several instances, the oxidation products have been characterized as those expected from normal cleavage,[25,108,109] a matter of some importance, because this solvent-modified oxidant may effect oxidation of single hydroxyl groups.[110,111]

Formation of a highly stable, tridentate complex may prevent cleavage when periodate is used above pH 7, although few classes of sugar compounds can satisfy the steric requirements for this kind of complexing. Thus far, the effect has been observed with α-D-ribopyranose, α-D-allopyranose, β-D-lyxopyranose, and several inositols—compounds affording a 1*a*,2*e*,3*a*-triol system—and with certain aldohexofuranose derivatives[24]; in glucofuranoses, the complex involves O-3, O-5, and O-6, and in galactofuranoses, O-2, O-5, and O-6.

There are several instances in which inductive effects undoubtedly suppress reactivity markedly. Thus, arabinono- and xylono-1,4-lactones are virtually unaffected by lead tetraacetate,[39] possibly because inductive electron withdrawal depresses the nucleophilicity of O-2 or destabilizes the Pb–O bond in the intermediate complex.[112] Inertness of a 2,3-*trans*-diol grouping within a lactone ring is evidenced in the observation that D-galactono-1,4-lactone is oxidized by periodate solely at C-5 and C-6.[113] Similarly, an electron-withdrawing substituent (such as a sulfonyloxy group) can markedly retard scission of an adjacent α-glycol grouping; for instance, the rate of oxidation of methyl 2-*O*-*p*-tolylsulfonyl-α-D-glucopyranoside is one thousandth that of the corresponding 2-*O*-methyl-D-glucoside,[94] although steric factors may also contribute substantially to the difference in rates. Hence, sulfonyl derivatives of unknown structure, or compounds containing similar deactivating (and, perhaps, bulky) substituents, may give a spurious, negative test, unless the reaction time is sufficiently long and the concentration of oxidant adequately high.

A combination of inductive and steric effects may also account for the fact that phenyl β-D-glucopyranoside (but not the α anomer) rapidly consumes only one molar equivalent of periodate, instead of the theoretical two; this oxidation is specific for the 3,4-diol grouping.[114] Methyl 6-*O*-trityl-α-D-mannopyranoside consumes only one molar equivalent of lead tetraacetate in benzene, but there is partial cleavage at both C-2,C-3 and C-3,C-4.[115] A pentose analogue, methyl β-L-arabinopyranoside, likewise reacts with *dilute*[115a] solutions of periodate in methyl sulfoxide to reduce one molar equivalent of oxidant; in the absence of the large substituent on C-5, however, cleavage occurs primarily at the C-3–C-4 bond.[115b] It was proposed[115b] that this under-oxidation arises because of intramolecular blocking of the 2-hydroxyl group of the initial dialdehyde by its incorporation into a 1,4-dioxane derivative that undergoes a second ring closure to afford the 3,6,8-trioxabicyclo[3.2.1]-octane derivative **27**; the later, unequivocal, spectroscopic characterization of the crystalline acetate of **27** provided support for this mechanism.[115c]

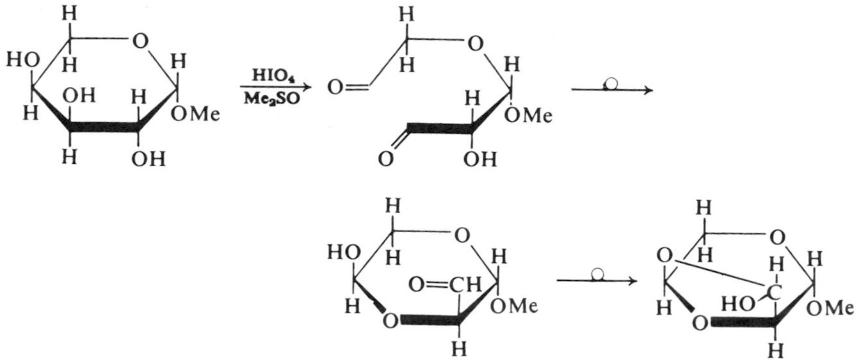

27

Even in highly polar solvents, the relatively low reactivity of hydroxy aldehydes[36-38] may lead to marked underoxidation of polyhydric alcohols. Thus, 4,6-O-ethylidene-D-mannose (and -D-galactose) and D-glucofuranose 5,6-carbonate each consume two moles of lead tetraacetate per mole in acetic acid when the concentration is 10 mM, but, in more-dilute solution (100 μM), take up only one mole at a significant rate.[38] This behavior was attributed to the minimal reactivity of the *aldehydo*-pentose formic esters that are produced by initial cleavage of the α-hydroxy hemiacetal group.

In other instances of apparent underoxidation, formate groups elaborated during the reaction serve as protecting substituents. Some examples are discussed in Section V,B in connection with reducing-sugar oxidations. An extreme example may be mentioned here, namely D-*erythro*-L-*galacto*-octose[116]; although this sugar has five *vic*-diol groupings, one mole reduces only two molar equivalents of lead tetraacetate readily, because stepwise degradation converts it into a 2,3-(or 2,4-)di-O-formyl-D-glucopyranose, which lacks a free 1,2-diol grouping.

Internal glycosyl residues of many oligosaccharides, particularly those linked to adjacent residues by (1→4) bonds and containing a 2,3-*trans*-diol grouping[117] (see Section V,C), are highly resistant to cleavage by lead tetraacetate in acetic acid, even under catalyzed conditions. A resistant 1,2-glycol was encountered[118] in the sophorosyl residue of a partially acetyl-ated glycolipid; although one mole of the compound consumed only one mole of periodate or lead tetraacetate under the usual conditions, a second *vic*-diol grouping was detected by oxidation with lead tetraacetate in pyridine. A di-D-fructofuranose 1,1′:2,2′-dianhydride[119] provides an unusual example of unreactivity in an oligosaccharide derivative; the 3,4-diol grouping of the β-D-glycosyl residue in the "disaccharide" is attacked at a markedly low rate by periodate, whereas the α-D-glycosyl residue is oxidized rapidly. This striking difference in reactivity is a consequence of local, structural rigidity, which constrains the diol grouping of the resistant residue to subtend an unfavorably large angle of about 150°. Oxidation of chondroitin and dermatan sulfates by aqueous periodic acid at pH 3 effects selective cleavage of 80% of the L-iduronic acid residues without significantly affecting the D-glucuronic acid residues.[119a]

In studies of polysaccharides, it is sometimes difficult to determine whether certain residues are resistant to oxidation for steric reasons, or because of the presence of (1→3) bonds or multiple linkages. This difficulty is compounded by the possibility of overoxidation, which militates against the use of greatly extended reaction periods. Poor solubility of certain polysaccharides (as well as of some compounds of low molecular weight) may strongly retard oxida-tion. Most probably this contributes to the relative inertness toward lead tetraacetate of suspensions of polysaccharides in acetic acid or pyridine[6]; however, as cellulose is readily oxidized in water by periodate under hetero-geneous conditions,[120,121] such surface effects as wetting or adsorption must play the more-important role. The underoxidation of alginates,[121a,b] xylans,[121c] and amylose[121b,d,e] was attributed to inter-residue hemiacetal formation as oxidation progresses; this would protect approximately every third glycosidyl residue in a linear (1→4)-linked polysaccharide (as in **28**

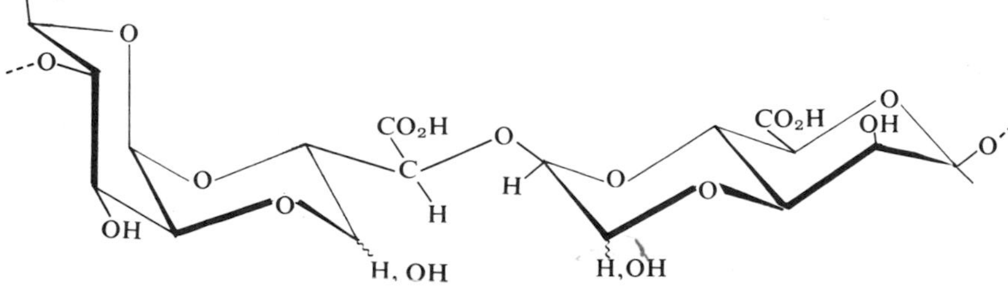

28

derived from alginic acid[121c]); the results of sequential applications of the Smith degradation (see Section VI,B) to amylose accord with this formulation.[121b] Solvent-induced, conformational effects have also been implicated as affecting the rates of such oxidations.[28b]

III. ACYCLIC ALDITOLS AND CYCLITOLS

Oxidative scission of 1,2-diol groupings in acyclic compounds is generally rapid and quantitative, providing an excellent means for structural examination of such partially substituted derivatives of alditols as esters[51,122,123] and acetals.[124–126] An early, highly fruitful application in this series was the finding that the common di-O-isopropylidene-D-mannitol is the 1,2:5,6-diacetal, shown by the fact that it consumes one mole of oxidant per mole and yields 2,3-O-isopropylidene-D-glyceraldehyde.[124] Reduction of this triose acetal affords 1,2-O-isopropylidene-L-glycerol, a starting compound for the synthesis of numerous, naturally occurring glycerides and phospholipids[127]; periodate oxidation of 1,6-dideoxy-1,6-difluoro-2,5-O-methylene-D-mannitol is a key step in the stereospecific synthesis of 1-deoxy-1-fluoro-L-glycerol.[127a] Similarly, glycol-cleavage oxidation was used to determine the structure of 1,3-O-benzylidene-D(or L)-arabinitol; through the product, 2,4-O-benzylidene-D(or L)-threose, it provides perhaps the best route to D- or L-threose and derivatives.[128]

In acyclic alditols, a vic-diol grouping consisting of secondary hydroxyl groups is usually oxidized more readily than one containing a primary and a secondary hydroxyl group,[22,129] and the $threo$ ($trans$) configuration is particularly vulnerable.[15,21,22,129] Consequently, initial attack of periodate on a polyhydric alcohol such as mannitol mainly cleaves the 3,4-$threo$-diol grouping and yields a glyceraldehyde, whereas galactitol is mainly split at C-2,C-3 and affords a threose.[22,129] Subsequent oxidation of one mole degrades these (as well as the primary products from other hexitols) to the expected, normal end products—two moles of formaldehyde and four moles of formic acid. When lead tetraacetate in acetic acid is used, underoxidation is sometimes observed.[130] This occurs most noticeably with galactitol, presumably because the intermediate tetrose is oxidized as a cyclic sugar in this medium and yields a stable formic ester. In these reactions, account must also be taken of the lead tetraacetate consumed in the (slower) oxidation of formic acid.[46,131]

Usually, although not invariably, vic-diol groupings including a tertiary hydroxyl group are relatively unreactive.[17,132]

The oxidation of an inositol differs notably from that of an acyclic hexitol; in principle, the main difference to be expected is the production of formic acid instead of formaldehyde. Thus, inositols show an *over-consumption* of periodate[133,134]; that is, one mole consumes ~ 6.7 moles of oxidant, not 6.0, and gives close to one mole of carbon dioxide, and only ~5 moles of formic acid, not 6. These results were accounted for in the following way[37,78,135]: the initial product is a hexodialdose (**29**), the glycol (but not hydroxy aldehyde) groupings of which are randomly split to yield glyoxal and tartronaldehyde (**30**). The latter then reacts as the tautomeric reductone (**18**, R = H) (as in the overoxidation route depicted in Section II,C), yielding glyoxylic acid (**20**, R = H) and, ultimately, carbon dioxide.

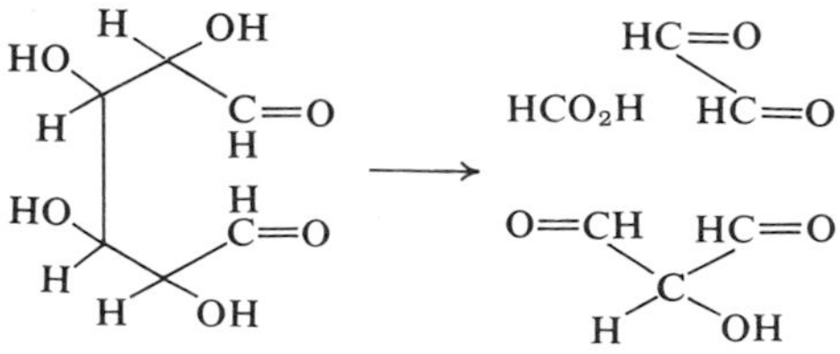

29   **30**

Although this pathway appears to be general for the inositols, the rates of oxidation vary widely; *cis*-inositol (all *cis*) reacts about two hundred times as fast as the *scyllo* isomer (all *trans*). For several of the most reactive isomers, the rates are higher than would be anticipated for a *cis*-diol of a cyclitol molecule, and this enhancement is attributed to excess steric strain caused by mutual repulsion of axially attached hydroxyl groups in these compounds.[136] The rule that, in the cyclohexane series, *cis*-diols react more rapidly than *trans* isomers with periodate and lead tetraacetate appears to hold generally for inositols and their derivatives,[136,137] including inosamines[138]; it has been utilized for assigning configuration.[137,138] Overoxidation of inositols is reportedly suppressed at 0° in 0.1M sulfuric acid solution.[23d]

As formic esters are not products of these periodate oxidations,[78] a hexodialdose, such as **29**, must react in an acyclic form; however, 1,3-di-*O*-methyl-*myo*-inositol shows exceptional behavior, in that it affords a high yield of a formic ester, which must be derived by cleavage of a furanose form of the intermediate hexodialdose.[139] Cyclization of the latter is regarded as being favored by the all-*trans* arrangement of groups on its furanose ring.[135]

The influence of cyclization of intermediate dialdehydes on the course of reaction is also evident during the oxidation of cyclopentanetetrols with lead tetraacetate at high dilution in acetic acid.[140] For example, the *cis*-diol of the (1,2,4/3) isomer (**31**) rapidly consumes one mole of oxidant per mole; subsequent attack on the intermediate pentodialdose **32** is slow, because it can involve only the relatively inactive 2-hydroxy aldehyde group of this acyclic form or a six-membered-ring hemialdal form. By contrast, the (1,2/3,5) isomer (33) rapidly consumes two moles of lead tetraacetate per mole, undoubtedly because the derived dialdehyde can be oxidized in the reactive furanose form 34. The 1,2- and 2,3-*cis*-diol groupings of the (1,2,3/4) isomer provide two possibilities for initial cleavage, one leading to 32, and the other to 34; hence, this isomer exhibits a rate of oxidant uptake that is intermediate between those for 31 and 33. A unique situation is afforded by the (1,2/3,4) isomer, which contains two highly reactive *vic*-diol groupings situated on the same ring, independent of each other; however, only one of these groupings can be cleaved rapidly, and reduction of a second molar equivalent of lead tetraacetate is slow, because, in both instances, the dialdose produced corresponds to an epimer of 32, which can form an unreactive, furanose ring to protect the hydroxyl group adjacent to the uncyclized aldehyde group.

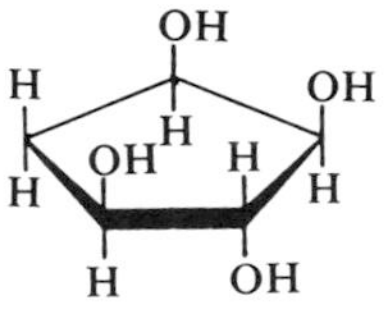

31 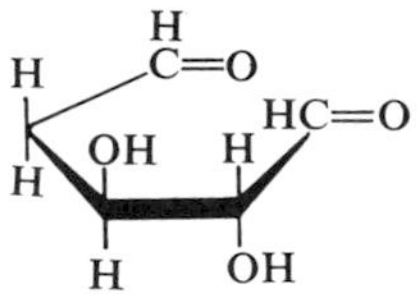   32 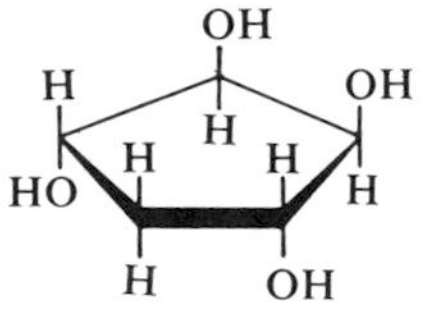   33

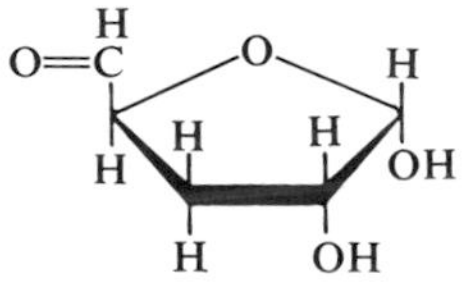

34

Of the cyclohexanetetrols, one mole of the *ortho* (1,2,3,4) isomers reduces the expected three moles of oxidant, whereas *para* (1,2,4,5) isomers show overoxidation, because of the concomitant formation of malonaldehyde[141] from C-2,C-3, and C-4. Although three of the *meta* (1,2,3,5) isomers undergo a normal uptake of oxidant, the (1,2,3/5) isomer is, as yet inexplicably, overoxidized.[142]

Oxidative interconversion of cyclitols and derivatives into acyclic, dicarboxylic acids has been widely used for structural elucidation in this series.[135] Early applications of the procedure helped determination of the constitution of shikimic and quinic acids[143]; for example, cleavage of the 1,2,3-triol grouping of methyl dihydroshikimate (35) produced a dialdehyde (36), which was oxidized[143] with bromine–water to the (known) degradation end product, tricarballylic acid (37). This approach to structural problems has proved particularly successful for determining the configuration of glycosides and related compounds.

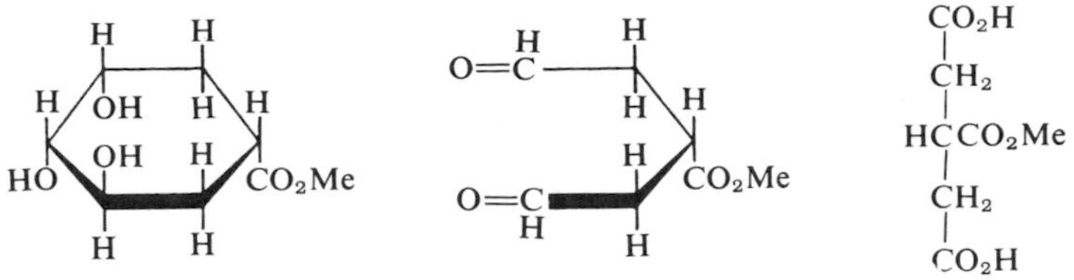

35   36   37

IV. GLYCOSIDES AND RELATED ALICYCLIC COMPOUNDS

A. REACTION CHARACTERISTICS; CONFIGURATIONAL RELATIONSHIPS

Criegee found that one mole of ethyl α-D-glucofuranoside rapidly reduces one mole of lead tetraacetate, and yields about one mole of formaldehyde, whereas one mole of methyl α-D-glucopyranoside gives no formaldehyde, even after two moles of oxidant have been reduced.[39] This difference constitutes a general means for distinguishing between five- and six-membered ring forms of various compounds. The exocyclic 5,6-diol grouping of a D-galactofuranoside is also cleaved more rapidly than the *trans* 2,3-diol grouping within the ring[144]; periodate shows the same effect, which has been utilized in descending from aldohexofuranosyl to aldopentofuranosyl derivatives.[145–147a] If further oxidation of the 5-*aldehydo* derivative (**38**, D-*gluco* isomer, or **39**, D-*galacto* isomer) is allowed to proceed, the trialdehyde **41** is obtained which, having a tartronaldehyde residue, is overoxidized (see Section II,C). Because of its highly reactive 2,3-*cis*-diol grouping, methyl α-D-mannofuranoside is mainly cleaved at the C-2–C-3 bond by lead tetraacetate.[39,144] Further oxidation is retarded, presumably because the resulting dialdehyde (**40**) adopts a stable, cyclic form (such as **42**). These stages are not evident in the oxidation of

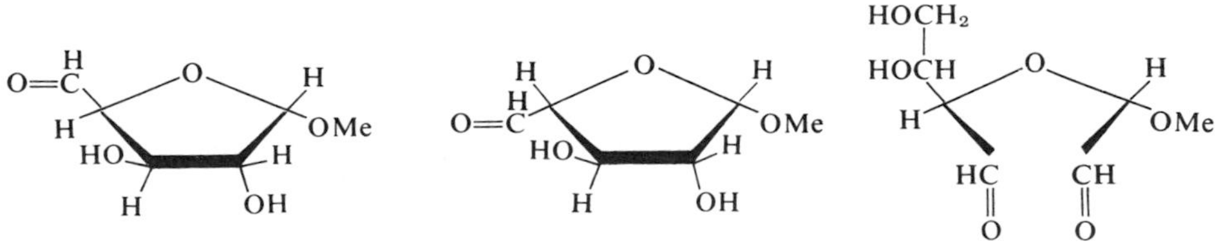

38   39   40

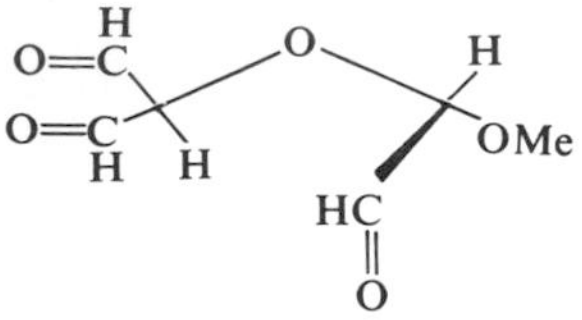

41 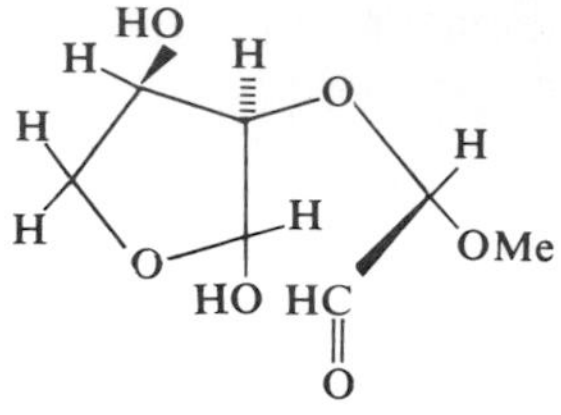   **42**

the D-mannoside by periodate, which appears to yield the trialdehyde **41** readily,[44] although the extreme reactivity of the 2,3-diol grouping probably accounts for the fact that this glycoside fails[24] to form a tridentate periodate complex at high pH.

Configurationally related 1,4-anhydrohexitols show closely analogous oxidation patterns, and this fact aided greatly in their characterization.[148,149]

Rates of cleavage of the 2,3-diol grouping of furanosides are, of course, very much higher for *cis* than for *trans* isomers. In addition, anomers are oxidized by lead tetraacetate at substantially different rates, the more reactive anomer having the 1-alkoxyl and 2-hydroxyl groups in *cis* relationship. For example, the rate for methyl β-D-threofuranoside is six times that of the α anomer,[150] whereas that for methyl α-D-erythrofuranoside is 3.5 times that of the β anomer.[66] The relative reactivities of the erythrosides toward lead tetraacetate contrast with their behavior toward tetraborate, as it is the β anomer that complexes the more extensively.[151] If these borate complexes serve as reasonable models for the cyclic lead complexes,[11] the enhanced rate of oxidation of the α anomer can be attributed to a relatively accelerated decomposition of its intermediate, because of greater internal repulsions, rather than to greater ease of complexing with the oxidant (see Section II,A).

Glycol-cleavage oxidation of glycopyranosides involves a well-defined stoichiometry under suitable conditions. Methyl α-D-glucopyranoside, for example, consumes two molar equivalents of periodate, yielding the syrupy dialdehyde **44** plus a mole of formic acid. The same reaction is accomplished by lead tetraacetate,[36] although due allowance must be made for the side-consumption of oxidant by the formic acid.[46] An elaborate study of oxidation products from glycosides permitted Jackson and Hudson[152–155] to correlate the ring size and the configuration of a whole series of such compounds with known structures and provided a general approach applicable to related types of compounds. Thus, one mole of certain methyl glycosides prepared from D-galactose, D-gulose, and D-mannose were all converted by oxidation with two moles of periodate into **44**. This was demonstrated by rotatory measurements and, more satisfactorily, by converting each product by hypobromite oxidation into a common, crystalline salt of the dicarboxylic acid **43**. The parent glycosides were thereby shown to be α-D-pyranosides. Similarly, a series of β-D-glycosides was correlated sterically by preparation[156] of the isomeric dialdehyde **45**, and a salt of the dicarboxylic acid **46**.

Oxidation of one mole of the anomeric methyl aldopentopyranosides with two moles of periodate[152] or lead tetraacetate[146,157] yields one of a pair of dialdehydes, a dextrorotatory product **48** being derived from α-D- or β-L-glycosides and levorotatory **49** from β-D or α-L isomers.[152] In turn, hypo-bromite oxidation affords either dicarboxylic acid **47** or **50**. Additional possibilities for configurational correlation accrue from the fact that a methyl aldotetrofuranoside similarly yields **48** or **49**, whereas a methyl aldopento-furanoside gives **44** or **45** as the final product.

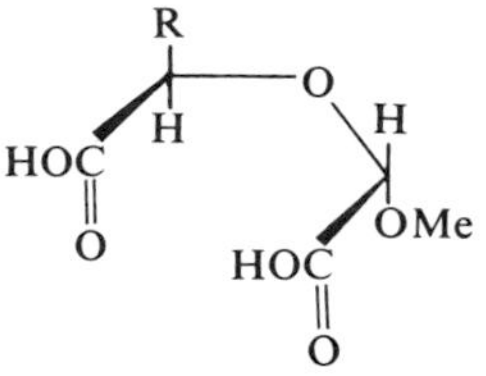

43 R = CH$_2$OH
47 R = H

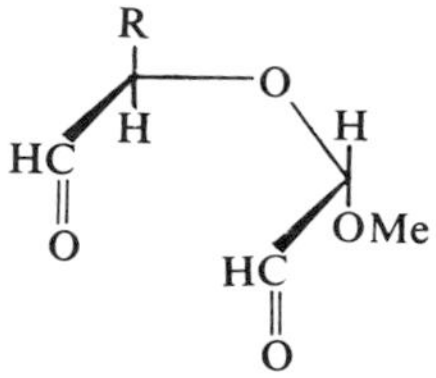

44 R = CH$_2$OH
48 R = H
51 R = CH$_2$OCHO

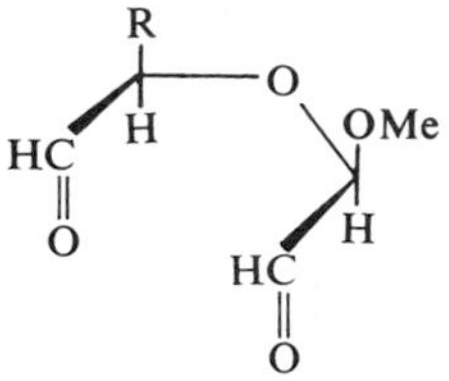

45 R = CH$_2$OH
49 R = H

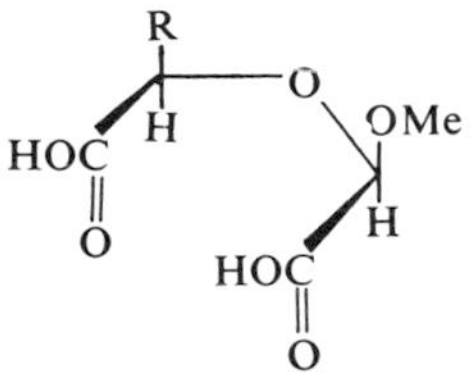

46 R = CH$_2$OH
50 R = H

Extension of Jackson and Hudson's approach is also feasible with 2-ketopyranosides, nucleosides,[147a,158] some types of anhydrides,[159,160] nonreducing and reducing disaccharides,[161] and glycosylamines.[161a]

Sometimes a better basis for such correlations is provided by reducing the dialdehyde,[162,163] which affords products containing fewer asymmetric centers; furthermore, these polyhydric alcohols frequently yield crystalline derivatives.

Configurational relationships may also be deduced by degrading the glycol-cleavage product to a known fragment. For example, noviose and mycarose were shown to be members of the L series, because glycosides of these anti-biotic sugars, on successive periodate cleavage, bromine oxidation, and hydrolysis, afforded (−)-3-hydroxy-2-methoxy-3-methylbutyric acid[164] and L-lactic acid,[165] respectively. A related application is found in the preparation[166] of (+)-[1-²H]ethanol, which was generated by a sequence of reactions initiated by periodate cleavage of the C-3–C-4 bond of butyl β-D-[5-²H]xylopyranoside of established stereochemistry, to give the deuterated dialdehyde.

Although stereoisomeric aldopyranosides are frequently oxidized to common end products, the rates at which these transformations take place differ in a substantial, but generally explicable, way.[36,167] When two hydroxyl groups of the 2,3,4-triol grouping have a *cis(a,e)* orientation, the first mole of oxidant per mole is reduced faster than when only *trans*-glycol groupings are present. Among the hexopyranosides, a *cis*-2,3-diol, as in methyl α-D-mannopyranoside, can give rise preferentially to a product (52) that contains an even more reactive, five-membered ring diol, causing an overall, rapid uptake of a second mole per mole. This second step is clearly evident in lead tetraacetate oxidations because the formic ester 51 is formed,[46] and not 44. Favored cleavage at C-3,C-4, such as would be expected with methyl α-D-galactopyranoside, results in a much slower reduction of the second mole per mole. In this instance, the initial cleavage product 53 cannot develop so reactive a type of *vic*-diol as is present in 52. However, as shown by the fact that the D-galactoside also gives 51 in high yield, reaction probably proceeds[46] largely via the hemiacetal 54.

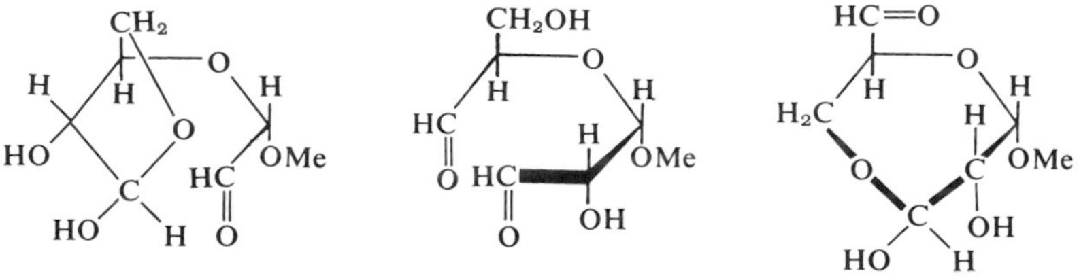

52　　　　　　　　53　　　　　　　　54

An understanding of the oxidation characteristics of hexopyranosides has been useful in establishing the stereochemistry of 1,5-anhydrohexitols.[167,168]

A diversity of conformational and configurational effects on the rate of periodate oxidation is evident with bicyclic derivatives of glycosides.[94,94a] In a series of 4,6-O-benzylidene aldohexopyranosides, for example, the higher reactivity associated with an *a,e* over an *e,e* orientation for the diol grouping is much enhanced when there is greater ring flexibility, as when the ring junction is *cis* (not *trans*). Also, since the inertness of a D-*altro* isomer (see Section II,D) can be attributed to a diaxial 2,3-diol grouping [and, hence, a $^4C_1(D)$ conformation], the observed oxidation of a *cis*-fused D-*ido* isomer implied a $^1C_4(D)$ conformation and a diequatorial 2,3-diol grouping in the latter. Anomeric configurational influences are more evident in compounds of this class[94] than among the parent glycosides.[36,167] Rate constants for the α anomers of the pairs examined are greater than those for the β anomers; this was attributed to greater steric interference with formation of the 2,3-periodate complex by an equatorial C-1 substituent.[94] Conformational information was also educed from the observation that 2,6-anhydro-1-deoxy-1,1-bis-(ethylsulfonyl)-D-allitol undergoes rapid oxidation by periodate; the $^4C_1(D)$ conformation, which would have the a,e,a-triol arrangement and, thus, would complex rather than oxidize, was logically excluded as a major contributor to the conformational equilibrium.[168a]

The fact that a glycopyranoside consumes more oxidant than a glycofuranoside, and also that it yields formic acid, finds routine application for the determination of ring size, not only of glycosides but of related classes of compounds. For example, a ready confirmation of the furanoid structure of the D-ribosyl group in uridine and cytidine was afforded by the observation[158,169] that only one mole of periodate per mole is consumed; thymidine and 2'-deoxycytidine, however, consume no oxidant, which is consistent with the presence of a furanoid ring in these 2'-deoxynucleosides.[170] Successful applications of these criteria of ring size (as well as of the production of formaldehyde) to naturally occurring glycosides, oligosaccharides, and polysaccharides (see Sections V,C and VI) are numerous. Although a number of interesting examples are found in the chemistry of C-glycosyl compounds, results in this series are sometimes inconclusive.[49,171]

A glycosyl residue of a disaccharide or higher oligosaccharide usually shows oxidation behavior similar to that of a structurally related simple glycoside. This fact has been utilized in determining the sequence of linkages in solanose.[172] Also, it is sometimes possible to open one particular glycosyl ring selectively in the presence of others in the same molecule. For example, the D-galactopyranosyl residues of raffinose and stachyose are oxidized by periodate much faster than the D-glucopyranosyl and D-fructofuranosyl groups of these oligosaccharides. This behavior permits selective removal of the oxidized fragments from the intact sucrose moiety.[173] The components of sucrose itself also show widely different reactivities, just as do other glycosides of D-glucopyranose and D-fructofuranose. Lead tetraacetate selectively oxidizes the D-fructofuranosyl group, whereas periodate attacks the D-glucopyranosyl group much more readily, affording two different partially oxidized sucroses.[174]

B. DIALDEHYDES

Dialdehydes formed by glycol cleavage of glycosides and other cyclic derivatives are capable of existing in a variety of modifications, depending on

References start on p. 1207.

such conditions as the solvent and the type of reaction to which they are subjected.[175] In water they may exist[175,176] as a hydrated, acyclic dialdehyde (such as **55**), as an internal hemiacetal (**56**), or as a hemialdal[177] (**57**); further cyclization of **56** is also possible in principle (see **28**, Section II,D). Formation of **57** requires the addition of the elements of a molecule of water,[176] and this has been envisaged[175] as proceeding via formation of an intermediate mono-aldehydic *gem*-diol.

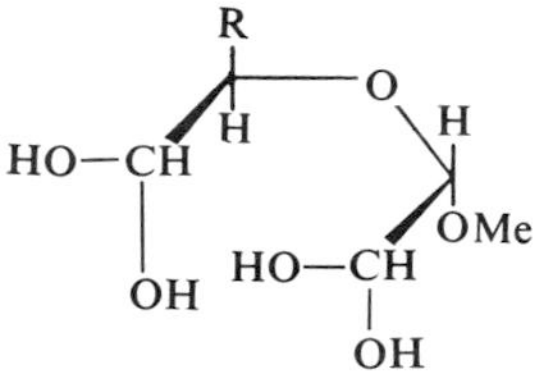

55 R = CH₂OH
58 R = H

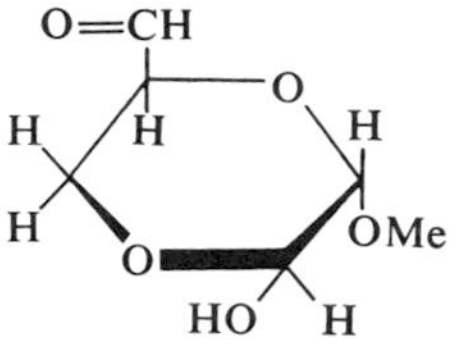

56

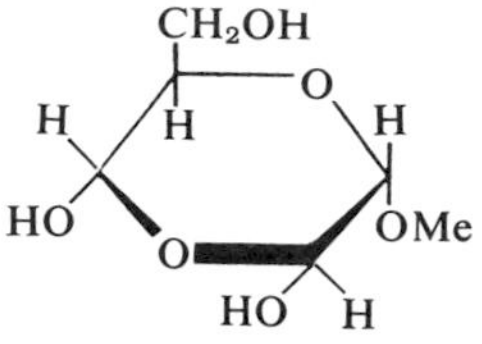

57

Relatively few dialdehydes are thus far known in crystalline form. The first reported example is that obtained from methyl α-L-rhamnopyranoside and related 6-deoxyglycosides[152]; its elemental composition is that of a dialdehyde monohydrate, but it has been found to possess a hemialdal structure, as it affords a bis(*p*-nitrobenzoate) and a dimethyl ether.[178] Similarly, n.m.r. spectroscopy showed that the compound contains two hydroxyl groups, and indicated that it exists almost exclusively as **59** in solution.[179]

For similar reasons, the dialdehydes **48** and **49**, obtained from methyl aldopentopyranosides or aldotetrofuranosides, can be formulated as hemi-aldals[175,180,181]; according to n.m.r.-spectral data, however, they exist in

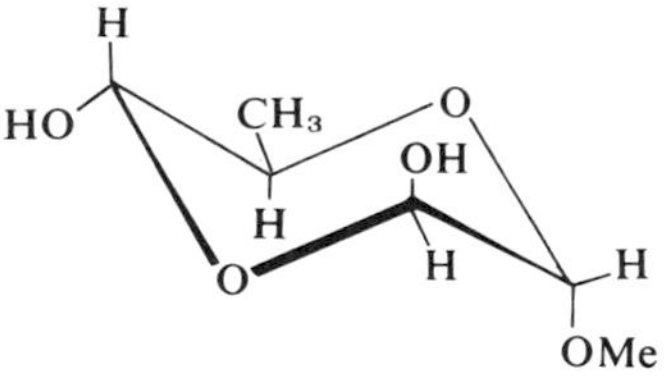

59

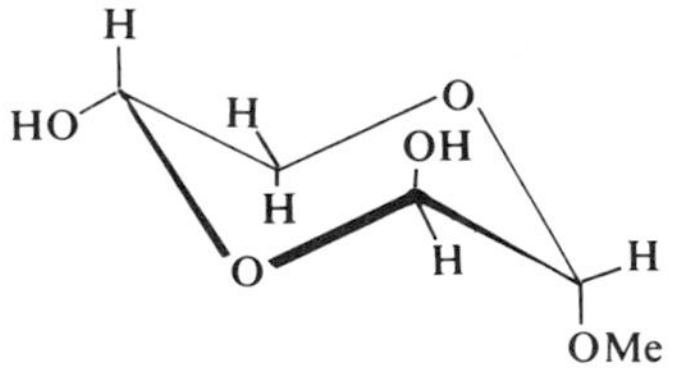

60

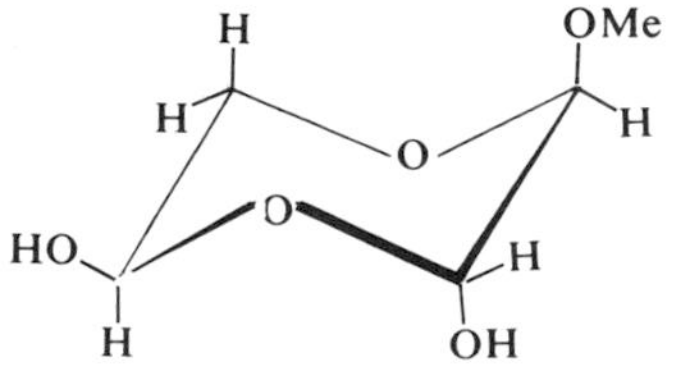

61

62 **63** **64** **65** **66**

deuterium oxide as an almost 1:1 mixture of the acyclic hydrated dialdehyde (**58** or as the β anomer) and the hemialdal (**60** or **61**) form. They occur almost exclusively as the latter form in such solvents as methyl sulfoxide or pyridine.[181]

The acyclic, hydrated form of the dialdehyde **62**, prepared from a 1,5-anhydropentitol (or 1,4-anhydrotetritol), is also moderately stable in aqueous solution, in which it exists in equilibrium with two hemialdal forms.[179] One hemialdal contains equatorially attached hydroxyl groups and may be represented as an equilibrium between **63** and **65**, whereas the other contains one equatorial and one axial hydroxyl group, and is depicted by the interconverting chair forms **64** and **66**. As was found with **55**, the acyclic form is displaced by the hemialdals in methyl sulfoxide or pyridine.

A seven-membered (1,4-dioxepan) ring hemialdal (**67**) is obtained by glycol cleavage of the 2,3-diol group of methyl 4,6-O-benzylidene-α-D-aldohexo-pyranosides.[182–185] In common with other hemialdals, this compound is readily converted into a monoalcoholate,[184] the favored location of the alkoxyl group being that shown (**67**, R = alkyl).[185] Oxidation of 1,2-O-cyclohexylidene-*myo*-inositol yields an *erythro*-tetrodialdose derivative that appears to exist as a mixture of isomeric, five-membered ring hemialdals. A crystalline hemialdal diacetate, prepared in high yield, was shown[186] to have structure **68**.

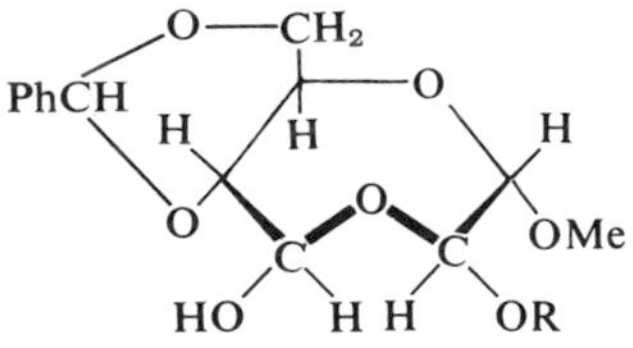

67 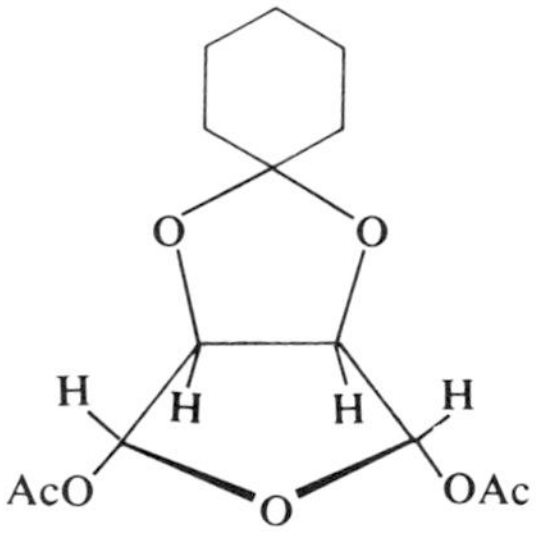   68 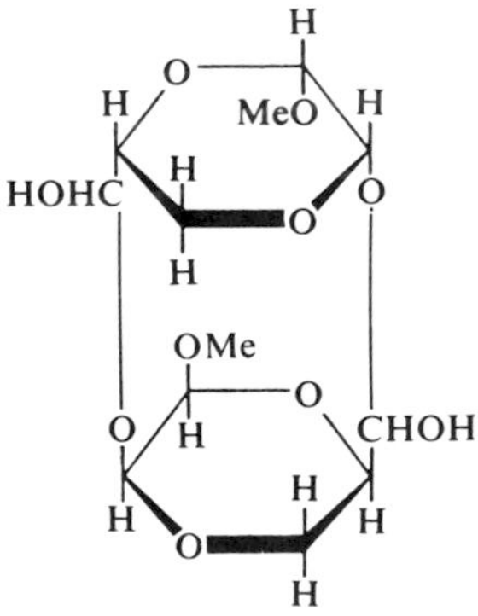   69

Dialdehydes obtained from methyl aldohexopyranosides and aldopento-furanosides (for instance, **44**, which is formed from α-D anomers) may assume a multitude of structures, because, in addition to acyclic and hemialdal forms, they can exist as internal hemiacetals.[115a,180,187–191] A crystalline, dimeric form of the latter type (**69**) has been characterized.[191] However, **44** affords different derivatives that represent various types of structures, and that can, therefore, best be described as a complex, equilibrium mixture of several forms.[175] Condensation of the dialdehyde derived from uridine with benzoyl-hydrazine effected replacement of the bridging oxygen atom of the hemialdal unit to afford an *N*-benzamidomorpholine derivative.[192]

Several other classes of dialdehydes prepared from monosaccharide derivatives, oligosaccharides, and polysaccharides are known.[175] A good deal of early interest centered on those obtained by periodate oxidation of cellulose and starch,[120,121,121b,193-198] as they constitute a novel, chemically modified form of these important polymers; the dialdehydic polymers themselves were later developed as materials of commercial significance.[175,199] Isolated examples of marginal activity in tests for inhibition of carcinoma have been reported for some periodate-oxidized hexose derivatives.[200] In addition, dialdehydes are starting materials for several useful syntheses; a particularly fruitful example is the synthesis of 3-amino-3-deoxyaldo-pentoses and -hexoses, in which the recyclization step is achieved by condensing *both* of the carbonyl groups with *one* molecule of nitromethane.[201] A related type of application is the synthesis of 1,4-oxathianes, which involves incorporation of a sulfur atom into the ring generated from the glycol-cleavage product.[202]

V. REDUCING SUGARS

A. INTRODUCTION

In the oxidation of reducing sugars, periodate and lead tetraacetate frequently act quite differently. These differences can usually be traced to differences in the rates of oxidation of (*a*) the various forms that the sugars assume in solution, and (*b*) the intermediates produced by the initial cleavage. In turn, these differences may arise merely because each of the oxidants is not normally used in the same solvent system as the other. Because anomeric pyranose–furanose and ring–aldehyde interconversions are faster in acetic acid than scissions of most kinds of *vic*-diols, a straightforward, overall course of reaction is sustained for most sugars; in water, however, the rates of oxidation often sufficiently exceed the rates of the tautomeric changes to produce a markedly different, sometimes highly complex outcome.

B. MONOSACCHARIDES AND PARTIALLY SUBSTITUTED DERIVATIVES

The sugars behave as their cyclic forms toward lead tetraacetate in acetic acid.[38,39,116] Oxidation primarily involves α-hydroxy hemiacetal groups, and it results in stepwise shortening of the carbon chain.[38,116,203,204,*] For example, the reaction of D-glucose (70) may be depicted as follows: (a) initial cleavage at the C-1–C-2 bond of α-D-glucofuranose (71) yields 3-O-formyl-D-arabinose (72), which, as the β-furanose (73), (b) is degraded to 2,3-di-O-formyl-D-erythrose (79); consecutively, steps (a) and (b) are much faster than the rate of oxidation of either α- or β-D-glucopyranose. Also, oxidation of 3-O-formyl-aldehydo-D-arabinose (72) is a slower process than its cyclization to a particularly reactive ring form, such as 73. Hence, several equilibrium displacements are maintained in this sequence in such a way that the reaction follows essentially a single pathway. Of the two steps (a) and (b), the former is implicated as rate controlling by the fact that changes in the concentration of reactants have relatively little effect on the oxidation rate.[38]

By contrast, periodate oxidation of D-glucose in water is initiated chiefly at C-1,C-2 of the pyranose form.[78,205–209] As anomerization is relatively slow, both anomers are oxidized, but the β anomer is the less reactive and is also cleaved at other glycol groupings.[78] Intermediate products, such as 4-O-formyl-D-arabinose (76), are probably oxidized faster than they can cyclize. The overall result is a moderately good yield of 2-O-formyl-D-glycerose (81) when three molar proportions of periodate are used,[205,208] although this ester is accompanied by smaller quantities of the other products possible, which range from formaldehyde to pentoses, accompanied by unoxidized D-glucose.[206]

These differences between the two oxidants and the influence of solvent are emphasized in the unimolar oxidation of 3-O-methyl-D-glucose (74), which, with each, yields a mono-O-formyl-2-O-methyl-D-arabinose[38,48,210]; as shown most clearly[48] with 3-O-methyl-D-[5-^{2}H]glucose, periodate cleavage gives the 4-formate (80, R' = H), whereas the 3-formate (75, R' = H) is produced by lead tetraacetate in acetic acid.[38,48] That the solvent is probably the main differentiating factor was found on comparing the oxidation of the 6-trityl ether of 74 in acetic acid with that in benzene[38]; in the latter medium, cleavage of the 1,2-diol of the pyranose is fast enough to obviate a furanose pathway almost completely, and the main product is 4-O-formyl-2-O-methyl-5-O-trityl-D-arabinose (80, R' = Tr), whereas in acetic acid the 3-formate (75, R' = Tr) is obtained.

The oxidation of one mole of D-mannose by lead tetraacetate in acetic acid, which involves rapid reduction of 2.8 molar proportions of oxidant,[38,116] is consistent with a principal reaction pathway in which the pyranose is degraded stepwise; 4-O-formyl-D-arabinose (76 ⇌ 77) is produced first, further

---

* Attack on the hydroxy hemiacetal should be favored by stabilizing of the developing carbonyl group in the transition state through electron release from the ring-oxygen atom,[6] and also by steric factors.[5,38]

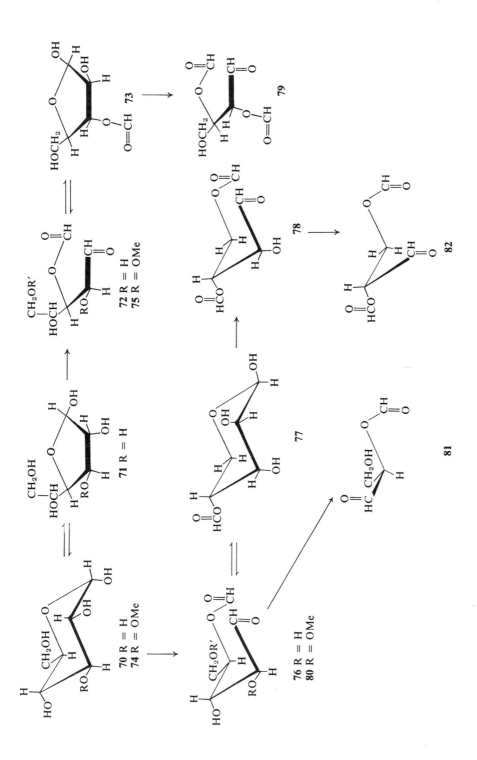

oxidized to 3,4-di-O-formyl-D-erythrose (78), and finally degraded to 2,3-di-O-formyl-D-glyceraldehyde (82). In aqueous periodate solution, however, cleavage at positions other than the α-hydroxy-hemiacetal grouping is a prominent reaction, and the behavior of D-mannose is very similar to that of D-glucose.[78]

The other aldohexoses, and the aldopentoses, are characterized by rapid values of uptake of lead tetraacetate ranging between those for D-glucose and D-mannose.[38,116] These intermediate levels of oxidation reflect differing proportions of furanose and pyranose pathways comprising the overall reactions. Periodate-oxidation data for these sugars are generally consistent with almost exclusive attack of pyranose forms, this behavior being most clearly evident with the aldopentoses.[208] Glycol-cleavage characteristics have, to some extent, been correlated with conformational[23] and configurational[38] properties of the sugars.

Gross differences between the action of periodate and lead tetraacetate are found also in the way in which they oxidize ketoses. The main pathway for periodate oxidation involves cleavage of the 1,2-diol to a glyoxylic ester[40,208,211]—for instance, 86 from D-fructose (83); the latter probably reacts in the pyranose form, whereas results for L-sorbose are more compatible with oxidation of a furanose form.[208] Lead tetraacetate cleaves the 2,3-α-hydroxy hemiacetal group almost exclusively, yielding a glycolic ester.[35,116,212,213] The reaction pathway for D-fructose may be depicted[116] by the sequence: D-fructofuranose (84) → 3-O-glycolyl-D-erythrose (85) → 3-O-formyl-2-O-glycolyl-D-glyceraldehyde (87). L-Sorbose gives the corresponding L-glyceraldehyde derivative,[212] and D-altro-heptulose affords the diester of D-erythrose.[35]

Even higher sugars that contain an exocyclic vic-diol or 1,2,3-triol group are specifically attacked by lead tetraacetate at the anomeric center and are degraded stepwise, as shown by the oxidation of heptoses,[116] an octose,[116] 2-octuloses,[213,214] and a 2-nonulose.[215] The products are always stable formates or formate-glycolates. Periodate, however, rapidly cleaves exocyclic diol groupings,[216] initiating a different overall reaction course that involves more-extensive oxidation.

As illustrated with 3-O-methyl-D-glucose (74), introduction of a substituent group at O-3 of an aldose results in a controlled, limited oxidation, regardless of the oxidant used.[48,210,217–221] Similar behavior is shown by 3-amino-3-deoxyaldoses in which the amino group is acetylated or permethylated.[92] A substituent on O-4 confines the oxidation largely to the 1,2,3-triol grouping. When an aldose contains a substituent on O-2, it is highly resistant toward lead tetraacetate[217,221] and relatively unreactive toward periodate; however, with each of these classes of derivative, the normal reaction may be obscured by the use of prolonged reaction periods or severe

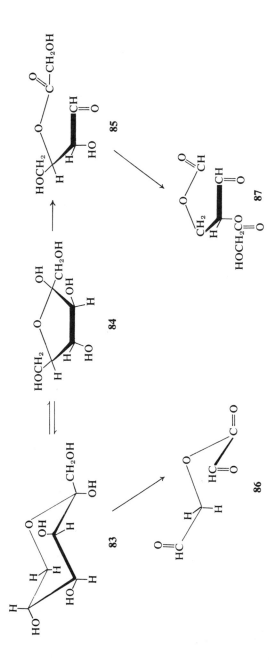

83 **84** **85**

86 **87**

conditions (high concentration, elevated temperature). Nonselective oxidation, which is thus promoted, may be further accentuated by the formation of substituted tartronaldehydes as intermediate products (see Section II,C).

The course of oxidation is essentially unaltered[220–223] for an aldohexose containing a substituent on O-6; however, a strongly electron-withdrawing group may promote overoxidation by destabilizing formic ester intermediates or by activating neighboring positions.[208,222,223] Thus, although D-*erythro*- and D-*threo*-tetruronic acids can be prepared by lead tetraacetate oxidation of D-glucuronic acid and D-galacturonic acid, respectively, overoxidation is more pronounced than in the reaction of the corresponding aldohexoses.[222] Similarly, D-erythrose 4-phosphate may be obtained by oxidizing D-glucose 6-phosphate with two molar proportions of lead tetraacetate,[223–225] but the stoichiometry of these reactions is less precise than that for neutral derivatives, and the experimental conditions can exert a profound effect on the composition of the products.[161,223–233]

C. OLIGOSACCHARIDES

The oxidation behavior of reducing oligosaccharides is essentially a combination of the patterns exhibited by monosubstituted derivatives of monosaccharides and by glycosides. In general, these merged patterns are sufficiently characteristic under suitable conditions to permit unambiguous characterization of oligosaccharides. Ideally, reducing disaccharides containing aldopentopyranose (R = H in **88**, **89**, and **G**) or aldohexopyranose (R = CH_2OH in **88**, **89** and **G**) residues show the following characteristics (moles per mole)[6,161,220,221,225–232]*:

(a) A (1→3)-linked disaccharide is converted into **88** with concomitant reduction of three moles of oxidant and liberation of one mole of formic acid.

(b) A (1→4)-linked disaccharide is degraded to **89**, which is accompanied by reduction of four moles of oxidant and liberation of two moles of formic acid; in the lead tetraacetate oxidation of a hexose (reducing) residue C-1 and C-2 both become formic ester groups, and only one equivalent of free acid is obtained.

(c) A (1→2)-linked disaccharide is degraded to **90**. Again, the (nonreducing) glycosyl group consumes two moles of oxidant, and yields one mole of formic acid; the reducing residue consumes two moles (pentose) or three

---

* These characteristics are observed when lead tetraacetate is used in aqueous acetic acid containing potassium acetate.[220,221] In glacial acetic acid, the degree of oxidation is much lower, because the reaction of the (reducing) glycose residues resembles that of monosaccharides (Section V,B).[221]

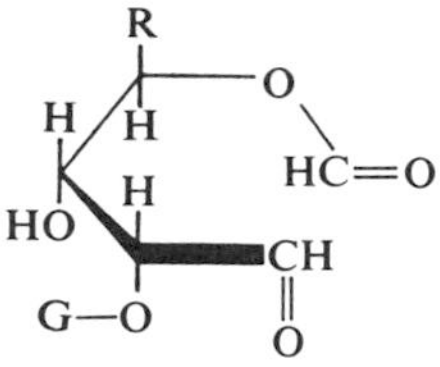

88 89 90 91

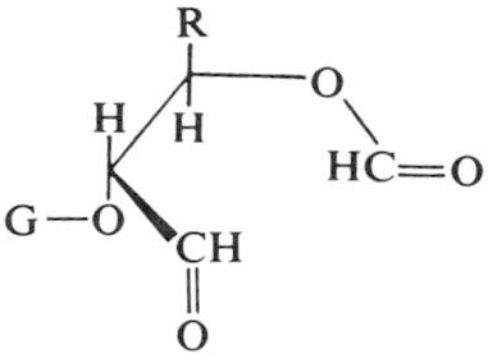

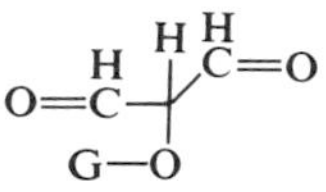

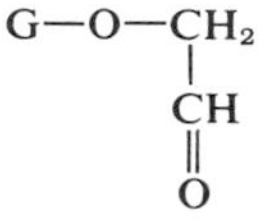

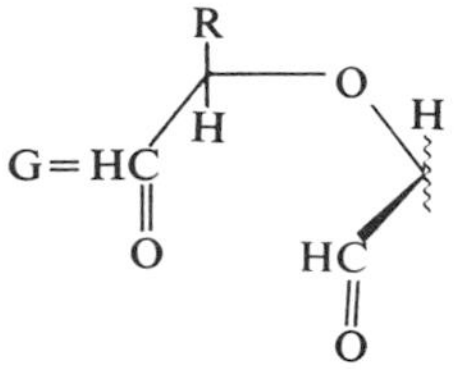

moles (hexose) of oxidant, and yields one mole (pentose) or two moles (hexose) of formic acid, *plus* one mole of formaldehyde from the primary alcohol group. When the reducing residue is that of a 6-deoxyaldohexose, acetaldehyde is liberated.[230]

(d) A (1→5)-linked (pentose) or (1→6)-linked (hexose) disaccharide is degraded to **91**, the uptake of oxidant being five, or six, moles, and the yield of formic acid four, or five moles, respectively.[233]

In practice, the results obtained are frequently complicated by over-oxidation[233] (see Section II,C). An early, comparative study of the periodate oxidation of maltose and isomaltose showed that the latter, a (1→6)-linked disaccharide, behaves as outlined in category (d), whereas the (1→4)-linked biose is extensively overoxidized; this information, however, differentiated between the two possible structures.[228] The overoxidation phase can be attributed to hydrolysis of the formate group of **89**, followed by cleavage of a 2-substituted tartronaldehyde, which is subject to further attack. Similarly, the ester group of product **88** can suffer hydrolysis, resulting in a spuriously high uptake of oxidant by (1→3)-linked disaccharides; however, over-oxidation is minimized when conditions are such (low temperature, controlled pH, low concentration of oxidant) as to stabilize formate groups. Because these groups are relatively stable in acetic acid, and because their formation is promoted under the reaction conditions, overoxidation of disaccharides is usually not a serious problem when lead tetraacetate is used.[161,220] When the 4-substituted reducing residue is a pentose, the product **89** is a glyceraldehyde moiety, which is not degraded further. The (1→2)-linked disaccharide is disposed toward overoxidation under all conditions because it yields a tartronaldehyde derivative (90) directly; this can be prevented, however, by initial reduction to the alditol, which is oxidized instead to a stable, 2-substituted glyceraldehyde.[161,230]

With disaccharides containing a 2-amino-2-deoxyaldose reducing residue, it is advantageous to use the N-acetyl derivative,[234,235] or perhaps better still, the N-acetyl derivative of the disaccharide alditol. The oxidation products formed from these derivatives do not become significantly overoxidized, and hence, the stoichiometry of the reaction is clearly indicative of the linkage position.[233–235] For example, one mole of a 3-substituted 2-amino-2-deoxy-aldohexose residue consumes two moles of oxidant, and yields one mole each of formic acid and formaldehyde, whereas the 4-substituted isomer consumes only one mole of oxidant and releases only one mole of formaldehyde.[234]

Oxidations involving 2-ketose or glycuronic acid residues also give satisfactory data,[161] the acids being best examined in the ester form[236] in order to minimize overoxidation.

A different approach to structural elucidation—namely, selective degradation—is made possible by the markedly disparate rates at which units may be oxidized[117,161,220,221]; usually, the reducing residue is the most susceptible. For example, treatment of cellobiose (92) with two moles of lead tetraacetate per mole yields 3,4-di-O-formyl-2-O-β-D-glucopyranosyl-D-erythrose, which has been characterized[161] by conversion into 2-O-β-D-glucopyranosyl-D-erythritol (93). A further, selective attack is feasible owing to the fact that the alditol residue of 93 is oxidized more rapidly than the glycosyl group. Hence, treatment of 93 with one molar proportion of oxidant produces 2-O-β-D-glucopyranosyl-L-glyceraldehyde, which is readily characterized[161] by reduction to 2-O-β-D-glucopyranosylglycerol (94). The same sequence of reactions converts maltose into the anomer of 94, 2-O-α-D-glucopyranosylglycerol,[221] and related sequences afforded anomeric pairs of

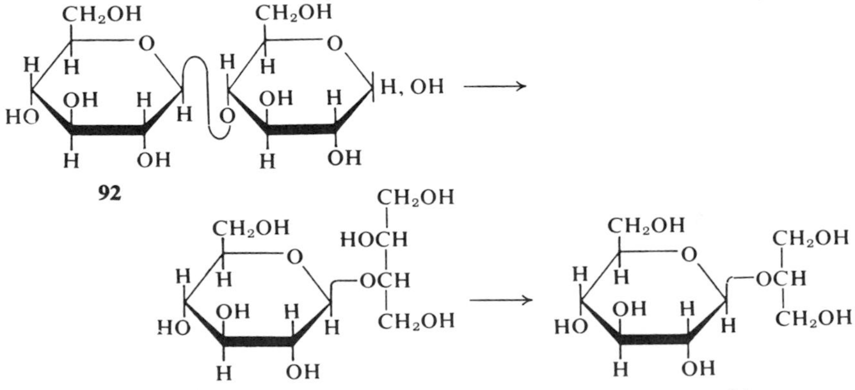

92

93 → 94

2-*O*-glycosylglycerols containing D-galactosyl, D-mannosyl, D-xylosyl, and L-arabinosyl groups.[221,237] Such compounds served as reference materials for establishing the configuration of a number of disaccharides.

This degradative technique is also applicable to higher oligosaccharides. For example, a trisaccharide was characterized as *O*-β-D-glucopyranosyl-(1→4)-*O*-β-D-glucopyranosyl-(1→3)-D-glucose (**95**) by the following series of reactions[238]:

β-D-Glc*p*-(1→4)-β-D-Glc*p*-(1→3)-D-Glc **95**

↓ $Pb(OAc)_4$ (1 equiv.)

β-D-Glc*p*-(1→4)-β-D-Glc*p*-(1→2)-D-Ara

↓ $NaBH_4$

β-D-Glc*p*-(1→4)-β-D-Glc*p*-(1→2)-D-Ara-itol

↓ $Pb(OAc)_4$ (2 equivs.)

β-D-Glc*p*-(1→4)-β-D-Glc*p*-(1→2)-D-glyceraldehyde

↓ $NaBH_4$

β-D-Glc*p*-(1→4)-β-D-Glc*p*-(1→2)-glycerol **96**

Identification of the final product as 2-*O*-β-cellobiosylglycerol (**96**) was accomplished by comparing it with **96** prepared from cellotriose by the sequence used for degradation of **92** to **94**. Similarly, the tetrasaccharide *O*-β-D-glucopyranosyl-(1→4)-*O*-β-D-glucopyranosyl-(1→4)-*O*-β-D-glucopyranosyl-(1→3)-D-glucose was identified by its conversion into 2-*O*-β-cellotriosylglycerol by the same sequence of reactions as used to degrade **95** to **96**. As the cellotriosylglycerol is also obtainable from authentic cellotetraose by selective degradation (as in **92**→**94**), the linkage positions and anomeric configurations for the tetrasaccharide were established simultaneously.[238]

Glycol-cleavage oxidation may sometimes provide a means for preferentially removing the nonreducing group of a trisaccharide or higher oligosaccharide[173,239,240]; this can be achieved when the end group is more readily oxidizable than the internal residues, a situation that is frequently encountered.[117] For example, oxidation of cellotriose (**97**) with four molar equivalents of lead tetraacetate, reduction, and partial hydrolysis[241,242] of the resultant polyhydric alcohol acetal **98** with acid affords[238] **93**. In the same way, *O*-β-D-glucopyranosyl-(1→3)-*O*-β-D-glucopyranosyl-(1→4)-*O*-β-D-glucopyranosyl-(1→4)-D-glucose was degraded to 2-*O*-β-D-cellobiosyl-D-erythritol (also prepared from cellotriose).[238] Selective removal of the nonreducing

References start on p. 1207.

group from trisaccharides was also achieved in other series[240,243,244] (sometimes by removal of the initially produced dialdehyde unit through the action
of phenylhydrazine or alkali[173,239,240]), illustrating the general utility of this
approach for structural studies.

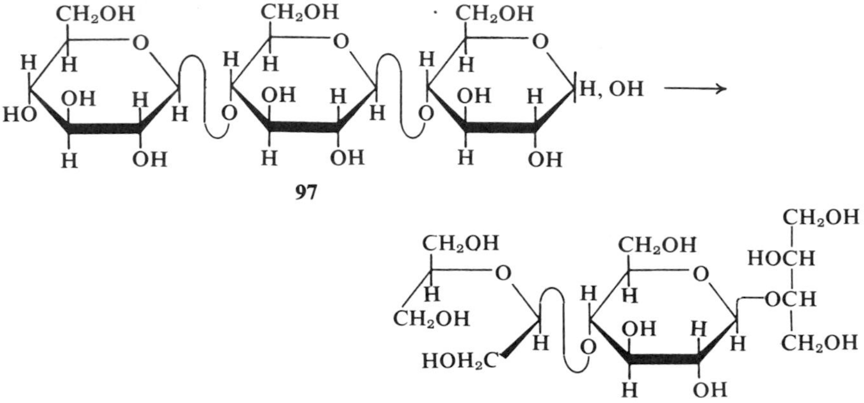

97

98

The reducing residue of the trisaccharide O-α-D-mannopyranosyl-$(1 \rightarrow 2)$-
O-α-D-mannopyranosyl-$(1 \rightarrow 2)$-D-mannose (**99**) was selectively detached after
lead tetraacetate oxidation[240]; the nonreducing group was converted (via
100, in several steps) into the 2-substituted triouronic acid **101**, which was
hydrolyzed with dilute acid to 2-O-α-D-mannopyranosyl-D-mannose. Another
example involves degradation of manninotriose to 6-O-α-D-galactopyranosyl-
D-galactose by oxidation with two molar proportions of lead tetraacetate and
subsequent alkaline hydrolysis of the 4-substituted D-erythrose.[245]

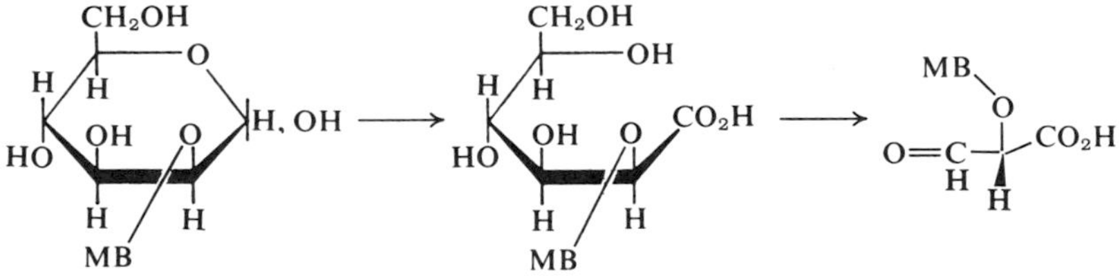

99   **100**   **101**

MB = 2-O-(α-D-mannopyranosyl)-α-D-mannopyranosyl

Mixed detecting reagents containing periodate are used in paper-chromatographic analysis of oligosaccharides. Periodate combined with alkaline silver
nitrate is employed to detect nonreducing sugars,[246] and a periodate–Schiff
reagent mixture is reported to produce specific color reactions from which
linkage positions may be determined.[247]

VI. POLYSACCHARIDES

A. OXIDATION PATTERNS; END-GROUP ANALYSIS

Periodate oxidation is a standard method for determining various structural features of polysaccharides; its earliest applications helped to define fundamental structures for cellulose,[120,121,193–195,197,198] starch,[121,196,248–250] glycogen,[44,45,251–253] and xylan.[195,254] It has been used to examine virtually every polysaccharide that has been studied, frequently in conjunction with classical methylation procedures.

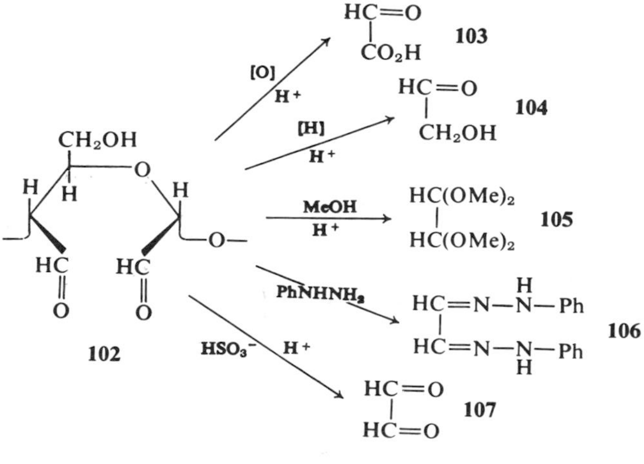

Characteristic patterns of oxidation are readily recognized for glycosyl residues that are joined by different types of linkage. An aldohexopyranosyl residue bonded to adjacent residues through O-1 and O-4 reduces one molar proportion of periodate.[120] Bromine oxidation of the resulting dialdehyde (102), followed by acid hydrolysis, yields D-erythronic and glyoxylic acids (103), showing that the C-2–C-3 bond is cleaved initially.[195] Similarly, reduction of 102, and acid hydrolysis, gives erythritol and glycolaldehyde (104).[196,255] More-direct demonstrations of the cleavage point in 102 are afforded by methanolysis, which converts[121] the glyoxal residue into the volatile tetramethyl acetal 105, or by treatment with phenylhydrazine, which affords glyoxal bis(phenylhydrazone) (106) and tetrulose phenylosazone (see Section VI,B).[256] Although the dialdehyde is relatively resistant to aqueous mineral acid, glyoxal (107) and D-erythrose are readily liberated by hydrolysis with sulfurous acid.[257] Also, 102 undergoes facile β elimination in the presence of alkali, a property that largely accounts for the marked alkali lability of periodate-oxidized cellulose and starch.[258–261] Several structures (of the kind considered in Sections II,D and IV,B for dialdehydes of low molecular weight) may be proposed for **102**, but the polymeric structure can also accommodate hemiacetal cross-links to an adjacent glycosyl residue or polymer molecule.[121b,d,e,175,175a]

A $(1 \rightarrow 2)$-linked aldohexopyranosyl residue also consumes one mole of periodate per mole of residues. However, this type of linkage is readily distinguished from the $(1 \rightarrow 4)$ by the fact that fragmentation of the dialdehyde by the various reactions applied to **102** affords two 3-carbon fragments, not a 2- and a 4-carbon fragment.

In the aldopentopyranosyl series, $(1 \rightarrow 4)$- and $(1 \rightarrow 2)$-linked residues also reduce only one mole of oxidant per mole, but the liberated products of cleavage are characteristically different; therefore, examination of the resultant dialdehydes, or of products derived from them, is necessary for differentiation between these two possibilities.

A $(1 \rightarrow 3)$-linked hexopyranosyl residue contains no *vic*-diol grouping and accordingly is not oxidized.

When the linkage to an aldohexopyranosyl residue is through O-6, three contiguous hydroxyl groups are available for oxidation. Consequently, such a residue is distinguishable from the others considered, as it consumes two molecules of oxidant and releases one molecule of formic acid per residue that reacts. A detailed kinetic analysis of such systems has been presented.[261a]

Although the stoichiometry of these oxidations is independent of the configuration, the rates of reaction differ widely.[262] In the $(1 \rightarrow 4)$-linked D-*gluco* series, for example, α anomers reduce periodate much faster than β anomers. In this respect, their behavior is related, not to that of simple glycosides (which show only small differences in rate between the anomers), but to that of conformationally rigid, fused-ring derivatives of the glycosides.[94] Within a given group, there may be large differences in the response to oxidation, indicative of variations in fine structure (as seen with amylopectins); glycogens, however, show remarkably uniform behavior. The $(1 \rightarrow 3)$-linked residues in nigeran or oat glucan are themselves unoxidized and appear to suppress the reactivity of the $(1 \rightarrow 4)$ residues present. As would be expected, residues having the *manno* configuration (2,3-*cis*-diol) oxidize at much higher rates than their *gluco* epimers. Periodate in aqueous perchloric acid oxidizes the glycuronic acid residues of glycosaminoglycuronans with sufficient selectivity that the reagent has been proposed[262a] for histochemical classification.

Nonreducing groups of polysaccharides, either in the aldo-pento- or -hexopyranose series (for instance, **108**, R = H) are oxidized by two moles of periodate per mole, yielding formic acid. This phenomenon provides the basis for a widely exploited method of estimating molecular weight[44,263]; similarly, branched polysaccharides yield formic acid in proportion to the ratio of terminal to nonterminal residues in the average repeating unit. The method has been of particular value in comparing various samples of glycogen,[44,45,252,264,265] amylopectin,[266,267] and dextran.[268-270] Application of this end-group analysis requires, however, a knowledge of the amount of formic acid released from other structural components of the molecule. A $(1\rightarrow6)$-linked hexopyranosyloxy residue (such as **108**, R = a glycosyl group) is one such additional source of formic acid requiring independent characterization, possibly by reduction of the dialdehyde **109** (R = a glycosyl group) and methylation of the derived polyhydric alcohol[271]; subsequent hydrolysis of the latter affords 1-*O*-methyl-D-glycerol (**110**), whereas the oxidized end group (**109**, R = H) simultaneously affords 1,3-di-*O*-methylglycerol (**111**).

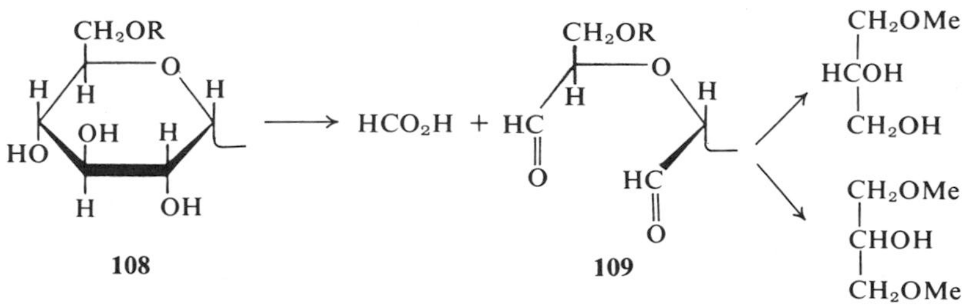

Formic acid is also produced from the reducing residue (as in the oxidation of **112** to **113**) in much the same way as for oligosaccharides (see Section V,C). The attendant problem of overoxidation is, therefore, encountered in oxidation of polysaccharides when the malonaldehyde-derived structures (**114**) are a product of the reaction of the reducing residue. Because the other residues are oxidized relatively slowly, overoxidation can proceed in the interim, exposing a succession of new, reducing residues (such as **115**) as the degradation proceeds along the chain. As the release of formaldehyde (**116**)

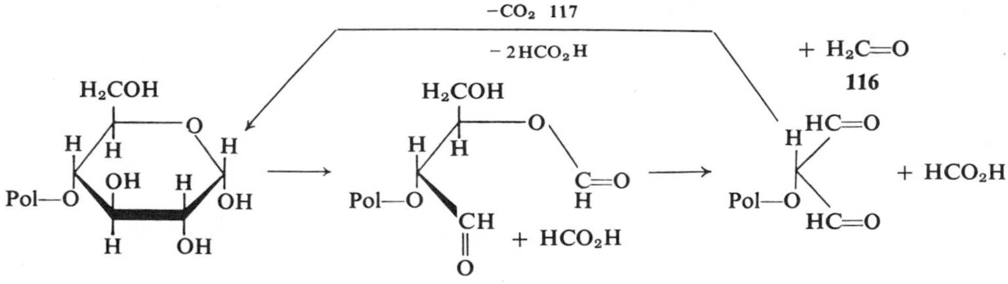

112 (Pol = POLYMER − *n* units)
115 (Pol = POLYMER − (*n* + 1) units)

References start on p. 1207.

and carbon dioxide (117) during these sequential steps approximates a reaction of zero order, overoxidation occurs at a linear rate, which can be corrected for by back-extrapolating the rate plot of the acid produced.

The amount of formaldehyde (116) liberated[233,272,273] provides a good index of the extent of overoxidation as the latter is derived from the primary alcohol group of degraded, reducing residues (112→114). The rate at which the formaldehyde is produced is proportional to the number of reducing residues,[274] so that a measure of the rate of overoxidation may itself provide an independent estimate of the degree of polymerization. Similarly, if the oxidative erosion process is arrested by formation of a nonoxidizable fragment (for instance, $ROCH_2CHO$, which would be produced from a 6-linked aldohexosyl group), the yield of formaldehyde at that stage determines the location of the stable structure in the polysaccharide chain.[272,273]

Under proper conditions, periodate oxidation of cellulose affords products rich in carboxylic acid groups[274a]; evaluation of the carboxyl content of oxycellulose can be complicated by processes of lactonization and by the presence of acidic, enediol groupings.[274b]

Lead tetraacetate has been used little for oxidation of polysaccharides.[6] Its ineffectiveness for this purpose probably stems mainly from the fact that the organic media commonly employed for the reagent are unable to dissolve polysaccharides. However, the use of methyl sulfoxide (containing about 10% of acetic acid), a good solvent for many polysaccharides, has permitted satisfactory oxidations to be conducted with lead tetraacetate.[275] Data thus obtained generally correspond closely to those obtained by periodate oxidation; however, the reaction rates are generally higher, and a more satisfactory recovery of the oxidized product appears to be feasible when methyl sulfoxide–acetic acid is the solvent.[275]

B. THE BARRY AND SMITH DEGRADATIONS

Many polysaccharides contain residues that do not have *vic*-diol groupings and that hence are inert to glycol-cleaving oxidants; such residues may be (1→3)-linked, function as branch points to which two or more other residues are joined, or be protected by substituents or by deoxygenation. Degradation of the partially oxidized polymers by removal of oxidized residues then facilitates study of the resistant parts of the molecules. Likewise, deliberate treatment of a polysaccharide with less than the theoretical amount of oxidant, and recovery of the unoxidized portions, affords a means of fragmenting the polymer.

Selective removal of oxidized residues (as in 118) may be effected in two principal ways: treatment with phenylhydrazine in hot, dilute acetic acid, according to Barry,[256] promotes rupture of the glycosidic bond with formation

118 106 119 120 121 93

of glyoxal bis(phenylhydrazone) (**106**), but leaves the unoxidized residue (**120**) intact; alternatively, reduction of **118** converts the dialdehyde residues into polyhydric alcohols (**119**) which, being acetals, are hydrolyzed by acid far faster than the glycopyranosyl group. The latter approach, developed by Smith and co-workers,[241,242] is the more satisfactory of the two (experimentally), but both procedures have provided many insights into the fine structure of many highly complex polysaccharides.

References start on p. 1207.

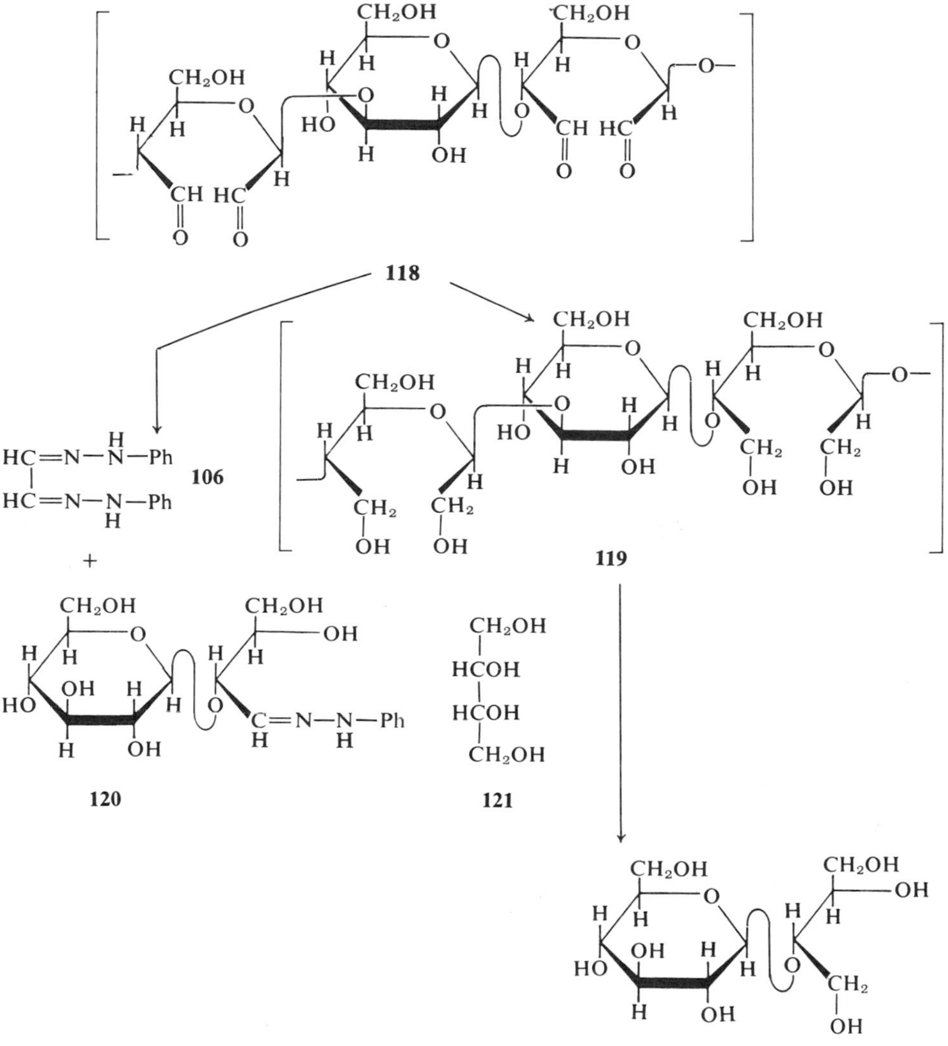

An early application[276] of the Barry procedure showed that snail galactan contains $(1\rightarrow3)$ and $(1\rightarrow6)$ linkages dispersed in a highly complex, dichotomous way. A succession of Barry degradations was applied, each degradation yielding a polymeric residue comprising only about half of the starting material, but still resembling the parent polysaccharide.

The value of the degradative approach is emphasized by the contrasting behavior of another type of $(1\rightarrow3)$, $(1\rightarrow6)$-linked D-glucopolysaccharide.[277,278] On the basis of the results of methylation analysis and periodate oxidation alone, the D-glucan could be regarded as structurally related to the snail galactan. However, degradation of the oxidized D-glucan afforded a polymeric residue that consumed very little periodate, and, hence, was affected little by a second degradation. Therefore, the $(1\rightarrow3)$ linkages are confined to a linear backbone of residues, whereas those of the $(1\rightarrow6)$ kind bind terminal glycosyl groups to the main chain.

Several other polysaccharides[279–284] show a stepwise response similar to that of snail galactan, in that polymeric products decreasing in size were isolated from successive Barry degradations. In these polysaccharides also, the periodate-resistant residues are assembled without interruption over large regions of a highly ramified molecule.

The degradative methods readily differentiate between these types of polysaccharides and those in which the nonoxidized residues are distributed more uniformly. For example, a xylan (from *Rhodomenia*), known to contain about equal proportions of $(1\rightarrow3)$ and $(1\rightarrow4)$ linkages, yields no polymeric products but gives D-*threo*-pentulose phenylosazone, showing that the linkages alternate uniformly along the polymer chain.[285] Similarly, nigeran was found to possess the same type of structure, but based on D-glucose; under the conditions of the Barry degradation it yielded D-*arabino*-hexulose phenylosazone,[286] whereas it was degraded[242] by the Smith method to 2-*O*-α-D-glucopyranosyl-D-erythritol (**93**). The arrangement of branching in beet arabinan and in certain arabinoxylans was determined in an analogous way. In these pentosans, the L-arabinofuranosyl residues are attached through O-3 (and sometimes O-2) of the residues in the main, pentosan chain, so that only residues constituting branch points are unoxidized by periodate. As the Barry degradation affords 3-*O*-α-L-arabinofuranosyl-1,3-dihydroxy-2-propanone phenylosazone from the arabinan, the branches of this polymer are on isolated arabinosyl residues;[287] the oxidized arabinoxylans, however, afford osazones of xylose, xylobiose, and xylotriose,[288] or, by the Smith procedure, 2-*O*-β-D-xylopyranosylglycerol and the corresponding di- and tri-D-xylosyl derivatives,[289] showing that the arabinosyl branches occur variously on isolated, adjacent, and three consecutive xylosyl residues.

Another example of the type of information obtainable is the degradation of a β-D-glucan from oat flour. Detection of 2-*O*-β-D-glucopyranosyl-D-erythritol (**93**) as a major product shows that $(1 \rightarrow 3)$- and $(1 \rightarrow 4)$-linked residues alternate, as in **118**. Erythritol (**121**) was another major product, which indicates that two (or more) adjacent, $(1 \rightarrow 4)$-linked D-glucosyl residues are also present. In addition, small proportions of 2-*O*-β-D-laminarabiosyl-D-erythritol and higher oligosaccharides of the series were formed, proving that the glucan contains occasional sequences of two, three, and more, consecutive, 3-*O*-substituted β-D-glucopyranosyl residues flanked by $(1 \rightarrow 4)$ links.[241,242]

Numerous other examples[241,242,289-291] further illustrate the great value of the Smith degradation in determinations of the fine structure of polysaccharides; however, minor variations on the classical procedure of Smith have also evolved. Dutton *et al.*[292] determined that gas–liquid chromatographic analyses of Smith degradation products [as per(trimethylsilyl) ethers] could be corrected, and calibrated, to provide a quantitative measure of the various products. A diminution in the degeneracy of fragments (various ring forms, anomers) was reported[293] to occur on oximation of the degraded product prior to etherification, and methanolysis prior to per(trimethylsilyl)ation has also been recommended.[294] Methylation of the oxidized product prior to reduction with borohydride differentiates free hydroxyl groups from reduced aldehydes.[295] Particular simplification derives from complete hydrolysis with acid, oximation, and forcing treatment with pyridine–acetic anhydride, which converts the degradation products into the well characterized aldononitrile peracetates.[296]

REFERENCES

1. L. Malaprade, *C.R. Acad. Sci.*, **186**, 382 (1928).
2. L. Malaprade, *Bull. Soc. Chim. Fr.*, [4] **43**, 683 (1928).
3. R. Criegee, *Ber.*, **64**, 260 (1931).
4. P. (F.) Fleury and J. Lange, *C.R. Acad. Sci.*, **195**, 1395 (1932).
5. J. M. Bobbitt, *Advan. Carbohyd. Chem.*, **11**, 1 (1956).
6. A. S. Perlin, *Advan. Carbohyd. Chem.*, **14**, 1 (1959).
6a. A. Haug, *Tidsskr. Kjemi, Bergvesen Met.*, **26**, 173 (1966); *Chem. Abstr.*, **66**, 46530 (1967); Z. Fialkiewiczowa, *Zesz. Nauk. Mat. Fiz. Chem. Wyzsza Szk. Pedagog. Gdansk.*, **7**, 179 (1967); *Chem. Abstr.*, **69**, 10625 (1968); Z. Fialkiewiczowa and J. Sokołowski, *ibid.*, **8**, 235 (1968); *Chem. Abstr.*, **71**, 61721 (1969); S. Fujibayashi, *Tampakushitsu Kakusan Koso, Bessatsu*, **11**, 151 (1968); *Chem. Abstr.*, **70**, 88138 (1969); G. Dryhurst, "Periodate Oxidation of Diol and Other Functional Groups: Analytical and Structural Applications," Pergamon, New York, 1970; A. J. Fatiadi, *Synthesis*, **4**, 229 (1974).
7. A comprehensive review is presented by C. A. Bunton, in "Oxidations in Organic Chemistry." Part A (K. Wiberg, ed.), Academic Press, New York, 1965, p. 367; see also, B. Sklarz, *Quart. Rev. Chem. Soc.*, **21**, 3 (1967).
8. L. J. Heidt, E. K. Gladding, and C. B. Purves, *Paper Trade J.*, **121**, 81 (1945).
9. G. J. Buist and C. A. Bunton, *J. Chem. Soc.*, 1406 (1954).
10. G. J. Buist, C. A. Bunton, and J. H. Miles, *J. Chem. Soc.*, 4575 (1957).

11. R. Criegee, L. Kraft, and B. Rank, *Ann.*, **507**, 159 (1933).

12. D. A. Rosenfeld, N. K. Richtmyer, and C. S. Hudson, *J. Amer. Chem. Soc.*, **70**, 2201 (1948).

13. S. J. Angyal and R. J. Young, *J. Amer. Chem. Soc.*, **81**, 5251, 5467 (1959).

14. R. Criegee, E. Büchner, and W. Walther, *Ber.*, **73**, 571 (1940).

15. R. E. Reeves, *J. Amer. Chem. Soc.*, **72**, 1499 (1950).

16. R. J. Dimler, H. A. Davis, and G. E. Hilbert, *J. Amer. Chem. Soc.*, **68**, 1377 (1946).

17. R. Criegee, E. Höger, G. Huber, P. Kruck, F. Marktscheffel, and H. Schellenberger, *Ann.*, **599**, 81 (1956).

18. C. C. Price and M. J. Knell, *J. Amer. Chem. Soc.*, **64**, 552 (1942).

19. G. J. Buist, C. A. Bunton, and J. H. Miles, *J. Chem. Soc.*, 743 (1959).

20. S. J. Angyal and C. G. MacDonald, *J. Chem. Soc.*, 686 (1952).

21. H. J. Backer, *Rec. Trav. Chim.* (Pays-Bas), **57**, 967 (1938).

22. J. C. P. Schwarz, *J. Chem. Soc.*, 276 (1957).

23. (*a*) G. R. Barker and D. F. Shaw, *J. Chem. Soc.*, 584 (1959); (*b*) K. Dijkstra, *Rec. Trav. Chim.* (Pays-Bas), **87**, 181 (1967); (*c*) P. Szabó, *Bull. Soc. Chim. Biol.*, **51**, 403 (1969); (*d*) S. R. Sarfati and P. Szabó, *Carbohyd. Res.*, **11**, 571 (1969); S. J. Angyal, D. Greeves, and V. A. Pickles, *ibid.*, **35**, 165 (1974).

24. A. S. Perlin and E. von Rudloff, *Can. J. Chem.*, **43**, 2071 (1965).

25. H. R. Goldschmid and A. S. Perlin, *Can. J. Chem.*, **38**, 2280 (1960).

26. C. A. Grob and P. W. Schiess, *Helv. Chim. Acta*, **43**, 1546 (1960).

27. R. Criegee and E. Büchner, *Ber.*, **73**, 563 (1940).

28. See R. Criegee, in "Oxidations in Organic Chemistry," Part A, (K. Wiberg, ed.), Academic Press, New York, 1965, p. 278.

28a. J. E. Scott and M. J. Tigwell, *Biochem. J.*, **123**, 46P (1971).

28b. T. P. Nevell and J. S. Shaw, *Chem. Ind.* (London), 772 (1968); R. G. Krylova, S. N. Ryadovskaya, L. I. Kostelian, and A. I. Usov, *Izv. Akad. Nauk SSSR, Ser. Khim.*, 2068 (1972).

29. Critical evaluations of methods commonly used for periodate and lead tetraacetate oxidations are presented in A. S. Perlin, *Methods Carbohyd. Chem.*, **1**, 427 (1962); R. D. Guthrie, *Methods Carbohyd. Chem.*, **1**, 432, 435, 445 (1962); J. C. Speck, Jr., *Methods Carbohyd. Chem.*, **1**, 441 (1962); A. S. Perlin, in "Oxidative Techniques and Applications in Organic Synthesis," Vol. 1 (R. L. Augustine, ed.), Dekker, New York, 1969, p. 189.

30. P. W. Clutterbuck and F. Reuter, *J. Chem. Soc.*, 1467 (1935).

31. E. Baer, R. Grosheintz, and H. O. L. Fischer, *J. Amer. Chem. Soc.*, **61**, 2607 (1939).

32. E. Baer, *J. Amer. Chem. Soc.*, **62**, 1597 (1940).

33. B. H. Nicolet and L. A. Shinn, *J. Amer. Chem. Soc.*, **61**, 1615 (1939).

34. P. F. Fleury, G. Poirot, and Y. Fievet, *C.R. Acad. Sci.*, **220**, 664 (1945).

35. C. Brice and A. S. Perlin, *Can. J. Biochem. Physiol.*, **35**, 7 (1957).

36. R. C. Hockett and W. S. McClenahan, *J. Amer. Chem. Soc.*, **61**, 1667 (1939).

37. J. C. P. Schwarz, *Chem. Ind.* (London), 1388 (1955).

38. A. S. Perlin, *Can. J. Chem.*, **42**, 2365 (1964).

38a. G. Dahlgren and K. L. Reed, *J. Amer. Chem. Soc.*, **89**, 1380 (1967).

39. R. Criegee, *Ann.*, **495**, 211 (1932).

40. D. B. Sprinson and E. Chargaff, *J. Biol. Chem.*, **164**, 433 (1946).

41. R. E. Reeves, *J. Amer. Chem. Soc.*, **63**, 1476 (1941).

42. D. A. MacFadyen, *J. Biol. Chem.*, **158**, 107 (1945).

42a. L. I. Kudryashov, M. A. Chlenov, P. N. Smirnov, and S. D. Kovacheva, *Zh. Obshch. Khim.*, **38**, 74 (1968).

43. G. Lindstedt, *Nature*, **156**, 448 (1945).

44. T. G. Halsall, E. L. Hirst, and J. K. N. Jones, *J. Chem. Soc.*, 1399, 1427 (1947).
45. M. Morrison, A. C. Kuyper, and J. M. Orten, *J. Amer. Chem. Soc.*, 75, 1502 (1953).
45a. J. F. Kennedy, *Methods Carbohyd. Chem.*, 6, 93 (1972).
46. A. S. Perlin, *J. Amer. Chem. Soc.*, 76, 5505 (1954).
47. A. S. Perlin, *Can. J. Chem.*, 42, 1365 (1964).
48. W. Mackie and A. S. Perlin, *Can. J. Chem.*, 43, 2645 (1965).
48a. A. M. Y. Ko and P. J. Somers, *Carbohyd. Res.*, 34, 57 (1974).
49. L. J. Haynes, *Advan. Carbohyd. Chem.*, 18, 227 (1963).
50. J. B. Pridham, *Advan. Carbohyd. Chem.*, 20, 371 (1965).
51. R. C. Hockett and H. G. Fletcher, Jr., *J. Amer. Chem. Soc.*, 66, 469 (1944).
52. A. S. Perlin, *Can. J. Chem.*, 41, 555 (1963).
53. P. J. Garegg, *Ark. Kemi*, 23, 255 (1964).
54. R. E. Reeves, *J. Amer. Chem. Soc.*, 71, 2868 (1949).
55. P. E. Verkade, *Rec. Trav. Chim.* (Pays-Bas), 57, 824 (1938).
56. R. C. Hockett and D. F. Mowery, Jr., *J. Amer. Chem. Soc.*, 65, 403 (1943).
57. C. Schöpf and A. Schmetterling, *Angew. Chem.*, 64, 591 (1952).
58. H. Fiedler, Doctoral Dissertation, Technischen Hochschule, Karlsruhe (1959), p. 38.
59. P. F. Fleury and J. Lange, *J. Pharm. Chim.*, [8] 17, 107 (1933).
60. H. H. Willard and L. H. Greathouse, *J. Amer. Chem. Soc.*, 60, 2869 (1938).
61. O. Dimroth and R. Schweizer, *Ber.*, 56, 1375 (1923).
61a. J. Elting, C.-C. Huang, and R. Montgomery, *Carbohyd. Res.*, 28, 387 (1973).
62. C. E. Crouthamel, H. V. Meek, D. S. Martin, and C. V. Banks, *J. Amer. Chem. Soc.*, 71, 3031 (1949).
63. J. S. Dixon and D. Lipkin, *Anal. Chem.*, 26, 1092 (1954).
64. G. O. Aspinall and R. J. Ferrier, *Chem. Ind.* (London), 1216 (1956); J. X. Khym, *Methods Carbohyd. Chem.*, 6, 87 (1972).
65. D. Benson, L. H. Sutcliffe, and J. Walkley, *J. Amer. Chem. Soc.*, 81, 4488 (1959).
66. A. S. Perlin and S. Suzuki, *Can. J. Chem.*, 40, 1226 (1962).
66a. G. Avigad, *Carbohyd. Res.*, 11, 119 (1969).
66b. R. D. Corlett, W. G. Breck, and G. W. Hay, *Can. J. Chem.*, 48, 2474 (1970).
67. F. S. H. Head, *Nature*, 165, 236 (1950).
68. C. F. Huebner, R. L. Lohmar, R. J. Dimler, S. Moore, and K. P. Link, *J. Biol. Chem.*, 159, 503 (1945).
69. C. F. Huebner, S. R. Ames, and E. C. Bubl, *J. Amer. Chem. Soc.*, 68, 1621 (1946).
70. P. (F.) Fleury and J. (E.) Courtois, *C.R. Acad. Sci.*, 223, 633 (1946).
71. M. Cantley, L. Hough, and A. O. Pittet, *J. Chem. Soc.*, 2527 (1963).
71a. J. E. Scott and M. J. Tigwell, *Carbohyd. Res.*, 28, 53 (1973).
72. R. W. Jeanloz, *Helv. Chim. Acta*, 27, 1509 (1944).
73. D. J. Bell, *J. Chem. Soc.*, 992 (1948).
74. G. D. Greville and D. H. Northcote, *J. Chem. Soc.*, 1945 (1952).
75. M. L. Wolfrom and J. M. Bobbitt, *J. Amer. Chem. Soc.*, 78, 2489 (1956).
75a. A. J. Fatiadi, *J. Res. Nat. Bur. Stand.*, 72A, 341 (1968).
76. J. C. P. Schwarz and M. MacDougall, *J. Chem. Soc.*, 3065 (1956).
77. J. L. Bose, A. B. Foster, and R. W. Stephens, *J. Chem. Soc.*, 3314 (1959).
78. S. J. Angyal and J. E. Klavins, *Aust. J. Chem.*, 14, 577 (1961).
79. B. G. Hudson and R. Barker, *J. Org. Chem.*, 32, 2101 (1967).
80. J. B. Lee, *J. Chem. Soc.*, 1474 (1960).
81. R. Criegee, *Ann.*, 481, 263 (1920).
82. P. F. Fleury, J. E. Courtois, R. Perlès, and L. Le Dizet, *Bull. Soc. Chim. Fr.*, 347 (1954).

82a. D. Charon and L. Szabó, *Carbohyd. Res.*, **34**, 271 (1974); *Eur. J. Biochem.*, **29**, 184 (1972).

83. F. Knoop, F. Ditt, W. Hecksteden, J. Maier, W. Merz, and R. Härle, *Z. Physiol. Chem.*, **239**, 30 (1936).

84. R. Criegee, *Angew. Chem.*, **53**, 321 (1940).

85. A. Neuberger, *J. Chem. Soc.*, 47 (1941).

86. C. Niemann and J. T. Hays, *J. Amer. Chem. Soc.*, **67**, 1302 (1945).

87. M. L. Wolfrom, R. U. Lemieux, and S. M. Olin, *J. Amer. Chem. Soc.*, **71**, 2870 (1949).

88. A. B. Foster and D. Horton, *J. Chem. Soc.*, 1890 (1958).

89. P. Karrer and J. Mayer, *Helv. Chim. Acta*, **20**, 407 (1937).

90. C. L. Stevens, K. Nagarajan, and T. H. Haskell, *J. Org. Chem.*, **27**, 2991 (1962).

91. E. H. Flynn, M. V. Sigal, Jr., P. F. Wiley, and K. Gerzon, *J. Amer. Chem. Soc.*, **76**, 3121 (1954).

92. F. A. Hochstein and P. P. Regna, *J. Amer. Chem. Soc.*, **77**, 3353 (1955).

93. M. J. Weiss, J. P. Joseph, H. M. Kissman, A. M. Small, R. E. Schaub, and F. J. McEvoy, *J. Amer. Chem. Soc.*, **81**, 4050 (1959).

94. J. Honeyman and C. J. G. Shaw, *J. Chem. Soc.*, 2454 (1959).

94a. C. B. Barlow and R. D. Guthrie, *Carbohyd. Res.*, **13**, 199 (1970).

94b. C. B. Barlow and R. D. Guthrie, *Carbohyd. Res.*, **11**, 53 (1969).

94c. J. M. J. Tronchet, F. Rachidzadeh, and J. Tronchet, *Helv. Chim. Acta*, **57**, 65 (1974).

95. W. A. Bonner and R. W. Drisko, *J. Amer. Chem. Soc.*, **73**, 3699 (1951).

96. L. Hough and M. I. Taha, *J. Chem. Soc.*, 2042 (1956).

97. L. Hough and M. I. Taha, *J. Chem. Soc.*, 3994 (1957).

98. D. Horton and D. H. Hutson, *Advan. Carbohyd. Chem.*, **18**, 123 (1963).

99. O. T. Schmidt and E. Wernicke, *Ann.*, **556**, 179 (1944).

100. E. J. Bourne, W. M. Corbett, M. Stacey, and R. Stephens, *Chem. Ind.* (London), 106 (1954).

101. W. M. Doane, B. S. Shasha, C. R. Russell, and C. E. Rist, *J. Org. Chem.*, **30**, 3071 (1965).

101a. C. J. Clayton and N. A Hughes, *Carbohyd. Res.*, **27**, 89 (1973).

102. G. V. Marinetti and G. Rouser, *J. Amer. Chem. Soc.*, **77**, 5345 (1955).

103. D. Grant and A. Holt, *J. Chem. Soc.*, 5026 (1960).

104. J. R. Turvey, M. J. Clancy, and T. P. Williams, *J. Chem. Soc.*, 1692 (1961).

104a. S. Peat, D. M. Bowker, and J. R. Turvey, *Carbohyd. Res.*, **7**, 225 (1968).

105. A. J. Cleaver, A. B. Foster, E. J. Hedgley, and W. G. Overend, *J. Chem. Soc.*, 2578 (1959).

105a. G. Just and A. Martel, *Tetrahedron Lett.*, 1517 (1973).

105b. G. Just and K. Grozinger, *Tetrahedron Lett.*, 4165 (1974).

106. B. H. Alexander, R. J. Dimler, and C. L. Mehltretter, *J. Amer. Chem. Soc.*, **73**, 4658 (1951).

107. R. J. Dimler, *Advan. Carbohyd. Chem.*, **7**, 37 (1952).

108. L. C. Stewart, E. Zissis, and N. K. Richtmyer, *J. Org. Chem.*, **28**, 1842 (1963).

109. S. J. Angyal and R. M. Hoskinson, *J. Chem. Soc.*, 2043 (1963).

110. R. Partch, *Tetrahedron Lett.*, 3071 (1964).

111. A. S. Perlin, unpublished (1965).

112. J. P. Cordner and K. H. Pausacker, *J. Chem. Soc.*, 102 (1953).

113. R. K. Hulyalkar and M. B. Perry, *Can. J. Chem.*, **43**, 3241 (1965).

114. E. F. Garner, I. J. Goldstein, R. Montgomery, and F. Smith, *J. Amer. Chem. Soc.*, **80**, 1206 (1958).

115. A. S. Perlin, unpublished (1966).
115a. J. J. M. Rowe, K. B. Gibney, M. T. Yang, and G. G. S. Dutton, *J. Amer. Chem. Soc.*, **90**, 1924 (1968).
115b. R. J. Yu and C. T. Bishop, *Can. J. Chem.*, **45**, 2195 (1967).
115c. J. Gelas, D. Horton, and J. D. Wander, *J. Org. Chem.*, **39**, 1946 (1974).
116. A. S. Perlin and C. Brice, *Can. J. Chem.*, **34**, 541 (1956).
117. A. S. Perlin and A. R. Lansdown, *Can. J. Chem.*, **34**, 451 (1956).
118. A. P. Tulloch, A. Hill, and J. F. T. Spencer, *Can. J. Chem.*, **46**, 3337 (1968).
119. R. U. Lemieux and R. Nagarajan, *Can. J. Chem.*, **42**, 1270 (1964).
119a. L.-Å. Fransson, *Carbohyd. Res.*, **36**, 339 (1974).
120. E. L. Jackson and C. S. Hudson, *J. Amer. Chem. Soc.*, **60**, 989 (1938).
121. D. H. Grangaard, E. K. Gladding, and C. B. Purves, *Paper Trade J.*, **115**, 75 (1942).
121a. B. Larsen and T. J. Painter, *Carbohyd. Res.*, **10**, 186 (1969).
121b. O. Smidsrød and T. (J.) Painter, *Carbohyd. Res.*, **26**, 125 (1973).
121c. T. J. Painter and B. Larsen, *Acta Chem. Scand.*, **24**, 813, 2366 (1970).
121d. T. J. Painter and B. Larsen, *Acta Chem. Scand.*, **24**, 2724 (1970); O. Smidsrød, B. Larsen, and T. (J.) Painter, *ibid.*, **24**, 3201 (1970).
121e. M. Fahmy Ishak and T. (J.) Painter, *Acta Chem. Scand.*, **25**, 3875 (1971); **27**, 1321 (1973).
122. G. Carrara, *Giorn. Chim. Ind. Ed. Appl.*, **14**, 236 (1932); *Chem. Abstr.*, **26**, 5069 (1932).
123. P. Brigl and H. Grüner, *Ber.*, **66**, 931 (1933).
124. H. O. L. Fischer and E. Baer, *Helv. Chim. Acta*, **17**, 622 (1934).
125. M. Steiger and T. Reichstein, *Helv. Chim. Acta*, **19**, 1016 (1936).
126. S. A. Barker and E. J. Bourne, *Advan. Carbohyd. Chem.*, **7**, 137 (1952).
127. E. Baer, *J. Amer. Oil Chem. Soc.*, **42**, 257 (1965).
127a. W. J. Lloyd and R. Harrison, *Carbohyd. Res.*, **26**, 91 (1973).
128. K. Gätzi and T. Reichstein, *Helv. Chim. Acta*, **21**, 195 (1938).
129. J. E. Courtois and M. Guernet, *Bull. Soc. Chim. Fr.*, 1388 (1957).
130. A. S. Perlin, unpublished (1954).
131. R. C. Hockett, M. T. Dienes, H. G. Fletcher, Jr., and H. E. Ramsden, *J. Amer. Chem. Soc.*, **66**, 467 (1944).
132. P. A. J. Gorin and A. S. Perlin, *Can. J. Chem.*, **36**, 480 (1958).
133. A. M. Stephen, *J. Chem. Soc.*, 738 (1952).
134. P. (F.) Fleury and L. Le Dizet, *Bull. Soc. Chim. Biol.*, **37**, 1099 (1955).
135. S. J. Angyal and L. Anderson, *Advan. Carbohyd. Chem.*, **14**, 135 (1959).
136. S. J. Angyal and D. J. McHugh, *J. Chem. Soc.*, 1423 (1957).
137. T. Posternak, *Helv. Chim. Acta*, **33**, 1597 (1950).
138. G. E. McCasland and D. A. Smith, *J. Amer. Chem. Soc.*, **73**, 516 (1951).
139. A. K. Kiang and K. H. Loke, *J. Chem. Soc.*, 480 (1956).
140. H. Z. Sable and A. S. Perlin, unpublished (1965).
141. G. E. McCasland and E. C. Horswill, *J. Amer. Chem. Soc.*, **76**, 2373 (1954).
142. G. E. McCasland, S. Furuta, L. F. Johnson, and J. N. Shoolery, *J. Org. Chem.*, **29**, 2354 (1964).
143. H. O. L. Fischer and G. Dangschat, *Helv. Chim. Acta*, **17**, 1200 (1934).
144. R. C. Hockett, M. H. Nickerson, and W. H. Reeder, III, *J. Amer. Chem. Soc.*, **66**, 472 (1944).
145. M. L. Wolfrom and K. Anno, *J. Amer. Chem. Soc.*, **75**, 1038 (1953).
146. O. Kjølberg, *Acta Chem. Scand.*, **14**, 1118 (1960).
147. E. Berner and O. Kjølberg, *Acta Chem. Scand.*, **14**, 909 (1960).
147a. P. Kohn, L. M. Lerner, and B. D. Kohn, *J. Org. Chem.*, **32**, 4076 (1967).

148. R. C. Hockett, M. Conley, M. Yusem, and R. I. Mason, *J. Amer. Chem. Soc.*, **68**, 922 (1946).
149. R. C. Hockett, H. G. Fletcher, Jr., E. Sheffield, R. M. Goepp, Jr., and S. Soltzberg, *J. Amer. Chem. Soc.*, **68**, 930 (1946).
150. J. N. Baxter and A. S. Perlin, *Can. J. Chem.*, **38**, 2217 (1960).
151. M. Mazurek and A. S. Perlin, *Can. J. Chem.*, **41**, 2403 (1963).
152. E. L. Jackson and C. S. Hudson, *J. Amer. Chem. Soc.*, **59**, 994 (1937).
153. W. D. Maclay and C. S. Hudson, *J. Amer. Chem. Soc.*, **60**, 2059 (1938).
154. E. L. Jackson and C. S. Hudson, *J. Amer. Chem. Soc.*, **61**, 959 (1939).
155. E. L. Jackson and C. S. Hudson, *J. Amer. Chem. Soc.*, **63**, 1229 (1941).
156. This sequence of oxidations may be used for determining the distribution of isotope in ^{14}C-labeled aldohexoses and aldopentoses; B. Boothroyd, S. A. Brown, J. A. Thorn, and A. C. Neish, *Can. J. Biochem. Physiol.*, **33**, 62 (1955); S. A. Brown, *ibid.* **33**, 368 (1955).
157. J. M. Grosheintz, *J. Amer. Chem. Soc.*, **61**, 3379 (1939).
158. J. Davoll, B. Lythgoe, and A. R. Todd, *J. Chem. Soc.*, 833 (1946).
159. L. Vargha and T. Puskaš, *Ber.*, **76**, 859 (1943).
160. R. C. Hockett, M. Zief, and R. M. Goepp, Jr., *J. Amer. Chem. Soc.*, **68**, 935 (1946).
161. A. J. Charlson and A. S. Perlin, *Can. J. Chem.*, **34**, 1804 (1956).
161a. A. B. Zanlungo, J. O. Deferrari, M. E. Gelpi, and R. A. Cadenas, *Carbohyd. Res.*, **35**, 33 (1974).
162. M. Abdel-Akher, J. E. Cadotte, B. A. Lewis, R. Montgomery, F. Smith, and J. W. Van Cleve, *Nature*, **171**, 474 (1953).
163. F. Smith and J. W. Van Cleve, *J. Amer. Chem. Soc.*, **77**, 3091 (1955).
164. C. H. Shunk, C. H. Stammer, E. A. Kaczka, E. Walton, C. F. Spencer, A. N. Wilson, J. W. Richter, F. W. Holly, and K. Folkers, *J. Amer. Chem. Soc.*, **78**, 1770 (1956).
165. D. M. Lemal, P. D. Pacht, and R. B. Woodward, *Tetrahedron*, **18**, 1275 (1962).
166. R. U. Lemieux and J. Howard, *Can. J. Chem.*, **41**, 308 (1963).
167. R. C. Hockett, M. T. Dienes, and H. E. Ramsden, *J. Amer. Chem. Soc.*, **65**, 1474 (1943); S. Honda, N. Hamajima, and K. Kaheki, *Carbohyd. Res.*, **68**, 77 (1979).
168. W. Freudenberg and E. F. Rogers, *J. Amer. Chem. Soc.*, **59**, 1602 (1937).
168a. B. Coxon and L. Hough, *Carbohyd. Res.*, **8**, 379 (1968).
169. D. M. Brown and B. Lythgoe, *J. Chem. Soc.*, 1990 (1950).
170. L. A. Manson and J. O. Lampen, *J. Biol. Chem.*, **191**, 87 (1951); for a review of applications involving nucleosides and nucleic acids, see G. Schmidt, *Methods Enzymol.*, **12**, 230 (1968).
171. L. J. Haynes, *Advan. Carbohyd. Chem.*, **20**, 357 (1965).
172. L. H. Briggs and L. C. Vining, *J. Chem. Soc.*, 2809 (1953).
173. A. K. Mitra and A. S. Perlin, *Can. J. Chem.*, **35**, 1079 (1957).
174. A. K. Mitra and A. S. Perlin, *Can. J. Chem.*, **37**, 2047 (1959).
175. R. D. Guthrie, *Advan. Carbohyd. Chem.*, **16**, 105 (1961).
175a. J. H. Pazur and L. S. Forsberg, *Carbohyd. Res.*, **58**, 222 (1977).
176. V. C. Barry and P. W. D. Mitchell, *J. Chem. Soc.*, 3631 (1953).
177. R. D. Guthrie and J. Honeyman, *Chem. Ind.* (London), 388 (1958).
178. I. J. Goldstein, B. A. Lewis, and F. Smith, *J. Amer. Chem. Soc.*, **80**, 939 (1958).
179. H. Greenberg and A. S. Perlin, *Carbohyd. Res.*, **35**, 195 (1974).
180. I. J. Goldstein and F. Smith, *J. Amer. Chem. Soc.*, **82**, 3421 (1960).
181. A. S. Perlin, *Can. J. Chem.*, **44**, 1757 (1966).
182. J. W. Rowen, F. H. Forziatti, and R. E. Reeves, *J. Amer. Chem. Soc.*, **73**, 4484 (1951).

183. I. J. Goldstein, B. A. Lewis, and F. Smith, *Chem. Ind.* (London), 595 (1958).
184. R. D. Guthrie and J. Honeyman, *J. Chem. Soc.*, 2441 (1959).
185. A. S. Perlin, *Can. J. Chem.*, **44**, 539 (1966).
186. S. J. Angyal and S. D. Gero, *Aust. J. Chem.*, **18**, 1973 (1965).
187. C. D. Hurd, P. J. Baker, Jr., R. P. Holysz, and W. H. Saunders, Jr., *J. Org. Chem.*, **18**, 186 (1953).
188. L. Mester and E. Moczar, *Chem. Ind.* (London), 761 (1957).
189. I. J. Goldstein and F. Smith, *J. Amer. Chem. Soc.*, **80**, 4681 (1958).
190. M. Guernet, A. Jurado Soler, and P. Malangeau, *Bull. Soc. Chim. Fr.*, 1188 (1963).
191. M. Cantley, J. R. Holker, and L. Hough, *J. Chem. Soc.*, 1555 (1965).
192. A. S. Jones and R. T. Walker, *Carbohyd. Res.*, **26**, 255 (1973).
193. D. H. Grangaard, J. H. Michell, and C. B. Purves, *J. Amer. Chem. Soc.*, **61**, 1290 (1939).
194. G. F. Davidson, *J. Text. Inst.*, **31**, T81 (1940).
195. G. Jayme, M. Sätre, and S. Maris, *Naturwissenschaften*, **29**, 768 (1941).
196. C. G. Caldwell and R. M. Hixon, *J. Biol. Chem.*, **123**, 595 (1938).
197. H. A. Rutherford, F. W. Minor, A. R. Martin, and M. Harris, *J. Res. Nat. Bur. Stand.*, **29**, 131 (1942).
198. F. S. H. Head, *J. Text. Inst.*, **44**, T209 (1953).
199. W. Dvonch and C. L. Mehltretter, *J. Amer. Chem. Soc.*, **74**, 5522 (1952).
200. W. Dvonch, H. Fletcher, III, F. J. Gregory, E. M. H. Healy, G. H. Warren, and H. E. Alburn, *Cancer Res.*, **26**, 2386 (1966).
201. H. H. Baer and H. O. L. Fischer, *Proc. Natl. Acad. Sci. U.S.*, **44**, 991 (1958); H. H. Baer, *Methods Carbohyd. Chem.*, **6**, 245 (1972).
202. K. W. Buck, F. A. Fahim, A. B. Foster, A. R. Perry, M. H. Qadir, and J. M. Webber, *Carbohyd. Res.*, **2**, 14 (1966).
203. A. S. Perlin, *J. Amer. Chem. Soc.*, **76**, 2595 (1954).
204. A. S. Perlin and C. Brice, *Can. J. Chem.*, **33**, 1216 (1955).
205. C. Schöpf and H. Wild, *Ber.*, **87**, 1571 (1954).
206. S. A. Warsi and W. J. Whelan, *Chem. Ind.* (London), 71 (1958).
207. F. S. H. Head, *Chem. Ind.* (London), 360 (1958).
208. L. Hough, T. J. Taylor, G. H. S. Thomas, and B. M. Woods, *J. Chem. Soc.*, 1212 (1958).
209. G. Hughes and T. P. Nevell, *Trans. Faraday Soc.*, **44**, 941 (1948).
210. G. R. Barker and D. C. C. Smith, *Chem. Ind.* (London), 1035 (1952).
211. Y. Khouvine and G. Arragon, *Bull. Soc. Chim. Fr.* [5] **8**, 676 (1941).
212. A. S. Perlin and C. Brice, *Can. J. Chem.*, **34**, 85, 541 (1956).
213. H. H. Sephton and N. K. Richtmyer, *J. Org. Chem.*, **28**, 1691 (1963).
214. A. J. Charlson and N. K. Richtmyer, *J. Amer. Chem. Soc.*, **81**, 1512 (1959).
215. H. H. Sephton and N. K. Richtmyer, *J. Org. Chem.*, **28**, 2388 (1963).
216. N. J. Antia and M. B. Perry, *Can. J. Chem.*, **38**, 1917 (1960).
217. J. F. Mahoney and C. B. Purves, *J. Amer. Chem. Soc.*, **64**, 9 (1942).
218. T. E. Timell, *Sv. Papperstidn.*, **51**, 52 (1948).
219. R. U. Lemieux and H. F. Bauer, *Can. J. Chem.*, **31**, 814 (1953).
220. A. S. Perlin, *Anal. Chem.*, **27**, 396 (1955).
221. A. J. Charlson, P. A. J. Gorin, and A. S. Perlin, *Can. J. Chem.*, **34**, 1811 (1956).
222. P. A. J. Gorin and A. S. Perlin, *Can. J. Chem.*, **34**, 693 (1956).
223. J. N. Baxter, A. S. Perlin, and F. J. Simpson, *Can. J. Biochem. Physiol.*, **37**, 199 (1959).
224. A. S. Sieben, A. S. Perlin, and F. J. Simpson, *Can. J. Biochem. Physiol.*, **44**, 663 (1966).

225. V. Klybas, M. Schramm, and E. Racker, *Arch. Biochem. Biophys.*, **80**, 229 (1959).
226. K. Ahlborg, *Sv. Kem. Tidskr.*, **54**, 205 (1942).
227. K. H. Meyer and P. Rathgeb, *Helv. Chim. Acta*, **32**, 1102 (1949).
228. M. L. Wolfrom, A. Thompson, A. N. O'Neill, and T. T. Galkowski, *J. Amer. Chem. Soc.*, **74**, 1062 (1952).
229. G. Neumüller and E. Vasseur, *Ark. Kemi*, **5**, 235 (1953).
230. A. N. O'Neill, *J. Amer. Chem. Soc.*, **76**, 5074 (1954).
231. L. Hough and M. B. Perry, *Chem. Ind.* (London), 768 (1956).
232. M. J. Clancy and W. J. Whelan, *Chem. Ind.* (London), 673 (1959).
233. R. W. Bailey and J. B. Pridham, *Advan. Carbohyd. Chem.*, **17**, 121 (1962).
234. G. Schiffman, E. A. Kabat, and S. Leskowitz, *J. Amer. Chem. Soc.*, **84**, 73 (1962).
235. S. A. Barker, A. B. Foster, M. Stacey, and J. M. Webber, *J. Chem. Soc.*, 2218 (1958).
236. A. B. Foster, R. Harrison, T. D. Inch, M. Stacey, and J. M. Webber, *J. Chem. Soc.*, 2279 (1963).
237. A. J. Charlson, P. A. J. Gorin, and A. S. Perlin, *Can. J. Chem.*, **35**, 365 (1957).
238. F. W. Parrish, A. S. Perlin, and E. T. Reese, *Can. J. Chem.*, **38**, 2094 (1960).
239. B. Lindberg, *Acta Chem. Scand.*, **9**, 1093 (1955).
240. P. A. J. Gorin and A. S. Perlin, *Can. J. Chem.*, **35**, 262 (1957).
241. I. J. Goldstein, G. W. Hay, B. A. Lewis, and F. Smith, *Abstr. Papers Amer. Chem. Soc. Meeting*, **135**, 3D (1959).
242. I. J. Goldstein, G. W. Hay, B. A. Lewis, and F. Smith, *Methods Carbohyd. Chem.*, **5**, 361 (1965).
243. N. K. Kochetkov, A. Ya. Khorlin, and V. J. Chirva, *Tetrahedron Lett.*, 2201 (1965).
244. P. A. J. Gorin, J. F. T. Spencer, and H. J. Phaff, *Can. J. Chem.*, **42**, 1341 (1964).
245. A. K. Mitra and A. S. Perlin, unpublished (1958), cited in ref. 5.
246. A. I. Usov and M. A. Rekhter, *Zh. Obshch. Khim.*, **39**, 912 (1969).
247. A. R. Archibald and J. G. Buchanan, *Carbohyd. Res.*, **11**, 558 (1969).
248. T. G. Halsall, E. L. Hirst, J. K. N. Jones, and A. Roudier, *Nature*, **160**, 899 (1947).
249. A. L. Potter and W. Z. Hassid, *J. Amer. Chem. Soc.*, **70**, 3488 (1948).
250. G. C. Gibbons and R. A. Boissonnas, *Helv. Chim. Acta*, **33**, 1477 (1950).
251. M. Schlamowitz, *J. Biol. Chem.*, **188**, 145 (1951).
252. M. Abdel-Akher and F. Smith, *J. Amer. Chem. Soc.*, **73**, 994 (1951).
253. D. J. Bell and D. J. Manners, *J. Chem. Soc.*, 1891 (1954).
254. S. K. Chanda, E. L. Hirst, J. K. N. Jones, and E. G. V. Percival, *J. Chem. Soc.*, 1289 (1950).
255. M. Abdel-Akher, J. K. Hamilton, R. Montgomery, and F. Smith, *J. Amer. Chem. Soc.*, **74**, 4970 (1952).
256. V. C. Barry, *Nature*, **152**, 538 (1943).
257. J. D. Moyer and H. S. Isbell, *Anal. Chem.*, **29**, 1862 (1957).
258. H. S. Isbell, *J. Res. Nat. Bur. Stand.*, **32**, 45 (1944).
259. G. F. Davidson and T. P. Nevell, *J. Text. Inst.*, **39**, T102 (1948).
260. D. O'Meara and G. N. Richards, *J. Chem. Soc.*, 4504 (1958).
261. R. L. Whistler, P. K. Chang, and G. N. Richards, *J. Amer. Chem. Soc.*, **81**, 3133 (1959); compare I.-L. Andresen, T. J. Painter, and O. Smisrød, *Carbohyd. Res.*, **59**, 563 (1977).
261a. M. F. Ishak and T. J. Painter, *Carbohyd. Res.*, **64**, 189 (1978).
262. F. W. Parrish and A. S. Perlin, unpublished (1960).
262a. J. E. Scott and R. J. Harbinson, *Histochemie*, **19**, 155 (1969); *Chem. Abstr.*, **72**, 3695 (1970).
263. F. Brown, S. Dunstan, T. G. Halsall, E. L. Hirst, and J. K. N. Jones, *Nature*, **156**, 785 (1945).

264. D. J. Manners and A. R. Archibald, *J. Chem. Soc.*, 2205 (1957).
265. O. Kjølberg, D. J. Manners, and A. Wright. *Comp. Biochem. Physiol.*, **8**, 353 (1963).
266. A. L. Potter, W. Z. Hassid, and M. A. Joslyn, *J. Amer. Chem. Soc.*, **71**, 4075 (1949).
267. I. C. MacWilliam and E. G. V. Percival, *J. Chem. Soc.*, 2259 (1951).
268. A. (R.) Jeanes and C. A. Wilham, *J. Amer. Chem. Soc.*, **72**, 2655 (1950).
269. J. C. Rankin and A. (R.) Jeanes, *J. Amer. Chem. Soc.*, **76**, 4435 (1954).
270. A. (R.) Jeanes, W. C. Haynes, C. A. Wilham, J. C. Rankin, E. H. Melvin, M. J. Austin, J. E. Clusky, B. E. Fischer, H. M. Tsuchiya, and C. E. Rist, *J. Amer. Chem. Soc.*, **76**, 5041 (1954).
271. I. J. Goldstein, J. K. Hamilton, and F. Smith, *J. Amer. Chem. Soc.*, **81**, 6252 (1959).
272. L. Hough and B. M. Woods, *Chem. Ind.* (London), 1421 (1957).
273. L. Hough, *Methods Carbohyd. Chem.*, **5**, 370 (1965).
274. F. W. Parrish and W. J. Whelan, *Nature*, **183**, 991 (1959).
274a. K. Kimov and V. Lateva, *Vysokomol. Soedin.*, *Ser. A*, **9**, 1646 (1967); *Chem. Abstr.* **67**, 101,149 (1967).
274b. G. M. Nabar and V. A. Shenai, *J. Appl. Polym. Sci.*, **14**, 1215 (1970); *Chem. Abstr.* **73**, 46,841 (1970).
275. V. Zitko and C. T. Bishop, *Can. J. Chem.*, **44**, 1749 (1966).
276. P. S. O'Colla, *Proc. Roy. Irish Acad.*, *Sec. B*, **55**, 165 (1952).
277. A. S. Perlin and W. A. Taber, *Can. J. Chem.*, **41**, 2278 (1963).
278. J. Johnson, S. Kirkwood, A. Misoki, T. E. Nelson, J. V. Scaletti, and F. Smith, *Chem. Ind.* (London), 820 (1963).
279. T. Dillon, D. F. O'Ceallachain, and P. S. O'Colla, *Proc. Roy. Irish Acad.*, *Sec. B*, **55**, 331 (1952); **57**, 31 (1954).
280. E. L. Hirst, J. J. O'Connell, and E. E. Percival, *Chem. Ind.* (London), 834 (1958).
281. J. J. O'Donnell and E. E. Percival, *J. Chem. Soc.*, 1739 (1959).
282. H. O. Bouveng, *Acta Chem. Scand.*, **15**, 78 (1961).
283. H. O. Bouveng and B. Lindberg, *Advan. Carbohyd. Chem.*, **15**, 53 (1960).
284. P. S. O'Colla, *Methods Carbohyd. Chem.*, **5**, 382 (1965).
285. V. C. Barry and P. W. D. Mitchell, *J. Chem. Soc.*, 4020 (1954).
286. S. A. Barker, E. J. Bourne, and M. Stacey, *J. Chem. Soc.*, 3084 (1953).
287. P. A. Finan, A. Nolan, and P. S. O'Colla, *Chem. Ind.* (London), 1404 (1958).
288. C. M. Ewald and A. S. Perlin, *Can. J. Chem.*, **37**, 1254 (1959).
289. G. O. Aspinall and K. M. Ross, *J. Chem. Soc.*, 1681 (1963).
290. G. G. S. Dutton and A. M. Unrau, *Can. J. Chem.*, **41**, 1417 (1963); **42**, 2048 (1964).
291. P. A. J. Gorin, K. Horitsu, and J. F. T. Spencer, *Can. J. Chem.*, **43**, 950 (1965); S. Hizukuri and S. Osaki, *Carbohyd. Res.*, **63**, 261 (1978); K. M. Aalmo, M. F. Ishak, and T. J. Painter, *ibid.*, **63**, C3 (1978).
292. G. G. S. Dutton, K. B. Gibney, G. D. Jensen, and P. E. Reid, *J. Chromatogr.*, **36**, 152 (1968).
293. H. Yamaguchi, T. Ikenaka, and Y. Matsushima, *J. Biochem.* (Tokyo), **63**, 553 (1968); **68**, 253 (1970).
294. K. W. Hughes and J. R. Clamp, *Biochim. Biophys. Acta*, **264**, 418 (1972).
295. P. Nánási and A. Lipták, *Carbohyd. Res.*, **29**, 193 (1973).
296. J. K. Baird, M. J. Holroyde, and D. C. Ellwood, *Carbohyd. Res.*, **27**, 464 (1973).

26. THE EFFECTS OF RADIATION ON CARBOHYDRATES

GLYN O. PHILLIPS

I. INTRODUCTION

Oxidative degradation is the most general consequence of the action of radiation on carbohydrates. The course and mechanism of degradation differ according to whether ultraviolet light[1,1a] or high-energy ionizing radiation is used.[2,2a] Investigations into the effects of ultraviolet and visible light on carbohydrates have derived much of their impetus from the desire to understand the photodegradation of cellulosic materials, for such reactions are of commercial significance. An understanding of the photosensitized degradation of carbohydrates may also be of value in the study of processes that operate in the photosynthetic cycle. As a result of developments in the nuclear-power industry, a large range of comparatively cheap high-energy radiation sources have become available. Interest has since this time been focused on the biological effects of these radiations, chemical utilization of radiations from fission products, and the use of radiation in the sterilization of food. In view of the important physiological role of carbohydrates and their wide occurrence in foods, detailed studies have been undertaken on the radiation chemistry of carbohydrates.

In this chapter, the primary processes by which radiations may initiate chemical change will first be discussed. Since the majority of investigations on carbohydrates have been undertaken on aqueous solutions, the conditions when photolysis or radiolysis of water may occur will be outlined. Distinction thereafter will be made between photodegradation, where ultraviolet and visible light are employed, and radiolysis, which may be promoted by high-energy radiations such as α-rays, β-rays, X-rays, and γ-rays, in aqueous solution or in the solid state.

II. PRIMARY PROCESSES

A. ULTRAVIOLET AND VISIBLE LIGHT[3–5]

1. *Excitation Processes*

The starting point in any photochemical interpretation is the long-established principle that only light that is absorbed by a molecule is effective in producing chemical change. Thus, the first step in any study of a photochemical process is to obtain the absorption spectrum of the compound being photolyzed. When light is absorbed, the energy of the molecule increases by an amount equal to the energy of the photon, which is

$$E = h\nu = hc/\lambda$$

where h is Planck's constant, c is the velocity of light, and ν and λ are the frequency and wavelength of the light. Changes may occur in the electronic, vibrational, or rotational energy of the molecule. Only visible or ultraviolet light can effect electronic changes in the molecule, and these will be accompanied by simultaneous variations in vibrational and rotational energy. During the formation of electronically excited states, an electron undergoes a transition from a lower to a higher level of energy in the molecule. Such changes in electronic energy are of the order of 100 kcal mole^{-1} and give rise to spectra in the visible or ultraviolet region at frequencies of about 10^{15} Hz.

The associated changes in vibrational energy are about 5 kcal mole^{-1}, and the rotational changes about 0.02 kcal mole^{-1}; these changes give rise to the coarse vibrational structure and the finer rotational structure of the electronic band. Electronic transitions of this type are governed by well-defined selection rules. One such rule (the Franck-Condon principle) states that, as the transition from the vibrational levels of one state to the vibrational levels of the other occur within about 10^{-5} sec, the nuclei of the molecule can be considered stationary. Since electronic absorption spectra arise from transitions of this kind, the maximum absorption will correspond to the Franck-Condon (vertical) transition.

a. Singlet and Triplet States.—Generally, in the ground state, molecules have two electrons in each orbital, and according to the Pauli principle the electrons must have opposed spins and hence "cancel" each other. Such a state is the singlet state. When one of the paired electrons is excited to a higher-energy orbital, the Pauli principle no longer governs its spin. When the two electrons have parallel spins, they confer paramagnetic properties on the molecule. Such a metastable state is the triplet (diradical) state.[6] The triplet states play an important part in certain photochemical processes. When, after excitation, the spins are in opposition, a diamagnetic molecule results (excited singlet state).

The natural lifetime of an excited state is inversely proportional to the strength of the absorption to that state, which is usually expressed as the molar extinction coefficient, ϵ, defined as

$$\epsilon = \frac{1}{cl} \log \frac{I_0}{I}$$

where c is the concentration in moles per liter, l is the length of the absorption path in centimeters, and (I/I_0) is the fraction of the incident light that is transmitted.

b. Types of Orbital Transition.—By reference to formaldehyde, it is possible to illustrate the various types of valence electrons and the promotions they may undergo.[7] Below are indicated the electrons forming single bonds (σ), the electrons involved in double bonds (π), and the unshared (nonbonding) electrons (n). The energy required to raise these electrons to antibonding

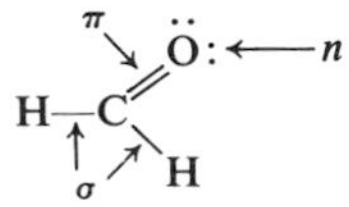

References start on p. 1287.

orbitals varies according to the type of bond. The n-electrons are the least strongly bound in a molecule, and, in the bonding levels, π-electrons have higher energies than the corresponding σ-electrons; in the antibonding levels, the order is reversed.[8]

	Level	Symbol
	Antibonding	σ^*
	Antibonding	π^*
E	Nonbonding	n
	Bonding	π
	Bonding	σ

Following absorption of light, the possible orbital promotions,[9] in order of increasing energy, are $n \to \pi^*$, $n \to \sigma^*$, $\pi \to \pi^*$, and $\sigma \to \sigma^*$. Consequently, $\sigma \to \sigma^*$ transitions of bonding electrons, as in the paraffins, are observed only in the far-ultraviolet region (<170 nm). Transitions involving bonding π-orbitals ($\pi \to \pi^*$) require less energy, and although some are observed in the far-ultraviolet region (for example, ethylene), many are found in the near-ultraviolet region. Excitation of an electron in a nonbonding orbital localized on an atom to an antibonding orbital (for example, $n \to \sigma^*$) is usually found in the far-ultraviolet region, as exhibited by saturated molecules (C—OH, C—NH$_2$, and C—halogen). The $n \to \pi^*$ transitions are found in molecules having a hetero atom multiply bound to another atom (C=O, C=S, and N=N), and they are generally found in the near-ultraviolet region. Acetone, for example, shows the $n \to \pi^*$ transition at 277 nm and the $n \to \sigma^*$ transition at 190 nm.

c. *Loss of Excitation Energy*.[3,4]—Excitation of a molecule by light does not necessarily lead to chemical change. Processes such as fluorescence, internal conversion of energy, nonradiative transition involving the triplet state, and energy transfer are all methods by which the excited state may lose its energy without chemical change occurring.[4]

Broadly speaking, permanent chemical change may arise as a result of two general types of processes. Where transitions lead to a repulsive state having a lifetime of less than one vibration, dissociation of the excited molecule may occur. Alternatively, the molecule may possess sufficient vibrational energy in either its upper or its lower energy state to cause it to dissociate. Generally, the bond that breaks is the bond responsible for the absorption of light. Chemical reaction may also be the result of intermolecular quenching of the excited state by a foreign molecule. The process may be primary—that is, one in which there is reaction between the quencher molecule and the excited molecule. It will be shown later that such reactions have a relevance to cellulose, where an excited molecule (Q*) may initiate reactions of the type

$$Q^* + RCH_2OH \longrightarrow \cdot QH + R\dot{C}HOH$$

Alternatively, a secondary reaction may lead to acquisition of sufficient energy by the quencher molecule from the excited molecule by an "inelastic" collision, to cause it to react. The excited molecule that becomes quenched is frequently referred to as a "photosensitizer." All reactions that require the absorption of light may be strictly termed photosensitized reactions, but the term has been expanded so that it now includes reactions initiated by the excited molecule in a molecule that does not absorb the light directly. Furthermore, truly sensitized reactions are those of the secondary type discussed above, where the absorbing substance acts as a catalyst and is not permanently altered. The term "photosensitizer" is, however, used loosely to cover both primary and secondary quenching effects outlined above.

2. Water

In aqueous solution, free radicals which may initiate the degradation of carbohydrates may be formed (1) by photolysis of water, by radiation of wavelength lower than 200 nm; (2) by sensitized decomposition of water to produce radicals.

a. Photolysis of Water.—The absorption spectrum of water vapor displays a continuum from 185 to 145 nm, with a maximum at about 165 nm. Below 145 nm there is a series of diffuse bands superimposed on a second continuum which extends below 120 nm. The molar extinction coefficient (ϵ) at the first maximum is 1480. The first broad continuum is to be expected[10-13] if the primary absorption results in a decomposition of a water molecule according to the equation $H_2O + h\nu \rightarrow \cdot H + \cdot OH$. Chen and Taylor[10] confirmed that such photolysis occurs when water vapor is irradiated in a flow system with 165-nm radiation from a hydrogen discharge. Chemical evidence demonstrates that liquid water undergoes photolysis by this path also.[14-16] For example, during photolysis of deaerated, aqueous solutions of methanol ($10^{-2} M$) with 184.9-nm radiation from a low-pressure mercury lamp, the formation of hydrogen and ethylene glycol was attributed to the following processes:

$$H_2O \longrightarrow \cdot H + \cdot OH$$

$$\cdot H + CH_3OH \longrightarrow H_2 + \cdot CH_2OH$$

$$\cdot OH + CH_3OH \longrightarrow H_2O + \cdot CH_2OH$$

$$2 \cdot CH_2OH \longrightarrow (CH_2OH)_2$$

Formaldehyde was also produced.

Other systems[16] also indicate that light of wavelength <200 nm can lead to the production of H atoms and OH radicals from water. Such reactive species could influence degradation of cellulosic materials considerably.

References start on p. 1287.

b. Photosensitized Decomposition of Water.—It has been suggested that, in the presence of compounds that can act as photosensitizers, the water molecule may be cleaved, although the energy of the radiation may be insufficient to induce photolysis of water. It has long been recognized that water may be decomposed by excited mercury,[17-19] but recently other types of photo-sensitizers have been recognized.

Riboflavin, for example, appears to promote the photochemical cleavage of water.[20-24] In deoxygenated aqueous solution, riboflavin is reduced to dehydroriboflavin on exposure to visible light. Addition of an "activator" increases the rate and permits photoreduction of riboflavin in the presence of oxygen. Materials that serve as activators include tertiary amines, thioethers, and other compounds containing an electronegative nitrogen or sulfur atom. Compounds in which the nitrogen atom can contribute to the ring, such as aryl amines, are not effective. The mechanism of photoreduction of riboflavin (R) in the presence of an activator (A) has been represented as shown. The

$$R + A + H_2O \longrightarrow (R—H_2O—A) \xrightarrow{h\nu} RH_2 + AO$$

water may be regarded as "suspended" between riboflavin and the activator, and "pulled apart" as the complex dissociates on absorption of radiant energy. The activator is not essential for the photoreduction. Through the formation of a riboflavin–water complex, it has been suggested that the bond strength of water may be lowered sufficiently to permit cleavage by light of wavelength 440 nm. At this wavelength, the energy absorbed per mole is 65 kcal, an amount clearly incapable of initiating direct photolysis of water. The splitting of water in this way, however, has been contested on energetic grounds.[24]

Mooney and Stonehill[25] have claimed that anthraquinone derivatives may also photosensitize the decomposition of water by light of wavelength greater than 400 nm. A mechanism was postulated in which the absorbing species is a complex of a molecule of hydration water, hydrogen-bonded to the carbonyl oxygen atom of a salt such as sodium 9,10-dihydro-9,10-dioxoanthracene-2-sulfonate ("silver salt"). The complex is supposedly cleaved during photo-excitation of the "silver salt" to give a hydroxyl radical, which then attacks the anthraquinone or anthraquinol ring.

Although the energy of the light used is only slightly more than half the energy of the H–OH bond, the authors claim that the validity of the mecha-nism can be substantiated on the basis of their calculations. If this claim proves to be correct, the proposed mechanism may have an important bearing on the photodegradation of carbohydrates that are sensitized by vat dyes. The hydroxyl radical produced might initiate abstraction processes in a molecule such as cellulose, and could lead to considerable chain scission and other oxidative processes, particularly in the presence of oxygen. Similar reactions involving methanol or water have been reported.[25,26] Subsequent experiments[26a] employing flash photolysis have indicated an alternative mechanism for the photodecomposition of anthraquinone salts. The mechanism accounts successfully for the observations of Mooney and Stonehill[25] and other studies on the products[26b] without introducing anomalous energetic considerations.

B. HIGH-ENERGY RADIATION

1. *Excitation and Ionization Processes*

The processes by which high-energy radiation and particles interact with matter have been described in detail.[27-31] Whether the radiation be electromagnetic (X-rays and γ-rays) or corpuscular (α-rays and β-rays), the final transfer of energy occurs via charged particles. With electromagnetic radiation, interaction of high-energy quanta with atoms of the medium through which they pass leads to ionization, since the energy of the quanta is substantially greater than the binding energy of an electron.

The behavior may be considered as a sequence of energy-transfer processes. First, the energy is transferred from the charged particle to the electronic system of the molecules of the medium, and for condensed systems, it is probable that energy deposition first extends over a domain. Ionization and excitation occur during this primary physical act. The ionization and excitation produced at this stage involve the electrons of a considerable volume of the medium, possibly 10^{-22} cm^3 or more, and last in their initial form for an extremely short period of time ($\sim 10^{-15}$ sec).[32] There is subsequent localization of the energy by way of a series of further energy-transfer processes to individual molecules, which undergo chemical change.

Little is known about the nature of the highly excited states initially produced, although photochemical and spectroscopic evidence provides some indication of their subsequent fate.[33] Excited states may also be formed by energy transfer from the first electrons (kinetic energy ~ 10 eV) formed during the primary act. These electrons can lose their energy by raising molecules in their paths to relatively low-lying excited states. Such excitations are not limited by the restrictions of optical transitions and may involve optically allowed transitions or produce triplet states (optically forbidden transitions). In this way, the excited states are formed some distance from the track of the primary electron. In polar media, particularly, the electrons may be trapped in potential wells, when the polarized aggregate has well-defined, quantized energy levels. Such solvated electrons have been detected in liquids and glasses by a variety of techniques.[34-37] Alternatively, the electrons, during their diffusion through

---

References start on p. 1287.

the matrix, may become attached to individual molecules.[38] For example, irradiation of glasses of naphthalene or tetracyanoethylene gives radical-anions, and halogen-containing molecules may react as shown. Such processes

$$RX + e^- \longrightarrow R\cdot + X^-$$

are well known in mass spectrometry.[39] A further possibility is that the electrons may recombine with positive holes after encounter with some 10^3 molecules, and so produce highly excited molecules by yet another energy-transfer process.

In contrast to liquids and glasses, crystals of nonpolar molecules, particularly those having few imperfections, contain relatively few trapping sites for electrons. Thus, although electrons may possess considerable mobility in such media, the chemical changes will depend in large measure on the charge-neutralization processes, which lead either to free radicals or to excited states. The net effect again is to form excited states at a considerable distance from the primary site of ionization. The subsequent migration of energy from the excited states, however the states are formed, has a considerable bearing on the chemical effects of radiation, and has been shown to be dependent to a marked extent on the state of aggregation in the condensed phase.[40] This behavior exerts a marked influence on the extent of degradation of crystalline carbohydrates.

In summary, it can be seen that energy deposition following high-energy irradiation is considerably less selective than during ultraviolet and visible irradiation, and that it gives rise to ionization and highly excited states.

2. Water

Fast-moving charged particles electronically excite and ionize water molecules. The secondary electrons released in this primary ionization can further ionize and dissociate water. In this way, the initially charged species H_2O^+ and e^- form in clusters or "spurs" and subsequently give rise to the dissociation products, $\cdot H$ and $\cdot OH$. A detailed account of these events has been given by Allen,[41] and only a brief outline will be given here. First, the fast-moving Compton recoil (photoelectric) electrons, and the secondary electrons that they produce, are rapidly moderated, initially to the subexcitation level of water (~ 6 eV), and then to thermal energies. Below 6 eV, the thermalization is slow and energy is dissipated mainly through vibrational and rotational energy transfers in about 10^{-13} sec. Gradually, the electron polarizes water molecules, which gradually restrict its movement until final orientation of adjacent water molecules is attained. At this stage ($\sim 10^{-10}$ sec) the electron becomes hydrated and possesses the properties of a normal univalent anion.[42,43] After hydration, e_{aq}^- reacts with water, liberating the hydrogen atom. This

$$e_{aq}^- + H_2O \longrightarrow \cdot H + OH^-$$

reaction is slow and is relatively unimportant in the presence of highly reactive e_{aq}^- scavengers. On the other hand, the electron-deficient H_2O^+ reacts according to the process:

$$H_2O^+ + H_2O \longrightarrow H_3O^+ + \cdot OH$$

Therefore, the final products of the ionization processes in pure liquid water are closely related to those obtained by direct photolysis of water. The chemical

$$H_2O^* \longrightarrow H\cdot + \cdot OH$$

species generated by the ionization and excitation acts are e_{aq}^-, $\cdot H$, and $\cdot OH$. According to the diffusion model, they are present in high concentration in the spur, which has an original radius of 1.5 nm.[44,45] During diffusion, pairwise recombination occurs, leading to the molecular products H_2 and H_2O_2, and to the homogeneously distributed e_{aq}^-, $\cdot H$, and $\cdot OH$. In recent years much attention has been directed toward measuring the yields of these free-diffusing species, characterizing their reactions, and determining their rate constants.[42,45a] The overall reaction may thus be represented:

$$H_2O \longrightarrow e_{aq}^-, \cdot H, \cdot OH, H_2O_2, H_2$$

III. PHOTODEGRADATION

Irradiation with ultraviolet and visible light can induce chemical changes in simple carbohydrates, although there have been many diverse views about the conditions necessary. One view is that photosensitizers such as iron or uranium salts must be present to initiate the decomposition of D-glucose.[46–48] On the other hand, the formation of carbon monoxide, carbon dioxide, and hydrogen[49] and changes in optical rotation[50] have also been reported in the absence of a sensitizer. Similarly, divergent views have been expressed concerning photodegradation of sucrose.[51–56] The historical aspects of this work have been reviewed in depth,[1] and a current article[1a] surveys photochemical reactions of sugars and their derivatives in general.

A. DIRECT PHOTOLYSIS

The first indication that light of wavelength below 280 nm is necessary for direct photolysis of D-glucose was given by Laurent et al.[57,58]; the yields of product were shown to be dependent on the concentration of D-glucose. Among the early products formed on irradiation of an alkaline solution of D-glucose is one having an absorption maximum at 265 nm, which shifts to 245 nm at pH values below 3.[58,59] Irradiated neutral solutions do not show the absorption maximum at 265 nm until alkali is added. It is significant that irradiation of $5 \times 10^{-2} M$ D-glucose solutions with high-energy radiation[60] also produces an absorption band at 265 nm. During irradiation of D-glucose solutions with unfiltered light from a medium-pressure mercury lamp,[61] there is a fall in pH, and carbon dioxide is evolved continuously. The rates of acid formation and of carbon dioxide evolution increase with the time period of irradiation and with increasing concentration of D-glucose. This indicates that *direct* absorption of radiation, and not indirect action, is the dominant process responsible for the degradation. Radioisotopic methods[60–62] have shown that stepwise degradation of the D-glucose occurs. The photolysis of D-glucitol with filtered and unfiltered light from a Hanovia medium-pressure mercury lamp has similarly been studied,[63] and the changes may be represented as follows:

CH₂OH	CHO	CO₂H	
CH_2OH	CHO	CO_2H	
HCOH	HCOH	HCOH	CHO
HOCH	HOCH	HOCH	HOCH
HCOH $\longrightarrow$	HCOH $\longrightarrow$	HCOH $\longrightarrow$	HCOH
HCOH	HCOH	HCOH	HCOH
CH_2OH	CH_2OH	CH_2OH	CH_2OH
D-Glucitol	D-Glucose	D-Gluconic acid	D-Arabinose

A similar degradation from the other extremity of the molecule leads to L-gulose, L-gulonic acid, and L-xylose. No significant decomposition is caused by the ozone formed.

For photochemical reaction to occur, it is necessary both that the light be *absorbed* by the sugars and that the absorbed light be sufficiently energetic to rupture bonds in the molecule. Unfortunately, owing to the general transparency of sugars, little attention has been given to details of the absorption spectra of sugars in neutral solution. An early claim[64] that there is an absorption band at about 280 nm has not been substantiated.[65–68] When acid is added to neutral solutions of D-glucose, an absorption maximum at 284 nm is observed, supposedly arising because of conversion of the nonabsorbing cyclic forms into open-chain carbonyl forms.[69]

It is possible to make some prediction about the nature of absorption spectra of sugars by considering the spectra of aliphatic alcohols and cyclic ethers. Aliphatic alcohols are transparent in the near-ultraviolet region; in the region of 200 nm[70,71] a continuum sets in, with a maximum occurring at about 180 nm.[70–74] Absorption becomes significant ($\epsilon = 50$) at 196.1 nm and, considering the width of the bands, the maxima for a series of alcohols fall in the narrow range 183.5 to 180.8 nm. The order of increasing wavenumber of the maxima and intensity of absorption is ethanol, isopropyl alcohol, methanol, and propyl alcohol.

Tetrahydrofuran, tetrahydropyran, and *p*-dioxane, three compounds that contain one or more oxygen atoms in rings that are otherwise saturated, show bands[74,75] in the vacuum ultraviolet between 200 and 166.7 nm. These compounds are practically transparent above 255 nm. There is a slight increase in absorption below this point, but to 227.3 nm the molar extinction coefficient remains below 1. An observation that may have some significance for the absorption spectra of carbohydrates in this region is that replacement of the 4-methylene group of tetrahydropyran by an oxygen atom, to give p-dioxane, causes the absorption band to move toward the visible region.

The spectrum of D-glucitol, measured in aqueous solution, accords with the observations. The solutions are almost transparent down to 300 nm; thereafter a continuum sets in which extends into the vacuum ultraviolet. After 217.2 nm the absorbancy rises sharply. In the region 200 to 253.7 nm the solutions obey Beer's law, and molar extinction coefficients have been calculated.[1] Since the straight-chain hydrocarbon corresponding to D-glucitol absorbs in the vacuum ultraviolet, the above spectrum may be attributed to electrons associated with oxygen atoms, with the strong probability that the nonbonding electrons are involved ($n \to \sigma^*$).

Absorption of radiation in the spectral region 230 to 240 nm is responsible for the major portion of the photodegradation when oxygenated solutions of D-glucitol are irradiated with a Hanovia medium-pressure mercury lamp. D-Glucitol absorbs the radiation directly, and, although the amount of light absorption is small, there is excitation of the nonbonding electrons to antibonding levels ($n \to \sigma^*$ transition), which initiates dissociation. It is unlikely that the active wavelengths are sufficiently energetic to excite electrons in nonbonding orbitals from the ground state of the molecule, or to induce direct ionization. Thus, the two requirements for photochemical reaction are fulfilled. Light of wavelength 230 to 253.7 nm, corresponding to an energy of 124 to 113 kcal mole^{-1}, is absorbed by the D-glucitol molecule, and it is sufficiently energetic to rupture a C–C bond or a C–O bond (80 to 90 kcal mole^{-1}) or to remove a hydrogen atom (100 kcal mole^{-1}).

Beélik and Hamilton[76] studied the ultraviolet (220 to 400 nm) irradiation of cellobiose, cellopentaose, methyl β-cellobioside, cellobiitol, and cellopentaitol as model compounds related to cellulose. The effects of irradiation were very similar. All of the compounds were fragmented, yielding comparable amounts of acidic compounds and products having absorption maxima near 260 nm. It was suggested that the degradation is initiated as a result of primary light absorption at an "acetal chromophore" near 265 nm which arises from the presence of the anomeric oxygen atom. The chemical changes induced in D-glucose and lower saccharides by radiation of wavelength 230 to 240 nm invariably take place initially at C-1 and so provide added evidence for the "acetal chromophore."[62]

In aqueous solution, ketoses are photolyzed more readily than aldoses.[77]

References start on p. 1287.

Products of the photolysis are the alditol having one carbon atom fewer than the parent sugar, carbon monoxide, and hydrogen. "Extreme" (presumably short wavelength) ultraviolet rays caused secondary reactions in which acid, carbon dioxide, and methane were produced. The behavior of a 10% aqueous solution of D-fructose was proposed as the basis for an actinometer to measure ultraviolet light.[78] Formaldehyde and methanol are other products that have been identified in solutions of D-fructose that had been irradiated with ultraviolet light.[50,79] It has been reported that the proportion of carbon dioxide in the evolved gases varies inversely with the intensity of the light. Acids, according to Cantieni,[80,81] are formed only in solutions heated to 80°, and the evolution of gas is accelerated by sodium chloride and retarded by potassium bromide, potassium iodide, acids, and bases. Photolysis occurs when the radiation is passed through glass but is much slower than when it is passed through quartz. At ordinary temperatures, a postirradiation release of gas was detected, but at high temperatures this effect was very small.

The pentose moieties of the nucleic acids are as susceptible as D-glucose to degradation by light from an unscreened, quartz lamp when the acids are irradiated[82] at pH 7.2.

On irradiation with light of 253.7-nm wavelength, L-ascorbic acid (AH_2) gives[83] dehydro-L-ascorbic acid (A). In the absence of air, the reaction is measurable only when the pH is greater than 2.5. Between pH 5 and 9, the reaction is independent of pH. In the presence of air, a greater amount of photooxidation occurs, and the product is further degraded, particularly at high pH values. A postirradiation oxidation is observed; this is accompanied by the formation of hydrogen peroxide, which is also formed in the main, irradiation-induced oxidation. It is suggested that irradiation causes the formation of an ascorbic radical $AH\cdot$, which can react with oxygen according to the following equations:

$$AH\cdot + O_2 \longrightarrow A + HO_2\cdot$$

$$2HO_2\cdot \longrightarrow H_2O_2 + O_2$$

Cleavage of the glycosidic link was observed[84,85] during the irradiation of aqueous solutions of aryl glycosides with monochromatic light of wavelength 254 nm. In this region, the aglycon is the absorbing center of the glycoside. For benzyl β-D-fructopyranoside, the photochemical efficiency for the photolysis varies by less than $\pm 5\%$ when: (1) the concentration of buffer is changed from 0 to 1 M acetate buffer and from 0 to 0.1 M phosphate at pH 3.5 to 6.5 and temperatures of 8° to 25°; (2) the light intensity is varied by a factor of over 100; and (3) the concentration of the glycoside is varied between 0.14 and 0.9 g in 50 ml of solution. The main reaction is, therefore, unimolecular, and its nondependence on temperature indicates that enough energy of the absorbed photon (110 kcal per einstein) enters the reactive center to produce the reaction without any measurable thermal activation being required. The products of the main photochemical reaction are the same as those produced by thermal hydrolysis, namely D-fructose and benzyl alcohol.

The reactive center is the acetal oxygen link between the aglycon and the glycosyl group. Very little of the main photolysis is caused, however, by photons directly absorbed by this group. This fact may be demonstrated by the lack of reaction when the benzyl group is replaced by the nonabsorbing methyl group, as in methyl β-D-fructopyranoside. When the aglycon is changed and benzyl β-D is replaced by benzyl α-D, phenyl α-D and β-D, and 2-(phenyl)ethyl α-D and β-D, there is no correlation between the ease of photochemical cleavage of the glycosidic link and its stability toward acid hydrolysis. In all cases, the changes in quantum yield are small, although the rates of hydrolysis differ by a factor of more than 10,000.

It appears that the energy of the absorbed photon, after absorption by the aglycon, must be transferred intramolecularly to the hemiacetal oxygen bridge (the reactive center). Although the effect is small, the efficiency of this transfer is greater for benzyl than for 2-(phenyl)ethyl, but phenyl D-glucosides are the least efficient of those studied. As observed during the photolysis of phenols in oxygen, it is probable that the primary process leads to the production of phenoxyl-type radicals (ArO·) and hydrated electrons.[86-88]

Photolysis of carbon–sulfur bonds in dithioacetals of aldoses can be effected in a stepwise manner, to give initially the 1-S-alkyl-1-thioalditol, and subsequently the 1-deoxyalditol.[87] Photochemical cleavage of carbon–sulfur bonds can be used as a route to deoxy sugars[88] (see Chapter 18).

Irradiation of polysaccharides in aqueous solution in the absence of a sensitizer leads to depolymerization.[89-91] Similar degradative changes have been observed when aqueous solutions of hyaluronic acid, sulfated hyaluronic acid, and heparin were irradiated with ultraviolet light and fast electrons.[92,93] Little is known, however, about the precise mechanisms of the direct photolysis of polysaccharides in solution.

B. PHOTOSENSITIZED DEGRADATION

Considerable information about the probable nature of the processes operative during fading and tendering of dyes absorbed on cellulosic fibers has been obtained by the study of the photosensitized oxidation of alcohols. Sensitizers initiating the photosensitized oxidation include benzophenone,[94,95] benzoquinone,[96] hydrogen peroxide,[97] and azo compounds.[98] For each of the sensitizers the pattern of chemical change is similar. In the absence of oxygen, primary and secondary alcohols are converted into the corresponding carbonyl compounds, and the sensitizer is hydrogenated. In oxygenated solutions, primary alcohols yield hydrogen peroxide, the corresponding aldehyde, and the carboxylic acid. Secondary alcohols yield, almost exclusively, the ketone and hydrogen peroxide. The mechanism has been examined in detail when the sensitizers are derivatives of anthraquinone because of their close structural relationship to the anthraquinonoid class of vat dyes.[99,100]

The reactivities of a variety of alcohols, carbohydrates, and ethers on photo-oxidation in the presence of these sensitizers have been studied by various workers.[100–102] The initial carbohydrate product from the hexitols appears to be the corresponding hexoses formed by oxidation of a primary alcohol group.[101] Hexonic acids are formed at a slower rate. Kinetic studies of such processes have suggested possible mechanisms.[102–109]

During the photosensitized oxidation of ethers and glycosides, Lock[110] observed that cleavage of the glycosidic link occurred also. When RCH_2OR' represents the glycoside benzyl β-D-glucopyranoside (R = phenyl and R' = β-D-glucopyranosyl), the following reactions occur:

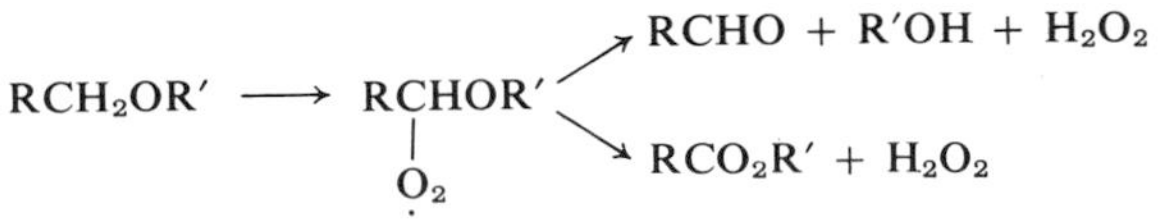

$$RCH_2OR' \longrightarrow \underset{\underset{O_2}{|}}{RCHOR'} \begin{cases} \nearrow RCHO + R'OH + H_2O_2 \\ \\ \searrow RCO_2R' + H_2O_2 \end{cases}$$

There is a close similarity in the behavior of N-alkylamides to that of alcohols, ethers, and glycosides.[102,110] A comparison is given in the following equations:

$$ROCH_2R' \longrightarrow ROCOR' + ROH + R'CHO$$

$$RCONHCH_2R' \longrightarrow RCONHCOR' + RCONH_2 + R'CHO$$

Heidt[111] considered that, since sucrose solutions are "as transparent as water" in the visible and near-ultraviolet to 200 nm, light in this region does not produce any significant effect. For inversion to occur, the presence of a photosensitizer is necessary. He investigated the quantum yields and kinetics of the production, by visible and ultraviolet light, of reducing sugars from aqueous solutions of sucrose containing uranyl salts.

The primary act is the capture of a photon by a uranyl salt, which, in its activated state, is sufficiently near the sucrose molecules and properly oriented to initiate the hydrolysis. The fraction of the activated uranyl molecules so disposed is proportional to the quantum yield (ϕ), defined as molecules of sucrose decomposed per photon absorbed, whence

$$\phi = B(UO_2SO_4)^n(\text{sucrose})^m(H_2O)^e \ldots /(UO_2SO_4) \tag{1}$$

where B is the proportionality constant, and the sum of the exponents in the numerator equals the number of reactant molecules entering the photochemically active cluster. When ϕ cannot exceed a finite value a,

$$\phi = a[1 - \exp(-gx)] \tag{2}$$

where x is directly proportional to the right-hand side of Eq. (1), and g includes B. At small value of gx, Eq. (2) becomes

$$\phi = agx \tag{3}$$

In this range of values, each exponent, except that of the photosensitizer, uranyl sulfate, equals the slope of the plot of log ϕ against the logarithm of the concentration of the corresponding reactant, when other variables are fixed. For the photosensitizer, however, $n = (1 + \text{the slope})$. The independence of ϕ, of the concentration of uranyl sulfate (between 0.008 and $0.0008\,M$), shows that only one uranyl molecule enters the reactive complex.

The points on a plot of log ϕ at 254 nm against log [sucrose] fell on a curved line that approached unit slope only when ϕ was less than 0.1. Therefore, the largest observed values of ϕ are too near to the maximum value for Eq. (3) to hold. The approach of the plot to unit slope when ϕ is small shows, however, that $m = 1$. Thus, one molecule of sucrose enters the reaction with each uranyl molecule.

The variation in ϕ with sucrose concentration may be expressed in the form $\phi = a\{1 - \exp(-g\,[\text{sucrose}])\}$. At 254 nm, $a = 0.3$ and $g = 0.6$, when [sucrose] is expressed in moles per liter and the solutions are buffered with acetate.

In the presence of titanium dioxide as photosensitizer, oxidation, not hydrolytic cleavage, appears to be the primary process during the irradiation of aqueous solutions of lactose, sucrose, maltose, or cellobiose.[112] An absorption band is produced at 267 nm, and the position of the maximum is dependent on pH.

Two distinct reactions promote the degradation of amylose when irradiated in oxygenated solution with 365-nm radiation in the presence of zinc oxide.[113,114] The first (reaction A), responsible for the release of reducing groups, is sensitized by zinc oxide. The second (reaction B), which causes the ultimate disappearance of reducing power, is not dependent on zinc oxide. Reaction A is a photooxidation, and reaction B is a hydrolytic type of process. (Shown on p. 1232.)

C. CELLULOSE

Photochemical investigations on cellulose have, in the main, been pursued independently of the main field of photochemistry, and therefore the results warrant separate consideration. It has been known since 1883 that the degradative effect of light on cotton and linen fabrics involves chemical processes.[115] Physical and chemical effects caused by the action of light lead to a loss in tensile strength of the fabric, and to formation of carboxyl and carbonyl

---

References start on p. 1287.

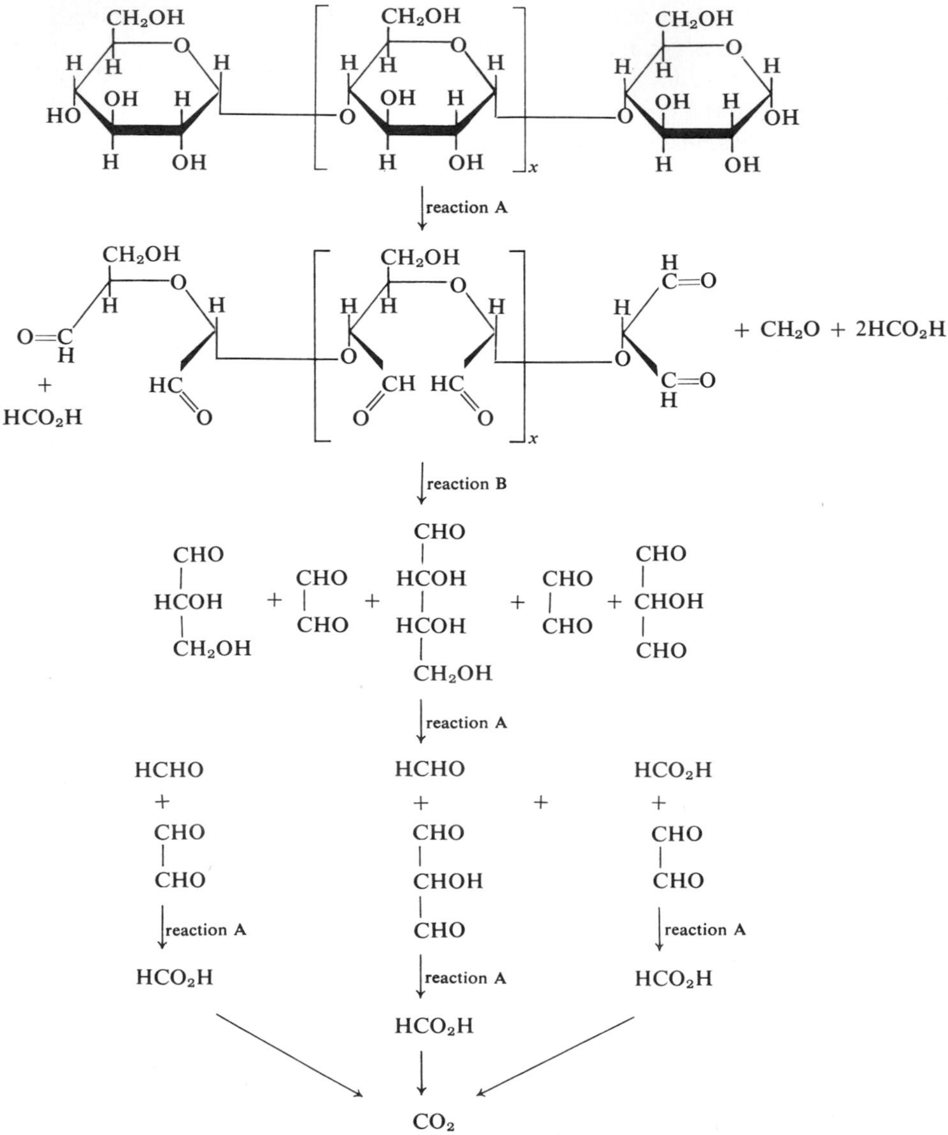

$$\text{reaction A} \quad \longrightarrow \quad \text{reaction B} \quad \longrightarrow \quad \text{reaction A}$$

$+ \; CH_2O + 2HCO_2H$

Scheme (top): cellulose chain units (CH_2OH, OH, O) with reaction A giving aldehyde and HCO_2H products $+ CH_2O + 2HCO_2H$.

reaction B:

$$\begin{array}{ccccccc}
\text{CHO} & & \text{CHO} & & \text{CHO} & & \text{CHO}\\
| & & | & & | & & |\\
\text{HCOH} & + & \text{CHO} & + & \text{HCOH} & + & \text{CHOH}\\
| & & & & | & & |\\
\text{CH}_2\text{OH} & & & & \text{HCOH} & & \text{CHO}\\
& & & & | & & \\
& & & & \text{CH}_2\text{OH} & &
\end{array}$$

reaction A:

$$\begin{array}{ccc}
\text{HCHO} & \text{HCHO} & \text{HCO}_2\text{H}\\
+ & + & +\\
\text{CHO} & \text{CHO} & \text{CHO}\\
| & | & |\\
\text{CHO} & \text{CHOH} & \text{CHO}\\
& | & \\
& \text{CHO} &
\end{array}$$

reaction A → HCO_2H (left) ; reaction A → HCO_2H (middle) ; reaction A → HCO_2H (right)

$$\longrightarrow \quad CO_2 \quad \longleftarrow$$

groups along the cellulose chain.[116–121] Much of the work in this field is technological and was carried out for the benefit of the textile industry.[122–124] Nevertheless, it is the work on cellulose and allied compounds that has provided the most reliable data on the effects of ultraviolet light on carbohydrates. Two general types of photochemical degradation will be considered—direct photolysis and photosensitized degradation.[125]

1. Direct Photolysis

Numerous investigators have reported chemical changes following the ultraviolet irradiation of cotton[116-134] and wood celluloses.[119,135-138] Considerable effort was applied to establishing whether a photosensitizer was required to promote the degradation of cellulose by light. The usual technique involves comparing the rates of degradation in oxygen-containing and oxygen-free atmospheres. Although it was initially thought that the presence of oxygen influenced the rate of degradation,[117,131,132] it is now clear that degradation by radiation of wavelength 253.7 nm is independent of the presence of oxygen.[119,121,125,129,130]

The initial disagreement probably arose because the outputs of the lamps used were substantially different at the critical region around 253.7 nm.[125] Stillings and Van Nostrand[132] irradiated cellulose under nitrogen, taking stringent precautions to remove and exclude oxygen during irradiation. When the oxygen content of the atmosphere was increased, enhanced degradation ensued. The mercury lamp used by these workers, according to Egerton,[125] emitted radiation from 248.3 to 578 nm, and of the total light produced, 58% had wavelengths greater than 350 nm, and only 22% had wavelengths below 300 nm; 18% had a wavelength of 365.4 nm. On this evidence, it is most unlikely that these investigators were observing the direct photolytic process, for it has been clearly demonstrated that no significant degradation occurs when pure cotton is exposed to light filtered through glass (which cuts off radiation below 340 nm), in the absence of oxygen.

With radiation from a Vitan quartz mercury-vapor lamp, from which 90% of the emitted radiation is of wavelength 253.7 nm, the loss of strength of cotton irradiated in atmospheres of carbon dioxide, nitrogen, or helium is about the same as in oxygen.[119] This lack of effect of oxygen was later confirmed.[125] For this type of photolytic breakdown, initiated by 253.7-nm radiation, it is significant that degradation, whether measured by loss in tensile strength or by increase in fluidity or reducing power, is greatly diminished by the presence of vat dyes. Examples include yellow and orange dyes, such as Caledon Yellow 5G and Cibanone Orange R, and green and blue dyes, such as Caledon Jade Green G and Caledon Blue R. This desensitizing effect of vat dyes was observed both in dry oxygen and in atmospheric air. This effect of dyes contrasts markedly with the way in which they accelerate degradation of textile materials irradiated by light of wavelengths greater than 340 nm.[139-141]

a. *Effect of Water Vapor and Temperature.*—Launer et al.[119] reported that water vapor exerts an inhibitory effect on the degradation initiated by light of

253.7 nm, and considered this to be an indication that some water molecules are in fundamental chemical combination with the cellulose. It was believed that, when water molecules enter the interchain spaces, the freedom of displacement for those carbon atoms in the vicinity of an H—O—H bond is decreased. Thus, when a far-ultraviolet photon is absorbed, the deformation of a given D-glucose residue is rendered less likely. Such an effect of deformability within the cellulose macromolecule is indicated by the results of Mason et al.[142] Parallel ramie fibers were irradiated with polarized ultraviolet light, and it was found that degradation was greater by about 25% with light vibrating crosswise to the fibers than with the vibrations lengthwise. By analogy with simpler compounds, it was concluded that photochemical degradation of cellulose chains is greatest if they are oriented so that the light energy vibrates in the direction in which the chains are most deformable.

The rates of photochemical degradation of undyed, bleached cotton yarn in oxygen, nitrogen, and air at 0% and 100% relative humidity were determined by using decrease in tensile strength as a measure of degradation. The rate of degradation was dependent on the presence of oxygen and was greater in atmospheres having a higher relative humidity.[117,140] These effects of moisture and oxygen on degradation of cellulose did not confirm the observations of Launer et al.[119] in the far-ultraviolet region, where increase in fluidity of cellulosic solutions was used as a measure of degradation of cellulose. For the near-ultraviolet region, they reported that the effect of moisture and oxygen on the degradation of cellulose was the reverse of that in the far-ultraviolet region. These apparent differences in effects of moisture and oxygen on the photochemical degradation of cellulose were attributed to the relationships between bond energies in the cellulosic chain and the energies of the photons involved. In the far-ultraviolet region, most of the photons involved would have sufficient energy to cleave the bonds. In the near-ultraviolet region, most photons would not have sufficient energy to cleave the bonds and would require the participation of oxygen.

b. Chemical Effects.—Generally, in evaluations of the effect of ultraviolet light, consideration has been given to modifications of the polymer system, including increases in solubility and reducing power, formation of carboxyl groups, and reduction in degree of polymerization.[119,125,132,142–147] Gaseous degradation products have also been identified, notably carbon dioxide,[126,132,142] carbon monoxide,[126,132] and hydrogen.[126] The latter was first identified during the irradiation of dry, purified cotton in vacuo at 40° with 253.7-nm radiation. It was suggested that the rigorous degassing procedure might have removed certain inhibiting factors which account for the previous nondetection of hydrogen. The rate of hydrogen evolution follows a parabolic rate law:

$$dx/dt = k/(1 + ax)$$

where t is the time; x, the number of moles of hydrogen formed; k, the initial rate of evolution; and a, a constant. A plot of log k against log (intensity of light) for any particular cellulose gives a straight line of unit slope. Thus $k = I_a\Phi_H$, where Φ_H is the initial quantum yield for the production of hydrogen, with a value of 10^{-2} mole/$Nh\nu$, and I_a is the intensity of the absorbed light. The rate of evolution of carbon monoxide and carbon dioxide remained relatively constant, and the zero-order rate equation showed a combined quantum yield of 10.9×10^{-4} mole/$Nh\nu$ for the first series of celluloses and 8×10^{-4} for the second series.

The quantum yield, calculated by Sippel,[148] is nearly unity for degradation by light of the middle-ultraviolet region (253.7 nm). The inefficiency of light in the near-ultraviolet region in the direct photolysis process is further confirmed by the low quantum yield (10^{-4} to 10^{-3}) calculated for degradation by light in this region.[119,148]

A postirradiation process, observed initially by Launer and Wilson,[119] was confirmed by Stillings and Van Nostrand[132] and by Egerton.[125] Degradation of cotton that had been irradiated under nitrogen or carbon dioxide was increased by storage in the dark for a period of several weeks. Cellulose that had been irradiated in oxygen also showed the postirradiation effect; however, the effect was smaller and became almost negligible when the time of irradiation in oxygen was extended from 4 to 10 days. The thermal nature of the post-irradiation process was indicated by the enhancing effect that followed when the temperature was raised to 70°.

Little consideration has been given to the mechanism of degradation. Most investigators were content to draw some type of analogy with acid hydrolysis and to assume cleavage of the glycosidic links. Launer and Wilson[119] mentioned the possibility of rupture in the vicinity of C-1, C-2, or C-3 of the D-glucose residues, rather than at the glycosidic linkage. If, for example, the bond between C-1 and C-2 were to be disrupted, this would not necessarily cause immediate rupture of the cellulose chain. But C-1, attached in this state to possibly three electronegative groups, could be sensitive to thermal oxidation, leading to chain disruption according to the postirradiation process mentioned above.

To account for the production of hydrogen, Flynn et al.[126,149] suggested that photolysis of the alcohol group occurs by the reaction shown. The increasing

$$\underset{\underset{H}{\diagdown}}{\overset{\overset{OH}{\diagup}}{\diagup\hspace{-0.3em}C\hspace{-0.3em}\diagdown}} + h\nu \longrightarrow \diagdown C{=}O + H_2$$

carbonyl-group concentration at the surface from this reaction might act as a nonvolatile filtering layer which would decrease the rate of photolysis of the alcohol groups and so explain the parabolic rate law that governs the evolution of hydrogen during irradiation.

References start on p. 1287.

A systematic attempt to establish a pattern of degradation was undertaken by Beélik and Hamilton,[76,136] who examined the smaller, water-soluble fragments produced during the initial stages of degradation of wood cellulose by ultraviolet light. The light source was a mercury lamp having 60% of its energy output in the ultraviolet region, with emission lines at 253.7, 265.2, 302.2, 313.1, and 365.4 nm. After irradiation, the ultraviolet absorption spectrum of the cellulose–water extract had a weak maximum at 260 nm, which was intensified and shifted to 270 nm at pH 12, but which disappeared at pH 2. Volatile and nonvolatile acids were formed, but 60% of the water-soluble portion was composed of neutral sugars. The following sugars were isolated and identified: D-xylose, xylo-biose, -triose, -tetraose, and -pentaose; D-glucose, cellobiose, and cellotriose; D-arabinose, 3-O-β-D-glucopyranosyl-D-arabinose, 3-O-β-cellobiosyl-D-arabinose, and D-mannose.

c. Primary Absorption Process.—Detailed studies on the absorption spectra of cellulose are as yet inconclusive. It has been suggested by Kujirai[144] that cellulose film shows a strong absorption below 200 nm and a weak absorption between 200 and 400 nm. Treiber,[150] in a review of the absorption properties of cellulose and its derivatives, concluded also that there is probably a selective absorption at about 278 nm which is "noticeable although weak." This absorption is attributed to the cyclic acetal ring and glycosidic oxygen atoms. It is clear, however, that more detailed experimental studies are necessary, by the refined techniques now available for the vacuum and far-ultraviolet regions, before the final nature of the absorption can be elucidated. Information concerning the primary absorption process must, therefore, be inferred from studies that give some indication of the effective spectral region that induces direct photolysis of cellulose.

It has now been demonstrated that radiation of wavelength greater than 340 nm is unable to photolyze pure cellulose directly.[76,136,144] For example, when Pyrex glass (6 mm thick) was interposed between the lamp and wood cellulose, the cellulose after irradiation showed no change in solubility, viscosity, or content of carboxyl and aldehyde groups. The formation of water-soluble acids was halved, and only traces of D-arabinose were detected. (The Pyrex filter transmitted less than 1% of light of 290 nm and only 10% of light of 310 nm.)

Kujirai[144] sought to distinguish the mechanisms of photolysis by 185-nm and 253.7-nm radiation on the basis of an oxygen effect. Apart from this isolated observation, however, there is no evidence that the direct photolytic process proceeds by a different mechanism at these two wavelengths. Direct photolysis of associated water by 185-nm radiation might account for this behavior.

No satisfactory reaction mechanisms have been advanced to explain all the products formed during direct photolysis. Although there can be little doubt that the wavelength of the light used in the various direct photolysis investigations is sufficiently energetic to rupture the carbon–carbon, carbon–oxygen, and carbon–hydrogen bonds in cellulose, another prerequisite is that the light must be *absorbed* by the molecule. From the available information, it does not appear that cellulose shows any significant absorption of light of wavelengths near 253.7 nm, and, therefore, it cannot be assumed that the latter condition is fulfilled.

To overcome this apparent contradiction, it has been postulated that a weak chromophoric group, the acetal group at C-1 of the D-glucose residues, is responsible for the absorption near 265 nm.[76,136] This suggestion appears to be the most probable mechanism at present. The experimental evidence

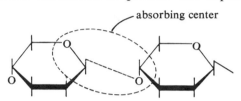

certainly does not justify an entirely new theory in which concurrent or alternate hydrolysis and oxidation are envisioned.[151–153]

2. Photosensitized Degradation

Light of wavelengths greater than 340 nm cannot induce degradation of cellulose directly. Certain dyes, however, and compounds such as zinc oxide and titanium dioxide, are capable of absorbing light in the near-ultraviolet or visible part of the spectrum and in their excited state can induce the degradation of cellulose. These processes may be collectively classified as photosensitized degradations of cellulose, although it must be emphasized that it is unlikely that all the sensitized reactions proceed by a common mechanism.

The photoinitiated tendering of cellulosic and synthetic fabrics that have been treated with vat dyes has been recognized for many years.[122,139–141,154–157] The technological aspects have been reviewed,[133] and it is clear that considerably more attention has been given to practical considerations than to fundamental studies of the mechanism of initiation of the tendering process. A conspicuous exception, however, is the series of studies into the mechanism of the auto-oxidation of cellulosic materials as photosensitized by quinonoid dyes. Here studies on model compounds in Section III,B allow a reasonably accurate extrapolation to the mechanism of tendering of cellulosic materials.[99–108,158]

a. Tendering by Vat Dyes.—In marked contrast to direct photolysis, it has been shown conclusively that the presence of oxygen is necessary for the photosensitized deterioration of cellulose. Dyes act as initiators and not as inhibitors. It was shown by Egerton[139-141] that moisture enhances the degradation and initiates further degradation, not only in the dyed fabric, but also by catalyzing the breakdown of adjacent, undyed fibers. Similar behavior is observed when either zinc oxide or titanium dioxide is employed as a sensitizer.

Two views were expressed about the detailed path of the reaction. Egerton[139-141] suggested that the light energy absorbed by the dye is transferred to oxygen in the surrounding atmosphere, with the production of activated oxygen. In the presence of water vapor, the activated oxygen is assumed to react to form hydrogen peroxide. The degradation of the textile material would then proceed by oxidation by activated oxygen and/or hydrogen peroxide, depending on the particular textile fiber and the experimental conditions.

Bamford et al.,[158] on the other hand, while associating photodegradation with the presence of hydrogen peroxide, considered that the primary step leading to the formation of hydrogen peroxide by the excited dye (D*) is the following:

$$D^* + OH^- \longrightarrow D^{\bar{\cdot}} + \cdot OH$$

Hydrogen peroxide would be formed subsequently by the combination of two hydroxyl radicals, or by the reaction of two $HO_2\cdot$ radicals formed according to the following equation:

$$D^{\bar{\cdot}} + H^+ \longrightarrow \cdot DH \xrightarrow{O_2} D + HO_2\cdot$$

The view of Bamford et al.[158,159] on the initiation of the sensitized photo-oxidation appears to be closer to current thought. They saw the initial reaction, in the absence of moisture, as occurring between the cellulose (R—H) and the excited dye (D*):

$$D^* + R—H \longrightarrow R\cdot + \cdot DH$$

Tendering may then take place by the following reactions:

$$R\cdot + O_2 \longrightarrow RO_2\cdot$$

$$RO_2\cdot + \cdot D—H \longrightarrow RO_2H + D$$

The sensitizer is unchanged, according to this explanation. When moisture is present, initiation may also take place by abstraction of hydrogen by hydroxyl radicals that may arise by photosensitized rupture of a water molecule.

$$R—H + \cdot OH \longrightarrow R\cdot + H_2O$$

The hydrogen atom formed could then lead to hydrogen peroxide.

$$HO_2\cdot + HO_2\cdot \longrightarrow H_2O_2 + O_2$$

The elegant investigations on model systems such as alcohols, glucosides, and disaccharides[99-107,158] now fully substantiate the view that the initial step in the tendering process is abstraction of a hydrogen atom from the cellulose by the photoexcited dye (rather than any kind of electron-transfer process), followed by addition of oxygen to give a peroxy radical. At least two reactions may follow. Decomposition of the peroxy radical may lead to chain scission and acid formation, or a carbonyl group may be formed.

Preliminary investigations with amylose have shown that changes in viscosity of amylose in different solvents during photodegradation can be correlated with the number of glycosidic bonds broken.[160] Oxygen absorbed during photodegradation also correlates with cleavage of glycosidic bonds, as required by the postulated mechanism, which seems, therefore, to offer an adequate explanation for photooxidation in the absence of moisture. When moisture is present, other processes, such as those described in relation to the sensitized photodecomposition of water, may contribute further to the initial abstractions.

Indirect confirmation of the above mechanism of phototendering comes from experiments where ultraviolet light was used to initiate grafting onto cellulose.[161,162] Following abstraction of a hydrogen atom from the cellulose by the photoexcited dye molecule, a free radical should be produced on the cellulose backbone. Evidence for the formation of such intermediate radicals is given by their ability to initiate vinyl polymerization and, although not certain, it is probable that a copolymer is grafted onto the cellulose backbone. The following mechanism is envisaged for grafting and the simultaneous production of homopolymer:

Absence of Oxygen

$$D + h\nu \longrightarrow D^* \tag{4}$$

$$\underset{\text{H}}{\overset{\text{OH}}{D^* + \text{cell—C—}}} \longrightarrow \cdot DH + \underset{\cdot}{\overset{\text{OH}}{\text{cell—C—}}} \tag{5}$$

$$\underset{\cdot}{\overset{\text{OH}}{\text{Cell—C—}}} + MH \longrightarrow \underset{\cdot MH}{\overset{\text{OH}}{\text{cell—C—}}} \longrightarrow \text{graft} \tag{6}$$

$$\underset{\cdot}{\overset{\text{OH}}{\text{Cell—C—}}} + MH \text{ (or SH)} \longrightarrow \underset{\text{H}}{\overset{\text{OH}}{\text{cell—C—}}} + M\cdot \text{ (or S}\cdot) \longrightarrow \text{homopolymer} \tag{7}$$

$$\underset{(MH)_n\cdot}{\overset{\text{OH}}{\text{Cell—C—}}} + SH \longrightarrow \underset{(MH)_nH}{\overset{\text{OH}}{\text{cell—C—}}} + S\cdot \tag{8}$$

$$D^* + MH \longrightarrow \cdot DH + M\cdot \longrightarrow \text{homopolymer} \tag{9}$$

In the above scheme MH represents the vinyl monomer, SH is the solvent—a possible hydrogen donor—and cell—$\overset{\text{OH}}{\underset{\text{H}}{\text{C}}}$— is a cellulose molecule. The process is initiated by the excited dye (reaction 4), and grafting occurs by reactions

(5) and (6). Homopolymerization preponderates under these conditions, by reactions (7), (8), and (9).

Presence of Oxygen

$$\cdot DH + O_2 \longrightarrow HO_2\cdot + D \tag{10}$$

$$\underset{\cdot}{\text{Cell—C—}}\overset{OH}{} + O_2 \longrightarrow \underset{O_2\cdot}{\text{cell—C—}}\overset{OH}{} \tag{11}$$

$$\underset{O_2\cdot}{\text{Cell—C—}}\overset{OH}{} + HO_2\cdot \longrightarrow \underset{|}{\text{cell—C}}\text{=O} + H_2O_2 + O_2 \tag{12}$$

$$\underset{H}{\text{Cell—C—}}\overset{OH}{} + \cdot OH \longrightarrow \text{cell—C—}\overset{OH}{} + H_2O \tag{13}$$

$$MH + HO_2\cdot \longrightarrow M\cdot + H_2O_2 \longrightarrow \text{homopolymer} \tag{14}$$
$$\text{(or }\cdot OH\text{)} \qquad \text{(or }H_2O\text{)}$$

Both grafting and homopolymerization are inhibited by oxygen, indicating that under these conditions reactions (11) and (12), which occur during photo-tendering, preponderate.

Although most of the sensitizer is regenerated in a cyclic process, there is evidence that side reactions also lead to degradation of the sensitizer and formation of oxidation products.[163-165] One view is that such side products are formed by attack of hydroxyl radicals produced during photodecomposition of water, leading to colored, hydroxylated derivatives,[26a] which have been shown to act as internal filters and to reduce the rate of the photoinduced reaction. Another possibility that cannot be excluded is that the semiquinone type of radical formed from the dye may undergo disproportionation.

The fate of the dye following excitation has been investigated by the use of rigid glass–solvent spectroscopy[166] and flash photolysis,[105-107] and has been reviewed in some detail.[167] The nature of the excited states has now been clarified.[167a]

b. *Sensitization by Metallic Oxides.*—Although it has long been recognized that metallic oxides may sensitize the photodegradation of cellulose,[122,124] little attempt has been made to elucidate the mechanism of the process. It has been shown, for example, that tungsten and molybdenum oxides,[168] when finely dispersed on dry cellulose (filter paper) or other polysaccharides, assume a blue color when exposed to light of 230 to 390 nm. Reduction to colored oxides of lower valence states was presumed to be responsible for the change.

The presence of moisture influenced the process. In the case of tungsten oxide, irradiation led to considerable degradation of the cellulose.

Consideration was given by Schurz et al.[112] to the role of titanium(IV) oxide as a catalyst in the light-damaging of cellulose. It was postulated that photo-oxidation in the presence of titanium oxide first causes oxidation of the D-glucose ring without splitting the D-glucosidic bonds, followed by secondary splitting of these bonds.

No information is available about the precise manner in which metallic oxides make the radiation energy available for photodegradation of cellulose. It is the author's view that the promotion of sensitized decomposition of water by the oxide, followed by direct abstraction of hydrogen atoms from the cellulose, may initiate photodegradation. For example, when Baur et al.[169] exposed suspensions of zinc oxide in water in the presence of air to sunlight for 10 to 15 hr, it was observed that the solutions liberated iodine from potassium iodide solution. When D-glucose was present during a similar exposure, hydrogen peroxide was formed; this result led these and later investigators[169-172] to conclude that hydrogen peroxide was the agent responsible for the oxidation of iodide and also for the photodegradation of cellulose. However, it is highly probable from later work in radiation studies of aqueous solutions that the processes involved in irradiations in the presence of zinc oxide are explained more satisfactorily in terms of precursors of the hydrogen peroxide.

When light is absorbed by zinc oxide, the initial reaction may involve transfer of an electron from zinc to an oxygen molecule absorbed on the surface. The resulting species, $\cdot O_2^-$, is a radical ion. It has been shown that organic alcohols[173,174] can be oxidized by way of this initiation step, followed either by transfer of electrons to the photopositive surface of the zinc oxide (after loss of a photoexcited electron by the crystal to O_2), or by hydrogen abstraction by the peroxide radical to form HO_2^-. The peroxide formed in the absence of organic material[172] may occur through reactions of the following type[175,176]:

$$\cdot O_2^- + H_2O \longrightarrow HO_2^- + \cdot OH$$

The $\cdot OH$ radical has been shown to be a powerful oxidizing entity and would find an ideal oxidizable substrate in cellulose. Similarly, the liberation of iodine from iodide solutions containing zinc oxide by the action of sunlight is probably due to processes such as the following:

$$\cdot OH + I^- \longrightarrow I\cdot + OH^-$$

$$\cdot I + \cdot I \longrightarrow I_2$$

For photodegradation of cellulose to occur, it is not necessary, therefore, to postulate the prior formation of hydrogen peroxide.

References start on p. 1287.

IV. RADIOLYSIS IN AQUEOUS SOLUTION[2,2a]

A. CHEMICAL CHANGES

Oxygen exerts an important influence on the nature of the products formed during high-energy irradiation of carbohydrates in aqueous solution. In radiation-chemical studies, it is, therefore, essential to maintain either evacuated or oxygenated conditions throughout a particular irradiation. Moreover, it must be emphasized that the method of deaerating a solution by passing nitrogen is not so satisfactory as the procedure of complete evacuation; Bourne *et al.*[177] have observed appreciable differences between sugar solutions irradiated under vacuum and under nitrogen. The use of argon to deoxygenate solutions of carbohydrates is now in common use and is satisfactory.

1. *Aldoses, Hexuloses, and Hexitols*

a. Aldoses.—Complex chemical changes follow the irradiation of hexose solutions with ^{60}Co γ-radiation and fast electrons,[178-189] and it was necessary to develop new radioisotopic methods to elucidate the nature of the products and the primary reactions.[190,190a] Yield–dose curves for the major products were obtained by carrier-dilution analysis at several dose levels, and in this way primary and secondary products were distinguished. The overall consumption of hexose during irradiation shows an initial *G* value* of 3.5 in oxygen and *in vacuo*. The main changes, given here for D-mannose,[182] are paralleled also with other hexoses.[187] The main processes in oxygen are as shown.

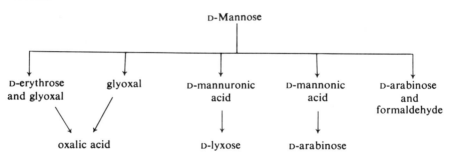

Under vacuum, the complications of secondary reactions involving oxygen are excluded, and the degradative pattern follows a rather different path.[187,188] The initial *G* values for the primary products are given in Table I.

A distinctive feature of irradiations under vacuum, not encountered in oxygen, is the formation of polymer at high doses. This behavior was first reported by Stacey *et al.*[191,192] A striking feature is the identical initial rate of degradation of D-glucose and D-mannose in oxygen and *in vacuo* (*G* = 3.5).

* Yield of radiation-induced change, molecules per 100 eV of energy.

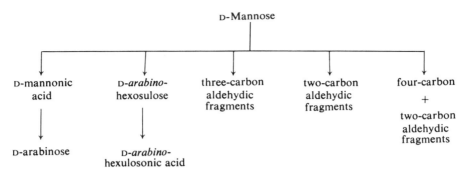

This indicates that the primary abstraction processes are similar and are independent of oxygen. On this basis, the difference between products found after irradiation in oxygen and *in vacuo* must be attributed to different secondary reactions of identical primary radicals. Certain products might, therefore, be expected to be the same under both conditions, and any differences should be rationalized by consideration of the fate of the radicals formed initially. The radiation stabilities of individual hexoses appear to be identical.[193]

Common processes observed are the formation of hexonic acids, and ring-scission reactions which lead to two- and three-carbon aldehydic fragments. The presence of oxygen does not appear to have any marked effect in controlling the extent of these processes. Oxygen influences the product arising from attack by a radical at C-2. D-*arabino*-Hexosulose is formed *in vacuo*, but ring

TABLE I

INITIAL G VALUES OF PRIMARY PRODUCTS FROM γ-IRRADIATED SOLUTIONS
($5 \times 10^{-2}M$) OF HEXOSES[187]

	D-*Glucose*		D-*Mannose*	
Constituent	In vacuo	*Oxygen*	In vacuo	*Oxygen*
Hexose ($-G$)	3.5	3.5	3.5	3.5
Hexonic acid	0.35	0.4–0.5	0.45	0.6–0.7
Hexuronic acid	Absent[a]	0.9	Absent[a]	0.8–1.0
Arabinose	S[b]		S[b]	0.5–0.6
Formaldehyde	S[b]			0.3
D-*arabino*-Hexosulose	0.4	Absent[a]	0.5	Absent[a]
Two-carbon fragments	0.85	0.8[c]	0.95	0.7
				(0.2)[d]
D-Erythrose		0.25	0.25	0.18
Three-carbon fragments	0.8		0.5	
Polymer	S[b]	Absent[a]	S[b]	Absent[a]

[a] Absent in significant proportion. [b] Secondary product. [c] Recalculated value. [d] Formed from C-1 and C-2.

scission occurs in oxygen to give D-arabinose and formaldehyde. Both re-actions occur to a comparable overall extent (G = 0.4 to 0.6). This behavior is analogous to that of aqueous glycolic acid during irradiation. The carbon–carbon scission that occurs in oxygen is diminished *in vacuo*.[194,195] However, the same primary radical may yield all the products.

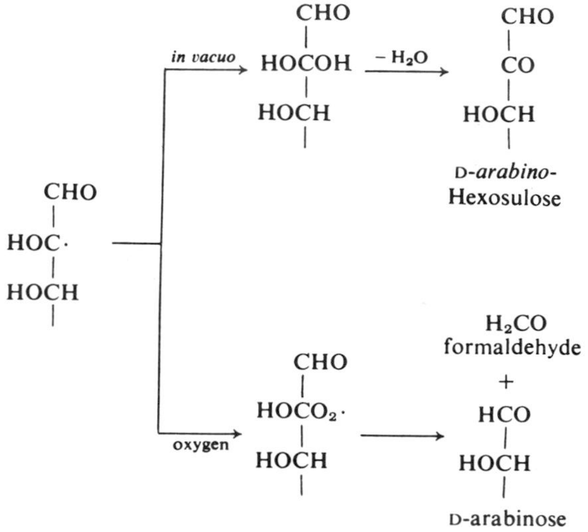

$$
\begin{array}{ccc}
 & \text{CHO} & \text{CHO} \\
 & | & | \\
\xrightarrow{\text{in vacuo}} \text{HOCOH} & \xrightarrow{-\text{H}_2\text{O}} & \text{CO} \\
 & | & | \\
 & \text{HOCH} & \text{HOCH} \\
 & | & | \\
 & & \text{D-}arabino\text{-} \\
 & & \text{Hexosulose}
\end{array}
$$

CHO
|
HOC·
|
HOCH
|

$$
\begin{array}{ccc}
 & & \text{H}_2\text{CO} \\
 & \text{CHO} & \text{formaldehyde} \\
 & | & + \\
\xrightarrow{\text{oxygen}} \text{HOCO}_2\cdot & \longrightarrow & \text{HCO} \\
 & | & | \\
 & \text{HOCH} & \text{HOCH} \\
 & | & | \\
 & & \text{D-arabinose}
\end{array}
$$

No uronic acid is found after irradiation *in vacuo*, and no polymer is pro-duced in oxygen. These observations may be rationalized by considering the fate of the primary radical RĊHOH, formed by hydrogen abstraction at C-6 in oxygen and *in vacuo*. After the initial dimerization step *in vacuo*, further radicals may be grafted onto the growing polymer chain. On the basis of such kinetics, simple competition between initial hexose and products for the primary species formed during radiolysis in water may be envisaged. From a practical point of view, it serves as a useful generalization to permit calculation of hexose concentration at any particular dose.

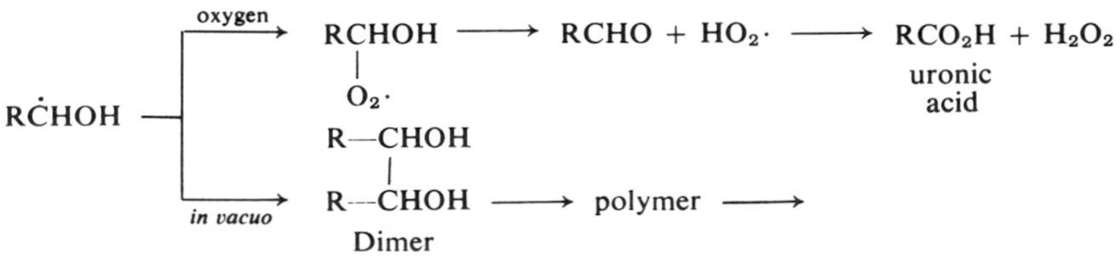

$$
\begin{array}{l}
\xrightarrow{\text{oxygen}} \text{RCHOH} \longrightarrow \text{RCHO} + \text{HO}_2\cdot \longrightarrow \text{RCO}_2\text{H} + \text{H}_2\text{O}_2 \\
\qquad\qquad | \qquad\qquad\qquad\qquad\qquad\qquad\qquad \text{uronic} \\
\text{RĊHOH} \qquad \text{O}_2\cdot \qquad\qquad\qquad\qquad\qquad\qquad \text{acid} \\
\qquad\qquad \text{R—CHOH} \\
\qquad\qquad | \\
\xrightarrow{\text{in vacuo}} \text{R—CHOH} \longrightarrow \text{polymer} \longrightarrow \\
\qquad\qquad \text{Dimer}
\end{array}
$$

Summarizing, therefore, it would appear that when hexoses are irradiated in dilute solution, attack is not confined to any particular part of the molecule. The nature of the products formed in oxygen and under vacuum demonstrate that all bonds are affected. Oxidation occurs at the extremities of the molecule, and ring scission simultaneously leads to lower fragments.

Pentoses are degraded similarly. When D-ribose is irradiated in oxygenated, aqueous solution, end-group oxidation and chain scission to yield lower aldehydic fragments are the main effects. This process has been studied in considerable detail.[195a]

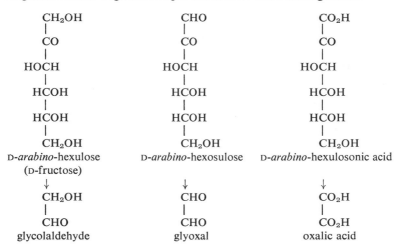

b. Hexuloses.—When D-fructose is irradiated in oxygenated solution, the following interrelated degradation processes have been distinguished.[181]

CH₂OH	CHO	CO₂H
CH_2OH	CHO	CO_2H
CO	CO	CO
HOCH	HOCH	HOCH
HCOH	HCOH	HCOH
HCOH	HCOH	HCOH
CH₂OH	CH₂OH	CH₂OH
D-*arabino*-hexulose (D-fructose)	D-*arabino*-hexosulose	D-*arabino*-hexulosonic acid
↓	↓	↓
CH₂OH	CHO	CO₂H
CHO	CHO	CO₂H
glycolaldehyde	glyoxal	oxalic acid

Evidence has also been educed for the formation of two fragments, reductone and 1,3-dihydroxy-2-propanone, by symmetrical scission of the molecules of D-fructose; these constituents are mainly responsible for the peak at 285 to

---

References start on p. 1287.

290 nm in the ultraviolet absorption spectrum of irradiated solutions of D-fructose.

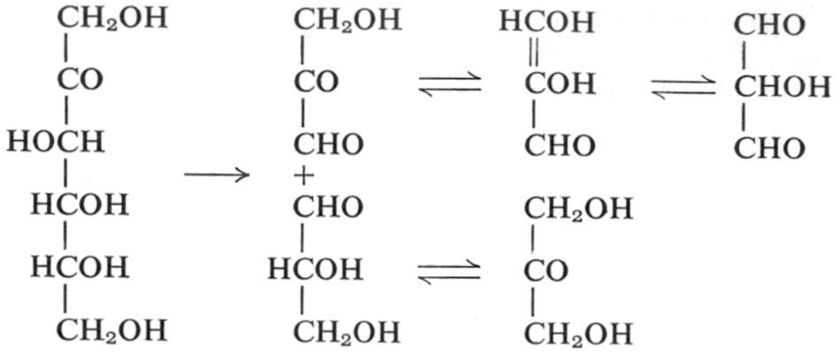

$$
\begin{array}{ccccc}
\text{CH}_2\text{OH} & \text{CH}_2\text{OH} & \text{HCOH} & & \text{CHO} \\
| & | & \| & & | \\
\text{CO} & \text{CO} \rightleftharpoons & \text{COH} \rightleftharpoons & & \text{CHOH} \\
| & | & | & & | \\
\text{HOCH} & \text{CHO} & \text{CHO} & & \text{CHO} \\
| \longrightarrow & + & & & \\
\text{HCOH} & \text{CHO} & \text{CH}_2\text{OH} & & \\
| & | & | & & \\
\text{HCOH} & \text{HCOH} \rightleftharpoons & \text{CO} & & \\
| & | & | & & \\
\text{CH}_2\text{OH} & \text{CH}_2\text{OH} & \text{CH}_2\text{OH} & &
\end{array}
$$

For the decomposition of D-fructose, the rate of consumption (G) is 4.0, close to that of D-glucose and D-mannose $(G = 3.5)$ under comparable conditions. Another primary process may involve oxidation of the primary alcohol group at C-6 to form D-*lyxo*-5-hexulosonic acid.

c. Hexitols.—The degradation induced in hexitols on irradiation in aqueous solution is more specific than that for hexoses under comparable conditions. When 1% solutions of D-mannitol are irradiated with fast electrons in an oxygenated medium, D-mannose is the main product; after prolonged irradiation, D-mannuronic acid is formed.[183] Similar results were obtained[179] with more-concentrated solutions (50%), and D-mannose was characterized in the product. D-Arabinose was formed as a secondary product, in addition to D-mannuronic acid.

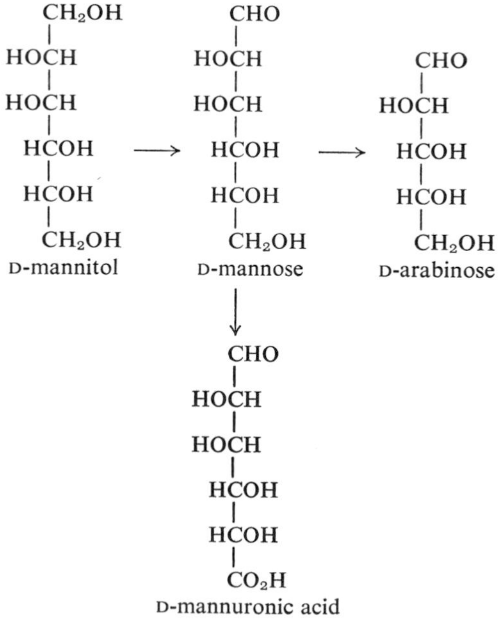

$$
\begin{array}{ccc}
\text{CH}_2\text{OH} & \text{CHO} & \\
| & | & \\
\text{HOCH} & \text{HOCH} & \text{CHO} \\
| & | & | \\
\text{HOCH} & \text{HOCH} & \text{HOCH} \\
| & | & | \\
\text{HCOH} \longrightarrow & \text{HCOH} \longrightarrow & \text{HCOH} \\
| & | & | \\
\text{HCOH} & \text{HCOH} & \text{HCOH} \\
| & | & | \\
\text{CH}_2\text{OH} & \text{CH}_2\text{OH} & \text{CH}_2\text{OH} \\
\text{D-mannitol} & \text{D-mannose} & \text{D-arabinose}
\end{array}
$$

$$
\begin{array}{c}
\downarrow \\
\text{CHO} \\
| \\
\text{HOCH} \\
| \\
\text{HOCH} \\
| \\
\text{HCOH} \\
| \\
\text{HCOH} \\
| \\
\text{CO}_2\text{H} \\
\text{D-mannuronic acid}
\end{array}
$$

Oxidation of either of the primary hydroxyl groups in D-mannitol would give D-mannose, and hence the characterization of D-mannose is easier than that of the products from irradiation of D-glucitol. Paper-chromatographic evidence reveals the presence of four main products from D-glucitol—glucose, gulose, xylose, and arabinose.[179,196] Carrier-dilution analyses have demonstrated that the stereochemical forms present are D-arabinose, L-xylose, and D-glucose.[185,196] On configurational grounds, therefore, it is deduced that gulose is present as the L enantiomorph. Oxidation of the primary alcohol group at one extremity of the D-glucitol molecule leads to D-glucose, and oxidation at the other end gives L-gulose. Similar considerations apply to the formation of pentoses. Simultaneously, the presence of formaldehyde was detected by isotope-dilution analysis. Thus, the degradation may be represented as shown. The yield–dose curves[185] for the hexoses and pentoses

```
                    CHO
                     |
                    HCOH            CHO
                     |               |
                    HOCH            HOCH
                     |               |          + H₂CO
    CH₂OH           HCOH            HCOH    formaldehyde
     |      ⟶       |       ⟶       |
    HCOH            HCOH            HCOH
     |               |               |
    HOCH            CH₂OH           CH₂OH
     |            D-glucose       D-arabinose
    HCOH
     |
    HCOH            CH₂OH           CHO
     |               |               |
    CH₂OH  ⟶        HCOH            HOCH           CH₂OH           CHO
  D-glucitol         |               |              |               |
                    HOCH            HOCH           HCOH            HOCH     + H₂CO
                     |       ≡       |              |               |     formalde-
                    HCOH            HCOH    ⟶      HOCH    ≡       HCOH      hyde
                     |               |              |               |
                    HCOH            HOCH           HCOH            HOCH
                     |               |              |               |
                    CHO             CH₂OH          CHO             CH₂OH
                  L-gulose                                       L-xylose
```

demonstrate that D-glucose and L-gulose are the primary products, formed at identical rates ($G = 1.2$), and D-arabinose and L-xylose are formed subsequently. Reactions proceeding with ring scission are also secondary. The rate of degradation is similar to that for hexoses; the G value for D-glucitol is 3.5 for $5 \times 10^{-2} M$ solutions.

Wolfrom et al.[179] commented on the similarity of the products from the irradiation of alditol solutions and from the action of Fenton's reagent (fer-

---

rous ions and hydrogen peroxide) thereon. D-Mannose was obtained from D-mannitol, in 40% yield, by the latter method,[197] through the agency of

$$Fe^{2+} + H_2O_2 \longrightarrow Fe^{3+} + \cdot OH + OH^-$$

hydroxyl radicals, which have now been shown to be the main species responsible for the radiolysis of carbohydrates in aqueous solution.[189]

When conditions of strict evacuation are maintained during irradiation of D-glucitol,[188] the yields of hexoses are lower than in oxygen ($G = 0.7$), although the rate of disappearance of D-glucitol is identical with the rate observed in oxygen ($G = 3.5$). Under vacuum, chain scission gains in significance, and glycolaldehyde and a tetrose are formed. The yield–dose curves for these fragments demonstrate that they are formed by primary processes, with an initial G value of about 1.0.

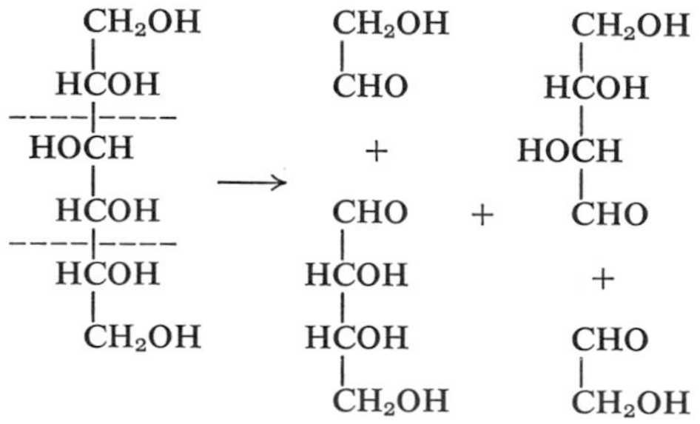

$$
\begin{array}{ccc}
CH_2OH & CH_2OH & CH_2OH \\
| & | & | \\
HCOH & CHO & HCOH \\
----|---- & & | \\
HOCH & + & HOCH \\
| & & | \\
HCOH & CHO \quad + & CHO \\
----|---- & | & \\
HCOH & HCOH & + \\
| & | & \\
CH_2OH & HCOH & CHO \\
& | & | \\
& CH_2OH & CH_2OH \\
\end{array}
$$

Gas analysis indicates that primary abstraction processes under vacuum involve hydrogen atoms as well as hydroxyl radicals. Under these conditions,

$$RCH_2OH + \cdot OH \longrightarrow R\dot{C}HOH + H_2O$$

$$RCH_2OH + \cdot H \longrightarrow R\dot{C}HOH + H_2$$

initial dimerization occurs as the primary step in the formation of the polymer. The polymer may be isolated after irradiation of D-glucitol in the absence of oxygen. Three primary processes have, therefore, been identified as occurring

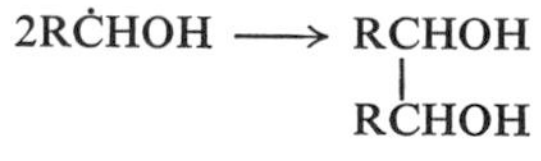

$$
2R\dot{C}HOH \longrightarrow \begin{array}{c} RCHOH \\ | \\ RCHOH \end{array}
$$

when solutions of D-glucitol are irradiated under vacuum.

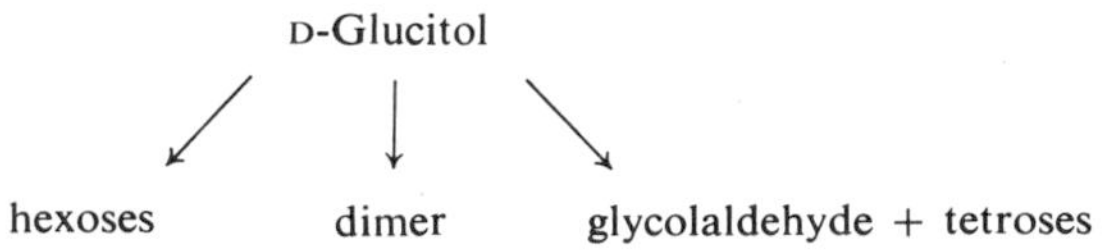

$$
\begin{array}{c}
\text{D-Glucitol} \\
\swarrow \quad \downarrow \quad \searrow \\
\text{hexoses} \quad \text{dimer} \quad \text{glycolaldehyde + tetroses}
\end{array}
$$

2. Disaccharides, Trisaccharides, Glycosides, and Lactones

Hydrolysis is the predominating reaction when maltose and cellobiose are irradiated with fast electrons in air-equilibrated solutions.[179,198] Hydrolysis at the glycosidic bond also occurs during irradiation of sucrose in aqueous solution, and acid is produced.[199,200] The accompanying overall change in optical rotation was proposed by Wright[201] for use as a pile-radiation dosimeter. In comparison with sucrose, methyl α-D-glucopyranoside is more resistant to radiation-induced hydrolysis.[202] Yield–dose curves obtained by carrier-dilution analysis reveal that D-glucose and D-fructose are primary products of the γ-irradiation of dilute, aqueous sucrose solution in the presence of oxygen, together with smaller amounts of D-*arabino*-hexosulose and D-gluconic acid.[203] D-Glucuronic acid, D-*arabino*-hexulosonic acid, D-arabinose, and two- and three-carbon aldehydic fragments arise by secondary processes. In the final stage, carbon dioxide and formic acid are formed. Hydrogen peroxide is produced continuously. The initial G values for D-glucose and D-fructose are 1.5 to 1.6; D-gluconic acid and D-*arabino*-hexosulose are formed with G values of 0.4 and 0.6, respectively.

The primary process leading simultaneously to D-gluconic acid, D-*arabino*-hexosulose, D-glucose, and D-fructose may be accounted for by two types of oxidative scission of the disaccharide linkage, at a and b. The former leads to D-fructose and D-gluconic acid, and the latter to D-glucose and D-*arabino*-hexosulose. A similar type of process was envisaged for the degradation of

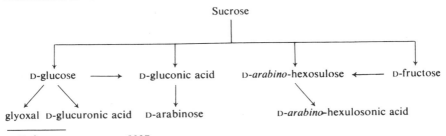

aqueous dextran with γ-radiation.[204] If the two types of scission occurred to a comparable extent, the amounts of the four main products would be of the same order. The results show, however, that although comparable amounts of D-gluconic acid and D-*arabino*-hexosulose are formed, the proportions of D-glucose and D-fructose are higher. It appears, therefore, that hydrolysis is the dominant process, but that it is accompanied to a smaller extent by the oxidative scission described. The overall degradation pattern for sucrose has been formulated as follows:

Sucrose

D-glucose → D-gluconic acid   D-*arabino*-hexosulose ← D-fructose

glyoxal D-glucuronic acid   D-arabinose   D-*arabino*-hexulosonic acid

The radiation-induced degradation of disaccharides in solution appears generally to proceed in this manner. The same two modes of cleavage of the glycosidic linkage were found also for lactose[205] and maltose.[206]

Hydrolysis is the predominating process when the trisaccharide raffinose is exposed to γ-rays in 2% aqueous solution.[179,207] The extent of hydrolysis increases with dose and, since two bonds are hydrolyzed, the reducing power–dose curve is nonlinear.

Schulte-Frohlinde, von Sonntag, and co-workers[190a,195a,207a,b] subsequently used coupled gas chromatographic–mass spectrometric techniques to identify many more products than previously found from the γ-irradiation of carbohydrates in aqueous solution and in the solid state. The behavior of cellobiose[207b,c] detailed in Scheme 1 is a typical example, which illustrates the types of chemical changes that occur in aqueous solution. The reactions of radicals produced by attack of $\cdot$OH at C-1' and C-1 are depicted below. With the advent of these powerful analytical methods, rapid new developments have been made in this area of product identification,[2a] for instance in the γ-radiolysis of polycrystalline cycloamylose hydrates.[207d] This approach is now transforming the field.[2a]

Radiation-induced hydrolysis is again the predominating process during irradiation of glycosides in aqueous solution. For methyl α-D-glucoside the $-G$ value is 2.6 for $10^{-2} M$ solutions.[208] Hydrolysis is accompanied by the formation of acid ($G = 0.36$) and, among the main radiolysis products, derivatives containing a deoxy group have been found.[209] With methyl, phenyl, and benzyl β-D-glucosides, it was shown that the nature of the aglycon does not exert a large effect on the overall decomposition, although the β-D-anomer is somewhat more labile than the α-D anomer. For the β-D derivatives, $-G$ values are 3.2 to 3.4, compared with 2.6 for the α-D derivatives. The mechanism proposed[210] for the breakdown is:

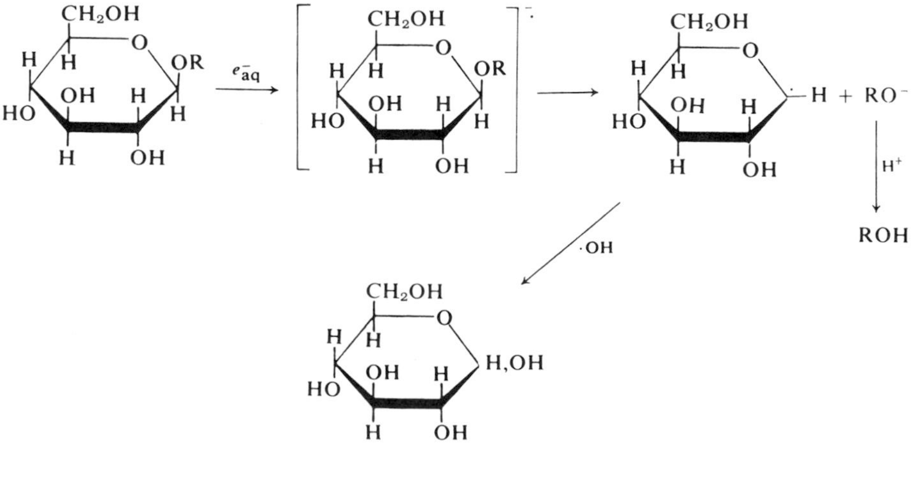

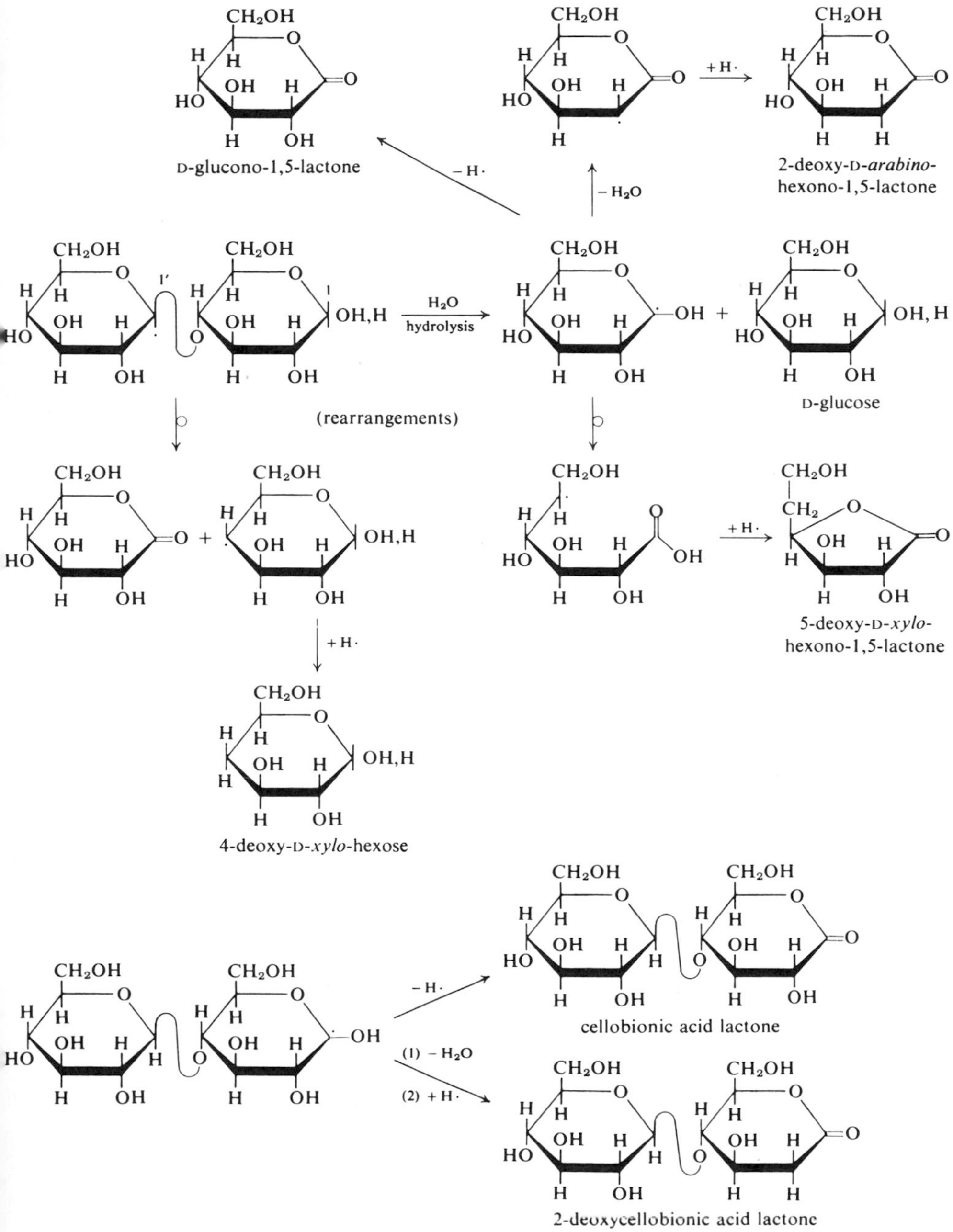
D-glucono-1,5-lactone
− H·
− H₂O
+ H·
2-deoxy-D-*arabino*-hexono-1,5-lactone
H₂O hydrolysis
D-glucose
(rearrangements)
+ H·
5-deoxy-D-*xylo*-hexono-1,5-lactone
4-deoxy-D-*xylo*-hexose
cellobionic acid lactone
(1) − H₂O
(2) + H·
2-deoxycellobionic acid lactone

SCHEME 1

References start on p. 1287.

Coleby[208] showed that when solutions of D-glucono-1,4-lactone, D-gulono-1,4-lactone, and L-gulono-1,4-lactone are irradiated with X-rays (210 kVp, 10 mA), γ-rays (^{60}Co), or fast electrons (4 MeV) under vacuum, the lactones are converted into the corresponding ascorbic acids.

$$
\begin{array}{ccc}
\left.\begin{array}{c}
\text{C=O} \\
| \\
\text{CHOH} \\
| \\
\text{CHOH} \\
| \\
\text{HCO}
\end{array}\right| & & \left.\begin{array}{c}
\text{C=O} \\
| \\
\text{COH} \\
\| \\
\text{COH} \\
| \\
\text{HCO}
\end{array}\right| \\
| & \longrightarrow & | \\
\text{CHOH} & & \text{CHOH} \\
| & & | \\
\text{CH}_2\text{OH} & & \text{CH}_2\text{OH}
\end{array}
$$

The route to the ascorbic acid from the lactone may involve abstraction of a hydrogen atom at C-2 or C-3, followed by enolization.

$$
\begin{array}{c}
| \\
\text{HCOH} \\
| \\
\text{HCOH} \\
|
\end{array}
\xrightarrow{\cdot\text{H or }\cdot\text{OH}}
\begin{array}{c}
| \\
\cdot\text{COH} \\
| \\
\text{HCOH} \\
|
\end{array}
\xrightarrow{\cdot\text{H or }\cdot\text{OH}}
\begin{array}{c}
| \\
\text{C=O} \\
| \\
\text{HCOH} \\
|
\end{array}
\rightleftharpoons
\begin{array}{c}
| \\
\text{COH} \\
\| \\
\text{COH} \\
|
\end{array}
$$

Irradiated solutions of carbohydrates have similar, characteristic ultraviolet absorption spectra.[178–181,184,208,211] A broad absorption occurs in the region of 240 to 300 nm, and the maxima, which may vary for individual carbohydrates, fall in the region of 260 to 290 nm. The intensity of the peak increases very markedly on addition of alkali, and the peak may shift to higher wavelengths. Several of the identified products absorb in this region, either in acid or in alkaline solution, and it is probable that more than one absorbing constituent is responsible for the spectrum observed. One compound which may contribute to the composite spectrum is 1,3-dihydroxy-2-propanone (λ_{max} 265 nm in neutral solution), formed by isomerization of glyceraldehyde, a change that occurs readily (particularly in alkali).[178] Reductone (λ_{max} in alkali, 287 nm, and in acid, 268 nm), D-arabino-hexosulose (λ_{max} in alkali, 265 nm) and the dienol form of D-arabino-hexulosonic acid (λ_{max} in alkali, 275 nm, and in acid, 230 nm), are other possible absorbing species. Particular attention has been drawn to the formation of a reductone during the irradiation of solutions of lactose.[212] However, all enediol structures, such as L-ascorbic acid (λ_{max} in neutral or acid solutions, 265 nm), would be expected to absorb strongly in this region, and, since "D-arabino-ascorbic" acid has been claimed to be a product from the irradiation of D-glucono-1,4-lactone in aqueous solution, secondary products may also contribute to the overall spectrum.

Evidence exists for the presence of slow postirradiation processes when sugar solutions are irradiated in oxygen and under vacuum. When solutions of D-glucose are irradiated under vacuum, the absorption at 265 nm increases steadily and attains a maximum 20 to 30 hr after irradiation has ceased.[178] Similarly, for D-fructose solutions irradiated in oxygen and under vacuum, the absorption maxima continue to increase for several days after irradiation has ceased. This process may be associated with the postirradiation decrease in concentration of hydrogen peroxide, at a rate of 3 to 4 $\times$ 10^{13} molecules min^{-1} ml^{-1} in irradiated D-fructose[181] and D-glucose[178] solutions. The occurrence of postirradiation reactions is further demonstrated in these systems by the observed liberation of gas for 24 to 30 hr after irradiation has been terminated.[178]

3. Polysaccharides

When polysaccharides are irradiated in solution, extensive degradation is observed. This behavior has been noted with dextran,[213,214] starch,[215,216] agar,[217] alginic acid, various gums,[218] pectins,[218–220] and hyaluronic acid.[221,222] The degradation is manifested in a decrease in viscosity and the formation of reducing substances.[216,218,222,223] The addition of sugars, such as sucrose, D-glucose, and D-fructose to a polysaccharide solution has a protective effect, presumably because the sugars are better scavengers of ·OH radicals than the polysaccharide.

When aqueous solutions of amylose or starch are irradiated, depolymerization occurs to give lower saccharides and "dextrins."[177,224] The process may, however, not be related entirely to simple hydrolysis, for acid is formed and an absorption maximum appears at 265 nm after irradiation.[207] Absorption in this region appears to be a feature characteristic of irradiated solutions of carbohydrates. The degradation of amylose[177] was followed by measuring the increase in reducing power and by the reduction in intensity of the color formed by adding iodine to the solution. According to the results obtained by both methods, oxygen inhibits the degradation. Although the products formed have not been identified with certainty, there are indications from paper chromatography that (in addition to D-glucose) maltose, maltotriose, glyoxal, and a tetrose are formed. The production of acid is independent of the extent of degradation of the amylose and depends only on the dose of radiation. The initial G value (acid) in oxygen is 1.5; under vacuum it is 1.4. It is thus evident that at least two primary processes take place simultaneously. One process leads to formation of acid and appears to be independent of oxygen, and the second process leads to degradation and is inhibited by the presence of oxygen.

Irradiation of solid potato-starch granules (moisture content 18.5%) leads to

---

References start on p. 1287.

changes similar to those observed with starch in solution.[225] After irradiation, the intrinsic viscosity and iodine-staining power of the starch specimens decreased, whereas the reducing power, lability to alkali, carboxyl value, and carbonyl number of the specimens increased with radiation dose. X-ray diffraction patterns showed that the crystalline part of the potato starch is damaged after doses of up to 1.5×10^7 roentgens. By paper chromatography, indications were obtained that D-glucose, maltose, D-arabinose, D-gluconic acid, D-glucuronic acid, and a series of saccharides of low molecular weight are formed by irradiation, with D-glucose and maltose preponderating. The gases produced during γ-irradiation are hydrogen, carbon monoxide, and carbon dioxide. When aqueous solutions of D-glucose are irradiated, hydrogen is the principal gas produced, with carbon dioxide and carbon monoxide being formed in smaller proportions. D-Arabinose, D-gluconic acid, and D-glucuronic acids are also among the products.[178] Therefore, it appears probable that, as for starch solutions, irradiation of starch granules leads, by oxidative and hydrolytic processes, to lower saccharides that subsequently undergo degradation.

When dextran is irradiated in aqueous solution with γ-radiation, one of the principal processes to result is hydrolysis to D-glucose, isomaltose, and isomaltotriose. Moreover, the yield–dose curve (increase in the concentration of product with the dose) for reducing substances indicates that formation of D-glucose and other lower saccharides is a primary process. Two other major products are D-gluconic acid and D-glucuronic acid, and the overall yield–dose curve for formation of acid indicates that these acids are also primary products, since they constitute the major acids produced. It is evident, therefore, that irradiation of dextran in solution causes degradation by at least two independent processes involving scission at the α-D-$(1 \to 6)$ linkage in the molecule. One leads to D-glucose and lower saccharides, and the other to D-gluconic acid and D-glucuronic acid. Glyoxal and D-erythrose, which are present in smaller proportions, are secondary products from D-glucose. Thus, random attack along the chain, as occurs with disaccharides, would lead to all the main products observed.

B. DEGRADATIVE MECHANISMS

During the action of ionizing radiation on dilute aqueous solutions, the energy is deposited almost exclusively in the water. As a result of the sequence of processes already described (Section II,B,2) several species are produced:

$$H_2O \xrightarrow{\quad\quad} e_{aq}^-,\ \cdot H,\ \cdot OH,\ H_2,\ H_2O_2$$

In aqueous solution, therefore, degradation is due largely to the effect of such reactive species. By the techniques of pulse radiolysis,[189] it has been possible to evaluate the reactivity of carbohydrates toward these species. Hexoses, pentoses, hexitols, and indeed any sugar that does not contain a reactive substituted group, are unreactive with solvated electrons (e_{aq}^-). For example, in a $5 \times 10^{-3} M$ solution of D-glucose, the second-order rate constant for the reaction e_{aq}^{-} + D-glucose is less than 10^7 mole^{-1} sec^{-1}. In neutral solution at solute concentrations $<5 \times 10^{-3} M$, it is the reaction with ·OH radicals that leads to degradation.[189,226] For the reaction ·OH + D-glucose, $k_2 = 1 \times 10^9$ mole^{-1} sec^{-1}. A transient species having an absorption maximum at 260 to 270 nm is produced as a result of initial abstraction by ·OH radicals. The transient species has been identified as an RĊHOH radical, produced as follows:

$$RCH_2OH + \cdot OH \longrightarrow R\dot{C}HOH + H_2O$$

This transient species disappears according to second-order kinetics:

$$2R\dot{C}HOH \longrightarrow RCHO + RCH_2OH$$

In oxygen, the primary abstraction process remains unaltered, and the $-G$ value for the disappearance of D-glucose remains the same. Different products result, however, because of the participation of oxygen in secondary processes. The overall stoichiometry is fulfilled by the mechanism shown:

$$R\dot{C}HOH + O_2 \longrightarrow \underset{\underset{\dot{O}_2}{|}}{RCHOH}$$

$$\underset{\underset{\dot{O}_2}{|}}{RCHOH} \longrightarrow RCHO + HO_2\cdot$$

$$2HO_2\cdot \longrightarrow H_2O_2 + O_2$$

Thus,

$$G_{(H_2O_2)} = \frac{(G_H + G_{OH})}{2} + G_p(H_2O_2)$$

where G_H, G_{OH}, and $G_p(H_2O_2)$ are the primary yields of H· and ·OH radicals and hydrogen peroxide. Similar reactions occur at positions other than C-6.

However, it is clear that abstraction by ·OH radicals cannot account satisfactorily for all features of carbohydrate degradation at concentrations above $5 \times 10^{-3} M$. Yields of product are dependent on solute concentration in the range 5×10^{-4} to $5 \times 10^{-2} M$, and the yields observed cannot be accounted for on the basis of accepted radical yields.[206,227] Typical behavior is shown by D-glucose, which on γ-irradiation in $5 \times 10^{-2} M$ solution at pH 7 gives a $-G$ (D-glucose) value of 3.5 in oxygen and also *in vacuo*. This value is greater than G_{OH}, which is 2.4 at pH 7. Furthermore, the radiolytic degradation of D-glucose in aqueous solution proceeds at $5 \times 10^{-4} M$ by a process that is chemically and kinetically different from that observed at $5 \times 10^{-2} M$. Whereas oxidative and chain-scission reactions occur at $5 \times 10^{-2} M$, only the former are found at $5 \times 10^{-4} M$. In $5 \times 10^{-4} M$ solution, the disappearance of D-glucose is first-order and the plot of log(D-glucose) against dose is linear.[187] In contrast, at $5 \times 10^{-4} M$, the disappearance of D-glucose is zero-order and the plot of D-glucose against dose is linear, at least up to 70% conversion of the initial D-glucose.[206] The difference in the behavior at the two concentrations has been attributed to transfer of excitation energy to the carbohydrate from water molecules bound to the carbohydrate. Excitation transfer from bound water is favorable as a result of the considerable overlap between the electrical-dipole fields of the donor and acceptor molecules.

Hyaluronic acid, the glycosaminoglycan derived from connective tissue, is particularly sensitive to radiation damage. This behavior has been related to its facile reaction with e_{aq}^- and $\cdot OH$ radicals.[226a] The radiation protection of hyaluronic acid by oxygen can, therefore, be attributed to the removal of e_{aq}^- by the reaction

$$e_{aq}^- + O_2 \longrightarrow \cdot O_2^-$$

When aryl glycosides are irradiated with γ-rays in aqueous solution, e_{aq}^- and $\cdot OH$ radicals attack the aromatic aglycon preferentially. A correlation

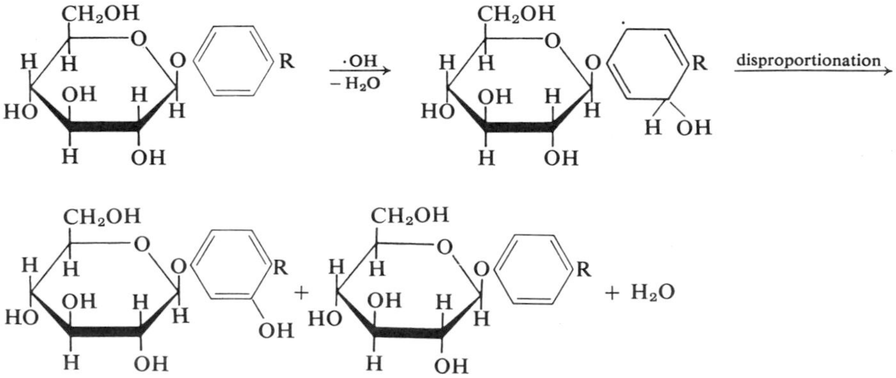

between the reaction rates of e_{aq}^- and $\cdot OH$ radicals, and the corresponding Hammett function of the substituent, demonstrates that e_{aq}^- acts in a manner analogous to that of nucleophilic reagents, and $\cdot OH$ in the manner of electrophilic reagents. Thus, electron-repelling groups decrease the rate of reaction of e_{aq}^- and facilitate the reaction of $\cdot OH$. Electron-attracting groups exert the opposite effect. Excitation transfer from excited water molecules bound to the carbohydrate leads to glycosidic scission.[228]

V. EFFECTS OF HIGH-ENERGY RADIATION IN THE SOLID STATE

A. CHEMICAL AND PHYSICAL CHANGES

The identification of the products of high-energy irradiation in the solid state has proved particularly intractable.[2a] By any method of analysis, the large excess of unchanged carbohydrate presents a serious obstacle. It is known that acids and other products are formed on γ-irradiation of solid α-D-glucose.[229,230] Isotope-dilution analysis was used to confirm the presence of D-gluconic and D-glucuronic acids among the products. Other acidic and neutral products have remained unidentified. These unidentified acids contain a high proportion of aldehyde groups and polymerize readily on standing.

Major attention in solid-state irradiations has been directed to the nature of the electron spin resonance (e.s.r.) spectra of the free radicals that are formed.[231-239] Changes with storage time and heat-treatment have been noted in the character of the e.s.r. spectrum of irradiated polycrystalline D-glucose, and changes in overall free-radical concentration have been found.[231,233] The free radicals produced from many mono- and disaccharides may be quite stable in air.[229] For a wide range of sugars, initial G (radical) values between 2 and 4 have been found.[229,240] The e.s.r. spectrum of the radical formed by γ-irradiation of D-glucose approximates to a three-line spectrum, with another single-line spectrum of slightly different g value superimposed.[234,240] Radicals of the type $-\dot{C}H_2$ at C-6 or $\cdot C-CH_2OH$ at C-5 could give rise to a three-line spectrum through interaction with the two methylene protons. The latter radical is the more probable, since it would derive some stabilization from the delocalization energy contributed by the neighboring hemiacetal oxygen atom. Furthermore, the splitting observed (about 17 gauss) is less than would be expected as a result of interaction with protons alone (about 22 gauss). The splitting may be attributed to a reduction in the spin density on the carbon atom owing to the mesomeric effect of the oxygen atom, such as occurs in radicals of the type $\cdot$CHROH. Only weak coupling between the unpaired electron and the OH proton would be expected, and doublets arising from strong interactions are resolved generally only in strongly acidic media.[241] The additional single line could be the result of a radical associated with the ring structure and is characteristic of most irradiated sugars. These radicals have been positively identified by studying γ-irradiated, frozen carbohydrates.[241a]

For α-D-glucopyranose monohydrate, D-glucitol, α-D-galactopyranose, and *myo*-inositol, the e.s.r. spectra of samples irradiated either with fast electrons or with X-rays were identical, indicating that the final, radical products produced by both types of radiation are the same.[237] Radicals produced in this way may possess half-lives of up to 12.5 weeks at 20° and approximately 8.5 hr at 50°. The formation of colors and fluorescence in the irradiated crystals is a further indication of ionization and excitation, producing trapped species within the crystal.

Several studies have also been made on the e.s.r. spectra of single crystals[233,234] of sucrose, L-sorbose, D-fructose, lactose hydrate, methyl

---

α-D-glucopyranoside,   D-glucono-1,4-lactone,   2-amino-2-deoxy-D-glucose hydrochloride, and 2,3:4,6-di-O-isopropylidene-α-L-sorbofuranose. It appears that any substituted group is selectively damaged by irradiation. For this reason the e.s.r. spectra of irradiated single crystals of sugar derivatives differ greatly from those found in the unsubstituted parent sugars. The free radicals formed in sugars by irradiation at 77°K are transformed by subsequent annealing. These transformations can be explained by a change in the configuration of the radical in most instances. However, for 2-amino-2-deoxy-D-glucose hydrochloride and 2,3:4,6-di-O-isopropylidene-α-L-sorbofuranose, the transformation includes migration of the free-radical site.

For sucrose crystals[233,234] it has been suggested that the radical is formed by loss of an OH group, whereby the remaining hydrogen atom on the same carbon atom has the strongest interaction with the unpaired electron. However, the same observation may also be consistent with the loss of a hydrogen atom.[236] In irradiated D-fructose and L-sorbose it has been concluded that the radical is formed by loss of H-5. Depending on the maximum temperature to which the sample is exposed, the free electron interacts either with a single hydrogen atom at C-6 or with all three hydrogen atoms at C-4 and C-6. The transition of the radical spectrum, which is due to a conformational change in the molecule, occurs at different temperatures for these two carbohydrates.[233]

Hydrolytic cleavage of the disaccharide bond was found to be the principal reaction during the action of ionizing radiation on methyl α-D-glucopyranoside and di-, tri-, and polysaccharides.[202,242,243]

However, disaccharides that are extremely sensitive to radiation[243a-d] experience relatively low conversion of the glycosidic bond.[243b] The major products are acids $(G > 40)$,[243a,b] which are formed primarily as a result of the modification of the reducing moiety of the disaccharide, as the hydrogen atom at C-1 is the most labile to radiolysis. The acid products have been identified[243] as 5-deoxylactobionic acid $(G = 40)$ and 2-deoxylactobionolactone $(G = 20)$; minor amounts of neutral products form by glycosidic bond scission $(G \sim 4.5)$. The efficient, radiolytic conversion into acid products has been rationalized in terms of facile, radical chain-reactions, which prevail in the solid state due to the existence of a peculiarly favorable crystal structure. The three types of transformation that have been recognized are presented in Scheme 2.

These transformations are also encountered in other systems. For example, 2-deoxy-D-*erythro*-pentose (route 2) yields 2,5-dideoxy-D-*erythro*-pentonic acid.[243e] Similarly, D-fructose yields 6-deoxy-D-*threo*-2,5-hexodiulose (M. Dizdaroglu, H. Leitich, and C. von Sonntag, personal communication) $(G = 60)$ at a dose of 0.8×10^{20} eV g^{-1}. As would be anticipated for radical chain-reactions, product yields are dose-rate dependent.

γ-Irradiation of polycrystalline cycloamylose hydrates[207d] leads to radical-

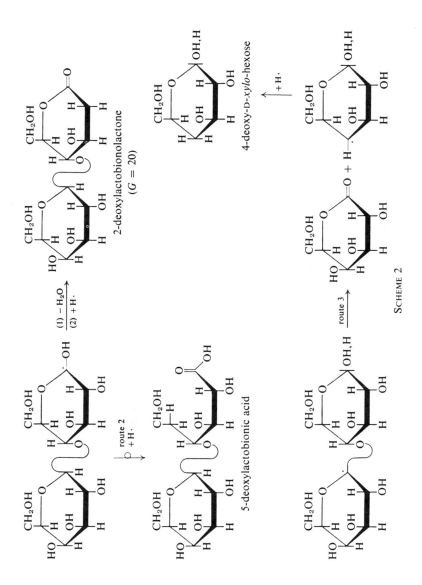

2-deoxylactobionolactone
(G = 20)

4-deoxy-D-xylo-hexose

5-deoxylactobionic acid

Scheme 2

induced opening of the cycloamylose ring and subsequent free-radical reactions giving products typical of those found with other carbohydrate systems.

On γ-irradiation of dry potato starch and its amylose and amylopectin fractions, there was an increase in reducing-group content that was proportional to dose.[244] Degradation of potato starch, amylose, and amylopectin was greatest at low water content (6%) and was considerably less at higher water content (20%).[245,246] Free radicals have been detected by e.s.r. after irradiation. The spectra, although poorly resolved, are typical of those obtained from γ-irradiated sugars.[233,246]

From the standpoint of product formation, the action of high-energy radiation on cellulose has received more study than radiation effects on any other carbohydrate in the solid state. Cellulose, in the form of wood, wood pulp, filter paper, cotton linters, and lint, has been irradiated with X-rays, β-rays, cathode rays, γ-rays, and neutrons, under various atmospheric conditions.[247-254] The overall results are similar. The physical and chemical properties of cellulose are not affected significantly until the cellulose has received a total dose of about 10^6 roentgens, after which the affected properties change rapidly with further increases in dose. Irradiation of cotton cellulose, as fibers or yarns, causes a decrease in tensile strength, elongation, elasticity, and tenacity.[253] The major chemical effects observed are chain cleavage and the formation of reducing groups and acid groups. The formation of reducing groups is the chemical change most readily apparent.

Evolution of gas occurs during the irradiation of cellulose.[255,256] When cellulose is irradiated in a vacuum or in an inert atmosphere the principal gaseous products are hydrogen, carbon monoxide, and carbon dioxide.[257,258] Dalton et al.[259] found that when the cellulose was efficiently outgassed before irradiation, the gas produced by irradiation is almost entirely hydrogen. The G (hydrogen) value rises from near 2 at high doses to $\sim$6 at 0.1 Mrads. Carbon monoxide and carbon dioxide are both produced at increasing rates as the total dose increases, and these gases appear, therefore, to be secondary products. The initial G values are 0.02 for carbon monoxide and 0.008 for carbon dioxide. The change in G value of hydrogen with dose is of considerable interest and may arise either (1) by the presence in cellulose of certain radiolytically weak links, or (2) by a diffusion-controlled repair mechanism. It was also observed by Dalton et al.[256] that the concentration of free radicals in cellulose, under conditions identical to those used for the work on gas yield, rises to a maximum within 10 Mrads of irradiation, and then remains constant. Thus radical–radical reactions appear to take place at an increasing rate as irradiation proceeds, leading to removal of hydrogen atoms and thus to a progressively lower yield of hydrogen. During irradiation of cellulose in vacuo, the appearance of a hydrogen molecule is associated with two chain fractures.[259] The G value for acid production is only about one-fifth of the

G value for chain fractures at 20 Mrads, but at 1 Mrad is about one-third; the ratio appears to approach one-half as zero dose is approached. The result for aldehyde production suggests that one aldehyde group is produced initially per chain fracture. It is likely, therefore, that radiation causes the breakdown of a D-glucose residue, to cause chain fracture and produce an aldehyde end group, a hydrogen molecule, and an acidic fragment. Subsequently, a second chain is broken by a mechanism that yields an aldehyde end group and causes a chain fracture, but gives no hydrogen or acid.

Free radicals have been detected by e.s.r. in purified cellulose after γ-irradiation *in vacuo*.[233,260,261] The spectrum is asymmetric, exhibiting five partially resolved peaks spanning about 50 gauss and having a spectroscopic splitting factor near $g = 2$. Differences in crystallinity have no obvious effect on the yield or on the nature of the e.s.r. spectrum of the radicals. The presence of about 8% of water during irradiation decreases the radical yield greatly and modifies the hyperfine structure. For dry cellulose the initial G value for radicals is 2.8. Thermal decay is imperfectly second-order and indicates the presence of several types of radicals having different lifetimes. A marked post-irradiation effect on the viscosity of cellulose that had been dried and irradiated in a nitrogen atmosphere was produced when the sample was stored in dry oxygen. The effect was terminated by the presence of water.[249,250] These changes reflect the decay of the free radicals produced during irradiation. The long life of the radicals is responsible for the success of graft polymerization with preirradiated cellulose.[262,263]

It is probable that breakdown of polymer chains occurs in the crystalline and amorphous regions equally.[255,264,265] Park[266] examined the effect of γ-radiation on cellulose crosslinked with formaldehyde and demonstrated that the reaction with formaldehyde, which occurs only in the amorphous regions of the cellulose, holds these regions together after irradiation.

The most precise information about the chemical course of the degradation of cellulose by ionizing radiation in oxygen comes from the work of Arthur *et al.*[247,267] Chain cleavage at C-1 gives D-*arabino*-hexulosonic acid on the reducing end of the chain, with uptake of 1 mole of oxygen, and an unaltered D-glucose residue is left on the nonreducing end of the other chain. Scission at C-4 results in chain cleavage, liberating the reducing end of the chain as an unaltered D-glucose residue and producing a keto group at C-4 of the non-reducing end of the other chain. For every two chain cleavages, one acid group is also produced. The products are D-glucose, cellobiose, and a homologous series of cello-oligosaccharides, together with D-*arabino*-hexulosonic acid and "2-keto-cellobionic" acid, and a homologous series of similar, oxidized cello-oligosaccharides.

References start on p. 1287.

The effects of radiation on the molecular properties of cellulose follow the relationship:

$$\ln P = k^1 \ln N_n + K^1$$

where P is the molecular property, N_n is the total number of ionizations or total dose absorbed, and k^1 and K^1 are constants.[247,267] For main-chain breaks, Charlesby[264] derived from Saeman's data[255] a G (scission) value equal to 10 and a G value of 10 for the destruction of D-glucose residues. Thus, it would appear that each scission of the main chain is accompanied by the destruction of one D-glucose residue.

Graft polymerization reactions on cellulose, as induced by radiation, will not be considered here. An account of this aspect of radiation interaction, with a detailed bibliography, has been given by Arthur et al.,[268] and a comprehensive account of the radiation chemistry of cellulose has been published.[268a]

B. Energy Transport

In this section, experiments will be summarized which indicate that energy transfer may occur efficiently in solid carbohydrates. Attention will be directed particularly to chemical manifestations of energy-transport processes with the object of demonstrating the occurrence of electronic energy transfer in high-energy experiments. Unequivocal evidence for such processes is scanty, not only in carbohydrate systems but generally.

1. Intermolecular Energy Transfer

a. D-*Glucose Crystals.*—Indications that energy-transfer processes might be important in the radiation chemistry of carbohydrates were obtained initially with D-glucose crystals. The physical state in which anhydrous α-D-glucose is irradiated with ionizing radiations profoundly influences the extent of radiation-induced decomposition.[269,269a] In particular, it was found that polycrystalline, anhydrous α-D-glucose, obtained from aqueous solution by freeze-drying, was more stable to ^{60}Co γ-radiation than was the initial α-D-glucose.

Polycrystalline α- and β-D-glucose behave similarly; from the yield–dose curves the initial $-G$ value is ∼20, and thereafter the $-G$ value for decomposition of D-glucose decreases. After a dose of 5.8×10^{21} eV/g, the $-G$ is ∼9. For the freeze-dried form, the $-G$ value is initially about 7. For α-D-glucose monohydrate the $-G$ value is initially about 11, and, as in the other examples studied, the value decreases as irradiation is continued. No differences were observed in the nature of the products produced on irradiation, although their overall yields for the various forms were examined. This may be illustrated be reference to acid yields (Table II), which show the same general variation as the $-G$ values of the individual forms of D-glucose. In contrast to acid yields, hydrogen and free-radical production are not appreciably influenced

TABLE II

G VALUES FOR ACID AND GAS FORMATION DURING IRRADIATION OF α-D-GLUCOSE[269]

Physical state of D-glucose	Acid			Hydrogen			Carbon dioxide (total)
	In vacuo	Air	CO_2	Total	Above solid	Retained by lattice	
Polycrystalline	13.2	13.3	13.0	3.8	1.06	2.74	
Freeze dried	3.7	3.7		4.0	2.0	2.0	
Monohydrate	5.2	5.0		4.1	1.23	2.87	0.6
Syrup	6.5		~5	4.75	4.65	0.1	0.3

by the state in which the solid is irradiated. For all the forms examined, the G (H_2) value is ~4 and the initial G (free radicals) value is also ~4. It is probable, therefore, that the processes leading to production of acid are different from those giving rise to hydrogen and long-lived free radicals that may be observed by electron spin resonance.

The high-radiation decomposition evident in polycrystalline anhydrous α- and β-D-glucose indicates a mechanism of energy transport in the crystal that is facilitated by a highly ordered structure. The existence of such an exciton-transport mechanism was suggested by Collinson et al.[270] to account for the high $-G$ ($FeCl_3$) values in crystalline solid solutions in various organic solvents. To account for the observations with sugar crystals, it was suggested that, on γ-irradiation, nonconducting excited (or exciton) states are produced by interaction of the allowed, excited electronic states of the identical molecules in the aggregate. This interaction is most probably associated with the extensive three-dimensional hydrogen-bonded network in these crystals. Chemical change will, on this view, be dependent on the efficiency of energy transfer, which will be at a maximum in a perfect crystal.

On this basis, the change in crystal regularity and the increased intermolecular distance (the syrup may contain up to 1% of water) would modify the exciton spectrum of the aggregate significantly. For the freeze-dried state no difference, compared with polycrystalline anhydrous α-D-glucose, was immediately apparent. From the X-ray powder diagrams of the two forms, freeze-dried D-glucose was crystallographically and chemically indistinguishable from anhydrous α-D-glucose.[269] However, examination of the surface characteristics by electron microscopy, with a replication technique, showed that the crystals produced after freeze-drying were highly dislocated. The characteristic river pattern was indicative of a textured solid,[271] where the imperfection and disorder in position reach such a degree that a space lattice is no longer a useful representation. It is highly probable, therefore, that the greater degree of imperfection in the freeze-dried state is the most plausible

References start on p. 1287.

explanation for the different behavior on irradiation.[269] The exciton model can, therefore, be extended to the results on the freeze-dried solid, for the crystal defects present can act as efficient exciton traps. Indeed, it is probable that the capture cross section for the reaction of excited states having defects would be larger than would arise from the presence of a substantial impurity.[272] Dissipation of the excitation, by interaction with a lattice imperfection, could thus lead to less-efficient localization of the exciton because of dissociation of a single molecule, which is the process responsible for chemical change. The freeze-dried state would thus be more stable to radiation.

For all of the crystal forms studied, the decomposition is unusually high in comparison with other groups of organic molecules.[273] High efficiency of energy transport and utilization in the crystal is indicated, which may be related to the hydrogen-bond network in the crystals. Such bonds have been shown to execute the transfer of energy efficiently.[274,275] For α-D-glucose monohydrate, there is a fully hydrogen-bonded structure present. Oxygen atoms at C-3, C-2, and C-6 are linked by hydrogen bonds through water to adjoining D-glucose molecules.[276] Therefore, the distance between atoms of adjoining D-glucose molecules will be greater in certain directions for the monohydrate than for the anhydrous D-glucose crystals, where four hydroxyl groups (at C-6, C-4, C-3, and C-2) are directly hydrogen-bonded to two oxygen atoms of other molecules. In addition, there appear to be bonds between the ring-oxygen atoms and hydroxyl groups at C-1 of neighboring molecules.[277] According to the resonance transfer mechanisms that have been envisaged,[278,279] the transfer efficiency would be expected to be dependent critically on intermolecular separations. On the basis of the exciton model, the transfer rate is proportional to R^{-3}, where R is the transfer distance. Thus, if $-G$ is an indication of transfer efficiency, the lower value for the monohydrate, as compared with α- and β-D-glucose, could result directly from the greater intermolecular separations in the monohydrate. The observed critical dependence of $-G$ on the degree of ordered crystallinity is not a requirement of the Förster mechanism and accords better with the exciton model.

b. Cycloamylose Complexes.—It has been suggested that, when crystals of D-glucose are irradiated, exciton states are produced in high yield over a domain in the crystal. On this view, chemical change will be dependent on whether the excitation energy is transformed into heat at lattice imperfections and other traps, or whether, in more perfect crystals, exciton transport becomes localized by dissociation of individual molecules of D-glucose. While the existence of such exciton states has been demonstrated in crystals of aromatic substances, there is no information about possible exciton states in carbohydrate crystals. If the above interpretation for the radiation-induced decomposition of the various forms of D-glucose is correct, and a collective and mobile excitation is produced, it should be possible to transfer this excitation to other molecules. In an attempt to demonstrate such energy transfer, systems were sought where close association between carbohydrates and a wide range of other organic molecules is possible. To this end, the cyclic oligosaccharides (the Schardinger dextrins) formed by the action of *Bacillus macerans* amylase on potato starch have been studied. These cycloamyloses form unique inclusion complexes with organic molecules.[280] Since the cyclo-amylose molecules are cyclic [joined by (1 → 4) α-D-glucosidic links], they do not contain a free reducing group and are thus not prone to radiation-induced oxidation reactions as a major, initial process.

For the purposes of the present discussion, two aspects of the effects of γ-irradiation on the pure, crystalline Schardinger dextrins are relevant.[281] First, the e.s.r. spectrum of γ-irradiated cycloheptaamylose (the cycloamylose containing seven D-glucose residues) shows mainly a hyperfine structure doublet having a splitting of 3 to 4 gauss and line width of 24 gauss.[282] To account for the small hyperfine splitting, a glycosyloxy radical of the type shown has been suggested, where there is only weak interaction with the proton at C-1. The G (radical) value for cycloheptaamylose, when irradiated

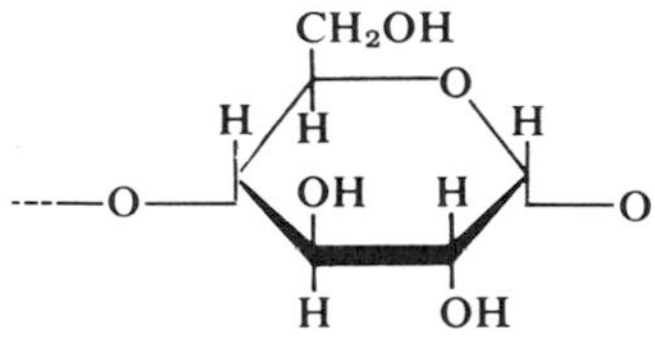

alone, is not less than 5. Furthermore, the major chemical change accompanying γ-irradiation is rupture of the macrocyclic system.[207d] No D-glucose is produced, and the main initial product is maltohexaose. D-Gluconic acid is the main acid produced, although another acid was also observed, particularly at high dose. The effects that have been noted may be represented as on p. 1266. The intermediate radical (2) would give rise to a carbonium ion (3) at C-4, which would react rapidly with water to produce the corresponding alcohol (4). The glycosyloxy site is the most probable intermediate for the formation of D-gluconic acid. The additional product would be maltohexaose, as observed experimentally.[282,283] For the pure cycloamyloses (six-, seven-, or eight-membered) the $-G$ value is about 15.

The effects of energy transfer in complexes of the cycloamyloses with organic molecules may be demonstrated in two ways.[282,284] First, an e.s.r. spectrum different in form and considerably diminished in intensity is found when a suitable complex is irradiated. Second, this behavior is accompanied by chemical protection of the cycloamylose. An example of the modification in the

---

References start on p. 1287.

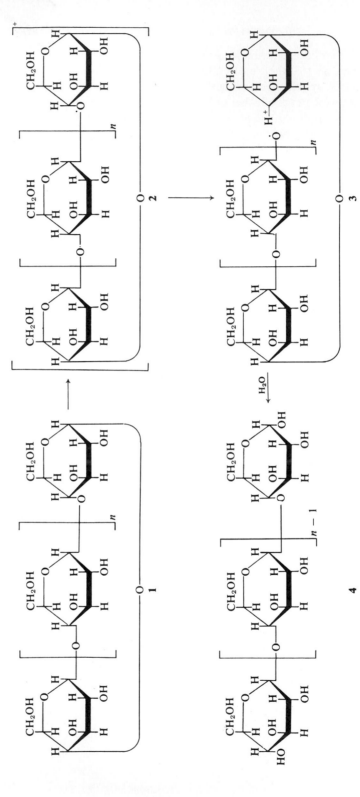

+ nonreducing acid

e.s.r. spectrum is shown in the spectrum of a γ-irradiated complex of benzene and cyclohexaamylose.[294] The doublet for the cycloamylose is replaced by a spectrum consisting of 36 lines, which may be assigned to a radical produced by the addition of a hydrogen atom to benzene. The following coupling constants may be assigned to this radical: for the protons of the methylene group,

49 $\pm$ 1 gauss; for the nuclei of the o- and p-hydrogen atoms, 10.9 $\pm$ 0.2 gauss; and for the nuclei of the m-hydrogen atoms, 2.7 $\pm$ 0.3 gauss. The three groups of lines have relative intensities of 1:2:1. A similar spectrum, giving almost identical coupling constants, has been observed[285,286] for the cyclohexadienyl radical produced when benzene is irradiated at temperatures below $-80°$. Detailed assignments have also been made for complexes with other aromatic molecules. In each case the radical produced can best be accounted for on the basis of addition of a hydrogen atom to the aryl group. The p-nitrophenol–cycloheptaamylose system provides a striking example of chemical protection following energy transfer. The $-G$ (cycloamylose) value, calculated on the basis of energy absorbed in the carbohydrate moiety only, is $\sim$1, compared with $\sim$15 for the uncomplexed cycloheptaamylose.

Table III shows results for a range of cycloamylose complexes. On the basis of these observations, charge or ionization transfer as an explanation for the observed effects can be excluded. If electron transfer occurred, the efficiency of chemical protection should be dependent on the electron affinity of the included molecule. However, no such dependence is shown. For example, although tetracyanoethylene and phenylacetylene differ in their electron affinities by a factor of at least 10^6, they confer the same degree of protection on the cycloamylose. Similarly, tetrachloroethylene and carbon tetrachloride, despite their high electron affinities and their facility in capturing electrons, confer negligible protection. Positive-hole migration is also unlikely, on the basis of similar reasoning, particularly since there is reliable mass spectrometric evidence that fragmentation of an excited cycloamylose molecule would accompany any significant charge separation.[287] Indeed, if the observed $-G$ value of 17 were to be the consequence of ion-pair formation, the mean energy for the production of an ion pair would be 12 eV, which is considerably less than the anticipated value of $\sim$ 30 eV. Finally, if there were to be appreciable electron migration followed by capture, the process would depend strongly on scavenger concentration. However, it has been shown that a more than tenfold variation in the 1-bromonaphthalene:cycloamylose ratio does not change the

References start on p. 1287.

TABLE III

RADIATION PROTECTION OF CYCLOAMYLOSES BY FORMATION OF COMPLEXES

Molecule included	Ratio molecule included: cyclo-amylose	Electron affinity[a]	$E_T (eV)$[b]	$E_1 (\beta)$[c]	$E_1 (eV)$[d]	G, radical	$-G$
Azobenzene	0.4	9			2.80		11.4
Anthracene	0.2	12	1.82	0.82	3.26		12.4
Tetracyanoethylene	0.3	7000			3.87		8.20
Methylnaphthalene	0.6		2.62	1.21	3.83	2.55	4.6
p-Nitrophenol					3.97		1.0
Nitrocyclohexane	1.0				4.45		2.3
Nitrobenzene	1.6	390	2.61		4.61	0.22	2.3
Toluene	0.7	0.01	3.58	1.86	4.60		2.6
Chlorobenzene	1.4	1			4.57		6.0
Iodobenzene	e	370			4.80		6.4
1-Bromonaphthalene	e		2.56		4.70		7.0
Phenylacetylene	1.2	0.01			4.76	1.65	8.9
Carbon tetrachloride	0.3	7000			~6.5		14.5
Tetrachloroethylene	1.3	~7000	~3.13		~7.0	3.4	16.0
Cyclohexane	0.4	0.01			7.5		16.0

[a] Values relative to chlorobenzene. [b] Calculated from the wavenumber of the phosphorescence band. [c] Calculated by L.C.A.O. molecular orbital method and quoted in terms of the resonance integral (β). [d] Calculated from absorption spectral data. [e] Various ratios used.

$-G$ (cycloamylose) value. Ionization transfer as the process responsible for the variation in the $-G$ (cycloamylose) value in the complexes (Table III), is therefore, most unlikely.

Transfer of electronic energy from the carbohydrate to the molecule included, however, accounts satisfactorily for the observations. Figure 1 shows the relation between $-G$ and the energy of the first singlet level (E_1) of the molecule included. Values of E_1 were calculated from measured excitation frequencies, and, where meaningful, for a series, by simple molecular-orbital theory. The results indicate that a resonance-transfer mechanism may be operative. Such transfer is most favorable when there is maximum overlap of the electric dipole fields of the emitter and acceptor molecules. Conditions for excitation transfer are optimal if the electronic level of the acceptor is somewhat lower than that of the emitter, and for energetic reasons, resonance transfer must occur to an acceptor having a lower energy gap. From Fig. 1, it is seen that transfer begins when the acceptor has $E_1 \sim 5$ eV and is a maximum at 4 eV, which sets limiting values on the energy of the excited states in the cycloamylose.

Triplet states do not appear to make any significant contribution to the transfer process, since there is no regular correlation between the $-G$ (cyclo-

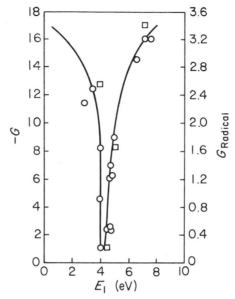

FIG. 1. Resonance-energy transfer leading to protection of cycloamyloses when complexed with organic molecules. ○ chemical yield; □ radical yield.

amylose) value and the first triplet energy level (Table III). Confirmation of this view comes from an experiment in which the 1-methyl group in the naphthalene nucleus was replaced by bromine. Although the change should enhance the $S \rightarrow T$ transition, it results in less efficient transfer of energy, since the $-G$ (cycloamylose) value for the complex with 1-bromonaphthalene is 7.0, as compared with 4.6 for 1-methylnaphthalene.

It was concluded, therefore, that on γ-irradiation of cycloamylose complexes, singlet–singlet resonance-energy transfer occurs, which leads to efficient protection of the carbohydrate.

A significant observation is that a similar correlation is found between the radical yield and E_1, and the chemical yield and E_1 (Fig. 1). If radical formation is due to hydrogen-atom attachment to the aromatic residue, this process may be related to energy transfer only as a result of a lower yield of primary hydrogen atoms in the carbohydrate moiety when efficient transfer occurs. Following rupture of a C—H bond in the carbohydrate, the excess energy would be carried as kinetic energy in the hydrogen atom. Since the activation energy of the process (H· + aromatic) is close to zero, attachment of a hydrogen atom could occur with hydrogen atoms that did not possess sufficient energy to

References start on p. 1287.

initiate further abstraction reactions in the carbohydrate. An excitation of 4 to 5 eV in the cycloamylose could be sufficient to cause bond rupture and simultaneous release of hydrogen atoms having excess energies of only about 0.2 to 0.5 eV. Capture would then be more favorable than further abstraction. Radical yields may be dependent, therefore, albeit indirectly, on the efficiency of energy transfer. By use of identical energy-transfer methods, amylose has been protected from radiation damage.[287a]

2. *Intramolecular Energy Transfer*

The carbohydrates are particularly suitable molecules for studying the effect of substituent groups on the chemical changes initiated by ionizing radiation, since the structure of the molecules is well established, and systematic variation in particular substituent groups is possible.

The radiation stabilization of glycosides was studied from this point of view.[288] During γ-irradiation of glycosides, the dominant radiation-induced process is scission at the glycosidic group. The extent to which this scission process occurs can be significantly reduced by substitution of aromatic groups into the molecule, and the behavior of these groups is closely analogous to the energy-transfer effects reported in the previous section. For example, γ-irradiation of methyl α-D-glucopyranoside gives D-glucose with a yield of $G = 2.3$. Replacement of the methyl group by a phenyl group protects the glycosidic bond, and for phenyl β-D-glucoside the G (D-glucose) value is 0.8. The hydrolytic nature of the cleavage was demonstrated by the observation that G (D-glucose) = 0.8 and G (phenol) = 0.8, with no other detectable products. The introduction of benzoyl groups into this molecule led to further radiation stabilization. For phenyl tetra-O-benzoyl-β-D-glucopyranoside, the production of phenol was further decreased ($G = 0.05$) (Table IV). It is interesting to observe that the stabilization caused by the aromatic groups is not due to a bond-strength effect. It is well established that aryl glycosides are more rapidly hydrolyzed than corresponding alkyl glycosides by aqueous acid.[289] By using phenyl β-maltoside heptaacetate, it can be demonstrated that protection extends beyond the length of one D-glucose molecule. When compared with the unprotected disaccharide, where scission at the glycosidic link occurs ($G = 2.1$), introduction of a phenyl group at C-1 stabilizes the glycosidic link at C-1 ($G = 0.05$) and also the disaccharide link ($G = 0.1$). Further examples are given in Table IV.

The criteria used to demonstrate energy transfer in cycloamylose complexes were chemical protection and changes in form and intensity of e.s.r. spectra. According to these same criteria, it would appear that intramolecular transfer of energy is responsible for the protective effects that have been described. Chemical protection has already been mentioned. The e.s.r. effects may be illustrated by reference to phenyl β-D-glycopyranoside. Disaccharides, and

TABLE IV

EFFECT OF AROMATIC GROUPS AFFORDING PROTECTION FROM RADIATION TO SOLID GLYCOSIDES

Glycoside irradiated	Products	G
Methyl α-D-glucopyranoside	D-Glucose	2.3
	Acid	1.5
Methyl tetra-O-acetyl-α-D-glucopyranoside	D-Glucose	2.3
	Deacetylated products	7.9
Methyl α-D-mannopyranoside	D-Mannose	2.0
	D-Glucose	0.8
Phenyl β-D-glucopyranoside	Phenol	0.8
	D-Glucose	a
Phenyl tetra-O-benzoyl-β-D-glucopyranoside	Phenol	0.05
Phenyl hepta-O-acetyl-β-maltoside	Disaccharide cleavage[b]	0.10
	Glycosidic cleavage[c]	0.05

[a] Not detectable. [b] Disaccharide link. [c] Glycosidic group.

glycosides in general, after γ-irradiation in the solid state, give e.s.r. spectra of a radical, consisting of a doublet[237] with hyperfine splitting of about 3 gauss and total line-width[288] of not more than 86 gauss; the G (radical) value is about 3. For phenyl β-D-glucoside, the e.s.r. spectrum consists of nine lines made up of a trio of triplets in the ratio of intensities of 1:2:1 and total line width of about 160 gauss; the G (radical) value is ~0.1. This e.s.r. spectrum may be accounted for on the basis of a radical formed by addition of an H atom to the phenyl group. By this assignment the coupling constants for the methylene protons would be 52 gauss, and for an aromatic ring proton, 12 gauss.

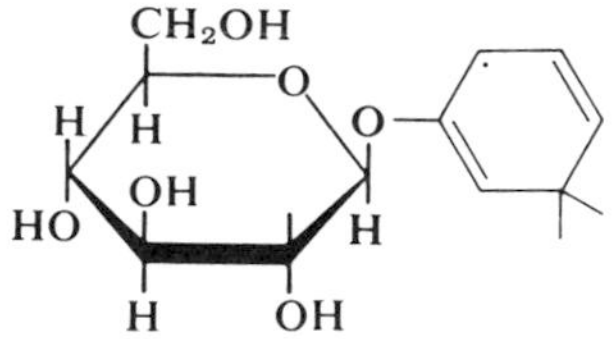

In the typical case of phenyl β-D-glucopyranoside, therefore, the protection provided by the phenyl group could be the result of intramolecular transfer of energy from the D-glucose ring to the aromatic group. The fluorescence spectrum of this substance has a maximum at 290 nm, whereas the absorption spectral maximum is at 271.7 nm. If the energy transferred were to be of these proportions, the excited molecule could be deactivated by fluorescence without any chemical change resulting. Therefore, to obtain estimates of the magnitude of the energy of the excited state in the sugar moiety, the aglycon was varied; Table V shows the relation between the G value for scission at the glycosidic linkage and E_1 for the aglycon. Although broadly the results show that

TABLE V

RADIATION PROTECTION OF GLYCOSIDES BY AROMATIC GROUPS[228]

Glycoside	E_1 $(eV)^a$	E_1 $(\beta)^b$	G (glycosidic scission)	G (radical)
Phenyl β-D-glucopyranoside	4.55	1.83	0.6	0.1
o-Cresyl β-D-glucopyranoside	4.53	1.79	0.75	
m-Cresyl β-D-glucopyranoside	4.54	1.82	0.83	
p-Cresyl β-D-glucopyranoside	4.43	1.72	1.2	
2,4-Dimethylphenyl β-D-glucopyranoside	4.42	1.73	1.1	
p-Nitrophenyl β-D-glucopyranoside	4.09		0.4	0.43
m-Nitrophenyl β-D-glucopyranoside	4.57		0.7	0.53
2-Naphthyl β-D-glucopyranoside	3.82	1.21	0.34	0.13
Benzyl β-D-arabinopyranoside	4.81		1.8	0.6
p-Methoxyphenyl β-D-glucopyranoside	4.39		0.8	
p-Chlorophenyl β-D-glucopyranoside	4.67		0.75	0.8
Anthranyl β-D-glucopyranoside	3.19	0.81	3.2	0.4

^{a} Calculated from absorption spectral data. ^{b} Calculated by an L.C.A.O. molecular-orbital method and expressed in terms of the resonance integral (β).

G (glycosidic scission) is critically dependent on E_1, anomalies are evident at $E_1 \sim 4.5$ eV (275.1 nm), where G values range from 0.6 to 1.2. Singlet–singlet transfer of resonance energy from the sugar moiety to the aromatic group appears to offer the best explanation for the results, and on this basis a discrete energy level of about 4.25 to 4.5 eV (291.2 to 275.1 nm) in the sugar would be indicated. It is noteworthy that the energy limits for the excitation level in the solid carbohydrate are consistent in the two chemical systems that have been examined—namely sugar glycosides, where the group is substituted in the molecule, and the inclusion complexes with cycloamylose, where the nature of the association is less definite.

Estimates of the distance over which energy transfer can occur in carbo-hydrates may be obtained by using fibrous cotton cellulose. Here the degree of substitution (d.s.) of aromatic groups can be varied and protection studied as a function of d.s. A d.s. of benzoyl groups on the cellulose molecule as low as 0.2 offered radiation protection of the fibrous cellulose after a dose of 1.2×10^{21} eV/g. If it is assumed that the benzoyl groups are substituted randomly on the cellulose molecule, the maximum distance between benzoyl groups may be calculated as a function of degree of substitution and as a binomial distribution with a probability of reaction ~ 0.95. At a d.s. of 0.2, the maximum spacing of benzoyl groups on the cellulose molecule was calculated to be 7.2 to 8.2 nm; at d.s. 0.5, 3.1 to 4.1 nm; at d.s. 1.1, 1.0 to 2.1 nm; and at d.s. 1.5, 0.2 to 0.1 nm. In the experimental case, the assumption that benzoyl groups are randomly distributed will not be exactly true, since the physical structure of the fibrous cellulose will limit the accessibility of highly ordered regions to reaction with benzoyl chloride. Therefore, in the experimental case, the maximum spacings of benzoyl groups will be greater than those indicated. Assuming that the breaking strength of the benzoylated cotton cellulose is related to its molecular weight,[247] the following relationships can be calculated: at a dose of 1.3×10^{21} eV/g of cellulose, the average estimated distances between molecular cleavages were increased from about 50 cellobiose units (50 nm) for unsubstituted cellulose to about 300 cellobiose units (300 nm) for benzoylated cellulose (d.s. 2.0). On the basis of the results for benzoylated cellulose (d.s. 0.2), therefore, energy transfer over at least 7 to 8 nm can be envisaged before protection ceases.

C. SELF-DECOMPOSITION

When carbon-14 techniques are applied to chemical problems, it is important that the ^{14}C-labeled compounds be chemically pure. It has frequently been assumed that such compounds are stable on storage. Evidence has accumulated that this assumption is not justified. Comprehensive reviews of this subject have appeared.[290,291] Attention here will be directed only to the self-decomposition of ^{14}C-carbohydrates.

The decomposition of a compound labeled with a radioactive isotope may be attributed to four main effects,[290,292] summarized in Table VI. For ^{14}C-labeled carbohydrates stored as freeze-dried samples under vacuum at room temperature, self-decomposition arises mainly by the primary (external) radiation effect and by the secondary radiation effect. However, it has been

TABLE VI

MODES OF DECOMPOSITION OF LABELED COMPOUNDS[a]

Mode of decomposition	Cause	Method of control
Primary (internal)	Natural isotopic decay	None, for a given specific activity
Primary (external)	Direct interaction of the α, β, or γ radiation with molecules of the compound	Dispersal of the labeled molecules
Secondary	Interaction of reactive species produced by the radiation with molecules of the compound	Dispersal of active molecules; cooling to low temperature, scavenging of free radicals
Chemical	Thermodynamic instability of compounds and poor choice of environment	Cooling to low temperatures, removal of harmful agents

[a] Secondary radiation and chemical effects can in particular circumstances supplement each other as, for example, during the self-decomposition of [U-^{14}C]dextran sulfate containing about 20 D-glucose residues per molecule, observed by Bayly and Weigel.[292,293]

observed that the alkalinity of normally washed Pyrex glass is detrimental to the stability of ^{14}C-labeled carbohydrate syrups, and it is desirable to store the samples at as low a temperature as possible, in order to decrease the rate of such unavoidable chemical reactions.

Secondary radiation and chemical effects can, under particular circumstances, supplement each other, as, for example, during the self-decomposition of [U-^{14}C]dextran sulfate containing about 20 D-glucose residues per molecule,[292,293] and having the relatively low specific radioactivity of 22.4 mCi per gram-atom of carbon (about 3 mCi/mmole of dextran sulfate). After three weeks in the freeze-dried form, the sample had charred and had become a total loss. The decomposition was attributed to the release of sulfuric acid by the secondary radiation effect, which had then released more sulfuric acid to destroy the material.

Carbohydrates are among the most susceptible of labeled compounds to decomposition under the agency of their own radiation. Arising from the decomposition is an enormous variety of products. For example, after a sample of D-[U-^{14}C]glucose (about 6 mg, having a specific activity of about 14.44 mCi/mM) had been stored as a freeze-dried sample in the dark for 26 months, 14.5% of the D-glucose had decomposed and 37 decomposition products were detected by two-dimensional paper chromatography–paper electrophoresis.[293] The self-decomposition of D-[U-^{14}C]mannose, D-[U-^{14}C]-ribose, and D-[U-^{14}C]fructose follows similar paths.[294] Certain of the products from D-[U-^{14}C]mannose were identified by paper chromatography as D-arabino-hexulosonic acid, D-lyxose, D-arabinose, and the lactones of D-mannonic acid. Two-, three-, and four-carbon aldehydic fragments were identified unequivocally by carrier dilution analysis.

Quantitative expression of radiochemical yields is provided by the G value (the number of molecules undergoing change per 100 eV input of energy). On a similar basis, the number of molecules decomposed per 100 eV liberated by the ^{14}C-sugar during storage may be calculated. This is referred to as the $G(-M)$ value and is related to the extent of self-decomposition by the expression[299]

$$P_d = f\bar{E}S_a \cdot 5.5 \times 10^{-9} G(-M)$$

where P_d is the initial percentage of decomposition per day, f is the fraction of the radiation energy absorbed by the system, $\bar{E}$ is the mean energy of the emission in electron volts, and S_a is the initial specific activity of the compound in millicuries per millimole. When the appropriate $\bar{E}$ value for carbon-14 is inserted, the equation reduces to

$$P_d = 2.65 \times 10^{-4} \cdot G(-M) \cdot S_a \cdot f$$

For ^{14}C-carbohydrates, $G(-M)$ values in the range of 4 to 900 have been reported[1]; the variation reflects the important influence exerted on the rate of decomposition by minor variations in the storage conditions. In many cases the value of the f term has been assumed to be unity, which is not a valid assumption for ^{14}C-carbohydrates. For a pure β-emitter, f is given by the expression

$$f = pl/2r(1.5 + \ln r/pl)$$

where l is the even thickness of deposit (cm), p is the density (mg cm^{-3}), and r is the mean range of the particles.[292] The self-decomposition of D-[U-^{14}C]-glucose has been measured under standardized conditions where f could be calculated.[295] The samples were deposited onto glass slides that had been stringently cleaned to prevent decomposition by chemical effects. The thickness of deposit was varied threefold to examine the effect of changing the proportion of energy absorbed on the amount of decomposition, and the samples were stored *in vacuo* at $-15°$. The extent of decomposition was measured by isotope-dilution analysis, and, from the total radioactivity recovered, the weight of deposit was determined. By assuming that the weight was deposited in the form of a cylinder of radius about 0.2 cm, which was measured for each deposit, the thickness was calculated. On this basis the value of f is 0.55 when three thicknesses are deposited and 0.28 when one thickness is deposited on the slide. By using the f values obtained in this way, the dose absorbed during storage may be calculated and $G (-M)$ values estimated. For the thicker deposit $G (-M)$ is 26, and for the thinner layer 29.

It has been suggested that the self-decomposition of ^{14}C-labeled sugars can be attributed, in the main, to reactive species produced by radiolysis of nonbonded water retained by the samples after freeze-drying.[292,294] However, a serious objection to this conclusion arises from the fact that high $G (-M)$ values have been observed, whereas only about 4 radical pairs per 100 eV are formed during the radiolysis of water. Therefore, to invoke $G (-M)$ values of $\sim$100, it is necessary to postulate subsequent chain reactions. No such chain reactions have been observed,[187] however, during irradiation of carbohydrates in aqueous solution, where $G (-M)$ values are 3.5 to 4.0. However, the more accurate $G (-M)$ values of 26 to 29 obtained for the self-decomposition under standardized conditions of geometry are of the same order as the $G (-M)$ values obtained with polycrystalline α-D-glucose when exposed to γ-radiation, and when the conditions are most efficient for energy transport. There is, therefore, a closer similarity between self-decomposition and γ-irradiation in the solid state than with the behavior of aqueous solutions on irradiation. The greater complexity of the products during self-decomposition of D-[U-^{14}C]glucose, compared with that of γ-irradiated aqueous solutions of D-glucose, is also an indication that some decomposition is due to effects of direct action. Further evidence that self-decomposition is due to a direct-action (or primary external radiation) effect comes from the observed greater stability of systems into which aromatic groups have been introduced.[296] Such groups

References start on p. 1287.

were found to act as energy sinks and hence also effect radiation protection during the γ-irradiation of solid glycosides. Under normal freeze-dried conditions it is, therefore, not necessary to invoke contributions from reactive species, formed by radiolysis of nonbonded water, to account for the observed high $G(-M)$ values, as was at one time suggested.[292–294]

Two methods for reducing the magnitude of decomposition from primary (external) radiation are (1) dispersion over a large area and (2) dilution. These procedures also decrease decomposition caused by secondary radiation effects, as is borne out by experimental results obtained[294] with D-[U-^{14}C]mannose. Aliquots (5 ml) of the pure sugar (about 100 μCi) in water (100 ml) were kept *in vacuo*. By isotope-dilution analysis, it was estimated that the rate of decomposition in the freeze-dried state was 7% a year, and in the frozen state 1% a year. The frozen state therefore appears to be the more satisfactory conditions for storage over long periods of time, an observation which is now becoming generally accepted.

For a detailed review of free-radical reactions of carbohydrates as studied by radiation techniques, see ref. 2a.

VI. OXIDATIVE–REDUCTIVE DEPOLYMERIZATION OF POLYSACCHARIDES*

In addition to the high-energy radiations and ultrasonics[297,298] described in earlier sections of this chapter, some compounds and metallic ions (autoxidants) are able to induce changes in carbohydrates that closely resemble those produced by ionizing radiations. It is now recognized that these autoxidants act through the formation of free radicals. It has long been known that hydrogen peroxide in conjunction with ferrous or ferric salts (Fenton's reagent)[299] causes degradation of starch, glycogen, and other glycans.[300] Fenton's reagent, aptly described as a nonspecific radiomimetic agent, generates the same type of radicals as those formed by the irradiation of water.[301,302] Reactions between hydrogen peroxide and Fe(II) and Fe(III) are envisaged as follows[303]:

$$Fe^{2+} + H_2O_2 \longrightarrow Fe^{3+} + OH\cdot + OH^-$$
$$OH\cdot + H_2O_2 \longrightarrow H_2O + HO_2\cdot$$
$$Fe^{2+} + HO_2\cdot \longrightarrow Fe^{3+} + HO_2^-$$

Ferric ions react with hydroperoxy radicals according to

$$Fe^{3+} + HO_2\cdot \longrightarrow Fe^{2+} + O_2 + H^+$$

The appearance of hydroperoxy intermediates in peroxide systems catalyzed by Cu(I, II), Fe(II, III), or Co(II) was detected by means of 2,6-di-*tert*-butyl-4-methylphenol,[304] and the presence of hydroxyl radicals in peroxide–Ti(III) solutions could be confirmed by electron spin resonance.[305]

* This section was prepared by A. Herp; the collaboration of T. Rickards and M. J. Harris is gratefully acknowledged.

In vivo, the metal ion–peroxide process is certainly one source of free radicals, as this is evidenced by the occurrence of such species in cellular metabolism.[306,307] Although in many instances radiomimetic chemicals may have damaging biological effects, free-radical intermediates produced endogenously are important in anabolic pathways. Thus, hydroxylation by hydroxyl free radicals is common in biological systems, a classical example being the conversion of tyramine into hydroxytyramine, which is mediated by L-ascorbic acid [308]; the latter compound, in the course of its autoxidation, seems to yield free-radical species of the type generated by the Fenton reagent.[309] Ascorbic acid in abundance, together with metal ions in the living organism, is an ideal system for the production of such free radicals. In the presence of copper ions, it is likely that the initial reaction is as follows[310]:

$$AH_2 + Cu^{2+} \longrightarrow Cu^+ + AH\cdot + H^+$$

(where AH_2 is the reduced form of L-ascorbic acid, and $AH\cdot$ is the organic free radical derived from the substrate). It has been claimed that such a free-radical mechanism is the main pathway in the ascorbic acid oxidase and peroxidase reactions.[311]

In vitro, such substances as ascorbic acid, cysteine, and ferrous salts, in the absence and in the presence of hydrogen peroxide, have been shown to degrade hyaluronic acid, a polysaccharide of high molecular weight which is responsible for the viscosity of synovial fluids.[312–316] Since the depolymerization by autoxidants takes place only when oxygen is present in the system, the phenomenon has been termed oxidative–reductive depolymerization, or the ORD reaction.[317] Because of its immediate biological interest, hyaluronic acid has been the substrate most extensively studied for this reaction.

A method widely used for following the course of oxidative–reductive depolymerization is based on measurements of the viscosity of aqueous solutions of a polysaccharide as a function of time, in the presence and in the absence of autoxidants and antioxidants. Measurements of viscosity have long been used for determining the degree of degradation of polysaccharides because it was found empirically that viscosity diminishes as degradation of the polymer proceeds. Viscosity data give an estimate of the molecular weight according to the Staudinger[318] equation (Eq. 15):

$$\eta_{sp}/c = kM^a \tag{15}$$

which relates the specific viscosity, η_{sp}, of homologous polymers of concentration c to the molecular weight M. The exponent a depends on the conformation of the polymer molecules in solution. Since the reduced viscosity, η_{sp}/c, of high polymers is not constant, the method of Elöd *et al.* is used,[319] namely

---

References start on p. 1287.

the extrapolation of the reduced viscosity to zero concentration. (For further details, see Chapter 35, Section II.)

According to experimental data established by Laurent *et al.*[320] for hyaluronic acid, Eq. (15) becomes

$$[\eta] = 0.036 M^{0.78}$$

from which the number of linkages broken during degradation can be calculated for different proportions of depolymerization, $[\eta]$ having the units of dl/g.

The intrinsic viscosity of hyaluronic acid may also be determined by use of the formula given by Sundblad[321]:

$$[\eta] = \frac{\eta_{sp}}{c(1 + 0.18\eta_{sp})}$$

where c is the concentration of hyaluronic acid in g/dl.

Since the specific fluidity ($\phi_{sp} = 1/\eta_{sp}$) values have been shown experimentally to be a linear function of time during the initial stages of degradation (Fig. 2), changes in specific viscosity may be related to the number of chain

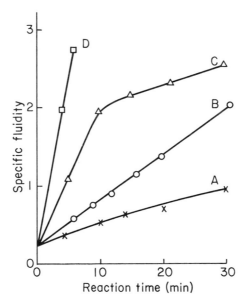

FIG. 2. Effect of copper and iron ions on the fluidity of hyaluronic acid in the presence of ascorbic acid. Reactions performed at 30° with 0.4 mg of hyaluronic acid per milliliter in 0.2 M phosphate buffer, pH 7.3. (A) Control (L-ascorbic acid, 0.33 mM). (B) 0.33 mM L-ascorbic acid + 3.3 μM FeSO$_4$. (C) 0.33 mM L-ascorbic acid + 3.3 μM CuSO$_4$. (D) 0.33 mM L-ascorbic acid + 3.3 μM FeSO$_4$ + 3.3 μM CuSO$_4$.

breaks, N, occurring in the polymer during depolymerization according to the following equation[322]:

$$d\phi_{sp}/dt \sim dN/dt$$

In addition to its simplicity, viscometry is a very sensitive tool because even a small proportion of chemical cleavages is detectable.

A. NATURE OF AUTOXIDANTS[323]

The most effective organic autoxidants in the oxidative–reductive depolymerization reaction are enediols, thiols, and quinones. Ascorbic acid, the enediol most widely studied, reduces the viscosity of hyaluronic acid and other acidic polysaccharides by about 90% within a few hours. *In vitro*, its activity does not seem to be stereospecific, because "D-*arabino*-ascorbic" acid is also active. The corresponding dehydroascorbic acids are similarly effective and decrease the specific viscosity of hyaluronic acid in phosphate buffer and at neutral pH to about the same degree as do the reduced forms.[323] In the process of depolymerization by L-ascorbic acid, the rate of change in fluidity of hyaluronic acid parallels the rate of disappearance of L-ascorbic acid, as measured by the combined formation of dehydro-L-ascorbic and L-*threo*-2,3-hexodiulosonic acids. As depolymerization continues to occur even after all of the L-ascorbic acid in the system has been oxidized, it must, therefore, be concluded that its oxidation products are also active; this interpretation is consistent with claims that L-*threo*-2,3-hexodiulosonic acid acts as a depolymerizing agent.

These observations suggest that both ascorbic and dehydroascorbic acids react by forming intermediates of the semiquinone type or monodehydroascorbic acid,[324] which play a role in the depolymerization, and that similar materials are formed in later stages of the oxidation of the ascorbic acids. If dehydroascorbic acid is formed during the degradation of the substrate when ascorbic acid is used as the autoxidant, its contribution in the phenomenon is presumably small, in view of the initial induction period of the fluidity/time curve when the oxidized compound is the autoxidant.[325] Quinones and hydroquinones also degrade hyaluronic acid, and, like ascorbic acid, the oxidized forms generally show effects of the same order as the reduced ones.[323] Folic acid and riboflavin, which have equally important biological functions, are both strongly active at physiological pH in the ORD reaction but require visible light. Compounds containing sulfhydryl groups range from moderate to good as depolymerizing agents, but corresponding disulfides are usually inactive. A few phenols are autoxidants, and, whereas the *meta*-substituted derivatives are fairly potent depolymerizing agents, the *ortho* and *para* derivatives induce no measurable degradation.[323]

References start on p. 1287.

Among inorganic ions, the most effective known are the cuprous, ferrous, and stannous cations, and the bisulfite anion. Hydrogen peroxide, by itself, is only slightly active in the system.[323]

The need for trace amounts of iron or copper ions for the oxidative–reductive depolymerization in the presence of organic autoxidants such as ascorbic acid is controversial. Some investigators believe that organic autoxidants require small amounts of metals for activity and that under conditions of rigorous "metal sterility" ascorbic acid and other reducing agents are not active [315,326]; others assume that ascorbic acid is capable of acting alone in degrading hyaluronic acid.[317]

The possibility that oxidative depolymerization by autoxidants requires trace amounts of metals is difficult to prove or to disprove. Weissberger and LuValle[327] found no evidence for the necessity of metal catalysis in the autoxidation of L-ascorbic acid. The strong suspicion that metal catalysts are involved in these autoxidation reactions has been reinforced by experiments in which a highly efficient iron-chelating resin (Chelex-100) was added to a reaction mixture buffered with phosphate and containing L-ascorbic acid as the reductant.[327a] In such a system, or in systems in which all of the reagents used had been prepurified by passage through Chelex-100, the depolymerization of hyaluronic acid was very slow; such systems exhibited rates of depolymerization proportional to the concentration of iron or copper ions added. These data clearly demonstrate that the L-ascorbic acid-induced degradation of the polysaccharide in phosphate buffer results from catalytic action of metal impurities that are encountered in commercial phosphate salts. The main function of the reductant in such systems is regeneration of the metal catalyst as shown:

$$\text{Ascorbic acid} + 2\text{Fe(III)} \longrightarrow \text{Dehydroascorbic acid} + 2\text{Fe(II)}$$

$$\text{Substrate} + n\text{Fe(II)} + O_2 \longrightarrow \text{Oxidized substrate} + n\text{Fe(III)}$$

The hydroxyl free-radical has specifically been implicated in these processes.[322]

An efficient system is, therefore, achieved by carrying out the reaction in the presence of micromolar quantities of metal catalyst in an electrolysis cell fitted with two platinum electrodes connected to a DC source. Depolymerization of the carbohydrates results from the cyclic reduction and autoxidation of the metal ions in solution. By using such a system, the degradation reaction can conveniently be followed by chemical means. In one such study, the free-radical-induced oxidation of a variety of sugar derivatives was about 20% during a 10-hour period.[327b]

B. Nature of Antioxidants

One feature of reactions involving free radicals is their general inhibition by organic compounds and by some inorganic ions when present at high concentrations. When hyaluronic acid is used as substrate, alcohols, acetate, bromide, chloride, and iodide ions strongly inhibit the oxidative–reductive depolymerization at $0.1 M$ concentration.[328] However, some substances inhibit markedly the degradation of hyaluronic acid even at millimolar concentrations. These compounds have been called antioxidants.[323] Some fairly potent antioxidants in the ascorbic acid–hyaluronic acid system are CNS^-, $Fe(CN)_6^{4-}$, alizarin (1,2-dihydroxyanthraquinone), catalase, 1,2-dithioglycerol (BAL), 8-quinolinol, DL-penicillamine (2-amino-3-methyl-3-thiobutyric acid), rutin, sodium N,N-diethyldithiocarbamate, and thiourea. Disulfuram, the dimer of N,N-diethyldithiocarbamic acid, is inactive in this system. Among dinitrophenols, only the *meta* isomer acts as an inhibitor. Generally, inorganic compounds are weaker inhibitors than organic ones. The most efficient organic compounds are alizarin, 8-quinolinol, sodium N,N-diethyldithiocarbamate, rutin, and the enzyme catalase. They all produced an inhibition of 80% or higher when L-ascorbic acid was used as the depolymerizing agent of hyaluronic acid. The inhibition of this reaction by catalase has been considered by some workers as evidence that peroxides participate in these degradations.[329]

Some compounds can serve both as autoxidants and as antioxidants, according to their concentration. Such a dual role is shown by 3,4-dinitrophenylalanine, 1,2-dithioglycerol, gallic acid (3,4,5-trihydroxybenzoic acid), and penicillamine.[323]

1. *Nature of Substrates Sensitive to the Oxidative–Reductive Depolymerization*

Although hyaluronic acid is the most widely studied substrate, the ORD reaction is observed with a large variety of polysaccharides and synthetic polymers. Alginate,[334,335] amylopectin,[330] cellulose,[331,332] dextran,[333] and pectins[336] are degraded by autoxidants. A comparison of the effect of L-ascorbic acid on the degradation of some polysaccharides is given in Table VII. However, as Smidsrød and co-workers[337] noticed, the criterion of changes in viscosity for following the oxidative–reductive depolymerization is valid only in the case of polymers for which the exponent a in the Staudinger equation (Eq. 1) is close to unity. By using the increase in reducing power instead of measuring decrease in viscosity for evaluating rates of depolymerization of polysaccharides for which the value of a in Eq. (15) is low, these authors concluded that neutral polysaccharides such as starch and dextran follow the same rate of degradation as do polysaccharides containing carboxyl groups. No linkage specificity was apparent,[333] but the limited data available in the literature do not permit definite conclusions concerning the relationship between linkage and degree of degradation.

References start on p. 1287.

TABLE VII

EFFECTS OF L-ASCORBIC ACID[a] ON THE SPECIFIC VISCOSITY OF SOME NATURAL AND SYNTHETIC POLYMERS

Polymer[b]	Concentration (%)	Initial η_{sp}	% Decrease in η_{sp} after 2 hr
Dextran	0.66	0.78	82
Hyaluronic acid	0.04	4.05	95
Copolymer of methyl vinyl ether and maleic anhydride	0.66	3.84	43
Pectin	0.66	8.48	56
Poly(acrylic acid)	0.73	0.25	54
Poly(methacrylic acid)	0.42	11.18	91
Propylene glycol alginate	0.3	2.18	71
Sodium alginate	0.33	8.81	82

[a] Concentration of L-ascorbic acid, 3.3 mM. [b] In 0.2 M phosphate buffer, pH 7.3, 30°.

C. GENERAL FEATURES OF THE ORD REACTION

Aside from the nature of the autoxidants and the polymer substrate, other factors, such as concentration, pH, ionic strength and type of salt, presence or absence of catalyst, light source, temperature, buffer components, and order of addition of various components to the reaction system, may influence the course of the reaction, either by modifying its rate, or by changing the reaction pathway. Free-radical systems often provide unexpected effects. The effects of some of these external factors on the oxidative–reductive depolymerization system will be briefly reviewed.

1. Concentration Dependence

Some autoxidants which promote depolymerization of oxygenated solutions of hyaluronic acid at low concentration exert an opposite effect when present at high concentrations. For example, with L-ascorbic acid as the autoxidant, it was observed[317] that, up to a concentration of 10 mM, the rate of degradation of the polymer increased, but at 40 mM and higher concentrations of autoxidant, the reaction rate began to decrease. In general, there seems to be an optimum concentration of autoxidant for maximal degradation to occur, and numerous compounds exhibit some protective effect at high concentration.

In the industrial application of autoxidants, this concentration effect is commonly used. Thus, L-ascorbic acid is employed as one of the most popular preservatives in foodstuffs[338,339]; copper ion is an inhibitor in certain types of washing processes, and an accelerator in the presence of other synergists.

2. Effect of pH

Maximum degradation of hyaluronic acid seems to take place in the pH range 4.5 to 7.4 when the autoxidant is ascorbic acid, cysteine, or a ferrous salt.[340] For other polysaccharides and related compounds, the pH optimum is comparable; for example, with sodium alginate as substrate, degradation by L-ascorbic acid is greatest[333] at pH 3.3 and 7.3. Optimum conditions of pH may vary according to the type of autoxidant. With hydroquinone, maximum degradation of hyaluronic acid and alginate is observed near pH 9. Dehydro-L-ascorbic acid has virtually no effect on the viscosity of hyaluronic acid solutions at pH 4.5 and below, but at pH 7.3 the diminution in viscosity is similar in magnitude to that induced by L-ascorbic acid.[341]

3. Effects of Catalysts

Addition of micromolar amounts of ferrous or cupric ions to L-ascorbic acid greatly enhances the rate of degradation of hyaluronic acid but apparently does not affect the extent of chemical scissions. Udenfriend et al.[308] have shown that the higher and lower valency forms of these ions act equivalently in the presence of ascorbic acids, which suggests that they be nonspecifically termed "iron" and "copper" ions; however, it must be stressed that, in the absence of ascorbic acid, the active ions are ferrous and cuprous ions. When acting alone, these ions are effective only in millimolar concentrations. When chelated with (ethylenedinitrilo)tetraacetic acid (EDTA), iron ions catalyze the autoxidation of hyaluronic acid by L-ascorbic acid. In contrast, EDTA abolishes the catalytic effect of copper in this system.[325]

4. Light Dependence

When the autoxidant is colored, visible light may be required for the reaction to take place. Known photosensitizers include eosin, folic acid, and riboflavin, all of which have been shown to degrade hyaluronic acid in the presence of light. Such compounds fail, however, to produce diminution in the viscosity of hyaluronic acid solutions as long as the substrate–photosensitizer system is kept in darkness.[323,326] The fabric and dyeing industries take advantage of this phenomenon by applying it to the process of graft polymerization of cellulose. In the process, the cellulose molecule is converted into a free radical by means of high-energy radiation or ultraviolet light in the presence of photosensitizers.[342] Fenton's reagent as well as other types of free-radical initiators have also been used.[343] (See also Chapter 36.)

5. Temperature

Degradation of polymers may also be brought about by thermal treatment, which seems to be pH-dependent.[344] The rate of degradation of polysaccharides by autoxidants is not significantly accelerated by a moderate rise in temperature.

References start on p. 1287.

6. Effect of Buffer System

The ORD reaction is affected in nature and extent by the type of buffer used in the reaction mixture. This may be attributable to the inhibitory effects of organic and some inorganic ions at high concentrations. Phosphate ions are seemingly the most suitable buffers, because they are not free-radical scavengers, and they also seem to be able to bind iron and copper ions. In phosphate buffer (0.2 M), the specific viscosity of a 0.04% solution of hyaluronic acid is lowered by about 90% within 10 minutes by ferrous sulfate (0.2 mM), but no noticeable change is observed when 0.2 M tris(hydroxymethyl)aminomethane (*Tris*) buffer is used instead. Also, L-ascorbic acid is much less efficient as a depolymerizing compound in some organic buffers than in inorganic buffers.[341]

D. Nature of Products Formed

When hyaluronic acid is incubated with autoxidizable compounds, changes (apparently irreversible) in this polymer substrate include a decrease in intrinsic viscosity, modified electrophoretic and ultracentrifugal patterns, and a loss of protein-precipitating ability. All these alterations are indicative of a decrease in molecular size of the polymer. The oxidative–reductive depolymerization, however, has a low efficiency in terms of chemical changes,[345] and usually not more than a few cleavages occur per molecule (Table VIII). Consequently,

TABLE VIII

Stoichiometry of ORD Reaction

Intrinsic viscosity[a] (dl/g)	Molecular weight[b] ($\times 10^{-6}$)	Number of linkages broken/molecule	Concentration autoxidant (mM)	Number of autoxidant molecules present/linkage broken[c]
		Ascorbic Acid as Autoxidant		
46	3.54		0.	
32	2.21	0.60	0.025	370
18	1.06	2.34	0.038	145
15	0.83	3.26	0.077	210
12	0.63	4.62	0.145	360
6	0.26	12.60	0.290	205
Average				260
		FeSO$_4$ as Autoxidant		
32	2.21	0.60	0.025	270
22	1.37	1.58	0.038	210
17	0.98	2.62	0.077	260
11	0.56	5.32	0.145	240
8	0.37	8.57	0.290	335
Average				280

[a] Calculated from the data of Pigman *et al.*[317] [b] $[\eta] = 36$ MW$^{0.78} \times 10^{-5}$.
[c] Initial concentration of hyaluronic acid was 0.113 μM with MW of 3.54 $\times 10^6$.

isolation of the final decomposition products of the oxidative–reductive depolymerization presents difficulties. Even after drastic and repeated treatment [by using a combination of L-ascorbic acid–Fe(II)–EDTA], not enough dialyzable material is obtained for the identification of breakdown products. However, chemical analyses show that the degradation is accompanied by the liberation of residues containing 2-acetamido-2-deoxyhexose at the termini and by a concomitant increase in apparent content of reducing sugars. Similar chemical changes were reported when hyaluronic acid was subjected to irradiation by X-rays[346] or deuterons.[347]

Other polysaccharides susceptible to oxidative–reductive depolymerizations were also examined for degradation products. Thus, thermal treatment of pectin or alginate solutions revealed in both instances an increase in reducing end groups and formation of 4,5-unsaturated, uronic acid derivatives. In neither case, however, was dialyzable material detectable.[344,348]

E. Mechanism of Degradation of Hyaluronic Acid and Other Polymers by Autoxidants

The requirement for oxygen, the nature of autoxidants and of antioxidants, and the general similarities between the effects of oxidative–reductive depolymerizations and radiation are strong evidence for the free-radical nature of the depolymerization observed. Peroxides may be involved, or they may be end products of free-radical recombinations. For some reason, not yet clarified, the chain cleavages are limited and the reaction is inefficient for both autoxidations and degradations by radiation.[323] The rather nonspecific character of organic substances in their ability to inhibit the oxidative–reductive depolymerization when used at high concentration, and the inhibitory effect shown in this system by the halogens in the increasing order F < Cl < Br < I, are also features that are characteristic of some free-radical systems.[345]

Anions of the lyotropic series show distinct variations in their effect in the ORD reaction, their inhibitory ability decreasing through the series $Fe(CN)_6^{4-}$, CNS^-, I^-, Br^-, NO_3^-, Cl^-, the classical order of Hofmeister. This may be attributable to competition between polymer and inhibitor for free-radical species as the rate-determining step.[345] The reactive species may be hydroxyl radicals, because a consistent and reasonable value has been obtained for the rate constant of the reaction step $HO\cdot$ + hyaluronic acid, by using data[348a,348b] previously recorded concerning the reaction rate of these anions with hydroxyl radicals.

The relative lack of influence of temperature on the rate of degradation of polysaccharides in the presence of autoxidants is also indicative of the

---

References start on p. 1287.

involvement of free-radical species, as is the reversal of depolymerization at high concentrations of autoxidant.

With the assumption that the hydroxyl free radical is the attacking agent, a possible mechanism for the degradation of polyacrylates in the oxidative-reductive depolymerization reaction is as follows:

$$\underset{\substack{| \\ \text{CO}_2^-}}{R-CH_2-\overset{\substack{\text{CO}_2^- \\ |}}{CH}}-CH_2-\overset{\substack{\text{CO}_2^- \\ |}}{CH}-CH_2-R' \longrightarrow R-CH_2-\overset{\substack{\text{CO}_2^- \\ |}}{CH}-\overset{\cdot}{CH}-\overset{\substack{\text{CO}_2^- \\ |}}{CH}-CH_2R' + HOH$$

$$+$$

$$HO\cdot$$

$$R-\overset{\cdot}{C}H_2 + CH=CH-\overset{\substack{\text{CO}_2^- \\ |}}{CH}-CH_2-R'$$

This proposed mechanism involves the preliminary withdrawal of a hydrogen atom by the hydroxyl free radical and subsequent β elimination. The presence of the carboxylate ion or its ester would assist the elimination by its tendency to attract electrons from the neighboring C—C linkage.

If the polysaccharides are degraded by a similar mechanism, the site of attack would be at C-5 for (1 → 4)-linked compounds, and it might then be expected that a carboxyl group at C-6 would enhance the β elimination.

In a study of the oxidative degradation of aldohexoses in the presence of hydroperoxides, Isbell et al.[348c] postulated a free-radical mechanism involving hydroxyl radicals similar to that already outlined. This process, according to the authors, would involve the abstraction of the hydrogen atom from the hydroxyl group at C-2 of the aldose with scission of the C-1–C-2 bond and a one-electron transfer to form the carbonyl group of the next lower aldose; the other electron forms the carbonyl group of formic acid, with regeneration of the hydroxyl radical. Significantly, the reaction is greatly accelerated by the addition of catalytic amounts of ferrous ions.

F. Significance of the ORD Reaction

The instability of polymeric carbohydrates in the presence of reducing agents, and the protective effect exerted by other compounds in the ORD reaction, may have practical applications as well as important biological implications.

From a practical point of view, the foregoing data may be used to advantage in isolating polysaccharides and for storing them without decomposition.[317,335] Thus, hyaluronic acid can be stabilized in solution by adding ethanol or a suitable organic buffer.

More important are the biological considerations, since the most active compounds in the oxidative–reductive depolymerization are commonly encountered in biological systems.[349] Autoxidations in the presence of ascorbic acid and copper and iron ions generate free-radical species which may be involved in normal and abnormal metabolic processes. Chemical systems that generate hydroxyl radicals have been shown to have bactericidal effects.[310] Aging,[350] oxygen toxicity,[351] mutation,[352] and hydroxylation mechanisms[308,309] are possible examples of such reactions. Dephosphorylation of sugar phosphates[353] or attack on the base moiety[354] of nucleic acids through participation of hydroxyl free radicals are possible pathways for the radiation-induced cleavage of DNA in biological systems. The hydroxylation of proline in procollagen requires ascorbic acid and Fe(II).[355] It has been suggested that the function of ascorbic acid depends on its ability to form the monodehydro-ascorbyl radical or to promote the formation of hydroxyl radicals.[309]

Inhibition of such reactions is of significance in that many of the compounds which exhibit protective (antioxidant) effects in the oxidative–reductive depolymerization are useful against radiation injuries in man and in animals.[356]

Hyaluronic acid, the basic component of the ground substance of soft connective tissue, is particularly susceptible to the action of autoxidants. It is, therefore, of interest that a general parallelism has been reported between the *in vitro* effects of hyaluronic acid and the *in vivo* effects on connective tissue permeability for several oxidation–reduction systems. Experiments on humans and on living animals have shown that substances that act as autoxidants *in vitro* increase interstitial permeability of skin, and those having antioxidant properties decrease it.[357,358]

REFERENCES (SECTIONS I–V)

1. G. O. Phillips, *Advan. Carbohyd. Chem.*, **18**, 9 (1963).
1a. R. W. Binkley, *Advan. Carbohyd. Chem. Biochem.*, **39**, in press (1980).
2. G. O. Phillips, *Advan. Carbohyd. Chem.*, **16**, 13 (1961).
2a. C. von Sonntag, *Advan. Carbohyd. Chem. Biochem.*, **37**, in press (1979).
3. R. M. Hochstrasser and G. B. Porter, *Quart. Rev.* (London), **14**, 146 (1960).
4. J. P. Simons, *Quart. Rev.* (London), **13**, 3 (1959).
5. J. G. Calvert and J. N. Pitts, Jr., "Photochemistry," Wiley, New York, 1966.
6. C. Reid, *Quart. Rev.* (London), **12**, 205 (1958).
7. M. Kasha, in "Comparative Effects of Radiation," M. Burton, J. S. Kirby-Smith, and J. L. Magee, Eds., Wiley, New York, 1960; *Radiat. Res. Suppl.*, **2**, 246 (1960).
8. S. F. Mason, *Quart. Rev.* (London), **15**, 287 (1961).
9. A. B. F. Duncan and F. A. Matsen, in "Technique of Organic Chemistry," A. Weissberger, Ed., Vol. 9, "Chemical Applications of Spectroscopy," W. West, Ed., Wiley (Interscience), New York, 1956, pp. 581, 629.
10. M. C. Chen and H. A. Taylor, *J. Chem. Phys.*, **27**, 857 (1957).
11. C. H. Giles and R. B. McKay, *Text. Res. J.*, **33**, 527 (1963).
12. W. C. Price, *J. Chem. Phys.*, **4**, 147 (1936).
13. M. Zelikoff and K. Watanabe, *J. Opt. Soc. Amer.*, **43**, 753 (1953).
14. J. Barrett and J. H. Baxendale, *Trans. Faraday Soc.*, **56**, 37 (1960).
15. H. Fricke and E. J. Hart, *J. Chem. Phys.*, **4**, 418 (1936).

16. M. C. Matheson, *Proc. Int. Conf. Peaceful Uses Atomic Energy*, Geneva, **29**, 387 (1958).
17. H. W. Melville, *Proc. Roy. Soc.* (London), *Ser. A*, **157**, 621 (1936).
18. J. R. Bates and H. S. Taylor, *J. Amer. Chem. Soc.*, **49**, 2436 (1927).
19. H. Senftleben and I. Rehren, *Z. Phys.*, **37**, 529 (1926).
20. W. J. Nickersen and J. R. Merkel, *Proc. Nat. Acad. Sci. U.S.*, **39**, 901 (1953).
21. W. J. Nickersen, *Biochim. Biophys. Acta*, **14**, 303 (1954).
22. W. J. Nickersen and G. J. Strauss, *J. Amer. Chem. Soc.*, **82**, 5007 (1960); **83**, 3187 (1961).
23. G. K. Oster, G. Oster, and G. Prati, *J. Amer. Chem. Soc.*, **79**, 595, 4836 (1957).
24. G. Oster, J. S. Bullin, and B. Holmstrom, *Experientia*, **18**, 249 (1962).
25. B. Mooney and H. I. Stonehill, *Chem. Ind.* (London), 1309 (1961).
26. H. C. A. van Beek and P. M. Heertjes, *J. Chem. Soc.*, 83 (1962).
26a. G. O. Phillips, N. Worthington, N. Sharpe, and J. F. McKellar, *Chem. Commun.*, 835 (1967); *J. Chem. Soc. (A)*, 767 (1969).
26b. A. D. Broadbent, *Chem. Commun.*, 382 (1967).
27. F. S. Dainton, *J. Phys. Colloid Chem.*, **52**, 490 (1948).
28. J. L. Magee, *Annu. Rev. Nucl. Sci.*, **3**, 171 (1954).
29. D. E. Lea, "Actions of Radiations on Living Cells," Cambridge University Press, London, 1965.
30. J. L. Magee and M. Burton, *J. Amer. Chem. Soc.*, **73**, 523 (1951).
31. M. Burton, *J. Chem. Educ.*, **28**, 404 (1951).
32. V. Fano, *in* "Comparative Effects of Radiation," M. Burton, J. S. Kirby-Smith, and J. L. Magee, Eds., Wiley, New York, 1961, p. 14.
33. F. S. Dainton, *in* "Chemical, Physical and Biological Aspects of Energy Transfer in Radiation Processes, G. O. Phillips, Ed., Elsevier, Amsterdam, 1966, p. 1.
34. G. Czapski and H. A. Schwarz, *J. Phys. Chem.*, **66**, 471 (1962).
35. E. Collinson, F. S. Dainton, D. R. Smith, and S. Tazuke, *Proc. Chem. Soc.*, 140 (1962).
36. E. J. Hart and J. W. Boag, *Nature*, **197**, 45 (1963).
37. J. P. Keene, *Nature*, **197**, 47 (1963).
38. W. H. Hammill, J. P. Guarino, M. R. Ronayne, and J. A. Ward, *Discussions Faraday Soc.*, **36**, 169 (1963).
39. W. M. Hickam and D. Berg, *J. Chem. Phys.*, **29**, 517 (1958).
40. E. Collinson, J. J. Conlay, and F. S. Dainton, *Discussions Faraday Soc.*, **36**, 153 (1963).
41. A. O. Allen, "The Radiation Chemistry of Water and Aqueous Solutions," Van Nostrand, Princeton, New Jersey, 1961.
42. E. J. Hart, *Annu. Rev. Nucl. Sci.*, **15**, 125 (1965).
43. *Symp. Solvated Electron*, Atlantic City, New Jersey, *Advan. Chem. Ser. No.* **50**, American Chemical Society, Washington, D.C., 1965.
44. A. Kuppermann, *Actions Chim. Biol. Radiat.*, *Ser. V*, 87–166 (1961).
45. H. A. Schwarz, *Radiat. Res. Suppl.*, **4**, 89 (1964).
45a. J. K. Thomas, *Radiation Rev.*, **1**, 183 (1968).
46. D. Berthelot and H. Gaudechon, *C. R. Acad. Sci.*, **155**, 831 (1912).
47. C. Neuberg, *Biochem. Z.*, **39**, 158 (1912).
48. L. M. Dillman, *J. Lab. Clin. Med.*, **17**, 236 (1961).
49. H. von Euler and E. Lindberg, *Biochem. Z.*, **39**, 410 (1912); *J. Chem. Soc.*, **102**, 407 (1912).

50. A. L. Bernoulli and R. Cantieni, *Helv. Chim. Acta*, **15**, 119 (1932).
51. H. von Euler and H. Ohlsen, *J. Chim. Phys.*, **9**, 416 (1911).
52. H. Bierry, V. Henri, and A. Rane, *C. R. Acad. Sci.*, **152**, 1629 (1911).
53. D. Berthelot and H. Gaudechon, *C. R. Acad. Sci.*, **155**, 1016 (1912); **156**, 468 (1913).
54. P. Beyersdorfer and W. Hess, *Ber.*, **57**, 1708 (1924).
55. A. K. Bhattacharya and N. R. Dhar, *J. Indian Chem. Soc.*, **6**, 879 (1929).
56. S. Szalay, *Magy. Biol. Kutatóintézet Munkái*, **8**, 417 (1935–6); *Chem. Abstr.*, **30**, 7041 (1936).
57. T. C. Laurent and E. M. Wertheim, *Acta Chem. Scand.*, **6**, 678 (1952).
58. T. C. Laurent, *J. Amer. Chem. Soc.*, **78**, 1875 (1956).
59. P. Holtz, *Arch. Exp. Pathol. Pharmakol.*, **182**, 141 (1936); P. Holtz and J. P. Becker, *ibid.*, **182**, 160 (1936).
60. G. O. Phillips, *Nature*, **173**, 1044 (1954); G. O. Phillips, G. J. Moody, and G. L. Mattock, *J. Chem. Soc.*, 3522 (1958).
61. G. O. Phillips and G. J. Moody, *J. Chem. Soc.*, 3398 (1960).
62. G. O. Phillips and T. Rickards, *J. Chem. Soc.* (*B*), 455 (1969).
63. G. O. Phillips and W. J. Criddle, *J. Chem. Soc.*, 3984 (1963).
64. J. E. Purvis, *J. Chem. Soc.*, **123**, 2519 (1923).
65. B. Tollens and H. Elsner, "Kurzes Handbuch der Kohlenhydrate," Barth, Leipzig, 1939, p. 213.
66. L. Kwiencinski, J. Meyer, and L. Marchlewski, *Hoppe-Seyler's Z. Physiol. Chem.*, **176**, 292 (1928).
67. W. G. Berl and C. E. Feazel, *J. Agr. Food Chem.*, **2**, 37 (1954).
68. F. Micheel, "Chemie der Zucker und Polysaccharide," Akademischer Verlag, Leipzig, 1939, p. 218.
69. E. Pacsu and L. A. Miller, *J. Amer. Chem. Soc.*, **70**, 523 (1948).
70. G. Herzberg and G. Scheibe, *Z. Phys. Chem.* (Leipzig), **41**, 390 (1930).
71. W. C. Price, *J. Chem. Phys.*, **3**, 256 (1935).
72. G. Fleming, M. M. Anderson, A. J. Harrison, and L. W. Pickett, *J. Chem. Phys.*, **30**, 351 (1959).
73. A. J. Harrison, B. J. Cederholm, and M. A. Terwilliger, *J. Chem. Phys.*, **30**, 355 (1959).
74. J. S. Lake and A. J. Harrison, *J. Chem. Phys.*, **30**, 361 (1959).
75. L. W. Pickett, M. Muntz, and E. M. McPherson, *J. Amer. Chem. Soc.*, **73**, 4862, 4865 (1951).
76. A. Beélik and J. K. Hamilton, *Papier*, **13**, 77 (1959).
77. D. Berthelot and H. Gaudechon, *C. R. Acad. Sci.*, **155**, 401, 1153 (1912).
78. D. Berthelot and H. Gaudechon, *C. R. Acad. Sci.*, **155**, 707 (1913).
79. H. Bierry, V. Henri, and A. Rane, *C. R. Acad. Sci.*, **151**, 316 (1910); *Biochem. Z.*, **64**, 257 (1914).
80. R. Cantieni, *Helv. Chim. Acta*, **18**, 473, 933, 1420 (1935).
81. R. Cantieni, *Helv. Chim. Acta*, **17**, 1528 (1934).
82. E. W. Rice, *Science*, **115**, 92 (1952).
83. P. Donzou, *Actions Chim. Biol. Radiations*, **4**, 86 (1958).
84. L. J. Heidt, *J. Amer. Chem. Soc.*, **61**, 2981 (1939); *J. Franklin Inst.*, **234**, 473 (1942).
85. G. Tanret, *C. R. Acad. Sci.*, **202**, 881 (1936); **201**, 1057 (1935).
86. G. Dobson and L. I. Grossweiner, *Trans. Faraday Soc.*, **61**, 708 (1965); H.-I. Joshek and L. I. Grossweiner, *J. Amer. Chem. Soc.*, **88**, 3261 (1966); H.-I. Joshek and S. I. Miller, *ibid.*, **88**, 3273 (1963).
87. D. Horton and J. S. Jewell, *J. Org. Chem.*, **31**, 509 (1966).

88. R. H. Bell, D. Horton, D. M. Williams, and E. Winter-Mihaly, *Carbohyd. Res.*, **58**, 109 (1977).
89. L. Massol, *C. R. Acad. Sci.*, **152**, 902 (1911).
90. J. Bielecki and R. Wurmser, *C. R. Acad. Sci.*, **154**, 1429 (1912).
91. F. Lieben, L. Lower, and B. Bauminger, *Biochem. Z.*, **271**, 209 (1934).
92. E. A. Balazs and T. C. Laurent, *J. Polymer Sci.*, **6**, 665 (1951).
93. E. A. Balazs, T. C. Laurent, A. F. Howe, and L. Varga, *Radiat. Res.*, **11**, 149 (1959).
94. A. Berthond, *Helv. Chim. Acta*, **16**, 592 (1933).
95. W. D. Cohen, *Rec. Trav. Chim.* (Pays-Bas), **40**, 433 (1920).
96. A. Berthond and D. Porret, *Helv. Chim. Acta*, **17**, 694 (1934).
97. H. S. Taylor and A. J. Gould, *J. Phys. Chem.*, **37**, 367 (1933).
98. B. E. Blaisdel, *J. Soc. Dyers Colour.*, **65**, 618 (1949).
99. J. L. Bolland and H. R. Cooper, *Proc. Roy. Soc.* (London), *Ser. A*, **225**, 405 (1954).
100. C. F. Wells, *Trans. Faraday Soc.*, **57**, 1703, 1719 (1961); *Discussions Faraday Soc.*, **29**, 219 (1960).
101. G. O. Phillips, P. Barber, and T. Rickards, *J. Chem. Soc.*, 3443 (1964).
102. H. R. Cooper, *Trans. Faraday Soc.*, **62**, 2865 (1966).
103. F. Wilkinson, *J. Phys. Chem.*, **66**, 2569 (1963).
104. J. Tickle and F. Wilkinson, *Trans. Faraday Soc.*, **61**, 1981 (1965).
105. N. K. Bridge and G. Porter, *Proc. Roy. Soc.* (London), *Ser. A*, **244**, 277 (1958).
106. N. K. Bridge, *Trans. Faraday Soc.*, **56**, 1001 (1960).
107. N. K. Bridge and M. Reed, *Trans. Faraday Soc.*, **56**, 1796 (1960).
108. P. J. Baugh, G. O. Phillips, and J. C. Arthur, Jr., *J. Phys. Chem.*, **70**, 3061 (1966).
109. P. J. Baugh, G. O. Phillips, and N. Worthington, *J. Soc. Dyers Colour.*, **85**, 241 (1969).
110. M. V. Lock, *Abstr. Papers Amer. Chem. Soc. Meeting*, **131**, 6D (1957); *Proc. Chem. Soc.*, 358 (1960).
111. L. J. Heidt, *J. Amer. Chem. Soc.*, **61**, 323 (1939).
112. J. Schurz and E. Kienzl, *Sv. Papperstidn.*, **61**, 844 (1958).
113. S. Peat, E. J. Bourne, and W. J. Whelan, *Nature*, **161**, 762 (1948).
114. W. J. Whelan and S. Peat, *J. Soc. Dyers Colour.*, **65**, 165 (1949).
115. G. Witz, *Bull. Soc. Ind. Rouen*, **11**, 188 (1883).
116. G. Barr and I. H. Hadfield, *J. Text. Inst.*, **18**, T 490 (1927).
117. P. W. Cunliffe and F. D. Farrow, *Shirley Inst. Mem.*, **7**, 1 (1928); *J. Text. Inst.*, **19**, T 169 (1928).
118. R. Haller and L. Wyszewianski, *Melliand Textilber.*, **10**, 951 (1929).
119. H. F. Launer and W. K. Wilson, *J. Res. Nat. Bur. Stand.*, **30**, 55 (1943); *J. Amer. Chem. Soc.*, **71**, 958 (1949).
120. W. Scharwin and A. Pakschwer, *Z. Angew. Chem.*, **40**, 1009 (1927).
121. A. J. Turner, *J. Soc. Dyers Colour.*, **36**, 165 (1920).
122. D. K. Appleby, *Amer. Dyestuff Reptr.*, **38**, 149 (1949).
123. P. W. Cunliffe, *J. Text. Inst.*, **15**, T 173 (1924).
124. H. M. Robinson and W. A. Reeves, *Amer. Dyestuff Reptr.*, **50**, 17 (1961).
125. For a full discussion of the two processes see G. S. Egerton, *J. Soc. Dyers Colour.*, **65**, 764 (1949).
126. J. H. Flynn, W. K. Wilson, and W. L. Morrow, *J. Res. Nat. Bur. Stand.*, **60**, 229 (1958).
127. E. Henser and G. N. Chamberlain, *J. Amer. Chem. Soc.*, **68**, 79 (1946).
128. E. Hibbert, *J. Soc. Dyers Colour.*, **43**, 261 (1927).
129. H. Kauffman, *Melliand Textilber.*, **7**, 617 (1916).

130. P. Neerman and H. Sommers, *Leipzig Monatschr. Textil-Ind.*, **40**, 161, 207 (1925).
131. S. Ogyri and T. Yamaguchi, *J. Soc. Chem. Ind. Jap.*, **40**, (Suppl. 300B), 356B (1937).
132. R. A. Stillings and R. J. Van Nostrand, *J. Amer. Chem. Soc.*, **66**, 753 (1944).
133. Symposium published in *J. Soc. Dyers Colour.*, **65**, 585–788 (1949).
134. P. Waentig, *Z. Angew. Chem.*, **36**, 357 (1923).
135. F. A. Abdie-Maumert, *Papeterie*, **77**, 593 (1955).
136. A. Beélik and J. K. Hamilton, *J. Org. Chem.*, **26**, 5074 (1961).
137. G. Centola, *Assoc. Tech. Ind. Papetiere Bull.*, **5**, 111 (1951).
138. L. Chadeyron, *Assoc. Tech. Ind. Papetiere Bull.*, **7**, 21 (1954).
139. G. S. Egerton, *J. Soc. Dyers Colour.*, **64**, 336 (1948).
140. G. S. Egerton, *J. Text. Inst.*, **39**, T 293 (1948).
141. G. S. Egerton, *Text. Res. J.*, **18**, 659 (1948).
142. C. W. Mason and F. B. Rosevear, *J. Amer. Chem. Soc.*, **61**, 2995 (1939).
143. V. L. Frampton, L. P. Foley, and H. H. Webber, *Arch. Biochem. Biophys.*, **18**, 345 (1948).
144. C. Kujirai, *Bull. Inst. Chem. Res. Kyoto Univ.*, **24**, 42 (1951).
145. C. Kujirai, *Bull. Inst. Chem. Res. Kyoto Univ.*, **31**, 228 (1953).
146. J. Schurz, *Sv. Papperstidn.*, **59**, 98 (1955).
147. H. Sihtola and B. C. Fogelberg, *Pap. Puu*, **36**, 430 (1954).
148. A. Sippel, *Faserforsch. Textiltech.*, **3**, 211 (1952); *Kolloid-Z.*, **127**, 79 (1952); *Melliand Textilber.*, **33**, 645 (1952).
149. J. H. Flynn, *J. Polymer Sci.*, **27**, 83 (1958).
150. E. Treiber, *Kolloid-Z.*, **130**, 39 (1953); *Sv. Papperstidn.*, **58**, 185 (1955).
151. B. C. Bera and P. N. Agarwal, "LABDEV" (Kanpur, India), **2**, 105 (1964).
152. B. C. Bera, S. D. Kapila, and A. L. Rao, *J. Sci. Ind. Res.*, **21D**, 444 (1962).
153. P. N. Agarwal, B. C. Bera, and L. K. Chaturvedi, *Indian J. Technol.*, **1**, 356 (1963).
154. D. Ashton and M. E. Robert, *J. Text. Inst.*, **44**, T 1 (1953).
155. A. D. Baskin and A. M. Kaplan, *Text. Res. J.*, **28**, 554 (1958).
156. W. N. Berard, S. G. Gremillion, Jr., and C. F. Goldthwait, *Text. Res. J.*, **26**, 81 (1956).
157. F. Dorr, *Z. Electrochem.*, **64**, 580 (1960).
158. C. H. Bamford and M. J. S. Dewar, *J. Soc. Dyers Colour.*, **65**, 674 (1949).
159. C. H. Bamford and M. J. S. Dewar, *Proc. Roy. Soc.* (London), *Ser. A*, **198**, 252 (1949).
160. G. A. Gilbert, H. Baum, and M. V. Lock, reported in *Chem. Eng. News*, **35**, 26 (1957) (June).
161. N. Geacintov, V. Stannett, E. W. Abrahamson, and J. J. Hermans, *J. Appl. Polymer Sci.*, **3**, 54 (1960).
162. N. Geacintov, V. Stannett, E. W. Abrahamson, and J. J. Hermans, *Makromol. Chem.*, **36**, 52 (1959).
163. R. Haller and G. Ziersch, *Melliand Textilber.*, **10**, 951 (1929).
164. E. Hibbert, *J. Soc. Dyers Colour.*, **43**, 261 (1927).
165. I. Rusznák and M. Fehérvári, *Magy. Textiltech.*, 127 (1959).
166. G. N. Lewis and M. Kasha, *J. Amer. Chem. Soc.*, **66**, 2100 (1944).
167. C. H. Giles and R. B. McKay, *Text. Res. J.*, **33**, 527 (1963).
167a. B. E. Hulme, E. J. Land, and G. O. Phillips, *Chem. Commun.*, 518 (1969).
168. A. Bollinger and N. T. Hinks, *Aust. J. Expt. Biol. Med. Sci.*, **27**, 569 (1949).
169. E. Baur and C. Neuweiler, *Helv. Chim. Acta*, **10**, 901 (1927).
170. C. F. Goodeve, *Trans. Faraday Soc.*, **33**, 340 (1937).
171. F. Scholefield and E. H. Goodyear, *Melliand Textilber.*, **10**, 867 (1929).

172. K. Yamafugi, N. Nishioeda, and H. Imagawa, *Biochem. Z.*, **301**, 404 (1939).
173. M. C. Markham, M. Hannan, and W. S. Evans, *J. Amer. Chem. Soc.*, **76**, 820 (1954).
174. M. C. Markham, M. Hannan, R. M. Paternosto, and C. B. Rose, *J. Amer. Chem. Soc.*, **80**, 5394 (1958).
175. M. C. Markham and K. J. Laidler, *J. Phys. Chem.*, **57**, 363 (1953).
176. N. Miller, *Rev. Pure Appl. Chem.*, **7**, 123 (1957).
177. E. J. Bourne, M. Stacey, and G. Vaughan, *Chem. Ind.* (London), 189 (1954).
178. G. O. Phillips, G. J. Moody, and G. L. Mattock, *J. Chem. Soc.*, 3522 (1958).
179. M. L. Wolfrom, W. W. Binkley, L. J. McCabe, T. M. Shen Han, and A. Michelakis, *Radiat. Res.*, **10**, 37 (1959).
180. C. T. Bothner-By and E. A. Balazs, *Radiat. Res.*, **6**, 302 (1959).
181. G. O. Phillips and G. J. Moody, *J. Chem. Soc.*, 754 (1960).
182. G. O. Phillips and W. J. Criddle, *J. Chem. Soc.*, 3404 (1960).
183. G. O. Phillips, *Nature*, **173**, 1044 (1954); *J. Chem. Soc.*, 297 (1963).
184. P. M. Grant and R. B. Ward, *J. Chem. Soc.*, 2871 (1959).
185. G. O. Phillips, G. L. Mattock, and G. J. Moody, *Proc. Int. Conf. Peaceful Uses Atomic Energy, Geneva*, **29**, 92 (1958).
186. G. O. Phillips and G. J. Moody, *Int. J. Appl. Radiat. Isotopes*, **6**, 78 (1959).
187. G. O. Phillips, *Radiat. Res.*, **18**, 446 (1963).
188. G. O. Phillips and W. J. Criddle, *J. Chem. Soc.*, 2733 (1962).
189. G. O. Phillips, W. Griffiths, and J. V. Davies, *J. Chem. Soc.* (*B*), 194 (1966).
190. G. O. Phillips and W. J. Criddle, *Proc. Int. Conf. Radioisotopes, Copenhagen*, p. 326 (1962).
190a. See, for example, A. G. W. Bradbury, and C. von Sonntag, *Carbohyd. Res.*, **62**, 223 (1978) for studies on radiolytic deamination of 2-amino-2-deoxy-D-glucose, and earlier papers there cited for general methodology and related work.
191. S. A. Barker, P. M. Grant, M. Stacey, and R. B. Ward, *J. Chem. Soc.*, 2648 (1959).
192. A. J. Bailey, S. A. Barker, I. R. Lloyd, and R. H. Moore, *Radiat. Res.*, **15**, 532, 538 (1961).
193. N. K. Kochetkov, L. I. Kudryashov, S. M. Yarovaya, and E. I. Bortsova, *Zh. Obshch. Khim.*, **35**, 270 (1965).
194. P. M. Grant, R. B. Ward, and D. H. Whiffen, *J. Chem. Soc.*, 4635 (1958).
195. P. M. Grant and R. B. Ward, *J. Chem. Soc.*, 2654 (1959).
195a. C. von Sonntag and M. Dizdaroglu, *Carbohyd. Res.*, **58**, 21 (1977).
196. G. O. Phillips and W. J. Criddle, *J. Chem. Soc.*, 3756, 3763 (1961).
197. F. Haber and W. Weiss, *Proc. Roy. Soc.* (London), *Ser. A*, **147**, 332 (1934).
198. M. A. Khenokh, E. A. Kuzicheva, and V. F. Evdokimov, *Dokl. Akad. Nauk SSSR*, **131**, 684 (1960).
199. G. L. Clark and K. R. Fitch, *J. Amer. Chem. Soc.*, **52**, 465 (1930).
200. G. L. Clark, L. W. Pickett, and E. D. Johnson, *Radiology*, **13**, 245 (1930).
201. J. Wright, *Discussions Faraday Soc.*, **12**, 60 (1952).
202. M. L. Wolfrom, W. W. Binkley, and L. J. McCabe, *J. Amer. Chem. Soc.*, **81**, 1442 (1959).
203. G. O. Phillips and G. J. Moody, *J. Chem. Soc.*, 155 (1960).
204. G. O. Phillips and G. J. Moody, *J. Chem. Soc.*, 3534 (1958).
205. S. Adachi, *J. Dairy Sci.*, **45**, 1427 (1962).
206. G. O. Phillips and K. W. Davies, *J. Chem. Soc.*, 205 (1964).
207. M. A. Khenokh, *Dokl. Akad. Nauk SSSR*, **104**, 746 (1955).
207a. V. Hartmann, C. von Sonntag, and D. Schulte-Frohlinde, *Z. Naturforsch.*, **25B**, 1394 (1970).

207b. M. Dizdaroglu and C. von Sonntag, *Z. Naturforsch.*, **28B**, 635 (1973).

207c. C. von Sonntag, M. Dizdaroglu, and D. Schulte-Frohlinde, *Z. Naturforsch.*, **31B**, 315 (1976).

207d. P. J. Baugh, J. I. Goodall, G. O. Phillips, C. von Sonntag, and M. Dizdaroglu, *Carbohyd. Res.*, **49**, 315 (1976).

208. N. K. Kochetkov, L. I. Kudryashov, and M. A. Chlenov, *Zh. Obshch. Khim.*, **35**, 897 (1965).

209. N. K. Kochetkov, L. I. Kudryashov, and M. A. Chlenov, *Izv. Akad. Nauk SSSR, Ser. Khim.*, 2115 (1964).

210. N. K. Kochetkov, L. I. Kudryashov, and M. A. Chlenov, *Zh. Obshch. Khim.*, **35**, 2246 (1965).

211. M. A. Khenokh, *Dokl. Akad. Nauk SSSR*, **104**, 746 (1955).

212. S. Adachi, *J. Dairy Sci.*, **45**, 962 (1962).

213. F. P. Price, W. D. Bellamy, and E. J. Lawton, *J. Phys. Chem.*, **58**, 821 (1954).

214. P. O. Kinell, K. A. Granath, and T. Vanngard, *Arkiv Fysik*, **13**, 272 (1958).

215. A. Brasch, W. Huber, and A. Waly, *Arch. Biochem. Biophys.*, **39**, 245 (1952).

216. M. A. Khenokh, *Zh. Obshch. Khim.*, **20**, 1560 (1950).

217. H. Kersten and C. H. Dwight, *J. Phys. Chem.*, **41**, 687 (1937).

218. R. N. Feinstein and L. L. Nejelski, *Radiat. Res.*, **2**, 8 (1955).

219. E. A. Roberts and B. E. Procter, *Food Res.*, **20**, 254 (1955).

220. Z. I. Kertesz, B. H. Morgan, L. W. Tuttle, and M. Lavin, *Radiat. Res.*, **5**, 372 (1956).

221. A. Caputo, *Nature*, **179**, 1133 (1957).

222. C. Ragan, C. P. Donlan, J. A. Cross, and A. F. Grubin, *Proc. Soc. Exp. Biol. Med.*, **66**, 170 (1947).

223. M. D. Schoenberg, R. E. Brookes, J. J. Hall, and M. Schneiderman, *Arch. Biochem. Biophys.*, **30**, 333 (1951).

224. H. A. Colwell and S. Russ, *Radium*, **9**, 230 (1912).

225. A. Mishina and Z. Nikuni, *Mem. Inst. Sci. Ind. Res. Osaka Univ.*, **17**, 215 (1960).

226. J. V. Davies, W. Griffiths, and G. O. Phillips, in "Pulse Radiolysis," M. Ebert, J. H. Baxendale, J. V. Keene, and A. J. Swallow, Eds., Academic Press, New York, 1965, p. 181.

226a. E. A. Balazs, J. V. Davies, G. O. Phillips, and M. Young, *Radiat. Res.*, **31**, 243 (1967).

227. G. O. Phillips and K. W. Davies, *J. Chem. Soc.*, 3981 (1964).

228. J. Moore, Ph.D. Thesis, University of Wales, 1966.

229. A. Ehrenberg, L. Ehrenberg, and G. Lofroth, *Acta Chem. Scand.*, **17**, 53 (1963); *Riso Rept. No.* **16**, 25 (1960).

230. G. J. Moody and G. O. Phillips, *Chem. Ind.* (London), 1247 (1959).

231. H. Veda, Z. Kuri, and S. Shida, *Nippon Kagaku Zasshi*, **82**, 8 (1961).

232. S. Dilli and J. L. Garnett, *Nature*, **198**, 984 (1963).

233. H. Veda, Z. Kuri, and S. Shida, *J. Chem. Phys.*, **35**, 2145 (1961); H. Veda, *J. Phys. Chem.*, **67**, 966, 2185 (1963).

234. H. Shields and P. Hamrick, *J. Chem. Phys.*, **37**, 202 (1962).

235. J. N. Herak, K. Adamic, and R. Blinc, *J. Chem. Phys.*, **42**, 2388 (1965).

236. G. Löfroth, Ph.D. dissertation, University of Stockholm, Stockholm, 1967.

237. D. Williams, J. E. Geusic, M. L. Wolfrom, and L. J. McCabe, *Proc. Nat. Acad. Sci. U.S.*, **44**, 1128 (1948); D. Williams, B. Schmidt, M. L. Wolfrom, A. Michelakis, and L. J. McCabe, *ibid.*, **47**, 1744 (1959).

238. J. Combrisson and J. Vebersfeld, *C. R. Acad. Sci.*, **268**, 397 (1954); J. Vebersfeld, *Ann. Phys.*, **1**, 395 (1956).

239. M. A. Collins, *Nature*, **193**, 984 (1963).

240. G. O. Phillips and P. J. Baugh, *J. Chem. Soc.* (*A*), 370 (1966).

241. J. R. Bolton, A. Carrington, and J. dos Santos-Viega, *Mol. Phys.*, **5**, 465 (1962).

241a. P. J. Baugh, K. Kershaw, and G. O. Phillips, *Nature*, **221**, 1138 (1969).

242. G. O. Phillips, F. A. Blouin, and J. C. Arthur, Jr., *Nature*, **202**, 1328 (1964); *Radiat. Res.*, **23**, 527 (1964).

243. G. O. Phillips, in "Current Topics in Radiation Research," M. Ebert and A. Howard, Eds., North-Holland Publ., Amsterdam, 1966, Chapt. 3.

243a. A. Ehrenberg, L. Ehrenberg, and G. Lofroth, *Acta Chem. Scand.*, **17**, 58 (1963); G. Lofroth, *ibid.*, **21**, 1997 (1967).

243b. N. U. Ahmed, P. J. Baugh, and G. O. Phillips, *J. Chem. Soc. Perkin II*, 1305 (1972).

243c. G. Lofroth, *Int. J. Radiat. Phys. Chem.*, **4**, 227 (1972).

243d. C. von Sonntag and M. Dizdaroglu, *Z. Naturforsch.*, **28B**, 365 (1973); M. Dizdaroglu, C. von Sonntag, and D. Schulte-Frohlinde, *Ann.*, 1592 (1973).

243e. C. von Sonntag, K. Neuwald, and M. Dizdaroglu, *Radiat. Res.*, **58**, 1 (1974).

244. Z. I. Kertesz, E. R. Schulz, G. Fox, and M. Gibson, *Food Res.*, **24**, 609 (1959).

245. L. Ehrenberg, M. Jaarma, and K. G. Zimmer, *Acta Chem. Scand.*, **11**, 950 (1957).

246. J. P. O'Meara and T. M. Sheen, *Food Technol.*, **9**, 132 (1957).

247. F. A. Blouin and J. C. Arthur, Jr., *Text. Res. J.*, **28**, 198 (1958); J. C. Arthur, Jr., *ibid.*, **28**, 204 (1958); J. C. Arthur, Jr., F. A. Blouin, and R. J. Dement, *Amer. Dyestuff Reptr.*, **49**, 383 (1960).

248. E. S. Gilfillan and L. Linden, *Text. Res. J.*, **25**, 773 (1955); **27**, 87 (1957).

249. R. E. Glegg, *Radiat. Res.*, **6**, 469 (1957).

250. R. E. Glegg and Z. I. Kertesz, *J. Polymer. Sci.*, **26**, 289 (1957); *Science*, **124**, 893 (1956).

251. D. J. Harmon, *Text. Res. J.*, **27**, 318 (1957).

252. H.-P. Pan, B. E. Proctor, S. A. Goldblith, H. M. Morgan, and R. Z. Naar, *Text. Res. J.*, **29**, 415 (1959).

253. O. Teszler, L. H. Kiser, P. W. Campbell, and H. A. Rutherford, *Text. Res. J.*, **28**, 456 (1958).

254. O. Teszler, H. Wiehart, and H. A. Rutherford, *Text. Res. J.*, **28**, 131 (1958).

255. J. F. Saeman, M. A. Millett, and E. J. Lawton, *Ind. Eng. Chem.*, **44**, 2848 (1952).

256. F. L. Dalton, M. R. Houlton, and J. A. Sykes, *Nature*, **200**, 862 (1963).

257. J. H. Flynn, W. K. Wilson, and W. L. Morrow, *J. Res. Nat. Bur. Stand.*, **60**, 2841 (1958).

258. S. Dilli and J. L. Garnett, *Chem. Ind.* (London), 409 (1963).

259. F. L. Dalton and M. R. Houlton, *Radiat. Res.*, **28**, 576 (1966).

260. R. E. Glaggand and Z. I. Kertesz, *J. Polymer Sci.*, **26**, 289 (1957).

261. J. Uebersfeld and J. Cambrisson, *C. R. Acad. Sci.*, **238**, 1397 (1954).

262. F. A. Blouin and J. C. Arthur, Jr., *Text. Res. J.*, **33**, 727 (1963).

263. Kh. Usmanov, B. I. Aikhodzhaev and V. O. Azizov, *Vysokomol. Soedin.*, **1**, 1570 (1959).

264. A. Charlesby, *J. Polymer Sci.*, **15**, 263 (1955).

265. H. Sobue and Y. Saito, *Bull. Chem. Soc. Japan*, **34**, 1344 (1961).

266. G. S. Park, *Polymer Lett.*, **1**, 617 (1963).

267. F. A. Blouin and J. C. Arthur, Jr., *J. Chem. Eng. Data*, **5**, 470 (1960).

268. J. C. Arthur, Jr., and F. A. Blouin, *Amer. Dyestuff Reptr.*, **51**, 1024 (1962).

268a. J. C. Arthur, Jr., in "Energetics and Mechanisms in Radiation Biology," G. O. Phillips, Ed., Academic Press, London, 1968, pp. 153–179.

269. G. O. Phillips and P. J. Baugh, *Nature*, **198**, 262 (1963); *J. Chem. Soc. (A)*, 370 (1966).

269a. G. Lofroth, *Acta Chem. Scand.*, **21**, 1997 (1967).

270. E. Collinson, J. J. Conlay, and F. S. Dainton, *Nature*, **194**, 1074 (1962); *Discussions Faraday Soc.*, **36**, 153 (1963).

271. A. Gilman, *Trans. Amer. Inst. Mining Met. Engrs.*, **203**, 1252 (1953); *J. Appl. Phys.*, **27**, 1018, 1262 (1965).

272. R. M. Hochstrasser, *Radiat. Res.*, **20**, 107 (1963).

273. B. M. Tolbert and R. M. Lemmon, *Radiat. Res.*, **3**, 52 (1955).

274. N. Riehl, *Naturwissenschaften*, **43**, 145 (1956).

275. H. N. Rexroad and W. Gordy, *Phys. Rev.*, **125**, 242 (1962).

276. R. C. G. Killean, W. G. Ferrier, and D. W. Young, *Acta Crystallogr.*, **15**, 911 (1962).

277. T. R. R. McDonald and C. A. Beevers, *Acta Crystallogr.*, **3**, 394 (1950).

278. T. Forster, *Naturwissenschaften*, **33**, 166 (1946); *Radiat. Res. Suppl.*, **2**, 326 (1960); *Ann. Phys.*, **2**, 55 (1948); *Discussions Faraday Soc.*, **27**, 7 (1959).

279. M. Kasha, *Radiat. Res.*, **20**, 55 (1963).

280. F. Cramer, *Angew. Chem.*, **68**, 115 (1956); *Rev. Pure Appl. Chem.*, **5**, 143 (1955).

281. G. O. Phillips and M. Young, *J. Chem. Soc. (A)*, 383 (1966).

282. G. O. Phillips and P. J. Baugh, *J. Chem. Soc. (A)*, 387 (1966).

283. U. K. Misra, J. C. Picken, and D. French, *Radiat. Res.*, **14**, 775 (1961).

284. G. O. Phillips and M. Young, *J. Chem. Soc. (A)*, 393 (1966).

285. H. Fisher, *J. Chem. Phys.*, **37**, 1094 (1962); *Kolloid-Z.*, **180**, 64 (1962).

286. R. W. Fessenden and R. H. Schuler, *J. Chem. Phys.*, **39**, 2147 (1963).

287. K. Biemann, D. C. DeJongh, and H. K. Schnoes, *J. Amer. Chem. Soc.*, **85**, 1763 (1963).

287a. G. O. Phillips and M. Young, *J. Chem. Soc. (A)*, 1240 (1968).

288. G. O. Phillips, F. A. Blouin, and J. C. Arthur, Jr., *Radiat. Res.*, **23**, 527 (1964).

289. A. B. Foster and W. G. Overend, *Chem. Ind.* (London), 566 (1955).

290. R. J. Bayly and E. A. Evans, *J. Label. Compounds*, **2**, 1 (1966).

291. P. Rochlin, *Chem. Rev.*, **65**, 685 (1965).

292. R. J. Bayly and H. Weigel, *Nature*, **188**, 384 (1960).

293. E. J. Bourne, D. H. Hutson, and H. Weigel, *J. Chem. Soc.*, 5153 (1960).

294. G. O. Phillips, W. J. Criddle, and G. J. Moody, *J. Chem. Soc.*, 4216 (1962).

295. G. O. Phillips and K. W. Davies, *J. Chem. Soc.*, 2654 (1965).

296. A. G. Lloyd, E. A. Balazs, G. Emberry, and F. S. Wusteman, *Biochem. J.*, **98**, 34 (1966).

REFERENCES (SECTION VI)

297. A. Weissler, *J. Amer. Chem. Soc.*, **81**, 1077 (1959).

298. C. Simionescu and V. Rusan, *Celul. Hirties* (Bucharest), **15**, 369 (1966); *Chem. Abstr.*, **66**, 66995 (1967).

299. H. J. H. Fenton, *J. Chem. Soc.*, **65**, 899 (1894).

300. C. Neuberg and S. Miura, *Biochem. Z.*, **36**, 37 (1911).

301. R. N. Feinstein and L. L. Nejelski, *Radiat. Res.*, **2**, 8 (1955).

302. F. Haber and J. Weiss, *Proc. Roy. Soc.* (London), *Ser. A*, **147**, 332 (1934).

303. I. M. Kolthoff and A. I. Medalia, *J. Amer. Chem. Soc.*, **71**, 3777, 3781 (1949).

304. G. M. Coppinger, *J. Amer. Chem. Soc.*, **79**, 2758 (1957).

305. W. T. Dixon and R. O. C. Norman, *Nature*, **196**, 891 (1962).
306. S. J. Leach, *Advan. Enzymol.*, **15**, 1 (1954).
307. I. Isenberg, *Physiol. Rev.*, **44**, 487 (1964).
308. S. Udenfriend, C. T. Clark, J. Axelrod, and B. B. Brodie, *J. Biol. Chem.*, **208**, 731 (1954).
309. W. B. Robertson, *Ann. N.Y. Acad. Sci.*, **92**, 159 (1961).
310. Y. Peloux, C. Nofre, A. Cier, and L. Colobert, *Ann. Inst. Pasteur*, **102**, 6 (1962).
311. I. Yamazaki and L. H. Piette, *Biochim. Biophys. Acta*, **50**, 62 (1961).
312. W. B. van Robertson, M. W. Ropes, and W. Bauer, *Biochem. J.*, **35**, 903 (1941).
313. J. Madinaveitia and T. H. H. Quibell, *Biochem. J.*, **35**, 453 (1941).
314. A. Pirie, *Brit. J. Exp. Pathol.*, **23**, 277 (1942).
315. B. Skanse and L. Sundblad, *Acta Physiol. Scand.*, **6**, 37 (1943).
316. C. W. Hale, *Biochem. J.*, **38**, 362 (1944).
317. W. Pigman, S. Rizvi, and H. L. Holley, *Arthritis Rheumat.*, **4**, 240 (1961).
318. H. Staudinger and W. Heuer, *Ber.*, **63**, 222 (1930).
319. E. Elöd and H. Schmid-Bielenberg, *Z. Phys. Chem.*, **B25**, 27 (1934).
320. T. C. Laurent, M. Ryan, and A. Pietruszkiewicz, *Biochim. Biophys. Acta*, **42**, 476 (1960).
321. L. Sundblad, *Acta Soc. Med. Upsalien*, **58**, 113 (1953).
322. T. Rickards, A. Herp, and W. Pigman, *J. Polymer Sci.*, **A1**, **5**, 931 (1967).
323. G. Matsumura, A. Herp, and W. Pigman, *Radiat. Res.*, **28**, 735 (1966).
324. N. G. Levandoski, E. M. Baker, and J. E. Canham, *Biochemistry*, **3**, 1465 (1964).
325. G. Matsumura and W. Pigman, *Arch. Biochem. Biophys.*, **110**, 526 (1965).
326. L. Sundblad and E. A. Balazs, in "The Amino Sugars," E. A. Balazs and R. W. Jeanloz, Eds., Vol. IIB, Academic Press, New York, 1966, p. 233.
327. A. Weissberger and J. E. LuValle, *J. Amer. Chem. Soc.*, **66**, 700 (1944).
327a. M. J. Harris, A. Herp, and W. Pigman, *J. Amer. Chem. Soc.*, **94**, 7570 (1972).
327b. M. J. Harris, A. Herp, and W. Pigman, *Arch. Biochem. Biophys.*, **142**, 615 (1971).
328. W. Pigman and S. Rizvi, *Biochem. Biophys. Res. Commun.*, **1**, 39 (1959).
329. K. Berneis, *Helv. Chim. Acta*, **46**, 57 (1963).
330. R. L. Whistler and R. Schweiger, *J. Amer. Chem. Soc.*, **71**, 3136 (1959).
331. T. S. Lawton and H. K. Nason, *Ind. Eng. Chem.*, **36**, 1128 (1944).
332. J. F. Haskins and M. S. Hogsed, *J. Org. Chem.*, **15**, 1264 (1950).
333. A. Herp, T. Rickards, G. Matsumura, L. B. Jakosalem, and W. Pigman, *Carbohyd. Res.*, **4**, 63 (1967).
334. D. L. Gilbert, R. Gerschman, J. Cohen, and W. Sherwood, *J. Amer. Chem. Soc.*, **79**, 5677 (1957).
335. O. Smidsrød, A. Haug, and B. Larsen, *Acta Chem. Scand.*, **17**, 2628 (1963).
336. P. Albersheim, H. Neukom, and H. Deuel, *Arch. Biochem. Biophys.*, **90**, 46 (1960).
337. O. Smidsrød, A. Haug, and B. Larsen, *Carbohyd. Res.*, **5**, 485 (1967).
338. G. Scott, "Atmospheric Oxidation and Antioxidants," Elsevier, Amsterdam, 1965, p. 361.
339. J. R. Chipault, in "Autoxidation and Antioxidants," Vol. II, W. O. Lundberg, Ed., Wiley (Interscience), New York, 1962, p. 508.
340. W. Pigman, W. Hawkins, E. Gramling, S. Rizvi, and H. L. Holley, *Arch. Biochem. Biophys.*, **89**, 184 (1960).
341. A. Herp, M. J. Harris, and T. Rickards, *Carbohyd. Res.*, **16**, 395 (1971).
342. J. H. Flynn and W. L. Morrow, *J. Polymer Sci.*, **A2**, 81 (1964); R. E. Florin and L. A. Wall, *ibid.*, **A1**, 1163 (1963).

343. G. N. Richards, *J. Appl. Polymer Sci.*, **5**, 539 (1961).
344. A. Haug, B. Larsen, and O. Smidsrød, *Acta Chem. Scand.*, **17**, 1466, 1473 (1963).
345. W. Pigman, G. Matsumura, and A. Herp, *Proc. 4th Int. Congr. Rheology, Providence, Rhode Island*, 1963, Wiley, New York, 1965, p. 505.
346. A. Caputo, *Nature*, **179**, 1133 (1957).
347. E. A. Balazs, T. C. Laurent, A. F. Howe, and L. Varga, *Radiat. Res.*, **11**, 149 (1959).
348. R. L. Whistler and J. N. BeMiller, *J. Amer. Chem. Soc.*, **82**, 457 (1960).
348a. J. K. Thomas, *Trans. Faraday Soc.*, **61**, 702 (1965).
348b. I. Krajlic and C. N. Trumbore, *J. Amer. Chem. Soc.*, **87**, 2547 (1965).
348c. H. S. Isbell, H. L. Frush, and E. T. Martin, *Carbohyd. Res.*, **26**, 287 (1973).
349. B. Commoner, J. Townsend, and G. E. Pake, *Nature*, **174**, 689 (1954).
350. L. Szilard, *Proc. Nat. Acad. Sci. U.S.*, **45**, 30 (1959); J. Bjorksten, *J. Amer. Geriatrics Soc.*, **25**, 396 (1977).
351. R. Gerschman, D. L. Gilbert, S. W. Nye, P. Dwyer, and W. O. Fenn, *Science*, **119**, 623 (1954).
352. D. Harman, *Radiat. Res.*, **16**, 753 (1963).
353. N. K. Kochetkov, L. I. Kudryashov, M. A. Chlenov, and L. P. Grineva, *Carbohyd. Res.*, **35**, 235 (1974).
354. L. S. Myers, Jr., *Federation Proc.*, **32**, 1882 (1973).
355. J. J. Hutton, Jr., A. L. Tappel, and S. Udenfriend, *Arch. Biochem. Biophys.*, **118**, 231 (1967).
356. P. Alexander, L. M. Bacq, S. F. Cousens, M. Fox, A. Herve, and T. Lazar, *Radiat. Res.*, **2**, 393 (1955).
357. J. Fabianek and A. Herp, *Excerpta Med. Int. Congr. Ser.*, **82**, 35 (1965).
358. A. Herp, J. Fabianek, A. Calick, and W. Pigman, *Ann. Rheumatic Diseases*, **25**, 345 (1966).

27. PHYSICAL METHODS FOR STRUCTURAL ANALYSIS

This chapter provides a survey of the principal physical methods that have been of major value in structural studies on carbohydrates and their derivatives. The emphasis in each section is on the principles of the method and general applications in the carbohydrate field. Further examples of detailed and specialized applications of various physicochemical techniques are to be found in individual chapters elsewhere in these volumes, notably in Chapters 1, 4, 5, 15, 28, 31, 35, and 45.

I. HIGH-RESOLUTION NUCLEAR MAGNETIC RESONANCE SPECTROSCOPY

LAURANCE D. HALL

A. GENERAL INTRODUCTION

In the two decades since the first[1] investigation of carbohydrate derivatives by proton (^{1}H) nuclear magnetic resonance (n.m.r.) spectroscopy, this technique has become firmly established as an extremely powerful physico-chemical tool for elucidating not only the static but also the dynamic structures of monosaccharides. Such applications have been the subject of several review articles,[2–8] and other reviews have emphasized studies of cyclitols,[9] instrumental methods,[10] and techniques for removing the unwanted overlap of individual proton resonances.[11] Additionally, each volume of the Chemical Society Specialist Periodical Report[12] entitled "Carbohydrate Chemistry" has a section devoted to n.m.r. studies.

Traditionally, most n.m.r. studies of carbohydrate derivatives have been concerned with protons, and the basic format of this chapter will follow that bias; however, the ever-increasing sophistication of n.m.r. spectrometers has made routine studies of many other magnetic nuclides ("heteronuclei"), of which ^{13}C, ^{15}N, ^{19}F, and ^{31}P have been of prime interest thus far. Accordingly, this chapter includes a *brief* introduction to a few, important, instrumental considerations, as well as a separate section devoted to studies of heteronuclei.

General aspects of the technique of high-resolution nuclear magnetic resonance spectroscopy have been described in numerous monographs[13–20] and in four series appearing on an annual basis.[21–24] In addition, several monographs[25–27] are concerned exclusively with the area of carbon-13 n.m.r. spectroscopy in organic chemistry.

1. *The Nuclear Magnet Model*

The simplest model on which a discussion of the n.m.r. technique can be based is that of the "nuclear magnet," in which any nucleus having spin $\frac{1}{2}$ (most commonly a proton) is considered to act as if it were a submicroscopic bar magnet. Such a magnet can adopt an orientation either parallel or anti-parallel to an external magnetic field, the two orientations constituting different energy states for the nucleus. Absorption of energy from the radio-frequency unit of the spectrometer enables resonating nuclear magnets to change from the lower to the higher energy state, and measurement of this process for each of the nuclei in the sample provides the histogram of instrument response versus radiofrequency that constitutes the n.m.r. spectrum.

The precise frequency at which any particular nucleus will resonate is determined by the effective magnetic field *at that nucleus*, and it is commonly presented as a displacement ("chemical shift") from the resonance frequency of a hypothetical, isolated nucleus. The chemical shift is influenced by a number of factors. Thus, for protons, the shielding effect of the surrounding electron cloud, which can be related to the electronegativity of nearby substituents, is generally dominant, whereas for nuclei having populated *p* or *d* orbitals, distortion from ideal geometry causes much larger shifts. A more subtle effect (spin coupling) arises from the mutual interaction between nonequivalent, magnetic nuclei that are proximal to one another in the molecule, which causes these resonances to exhibit multiplet structure, whereas other nuclei that do not experience such interactions give single, narrow resonances.

Although the nuclear magnet model is somewhat naive, it can be used to rationalize much of the nuclear magnetic resonance phenomenon, and, moreover, it can be extended readily to encompass discussions of spin–spin coupling and of the double-resonance technique.

2. Description of the N.M.R. Spectrum

Any n.m.r. spectrum can be fully described by five sets of parameters, namely:

1. The relative areas, or *integrals*, of each separate resonance signal; these areas are proportional to the relative numbers of nuclei resonating at each particular frequency (or magnetic field, since the frequency and field values corresponding to a resonance condition are directly related).

2. The resonance positions, or *chemical shifts*, of signals in the spectrum, which are defined with respect to the resonance position of an arbitrarily selected, internal standard; the normal reference signal for proton and carbon-13 spectra is tetramethylsilane (Me_4Si). For convenience, these values are quoted in dimensionless units of parts per million (p.p.m.) on either the δ scale, in which:

$$\delta = \frac{\text{frequency separation (in Hz) between the resonance and } Me_4Si \times 10^6}{\text{spectrometer radiofrequency (in Hz)}}$$

or on the τ scale, in which $\tau = (10 - \delta)$.

3. The magnitudes of the various *spin couplings*; such couplings, which cause many resonances to have resolvable fine structure, are reported in frequency units [Hertz (Hz), formerly cycles per second (c.p.s.)].

4. *The spin–lattice relaxation times*, or T_1 values; these are the time constants for the transfer of absorbed energy from the resonating nuclei to their surrounding environment; a T_1 value has dimensions of time (seconds).

5. *The spin–spin relaxation times*, or T_2 values; these are time constants for loss of phase coherence of the resonating nuclei, and this type of relaxation time constant has been little studied by organic chemists.

P.m.r. spectroscopy can be applied to problems in the carbohydrate field at any one of three different levels, which are, in order of increasing complexity: (1) the identification of substituent groups present in the molecule;

(2) the assignment of configuration and of ring size to cyclic derivatives; and (3) the elucidation of the conformations (and configurations) of both cyclic and acyclic derivatives (see also Vol. IA, Chapter 5). Investigations in the first two categories are essentially routine, especially for pyranose derivatives, but precise assignments of conformation to furanose derivatives are still a difficult task for which there is a dearth of reference data.

B. Instrumental Considerations

In a conventional ^{1}H-n.m.r. spectrometer operating in the frequency-sweep mode, the magnetic field is kept constant, and the individual protons are sequentially excited into resonance by a rather weak, "observing" radiofrequency field that is scanned through an appropriate frequency range; typically, a sweep width of 1000 Hz is scanned during 250 seconds. This mode of operation gives entirely acceptable ^{1}H spectra for solutions that are 0.1 to 0.4 molar; it can also be used for studies of other heteronuclear species, in which repeated scans of the same spectral region are generally required to overcome the lower relative sensitivity (Table I) of those nuclei. Advantage is taken here of the fact that the signal-to-noise ratio of an averaged spectrum is proportional to $\sqrt{N}$, where N is the number of scans that are averaged; however, if each scan takes 250 seconds, such time-averaging experiments can be extremely time-consuming.

The development[7,20,28,29] of Fourier-transform n.m.r. spectrometers has substantially shortened the amount of time required for time-averaging experiments. In this technique, all the nuclei of a particular isotope are simultaneously excited into resonance by application of a short (20 to 60 × 10^{-6} second), strong (100 to 1000 watts) pulse of radiofrequency power whose

TABLE I

N.M.R. Parameters for Common Nuclei[a]

Isotope	N.m.r. frequency at 23.487 T	Natural abundance (%)	Relative sensitivity[b]
^{1}H	100.00	99.985	1.00
^{13}C	25.144	1.108	1.59×10^{-2}
^{15}N	10.133	0.37	1.04×10^{-3}
^{19}F	94.077	100.	8.33×10^{-1}
^{31}P	40.481	100.	6.63×10^{-2}

[a] Data published by Varian Associates, Palo Alto, California, 1968.

[b] For equal numbers of nuclei at constant magnetic field. This figure is the product of the natural abundance and the intrinsic sensitivity of the nuclide in an n.m.r. experiment.

frequency lies to one side of the spectral region to be measured. Subsequent to this pulse, the receiver section of the spectrometer is switched on for 1 to 10 seconds, during which time the n.m.r. spectrometer records the re-emission of absorbed energy in the form of a free-induction decay (f.i.d.) signal; in the time domain the sample may be excited repeatedly and the individual f.i.d. signals summed coherently in the memory of a digital computer. The f.i.d. signal consists of a plot of detector response versus time, which is converted, by the mathematical manipulation known as Fourier transformation, into the more familiar, frequency-domain n.m.r. spectrum, which is a plot of peak intensity versus frequency. The spectrum obtained in this way is essentially identical to that obtained from a conventional spectrometer, except that it is obtained more quickly. Although this saving of time is of minor importance for most proton n.m.r. studies of adequate (10 to 100 mg) samples, it is of major significance in the acquisition of heteronuclear n.m.r. spectra, for which the spectrum must be sampled repeatedly and time-averaged to overcome the unfavorably low sensitivity (Table I) of most heteronuclei. A further advantage, which applies equally to studies of both proton and heteronuclear species, is that the Fourier-transform method opens up an extensive range of other forms of data manipulation, including the direct measurement of relaxation times, as discussed in Section E.

Of the many ancillary n.m.r. experiments, the most useful is the technique of double irradiation ("spin decoupling"). For example, this technique can be used to bring a single proton into resonance, thereby eliminating its couplings from the other proton resonances and identifying which protons are spin-coupled.[5,7,30] Heteronuclear, double-irradiation experiments may also be performed,[30] in which the resonance signal of one nuclide is observed in the customary way while different nuclear isotopes are irradiated at their resonance frequency with power derived from a separate radiofrequency source. This approach is widely used in carbon-13 studies, in which simultaneous irradiation of the resonances of all the protons in the sample not only narrows every ^{13}C resonance to a singlet, thereby simplifying the spectrum, but also improves the effective signal-to-noise of the ^{13}C-n.m.r. spectrum (see Section G) by a cross-relaxation process known as the nuclear Overhauser effect.

C. CHEMICAL SHIFTS

Provided that suitable reference values are available, measurements of chemical shift may be used for the direct identification of substituent groups in a carbohydrate derivative. Table II lists the proton chemical shifts of some of the substituents most commonly encountered in the carbohydrate field.

References start on p. 1320

TABLE II

Proton Chemical Shifts (p.p.m., δ-Scale) for Substituents Attached to a Carbohydrate

Substituent	Range	Substituent	Range
C_6H_5CO	7.0–8.4	CH_3N-	2.8–3.0
C_6H_5C	7.2–7.4	CH_3S-	2.4–2.7
C_6H_5CH—O | O	5.3–5.8	CH_3COS or p-$CH_3C_6H_4SO_3$-	2.4–2.5
$C_6H_5CH_2O$-	4.5–4.3	CH_3CO_2-	2.0–2.2
CH_3O-	3.3–3.5	CH_3CONH-	1.8–2.1
CH_3SO_3-	2.8–3.1	$(CH_3)_2C\begin{smallmatrix}O\\ \\O\end{smallmatrix}$	1.3–1.6

Measurements of relative peak areas in integrated spectra give the relative numbers of protons in each of the various substituents. It should be recognized that the chemical shifts of many of these substituents are solvent-dependent (see Section F), and so comparisons between different solvents may not be made indiscriminately. Specific intramolecular shifts may also occur, especially when compounds have carbonyl, aromatic, or other unsaturated substituents.

In many instances, the chemical shifts of individual ring protons can be determined, and these values may also facilitate structural assignments. As a general rule,[1] anomeric protons resonate at lower field than methine protons, whereas methylene protons resonate at somewhat higher field; however, all these shifts depend to a certain extent on the nature of the geminal substituent, and occasionally the foregoing orders are reversed. A characteristic shift of ring-proton resonances to low field accompanies[31] esterification of a geminal hydroxyl group, and this displacement is often useful as an aid in the assignment of ring-proton resonances.

The chemical shifts of certain groups, notably those of acetyl groups and of anomeric and methine ring protons in pyranose sugars, exhibit regular, stereospecific dependencies, which may be correlated with configuration. In spectra measured in chloroform, the methyl resonance of an axially oriented acetoxyl substituent is usually observed[1] at lower field than that of an equatorially oriented substituent. Equatorial ring protons of pyranose sugars in water[32–34] and of their substituted derivatives in chloroform[1,35,36] generally resonate at lower field than their axial counterparts. Many of the *apparent* exceptions to this "rule" have been rationalized by Lemieux and Stevens in terms of long-range shielding and deshielding by hydroxyl[32] and acetoxyl substituents.[35] The essential content of this empirically determined rule is summarized in the accompanying two formulas, which depict the *downfield* chemical-shift changes (in p.p.m.) that would accompany the

$$
\begin{array}{cc}
\text{H} & \text{H} \\
\text{AcO} \quad \text{CH}_2\text{OAc} \quad \text{O} & \text{HO} \quad \text{CH}_2\text{OH} \quad \text{O} \\
\text{H}{-}(0.2) & \text{H}{-}(0.3) \\
(0.25){-}{-}\text{H} & (0.35){-}{-}\text{H} \\
\text{AcO} \qquad \text{AcO} \quad \text{OAc} & \text{HO} \qquad \text{HO} \quad \text{OH} \\
\text{H} \qquad \text{H} & \text{H} \qquad \text{H} \\
(0.25) \qquad (0.6) & (0.35) \qquad (0.6)
\end{array}
$$

inversion from an equatorial to an axial orientation of the substituent attached to carbon-1; for a more detailed appraisal of the accuracy of those rules, the reader may consult refs. 8, 32, and 35.

Considerably less detailed information is available concerning chemical shifts in furanoses; two extensive documentations are those of Stevens and Fletcher[37] for pentofuranose derivatives and of Hall, Slessor, and co-workers[38] for the isomeric 1,2:5,6-di-*O*-isopropylidene-D-hexofuranoses.

It is noteworthy[32–34,39] that the anomeric protons of furanose sugars often resonate downfield of the pyranose forms. This factor, coupled with the pronounced shift differential between axial and equatorial protons attached to a pyranose ring and the deshielding of anomeric protons compared with other ring protons, has made ¹H-n.m.r. spectroscopy a powerful tool for studying the composition of free sugars in solution.[32–34,40–42] Measurement by integration of the relative areas beneath the resonances of the individual ring protons of different anomeric and tautomeric forms gives a fairly accurate (about ± 3%) estimate of the relative proportions of the individual species present in the mixture at mutarotational equilibrium. Similar measurements of solutions that are undergoing mutarotation allow the process to be monitored and its rate determined; this approach is especially valuable for sugars that exhibit complex mutarotations through involvement of substantial proportions of more than two tautomers (see Chapters 4 and 5).

Since the pioneering work of C. D. Jardetzky[43] and Lemieux,[44] there have been numerous studies[45,46] of nucleoside and nucleotide derivatives that provide some information concerning structures of D-ribofuranose and analogous 2-deoxy sugars.

The ¹H chemical shifts of derivatives of oligo-[47] and poly-saccharides have also been studied.[48] Insofar as reasonably dispersed spectra can be obtained

---

References start on p. 1320.

for such derivatives, the chemical shifts appear to follow the same trends as those of monosaccharides. In such series as the acetylated cello-oligosaccharides, it is possible to observe discrete signals for protons in the chain terminal residue and thereby determine such features as chain length and anomeric configuration in the terminus.[49]

D. COUPLING CONSTANTS

It is indeed fortunate for the carbohydrate chemist that vicinal proton–proton coupling constants depend to a marked extent on the dihedral angle (ϕ) separating the two protons, as this dependence not only provides the soundest criterion for making configurational assignments, but also opens

the possibility of specifying the conformation in solution. Following early experimental work,[1] valence-bond calculations suggested[50] that the magnitude of such vicinal couplings (J) exhibits an angular dependence of the approximate form:

$$J = J_0 \cos^2 \phi + C$$

where J_0 and C are constants. Originally[50] these were allocated values of $J_0 = 8.5$ for $0° \leqslant \phi \leqslant 90°$, $J_0 = 9.5$ for $90° \leqslant \phi \leqslant 180°$, and $C = -0.28$. Although some of the early workers[40,51,52] suggested (largely on intuitive grounds) somewhat larger values for these constants to allow for undefined "substituent dependencies," many investigators overlooked the cautionary statements that were included in Karplus' original publication. This oversight caused sufficient confusion that Karplus published[53] a second paper, reaffirming his original, cautionary statements emphasizing that the dihedral, angular dependence is only one of several factors influencing the magnitude of vicinal couplings. Some of those dependencies are discussed here in more detail, but it should be pointed out at the onset that there seems little reason to doubt the broad qualitative validity of the Karplus dependence—namely, that large (~ 10 Hz) vicinal couplings indicate antiparallel protons, whereas small values (~ 3 Hz) are typical of *gauche*-disposed protons. For most pyranose sugars, measurement of the vicinal $^1H-^1H$ coupling constants permits determination of the overall conformation and the relative configuration at individual carbon atoms. However, for furanose derivatives, conformational and even configurational assignments may be difficult. Detailed conformational assignments for both pyranose and furanose sugars, with actual values for dihedral angles, require a more-precise knowledge of the other factors that influence vicinal couplings.

The work of Williamson,[54] together with that of Laszlo and Schleyer,[55] illustrates quite convincingly the dependence of vicinal couplings on the electronegativities of other substituents attached to the carbon atoms. For the carbohydrate chemist, the most important feature of these studies lies in the implication that, although the couplings required for the "anomeric"

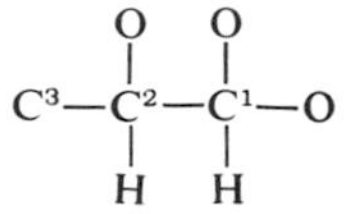

$$C^3-C^2-C^1-O \qquad\qquad C-C-C-C$$

"Anomeric" fragment 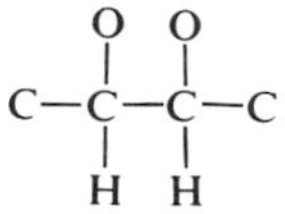   "Nonanomeric" fragment

fragment should differ from those of a similarly disposed "nonanomeric" fragment, no allowance was made in the Karplus equation for a change in the O-substituent as, for example, in a change from –OH to –OCOCH$_3$. Subsequent treatments[56] that take into account such effective changes in electronegativity afford somewhat better quantitative agreement. A rather more serious obstacle to quantitative conformational studies than the foregoing electronegativity dependence is the configurational effect first enunciated by Williams and Bhacca[57] for steroidal derivatives. They found that, for a series of steroidal alcohols and their acetates, the values of $J_{a,e}$ differed significantly from the values of $J_{e,a}$, even though a dihedral angle of 60° should be present in both relationships. Although this individual set of

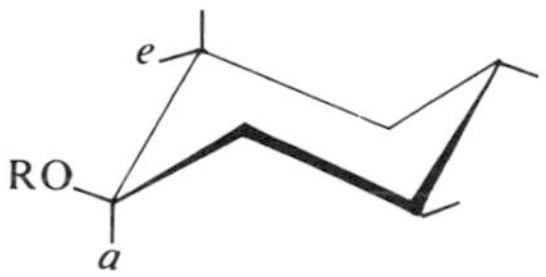

$$J_{a,e} = 5.5 \pm 1.0 \text{ Hz} \qquad\qquad J_{e,a} = 2.5\text{-}3.2 \text{ Hz}$$

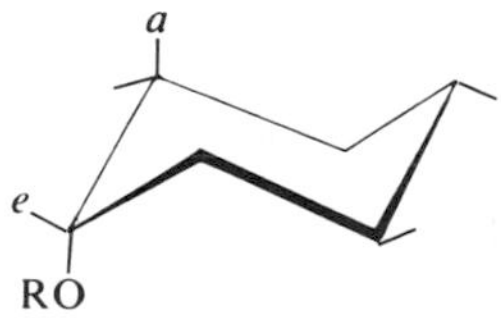

observations might be explained on the basis of some conformational deformation, it can scarcely be coincidental that similar inequalities have been observed in a variety of other systems,[58] including carbohydrates.[5,59] Clearly, a dependence of this nature severely limits the applicability of the Karplus approach for quantitative determinations of conformation. A further, nonsystematic dependence has been observed by Abraham and Thomas,[60] but this does not appear to be directly applicable with carbohydrate systems.

Configurational and conformational studies of pyranose derivatives based on vicinal H–C–C–H coupling constants are too numerous to mention separately here. An excellent review article[61] has summarized much of this work, with special reference to systems that are conformationally mobile; temperature-dependent effects in p.m.r. spectra of molecules in the latter

---

group sometimes permit the spectroscopic isolation of both constituents of a conformational equilibrium.

Studies of furanose derivatives are hindered both by the intrinsic conformational mobility of five-membered rings (see Chapter 5) as well as by the paucity of reference data.[37,38,51] The fact that the energy barrier to conformational inversion in such systems is low, coupled with the small free-energy difference between individual conformers, means that all observed n.m.r. parameters are time-averaged. One approach[62] to the uncertainties of the conformational mobility uses individual coupling constants to define those segments of the pseudorotational cycle that are *not* populated. Acyclic derivatives exhibit even greater conformational mobility, but favored conformations have been educed[63] from spin-coupling data in a variety of systems. As in studies of cyclic molecules, investigators are well advised[64] not to overinterpret the quantitative significance of coupling constants measured in flexible molecules, particularly when a number of conformational variables operate simultaneously.

Other classes of vicinal proton couplings that have well-defined, stereospecific dependencies include those of epi-derivatives[65] (H–C—C–H, where X = O,S,NH), H–C–N–H (ref. 66), and H–C–O–H (ref. 67). The last of these in particular is of considerable importance for carbohydrate derivatives,[67,78] and Fraser and co-workers[69] have described a quantitative rationale for this dependence.

Studies[5,56,70] of pentopyranose and 2-deoxy-D-hexopyranose derivatives have shown that the coupling constants of geminal protons (^2J) are dependent on the relative orientation of the substituent at an adjacent carbon atom. A theoretical rationale has been offered[71] for this dependence, which should find considerable utility in configurational assignments.

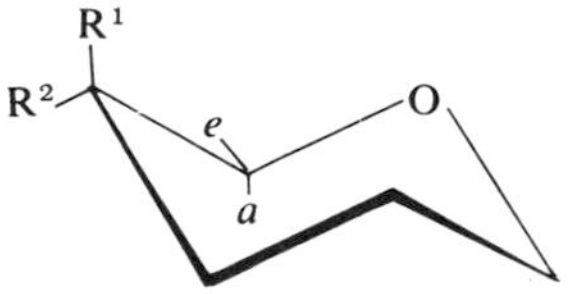

$R^1 = H, R^2 = OAc, J_{a,e} = -11$ Hz

$R^1 = OAc, R^2 = H, J_{a,e} = -13$ Hz

The second generation of high-resolution n.m.r. spectrometers has uncovered an ever-increasing variety of so-called, long-range couplings—that is, interactions between protons separated by four (^4J) or more bonds.[72] Insofar as these may occur in both saturated and unsaturated systems, and across carbon or hetero atoms, and since they also exhibit stereospecific dependencies, such couplings can afford a wealth of stereochemical information. The absolute signs of many of these coupling constants vary with stereochemistry, and this factor further adds to their utility as a structural tool.

The first, well-defined stereospecificity for ^4J couplings in a saturated sugar was observed[73] for 1,6-anhydro-D-hexopyranose derivatives, where $^4J_{e,e}$ couplings ranging between 1.1 and 1.8 Hz were noted.[73a] Other workers have also observed such couplings,[74] and Barfield[75] has made theoretical calculations of their stereochemical dependencies; the variations in sign predicted by those calculations have been experimentally confirmed.[76] Several examples of couplings across five bonds (^5J) have also been observed.[77]

Long-range couplings (^4J and ^5J) for unsaturated carbohydrate derivatives have also been observed.[78,79] Theoretical calculations have resulted in the formulation[72,80] of quantitative rationales, and the signs of some allylic couplings have been determined experimentally.[5,76,81]

In concluding this section, it is worthwhile commenting that the majority of the studies of carbohydrates have used first-order coupling constants—that is to say, the spacings between individual lines in the n.m.r. spectrum have been assumed to be numerically identical to the coupling constants. This is strictly true only when the chemical shift between the resonance is large compared with their mutual couplings; a ratio of 10:1 is generally considered sufficient. As this ratio decreases, so the error in a first-order coupling value also increases. Such errors are generally not *chemically* significant for configurational studies, but they would be serious if quantitative evaluations of conformations were attempted. Several computer programs are now available that can be used either to simulate n.m.r. spectra or to fit a measured spectrum by iteration, and there is no justification for using inaccurate coupling constants in predicting conformations. Coxon has given an excellent account of this topic.[7]

E. Spin–Lattice Relaxation

Although the phenomenon of spin–lattice relaxation has long attracted the attention of physicists,[13,20] it has been neglected by most practicing organic chemists, largely because the earlier experimental methods used by physicists are incompatible with the complex n.m.r. spectra of organic molecules. However, the advent of Fourier-transform, pulsed n.m.r. spectrometers has made routine[20,26,28,29] the measurement of individual spin–lattice relaxation times (T_1 values) for even quite complex molecules, so that it is appropriate for carbohydrate chemists to utilize such data.

Spin–lattice relaxation involves a transfer of magnetization between the nuclei of interest (the spins) and their surrounding environment (the lattice); the rate at which this energy transfer occurs is the spin–lattice relaxation time

---

References start on p. 1320.

(T_1 value). In a typical T_1 measurement for protons, all the protons are simultaneously irradiated with a pulse of radiofrequency power sufficient to tip their magnetization through $180°$; a selected period of time (t) is then allowed to elapse, during which some of the magnetization relaxes into the lattice, and then the residual magnetization is measured by application of a $90°$ pulse of radiofrequency, measurement of the free-induction decay signal, and subsequent Fourier transformation to the required n.m.r. spectrum. The overall outcome of this two-pulse sequence is a "partially relaxed" spectrum, in which the relative intensities of the individual peaks are determined by their relative relaxation rates. After a long delay period to allow all the protons to return to an equilibrium condition, another two-pulse sequence is applied, with a different pulse-delay period (t), to give a second, partially relaxed spectrum. This process is continued a number of times, typically between ten and twenty, and the peak heights of the individual resonances of each spectrum are measured. The T_1 value of each resonance is then obtained as the slope of a plot of "ln (peak height)" versus "pulse-delay time," expressed in seconds.

In principle, a number of different mechanisms can contribute to the spin–lattice relaxation of a proton, but in practice it is possible to make measurements under conditions such that one mechanism, intramolecular dipole–dipole (i.d.d.) relaxation, is dominant or even exclusive; this merely[82,83] requires that measurements be made using dilute solutions (concentration less than 0.1 molar) in a magnetically inert liquid (typically a fully deuterated solvent) and in the absence of any paramagnetic impurities (dissolved oxygen or metal ions). The importance of this situation becomes apparent on consideration of the form of this relaxation mechanism, which is given here for relaxation between two arbitrary nuclei, R and S:

$$\left(\frac{1}{T_1}\right)_{\text{(i.d.d.)}} \propto \frac{\gamma_R{}^2\gamma_S{}^2}{(r_{R-S})^6} \cdot \tau_c \, (R - S)$$

Here, γ_R and γ_S are the magnetogyric ratios of R and S, respectively, (r_{R-S}) is the internuclear separation between these two nuclei, and $\tau_c \, (R - S)$ is a motional correlation time that is related to the rate of motion of the nuclei in solution. Because γ_H has a far larger value than that of any other commonly occurring nucleus, it follows that the relaxation of a proton in a carbohydrate molecule generally involves only the other protons in that same molecule. Furthermore, as the efficiency with which one proton may cause another to relax decreases so rapidly with increase in internuclear separation, it follows that the relaxation of any proton is dominated by its nearest neighbors.

As a result of this dependence, measurements of proton spin–lattice relaxation times offer a practical method for relating interproton distances and, eventually, measuring their absolute values.[81a]

The proton T_1 values of the anomeric protons of D-glucose (**1**), D-galactose (**2**), and D-mannose (**3**) illustrate[82] the type of configurational dependence observed in pyranose sugars. The relaxation of an axially oriented, anomeric proton is largely influenced by contributions from the axial protons at positions 3 and 5; these contributions are substantially decreased when the anomeric proton is equatorially oriented. This type of dependence has also been demonstrated in 2-deoxy-2-halohexopyranose derivatives,[83] and in a wide selection of monosaccharide[82] and oligosaccharide[84] derivatives.

1(β)

T_1 (sec) for H-1: 2.3

1(α)

T_1 (sec) for H-1: 4.4

2(β)

2.4

2(α)

4.0

3(β)

1.8

3(α)

5.9

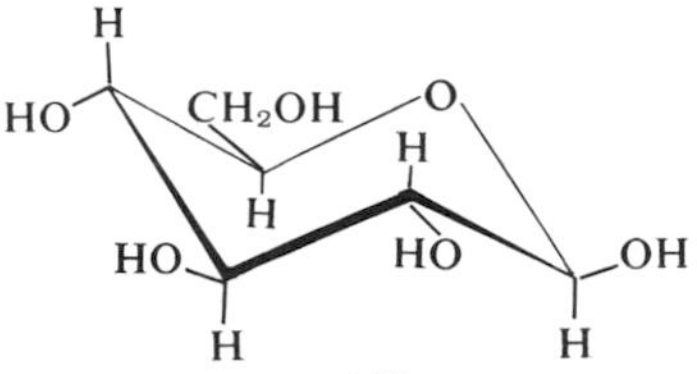

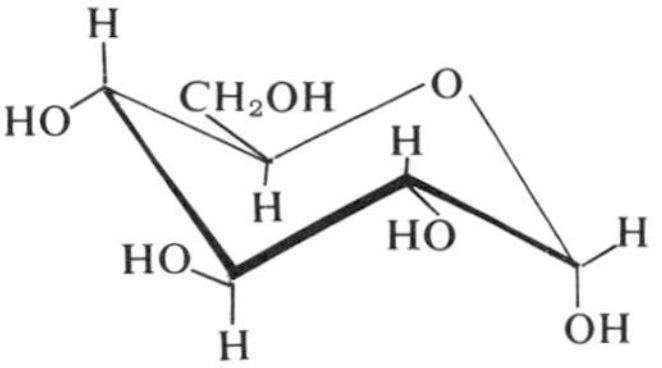

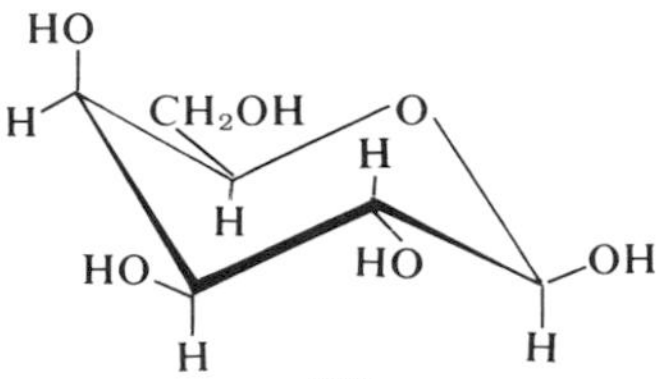

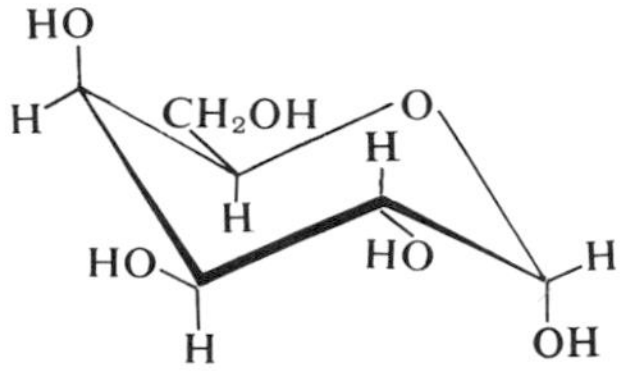

The practical potential of spin–lattice relaxation studies for conformations in solution has benefited from the advent of simpler, less-expensive Fourier-transform spectrometers, which has made possible the routine measurement of T_1 values for carbon-13 signals.[84a]

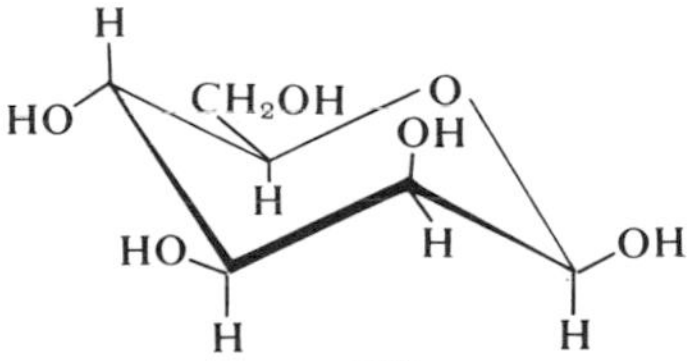

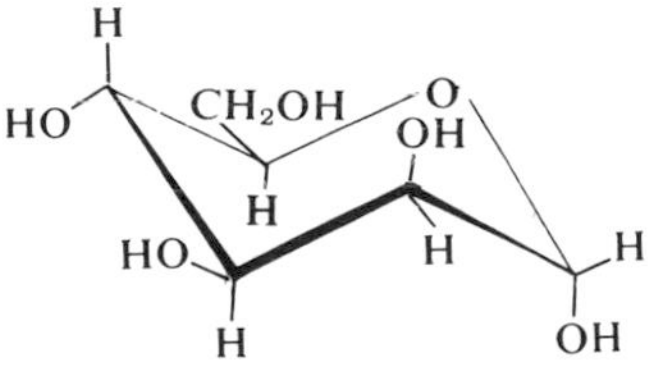

F. Aids to Spectral Assignments

1. *Chemical Techniques*

The principal limitations of proton n.m.r. spectroscopy are associated with the low intrinsic sensitivity of proton chemical shifts with respect to changes in chemical environment; as a result, the proton spectra, even of simple

References start on p. 1320.

monosaccharides, are often complicated by the overlap of several proton resonances, and the spectra of oligosaccharides are generally dominated by this complication. Possible solutions to this "hidden resonance[11] problem" may be categorized into two broad groups—namely, chemical and instrumental methods.

Of the chemical methods, change of solvent is the simplest way to obtain small, but useful changes in shifts, as, for example,[85] in the change from chloroform to benzene, acetone, or acetonitrile, or from water to pyridine or methyl sulfoxide. It is frequently necessary to use a deuterated solvent to avoid interfering solvent resonances, and a mixture of solvents often affords a well-dispersed proton spectrum. Many of these solvent-induced shifts appear to arise from favored solvation of specific sites of the solute molecule[86,87]; this is especially true for aromatic solvents.[88] Thus, the chemical shifts of acetate methyl resonances, which exhibit a reasonably predictable dependence on stereochemical disposition for solutions in chloroform, are substantially[89] altered when benzene is the solvent.

Profound changes in chemical shift can be induced by addition of small quantities, typically less than 10^{-2} molar equivalents, of any one of a variety of paramagnetic species, generally metal cations. Paramagnetic cations of transition metals and of many of the lanthanides induce relatively large changes in chemical shifts of resonances in other molecules in their presence, but this effect is generally of little value because of an accompanying, extreme line broadening of these resonances caused by enhanced relaxation. Europium^{3+} and, to a smaller extent, praseodymium^{3+} are exceptional in inducing fairly large chemical-shift displacements before the multiplet structure is lost through increased line widths. Reuben and Fiat[90] demonstrated induced shifts of water protons complexed to lanthanide cations. Pursuing this line, Williams and co-workers[91] complexed lysozyme to Gd^{3+} and to Eu^{3+}; they then added methyl 2-acetamido-2-deoxy-β-D-glucopyranoside to the modified enzyme and observed induced shifts of the methyl resonances in the sugar, demonstrating that the site of coordination relative to the cation may be deduced from selective broadening and displacement, respectively, of resonances due to protons in the substrate. Later studies[63,92–94] have revealed that such free cations in hydroxylic solvents are of utility in enhancing spectral dispersion of unprotected sugar derivatives. Carboxylate groups[94] are particularly good sites for strong complexation with Pr^{3+} and Eu^{3+} cations, and excellent spectral dispersion can be achieved with such compounds as monolactones of aldaric acids.[95]

Paramagnetically induced shifts became a practical, general tool for enhancing spectral dispersion with the discovery by Hinckley[96] in 1969 that a hindered Lewis acid complex of Eu^{3+} coordinates with cholesterol to move a number of resonances out of the high-field envelope of signals. The active principle of the complex,[97] tris[2,2,6,6-tetramethyl-3,5-heptanedionato]-europium(III) [Eu(dpm)$_3$], and a less-hindered, stronger-coordinating complex, tris[6,6,7,7,8,8,8-heptafluoro-2,2-dimethyl-3,5-octanedionato]-europium(III)[98] [Eu(fod)$_3$], have been shown to possess broad utility in simplifying overlapped n.m.r. spectra of partially protected sugar derivatives, since their initial application to carbohydrates by Armitage and Hall[99] and by David and co-workers[100] in 1970.

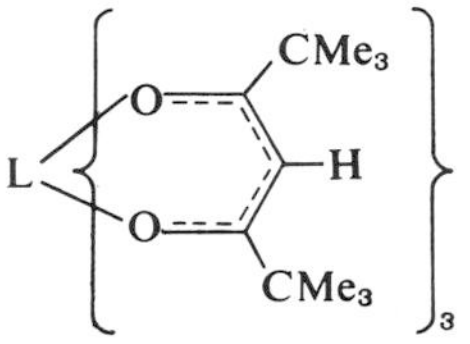

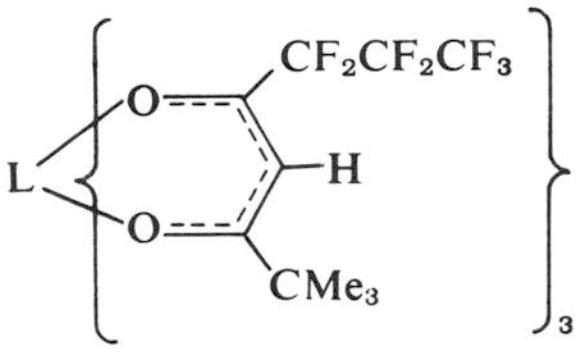

The dpm and fod derivatives of europium generally induce shifts to low field, whereas the corresponding derivatives of praseodymium cause shifts to high field. Other lanthanides are not generally practical[101] for use as proton-shift reagents.

The operation of these shift reagents depends on the occurrence of a rapid, reversible equilibrium between the carbohydrate derivative and the shift re-agent, and as the latter are "hard acids," such oxygen-donor substituents as OH or OCOR are particularly effective in promoting the necessary association.

If the induced change of chemical shift ($\Delta\delta$) arises solely from dipolar (pseudo-contact) interactions,[102] then its magnitude is given by the expression

$$\Delta\delta = k[Ln](3\cos^2\theta - 1)/r^3,$$

where [Ln] is the concentration of lanthanide ion, and r is the distance between the ion and the nucleus under consideration. As this relationship includes dependence on distance, it provides (at least in principle) a method for determining the geometry of the lanthanide–substrate complex and, by extrapolation, that of the substrate itself. Unfortunately, the ease and general-ity of this approach is seriously complicated by a variety of problems, especially those associated with an adequate definition[103] of the geometric term involving "θ". Nevertheless, this approach has been valuable in a number of applications, as in determining the stereochemistry of branched-chain sugars containing tertiary alcohol groups, where the interpretation may be consider-ably simplified by comparing a series of closely related compounds.[104]

Now that it is routinely feasible to measure spin–lattice relaxation times, use can also be made of paramagnetic species such as Gd^{3+} and Gd(dpm)$_3$, which induce changes in T_1 values without affecting chemical shifts. These

References start on p. 1320.

changes in relaxation, which are proportional to the inverse sixth power of the distance between the metal and the probe nucleus, may result in the selective broadening, and hence elimination, of individual proton[91,105] or ^{13}C resonances.[106] This behavior has been proposed[106] for differentiation of anomeric forms of uronic acids.

As deuterium and hydrogen have different resonance frequencies, specific deuteriation can effectively remove selected resonances from a proton n.m.r. spectrum; indeed, this is the reason why specifically deuteriated solvents are used. Although deuterium has a nuclear spin of unity, its spin coupling with suitably disposed protons is only $\sim 15\%$ that of the corresponding proton–proton coupling, and this coupling can either be ignored or be removed by double irradiation at the deuterium frequency. The replacement of specific ring protons by deuterium atoms has frequently been used[107–110] as a means of investigating the stereospecificity of certain reaction mechanisms, and Horton and co-workers[89,111] have used specifically N- and O-trideuterio-acetylated ($OCOCD_3$) derivatives to study the chemical shifts of acetate groups. Other investigators[112] have used selective O-trideuteriomethylation to identify O-methylated methyl glycopyranosides.

Specific O-deuteriation of alcohols is readily achieved by dissolution of the alcohol in deuterium oxide, and N-deuteriation of most amines is also effected readily. For alkylamides, N-deuteriation takes place relatively slowly (several minutes at room temperature) by direct exchange, whereas deuteriation of some arylamines may occur only in the presence of a base—for example, tributylamine.[113] Selective deuteriation of individual ring protons is necessarily more difficult to accomplish, but the widespread availability of deuteriated reducing reagents and specifically oxidized sugars allows for a wide diversity of syntheses.[114]

2. Instrumental Techniques

A variety of instrumental methods are available both for simplifying n.m.r. spectra and for confirming spectral assignments. The simplest, but also the most expensive, is to measure the n.m.r. spectrum with a spectrometer that operates at a higher magnetic field, using the fact that the dispersion of a spectrum increases proportionally with the field strength, whereas the width of a spin-coupled multiplet remains constant. In 1980, the highest field available is equivalent to a proton resonance frequency of 600 MHz.

A more-generally available method involves the use of the double-resonance (spin-decoupling) technique.[5,7,30] In essence, this technique allows for the spin coupling of one, or more, sets of nuclei to be effectively removed from a spectrum, causing the concomitant simplification of other resonances that would normally have been subject to the coupling. These experiments can be performed between nuclei of the same species (homonuclear), for example, ^{1}H-[^{1}H],[115] or of different species (heteronuclear), for example, ^{1}H-[^{13}C].[116] A useful extension of spin decoupling, referred to by the acronym INDOR[117] (INternuclear DOuble Resonance), uses[30,118] a weak, observing radiofrequency field to monitor the intensity of a single transition while a weak irradiating field is scanned through the spectrum; each time the irradiating field passes through a transition that is spin-coupled to the line that is being monitored, a peak is produced. The unique feature of this experiment is that it effectively removes from the final spectrum all resonances other than those that are spin-coupled to the monitor resonance; hence it provides a powerful method for solving some hidden-resonance problems.[11]

Developments in Fourier-transform n.m.r. spectroscopy have opened up an extensive variety of data-manipulation experiments, many of which have relevance to the carbohydrate chemist. Partially relaxed n.m.r. spectroscopy, for example, may be used to discriminate against a signal that is characterized by a T_1 value different from that of a proximal resonance. This technique depends on the fact that, during a Fourier-transform determination of spin-lattice relaxation (Section E), the magnetization of each individual resonance decays to zero intensity independently of the others. If a second, observing pulse is applied at precisely the time that the net magnetization of a specific resonance attains zero intensity, that resonance will then be removed from the final n.m.r. spectrum. Several workers[77,79,119] independently used this approach to remove undesired solvent resonances; individual resonances of overlapping ring protons having significantly different T_1 values have also been so separated,[83] but 2-dimensional J spectroscopy (a two-pulse, two-transform technique) is superior for this purpose.[120]

G. STUDIES OF NUCLEI OTHER THAN PROTONS

Although the vast majority of n.m.r. investigations of carbohydrates have been, and presumably will continue to be, concerned principally with protons, a number of other naturally occurring magnetic nuclei are suitable for study. Of these, carbon-13 is of prime importance,[25,26,121] although nitrogen-15 (ref. 122), fluorine-19 (ref. 123), and phosphorus-31 (ref. 124) have also been studied; all of these have nuclear spin 1/2, and the latter two are 100% naturally abundant. The principal barriers to the study of such nuclei are associated with their low intrinsic sensitivity (Table I); however, the advent of Fourier-transform n.m.r. spectroscopy and heteronuclear double-resonance methods has ameliorated these difficulties. Reasons for studying heteronuclear spectra rather than proton spectra are numerous, the most general

---

References start on p. 1320.

TABLE III

COUPLING CONSTANTS FOR SPECIFICALLY
FLUORINATED PYRANOSE DERIVATIVES
(IN Hz)

Coupling constant		Range
Fluorine–proton		
^2J		+48 to +53
^3J	a,a	+24 to +38
	a,e	+ 1.5 to +12
^4J	e,e	+ 4
	e,a	− 1
Fluorine–fluorine		
^3J	a,a	−20.0
	a,e	−13 to −19
^4J	e,e	− 3
	e,a	+ 1
	a,a	+10

being the enhanced sensitivity of heteronuclear chemical shifts to environmental changes as compared with an equivalent proton substituent. The chemical shifts of heteronuclear resonances are commonly expressed relative to the signals of the following compounds: ^{13}C, $(^{13}CH_3)SiMe_3$; ^{19}F, $CFCl_3$; ^{31}P, P_4O_6. There appear to be no well-defined reference compounds for other nuclear species.

The ^{19}F-n.m.r. spectra of specifically fluorinated carbohydrate derivatives have been extensively studied by the groups of Hall,[125] of Foster,[126] and of Kent.[127] Tables III and IV summarize pertinent chemical-shift and coupling-constant data. It may be noted that many of the dependencies illustrated by these parameters parallel those of the corresponding proton parameters, except that those for fluorine are generally larger; for example, the intrinsic chemical-shift difference between axial and equatorial protons is ∼0.5 p.p.m., whereas the corresponding shift for fluorine substituents is ∼11 p.p.m.

TABLE IV

CHEMICAL SHIFTS FOR SPECIFICALLY
FLUORINATED CARBOHYDRATES (IN
P.P.M. FROM $CFCl_3$)

Substituent	Range
Glycosyl fluoride	113–150
Secondary fluoride	195–207
Primary fluoride	213–235

Consequently, ^{19}F-n.m.r. spectroscopy provides a powerful tool for elucidating the structures of specifically fluorinated carbohydrates. Fluorine may also be used as a structural probe for studying the interaction of carbohydrate derivatives with enzymes and other biopolymers, and several investigators[128] have reported studies on the interaction of lysozyme with specifically fluorinated derivatives of 2-acetamido-2-deoxy-D-glucopyranose.

Despite the great biological importance of phosphorylated sugars, relatively few studies have been made of the phosphorus-31 spectra of such derivatives. Hall and Malcolm[129,130] early established the angular dependence of vicinal ^{31}P–O–C–[1]H coupling constants in phosphoric esters to be that depicted. This

$$\phi \sim 60°, J \sim 2 \text{ Hz}$$

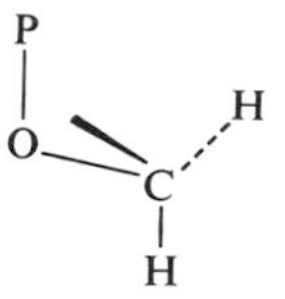

$$\phi \sim 180°, J \sim 22 \text{ Hz}$$

observation has been corroborated by others,[131] and several groups, particularly those of Smith[132] and of Hruska,[133] subsequently used these results to study the stereochemistry of nucleotide derivatives. A similar angular dependence of the ^{31}P–C–C–[1]H couplings of sugar phosphonates has also been firmly established.[134] Narrow ^{31}P resonance signals are observed in high polymer derivatives of regular structure, and they have been used to assign anomeric configurations to the hexosyl phosphate residues in yeast O-phosphonohexoglycans.[134a]

Coxon[135] has examined amino sugar derivatives containing nitrogen-15, enriched to $> 99\%$. Couplings between ^{15}N and ^{1}H were observed as follows: [1]J 91.3, [2]J 0.5, and [3]J 1.5 Hz. Mester and co-workers used the same method to study sugar osazones[136] and formazans.[137]

Perlin[120] has provided a survey of many studies of the carbon-13 n.m.r. spectra of carbohydrates. The first reports[138–140] of the pronounced, stereospecific dependence of ^{13}C chemical shifts were made in 1969, and the data from ref. 138 are summarized in Fig. 1. The range of the corresponding ^{1}H shifts is given in the insert; it clearly illustrates the very substantial (~ 30-fold) increase in the dispersion of the ^{13}C spectra. Definitive studies soon followed from the groups of Perlin,[141] of Roberts,[142] and of others.[143] Data for these early studies were obtained by using frequency-swept, continuous-wave spectrometers; noise-modulated irradiation of the entire proton spectrum was used to improve the signal-to-noise ratio both indirectly, by collapsing

---

References start on p. 1320.

each carbon resonance to a singlet, and directly, by a cross-relaxation (nuclear Overhauser) enhancement of the signal intensity. That level of technology afforded adequate chemical shifts, but proton decoupling eliminated all the coupling constants. Coupling data were obtained by Perlin[139] using ^{13}C-enriched compounds, and later, with the advent of Fourier-transform spectrometers, Bock and Pedersen[144] measured coupling values of natural-abundance molecules using gated proton decoupling. In this important experiment (compare Section F,1), the proton irradiation frequency is switched off just before the free-induction decay is measured; in this way, the ^{13}C-n.m.r. spectrum exhibits all the carbon–proton couplings and still retains most of the intensity enhancement associated with the nuclear Overhauser effect.

Carbon-13 shifts of the anomeric center in pyranose sugars display a number of important dependencies. The overall dependence with regard to structural environment is summarized[138] in Fig. 1, where it may be seen that the total range of shifts is ~ 30 times as great as that for ^{1}H shifts. These shifts also show a marked dependence on the orientation and number of substituents attached to the same carbon atom, and to the adjacent (α) and next (β) atoms; this is illustrated[138] in Fig. 2. These generalizations are fully documented for inositols, for pentopyranoses, and for hexopyranoses in the review article by Kotowycz and Lemieux.[8] These observations have also been extended to various oligosaccharides,[145] affording direct stereochemical information.

It should be borne in mind that specific attribution of individual ^{13}C resonances is fully secure only when the different carbon atoms differ considerably in their structural environment, or when direct methods, such as specific ^{13}C enrichment or heteronuclear decoupling techniques, are used as an aid in signal identification.

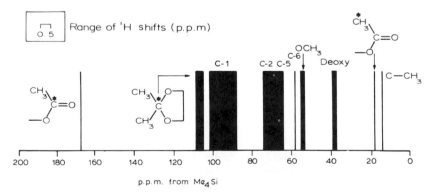

FIG. 1. Range of ^{13}C chemical shifts. The insert shows the spectral range of the corresponding ^{1}H chemical shifts.

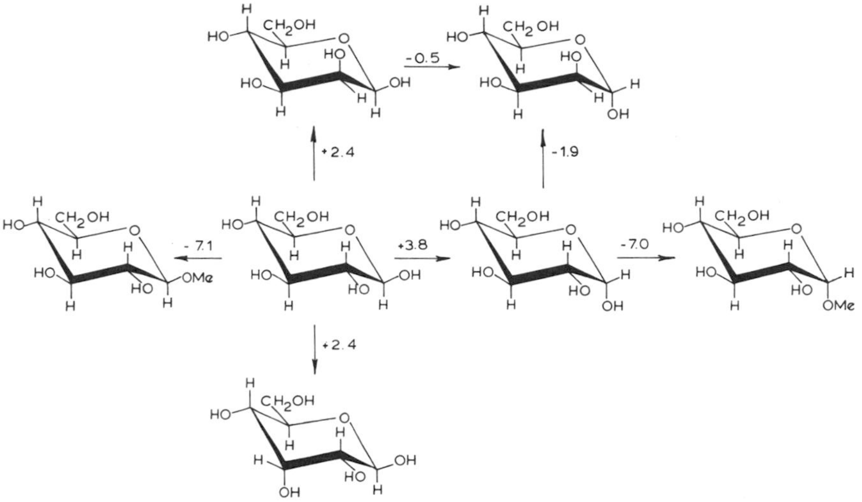

FIG. 2.   Relative ^{13}C n.m.r. chemical shifts (p.p.m.) as a function of structural change in pyranoid sugar derivatives. The shift values refer to the carbon atom undergoing stereochemical or substitional modification (after ref. 138).

Although most of the foregoing studies have been concerned with measurements of chemical shifts, chemical[139] and instrumental[144] advances have made feasible the measurement of ^{1}H–^{13}C coupling constants. Study of this nucleus is fast becoming one of the major arenas of research in the sugars. Bock and Pedersen[144] have shown that the magnitude of the $^1J_{^{13}C-^1H}$ coupling of the anomeric carbon atom of pyranose derivatives depends on the anomeric configuration, being ~ 160 Hz for β anomers and ~ 170 Hz for α anomers. Three accounts[146–148] of the stereospecific dependencies of ^{13}C–^{1}H coupling constants appeared simultaneously; that by Lemieux[146] contains a number of fundamentally important dependencies related to nucleosides and to the orientation of the aglycon of methyl hexopyranosides.

H. CONCLUDING REMARKS:
APPLICATIONS TO MACROMOLECULAR CARBOHYDRATES

It was observed at the beginning of this chapter that most n.m.r. studies of carbohydrates have used proton substituents as the nuclear probe. A further bias has been toward studies of monosaccharides, and, as that area now seems to have been rather thoroughly explored, much more attention in the future will be invested in the area of oligo- and poly-saccharide chemistry.

References start on p. 1320.

This is a formidable challenge; all the problems of monosaccharide studies become more serious, and, for proton n.m.r. studies, the hidden resonance problem may be overwhelming. However, such studies as those of Dutton's group[149] on intact capsular polysaccharides of *Klebsiella* having molecular weights of 10^5 to 10^6 show that even proton studies have future potential.

Still more exciting is the broad potential of heteronuclear n.m.r. spectroscopy, particularly with carbon-13. Perlin and his co-workers[150] have used ^{13}C-n.m.r. spectroscopy for structural assignments in heparin and related glycosaminoglycans, and a series of reports from Jennings, Smith, and their colleagues have demonstrated the great potential of the technique for structural work with both simple and complex polysaccharides. They have made full assignments[151] for several linear D-glucans and related oligosaccharides,[152] as well as some cycloamyloses,[153] and they have shown the value of methylation as an aid in attribution of signals. As a prelude to examination of meningococcal polysaccharide antigens, they[154] have assigned the signals in various N-acetylhexosamines and their phosphates, using specific ^2H-labeling to attribute certain ^{13}C signals, and have used these data to obtain information[155] on the composition, sequence, anomeric configuration, linkage position, mode of substitution, and conformation of these phosphorylated polysaccharides, which are built up from hexosamine residues joined through phosphoric diester linkages. Neuraminic acid-containing polysaccharides have been similarly examined,[156] and Gorin[157] has discussed the methodology of ^{13}C assignment with specific reference to bacterial D-mannans. A comprehensive article by Gorin[158] discusses ^{13}C-n.m.r. spectroscopy of carbohydrates with specific reference to polysaccharides.

A further aspect awaiting additional development is the use of n.m.r. spectroscopy to study the molecular association of carbohydrates with such biopolymers as enzymes and biomembranes. The work with lysozyme referred to earlier,[128] and the study[159] of the binding of methyl D-glucopyranosides to concanavalin A are clear examples of the portent of such studies.

REFERENCES

1. R. U. Lemieux, R. K. Kullnig, H. J. Bernstein, and W. G. Schneider, *J. Amer. Chem. Soc.*, **80**, 6098 (1958).
2. R. U. Lemieux and D. R. Lineback, *Annu. Rev. Biochem.*, **32**, 155 (1963).
3. L. D. Hall, *Advan. Carbohyd. Chem.*, **19**, 51 (1964).
4. L. Hough and A. C. Richardson, in "Rodd's Chemistry of Carbon Compounds," 2nd ed., S. Coffey, Ed., Vol. 1, Part F, Elsevier, London, 1967, pp. 139–165.
5. L. D. Hall and J. F. Manville, *Advan. Chem. Ser.*, **74**, 228 (1968).
6. T. D. Inch, *Annu. Rep. N.M.R. Spectrosc.*, **2**, 35 (1969); *ibid,*, **5A**, 305 (1972).
7. B. Coxon, *Methods Carbohyd. Chem.*, **6**, 513 (1972).
8. G. Kotowycz and R. U. Lemieux, *Chem. Rev.*, **73**, 669 (1973).
9. G. E. McCasland, *Advan. Carbohyd. Chem.*, **20**, 11 (1965).

10. B. Coxon, *Advan. Carbohyd. Chem. Biochem.*, **27**, 7 (1972).
11. L. D. Hall, *Advan. Carbohyd. Chem. Biochem.*, **29**, 11 (1974).
12. "Carbohydrate Chemistry," Vols. 1–10, Specialist Periodical Reports, The Chemical Society, London, 1968–1978.
13. J. A. Pople, W. G. Schneider, and H. J. Bernstein, "High Resolution Nuclear Magnetic Resonance," McGraw-Hill, New York, 1959.
14. J. W. Emsley, J. Feeney, and L. H. Sutcliffe, "High Resolution Nuclear Magnetic Resonance Spectroscopy," Vols. I–II, Pergamon, Oxford, 1965.
15. A. Carrington and A. D. McLachlan, "Introduction to Magnetic Resonance," Harper, New York, 1967.
16. L. M. Jackman and S. Sternhell, "Applications of Nuclear Magnetic Resonance Spectroscopy in Organic Chemistry," 2nd ed., Pergamon, Oxford, 1969.
17. F. A. Bovey, "Nuclear Magnetic Resonance Spectroscopy," Academic Press, New York, 1969.
18. E. D. Becker, "High Resolution N.M.R.," Academic Press, New York, 1969.
19. R. M. Lynden-Bell and R. K. Harris, "Nuclear Magnetic Resonance Spectroscopy," Nelson, London, 1969.
20. T. C. Farrar and E. D. Becker, "Pulse and Fourier Transform N.M.R. Introduction to Theory and Methods," Academic Press, New York, 1971.
21. *Advan. Magn. Resonance*, Vols. 1–9, Academic Press, New York, 1965–1977.
22. "Annual Report on N.M.R. Spectroscopy," Vols. 1–6C, Academic Press, New York, 1968–1977.
23. "Progress in Nuclear Magnetic Resonance Spectroscopy," Vols. 1–11, Pergamon, Oxford, 1966–1977.
24. "Nuclear Magnetic Resonance," Vols. 1–7, Specialist Periodical Reports, The Chemical Society, London, 1972–1978.
25. J. B. Stothers, "Carbon-13 N.M.R. Spectroscopy. Organic Chemistry—A Series of Monographs," Vol. 24, A. T. Blomquist and H. Wasserman, Eds., Academic Press, New York, 1972.
26. G. C. Levy and G. L. Nelson, "Carbon-13 Nuclear Magnetic Resonance for Organic Chemists," Wiley–Interscience New York, 1972.
27. L. F. Johnson and W. C. Jankowsky, "Carbon-13 N.M.R. Spectra," Wiley–Interscience, New York, 1972.
28. R. R. Ernst and W. A. Anderson, *Rev. Sci. Instrum.*, **37**, 93 (1966); R. R. Ernst, *Advan. Magn. Resonance*, **2**, 1 (1966).
29. H. D. W. Hill and R. Freeman, "Introduction to Fourier Transform N.M.R.," Varian Associates, Palo Alto, 1970.
30. J. D. Baldeschweiler and E. W. Randall, *Chem. Rev.*, **63**, 91 (1963); R. A. Hoffman and S. Forsén, *Progr. N.M.R. Spectrosc.*, **1**, 15 (1966); W. McFarlane, *Annu. Rev. N.M.R. Spectrosc.*, **1**, 135 (1968).
31. L. M. Jackman, "Applications of Nuclear Magnetic Resonance Spectroscopy in Organic Chemistry," Pergamon, Oxford, 1959.
32. R. U. Lemieux and J. D. Stevens, *Can. J. Chem.*, **44**, 249 (1966).
33. S. J. Angyal, *Aust. J. Chem.*, **21**, 2737 (1968).
34. S. J. Angyal, *Angew. Chem. Int. Ed.*, **8**, 157 (1969), and references therein.
35. R. U. Lemieux and J. D. Stevens, *Can. J. Chem.*, **43**, 2059 (1965).
36. P. L. Durette and D. Horton, *Carbohyd. Res.*, **18**, 403 (1971); see also ref. 61.
37. J. D. Stevens and H. G. Fletcher, Jr., *J. Org. Chem.*, **33**, 1799 (1968).
38. L. D. Hall, S. A. Black, K. N. Slessor, and A. S. Tracey, *Can. J. Chem.*, **50**, 1912 (1972).

39. B. Capon and D. Thacker, *Proc. Chem. Soc.*, 369 (1964).
40. R. W. Lenz and J. P. Heeschen, *J. Polymer Sci.*, **51**, 247 (1961).
41. M. Rudrum and D. F. Shaw, *J. Chem. Soc.*, 52 (1965).
42. S. J. Angyal, *Advan. Chem. Ser.*, **117**, 106 (1972).
43. C. D. Jardetzky, *J. Amer. Chem. Soc.*, **82**, 229 (1960).
44. R. U. Lemieux, *Can. J. Chem.*, **39**, 116 (1961).
45. C. Altona and M. Sundaralingam, *J. Amer. Chem. Soc.*, **95**, 2333 (1973), and references cited therein.
46. L. B. Townsend, in "Synthetic Procedures in Nucleic Acid, Chemistry," W. W. Zorbach and R. S. Tipson, Eds., Vol. 2, Wiley–Interscience, New York, 1973, p. 267.
47. W. W. Binkley, D. Horton, and N. S. Bhacca, *Carbohyd. Res.*, **10**, 245 (1969); B. Casu, M. Reggiani, G. G. Gallo, and A. Vigevani, *ibid.*, **12**, 157 (1970); M. Anteunis, A. de Bruyn, and G. Verhegge, *ibid.*, **44**, 101 (1975).
48. H. Friebolin, G. Keilich, and E. Siefert, *Angew. Chem. Int. Ed.*, **8**, 766 (1969); *Org. Magn.Resonance*, **2**, 457 (1970); *ibid.*, **3**, 31 (1971); P. A. J. Gorin, M. Mazurek, and J. F. T. Spencer, *Can. J. Chem.*, **46**, 2305 (1968); A. S. Perlin, B. Casu, G. R. Sanderson, and L. F. Johnson, *Can. J. Chem.*, **48**, 2260 (1970).
49. D. Gagnaire and M. Vincendon, *Carbohyd. Res.*, **31**, 367 (1973).
50. M. Karplus, *J. Chem. Phys.*, **30**, 11 (1959).
51. R. J. Abraham, L. D. Hall, L. Hough, and K. A. McLauchlan, *J. Chem. Soc.*, 3699 (1962).
52. R. U. Lemieux, J. D. Stevens, and R. R. Fraser, *Can. J. Chem.*, **40**, 1955 (1962).
53. M. Karplus, *J. Amer. Chem. Soc.*, **85**, 2870 (1963).
54. K. L. Williamson, *J. Amer. Chem. Soc.*, **85**, 516 (1963); K. L. Williamson, C. A. Lanford, and C. R. Nicholson, *ibid.*, **86**, 762 (1964).
55. P. Laszlo and P. von R. Schleyer, *J. Amer. Chem. Soc.*, **85**, 2709 (1963).
56. P. L. Durette and D. Horton, *Org. Magn. Resonance*, **3**, 417 (1971); L. G. Vorontsova and A. F. Bochkov, *ibid.*, **6**, 654 (1974).
57. D. H. Williams and N. S. Bhacca, *J. Amer. Chem. Soc.*, **86**, 2742 (1964).
58. H. Booth, *Tetrahedron Lett.*, 411 (1965).
59. B. Coxon, *Tetrahedron*, **21**, 3481 (1965).
60. R. J. Abraham and W. A. Thomas, *Chem. Commun.*, 431 (1965).
61. P. L. Durette and D. Horton, *Advan. Carbohyd. Chem. Biochem.*, **26**, 49 (1971).
62. L. D. Hall, P. R. Steiner, and C. Pedersen, *Can. J. Chem.*, **48**, 1155 (1970).
63. D. Horton and J. D. Wander, *J. Org. Chem.*, **39**, 1859 (1974), and references cited therein; compare M. Blanc-Muesser, J. Defaye, and D. Horton, *Carbohyd. Res.*, **68**, 175 (1979).
64. H. El Khadem, D. Horton, and J. D. Wander, *J. Org. Chem.*, **37**, 1630 (1972); D. Horton, P. L. Durette, and J. D. Wander, *Ann. N. Y. Acad. Sci.*, **222**, 884 (1973).
65. D. H. Buss, L. D. Hall, L. Hough, and J. F. Manville, *Tetrahedron*, **21**, 69 (1965); F. Sweet and R. K. Brown, *Can. J. Chem.*, **46**, 1481 (1969).
66. V. F. Bystrov, S. L. Portnova, V. I. Tsetlin, V. T. Ivanov, and Yu. A. Ovchinnikov, *Tetrahedron*, **25**, 493 (1969).
67. O. L. Chapman and R. W. King, *J. Amer. Chem. Soc.*, **86**, 1256 (1964); C. P. Rader, *ibid.*, **88**, 1713 (1966).
68. B. Casu, M. Reggiani, G. G. Gallo, and A. Vigevani, *Tetrahedron*, **22**, 3061 (1966); A. S. Perlin, *Can. J. Chem.*, **44**, 539 (1966); J. C. Jochims, G. Taigel, A. Seeliger, P. Lutz, and H. E. Driesen, *Tetrahedron Lett.*, 4363 (1967); D. B. Davies and S. S. Danyluk, *Can. J. Chem.*, **48**, 3112 (1970).
69. R. R. Fraser, M. Kaufman, P. Morand, and G. Goril, *Can. J. Chem.*, **47**, 403 (1969).

70. B. Coxon, *Tetrahedron*, **22**, 2281 (1966); P. L. Durette and D. Horton, *Carbohyd. Res.*, **18**, 57 (1971); J. M. J. Tronchet, F. Barbalat-Rey, and J. Tronchet, *ibid.*, **41**, 1 (1975).
71. J. A. Pople and A. A. Bothner-By, *J. Chem. Phys.*, **42**, 1339 (1965); M. Barfield and D. M. Grant, *Advan. Magn. Resonance*, **1**, 149 (1965).
72. S. Sternhell, *Rev. Pure Appl. Chem.*, **14**, 15 (1964); M. Barfield and B. Chakrabarti, *Chem. Rev.*, **69**, 757 (1969).
73. L. D. Hall and L. Hough, *Proc. Chem. Soc.*, 382 (1962).
73a. See also M. Černý and J. Staněk, Jr., *Advan. Carbohyd. Chem. Biochem.*, **34**, 23 (1977), especially pp. 56–62 for a detailed discussion of n.m.r. studies on 1,6-anhydrohexopyranose systems.
74. D. Horton and J. S. Jewell, *Carbohyd. Res.*, **5**, 149 (1967).
75. M. Barfield, *J. Chem. Phys.*, **41**, 3825 (1964); *J. Amer. Chem. Soc.*, **93**, 1066 (1971).
76. L. D. Hall and J. F. Manville, *Carbohyd. Res.*, **8**, 295 (1968).
77. B. Coxon, *Carbohyd. Res.*, **12**, 313 (1970).
78. R. U. Lemieux, K. A. Watanabe, and A. A. Pavia, *Can. J. Chem.*, **47**, 4413 (1969), and references therein.
79. R. J. Ferrier and M. M. Ponpipom, *J. Chem. Soc. (C)*, 560 (1971).
80. E. W. Garbisch, Jr., *J. Amer. Chem. Soc.*, **86**, 5561 (1964).
81. B. Coxon, H. J. Jennings, and K. A. McLauchlan, *Tetrahedron*, **23**, 2395 (1967).
81a. J. M. Berry, L. D. Hall, D. G. Welder, and K. F. Wong, *Amer. Chem. Soc. Symp. Ser.*, **87**, 30 (1979).
82. L. D. Hall and C. M. Preston, *Chem. Commun.*, 1319 (1972); *Carbohyd. Res.*, **37**, 267 (1974).
83. C. W. M. Grant, L. D. Hall, and C. M. Preston, *J. Amer. Chem. Soc.*, **95**, 7742 (1973); R. Freeman, H. D. W. Hill, B. L. Tomlinson, and L. D. Hall, *J. Chem. Phys.*, **61**, 4466 (1974).
84. L. D. Hall and C. M. Preston, *Carbohyd. Res.*, **27**, 332 (1973); **49**, 3 (1976).
84a. K. Bock and L. D. Hall, *Carbohyd. Res.*, **40**, C3 (1975).
85. C. V. Holland, D. Horton, M. J. Miller, and N. S. Bhacca, *J. Org. Chem.*, **32**, 3077 (1967).
86. P. Laszlo, *Progr. N.M.R. Spectrosc.*, **3**, 231 (1967).
87. J. Ronayne and D. H. Williams, *Annu. Rev. N.M.R. Spectrosc.*, **2**, 83 (1969).
88. M. H. Freemantle and W. G. Overend, *Chem. Commun.*, 503 (1968); *J. Chem. Soc. (B)*, 547, 551 (1969).
89. D. Horton and J. H. Lauterbach, *J. Org. Chem.*, **34**, 86 (1969); *Carbohyd. Res.*, **43**, 9 (1975).
90. J. Reuben and D. Fiat, *Chem. Commun.*, 729 (1967).
91. K. G. Morallee, E. Nieboer, F. J. C. Rossotti, R. J. P. Williams, A. V. Xavier, and R. A. Dwek, *Chem. Commun.*, 1132 (1970).
92. C. D. Barry, A. C. T. North, J. A. Glasel, R. J. P. Williams, and A. V. Xavier, *Nature*, **323**, 236 (1971); I. M. Armitage, G. Dunsmore, L. D. Hall, and A. G. Marshall, *Can. J. Chem.*, **50**, 2119 (1972); A. Arduini, I. M. Armitage, L. D. Hall, and A. G. Marshall, *Carbohyd. Res.*, **26**, 255 (1973).
93. S. J. Angyal and K. P. Davies, *Chem. Commun.*, 500 (1971).
94. J. K. M. Sanders and D. H. Williams, *Tetrahedron Lett.*, 2813 (1971); F. A. Hart, G. P. Moss, and M. L. Staniforth, *ibid.*, 3389 (1971).
95. D. Horton and Z. Wałaszek, *Abstr. Pap. Amer. Chem. Soc. Meeting*, **170**, CARB-10 (1975).
96. C. C. Hinckley, *J. Amer. Chem. Soc.*, **91**, 5160 (1969).

97. J. K. M. Sanders and D. H. Williams, *Chem. Commun.*, 422 (1970); *J. Amer. Chem. Soc.*, **93**, 641 (1971).

98. R. E. Rondeau and R. E. Sievers, *J. Amer. Chem. Soc.*, **93**, 1522 (1971).

99. I. Armitage and L. D. Hall, *Chem. Ind.* (London), 1537 (1970); *Can. J. Chem.*, **49**, 2770 (1971).

100. P. Girard, H. Kagan, and S. David, *Bull. Soc. Chim. Fr.*, 4515 (1970).

101. N. Ahmad, N. S. Bhacca, J. Selbin, and J. D. Wander, *J. Amer. Chem. Soc.*, **93**, 2564 (1971).

102. H. M. McConnell and R. E. Robertson, *J. Chem. Phys.*, **29**, 1361 (1958); B. Bleaney, *J. Magn. Resonance*, **8**, 91 (1972).

103. I. M. Armitage, L. D. Hall, A. G. Marshall, and L. Werbelow, *J. Amer. Chem. Soc.*, **95**, 1437 (1973); see also, for example, P. McArdle, J. O. Wood, E. E. Lee, and M. J. Conneely, *Carbohyd. Res.*, **69**, 39 (1979) for approaches to quantitative treatment of shift-reagent data.

104. S. D. Gero, D. Horton, A. M. Sepulchre, and J. D. Wander, *Tetrahedron*, **29**, 2963 (1973); *J. Org. Chem.*, **40**, 1061 (1975); D. Horton and J. D. Wander, *Carbohyd. Res.*, **39**, 141 (1975).

105. L. D. Hall and C. M. Preston, *Carbohyd. Res.*, **41**, 53 (1975).

106. B. Casu, G. Gatti, N. Cyr, and A. S. Perlin, *Carbohyd. Res.*, **41**, C6 (1975).

107. R. U. Lemieux and S. Levine, *Can. J. Chem.*, **42**, 1473 (1964); R. U. Lemieux, A. A. Pavia, J. C. Martin, and K. A. Watanabe, *ibid.*, **47**, 4427 (1969).

108. A. Rosenthal and D. Abson, *Can. J. Chem.*, **42**, 1811 (1964); **43**, 1318 (1965); A. Rosenthal and H. J. Koch, *ibid.*, **43**, 1375 (1965).

109. D. Horton, J. S. Jewell, E. K. Just, and J. D. Wander, *Carbohyd. Res.*, **18**, 49 (1971); D. Gagnaire, D. Horton, and F. R. Taravel, *Carbohyd. Res.*, **27**, 363 (1973).

110. W. Mackie and A. S. Perlin, *Can. J. Chem.*, **43**, 2921 (1965).

111. D. Horton, J. B. Hughes, J. S. Jewell, K. D. Philips, and W. N. Turner, *J. Org. Chem.*, **32**, 1973 (1967); D. Horton, W. E. Mast, and K. D. Philips, *J. Org. Chem.*, **32**, 1471 (1967).

112. D. Gagnaire and L. Odier, *Carbohyd. Res.*, **11**, 33 (1969); E. B. Rathbone and A. M. Stephen, *Tetrahedron Lett.*, 1339 (1970).

113. A. E. El Ashmawy and D. Horton, *Carbohyd. Res.*, **3**, 191 (1966).

114. J. E. G. Barnett and D. L. Corina, *Advan. Carbohyd. Chem. Biochem.*, **27**, 128 (1972).

115. R. Burton, L. D. Hall, and P. R. Steiner, *Can. J. Chem.*, **48**, 2679 (1970); B. Coxon, *Carbohyd. Res.*, **18**, 427 (1971).

116. R. Burton, L. D. Hall, and P. R. Steiner, *Can. J. Chem.*, **49**, 588 (1971).

117. E. B. Baker, *J. Chem. Phys.*, **37**, 911 (1962).

118. V. J. Kowalewski, *Progr. Nucl. Magn. Resonance Spectrosc.*, **5**, 1 (1969).

119. A. G. Redfield and R. K. Gupta, *J. Chem. Phys.*, **54**, 1418 (1971); S. L. Patt and B. D. Sykes, *ibid.*, **56**, 2182 (1972); F. W. Benz, J. Feeney, and G. C. K. Roberts, *J. Magn. Resonance*, **8**, 114 (1972).

120. R. Freeman and G. A. Morris, *Bull. Magn. Resonance*, **1**, 5 (1979).

121. A. S. Perlin, *MTP Intern. Rev. Sci., Org. Chem.*, Ser. 2, Vol. 7, Butterworth, London, 1976, pp. 1–34.

122. M. Witanowski and G. A. Webb (eds.), "Nitrogen N.M.R.," Plenum, New York, 1973.

123. R. F. Fields, *Annu. Rep. N.M.R. Spectrosc.*, **5A**, 99 (1972).

124. M. M. Crutchfield, C. H. Dungan, J. H. Letcher, V. Mark, and J. R. van Wazer, "^{31}P Nuclear Magnetic Resonance," Vol. 5 of "Topics in Phosphorus Chemistry," (M. Grayson and E. J. Griffith, eds.), Wiley–Interscience, New York, 1967; G. Mavel, *Annu. Rep. N.M.R. Spectrosc.*, **5B**, 1 (1973).

125. L. D. Hall, R. N. Johnson, A. B. Foster, and J. H. Westwood, *Can. J. Chem.*, **49**, 236 (1971).

126. A. B. Foster, J. H. Westwood, B. Donaldson, and L. D. Hall, *Carbohyd. Res.*, **25**, 228 (1972), and references therein.

127. P. W. Kent and R. A. Dwek, *Biochem. J.*, **211**, 11 (1971), and references therein.

128. M. A. Raftery, W. H. Huestis, and F. Miller, *Cold Spring Harbor Symp. Quant. Biol.*, **36**, 541 (1971); C. G. Butchard, R. A. Dwek, P. W. Kent, R. J. P. Williams. and A. V. Xavier, *Eur. J. Biochem.*, **27**, 548 (1972); C. W. M. Grant and L. D. Hall, *Carbohyd. Res.*, **24**, 218 (1972).

129. L. D. Hall and R. B. Malcolm, *Chem. Ind.* (London), 92 (1968).

130. L. D. Hall and R. B. Malcolm, *Can. J. Chem.*, **50**, 2092, 2111 (1972).

131. W. G. Bentrude and H. W. Tan, *J. Amer. Chem. Soc.*, **94**, 8222 (1972); J. A. Mosbo and J. G. Verkade, *ibid.*, **94**, 8224 (1972), and references therein.

132. H. H. Mantsch and I. C. P. Smith, *Biochem. Biophys. Res. Commun.*, **46**, 808 (1972), and references therein.

133. D. J. Wood, F. E. Hruska, and K. K. Ogilvie, *Can. J. Chem.*, **52**, 3353 (1974), and references therein; compare P. J. Cozzone and O. Jardetzky, *Biochemistry*, **15**, 4853, 4860 (1976).

134. L. Evelyn, L. D. Hall, P. R. Steiner, and D. H. Stokes, *Chem. Commun.*, 576 (1969); *Org. Magn. Resonance*, **5**, 141 (1973); L. Evelyn, L. D. Hall, L. Lynn, P. R. Steiner, and D. H. Stokes, *Carbohyd. Res.*, **27**, 21 (1973), and references therein.

134a. A. J. R. Costello, T. Glonek, M. E. Slodki, and F. R. Seymour, *Carbohyd. Res.*, **42**, 23 (1975).

135. B. Coxon, *Carbohyd. Res.*, **11**, 153 (1969).

136. L. Mester, G. Vass, A. Stephen, and J. Parello, *Tetrahedron Lett.*, 4053 (1968); L. Mester, A. Stephen, and J. Parello, *ibid.*, 4119 (1968).

137. L. Mester and G. Vass, *Tetrahedron Lett.*, 3847 (1969).

138. L. D. Hall and L. F. Johnson, *Chem. Commun.*, 509 (1969).

139. A. S. Perlin and B. Casu, *Tetrahedron Lett.*, 2921 (1969).

140. D. E. Dorman, S. J. Angyal, and J. D. Roberts, *Proc. Nat. Acad. Sci. U.S.*, **63**, 612 (1969).

141. A. S. Perlin, B. Casu, and H. J. Koch, *Can. J. Chem.*, **48**, 1355 (1970); H. J. Koch and A. S. Perlin, *Carbohyd. Res.*, **15**, 403 (1970); A. S. Perlin and H. J. Koch, *Can. J. Chem.*, **48**, 2639 (1970), and references therein.

142. D. E. Dorman and J. D. Roberts, *J. Amer. Chem. Soc.*, **92**, 1355 (1970), and references therein.

143. W. Voelter, E. Breitmaier, G. Jung, T. Keller, and D. Hiss, *Angew. Chem. Int. Ed.*, **9**, 803 (1970); A. Allerhand, D. Doddrell, and R. Komoroski, *J. Chem. Phys.*, **55**, 189 (1971); W. A. Szarek, D. M. Vyas, S. D. Gero, and G. Lukacs, *Can. J. Chem.*, **52**, 3394 (1974), and references therein.

144. K. Bock and C. Pedersen, *J. Chem. Soc. Perkin II*, 293 (1973).

145. A. Allerhand and D. Doddrell, *J. Amer. Chem. Soc.*, **93**, 2777 (1971); D. Doddrell, and A. Allerhand, *ibid.*, 2779 (1971); W. W. Binkley, D. Horton, N. S. Bhacca, and J. D. Wander, *Carbohyd. Res.*, **23**, 301 (1972); P. Colson and R. R. King, *ibid.*, **47**, 1 (1976); F. R. Seymour, R. D. Knapp, J. E. Zweig, and S. H. Bishop, *ibid.*, **72**, 57 (1979).

146. R. U. Lemieux, *Ann. N.Y. Acad. Sci.*, **222**, 915 (1973).

147. A. S. Perlin, N. Cyr, H. J. Koch, and B. Korsch, *Ann. N.Y. Acad. Sci.*, **222**, 935 (1973).

148. D. E. Dorman, *Ann. N.Y. Acad. Sci.*, **222**, 943 (1973).

149. Y. M. Choy and G. G. S. Dutton, *Can. J. Chem.*, **51**, 198 (1973); J.-P. Joseleau, M. Lapeyre, M. Vignon, and G. G. S. Dutton, *Carbohyd. Res.*, **67**, 197 (1978), and references therein to other studies of *Klebsiella* polysaccharides.

150. A. S. Perlin, N. M. K. Ng Ying Kin, S. S. Bhattacharjee, and L. F. Johnson, *Can. J. Chem.*, **50**, 2437 (1972); compare G. Gatti, B. Casu, G. Torri, and J. R. Vercellotti, *Carbohyd. Res.*, **68**, C3 (1979).

151. P. Colson, H. J. Jennings, and I. C. P. Smith, *J. Amer. Chem. Soc.*, **96**, 8081 (1974).

152. Compare P. Colson, K. N. Slessor, H. J. Jennings, and I. C. P. Smith, *Can. J. Chem.*, **53**, 1030 (1975); P. Colson, H. C. Jarrell, B. L. Lamberts, and I. C. P. Smith, *Carbohyd. Res.*, **71**, 265 (1979); see also B. Capon, D. S. Rycroft, and J. W. Thomson, *ibid.*, **70**, 145 (1979) for ^{13}C studies on acetylated cello-oligosaccharides.

153. Compare T. Usui, N. Yamaoka, K. Matsuda, K. Tuzimura, H. Sugiyama, and S. Sato, *J. Chem. Soc. Perkin I*, 2425 (1973).

154. D. R. Bundle, H. J. Jennings, and I. C. P. Smith, *Can. J. Chem.*, **51**, 3812 (1973).

155. D. R. Bundle, H. J. Jennings, and I. C. P. Smith, *J. Biol. Chem.*, **249**, 2275 (1974).

156. A. K. Bhattacharjee, H. J. Jennings, C. P. Kenny, A. Martin, and I. C. P. Smith, *J. Biol. Chem.*, **250**, 1926 (1975).

157. P. A. J. Gorin, *Carbohyd. Res.*, **39**, 3 (1975).

158. P. A. J. Gorin, *Advan. Carbohyd. Chem. Biochem.*, **38**, in press (1980).

159. C. F. Brewer, D. Marcus, A. P. Grollman, and H. Sternlicht, *Ann. N. Y. Acad. Sci.*, **222**, 978 (1973).

II. MASS SPECTROMETRY

Don C. DeJongh

A. Introduction

A mass spectrum consists of a plot of the relative intensities of gaseous ions, formed by ionization and subsequent fragmentation of gaseous molecules, against their mass-to-charge ratio (m/e).[1] The emergence of mass spectrometry as an analytical technique of wide applicability in natural-product chemistry is a result of the recognition of correlations existing between the structure of a complex organic molecule and its mass spectrum.[2] The versatility of mass spectrometry has been greatly increased by advances in instrumentation and inlet techniques that make it possible to obtain directly the elemental composition of all ions with 1 microgram or less of sample.[3]

The need to volatilize compounds before ionization, at temperatures and pressures at which they are thermally stable, initially limited the application of mass spectrometry in the field of carbohydrate chemistry. Another limiting factor has been the extensive fragmentation of ions arising from the highly substituted and oxygenated carbohydrates and their derivatives, producing complex mass spectra with few, if any, intense peaks in the high m/e region. This latter limitation has led to pioneering studies of many classes and derivatives of carbohydrates with the object of systematically determining what structural information is obtainable from the mass spectrum of each.[4] Since the mass spectra of these classes and derivatives have different characteristics, the choice among them depends on their availability and the information desired. Also, alternative methods of effecting ionization, discussed in the final section, have greatly extended the application of mass spectrometry in this field.

Since only minute samples are required, the combination of mass spectrometry, directly[5] and indirectly[6] with gas chromatography, and indirectly[7] with thin-layer chromatography, has dramatically increased the scope of mass spectrometry as an analytical tool. These combination techniques are particularly important in carbohydrate chemistry, since it is often difficult to obtain samples in a pure crystalline state.

In the following sections, the general applications of mass spectrometry to carbohydrate chemistry will be discussed in terms of free carbohydrates and their glycosides, and their permethylated, per(trimethylsilylated), and peracetylated derivatives, their dithioacetals, and their isopropylidene acetals. Deoxy, amino, and anhydro sugars will be treated under the derivatives through which they were studied. The mass spectra of nucleosides, steroid glycosides, and carbohydrates having nitrogen or sulfur in the ring will be covered separately. Unless stated otherwise, the mass spectra discussed are positive-ion spectra obtained with an electron-impact ion source. Specific applications of mass spectrometry, where relevant, are also mentioned elsewhere in this book.

B. THE MASS SPECTRA OF FREE CARBOHYDRATES
AND THEIR GLYCOSIDES

Extensive fragmentation to ions of low m/e ratios makes the detection of molecular-ion peaks very difficult in the mass spectra of free carbohydrates and glycosides. Structural ambiguity of many of these peaks at low m/e values limits the general usefulness of these mass spectra for structural studies. For example, the mass spectrum of D-ribose has no detectable peak corresponding to the molecular ion at m/e 150, but peaks of relatively low intensity appear at m/e 132 and m/e 119, from the loss of water and of a hydroxymethyl radical from C-4, respectively.[8] Also, no molecular-ion peak is present in the spectrum of methyl β-D-glucopyranoside; the first fragment corresponds to loss of a methoxyl group from C-1 of the molecular ion.[9]

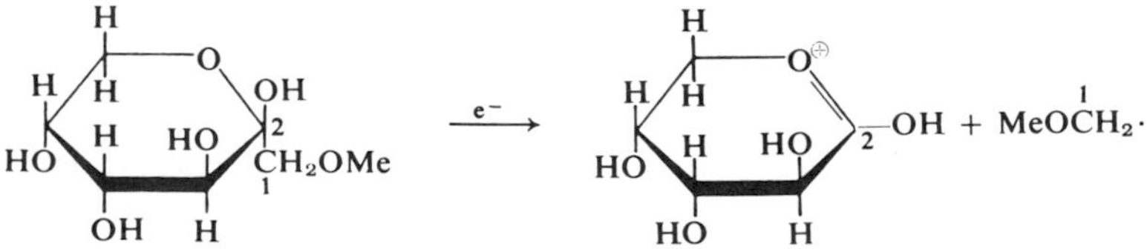

Information can be obtained as to the site of methylation, as can be seen in Eqs. (1) and (2), from a comparison of the mass spectra of 1-O-methyl-β-D-fructopyranose (**1**) and 3-*O*-methyl-β-D-fructopyranose[10] (**2**). Both compounds undergo fragmentation of the C-1–C-2 bond with charge retention on C-2, but 45 mass units are lost in one case, whereas 31 are lost in the other.

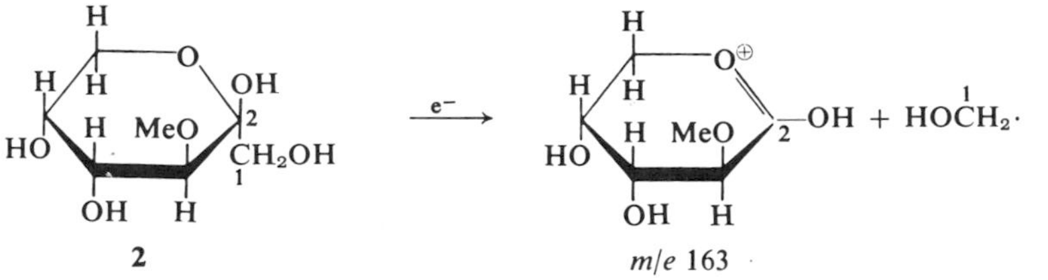

$$2 \qquad\qquad m/e\ 163$$

A study of rearrangements of hexose 4- and 5-sulfonates demonstrates a situation in which mass spectrometry played an important role.[11] Ring size and the nature and position of substitution in compound **3**, formed from a rearrangement product, were elucidated with the aid of the mass spectrum.

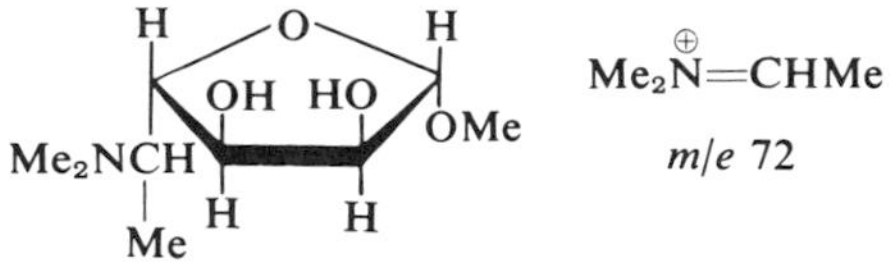

3

The most intense peak was at m/e 72, corresponding to cleavage of the side chain attached at C-4; a peak at m/e 174 resulted from the loss of the methoxyl group at C-1. Thus, from its mass spectrum, compound **3** could be recognized as a methyl furanoside having an *N,N*-dimethylamino substituent at C-5.

The mass spectrum of aldgamycin E, a neutral macrolide antibiotic (see Vol. IIA, Chapter 31), provided a molecular weight and evidence of the presence of sugars identified as mycinose and aldgarose: molecular-ion peak, m/e 742; loss of mycinosyl, m/e 567; loss of aldgarosyl, m/e 540; and loss of both mycinosyl and aldgarosyl,[12] m/e 333. The mass spectrum of the methyl glycoside of aldgarose provided data, in conjunction with infrared and n.m.r. spectrometry, leading to its identification as a cyclic carbonate derivative of a sugar.

An attempt has been made to correlate appearance-potential measurements on the $C_6H_{11}O_5{}^+$ ion from methyl α- and β-D-glucopyranosides, and from four disaccharides, with the anomeric configuration.[13] A slightly lower appearance-potential for the ion from α-glycosides is observed. The differences are so small and accurate determinations so inherently difficult with conventional instruments that the technique has not found further application.

Partial methylation enhances volatility of glycosides and produces fragmentation patterns that correspond to those of fully methylated analogues (Section

References start on p. 1351.

II,C,2), except that cleavages involving —OCH_3 substituents are quantitatively favored over those involving —OH substituents.[14]

The mass spectra of the stereoisomeric inositols (see Vol. IB, Chapter 15) also have no molecular-ion peaks or intense peaks in the high m/e region.[14a] The isomeric compounds fragment similarly, but their mass spectra do show differences of relative intensity that have been interpreted in terms of stereochemistry.

C. THE MASS SPECTRA OF PERMETHYLATED, PER(TRIMETHYLSILYLATED), AND PERACETYLATED CARBOHYDRATES

1. Introduction

Peracetylated, permethylated, and per(trimethylsilylated) carbohydrates have been widely studied by mass spectrometry because they are volatile derivatives, easily prepared on a small scale, which can be separated and purified by thin-layer and gas chromatography. Their fragmentations upon electron impact are sensitive to ring size and substitution, but stereochemical variations lead only to differences in relative intensities of peaks rather than to differences in fragmentation pathways. The mass spectra of these derivatives generally do not exhibit molecular-ion peaks, and peaks in the high m/e region are of relatively low intensity.

2. Permethylated Carbohydrates

A most valuable[15] analytical application of the mass spectra of methyl ethers is the identification of partially methylated monosaccharides.[16] Methylation is a commonly used procedure for analyzing the structure of naturally occurring compounds, such as polysaccharides, containing carbohydrate monomer units (see Vol. IA, Chapter 12, and Vol. IIA, Chapter 34). A partially methylated monosaccharide, obtained from hydrolysis of a methylated polysaccharide, is deuteriomethylated, and the mass spectrum of the product is compared with the mass spectra of authentic samples. The shifts in location of peaks due to the presence of deuterium in place of hydrogen provide a basis for locating the position of the trideuteriomethyl groups and, thus, of the free hydroxyl groups in the starting compound.[4(d)]

Some of the major peaks in the mass spectrum of methyl 2,3,4-tri-O-methyl-β-D-arabinopyranoside (4) have been assigned the structures shown in Fig. 1 on the basis of extensive deuterium labeling.[17] The mass spectrum does not exhibit a molecular-ion peak, the peak of highest mass being m/e 176. From the deuterium labeling studies, the m/e 101 fragment was found to retain carbon atoms 1-2-3 and 2-3-4 and the m/e 88 fragment to retain carbon atoms 1-2, 3-4, and 2-3. The m/e 75 fragment, which contains two methoxyl groups on C-1, is

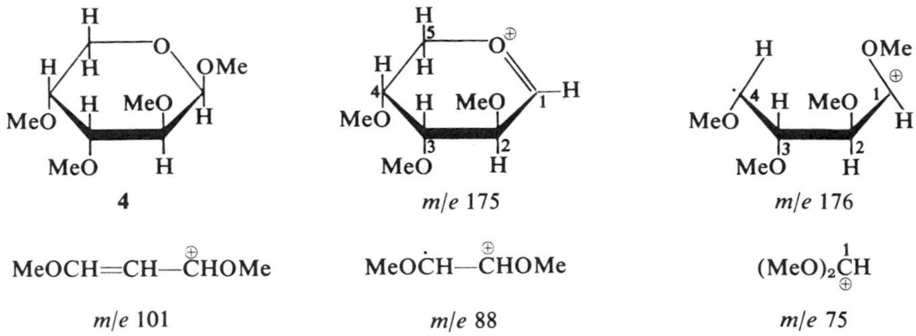

FIG. 1. Peaks and assigned structures from the mass spectrum of methyl 2,3,4-tri-*O*-methyl-β-D-arabinopyranoside (**4**).

apparently formed by migration of the methoxyl group on C-3 to C-1. Furanosides can readily be differentiated from pyranosides, as can be seen in the characteristic cleavage of the C-4–C-5 bond, shown in Eq. (3) for methyl 2,3,5,6-tetra-*O*-methyl-α-D-galactofuranoside[18] (**5**).

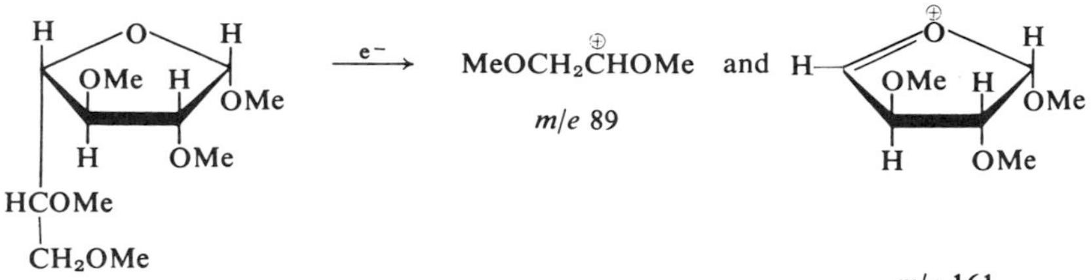

(3)

5

Permethylated amino sugars,[19] deoxy sugars,[20] anhydro sugars,[21] keto sugars,[22] and partially methylated alditol acetates[23] also fragment in characteristic manners which can be correlated with substitution and structural variations.[23a] For example, methyl 1,3,4,5-tetra-*O*-methyl-β-D-fructopyranoside (**6**) easily loses a methoxyl radical from C-2, or loses C-1 and its substituent, upon electron impact; it also gives an ion at *m/e* 119 from transfer of a methoxyl group from C-4 to C-2, as illustrated[22] in Eq. (4).

The peak containing C-1 and a methoxyl group rearranged from C-3, *m/e* 75 in Fig. 1, shifts characteristically in the mass spectra of various permethylated acetamido sugars.[19] This fragment is found at *m/e* 102 in the spectrum of *N*-acetyl-2,3,4,6-tetra-*O*-methyl-β-D-glucopyranosylamine and at *m/e* 75 in

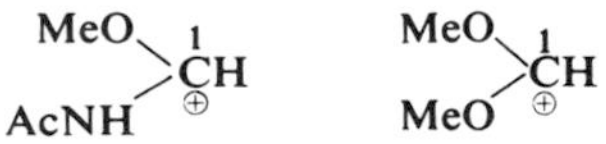

m/e 102 *m/e* 75

References start on p. 1351.

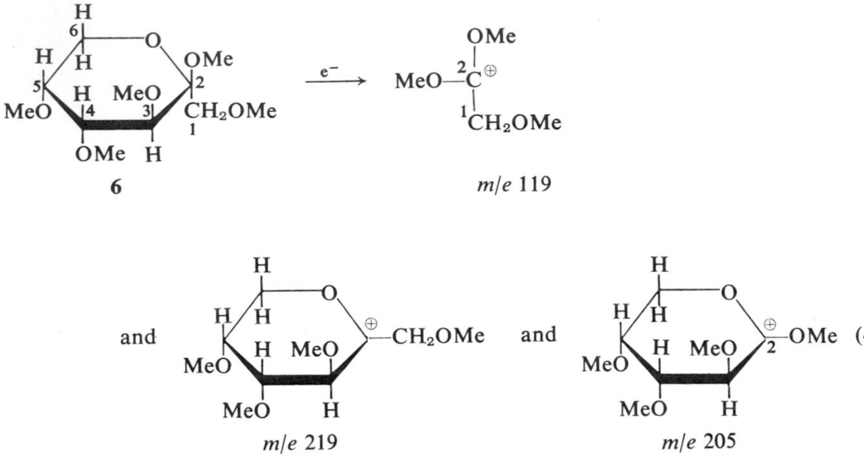

6   *m/e* 119

m/e 219   *m/e* 205   (4)

the spectra of the 5- and 6-acetamido isomers. A 3-acetamido group does not migrate to C-1 as does a 3-methoxyl group. A mass spectrum of methyl 2-amino-2-deoxy-3,4,6-tri-O-methyl-α-D-galactopyranoside hydrochloride has been reported, although it is likely that the hydrogen chloride separated thermally from the sugar molecule in the ion source (120°) before ionization.[24] Detailed analyses have been reported of the fragmentation of methyl (methyl tri-O-methylhexopyranosiduronates) and their amides.[24a]

It is possible to recognize disaccharides connected by (1 → 2), (1 → 4), and (1 → 6) linkages on the basis of fragmentation in the mass spectrometer,[25] although differentiation between (1 → 4)- and (1 → 2)-linked disaccharides is rather difficult.

3. *Per(trimethylsilylated) Carbohydrates*

The ease with which sugars and their derivatives are converted into their per(trimethylsilyl) ethers, the facility with which the products can be separated by gas–liquid chromatography, and the fact that much structural information can be derived from the mass spectra of these ethers have made the mass spectrometry of the per(trimethylsilylated) sugars one of the most fruitful newer techniques in carbohydrate chemistry. By use of a coupled gas chromatograph–mass spectrometer it is possible to examine the separated components of complex mixtures with very small samples. The fragmentation patterns of the per(trimethylsilyl) ethers of D-glucose, various glycosides, and some amino sugars have been analyzed in detail with the use of deuterium labeling and exact-mass data from high-resolution measurements.[26] It was shown that the spectra permit recognition of ring size and also degree and position of substitution by various groups in the original sugar.[26]

Mass spectrometry of per(trimethylsilyl) ethers has been used in studies on carbohydrate-containing antibiotics[27] (see also Vol. IIA, Chapter 31), and on various glycolipid derivatives (Vol. IIB, Chapter 44), including cerebrosides[28] and glycosphingolipids.[29] Applications with disaccharides, including cellobiose, lactose, laminarabiose, and maltose, showed[30,31] that differences in the spectra characteristic of the disaccharide link can be recognized. Applications with sucrose and enzymically produced homologues such as the trisaccharide 1-kestose and the tetrasaccharide nystose showed that the spectra allow determination of the presence and location of hexulofuranose (ketohexofuranose) residues; the per(trimethylsilyl) ethers are superior to the peracetates for this purpose.[32]

Molecular-ion peaks are usually either absent or of very low intensity in the per(trimethylsilyl) ethers, but a weak peak for the loss of $CH_3 \cdot$ from the molecular ion is generally observed.[26,30] However, even in a molecule as large as the tetradeca(trimethylsilyl) ether of stachyose (mol. wt. 1674) a weak molecular-ion peak can still be detected.[32]

4. Peracetylated Carbohydrates

The mass spectra of peracetylated hexoses and pentoses in their furanose,[33] pyranose,[33] and acyclic[34] forms can be interpreted in terms of structural variations such as ring size and substitution differences, but not particularly in terms of stereochemistry. Similar studies have been reported with peracetylated 1,6-anhydrohexopyranoses,[35] and peracetylated anhydrodeoxyalditols, unsaturated sugars, and branched-chain sugars.[36] As with the permethylated carbohydrates, stereoisomers of peracetylated carbohydrates give mass spectra differing in relative intensities rather than in fragmentation paths. Noticeable

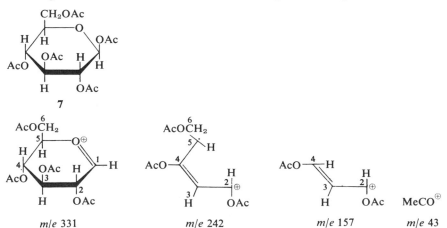

FIG. 2. Representative fragments in the mass spectrum[33] of β-D-glucopyranose penta-acetate.

References start on p. 1351.

quantitative differences, such as in the intensities of high-mass-range peaks, have been observed in the mass spectra of peracetylated carbohydrates taken on different instruments under different inlet conditions.[36a]

No molecular-ion peaks are observed in the mass spectra of these derivatives. Series of fragments can be recognized within which the individual peaks differ by 60 and 42 mass units, corresponding to successive eliminations of acetic acid and ketene, respectively, from the acetoxyl substituents. Representative fragments from each series are shown in Fig. 2 from the mass spectrum[33] of β-D-glucopyranose pentaacetate (7). These assignments are supported by peak shifts in the mass spectra of the corresponding compounds prepared with acetic anhydride-d_6. A similar study has been made with the pentaacetate of 2-amino-2-deoxy-D-glucose.[37]

In contrast to the minor relative-intensity differences among the mass spectra of epimers and anomers, the mass spectra of hexulopyranoses (ketohexopyranoses) differ appreciably from those of their hexopyranose (aldohexopyranose) isomers, as can be seen from a comparison of the representative fragments in Figs. 2 and 3.

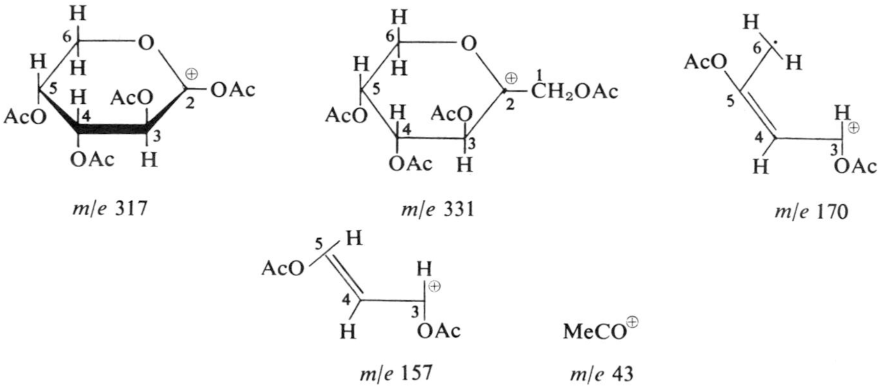

m/e 317   *m/e* 331   *m/e* 170

m/e 157   *m/e* 43

FIG. 3. Representative fragments in the mass spectrum[33] of β-D-fructopyranose penta-acetate

The mass spectra of the corresponding acyclic peracetates exhibit fragmentation involving the carbonyl group[34]:

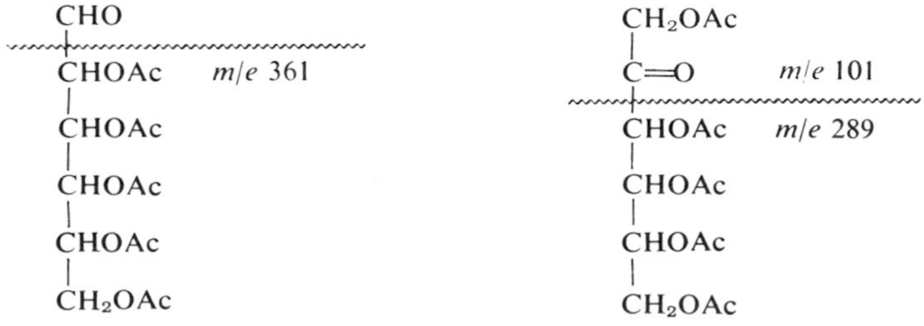

In the mass spectra of peracetylated carbohydrates in the furanose form, peaks are found corresponding to fragments formed by cleavage of the C-4–C-5 bond.[33] For example, mass spectrometry played a major role in identifying a furanose product from a rearrangement of 2-(N-acetylbenzamido)-2-deoxy-D-glucose.[38] Upon heating this compound in chloroform solution and subsequently acetylating the product mixture, four products were isolated, of which three were readily identified from previously known compounds. The fourth product, formed in substantial quantity, was identified as 2-acetamido-3,5,6-tri-O-acetyl-1-O-benzoyl-2-deoxy-α-D-glucofuranose (8) by its n.m.r. and mass spectra; Eq. (5) shows one fragmentation that was used to assign the furanose structure.

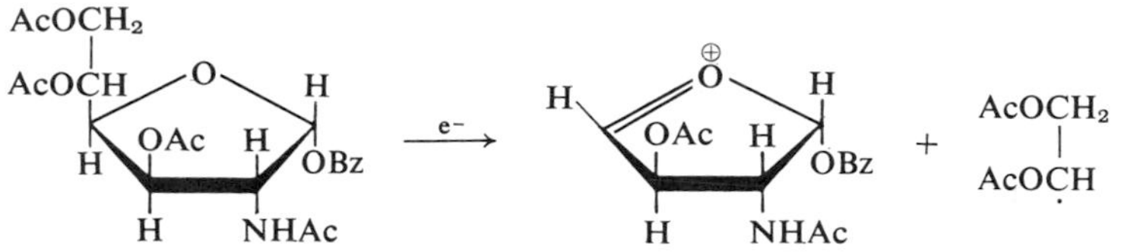

The position of the deoxy function can be recognized from the mass spectra of 6-deoxy-α-D-glucopyranose tetraacetate and 2-deoxy-β-D-*arabino*-hexopyranose tetraacetate.[33] Caution must be exercised in assigning ring size as the peracetates become less substituted. The mass spectra of isomeric *cis*- and *trans*-2,3-diacetoxytetrahydropyran (9 and 10, respectively) and the acetylated tetrahydrofuran-2-yl-carboxaldehyde (11) are nearly identical.[39] This similarity

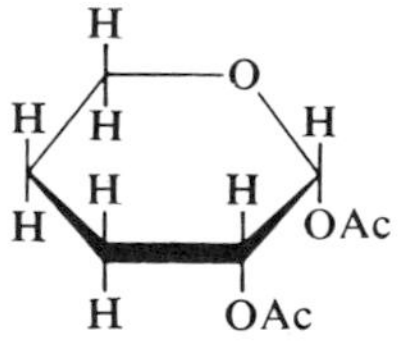

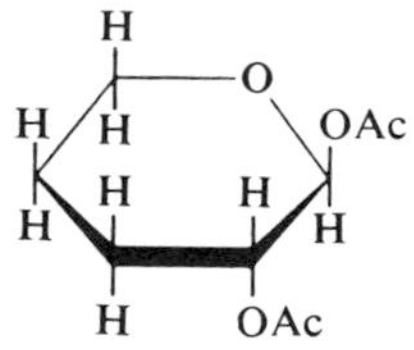

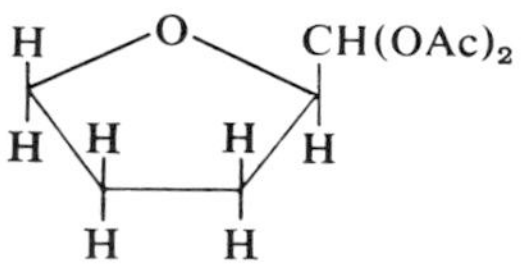

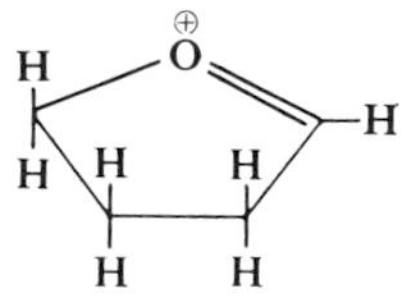

References start on p. 1351.

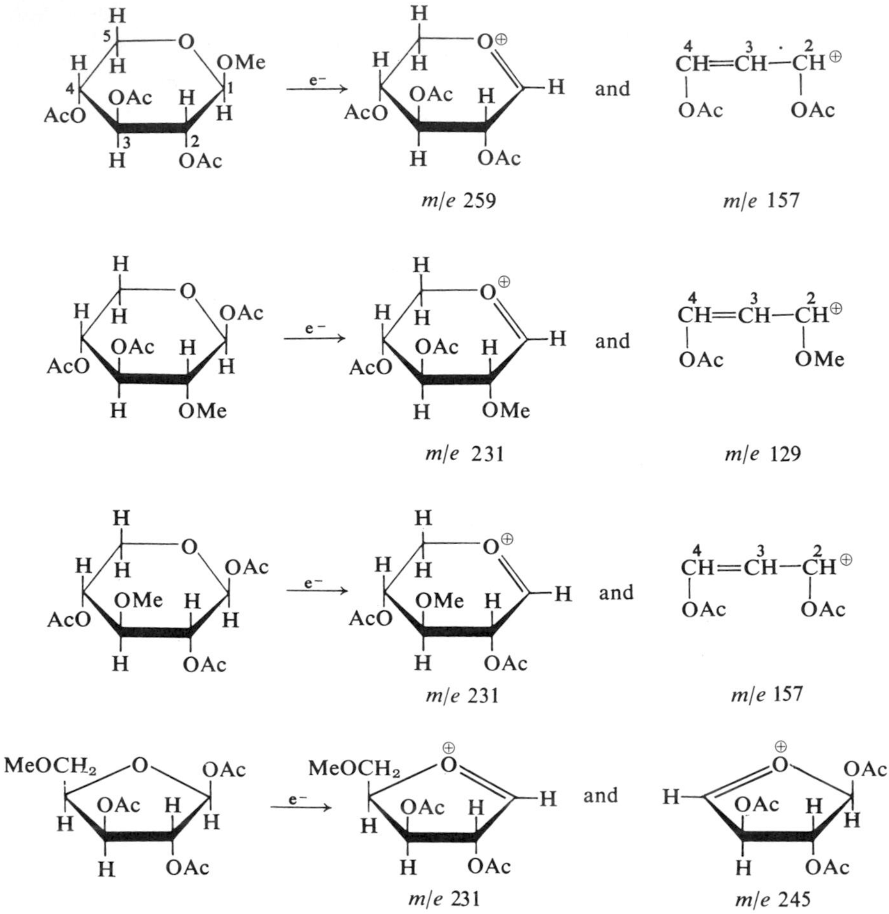

FIG. 4. Characteristic fragments in the mass spectra of isomeric mono-O-methyl-D-xylose triacetates.

has been explained by an electron-impact-induced ring contraction and fragmentation to a common ion (12).

To illustrate the differences that occur in location of peaks when substituents are interchanged, representative fragments from the isomeric 1-, 2-, 3-, and 5-O-methyl-D-xylose triacetates are compared in Fig. 4.[40]

Peracetylated aryl glycosides—for example, arbutin pentaacetate—readily lose the aryloxy radical from the molecular ion, and the latter is rarely observed. Thus, independent of the nature of the aglycon, the acetylated aryl glycosides give very similar fragmentation patterns.[41] The identity of the aryl group is readily determined from the mass spectrum.

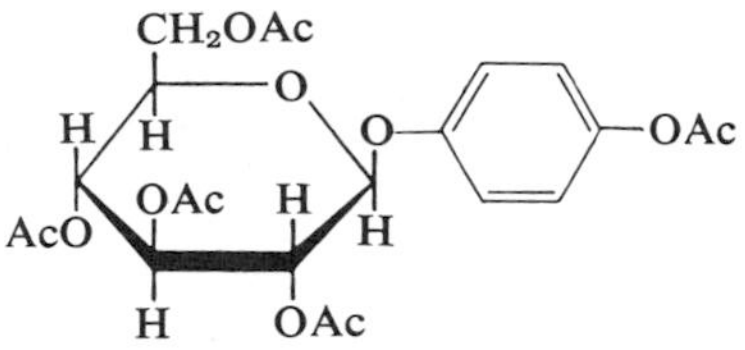

Arbutin pentaacetate

D. The Mass Spectra of Dithioacetals of Monosaccharides

A very important feature of the mass spectra of diethyl dithioacetals of monosaccharides is the presence of molecular-ion peaks, from which the molecular weight can be readily obtained. For example, the molecular-ion peak from D-arabinose diethyl dithioacetal (13) is of intensity 20% of the most intense peak in the spectrum.[42] No molecular-ion peak is found in the mass spectrum of the corresponding ethylene dithioacetal, on the other hand.[42]

The electron-impact-induced fragmentation paths of these diethyl dithioacetals can be correlated with structure without the complications associated with cyclic forms. Two characteristic fragment ions are the diethyl dithioacetal fragment, m/e 135, and a fragment found 61 mass units lower than the molecular ion, as illustrated in Eq. (6) for compound 13. The 135-fragment is generally the most intense peak in the spectrum, except when a 2-deoxy function is present.

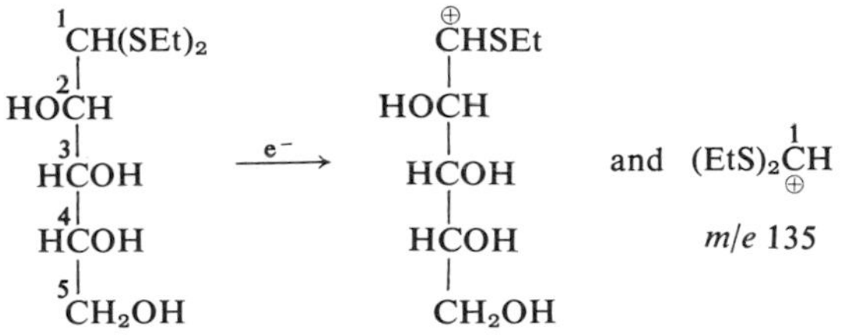

$$\begin{array}{ccc} \overset{1}{\text{CH(SEt)}_2} & \overset{\oplus}{\text{CHSEt}} & \\ \overset{2}{\text{HOCH}} & \text{HOCH} & \\ \overset{3}{\text{HCOH}} \xrightarrow{e^-} & \text{HCOH} & \text{and } (\text{EtS})_2\overset{1}{\underset{\oplus}{\text{CH}}} \\ \overset{4}{\text{HCOH}} & \text{HCOH} & m/e\ 135 \\ \overset{5}{\text{CH}_2\text{OH}} & \text{CH}_2\text{OH} & \end{array} \qquad (6)$$

13   M − 61, m/e 195

The (M − 61) ion, formed by loss of a C-1 substituent, fragments further by elimination of a molecule of water.

Three other important fragments, referred to as A, B, and C, are present in the mass spectra of diethyl dithioacetals[42]; the portions of the original molecule remaining in these fragments are shown in Fig. 5. A cyclic structure has been suggested for fragment B, since both substituents have been lost from C-1. Data are not available to specify the origin of the hydrogen atom incorporated into fragment C from elsewhere in the molecule.

Shifts in the locations of these fragments in O-methyl-substituted diethyl dithioacetals provide a means of determining the extent, position, and nature

References start on p. 1351.

$$
\begin{array}{c}
{}^1\text{C(SEt)}_2 \\
\parallel \\
{}^2\text{CH} \\
\mid \\
{}^3\underset{\oplus}{\text{CHOH}}
\end{array}
\qquad\qquad
[\text{M} - \text{EtS}\cdot - \text{EtSH}]^{\oplus}
\qquad\qquad
\left[H + \begin{array}{c} {}^1\text{CHSEt} \\ \parallel \\ {}^2\text{CHOH} \end{array} \right]^{\oplus}
$$

fragment A, m/e 177   fragment B, M − 123   fragment C, m/e 105

FIG. 5. Fragments A, B, and C in the mass spectra of diethyl dithioacetals of mono-saccharides.

of substitution.[42,43] For example, the shift of fragment A from m/e 177 (Fig. 5) to m/e 191 in the mass spectrum[43] of mycinose diethyl dithioacetal (**14**) is due to the methoxyl group on C-3. Fragment C at m/e 119 and a fragment at m/e 179 are characteristic of the 2-methoxyl group. The molecular-ion peak has a relative intensity of 25%, and fragment B is found at m/e 175. The principal ions found in the mass spectrum of compound **14** are summarized in Fig. 6, together with their relative intensities.

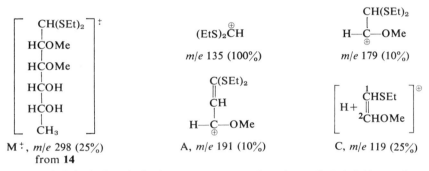

M$\overset{+}{\cdot}$, m/e 298 (25%)
from **14**
          A, m/e 191 (10%)   C, m/e 119 (25%)

FIG. 6. Principal peaks in the mass spectrum of mycinose diethyl dithioacetal.

N-Acetylated amino sugar diethyl dithioacetals (see Vol. IB, Chapter 16) have been investigated also, including derivatives of 2-amino-2-deoxy-, 2,6-diamino-2,6-dideoxy-, 3,6-diamino-3,6-dideoxy-, 3-amino-3,6-dideoxy-, and 6-amino-6-deoxyhexoses, and 3-amino-3-deoxy-, 4-amino-4,5-dideoxy-, 5-amino-5-deoxy-, and 4,5-diamino-4,5-dideoxypentoses.[44] Fragmentation is similar to that shown in Eq. (6) and Fig. 5, except for the m/e location of peaks containing the acetamido functions. These shifts in location of key peaks are characteristic of the position of the acetamido function and can be used for identification. For example, Eq. (7) shows the key fragments that identify the position of the acetamido functions in the mass spectrum of 3,6-diacetamido-3,6-dideoxy-D-altrose diethyl dithioacetal (**15**). In the mass spectrum of the isomeric 2,6-diacetamido-2,6-dideoxy-L-idose diethyl dithioacetal, fragment A is found at m/e 177 because the hydroxyl group is now at C-3; the C-6 fragment is present again at m/e 72.

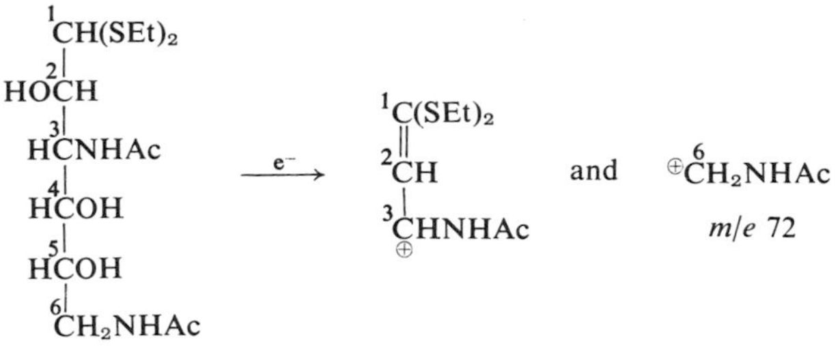

$$
\begin{array}{c}
\overset{1}{C}H(SEt)_2 \\
\overset{2}{HOCH} \\
\overset{3}{HCNHAc} \\
\overset{4}{HCOH} \\
\overset{5}{HCOH} \\
\overset{6}{CH_2NHAc}
\end{array}
\quad \xrightarrow{e^-} \quad
\begin{array}{c}
\overset{1}{C}(SEt)_2 \\
\overset{2}{CH} \\
\overset{3}{CHNHAc} \\
\oplus
\end{array}
\quad \text{and} \quad \overset{6}{\oplus}CH_2NHAc \qquad (7)
$$

15   A, m/e 218 m/e 72

The deoxy sugars (see Vol. IB, Chapter 17) are an equally important class of carbohydrates, long known as components of biological materials. Characteristic fragmentations are also observed in the mass spectra of their diethyl dithioacetals which permit the location of deoxy functions and allow definitive structure assignments to these compounds. Monodeoxyhexoses[42,43] and pentoses[43] and 2,3-, 2,6-, 3,6-, and 4,6-dideoxyhexoses[43] have been investigated.

The mass spectra of these deoxy derivatives are characterized by an apparent reluctance of ions to fragment by cleavage of bonds leading to primary radicals on the methylene group. For example, m/e 135, the diethyl dithioacetal fragment, is the most intense peak in the mass spectrum of 3,6-dideoxy-L-*lyxo*-hexose diethyl dithioacetal (**16**) (Eq. 8), whereas its relative intensity is only 8% in the mass spectrum of 2,6-dideoxy-D-*ribo*-hexose diethyl dithioacetal (**17**)

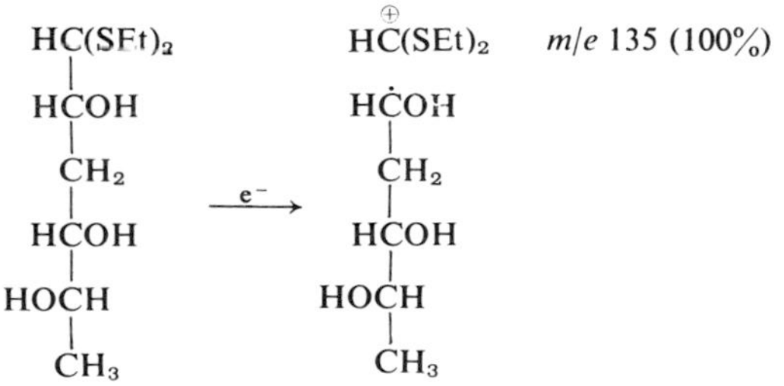

$$
\begin{array}{c}
HC(SFt)_2 \\
HCOH \\
CH_2 \\
HCOH \\
HOCH \\
CH_3
\end{array}
\quad \xrightarrow{e^-} \quad
\begin{array}{c}
\overset{\oplus}{H}C(SEt)_2 \\
HCOH \\
CH_2 \\
HCOH \\
HOCH \\
CH_3
\end{array}
\quad m/e\ 135\ (100\%) \qquad (8)
$$

16

$$
\begin{array}{c}
HC(SEt)_2 \\
CH_2 \\
HCOH \\
HCOH \\
HCOH \\
CH_3
\end{array}
\quad \xrightarrow{e^-} \quad
\begin{array}{c}
\overset{\oplus}{H}C(SEt)_2 \\
CH_2 \\
HCOH \\
HCOH \\
HCOH \\
CH_3
\end{array}
\quad m/e\ 135\ (8\%)
\quad
\begin{array}{c}
HC(SEt)_2 \\
CH_2 \\
CH_2 \\
HCOH \\
HCOH \\
CH_2OH
\end{array}
\quad \xrightarrow{e^-} \quad m/e\ 135\ (3\%) \quad (9)
$$

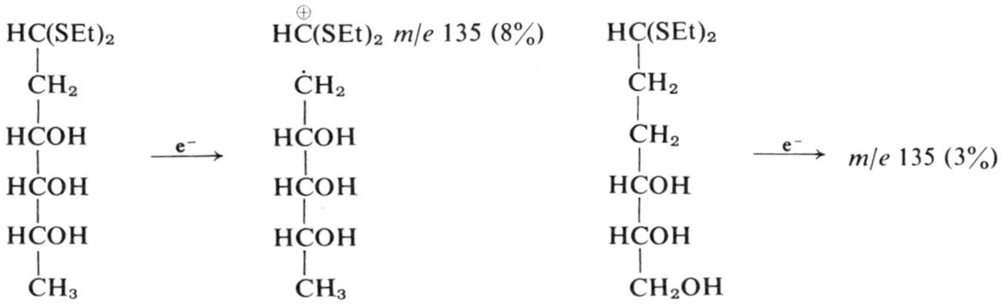

17   **18**

(Eq. 9) and 3% in the spectrum of 2,3-dideoxy-D-*erythro*-hexose diethyl dithioacetal (**18**).

Fragment C, which retains C-1 and C-2, is found at m/e 89 in the spectrum of compound **17** rather than at m/e 105, as shown in Fig. 5; it is not found in the mass spectra of compounds **16** and **18** because of the methylene group at C-3. Fragment A is found 16 mass units lower than shown in Fig. 5 in the mass spectrum of the 3-deoxy compound **16**, but does not appear in the spectra of 4-deoxy derivatives. The presence of a deoxy function at C-5 or C-6 does not affect the formation of fragments A, B, or C because of its remote position with respect to the bonds most susceptible to cleavage.

$$\left[H + \overset{\overset{\displaystyle 1}{\underset{\displaystyle \|}{CHSEt}}}{\underset{\displaystyle 2CH_2}{}} \right]^{\oplus} \qquad\qquad \overset{\underset{\displaystyle \|}{C(SEt)_2}}{\underset{\underset{\displaystyle \oplus}{\displaystyle CH_2}}{\underset{\displaystyle |}{CH}}}$$

C from compound **17**   A from compound **16**

m/e 89 m/e 161

Peracetylated dithioacetals also have been investigated.[45,46] Their fragmentation paths are similar to those of their unacetylated precursors, except for the complicating factor of the acetoxyl groups. The acetates can be introduced through conventional reservoir-type heated inlet systems, but the unacetylated analogues must be introduced directly into the ion source. The peak at m/e 135, the diethyl dithioacetal fragment, is absent from the mass spectrum of D-fructose diethyl dithioacetal pentaacetate because of the highly substituted acetal carbon atom.[45]

Crystalline diethyl dithioacetals of monosaccharides can be conveniently prepared on a small scale (see Vol. IA, Chapter 10). Mercaptolysis of glycosidic bonds with a thiol and a strong acid is frequently employed for isolation of carbohydrate residues from complex molecules. A complementary procedure to the structure determination of carbohydrates from the mass spectra of their diethyl dithioacetals is the deduction of stereochemistry and partial or total configuration through the MacDonald–Fischer degradation of the corresponding sulfones.[47] Thus, by a combination of mass spectrometry and a two-step chemical reaction, information regarding both the structure and the configuration of a sugar can be obtained from a small amount of a common intermediate.

E. THE MASS SPECTRA OF ISOPROPYLIDENE ACETALS OF CARBOHYDRATES

The mass spectra of per-*O*-acetyl, per-*O*-methyl, and dithioacetal derivatives of carbohydrates are quite insensitive with respect to stereochemistry. The differences in spatial relationships between the substituents on the pyranose or furanose rings or on the acyclic chains do not cause appreciable differences in fragmentation patterns upon electron impact. Differences that are observed in the mass spectra are variations in relative intensity rather than differences in fragmentation paths, and are too small to permit interpretation in terms of stereochemistry, except for direct comparison with the mass spectrum of an authentic sample run under identical conditions (but see p. 1349).

Since the formation of isopropylidene acetals (Vol. IA, Chapter 11) depends on the presence of *cis*-1,2- or *cis*-1,3-diol groups, epimers may give derivatives that are structural isomers, to which mass spectrometry is very sensitive, rather than stereoisomers, to which it is not. For example, D-glucose forms 1,2:5,6-di-O-isopropylidene-D-glucofuranose (**19**), whereas D-galactose forms 1,2:3,4-di-O-isopropylidene-D-galactopyranose (**20**). Cleavage of the C-4–C-5 bond

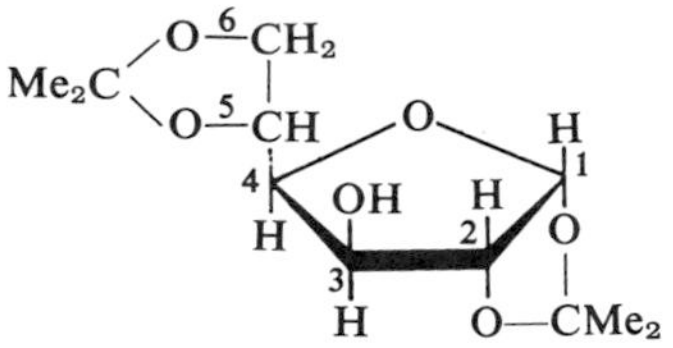

19 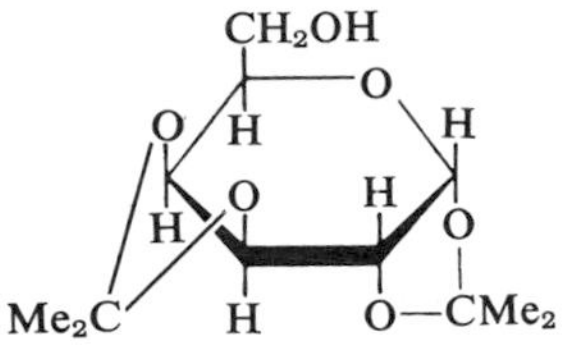   **20**

in compound **19** is especially favorable upon electron bombardment (Eq. 10) and is mainly responsible for the major differences [48] between the mass spectra of compounds **19** and **20**.

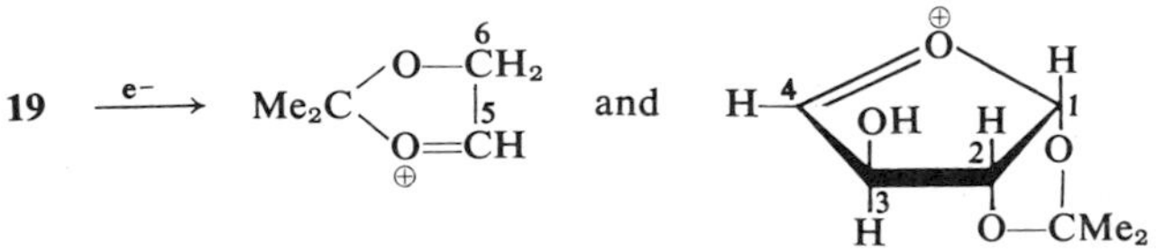

$$m/e\ 101 \qquad\qquad m/e\ 159$$

D-Fructose forms two isomeric di-O-isopropylidene-D-fructopyranoses, 2,3:4,5 (**21**) and 1,2:4,5 (**22**), which also fragment in characteristic manners (Eqs. 11 and 12), distinguishing them from one another and from other hexose derivatives. [48]

D-Ribose forms a mono-O-isopropylidene-D-ribofuranose, whereas L-

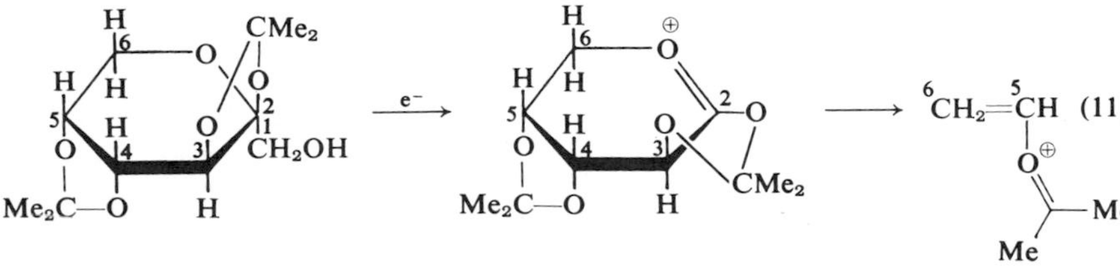

21 $m/e\ 229$ $m/e\ 85$

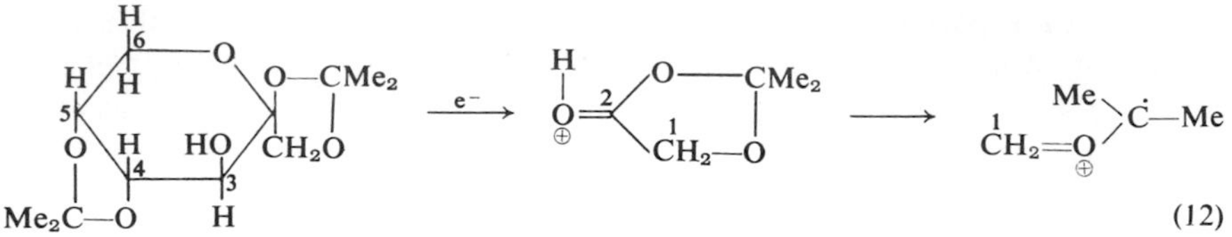

(12)

22 m/e 117 m/e 72

arabinose and D-xylose form di-O-isopropylidene L-arabinopyranose and D-xylofuranose, respectively. These three derivatives can be readily differentiated from one another on the basis of their mass spectra.[48]

No molecular-ion peaks are observed in the mass spectra of these isopropylidene acetals, but molecular weights can be determined readily from the relatively intense peak 15 mass units lower than the molecular ion, which results from loss of a methyl radical from a 1,3-dioxolane ring (Eq. 13). The

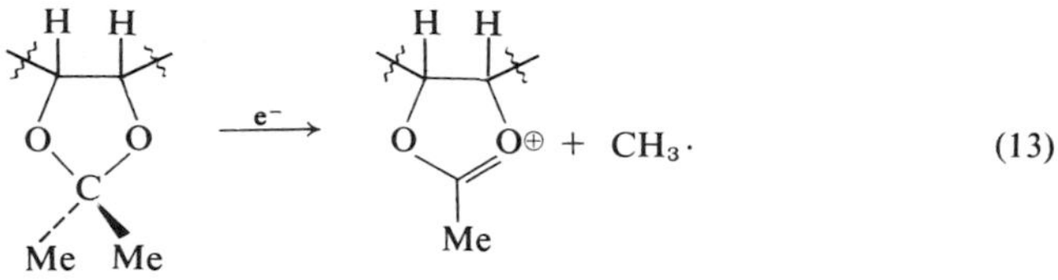

(13)

(M − 15)

fragmentation paths reported were corroborated by accurate mass measurement and deuterium labeling with derivatives prepared[48] with acetone-d_6. For example, the peak at m/e 72 in Eq. (12) was found to have a mass of 72.0585, only 0.0010 mass unit from the calculated mass of C_4H_8O, and was shown to contain two methyl groups because it appeared at m/e 78 in the mass spectrum of the analogue prepared with acetone-d_6.

The structure of a previously unreported di-O-isopropylidene-D-galactose was elucidated from its mass spectrum.[48] When acetone, cupric sulfate, and D-galactose are heated for 18 hours at 100°, a mixture of isopropylidene acetals forms which was shown by gas chromatography to be 80% of compound **20** and 20% of another product. The second product was shown to be a 1,2:5,6-di-O-isopropylidenehexofuranose from its mass spectrum. Thus, the second di-O-isopropylidene-D-galactose has the structure **23**.

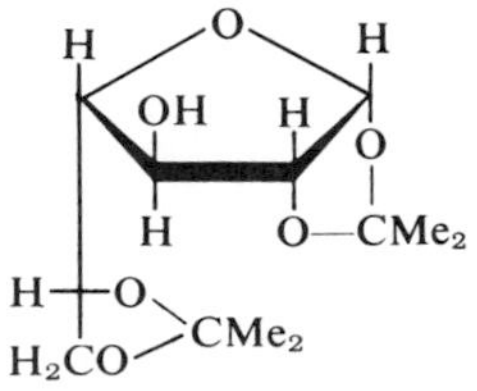

23

The mass spectrum of methyl 5-acetamido-5,6-dideoxy-2,3-O-isopropyli-dene-α-L-talofuranoside (**24**) was used to support a structure assignment.[49] Characteristic peaks are found at m/e 228 from loss of the C-1 methoxyl group, and at m/e 244 from loss of a methyl group (Eq. 14). The two most intense peaks in the spectrum are m/e 86 and m/e 44 from cleavage of the C-4–C-5 bond (Eq. 15).

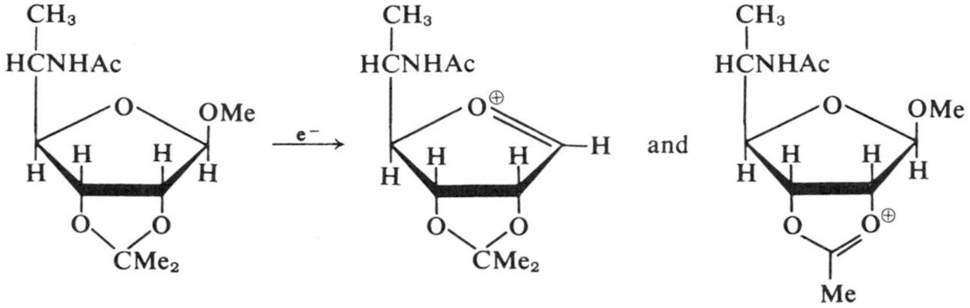

$$\text{(14)}$$

24 m/e 228 m/e 244

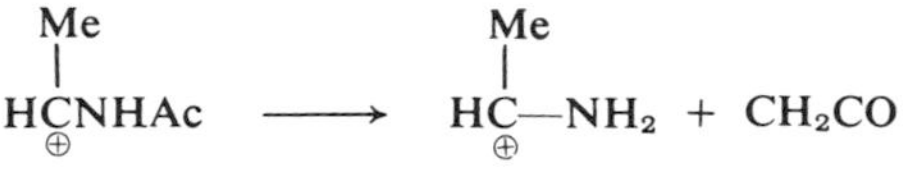

$$\text{(15)}$$

m/e 86 m/e 44

F. THE MASS SPECTRA OF NUCLEOSIDES

Even though nucleosides (Vol. IIA, Chapter 29) have very low volatilities, it is possible to sublime them directly into the electron beam and thereby obtain their mass spectra. The mass spectra of a number of nucleosides of known structures have been studied[50] and found to yield information permitting identification of the base. Prominent peaks one and two mass units higher than the mass of the base moiety (B) are formed from rearrangement processes, with the transferred hydrogen atoms originating largely from the hydroxyl groups of the sugar moiety (S).

A peak is also found for the sugar moiety, although its intensity is much lower if the base is a purine than if it is a pyrimidine.[50] The mass spectra of nucleosides epimeric in the sugar portion show variations in the relative intensities of some peaks, but not in fragmentation paths. Molecular-ion peaks are observed, in contrast to the spectra of free monosaccharides and methyl glycosides.

In the mass spectrum of uridine (**25**), the peak at m/e 113 from the base portion and two rearranged hydrogen atoms (B + 2H) is the most intense (100%), and the peak at m/e 133 from the sugar portion (S) is of 75% relative intensity.[50] On the other hand, adenosine (**26**) gives a (B + H) peak at m/e 135 which is the most intense and an (S) peak at m/e 133 of about 2.5% relative intensity.

References start on p. 1351.

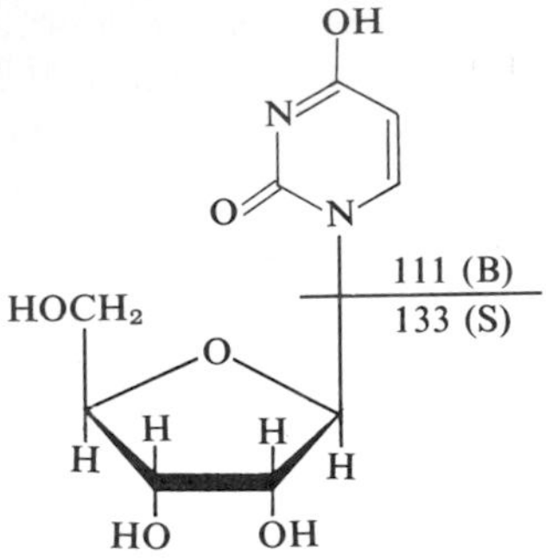

OH

111 (B)
HOCH₂   133 (S)

H H
H   H
HO   OH

25

NH₂

5′   134 (B)
HOCH₂   133 (S)

4′   H H   1′
H 3′   H
2′
HO   OH

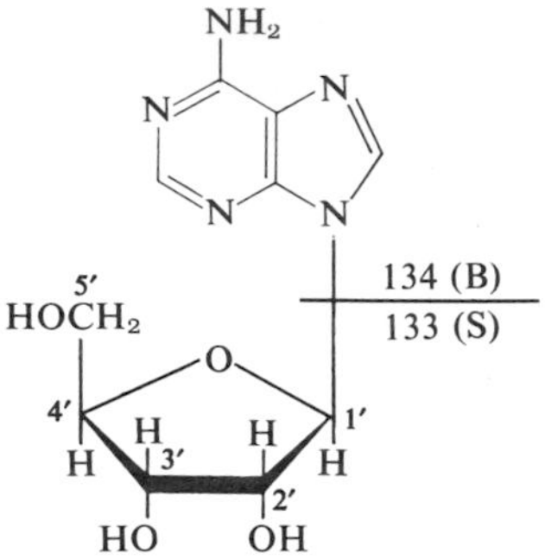

26

Another important fragmentation of both uridine and adenosine is the formation of a fragment 89 mass units lower than the molecular ion. This fragment does not form in the fragmentation of 5′-deoxynucleosides and is found 2 mass units higher in the mass spectra of the nucleosides that have been equilibrated with deuterium oxide. The structure proposed [50] for this fragment contains the base moiety, C-1′, and C-2′, together with a hydrogen atom transferred from the C-5′ hydroxyl group to C-1′ (Eq. 16).

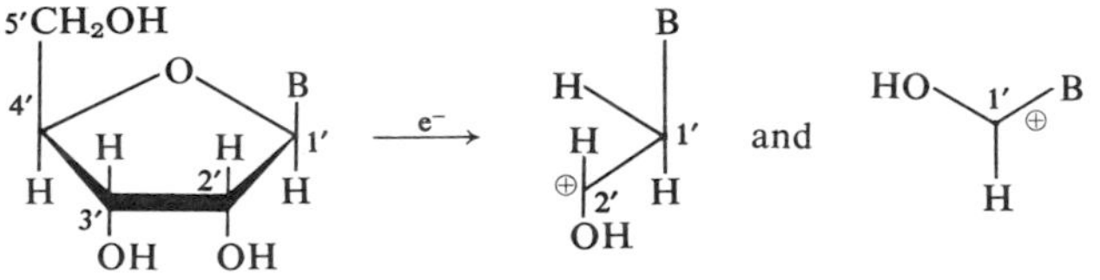

5′CH₂OH

4′   B
H H 1′
H 2′ H
3′
OH OH

$\xrightarrow{e^-}$

B
H
H 1′
2′ H
OH

M − 89

and

HO 1′ B
H

B+30

(16)

A peak is also found 30 mass units higher than the base moiety (B + 30 in Eq. 16).[50] It is of very low intensity in the mass spectrum of 2′-deoxyadenosine, indicating that a 2′-hydroxyl group is essential for it to be a major peak. This conclusion is substantiated by the shift of the peak to one mass unit higher in the mass spectra of the nucleosides equilibrated with deuterium oxide.

The fragmentations shown in Eq. (16) were especially useful for providing further evidence for the 3′-deoxyadenosine structure (3′-deoxy analogue of **26**) of the antibiotic cordycepin.[51] Furthermore, the mass spectrum of cordycepin is identical to the mass spectrum of synthetic 3′-deoxyadenosine. The molecular-ion peak at m/e 251 and the (B + H) peak at m/e 135 correspond to a deoxypentoside of adenine. The formation of a (B + 30) peak agrees with the presence of a 2′-hydroxyl group, and the shift of the (M − 89) peak in Eq. (16) to M-73 in the mass spectrum of cordycepin suggests that a 3′-deoxy function and a C-5′ hydroxyl group are contained in the sugar portion.

Mass spectrometry aided in the identification of a nucleoside from the soluble ribonucleic acid of yeast.[52,53] An empirical formula of $C_{15}H_{21}H_5O_4$ was obtained from a high-resolution mass spectrum.[52] The abundance of ions containing five nitrogen and no oxygen atoms pointed toward a pentosyladenine substituted with a C_5 fragment. The appearance in the high resolution mass spectrum of ions of the type C_nH_mN up to $C_5H_{10}N$ suggested a five-carbon alkenyl or a cyclopentyl substituent on the amino group at the 6-position of adenine. On the basis of fragmentation the cyclopentyl possibility was excluded. Nuclear magnetic resonance spectroscopy permitted location of the double bond in the alkenyl substituent. The 6-(3-methyl-2-butenylamino)-9-β-D-ribofuranosylpurine structure (**27**) arrived at by high-resolution mass spectrometry was subsequently shown to be correct by synthesis.[53]

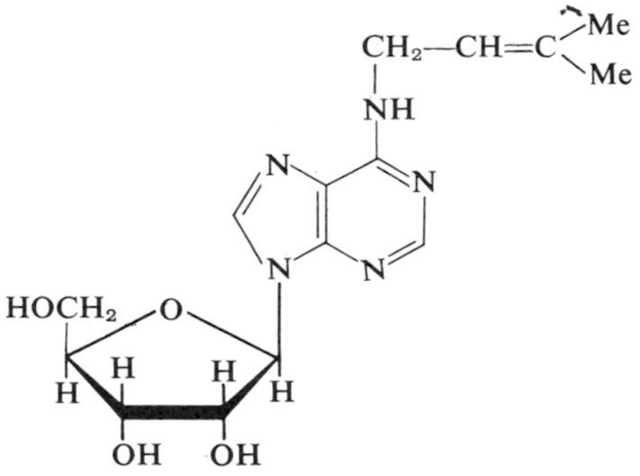

27

The mass spectra of puromycin (**28**) and puromycin nucleoside (**29**) have been reported, together with spectra of some of their derivatives.[54] Fragmentation of the nucleoside is directed by the purine moiety.

The mass spectrum of puromycin (**28**) has peaks arising from fragmentation

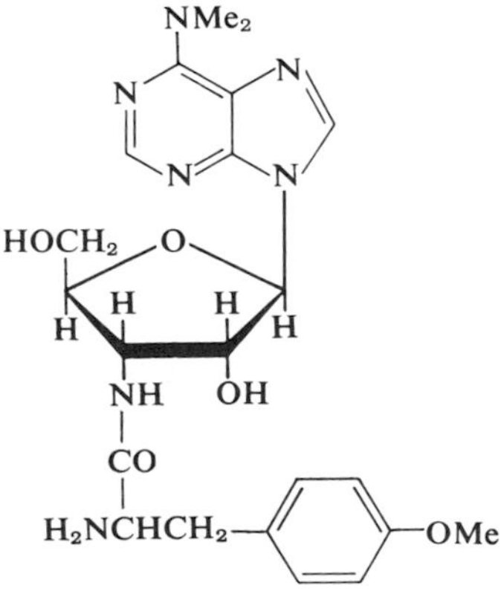

28

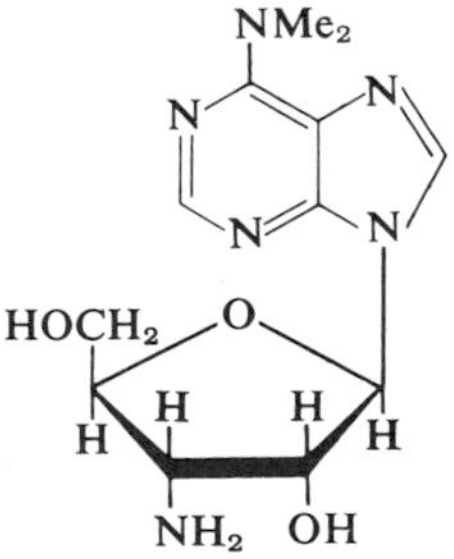

29

References start on p. 1351.

of bonds on either side of the amide carbonyl group, and also from fragmentation of the benzylic bond. Peaks characteristic of the puromycin nucleoside are also present. The superiority of the (volatile) per(trimethylsilyl)[54a] and permethyl[54b] ethers for analysis of nucleosides has been emphasized.

G. THE MASS SPECTRA OF STEROID GLYCOSIDES

The mass spectra of three cardiac glycosides (see Vol. IIA, Chapter 32, for detailed discussion of these glycosides) have been studied as models in order to explore the potentiality of mass spectrometry in structural studies in that area. Somalin[55,56] (30), cymarin,[55] and helveticoside[55] are composed of a steroid moiety linked to a carbohydrate by a glycosidic bond. In the positive ion mass spectrum of somalin a very minute molecular-ion peak is observed; a major

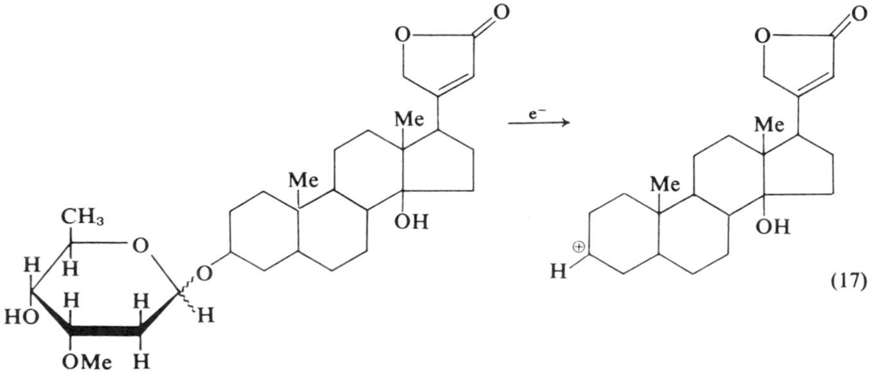

$$30 \qquad\qquad m/e\ 357 \qquad (17)$$

fragmentation is the cleavage of the glycosidic bond as shown[55,56] in Eq. (17). In the region of m/e 338–190, the mass spectrum of somalin is nearly identical with that[55] of digitoxigenin, the steroid aglycon of somalin.

A negative-ion mass spectrum is obtained by accelerating and separating the negative ions formed by electron capture, rather than the positive ions formed by electron expulsion as in a positive-ion mass spectrum.[56a] Negative-ion mass spectra show a low yield of negative ions and a lack of extensive fragmentation. The most characteristic feature of the negative-ion mass spectra of alcohols is the absence of a molecular-ion peak and the presence of a strong anion one mass unit lower than the molecular weight, caused by loss of an oxygen-bonded hydrogen atom[57] (Eq. 18). Whereas the positive-ion mass spectrum of somalin

$$R\!-\!O\!-\!H + e^- \longrightarrow R\!-\!O^- + H\cdot \qquad (18)$$

contains a minute molecular-ion peak, the negative-ion mass spectrum has an intense peak one mass unit lower, and an intense peak at m/e 373 which was

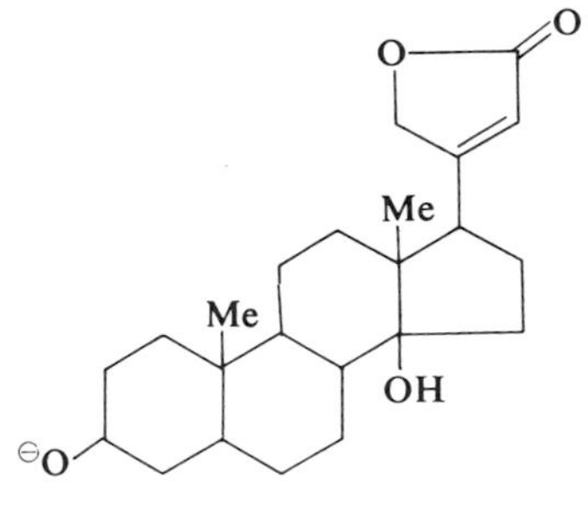

m/e 373

31

assigned to alkoxyl structure[55] **31**. These results illustrate how complementary information is available from both types of mass spectra.

Interpretation of the positive-ion mass spectrum of deacyl-kondurango-genin-A-monoside (**32**) is similar to that of somalin.[58] Thermal degradation appears to occur to a certain extent during volatilization.

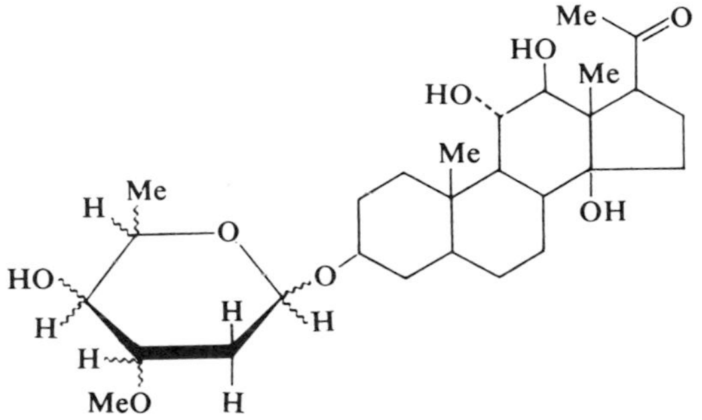

32

H. THE MASS SPECTRA OF CARBOHYDRATES HAVING NITROGEN OR SULFUR IN THE RING

The mass-spectral fragmentation[19] of methyl 5-acetamido-5-deoxy-2,3,4-tri-*O*-methyl-β-D-xylopyranoside (**33**) is similar to the fragmentation of methyl 2,3,4-tri-*O*-methyl-β-D-arabinopyranoside (Fig. 1). The peaks at *m/e* 75, 88,

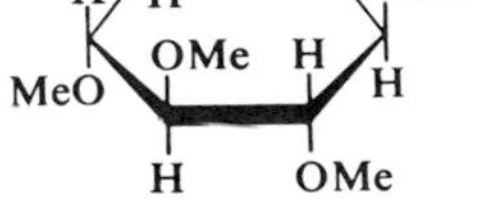

33

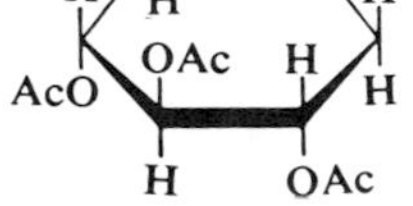

34

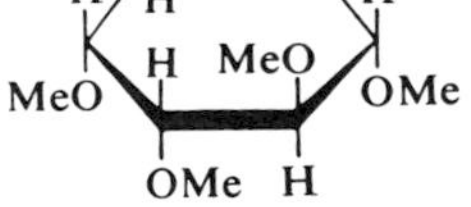

35

References start on p. 1351.

and 101 are prominent, although no peak is found at m/e 176 from loss of C-5 and the ring nitrogen and its substituent. No molecular-ion peak is observed, but a series of peaks beginning with the elimination of the C-1 methoxyl group, followed by successive losses of methanol, occurs in the high-mass range. Compound **33** undergoes thermal degradation to 3-methoxypyridine at temperatures above 200°, so that caution must be exercised in choosing inlet and ion–source conditions.

The mass spectrum of N-acetyl-2,3,4-tri-O-acetyl-1,5-dideoxy-1,5-iminoxylitol (**34**) has a detectable molecular-ion peak of about 1% relative intensity.[59] The major fragmentation leads to an ion at m/e 139 by successive losses of two molecules of acetic acid and one of ketene.

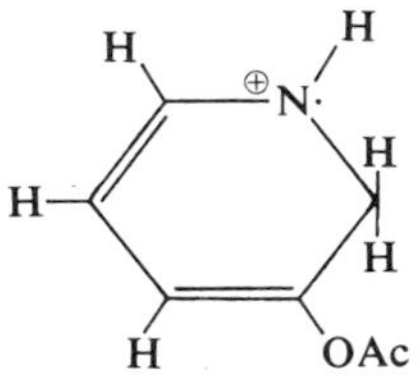

m/e 139

The presence of sulfur in methyl 2,3,4-tri-O-methyl-5-thio-α-D-xylopyranoside (**35**) stabilizes the charge in the molecular ion so that it is of about 12% relative intensity.[19] Fragmentation is similar to fragmentation of compound **33**. Peaks are present for the successive loss of three molecules of methanol from the molecular ion.

I. ALTERNATIVE METHODS OF EFFECTING IONIZATION

The electron-impact technique, as described in the preceding sections, is the standard method of producing ions for mass-spectrometric identification; it generally renders information from extensive fragmentation of the carbohydrate molecule. Decreasing the energy of electrons used in direct bombardment does not cause a significant increase in the population of large-mass ions resultant from a given sample, but other techniques, particularly chemical ionization and field desorption, have been developed that favor minimal fragmentation and often afford prominent, intact molecular ions. In general, information obtained by these newer methods is complementary to, rather than a replacement for, data obtained from electron impact.

1. Chemical Ionization

For the technique of chemical ionization (c.i.), the sample is volatilized into an ion source containing a so-called "reagent gas." Electron bombardment of the source contents produces mainly ions derived from the reagent gas (most commonly methane, isobutane, or ammonia), which undergo a series of ion–molecule reactions to produce lower energy, proton-donating ions (principally CH_5^+, $C_4H_9^+$, or NH_4^+, respectively) that transfer a charged particle (commonly H^+) to or from the sample molecule.[60] As carbohydrates invariably behave as Lewis bases, capture ions are formed initially by c.i. and may account for a sizeable proportion of the total ionization. The c.i. (isobutane) mass spectrum of 1,6-anhydro-β-D-galactopyranose features a proton-capture (MH$^+$, m/e 163) ion as the base peak, and the corresponding c.i. (ammonia) mass spectrum consists essentially of a single peak at m/e 180, corresponding to an intact molecule plus an ammonium ion $(M + NH_4)^+$. In contrast, however, D-glucose diethyl dithioacetal affords only a weak (10% relative abundance) $(M + NH_4)^+$ ion from c.i. (ammonia), and no detectable (MH)$^+$ ion from c.i. (isobutane).[61]

Fragmentation subsequent to chemical ionization is much simpler than under electron impact and generally consists of elimination reactions of substituents as small, neutral fragments. Chemical ionization (isobutane) of D-glucose diethyl dithioacetal pentaacetate[61] initially produces an unstable (MH)$^+$ ion that loses the elements of ethanethiol; the ion thereby produced subsequently decomposes by losses of acetic acid, acetic anhydride, and ketene in various combinations, from which the presence of five acetate groups may be deduced. No carbon–carbon bonds are broken. Fragment intensities display recognizable differences as a function of stereochemistry, in contrast to e.i.–m.s.; a detailed analysis of this aspect, as it applies to the ammonia-mediated c.i.–m.s. of O-trimethylsilyl derivatives of aldoses, has been published.[61a]

The thioglycoside antibiotics lincomycin and clindamycin[62] also afford prominent MH$^+$ ions from c.i. (isobutane), whereas the glycosyl cleavage reaction of β-D-fructofuranose pentaacetate is so strongly favored that, even under c.i. (ammonia) conditions, the largest-mass ion is the base peak at m/e 331; this ion corresponds to loss of the anomeric acetoxyl group plus the cationic species transferred.[61]

Partial surveys are available that give consideration to the effect of functional groups[61] and of reagent gases[63] upon decomposition, and Foltz[64] has reviewed applications of c.i.–m.s. to carbohydrates and related molecules.

2. Field Ionization

As an alternative to bombardment with fast electrons, ionization may also be effected by extraction of an electron into an intense ($\sim 10^8$ V/cm) electric field near the surface of a "field emitter," a metal surface having microscopically narrow edges. Ions thus formed suffer only a minimal increase of internal energy beyond that required to effect volatilization, so that molecular ions or intact capture ions resulting from intermolecular proton transfer are a common feature of field-ionization mass spectra of thermally stable molecules. Fragmentation processes, which are usually of minor significance, are mainly restricted to such strongly favored decomposition modes as loss of water or glycosyl cleavage.

The base and second-largest peaks of the f.i. mass spectrum of D-ribose[65] are the $(M + 1)^+$ and M^+ ions, which account for over 30% of the net ions formed. The molecular ion of D-ribose is essentially undetectable under conditions of electron-impact ionization at 70 eV, and the generalization that lowering the electron energy will not increase the intensity of M^+ is held, so that some value attaches to this observation. However, the fundamental problem of decomposition prior to ionization remains, and the efficiency of ionization (number of ions per gram of sample) is poor, so that applications of f.i.–m.s. in carbohydrate chemistry will probably be less important than those of the related technique of field-desorption mass spectrometry. The relationship of peak width to kinetics of decomposition may find use in theoretical studies of modes of decomposition. The principle is discussed in detail in a review of the technique of f.i.–m.s. by Beckey.[66]

3. *Field Desorption*

This technique,[67] clearly the most important advance in mass spectrometry of carbohydrates since the introduction of c.i.–m.s., requires the preparation of a special field anode, to which is applied a film of sample. Ionization occurs on the surface of the coated anode, in an electric field gradient, and precedes the separation of ions into the vapor phase, so that nonvolatile molecules may thus be studied. Ionization efficiency approaches that of electron impact, and for each sample there exists an optimal temperature (T*) of the field anode, at which the intensity of the molecular ion is both a relative and an absolute maximum. The molecular ion is not uncommonly the only ion produced; thermally induced fragmentation increases as the emitter temperature is raised, although the normal procedure is to record the spectrum at T*.

Because of the difficulty of measuring mass spectra of underivatized carbohydrates by conventional ionization methods, Beckey et al. selected, among their earliest examples, D-glucose[68] and cellobiose,[69] both of which afforded prominent ions that constituted the intact molecule. More-complex substances, including nucleosides,[70] and aryl and steroidal glycosides,[71] were subsequently shown to afford M^+ or $M + 1$ ions in abundance, together with ions characteristic both of the sugar portion and the aglycon. The presence of alkali-metal cations (Na^+ or K^+) results in the erratic formation[70,72] of $(M + Na)^+$ or $(M + K)^+$ ions; this process randomizes quantitative relationships within the spectrum. 16-O-(α-D-Glucopyranosyluronic acid)estriol isolated from urine has been determined[72] by this technique, indicating the potential for direct clinical applications.

Among the various ionization techniques, field desorption introduces the smallest amount of thermal stress; in an attempt to distinguish diastereo-isomers, Beckey and co-workers[73] examined 6-bromo-2-naphthyl α- and β-D-glucopyranosides and found that the intensity ratios of ions corresponding to the sugar portion and to the aglycon differed by a factor of only 2 for these presumably favorable examples. A detailed description of the technique and several applications is contained in a review by Beckey and Schulten.[67]

REFERENCES

1. For reviews of mass spectrometry and its applications to organic chemistry see (a) J. H. Beynon, "Mass Spectrometry and Its Applications to Organic Chemistry," Elsevier, Amsterdam, 1960; (b) K. Biemann, "Mass Spectrometry," McGraw–Hill, New York, 1962; F. W. McLafferty, Ed., "Mass Spectrometry of Organic Ions," Academic Press, New York, 1963; and R. W. Kiser, "Introduction to Mass Spectrometry and Its Applications," Prentice-Hall, Englewood Cliffs, New Jersey, 1965.

2. See H. Budzikiewicz, C. Djerassi, and D. H. Williams, (a) "Interpretation of Mass Spectra of Organic Compounds," Holden–Day, San Francisco, California, 1964; (b) "Structure Elucidation of Natural Products by Mass Spectrometry," Vol. I: "Alkaloids," Holden–Day, San Francisco, California, 1964; Vol. II: "Steroids, Sugars, and Miscellaneous Classes," Holden–Day, San Francisco, California, 1964; see also "Biochemical Applications of Mass Spectrometry," G. R. Waller, Ed., Wiley, New York, 1972.

3. For a review see F. W. McLafferty, *Science*, **151**, 641 (1966).

4. For reviews see (a) reference 2b, Vol. II, chap. 27; (b) K. Heyns, H. F. Grützmacher, H. Scharmann, and D. Müller, *Fortschr. Chem. Forsch.*, **5**, 448 (1966); (c) N. K. Kochetkov and O. S. Chizhov, *Advan. Carbohyd. Chem.*, **21**, 39 (1966); *Methods Carbohyd. Chem.*, **6**, 540 (1972); (d) J. Lönngren and S. Svensson, *Advan. Carbohyd. Chem. Biochem.*, **29**, 41 (1974).

5. See F. A. J. M. Leemans and J. A. McCloskey, *J. Amer. Oil Chem. Soc.*, **44**, 11 (1967).

6. J. W. Amy, E. M. Chait, W. E. Baitinger, and F. W. McLafferty, *Anal. Chem.*, **37**, 1265 (1965).

7. K. Heyns and H. F. Grützmacher, *Angew. Chem. Int. Ed.*, **1**, 400 (1962).

8. Ref. 1b, p. 350.

9. R. I. Reed, W. K. Reid, and J. M. Wilson, in "Advances in Mass Spectrometry," R. M. Elliott, Ed., Vol. II, Macmillan, New York, 1963, p. 420.

10. K. Biemann, H. K. Schnoes, and J. A. McCloskey, *Chem. Ind.* (London), 448 (1963).

11. C. L. Stevens, R. P. Glinski, K. G. Taylor, P. Blumbergs, and F. Sirokman, *J. Amer. Chem. Soc.*, **88**, 2073 (1966).

12. M. P. Kunstman, L. A. Mitscher, and N. Bohonos, *Tetrahedron Lett.*, 839 (1966).

13. P. A. Finan, R. I. Reed, and W. Snedden, *Chem. Ind.* (London), 1172 (1958).

14. K. Heyns, G. Kissling, and D. Müller, *Carbohyd. Res.*, **4**, 452 (1967); V. Kováčik and P. Kováč, *ibid.*, **24**, 23 (1972).

14a. Ref. 9, p. 421.

15. See, for example, O. Larm and B. Lindberg, *Advan. Carbohyd. Chem. Biochem.*, **33**, 295 (1976) and references cited, for applications in methylation analysis of pneumococcal polysaccharides.

16. N. K. Kochetkov and O. S. Chizhov, *Tetrahedron*, **21**, 2029 (1965); K. Heyns, K. R. Sperling, and H. F. Grützmacher, *Carbohyd. Res.*, **9**, 79 (1969).

17. K. Heyns and D. Müller, *Tetrahedron*, **21**, 55 (1965).

18. Ref. 4b, p. 465; compare E. G. de Jong, W. Heerma, B. Dujardin, J. Haverkamp, and J. Vliegenthart, *Carbohyd. Res.*, **60**, 229 (1978).

19. K. Heyns and D. Müller, *Tetrahedron*, **21**, 3151 (1965); N. K. Kochetkov, O. S. Chizhov, and B. M. Zolotarev, *Carbohyd. Res.*, **2**, 89 (1966).

20. N. K. Kochetkov, O. S. Chizhov, and B. M. Zolotarev, *Dokl. Akad. Nauk SSSR*, **165**, 569 (1965).

21. K. Heyns and H. Scharmann, *Carbohyd. Res.*, **1**, 371 (1966).

22. Ref. 4b, p. 473.

23. H. Björndal, B. Lindberg, and S. Svensson, *Carbohyd. Res.*, **5**, 433 (1967); compare L. S. Golovkina, O. S. Chizhov, and N. S. Vulf'son, *Izv. Akad. Nauk SSSR Ser., Khim.*, **11**, 1915 (1966).

23a. Peracetylated, partially methylated aldononitrile acetates are also useful in this regard; see, for example, F. R. Seymour, E. C. M. Chen, and S. H. Bishop, *Carbohyd. Res.*, **68**, 113 (1979) for applications in linkage analysis of dextrans and for leading references.

24. K. Heyns and D. Müller, *Tetrahedron Lett.*, 617 (1966).

24a. V. Kováčik and P. Kováč, *Carbohyd. Res.*, **38**, 25 (1974); *Org. Mass Spectrom.*, **9**, 172 (1974); see also J. F. Kennedy and S. M. Robertson, *Carbohyd. Res.*, **67**, 1 (1978).

25. O. S. Chizhov, L. A. Poljankova, and N. K. Kochetkov, *Dokl. Acad. Nauk SSSR*, **158**, 685 (1964); cf. J. Karliner, *Tetrahedron Lett.*, 3545 (1968).

26. D. C. DeJongh, T. Radford, J. D. Hribar, S. Hanessian, M. Bieber, G. Dawson, and C. C. Sweeley, *J. Amer. Chem. Soc.*, **91**, 1728 (1969).

27. D. C. DeJongh, J. D. Hribar, S. Hanessian, and P. W. K. Woo, *J. Amer. Chem. Soc.*, **89**, 3364 (1967).

28. K. Samuelsson and B. Samuelsson, *Biochem. Biophys. Res. Commun.*, **37**, 15 (1969).

29. C. C. Sweeley and G. Dawson, *Biochem. Biophys. Res. Commun.*, **37**, 6 (1969).

30. O. S. Chizhov, N. V. Molodtsov, and N. K. Kochetkov, *Carbohyd. Res.*, **4**, 273 (1967).

31. N. K. Kochetkov, N. V. Molodtsov, and O. S. Chizhov, *Tetrahedron*, **24**, 5587 (1968).

32. W. W. Binkley, R. C. Dougherty, D. Horton, and J. D. Wander, *Carbohyd. Res.*, **17**, 127 (1971).

33. K. Biemann, D. C. DeJongh, and H. K. Schnoes, *J. Amer. Chem. Soc.*, **85**, 1763 (1963).

34. D. C. DeJongh, *J. Org. Chem.*, **30**, 453 (1965).

35. A. Rosenthal, *Carbohyd. Res.*, **8**, 61 (1968).

36. K. Heyns and H. Scharmann, *Ber.*, **99**, 3461 (1966).

36a. Ref. 2b, Vol. II, p. 204.

37. R. C. Dougherty, D. Horton, K. D. Philips, and J. D. Wander, *Org. Mass Spectrom.*, **7**, 805 (1973).

38. T. D. Inch and H. G. Fletcher, Jr., *J. Org. Chem.*, **31**, 1821 (1966).

39. M. Venugopalan and C. B. Anderson, *Chem. Ind.* (London), 370 (1964).

40. D. C. DeJongh and K. Biemann, *J. Amer. Chem. Soc.*, **85**, 2289 (1963).

41. E. Haslem, *Carbohyd. Res.*, **5**, 161 (1967); compare K. Heyns and D. Müller, *Tetrahedron Lett.*, 6061 (1966).

42. D. C. DeJongh, *J. Org. Chem.*, **30**, 1563 (1965).

43. D. C. DeJongh and S. Hanessian, *J. Amer. Chem. Soc.*, **88**, 3114 (1966).

44. D. C. DeJongh and S. Hanessian, *J. Amer. Chem. Soc.*, **87**, 3744 (1965).

45. D. C. DeJongh, *J. Amer. Chem. Soc.*, **86**, 3149 (1964).

46. D. C. DeJongh, *J. Amer. Chem. Soc.*, **86**, 4027 (1964).

47. D. L. MacDonald and H. O. L. Fischer, *J. Amer. Chem. Soc.*, **74**, 2087 (1952); compare J. D. Wander and D. Horton, *Advan. Carbohyd. Chem. Biochem.*, **31**, 16 (1976).

48. D. C. DeJongh and K. Biemann, *J. Amer. Chem. Soc.*, **86**, 67 (1964).

49. S. Hanessian, *Chem. Commun.*, 796 (1966).

50. K. Biemann and J. A. McCloskey, *J. Amer. Chem. Soc.*, **84**, 2005 (1962).

51. S. Hanessian, D. C. DeJongh, and J. A. McCloskey, *Biochim. Biophys. Acta*, **117**, 480 (1966).

52. K. Biemann, S. Tsunakawa, J. Sonnenbichler, H. Feldmann, D. Dütting, and H. G. Zachau, *Angew. Chem.*, **78**, 600 (1966).

53. R. H. Hall, M. J. Robins, L. Stasiuk, and R. Thedford, *J. Amer. Chem. Soc.*, **88**, 2614 (1966).

54. S. H. Eggers, S. I. Biedron, and A. O. Hawtrey, *Tetrahedron Lett.*, 3271 (1966).

54a. J. A. McCloskey, A. M. Lawson, K. Tsuboyama, I. M. Kreuger, and R. L. Stillwell, *J. Amer. Chem. Soc.*, **90**, 4182 (1968).

54b. D. L. von Minden and J. A. McCloskey, *J. Amer. Chem. Soc.*, **95**, 7480 (1973).

55. M. v. Ardenne, R. Tümmler, E. Weiss, and T. Reichstein, *Helv. Chim. Acta*, **47**, 1032 (1964).

56. G. Spiteller, *Z. Anal. Chem.*, **197**, 1 (1963).

56a. D. F. Hunt, G. C. Stafford, F. W. Crow, and J. W. Russell, *Anal. Chem.*, **48**, 2098 (1976).

57. C. E. Melton and P. S. Rudolph, *J. Chem. Phys.*, **31**, 1485 (1959).

58. R. Tschesche, P. Welzel, and H.-W. Fehlhaber, *Tetrahedron*, **21**, 1797 (1965).

59. H. Paulsen, *Ann.*, **683**, 187 (1965).

60. F. H. Field, *Accounts Chem. Res.*, **1**, 42 (1968).

61. D. Horton, J. D. Wander, and R. L. Foltz, *Carbohyd. Res.*, **36**, 75 (1974).

61a. T. Murata and S. Takahashi, *Carbohyd. Res.*, **62**, 1 (1978).

62. D. Horton, J. D. Wander, and R. L. Foltz, *Anal. Biochem.*, **59**, 452 (1974).

63. M. S. Wilson and J. A. McCloskey, *J. Amer. Chem. Soc.*, **97**, 3436 (1975).

64. R. L. Foltz, *Chem. Tech.*, **5**, 39 (1975).

65. H. D. Beckey, H. Knöppel, H. G. Metzinger, and P. Schulze, *Advan. Mass Spectrom.*, **3**, 35 (1966).

66. H. D. Beckey, *Angew. Chem.*, **81**, 662 (1969); "Field-Ionization Mass Spectrometry". Pergamon Press, New York, 1971.

67. H. D. Beckey and H.-R. Schulten, *Angew. Chem.*, **87**, 425 (1975); *Angew. Chem. Int. Edn. Engl.*, **14**, 403 (1975).

68. H. D. Beckey, *Int. J. Mass Spectrom. Ion Phys.*, **2**, 500 (1969).

69. H. Krone and H. D. Beckey, *Org. Mass Spectrom.*, **2**, 427 (1969).

70. H.-R. Schulten and H. D. Beckey, *Org. Mass Spectrom.*, **7**, 861 (1973).

71. H.-R. Schulten and D. E. Gaines, *Biomed. Mass Spectrom.*, **1**, 120 (1974).

72. H. Adlercrutz, B. Soltmann, and M. J. Tikkanen, *J. Steroid Biochem.*, **5**, 163 (1974).

73. W. D. Lehmann, H.-R. Schulten, and H. D. Beckey, *Org. Mass Spectrom.*, **7**, 1103 (1973).

III. POLARIMETRY

R. J. Ferrier

A. Introduction

Although optical activity is one of the physical properties of carbohydrates most readily and often measured, it probably is more complex and fundamentally less well understood than any other. This is frequently—in analytical work, for example—of small account, but, nevertheless, a full appreciation of the interrelationships between optical activity and structure, and of solvent, concentration, and temperature dependencies, is to be aimed at. Despite the basic difficulties involved, polarimetry has played a notable part throughout the history of carbohydrate chemistry in providing empirical means for analyzing certain structural features of sugars; Hudson's Isorotation Rules had, for example, until the advent of nuclear magnetic resonance spectroscopy, served as the main method for anomeric classification. In the vast majority of instances, measurements have unfortunately been made only at 589 nm (sodium D line), and optical rotatory dispersion (o.r.d.) studies have played a surprisingly minor role. In recent years, however, the significance of o.r.d. and circular dichroism (c.d.) has increased appreciably in this field.

An ultraviolet-absorbing chromophore in an organic molecule can acquire induced asymmetry from its molecular environment; the electronic transition associated with the absorption thus can become "optically active," and the o.r.d. curve will show a Cotton effect, and the circular dichroism curve a maximum (or minimum) near the center of the absorption band (Fig. 1). Away from the region of absorption, an o.r.d. curve simplifies to a plain curve of the form $[\phi] = K/(\lambda^2 - \lambda_0^2)$ (Drude equation), where $[\phi]$ is the molecular rotation at wavelength λ, K is a constant, and λ_0 is the wavelength of the absorption maximum. The rotation therefore diminishes with distance from the maximum. The observed rotation of a compound at any wavelength is consequently the sum of the partial rotations associated with all the optically active chromophores present, and on the "red side" of the absorption bands the asymmetric chromophore having the highest absorption maximum contributes most to the overall optical activity, its relative significance increasing as its maximum is approached.[1]

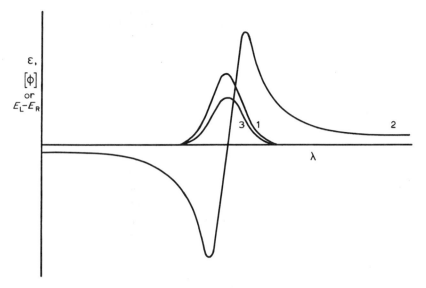

FIG. 1. Absorption spectrum (curve 1) of a typical optically active chromophore (ϵ versus λ). Curve 2, rotatory dispersion ([ϕ] versus λ; positive Cotton effect) associated with it. Curve 3, circular dichroism ($E_L - E_R$ versus λ) associated with it.

In simple carbohydrates, the absorption bands at the longest wavelength are associated with the oxygen atoms, none of which absorb in the readily accessible part of the ultraviolet (> 200 nm). Consequently, plain o.r.d. curves are generally observed; at 589 nm no single chromophore predominates completely, and the rotations measured comprise the sum of several significant partial rotations. Although this approach to the fundamental analysis of composite rotations is soundest, the alternative method of subdividing them into components based on contributions from individual asymmetric atoms or conformational "elements" is more readily applied, and is the basis of several approaches which have led to useful methods of configurational and conformational analysis and which are outlined in the next section.

B. MONOCHROMATIC POLARIMETRY

1. Empirical Methods of Configurational Analysis Based on the Consideration of Isolated Asymmetric Centers

Although it would normally be advantageous to measure optical rotations at short wavelengths, the sodium D line (589 nm) is conventionally employed, and the empiricisms to be discussed were established from results obtained at

---

* *References start on p. 1373.*

this wavelength. Many of these, particularly those concerned with quantitative aspects, may apply only in this region of the spectrum, and, conceivably, observations made at shorter wavelengths could reveal useful relationships as yet undetected.

a. The Anomeric Center.—The simplest approach involves application of Van't Hoff's Principle of Optical Superposition, which proposes the additivity of the rotational contributions of different asymmetric centers in a complex molecule. Although the principle is known to be fundamentally unsound (except when interaction between such centers is impossible), it was found by Hudson to be applicable with great profit to the determination of configuration at the anomeric center of free sugars and glycosides. His celebrated Isorotation Rules [2-4] have their basis in the Principle and still retain their value and popularity, although it is now clear that they are not valid under all conditions.

According to these Rules, the rotation of a glycosyl derivative may be considered to be composed of a value A, the contribution of the anomeric center, and a value B, that of the remainder of the molecule. The A value is positive for α-D (and β-L) compounds and negative for β-D (and α-L) compounds, so that, by the Principle of Optical Superposition, the rotations of

X = OH for free sugars, *O*-alkyl or *O*-aryl for glycosides,
halogen for glycosyl halides, and so on.

α-D and β-D derivatives are $+A + B$ and $-A + B$, respectively. Thus, knowledge of the molecular rotations* $[\phi]$, ($[\alpha]_D \times$ mol. wt./100) of any anomeric pair of derivatives can be used to determine the partial rotational contributions: $[\phi]_\alpha + [\phi]_\beta = +A + B - A + B = 2B$ and $[\phi]_\alpha - [\phi]_\beta = +A + B + A - B = 2A$. After examining the rotation of many anomeric pairs of compounds, Hudson formulated his two Isorotation Rules: (I) "The rotation of carbon 1 (A value) in the case of many substances of the sugar group is affected in only a minor degree by changes in the structure of the remainder of the molecule"; and (II) "Changes in the structure of carbon 1 in the case of many substances of the sugar group affect in only a minor degree the rotation of the remainder of the molecule" (B value).

Values of $2A$ for the anomers of D-xylose, D-glucose, and D-galactose are $+170°$, $+168°$, and $+177°$, which support the first Rule, but the corresponding figures for D-lyxose, D-mannose, D-talose, and D-rhamnose are $+117°$, $+84°$,

---

* Molecular rotations are also frequently expressed as $[M]_D = [\alpha]_D \times$ mol. wt., especially in the earlier literature; the practice of dividing by 100 became current with the advent of recording spectropolarimeters for optical rotatory dispersion measurements.

+98°, and +79°, which suggest that A values fall into two categories according to the configuration at C-2. D-Arabinose, however, gives $2A = +170°$ and is thus anomalous as it has the L configuration at C-2. When classification is carried out in conformational terms (see Vol. IA, Chapter 5) the anomaly is explicable: those sugars in which the hydroxyl group at C-2 is equatorial give $2A$ values near +170°, and those having axial C-2 hydroxyl groups have $2A$ values near +100°. Closely parallel results come from examination of the molecular rotations of glycosides; again there are two sets of A values, and again the arabinosides (conformationally unusual) are exceptional on a configurational basis. The first Rule holds therefore within each class of compound. The $2A$ values[4a] for mannose and glucose derivatives become much closer at 80° than at 20°, presumably because of changes in conformational equilibria.

The second Rule requires that the B value for any particular glycosyl ring be independent of the nature of the anomeric group; for example, in agreement with this, the $2B$ values for the anomerically pure glucopyranoses, glucopyranosyl fluorides, and methyl, ethyl, and propyl glucopyranosides are 236, 220, 242, 240, and 227, respectively. When, however, polarizable groups such as halogens or certain alkoxy or aryloxy groups are attached at C-1 this second rule breaks down, but new linear correlations are observed between molecular rotations and refractions of the bonds connecting the substituents to the anomeric centers.[5] Again, therefore, Hudson's Rule applies only under restricted conditions.

Hudson's Isorotation Rules can be applied as in the following example: a reducing disaccharide ($[\alpha]_D + 65° \rightarrow +37°$), isolated from natural sources, was characterized by conventional chemical procedures as a 2-O-D-xylopyranosyl-L-arabinose. From its mutarotation it was seen to have the β configuration at the reducing center, but the anomeric configuration of the D-xylopyranosyl group was undetermined. If the rotations of the four possible forms (furanose modifications being ignored) are taken as being approximately $B_{\text{methyl D-xylopyranoside}} \mp A_{\text{methyl D-xylopyranoside}} + B_{\text{L-arabinose}} \mp A_{\text{L-arabinose}}$, and $[\phi]_\alpha$ and $[\phi]_\beta$ for methyl D-xylopyranoside are taken as +252° and −107°, respectively, and $[\phi]_\alpha$ and $[\phi]_\beta$ for L-arabinose as +133° and +303°, respectively, then $A_{\text{methyl D-xylopyranoside}} = 359/2 = +180°$, $B_{\text{methyl D-xylopyranoside}} = 145/2 = 72°$, $A_{\text{L-arabinose}} = 170/2 = 85°$, and $B_{\text{L-arabinose}} = 436/2 = +218°$, and the calculated molecular rotations are:

O-α-D-xylopyranosyl-α-L-arabinopyranose $72 + 180 + 218 - 85 =$
$$+385° \ ([\alpha]_D + 136°)$$

O-α-D-xylopyranosyl-β-L-arabinopyranose $72 + 180 + 218 + 85 =$
$$+ 555° \ ([\alpha]_D + 196°)$$

References start on p. 1373.

O-β-D-xylopyranosyl-α-L-arabinopyranose $72 - 180 + 218 - 85 =$
$$+ 25° \; ([\alpha]_D + 9°)$$

O-β-D-xylopyranosyl-β-L-arabinopyranose $72 - 180 + 218 + 85 =$
$$+ 195° \; ([\alpha]_D + 69°)$$

The error involved in the use of $A_{\text{methyl xylopyranoside}}$ instead of a value obtained from xylosyl disaccharides is believed to be small ($A_{\text{methyl glucopyranoside}} = +186°$, $A_{\text{2-}O\text{-D-glucopyranosylglycerol}} = +191°$, and the A value for the non-reducing moieties in β-maltose and β-cellobiose is $+184°$), and similarly, the inaccuracies associated with the use of $B_{\text{L-arabinose}}$ instead of $B_{\text{substituted L-arabinose}}$ are neglected ($B_{\text{D-glucose}} = +118°$, $B_{\text{3-}O\text{-methyl-D-glucose}} = +131°$). It was concluded, therefore, that the isolated disaccharide is the β,β isomer (**1**), and synthetic work corroborated this.[6] In contrast, a 3-linked D-xylopyranosyl-L-arabinose obtained from a different natural source has $[\alpha]_D + 166° \rightarrow +182°$ and it can safely be assigned the α,α configuration.

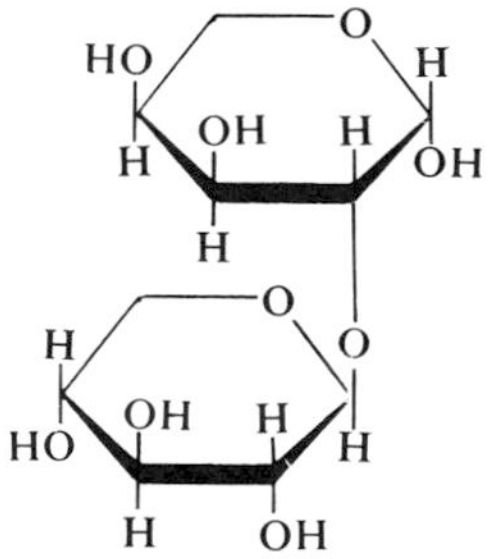

1

As calculations of this type must involve the use of rotations of pure anomers of free sugars, appreciable inaccuracies are entailed, but nevertheless valuable results have been obtained. Errors are minimized by early examination of mutarotating solutions, and by the use of A and B values derived from reference compounds as similar in structure as possible to those under examination.[7]

A further application of the Rules reveals that the known isomeric glucoses have pyranoid rings, as their $2B$ value is $236°$, whereas those of the methyl glucopyranosides and glucofuranosides are $243°$ and $25°$, respectively. Related considerations following observations of the optical rotations of arabinose derivatives have led to empiricisms from which ring sizes of arabinosides and positions of substitution of free arabinoses can be deduced. Thus, for example, 4-O-substituted L-arabinoses, which can adopt only the pyranoid ring form, have $[\alpha]_D$ values of $+125°$ to $+140°$, whereas 5-O-substituted derivatives have values of $-25°$ to $-50°$. Compounds that are free to adopt either five- or six-membered ring forms have[8] intermediate values of $+90°$ to $+110°$.

Although β-D-glucopyranosides of the primary and secondary alcohols have molecular rotations usually falling in the interval $-65°$ to $-100°$, those derived from phenols exhibit molecular rotations greater than $-170°$. The corresponding derivatives of tertiary alcohols have molecular rotations near $-40°$. For the aromatic glucosides, an interesting correlation exists between the effect of substituent groups present in the phenyl nucleus on the rotations and the influence of the same groups on substitution reactions of benzene derivatives. The "ortho–para directing groups," when substituted in the aromatic nucleus of phenyl β-D-glucopyranoside, have little or no effect on the rotation. However, "meta-directing groups" in positions meta and para to the glucosidic connection cause the rotation of the glucoside to become appreciably more negative than for phenyl β-glucopyranoside. Thus, the value of $-310°$ for p-nitrophenyl β-glucopyranoside may be compared with that of $-182°$ for phenyl β-glucopyranoside. Di-ortho-substituted derivatives have anomalously low molecular rotations near those of the tertiary alkyl β-glucosides ($-40°$ to $-50°$). The influence of substituents in the aromatic nucleus of phenyl β-D-glucopyranoside parallels the effect of the same groups on the acidity of the corresponding substituted phenols.[8a]

Some of the ortho-substituted phenyl β-glucopyranoside tetraacetates have anomalous positive rotations. Thus, o-nitrophenyl β-D-glucopyranoside tetraacetate has a molecular rotation of $+211°$ ($[\alpha]_D = 45°$), as compared with the negative values $-174°$ and $-192°$ for the m- and p-isomers. The positively rotating derivatives have a very large temperature coefficient, and their rotations become negative at higher temperatures, although the rotations of the m- and p-isomers are affected only to a minor degree by an increase of temperature. This and other evidence makes it probable that the positive rotation of certain of the o-substituted phenyl β-glucopyranoside tetraacetates is due to a bonding of the group at the ortho position with an acetyl group in the sugar portion of the molecule.[8b] Other workers have shown that the large temperature coefficients exhibited by o-nitrophenyl compounds result from interaction between the nitro group and the carbohydrate.[8c]

Hudson's proposal that, in the D series, the more dextrorotatory of an anomeric pair (namely, that having A positive) be termed α has been replaced by that of Freudenberg who assigned the designation α to the anomer having the hydroxyl (or equivalent) group at the anomeric center and that at the highest-numbered asymmetric carbon atom in the cis relationship (see Vol. IA, Chapter 1). Until recently, the two proposals were found to be consistent. There are now, however, several examples known of D anomeric pairs in which the α isomer (Freudenberg convention) is the less dextrorotatory, and, in each instance, a chromophore, the asymmetry of which controls rotational characteristics, is present near the anomeric center. Several pairs of pyrimidine nucleosides [such as thymidine (2) and its anomer] which have been examined,[9] the anomeric tetra-*O*-acetyl-3-deoxy-D-*erythro*-hex-2-enopyranoses[10] (3), and the 1,3,4,6-tetra-*O*-acetyl-2-deoxy-2-(2,4-dinitroanilino)-D-glucopyranoses[11] (4) are exceptional.

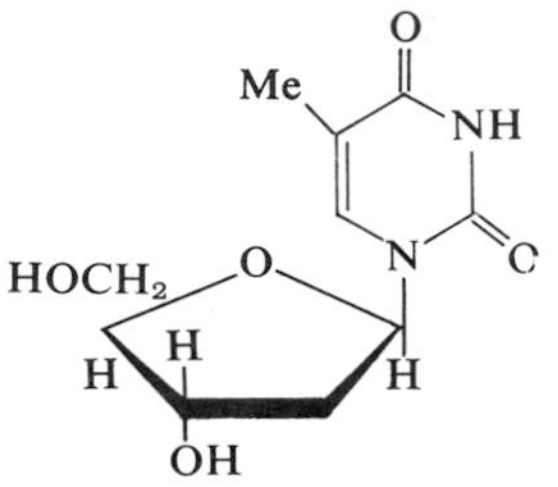

2 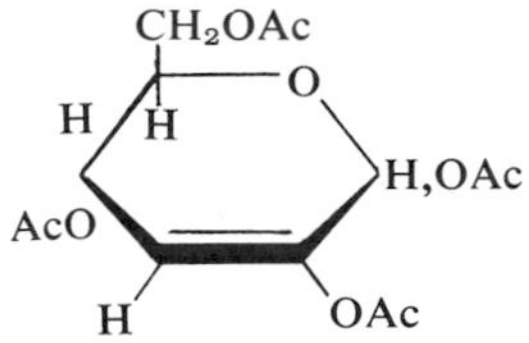   3 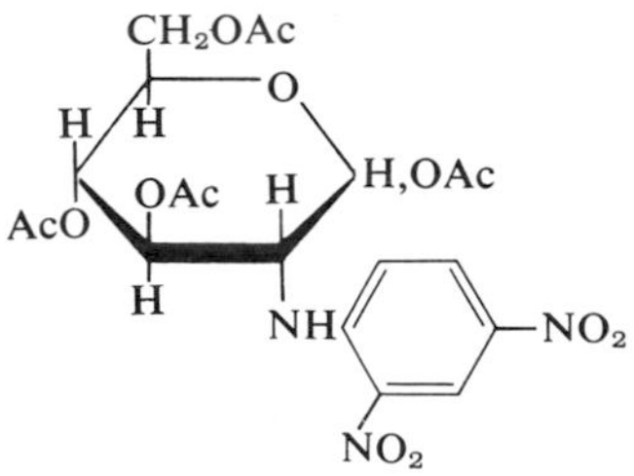   4

b. Centers Other than the Anomeric.—Although Hudson's Isorotation Rules relate to the anomeric center alone, further empirical generalizations have been enunciated to permit configurational assignment at other positions. It has been recognized, for example, that dextrorotatory amides,[12] phenylhydrazides,[13] and benzimidazoles[14] are produced from aldonic acids having the D configuration at C-2, and that aldoses having the D configuration at C-3 afford dextrorotatory phenylosotriazoles on oxidation of their phenylosazones[15]: D-*arabino*-hexulose phenylosotriazole (5), for example, has $[\alpha]_D$ $-82°$. Similarly, the (polyhydroxyalkyl)-pyrroles and -furans obtained on reaction of 2-amino-2-deoxyaldoses and aldoses with β-dicarbonyl compounds are dextrorotatory[16] when C-3 is D. In each of these examples, the chiral center adjacent to the unsaturated chromophore provides the greatest contribution to the net rotation, so that the sign of rotation is determined by the configuration at this position. (Generalized Heterocycle Rule[16a].)

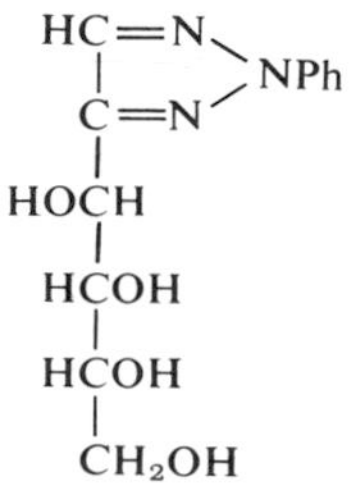

5

Lastly, configurations at C-4 and C-5 may be determined by application of Hudson's Lactone Rule,[17] which specifies that, when lactonization of an aldonic acid is accompanied by a dextrorotatory change, the hydroxyl group involved has the D configuration, and vice versa. As a corollary, the Lactone Rule provides a means of determining ring size in those compounds having a *threo*-diol at C-4 and C-5, but infrared spectroscopy offers a more general and reliable method for such determination.

Double bonds in pyranoid rings may invalidate Hudson's Isorotation Rules (see compounds **3**), and it appears that Mills' observation[18] [that an allylic hydroxyl or substituted hydroxyl group present on a cyclohexene ring in the configuration depicted (**6**) contributes negatively to optical rotation] applies. The rotations of glycals that are epimeric at C-3 (see Section B,1,c), and of pairs of isomeric hexopyranosides possessing double bonds at C-2–C-3 and differing in configuration at C-4 bear this out.[19]

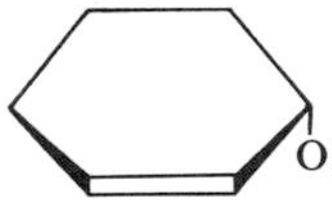

6

c. Generalizations.—Bose and Chatterjee[20] have formulated a useful but still completely empirical generalization that can be used to assess the qualitative contribution of various asymmetric centers to the rotation of cyclic compounds, and have incidentally provided a correlation between the Isorotation and Lactone Rules. They postulated that, if any asymmetrically substituted atom is viewed from the center of the ring so that the bulkier substituent **B** is below the ring and the less bulky one **A** is above, then if the larger adjacent ring group (**L**) is on the right of **S** (the smaller adjacent group) this epimer (**7**) will be more dextrorotatory. Relative "sizes" of common ring groups are:

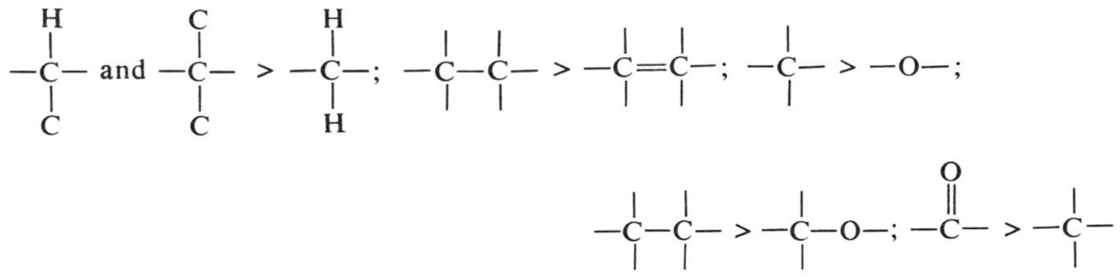

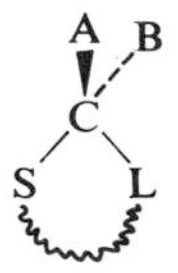

7

Although polarizability rather than "size" of neighboring groups is of more fundamental significance, the empiricisms correlate with observation with respect to the configurations at C-5 of hexopyranoses, pyranosides, and δ-lactones; at C-4 of pyranosides, furanosides and δ-lactones; at C-3 of glycals and γ-lactones; at C-2 of glycosides and lactones; and at C-1 of free sugars and glycosides. Thus, for example, the isomers **8** (α-D-aldose; compare Isorotation Rules), **9** (γ-lactone or pentofuranoside having the D configuration at C-4; compare Lactone Rule), **10** (glycal having the D configuration at C-3), **11** (sugar, glycoside, or lactone having the D configuration at C-2) are all more dextrorotatory than their epimers.

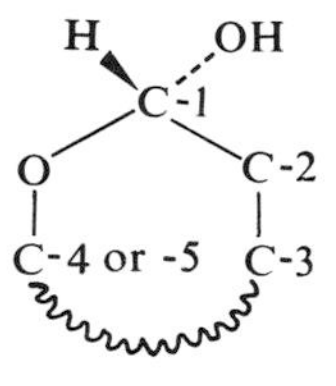

$$\textbf{8} \qquad\qquad \textbf{9} \qquad\qquad \textbf{10} \qquad\qquad \textbf{11}$$

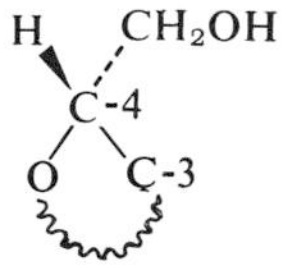

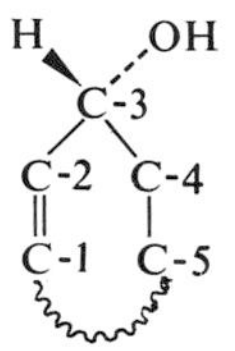

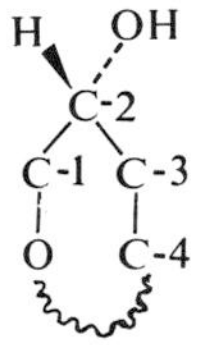

In a similar but less-general study of acylated aldopento-furanoses, -pyranoses, and glycosyl halides, Fletcher and co-workers[21] showed that 1,4- and 1,5-anhydropentitol derivatives had rotations intermediate between those of the corresponding anomeric peracylated aldoses or acylglycosyl halides. Similarly, the rotations of 2-deoxy and 1,2-dideoxy compounds lie between those of the corresponding C-2 epimers and the 2-deoxy anomers, respectively. With this generalization it is therefore possible, for example, to determine the anomeric configuration of a glycosyl derivative from knowledge of its optical rotation, and that of the equivalent anhydroalditol. This study also revealed that acylated pentofuranosyl halides and esters, and corresponding 1- and 2-deoxy and 1,2-dideoxy derivatives, are all more dextrorotatory than their pyranosyl counterparts, and that A values (Hudson's) for furanosyl esters are appreciably less than those for the pyranosyl analogues.

2. Empirical Methods of Configurational and Conformational Analysis Based on the Consideration of Complete Molecules

A later and more comprehensive, but still empirical, approach to the correlation of optical rotational properties and structures of pyranoid carbohydrates was introduced by Whiffen.[22] In applying superposition rules, he considered resultant rotation to be constituted of terms associated with individual asymmetric centers together with those associated with asymmetric disposition of substituents along chemical bonds. In compounds containing only carbon, hydrogen, and oxygen and no multiple bonds, he assumed that contributions of the first type are small and can be ignored, and that observed rotation can be taken therefore as the sum of the terms of the second type which, for a bond unit UVWC–CXYZ, can be expressed as UC–CX (U/X in abbreviated form) taken together with all similar elements U/Y, U/Z, V/X, V/Y, V/Z, W/X, W/Y, and W/Z. In a pyranoid chair conformation, the angles subtended by the U,V,W–C and the X,Y,Z–C bonds must be close to either $\pm 60°$ or $180°$, and as a sine function relates those angles to the rotations of the units, the three planar (antiparallel) elements do not contribute, and the rotation of the general unit is thus $U/X - U/Z + V/Y - V/X + W/Z - W/Y$—namely, $U/X - X/V + V/Y - Y/W + W/Z - Z/U$. ($U/Z \equiv Z/U$, because this change merely involves observation from the other end of the C–C bond and has no effect upon the rotation; the sign convention assigns a $+$ value to general units whose substituents advance into each other as a right-hand screw.)

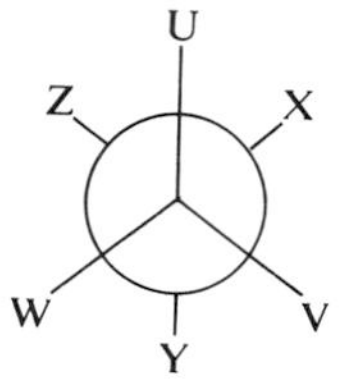

Whiffen analyzed complete cyclitols, pyranoses, and pyranosides of known structure, conformation, and rotation in this way, by considering each bond in the ring in turn. After summation, he was able to express the totals in terms of the six parameters (F–K) shown, which were evaluated by solving a set of simultaneous equations.

Rotational parameter	Symbol	Value (degrees)
$O/O - 2\,O/H + H/H$	F	$+\ 45$
$O_G/O - O_G/H - O/H + H/H$	G	$+\ 32$
$O/C - C/H - O/H + H/H$	H	$+\ 34$
$O_R/O - C_R/O - O_R/H + C_R/H$	I	$+\ 43$
$\bar{C}_R/O_G - \bar{C}_R/H + C_R/H - C_R/O_G + O_G/H - H/H$	J	$+113$
$\bar{C}_R/C - \bar{C}_R/H + C_R/H - C_R/C + C/H - H/H$	K	$ca.\ -\ 29$

For D-hexoses, a value of $+30°$ is allowed to account for the asymmetry associated with the C-6–oxygen bond (this complication does not arise with the 6-deoxyhexoses), and for L-hexoses $-30°$ is to be added. Similarly, $+100$ is added for α-D-glycosides to compensate for restriction of rotation imposed about the C-1–O-1 bond, and $-100°$ is allowed for β-D-glycosides. In the L series the signs of these values are reversed.

The utility of the method is illustrated by the following calculation of the rotation (in water) of methyl 2,3-dideoxy-α-D-*erythro*-hexopyranoside in the

$^4C_1(D)$ conformation (**12**). Whiffen's terminology is used; the glycosidic oxygen atom (O_G) is distinguished from the ring oxygen atom (O_R) and from hydroxyl oxygen atoms (O), and ring carbon atoms ($\bar{C}_R$) corresponding to U in the element UO–CX ($\bar{U}/X$) are differentiated from other ring carbon atoms (C_R) and from exocyclic carbon atoms (C).

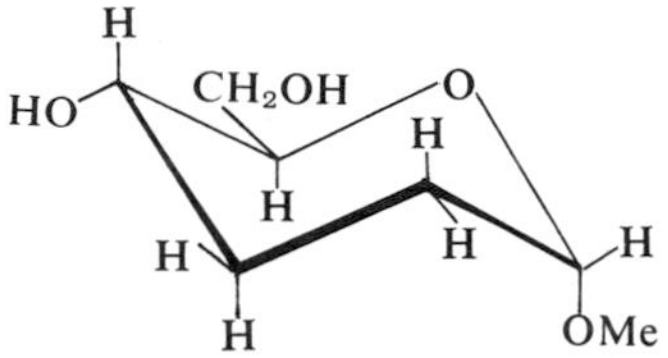

12

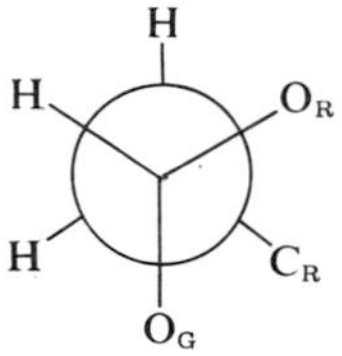

C-1–C-2 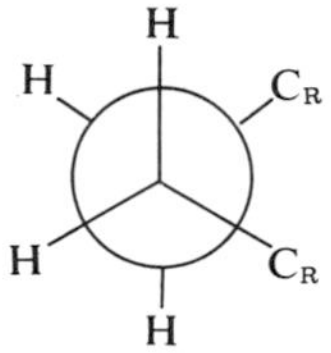 C-2–C-3 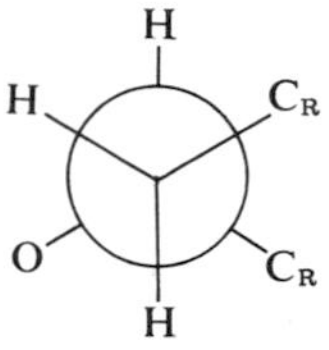 C-3–C-4

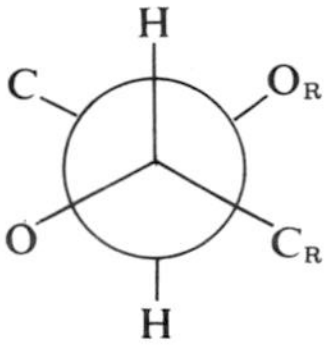

C-4–C-5 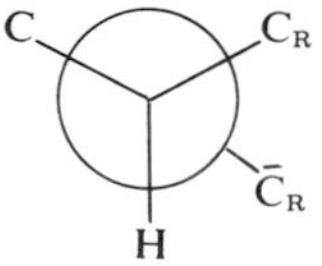 C-5–O_R 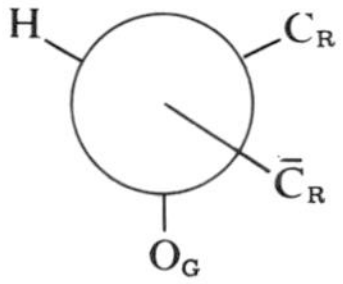 O_R–C-1

Summing the elements: $-H/O_R + O_R/C_R - C_R/O_G + O_G/H$

$$H/H + H/H \qquad \ldots \qquad \text{(C-1–C-2)}$$

$$+H/C_R - C_R/C_R + C_R/H - H/H +$$
$$H/H - H/H \qquad \ldots \qquad \text{(C-2–C-3)}$$

$$-H/C_R + C_R/C_R - C_R/H + H/O -$$
$$O/H + H/H \qquad \ldots \qquad \text{(C-3–C-4)}$$

$$+H/O_R - O_R/C_R + C_R/H - H/O +$$
$$O/C - C/H \qquad \ldots \qquad \text{(C-4–C-5)}$$

$$+C_R/\bar{C}_R - \bar{C}_R/H \qquad \ldots \qquad \text{(C-5–}O_R\text{)}$$

$$-C_R/\bar{C}_R + \bar{C}_R/O_G \qquad \ldots \qquad \text{(}O_R\text{–C-1)}$$

$$= \bar{C}_R/O_G - \bar{C}_R/H + C_R/H - C_R/O_G + O_G/H + O/C - C/H - O/H = J + H$$

Additionally, $+30°$ and $+100°$ have to be allowed for the asymmetric contribution from the C-6 group and the aglycon, respectively, and so the final molecular rotation is $113° + 34° + 30° + 100° = 277°$. Calculation of the rotation of the compound in the $^1C_4(D)$ conformation gives $+58°$ ($K - I +$

130°), and, as the observed value is $+286°$, it can be concluded that the $^4C_1(D)$ chair is adopted in aqueous solution.

A method similar in its empirical approach but applicable to any optically active cyclic or acyclic compound has been advanced by Brewster,[23] but Lemieux and Martin's development[24] of the Whiffen approach promises to be appreciably more useful in carbohydrate chemistry. These authors re-emphasize the basically primitive character of all these empirical approaches, but also stress their potential usefulness not only for determining rotations of molecules in fixed conformations but also for studying conformational equilibria. This extension follows from the observation[25] that the different optical rotations exhibited by certain compounds in different solvents result from alterations in molecular conformations and not, for example, from changes in direct effects of solvation.

Their method to some degree both simplifies that of Whiffen and extends it. The rotational parameters are reduced to four, and they are based simply on pair-wise, vicinal interactions between *gauche*-related oxygen and carbon atoms attached to directly bonded oxygen or carbon atoms; the treatment does not consider related interactions involving hydrogen atoms, nor does it differentiate between different types of carbon and oxygen atoms. Their parameters, following the notation on p. 1363, are thus OC–CC (designated O/C; evaluated as 10°), OC–CO (O/O; 45°); OC–OC (O/C$_0$; 115°) and CC–OC (C/C$_0$; 60°). The values assigned were established by examination of simple model compounds such as 2-methyltetrahydropyran and 3-hydroxy-2-methyltetrahydropyran. By this procedure, the molecular rotation of methyl 2,3-dideoxy-α-D-*erythro*-hexopyranoside (p. 1364) is computed as follows: O/C$_0$ (O-1–C-5) + O/C$_0$ (Me–O-5, assuming that the methyl group is *gauche* to H-1 and the ring oxygen atom) + O/O (O-5–O-6, assuming that O-6 is *gauche* to H-5 and the ring oxygen atom) = $+275°$. For this approach to be applicable to complex molecules, information on favored rotamer states for complex groups (such as OCH_3 and CH_2OH) attached to ring carbon atoms is required, but of course these required data can be provided by polarimetric examination of simple model compounds.

Although each of these approaches may be applied to advantage, (see for example, Vol. IA, p. 427), none offers a general explanation, based on sound theory, of the interrelationship between measured optical rotation and structure, and there is still appreciable room for development in the field. Any comprehensive treatment must take into account, in addition to gross structural and conformational features, factors relating to the steric dispositions of exocyclic groups and the effects of solvation. Using Kirkwood's polariz-ability theory,[26] Yamana[27] has initiated such an approach and has established, for example, the validity of the absolute configuration of sugars initially assumed by Fischer, and of the Isorotation Rules. In addition, he has concluded that glucose, mannose, and galactose adopt the $^4C_1(D)$ or $^1C_4(L)$ conformation in solution, and has calculated the probable orientation of the C-6–oxygen bonds in hexoses and of the O-1–aglycon bonds in glycosides.

C. Optical Rotatory Dispersion and Circular Dichroism

Despite the fact that measurements of o.r.d. spectra of carbohydrates were conducted[28,29] in the 1930s, the technique has played a surprisingly unimportant role, even during recent years when instrumentational advances have been extensive and its utility in natural-product studies has been demonstrated convincingly.[30] However, the successes achieved in other fields and the availability of commercial spectropolarimeters which allow automatic measurement down to 200 nm and below have stimulated a revival in interest.

In free sugars and glycosides, the asymmetric chromophores of greatest significance in determining rotatory characteristics are the ring and anomeric oxygen atoms, which absorb below 200 nm. As anticipated, therefore, these simple compounds normally display plain o.r.d. curves[31,32] down to this wavelength (Fig. 2). D-Galactose and 6-deoxy-D-galactose are, however,

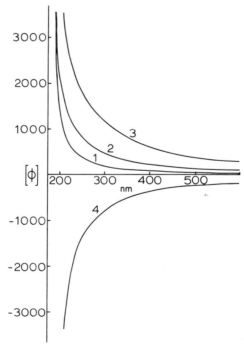

FIG. 2. Optical rotatory dispersion curves in water of (1) D-mannose, (2) D-glucose (both equilibrated), (3) methyl α-D-xylopyranoside, and (4) methyl β-D-xylopyranoside.

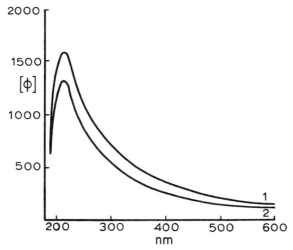

FIG. 3. Optical rotatory dispersion curves in water of (1) D-galactose and (2) 6-deoxy-D-galactose (both equilibrated).

exceptional in giving plain curves immediately on dissolving, but after equilibration their curves are anomalous, and show an extremum near 210 nm (Fig. 3) which appears to be the first extremum of a positive Cotton effect.[31,32] Listowsky *et al.*[31] have shown that these anomalies result not from the presence of an asymmetric chromophore but arise mainly by simple superposition of the o.r.d. curves of the two pyranoid forms present at equilibrium (approximately 30% α and 70% β). The possibility of this superposition is indicated by the "synthetic spectrum" of a 3:7 mixture of methyl α- and β-D-galactopyranosides (Fig. 4).

By comparing the o.r.d. curve of pairs of pyranoid compounds differing in one individual feature—for example, the configuration at one asymmetric center or the presence of a C-5 substituent—Listowsky *et al.*[31] determined the qualitative contribution of these features to o.r.d. curves, distinguishing between long-wavelength effects and those operating in the 200 nm region—namely, close to the first Cotton effect. They arrived at the conclusions given in Table I, from which it can be seen that the sign of the contribution of some features can alter with wavelength.

The long-wavelength results concur with the postulates of Bose and Chatterjee[20] and with those obtainable by the Whiffen[22] or Brewster[23] procedures. Those measured at short wavelengths allow the generalization that, in the $^4C_1(D)$ conformation, axial components at C-1 and C-2 contribute

References start on p. 1373.

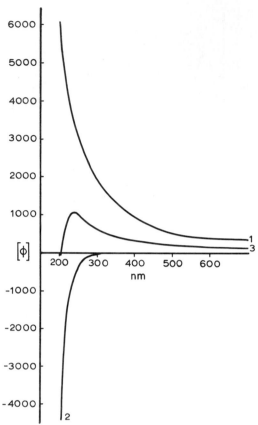

Fig. 4. Optical rotatory dispersion curves in water of (1) methyl α-D-galactopyranoside, (2) methyl β-D-galactopyranoside, and (3) a 3:7 mixture of these (curve derived algebraically).

positively and those at C-4 and C-5 contribute negatively. Measurements at wavelengths down to ~190 nm have shown[33] that the signs of the longest-wavelength c.d. curves of free sugars and simple pyranosides can be determined and, somewhat surprisingly, that these are dependent more on the configuration at C-4 than at C-1.

Carbohydrate derivatives that contain suitable dissymmetric chromophores do show "anomalous" o.r.d. spectra and do exhibit c.d. maxima. Compounds possessing free carbonyl groups—for example, 2,3,4,5,6-penta-O-acetyl-D-glucose[29b] (13), methyl 3,4-O-isopropylidene-β-D-*erythro*-pentopyranosid-2-ulose[34] (14, Fig. 5), and 1,6-anhydrohexopyranosuloses[35]—show characteristic curves centered in the 290-nm region, but this figure may be as high as 330 nm

TABLE I

SIGNS OF ROTATORY CONTRIBUTIONS OF INDIVIDUAL CENTERS IN THE PYRANOSE SERIES

Position and configuration of hydroxyl group	*Contribution at long wavelength*	*Contribution near 200 nm*
Carbon atom 1,D	+	+
1,L	−	−
2,D	+	−
2,L	−	+
3,D	−	−
4,L	+	−
5,L	−	−
5,D (with 4,D)	+	+
5,D (with 4,L and 1,D)	+	−
5,D (with 4,L and 1,L)	−	−

for strained-ring, cyclic carbonyl compounds.[35a] For the last class of compound, the octant rule has been found to apply.[35b] The configuration (D or L) of a sugar may be determined conveniently by c.d. measurements at 213 nm on the acetylated alditol or a methylated derivative of it; the acetoxyl group functions as a chromophore. Only milligram quantities of material are required, and the method is well suited to analyzing fractions obtained by gas–liquid chromatography.[36]

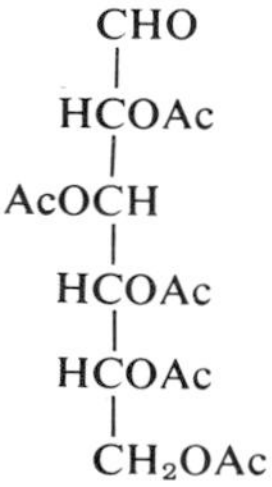
CHO
HCOAc
AcOCH
HCOAc
HCOAc
CH₂OAc

13

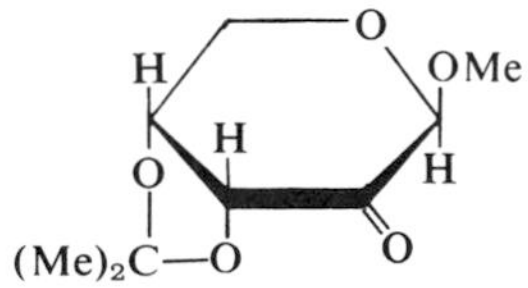
O
H
OMe
H
O
H
(Me)₂C—O
O

14

The $n \rightarrow \pi^*$ transition of lactones giving an absorption near 225 nm is accessible with modern spectropolarimeters, and o.r.d. studies of carbohydrate γ-lactones have revealed that compounds having the D configuration at C-2 show negative Cotton effects.[37a] As Hudson's Lactone Rule (Section II,A,2) is applicable to the determination of configuration at the asymmetric center involved in the ring junction, complete o.r.d. measurements may thus be used to allow configurational assignment to the asymmetric carbon atoms on either

References start on p. 1373.

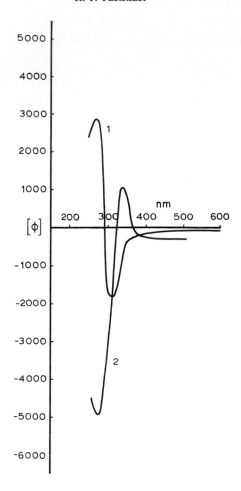

Fig. 5. Optical rotatory dispersion curves of (1) 2,3,4,5,6-penta-*O*-acetyl-D-glucose (**13**) in chloroform, and (2) methyl 3,4-*O*-isopropylidene-β-D-*erythro*-pentopyranosid-2-ulose (**14**) in methanol.

side of the carbonyl group. The similarity of the o.r.d. curves of D-ribono-1,4-lactone (**15**) ($[\alpha]_D$ + 43°) and its 2,3-isopropylidene acetal ($[\alpha]_D$ − 80°) allowed the conclusion, which was confirmed by n.m.r.-spectroscopic analysis, that fusion of the acetal ring causes only minor changes in ring conformation, and that the apparently inconsistent $[\alpha]_D$ values result from change in the electronic characteristics of the groups near the carbonyl chromophore.[37b]

Interpretation of o.r.d. curves of compounds bearing most other chromophores having suitable absorbing characteristics are complicated by the sensitivity of the rotation curves to the conformations adopted by the chromo-

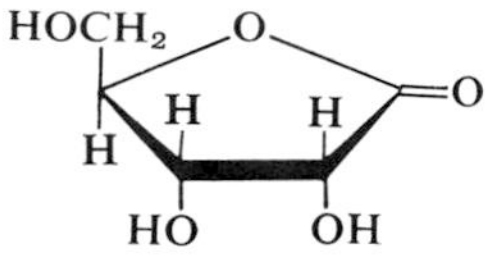

15

phores themselves and to their environments, so that, for example, the Cotton effects given by xanthates (λ_{max} 354 nm)[38] may not reflect the absolute configuration of the alcoholic centers involved. However, an extensive study of the c.d. spectra of carbohydrate azido derivatives has led to the consideration of an octant rule for these compounds,[39a] and, specifically, 4-azido-4-deoxy-pentopyranosides have been found to give positive Cotton effects when the ring conformation is $^4C_1(D)$ and the azide group is axial, or when the conformation is $^1C_4(D)$ and the azide is equatorial.[39b] Despite this type of possible generalization, the sensitivity of o.r.d. and c.d. behavior in compounds of this general class is illustrated by the observations that acetylation of N-(β-D-glucopyranosyl)-3-carboxamidopyridinium bromide[40] (**16**), methyl 3-deoxy-3-phenylazo-α-D-glucopyranoside[41] (**17**), and aldose benzylphenyl-hydrazones[42] causes complete inversion of the signs of the observed Cotton effects.

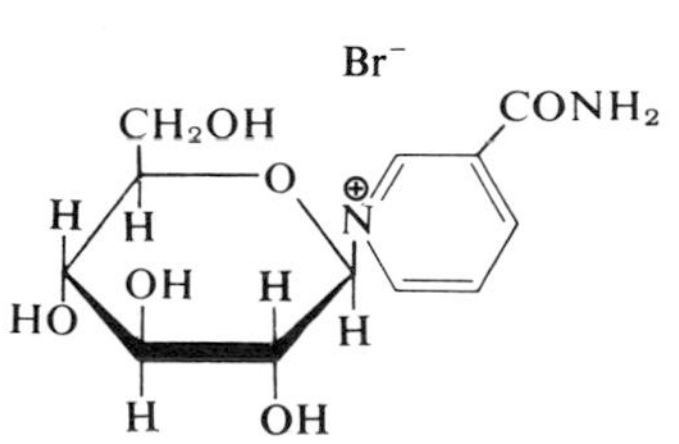

16 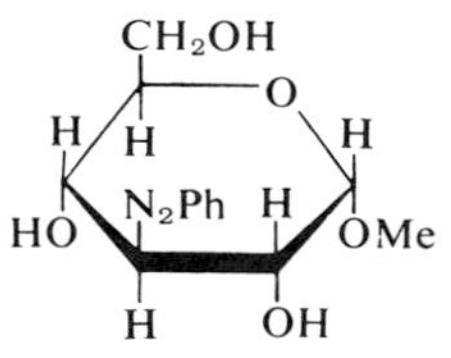   **17**

It has been proposed that the signs of Cotton effects near 275 nm in the o.r.d. spectra of the 1-deoxy-1-nitroalditols may be related to the configurations at C-2 (D hydroxyl groups lead to positive curves),[43] but with the aldose diethyl dithioacetals and ethylene dithioacetals, apparently configuration at C-3 is the most important in determining the nature of the Cotton effects (near 235 to 240 nm).[44] In a study intended to demonstrate the versatility of this approach to configurational analysis of acyclic sugar derivatives, o.r.d. curves of aldose benzimidazoles, quinoxalines, flavazoles, and anhydro-osazones were examined and shown to be relatable to configurations at C-2, C-3, C-4, and C-5, respectively, these being, in each instance, the asymmetric centers adjacent to the heterocyclic chromophore.[45] More specifically, 2-benzyl-2-phenyl-hydrazones of aldoses that have the L configuration at C-2 show positive

References start on p. 1373.

Cotton effects near 300 nm, and this observation[42] probably affords the most reliable of the polarimetric methods for determining configurations at C-2.

In the nucleoside field, extensive studies have shown that o.r.d. properties of these compounds are dependent not only on anomeric configuration but markedly on the relative orientations of the sugar and heterocyclic rings.[46] In consequence, it has been concluded that early generalizations[9,47] indicating that α- and β-D-glycosyl pyrimidines give negative and positive curves, respectively (with the reverse situation for the analogous purine compounds), are too restrictive, although they do apply in many instances.

As with many pyrimidine nucleosides, the 1,2,4,6-tetra-O-acetyl-3-deoxy-D-erythro-hex-2-enopyranoses (Section B,1,a) are anomalous with respect to Hudson's Rules over the entire spectral range, but the anomeric p-nitrophenyl 4,6-di-O-acetyl-2,3-dideoxy-D-erythro-hex-2-enosides (18) display curves that cross at 290 nm near the center of Cotton effects, and with these anomers the α form is the more dextrorotatory only above this wavelength.[10] Consequently the Isorotation Rules are shown to be unreliable for certain compounds, and they are in some instances even wavelength-dependent.

18

In his classic studies on pyranoside conformations (see Vol. IA, Chapter 5), Reeves used direct polarimetry to examine the orientation and chirality of vicinal diols, and extensions of this approach have shown that o.r.d. or c.d. studies are more reliable for the purpose[48] because chirality can be directly related to Cotton effects or c.d. maxima of cuprammonium complexes at 300 and 580 nm. By this means the structure of the 2,3-dihydroxyoctalin derivative rishitin has been determined,[49a] and the conformations of furanosides having 2,3-cis related diols have been examined.[49b] In related manner, the chirality of vicinal diols can be determined by examination of the circular dichroism of derived p-chlorobenzoates[50] or thionocarbonates.[51] α-Amino alcohols may also be configurationally correlated either by direct polarimetry[52] or by c.d. examination[53] of cuprammonium complexes.

Although circular dichroism provides essentially the same information as does optical rotatory dispersion, it can, on occasions, be applied more profitably, and the technique is finding increased use. Thus, the negative Cotton effect of methyl α-D-glucopyranoside 6-(S-benzylxanthate) is masked by a strong, positive base rotation, whereas the circular dichroism, which is manifested only in the region of an asymmetric absorption band, is unaffected by

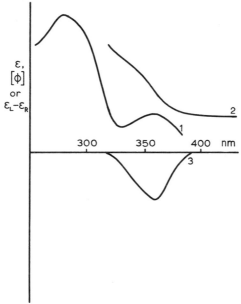

FIG. 6. Ultraviolet absorption spectrum (1), optical rotatory dispersion curve (2), and circular dichroism curve (3), of methyl α-D-glucopyranoside 6-(S-benzylxanthate) in 1,4-dioxane.

the influence of the remainder of the molecule (Fig. 6).[54] Several trithiocarbonates studied by Djerassi *et al.* provide other instances in which circular dichroism has clear advantages over optical rotatory dispersion.[55]

Magnetic circular dichroism has not as yet been extensively exploited in carbohydrate chemistry, but studies with nucleosides[56] have shown the technique to be applicable with minute samples and to be utilizable as a means of distinguishing between purine and pyrimidine derivatives. In addition, it can be used to resolve overlapping c.d. bands and, in conjunction with natural c.d. studies, to provide a sensitive means of "fingerprinting" optically active compounds.

REFERENCES

1. For a fuller general discussion of the theory of optical rotation, see J. C. P. Schwarz, in "Physical Methods in Organic Chemistry" (J. C. P. Schwarz, ed.), Chapter 6, Oliver & Boyd, Edinburgh, 1964.
2. C. S. Hudson, *J. Amer. Chem. Soc.*, **31**, 66 (1909); **38**, 1566 (1916).
3. F. J. Bates and Associates, *Nat. Bur. Stand. Circ. C440*, 411 (1942).
4. J. Staněk, M. Černý, J. Kocourek, and J. Pacák, "The Monosaccharides," Academic Press, New York, 1963.
4a. W. Kauzmann, *J. Amer. Chem. Soc.*, **64**, 1626 (1942).

5. S. Yamana, *J. Org. Chem.*, **31**, 3698 (1966); **33**, 185, 1819 (1968); *Tetrahedron*, 1559 (1968); D. D. Davis and F. R. Jensen, *J. Org. Chem.*, **35**, 3410 (1970).

6. G. O. Aspinall and R. J. Ferrier, *J. Chem. Soc.*, 1501 (1958).

7. Compare W. Korytnyk, *J. Chem. Soc.*, 650 (1959).

8. S. C. Williams and J. K. N. Jones, *Can. J. Chem.*, **45**, 275 (1967).

8a. W. Pigman and H. S. Isbell, *J. Res. Nat. Bur. Stand.*, **27**, 9 (1941).

8b. W. Pigman, *J. Res. Nat. Bur. Stand.*, **33**, 129 (1944).

8c. B. Capon, W. G. Overend, and M. Sobell, *J. Chem. Soc.*, 5172 (1961).

9. T. R. Emerson and T. L. V. Ulbricht, *Chem. Ind.* (London), 2129 (1964).

10. R. J. Ferrier, W. G. Overend, and G. H. Sankey, *J. Chem. Soc.*, 2830 (1965).

11. D. Horton, *J. Org. Chem.*, **29**, 1776 (1964); S. Guberman and D. Horton, *ibid.*, **32**, 294 (1967).

12. C. S. Hudson, *J. Amer. Chem. Soc.*, **40**, 813 (1918).

13. C. S. Hudson, *J. Amer. Chem. Soc.*, **39**, 462 (1917); P. A. Levene and G. M. Meyer, *J. Biol. Chem.*, **31**, 623 (1917).

14. N. K. Richtmyer and C. S. Hudson, *J. Amer. Chem. Soc.*, **64**, 1612 (1942).

15. H. El Khadem, *J. Org. Chem.*, **28**, 2478 (1963).

16. F. García González and A. Gómez Sánchez, *Advan. Carbohyd. Chem.*, **20**, 303 (1965).

16a. H. S. El Khadem and Z. M. El-Shafei, *Tetrahedron Lett.*, 1887 (1963); J. A. Mills, *Aust. J. Chem.*, **17**, 277 (1964); compare H. El Khadem, *Carbohyd. Res.*, **59**, 11 (1977) for extension to a generalized circular-dichroism rule for nitrogen heterocycles attached to alditol residues.

17. C. S. Hudson, *J. Amer. Chem. Soc.*, **32**, 338 (1910).

18. J. A. Mills, *J. Chem. Soc.*, 4976 (1952).

19. D. M. Ciment, R. J. Ferrier, and W. G. Overend, *J. Chem. Soc.* (*C*), 446 (1966).

20. A. K. Bose and B. G. Chatterjee, *J. Org. Chem.*, **23**, 1425 (1958).

21. A. K. Bhattacharya, R. K. Ness, and H. G. Fletcher, Jr., *J. Org. Chem.*, **28**, 428 (1963).

22. D. H. Whiffen, *Chem. Ind.* (London), 964 (1956).

23. J. H. Brewster, *J. Amer. Chem. Soc.*, **81**, 5475, 5483, 5493 (1959).

24. R. U. Lemieux and J. C. Martin, *Carbohyd. Res.*, **13**, 139 (1970).

25. R. U. Lemieux, A. A. Pavia, J. C. Martin, and K. A. Watanabe, *Can. J. Chem.*, **47**, 4427 (1969).

26. J. G. Kirkwood, *J. Chem. Phys.*, **5**, 479 (1937).

27. S. Yamana, *Bull. Chem. Soc. Japan*, **30**, 203, 207, 916, 920 (1957); **31**, 558, 564 (1958); **32**, 597 (1959); **35**, 1269, 1421, 1950 (1962); **36**, 473 (1963).

28. T. L. Harris, E. L. Hirst, and C. E. Wood, *J. Chem. Soc.*, 2108 (1932); 1825 (1934); 1658 (1935); 848 (1937); R. W. Herbert, E. L. Hirst, and C. E. Wood, *ibid.*, 1564 (1933); 1151 (1934); R. W. Herbert, E. L. Hirst, H. Samuels, and C. E. Wood, *ibid.*, 295 (1935); T. L. Harris, R. W. Herbert, E. L. Hirst, C. E. Wood, and H. Woodward, *ibid.*, 1403 (1936).

29. (a) M. L. Wolfrom and W. R. Brode, *J. Amer. Chem. Soc.*, **53**, 2279 (1931); (b) H. Hudson, M. L. Wolfrom, and T. M. Lowry, *J. Chem. Soc.*, 1179 (1933); (c) W. C. G. Baldwin, M. L. Wolfrom, and T. M. Lowry, *ibid.*, 696 (1935).

30. C. Djerassi, "Optical Rotatory Dispersion: Applications to Organic Chemistry," McGraw-Hill, New York, 1960; P. Crabbé, "Optical Rotatory Dispersion and Circular Dichroism in Organic Chemistry," Holden-Day, San Francisco, 1965.

31. I. Listowsky, G. Avigad, and S. Englard, *J. Amer. Chem. Soc.*, **87**, 1765 (1965).

32. N. Pace, C. Tanford, and E. A. Davidson, *J. Amer. Chem. Soc.*, **86**, 3160 (1964).

33. I. Listowsky and S. Englard, *Biochem. Biophys. Res. Comm.*, **30**, 329 (1968).

34. J. S. Burton, W. G. Overend, and N. R. Williams, *J. Chem. Soc.*, 3433 (1965).

35. (a) D. Horton and J. S. Jewell, *Carbohyd. Res.*, 5, 149 (1967); (b) K. Heyns, J. Weyer, and H. Paulsen, *Ber.*, 100, 2317 (1967).

36. G. M. Bebault, J. M. Berry, Y. M. Choy, G. G. S. Dutton, N. Funnell, L. D. Hayward, and A. M. Stephen, *Can. J. Chem.*, 51, 324 (1973).

37. (a) T. Okuda, S. Harigaya, and A. Kiyomoto, *Chem. Pharm. Bull.* (Tokyo), 12, 504 (1964); (b) R. J. Abraham, L. D. Hall, L. Hough, K. A. McLauchlan, and H. J. Miller, *J. Chem. Soc.*, 748 (1963).

38. D. Horton and M. L. Wolfrom, *J. Org. Chem.*, 27, 1794 (1962).

39. (a) H. Paulsen, *Ber.*, 101, 1571 (1968); (b) T. Sticzay, P. Šipoš, and Š. Bauer, *Carbohyd. Res.*, 10, 469 (1969).

40. R. U. Lemieux and J. W. Lown, *Can. J. Chem.*, 41, 889 (1963).

41. E. O. Bishop, G. J. F. Chittenden, R. D. Guthrie, A. F. Johnson, and J. F.McCarthy, *Chem. Commun.*, 93 (1965).

42. W. S. Chilton, *J. Org. Chem.*, 33, 4459 (1968).

43. C. Satoh, A. Kiyomoto, and T. Okuda, *Chem. Pharm. Bull.* (Tokyo), 12, 518 (1964); C. Satoh and A. Kiyomoto, *Carbohyd. Res.*, 23, 450 (1972).

44. M. K. Hargreaves and D. L. Marshall, *Carbohyd. Res.*, 29, 339 (1973); B. Capon, N. Dennis, R. J. Ferrier, and D. L. Marshall, unpublished results, 1965; J. D. Wander and D. Horton, *Advan. Carbohyd. Chem. Biochem.*, 32, 15 (1976).

45. W. S. Chilton and R. C. Krahn, *J. Amer. Chem. Soc.*, 89, 4129 (1967), but see also ref. 16a.

46. T. R. Emerson, R. J. Swan, and T. L. V. Ulbricht, *Biochemistry*, 6, 843 (1967); G. T. Rogers and T. L. V. Ulbricht, *Biochem. Biophys. Res. Commun.*, 39, 414, 419 (1970); D. W. Miles, M. J. Robins, R. K. Robins, M. W. Winkley, and H. Eyring, *J. Amer. Chem. Soc.*, 91, 831 (1969); T. L. V. Ulbricht, in "Synthetic Procedures in Nucleic Acid Chemistry" (W. W. Zorbach and R. S. Tipson, eds.), Vol. 2, Wiley–Interscience, New York, 1973, p. 177.

47. T. L. V. Ulbricht, J. P. Jennings, P. M. Scopes, and W. Klyne, *Tetrahedron Lett.*, 695 (1964).

48. W. Voelter, H. Bauer, and G. Kuhfittig, *Chimia*, 27, 274 (1973).

49. (a) S. T. K. Bukhari and R. D. Guthrie, *J. Chem. Soc.* (*C*), 1073 (1969); (b) S. T. K. Bukhari and R. D. Guthrie, *Carbohyd. Res.*, 12, 469 (1970).

50. N. Harada, H. Sato, and K. Nakanishi, *Chem. Commun.*, 1691 (1970); compare J. Dillon and K. Nakanishi, *J. Amer. Chem. Soc.*, 97, 5409, 5417 (1975).

51. A. H. Haines and C. S. P. Jenkins, *Chem. Commun.*, 350 (1969).

52. C. B. Barlow and R. D. Guthrie, *J. Chem. Soc.* (*C*), 1194 (1967).

53. S. T. K. Bukhari, R. D. Guthrie, A. I. Scott, and A. D. Wrixon, *Chem. Commun.*, 1580 (1968).

54. C. Djerassi, H. Wolf, and E. Bunnenberg, *J. Amer. Chem. Soc.*, 84, 4552 (1962).

55. C. Djerassi, H. Wolf, D. A. Lightner, E. Bunnenberg, K. Takeda, T. Komeno, and K. Kuriyama, *Tetrahedron*, 19, 1547 (1963).

56. W. Voelter, R. Records, E. Bunnenberg, and C. Djerassi, *J. Amer. Chem. Soc.*, 90, 6163 (1968); W. Voelter, G. Barth, R. Records, E. Bunnenberg, and C. Djerassi, *ibid.*, 91, 6165 (1969).

IV. ELECTRONIC (ULTRAVIOLET) SPECTROSCOPY

H. S. El Khadem and Frank S. Parker

A. Principles and Techniques

When organic molecules absorb energy in the ultraviolet and visible regions, electrons in sigma (σ), pi (π), or nonbonding (n) orbitals undergo promotion from the ground state to higher-energy, antibonding levels. The antibonding orbital related to the σ bond is called a σ^* orbital, and that involving a π bond is called a π^* orbital. There are no antibonding orbitals associated with the nonbonding electrons.

In general, the bands associated with transitions from nonbonding states to antibonding π orbitals ($n \rightarrow \pi^*$) are weaker and appear at longer wavelength than those arising from transitions of π electrons to their antibonding orbitals ($\pi \rightarrow \pi^*$). The $n \rightarrow \pi^*$ bands undergo a blue (hypsochromic) shift on changing from a nonpolar solvent to a hydroxylic one and in nitrogen derivatives may show vibrational fine structure in nonpolar solvents that is lost in the hydroxylic solvents. The nonbonding lone pair of electrons is particularly subject to hydrogen-bond formation with the solvent molecules, so that these electrons require additional energy for their promotion to π^* orbitals, thus causing the shift of the absorption band toward shorter wavelength. The $n \rightarrow \pi^*$ transitions are slightly affected by conjugation, which lowers the energy of the π^* orbital but does not affect the n orbital; the transitions may undergo a blue shift by inductive or electromeric effects of charged centers in the molecule.

The $\pi \rightarrow \pi^*$ bands are more strongly affected by conjugation, which causes a marked bathochromic shift (shift to longer wavelength). This is because the energy of the transition is lowered, when the conjugated system is extended, by both the raising of the donor orbitals and the lowering of the acceptor orbitals, so that the $\pi \rightarrow \pi^*$ bands may extend as far as the $n \rightarrow \pi^*$ bands and even overlap them. The $\pi \rightarrow \pi^*$ bands may also possess vibrational fine structure, but this structure is retained in hydroxylic solvents and permits differentiation of $\pi \to \pi^*$ bands from the $n \to \pi^*$ bands.

The transitions between nonbonding and antibonding σ electrons ($n \to \sigma^*$) involving a heteroatom give rise to strong bands in the 200-nm region, usually below the measured spectral range. They occur at somewhat longer wavelength in the case of nitrogen than with oxygen because the lone pair of electrons in the latter occupy a lower energy level than those of nitrogen, so that the transition to the σ^* level requires greater energy. In the case of sulfur, the electrons are more weakly bound, and excitation of the nonbonding electrons requires less energy; the $n \to \sigma^*$ bands of saturated sulfur compounds appear at longer wavelength (~ 220 nm) and thus may be detected with ordinary spectrophotometers.

The bands arising from transition of σ electrons to their antibonding orbitals ($\sigma \to \sigma^*$) occur in the vacuum ultraviolet and can be detected only with special spectrophotometers. Figure 1 summarizes the general relationships between electronic energy levels and the associated spectral bands.

An ultraviolet spectrum is a continuous record of absorbance A ("optical density") versus wavelength. A is defined as log I_0/I, where I_0 is the intensity of incident light and I is the intensity of transmitted light. The Beer–Lambert law states that $A = \epsilon c l$, where ϵ is the molar extinction coefficient, c is the molar concentration, and l is the path length in centimeters. Another term, $E_{1\text{cm}}^{1\%}$, the absorbance of a 1% solution of a substance of unknown

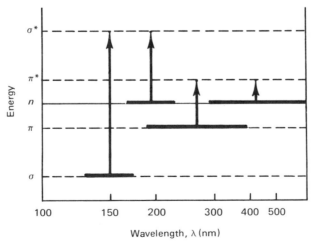

FIG. 1. Schematic relationship between electronic energy levels and the corresponding spectral bands.

* References start on p. 1390.

molecular weight in a 1.0-cm. cell, is also used. Percent transmission ($I/I_0 \times 100$) versus wavelength is frequently recorded.

The apparatus and technique of absorption spectrophotometry have been described in many texts; it suffices here to mention that most spectrophotometers record absorbance or transmission curves between 190 and 1000 nm and are capable of repeating the spectrum or a portion thereof at fixed intervals to monitor reaction kinetics. Because of the experimental difficulty involved, and the need for special apparatus, carbohydrate derivatives have seldom been studied in the vacuum ultraviolet or in the region beyond the visible (near infrared), both of which merit greater attention by the spectroscopist and the carbohydrate section.

The lower wavelength cut-off for solvents commonly used for carbohydrates and their derivatives are: water, 191 nm; ethanol, 204 nm; chloroform, 237 nm, and cyclohexane, 195 nm, for 1-cm cells. It is advantageous to measure the spectrum of a substance in more than one solvent and, whenever possible, in a hydroxylic, a polar, and a nonpolar solvent, to observe shifts, if any, in the absorption maxima. This information, together with a study of the effects of substitution in the neighborhood of a chromophore, facilitates the assignment of the absorption bands to the various transitions.

More details may be found in the general references[1–5] on ultraviolet spectroscopy given at the end of this section.

B. Electronic Spectra of Simple Carbohydrates and Their Derivatives

Free sugars are not normally found in the aldehydo or keto forms to any appreciable extent, but exist mainly in cyclic hemiacetal forms, which do not have π electrons. They are not, therefore, expected to show any absorption beyond 190 nm; this is because their absorption at longest wavelength would arise from promotion of a nonbonding electron from the lone pair on oxygen, to an antibonding σ orbital ($n \rightarrow \sigma^*$). Such transitions, as stated earlier, lie below the measured range of conventional spectrometers. This proposition was confirmed experimentally by studying the spectra of 4-oxa-5α-cholestan-3α-ol, a compound prepared as a model for pyranoid sugars, which shows no ultraviolet absorption above 200 nm and no carbonyl bands in the infrared.[6]

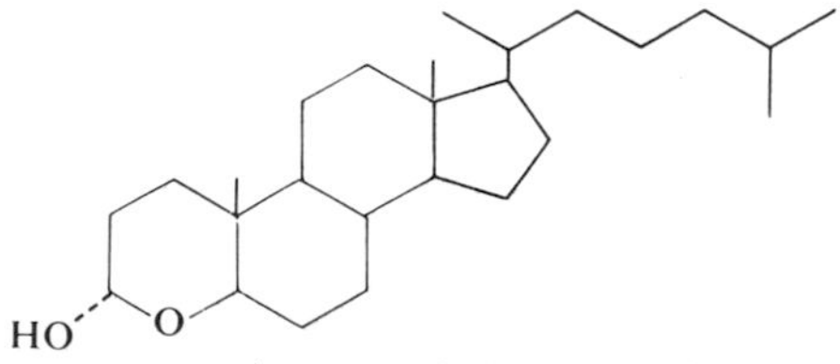

4-oxa-5α-cholestan-3α-ol

Of course this does not mean that such compounds do not absorb in the far ultraviolet, and if their spectra were measured under vacuum and in the

TABLE I

ABSORPTION CHARACTERISTICS OF TYPICAL CARBOHYDRATE DERIVATIVES

Compound type	λ_{max} (nm)	Ref.
Aldoses	< 200	50
Ketoses	280	50, 51
Aldosuloses	228	53
Aldonic acids and their lactones	< 200	54, 55
Aldonic acid amides	< 200	56
Nitriles	< 200	39
Azides	287	37
Oximes	< 200	38
Acetyl derivatives and methyl ethers	< 200	4
Benzoyl derivatives	~ 230, 275, 280	4
Benzylidene acetals	210, 250, 256, 260, 266	11
p-Nitrobenzoates	~ 261, 270, 280	4
3,5-Dinitrobenzoates	260, 270, 280, 290	4
2,4-Dinitroanilino derivatives	208, 264, 333	59

absence of solvents, strong $\sigma \to \sigma^*$ and $n \to \sigma^*$ bands would be observed in that region. To possess characteristic absorptions measurable with the ordinary u.v. spectrophotometers, organic molecules usually require π electrons (see Fig. 1). These are found in unsaturated groups often referred to as "chromophores," a name introduced by Witt in 1876 at a time when only absorptions above 220 nm could be detected.

The common chromophores encountered in carbohydrate derivatives are the C=C, C=N, N=N, C=O, C=S, and S=O groups, and also aromatic and heteroaromatic rings (Table I). To expedite the examination of their ultraviolet spectra, carbohydrate derivatives have been grouped in this section according to the chromophores present in their molecules. The references given have been selected from the voluminous literature to demonstrate specific points of interest rather than to cite all pertinent ultraviolet data. Particularly useful were modern papers in which the authors assigned the absorptions to the different orbital transitions, but in some obvious cases where this was not done, probable assignments have been suggested. Excluded here are the chromogens produced by the breakdown of sugars by heat, acids, alkali, nitrous acid, or other processes, as well as colored products formed by the addition of other reagents. Such color-producing reactions are the basis of various colorimetric analyses and reducing-sugar determinations, discussed in detail by Aminoff and by Binkley, respectively (see Vol. IIB, Chapter 45); they have also been reviewed by Dische.[5] Another source of reference for color reactions of sugars is the three volumes of Hershenson's index,[7] where

* References start on p. 1390.

the sugars are listed together with the color-forming reagents used; reference is also made to original published spectra of the colored products. The effect of ultraviolet radiation on carbohydrates is discussed by Phillips in Chapter 26.

1. Derivatives Having C=C Groups

The $\pi \to \pi^*$ transitions of an isolated C=C group fall below the limit of the spectrophotometer range at $\lambda_{max} \sim 180$ nm. Only the "tail" of this absorption is normally observed for simple alkenic groups in saccharides; thus compounds **1** and **2** show "tail" absorption near 205 nm.[8] When the double bond is conjugated with a carbonyl double bond, the absorption band undergoes a bathochromic shift, and thus compounds **3** and **4** absorb maximally at 228 nm.[9] Ascorbic acid (**5**) absorbs at 245 nm in the nondissociated form and at 265 nm in the monodissociated form.[10] Furanose and pyranose derivatives having two conjugated, annular, double bonds display the absorptions characteristic of furans and pyrans; for example, compound **6** shows λ_{max} 218 nm,[11,12] and 2,4,6-trimethylpyran has λ_{max} 221 and 278 nm.[13]

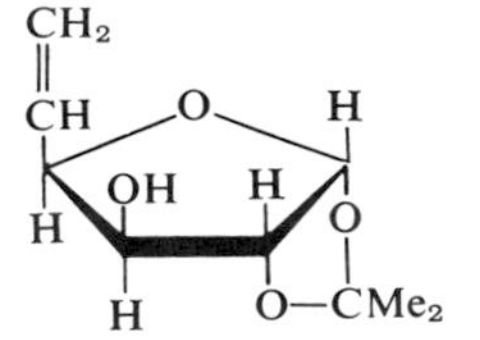

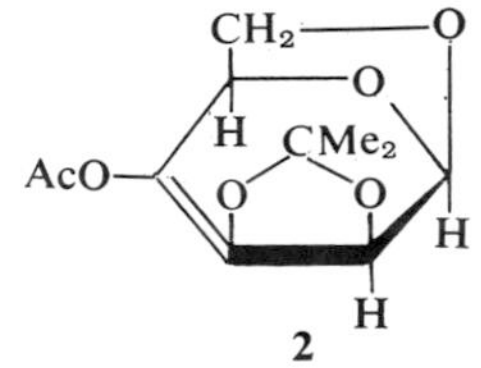

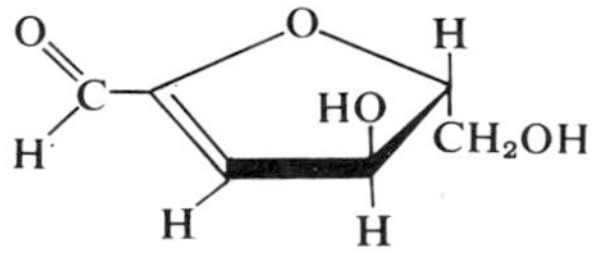

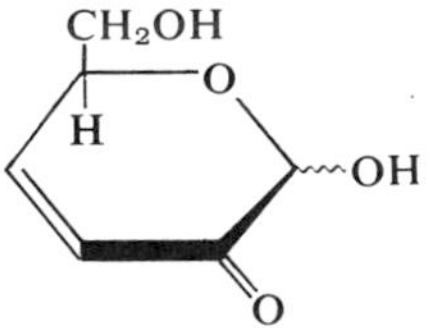

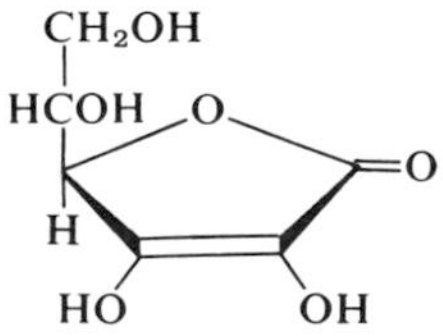

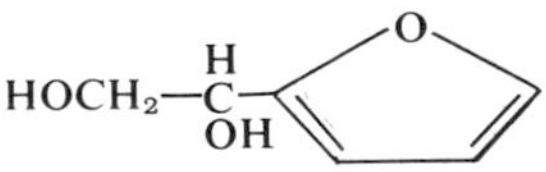

2. Derivatives Having C=N and N=N Groups

The true aldimines and Schiff bases expected from the condensation of aldoses with ammonia or amines do not generally exist as such in the sugar series. They rearrange to cyclic products and Amadori compounds, as discussed in Chapter 20. Some hydrazones, and most bis(hydrazones), exist in the acyclic form. Their spectra are, however, more complicated than those of simple hydrazones because the C=N groups in these compounds are often involved in prototropic rearrangements with adjacent arylamino groups (in hydrazono–azo tautomerism), and with the various groups of the sugar side chain, to give cyclic products that may, in turn, give rise to equilibria between the hydrazono and hydrazino forms. The C=N group may also be involved in the formation of hydrogen-bonded chelates with adjacent OH or NH groups.

In general, the hydrazono forms are favored over the azo forms; this generalization applies to simple hydrazones as well as to the hydrazones and osazones of the sugars. There are, however, examples where chelation is known to stabilize a phenylazo group, as in compound 7 (λ_{max} 251, 286, and 487 nm), which is the red, enolic, phenylhydrazono–phenylazo tautomer of the keto bis(phenylhydrazono) form[14] 8. Another example of the conversion of

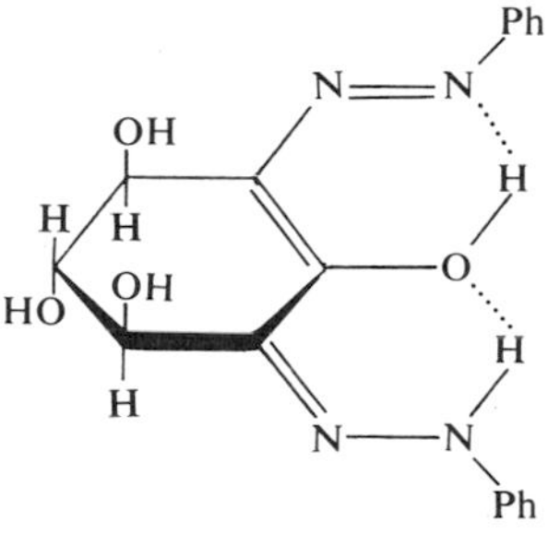

7 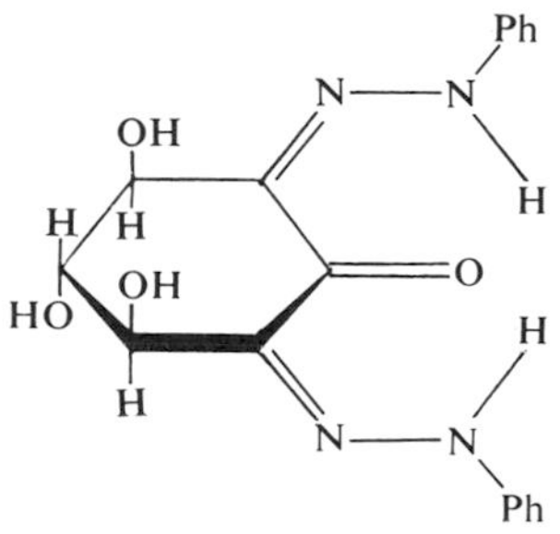   8

hydrazono derivatives into azo forms is observed when acetylated saccharide hydrazones are warmed in the presence of catalytic amounts of pyridine to produce azoethylenes (for instance, compound 9). These compounds are yellow and have a weak absorption maximum at ∼425 nm, which is absent in spectra of the parent hydrazone acetates[14–17] (such as 10), and which may be assigned to $n \rightarrow \pi^*$ transitions.

Cyclized hydrazones of sugars are examples of derivatives existing in the hydrazino forms. Their spectra are somewhat different from those of the acyclic forms,[18,19] a property used by Tipson[20] to assign an open-chain structure to D-ribose hydrazone. Another useful tool to differentiate between the acyclic and cyclic hydrazones is the formazan reaction[21–24] (see Chapter 21). The bright red formazans, with their unmistakable spectra, are obtained only from the acyclic hydrazones.

---

* References start on p. 1390.

ULTRAVIOLET DATA FOR 9	
λ_{max}(nm)	ϵ
222	11,000
227	12,000
274	10,800
420	200

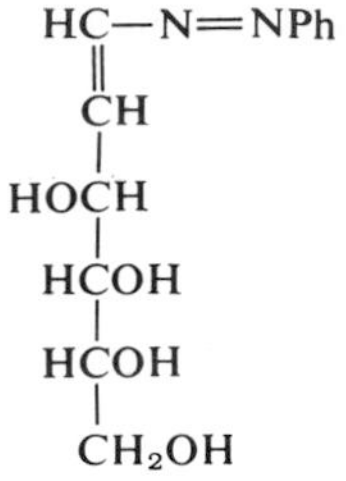

```
HC—N=NPh
  ‖
  CH
  |
HOCH
  |
 HCOH
  |
 HCOH
  |
 CH₂OH
   9
```

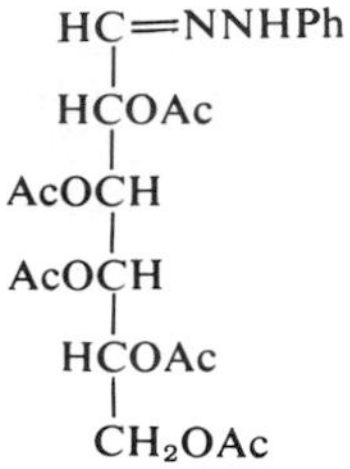

```
HC=NNHPh
  |
 HCOAc
  |
 AcOCH
  |
 AcOCH
  |
 HCOAc
  |
 CH₂OAc
   10
```

Chelation exerts a marked influence on the absorption spectra of osazones; a study by Chapman[25] of the n.m.r. spectra of various bis(phenylhydrazones) and phenylosazones of sugars revealed that the differences previously[26–30] observed between the electronic spectra of the latter and, for instance, glyoxal bis(phenylhydrazone) (11) were due to the fact that the sugar derivatives possess chelated rings, whereas the glyoxal derivative exists in the unchelated form. Similarly, the spectra of N,N-disubstituted osazones (12),[30] which cannot form hydrogen-bonded chelates, differ from those of monosubstituted phenylosazones 13 (which are chelated), in that their lower-energy bands are shifted to shorter wavelength. Because of the increased conjugation, the $n \to \pi^*$ bands in most of the foregoing compounds become swamped by the more intense $\pi \to \pi^*$ bands, so that the longer-wavelength absorptions are probably due to both types of transition.

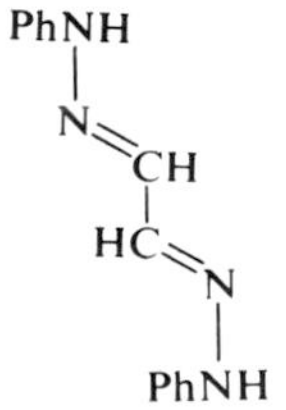

11 (λ_{max} 303,378 nm) 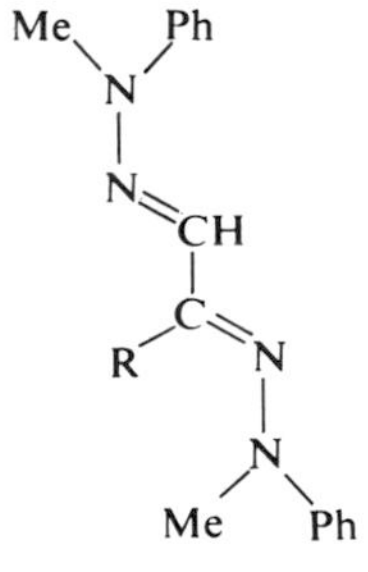   12 (λ_{max} 335 nm)   13 (λ_{max} 256, 310, 396 nm)
(R = polyhydroxyalkyl)

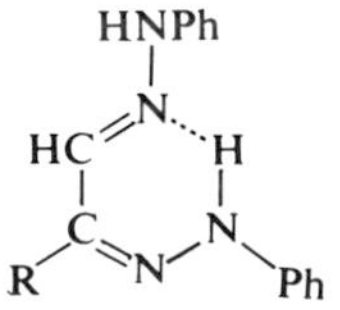

The position of the longer-wavelength absorption band undergoes a bathochromic shift in proceeding from hydrazones[18] to osazones[30,31] and from alkyl[19] (or acyl[32]) to aryl[30,31] derivatives. Consequently, the hydrazones, whether acyl- or aryl-substituted, are colorless except for the nitro- and polynitro-phenyl derivatives; the acylosazones are likewise colorless, but the arylosazones are golden yellow, and their nitro derivatives are red; mixed 1-aryl-2-acyl-bis(hydrazones) have intermediate colors.[33] Also red are the bis(arylhydrazones) of dehydro-L-ascorbic acid,[31] which have their hydrazone residues in conjugation with the lactone carbonyl group; this conjugation also influences the monohydrazones of dehydro-L-ascorbic acid.[34] Interestingly, such cyclic α-diketo bis(hydrazones) as **7** and **14** absorb at longer wavelength than similar acyclic compounds such as mesoxalaldehyde bis(phenylhydrazone)[35] (**15**). This is probably because the keto group and the two hydrazone residues are nearly coplanar in the cyclic compounds, and the increased conjugation causes a shift of the $\pi \rightarrow \pi^*$ transitions to longer wavelength.

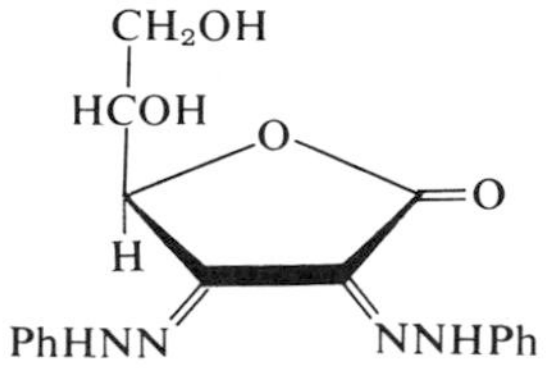

CH$_2$OH
|
HCOH
|
H

PhHNN   NNHPh
14 (λ_{max} 261, 274, 352, 444)

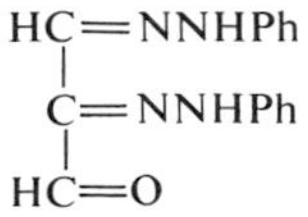

HC=NNHPh
|
C=NNHPh
|
HC=O

15 (λ_{max} 270, 365)

The osazone formazans, which have phenylazo groups conjugated with the bis(phenylhydrazone) residues, absorb in the same region: λ_{max} 335 and 440 nm.[36]

Sugar azides show a weak absorption near 287 nm, which has been assigned to $\pi \rightarrow \pi^*$ transitions[37]; on the other hand, diazo sugar derivatives, such as **16**, absorb at λ_{max} 242 nm (ϵ 10,620).[38]

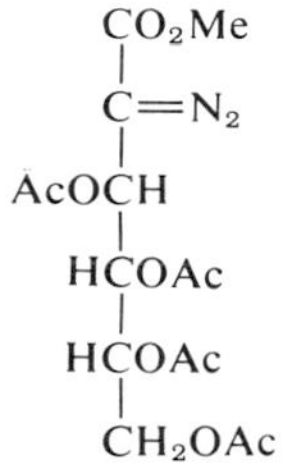

CO$_2$Me
|
C=N$_2$
|
AcOCH
|
HCOAc
|
HCOAc
|
CH$_2$OAc

16

The spectra of certain oximes[38a] and nitriles[39] have been recorded, but these are of little significance, as they exhibit no characteristic absorptions, probably because of cyclization.

3. Derivatives Having C=O Groups

As mentioned earlier, the spectra of the reducing sugars are affected by their existence in equilibria between several cyclic forms that lack the C=O chromophore, and by variation from one sugar to another in the position of this equilibrium. Furthermore, reducing sugars are difficult to purify, and often contain contaminants and degradation products possessing strong absorbances that may be quite misleading. This was early pointed out by

---

* References start on p. 1390.

Sørensen[40] in a review covering the period up to the late 1930s, where he showed how impurities in samples had led earlier investigators to erroneous conclusions regarding the amount of aldehyde form present in reducing sugars in aqueous solutions. Although the parent compound of furanoses (tetrahydrofuran-2-ol) and that of pyranoses (tetrahydropyran-2-ol) contain considerable proportions of the aldehyde form in aqueous solutions,[41] only very small proportions of the open-chain aldehyde forms[42] are present in equilibrated solutions of pentoses and hexoses (estimated at 0.0026% for D-glucose), and substitution tends to cause the cyclic forms to be favored over open-chain ones.[43]

There are instances, however, where somewhat larger proportions of aldehyde may be present in aqueous solutions of sugars. For example, D-idose is unlike most sugars in that it gives a positive Schiff reaction for aldehydes,[44] and 5-O-methyl-D-glucose, and 5,6-di-O-methyl-D-glucose, which cannot form pyranose rings, react with both Schiff's reagent and cold Fehling's solution.[45] Other examples are 3,6-anhydro-2,4-di-O-methyl-D-galactose,[46] and the corresponding D-glucose[47] and D-mannose[48] derivatives; these appear to exist mainly in the aldehyde form because of the instability of their pyranoid tautomers.

Aliphatic aldehydes and ketones show a very weak absorption band in the region of 300 nm (ϵ ~10) arising from transitions from nonbonding orbitals to antibonding π orbitals. This band has been observed in sugar derivatives; for instance 1,6-anhydro-2,3-O-isopropylidene-β-D-*lyxo*-hexo-pyranos-4-ulose (**17**), a ketone that cannot tautomerize to a hemiacetal, shows[49] λ_{max} 300 nm (ϵ 50) in chloroform. In water there is no absorption between 250 and 500 nm, presumably because the ketone is largely converted into the hydrated form, a *gem*-diol. Absorption spectra have been presented

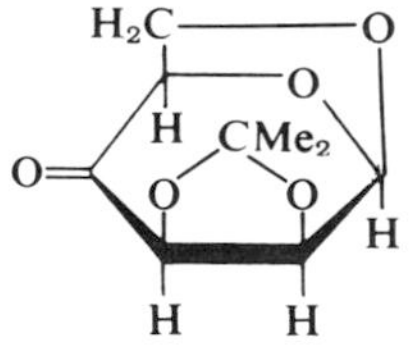

17

for various D-fructose preparations that show absorption maxima near 280 nm, supposedly due to the keto group.[50] It is possible, however, that they might be caused by impurities and, as in the case of D-glucose, might disappear upon purification.[51] Subsequently, 2,7-anhydro-β-D-*altro*-heptulofuranose and D-*altro*-heptulose were reported[52] to have absorptions at 245 and 280 nm, respectively.

Aldosuloses (*osones*) generally exist in cyclic forms; whereas simple α,β-diketones and keto aldehydes are yellow, osones are colorless, and they show no absorption above 240 nm. Thus, for example, 3-deoxy-D-*erythro*-hexosulose[53] shows maximal absorption at 228 nm and its 3,4-unsaturated derivative[9] at 233 nm; these values are hardly compatible with acyclic α-keto aldehyde structures, which would be expected to absorb strongly above 400 nm.

In lactones, the annellation of the oxygen atom would tend to increase the energy interval between the nonbonding level and the lowest unoccupied π^* orbital, so that the $n \rightarrow \pi^*$ absorptions of these compounds would be expected to undergo a very large blue shift (about 90 nm) as compared with the unperturbed carbonyl absorption. It is not surprising, therefore, that the spectra of aldonic acids and their γ- and δ-lactones do not show any discrete bands in their ultraviolet spectra.[54,55] For the same reason, the corresponding amides[56] likewise lack any characteristic absorption. It should be noted that esters and amides of simple aliphatic carboxylic acids absorb below 210 nm.

4. *Sulfur-containing Derivatives*

The nonbonding electrons of sulfur are more weakly bound than those of oxygen, and thus require less energy for their excitation. Accordingly, sulfur compounds, whether saturated or unsaturated, absorb at longer wavelengths than their oxygen counterparts. The absorptions of many saturated sulfur compounds lie at sufficiently long wavelength that they may be detected with the usual instruments. Thus 1-thio-β-D-galactopyranose pentaacetate[57] (**18**) absorbs at 226 nm (ε 3500). This absorption is quite characteristic of acetylthio groups and may be used to differentiate them from acetoxyl groups.

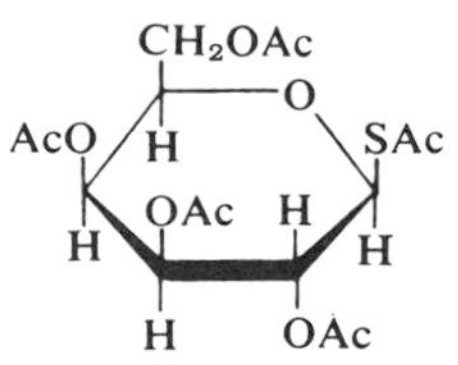

CH₂OAc
O
AcO
H
SAc
OAc
H
H
H
H
OAc

18

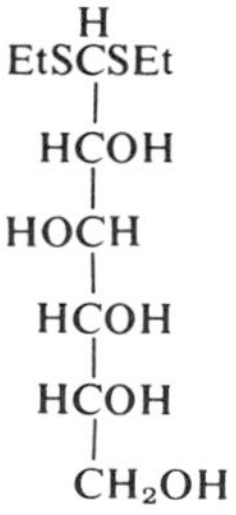

H
EtSCSEt
HCOH
HOCH
HCOH
HCOH
CH₂OH

19

Aldose dialkyl dithioacetals—for example, compound **19**—exhibit an ultraviolet absorption maximum near 213 nm (ε 1630) and a second maximum at 232 (ε 600), as do the dialkyl dithioacetals of simple aldehydes. The absorption at shorter wavelength corresponds to photoexcitation of the C–S bond,

* *References start on p. 1390.*

whereas the longer-wavelength absorption is indicative of a photoexcited state involving direct geminal interaction of the sulfur atoms. The latter state is probably[58] stabilized by extra delocalization of the electrons, made possible by expansion of the sulfur valence shell beyond eight electrons through the use of the 3d orbitals in a structure of the type of **20**.

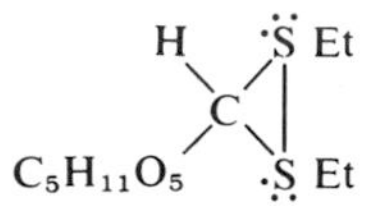

$$\text{H} \quad \ddot{\text{S}} \text{ Et}$$
$$\diagdown \diagup$$
$$\text{C}$$
$$\diagup \diagdown$$
$$C_5H_{11}O_5 \quad \cdot\ddot{\text{S}} \text{ Et}$$

20

For sulfur compounds having C=S groups, $\pi \to \pi^*$ and $n \to \pi^*$ transitions may be expected. Thus, in glycosyl ethylxanthates,[59] for example compound **21**, two bands were observed in ethanol; the first, at 244 nm (ϵ 2800), may be assigned to the $\pi \to \pi^*$ transition of the C=S bond, and the second, at 359 nm (ϵ 200), to the $n \to \pi^*$ transition of the S—C group.

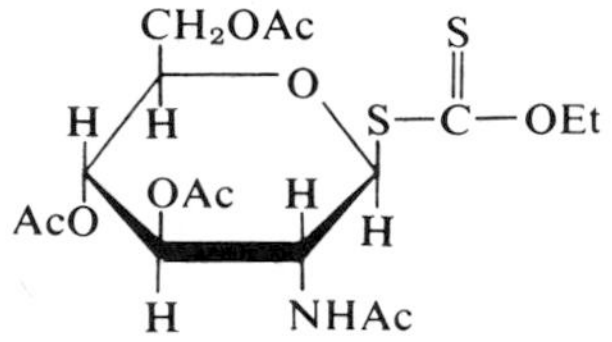

21

Thiocarbamates[60] show strong absorption at about 249 nm (ϵ 13,000 to 19,000), moderate near 204 nm (ϵ 4000 to 9000), and weak between 285 and 300 nm, with no distinct maximum in the last region. These bands appear to be characteristic of the O-(N,N-dialkylthiocarbamoyl) group; the $n \to \pi^*$ transition of the thiocarbonyl group is certainly in the last region. A typical example is 1-O-(N,N-dimethylthiocarbamoyl)-2,3:5,6-di-O-isopropylidene-α-D-mannofuranose (**22**), which has λ_{max}^{EtOH} 205 nm (ϵ 9,000), 249 nm (ϵ 19,000), and shoulder absorption at 285 to 300 nm. A third example of absorption by the thiocarbonyl group is afforded by 1,2-O-isopropylidene-α-D-glucofuranose 5,6-thionocarbonate (**23**). This compound shows[61] a strong absorption in ethanol at 234 nm (ϵ 12,300) and a weak absorption at 306 nm (ϵ 34), evidently owing to $\pi \to \pi^*$ and $n \to \pi^*$ transitions, respectively.

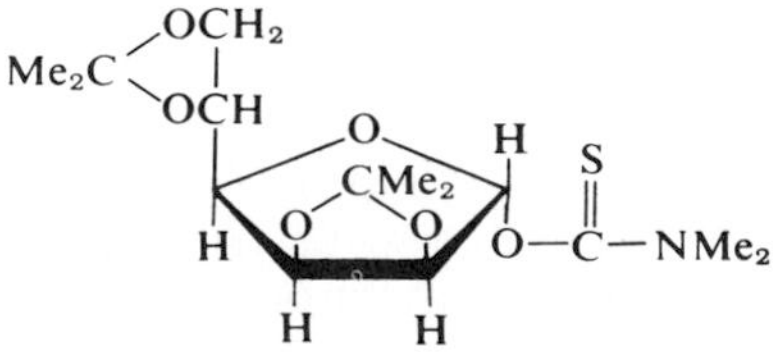

22

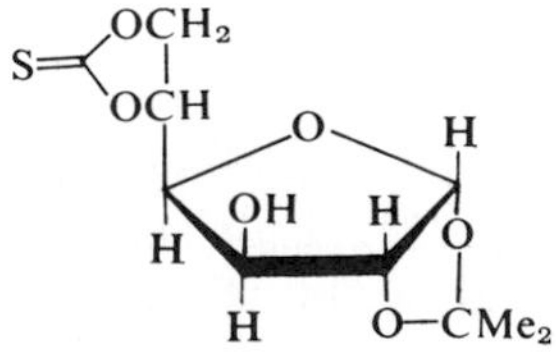

23

The u.v. absorption spectra[62] of two unsaturated disulfones were shown to exhibit broad maxima in the 270-nm region in ethyl alcohol. These compounds were 3,4,5,6-tetra-O-acetyl-1,2-dideoxy-1,1-bis(ethylsulfonyl)-D-*arabino*-hex-1-enitol and 3,4,5,6-tetra-O-acetyl-1,2-dideoxy-1,1-bis(ethylsulfonyl)-D-*lyxo*-hex-1-enitol. The former shows a shoulder near 280 nm.

5. Derivatives Having Aromatic and Heteroaromatic Rings

The characteristic absorptions of aromatic and heteroaromatic rings are so strong that they dominate the spectra of carbohydrate derivatives containing them. Thus O-benzoyl and O-benzylidene derivatives of methyl glycosides show hardly any absorptions other than those of the phenyl ring. The same applies to the N-(2,4-dinitrophenyl) derivatives of amino sugars and to the phenylazophenyl glycosides.[63,64] Of course, when stronger chromophores are present in the molecule, as in osazone benzoates,[65] the absorptions of these groups can be observed in addition to those of the phenyl groups.

Similarly, phenyl glycosides, nucleosides, and hydroxyalkyl heteroaromatic compounds arising from the cyclization of sugar derivatives usually show absorptions analogous to those of the methyl-substituted parent aromatic or heterocyclic compound. This correspondence was utilized in the determination of the position of linkage of D-ribose to the heterocyclic nucleus in purine nucleosides. The spectra of these compounds were compared with those of the 7- and 9-methyl aglycons and found to be identical with the latter[66,67] (see also Vol. IIA, p. 57).

It is evident that, in aromatic compounds, the electron-transfer bands are responsible for the absorptions above 210 nm, and in heteroaromatic compounds these absorptions will be supplemented by $n \rightarrow \pi^*$ transitions. In six-membered aromatic and heteroaromatic rings, the observed bands are usually classified according to Clar[68] into α, p, and β bands, with the α band at the longer-wavelength side of the spectrum and the β band at the shorter. In addition to these, a weak triplet absorption (t band) is sometimes found on the long-wavelength side of the spectra of these compounds. The electron-transfer absorptions of phenyl derivatives and of monoheteroaromatic, six-membered rings are more intense than the α band of the parent benzene, in which the component transition moments cancel each other. In p-substituted phenyl derivatives, the intensity is increased fourfold compared with the local excitation in benzene, as the component charge is doubled and the increase in absorption is proportional to the square of the resultant transition moments. The presence of electronegative groups in the p-position also causes a shift of the α bands toward the red. Accordingly, it may be expected that N-phenyl derivatives, and annellated azines obtained from saccharide nitrogen deriva-

---

tives, would show a strong $\pi \rightarrow \pi^*$ band overlapping the weaker $n \rightarrow \pi^*$ bands. It is, however, possible to find $n \rightarrow \pi^*$ bands in some instances, as in sugar-derived quinoxalines[69-72] where the parent heterocycle[73] has an $n \rightarrow \pi^*$ band at 330 nm (ϵ 200).[72]

The monocyclic, five-membered, nitrogen heterocycles do not usually show $n \rightarrow \pi^*$ bands, and the longer-wavelength absorptions of these compounds have been attributed to $\pi \rightarrow \pi^*$ transitions.[74] Accordingly, the absorption bands observed with saccharide pyrroles,[75,76] pyrazoles,[77-81] imidazoles,[82-86] and triazoles[87-90] are almost certainly $\pi \rightarrow \pi^*$ bands. An exception to this generalization is with the purines; their 9-methyl derivatives have been found to give $n \rightarrow \pi^*$ absorptions at the long-wavelength edge of the lowest-energy $\pi \rightarrow \pi^*$ band. Similarly, the u.v. spectra of nucleotides have been found to show $n \rightarrow \pi^*$ bands at the limit of the long-wavelength absorption, submerged under the stronger $\pi \rightarrow \pi^*$ bands.[91]

C. Spectra of Glycoproteins, Nucleic Acids, and Related Substances

Polysaccharides unsubstituted by chromophores do not exhibit significant ultraviolet absorption. An almost flat curve was reported for neuraminic acid from 240 to 310 nm, with slight maxima near 260 and 280 nm.[92] On the other hand, glycoproteins display characteristic spectra having well-defined maxima, because of their protein component.

The ultraviolet absorption of proteins[93,94] has been studied extensively in solutions and in histological specimens.[95] Most proteins show a principal band near 280 nm, a minimum near 250 nm, and a steeply rising end-absorption below 250 nm. A few show absorptions of uncertain origin near 250 nm. The absorption in the 220-nm region has been ascribed to the peptide linkage.[96] Upon removal of the protein portion of a glycoprotein, the 280-nm band disappears. Thus when a corneal glycosaminoglycan–protein complex, having λ_{max} 280 nm indicative of protein, was subjected to the action of trypsin, the absorption disappeared.[97] Similarly, treatment of crude potassium chondroitin sulfate (from cartilage) with kaolin removed a protein contaminant that caused shoulder absorption near 280 nm.[98]

An ethanol-precipitated hyaluronic acid preparation from synovial fluid, containing 30% protein, showed a maximum in the region of 275 nm in neutral or acid solution, with a shift to about 285 nm in $0.1 M$ sodium hydroxide.[99] This property is typical of proteins and is due mainly to the tyrosine and tryptophan that they contain.

The ultraviolet spectrum of a nucleoprotein is composed of contributions from the nucleic acid and the protein moieties. As proteins absorb strongly at 280 nm and nucleic acids absorb strongly at 260 nm, a correction for nucleic acid content may be made by using an empirical equation[100]:

$$\text{Protein concentration (mg/ml)} = 1.55 \, D_{280nm} - 0.76 \, D_{260nm}$$

A difficulty with this method is its lack of specificity, owing to possible interference by other substances that absorb in the region 250 to 290 nm.

The ultraviolet absorptions of nucleosides, nucleotides, and nucleic acids are due essentially to the purine and pyrimidine moieties.[67] Many nucleotides are barely distinguishable by their u.v. spectra from their parent nucleosides. Absorption curves for the purines, pyrimidines, nucleosides, and nucleotides have been recorded over the range 210 to 300 nm, and absorption maxima for the various compounds are well established.[101] It should be pointed out that pH is an important factor to consider when investigating the absorption maxima and their intensities for these compounds, as the spectra vary with pH in a manner that may be usefully correlated with structure.[102]

A practicable convention for reporting molar absorptivities of nucleic acids in aqueous solution is based on phosphorus content[103]:

$$\epsilon(P) = A/C \cdot d$$

where $\epsilon(P)$ is the molar absorptivity, A is absorbance, d is internal cell length in centimeters, and C is gram-atoms of phosphorus per liter.

When native DNA is altered by heat, there is a marked hyperchromic effect,[104] denoting a decrease in hydrogen bonding in the DNA. Rich and Kasha[91] have shown that, although there is a strong hypochromism in the main absorption bands of polyadenylic acid and polyuridylic acid when they form a 1:1 helical complex, the long-wavelength "tail" of the absorption band shows a hyperchromic effect. This is expected from $n \rightarrow \pi^*$ transitions hidden in the "tail" of the main $\pi \rightarrow \pi^*$ bands mentioned earlier.

The ultraviolet spectrum of flavine adenine dinucleotide (FAD), has maxima at 960 nm and 450 nm,[105] whereas for riboflavin and D-glucosyl riboflavin the larger wavelength maximum lies at 510 nm.[106] Nicotinamide adenine dinucleotide (NAD$^+$) in the reduced state (NADH) has an absorption maximum at 340 nm, but the oxidized form does not have a maximum at this wavelength[107,108]; this difference is the basis of numerous spectrophotometric assays for monitoring enzyme-catalyzed reactions.

The cobamide coenzyme form of vitamin B_{12} contains two carbohydrates.[109] A D-ribofuranosyl 3-phosphate group is linked through nitrogen to 5,6-dimethylbenzimidazole, and 5'-deoxyadenosine is also present, linked through the 5'-carbon atom to the cobalt atom of the vitamin. The cobamide coenzyme form has no absorption maximum near 360 nm, but in the presence of cyanide and light the cobalt is oxidized to the trivalent state, and the 5'-deoxyadenosine is replaced by cyanide, whereupon the resultant cyanocobalamin exhibits a maximum at 361 nm ($E_{1cm}^{1\%}$ 107) that may be used to monitor this reaction. Cyanocobalamin also has maxima at 378 and 550 nm in water, with extinction coefficients ($E_{1cm}^{1\%}$) of 115 and 64, respectively.

---

* References start on p. 1390.

An "ultraspecific" micromethod for the determination of D-glucose, which is advantageous when D-glucose must be determined in the presence of relatively high concentrations of other sugars such as D-mannose, uses a highly stereospecific D-glucokinase obtained from *Aerobacter aerogenes*. The change in absorbance of reduced nucleotide coenzyme (NADPH) at 340 nm provides a measure of the D-glucose present. Nanomole quantities of D-glucose may be determined in the presence of relatively high concentrations of D-mannose, D-fructose, 2-deoxy-D-*arabino*-hexose, maltose, and many other sugars.[110]

Basic metachromatic dyes are widely used for the histological localization of such substances as sulfated glycosaminoglycans, nucleic acids, and hyaluronic acid[111–114] (see Vol. IIB, Chapter 45, Section IV). For example, Toluidine Blue alone in water displays an absorption maximum near 620 nm.[111] When a metachromatic substance is added to a solution containing Toluidine Blue, a new, broader peak of lower intensity appears at 540 nm with a concurrent lowering of the band at 620 nm, which becomes a shoulder. Metachromasy is discussed in greater detail by Swift.[115]

It should be stressed here that many of the spectrophotometric procedures that have been used for particular sugars have limited specificity, and their application in biological assays requires many precautions.

An excellent spectrophotometric reagent must finally be mentioned because of its importance in analytical chemistry, and that is the iodine–starch complex. The amylose component of starch interacts with iodine to give a characteristic deep-blue complex having λ_{max} near 625 nm ($\epsilon \sim 40,000$), in which the iodine molecules are situated in the core of a helical (1→4)-α-D-glucan chain (see Vol. IIB, Chapter 38). For full development of this complex, the glucan chain must be sufficiently long and unbranched; otherwise the typical "blue values" are not observed.[116] "Blue values" have been used[117] to compare certain varieties of starch, amylose, and amylopectin. The absorption curves of the polysaccharide–iodine complexes were determined between 400 and 700 nm, and the observed maxima range from about 550 to about 660 nm. The curves have been used to obtain evidence regarding the linearity of amylose, the chain length of degraded amyloses, and the structure of amylopectin.

REFERENCES

1. H. H. Jaffé and M. Orchin, "Theory and Applications of Ultraviolet Spectroscopy," Wiley, New York, 1962.
2. A. Gillam and E. S. Stern, "An Introduction to Electronic Absorption Spectroscopy in Organic Chemistry," Arnold, London, 1957.
3. R. P. Bauman, "Absorption Spectroscopy," Wiley, New York, 1962.
4. A. I. Scott, "Interpretation of the Ultraviolet Spectra of Natural Products," Pergamon, London, 1964.
5. Z. Dische, *Methods Carbohyd. Chem.*, 1, 475 (1962).
6. J. T. Edward, P. F. Morand, and I. Puskas, *Can. J. Chem.*, 39, 2069 (1961).

7. H. M. Hershenson, "Ultraviolet and Visible Spectra Index for 1930–1954," Academic Press, New York, 1956; "Ultraviolet and Visible Spectra Index for 1955–1959," published 1961; "Ultraviolet and Visible Spectra Index for 1960–1963," published 1966.
8. D. Horton, personal communication.
9. M. L. Mednick, *J. Org. Chem.*, **27**, 398 (1962); compare H. Esterbauer, E. B. Sanders, and J. Schubert, *Carbohyd. Res.*, **44**, 126 (1975).
10. J. S. Lawendel, *Nature*, **180**, 434 (1957).
11. E. L. Albano, D. Horton, and T. Tsuchiya, *Carbohyd. Res.*, **2**, 349 (1966).
12. D. Horton and T. Tsuchiya, *Carbohyd. Res.*, **3**, 257 (1966).
13. R. Gompper and O. Christmann, *Ber.*, **94**, 1784 (1961).
14. A. J. Fatiadi and H. S. Isbell, *Carbohyd. Res.*, **5**, 302 (1967).
15. M. L. Wolfrom, A. Thompson, and D. R. Lineback, *J. Org. Chem.*, **27**, 2563 (1962).
16. M. L. Wolfrom, G. Fraenkel, D. R. Lineback, and F. Komitsky, Jr., *J. Org. Chem.*, **29**, 457 (1964).
17. H. El Khadem, M. L. Wolfrom, Z. M. El-Shafei, and S. H. El Ashry, *Carbohyd. Res.*, **4**, 225 (1967).
18. H.-H. Stroh and B. Ihlo, *Ber.*, **96**, 658 (1963).
19. H.-H. Stroh and H. G. Scharnow, *Ber.*, **98**, 1588 (1965).
20. R. S. Tipson, *J. Org. Chem.*, **27**, 2272 (1962).
21. G. Zemplén and L. Mester, *Acta Chim. Acad. Sci. Hung.*, **2**, 9 (1952).
22. L. Mester and A. Major, *J. Amer. Chem. Soc.*, **77**, 4297 (1955).
23. L. Mester and A. Messmer, *Methods Carbohyd. Chem.*, **2**, 119 (1963).
24. G. V. D. Tiers, S. Ploven, and S. Searles, *J. Org. Chem.*, **25**, 285 (1960).
25. O. L. Chapman, R. W. King, W. J. Welstead, Jr., and T. J. Murphy, *J. Amer. Chem. Soc.*, **86**, 4968 (1964).
26. L. L. Engel, *J. Amer. Chem. Soc.*, **57**, 2419 (1935).
27. P. Grammaticakis, *C.R. Acad. Sci.*, **223**, 1139 (1946).
28. V. C. Barry, J. E. McCormick, and P. W. D. Mitchell, *J. Chem. Soc.*, 222 (1955).
29. G. Henseke, G. Hanisch, and H. Fischer, *Ann.*, **643**, 161 (1961).
30. G. Henseke and H. J. Binte, *Chimia*, **12**, 103 (1958).
31. H. El Khadem and S. H. El Ashry, *Carbohyd. Res.*, **7**, 501 (1968).
32. H. El Khadem and M. A. E. Shaban, *Carbohyd. Res.*, **2**, 178 (1966); **3**, 416 (1967).
33. H. El Khadem, M. M. A. Abdel Rahman, and M. A. E. Shaban, *Carbohyd. Res.*, **6**, 465 (1968).
34. J. H. Roe and C. A. Kuether, *J. Biol. Chem.*, **147**, 399 (1943).
35. H. El Khadem, M. A. E. Shaban, and M. A. M. Nassr, *Carbohyd. Res.*, **8**, 113 (1968).
36. L. Mester, *Advan. Carbohyd. Chem.*, **13**, 105 (1958).
37. S. Hanessian and N. R. Plessas, *Chem. Commun.*, 706 (1968).
38. D. Horton and K. D. Philips, *Carbohyd. Res.*, **22**, 151 (1972).
38a. L. Anderson and H. A. Lardy, *J. Amer. Chem. Soc.*, **72**, 3141 (1950).
39. P. E. Papadakis, *J. Amer. Chem. Soc.*, **64**, 1950 (1942).
40. N. A. Sørensen, *Kgl. Norske Videnskabers Selskabs Skrifter*, No. 2, 160 (1937).
41. C. D. Hurd and W. H. Saunders, *J. Amer. Chem. Soc.*, **74**, 5324 (1952).
42. J. M. Los, L. B. Simpson, and K. Wiesner, *J. Amer. Chem. Soc.*, **78**, 1564 (1956).
43. G. S. Hammond, "Steric Effects on Equilibrated Systems," in "Steric Effects in Organic Chemistry" (M. S. Newman, ed.), Wiley, New York, 1956, p. 460; for an explanation, see N. L. Allinger and V. Zalkow, *J. Org. Chem.*, **25**, 701 (1960).
44. L. von Vargha, *Ber.*, **87**, 1351 (1954).
45. L. von Vargha, *Ber.*, **69**, 2098 (1936); M. R. Salmon and G. Powell, *J. Amer. Chem. Soc.*, **61**, 3507 (1939).

46. W. N. Haworth, J. Jackson, and F. Smith, *J. Chem. Soc.*, 620 (1940).
47. W. N. Haworth, L. N. Owen, and F. Smith, *J. Chem. Soc.*, 88 (1941).
48. A. B. Foster, W. G. Overend, M. Stacey, and G. Vaughan, *J. Chem. Soc.*, 3367 (1954).
49. D. Horton and J. S. Jewell, *Carbohyd. Res.*, **2**, 251 (1966).
50. I. Kwiecinski, J. Meyer, and L. Marchlewski, *Hoppe-Seyler's Z. Physiol. Chem.*, **176**, 292 (1928); see L. A. T. Verhaar and J. M. H. Dirkx, *Carbohyd. Res.*, **62**, 197 (1978) for use of absorption at ~200 nm for monitoring separations of free sugars.
51. K. S. Dodgson and B. Spencer, *Biochem. J.*, **57**, 310 (1954).
52. I. P. Zill and N. E. Tolbert, *J. Amer. Chem. Soc.*, **76**, 2929 (1954).
53. M. L. Wolfrom, R. D. Schuetz, and L. F. Cavalieri, *J. Amer. Chem. Soc.*, **70**, 514 (1948); compare A. A. El-Dash and J. E. Hodge, *Carbohyd. Res.*, **18**, 259 (1971).
54. T. L. Harris, E. L. Hirst, and C. E. Wood, *J. Chem. Soc.*, 848 (1937).
55. R. W. Herbert, E. L. Hirst, H. Samuels, and C. E. Wood, *J. Chem. Soc.*, 295 (1935).
56. T. L. Harris, E. L. Hirst, and C. E. Wood, *J. Chem. Soc.*, 1660 (1935).
57. C. V. Holland, D. Horton, M. J. Miller, and N. S. Bhacca, *J. Org. Chem.*, **32**, 3077 (1967).
58. D. Horton and J. S. Jewell, *J. Org. Chem.*, **31**, 509 (1966).
59. D. Horton and M. L. Wolfrom, *J. Org. Chem.*, **27**, 1794 (1962).
60. D. Horton and H. S. Prihar, *Carbohyd. Res.*, **4**, 115 (1967).
61. D. Horton and W. N. Turner, *Carbohyd. Res.*, **1**, 444 (1966).
62. D. L. MacDonald and H. O. L. Fischer, *J. Amer. Chem. Soc.*, **74**, 2087 (1952).
63. R. P. Zelinski and W. A. Bonner, *J. Amer. Chem. Soc.*, **71**, 1791 (1949).
64. J. N. Smith and R. T. Williams, *Biochem. J.*, **42**, 538 (1948).
65. H. El Khadem, M. L. Wolfrom, and D. Horton, *J. Org. Chem.*, **30**, 838 (1965).
66. J. M. Gulland and L. F. Story, *J. Chem. Soc.*, 625 (1938).
67. A. Albert, in "Synthetic Procedures in Nucleic Acid Chemistry" (W. W. Zorbach and R. S. Tipson, eds.), Vol. 2, Wiley–Interscience, New York, 1973, p. 47.
68. E. Clar, "Aromatische Kohlenwasserstoffe," Springer Verlag, 1952.
69. R. Kuhn and F. Bär, *Ber.*, **67**, 898 (1934).
70. G. Henseke and R. Jakobi, *Ann.*, **684**, 146 (1965).
71. G. Henseke and K. Dietrich, *Ber.*, **92**, 1550 (1959).
72. For references see G. Henseke, *Z. Chem.*, **6**, 329 (1966).
73. S. F. Mason, *J. Chem. Soc. Spec. Publ.*, **3**, 139 (1955).
74. S. F. Mason, in "Physical Methods in Heterocyclic Chemistry" (A. R. Katritzky, ed.), Vol. 2, Academic Press, New York, 1963, p. 1.
75. A. Gómez Sánchez, L. Rey Romero, and F. García González, *An. Real Soc. Espan. Fis. Quim.* (Madrid), **60B**, 505 (1964).
76. F. García González and A. Gómez Sánchez, *Advan. Carbohyd. Chem.*, **20**, 303 (1965).
77. H. Ohle, *Ber.*, **67**, 1750 (1934).
78. H. El Khadem and S. H. El Ashry, *J. Chem. Soc.* (*C*), 2248 (1968).
79. G. Hanisch and G. Henseke, *Ber.*, **101**, 4170 (1968).
80. H. El Khadem and M. M. Mohammed-Ali, *J. Chem. Soc.*, 4929 (1963).
81. H. El Khadem, *J. Org. Chem.*, **29**, 3072 (1964).
82. W. S. Chilton and R. C. Krahn, *J. Amer. Chem. Soc.*, **89**, 4129 (1967).
83. C. F. Hueber, R. Lohmar, R. J. Dimler, S. Moore, and K. P. Link, *J. Biol. Chem.*, **159**, 503 (1945).
84. N. K. Richtmyer, *Advan. Carbohyd. Chem.*, **6**, 175 (1951).

85. F. Fernández-Bolaños, F. García González, J. Gasch Gómez, and M. Menéndez Gallego, *Tetrahedron*, **19**, 1883 (1963).
86. F. García González, M. Menéndez Gallego, F. Aríza Toro, and C. Alvarez González, *An. Real Soc. Españ. Fís. Quím.* (Madrid), **64B**, 407 (1968).
87. H. El Khadem and Z. M. El-Shafei, *J. Chem. Soc.*, 3117 (1958); 1655 (1959).
88. H. El Khadem, *Advan. Carbohyd. Chem.*, **18**, 99 (1963).
89. H. El Khadem and M. A. E. Shaban, *J. Chem. Soc.* (*C*), 519 (1967).
90. H. El Khadem, M. A. M. Nassr, and M. A. E. Shaban, *J. Chem. Soc.* (*C*), 1465 (1968).
91. A. Rich and M. Kasha, *J. Amer. Chem. Soc.*, **82**, 6197 (1960).
92. R. E. Trucco and R. Caputto, *J. Biol. Chem.*, **206**, 901 (1954).
93. G. H. Beaven and E. R. Holiday, *Advan. Protein Chem.*, **7**, 319 (1952); G. H. Beaven, *Advan. Spectrosc.*, **2**, 331 (1961).
94. E. Schauenstein, *Acta Histochem.*, **4**, 208 (1957).
95. W. Sandritter, *Acta Histochem.*, **4**, 276 (1957).
96. R. B. Setlow and W. R. Guild, *Arch. Biochem. Biophys.*, **34**, 223 (1951).
97. A. M. Woodin, *Biochem. J.*, **58**, 50 (1954).
98. J. Einbinder and M. Schubert, *J. Biol. Chem.*, **185**, 725 (1950).
99. A. G. Ogston and J. E. Stanier, *Biochem. J.*, **46**, 364 (1950).
100. E. Layne, *Methods Enzymol.*, **3**, 454 (1957).
101. G. H. Beaven, E. R. Holiday, and E. A. Johnson, "Optical Properties of Nucleic Acids and Their Components," in "The Nucleic Acids" (E. Chargaff and J. N. Davidson, eds.), Vol. I, Academic Press, 1955.
102. L. B. Townsend, R. K. Robins, R. N. Loeppky, and N. J. Leonard, *J. Amer. Chem. Soc.*, **86**, 5320 (1964); J. J. Fox and I. Wempen, *Advan. Carbohyd. Chem.*, **14,** 307 (1959).
103. E. Chargaff and S. Zamenhof, *J. Biol. Chem.*, **173**, 327 (1948).
104. P. Doty in "The Structure and Biosynthesis of Macromolecules," (D. J. Bell and J. K. Grant, eds.), Biochemical Society Symposia, No. 21, Cambridge University Press, New York, 1962, p. 8.
105. H. R. Mahler, *J. Biol. Chem.*, **206**, 13 (1954).
106. L. G. Whitby, *Biochem. J.*, **54**, 437 (1953).
107. J. F. Taylor, S. F. Velick, G. T. Cori, C. F. Cori, and M. W. Slein, *J. Biol. Chem.*, **173**, 619 (1948).
108. E. S. West, W. R. Todd, H. S. Mason, and J. T. Van Bruggen, "Textbook of Biochemistry," 4th ed., Macmillan, New York, 1966, p. 924.
109. A. White, P. Handler, and E. L. Smith, "Principles of Biochemistry," 3rd ed., McGraw-Hill, 1964.
110. M. Y. Kamel, R. R. Hart, and R. L. Anderson, *Anal. Biochem.*, **18**, 270 (1967).
111. J. W. Kelly, *Arch. Biochem. Biophys.*, **55**, 130 (1955).
112. N. Weissman, W. H. Carnes, P. S. Rubin, and J. Fisher, *J. Amer. Chem. Soc.*, **74**, 1423 (1952).
113. B. Sylven and H. Malingren, *Lab. Invest.*, **1**, 413 (1952).
114. L. Hansen, B. Hegner, and C. E. Jensen, *Acta Chem. Scand.*, **13**, 1167 (1959).
115. H. Swift, "Cytochemical Techniques for Nucleic Acids," in "The Nucleic Acids" (E. Chargaff and J. N. Davidson, eds.), Vol. II, Academic Press, 1955.
116. R. R. Baldwin, R. S. Bear, and R. Rundle, *J. Amer. Chem. Soc.*, **66**, 111 (1944).
117. R. M. McCready and W. Z. Hassid, *J. Amer. Chem. Soc.*, **65**, 1154 (1943); I. A. Wolff, B. T. Hofreiter, P. R. Watson, W. L. Deatherage, and M. M. MacMasters, *ibid.*, **77**, 1654 (1955).

V. INFRARED SPECTROSCOPY*

R. STUART TIPSON AND FRANK S. PARKER

A. PRINCIPLES AND INSTRUMENTATION [1]

A molecule can exist in a number of energy states, and a change from one level to another can result from absorption or in emission of energy, if selection rules permit the transition. One of the three kinds of molecular spectra, the vibrational spectrum, is obtained when absorption of radiant energy causes changes in the energy of molecular vibration. Most of the vibrational absorption bands[1] are found in the range† of 2–100 μm (5000–100 cm^{-1}); however, as the spectrometers usually available to the organic chemist cover the range of 2–40 μm (5000–250 cm^{-1}), discussion of infrared spectra will be confined to this range.

The range of traditional interest to the carbohydrate chemist is 2–15 μm (5000–667 cm^{-1}), and accurate recordings of spectra may be obtained with a recording, double-beam, infrared spectrophotometer equipped with a prism of sodium chloride or a diffraction grating. In such instruments, the sample is exposed to infrared radiation, and that passing through the sample is resolved with the prism or grating; the spectrum is then scanned, and the frequencies (bands) at which radiation is absorbed are recorded, providing an infrared

---

* Contribution, in part, from the National Bureau of Standards.

† For this region, the position of a band in the spectrum is expressed in two ways: (1) as the wavelength (λ) in micrometers (formerly called microns; 1 μm $= 10^{-6}$ m), or (2) as the wavenumber (ν) in reciprocal centimeters (kaysers). The wavenumber is the reciprocal of the wavelength; $\nu = 1/\lambda$; thus, ν (in cm^{-1}) equals $10^4/\lambda$ (in μm). Wave-numbers have the advantage of being directly proportional to energy. As instruments are now available that record linearly either in wavelength or in wavenumber, bands will be expressed in both ways. For wavelengths in the range of 2.00–9.99 μm, the corresponding values of wavenumber have been given to the nearest 5 or 0 for the last figure.

absorption spectrogram. For measurement of bands in the region of 2.7–3.6 μm (3700–2780 cm^{-1}), the greater dispersion given by a lithium fluoride prism or a comparable diffraction-grating is an asset. A prism of potassium bromide may be used for the range of 2–25 μm (5000–400 cm^{-1}), and a prism of cesium bromide for 15–40 μm (667–250 cm^{-1}).

Smaller (and cheaper) prism instruments are available as an analytical tool for rapid, routine recording of useful but less accurate spectra. In the range of 10–15 μm (1000–667 cm^{-1}), such an instrument can routinely measure wavenumbers accurate to within ± 2 cm^{-1}. In the range of 2.5–10 μm (4000–1000 cm^{-1}), the inaccuracy is increasingly greater with increasing wavenumber, but the reproducibility is satisfactory.

B. SAMPLING TECHNIQUES

1. *Phase*

A carbohydrate or derivative is normally examined as a film of an anhydrous syrup, a solution in an organic solvent or water (in a suitable, water-insoluble cell), a mull in a hydrocarbon oil (Nujol) or an appropriate, perhalogenated hydrocarbon, or a pellet of compressed alkali-metal halide. Although recording is usually done with the sample at room temperature, bands may be sharper at lower temperatures.[1a]

The spectrum of a compound may be different for different physical states. Polymorphs of a compound may show differences in a few features of their spectra. An example[2] is *N*-benzoyl-2,3,4,6-tetra-*O*-benzoyl-β-D-glucosylamine; this exists in a form of m.p. 113–115° which, when heated to 117–120° and allowed to crystallize from the melt, gives a form of m.p. 184° having a somewhat different spectrum (Nujol mull). Also, different crystal habits (same melting point) of a compound (resulting, for example, from crystallization of samples from different solvents) may give somewhat different spectra if examined as mulls. Shifts of up to 20 cm^{-1} for certain bands have been noted[3] between crystalline and amorphous forms of some carbohydrates. In all such instances, the molten materials, or solutions in the same solvent, give identical spectra.

The spectra of crystalline materials often show more bands than those of the same compounds in solution.[4] Moreover, band positions may be shifted, or band intensities altered, if the compound forms hydrogen bonds with the solvent. If the concentration is below 5 mM, intermolecular hydrogen-bonding is negligible for solutions in carbon tetrachloride and similar solvents[5,6]; this effect has been used in studying intramolecular hydrogen-bonding in model compounds related to sugars.[7]

Pelleting of certain free sugars with an alkali halide changes a crystalline sample to an amorphous form.[8,9] Moreover, some sugars (for example α-D-glucose) may form a complex with traces of sodium bromide or sodium iodide present in pellets of the corresponding potassium halide.[9] If the pelleting halide is not dry,[10] or if it acquires moisture, the sugar may mutarotate or form a hydrate should the pellet be stored.[11] Consequently, such spectra should always be compared with those obtained for a Nujol mull of the compound, or the pellets should not be stored. However, where examined immediately after preparation of the pellets, eight of twenty-four aldopyranosides[12] show interaction with the pelleting halide, and six of twenty-seven free sugars[13] give spectrograms that are different in Nujol and in potassium iodide.

2. *Comparison of Samples*

As the result of recording and comparing the infrared spectra of a large number of organic compounds, it has been found that every compound having a unique structure gives a unique spectrum. Consequently, for an enantio-morphic pair, every detail of whose structure is identical (but in mirror image), the spectra of the solids, in the same polymorphic modification, are identical.[3] The spectra may differ slightly because of polymorphism, but for the gaseous or liquid (melt or solution) phases, the spectra are identical.

Also, for a large molecule, a slight change in structure may change the spectrum only slightly. Thus, proceeding up a series of oligosaccharides containing the same anomeric form of the same monosaccharide residue only, the spectra for the tetraose, pentaose, hexaose, etc., become more and more alike.[14]

Consequently, with these exceptions, pure compounds can be identified unequivocally by their infrared spectra. For a carbohydrate, the infrared spectrum is usually a much more specific and characteristic property than the ultraviolet spectrum, melting point, boiling point, density, or refractive index. Thus, although the anomers of a sugar or glycoside differ only in the configuration at one carbon atom, their infrared spectra are quite different.[14] Also, the spectrum of the furanoid form of a sugar derivative differs from that of the pyranoid form in the same anomeric modification, as with methyl β-D-ribo-furanoside and methyl β-D-ribopyranoside.[15]

Hence, the simplest use of infrared spectra is for the identification of a compound. The spectrum of a sample to be identified is compared with the spectra of pure compounds of known structure, measured with the same sampling technique. In this use, it is not necessary that the spectra be analyzed for characteristic absorption bands. However, reference spectra[16–20] of related, pure compounds are needed.

Another use is the precise identification of spectra already in the literature, where the compound studied had been inadequately described. For example, in 1950, Kuhn[14] recorded the spectra of ten of the crystalline free sugars, each

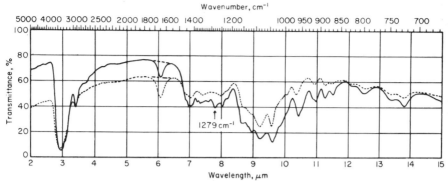

FIG. 1. Infrared spectra of commercial (---) and purified (——) β-D-ribose.

in a Nujol mull, but except for one (α-D-glucose), he did not mention which anomer had been employed. Similarly, Urbański et al.[21] recorded the spectra of six sugars, without specifying the anomeric form. These have since been identified[13] by comparison with the spectra of authentic forms of those sugars.

In another application, infrared spectra can be used for estimating the purity of a sample. Thus, Fig. 1 shows the spectrum of pure β-D-ribose[13] and that of a commercial specimen. It is obvious that the commercial material is impure; for example, its spectrum does not display a band at 7.82 μm (1280 cm^{-1}) that is present in the spectrum of the pure sugar. In other instances, a crude material may show a band absent from the spectrum of the pure compound; thus, samples of commercial D-mannitol showed a band at 12.80 μm (781 cm^{-1}), due to D-glucitol, that could be removed by recrystallization from aqueous ethanol.[22]

The same figure also serves to illustrate that the spectra may be used in monitoring the isolation or purification of a desired product obtained, for example, by distillation or chromatography. It is not necessary to know what the compound is (nor, initially, to have an infrared spectrum of the pure compound); the concentration or purification procedure can be monitored by observing some characteristic infrared band. For example, by recrystallizing β-D-ribose until the band at 7.82 μm (1280 cm^{-1}) appears and remains of constant intensity, a pure sample may be obtained without any knowledge of the actual significance of this band. However, the minimum detectable amount of impurity (which may also, of course, contribute bands absent from the spectrum of the pure material) varies enormously from one compound to another.

Nuclear magnetic resonance spectra are often superior to infrared spectra for comparisons to establish identity and purity of carbohydrate derivatives (see Chapter 27, Section I).

C. INTERPRETATION OF SPECTRA

1. General Remarks

A more interpretive use of infrared spectra permits qualitative analysis for specific groups of atoms. The motions constantly undergone by the atoms and bonds of a molecule may be examined with a model consisting of springs and balls. A diatomic molecule A—B may be represented by two balls, each of a mass proportional to those of atoms A and B, respectively, connected by a spring of strength proportional to the force binding atoms A and B. Molecule A—B can undergo a stretching vibration only along the bond between atoms A and B; this vibration occurs if the spring is compressed and then the whole molecule is permitted to relax on its own. If the resulting, periodic oscillation is moderate, the system follows Hooke's law approximately, and the frequency (ν) of the stretching vibration is given by the equation for simple harmonic vibrations, namely $\nu = (f/\mu)^{1/2}/2\pi c$, where f is the force constant of the bond, c is the velocity of light, and μ is the reduced mass, defined as $\mu = m_A m_B/(m_A + m_B)$, where m_A and m_B are the masses of atoms A and B, respectively. The equation for ν applies only to two atoms joined by a bond of a particular strength.

For a polyatomic molecule, there are many ways in which pairs of atoms may vibrate, bend, rock, or twist, relative to each other and to other pairs, or larger groups, of atoms. Although a mathematical treatment of the infrared spectrum of any pure compound could provide information on the structure through checking of consistency with possible models, the degree of difficulty of such a calculation is a function of the number of atoms and of the symmetry of their geometrical arrangement. Thus, for most sugars, such a treatment is very difficult.

Consequently, in attempting to correlate the structure of the molecule with the frequencies observed, use has been made of the empirical approach. The method is based on the assumption (not necessarily justified) that the vibrations of a certain group are fairly independent of the rest of the molecule. Then, the spectra (recorded under like conditions) of a large number of compounds having this feature are examined, to find out which bands the spectra have in common. For example, the spectra of twenty-four acetylated glycosides were studied,[18] together with those for twenty-one nonacetylated glycosides.[12] It was then obvious that (1) all of the acetates showed a band at 5.64–5.76 μm (1775–1735 cm^{-1}; C=O stretching frequency) not shown by any of the non-acetylated compounds; and (2) all of the nonacetylated glycosides, but not the acetates, showed a band at 2.93–3.05 μm (3415–3280 cm^{-1}; —O—H stretching). All of the hydrates (but no anhydrous compounds) showed a band at

6.01–6.12 μm (1665–1635 cm^{-1}). Here, attention was paid only to the positions of absorption bands, regardless of their relative intensities.

It has been found that, in the absence of disturbing effects, all compounds containing the same group will absorb infrared radiation at almost the same wavelength. Bands in the region from 2.0 to *ca.* 7 μm (5000 to *ca.* 1430 cm^{-1}) are usually characteristic group frequencies; for these, the associated vibration is well localized, as in C=O and X—H stretching vibrations, and the precise positions of the bands are more reliable than for those in the region of 7–15 μm (1430–667 cm^{-1}). The latter interval, or that of 7–11 μm (1430–910 cm^{-1}), is often known as the "fingerprint region," as it is usually rich in bands that collectively provide a characteristic of the compound; however, because the origin of these bands is often not readily determined, a detailed study of this part of the spectrum is generally deferred until the rest of the spectrum has been examined. For carbohydrates and derivatives, the region of 8–10 μm (1250–1000 cm^{-1}) contains bands for —C—O— of esters, ethers, and hydroxyl groups and may not show highly individual bands for such structures. The bands in the region of 7–15 μm (1430–667 cm^{-1}) may arise from single-bond, skeletal stretching vibrations (between atoms of similar, or the same, mass) or from bending vibrations. The latter occur in this region because less energy is required to produce them; certain of these bands have diagnostic value for groups, but most of them are greatly influenced by structural changes in the molecule.

Presence of a band at a position expected for a certain group is not conclusive evidence that that group is present in the compound, because of the possibility of interference by other absorptions. However, absence of a group absorption usually indicates absence of that group from the compound, provided that effects (such as hydrogen bonding) that could shift or even remove the band are not operative. For example, the bands for carbonyl groups are always very strong; if the spectrum has only a weak band at 5.88 μm (1700 cm^{-1}), (1) this band is not due to carbonyl, (2) the molecule is very large and has, perhaps, only one carbonyl group, or (3) a carbonyl compound is present in small proportion. Some group absorptions may be quite different in intensity from one compound to another, even though the wavelength of the absorption is about the same.

2. *Functional Groups in Carbohydrates and Their Derivatives*

Except for the electrical or steric effects mentioned, every organic compound that possesses a particular group will show the corresponding characteristic group frequency in its spectrum. Many compilations of such group frequencies are available.[1] Table I lists, and provides an estimate of the relative intensities of, the group frequencies in which the sugar chemist is likely to be interested;

TABLE I
CHARACTERISTIC INFRARED BANDS SHOWN BY VARIOUS GROUPS

Range μm	Range cm⁻¹	Intensity[a]	Group	Type of vibration[b]	Remarks
2.22–2.38	4505–4200	w	C—H	str.	Aliphatic (combination)
2.35–2.50	4255–4000	w	C—H	str.	Aromatic (combination)
2.74–2.86	3650–3500	var	O—H	str.	Free OH, oxime
2.75–2.76	3640–3625	m(sharp)	O—H	str.	Free OH, alcohols
2.78–3.23	3600–3100	m	O—H	str.	Water of crystallization
2.79–2.92	3590–3425	var(sharp)	O—H	str.	Intramolecularly bonded OH
2.82–2.86	3550–3500	m	O—H	str.	Free OH, carboxylic acid (v. dil. soln.)
2.82–2.90	3550–3450	var(sharp)	O—H	str.	Intermolecularly bonded OH (dimeric)
2.82–3.13	3550–3195	w	C=O	str.	Carbonyl (first overtone)
~2.84	~3520	s	N—H	str.	Primary amide (free)
~2.86	~3500	m	N—H	asym. str.	Primary amine, free NH (dil. soln.)
2.86–3.03	3500–3300	m	N—H	str.	Secondary amine, free NH
2.86–3.27	3500–3060	m	N—H	str.	Associated NH, amine, or amide
~2.94	~3400	s	N—H	str.	Primary amide (free)
~2.94	~3400	m	N—H	sym. str.	Primary amine, free NH (dil. soln.)
2.94–3.10	3400–3225	s(broad)	O—H	str.	Intermolecularly bonded OH (polymeric)
~2.96	~3380	m	NH_3^+	str.	Amine salt (soln.)
2.98–3.18	3355–3145	m	NH_3^+	str.	Amine salt (solid); several bands
~2.99	~3350	m	N—H	str.	Primary amide (bonded)
~3.03	~3300	s	C—H	str.	≡C—H, acetylenes
3.03–4.00	3300–2500	w(v broad)	O—H	str.	H-bonded carboxylic acid dimers
~3.05	~3280	m	NH_3^+	str.	Amine salt (soln.)
~3.15	~3175	m	N—H	str.	Primary amide (bonded)
3.17–3.28	3155–3050	w	C—H	str.	—CH=C—O— and —C=CH—O—

μm	cm⁻¹	Intensity	Bond	Vibration	Assignment
3.23–3.25	3095–3075	m	C—H	str.	RCH=CH₂, olefin
3.25–3.30	3075–3030	w-m	C—H	str.	C—H of aromatic ring
3.28–3.34	3050–2995	w	C—H	str.	Of epoxide (shifts to 3040–3030 if ring strain increases)
3.29–3.32	3040–3010	s; m	C—H	str.	>C—H; RCH=CH₂, RCH=CHR' (cis or trans), RCR'=CHR₂, olefin
~3.38	~2960	s	C—H	asym. str.	C-methyl
~3.42	~2925	s	C—H	asym. str.	>CH₂, methylene, Ar—CH₃
3.45–3.47	2900–2880	w	C—H	str.	C—H, methine
3.45–3.70	2900–2705 (two)	w	C—H	str.	—C(=O)H, aldehyde
3.45–4.35	2900–2300 (several)	w	N—H	str.	Quaternary amine salt, bonded
~3.48	~2875	s	C—H	sym. str.	C-methyl
~3.51	~2850	s	C—H	sym. str.	>CH₂, methylene
3.53–3.55	2835–2815		C—H	str.	O-methyl
~3.54	~2825	m	C—H	str.	—CH〈OCH₂– / OCH₂–, alkyl acetal
~3.60	~2780		C—H	str.	—O—CH₂—O—
3.70–3.91	2705–2560	w(broad)	P—OH	str.	Phosphoric ester, H-bonded (May be several bands)
3.70–4.35	2705–2300	s	NH₂⁺, NH⁺	str.	(May be several bands)
~3.88	~2580	w	S—H	str.	Thiol, free
~4.17	~2400	w	S—H	str.	Thiol, H-bonded
~4.41	~2270	vs	N=C=O	asym. str.	Isocyanate
4.42–4.46	2260–2240	w	—C≡N	str.	Satd. nitrile
4.42–4.57	2260–2190	var.	C≡C	str.	RC≡CR', acetylenes
4.48–4.51	2230–2215	s	—C≡N	str.	Unsaturated, conjugated nitrile
4.55–4.88	2200–2050	vs	C=S	asym. str.	—N=C=S, isothiocyanate (two or more bands)

^a Key: m, moderate; s, strong; v, very; var, variable; w, weak.
^b asym., asymmetrical; def., deformation; str., stretching; sym., symmetrical.

TABLE I—continued

Range (μm)	Range (cm⁻¹)	Intensity[a]	Group	Type of vibration[b]	Remarks
4.55–5.00	2200–2000	s	C≡N	str.	Cyanide, thiocyanate, cyanate
4.59–4.72	2180–2120		N≡C	str.	R—$\overset{+}{N}\!\equiv\!\overset{-}{C}$
4.63–4.72	2160–2120	s	N≡N	str.	Azide
4.67–4.76	2140–2100	w	C≡C	str.	RC≡C—H; acetylenes
~5.52	~1810	s	C=O	str.	—COCl, aliphatic acid chloride
5.62–5.75	1780–1740	s	C=O	str.	—O—(C=O)—O—, carbonate
~5.65	~1770	s	C=O	str.	γ-Lactone
5.73–5.76	1745–1735	s	C=O	str.	Saturated esters
~5.75	~1740	s	C=O	str.	δ-Lactone
5.75–5.81	1740–1720	s	C=O	str.	—C(=O)H, aldehyde
~5.80	~1725	s	C=O	str.	Formic ester
5.80–5.87	1725–1705	s	C=O	str.	Ketone
~5.81	~1720	s	C=O	str.	Benzoic ester
5.81–5.88	1720–1700	s	C=O	str.	—CO₂H; aliphatic carboxylic acid (dimer)
5.88–5.99	1700–1670	s	C=O	str.	—CONHR, secondary amide, free (dil. soln.): Amide I
5.92–5.99	1690–1670	s	C=O	str.	—CONH₂, primary amide, free (dil. soln.): Amide I
5.95–6.13	1680–1630	s	C=O	str.	Secondary amide (solid)
5.95–6.17	1680–1620	var	C=C	str.	Nonconjugated C=C
5.96–6.00	1680–1670		C=C	str.	trans-Olefin; RHC=CHR'
~5.97	~1675	s	C=S	str.	Thioester
~5.97	~1675	s	C=O	str.	Thioester
~5.99	~1670	w	C=N	str.	Aliphatic oxime
5.99–6.17	1670–1620	s	C=O	str.	Primary amide (solid), H-bonded, two bands: Amide I
6.02–6.05	1660–1650		C=C	str.	cis-Olefin; RHC=CHR'
6.03–6.07	1660–1650	m	C=C	str.	Terminal olefin; RR'C=CH₂

μm	cm⁻¹		Bond	Mode	Assignment
6.06–6.17	1650–1620	s	N—H	def.	Primary amide (solid): Amide II band
6.06–6.25	1650–1600	s	NO_2	asym. str.	$-O-NO_2$, nitrate
6.06–6.33	1650–1580	m–s	N—H	def.	NH_2; primary amine
6.06–6.45	1650–1550	w	N—H	def.	NHR; secondary amine
6.07–6.11	1650–1640	var	C=C	str.	Terminal olefin; $RHC=CH_2$
~6.15	~1625	s	C=C	str.	Ph-conjugated C=C
6.15–6.31	1625–1585	m	C=C	skeletal, in-plane	Aromatic C=C
6.17–6.29	1620–1590	s	N—H	def.	Primary amide (dil. soln.)
6.17–6.41	1620–1560	m–s	N—H	def.	Amine
6.21–6.49	1610–1540	vs	C—O	asym. str.	$-CO_2^-$, carboxylate
~6.25	~1600	s	C=C	str.	C= or C=C conjugated with C=C
~6.31	~1585	m	NH_3^+	asym. def.	Amine salt
6.33–6.58	1580–1520	m	C=N (plus C=C)	asym. def. interaction effects	Pyrimidines
6.37–6.60	1570–1515	s	N—H	def.	Secondary amide (solid): Amide II band
6.45–6.62	1550–1510	s	N—H	def.	Secondary amide (dil. soln.)
~6.67	~1500	var	C=C	skeletal, in-plane	Aromatic C=C
6.67–6.80	1500–1470	s	C=S	str.	$-N=C=S$
6.67–7.69	1500–1300	m	NH_3^+	sym. def.	Amine salt
~6.81	~1470	s	C—H	scissoring	Alkane $-CH_2-$
~6.85	~1460	m	C—H	asym. bend	$-CH_3$
6.85–7.14	1460–1400	s	C—O	sym. str.	$-CO_2^-$, carboxylate
~6.87	~1455	s	C—H	scissoring	Alicyclic $-CH_2-$
6.90–7.14	1450–1400	w	$-N=N-$	str.	Azo
6.94–7.17	1440–1395	w	C—O	str. (plus OH def.)	Carboxylic acid
6.94–7.41	1440–1350	s	S=O	str.	$(RO)_2SO_2$, sulfuric ester
6.94–7.55	1440–1325	m	C—C	str.	Aliphatic aldehyde

[a] Key: m, moderate; s, strong; v, very; var, variable; w, weak.
[b] asym., asymmetrical; def., deformation; str., stretching; sym., symmetrical.

TABLE I—continued

Range		Intensity^a	Group	Type of vibration^b	Remarks
μm	cm⁻¹				
7.04–7.11	1420–1405	w	C—H	in-plane bend	C=CH₂
7.04–7.52	1420–1330	s	S=O	str.	ROSO₂R', sulfonic ester
7.05–7.14	1420–1400	m	C—N	str.	Primary amide
~7.09	~1410	w	C—N	str.	Aliphatic amine
7.19–7.35	doublet, 1390–1360	m	C—H	sym. bend	gem-Dimethyl
7.22–7.27	1385–1375	m	C—H	sym. bend	—CH₃
7.30–8.00	1370–1250	s	C—O	str.	Lactone
~7.46	~1340	w	C—H	bend	Alkane C—H
7.46–7.81	1340–1280	s	S=O	sym. str.	R₂SO₂, sulfone
7.46–8.47	1340–1180	w	N=N	str.	Azide
7.58–8.26	1320–1210	s	C—O	str.	Carboxylic acid
7.63–8.00	1310–1250	s	C—O	str.	Benzoic ester, phthalic ester
7.66–8.33	1305–1200	m	N—H	def.	Secondary amide, Amide III band
7.69–8.00	1300–1250	s	NO₂	sym. str.	—O—NO₂, nitrate
7.69–8.33	1300–1200	s	P=O	str.	Phosphoric ester, free P=O
7.87–8.70	1270–1150	s	C—O	str.	—(O=)C—O—R in esters
7.96–8.12	1255–1230	s	C—O	str.	CH₃COOR, acetic ester
~8.00	~1250	s	C—O	str.	Methylene acetal
~8.00	~1250	s	C—O	str.	Epoxide
~8.00	~1250	vs	Si—CH₃	sym. CH₃ def.	Si(CH₃)₃, trimethylsilyl
8.00–8.70	1250–1150	vs	P=O	str.	Phosphoric ester, H-bonded P=O
8.10–8.25	1235–1210	s	C=S	str.	R₂C=S, thioketone
8.13–8.70	1230–1150	s	S=O	str.	(RO)₂SO₂, sulfuric ester

μm	cm⁻¹	Intensity	Bond	Mode	Assignment
8.16–8.51, 8.89–9.17, 9.35–10.00	1225–1175, 1125–1090, 1070–1000 (two)	w	C—H	in-plane bend	p-Substituted phenyl
8.20–9.80	1220–1020	m	C—N	str.	Aliphatic amine
8.33–8.73	1200–1145	s	S=O	str.	$ROSO_2R'$, sulfonic ester
8.33–9.62	1200–1040	s	C—O	str.	C—O—C—O—C, cyclic acetal (four to five bands)
8.33–8.55	1200–1170	s	C—O	str.	Propionic and higher esters
8.33–10.00	1200–1000	s	C—OH	str.	Alcohols
8.44–8.51	1185–1175	s	C—O	str.	Formic ester
8.51–8.58, 8.55–8.77	1175–1165, and 1170–1140	s	C—H	skeletal	$(CH_3)_2C<$, isopropyl
8.51–8.89, 9.01–9.35, 9.35–10.00	1175–1125, 1110–1070, 1070–1000	w	C—H	in-plane bend	Unsubstituted phenyl
8.70–9.09	1150–1100	s	S=O	asym. str.	R_2SO_2, sulfone
8.70–9.09	1150–1100	s	C—O	str.	Benzoic ester, phthalic ester
8.70–9.35	1150–1070	s	C—O—C	asym. str.	Aliphatic ether
~8.93	~1120	s	C=S	str.	—NH—(C=S)—, thioamide
9.01–10.00	1110–1000	s	C—F	str.	Monofluoro derivatives
9.17–9.71	1090–1030	vs	P—O—C	str.	Phosphoric ester
9.17–9.80	1090–1020	vs	Si—O	str.	Si—O—C, trimethylsilyl
9.45–9.50	1060–1055	s	C=S	str.	(RS)$_2$C=S, trithiocarbonate
9.52–9.80	1050–1020	s	S=O	str.	>S=O, sulfoxide
~9.62	~1040		C—O	str.	Methylene acetal
9.95–10.10, 10.93–10.99	1005–990 and 915–910	vs	C—H	bend	C=C—H, vinyl
10.05–10.15, 10.99–11.05	995–985 and 910–905		C—H	out-of-plane bend	$RCH=CH_2$

[a] Key: m, moderate; s, strong; v, very; var, variable; w, weak.
[b] asym., asymmetrical; def, deformation; str., stretching; sym., symmetrical.

References start on p. 1433.

TABLE I—continued

Range		Intensity^a	Group	Type of vibration^b	Remarks
μm	cm⁻¹				
10.20–10.36	980–965		C—H	out-of-plane bend	*trans*-RHC=CHR'
10.31–10.64	970–940	broad	P—O—P		Pyrophosphate
10.36–10.42, 10.58–10.64	965–960 and 945–940	s	C—H	bend	Vinyl ether
10.42–10.75	960–930		N—O	str.	Oxime
10.53–12.35	950–810		C—O	str.	Epoxide
~10.55	~948	s	C—H	bend	Vinyl ester
~10.81	~925		C—O	str.	Methylene acetal
11.17–11.30	895–885		C—H	out-of-plane bend	RR'C=CH₂
~11.90	~840	vs	Si—CH₃	str.	Si(CH₃)₃, trimethylsilyl
11.90–12.66	840–790		C—H	out-of-plane bend	RR'C=CHR"
11.90–12.66	840–790	m	C—H	skeletal	(CH₃)₂C<, isopropyl
11.90–13.33	840–750		C—O	str.	Epoxide

12.00–12.35	833–810	C—H	out-of-plane bend	vs	p-Substituted phenyl
~12.50	~800	NH_3^+	rocking	w	Amine salt
~12.50	~800	NH_2^+	rocking	w	
12.99–13.70 14.08–14.49	770–730, 710–690	C—H	out-of-plane bend	s	Unsubstituted phenyl
~13.25	~755	Si—CH_3	str.	vs	Si(CH_3)₃, trimethylsilyl
13.33–14.29	750–700	C—Cl	str.	s	Monochloro derivatives
~13.89	~720	N—H	def.	m(broad)	Secondary amide, bonded: Amide V band
14.18–17.54	705–570	C—S	str.	w	Thiol, sulfide
~14.49	~690	C—H	out-of-plane bend		cis-RHC=CHR'
~15.38	~650	C—Br	str.	s	Bromo derivatives
16.67–20.83	600–480	C—I	str.	s	Iodo derivatives
18.18–22.22	550–450	S—S	str.	vw	Disulfide

[a] Key: m, moderate; s, strong; v, very; var, variable; w, weak.
[b] asym., asymmetrical; def., deformation; str., stretching; sym., symmetrical.

References start on p. 1433.

about half of them lie in the range of 2–7 μm (5000–1430 cm^{-1}) and the rest above 7 μm (below 1430 cm^{-1}).

Description of a group vibration as A—B stretching or A—B—C bending is, at best, a first approximation. All of the bonds in a group are usually changing to some extent during a vibration. One bond may dominate, and it is then reasonable to describe the vibration as the motion of that bond, but this is by no means always justifiable. Thus, for the amide group vibrations, the motions of C=O, C—N, and N—H stretching and the corresponding bending motions are often highly mixed; consequently, the amide group vibrations are described as Amide I, II, III, and so on.

In the following discussion of some of the more important bands, all values for their positions are approximate unless a range is stated; even so, although frequencies generally fall within the limits indicated, they may, in special cases, lie beyond these ranges. Consequently, when the correlations are used, all other available evidence should also receive consideration.

a. C—H Bands.—All sugars and their derivatives possess the methylidyne (methine) grouping —C—H, which shows a band at about 3.45 μm (2900 cm^{-1}), assigned to C—H stretching. Thus, a group of twenty-eight cyclic acetals of various sugars all showed[4] a band at 3.32–3.37 μm (3010–2970 cm^{-1}).

Acetylenic compounds give a strong band for C—H stretching of ≡C—H at about 3.07 μm (3255 cm^{-1}); this occurs at 3.03 μm (3300 cm^{-1}) for 1,2,3,4-tetra-*O*-acetyl-5,6-dideoxy-D-*lyxo*-hex-5-ynitol[23] and at 3.11 μm (3215 cm^{-1}) for 6,7-dideoxy-1,2:3,4-di-*O*-isopropylidene-D-*gulo*-hept-6-ynitol[24]; the L-*manno* isomer of the latter shows the band at 3.06 μm (3270 cm^{-1}).

Other C—H stretching frequencies encountered in sugar derivatives are: olefinic =CH$_2$ at 3.25 μm (3075 cm^{-1}) and 3.36 μm (2975 cm^{-1}); Ar—H at 3.28 μm (3050 cm^{-1}); olefinic =C—H at 3.31 μm (3020 cm^{-1}); —CH$_2$— at 3.42 μm (2925 cm^{-1}) and 3.51 μm (2850 cm^{-1}); and *C*-methyl at 3.38 μm (2960 cm^{-1}) and 3.49 μm (2865 cm^{-1}). All of twenty-one methyl aldopyranosides studied[12] show a characteristic C—H stretching band at 3.47–3.52 μm (2880–2840 cm^{-1}), not shown by *C*-methyl or ethoxyl groups,[25] that permits detection of the glycosidic methoxyl group, regardless of the anomeric disposition or of substitution (or nonsubstitution) at C-5.

Some deformation frequencies for C—H are: —CH$_2$— at 6.89 μm (1450 cm^{-1}), and *C*-methyl at 7.27 μm (1375 cm^{-1}). The *C*-methyls of the isopropyl group show bands at 7.25 and 7.30 μm (1380 and 1370 cm^{-1}) that are particularly useful for indicating the presence of the isopropylidene acetal structure.

Weak bands for C—H in-plane deformations of the unsubstituted phenyl group are found at 8.51–8.89 μm (1175–1125 cm^{-1}), 9.01–9.35 μm (1110–1070 cm^{-1}), and 9.35–10.00 μm (1070–1000 cm^{-1}); strong bands for C—H out-of-plane deformations occur at 12.99–13.70 μm (770–730 cm^{-1}) and at 14.10–14.50 μm (710–690 cm^{-1}).

For substituted phenyl groups, the bands for C—H in-plane and out-of-plane deformation differ according to the position and degree of substitution. For the p-substituted phenyl group, most commonly encountered in sugar chemistry, weak bands (in-plane deformations) are found at 8.17–8.51 μm (1225–1175 cm^{-1}), 8.89–9.17 μm (1125–1090 cm^{-1}), and 9.35–10.00 μm (1070–1000 cm^{-1}), and a strong band (out-of-plane deformation) at 11.63–12.50 μm (860–800 cm^{-1}). Thus, the p-substituted phenyl group of p-toluenesulfonic esters of alditols[26] and sugars[27] shows a hydrogen out-of-plane deformation at 12.05–12.35 μm (830–810 cm^{-1}), not shown by methanesulfonates.

The vinyl group CHR=CH$_2$ shows a weak band at 7.04–7.09 μm (1420–1410 cm^{-1}), a band at 7.69–7.75 μm (1300–1290 cm^{-1}), a medium-strength band at 10.05–10.15 μm (995–985 cm^{-1}), and a strong band at 10.93–11.05 μm (915–905 cm^{-1}). A cis double bond shows a weak band at 7.04–7.14 μm (1420–1400 cm^{-1}) and a strong band at 13.70–15.04 μm (730–665 cm^{-1}); a $trans$ double bond shows a weak band at 7.55–7.75 μm (1325–1290 cm^{-1}) and a strong band at 10.20–10.42 μm (980–960 cm^{-1}). For example, (E)-3,4-dideoxy-D-$threo$-hex-3-enitol[28] shows bands at 7.55 and 10.25 μm (1325 and 976 cm^{-1}); its 1,2:5,6-di-O-isopropylidene derivative shows bands at 7.65 μm (1305 cm^{-1}) and 10.30 μm (971 cm^{-1}).

Stacey et $al.$[29] have succeeded in identifying the C—H deformation vibrations at the anomeric carbon atom of various sugars by replacing the hydrogen atom on C-1 with deuterium. For example, to prepare α-D-[1-^{2}H]glucopyranose, they dissolved D-glucono-1,5-lactone in deuterium oxide, reduced the carbonyl group to >CD(OD) with sodium amalgam in deuterium oxide, and then converted the OD groups into OH groups by dissolving three times in water and evaporating; the C—D bond at C-1 remained unchanged. The spectrum of the α-D-[1-^{2}H]glucopyranose was then compared with that of α-D-glucopyranose. By theory, ^{1}H deformation frequencies are approximately $\sqrt{2}$ times the corresponding deuterium frequencies (in cm^{-1}) if the deformation corresponds to a pure bending or stretching mode. They found C-1—H at 7.28 μm (1375 cm^{-1}) and C-1—D at 9.13 μm (1095 cm^{-1}) (frequency ratio 1.26); and C-1—H at 7.79 μm (1285 cm^{-1}) and C-1—D at 10.36 μm (965 cm^{-1}) (ratio 1.33), as compared to the theoretical ratio of 1.414. Similar assignments were made for the β anomer and for the two anomers of other sugars.

$b.$ $N—H$ $Bands.$—In dilute solution in a nonpolar solvent, primary amines show two bands in the region of 2.86–3.03 μm (3495–3300 cm^{-1}) due to stretching vibrations of the NH$_2$ group. If hydrogen bonding occurs, or if the solid is examined, the range is shifted to 2.86–3.23 μm (3495–3095 cm⁻¹). Secondary amines in dilute solution show only one N—H stretching band, at 2.94–3.03 μm (3400–3300 cm⁻¹).

An N—H deformation frequency is shown by primary amines at 6.08–6.45 μm (1645–1550 cm⁻¹); thus, D-glucosylamine shows a band at 6.17 μm (1620 cm⁻¹), and 2-amino-2-deoxy-D-glucopyranose at 6.25 μm (1600 cm⁻¹).[3] A band at 6.33–6.62 μm (1580–1510 cm⁻¹) is shown by —NH—.

The NH₂ deformation frequency of primary amides occurs at 6.06–6.17 μm (1650–1620 cm⁻¹) for the solid, and at 6.17–6.29 μm (1620–1590 cm⁻¹) for solutions; it is called the Amide II band. Secondary amides, having an NH group, show the Amide II band at 6.37–6.60 μm (1570–1515 cm⁻¹) for the solid, and at 6.45–6.62 μm (1550–1510 cm⁻¹) for solutions. Hydrogen-bonded secondary amides show an NH deformation mode near 13.89 μm (720 cm⁻¹), called the Amide V band.

The spectra of sixteen 1-acetamido derivatives of sugars[19] show at least one band at 2.98–3.09 μm (3355–3235 cm⁻¹), attributed to N—H stretching; and, at 6.35–6.49 μm (1575–1540 cm⁻¹), the Amide II band. In a study of the spectra of sixty 1-acylamido derivatives of aldofuranoid, aldopyranoid, and acyclic sugars,[2] all of the compounds were secondary amides, and all of them show at least one band at 2.89–3.10 μm (3460–3225 cm⁻¹) due to N—H stretching; in this region, completely esterified compounds could not be distinguished from those having free hydroxyl groups that would show O—H stretching in the same region. All of the compounds show a band at 6.35–6.65 μm (1575–1505 cm⁻¹; Amide II).

The acyclic 1,1-bis(acylamido)-1-deoxyalditols show[2] two Amide II bands, suggesting that the two acylamido groups on C-1 of these compounds are not equivalent. They may have a hydrogen-bonded structure, possibly of the following type:

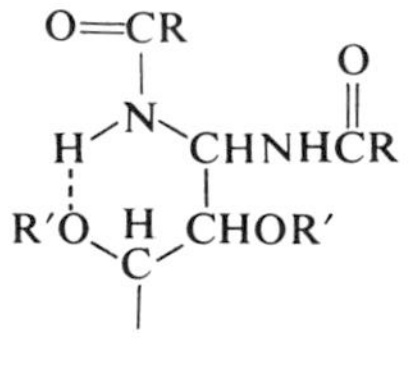

$$
\begin{array}{c}
O{=}CR \\
|\\
\underset{H}{}\;\;N\!\!-\!\!CHNHCR{=}O \\
R'O\;\;H\;\;CHOR' \\
\diagdown C \diagup \\
|
\end{array}
$$

where R is Me, Et, or Ph; and R' is H, Ac, EtCO, or Bz.

c. O—H Bands.—Compounds having a free hydroxyl group show a band for O—H stretching at 2.68–2.84 μm (3730–3520 cm⁻¹). The O—H bond is weakened if the hydroxyl group is hydrogen-bonded, and the band is broadened, with a shift to longer wavelength, 2.84–3.22 μm (3520–3105 cm⁻¹). The band for O—H deformation lies at 9.26–9.71 μm (1080–1030 cm⁻¹). For sugars, differentiation of primary from anomeric and secondary hydroxyl groups by

OH frequencies is not feasible, because of frequency shifts caused by hydrogen bonding.

Infrared spectra may be used for the detection of hydrogen bonding, which may be between one molecule and another (intermolecular) or in one molecule (intramolecular). For compounds soluble in carbon tetrachloride, intermolecular hydrogen-bonding is negligible at concentrations below 5 mM, and absorptions at *ca.* 2.79 μm (*ca.* 3585 cm^{-1}) and *ca.* 2.76 μm (*ca.* 3625 cm^{-1}) may be associated with bonded and free hydroxyl groups, respectively.[5] In this way, the extent of intramolecular hydrogen-bonding may be determined.[7,30] In pyranoid (and *m*- or *p*-dioxane) derivatives, a suitably located hydroxyl group can form a hydrogen bond with the ring-oxygen atom, as shown in Fig. 2. The spectra are consistent with the equilibria shown, assuming that the molecules exist preponderantly in the chair conformations (as against non-chair conformations). For compounds that are insufficiently soluble in carbon tetrachloride, Åkermark[31] has suggested the use of *p*-dioxane, which disrupts intermolecular hydrogen-bonds but does not affect strong intramolecular hydrogen-bonds.

In solution, both penta-*O*-acetyl-*aldehydo*-D-galactose aldehydrol and the

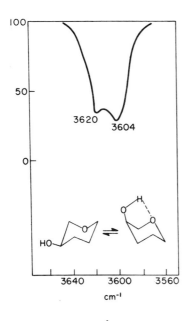
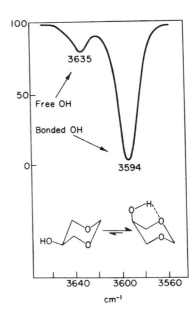

A   B

FIG. 2. Infrared spectra of tetrahydropyran-3-ol (*A*) and of 1,3-*O*-methyleneglycerol (*B*) in carbon tetrachloride; both solutions 5 mM. From ref. 7.

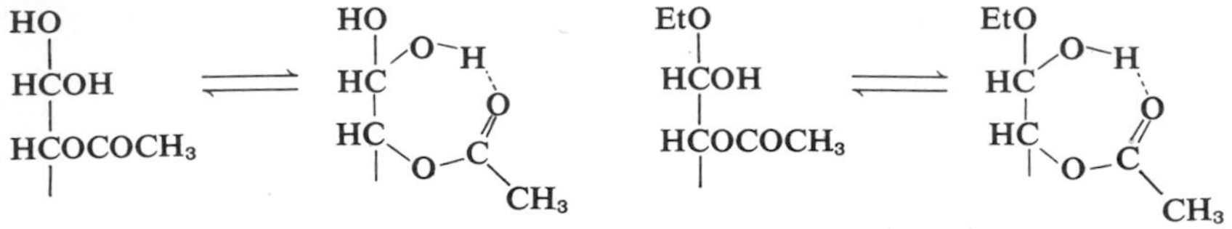

Aldehydrol   Hemiacetal

corresponding ethyl hemiacetal show [17] a band at 2.78 μm (3595 cm^{-1}) for free hydroxyl, and a band at 2.87 μm (3485 cm^{-1}) for hydrogen-bonded hydroxyl.

The OD bands in the spectra of fully O-deuterated sugars have been examined.[32] The OH band at 2.95 μm (3390 cm^{-1}) of α,β-D-glucose was shifted, by deuteration, to 4.05 μm (2470 cm^{-1}), a wavenumber ratio of 1.37, and other new bands appeared.

d. C≡C and C=C Bands.—The C≡C stretching absorption is weak and has limited diagnostic value; it is not shown by symmetrically disubstituted acetylenes. A weak stretching band for H—C≡C—R lies at 4.65–4.76 μm (2150–2100 cm^{-1}); thus, 1,2,3,4-tetra-O-acetyl-5,6-dideoxy-D-*lyxo*-hex-5-ynitol shows [23] a band at 4.65 μm (2150 cm^{-1}), and two 6,7-dideoxyhept-6-ynitol derivatives show [24] the band at 4.72–4.74 μm (2120–2110 cm^{-1}). For R—C≡C—R', the band is at 4.42–4.57 μm (2260–2190 cm^{-1}).

Stretching frequencies for (phenyl) conjugated C=C bonds give a strong band at about 6.15 μm (1625 cm^{-1}); for C=O or C=C conjugation, the band occurs at about 6.25 μm (1600 cm^{-1}). Bands for nonconjugated C=C bonds occur at 5.95–6.17 μm (1680–1620 cm^{-1}); bands are found for a *cis* double bond in RCH=CHR' at 6.02–6.05 μm (1660–1655 cm^{-1}), and for the *trans* form at 5.96–6.05 μm (1680–1655 cm^{-1}). For example, (*E*)-3,4-dideoxy-D-*threo*-hex-3-enitol shows [28] a band at 6.05 μm (1655 cm^{-1}). A terminal, exocyclic double bond, as in H$_2$C=CRR', shows a band at 6.03–6.07 μm (1660–1645 cm^{-1}), and H$_2$C=CHR absorbs at 6.07–6.11 μm (1645–1635 cm^{-1}).

The unsubstituted phenyl ring shows bands of moderate, variable intensity for skeletal, in-plane, stretching vibrations of aryl C=C at 6.16–6.35 μm (1625–1575 cm^{-1}), 6.29–6.36 μm (1590–1570 cm^{-1}), 6.38–6.94 μm (1565–1440 cm^{-1}), and 6.56–6.78 μm (1525–1475 cm^{-1}). Thus, the phenyl ring of the benzoyl group shows bands at 6.25 and 6.32 μm (1600 and 1580 cm^{-1}). For example, a group of 1-acylamido sugar derivatives having N-benzoyl or O-benzoyl groups, or both, show [2] bands at 6.20–6.25 μm (1615–1600 cm^{-1}), 6.30–6.38 μm (1585–1565 cm^{-1}), and 6.64–6.77 μm (1505–1475 cm^{-1}). The p-substituted phenyl ring of p-toluenesulfonic esters of alditols [26] and sugars [27] shows an aryl C=C band at 6.23–6.25 μm (1605–1600 cm^{-1}) that differentiates them from methanesulfonates.

e. C≡N, C═N, and C—N Bands.—The stretching band for C≡N lies in the range of 4.17–4.76 μm (2400–2100 cm^{-1}); thus, for R—C≡N, it is found at 4.43–4.47 μm (2260–2240 cm^{-1}), and, for conjugated R—C≡N, at 4.47–4.52 μm (2240–2215 cm^{-1}). For isocyanides, R—N≡C, the band is at 4.55–4.76 μm (2200–2100 cm^{-1}). For —S—C≡N, the band is found at *ca.* 4.63 μm (*ca.* 2160 cm^{-1}).

Compounds containing C═N— show a band at 6.02–6.21 μm (1660–1610 cm^{-1}) that has been used in determining whether such compounds as *N*-substituted glycosylamines are cyclic or acyclic (see Chapter 20). However, the compound examined must be scrupulously dry, as moisture shows a band at 6.06–6.25 μm (1650–1600 cm^{-1}). Moreover, hydrates cannot be employed, as water of crystallization shows a band at 6.06–6.10 μm (1650–1640 cm^{-1}).

For —N═C═S (isothiocyanate) and —N═C═N— (carbodiimides), a strong band is shown at *ca.* 4.76 μm (*ca.* 2100 cm^{-1}).

Aliphatic amines show a band of moderate intensity for C—N stretching at 8.20–9.80 μm (1220–1020 cm^{-1}) and a weak band at about 7.09 μm (1410 cm^{-1}). Nitro compounds show a band (moderate intensity) for C—N stretching at 10.87–11.76 μm (920–850 cm^{-1}), and primary amides at 7.05–7.15 μm (1420–1400 cm^{-1}). Of less diagnostic value are a C—N band at 7.30–7.63 μm (1370–1310 cm^{-1}) for the *N*-methyl group, and the Ph—N stretching band, observed[33] at 8.70–8.84 μm (1150–1130 cm^{-1}) or[34] 8.62–8.85 μm (1160–1130 cm^{-1}) for phenylhydrazones and phenylazo derivatives.

f. C═O Bands.—*Aldehydes and ketones.* The C═O stretching frequency of the carbonyl group in aldehydes and ketones lies at 5.78–6.00 μm (1730–1665 cm^{-1}). Thus, for the acyclic form of certain aldoses and ketoses (in a lyophilizate of the mutarotational equilibrium mixture), an extremely weak band is detectable[13] at 5.82 μm (1720 cm^{-1}). Kuhn[14] attributed a band at 6.2 μm (1615 cm^{-1}) shown by periodate-oxidized methyl α-D-glucopyranoside to aldehydic carbonyl. Periodate-oxidized cellulose shows only a very weak band[35] and has been shown[36] to exist mainly as the hemialdal —CH(OH)—O—CH(OH)—, formed by hydration of two aldehyde groups per oxidized residue.

If the bands given by two different kinds of groups lie in close proximity in the spectrum (see Table I), the bands may be separate, but often one appears as a shoulder on the other, or one may completely obscure the other. For example, a number of anhydrous, monomeric, aldehydo sugar acetates show no strong band[17] for the aldehyde carbonyl group, presumably because it is obscured by the acetate carbonyl band. As a corollary, in the absence of other information, the CHO band could be mistaken for OAc or OBz. If such interference is absent, aldehydo and keto sugars show the C═O band; for example, 3-*O*-benzyl-1,2-*O*-isopropylidene-α-D-*xylo*-pentodialdo-1,4-furanose shows[37] a

---

band at 5.8 μm (1725 cm^{-1}), and the anomers of methyl D-*erythro*-pento-pyranosid-3-ulose show[37a] a strong band at 5.82 μm (1720 cm^{-1}).

Un-ionized carboxylic acids. The C=O stretching frequency appears at 5.76 μm (1735 cm^{-1}) for —C(=O)OH of un-ionized carboxylic acids,[38,39] including alginic acid, chondroitin sulfate, hyaluronic acid, and the pneumococcal polysaccharides.

Lactones. In 1958, Barker et al.[40] found that twenty-two out of twenty-four aldono-1,4-lactones show a band at 5.59–5.67 μm (1790–1765 cm^{-1}), and all of eleven aldono-1,5-lactones show a band at 5.68–5.79 μm (1760–1725 cm^{-1}). Consequently, if the spectrum shows a strong band at 5.60 μm (1785 cm^{-1}), there is a strong possibility that the aldonolactone is 1,4, and if it shows a band at 5.78 μm (1730 cm^{-1}), there is a possibility that it is 1,5. However, if it shows a band at 5.65–5.70 μm (1770–1755 cm^{-1}), some other method for distinguishing between the two should be used. The 6,3-lactones of 1,2-O-isopropylidene-α-D-gluco(and β-L-ido)furanuronic acid show[4,41] C=O stretching at 5.59–5.67 μm (1790–1765 cm^{-1}).

Acetates and other esters. The C=O stretching vibration of the O-acetyl group gives rise to strong absorption at 5.72–5.80 μm (1750–1725 cm^{-1}); thus, the octaacetates of α-cellobiose, α-gentiobiose, and β-maltose show[42] a strong band at 5.79 μm (1725 cm^{-1}), 5.72 μm (1750 cm^{-1}), and 5.76 μm (1735 cm^{-1}), respectively. Similarly, six acetates of cyclic acetals of sugars show[4] a band at 5.72–5.75 μm (1750–1740 cm^{-1}); all of twenty-four acetylated aldopyranosides show[18] at least one band in the region of 5.67–5.76 μm (1765–1735 cm^{-1}); all of eight reducing, pyranose acetates show[19] a band at 5.71–5.76 μm (1750–1735 cm^{-1}); all of twenty fully acetylated pyranoses[20] show a band at 5.69–5.74 μm (1755–1740 cm^{-1}); and fourteen acetates (and a tetrapropionate) of 1-acyl-amido derivatives of sugars all show[2] a band at 5.68–5.74 μm (1760–1740 cm^{-1}), except for N-acetyl-2,3,4-tri-O-acetyl-β-D-ribosylamine, which shows[43] a band at 5.82 μm (1720 cm^{-1}). Benzoates of the same group of 1-acylamido derivatives show[2] a band at 5.73–5.79 μm (1745–1725 cm^{-1}), except for 1,1-bis-(benzamido)-6-O-benzoyl-1-deoxy-D-glucitol, which shows a band at 5.89 μm (1700 cm^{-1}). Mixed esters (acetate–benzoates) show two bands in this region.

The five-membered, cyclic carbonate group in sugar carbonates shows a higher C=O stretching frequency than mixed-ester carbonates of sugars, average values lying at 5.49 μm (1820 cm^{-1}) and 5.68 μm (1760 cm^{-1}), respectively.[44] If the cyclic carbonate is *trans*-fused in sugar derivatives, a strong C=O stretching band is shown[45] at 5.43 μm (1840 cm^{-1}); the band is not shown by *cis*-fused carbonates.

Primary amides. The C=O stretching frequency for —C(=O)NH$_2$ lies near 6.06 μm (1650 cm^{-1}) for solids, and near 5.92 μm (1690 cm^{-1}) for dilute solutions; known as the Amide I band, it is also shown by secondary and tertiary amides. Three glycopyranuronamide derivatives show[19] this band at 6.00–6.02 μm (1665–1660 cm^{-1}).

N-*Acetyl and* S-*acetyl*. The Amide I band for the monosubstituted amide group, as in —NH—(C=O)—CH₃, occurs at *ca.* 6.04 μm (1655 cm⁻¹) for solids, and at 5.88–5.99 μm (1700–1670 cm⁻¹) for dilute solutions. Thus, the anomers of 2-acetamido-2-deoxy-D-glucopyranose and their tetraacetates show[46] the band at 5.97–6.19 μm (1675–1615 cm⁻¹). For all of sixteen 1-acetamido derivatives of sugars, the band is shown[19] at 5.85–6.02 μm (1710–1660 cm⁻¹). Hydrates show bands at 6.01–6.09 μm (1665–1640 cm⁻¹) that are partly or completely obscured by Amide I bands in the same region. Of sixty 1-acylamido derivatives of sugars, all show[2] a band at 5.95–6.15 μm (1680–1625 cm⁻¹). The Amide I band occurs near 6.07 μm (1650 cm⁻¹) for such polysaccharides as chitin,[47] but at 6.18 μm (1620 cm⁻¹) for 2-acetamido-2-deoxy-α-D-glucopyranose, and at 6.01 μm (1665 cm⁻¹) for its tetraacetate, probably because of differences in the hydrogen bonding of the C=O group to OH and NH groups.[3] Because an ionized carboxyl group will absorb in this region, the spectra of such polysaccharides as chondroitin sulfates A, B, and C should be recorded for films cast from an acid solution.

The carbonyl group in the *S*-acetyl group, —S—(C=O)—CH₃, shows[48] a band near 5.95 μm (1680 cm⁻¹). Thus, this group can be distinguished from the *O*-acetyl group, which absorbs near 5.75 μm (1740 cm⁻¹) and the *N*-acetyl group, which absorbs near 6.10 μm (1640 cm⁻¹).

g. C—O Bands.—Esters. Strong bands for the C—O stretching vibrations are shown by esters: for example, by acetates at 8.00–8.13 μm (1250–1230 cm⁻¹), by formates at 8.33–8.48 μm (1200–1180 cm⁻¹), and by propionates at 8.33–8.55 μm (1200–1170 cm⁻¹). Esters of aromatic acids show two, strong, C—O stretching bands at 7.69–8.00 μm (1300–1250 cm⁻¹) and at 8.70–9.09 μm (1150–1000 cm⁻¹).

Carboxylate ions. Salts of carboxylic acids, such as the lithium and barium salts of 1,2-*O*-isopropylidene-α-D-glucofuranuronic acid, show[4] a carbon—oxygen stretching band (strong) at 6.11–6.25 μm (1635–1600 cm⁻¹) that distinguishes carboxylate anions from esters, and a band (medium strength) at 7.04–7.69 μm (1420–1300 cm⁻¹).

h. N≡N, N=N, and NO₂ Bands.—The N≡N stretching vibration of azides gives a strong band at 4.63–4.72 μm (2160–2120 cm⁻¹) and a weak band at 7.46–8.48 μm (1340–1180 cm⁻¹). The N=N group in aromatic diazo compounds gives bands at 6.31–6.37 μm (1585–1570 cm⁻¹) and at 7.04–7.24 μm (1420–1380 cm⁻¹). According to Bassignana and Cogrossi,[34] the bands lie at 6.32–6.42 μm (1580–1560 cm⁻¹) and at 6.95–7.25 μm (1440–1380 cm⁻¹).

Nitric esters of sugars show[27] strong bands for asymmetrical NO₂ stretching at 6.00–6.20 μm (1665–1615 cm⁻¹) and for symmetrical NO₂ stretching at

---

References start on p. 1433.

7.78–7.89 μm (1285–1265 cm^{-1}); a broad band (for O—NO$_2$ stretching) at 11.48–12.08 μm (871–828 cm^{-1}) is useful only for confirmation, as it occurs in a region that contains bands (Types 2a, 2c, and C) for sugars and the C—H out-of-plane deformation of the p-substituted phenyl group.

i. S=O, —SO$_2$—, and C=S Bands.—Esters of sulfuric acid such as the 6-sulfates of D-glucose, D-galactose, and 2-acetamido-2-deoxy-D-glucose show[49] an intense band at 8.07 μm (1240 cm^{-1}), due to S=O stretching vibrations. The position of a band at 11.76–12.19 μm (850–820 cm^{-1}) for the C—O—S frequency of the sulfate group has been correlated with its attachment as a primary, axial secondary, or equatorial secondary substituent (see Section V,C,4).

Sulfones show strong bands at 7.38–7.64 μm (1355–1310 cm^{-1}) and at 8.62–9.01 μm (1160–1110 cm^{-1}) for S=O stretching; thus, the monosulfone obtained by oxidizing penta-O-acetyl-D-glucose dibenzyl dithioacetal shows a band at 7.66 μm (1305 cm^{-1}).[22]

p-Toluenesulfonic esters of alditols[26] show bands at 7.3–7.4 μm (1370–1350 cm^{-1}) and at 8.4–8.5 μm (1190–1175 cm^{-1}) for the asymmetrical and symmetrical stretching modes of the —SO$_2$— group.[50,51] Sulfonic esters of sugars usually show two bands in each region.[27] Bands for C—O—S at 11.8–11.9 μm (847–840 cm^{-1}) and at 11.2–11.4 μm (893–877 cm^{-1}) have been correlated with the axial and equatorial disposition, respectively, of the sulfonic ester group (see Section V,C,4), although this has been disputed.[51a]

The C=S group of thionocarbonates shows[52] bands at 7.52 μm (1330 cm^{-1}) and at 7.65 μm (1305 cm^{-1}); a band at 8.40 μm (1190 cm^{-1}) is also found.[51]

Dimethylthiocarbamates of sugars show, for OC(=S)NMe$_2$, a strong band[53] at 6.5–6.6 μm (1540–1515 cm^{-1}). N,N-Dialkyldithiocarbamates show, for —SC(=S)NMe$_2$, a strong band at 6.71–6.80 μm (1490–1470 cm^{-1}).

j. Miscellaneous.—The following bands are useful. (1) *Polystyrene* (for calibration): 3.51 μm (2850 cm^{-1}), 6.24 μm (1600 cm^{-1}), and 11.03 μm (907 cm^{-1}). (2) *Liquid indene* (for calibration): 6.21 μm (1610 cm^{-1}), 6.44 μm (1565 cm^{-1}), 7.18 μm (1395 cm^{-1}), 7.34 μm (1360 cm^{-1}), 9.81 μm (1020 cm^{-1}), 11.61 μm (861 cm^{-1}), and 13.69 μm (731 cm^{-1}). (3) *Nujol*: 3.43 μm (2920 cm^{-1}), 3.50 μm (2861 cm^{-1}), 6.86 μm (1460 cm^{-1}), and 7.26 μm (1380 cm^{-1}) [and a weak band at 13.89 μm (720 cm^{-1})]. (4) *Water* (as moisture): 2.92 μm (3430 cm^{-1}) and 6.06–6.25 μm (1650–1600 cm^{-1}).

Spurious bands may be introduced from a variety of sources, including silicone grease used on stopcocks, polyethylene powder from laboratory ware, and phthalic esters from the plasticizer in plastic tubing. About thirty-five spurious bands have been listed.[54]

3. Correlations for the Fingerprint Region and Beyond

a. The Fingerprint Region.—That part of the fingerprint region from 7 to 11 μm (1430–910 cm^{-1}) is rich in absorption bands, because it contains bands caused both by certain stretching vibrations and by bending vibrations. This region is particularly characteristic of a compound because, here, small differences in structure cause large spectral effects. Consequently, the region is valuable for identification of a compound. Whereas similar compounds may show quite similar spectra in the range of 2–7 μm (5000–1430 cm^{-1}), their fingerprint regions will often be different; the region is therefore useful in differentiating between isomers, including anomers.

On the other hand, the fingerprint region is often less useful for recognizing the presence of particular organic groups. Thus, in the region below 7 μm (above 1430 cm^{-1}), the presence or absence of bands for characteristic groups may give valuable, reliable information, whereas bands in the fingerprint region do not necessarily afford the same kind of help. For example, the characteristic group frequencies for α-(D or L)-glucopyranose are essentially the same as those for β-(D or L)-glucopyranose, but, for crystals in which the molecules have the favored chair conformation, the interactions of, for example, the hydroxyl group at C-1 with the ring-oxygen atom of a neighboring molecule obviously are different for the axially attached group of the α-(D or L) anomer and the equatorially attached group of the β-(D or L) anomer, and so the fingerprint regions are different. Figure 3 shows a small part of the hydrogen-bonded structure proposed[55] for crystalline α-D-glucopyranose; obviously, the bonding between the C-1 hydroxyl group of one molecule and the ring-oxygen atom of a neighbor would be different for the β-D anomer. Analysis of such differences by a combination of X-ray crystal-structure analysis, nuclear magnetic resonance spectroscopy, and infrared spectroscopy should eventually lead to improved interpretation of the bands in the fingerprint region and beyond, but this goal has not yet been attained.

b. Correlations for 10–15 μm (1000–667 cm^{-1}).—*Correlations for certain aldopyranose derivatives.* For part of this range, namely, 10.42–13.70 μm (960–730 cm^{-1}), Barker and co-workers[56–58] have sought to identify infrared bands characteristic of several aldopyranoses and their derivatives. For D-glucopyranose derivatives, they identified[56] three principal sets of bands, given in Table II. If the bands are to have diagnostic value for (D or L)-glucopyranose derivatives, α anomers should not show type 2b absorption, and β anomers should not show type 2a absorption. However, they found[56] that (1) some α anomers exhibit type 1 absorption in the range of type 2b bands, and (2) some β anomers show "weak peaks of type 2a," which they tentatively attributed to

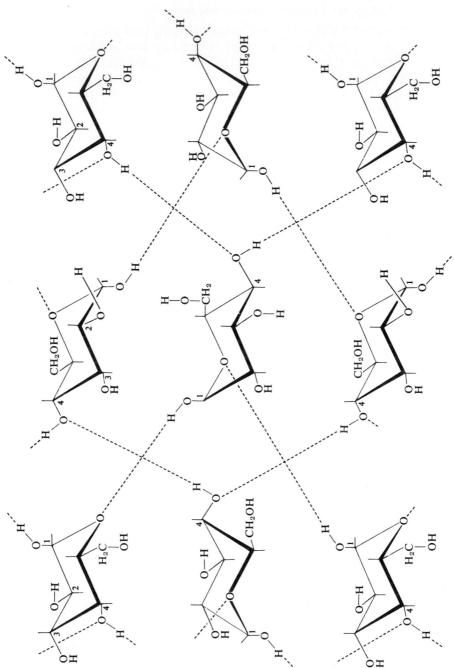

Fig. 3. Some hydrogen bonds between molecules in different layers of crystalline α-D-glucopyranose-$^4C_1$. From results in ref. 55.

TABLE II

POSITIONS (MEAN VALUES) OF VARIOUS TYPES OF BANDS FOR
D-GLUCOPYRANOSE DERIVATIVES[56]

Linkage	Type 1		Type 2a		Type 2b		Type 3	
	μm	cm^{-1}	μm	cm^{-1}	μm	cm^{-1}	μm	cm^{-1}
α Anomeric								
Monosaccharides	10.93	(915)	11.81	(847)			13.04	(767)
	11.11	(900)	11.88	(842)			13.32	(751)
Higher saccharides	10.75	(930)	11.86	(843)			13.14	(761)
	10.90	(917)	11.92	(839)			13.02	(768)
β Anomeric								
Monosaccharides	10.94	(914)			11.16	(896)	—	—
	10.89[a]	(918)[a]			11.22	(891)	12.95[b]	(772)[b]
Higher saccharides	10.86	(921)			11.24	(890)	12.92[c]	(774)[c]

[a] Six of ten compounds did not show this band. [b] Eleven of sixteen compounds did not show this band. [c] Five of sixteen compounds did not show this band.

traces of the α anomers. They found that the type 2a band can be used with considerable confidence for diagnosing the α anomeric form, particularly in polymers of glucopyranose. The type 2b band was not found useful for diagnosing the β anomeric form, but the *absence* of the type 2a band was considered very useful for diagnosing the β anomeric form. They regarded the bands of types 1 and 3 as useful only for determining points of linkage in polymers of α-glucopyranose.

The infrared spectra of some more glucopyranose derivatives were examined,[57] and slightly different positions were found for bands of type 2a and type 3 (see Table III). As before, some of the α anomers were found to show type 1 absorption in the range of type 2b bands. Also, they pointed out that derivatives containing a benzene ring may show absorption in the region of the type 2a band, and that acetates absorb in the region of the type 2b band. Their results[57] for bands characteristic of four other aldopyranoses and their derivatives are also summarized in Table III. For manno- and galacto-pyranose derivatives, the type 2a band can be used for diagnosing α anomers; absence of the type 2a band is useful for diagnosing the β anomeric form. A band at 11.30–11.53 μm (876 $\pm$ 9 cm^{-1}), designated type 2c, was found characteristic[57] of mannopyranose derivatives; a band of type 2c, at 11.39–11.57 μm (871 $\pm$ 7 cm^{-1}), was considered characteristic of galactopyranose derivatives. The mean frequency for a given type of absorption may change with the group configuration; thus, the mean for type 3 absorption is at 12.64 μm (791 cm^{-1}) for the

References start on p. 1433.

TABLE III

Infrared Bands (Mean Values) Shown by Five (D or L)-Aldopyranoses and Their Derivatives[56,57]

Type of band	Xylose μm	Xylose (cm^{-1})	Arabinose μm	Arabinose (cm^{-1})	Glucose μm	Glucose (cm^{-1})	Mannose μm	Mannose (cm^{-1})	Galactose μm	Galactose (cm^{-1})
Both anomers										
1	?	?	?	?	10.90	(917)	?	?	?	?
2c	—	—	—	—	—	—	11.42	(876)	11.48	(871)
3			13.11	(763)	12.99[a] 13.28[a]	(770)[a] (753)[a]	12.64[a]	(791)[a]	13.30[a]	(752)[a]
α anomers only										
2a	—	—	—	—	11.85[a] 11.86[a]	(844)[a] (843)[a]	12.00[a]	(833)[a]	12.12[a]	(825)[a]
3	13.35	(749)								
β anomers only										
2a	—	—	11.86	(843)						
2b			11.82	(846)	11.22[b] 11.24[b]	(891)[b] (890)[b]	11.20[c]	(893)[c]	11.17[c]	(895)[c]

[a] Many derivatives containing a benzene ring absorb here. [b] Must be confirmed by absence of absorption at ca. 11.85 μm (ca. 844 cm^{-1}). [c] Bands for other types of vibration also occur here.

TABLE IV

BANDS POSSIBLY CHARACTERISTIC OF VARIOUS FEATURES OF SOME
ALDOPYRANOSE DERIVATIVES

Type of band	Structural feature	Bands[a]		References
		μm	cm⁻¹	
	Terminal C-methyl-group rocking[b]	10.34	967	58
1	Antisymmetrical ring-vibration[c]	10.90	917	56
2b	Anomeric C—H axial bond	11.22	891	56
2c	Equatorial C—H deformation (other than anomeric C—H)	11.36	880	57
	Ring-methylene rocking vibration (if not adjacent to the ring-oxygen atom)	11.53	867	58
2a	Anomeric C—H equatorial bond	11.85	844	56
3	Symmetrical ring-breathing vibration	12.99	770	56

[a] Mean value. [b] This band may not have diagnostic value. [c] For glucopyranose derivatives.

manno configuration, but at 13.30 μm (752 cm⁻¹) for the *gluco* and *galacto* configurations.

In addition, Barker and co-workers[58] found that 2- and 3-deoxy derivatives of gluco-, manno-, and galacto-pyranose show absorption at 11.51–11.56 μm (869–865 cm⁻¹); seven 6-deoxy derivatives of mannopyranose or galacto-pyranose show a band near 10.34 μm (967 cm⁻¹).

Application of these correlations[56–58] has proved useful[3] in the study of many related compounds, including oligosaccharides and polysaccharides. Assignments suggested[3,56–58] for the bands are given in Table IV. In the range of 10–15 μm (1000–667 cm⁻¹), methyl β-D-xylopyranoside shows only three bands,[12] at 10.25 μm (976 cm⁻¹), 10.38 μm (963 cm⁻¹), and 11.14 μm (898 cm⁻¹); indeed, β-(D or L)-xylopyranose derivatives are not characterizable by *any* of the bands listed in Table III. This means that none of the bands listed in Tables III and IV can be regarded as characteristic of the pyranoid ring, per se, of aldopyranoid derivatives. Theoretical justification for these and some other empirical assignments have been provided by normal coordinate analysis.[58a]

Correlations for furanoid and pyranoid forms of aldose and ketose derivatives. For aldo- and keto-furanose derivatives, Barker and Stephens[59] noted absorption bands at the following mean values: type A, at 10.82 μm (924 cm⁻¹), and type D, at 12.52 μm (799 cm⁻¹). Type A absorption could not be distinguished from types 1 or 2b of aldopyranoses, and therefore has no diagnostic value in differentiating between furanoid and pyranoid aldoses. In addition, most of the furanoid compounds also show type B absorption at 11.38 μm (879 cm⁻¹) and type C absorption at 11.66 μm (858 cm⁻¹). It has been found[2,4,19] that these correlations are, in most instances, restricted to the compounds originally studied,[59] and cannot be extended to have a wider diagnostic applicability to related compounds.

In 1962, the infrared spectra of most of the readily available, unsubstituted aldo- and keto-pentoses and aldo- and keto-hexoses were published.[13] In 1964, Verstraeten[60] made a study of these spectra, together with those of some additional 2-ketoses, and obtained evidence that most of the common sugars having a cyclic structure, and their derivatives, display type 1 absorption at a mean value of 10.76 μm (929 cm^{-1}). Hence, the type 1 (type A) band has no diagnostic value for distinguishing between aldoses and ketoses, and between glycofuranoses and glycopyranoses. Moreover, as it is shown[2] by acyclic 1-acyl-amido derivatives of sugars, it has no diagnostic value for distinguishing between cyclic and acyclic forms of such compounds.

The same author[60] observed that some ketopyranoses, as well as aldopyranoses, show a type 3 band at 12.80 μm (781 cm^{-1}). Hence, this band, too, has limited diagnostic value. He concluded that type 3 absorption is shown provided that two conditions are met: (1) the sugar must have a pyranoid ring, and (2) this pyranoid form must assume a conformation having at least one axial hydroxyl group. If the number of axial hydroxyl groups is increased (thereby decreasing the conformational stability), type 3 absorption becomes manifest. For example, β-(D or L)-xylopyranose, which shows no type 3 absorption, is devoid of axial hydroxyl groups, whereas the α anomer in the favored conformation, which has an axial hydroxyl group at C-1, shows absorption at 13.16 μm (760 cm^{-1}).

It was found[60] that 2-ketoses display "type I" bands at 11.44 μm (874 cm^{-1}) and "type IIA" bands at 12.24 μm (817 cm^{-1}), regardless of whether the 2-ketoses are pyranose or furanose. These bands were ascribed to the presence of the following structural feature:

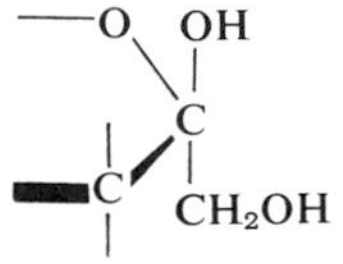

and were tentatively assigned to a skeletal vibration. However, six aldoses also show these bands. The type I band, which appears to be the same as Barker's type B band for aldo- and keto-furanoses at 11.38 μm (879 cm^{-1}), has[2] no diagnostic value for 60 aldofuranoid, aldopyranoid, and acyclic 1-acylamido derivatives. The type IIA band lies in about the same range as Barker's type D band for aldo- and keto-furanose derivatives, which is at 12.52 μm (799 cm^{-1}). If the hydroxyl groups of a 2-ketofuranose are substituted, or if C-2 of the 2-ketofuranose is joined to a pyranoid or furanoid structure, a type IIB band appears at 11.99 μm (834 cm^{-1}), in addition to or instead of the type IIA band.

Verstraeten[60] found that only furanoses show "type 2" absorption at 11.76

μm (850 cm^{-1}). He stated that his type 2 absorption is the same as the type C absorption of Barker and Stephens,[59] and to avoid confusion, it should be referred to as the latter. The type C band is shown by both aldo- and keto-furanoses, and therefore cannot be used for distinguishing between them.

It has been found[2] that, if an N-acetyl group (but no ester group) is present, the bands of types C, 3, IIA, and IIB may have diagnostic value; if an N-benzoyl group (but no ester group) is present, the bands of types 3, IIA, and IIB may have diagnostic value. For N-acetyl-O-acetyl derivatives of sugars, the bands of types IIA and IIB may differentiate between ketoses and nonketoses, but not between cyclic and acyclic compounds.

4. Conformational Studies

In studying the conformations of sugar derivatives, the most direct information is obtained by n.m.r. spectroscopy. However, the empirical correlation of infrared spectra has been used[12] to give conformational information. Suppose that the spectra of the α and β anomers of the methyl pyranosides of the four aldopentoses and eight aldohexoses were to be recorded. This would provide twenty-four spectra of closely related compounds. Each compound has C—H, C—OH, C—OCH$_3$, and a pyranoid ring, and yet the spectrum of each is unique because the precise positions of the various bands change from one compound to another, depending on interactions arising from configuration and conformation and on the presence or absence of the CH$_2$OH group.

As an example, the spectra of the α and β anomers of methyl D-xylopyranoside and methyl L-arabinopyranoside were studied.[12] All bands shown in common by the four glycosides were ignored. All bands then shown in common by the two xylosides were regarded as characteristic of the *xylo* configuration and were ignored; similarly, all bands shown in common by the two arabinosides were ignored. This left a set of bands differentiating between the anomers of the xylosides, on the one hand, and between the arabinoside anomers, on the other (see Table V). This indicated a similarity between the β-D-xylopyranoside and the α-L-arabinopyranoside. Since the conformation of methyl β-D-xylopyranoside has been shown[61] by X-ray studies to be that shown in Fig. 4, the conformational correlations are as indicated. These formulas are in agreement with the conformations predicted by considerations of interaction energies (see Volume IA, Chapter 5).

This correlation is purely empirical, but the same kind of comparison has been made for other pairs of anomers of (1) methyl aldopyranosides,[12] (2) acetylated methyl aldopyranosides,[18] and (3) fully acetylated aldopyranoses.[20] In every instance, the empirical observations made on the infrared spectra agree with the predicted conformations. Those sugar derivatives for which one chair conformation is not predicted to be strongly favored over the other give

---

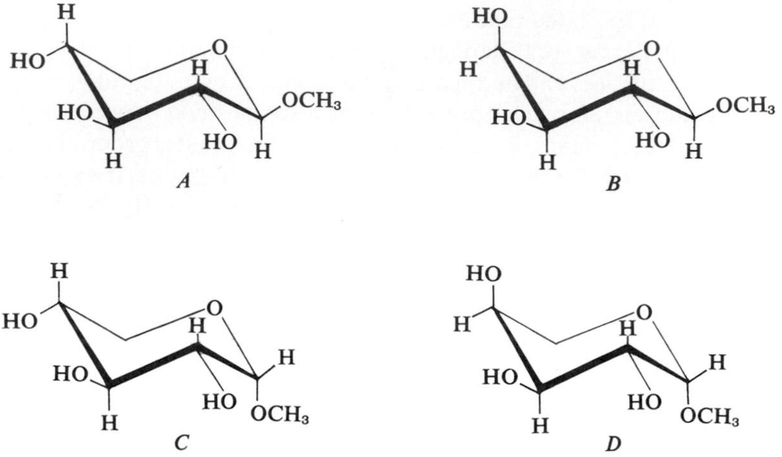

FIG. 4. The structures of methyl β-D-xylopyranoside-$^4C_1$ (A) and methyl α-L-arabino-pyranoside-$^4C_1$ (B) compared with those of methyl α-D-xylopyranoside-$^4C_1$ (C) and methyl β-L-arabinopyranoside-$^4C_1$ (D).

TABLE V

ANOMER-DIFFERENTIATING BANDS (CM^{-1}) SHOWN
BY FOUR METHYL PYRANOSIDES[12]

D-xylo		L-arabino	
β	α	α	β
3448	—	3460	—
1385	—	1395	—
1295	—	1295	—
1218	—	1227	—
1060	—	1058	—
976	—	973	—
645	—	646	—
473	—	487	—
—	3333	—	3322
—	2710	—	2695
—	741	—	744
—	437	—	433

data that do not fit in the correlations. Examples are: methyl α- and β-D-lyxo-pyranoside and their triacetates, methyl α-D-gulopyranoside and its tetra-acetate, and penta-O-acetyl-α-D-gulopyranose.

Among fully acetylated monosaccharides, those having an axial OAc at C-1 show[17] a band, possibly owing to a C—O stretching vibration, at 8.59–8.67 μm (1175–1155 cm^{-1}); if the group is equatorial, a band is shown at 8.87 μm (1125 cm^{-1}). For each region, the other anomer shows the absorption only weakly or not at all. The results with compounds having the *gulo*, *ido*, and *talo* configurations suggest that they exist in the $^1C_4$(D) or $^4C_1$(L) conformation, or as a mixture of chair conformers.

Acetylated methyl glycosides[17] having an axial OMe at C-1 show bands at 8.31–8.35 μm (1205–1200 cm^{-1}) and at 8.75–8.85 μm (1145–1130 cm^{-1}); those having the group equatorial show no absorption in either region.

Because polysulfated hyaluronic acid, which has only equatorial sulfate groups, shows a band at 12.19 μm (820 cm^{-1}), Orr[39] decided that the sulfate group of chondroitin sulfate C, showing a band at 12.12 μm (825 cm^{-1}), is equatorially attached, and that that of chondroitin sulfate A, giving a band at 11.70 μm (855 cm^{-1}), is axially attached. He ascribed the bands to the C—O—S vibration. It was then found[62] that the equatorial sulfate group in D-glucose 3-sulfate gives a band at 12.02 μm (832 cm^{-1}), and that a band at 12.19 μm (820 cm^{-1}) is shown by the 6-sulfates of D-galactose, D-glucose, and 2-acetamido-2-deoxy-D-glucose, in which the ester group is on the equatorial, primary hydroxyl group. Hence, chondroitin sulfates C and D probably have an equatorial sulfate group on C-6, and chondroitin sulfates A and B have an axial sulfate group on C-4 of the 2-acetamido-2-deoxy-D-galactose residues. Chondroitin polysulfate has sulfate groups at both C-4 and C-6.

Similarly, λ-carrageenan shows[63] a broad band at 11.62–12.35 μm (860–810 cm^{-1}), with a maximum at 12.09 μm (827 cm^{-1}) compatible with the presence of residues of D-galactose 2,6-disulfate (diequatorial) and D-galactose 4-sulfate (axial).

A strong band at 11.8–11.9 μm (848–840 cm^{-1}) and a weak one at 11.2–11.4 μm (893–877 cm^{-1}) are shown by sulfonic esters of pyranoid sugars, and these absorptions have been attributed[50] to the C—O—S vibration of an equatorial and an axial sulfonic ester group, respectively, on a pyranoid ring.

D. EXAMPLES OF USE OF INFRARED SPECTRA

In addition to those already mentioned, the following are some examples of the uses of infrared spectra.

1. *Qualitative Uses*

Infrared spectra may be employed in monitoring the course of a reaction. For example, if a compound requires several methylations to give a completely methylated product, the extent of methylation may be determined by observing the disappearance of OH absorption from the infrared spectrum, so that quantitative determination for methoxyl need be made only on the final product.

References start on p. 1433.

The elementary approach has been applied [13] in studying the mutarotation of sugars. For a number of monosaccharides, an aqueous solution of one crystalline anomer was kept until mutarotation was complete; the solution was then lyophilized, and the spectrum of the lyophilizate was recorded and compared with those of the two crystalline anomers. For D-glucose, L-rhamnose, and D-mannose, all the bands in the spectrum of the equilibrium mixture that are not shown by the α-pyranose anomer either (1) are shown by the β-pyranose anomer or (2) could be due to overlapping and summation of closely situated bands of the two anomers, indicating that the equilibrium mixture consists of the anomers of the pyranose form; this conclusion is in agreement with the results of earlier optical rotation studies by Isbell and Pigman.[64] In contrast, for D-talose, the spectrum of the lyophilizate showed bands absent from the spectrum of either anomer of the pyranose form. These results also agree with those of the earlier work [65] (namely, that the equilibrium mixture contains the α- and β-pyranose and the α- and β-furanose forms), and have since been confirmed, and quantified, by n.m.r. spectroscopy.[66]

2. Quantitative Uses

Mutarotation of sugars has also been studied quantitatively. Parker [67a] recorded the spectra, for the range 6.00–11.00 μm (1665–909 cm^{-1}), of 20% aqueous solutions of α-D-glucose, β-D-glucose, and β-D-mannose (1) 2.5 minutes after dissolution, and (2) at the end of mutarotation. By monitoring the change in percent transmittance of 10% solutions [at 8.75 μm (1145 cm^{-1}) for α- and β-D-glucose, and at 8.60 μm (1165 cm^{-1}) for β-D-mannose] with time, he was able to determine the mutarotation constants; these agreed well with those determined from measurements of change in optical rotation. The kinetics of the action of β-D-glucosidase on phenyl β-D-glucopyranoside at pH 5.2 were determined [67b] by monitoring the increase in absorbance at 8.70 μm (1150 cm^{-1}) with time; this band is shown by β-D-glucose, but not by the glucoside.

The intensity of a group frequency may be used for quantitative analysis, because it depends on the magnitude of the change in dipole moment that is associated with the molecular vibration. Consequently, strong bands are usually caused by the vibrations of polar linkages, such as O—H, N—H, C=O, C—N, etc. To a first approximation, the intensity of an absorption band characteristic of a specific group is proportional to the amount of that group present. Figure 5 shows part of the infrared spectra of equimolar solutions of three acetylated methyl glycopyranosides.[17] It may be seen that the area under the curve increases—from three, to four, to five acetyl groups. The procedure could be adapted for use as an analytical method—in this case, for acetyl; but any other group giving a characteristic band could similarly be analyzed for. In most cases, however, n.m.r. spectroscopy is a more convenient tool for quantitative analysis.

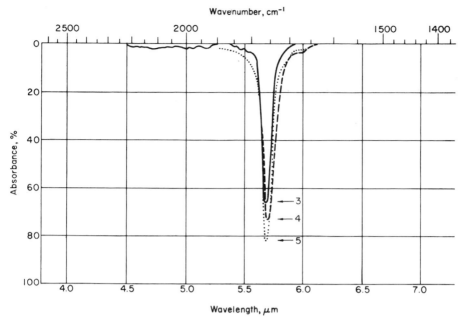

FIG. 5. Infrared spectra of tri-*O*-acetyl-α-L-rhamnopyranoside (**3**), tetra-*O*-acetyl-β-D-mannopyranoside (**4**), and penta-*O*-acetyl-D-*glycero*-β-D-*gulo*-heptopyranoside (**5**). All solutions 19.7 m*M* (**3** and **5** in carbon tetrachloride; **4** in chloroform).[17]

3. *Determination of Structure*

An example of solution of a structural problem involves the D-talose mono-benzoate (**1**) obtained[65] as a by-product from the action of peroxybenzoic acid on D-galactal. There seemed a possibility that it might be a 1,2-(orthobenzoate) (**1a**), so its spectrum was compared[68] with that of 1,2-*O*-(1-methoxyethylidene)-L-rhamnose. As may be seen from Fig. 6, the latter has no band at 5.77 μm (1735 cm⁻¹), whereas **1** has a strong ester-carbonyl absorption there. Hence, **1** is not an orthobenzoate but a benzoate; it was thought to be 1-*O*-benzoyl-α-D-talose but was later[69] shown to be the β-D anomer (**2**).

$1735 cm^{-1}$

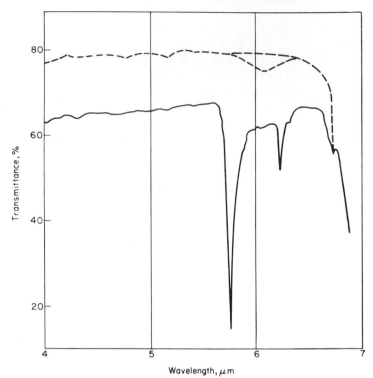

FIG. 6. Infrared spectra of L-rhamnose 1,2-(methyl orthoacetate) (---) and 1-O-benzoyl-β-D-talopyranose (——).[68]

For sugars in which the hetero atom of the ring may be nitrogen, the infrared spectra immediately show which form has this structure. For example, of the two ring-forms of 5-acetamido-5-deoxy-L-arabinose,[70] one shows bands at 3.03 μm (3300 cm⁻¹), 6.14 μm (1630 cm⁻¹), and 6.43 μm (1555 cm⁻¹), for OH and NH, NAc, and NH, respectively, and is therefore the furanose form (3), whereas the other shows bands at 2.96 μm (3380 cm⁻¹) for OH, and 6.19 and

6.27 μm (1615 and 1595 cm^{-1}) for NAc, but no —NH— absorption near 6.43 μm (1555 cm^{-1}), and is therefore the pyranose form (4).

In another example, 4-acetamido-4,5-dideoxy-2,3-O-isopropylidene-*alde-hydo*-L-xylose (5) shows bands[71] at 3.10 μm (3225 cm^{-1}; NH), 5.81 μm (1720 cm^{-1}; HC=O), 6.10 and 6.50 μm (1640 and 1540 cm^{-1}; NHAc), and 7.30 μm (1370 cm^{-1}; CMe$_2$). On treatment of compound 5 with acetic acid, 4-acet-amido-4,5-dideoxy-α,β-L-xylofuranose (6) was obtained; this compound

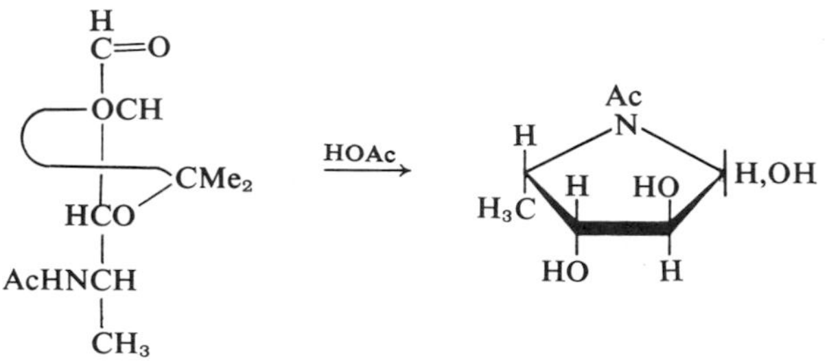

5   6

(which cannot exist in a pyranoid form) shows C=O absorption at 6.12 μm (1635 cm^{-1}; Amide I).

The Schiff-base structure 7 was proposed[72] for N-o-tolyl-D-glucosylamine, because it showed a C=N band at 6.05 μm (1655 cm^{-1}); however, the pure

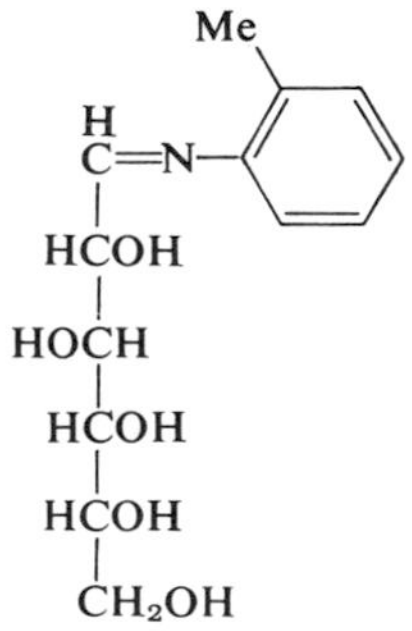

7 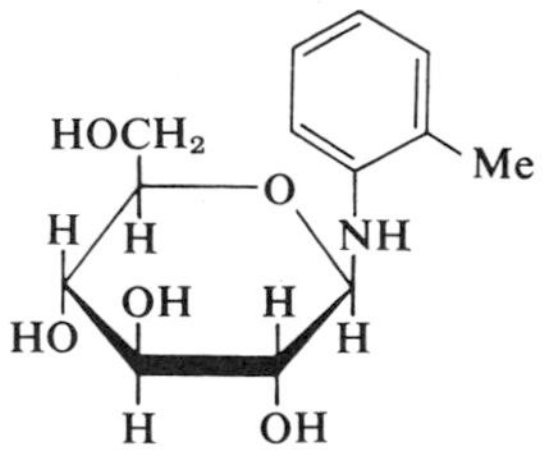   8

compound shows[15] no band at 6.05 μm (1655 cm^{-1}), indicating that the structure is cyclic, probably the pyranoid form (8). All of the N-substituted glycosylamines examined by Ellis[15] were found, from their spectra, to have cyclic structures.

References start on p. 1433.

The oximes of arabinose,[72] rhamnose,[72] and fructose[73] show a weak band at 6.05 μm (1655 cm⁻¹), which may indicate the acyclic form, but it is possible

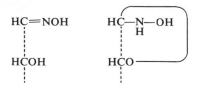

that the N—H of the cyclic form might show a weak band of about the same frequency. D-Glucose oxime does not show a band in this region[72,73] and is, presumably, cyclic. The acetyl derivatives of sugar oximes are undoubtedly cyclic, as they show the characteristic bands for the N-acetyl group.[73]

Infrared spectroscopy has proved to be inappropriate for determining whether N^2-substituted hydrazones of sugars are cyclic or acyclic. Even for N^2-substituted hydrazones that are known to be acyclic, the intensity of the C=N band is so weak as to be unobservable.[74] Phenylosazones of sugars are known to exist preponderantly in the acyclic form (see Chapter 21), and they show[75] a C=N band at 6.30 μm (1585 cm⁻¹).

Acyclic semicarbazones usually show a band at 5.97–6.08 μm (1675–1645 cm⁻¹) for C=N. Acyclic thiosemicarbazones show a weak band at 6.06–6.14 μm (1650–1630 cm⁻¹); the thiosemicarbazones of seven aldoses that do not show this band are therefore considered to have a cyclic structure.[76]

It has been suggested[76a] that inductive effects of the substituents at each end of the C=N bond may coincidentally give rise to little net polarization of the bond, causing its infrared absorption to be weak or unobservable. Consequently, for all compounds that might conceivably contain a C=N bond but that show no C=N band in the infrared spectrum, the result should be verified by laser-Raman spectroscopy (see Section V,E).

Cellulose I, II, and III differ in the amorphous and crystalline regions, and their spectra show differences in the O—H stretching range. Thus, cellulose I shows[77] five bands in the OH region, and, when cellulose film is treated[78,79] with deuterium oxide vapor, the intensity of one of these, at 2.78–3.03 μm (3600–3000 cm⁻¹), rapidly decreases while an O—D band at 3.70–4.17 μm (2700–2400 cm⁻¹) appears; the band affected is the OH stretching of the amorphous regions. The four bands that remain are those of hydrogen-bonded hydroxyl groups in the crystalline region. The ratio of the intensities of the OH and OD bands then gives an indication of the proportion of hydroxyl groups that are hydrogen-bonded in a crystalline manner, and this ratio thus provides a measure of the crystallinity.[80]

The directions of the hydroxyl groups were determined by using plane-polarized, infrared radiation[81] (see Section V,E). It was shown that the OH band at 3.02 μm (3310 cm^{-1}) is "perpendicular," and it was suggested that some of the hydroxyl bonds lie along the chain direction and form intramolecular hydrogen-bonds.

The spectrum of starch indicates[82] that the hydroxyl groups are extensively hydrogen-bonded. The spectrum of potato starch differs from that of corn starch, especially in absorption regions for oxygen-containing groups. Thus, absorption is stronger[83] for the corn starch at 5.95 μm (1680 cm^{-1}), 9.5–10.5 μm (1055–952 cm^{-1}), and 11.7 μm (855 cm^{-1}), whereas it is stronger for potato starch at 10.8 μm (926 cm^{-1}). Arrowroot, corn, potato, rice, and wheat starches show bands at 3 μm (3335 cm^{-1}), 4.75 μm (2105 cm^{-1}), and 6.15 μm (1625 cm^{-1}). When the water content of the starch is changed, the band at 3 μm (3335 cm^{-1}) undergoes a change characteristic of the particular starch, and this property may be used for identifying and classifying starches.[84]

E. Special Techniques

Plane-polarized radiation has proved useful in studying oriented films of such polysaccharides as cellulose, chitin, and xylan, providing information[81,85] not given by other techniques. It has been used mainly in studying the crystal structure of samples having uniaxial orientation, as in fibers, in which the polymer chains are aligned along the fiber axis. The spectrum is measured with the radiation vector (1) parallel and (2) perpendicular to the chain direction. The respective intensity of the "parallel" and the "perpendicular" bands depends on the direction of the transition moment of the vibration responsible for the band—that is, whether it is mainly parallel or perpendicular to the chain direction. Thus, in the C—H stretching region of 3.33–3.57 μm (3000–2800 cm^{-1}) of the spectrum of cellulose, there are a number of partially overlapping bands, some of which cannot be assigned on the basis of group frequency. The C—H bonds of the ring are known to be approximately perpendicular to the chain axis, and so the dichroism of the bands associated with them must also be perpendicular; consequently, the perpendicular bands at ca. 3.45 μm (2900 cm^{-1}) have been assigned[86] to those vibrations. The cellulose band at 3.51 μm (2855 cm^{-1}) was assigned to the symmetrical stretch for CH$_2$, and a band at 7.00 μm (1430 cm^{-1}) to the CH$_2$ symmetrical bending mode.

The direction of the bonds in a crystal may also be determined. α-D-Glucopyranose shows one band at 2.94 μm (3405 cm^{-1}) and four others in the region of 2.99–3.12 μm (3345–3205 cm^{-1}). These have been correlated[79] with the results of X-ray diffraction studies, which show[55] that there is one

$$O—H \cdots O \overset{\displaystyle /C}{\underset{\displaystyle \backslash C}{}} \quad \text{bond (between two molecules), and that there are four}$$

O—H····OH bonds (see Fig. 3 for part of the structure). Similarly, the directions of the hydroxyl groups in sucrose have been determined.[87]

The technique of attenuated total reflection[88] is useful for samples not amenable to examination by transmission spectroscopy; the radiation penetrates a short distance into the sample and is attenuated by absorption, and the extent of attenuation is independent of the thickness of the sample. For measurements of aqueous solutions, an attachment is placed in both beams of the spectrometer to compensate for absorption by water, making unnecessary the use of very thin, accurately matched cells to avoid interference from infrared absorption by water. The method has been applied[89] in the region of 14.29–40.00 μm (700–250 cm^{-1}) to the study of twenty carbohydrates. For those compounds whose transmittance spectra in this region had previously been recorded,[12,13] the attenuated total reflection spectra were in good agreement. In this range, the anomers of a sugar afford different spectra.

The use of a micro-die for preparing alkali halide pellets, and of lead thiocyanate [which is water-soluble and gives a single band at 4.85 μm (2060 cm^{-1})] as an internal reference standard, permits identification and quantitative determination of water-soluble carbohydrates in the micro quantities typically obtained by paper chromatography.[90]

Raman spectra, measurements of inelastic scattering of visible, monochromatic light, give information similar to that given by infrared spectra. Band shifts ($\Delta\nu$, in cm^{-1} from the incident frequency), instead of absolute frequencies of absorption, are the important characteristics of the Raman effect. Although these band shifts correspond to infrared group frequencies, their respective values often do not precisely coincide. The band shifts are characteristic of the molecule irradiated. For example, in the spectra of carboxylic acid dimers, the band for the C=O symmetrical stretching mode occurs strongly at *ca.* 6.02 μm (1660 cm^{-1}) in Raman spectra, but is very weak (not normally observed) in infrared spectra; in contrast, the band for the C=O asymmetrical stretching mode, at *ca.* 5.85 μm (1710 cm^{-1}), is weak in Raman, but very strong in infrared, spectra. Thus, Raman spectra supplement infrared spectra. Raman spectra have been studied for some free sugars[91] and some 1,3-dioxolane compounds related to sugar acetals.[92] The technique of laser-Raman spectroscopy has led to greater activity in this field. The laser-Raman spectrum of D-glucose and other sugars have been recorded,[93] and it has been found that derivatives of two hexopyranos-4-ulose 4-oximes and of a 4-C-formylhexopyranose 4^1-oxime (all containing the C=NOH group), which, in their infrared spectra show weak or negligible absorption for C=N near 6.02 μm (1660 cm^{-1}), have a moderate-to-strong band at $\Delta\nu$ 1665–1650 cm^{-1} in their laser-Raman spectra.[94] Koenig[95] has discussed the Raman spectra of carbohydrates in aqueous solution and in the solid state, including such examples as D-glucose, maltose, cellobiose, dextran, amylose, amylopectin, glycogen, and cellulose; he assigned the various vibrations observed to specific motions, and identified bands characteristic of α and β configurations at the anomeric center. By deuterium-substitution methods, Vasko *et al.*[96] have identified vibrational modes for D-glucose, maltose, cellobiose, and dextran, which they related to O—H and C—H bonds; they further examined the vibrational spectra of α-D-glucopyranose by normal-coordinate analysis. Cael *et al.*[97] have applied normal-coordinate analysis in studying the Raman spectra of the polymorphic forms of amylose, and they have identified configuration- and conformation-sensitive modes for D-glucose.

REFERENCES

1. For extensive treatments of the theory and practice of infrared spectroscopy, see: F. A. Miller, *in* "Organic Chemistry, An Advanced Treatise" (H. Gilman, ed.), Vol. 3, Chapter 2, Wiley, New York, 1953; R. N. Jones and C. Sandorfy, *in* "Chemical Applications of Spectroscopy; Technique of Organic Chemistry," (W. West, ed.), Vol. 9, Chapter 4, Interscience, New York, 1956; L. J. Bellamy, "The Infra-red Spectra of Complex Molecules," 2nd ed., Wiley, New York, 1958; G. K. T. Conn and D. G. Avery, "Infrared Methods: Principles and Applications," Academic Press, New York, 1960; K. Nakanishi, "Practical Infrared Absorption Spectroscopy," Holden–Day, San Francisco, 1962; W. Brügel, "An Introduction to Infrared Spectroscopy" (translated from the German), Wiley, New York, 1962; W. J. Potts, Jr., "Chemical Infrared Spectroscopy," Wiley, New York, 1963; C. N. R. Rao, "Chemical Applications of Infrared Spectroscopy," Academic Press, New York, 1963; R. G. White, "Handbook of Industrial Infrared Analysis," Plenum Press, New York, 1964; N. B. Colthup, L. H. Daly, and S. E. Wiberly, "Introduction to Infrared and Raman Spectroscopy," Academic Press, New York, 1964; H. A. Szymanski, "IR Theory and Practice of Infrared Spectroscopy," Plenum Press, New York, 1964; J. R. Dyer, "Applications of Absorption Spectroscopy of Organic Compounds," Prentice–Hall, Englewood Cliffs, N.J., 1965, H. A. Szymanski, "Raman Spectroscopy: Theory and Practice," Plenum Press, New York, 1967; L. J. Bellamy, "Advances in Infrared Group Frequencies," Methuen, London, 1968; A. D. Cross and R. A. Jones, "Introduction to Practical Infrared Spectroscopy," 3rd ed., Plenum Press, New York, 1969; H. A. Szymanski, "Correlation of Infrared and Raman Spectra of Organic Compounds," Hertillon Press, Cambridge Springs, Pa., 1969; F. S. Parker, "Applications of Infrared Spectroscopy in Biochemistry, Biology, and Medicine," Plenum Press, New York, 1971; M. Avram and G. Mateescu, "Infrared Spectroscopy," Wiley (Interscience), New York, 1972.

1a. J. E. Katon, J. T. Miller, Jr., and F. F. Bentley, *Arch. Biochem. Biophys.*, **121**, 798 (1967); *Carbohyd. Res.*, **10**, 505 (1969).

2. R. S. Tipson, A. S. Cerezo, V. Deulofeu, and A. Cohen, *J. Res. Nat. Bur. Stand.*, *A*, **71**, 53 (1967).

3. S. A. Barker, E. J. Bourne, and D. H. Whiffen, *Methods Biochem. Anal.*, **3**, 213 (1956).

4. R. S. Tipson, H. S. Isbell, and J. E. Stewart, *J. Res. Nat. Bur. Stand.*, **62**, 257 (1959).

5. L. P. Kuhn, *J. Amer. Chem. Soc.*, **74**, 2492 (1952); **76**, 4323 (1954).

6. A. R. H. Cole and P. R. Jefferies, *J. Chem. Soc.*, 4391 (1956).

7. S. A. Barker, J. S. Brimacombe, A. B. Foster, D. H. Whiffen, and G. Zweifel, *Tetrahedron*, **7**, 10 (1959).

8. V. C. Farmer, *Spectrochim. Acta*, **8**, 374 (1957).

9. V. C. Farmer, *Chem. Ind.* (London), 1306 (1959).
10. S. A. Barker, E. J. Bourne, H. Weigel, and D. H. Whiffen, *Chem. Ind.* (London), 318 (1956).
11. S. A. Barker, E. J. Bourne, W. B. Neely, and D. H. Whiffen, *Chem. Ind.* (London), 1418 (1954).
12. R. S. Tipson and H. S. Isbell, *J. Res. Nat. Bur. Stand.*, *A*, **64**, 239 (1960).
13. R. S. Tipson and H. S. Isbell, *J. Res. Nat. Bur. Stand.*, *A*, **66**, 31 (1962).
14. L. P. Kuhn, *Anal. Chem.*, **22**, 276 (1950).
15. G. P. Ellis, *J. Chem. Soc.* (*B*), 572 (1966).
16. "American Petroleum Institute Project 44, Infrared Absorption Spectra," Data Distribution Office, Agricultural and Mechanical College of Texas, College Station, Texas 77843 (2400 spectra); American Society for Testing and Materials, 1916 Race St., Philadelphia, Pennsylvania 19103 (IBM cards; index to over 55,500 published spectra); "The Documentation of Molecular Spectroscopy (DMS)," Butterworth, Bethesda, Maryland 20014 (almost 200 sugar derivatives); H. M. Hershenson, "Infrared Absorption Spectra, Indices for 1945–1957, and 1958–1962," Academic Press, New York, 1959 and 1964; "National Research Council—National Bureau of Standards Catalog of Infrared Spectra and Bibliography," National Bureau of Standards, Washington, D.C. 20234 (punch cards); Ref. 2 [spectra of 23 *N*-glycosylacylamines and 21 1,1-bis(acylamido)-1-deoxyalditols]; Ref. 4 (28 cyclic acetals of sugars); Ref. 12 (24 aldopyranosides); Ref. 13 (6 anomeric pairs of sugars and 12 single anomers); Ref. 14 (79 carbohydrates); Sadtler Research Laboratories, 3316 Spring Garden St., Philadelphia, Pennsylvania 19104 (about 200 sugar derivatives); H. A. Szymanski, "Infrared Band Handbook," 2nd ed., Plenum Press, New York, (1970); R. L. Whistler and L. R. House, *Anal. Chem.*, **25**, 1463 (1953) (35 monosaccharides); J. W. White, Jr., C. R. Eddy, J. Petty, and N. Hoban, *Anal. Chem.*, **30**, 506 (1958) (10 disaccharides and their acetates).
17. H. S. Isbell, F. A. Smith, E. C. Creitz, H. L. Frush, J. D. Moyer, and J. E. Stewart, *J. Res. Nat. Bur. Stand.*, **59**, 41 (1957); 56 sugar acetates.
18. R. S. Tipson and H. S. Isbell, *J. Res. Nat. Bur. Stand.*, *A*, **64**, 405 (1960); 24 acetylated aldopyranosides.
19. R. S. Tipson and H. S. Isbell, *J. Res. Nat. Bur. Stand.*, *A*, **65**, 31 (1961); 8 reducing pyranose acetates and 11 *N*-glycopyranosylacetamides.
20. R. S. Tipson and H. S. Isbell, *J. Res. Nat. Bur. Stand.*, *A*, **65**, 249 (1961); 20 peracetylated pyranoses.
21. T. Urbański, W. Hofmann, and M. Witanowski, *Bull. Acad. Polon. Sci. Ser. Chim. Geol. Geograph.*, **7**, 619 (1959).
22. D. H. Whiffen, *Chem. Ind.* (London), 129 (1957).
23. C. D. Hurd and H. Jenkins, *Carbohyd. Res.*, **2**, 240 (1966).
24. D. Horton and J. M. J. Tronchet, *Carbohyd. Res.*, **2**, 351 (1966).
25. H. B. Henbest, G. D. Meakins, B. Nicholls, and A. A. Wagland, *J. Chem. Soc.*, 1462 (1957).
26. R. S. Tipson, *J. Amer. Chem. Soc.*, **74**, 1354 (1952).
27. R. D. Guthrie and H. Spedding, *J. Chem. Soc.*, 953 (1960).
28. R. S. Tipson and A. Cohen, *Carbohyd. Res.*, **1**, 338 (1966).
29. M. Stacey, R. H. Moore, S. A. Barker, H. Weigel, E. J. Bourne, and D. H. Whiffen, *Proc. U.N. Intern. Conf. Peaceful Uses At. Energy*, 2nd, *Geneva*, **20**, 251 (1958).
30. J. S. Brimacombe, A. B. Foster, M. Stacey, and D. H. Whiffen, *Tetrahedron*, **4**, 351 (1958).
31. B. Åkermark, *Acta Chem. Scand.*, **15**, 985 (1961).

32. R. S. Tipson, W. S. Layne, and A. Cohen, unpublished results (1962).
33. R. Kübler, W. Lüttke, and S. Weckberlin, *Z. Elektrochem.*, **64**, 650 (1960); A. Rosenthal and M. R. S. Weir, *J. Org. Chem.*, **28**, 3025 (1963).
34. P. Bassignana and C. Cogrossi, *Tetrahedron*, **20**, 2361 (1964).
35. J. W. Rowen, F. H. Forziati, and R. E. Reeves, *J. Amer. Chem. Soc.*, **73**, 4484 (1951).
36. H. Spedding, *J. Chem. Soc.*, 3147 (1960).
37. M. L. Wolfrom and S. Hanessian, *J. Org. Chem.*, **27**, 1800 (1962).
37a. B. Lindberg and K. N. Slessor, *Carbohyd. Res.*, **1**, 493 (1966).
38. S. F. D. Orr, R. J. C. Harris, and B. Sylvén, *Nature*, **169**, 544 (1952); H. J. R. Stevenson and S. Levine, *Science*, **116**, 705 (1952); S. Levine, H. J. R. Stevenson, and P. W. Kabler, *Arch. Biochem. Biophys.*, **45**, 65 (1953).
39. S. F. D. Orr, *Biochim. Biophys. Acta*, **14**, 173 (1954).
40. S. A. Barker, E. J. Bourne, R. M. Pinkard, and D. H. Whiffen, *Chem. Ind.* (London), 658 (1958).
41. See also, Y. Nitta, J. Ide, A. Momose, and M. Kawada, *Yakugaku Zasshi*, **82**, 790 (1962); *Chem. Abstr.*, **57**, 14601 (1962).
42. R. Stephens, Ph.D. Thesis, University of Birmingham, England (1954).
43. R. S. Tipson, *J. Org. Chem.*, **26**, 2462 (1961).
44. L. Hough, J. E. Priddle, R. S. Theobald, G. R. Barker, T. Douglas, and J. W. Spoors, *Chem. Ind.* (London), 148 (1960).
45. W. M. Doane, B. S. Shasha, E. I. Stout, C. R. Russell, and C. E. Rist, *Abstr. Papers Amer. Chem. Soc. Meeting*, **153**, 14C (1967).
46. F. Micheel, F. P. van de Kamp, and H. Wulff, *Ber.*, **88**, 2011 (1955).
47. S. E. Darmon and K. M. Rudall, *Discussions Faraday Soc.*, **9**, 251 (1950).
48. D. Horton and M. L. Wolfrom, *J. Org. Chem.*, **27**, 1794 (1962).
49. A. G. Lloyd and K. S. Dodgson, *Nature*, **184**, 548 (1959).
50. K. Onodera, S. Hirano, and N. Kashimura, *Carbohyd. Res.*, **1**, 208 (1965).
51. E. Albano, D. Horton, and T. Tsuchiya, *Carbohyd. Res.*, **2**, 349 (1966).
51a. R. C. Chalk, M. E. Evans, F. W. Parrish, and J. A. Sousa, *Carbohyd. Res.*, **61**, 549 (1978).
52. F. A. Carey, D. H. Ball, and L. Long, Jr., *Carbohyd. Res.*, **3**, 205 (1966).
53. D. Horton and H. S. Prihar, *Carbohyd. Res.*, **4**, 115 (1967).
54. P. J. Launer, *Perkin–Elmer Instr. News*, **13** (3), 10 (1962).
55. T. R. R. McDonald and C. A. Beevers, *Acta Crystallogr.*, **5**, 654 (1952).
56. S. A. Barker, E. J. Bourne, M. Stacey, and D. H. Whiffen, *J. Chem. Soc.*, 171 (1954).
57. S. A. Barker, E. J. Bourne, R. Stephens, and D. H. Whiffen, *J. Chem. Soc.*, 3468 (1954).
58. S. A. Barker, E. J. Bourne, R. Stephens, and D. H. Whiffen, *J. Chem. Soc.*, 4211 (1954).
58a. J. J. Cael, J. L. Koenig, and J. B. Blackwell, *Carbohyd. Res.*, **32**, 79 (1974).
59. S. A. Barker and R. Stephens, *J. Chem. Soc.*, 4550 (1954).
60. L. M. J. Verstraeten, *Anal. Chem.*, **36**, 1040 (1964); see also, *idem*, *Carbohyd. Res.*, **1**, 481 (1966).
61. C. J. Brown, *Acta Crystallogr.*, **13**, 1049 (1960).
62. A. G. Lloyd and K. S. Dodgson, *Biochim. Biophys. Acta*, **46**, 116 (1961).
63. D. A. Rees, *J. Chem. Soc.*, 1821 (1963).
64. H. S. Isbell and W. W. Pigman, *J. Res. Nat. Bur. Stand.*, **18**, 141 (1937).
65. W. W. Pigman and H. S. Isbell, *J. Res. Nat. Bur. Stand.*, **19**, 189 (1937).
66. S. J. Angyal and V. A. Pickles, *Carbohyd. Res.*, **4**, 269 (1967).
67. (a) F. S. Parker, *Biochim. Biophys. Acta*, **42**, 513 (1960); (b) *Nature*, **203**, 975 (1964).

68. H. S. Isbell, J. E. Stewart, H. L. Frush, J. D. Moyer, and F. A. Smith, *J. Res. Nat. Bur. Stand.*, **57**, 179 (1956).
69. H. B. Wood, Jr., and H. G. Fletcher, Jr., *J. Amer. Chem. Soc.*, **79**, 3234 (1957).
70. J. K. N. Jones and J. C. Turner, *J. Chem. Soc.*, 4699 (1962).
71. A. E. El-Ashmawy and D. Horton, *Carbohyd. Res.*, **1**, 164 (1965).
72. F. Legay, *C. R. Acad. Sci.*, **234**, 1612 (1952).
73. H. Bredereck, A. Wagner, D. Hummel, and H. Kreiselmeier, *Ber.*, **89**, 1532 (1956).
74. J. Fabian and M. Legrand, *Bull. Soc. Chim. Fr.*, 1461 (1956); J. Fabian, M. Legrand, and P. Poirier, *ibid.*, 1499 (1956); H. S. Blair and G. A. F. Roberts, *J. Chem. Soc. (C)*, 2425 (1967).
75. W. Otting, *Ann.*, **640**, 44 (1961).
76. J. R. Holker, cited in H. Spedding, *Advan. Carbohyd. Chem.*, **19**, 23 (1964); see p. 40 and footnote 51 on p. 37 thereof.
76a. E. L. Albano and D. Horton, *Carbohyd. Res.*, **11**, 485 (1969).
77. L. Brown, P. Holliday, and I. F. Trotter, *J. Chem. Soc.*, 1532 (1951).
78. K. E. Almin, *Sv. Papperstidn.*, **55**, 767 (1952).
79. H. J. Marrinan and J. Mann, *J. Appl. Chem.* (London), **4**, 204 (1954).
80. J. Mann and H. J. Marrinan, *Trans. Faraday Soc.*, **52**, 481, 487, 492 (1956).
81. J. Mann and H. J. Marrinan, *J. Polymer Sci.*, **27**, 595 (1958); **32**, 357 (1958); M. Tsuboi, *ibid.*, **25**, 159 (1957).
82. M. Samec, *Staerke*, **5**, 105 (1953).
83. M. Samec, *J. Polymer Sci.*, **23**, 801 (1957).
84. O. Yovanovitch, *C. R. Acad. Sci.*, **252**, 2884 (1961).
85. C. Y. Liang and R. H. Marchessault, *J. Polymer Sci.*, **37**, 385 (1959); **39**, 269 (1959); **43**, 85 (1960); F. G. Pearson, R. H. Marchessault, and C. Y. Liang, *ibid.*, **43**, 71, 101 (1960); **59**, 357 (1962); R. H. Marchessault, F. G. Pearson, and C. Y. Liang, *Biochim. Biophys. Acta*, **45**, 499 (1960); see also D. H. Isaac, J. Blackwell, J. L. Koenig, E. D. T. Atkins, and J. K. Sheehan, *Carbohyd. Res.*, **50**, 169 (1976) for applications with glycosaminoglycans.
86. R. H. Marchessault, *Pure Appl. Chem.* (London), **5**, 107 (1962).
87. J. W. Ellis and J. Bath, *J. Chem. Phys.*, **6**, 221 (1938).
88. J. Fahrenfort, *Spectrochim. Acta*, **17**, 698 (1961).
89. F. S. Parker and R. Ans, *Appl. Spectrosc.*, **20**, 384 (1966).
90. F. E. Resnik, L. S. Harrow, J. C. Holmes, M. E. Bill, and F. L. Greene, *Anal. Chem.*, **29**, 1874 (1957).
91. F. H. Spedding and R. F. Stamm, *J. Chem. Phys.*, **10**, 176 (1942).
92. S. A. Barker, E. J. Bourne, R. M. Pinkard, and D. H. Whiffen, *J. Chem. Soc.*, 802, 807 (1959).
92a. M. C. Tobin, "Laser Raman Spectroscopy," in "Physical Techniques in Biological Research" (G. Oster, ed.), Vol. 1, Part A, Academic Press, New York, 1971; F. S. Parker, *Appl. Spectrosc.*, **29**, 129 (1975).
93. I. R. Beattie, *Chem. Brit.*, **3**, 347 (1967); C. Y. She, N. D. Dinh, and A. T. Tu, *Biochim. Biophys. Acta*, **372**, 345 (1974).
94. D. Horton, E. K. Just, and B. Gross, *Carbohyd. Res.*, **16**, 239 (1971).
95. J. L. Koenig, *J. Polymer Sci. (D)*, **6**, 59 (1972).
96. P. D. Vasko, J. Blackwell, and J. L. Koenig, *Carbohyd. Res.*, **19**, 297 (1971); **23**, 407 (1972).
97. J. J. Cael, J. L. Koenig, and J. Blackwell, *Carbohyd. Res.*, **29**, 123 (1973); **32**, 79 (1974).

VI. X-RAY AND NEUTRON DIFFRACTION

R. J. Ferrier

A. Scope and Limitations of the Techniques

Diffraction techniques are unique among the physical methods used for structural analysis of organic compounds in providing, at least in principle, the means for accurately locating the position of every atom within a molecule. However, the formidable problems associated with the measurement and analysis of diffraction data had, until *ca.* 1970, set these procedures apart as a province of specialists. In subsequent years, however, the advent of automatic diffractometers and computerized methods for data reduction have greatly decreased the effort and difficulty associated with X-ray crystallographic analyses. In consequence, X-ray crystal structures of carbohydrates and their derivatives are now being solved almost routinely. A "heavy atom" is no longer essential, although its presence greatly facilitates solution of structures.

The allied technique of neutron diffraction is potentially of great interest, as it provides an accurate means of locating hydrogen atoms, which diffract X-rays only weakly, but this technique has not yet contributed a great deal to carbohydrate chemistry. Nevertheless, very accurate structural parameters for sucrose[1] and α-D-glucopyranose[2] have been obtained by this procedure, thus effectively demonstrating its potential in the field.

In principle at least, provided that a suitable single crystal of the compound under investigation can be obtained (it need weigh only about 0.1 mg), full information can be determined on *relative* molecular configuration and other features of general structural significance, on conformation, on bond lengths and bond angles, and on details of intermolecular associations in the lattice. Specialized techniques also permit the determination of *absolute* configurations. With information on the crystal structures of several hundred carbohydrates and their derivatives at hand, generalizations can be considered on each of these points.

Because of the greater convenience of other powerful instrumental methods—particularly n.m.r. spectroscopy—the need to resort to crystallography for configurational determination of a compound has been infrequent. However, in some situations as, for example, with compounds **1** (ref. 3) or **2** (ref. 4) other

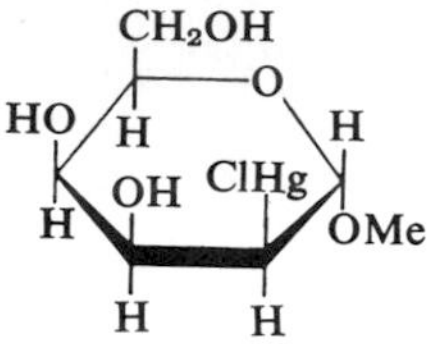

methods are inadequate, and X-ray diffraction has been applied with unique success. On occasion, desired structural detail is readily obtainable only by this procedure; the (*p*-bromophenyl)hydrazone of D-arabinose, for example, has been shown to exist in the crystal in the α-pyranoid ring form,[5] whereas the corresponding derivative of D-ribose exists in the acyclic form.[6]

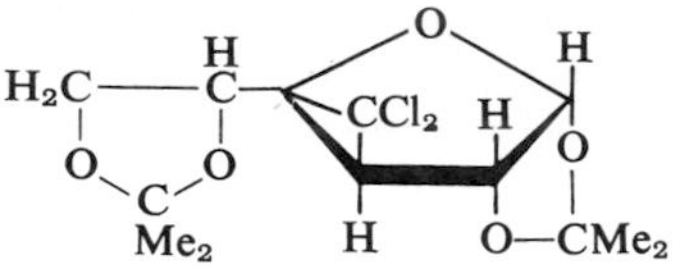

As well as contributing detail of this type to the field, crystallography has also furnished stereochemical correlations of broad, general significance. For example, by ascertaining the absolute configuration of a salt of (+)-tartaric acid (the aldaric acid derivable from L-threose), Bijvoet and co-workers[7] showed that Fischer's relative configurational assignments to the sugars were fortuitously correct in the absolute sense (see Vol. IA, Chapter 1). Furthermore, many crystallographic analyses have proved, in accordance with Freudenberg's postulate and Böeseken's findings, that cyclic sugars and their derivatives designated α have the same configuration at the anomeric center as at the highest numbered asymmetric carbon atom in the backbone chain (see Vol. IA, Chapter 1).

When crystallographic data allow hydrogen atoms to be located, it becomes possible to make direct comparisons between proton–proton dihedral angles estimated by n.m.r. spectroscopy for molecules in solution with actual angles in the crystal; such a determination has been made[8] for tri-*O*-acetyl-β-D-arabinopyranosyl bromide (**3**), a pyranoid sugar derivative that adopts the $^1C_4$(D) conformation both in the crystal and in solution.

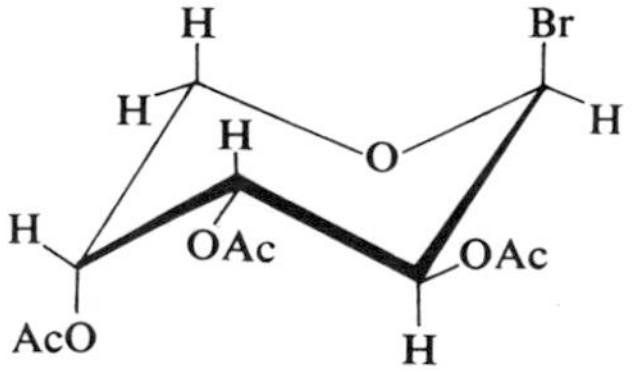

A surprisingly close correlation is observed between the conformations of carbohydrate derivatives determined crystallographically in the solid state and those determined by other techniques for the molecules in solution[9]; such correspondence is not necessarily to be expected, because lattice energies in the crystal could conceivably be considerably in excess of the energies required to effect conformational change in solution. This correlation is true not only for pyranoid derivatives and furanoid derivatives (in which 1,2-*cis* nonbonded interactions appear to be the most significant and are minimized through displacements of C-2 or C-3 away from a plane defined by the remaining atoms of the ring), but also for acyclic derivatives (see Vol. IA, Chapter 5). This is best illustrated by the work of Jeffrey and his colleagues,[10] who have shown that alditols exist in the crystal, as in aqueous solution (see Vol. IA, Chapter 5), preferentially in the planar zigzag conformation, except when this arrangement would bring 1,3-*cis*-related C–O bonds into an eclipsed relationship. In this event, interactions analogous to axial–axial interactions between oxygen atoms on six-membered rings appear to cause distortions that result, for example, in ribitol's adopting the "sickle" conformation (**4**), in which there are no such eclipsing interactions, rather than the extended, planar form (**5**). In contrast,

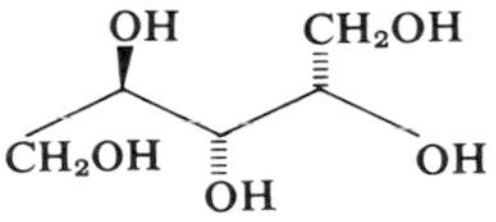

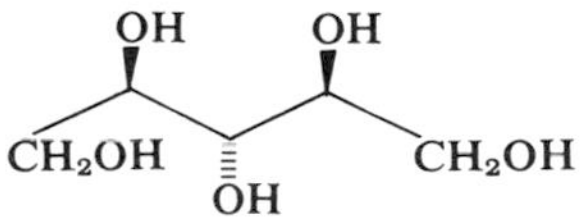

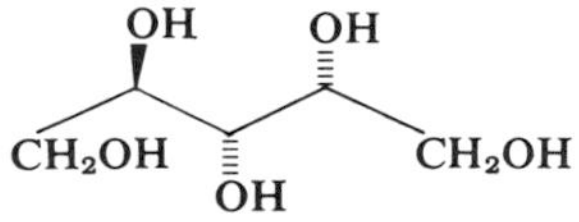

D-arabinitol favors the extended conformation (**6**), which has no such interactions.

An interesting point relating to bond lengths has been noted by the same group of workers.[11] They have observed that the C-1–O-1 bonds of pyranoid derivatives are appreciably shorter than the other C–O bonds, regardless of their axial or equatorial orientation, but that the shortening is less when the bonds are axial, as might have been expected from the operation of dipolar repulsions connected with the anomeric effect (Vol. IA, Chapter 5). This phenomenon apparently persists in furanoid derivatives and also in thio-

References start on p. 1444.

glycosides, and may be consequent upon regularities of intermolecular hydrogen-bonding patterns.

Intermolecular association of free sugars in the crystal lattice usually involves the maximum possible number of hydrogen bonds; the ring oxygen atom is invariably an acceptor of a hydrogen bond (this would tend to extend the C-1–O-5 and C-5–O-5 bonds in pyranoid derivatives), and most commonly each hydroxyl group is involved in hydrogen-bond donation (donation by OH-1 would lead to the observed decrease in the length of C-1–O-1) or acceptance.

The subject of X-ray crystallography of simple carbohydrates and their derivatives has been treated in two detailed articles,[12,13] and in a series[14,15] of annotated bibliographies, the latter appearing regularly in *Advances in Carbohydrate Chemistry and Biochemistry*; uncritical tabulations of individual structures that have been solved are collected in the series of Specialist Periodical Reports on Carbohydrate Chemistry.[16]

In addition to its use with crystalline compounds of low molecular weight, X-ray crystallography is a most valuable tool for furnishing information on the molecular architecture of crystalline, polymeric carbohydrates. Indeed, the classic work of Meyer and Misch[17] on the structure of crystalline cellulose provided an early, direct illustration of the β-D-glucopyranose ring system in the $^4C_1(D)$ chair conformation. The sharp resolution of diffraction patterns observed with crystalline compounds of low molecular weight is not attainable with crystalline polymers; nevertheless, a wealth of structural information has been achieved with polymeric carbohydrates by the crystallographic technique. Detailed discussions of this subject have been provided by Marchessault and co-workers[18,19]; individual applications to particular polysaccharides are discussed in the respective chapters in Part II of this work (see especially Vol. IIA, pp. 384, 419, 452; Vol. IIB, p. 492), and an independent series of annotated bibliographies of X-ray crystallographic studies of polysaccharides has been instituted[19] in *Advances in Carbohydrate Chemistry and Biochemistry*. In crystallographic work on the amylose–iodine complex, Rundle and co-workers[20] introduced the concept of a helical conformation for a biopolymer long before the idea attained wide popularity through subsequent work on proteins and deoxyribonucleic acid. Later studies of polysaccharides have emphasized correlation of empirically determined conformations with those predicted on the basis of energy minima and maxima.

B. List of Key Structures Determined by Crystallography

The following illustrative list includes some important carbohydrate derivatives that have been examined crystallographically. The original literature references are not included; however, these can be obtained from the review sources cited. In some instances the compounds were examined as hydrates, but this fact is not specifically noted.

MONOSACCHARIDES[12–16]

Pentoses

β-D- and β-L-Arabinopyranose
2-Deoxy-β-D-*erythro*-
　pentopyranose
β-D-Lyxopyranose

β-D-Ribofuranose tetraacetate
α- and β-D-Arabinopyranose
　tetraacetate
α-D-Xylopyranose

Hexoses

α- and β-D-Galactopyranose
2-Acetamido-2-deoxy-α-D-
　galactopyranose
2-Amino-2-deoxy-β-D-galacto-
　pyranose hydrochloride
α- and β-D-Glucopyranose
2-Acetamido-2-deoxy-α-D-
　glucopyranose

2-Amino-2-deoxy-α-D-glucopyranose
　hydrochloride and
　hydrobromide
α-D,L-Mannopyranose
α-L-Rhamnopyranose
β-D-Fructopyranose
α-L-Sorbopyranose
α-D-Tagatopyranose

Heptoses

α-D-*altro*-3-Heptulofuranose

β-D-*manno*-3-Heptulopyranose

OLIGOSACCHARIDES[12–16]

β-Cellobiose
Methyl β-cellobioside–methanol
β-Cellotetraose
Cyclohexaamylose inclusion
　complexes
1-Kestose
α- and β-Lactose

β-Maltose
Methyl β-maltoside
Melezitose
Planteose
Raffinose
Sucrose
α,α-Trehalose

POLYSACCHARIDES[18,19]

Alginic acid
B-Amylose
V-Amylose
Cellulose
Chitin
L-Guluronan

(1 → 4)-β-D-Mannan
D-Mannuronan
Nigeran
Pectic acid
(1 → 3)-β-D-Xylan
(1 → 4)-β-D-Xylan

References start on p. 1444.

GLYCOSIDES AND THIOGLYCOSIDES[12-16]

Methyl α-D-lyxofuranoside
Methyl β-D-ribopyranoside
Methyl 1-thio-α-D-ribopyranoside
Methyl 5-thio-α- and -β-D-ribopyranoside
Methyl 1,5-dithio-α- and -β-D-ribopyranoside
Methyl β-D-xylopyranoside
Methyl α-D-xylopyranoside triacetate
Methyl 1-thio-β-D-xylopyranoside
Methyl α-D-altropyranoside
Methyl α-D-galactopyranoside
Methyl 6-bromo-6-deoxy-α-D-galactopyranoside

Methyl 2-chloro-2-deoxy-α,β-D-galactopyranoside
Methyl α-D-glucopyranoside
Sinigrin
Methyl α-D-glucoseptanoside tetraacetate
Ethyl 1-thio-α-D-glucofuranoside
Methyl α-D-mannopyranoside
Ethyl 2-S-ethyl-1,2-dithio-α-D-mannofuranoside
Methyl 4,6-dideoxy-4-(dimethylamino)-α-D-talo-pyranoside methiodide

ALDITOLS[13-16]

Erythritol
D,L-Arabinitol
Ribitol
Xylitol
Allitol

Galactitol
D-Glucitol
D-Iditol
D-Mannitol

ACIDS AND DERIVATIVES[12-16]

Calcium (and strontium) L-arabinonate
L-Ascorbic acid
Calcium (and sodium) L-ascorbate
D-Galactonic acid
D-Galactono-1,4-lactone
Methyl (methyl α-D-galactopyranosid)uronate
Potassium D-gluconate
D-Gluconic acid 6-phosphate, trisodium salt

D-Glucono-1,5-lactone
Potassium β-D-gluco-pyranuronate
α-D-Glucopyranuronamide
β-D-Glucopyranurono-6,3-lactone
β-D-Glucofuranurono-6,3-lactone
D-Glucaric acid
Calcium D-xylo-5-hexulosonate
Calcium and strontium α-D-glucoisosaccharinate

ANHYDRO SUGARS [13–16]

Methyl 3,6-anhydro-α-D-galactopyranoside

Methyl 3,6-anhydro-α-D-glucopyranoside

1,6-Anhydro-β-D-glucopyranose

1,6-Anhydro-β-D-glucopyranose triacetate

1,6:2,3-Dianhydro-β-D-gulopyranose

1,6-Anhydro-β-D-mannofuranose

2,6-Anhydro-β-D-fructofuranose

2,7-Anhydro-β-D-*altro*-heptulopyranose

CARBOHYDRATE-CONTAINING ANTIBIOTICS [13–16]

Blasticidin S bromide

Cytosamine triacetate

Erythromycin A hydroiodide

Formycin hydrobromide

Isoquinocycline A

Kanamycin

Kasugamycin hydrobromide

Lincomycin hydrochloride

Polyoxin C

Puromycin

Showdomycin

Tubercidin

MISCELLANEOUS DERIVATIVES [12–16]

D-Arabinose p-bromophenyl-hydrazone

2,3,4-Tri-O-acetyl-β-D-arabino-pyranosyl bromide

D-Ribose 5-phosphate, barium salt

D-Ribose p-bromophenyl-hydrazone

2,3,4-Tri-O-benzoyl-β-D-xylopyranosyl bromide

α-D-Glucopyranosyl phosphate, dipotassium salt, potassium sodium salt

D-Glucose p-bromophenyl-hydrazone

2-S-Ethyl-2-thio-D-mannose diethyl dithioacetal

1,2:4,5-Di-O-isopropylidene-β-D-fructopyranose

N-Acetylneuraminic acid

epi-Inositol

myo-Inositol

myo-Inositol 2-phosphate

OTHER CARBOHYDRATE-CONTAINING COMPOUNDS

Nearly all of the D-ribofuranosyl nucleosides and their 5'-phosphates, as well as several D-arabinofuranosyl nucleosides, have been examined by X-ray crystallography.[14–16] Structures of a large number of sugars having partial or mixed substitution are listed in refs. 14–16, 18, and 19.

REFERENCES

1. G. M. Brown and H. A. Levy, *Science*, **141**, 921 (1963).
2. G. M. Brown and H. A. Levy, *Science*, **147**, 1038 (1965).
3. J. Bain and M. M. Harding, *J. Chem. Soc.*, 4025 (1965).
4. J. S. Brimacombe, P. A. Gent, and T. A. Hamor, *J. Chem. Soc.* (*B*), 1566 (1968).
5. S. Furberg and C. S. Petersen, *Acta Chem. Scand.*, **16**, 1539 (1962).
6. K. Bjamer, S. Furberg, and C. S. Petersen, *Acta Chem. Scand.*, **18**, 587 (1964).
7. A. F. Peerdeman, A. J. van Bommel, and J. M. Bijvoet, *Koninkl. Ned. Akad. Wetenschap. Proc.*, **54**, 3 (1951).
8. P. W. R. Corfield, J. D. Mokren, P. L. Durette, and D. Horton, *Carbohyd. Res.*, **23**, 161 (1972).
9. P. L. Durette and D. Horton, *Advan. Carbohyd. Chem. Biochem.*, **26**, 49 (1971); D. Horton, P. L. Durette, and J. D. Wander, *Ann. N.Y. Acad. Sci.*, **222**, 884 (1973).
10. G. A. Jeffrey and H. S. Kim, *Carbohyd. Res.*, **14**, 207 (1970).
11. H. M. Berman, S. S. C. Chu, and G. A. Jeffrey, *Science*, **157**, 1576 (1967); G. A. Jeffrey, *Amer. Chem. Soc. Symp. Ser.*, **87**, 50 (1979).
12. G. A. Jeffrey and R. D. Rosenstein, *Advan. Carbohyd. Chem.*, **19**, 7 (1964).
13. G. Strahs, *Advan. Carbohyd. Chem. Biochem.*, **25**, 53 (1970).
14. G. A. Jeffrey and M. Sundaralingam, *Advan. Carbohyd. Chem. Biochem.*, **30**, 445 (1974).
15. G. A. Jeffrey and M. Sundaralingam, *Advan. Carbohyd. Chem. Biochem.*, **31**, 347 (1975); **32**, 353 (1976); **34**, 345 (1977); **37**, in press (1979).
16. *Carbohydrate Chemistry*, Vols. 2–10; Specialist Periodical Reports, The Chemical Society, London, 1969–1978.
17. K. H. Meyer and L. Misch, *Helv. Chim. Acta*, **20**, 232 (1937).
18. R. H. Marchessault and A. Sarko, *Advan. Carbohyd. Chem.*, **22**, 421 (1967).
19. R. H. Marchessault and P. R. Sundararajan, *Advan. Carbohyd. Chem. Biochem.*, **33**, 387 (1976); **35**, 377 (1978); **36**, 315 (1979).
20. R. E. Rundle and D. French, *J. Amer. Chem. Soc.*, **65**, 1707 (1943); R. E. Rundle and F. C. Edwards, *ibid.*, **65**, 2200 (1943); compare R. S. Bear, *ibid.*, **64**, 1388 (1942).

28. SEPARATION METHODS: CHROMATOGRAPHY AND ELECTROPHORESIS

MARTIN I. HOROWITZ*

I. INTRODUCTION

Chromatographic and electrophoretic methods afford information on the homogeneity, molecular size, and structure of a carbohydrate and provide useful data—the R_f, R_{Glc}, and M_{Glc} values—which are helpful in the identification of the compound in question. These separation methods also provide a means for isolation of carbohydrates and their derivatives on a preparative scale. Applications to the field of carbohydrate chemistry of the various

* This work was supported by PHS Research Grant No. AM–15475–07 from the National Institute of Arthritis and Metabolic Disease.

separation methods—namely, partition (including gas–liquid partition), adsorption, and ion-exchange chromatography, gel filtration, and zone electrophoresis—will be discussed here. A more detailed description of the theory and practice of these methods may be found in various standard text books[1-4] and in *Methods in Carbohydrate Chemistry*, Vols. I–VIII (1962–1979).

II. PARTITION CHROMATOGRAPHY

In partition chromatography, the substance being chromatographed is distributed between two solvent phases, a stationary phase and a moving phase, according to the partition coefficient of the substance between the two phases. The water-rich phase usually is kept stationary while the phase rich in organic solvent moves over it. In paper partition chromatography, the sample is applied along a base line on a sheet of filter paper, either as a small circular zone or as a narrow streak, and the solvent is allowed to flow over the paper by capillarity in a closed chamber maintained at a reasonably constant temperature. Ascending, descending, and circular modifications of flow movement have been described, although ascending and descending flow are the most widely used. After a suitable development time, the paper is removed from the chamber, dried, and sprayed with a reagent that reveals the separated components of the sample as colored spots.

The R_f value of a compound is defined as the distance traveled by the compound relative to the distance traveled by the solvent phase.

Other R values are used when it is desired to express the rate of movement of the compound relative to the rate of movement of a reference compound. For example, the ratios are designated R_{Glc}, R_{Lac}, and R_{Xyl} when glucose, lactose, and xylose, respectively, are used as the reference compounds.

The ratio of the distribution of a compound between the two phases (the partition coefficient, α) is related to the amount of free energy needed to transport one mole of the compound from one phase to the other. The R_f value is related to α by the expression

$$\alpha = \frac{A_L}{A_S}\left(\frac{1}{R_f} - 1\right)$$

where A_L/A_S is the ratio of volumes of the organic solvent phase to water phase.[5]

A. CORRELATION OF STRUCTURE WITH MIGRATION ON PAPER

The R_f value of a sugar increases as the water content of the mobile phase increases. Isherwood and Jermyn[6] have interpreted this fact as indicating that sugars are strongly hydrated in aqueous solution. They observed that a plot of log $(1/R_f - 1)$ or R_M, against log N, where N is the mole fraction of water in the solvent, gave a straight line. Phenolic solvents were exceptions. Isherwood and Jermyn[6] also formulated a number of empirical rules, including the following: (1) Sugars having a furanose ring exhibit higher R_f values than those having a pyranose ring. (2) The R_f value is dependent mainly on the configuration of the hydroxyl groups.

Jaeger et al.[7] developed the concept that the R_f value is related to the interaction of the hydroxyl groups of the sugar through hydrogen bonding with water as the stationary phase. They observed that, the greater the number of equatorial hydroxyl groups (of the favored pyranose conformation of the sugar in solution), the lower is the R_f value of the sugar; this observation is interpreted as reflecting a low relative solubility of the sugar in the organic solvent (or moving phase) owing to the hydration of the sterically accessible, equatorial hydroxyl groups. Axial hydroxyl groups are hydrated less easily for steric reasons and lead to an increase in the solubility of the sugar in organic solvents, with increase in the R_f value. The R_f values decrease in a regular fashion with increasing number of sugar residues in a homologous series of oligosaccharides. French and Wild[8] plotted the logarithm of a partition function, $\alpha = R_f/(1 - R_f)$, against the molecular size for each oligosaccharide in a regular homologous series and obtained characteristic straight lines for each series. The R_f values were lower for $(1 \to 6)$-linked sugars than for $(1 \text{-} > 4)$-linked sugars. The decrease in R_f value with increase in the number of monosaccharide residues also holds for homologous series of the N-benzylglycosylamine derivatives of oligosaccharides.[9]

B. SOLVENT SYSTEMS

Usually a compromise must be made between a mobile phase which gives an optimum R_f value and one which gives the greatest distance between spots. Jermyn and Isherwood[10] observed that the best resolution of pairs of sugars is obtained with systems in which the sugars exhibit R_f values of about 0.2–0.3; optimal separation is effected by allowing the solvent to run off the end of the paper, or by multiple development. Since it is difficult, in practice, to get two-component solvent systems that give R_f values of 0.2–0.3 for pairs of sugars, a third component, soluble in both phases, may be added; this usually increases the solubility of the sugar and allows greater flexibility in obtaining R_f values in the desired range. Multiple development has been used to improve the resolution of compounds having low R_f values. In this procedure, the distance between spots having low R_f values increases with the number of developments.

References start on p. 1470.

TABLE I

SOME SOLVENT SYSTEMS COMMONLY USED FOR PAPER CHROMATOGRAPHY

Substance	Solvents	Proportions (v/v)
N-Acetylated amino sugars[16]	Borated paper; 0.2 M borate pH 8; ethyl acetate–pyridine–water	2:1:2
Alditols[19,20]	Propyl alcohol–ethyl acetate–water	7:1:2
	Ethyl acetate–pyridine–water	14:5:1
Amino sugars[15]	Pyridine–ethyl acetate–acetic acid–water	5:5:1:3
Lactones[13]	Pyridine–ethyl acetate–water	11:40:6
Methylated sugars[14]	Butyl alcohol–ethanol–water	5:1:4
Neutral sugars[4]	Ethyl acetate–pyridine–water	8:2:1 or 2:1:2 or 2:1:5 or 10:4:3
Neutral sugars, sugar acids[12]	Butyl alcohol–acetic acid–water	4:1:5
Oligosaccharides[21,22]	Butyl alcohol–pyridine–benzene–water	5:3:1:3
	Propyl alcohol–ethyl acetate–water	6:1:3
Sialic acids[17,18]	Butyl alcohol–acetic acid–water	3:2:1
	Butyl alcohol–propyl alcohol–0.1 N HCl	1:2:1

Durso and Mueller[11] have described a systematic approach to the formulation of solvent systems.

Some solvent systems useful in partition chromatography are listed in Table I. A more-extensive list of solvent systems may be found elsewhere.[3,5,11a]

The solvent system: nitromethane–acetic acid–ethanol–water saturated with boric acid 8:1:1:1 (v/v/v/v) resolves the common monosaccharides from each other and from the corresponding alditols.[11b]

Reversed phase chromatography employs nonpolar solvents as the fixed phase and nonpolar or polar solvents as the mobile phase. This system is advantageous for the separation of compounds that are insoluble in water, such as methylated[23,24] and acetylated sugars[23] and acetylated sugar alcohols.[23] In the procedure developed by Wickberg,[23] paper is impregnated with a solution (25%, v/v) of methyl sulfoxide in toluene. The toluene is removed, and the chromatogram is developed with light petroleum ether, isopropyl ether, ethyl ether–methyl sulfoxide (25:1, v/v), or benzene–methyl sulfoxide (20:1, v/v).

C. PREPARATION OF A SAMPLE FOR CHROMATOGRAPHY

Inorganic salts (at concentrations greater than 5%) interfere with the chromatographic process by moving separately as cations and anions and by creating local zones of acidity and basicity.[25] This action affects the water content of the two phases and also interferes with the coloration of the zones of the solute. Electrolytic desalting,[3] solvent extraction, and selective precipitation—for example, sulfate ions as barium sulfate—have been used to reduce the salt content of sugar samples, but these techniques have largely been superseded by ion-exchange[25a] and gel-filtration methods.

D. DETECTION AND QUANTITATION

Detection of sugars on paper may be effected by spraying or by dipping the developed chromatogram in a solution of an appropriate reagent. Color reagents may be grouped into three types—those that react with both reducing and nonreducing carbohydrates, those that react with reducing sugars alone, and those that react with a sugar by virtue of a special functional group of the sugar, such as an amino group or a carboxyl group. Many of these reagents are highly sensitive and can reveal 1 to 2 μg of a carbohydrate on paper and even smaller amounts on thin-layer plates. Ammonium molybdate,[26] periodate–benzidine,[27] periodate–permanganate–benzidine,[28] periodate–permanganate,[29] and periodic acid–fuchsin reagents (for protein-bound carbohydrates)[30] are used to reveal both reducing and nonreducing sugars. Reducing sugars may be revealed by alkaline silver nitrate,[31] aniline hydrogen phthalate,[32] aniline oxalate,[33] aniline trichloroacetate, aniline phosphate,[34] or aniline citrate,[21] triphenyltetrazolium chloride,[10,35,36] p-anisidine hydrochloride,[34,37] and phenolic reagents.[38] Alkaline silver nitrate also may be used for nonreducing sugars containing vicinal diols, but this application requires a longer exposure to the silver nitrate reagent, and borate and salts interfere with the staining of carbohydrates by the silver nitrate spray. Paper developed in borate solution may be dipped in a solution of acetone containing hydrofluoric acid[39] and then stained with silver nitrate solution. Detection of carbohydrates in the presence of boric acid may be facilitated by substitution of 0.5 M NaOH in 80% ethanol containing 4% pentaerythritol[11b,39a] for methanolic NaOH as used in the conventional silver nitrate dipping technique. Borate does not interfere with the lead acetate reagent[40] or with the periodate–benzidine reagents, and these reagents may be used instead of silver nitrate when the material is chromatographed on borate paper. Enzyme preparations such as invertase,[41] D-glucose oxidase,[42] and D-galactose oxidase[43] may be used before or after chromatography, in conjunction with staining reagents, for various special applications. When fiber-glass paper is employed, the sheets may be sprayed with concentrated sulfuric acid and heated to produce brown spots by charring[44]; this procedure is advantageous for nonreducing, methylated sugars. Isotopically labeled sugars can be detected by an appropriate isotope scanning device or by autoradiography. Reducing sugars may be reduced with [^{3}H]borohydride, and the corresponding alditols separated by paper chromatography as their borate complexes. Quantitative determinations may be performed by counting the tritiated compounds, provided tritium of known specific activity is used and an internal standard such as [^{14}C]glucose is included.[44a]

More-specific reagents may be employed to detect individual structural features. For example, amino sugars are detected[45,46] as purple spots by ninhydrin or as red spots by the p-aminobenzaldehyde–2,4-pentanedione reagent of Elson and Morgan; 2-acetamido–2-deoxyaldoses are detected as purple spots by the use of p-aminobenzaldehyde alone. Sugar acids may be detected by acid-base indicators, such as Bromophenol Blue, or they may be converted into their esters by diazomethane and detected by means of hydroxamic acid.[3] Sugar sulfates may be visualized with ethanolic solutions of barium chloride and sodium rhodizonate,[4,47] and sugar phosphates with a molybdate reagent.[48] Directions for the preparation and use of these reagents have been compiled by Macek.[3] The monograph by Montreuil et al.[49] should be consulted for an evaluation of the various color reagents used for the determination of sugars present in glycoproteins by quantitative chromatographic methods.

After detection of the sugar by means of guide strips, it is possible to elute the compound from the paper for analysis or characterization. Quantitative analysis may be performed either in situ on the filter paper or on solutions containing the eluted sugars. In the former method, the areas of the spots of the sugars being determined are measured directly on the paper and compared to areas of a graded concentration of standards.[50] Photometric determination of color densities[51] may also be used. Superior results are claimed for the technique whereby the sugar samples and standards are eluted from the paper and subsequently analyzed by a suitable reagent [52,53]; the accuracy of this method is ±5%.

It is possible to chromatograph 0.5 mg of sugar per square centimeter of Whatman No. 1 paper; this amounts to a maximum of 25 mg on a 50-cm-wide sheet.[54] Even larger amounts of sugar can be carried on thicker sheets, such as Whatman No. 3, but when more than 100 mg of sugar is to be isolated, a column method is more convenient. Satisfactory preparative-scale separations have been obtained by partition on cellulose powder,[55] hydrocellulose,[55,56] microcrystalline cellulose (Avicel),[57] starch,[58] and Celite columns.[5,59] Cellulose powder has a relatively high capacity, and separations obtained on cellulose columns are similar to those obtained on paper. Hydrocellulose has a higher capacity and better resolving power for methylated derivatives.[56] Starch affords excellent resolution of monosaccharides, but the shortcomings of starch include need for pretreatment, low capacity, and low flow-rate of solvent. Celite gives a more rapid flow rate than cellulose, good separation of monosaccharides and other sugar derivatives, and minimal elution of column impurities. Monosaccharides,[55,56,59] methylated sugars[55,56,59] and other derivatives,[59] and oligosaccharides have been separated on a several-gram scale by column methods.

Ion-exchange resins have been used as the support phase for partition chromatography of carbohydrates. In this technique, a solution of carbohydrates (pentitols, hexitols, pentoses, and hexoses) in 86% ethanol (v/v) is added to a column of anion-exchange resin beads in the sulfate form, and the column is developed with this solvent at an elevated temperature. Excellent quantitative separations have been achieved, and a precision of $\pm 2\%$ was reported in the analysis of the effluent.[60]

Partition chromatography may also be performed on thin-layer plates. Thin-layer chromatography (t.l.c.) is a technique which employs a thin coating of a porous medium (generally 0.25 mm to 2.0 mm thick) on a glass plate or a foil. Capillary migration of the solvent through the coating achieves rapid development of chromatograms (1 to 3 hours for t.l.c. compared with 16 to 48 hours required for most paper-chromatographic separations). The technique gives highly compact zones and requires smaller amounts of sample than those used in paper chromatography. It can also be used for quantitative analysis.[61-63]

Although t.l.c. was originally introduced as a method of adsorption chromatography, the technique has also been used for partition chromatography with silica gel, kieselguhr, or cellulose as the support for the stationary (water) phase.[47] Usually, calcium sulfate is incorporated as a binder with silica gel and with kieselguhr. The binder helps to fix the supporting medium to the glass plate. The letter G (for gypsum) is added after the name of the adsorbent to indicate that it contains calcium sulfate as binder—for example, silica gel G or kieselguhr G. Admixture of silica gel G with salts[63] and the introduction of water into the system are usually required to achieve good separation of unsubstituted sugars and polar derivatives on silica gel. Such modifications are believed to change the mechanism from adsorption to partition or to a combination of the two mechanisms. Silica gel G impregnated with boric acid[63] or phosphate buffer,[64] and kieselguhr G impregnated with salts,[65-67] have been used for the separation of neutral sugars. Silica gel accommodates a larger load of sample and affords better resolution of mixtures of sugars than kieselguhr.[47] Mixtures of silica gel G and kieselguhr G,[65,68] Celite and starch,[69] and silica gel G and alumina (Alusil)[67,70] have also been used for the resolution of mixtures of neutral sugars. Aqueous solvents on silica gel G impregnated with salts have been used to separate mixtures of sugar acids,[71] malto-oligosaccharides,[72] oligosaccharides containing sialic acid,[73] phenolic glycosides,[74] and dansylated amino sugars.[75]

Cellulose MN 300 has been widely used[49,76] for t.l.c. to obtain separations similar to those obtained by paper chromatography. Vomhof and Tucker[77] described the chromatography of sugars in nine solvent systems using cellulose MN 300; separation of sugar phosphates on this support also has been

---

reported.[78] Cellulose-coated plates have also afforded separations of amino sugars[79] and sialic acids.[80] Wolfrom and co-workers[57,61] have used micro-crystalline cellulose (Avicel, a product of American Viscose Co., Marcus Hook, Pennsylvania) as a support for partition t.l.c. Free sugars, glycosides, methyl ethers, hydroxy acids, lactones, amino sugars, and amino acids[61] have been effectively separated on Avicel. This material has a high physical strength, and the plates are available commercially. Polycarbonate sheets[81] containing silica gel G provide fair separation of hexoses, pentoses, and some sugar alcohols.

Ghebregzabher et al.[82] have written a comprehensive review on thin-layer chromatography of carbohydrates.

III. GEL FILTRATION AND AFFINITY CHROMATOGRAPHY

The technique of gel filtration on columns of cross-linked dextrans has been employed successfully to desalt and to fractionate mixtures of low- and high-molecular-weight carbohydrates,[83] glycopeptides,[84] and glycoproteins.[85] The mechanism of fractionation here entails exclusion of molecules above a certain size from the interior of the three-dimensional network of dextran chains, partial accessibility of the interior network to smaller molecules, and a greater or complete accessibility of the interior to solutes and salts of still smaller size.[86] Complete exclusion results in a shorter path for the molecules to traverse in their journey from the top to the bottom of the column, whereas inclusion results in a longer, more tortuous path down the column; consequently, larger molecules leave the column faster than smaller molecules. A graded series of cross-linked dextrans, known as the Sephadex G series, is commercially available (Pharmacia Co.). Passage through columns of polyacrylamide beads (Bio-Rad Laboratories, Richmond, California) and agarose beads[83] also has been used to separate molecules of different molecular size.[87] Agarose columns have been used successfully for the fractionation of glycosaminoglycans and proteoglycans.[88]

Some deviations from a pure sieving mechanism arise when cyclic unsaturated compounds or charged compounds are used. Cyclic unsaturated compounds are adsorbed to some extent on Sephadex columns and, accordingly, are retarded in their passage down the column. Negatively charged substances are accelerated, and positively charged substances are retarded, because of interaction with some negative sites, presumably carboxyl groups, on Sephadex gels described as having 10 microequivalents of acidic groups per gram of dry gel.[86] These interactions can be minimized by sieving in the presence of electrolytes or weak acids.[89]

Gel filtration may be used for estimating molecular-weight distributions of polysaccharides from their elution patterns. It is important that appropriate standards be used to calibrate the columns; for example, in estimating the molecular weight of a dextran fraction a column calibrated with the homologous series of dextran compounds should be used.[90] The commonly employed method of estimating molecular weight by gel filtration is based on the linear relation between elution volume, V_e, and log of molecular weight found for structurally related compounds.

Differences in the distribution coefficients (K_d) are usually too small to afford separations of monosaccharides from one another; if a difference in size exists between two monosaccharides, for example, D-glucose and D-ribose, then it may be possible to effect a separation with a gel that is highly cross-linked.[91] Resolution of mono-, di-, and tri-saccharides into the corresponding saccharide groups may be accomplished with either Bio-Gel P-2[91] or Sephadex G-10 or G-15.[92] Chitin oligosaccharides have been fractionated on Bio-Gel P-2,[93] and oligosaccharides from enzymic digestion of hyaluronic acid, chondroitin 4-sulfate, and heparin[94] have been separated on columns of Sephadex G-25; gluco-oligosaccharides of d.p. 2–15 may be resolved[94a] on Bio-Gel P-4.

Gel filtration of acidic glycosaminoglycans[95] using Sephadex G-200 columns was described; gel filtration of these compounds on thin layer plates has also been used for molecular weight estimation,[96] but this method is difficult to apply for accurate estimation when the polymers are very polydisperse. Proteoglycans may be chromatographed on columns of agarose (Bio-Gel A-150m)[97] as an aid to investigating their heterogeneity. The molecular sieve chromatography of proteoglycans on columns of controlled-pore glass beads (CPG) or 1% agarose has been compared; the resolving power and recovery were comparable, but CPG columns could be operated at much higher flow rates.[97]

Gel filtration on the lipophilic gel Sephadex LH-20 may be used to fractionate methylated and acetylated carbohydrates and oligosaccharides[96] and to separate ethylated carbohydrate from methyl sulfoxide and other reagents.[98]

Affinity chromatography is a separation technique that takes advantage of the affinity of (usually) proteins for specific ligands, for instance, enzyme–substrate, antibody–antigen or antibody–hapten, glycoprotein–lectin, receptor–hormone, and others. Low- or high-molecular-weight ligands are coupled to derivatives of an inert support such as agarose or poly(acrylamide). Spacer arms, to increase the accessibility of immobilized ligands, are sometimes employed. An aqueous mixture containing the desired constituent is filtered through the "affinity" column, which selectively retains substances interacting with the ligand. The column is then washed and developed (a) with buffers, varying the ionic strength, pH, or both, or (b) with solutions containing substances with an affinity for the ligand. Thus, where concanavalin A (which binds α-D-mannopyranosyl or α-D-glucopyranosyl residues having unmodified

---

References start on p. 1470.

hydroxyl groups at the C-3, C-4, and C-6 positions[99]) is the ligand and a glyco-protein containing α-mannosyl groups is bound to the column via the ligand, the glycoprotein may be desorbed from the column by an eluent containing methyl α-D-mannopyranoside.

Affinity chromatography is the subject of a complete volume in "Methods in Enzymology"[100] and of a Biochemical Society Symposium,[100] and individuals interested in theory and applications of affinity chromatography should consult these sources for orientation and for varied applications. The following are some selected examples of applications to carbohydrates: purification of glycosyl transferases,[101] lectins,[102] lectin–glycoprotein interactions,[103] and antibody–monosaccharide or antibody–oligosaccharide interactions.[99]

IV. ADSORPTION CHROMATOGRAPHY

Adsorption chromatography on columns or on thin-layer plates is a very useful method for the separation and purification of simple sugars, sugar derivatives, and oligosaccharides. Column separations can be performed either by extrusion or by elution. In the elution technique, the components of a mixture, dissolved in a solvent, are added to the top of a column previously equilibrated with the solvent. On carbon columns, some of the constituents of the mixture, notably inorganic salts, are not adsorbed and appear in the initial effluent. The adsorbed constituents are selectively desorbed by the sequential introduction, either discontinuously or by gradient, of solvent or solvent mixtures. The solvent molecules displace the adsorbed materials and are adsorbed themselves in turn. Frontal analysis[103a] and displacement chromatography[104] have been used at times, but zonal or elution chromatography, the method described above, is most commonly used.

Adsorption, or retention of solute on the adsorbent surface, is generally considered in terms of the surface tension of the solid adsorbent. Molecules in the interior of the adsorbent are subject to equal forces in all directions, whereas molecules at the surface are subject to unbalanced forces which result in a net inward force, causing the molecules to be adsorbed at the surface. London dispersion forces or van der Waals forces are believed to be the principal forces operative in adsorption by nonpolar adsorbents such as carbon,[1] although some ion-exchange and polar sites also are present in most samples of carbon. The larger the number of atoms in the solute molecule, the larger is the number of interactions between the nonpolar adsorbent and the solute, and therefore solutes of higher molecular weight are usually adsorbed more tightly on activated carbons than are those of low molecular weight. On the other hand, dipole interactions and hydrogen bonding predominate in the adsorption of solute by polar adsorbents, such as silica gel (partially dehydrated silicic acid). The more polar the functional grouping on the solute, the tighter is the adsorption. An increase in molecular weight of the solute by addition of nonpolar groups decreases the polar character of the solute.[1] Accordingly, polar compounds of high molecular weight may be adsorbed less strongly on polar adsorbents than are compounds of lower molecular weight carrying the same polar functional group.

Chromatography on activated carbon is commonly used for column separations of unsubstituted carbohydrates. Such carbons are useful for desalting and for fractionation of oligosaccharides, but they usually will not resolve mixtures of monosaccharides. A successful application of carbon chromatography is shown in Fig. 1, which illustrates the zonal separation of the components present in an acid hydrolyzate of cyclohexaamylose by gradient elution with aqueous ethanol.[105] The carbon was pretreated with 1% ethanolic stearic acid to block the very active sites and capillary pores which, if accessible, lead to poor recovery and erratic results. Activated carbons are noted for their high adsorptive capacity, ready availability, cheapness, and suitability for use with aqueous and polar organic solvents. The flow of solvent through the beds of carbon is extremely slow, and Celite is usually mixed with the carbon to accelerate flow.

Polar adsorbents, such as silica gel, hydrated magnesium oxide, fuller's earths (Floridins), and alumina,[106] have been used for the separation of substituted sugars. Benzene–petroleum ether and benzene–ethanol, and binary mixtures of adjacent solvents in the eluotropic series, are solvent combinations frequently employed here. The colored p-phenylazobenzoic esters of sugars were separated successfully on silicic acid,[107] on Magnesol,[108] and on calcium acid silicates.[109] Tetra- and penta-acetates of D-glucose were resolved on Magnesol,[110] and this adsorbent is extremely effective for the separation of isomeric oligosaccharides as their peracetates.[111] Sugar acids are separated

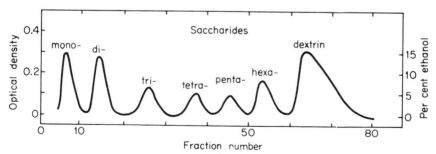

FIG. 1. Carbon-column chromatography of a mixture of mono- to hexa-saccharides from an acid hydrolyzate of cyclohexaamylose (courtesy of W. W. Binkley[105] and Academic Press).

References start on p. 1470.

from the parent sugars on fuller's earth clays,[112] in keeping with the behavior of polar solutes on polar adsorbents. Colorless derivatives may be detected on columns by brushing alkaline permanganate along the outside of the extruded column.[111]

Polar adsorbents, especially silica gel, are extensively used for thin-layer chromatography of substituted sugars of low polarity, particularly in synthetic work. It is often possible to examine reaction mixtures for the appearance of intermediate products and to monitor the course of a reaction with time. Thin-layer chromatography on silica gel often may be performed without the need of desalting the sample.[113] Separations by thin-layer adsorption chromatography require the use of heat-activated adsorbents and nonaqueous solvents. R_f values are less reproducible and thus less significant in adsorption than in partition or in reversed-phase-partition thin-layer chromatography. The rate of migration of a compound in thin-layer chromatography is affected by a number of operational variables, such as the amount of sample applied, grain size and activity of the adsorbent, and the thickness of the adsorbent layer. The rate of migration also is affected by the nature of the binder, the composition of the solvent, saturation of the chamber with vapor of the solvent, and the temperature of the system. Accordingly, expressing R_f values relative to a reference substance applied in similar amount is the preferred method for reporting migration data in thin-layer adsorption chromatography. Silica gel G plates have been used for resolving mixtures of sugar derivatives such as trimethylsilyl ethers,[114] methyl ethers,[115,116] esters (Fig. 2),[117,118] dithioacetals,[66] hydrazones,[119,120] and acetals.[121] Dansylated amino sugars,[75] deoxy sugars,[122] substituted cerebrosides,[123] glycolipids,[124] and gangliosides[125] also have been separated on this adsorbent. Alumina plates have been employed for thioacetals, halogenodeoxy sugar derivatives, and acetylated or isopropylidene and benzylidene derivatives,[126] and for the resolution of intermediates in the synthesis of serine-O-glycosides.[127] Preparative separations, on a gram scale, are readily achieved by thin-layer chromatography in either adsorption or partition systems.[128]

Thin-layer chromatography of glycolipids has been extensively used for quantitative analysis and heterogeneity studies, both on an analytical scale and also on a preparative scale. The monograph "Glycolipid Methodology" should be consulted for applications.[129]

V. ION EXCHANGE

Ion-exchange chromatography, when applicable, is particularly suitable for the separation of large amounts of solute and may be adapted to automatic operation for analytical determinations. Techniques for the quantitative separation of neutral sugars based on ion exchange have been developed by Hallén,[130] Walborg et al.,[131,132] Syamananda et al.,[133] and Lee et al.[134]

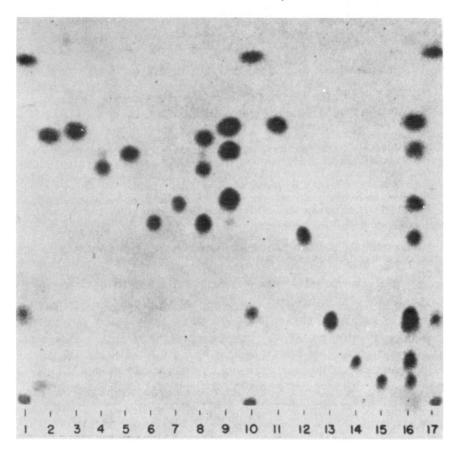

FIG. 2. Thin-layer chromatogram of various carbohydrate acetates. Adsorbent, silica gel G; solvent, 4% methanol in benzene (v/v); double development (2 × 17 cm). Components in decreasing order of mobility: (1, 10, 17) (dyes as controls) Sudan III, methyl red, methyl orange; (2) α-D-glucopyranose pentaacetate; (3, 11) β-D-glucopyranose pentaacetate; (4) α-maltose octaacetate; (5) β-maltose octaacetate; (6) α-cellobiose octaacetate; (7) β-cellobiose octaacetate; (8) compounds 2, 4, and 6; (9) compounds 3 (11), 5, and 7; (12) β-laminaribiose octaacetate; (13) β-laminaritriose undecaacetate; (14) β-laminaritetraose tetradecaacetate; (15) β-laminaripentaose heptadecaacetate; (16) compounds 3 (11), 5, 7, 12-15. (Trace impurities in 4 and 7 were anomers of the compounds cited.) (Courtesy of Drs. M. E. Tate and C. T. Bishop,[117] and reproduced by permission of the National Research Council of Canada from *Can. J. Chem.*)

Anion-exchange resins in the borate form are used to immobilize the sugars, and the sugar borate complexes are eluted from the columns with borate buffers. Quantitative analysis by this ion-exchange technique affords a precision comparable to that obtained with the partition method on an ion-exchange support mentioned earlier (p. 1451). Retention data for anion-exchange chromatography of 33 alditols and deoxyalditols were reported by Larsson and Samuelson.[135] They found that the distribution coefficient increases with increasing stability and charge of the borate complex. Vicinal hydroxyl groups in a *gauche* conformation impart the ability to form strong complexes.

Ion-exchange resins have been used successfully for the separations of amino sugars,[136–138] sialic acids,[139] sialyl oligosaccharides,[140] uronic acids,[141] sugar phosphates,[142] nucleoside glycosyl diphosphates,[143] lipopolysaccharides,[144] acidic glycosaminoglycans,[145,146] sugar alcohols,[147] and Amadori rearrangement products.[148] Paper strips impregnated with ion-exchange resins also have been used for the separation of carbohydrates.[149]

Ion-exchange resins can be used for desalting hydrolyzates and extracts.[25a] Such resins should be used with caution in the hydroxyl form, since degradation and isomerization of sugars may occur.[150–152] The use of resins in the hydrogen carbonate, formate, or acetate forms will minimize such changes. Excessive amounts of resins should be avoided because losses of sugars by adsorption may occur.

Ion-exchange celluloses, most notably diethylaminoethyl (DEAE) cellulose[153] and DEAE Sephadex,[154] have been employed extensively for the purification of glycoproteins,[155,156] glycopeptides,[87] and acidic polysaccharides.[157] Glycolipids have been separated from gangliosides on columns of DEAE cellulose using nonaqueous solvent mixtures as elutants.[158] Glycosaminoglycans have been fractionated on columns of ECTEOLA cellulose (a product of reaction of cellulose with epichlorohydrin and triethanolamine in the presence of strong alkali),[146] but this technique is less convenient and not so widely used as detergent fractionation techniques employing cetylpyridinium or cetyltrimethylammonium cations to form insoluble complexes with the polyanions, which are then selectively solubilized by varying the ionic strength of the solvent.[159] For obtaining preparations of high purity, the two methods usually are used sequentially—first fractionation with detergents and then column chromatography on anion exchangers, usually Dowex-1 or, occasionally, ECTEOLA.

VI. HIGH PERFORMANCE LIQUID CHROMATOGRAPHY

High performance liquid chromatography (HPLC), also often called high pressure liquid chromatography, has been used to facilitate more-rapid separations and analytical procedures than those obtained by the conventional application of column chromatography. HPLC is similar in principle to gas–liquid chromatography in utilizing a sample injection system, a separating column, an eluent under pressure, a solute detector, and a sample collection system. The liquid chromatography systems, however, use liquid instead of gas as the driving eluent and use very high pressures to move the liquid owing to the high viscosity of liquids. The principal detection systems in use are based either on U.V. absorption or refractive index change (for example, Waters Associates ALC-100 liquid chromatograph, Framingham, Mass.). Elevated or wide-ranging temperature changes are not often necessary during HPLC since many of the applications work well at ambient temperature. Separations may occur by liquid–liquid partition or adsorption using an organic mobile phase or by ion exchange or liquid–liquid partition using an aqueous mobile phase.[160] Supports are either "reversed phases" such as octadecyl groups bonded to silica, silica or alumina, or ion-exchangers.[160] Separations of 3000 to 5000 theoretical plates can be achieved with columns using pressure differences less than 150 atm. A monograph has been published on liquid chromatography including applications to carbohydrates,[161] and a symposium has been held on separations of carbohydrates.[161a] A typical application[162] was the fractionation of acetolysis products of extracellular yeast D-mannans using the Waters Associates μ-Bondapak-NH$_2$ "Carbohydrate" column and the solvent system water–acetonitrile (7:13 v/v). Preparative separations of ketoses in admixture are readily achieved.[162a] Perbenzoylated derivatives of monosaccharides and disaccharides have been separated by HPLC,[163] and separations of monoderivatized carbohydrates on Partisil-10PAC and the effect of a number of acids and buffer salts on the separation of mono-, di-, and tri-saccharides have also been examined.[164] Examples of the applications of HPLC in ion-exchange chromatography[165] of carbohydrates and to gel filtration of D-glucose oligomers have also been reported.[166]

VII. ELECTROPHORESIS

A. Moving-Boundary Electrophoresis

Moving-boundary or free electrophoresis[167–169] is a method used for the study of the migration of charged particles in solution, under the influence of an electric field, in the absence of a supporting medium. The number of boundaries formed may give information concerning the number of differently charged components present in a complex mixture. The method does not have much value for isolation of components. Moving-boundary electrophoresis is one of the methods of choice for determining mobility, homogeneity, and isoelectric point of a macromolecule, and for studying the interactions of high-molecular weight substances. Indeed, moving-boundary electrophoresis may be the only electrophoretic method suitable for the study of substances which

References start on p. 1470.

occlude the pores of supporting media, adhere to the support, or aggregate at the high concentrations sometimes employed in zone electrophoresis. The moving-boundary method has been applied in investigations of hyaluronic acid,[170] heparin,[171] "chondromucoprotein,"[172] mucins,[173] and polysaccharides.[174]

B. Zone Electrophoresis

In zone electrophoresis, a supporting matrix, such as paper, cellulose acetate, agar, agarose, poly(acrylamide), starch, poly(vinyl chloride), or Celite, is used to achieve separations of individual components of a mixture based on their electrical charge.[4,175] This method can be used not only for analysis but also for isolation of components. Equipment for zone electrophoresis is easier to operate and less expensive than equipment for moving-boundary electrophoresis. Other features which make zone electrophoresis popular are the variable methods of analysis that can be utilized, including differential staining, immunodiffusion,[176] and elution analysis; the small amounts of sample required (micrograms); and its applicability to the study of low-molecular weight solutes for which the boundaries would be unstable in moving-boundary electrophoresis because of diffusion.

In zone electrophoresis the sample is applied as a narrow band onto the support, which may be in the form of a thin layer, slab, or column. The support is wet with buffer and makes contact with buffer reservoirs that are connected in series to the electrode vessels. An electric potential is applied, and, after a suitable time, the current flow is stopped. The electrophoresis pattern is revealed either by staining the support with a suitable reagent or by analyzing the extracts from segments of the slab or from fractions eluted from the column. Details of the technique have been published.[177]

Low-voltage gradients, in the range of 1 to 5 volts/cm, may be used for separations on paper or on cellulose acetate of charged, high-molecular weight compounds, such as heparin, hyaluronic acid, chondroitin sulfate,[178,179] as shown in Fig. 3 (ref. 180), and glycoproteins.[87,181] High-voltage gradients (over 20 volts/cm) are preferred for the separation of low-molecular weight solutes because separation is accelerated and diffusion is kept to a minimum, provided that cooling is employed to dissipate the heat generated.

The electrophoretic technique is especially useful for samples containing inorganic salts. Contamination of a sample by salts can vitiate separations by paper-chromatography and interfere with detection, but it presents no problem in zone electrophoresis, unless very high concentrations of salt are present in the sample. However, zone electrophoresis fails to resolve some mixtures, such as uncomplexed, neutral sugars, readily separable by partition chromatography. Zone electrophoresis should be regarded as a technique complementary to paper chromatography.

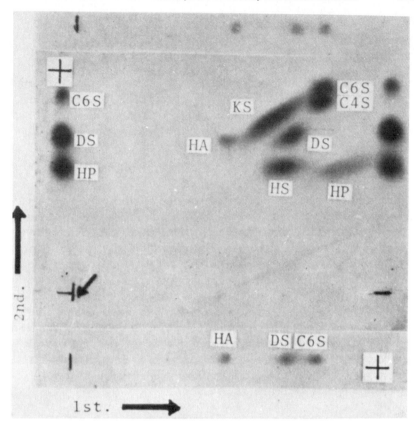

FIG. 3. Two-dimensional electrophoresis of a mixture of seven acidic glycosamino-glycans plus reference standards on cellulose acetate (10 × 10 cm). The mixture is composed of hyaluronic acid (HA, 3.7 µg), chondroitin 6-sulfate (C6S, 1.7 µg), dermatan sulfate (DS, 1.3 µg), keratan sulfate (2 µg), heparan sulfate (1.6 µg), and heparin (HP, 1.1 µg) in 4 µl. Reference standards for the run in the first dimension are C6S, DS, and HA, and in the second dimension are C6S, DS, and HP, respectively. Directions of migration are indicated by large arrows. A small arrow indicates the origin of the mixture of seven glycosaminoglycans. Electrophoretic systems; 0.1 M pyridine/0.47 M formic acid, pH 3, at 1 mA/cm for 1.7 hr in the first dimension and 0.1 M barium acetate, pH 8, at 1 mA/cm for 4.5 hr in the second. (Courtesy of Drs. R. Hata and Y. Nagai,[180] and reproduced by permission of *Anal. Biochem.*)

An index of relative mobility commonly employed in zone electrophoresis of carbohydrates is the M_{Glc} value, where

$$M_{\text{Glc}} = \frac{\text{True distance of migration of a substance}}{\text{True distance of migration of D-glucose}}$$

The true distance of migration is obtained by correction for the electroendosmotic flow or movement of the buffer solution under the influence of the electric field. Electroendosmotic flow is toward the cathode on electrophoresis in borate buffer at alkaline pH values, and the true distance moved is that measured relative to the position of an uncharged marker—for example, 2,3,4,6-tetra-O-methyl-α-D-glucose. The mobilities of oligosaccharides are frequently given relative to that of a standard disaccharide such as lactose (M_{Lac}). Absolute mobilities (cm $\times$ V^{-1} $\times$ sec^{-1}) are useful for indicating the time required to obtain a given separation.

1. Mobility of Carbohydrate Complexes

Neutral polysaccharides, neutral sugars, and alditols remain stationary or move only by electroendosmosis unless a complexing agent is present. This property may be of use when separation of neutral sugars from amino acids, amino sugars, sugar acids, and other charged substances is desired.[182] However, migration of uncharged carbohydrates [183,184] and alditols [40] may be achieved by complexing with borate buffers at an alkaline pH or with other suitable agents.

Maximum mobilities of carbohydrates are observed with borate buffers at pH 9 to 10. Table II shows the mobilities of carbohydrates in borate buffers as a function of pH. Best-defined zones occur at high pH values. The relative mobilities of some pairs of sugars change with the pH of the buffer; for example, D-glucose moves faster than D-fructose at pH 9 to 10, but the reverse occurs at pH 7 to 8. Therefore, attention must be given to the pH of the complexing buffer in reporting M_{Glc} values.

TABLE II

Mobilities of Sugars[a] (cm V^{-1} sec^{-1} $\times$ 10^5) at 20° in Borate Buffer at Various pH Values, on Whatman No. 1 Paper[b]

pH:	7.0	8.0	8.6	9.2	9.7
Arabinose	3.2	6.5	10.3	13.3	13.9
Cellobiose	<0.5	0.5	1.5	3.2	4.5
Fructose	8.2	9.7	11.4	12.5	13.1
Galactose	2.8	5.8	9.6	13.0	13.1
Glucose	2.4	6.5	11.4	14.5	14.6
Mannose	2.6	4.9	7.8	9.8	10.0
Raffinose	0.5	0.9	1.7	3.6	4.8
Rhamnose	1.3	2.4	4.4	7.1	7.8
Ribose	7.0	9.1	10.2	10.9	11.0
Sorbose	8.7	10.4	12.2	14.1	14.3

[a] D and L designations were not specified. The configuration of a sugar is omitted here and in work cited elsewhere in this chapter where the investigators failed to specify the D or L configuration.

[b] Courtesy of Drs. R. Consden and W. M. Stanier[183] and Nature.

Electrophoresis of disaccharides and disaccharide alditols in borate or molybdate buffers can give information on the position of the interglycosidic linkage. For example, 2-O-β-D-glucopyranosyl-D-glucitol is readily differentiated from 3-O-β-D-glucopyranosyl-D-glucitol, since the M_{Glc} value in borate is 1.2 for the former and 0.1 for the latter.

The stereochemical requirements for formation of borate complexes have been reviewed.[185] An oxygen-to-oxygen distance of 2.4 Å or the possibility of approaching this distance in one, or preferably more than one, of the conformations of the polyhydroxy compounds is a requirement for complex formation. However, nonbonded interactions, such as those between axial hydroxyl groups or between axial methyl and axial hydroxyl groups, may prevent the formation of borate complexes or markedly diminish the stability of the borate complex. Paper electrophoresis in sulfonate, benzeneboronic acid,[186] germanate,[187,188] stannate,[185] arsenite,[39a,189] molybdate,[190] tungstate,[185] basic lead acetate,[39a,189] vanadate,[185] and hydrogensulfite[191] have been used to advantage in special situations. The mobility values obtained for some of these buffers differ greatly from those observed in borate buffer. The complexing pattern of alditol derivatives with molybdate is especially sensitive to the position of substitution and is useful for the characterization of alditols of oligosaccharides.[190]

2. Effect of Molecular Weight on the Mobility of Carbohydrates

The electrophoretic mobility of sugars in the presence of a complexing agent such as borate is dependent on structure rather than on molecular size, except for sieving effects, which may become pronounced with macromolecules. This behavior is markedly different from that observed in paper chromatography, carbon chromatography, and Sephadex gel filtration, for which molecular size is an important factor in determining the order of migration or elution.

Treatment of reducing sugars with benzylamine, followed by electrophoresis of the resultant N-benzylglycosylamine in acidic, noncomplexing buffers, give M_{Glc} values that decrease with increasing molecular size. Molecular sizes of aldoses, up to hexasaccharides at least, may be estimated by using an appropriate set of reference standards.[192] Similar observations regarding mobility as a function of molecular size have been made for the sulfite addition compounds.[185,191]

Acrylamide gel electrophoresis, especially in the presence of sodium dodecyl sulfate (SDS), has been used for evaluating homogeneity and for resolving mixtures of proteins, glycoproteins, and other membrane constituents.[193,194] The SDS acrylamide gel electrophoresis technique has been adapted successfully for the estimation of molecular weights of proteins,[195,196] but this technique has failed with certain glycoprotein preparations.[194] Segrest et al.[194] have suggested that glycoproteins do not bind sufficient SDS and hence exhibit a lower charge-density and consequently an anomalously high, apparent molecular weight.

Hsu et al.[197] fractionated and determined the molecular weights of chondroitin sulfate preparations by poly(acrylamide) gel electrophoresis; Mathews[198] refined this technique further for application to other glycos-aminoglycans as well by maintaining a high salt concentration of $\sim 0.2\,M$, thus favoring the random-coil conformation and minimizing viscous interaction of the polyanion with the solvent.

VIII. GAS–LIQUID CHROMATOGRAPHY OF CARBOHYDRATES

A. GENERAL REMARKS

Gas chromatography is a separation process in which the components to be separated are volatilized and distributed between a moving gas phase and a stationary support phase.[199,200] The stationary phase may be a solid (gas–solid chromatography) or a liquid adsorbed on an inert support (gas–liquid chromatography). This discussion will be restricted to gas–liquid chromatography, since separation of carbohydrates has been studied by this technique. Gas–solid chromatography has been utilized largely for separations of gases and low-boiling, nonpolar solutes. Helium or nitrogen gas serves as an inert carrier of the vaporized components. The sample, usually a solution of volatile derivatives of the carbohydrates, is injected by a syringe into a heated chamber of small void-volume in which the sugar derivatives are volatilized, and the components of the sample are swept along in a stream of the carrier gas. The volatile components move through the column and are distributed between the gas phase and the liquid-coated solid phase. The column is heated uniformly to maintain the components in the vapor phase and to afford a suitable partition ratio (K) of the sample between liquid and gas phases. The temperature of the column may be held constant, increased in discrete steps, or increased regularly. Exit of the various components from the column is monitored and determined by a suitable detector, such as a flame-ionization detector. Modifications, which include coupling of the instrument to a mass spectrometer or to a liquid-scintillation spectrometer, are commonly used. Of these three, ancillary systems, mass spectrometry has been the subject of the most intensive investigations. Examination of the fragmentation patterns resulting from the impact of electrons on components in the g.l.c. effluent has facilitated structural elucidation of mono-,[201] di-,[202] oligo-,[203] and poly-saccharides,[204] sugar acids,[205] amino sugars,[206] glycolipids,[207] and nucleosides.[208] (See Chapter 27, Section II.)

B. SUPPORTS AND LIQUID PHASES

Golay[209,210] has developed columns which are not filled with solid support but which contain the liquid phase coated as a thin film on the walls of the column. These columns, referred to as Golay-type columns, open-tubular columns, or capillary columns, exhibit greater efficiency (greater number of theoretical plates) and are generally used at a lower temperature. These columns are superior for analytical work; however, they cannot be adapted readily for preparative-scale work because the small amount of stationary liquid used restricts the amount of sample that can be injected to about 1 μg or less.

Both polar and nonpolar liquid phases have been employed successfully. Polar phases exert strong attractive forces whereas nonpolar phases exert weak attractive forces.[211] Usually, components are more likely to be eluted from the column in order of their vapor pressures or boiling points when chromatographed on a nonpolar liquid phase than when chromatographed on a polar phase. Structural differences and electronic interactions with the liquid phase are the determining factors in governing the rate of elution from a polar column.

Nonpolar coatings perform extremely well for separations based on "carbon number," but they also have worked (though not quite so well as polar phases) for separations within a group of substituted sugars containing the same number of carbon atoms. Examples of nonpolar liquid coatings are the methyl-silicone SE-30, the methylphenylsilicone SE-52, the fluorinated hydrocarbon QF-1, and the hydrocarbon polymers Apiezon M and Apiezon L. The silicone gums SE-52 and SE-30 are useful for scanning a wide array of carbohydrates, ranging from trioses to oligosaccharides, as their per(trimethylsilyl) ethers. Separation of per(trimethylsilyl) ethers of a variety of carbohydrates on SE-52 is shown[212] in Fig. 4. The silicone gums are heat-resistant, have a high boiling point, and do not "bleed" easily. Accordingly, after the removal of low-boiling derivatives, the high-boiling per(trimethylsilyl) ethers of oligosaccharides, such as that of stachyose with a molecular weight of 1676, may be eluted by elevating the temperature to 250° without harming the column coating.

Polar coatings such as butanediol succinate, polyethylene glycol succinate, polyethylene glycol succinate linked to a silicone of the cyanoethyl type (ECNSS-M), and Carbowax 20M terminated with terephthalic acid afford more-complete separation of the monosaccharides and of compounds within a narrow and low-molecular weight range than the nonpolar coatings. Polar polyesters do not withstand elevated temperatures, and their use is limited largely to derivatives of monosaccharides. OV-225 has generally replaced

---

References start on p. 1470.

ECNSS-M as the coating of choice, especially in mass-spectrometric applications, because of the greater stability of the former.[213]

C. DERIVATIVES

1. *Silyl Ethers*

The carbohydrate derivatives used in gas–liquid chromatography must be volatile yet stable at the operating temperature of the column, and must not be adsorbed irreversibly on the stationary phase. The per(trimethylsilyl) ethers meet these requirements, are easy to prepare, and have been used for a wide variety of compounds, including glycosides, deoxy sugars, cyclitols, hexoses, amino sugars, and sialic acids (Fig. 4).

The usual conditions for trimethylsilylation are as follows: The carbohydrate (generally 1 to 10 mg of dry powder) is treated with 1 ml of anhydrous pyridine containing 0.2 ml of hexamethyldisilazane and 0.1 ml of chlorotrimethylsilane. The reaction is essentially completed within 5 minutes at room temperature.

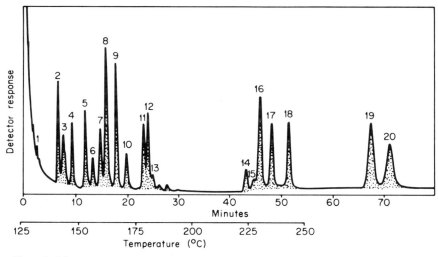

FIG. 4. Linear temperature-programmed separation of trimethylsilyl ethers. The column used was SE-52, with an initial temperature of 125°. The instrument was programmed for a temperature increase of 2.3° per minute, with holding under isothermal conditions when 250° was attained. The components are: (1) erythrose; (2) β-arabinose; (3) ribose; (4) α-xylose; (5) methyl α-mannopyranoside; (6) α-gulose; (7) α-galactose; (8) α-glucose; (9) L-ascorbic acid; (10) β-glucose; (11) 2-acetamido-2-deoxyglucose; (12) 2-acetamido-2-deoxygalactose; (13) D-*glycero*-D-*gulo*-heptonolactone; (14) sucrose; (15) α-maltose; (16) β-maltose; (17) β-cellobiose; (18) gentiobiose; (19) raffinose; (20) melezitose. (Reproduced by courtesy of C. C. Sweeley *et al.*[212] and *J. Amer. Chem. Soc.*)

When only very small amounts of sugars (less than 1 mg) are available, the sample may be triturated with a small volume of the reagent mixture.[214]

Per(dimethylsilyl) ethers of carbohydrates may be prepared in a similar manner by substituting chlorodimethylsilane and tetramethyldisilazane for the above silylating reagents.[199] The retention times for the dimethylsilyl ethers may be almost half those of the corresponding trimethylsilyl ethers, and the relative retention times for two anomers as the dimethylsilyl ethers may differ from those of the trimethylsilyl ethers. Sometimes, separation of sugars may be improved if the dimethylsilyl, rather than the trimethylsilyl, ethers are employed, as in the separation of arabinitol from ribitol. For other sugars, such as the aldohexoses, the trimethylsilyl ethers are separated better than the dimethylsilyl ethers.[199] The dimethylsilyl ethers of sugars may be particularly useful for compounds of high molecular weight for which the volatility of the trimethylsilyl ether derivatives are low.

Quantitative analysis of the trimethylsilyl ethers of monosaccharides have been performed on mixtures of pure sugars,[215–217] on hydrolyzates of crude wood cellulose,[218] on glycolipids, glycoproteins, and glycopeptides, on gastric juice,[221] and on other body fluids.[199] Internal standards are employed to eliminate errors. Calibration curves[222] for each sugar are obtained by chromatography of various amounts of the sugar in the presence of constant amounts of an internal standard. The ratio of the total area (sum of areas of the anomeric forms) of the sugar to that of the internal standard is plotted against the ratio of the weight of the sugar to that of the standard. By adding the same amount

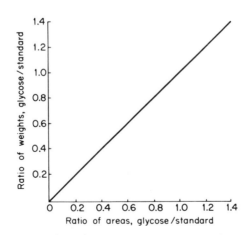

FIG. 5. Calibration curve for sugar derivatives by using an internal standard. Various amounts of each sugar derivative are chromatographed with constant amounts of internal standard.

References start on p. 1470.

of internal standard to the mixture of unknown sugars, silylating, and chromatographing the mixture, it is possible to determine the concentration of solute in the sample by referring to a standard curve (Fig. 5).

Since a mixture of three monosaccharides such as D-glucose, D-mannose, and D-galactose may give seven or more peaks (representing the α-D- and β-D-pyranoses, the furanoses, the aldehyde, and the aldehydrol forms), difficulties can arise in interpreting patterns of complex mixtures. Moreover, overlap by an anomer of a second sugar may occur. Reduction of the sugars to the corresponding alditols with borohydride, followed by acetylation, may facilitate resolution of some mixtures.[223] Mixtures of the per-O-trimethylsilyl derivatives of various inositols have been resolved successfully.[224]

Quantitative analysis of 2-acetamido-2-deoxy-D-glucose and 2-acetamido-2-deoxy-D-galactose as their trimethylsilyl ethers, by using 2-acetamido-2-deoxy-D-glucitol as internal standard, has been achieved. As little as 0.5 mg of these components in a sample may be determined.[217,225–227] Separations of the N-trimethylsilyl-per-O-trimethylsilyl derivatives of 2-amino-2-deoxy-D-glucose and 2-amino-2-deoxy-D-galactose also have been obtained.[227,228] The various 2-amino-2,6-dideoxyhexoses also have been successfully resolved as their trimethylsilyl derivatives.[229]

A procedure for the analysis of hexuronic acids was described by Perry and Hulyalkar.[217,230] The hexuronic acids were converted into aldonic acids by sodium borohydride and were further converted into the aldono-1,4-lactones. Trimethylsilylation of the aldonolactones yielded the volatile, stable 2,3,5,6-tetrakis-O-trimethylsilylaldono-1,4-lactones, which were resolved well by gas–liquid chromatography on a column of neopentylglycol sebacate. Analysis of synthetic mixtures showed errors of only a few tenths of 1%, but the quantitative analysis of hexuronic acids in polymers was less accurate because of destruction during hydrolysis. The analysis of glycosaminoglycans and their uronic acids by gas–liquid chromatography has been described by Berenson and co-workers.[231]

2. Methyl Ethers

Gas–liquid chromatography of carbohydrate derivatives was first performed for methylated methyl glycosides.[232] The derivatives were separated on columns of Apiezon M or Celite-545 and were recovered quantitatively. Separations of many fully methylated and partially methylated methyl glycosides have been achieved. The α-D and β-D anomers may appear as separate peaks, and caution must be exercised in the interpretation of the patterns. Reduction of partially methylated sugars to the alditol ethers, followed by acetylation of the unsubstituted hydroxyl groups, affords derivatives of higher volatility, and the resolution of the partially methylated sugars is increased.[233] Bishop[233] has reported relative retention volumes for many methylated sugars and other derivatives, and a compendious review by Dutton,[217] compares the behavior

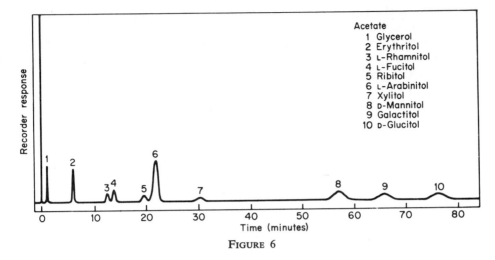

FIGURE 6

in g.l.c. of the methylated derivatives of each sugar and contains an extensive listing of column coatings, temperatures for elution of the methylated sugar, natural product or model compound analyzed, and appropriate references.

3. Acetates

Separations obtained by Jeanes and co-workers[223] with complex mixtures of alditol acetates (Fig. 6) are superior to those obtained for the corresponding mixtures of the trimethylsilyl ethers of aldoses or alditols. The carbohydrates present in glycoproteins were quantitatively determined by g.l.c. of their alditol acetates,[234] and this procedure, or a variant of it employing the trifluoroacetyl derivatives,[235] is widely used in the analysis of glycoproteins and oligosaccharides. When the mixture is not too complex, analysis of the trimethylsilyl ethers is preferred, since the reaction requires only 5 to 15 minutes, as compared to 4 hours for acetylation. Kim et al.[236] have described the application of g.l.c. of alditol acetates to the analysis of neutral sugars in glycoproteins; the acetylated aldononitriles are also useful in identification of aldoses by g.l.c.[236a]

Gas–liquid chromatography of ethylene glycol and glycerol acetates has been used for the identification and quantitation of the cleavage products obtained from sugars by periodate oxidation followed by reduction.[237]

4. Miscellaneous

Sugars have been separated by gas–liquid chromatography of the ethylidene, isopropylidene, and benzylidene acetals.[199,238] Nucleosides and esters of sugar

---

References start on p. 1470.

phosphates also have been separated.[239] The procedure has also been used for the quantitative determination of acetyl groups.[240] Alcohols and various glycols have been separated readily on columns of uncoated polyaromatic resins.[241]

The use of ethylation for the analyses of glycosyl linkages of polysaccharides was reported by Sweet et al.[242] Ethylation was followed by hydrolysis, reduction, formation of ethylated alditol acetates, and gas–liquid chromatographic analysis. Analysis by gas–liquid chromatography of the sugars of methylated sugar gum was reported by Aspinall et al.[243] and of the alditol acetates from the pectic polysaccharides of rapeseed by Siddiqui and Wood.[244] Combined gas–liquid chromatographic–mass spectrometric analyses were reported for various methylated derivatives of 2-acetamido-2-deoxy-D-glucose[245] and 2-deoxy-2-(N-methylacetamido)-D-galactose.[246] Kennedy et al.[247] reported the separation by gas–liquid chromatography of O-trimethylsilylated derivatives of D-galacturonic, D-glucuronic, D-iduronic, and D-mannuronic acids and recommended silylation of these compounds in methyl sulfoxide. Dudman and Whittle[248] discussed the presence of phthalic esters on gas–liquid chromatograms as contaminants of sugar peaks and, accordingly, the need for other methods of identification. Mass spectrometric analysis is useful in this respect because phthalates and sugar derivatives have different mass spectra. Phthalic esters (except for the dimethyl esters) give a characteristic[249] ion at m/e 149.

The separation and quantitative determination of neutral sugars and hexosamines from glycoproteins and mucopolysaccharides by gas–liquid chromatography of the aldononitrile acetate derivatives was reported by Varma and Varma.[250] In this application, the hexosamines were subjected to nitrous acid deamination, and the resultant anhydroaldoses were derivatized. This technique was applied to urinary and brain glycoproteins, thyroglobulin, chondroitin sulfates, and hyaluronic acid.

REFERENCES

1. C. J. O. R. Morris and P. Morris, in "Separation Methods in Biochemistry," Wiley (Interscience), New York, 1964.
2. H. P. Burchfield and E. E. Storrs, in "Biochemical Applications of Gas Chromatography," Academic Press, New York, 1962.
3. K. Macek, in "Paper Chromatography," I. M. Hais and K. Macek, Eds., Academic Press, New York, 1963.
4. A. B. Foster, Advan. Carbohyd. Chem., 12, 81 (1957).
5. G. N. Kowkabany, Advan. Carbohyd. Chem., 9, 304 (1954).
6. F. A. Isherwood and M. A. Jermyn, Biochem. J., 48, 515 (1951).
7. H. Jaeger, A. Ramel, and O. Schindler, Helv. Chim. Acta, 40, 1310 (1957).
8. D. French and G. M. Wild, J. Amer. Chem. Soc., 75, 2612 (1953).
9. E. J. Bourne, Nature, 171, 385 (1953).

10. M. A. Jermyn and F. A. Isherwood, *Biochem. J.*, **44**, 402 (1948).
11. D. F. Durso and W. A. Mueller, *Anal. Chem.*, **28**, 1366 (1956).
11a. A. Jeanes, C. S. Wise, and R. J. Dimler, *Anal. Chem.*, **23**, 415 (1957).
11b. J. F. Robyt, *Carbohyd. Res.*, **40**, 373 (1975).
12. S. M. Partridge, *Nature*, **158**, 270 (1946).
13. F. G. Fischer and Z. Dörfel, *Hoppe-Seyler's Z. Physiol. Chem.*, **297**, 164 (1954).
14. E. L. Hirst and J. K. N. Jones, *J. Chem. Soc.*, 1659 (1949).
15. F. G. Fischer and H. J. Nebel, *Hoppe-Seyler's Z. Physiol. Chem.*, **302**, 10 (1955).
16. E. Cabib, L. F. Leloir, and C. E. Cardini, *J. Biol. Chem.*, **203**, 1055 (1953).
17. R. G. Spiro, *J. Biol. Chem.*, **235**, 2860 (1960).
18. E. Svennerholm and L. Svennerholm, *Nature*, **181**, 1154 (1958).
19. J. Cerbulis, *Anal. Chem.*, **27**, 140 (1955).
20. D. Waldi and F. Munter, *Naturwissenschaften*, **37**, 393 (1962).
21. A. A. White and W. C. Hess, *Arch. Biochem. Biophys.*, **64**, 57 (1956).
22. J. L. Buchan and R. I. Savage, *Analyst*, **77**, 401 (1952).
23. B. Wickberg, *Acta Chem. Scand.*, **12**, 615 (1958); B. Wickberg, *Methods Carbohyd. Chem.* **1**, 31 (1962).
24. K. Wallenfels, G. Bechtler, R. Kuhn, H. Trischmann, and H. Egge, *Angew. Chem. Int. Ed.*, **2**, 515 (1963).
25. S. M. Partridge, *Biochem. Soc. Symp.*, **3**, 52 (1949).
25a. V. Vitek and K. Vitek, *J. Chromatog.*, **60**, 381 (1971).
26. S. Aronoff and L. Vernon, *Arch. Biochem. Biophys.*, **28**, 424 (1950).
27. J. A. Cifonelli and F. Smith, *Anal. Chem.*, **26**, 1132 (1954).
28. M. L. Wolfrom and J. B. Miller, *Anal. Chem.*, **28**, 1037 (1956).
29. R. U. Lemieux and H. F. Bauer, *Anal. Chem.*, **26**, 920 (1954).
30. E. Köiw and A. Gronwall, *Scand. J. Clin. Lab. Invest.*, **4**, 244 (1952).
31. W. E. Trevelyan, C. P. J. Glaudemans, and A. L. Currie, *Nature*, **166**, 444 (1952).
32. S. M. Partridge, *Nature*, **164**, 443 (1949); G. Caldes and B. Prescott, *J. Chromatog.*, **84**, 220 (1973).
33. S. M. Partridge, *Biochem. Soc. Symp.*, **3**, 52 (1949).
34. L. Hough, J. K. N. Jones, and W. H. Wadman, *J. Chem. Soc.*, 1702 (1950).
35. K. Wallenfels, *Naturwissenschaften*, **37**, 491 (1950); K. Wallenfels, E. Berndt, and G. Limberg, *Angew. Chem.*, **65**, 581 (1953);
36. W. E. Trevelyan, D. P. Procter, and J. S. Harrison, *Nature*, **166**, 444 (1950); F. G. Fischer and H. Dörfel, *Hoppe-Seyler's Z. Physiol. Chem.*, **297**, 164 (1954); M. Trojna and J. Hubacek, *Collect. Czech. Chem. Commun.*, **37**, 3595 (1972).
37. J. B. Pridham, *Anal. Chem.*, **28**, 1967 (1956).
38. W. G. C. Forsyth, *Nature*, **161**, 239 (1948); S. M. Partridge and R. G. Westall, *Biochem. J.*, **42**, 238 (1948).
39. J. Britton, *Biochem. J.*, **73**, 19P (1959).
39a. J. L. Frahn and J. A. Mills, *Aust. J. Chem.*, **12**, 65 (1959).
40. J. G. Buchanan, C. A. Dekker, and A. G. Long, *J. Chem. Soc.*, 3162 (1950); D. Gross, *Nature*, **176**, 362 (1955).
41. K. F. Williams and A. Bevenue, *Science*, **113**, 582 (1951); F. J. Bealing and J. S. D. Bacon, *Biochem. J.*, **53**, 277 (1953).
42. J. H. Pazur and K. Kleppe, *Biochemistry*, **3**, 578 (1964).
43. G. Avigad, D. Amaral, C. Asensio, and B. L. Horecker, *J. Biol. Chem.*, **237**, 2736 (1962).
44. O. Gabriel and D. Igals, *Biochem. Biophys. Res. Commun.*, **2**, 358 (1960).

44a. C. A. Knutson, *Carbohyd. Res.*, **43**, 225 (1975); H. E. Conrad, *Methods Carbohyd. Chem.*, **7**, 71 (1976).

45. S. M. Partridge, *Biochem. J.*, **42**, 238 (1948).

46. D. Horton, J. R. Vercellotti, and M. L. Wolfrom, *Biochem. Biophys. Acta*, **50**, 358 (1961).

47. E. Stahl and U. Kaltenbach, *in* "Thin Layer Chromatography," E. Stahl, Ed., Academic Press, New York, 1965, p. 461.

48. C. H. Hanes and F. A. Isherwood, *Nature*, **164**, 1107 (1949).

49. J. Montreuil, G. Spik, and A. Konarska, "Microdosage des Glucides; Méthodes Chromatographiques de Dosage des Oses 'Neutres'," Faculté des Sciences de Lille, Lille, 1967.

50. R. B. Fisher, D. S. Parsons, and G. A. Morrison, *Nature*, **161**, 764 (1948).

51. E. F. McFarren, K. Brand, and H. R. Rutkowski, *Anal. Chem.*, **23**, 1146 (1951).

52. M. Dubois, K. Gilles, J. K. Hamilton, P. A. Rebers, and F. Smith, *Nature*, **168**, 167 (1951).

53. M. Dubois, K. Gilles, J. K. Hamilton, P. A. Rebers, and F. Smith, *Anal. Chem.*, **28**, 1366 (1956).

54. H. H. Brownell, J. G. Hamilton, and A. A. Casselman, *Anal. Chem.*, **29**, 550 (1957).

55. L. Hough, J. K. N. Jones, and W. H. Wadman, *J. Chem. Soc.*, 2511 (1949).

56. J. D. Geerdes, B. A. Lewis, R. Montgomery, and F. Smith, *Anal. Chem.*, **26**, 2646 (1954).

57. M. L. Wolfrom, D. L. Patin, and R. M. de Lederkremer, *J. Chromatog.*, **17**, 488 (1965).

58. S. Gardell, *Acta Chem. Scand.*, **7**, 201 (1953).

59. H. M. Kissman, C. Pidacks, and B. R. Baker, *J. Amer. Chem. Soc.*, **77**, 18 (1955); R. U. Lemieux, C. T. Bishop, and G. E. Pelletier, *Can. J. Chem.*, **34**, 1365 (1956); R. H. Hall, *Anal. Biochem.*, **4**, 395 (1962).

60. O. Samuelson and H. Stroemberg, *Carbohyd. Res.*, **3**, 89 (1966); B. Arividi and O. Samuelson, *Sv. Kem. Tidskr.*, **77**, 84 (1965); I. L. Larson and O. Samuelson, *Acta Chem. Scand.*, **19**, 1357 (1965).

60a. R. L. Munier and A. M. Drapier, *C. R. Acad. Sci., Ser. C*, **274**, 1005 (1972).

61. M. L. Wolfrom, R. M. de Lederkremer, and G. Schwab, *J. Chromatog.*, **22**, 474 (1966); D. Horton, A. Tanimura, and M. L. Wolfrom, *ibid.*, **23**, 309 (1966).

62. W. M. Lamkin, D. N. Ward, and E. F. Walborg, *Anal. Biochem.*, **17**, 485 (1966).

63. G. Pastuska, *Z. Anal. Chem.*, **179**, 427 (1961); R. Ohman, *Acta Chem. Scand.*, **21**, 1670 (1967); F. Kraffczyk, R. Helges, and H. J. Bremer, *Clin. Chim. Acta*, **42**, 303 (1972); A. E. Gal, *Anal. Biochem.*, **24**, 452 (1968).

64. E. Ragazzi and G. Veronese, *Farmaco Ed. Prat.*, **18**, 152 (1963); Y. S. Ovadov, E. V. Evtushenko, V. E. Vaskofsky, R. G. Ovodova, and J. F. Soloreva, *J. Chromatog.*, **26**, 111 (1967).

65. E. Stahl and V. Kaltenbach, *J. Chromatog.*, **5**, 351 (1961).

66. K. Kringstad, *Acta Chem. Scand.*, **18**, 2399 (1964); V. Paces and J. Frgala, *J. Chromatog.*, **79**, 373 (1973).

67. D. Waldi, *J. Chromatog.*, **18**, 417 (1965).

68. D. W. Bowker and J. R. Turvey, *J. Chromatog.*, **22**, 486 (1966).

69. B. Shasha and R. L. Whistler, *J. Chromatog.*, **14**, 532 (1964).

70. E. Waldi, *in* "Thin Layer Chromatography," E. Stahl, Ed., Academic Press, New York, 1965, p. 466.

71. E. Bancher, H. Scherz, and K. Kaindel, *Microchim. Acta*, 1043 (1964).

72. C. N. Huber, H. D. Scobell, H. Tai, and E. E. Fisher, *Anal. Chem.*, **40**, 207 (1968); see also D. Nurok and A. Zlatkis, *Carbohyd. Res.*, **65**, 265 (1978).
73. E. Granzer, *Hoppe-Seyler's Z. Physiol. Chem.*, **328**, 277 (1962).
74. R. C. S. Audette, G. Blunder, J. W. Steele, and C. S. C. Wong, *J. Chromatog.*, **25**, 367 (1966).
75. A. A. Galoyan, B. K. Mesrob, and V. Holeysovsky, *J. Chromatog.*, **24**, 440 (1966).
76. A. Schweiger, *J. Chromatog.*, **9**, 374 (1962); R. Spitschan, *ibid.*, **61**, 169 (1971).
77. D. W. Vomhof and T. C. Tucker, *J. Chromatog.*, **17**, 300 (1965).
78. J. W. F. Davidson and W. G. Drew, *J. Chromatog.*, **21**, 319 (1966).
79. M. Günther and A. Schweiger, *J. Chromatog.*, **17**, 702 (1965).
80. P. Mendl and H. Tuppy, *Monatsh. Chem.*, **96**, 802 (1965); E. Klenk, L. Hof, and L. Georgias, *Hoppe-Seyler's Z. Physiol. Chem.*, **348**, 149 (1967).
81. D. M. W. Anderson and J. F. Stoddart, *Carbohyd. Res.*, **1**, 417 (1966).
82. M. Ghebregzabher, S. Rufini, B. Monaldi, and M. Lato, *J. Chromatog.*, **127**, 133 (1976).
83. J. Poráth, *Nature*, **183**, 1657 (1959); Y. C. Lee and R. Montgomery, *Methods Carbohyd. Chem.*, **5**, 28 (1965); S. Hjerten and R. Mosbach, *Anal. Biochem.*, **3**, 109 (1962); A. Polson, *Biochim. Biophys. Acta*, **50**, 565 (1961); S. Hjerten, *Arch. Biochem. Biophys.*, **99**, 466 (1962).
84. R. G. Spiro, *J. Biol. Chem.*, **242**, 1923 (1967).
85. A. Fiori, G. V. Giusti, and G. Panari, *J. Chromatog.*, **55**, 365 (1971); A. Fiori, G. Panari, G. V. Giusti, and G. Brandi, *ibid.*, **84**, 335 (1973).
86. P. Andrews, *Brit. Med. Bull.*, **22**, 112 (1966).
87. T. Pamer, G. B. J. Glass, and M. I. Horowitz, *Biochemistry*, **7**, 3821 (1968).
88. H. U. Choi, K. Meyer, and R. Swarm, *Proc. Nat. Acad. Sci. U.S.*, **68**, 877 (1971); K. D. Brandt, C. P. Tsiganos, and H. Muir, *Biochim. Biophys. Acta*, **320**, 453 (1973).
89. M. Janado and K. Onodera, *Agr. Biol. Chem.* (Tokyo), **36**, 2027 (1972).
90. D. M. W. Anderson and J. F. Stoddart, *Carbohyd. Res.*, **2**, 104 (1966).
91. M. John, G. Trenel, and H. Dellweg, *J. Chromatog.*, **42**, 476 (1969).
92. J. M. Goodson and V. Distefano, *J. Chromatog.*, **45**, 139 (1969); J. M. Goodson, V. Distefano, and J. C. Smith, *ibid.*, **54**, 43 (1971).
93. M. A. Raftery, T. Rand-Meir, F. W. Dahlquist, S. M. Parsons, C. L. Borders, Jr., R. G. Wolcott, W. Beranek, Jr., and L. Jao, *Anal. Biochem.*, **30**, 427 (1969).
94. P. Flodin, J. D. Gregory, and L. Rodén, *Anal. Biochem.*, **8**, 424 (1964); C. P. Dietrech, *Biochem. J.*, **108**, 647 (1968).
94a. F. Schmidt and B. S. Enevoldsen, *Carbohyd. Res.*, **61**, 197 (1978).
95. A. Wasteson, *Biochim. Biophys. Acta*, **177**, 152 (1969); *Biochem. J.*, **122**, 477 (1971); P. A. Revell and H. Muir, *ibid.*, **130**, 597 (1972); J. J. Hopwood and H. C. Robinson, *ibid.*, **135**, 63 (1973).
96. G. Tortolani and C. Romagloni, *Anal. Biochem.*, **66**, 29 (1975).
97. R. H. Pearce and B. J. Grimmer, *Biochem. J.*, **157**, 753 (1976); P. L. Lever and P. F. Goetnik, *Anal. Biochem.*, **75**, 67 (1976).
98. D. P. Sweet, P. Albersheim, and R. H. Shapiro, *Carbohyd. Res.*, **40**, 199 (1975).
99. R. F. Murphy, *Biochem. Soc. Trans.*, **2**, 1298 (1974); A. M. Jeffrey, D. A. Zopf, and V. Ginsburg, *Biochem. Biophys. Res. Commun.*, **62**, 608 (1975); I. J. Goldstein, C. M. Reichkert, A. Misaki, and A. J. Gorin, *Biochim. Biophys. Acta*, **317**, 500 (1973); M. E. A. Periera and E. A. Kabat, *J. Exptl. Med.*, **143**, 422 (1976).
100. W. B. Jakoby and M. Wilchek, Eds., "Methods in Enzymology," Vol. XXXIV, Academic Press, New York, 1974; *Biochem. Soc. Trans.*, **2**, 1289 (1974).

101. N. B. Schwartz and L. Rodén, *Carbohyd. Res.*, **37**, 167 (1974); *J. Biol. Chem.*, **250**, 5200 (1975).

102. V. Horejsi and J. Kocourek, *Biochim. Biophys. Acta*, **297**, 346 (1973); H. Lis and N. Sharon, *Annu. Rev. Biochem.*, **42**, 541 (1973); see also I. J. Goldstein and C. E. Hayes, *Advan. Carbohyd. Chem. Biochem.*, **35**, 128 (1978).

103. M. Tomana, W. Neidermeier, J. Mestecky, R. E. Schrohenloher, and S. Porch, *Anal. Biochem.*, **72**, 389 (1976).

103a. L. Hagdahl, *Acta Chem. Scand.*, **2**, 574 (1948).

104. C. Weibull and A. Tiselius, *Ark. Kemi Mineral Geol.*, **19A** [19], 22 (1945).

105. W. W. Binkley, *Advan. Carbohyd. Chem.*, **10**, 55 (1955).

106. J. K. N. Jones, *J. Chem. Soc.*, 333 (1944); R. U. Lemieux and G. Huber, *J. Amer. Chem. Soc.*, **75**, 4118 (1953).

107. G. H. Coleman and C. M. McCloskey, *J. Amer. Chem. Soc.*, **65**, 1588 (1943).

108. G. H. Coleman, A. G. Farnham, and A. Miller, *J. Amer. Chem. Soc.*, **64**, 1501 (1942); G. Constantopoulos, A. S. Dekabon, and W. R. Carroll, *Anal. Biochem.*, **31**, 59 (1969).

109. E. E. Dickey and M. L. Wolfrom, *J. Amer. Chem. Soc.*, **71**, 825 (1949).

110. W. W. Binkley and M. L. Wolfrom, "Chromatography of Sugar and Related Substances," Scientific Report, No. 10, Sugar Research Foundation, Inc., New York, August, 1948.

111. A. Thompson, *Methods Carbohyd. Chem.*, **1**, 36 (1962).

112. B. W. Lew, M. L. Wolfrom, and R. M. Goepp, Jr., *J. Amer. Chem. Soc.*, **68**, 1449 (1946).

113. E. Guilloux and S. Beaugiraud, *Bull. Soc. Chim. Fr.*, 259 (1965).

114. J. E. Kärkkainen, E. O. Haahti, and A. A. Lehtonen, *Anal. Chem.*, **36**, 1316 (1966).

115. P. J. Stoffyn, *J. Amer. Oil Chem. Soc.*, **43**, 4813 (1966); R. G. Spiro, *J. Biol. Chem.*, **242**, 4813 (1967); V. Prey, H. Berbalk, and M. Kauz, *Microchim. Acta*, **449** (1962).

116. B. Fournet, Y. Leroy, and J. Montreuil, *Colloq. Int. C.N.R.S.*, **221**, 111 (1973).

117. R. M. Powell and M. S. Feather, *Carbohyd. Res.*, **4**, 486 (1967); M. E. Tate and C. T. Bishop, *Can. J. Chem.*, **40**, 1043 (1962).

118. Y. Yoshioka, M. Emori, T. Ikekawa, and F. Fukuoka, *Carbohyd. Res.*, **43**, 305 (1975).

119. Z. Kowalewski, O. Schindler, H. Jaeger, and T. Reichstein, *Helv. Chim. Acta*, **43**, 1280 (1960)

120. E. F. L. J. Anet, *J. Chromatog.*, **1**, 291 (1962).

121. H. R. Sinclair and W. J. Wheadon, *Carbohyd. Res.*, **4**, 292 (1967).

122. S. N. McNally and W. G. Overend, *J. Chromatog.*, **21**, 160 (1966).

123. H. M. Flowers, *Carbohyd. Res.*, **5**, 126 (1967).

124. S. Hakomori and G. D. Strycharz, *Biochemistry*, **7**, 1279 (1968); B. L. Slomiany and M. I. Horowitz, *Biochim. Biophys. Acta*, **218**, 278 (1970); E. L. Smith and J. M. McKibbin, *Anal. Biochem.*, **45**, 608 (1972); W. J. Esselman, J. R. Ackerman, and C. C. Sweeley, *J. Biol. Chem.*, **248**, 7310 (1973); Y.-T. Li, M. Y. Mazzotta, C.-C. Wan, R. Orth, and S.-C. Li, *ibid.*, **248**, 7512 (1973).

125. R. K. Yu and R. W. Ledeen, *J. Lipid Res.*, **13**, 680 (1972); G. Yogeeswaran, J. R. Wherrett, S. Chatterjee, and R. K. Murray, *J. Biol. Chem.*, **245**, 6718 (1970); G. Tettamanti, F. Bonali, S. Marchesini, and V. Zambotti, *Biochim. Biophys. Acta*, **296**, 160 (1973); L. Svennerholm, *Methods Carbohyd. Chem.*, **7**, 464 (1972).

126. N. K. Kochetkov, B. A. Dmitriev, and A. I. Usov, *Dokl. Akad. Nauk SSSR*, **143**, 863 (1962).

127. V. A. Derevitskaya, M. G. Vafina, and N. K. Kochetkov, *Carbohyd. Res.*, **3**, 377 (1967).
128. D. Horton and T. Tsuchiya, *Carbohyd. Res.*, **5**, 426 (1967).
129. L. A. Witting, Ed., "Glycolipid Methodology," Amer. Oil Chem. Soc., Champaign, Illinois, 1976, pp. 1–438.
130. A. Hallén, *Acta Chem. Scand.*, **14**, 2249 (1960).
131. E. F. Walborg, L. Christensson, and S. Gardell, *Anal. Biochem.*, **13**, 177 (1965).
132. E. F. Walborg, Jr., D. B. Ray, and L. E. Öhrberg, *Anal. Biochem.*, **29**, 433 (1969).
132a. E. F. Walborg, Jr., and R. S. Lantz, *Anal. Biochem.*, **22**, 123 (1968).
133. R. Syamananda, R. C. Staples, and R. J. Block, *Contrib. Boyce Thompson Inst.*, **21**, 363 (1962).
134. Y. C. Lee, G. S. Johnson, B. White, and A. Scocca, *Anal. Biochem.*, **43**, 640 (1971); Y. C. Lee, *in* "Methods in Enzymology," V. Ginsburg, Ed., Vol. 28, Part B, Academic Press, New York, 1972, p. 63.
134a. J. I. Ohms, J. V. Benson, Jr., and J. A. Patterson, *Anal. Biochem.*, **20**, 51 (1967).
135. K. Larsson and O. Samuelson, *Carbohyd. Res.*, **50**, 1 (1976).
136. N. F. Boas, *J. Biol. Chem.*, **204**, 553 (1953); S. Gardell, *Acta Chem. Scand.*, **7**, 207 (1953).
137. Y. Hashimoto and W. Pigman, *Ann. N.Y. Acad. Sci.*, **93**, 541 (1962); J. D. Gregory and L. V. Lenten, *in* "Automation in Analytical Chemistry," L. T. Skeggs, Jr., Ed., Medaid Inc., New York, 1966, p. 620.
138. M. J. Crumpton, *Biochem. J.*, **72**, 479 (1959).
139. L. Svennerholm, *Acta Soc. Med. Upsalien*, **61**, 75 (1956).
140. H. S. Barra and R. Caputto, *Biochim. Biophys. Acta*, **101**, 367 (1965).
141. L. Rodén and G. Armand, *J. Biol. Chem.*, **241**, 65 (1966); L.-Å. Fransson, L. Rodén, and M. L. Spach, *Anal. Biochem.*, **21**, 317 (1968).
142. J. X. Khym and W. E. Cohn, *J. Amer. Chem. Soc.*, **75**, 1153 (1953).
143. S. M. Moreno, M. Cookier, G. Guerrero, M. E. J. De Recondo, and E. F. Recondo, *Carbohyd. Res.*, **47**, 275 (1976).
144. Y. C. Lee, *J. Biol. Chem.*, **241**, 1899 (1966).
145. S. Schiller, G. A. Glover, and A. Dorfman, *J. Biol. Chem.*, **236**, 989 (1961).
146. N. R. Ringertz and P. Reichard, *Acta Chem. Scand.*, **14**, 303 (1960).
147. S. A. Barker, M. J. How, P. V. Peplow, and P. J. Somers, *Anal. Biochem.*, **26**, 219 (1968); S. J. Angyal, G. S. Bethell, and R. J. Beveridge, *Carbohyd. Res.*, **73**, 9 (1979).
148. S. Adachi, *Anal. Biochem.*, **9**, 150 (1964).
149. H. M. Howell, H. E. Conrad, and E. W. Voss, Jr., *Immunochem.*, **10**, 761 (1973).
150. J. D. Phillips and A. Pollard, *Nature*, **171**, 41 (1953).
151. C. N. Turton and E. Pascu, *J. Amer. Chem. Soc.*, **77**, 1059 (1955).
152. J. C. Sowden, *J. Amer. Chem. Soc.*, **76**, 4487 (1954).
153. C. M. Thompson, *in* "Methods in Medical Research," R. E. Olson, Ed., Year Book Medical Publishers, Inc., Chicago, 1970, Vol. 12, p. 119.
154. C. A. Male, *in* "Methods in Medical Research," R. E. Olson, Ed., Year Book Medical Publishers, Inc., Chicago, 1970, Vol. 12, p. 221; M. Höök, U. Lindahl, A. Hallén, and G. Bäckström, *J. Biol. Chem.*, **250**, 6065 (1975).
155. J. R. Dunstone and W. T. J. Morgan, *Biochim. Biophys. Acta*, **101**, 300 (1965).
156. J. Goa, *Acta Chem. Scand.*, **15**, 1975 (1961).
157. H. Meier, *Acta Chem. Scand.*, **15**, 1381 (1961).
158. G. Rouser, G. Kritchevsky, A. Yamamoto, G. Simon, C. Galli, and A. J. Bauman, *in* "Methods in Enzymology," J. M. Lowenstein, Ed., Academic Press, New York, 1969, Vol. 14, p. 272.

159. J. E. Scott, *Methods Carbohyd. Chem.*, **5**, 38 (1965).
160. I. Halasz, *Biochem. Soc. Trans.*, **3**, 860 (1975); F. Bailey, *ibid.*, **3**, 861 (1975).
161. Z. Deyl, K. Macek, and J. Janak, Eds., "Liquid Column Chromatography. A Survey of Modern Techniques and Applications," Elsevier Scientific Publishing Co., Amsterdam, 1975.
161a. G. D. McGinnis (presiding), *Abstracts Papers Amer. Chem. Soc. Meeting*, **171**, CARB-17–CARB-26 (1976); see also G. D. McGinnis, *Methods Carbohyd. Res.*, **8**, in press (1979).
162. F. R. Seymour, M. E. Slodki, R. D. Plattner, and R. M. Stodola, *Carbohyd. Res.*, **48**, 225 (1976).
162a. L. W. Doner, *Carbohyd. Res.*, **70**, 209 (1979).
163. J. Lehrfeld, *J. Chromatog.*, **129**, 141 (1976).
164. F. M. Rabel, A. G. Caputo, and E. T. Butts, *J. Chromatog.*, **126**, 731 (1976).
165. W. Voelter and H. Bauer, *J. Chromatog.*, **126**, 731 (1976).
166. N. K. Sabbagh and I. S. Fargeson, *J. Chromatog.*, **120**, 55 (1976); see also Ref. 94a.
167. A. Tiselius, *Trans. Faraday Soc.*, **33**, 524 (1937).
168. L. C. Longsworth, *in* "Methods in Medical Research," A. C. Corcoran, Ed., Yearbook Publ., Chicago, Illinois, 1952, Vol. 5, pp. 63–106.
169. L. C. Longsworth, *in* "Electrophoresis," M. Bier, Ed., Academic Press, New York, 1959, Chapters 3 and 4.
170. L. Varga, *J. Biol. Chem.*, **217**, 651 (1955).
171. R. Jensen, O. Snellman, and B. Sylven, *J. Biol. Chem.*, **174**, 265 (1948).
172. R. C. Warner and M. Schubert, *J. Amer. Chem.*, **80**, 5166 (1958).
173. S. Tsuiki, Y. Hashimota, and W. Pigman, *J. Biol. Chem.*, **236**, 2172 (1961).
174. R. L. Whistler and C. S. Campbell, *Methods Carbohyd. Chem.*, **5**, 201 (1965).
175. T. Wieland, *in* "Electrophoresis," M. Bier, Ed., Academic Press, New York, 1959, pp. 493–530.
176. J. Uriel, *in* "Immunoelectrophoretic Analysis," P. Grabar and P. Burtin, Eds., Elsevier, Amsterdam, 1960, Chapter 2.
177. H. G. Kunkel and R. Trautman, *in* "Electrophoresis," M. Bier, Ed., Academic Press, New York, 1959, pp. 225–262.
178. N. Seno, K. Anno, K. Kondo, S. Nagase, and S. Saito, *Anal. Biochem.*, **37**, 197 (1968); A. Gardais, J. Picard, and C. Tarasse, *J. Chromatog.*, **42**, 396 (1969).
179. M. Breen, H. G. Weinstein, M. Anderson, and A. Veis, *Anal. Biochem.*, **35**, 146 (1970); E. Wessler, *ibid.*, **41**, 67 (1971); D. Hsu, P. Hoffman, and T. A. Mashburn, Jr., *ibid.*, **46**, 156 (1972); C. P. Dietrich and S. M. C. Dietrich, *ibid.*, **46**, 209 (1972).
180. R. Hata and Y. Nagai, *Anal. Biochem.*, **45**, 462 (1972).
181. K. Arai and H. Wallace, *Anal. Biochem.*, **31**, 71 (1969).
182. D. H. Northcote and J. D. Pickett-Heaps, *Biochem. J.*, **82**, 509 (1966).
183. R. Consden and W. M. Stanier, *Nature*, **169**, 783 (1952).
184. D. M. Carlson, *Anal. Biochem.*, **20**, 195 (1967).
185. H. Weigel, *Advan. Carbohyd. Chem.*, **18**, 61 (1963).
186. P. J. Garegg and B. L. Lindberg, *Acta Chem. Scand.*, **15**, 1913 (1961).
187. B. Lindberg and B. Swan, *Acta Chem. Scand.*, **14**, 1043 (1960).
188. W. J. Popiel, *Chem. Ind.* (London), 434 (1961).
189. J. L. Frahn and J. A. Mills, *Chem. Ind.* (London), 578 (1956).
190. E. J. Bourne, D. H. Hutson, and H. Weigel, *J. Chem. Soc.*, 4252 (1960).
191. J. L. Frahn and J. A. Mills, *Chem. Ind.* (London), 1137 (1956).
192. S. A. Barker, E. J. Bourne, and O. Theander, *J. Chem. Soc.*, 4276 (1955).

193. G. Fairbanks, T. L. Steck, and D. F. H. Wallach, *Biochemistry*, **10**, 2606 (1971).
194. J. P. Segrest, R. L. Jackson, E. P. Andrews, and V. T. Marchesi, *Biochem. Biophys. Res. Commun.*, **44**, 390 (1971).
195. A. L. Shapiro, E. Venuela, and J. V. Maisel, Jr., *Biochem. Biophys. Res. Commun.*, **28**, 815 (1967)
196. A. L. Shapiro and J. V. Maizel, *Anal. Biochem.*, **29**, 505 (1969).
197. D.-S. Hsu, P. Hoffman, and T. A. Mashburn, Jr., *Biochim. Biophys. Acta*, **338**, 254 (1974).
198. M. B. Mathews, *Methods Carbohyd. Chem.*, **7**, 116 (1976); M. B. Mathews and L. Decker, *Biochim. Biophys. Acta*, **244**, 30 (1971).
199. W. W. Wells, C. C. Sweeley, and R. Bentley, *in* "Biomedical Applications of Gas Chromatography," H. A. Szymanski, Ed., Plenum Press, New York, 1964, pp. 169–224.
200. L. S. Ettre, "Open Tubular Columns in Gas Chromatography," Plenum Press, New York, 1965.
201. G. Petersson and O. Samuelson, *Sv. Papperstidn.*, **71**, 77, 737 (1968).
202. O. S. Chizhov, N. V. Molodtsov, and N. K. Kochetkov, *Carbohyd. Res.*, **4**, 273 (1967).
203. N. K. Kochetkov, O. S. Chizhov, and N. V. Molodtsov, *Tetrahedron*, **24**, 5587 (1968); B. Lindberg, J. Lönngren, and W. Nimmich, *Acta Chem. Scand.*, **26**, 2231 (1972).
204. B. Lindberg, *in* "Methods in Enzymology," V. Ginsburg, Ed., Academic Press, New York, 1972, Vol. 28, Part B, pp. 178–195.
205. G. Petersson, H. Reidl, and O. Samuelson, *Sv. Papperstidn.*, **70**, 371 (1967).
206. J. Kärkkäinen and R. Vikho, *Carbohyd. Res.*, **10**, 113 (1969).
207. H.-J. Yang and S. Hakomori, *J. Biol. Chem.*, **246**, 1192 (1971); K. Stellner, H. Saito, and S. Hakomori, *Arch. Biochem. Biophys.*, **155**, 464 (1973).
208. W. A. Koenig, L. C. Smith, P. F. Crain, and J. A. McCloskey, *Biochemistry*, **10**, 3968 (1971).
209. M. J. E. Golay, *in* "Gas Chromatography," V. J. Coates, H. J. Nebels, and I. S. Fagerson, Eds., Academic Press, New York, 1958, pp. 1–13.
210. J. Szafranek, C. D. Pffafenberger, and E. C. Horning, *Anal. Lett.*, **6**, 479 (1973).
211. H. H. Hausdorff and N. Brenner, *Oil Gas J.*, **21**, 1958.
212. C. C. Sweeley, R. Bentley, M. Makita, and W. W. Wells, *J. Amer. Chem. Soc.*, **85**, 2497 (1963).
213. J. Lönngren and Å. Pilotti, *Acta Chem. Scand.*, **25**, 1144 (1971); H. Bjorndal, B. Lindberg, J. Lönngren, K. Nilsson, and W. Nimmich, *ibid.*, **26**, 1269 (1972); S. K. Kundec and R. W. Ledeen, *Carbohyd. Res.*, **39**, 179 (1975).
214. C. H. Bolton, J. R. Clamp, and L. Hough, *Biochem. J.*, **96**, 5C (1965); C. C. Sweeley and R. V. P. Tao, *Methods Carbohyd. Chem.*, **6**, 8 (1972).
215. J. M. Rickey, M. G. Rickey, Jr., and R. Shraer, *Anal. Biochem.*, **9**, 272 (1964).
216. J. S. Sawardeker and J. H. Sloneker, *Anal. Chem.*, **37**, 945 (1965).
217. G. G. S. Dutton, *Advan. Carbohyd. Chem. Biochem.*, **28**, 11 (1973); **30**, 9 (1974).
218. H. E. Brower, J. E. Jeffrey, and M. W. Folsom, *Anal. Chem.*, **38**, 362 (1966).
219. C. C. Sweeley and B. Walker, *Anal. Chem.*, **36**, 1461 (1964).
220. G. Dawson and J. R. Clamp, *in* "Methods in Medical Research," R. E. Olson, Ed., Year Book Medical Publishers, Inc., Chicago, 1970, Vol. 12, pp. 131–136.
221. M. D. G. Oates and J. Schrager, *Biochem. J.*, **97**, 697 (1965).
222. J. S. Sawardeker, J. H. Sloneker, and R. J. Dimler, *J. Chromatog.*, **20**, 260 (1965).
223. J. S. Sawardeker, J. H. Sloneker, and A. Jeanes, *Anal. Chem.*, **37**, 1602 (1965).

224. Y. C. Lee and C. E. Ballou, *J. Chromatog.*, **18**, 147 (1965).
225. M. B. Perry, *Can. J. Biochem.*, **42**, 451 (1964); M. I. Horowitz and M. R. Delman, *J. Chromatog.*, **21**, 300 (1966).
226. M. B. Perry and A. C. Webb, *Can. J. Biochem.*, **46**, 1163 (1968).
227. R. E. Hurst, *Carbohyd. Res.*, **30**, 143 (1973).
228. B. Radhakrishnamurthy, E. R. Dalferes, Jr., and G. S. Berenson, *Anal. Biochem.*, **17**, 545 (1966).
230. M. B. Perry and R. K. Hulyalkar, *Can. J. Biochem.*, **43**, 573 (1965).
231. B. Radhakrishnamurthy, E. R. Dalferes, Jr., and G. S. Berenson, *Anal. Biochem.*, **24**, 397 (1968).
232. A. G. McInnes, D. H. Ball, F. P. Cooper, and C. T. Bishop, *J. Chromatog.*, **1**, 556 (1958).
233. C. T. Bishop, *Advan. Carbohyd. Chem.*, **19**, 95 (1964).
234. W. F. Lehnhardt and R. J. Winzler, *J. Chromatog.*, **34**, 471 (1968); W. Niedermeier, *Anal. Biochem.*, **40**, 465 (1971); W. Niedermeier, T. Kirkland, R. T. Acton, and J. C. Bennett, *Biochim. Biophys. Acta*, **237**, 442 (1971); L. J. Griggs, A. Post, E. R. White, J. A. Finkelstein, W. E. Moeckel, K. G. Holden, J. E. Zarembo and J. A. Weisbach, *Anal. Biochem.*, **43**, 369 (1971).
235. Z. P. Zanetta, W. C. Breckenridge, and G. Vincendon, *J. Chromatog.*, **69**, 291 (1972).
236. J. H. Kim, B. Shome, T. H. Liao, and J. G. Pierce, *Anal. Biochem.*, **20**, 258 (1967).
236a. F. R. Seymour, E. C. M. Chen, and S. H. Bishop, *Carbohyd. Res.*, **73**, 19 (1979).
237. C. T. Bishop and F. P. Cooper, *Can. J. Chem.*, **38**, 793 (1960); J. K. Baird, M. J. Holroyd, and D. C. Ellwoode, *Carbohyd. Res.*, **27**, 464 (1973).
238. H. W. Kircher, *Anal. Chem.*, **32**, 1103 (1960); C. T. Bishop and F. P. Cooper, *Can. J. Chem.*, **38**, 388 (1960).
239. W. W. Wells, T. Katagi, R. Bentley, and C. C. Sweeley, *Biochim. Biophys. Acta*, **82**, 408 (1964).
240. D. N. Ward, J. A. Coffee, D. B. Ray, and W. M. Lamkin, *Anal. Biochem.*, **14**, 243 (1966); G. R. Shepherd and B. J. Noland, *ibid.*, **26**, 325 (1968); B. Radhakrishnamurthy, E. R. Dalferes, Jr., and G. S. Berenson, *ibid.*, **26**, 61 (1968).
241. O. L. Hollis, *Anal. Chem.*, **38**, 309 (1966).
242. D. P. Sweet, P. Albersheim, and R. H. Shapiro, *Carbohyd. Res.*, **40**, 199, 217 (1975).
243. G. O. Aspinall, A. S. Chaudhari, and C. C. Whitehead, *Carbohyd. Res.*, **47**, 119 (1976).
244. I. R. Siddiqui and P. J. Wood, *Carbohyd. Res.*, **50**, 97 (1976).
245. S. K. Kunda and R. W. Ledeen, *Carbohyd. Res.*, **39**, 329 (1975); N.-V. Din and R. W. Jeanloz, *ibid.*, **47**, 245 (1976).
246. S. K. Kunda and R. W. Ledeen, *Carbohyd. Res.*, **39**, 179 (1975).
247. J. F. Kennedy, S. M. Robertson, and M. Stacey, *Carbohyd. Res.*, **49**, 243 (1976).
248. W. F. Dudman and C. P. Whittle, *Carbohyd. Res.*, **46**, 267 (1976).
249. H. M. Fales, G. W. A. Milne, and R. S. Nicholson, *Anal. Chem.*, **43**, 1785 (1971).
250. R. Varma and R. S. Varma, *J. Chromatog.*, **128**, 45 (1976).

AUTHOR INDEX
Volume IB

Numbers in parentheses are reference numbers and indicate that an author's work is referred to although his name is not cited in the text. Numbers in italics show the page on which the complete reference is listed. This index covers Volume IB only; authors of papers cited in Volume IA are listed in the Author Index at the end of that volume.

SUBJECT INDEX
Volumes IA and IB

C

Cancer, bladder, D-glucaro-1,4-lactone in treatment of, 1040

Carbamates, thio-, ultraviolet absorption spectra, 1386

Carbamic acid, *N,N*-dimethylthio-, esters with sugars, 232

Carbamide, *N*-bromo-, as oxidizing agent, 1016

Carbanilates, sugar, preparation and properties of, 226

Carbohydrates, *see also* Sugars

  carbon-13 labeled, nuclear magnetic resonance spectroscopy of, 1317–1319

  carbon-14 labeled, self-destruction of, 1273–1276

  carbon-14 labeling, 154–156

  chromatography of, 1445–1459, 1464–1470

  definition, 2, 7

  electrophoresis of, 1459–1464

  halogen derivatives of cyclic, 239–251

  isotopically labeled, preparation of, 120, 151, 154–158

  nitrogen-containing, mass spectra of, 1347

  occurrence and uses of, 1

  oxidation by nitric acid, 1136

  peracetylated, mass spectra of, 1333

  permethylated, mass spectra of, 1330–1332

  per(trimethylsilylated), mass spectra of, 1332

  phosphates and other inorganic esters, 253–277

  radiation effect on, 1217–1297

  reduction of, 989–1011

  structural analysis, high-resolution nuclear magnetic resonance spectroscopy, 1299–1326

    infrared spectroscopy, 1394–1436

    mass spectrometry, 1321–1353

    polarimetry, 1354–1375

    ultraviolet spectroscopy, 1376–1393

    X-ray and neutron diffraction, 1437–1444

  sulfur-containing, mass spectra of, 1347

Carbonates

  of carbohydrates, preparation of, 229

  thio-, of carbohydrates, preparation of, 229

Carbon-14 labeled compounds, self-destruction of, 1273–1276

Carbonyl compounds

  1,2-di-, oxidative cleavage of, 1171–1173

  ultraviolet absorption spectra, 1383

Carbonyl groups, determination of, 1057

Carboxylic acids, infrared spectroscopy of, 1414

Carcinoma cells, nontoxic inhibitors of Ehrlich ascites, 807

Carrageenan, mercaptolysis and structure of, 356

Carubinose, *see* Mannose, D-

Casiniroedine, isolation and structure of, 892

Catalysis, acid-base, of enolization, 175–178

Catalysts

  for acetylation and anomerization of sugars, 220

  for aryl glycoside synthesis by Helferich reaction, 293

  for mutarotation, enzymes as, 174

Celesticetin, constituent of, 735

Cellobiitol, ultraviolet irradiation of, 1227

Cellobiose

  irradiation of, 1231

    in aqueous solution, 1248, 1250, 1251

  microbial oxidation of, 1156

  oxidation by permanganate, 1147

  oxidative cleavage of, 1198

  semicarbazone, structure of, 936

  ultraviolet irradiation of, 1227

Cellobioside

  ethyl hepta-*O*-acetyl-α-, preparation by Helferich reaction, 294

  methyl, oxidation of, by chlorine, 1115

β-Cellobioside, methyl, ultraviolet irradiation of, 1227

Cellobiouronic acid, occurrence and synthesis of, 1028

Cellopentaitol, ultraviolet irradiation of, 1227

Cellopentaose, ultraviolet irradiation of, 1227

Cellotriose, oxidative cleavage of, 1199

Cellulose

  determination of carbonyl groups in, 1057

  dialdehyde from periodate oxidation of, 1191

  high-energy radiation of, 1260–1262

  history, 6

T

Z